Mechanical Engineers' Handbook

Mechanical Engineers' Handbook
Third Edition

Instrumentation, Systems, Controls, and MEMS

Edited by
Myer Kutz

JOHN WILEY & SONS, INC.

This book is printed on acid-free paper. ♾

Published by John Wiley & Sons, Inc., Hoboken, New Jersey.
Published simultaneously in Canada.

Library of Congress Cataloging-in-Publication Data:
Mechanical engineers' handbook/edited by Myer Kutz.—3rd ed.
p. cm.
Includes bibliographical references and index.
ISBN-13 978-0-471-44990-4
ISBN-10 0-471-44990-3 (cloth)
1. Mechanical engineering—Handbooks, manuals, etc. I. Kutz, Myer.
TJ151.M395 2005
621—dc22

2005008603

Printed in the United States of America.

10 9 8 7 6 5 4 3 2 1

To Bill and Judy, always there

Contents

Preface

The second volume of the third edition of the *Mechanical Engineers' Handbook* ("ME3") is comprised of two major parts: Part 1, Instrumentation, with eight chapters, and Part 2, Systems, Controls, and MEMS, with 13 chapters. The two parts are linked in the sense that most feedback control systems require measurement transducers. Most of the chapters in this volume originated not only in earlier editions of the *Mechanical Engineers' Handbook* but also in a book called *Instrumentation and Control,* which was edited by Chester L. Nachtigal and published by Wiley in 1990. Some of these chapters have been either updated or extensively revised. Some have been replaced. Others, which present timeless, fundamental concepts, have been included without change.[1] In addition, there are chapters that are entirely new, including Digital Integrated Circuits: A Practical Application (Chapter 8), Neural Networks in Control Systems (Chapter 19), Mechatronics (Chapter 20), and Introduction to Microelectromechanical Systems (MEMS): Design and Application (Chapter 21).

The instrumentation chapters basically are arranged, as they were in the Nachtigal volume, in the order of the flow of information in real measurement systems. These chapters start with fundamentals of transducer design, present transducers used by mechanical engineers, including strain gages, temperature transducers such as thermocouples and thermistors, and flowmeters, and then discuss issues involved in processing signals from transducers and in acquiring and displaying data. A general chapter on measurement fundamentals, updated from the second edition of *Mechanical Engineers' Handbook* ("ME2"), as well as the chapter on digital integrated circuits have been added to the half-dozen *Instrumentation and Control* chapters in this first part.

The systems and control chapters in the second part of this volume start with three chapters from ME2, two of which have been updated, and move on to seven chapters from Nachtigal, only two of which required updating. These ten chapters present a general discussion of systems engineering; fundamentals of control system design, analysis, and performance modification; and detailed information about the design of servoactuators, controllers, and general-purpose control devices. This second part of Vol. II concludes with the chapters, all of them new to the handbook, on what are termed "new departures"—neural networks, mechatronics, and MEMS. These topics have become increasingly important to mechanical engineers in recent years.

---

[1] A new edition of *Instrumentation and Control* has been sought after but has never appeared. Because several chapters had numerous contributors, it proved impossible to update or revise them or even to find anyone to write new chapters on the same topics on the schedule that other contributors could meet. Because the material in these chapters was outdated, they have been dropped from this edition, but may be revised for future editions.

Vision for the Third Edition

Basic engineering disciplines are not static, no matter how old and well established they are. The field of mechanical engineering is no exception. Movement within this broadly based discipline is multidimensional. Even the classic subjects on which the discipline was founded, such as mechanics of materials and heat transfer, continue to evolve. Mechanical engineers continue to be heavily involved with disciplines allied to mechanical engineering, such as industrial and manufacturing engineering, which are also constantly evolving. Advances in other major disciplines, such as electrical and electronics engineering, have significant impact on the work of mechanical engineers. New subject areas, such as neural networks, suddenly become all the rage.

In response to this exciting, dynamic atmosphere, the *Mechanical Engineers' Handbook* is expanding dramatically, from one volume to four volumes. The third edition not only is incorporating updates and revisions to chapters in the second edition, which was published in 1998, but also is adding 24 chapters on entirely new subjects as well, incorporating updates and revisions to chapters in the *Handbook of Materials Selection,* which was published in 2002, as well as to chapters in *Instrumentation and Control,* edited by Chester Nachtigal and published in 1990.

The four volumes of the third edition are arranged as follows:

Volume I: *Materials and Mechanical Design*—36 chapters
- Part 1. Materials—14 chapters
- Part 2. Mechanical Design—22 chapters

Volume II: *Instrumentation, Systems, Controls, and MEMS*—21 chapters
- Part 1. Instrumentation—8 chapters
- Part 2. Systems, Controls, and MEMS—13 chapters

Volume III: *Manufacturing and Management*—24 chapters
- Part 1. Manufacturing—12 chapters
- Part 2. Management, Finance, Quality, Law, and Research—12 chapters

Volume IV: *Energy and Power*—31 chapters
- Part 1: Energy—15 chapters
- Part 2: Power—16 chapters

The mechanical engineering literature is extensive and has been so for a considerable period of time. Many textbooks, reference works, and manuals as well as a substantial number of journals exist. Numerous commercial publishers and professional societies, particularly in the United States and Europe, distribute these materials. The literature grows continuously, as applied mechanical engineering research finds new ways of designing, controlling, measuring, making and maintaining things, and monitoring and evaluating technologies, infrastructures, and systems.

Most professional-level mechanical engineering publications tend to be specialized, directed to the specific needs of particular groups of practitioners. Overall, however, the mechanical engineering audience is broad and multidisciplinary. Practitioners work in a variety of organizations, including institutions of higher learning, design, manufacturing, and consulting firms as well as federal, state, and local government agencies. A rationale for an expanded general mechanical engineering handbook is that every practitioner, researcher, and bureaucrat cannot be an expert on every topic, especially in so broad and multidisciplinary a field, and may need an authoritative professional summary of a subject with which he or she is not intimately familiar.

Starting with the first edition, which was published in 1986, our intention has always been that the *Mechanical Engineers' Handbook* stand at the intersection of textbooks, research papers, and design manuals. For example, we want the handbook to help young engineers move from the college classroom to the professional office and laboratory where they may have to deal with issues and problems in areas they have not studied extensively in school.

With this expanded third edition, we have produced a practical reference for the mechanical engineer who is seeking to answer a question, solve a problem, reduce a cost, or improve a system or facility. The handbook is not a research monograph. The chapters offer design techniques, illustrate successful applications, or provide guidelines to improving the performance, the life expectancy, the effectiveness, or the usefulness of parts, assemblies, and systems. The purpose is to show readers what options are available in a particular situation and which option they might choose to solve problems at hand.

The aim of this expanded handbook is to serve as a source of practical advice to readers. We hope that the handbook will be the first information resource a practicing engineer consults when faced with a new problem or opportunity—even before turning to other print sources, even officially sanctioned ones, or to sites on the Internet. (The second edition has been available online on knovel.com.) In each chapter, the reader should feel that he or she is in the hands of an experienced consultant who is providing sensible advice that can lead to beneficial action and results.

Can a single handbook, even spread out over four volumes, cover this broad, interdisciplinary field? We have designed the third edition of the *Mechanical Engineers' Handbook* as if it were serving as a core for an Internet-based information source. Many chapters in the handbook point readers to information sources on the Web dealing with the subjects addressed. Furthermore, where appropriate, enough analytical techniques and data are provided to allow the reader to employ a preliminary approach to solving problems.

The contributors have written, to the extent their backgrounds and capabilities make possible, in a style that reflects practical discussion informed by real-world experience. We would like readers to feel that they are in the presence of experienced teachers and consultants who know about the multiplicity of technical issues that impinge on any topic within mechanical engineering. At the same time, the level is such that students and recent graduates can find the handbook as accessible as experienced engineers.

Contributors

Adam C. Bell
Dartmouth, Nova Scotia, Canada

Sujeet Chand
Rockwell Automation
Milwaukee, Wisconsin

James H. Christensen
Holobloc, Inc.
Cleveland Heights, Ohio

Shane Farritor
University of Nebraska–Lincoln
Lincoln, Nebraska

Keith Folken
Peoria, Illinois

Shuzhi Sam Ge
National University of Singapora
Singapore

Jerry Lee Hall
Hall-Wade Engineering Services
and
Iowa State University
Ames, Iowa

Syed Hamid
Halliburton Services
Duncan, Oklahoma

E. L. Hixson
University of Texas
Austin, Texas

Suhada Jayasuriya
Texas A&M University
College Station, Texas

Robert J. Kretschmann
Rockwell Automation
Mayfield Heights, Ohio

F. L. Lewis
University of Texas at Arlington
Fort Worth, Texas

Philip C. Milliman
Weyerhaeuser Company
Federal Way, Washington

Robert J. Moffat
Stanford University
Stanford, California

Mahmood Naim
Union Carbide Corporation
Indianapolis, Indiana

Thomas Peter Neal
Lake View, New York

William J. Palm III
University of Rhode Island
Kingston, Rhode Island

Karl N. Reid
Oklahoma State University
Stillwater, Oklahoma

Todd Rhoad
Austin, Texas

E. A. Ripperger
University of Texas
Austin, Texas

Andrew P. Sage
George Mason University
Fairfax, Virginia

Krishnaswamy Srinivasan
The Ohio State University
Columbus, Ohio

Sriram Sundararajan
Iowa State University
Ames, Iowa

John Turnbull
Case Western Reserve University
Cleveland, Ohio

Patrick L. Walter
Texas Christian University
Fort Worth, Texas

K. Preston White, Jr.
University of Virginia
Charlottesville, Virginia

Kazuhiko Yokoyama
Yaskawa Electric Corporation
Tokyo, Japan

M. E. Zaghloul
The George Washington University
Washington, D.C.

Mechanical Engineers' Handbook

PART 1

INSTRUMENTATION

CHAPTER 1

INSTRUMENT STATICS

Jerry Lee Hall
Department of Mechanical Engineering
Iowa State University
Ames, Iowa

Sriram Sundararajan
Department of Mechanical Engineering
Iowa State University
Ames, Iowa

Mahmood Naim
Union Carbide Corporation
Indianapolis, Indiana

1 TERMINOLOGY

1.1 Transducer Characteristics

A *measurement system* extracts information about a measurable quantity from some medium of interest and communicates this measured data to the observer. The measurement of any variable is accomplished by an instrumentation system composed of *transducers*. Each transducer is an energy conversion device and requires energy transfer into the device before the variable of interest can be detected.

The Instrument Society of America (ISA) defines *transducer* as "a device that provides usable output in response to a specified measurand." The *measurand* is "a physical quantity, property or condition which is measured." The *output* is "the electrical quantity, produced by a transducer, which is a function of the applied measurand."[1]

It should be made very clear that the act of measurement involves transfer of energy between the measured medium and the measuring system and hence the measured quantity is disturbed to some extent, making a perfect measurement unrealistic. Therefore, pressure cannot be measured without an accompanying change in volume, force cannot be measured without an accompanying change in length, and voltage cannot be measured without an accompanying flow of charge. Instead measures must be taken to minimize the energy transfer from the source to be measured if the measurement is to be accurate.

There are several categorical characteristics for a transducer (or measurement system). When the measurand maintains a steady value or varies very slowly with time, transducer performance can be described in terms of *static characteristics*. Instruments associated with rapidly varying measurands require additional qualifications termed *dynamic characteristics*. Other performance descriptors include *environmental characteristics* (for situations involving varying environmental operating conditions), *reliability characteristics* (related to the life expectancy of the instrument under various operating conditions), *theoretical characteristics* (describing the ideal behavior of the instrument in terms of mathematical or graphical relationships), and *noise characteristics* (external factors that can contribute to the measurement process such as electromagnetic surroundings, humidity, acoustic and thermal vibrations, etc.). In this chapter, we will describe the considerations associated with evaluating numerical values for the static characteristics of an instrument.

1.2 Definitions

The description of a transducer and its role in a measuring system is based on most of the definitions that follow. Further details of these definitions can be found in other works.[2–4]

Static calibration is the process of measuring the static characteristics of an instrument. This involves applying a range of known values of static input to the instrument and recording the corresponding outputs. The data obtained are presented in a tabular or graphical form.

Range is defined by the upper and lower limits of the measured values that an instrument can measure. Instruments are designed to provide predictable performance and, often, enhanced linearity over the range specified.

Sensitivity is defined as the change in the output signal relative to the change in the input signal at an operating point. Sensitivity may be constant over the range of the input signal to a transducer or it can vary. Instruments that have a constant sensitivity are called "linear."

Resolution is defined as the smallest change in the input signal that will yield a readable change in the output of the measuring system at its operating point.

Threshold of an instrument is the minimum input for which there will be an output. Below this minimum input the instrument will read zero.

Zero of an instrument refers to a selected datum. The output of an instrument is adjusted to read zero at a predefined point in the measured range. For example, the output of a Celsius thermometer is zero at the freezing point of water; the output of a pressure gage may be zero at atmospheric pressure.

Zero drift is the change in output from its set zero value over a specified period of time. Zero drift occurs due to changes in ambient conditions, changes in electrical conditions, aging of components, or mechanical damage. The error introduced may be significant when a transducer is used for long-term measurement.

Creep is a change in output occurring over a specific time period while the measurand is held constant at a value other than zero and all environmental conditions are held constant.

Accuracy is the maximum amount of difference between a measured variable and its true value. It is usually expressed as a percentage of full-scale output. In the strictest sense, accuracy is never known because the true value is never really known.

Precision is the difference between a measured variable and the best estimate (as obtained from the measured variable) of the true value of the measured variable. It is a measure of repeatability. Precise measurements have small dispersion but may have poor accuracy if they are not close to the true value. Figure 1*a* shows the differences between accuracy and precision.

Linearity describes the maximum deviation of the output of an instrument from a best-fitting straight line through the calibration data. Most instruments are designed so that the output is a linear function of the input. Linearity is based on the type of straight line fitted to the calibration data. For example, least-squares linearity is referenced to that straight line for which the sum of the squares of the residuals is minimized. The

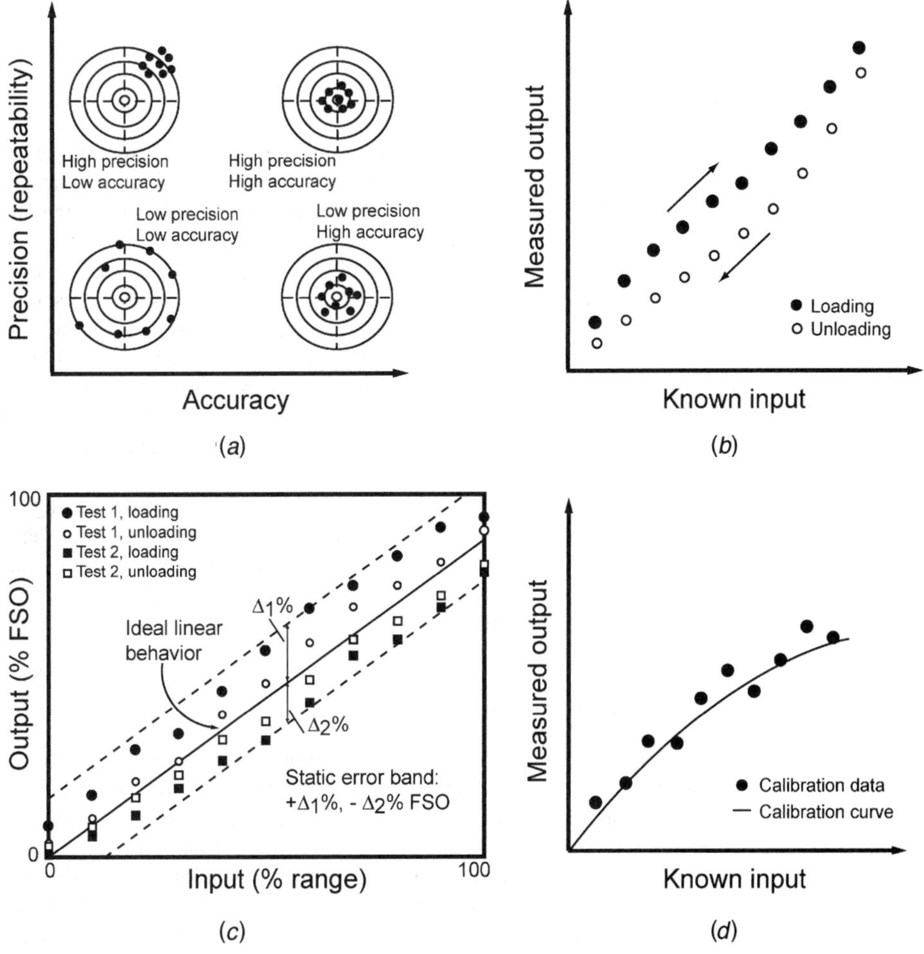

Figure 1 Schematics illustrating concepts of (*a*) accuracy and precision, (*b*) hysteresis, (*c*) a static error band, and (*d*) fitting a curve to the calibration data.

term "residual" refers to the deviations of output readings from their corresponding values on the straight line fitted through the data.

Hysteresis is the maximum difference in output, at any measured value within the specified range, when the value is approached first with increasing and then with decreasing measurand. Hysteresis is typically caused by a lag in the action of the sensing element of the transducer. Loading the instrument through a cycle of first increasing values, then decreasing values, of the measurand provides a hysteresis loop, as shown in Fig. 1*b*. Hysteresis is usually expressed in percent of full-scale output.

Error band is the band of maximum deviation of output values from a specified reference line or curve. A static error band (see Fig. 1*c*) is obtained by static calibration. It is determined on the basis of maximum deviations observed over at least two consecutive calibration cycles so as to include repeatability. Error band accounts for deviations that may be due to nonlinearity, nonrepeatability, hysteresis, zero shift, sensitivity shift, and so forth. It is a convenient way to specify transducer behavior when individual types of deviations need not be specified nor determined.[5]

2 STATIC CALIBRATION

2.1 Calibration Process

Calibration is the process of comparison of the output of a measuring system to the values of a range of known inputs. For example, a pressure gage is calibrated by a device called a "dead-weight" tester, where known pressures are applied to the gage and the output of the gage is recorded over its complete range of operation.

The calibration signal should, as closely as possible, be the same as the type of input signal to be measured. Most calibrations are performed by means of static or level calibration signals since they are usually easy to produce and maintain accurately. However, a measuring system calibrated with static signals may not read correctly when subjected to the dynamic input signals since the natural dynamic characteristics and the response characteristics of the measurement system to the input forcing function would not be accounted for with a static calibration. A measurement system used for dynamic signals should be calibrated using known dynamic inputs.

A static calibration should include both increasing and decreasing values of the known input signal and a repetition of the input signal.[6] This allows one to determine hysteresis as well as the repeatability of the measuring system, as shown in Fig. 1*c*. The sensitivity of the measuring system is obtained from the slope of a suitable line or curve plotted through the calibration points at any level of the input signal.

2.2 Fitting Equations to Calibration Data

Though linear in most cases, the calibration plot of a specific measurement system may require a choice of a nonlinear functional form for the relationship that best describes the calibration data, as shown in Fig. 1*d*. This functional form (or curve fit) may be a standard polynomial type or may be one of a transcendental function type. Statistics are used to fit a desired function to the calibration data. A detailed description of the mathematical basis of the selection process used to determine the appropriate function to fit the data can be found elsewhere.[7] Most of today's graphing software allow the user to select the type of fit required. A very common method used to describe the quality of "fit" of a chosen functional form is the "least-squares fit." The principle used in making this type of curve fit is to minimize the sum of the squares of the deviations of the data from the assumed curve. These deviations from the assumed curve may be due to errors in one or more variables. If the error is in one variable, the technique is called linear regression and is the common case encountered in engineering measurements. If several variables are involved, it is called multiple regression. Two assumptions are often used with the least-squares method: (i) the x variable (usually the input to the calibration process) has relatively little error as compared to the y (measured) variable and (ii) the magnitude of the uncertainty in y is not dependent on the magnitude of the x variable. The methodology for evaluating calibration curves in systems where the magnitude of the uncertainty in the measured value varies with the value of the input variable can be found elsewhere.[8]

Although almost all graphing software packages include the least-squares fit analysis, thus enabling the user to identify the best-fit curve with minimum effort, a brief description of the mathematical process is given here. To illustrate the least-squares technique, assume that an equation of the following polynomial form will fit a given set of data:

$$y = a + bx^1 + cx^2 + \cdots + mx^k \tag{1}$$

If the data points are denoted by (x_i, y_i), where i ranges from 1 to n, then the expression for summation of the residuals is

$$\sum_{i=1}^{n} (y_i - y)^2 = R \tag{2}$$

The least-squares method requires that R be minimized. The parameters used for the minimization are the unknown coefficients $a, b, c, \ldots, m$ in the assumed equation. The following differentiation yields $k + 1$ equations called "normal equations" to determine the $k + 1$ coefficients in the assumed relation. The coefficients $a, b, c, \ldots, m$ are found by solving the normal equations simultaneously:

$$\frac{\partial R}{\partial a} = \frac{\partial R}{\partial b} = \frac{\partial R}{\partial c} = \cdots = \frac{\partial R}{\partial m} = 0 \tag{3}$$

For example, if $k = 1$, then the polynomial is of first degree (a straight line) and the normal equations become

$$\sum y_i = a(n) + b\sum x_i \qquad \sum x_i y_i = a\sum x_i + b\sum x_i^2 \tag{4}$$

and the coefficients a and b are

$$a = \frac{\sum x^2 \sum y - \sum x \sum xy}{n\sum x^2 - (\sum x)^2} \qquad b = \frac{n\sum xy - \sum x \sum y}{n\sum x^2 - (\sum x)^2} \tag{5}$$

The resulting curve ($y = a + bx$) is called the regression curve of y on x. It can be shown that a regression curve fit by the least-squares method passes through the centroid $(\bar{x}, \bar{y})$ of the data.[9] If two new variables X and Y are defined as

$$X = x - \bar{x} \qquad \text{and} \qquad Y = y - \bar{y} \tag{6}$$

then

$$\sum X = 0 = \sum Y \tag{7}$$

Substitution of these new variables in the normal equations for a straight line yields the following result for a and b:

$$a = 0 \qquad b = \frac{\Sigma XY}{\Sigma X^2} \tag{8}$$

The regression line becomes

$$Y = bX \tag{9}$$

The technique described above will yield a curve based on an assumed form that will fit a set of data. This curve may not be the best one that could be found, but it will be the best based on the assumed form. Therefore, the "goodness of fit" must be determined to check that the fitted curve follows the physical data as closely as possible.

Example 1 Choice of Functional Form. Find a suitable equation to represent the following calibration data:

x = [3, 4, 5, 7, 9, 12, 13, 14, 17, 20, 23, 25, 34, 38, 42, 45]

y = [5.5, 7.75, 10.6, 13.4, 18.5, 23.6, 26.2, 27.8, 30.5, 33.5, 35, 35.4, 41, 42.1, 44.6, 46.2]

Solution: A computer program can be written or graphing software (e.g., Microsoft Excel) used to fit the data to several assumed forms, as given in Table 1. The data can be plotted and the best-fitting curve selected on the basis of minimum residual error, maximum correlation coefficient, or smallest maximum absolute deviation, as shown in Table 1.

The analysis shows that the assumed equation $y = a + (b)\log(x)$ represents the best fit through the data as it has the smallest maximum deviation and the highest correlation coefficient. Also note that the equation $y = 1/a + bx$ is not appropriate for these data because it has a negative correlation coefficient.

Example 2 Nonlinear Regression. Find the regression coefficients a, b, and c if the assumed behavior of the (x, y) data is $y = a + bx + cx^2$:

x = [2, 3, 4, 5, 6, 7, 8, 9, 10, 11, 12, 13, 14, 15]

y = [0.26, 0.38, 0.55, 0.70, 1.05, 1.36, 1.75, 2.20, 2.70, 3.20, 3.75, 4.40, 5.00, 6.00]

Table 1 Statistical Analysis for Example 1

Assumed Equation	Regression Coefficient a	Regression Coefficient b	Residual Error, R	Maximum Deviation	Correlation Coefficient[a]
$y = bx$	—	1.254	56.767	10.245	0.700
$y = a + bx$	9.956	0.907	20.249	7.178	0.893
$y = ae^{bx}$	10.863	0.040	70.274	18.612	0.581
$y = 1/(a + bx)$	0.098	−0.002	14257.327	341.451	−74.302
$y = a + b/x$	40.615	−133.324	32.275	9.326	0.830
$y = a + b \log x$	−14.188	15.612	1.542	2.791	0.992
$y = ax^b$	3.143	0.752	20.524	8.767	0.892
$y = x/(a + bx)$	0.496	0.005	48.553	14.600	0.744

[a]Defined in Section 4.3.7.

From Eqs. (1)–(3)

$$y = a + bx + cx^2$$
$$xy = ax + bx^2 + cx^3 \qquad (10)$$
$$x^2y = ax^2 + bx^3 + cx^4$$

A simultaneous solution of the above equations provides the desired regression coefficients:

$$a = 0.1959 \qquad b = -0.0205 \qquad c = 0.0266 \qquad (11)$$

As was mentioned previously, a measurement process can only give you the best estimate of the measurand. In addition, engineering measurements taken repeatedly under seemingly identical conditions normally show variations in measured values. A statistical treatment of measurement data is therefore a necessity.

3 STATISTICS IN THE MEASUREMENT PROCESS

3.1 Unbiased Estimates

Data sets typically have two very important characteristics: *central tendency* (or most representative value) and *dispersion* (or scatter). Other characteristics such as skewness and kurtosis (or peakedness) may also be of importance but will not be considered here.[10]

A basic problem in every quantitative experiment is that of obtaining an unbiased estimate of the true value of a quantity as well as an unbiased measure of the dispersion or uncertainty in the measured variable. Philosophically, in any measurement process a deterministic event is observed through a "foggy" window. If so, ultimate refinement of the measuring system would result in all values of measurements to be the true value μ. Because errors occur in all measurements, one can never exactly measure the true value of any quantity. Continued refinement of the methods used in any measurement will yield closer and closer approximations, but there is always a limit beyond which refinements cannot be made. To determine the relation that a measured value has with the true value, we must specify the unbiased estimate $\bar{x}$ of the true value μ of a measurement and its uncertainty (or precision) interval W_x based on a desired confidence level (or probability of occurrence).

An *unbiased estimator*[9] exists if the mean of its distribution is the same as the quantity being estimated. Thus, for sample mean $\bar{x}$ to be an unbiased estimator of population mean μ, the mean of the distribution of sample means, $\bar{\bar{x}}$, must be equal to the population mean.

3.2 Sampling

Unbiased estimates for determining population mean, population variance, and variance of the sample mean depend on the type of sampling procedure used.

Sampling with Replacement (Random Sampling)

$$\hat{\mu} = \bar{x} \qquad (12)$$

where $\bar{x}$ is the sample mean and $\hat{\mu}$ is the unbiased estimate of the population mean, μ;

$$\hat{\sigma}^2 = S^2\left(\frac{n}{n-1}\right)$$

where S^2 is the sample variance

$$S^2 = \frac{\sum(x_i - \bar{x})^2}{n} \tag{13}$$

and

$$\hat{\sigma}_{\bar{x}}^2 = \frac{\hat{\sigma}^2}{n} \tag{14}$$

where $\hat{\sigma}_{\bar{x}}^2$ is the variance of the mean.

Sampling without Replacement (Usual Case)

$$\hat{\mu} = \bar{x} \tag{15}$$

$$\hat{\sigma}^2 = S^2\left(\frac{n}{n-1}\right)\left(\frac{N-1}{N}\right) \tag{16}$$

where N is the population size and n the sample size, and

$$\hat{\sigma}_{\bar{x}}^2 = \frac{\hat{\sigma}^2}{n}\left(\frac{N-n}{N-1}\right) \tag{17}$$

Note that sampling without replacement from an extremely large population is equivalent to random sampling.

3.3 Types of Errors

There are at least three types of errors that one must consider in making measurements. They are systematic (or fixed) errors, illegitimate errors (or mistakes), and random errors:

Systematic errors are of consistent form. They result from conditions or procedures that are correctable. This type of error may generally be eliminated by calibration.

Illegitimate errors are mistakes and should not exist. They may be eliminated by using care in the experiment, proper measurement procedures, and repetition of the measurement.

Random errors are accidental errors that occur in all measurements. They are characterized by their inconsistent nature, and their origin cannot be determined in the measurement process. These errors are estimated by statistical analysis.

If the illegitimate errors can be eliminated by care and proper measurement procedures and the systematic errors can be eliminated by calibrating the measurement system, then the random errors remain to be determined by statistical analysis to yield the precision of the measurement.

3.4 Propagation of Error or Uncertainty

In many cases the desired quantity cannot be measured directly but must be calculated from the most representative value (e.g., the mean) of two or more measured quantities. It is desirable to know the uncertainty or precision of such calculated quantities.

Precision is specified by quantities called *precision indexes* (denoted by W_x) that are calculated from the random errors of a set of measurements. A $\pm W_x$ should be specified for every measured variable. The confidence limits or probability for obtaining the range $\pm W_x$ is generally specified directly or is implied by the particular type of precision index being used.

The precision index of a calculated quantity depends on the precision indexes of the measured quantities required for the calculations.[9] If the measured quantities are determined independently and if their distribution about a measure of central tendency is approximately symmetrical, the following "propagation-of-error" equation is valid[11]:

$$W_R^2 = \Sigma\left(\frac{\partial R}{\partial x_i}\right)^2 W_{x_i}^2 \tag{18}$$

In this equation, R represents the calculated quantity and $x_1, x_2, \ldots, x_n$ represent the measured independent variables so that mathematically we have $R = f(x_1, x_2, \ldots, x_n)$. The precision index is a measure of dispersion about the central tendency and is denoted by W in Eq. (18). The standard deviation is often used for W; however, any precision index will do as long as the same type of precision index is used in each term of the equation.

A simplified form of this propagation-of-error equation results if the function R has the form

$$R = kx_1^a x_2^b x_3^c \cdots x_n^m \tag{19}$$

where the exponents $a, b, \ldots, m$ may be positive or negative, integer or noninteger. The simplified result for the precision W_R in R is

$$\left(\frac{W_R}{R}\right)^2 = a^2\left(\frac{W_{x1}}{x_1}\right)^2 + b^2\left(\frac{W_{x2}}{x_2}\right)^2 + \cdots + m^2\left(\frac{W_{xn}}{x_n}\right)^2 \tag{20}$$

The propagation-of-error equation is also used in planning experiments. If a certain precision is desired on the calculated result R, the precision of the measured variables can be determined from this equation. Then, the cost of a proposed measurement system can be determined as it is directly related to precision.

Example 3 Propagation of Uncertainty. Determine the resistivity and its uncertainty for a conducting wire of circular cross section from the measurements of resistance, length, and diameter. Given

$$R = \rho\frac{L}{A} = \rho\frac{4L}{\pi D^2} \qquad \text{or} \qquad \rho = \frac{\pi D^2 R}{4L} \tag{21}$$

$R = 0.0959 \pm 0.0001\ \Omega \qquad L = 250 \pm 2.5$ cm $\qquad D = 0.100 \pm 0.001$ cm

where R = wire resistance, Ω
L = wire length, cm
A = cross-sectional area, $= \pi D^2/4$, cm²
ρ = wire resistivity, Ω·cm

Solution: Thus the resistivity is

$$\rho = \frac{(\pi)(0.100)^2(0.0959)}{4(250)} = 3.01 \times 10^{-6}\ \Omega\cdot\text{cm}$$

The propagation of variance (or precision index) equation for ρ reduces to the simplified form, that is,

$$\left(\frac{W\rho}{\rho}\right)^2 = 4\left(\frac{W_D}{D}\right)^2 + \left(\frac{W_R}{R}\right)^2 + \left(\frac{W_L}{L}\right)^2$$
$$= 4\left(\frac{0.001}{0.10}\right)^2 + \left(\frac{0.0001}{0.0959}\right)^2 + \left(\frac{2.5}{250}\right)^2$$
$$= 4.00 \times 10^{-4} + 1.09 \times 10^{-6} + 1.00 \times 10^{-4}$$
$$= 5.01 \times 10^{-4}$$

The resulting resistivity ρ and its precision W_ρ are

$$W_\rho = \rho\sqrt{(5.01)10^{-4}} = \pm 6.74 \times 10^{-8}$$
$$\rho = (3.01 \pm 0.07) \times 10^{-6}\ \Omega\cdot\text{cm}$$

3.5 Uncertainty Interval

When several measurements of a variable have been obtained to form a data set (multisample data), the best estimates of the most representative value (mean) and dispersion (standard deviation) are obtained from the formulas in Section 3.2. When a single measurement exists (or when the data are taken so that they are equivalent to a single measurement), the standard deviation cannot be determined and the data are said to be "single-sample" data. Under these conditions the only estimate of the true value is the single measurement, and the uncertainty interval must be estimated by the observer.[12] It is recommended that the precision index be estimated as the maximum reasonable error. This corresponds approximately to the 99% confidence level associated with multisample data.

Uncertainty Interval Considering Random Error

Once the unbiased estimates of mean and variance are determined from the data sample, the uncertainty interval for μ is

$$\mu = \hat{\mu} \pm \hat{W} = \hat{\mu} \pm k(\nu, \gamma)\hat{\sigma} \tag{22}$$

where $\hat{\mu}$ represents the most representative value of μ from the measured data and $\hat{W}$ is the uncertainty interval or precision index associated with the estimate of μ. The magnitude of the precision index or uncertainty interval depends on the confidence level γ (or probability chosen), the amount of data n, and the type of probability distribution governing the distribution of measured items.

The uncertainty interval $\hat{W}$ can be replaced by $k\hat{\sigma}$, where $\hat{\sigma}$ is the standard deviation (measure of dispersion) of the population as estimated from the sample and k is a constant that depends on the probability distribution function, the confidence level γ, and the amount of data n. For example, with a Gaussian distribution the 95% confidence limits are $\hat{W} = 1.96\sigma$, where $k = 1.96$ and is independent of n. For a t-distribution, $k = 2.78$, 2.06, and 1.96 with a sample size of 5, 25, and ∞, respectively, at the 95% level of confidence probability. Note that $\nu = n - 1$ for the t-distribution. The t-distribution is the same as the Gaussian distribution as $n \rightarrow \infty$.

Uncertainty Interval Considering Random Error with Resolution, Truncation, and Significant Digits

The uncertainty interval $\hat{W}$ in Eq. (22) assumes a set of measured values with only random error present. Furthermore, the set of measured values is assumed to have unbounded significant digits and to have been obtained with a measuring system having infinite resolution. When finite resolution exists and truncation of digits occurs, the uncertainty interval may be larger than that predicted by consideration of the random error only. The uncertainty interval can never be less than the resolution limits or truncation limits of the measured values.

Resolution and Truncation

Let $\{s_n\}$ be the theoretically possible set of measurements of unbound significant digits from a measuring system of infinite resolution and let $\{x_n\}$ be the actual set of measurements expressed to m significant digits from a measuring system of finite resolution. Then the quantity $s_i - x_i = \pm e_i$ is the resolution or truncation deficiency caused by the measurement process. The unbiased estimates of mean and variance are

$$\hat{\mu} = \frac{\sum s_i}{n} = \bar{s} \qquad \text{and} \qquad \hat{\sigma}^2 = \frac{\sum(s_i - \bar{s})^2}{n - 1} \tag{23}$$

Noting that the set $\{x_n\}$ is available rather than $\{s_n\}$, the required mean and variance are

$$\hat{\mu} = \frac{\sum x_i}{n} \pm \frac{\sum e_i}{n} = \bar{x} \pm \frac{\sum e_i}{n} \qquad \text{and} \qquad \hat{\sigma}^2 = \frac{\sum(x_i - \bar{x})^2}{n - 1} \tag{24}$$

The truncation or resolution has no effect on the estimate of variance but does affect the estimate of the mean. The truncation error e_i is not necessarily distributed randomly and may all be of the same sign. Thus $\bar{x}$ can be biased as much as $\sum e_i / n = \bar{e}$ high or low from the unbiased estimate of the value of μ so that $\hat{\mu} = \bar{x} \pm \bar{e}$.

If e_i is a random variable observed through a "cloudy window" with a measuring system of finite resolution, the value of e_i may be plus or minus but its upper bound is R (the resolution of the measurement). Thus the resolution error is no larger than R and $\hat{\mu} = \bar{x} \pm Rn/n = \bar{x} \pm R$.

If the truncation is never more than that dictated by the resolution limits (R) of the measurement system, the uncertainty in $\bar{x}$ as a measure of the most representative value of μ is never larger than R plus the uncertainty due to the random error. Thus $\hat{\mu} = \bar{x} \pm (\hat{W} + \text{R})$. It should be emphasized that the uncertainty interval can never be less than the resolution bounds of the measurement. The resolution bounds cannot be reduced without changing the measurement system.

Significant Digits

When x_i is observed to m significant digits, the uncertainty (except for random error) is never more than $\pm 5/10^m$ and the bounds on s_i are equal to $x_i \pm 5/10^m$ so that

$$x_i - \frac{5}{10^m} < s_i < x_i + \frac{5}{10^m} \tag{25}$$

The relation for $\hat{\mu}$ for m significant digits is then from Eq. 24.

$$\hat{\mu} = \bar{x} \pm \frac{\sum e_i}{n} = \bar{x} \pm \frac{\sum(5/10^m)}{n} = \bar{x} \pm \frac{5}{10^m} \tag{26}$$

The estimated value of variance is not affected by the constant magnitude of $5/10^m$. When the uncertainty due to significant digits is combined with the resolution limits and random error, the uncertainty interval on $\hat{\mu}$ becomes

$$\hat{\mu} = \bar{x} \pm \left(\hat{W} + R + \frac{5}{10^m}\right) \tag{27}$$

This illustrates that the number of significant digits of a measurement should be carefully chosen in relation to the resolution limits of the measuring system so that $5/10^m$ has about the same magnitude as R. Additional significant digits would imply more accuracy to the measurement than would actually exist based on the resolving ability of the measuring system.

3.6 Amount of Data to Take

Exactly what data to take and how much data to take are two important questions to be answered in any experiment. Assuming that the correct variables have been measured, the amount of data to be obtained can be determined by using the relation

$$\mu = \bar{\bar{x}} \pm \left(\hat{W}_{\bar{x}} + R + \frac{5}{10^m} \right) \tag{28}$$

where it is presumed that several sample sets may exist for estimation of μ and that the mean of means of the sample sets is denoted by $\bar{\bar{x}}$. This equation can be rewritten using Eqs. (13) and (14) (assuming random sampling):

$$\begin{aligned} \mu &= \bar{\bar{x}} \pm \left(k(\nu,\gamma)\hat{\sigma}_{\bar{x}} + R + \frac{5}{10^m} \right) \\ &= \bar{\bar{x}} \pm \left(k(\nu,\gamma)\frac{\hat{\sigma}}{\sqrt{n}} + R + \frac{5}{10^m} \right) \end{aligned} \tag{29}$$

The value of n to achieve the difference in $\mu - \bar{\bar{x}}$ within a stated percent of μ can be determined from

$$n = \left[\frac{k(\upsilon,\gamma)\hat{\sigma}}{(\%/100)\hat{\mu} - R - (5/10^m)} \right]^2 \tag{30}$$

This equation can only yield valid values of n once valid estimates of $\hat{\mu}$, $\hat{\sigma}$, k, R, and m are available. This means that the most correct values of n can only be obtained once the measurement system and data-taking procedure have been specified so that R and m are known. Furthermore, either a preliminary experiment or a portion of the actual experiment should be performed to obtain good estimates of $\hat{\mu}$ and $\hat{\sigma}$. Because k depends not only on the type of distribution the data follows but also on the sample size n, the solution is iterative. Thus, the most valid estimates of the amount of data to take can only be obtained after the experiment has begun. However, the equation can be quite useful for prediction purposes if one wishes to estimate values of $\hat{\mu}$, $\hat{\sigma}$, k, R, and m. This is especially important in experiments for which the cost of a single run may be relatively high.

Example 4 Amount of Data to Take. The life for a certain type of automotive tire is to be established. The mean and standard deviation of the life estimated for these tires are 84,000 and ±7,230 km, respectively, from a sample of nine tires. On the basis of the sample, how much data are required to establish the life of this type of tire to within ±10% with 90% confidence and a resolution of 5 km?

Solution: Confidence limits are as follows:

$$\mu = \hat{\mu} \pm \left(t\hat{\sigma}_{\bar{x}} + R + \frac{5}{10^m} \right)$$

$$\mu - \hat{\mu} = (0.10)\bar{x} = t\hat{\sigma}_{\bar{x}} + R + \frac{5}{10^m} = t\frac{\hat{\sigma}_x}{\sqrt{n}} + R + \frac{5}{10^m}$$

$$\frac{t}{\sqrt{n}} = \frac{(0.10)\bar{x} - R - (5/10^m)}{\hat{\sigma}_x} = \frac{(0.10)(84000) - 5}{7230} = 1.6$$

n	ν	$t(\nu,0.10)^a$	$t/\sqrt{n}$
2	1	6.31	3.65
3	2	2.92	1.46
5	4	2.13	0.87

[a]From a t-statistic table.[9]

Thus a sample of three tires is sufficient to establish the tire life within ±10% at a 90% level of confidence.

3.7 Goodness of Fit

Statistical methods can be used to fit a curve to a given data set. In general, the least-squares principle is used to minimize the sum of the squares of deviations away from the curve to be fitted. The deviations from an assumed curve $y = f(x)$ are due to errors in y, in x, or in both y and x. In most cases the errors in the independent variable x are much smaller than the dependent variable y. Therefore, only the errors in y are considered for the least-squares curve. The goodness of fit of an assumed curve is defined by the correlation coefficient r, where

$$r = \pm\sqrt{\frac{\sum(y - \bar{y})^2}{\sum(y_i - \bar{y})^2}} = \pm\sqrt{1 - \frac{\sum(y_i - y)^2}{\sum(y_i - \bar{y})^2}}$$

$$= \pm\sqrt{1 - \frac{\hat{\sigma}_{y,x}^2}{\hat{\sigma}_y^2}} \tag{31}$$

where $\sum(y_i - \bar{y})^2$ = total variation (variation about mean)
$\sum(y_i - y)^2$ = unexplained variation (variation about regression)
$\sum(y - \bar{y})^2$ = explained variation (variation based on assumed regression equation)
$\hat{\sigma}_y$ = estimated population standard deviation of y variable
$\hat{\sigma}_{y,x}$ = standard error of estimate of y on x

When the correlation coefficient r is zero, the data cannot be explained by the assumed curve. However, when r is close to ±1, the variation of y with respect to x can be explained by the assumed curve and a good correlation is indicated between the variables x and y.

The probabilistic statement for the goodness-of-fit test is given by

$$P[r_{\text{calc}} > r] = \alpha = 1 - \gamma \tag{32}$$

where r_{calc} is calculated from Eq. (31) and the null and alternate hypotheses are as follows:

H_0: No correlation of assumed regression equation with data.

H_1: Correlation of regression equation with data.

The goodness-of-fit for a straight line is determined by comparing r_{calc} with the value of r obtained at $n - 2$ degrees of freedom at a selected confidence level γ from tables. If $r_{\text{calc}} > r$, the null hypothesis is rejected and a significant fit of the data within the confidence level specified is inferred. However, if $r_{\text{calc}} < r$, the null hypothesis cannot be rejected and no correlation of the curve fit with the data is inferred at the chosen confidence level.

Example 5 Goodness-of-Fit Test. The given x–y data were fitted to a curve $y = a + bx$ by the method of least-squares linear regression. Determine the goodness of fit at a 5% significance level (95% confidence level):

$$x = [56, 58, 60, 70, 72, 75, 77, 77, 82, 87, 92, 104, 125]$$

$$y = [51, 60, 60, 52, 70, 65, 49, 60, 63, 61, 64, 84, 75]$$

At $\alpha = 0.05$, the value of r is 0.55. Thus $P[r_{\text{calc}} > 0.55] = 0.05$. The least-squares regression equation is calculated to be $y = 39.32 + 0.30x$, and the correlation coefficient is calculated as $r_{\text{calc}} = 0.61$. Therefore, a satisfactory fit of the regression line to that data is inferred at the 5% significance level (95% confidence level).

3.8 Probability Density Functions

Consider the measurement of a quantity x. Let

x_i = i measurement of quantity
$\bar{x}$ = most representative value of measured quantity
d_i = deviation of ith measurement from $\bar{x}$, $= x_i - \bar{x}$
n = total number of measurements
Δx = smallest measurable change of x, also known as "least count"
m_j = number of measurements in x_j size group

A *histogram* and the corresponding *frequency polygon* obtained from the data are shown in Fig. 2, where the x_j size group is taken to be inclusive at the lower limit and exclusive at the upper limit. Thus, each x_j size group corresponds to the following range of values:

$$x_j - \left(\frac{1}{2}\Delta x_j\right) \leq x < x_j + \left(\frac{1}{2}\Delta x_j\right) \tag{33}$$

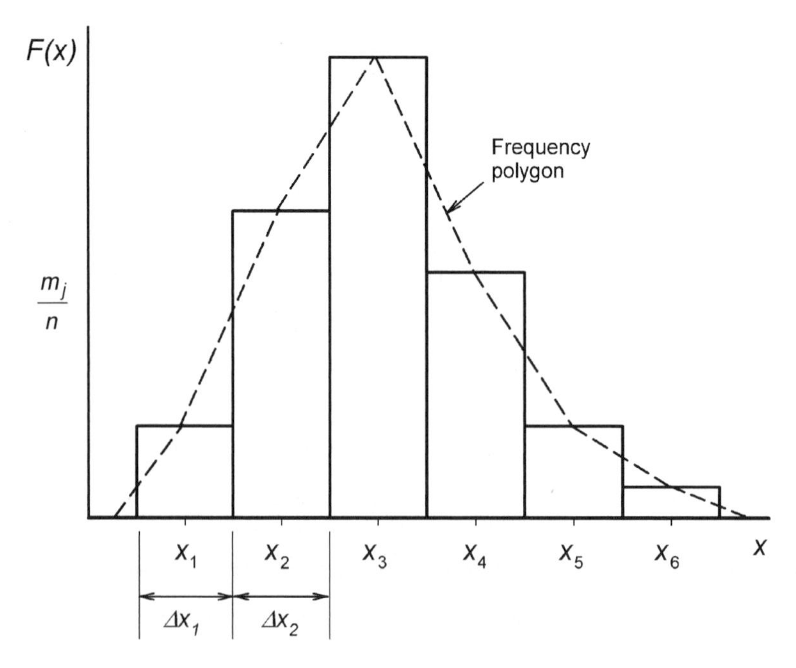

Figure 2 Histogram of a data set and the corresponding frequency polygon.

The height of any rectangle of the histogram shown is denoted as the relative number m_j/n and is equal to the statistical probability $F_j(x)$ that a measured value will have the size $x_j \pm (\Delta x_j/2)$. The area of each rectangle can be made equal to the relative number by transforming the ordinate of the histogram in the following way: Area = relative number = $F_j(x) = m_j/n = p_j(x)\ \Delta x$. Thus

$$p_j(x) = \frac{m_j}{n\ \Delta x} \tag{34}$$

The shape of the histogram is preserved in this transformation since the ordinate is merely changed by a constant scale factor ($1/\Delta x$). The resulting diagram is called the *probability density diagram.* The sum of the areas underneath all rectangles is then equal to 1 [i.e., $\sum p_j(x)\ \Delta x_j = 1$]. In the limit, as the number of data approaches infinity and the least count becomes very small, a smooth curve called the *probability density function* is obtained. For this smooth curve we note that

$$\int_{-\infty}^{+\infty} p(x)\ dx = 1 \tag{35}$$

and that the probability of any measurement, x, having values between x_a and x_b is found from

$$P(x_a \leq x \leq x_b) = \int_{x_a}^{x_b} p(x)\ dx \tag{36}$$

To integrate this expression, the exact probability density function $p(x)$ is required. Based on the assumptions made, several forms of frequency distribution laws have been obtained. The distribution of a proposed set of measurements is usually unknown in advance. However, the Gaussian (or normal) distribution fits observed data distributions in a large number of cases. The Gaussian probability density function is given by the expression

$$p(x) = (1/\sigma\sqrt{2\pi})e^{-(x-\bar{x})^2/2\sigma^2} \tag{37}$$

where σ is the standard deviation. The standard deviation is a measure of dispersion and is defined by the relation

$$\sigma = \frac{\int_{-\infty}^{+\infty} x^2 p(x)\ dx}{\int_{-\infty}^{+\infty} p(x)\ dx} = \sqrt{\frac{\sum(x_i - \mu)^2}{n}} \tag{38}$$

3.9 Determination of Confidence Limits on μ

If a set of measurements is given by a random variable x, then the central limit theorem[13] states that the distribution of means, $\bar{x}$, of the samples of size n is Gaussian (normal) with mean μ and variance $\sigma_{\bar{x}}^2 = \sigma^2/n$, that is, $\bar{x} \sim G(\mu, \sigma^2/n)$. [Also, the random variable $z = (\bar{x} - \mu)/(\sigma/\sqrt{n})$ is Gaussian with a mean of zero and a variance of unity, that is, $z \sim G(0,1)$.] The random variable z is used to determine the confidence limits on μ due to random error of the measurements when σ is known.

The confidence limit is determined from the following probabilistic statement and the Gaussian distribution for a desired confidence level γ:

$$P\left[-z < \frac{\bar{x} - \mu}{\sigma/\sqrt{n}} < z\right] = \gamma \tag{39}$$

It shows a γ probability or "confidence level" that the experimental value of z will be between $\pm z$ obtained from the Gaussian distribution table. For a 95% confidence level, $z = \pm 1.96$ from the Gaussian table[9] and $P[-1.96 < z < 1.96] = \gamma$, where $z = (\bar{x} - \mu)/(\sigma/\sqrt{n})$. Therefore, the expression for the 95% confidence limits on μ is

$$\mu = \bar{x} \pm 1.96 \frac{\sigma}{\sqrt{n}} \tag{40}$$

In general

$$\mu = \bar{x} \pm k \frac{\sigma}{\sqrt{n}} \tag{41}$$

where k is found from the Gaussian table for the specified value of confidence, γ.

If the population variance is not known and must be estimated from a sample, the statistic $(\bar{x} - \mu)/(\sigma/\sqrt{n})$ is not distributed normally but follows the t-distribution. When n is very large, the t-distribution is the same as the Gaussian distribution. When n is finite, the value of k is the "t" value obtained from the t-distribution table.[9] The probabilistic statement then becomes

$$P\left[-t < \frac{\bar{x} - \mu}{\sigma/\sqrt{n}} < +t\right] = \gamma \tag{42}$$

and the inequality yields the expression for the confidence limits on μ:

$$\mu = \bar{x} \pm t \frac{\sigma}{\sqrt{n}} \tag{43}$$

If the effects of resolution and significant digits are included, the expression becomes as previously indicated in Eq. (29):

$$\mu = \bar{x} \pm \left(t \frac{\sigma}{\sqrt{n}} + R + \frac{5}{10^m}\right) \tag{44}$$

3.10 Confidence Limits on Regression Lines

The least-squares method is used to fit a straight line to data that are either linear or transformed to a linear relation.[9,14] The following method assumes that the uncertainty in the variable x is negligible compared to the uncertainty in the variable y and that the uncertainty in the variable y is independent of the magnitude of the variable x. Figure 3 and the definitions that follow are used to obtain confidence levels relative to regression lines fitted to experimental data:

a = intercept of regression line
b = slope of regression line, $= \Sigma XY/\Sigma X^2$
y_i = value of y from data at $x = x_i$
$\hat{y}_i$ = value of y from regression line at $x = \hat{x}_i$; note that $\hat{y}_i = a + b\hat{x}_i$ for a straight line

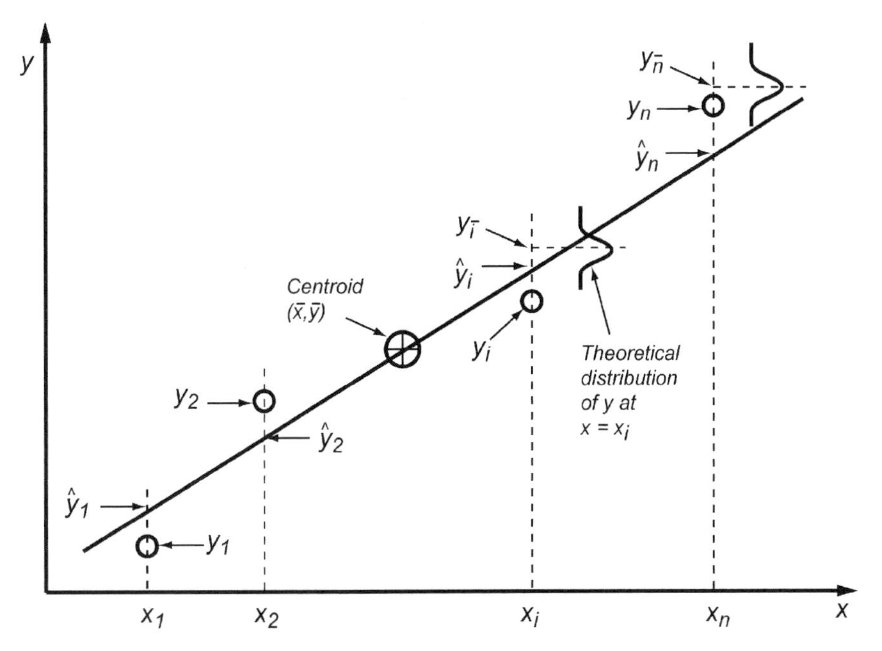

Figure 3 Schematic of a regression line.

$y_{\bar{i}}$ = mean estimated value of y_i at $x = x_i$; mean of distribution of y values at $x = x_i$; if there is only one measurement of y at $x = x_i$, then that value of y (i.e., y_i) is best estimate of $y_{\bar{i}}$

ν = degrees of freedom in fitting regression line to data ($\nu = n - 2$ for straight line)

$\sum(y_i - \hat{y}_i)^2/v = \hat{\sigma}^2_{y,x}$ = unexplained variance (for regression line) where $\hat{\sigma}_{y,x}$ is standard deviation of estimate

$\sigma^2_{\bar{y},x} = \sigma^2_{y,x}/n$ from central limit theorem

$\sigma^2_b = \sigma^2_{y,x}/\sum X^2$ = estimate of variance on slope

Slope-Centroid Approximation

This method assumes that the placement uncertainty of the regression line is due to uncertainties in the centroid ($\bar{x}$, $\bar{y}$) of the data and the slope b of the regression line passing through this centroid. These uncertainties are determined from the following relations and are shown in Fig. 4:

$$\text{Centroid:} \qquad \mu_{\bar{y}} = \hat{\mu}_{\bar{y}} \pm t(\nu,\gamma)\hat{\sigma}_{\bar{y},x} = \bar{y} \pm t\frac{\hat{\sigma}_{y,x}}{\sqrt{n}} \tag{45}$$

$$\text{Slope:} \qquad \mu_b = \hat{b} \pm t(\nu,\gamma)\hat{\sigma}_b = \hat{b} \pm t\frac{\hat{\sigma}_{y,x}}{\sum X^2} \tag{46}$$

Point-by-Point Approximation

This is a better approximation than the slope-centroid technique. It gives confidence limits of the points $y_{\bar{i}}$, where

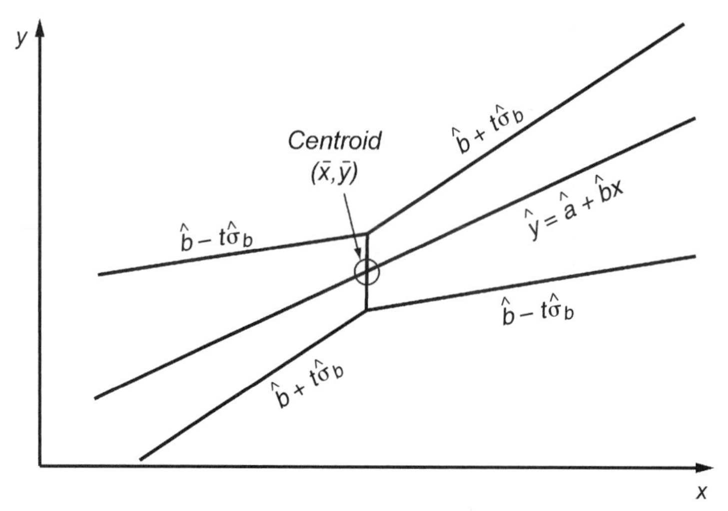

Figure 4 Schematic illustrating confidence limits on a regression line.

$$\mu_{y\bar{i}} = \hat{y}_{\bar{i}} \pm t(\nu,\gamma)\hat{\sigma}_{y\bar{i}} \tag{47}$$

and

$$\hat{\sigma}^2_{y_i} = \hat{\sigma}^2_{y,x}\left(\frac{1}{n} + \frac{X_i^2}{\Sigma X_i^2}\right) \tag{48}$$

At the centroid where $X_i = 0$, Eq. (47) reduces to the result for $\mu_{\bar{y}}$ given in Eq. (45).

Line as a Whole

More uncertainty is involved when all points for the line are taken collectively. The "price" paid for the additional uncertainty is given by replacing $t(\upsilon,\gamma)$ in the confidence interval relation by $\sqrt{2F}$, where $F = F\,(2,\ \nu,\ \gamma)$ obtained from an "F" table statistical distribution. Thus the uncertainty interval for placement of the "line as a whole" confidence limit at $x = x_i$ is found by the formula

$$\mu_{\text{Line}} = \hat{y}_i \pm \sqrt{2F}\hat{\sigma}_{y,x}\sqrt{\frac{1}{n} + \frac{X_i}{\Sigma X_i^2}} \tag{49}$$

Future Estimate of a Single Point at $x = x_i$

This gives the expected confidence limits on a future estimate value of y at $x = x_i$ in a relation to a prior regression estimate. This confidence limit can be found from the relation

$$\mu_y = \hat{y}_i \pm t(\nu,\gamma)\hat{\sigma}_{yi} \tag{50}$$

where $\hat{y}_i$ is the best estimate of y from the regression line and

$$\hat{\sigma}^2_{y_i} = \hat{\sigma}^2_{y,x}\left(1 + \frac{1}{n} + \frac{X_i^2}{\Sigma X_i^2}\right) \tag{51}$$

If one uses $\gamma = 0.99$, nearly all (99%) of the observed data points should be within the uncertainty limits calculated by Eq. (51).

Example 6 Confidence Limits on a Regression Line. Calibration data for a copper–constantan thermocouple are given:

Temperature: $x = [0, 10, 20, 30, 40\ 50, 60, 70, 80, 90, 100]$ °C

Voltage: $y = [-0.89, -0.53, -9.15, -0.20, 0.61, 1.03, 1.45, 1.88, 2.31, 2.78, 3.22]$ mV

If the variation of y is expected to be linear over the range of x and the uncertainty in temperature is much less than the uncertainty in voltage, then:

1. Determine the linear relation between y and x.
2. Test the goodness of fit at the 95% confidence level.
3. Determine the 90% confidence limits of points on the line by the slope-centroid technique.
4. Determine the 90% confidence limits on the intercept of the regression line.
5. Determine the 90% confidence limits on a future estimated point of temperature at 120°C.
6. Determine the 90% confidence limits on the whole line.
7. How much data would be required to determine the centroid of the voltage values within 1% at the 90% confidence level?

The calculations are as follows:

$$\bar{x} = 50.0 \qquad \bar{y} = 1.0827 \qquad y = a + bx$$

$$\sum x_i = 550 \qquad \sum y_i = 11.91 \qquad \sum x_i y_i = 1049.21$$

$$\sum (x_i)^2 = 38{,}500.0 \qquad \sum (y_i)^2 = 31.6433$$

$$\sum (x_i - \bar{x})^2 = \sum X_i^2 = 11{,}000.00 \qquad \sum (y_i - \bar{y})^2 = \sum Y_i^2 = 18.7420$$

$$\sum (x_i - \bar{x})(y_i - \bar{y}) = \sum X_i Y_i = 453.70$$

$$\sum (y_i - \hat{y}_i)^2 = 0.0299$$

1. $b = \sum XY / \sum X^2 = 453.7/11{,}000 = 0.0412$

$a = \bar{y} - b\bar{x} = 1.0827 - (0.0412)(50.00) = 1.0827 - 2.0600$

$= -0.9773$

2. $r_{\text{exp}} = \sqrt{\sum (\hat{y} - \bar{y})^2 / \sum (y_i - \bar{y})^2} = \sqrt{1 - \sum (y_i - \hat{y}_i)^2 / \sum (y_i - \bar{y})^2}$

$= \sum XY / \sqrt{\sum X^2 \sum Y^2} = 0.998$

$r_{\text{Table}} = r(\alpha, \nu) = r(0.05,9) = 0.602$

$P[r_{\text{exp}} > r_{\text{Table}}] = \alpha$ with H_0 of no correlation; therefore reject H_0 and infer significant correlation

3. $t = t(\gamma, \upsilon) = t(0.90,9) = 1.833$ (see Ref. 6)

$\hat{\sigma}^2_{y,x} = \sum (y_i - \hat{y}_i)^2 / \nu = 0.0299/9 = 0.00333$

$$\hat{\sigma}_{y,x} = \pm 0.05777, \ \hat{\sigma}_{\bar{y},x} = \pm 0.0173$$

$$\hat{\sigma}_b = \hat{\sigma}_{y,x}/\sqrt{\Sigma X^2} = \pm 0.000550$$

$$\mu_{\bar{y}} = \hat{\mu}_{\bar{y}} \pm t\hat{\sigma}_{\bar{y},x} = y \pm t\hat{\sigma}_{\bar{y},x}$$

$$= 1.0827 \pm (1.833)(0.01737) = 1.0827 \pm 0.0318$$

$$\mu_b = \hat{b} \pm t\hat{\sigma}_b = 0.0412 \pm (1.833)(0.00055)$$

$$= 0.0412 \pm 0.00101$$

4. $\mu_{y\bar{i}} = \hat{\mu}_{y\bar{i}} \pm t\hat{\sigma}_{y\bar{i}} = y_{\bar{i}} \pm t\hat{\sigma}_{y,x}\sqrt{1/n + X^2/\Sigma X^2}$

$$= -0.9773 \pm (1.833)(0.0325)$$

$$= -0.9773 \pm 0.0596$$

5. $\mu_{yi} = \hat{\mu}_{yi} \pm t\hat{\sigma}_{yi} = \hat{y}_i + t\hat{\sigma}_{y,x}\sqrt{1 + 1/n + X^2/\Sigma X^2}$

$$= (-0.9773 + 4.9500) \pm (1.833)(0.0715)$$

$$= 3.9727 \pm 0.1310$$

6. $\mu_{y\bar{i}\,\text{Line}} = \hat{\mu}_{y\bar{i}} \pm \sqrt{2F(2, v)}\hat{\sigma}_{y\bar{i}}$; for any given point compare with 4 above

$$= \hat{y}_{\bar{i}} \pm \sqrt{(2)(3.01)}\hat{\sigma}_{y\bar{i}}$$

$$= \hat{y}_{\bar{i}} \pm (2.46)\hat{\sigma}_{y\bar{i}}$$

$= -0.9773 \pm 0.0799$; for point of 4 above

7. $\mu_{\bar{y}i} = \hat{\mu}_y \pm t\hat{\sigma}_{\bar{y},x} = \bar{y} \pm t\dfrac{\hat{\sigma}_{y,x}}{\sqrt{n}}$

$$\sigma_{\bar{y},x} = \sigma_{\bar{y},x} = \sigma_{y,x}/\sqrt{n}$$

$$\hat{\sigma}_{y,x} = \pm 0.0577$$

$$\mu_{\bar{y}} - \bar{y} = (1\%)(\mu_y) \cong (1\%)(\bar{y}) = 0.010827$$

$$t/\sqrt{n} = (1\%)\bar{y}/\hat{\sigma}_{y,x} = 0.010827/0.05770 = 0.188$$

From the t table[9] at $t/\sqrt{n} = 0.188$ with $\nu = n - 2$ for a straight line (two constraints).

ν	$t(0.10,\nu)$	$t/\sqrt{n}$
60	1.671	0.213
90	1.663	0.174
75	1.674	0.188

Therefore $n = \nu + 2 = 77$ is the amount of data to obtain to assure the precision and confidence level desired.

3.11 Inference and Comparison

Events are not only deterministic but probabilistic. Under certain specified conditions some events will always happen. Some other events, however, may or may not happen. Under the same specified conditions, the latter ones depend on chance and, therefore, the probability of occurrence of such events is of concern. For example, it is quite certain that a tossed unbiased die will fall down. However, it is not at all certain which face will appear on top when the die comes to rest. The probabilistic nature of some events is apparent when questions of the following type are asked. Does medicine A cure a disease better than medicine B? What is the ultimate strength of 1020 steel? What total mileage will brand X tire yield? Which heat treatment process is better for a given part?

Answering such questions involves designing experiments, performing measurements, analyzing the data, and interpreting the results. In this endeavor two common phenomena are observed: (1) repeated measurements of the same attribute differ due to measurement error and resolving capability of the measurement system and (2) corresponding attributes of identical entities differ due to material differences, manufacturing tolerances, tool wear, and so on.

Conclusions based on experiments are statistical inferences and can only be made with some element of doubt. Experiments are performed to make statistical inferences with minimum doubt. Therefore, experiments are designed specifying the data required, amount of data needed, and the confidence limits desired in drawing conclusions. In this process an instrumentation system is specified, a data-taking procedure is outlined, and a statistical method is used to make conclusions at preselected confidence levels.

In statistical analysis of experimental data, the descriptive and inference tasks are considered. The descriptive task is to present a comprehensible set of observations. The inference task determines the truth of the whole by examination of a sample. The inference task requires sampling, comparison, and a variety of statistical tests to obtain unbiased estimates and confidence limits to make decisions.

Statistical Testing

A statistical hypothesis is an assertion relative to the distribution of a random variable. The test of a statistical hypothesis is a procedure to accept or reject the hypothesis. A hypothesis is stated such that the experiment attempts to nullify the hypothesis. Therefore, the hypothesis under test is called the *null* hypothesis and symbolized by H_0. All alternatives to the null hypothesis are termed the *alternate* hypothesis and are symbolized by H_1.[15]

If the results of the experiment cannot reject H_0, the experiment cannot detect the differences in measurements at the chosen probability level.

Statistical testing determines if a process or item is better than another with some stated degree of confidence. The concept can be used with a certain statistical distribution to determine the confidence limits.

The following procedure is used in statistical testing:

1. Define H_0 and H_1.
2. Choose the confidence level γ of the test.
3. Form an appropriate probabilistic statement.
4. Using the appropriate statistical distribution, perform the indicated calculation.
5. Make a decision concerning the hypothesis and/or determine confidence limits.

Two types of error are possible in statistical testing. A Type I error is that of rejecting a true null hypothesis (rejecting truth). A Type II error is that of accepting a false null hypothesis (embracing fiction). The confidence levels and sample size are chosen to minimize the probability of making a Type I or Type II error. Figure 5 illustrates the Type I (α) and Type II (β) errors, where

H_0 = sample with $n = 1$ comes from $N(10,16)$
H_1 = sample with $n = 1$ comes from $N(17,16)$

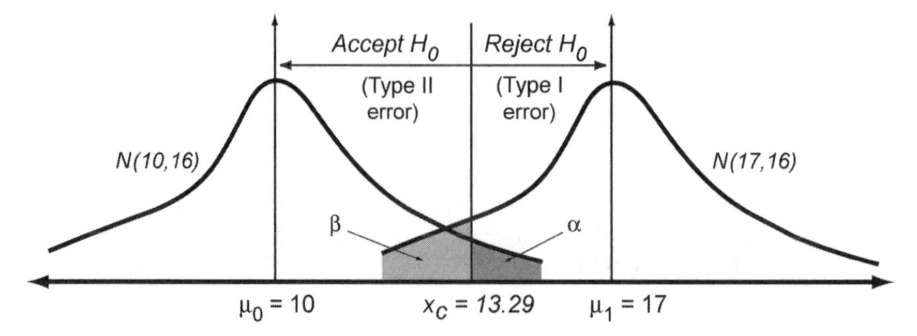

Figure 5 Probability of Type I and Type II errors.

α = probability of Type I error [concluding data came from $N(17,16)$ when data actually came from $N(10,16)$]
β = probability of a Type II error [accepting that data come from $N(10,16)$ when data actually come from $N(17,16)$]

When $\alpha = 0.05$ and $n = 1$, the critical value of x is 16.58 and $\beta = 0.458$. If the value of x obtained from the measurement is less than 16.58, accept H_0. If x is larger than 16.58, reject H_0 and infer H_1. For $\beta = 0.458$ there is a large chance for a Type II error. To minimize this error, increase α or the sample size. For example, when $\alpha = 0.05$ and $n = 4$, the critical value of x is 13.29 and $\beta = 0.032$ as shown in Fig. 5. Thus, the chance of a Type II error is significantly decreased by increasing the sample size. Various statistical tests are summarized in the flow chart shown in Fig. 6.

Comparison of Variability
To test whether two samples are from the same population, their variability or dispersion characteristics are first compared using F-statistics.[9,15]

If x and y are random variables from two different samples, the parameters $U = \sum(x_i - \bar{x})^2/\sigma_x^2$ and $V = \sum(y_i - \bar{y})^2/\sigma_y^2$ are also random variables and have chi-square distributions (see Fig. 7*a*) with $\nu_1 = n_1 - 1$ and $\nu_2 = n_2 - 1$ degrees of freedom, respectively. The random variable W formed by the ratio $(U/\nu_1)/(V/\nu_2)$ has an F-distribution with ν_1 and ν_2 degrees of freedom [i.e., $W \sim F(\gamma, \nu_1, \nu_2)$]. The quotient W is symmetric; therefore, $1/W$ also has an F-distribution with ν_2 and ν_1 degrees of freedom [i.e., $1/W \sim F(\alpha, \nu_2, \nu_1)$]. Figure 7*b* shows the F-distribution and its probabilistic statement:

$$P[F_L < W < F_R] = \gamma \tag{52}$$

where $W = (U/\nu_1)/(V/\nu_2) = \hat{\sigma}_1^2/\hat{\sigma}_2^2$ with

$$H_0 \text{ as } \sigma_1^2 = \sigma_2^2 \qquad \text{and} \qquad H_1 \text{ as } \sigma_1^2 \neq \sigma_2^2 \tag{53}$$

Here, W is calculated from values of $\hat{\sigma}_1^2$ and $\hat{\sigma}_2^2$ obtained from the samples.

Example 7 Testing for Homogeneous Variances. Test the hypothesis at the 90% level of confidence that $\sigma_1^2 = \sigma_2^2$ when the samples of $n_1 = 16$ and $n_2 = 12$ yielded $\hat{\sigma}_1^2 = 5.0$ and $\hat{\sigma}_2^2 = 2.5$. Here H_0 is $\sigma_1 = \sigma_2$ and H_1 is $\sigma_1 \neq \sigma_2$

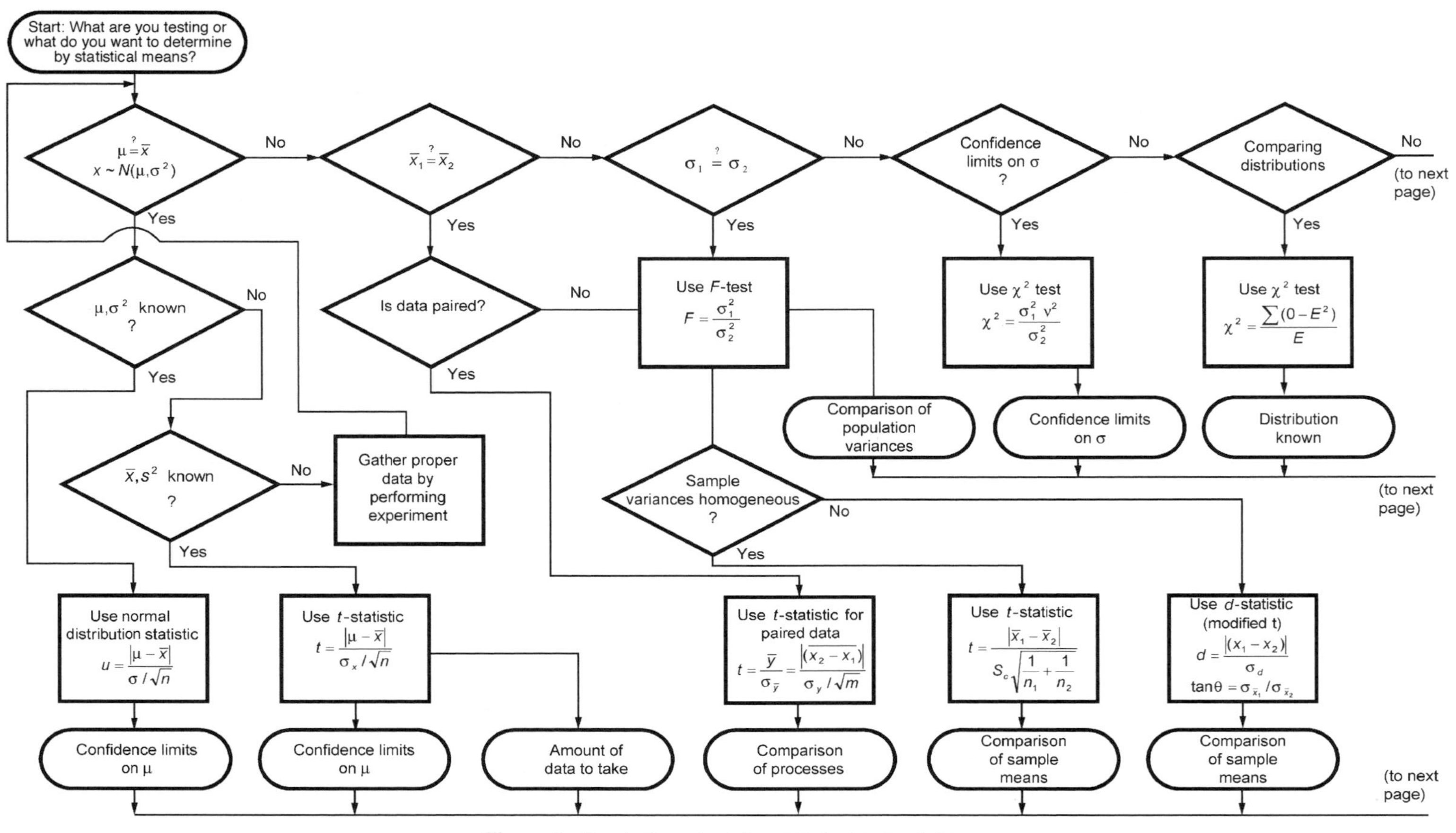

Figure 6 User's flow chart for statistical tests of data.

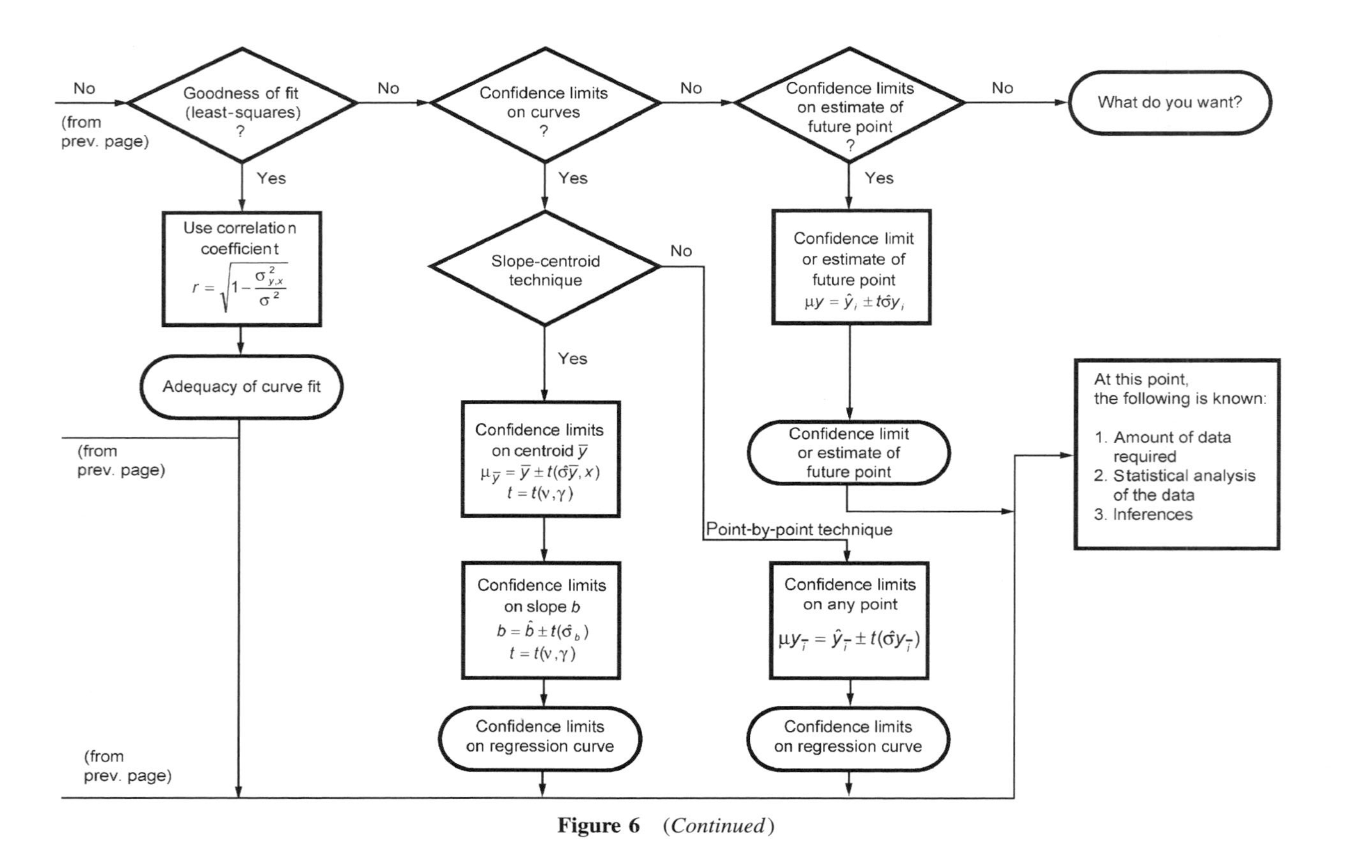

Figure 6 (*Continued*)

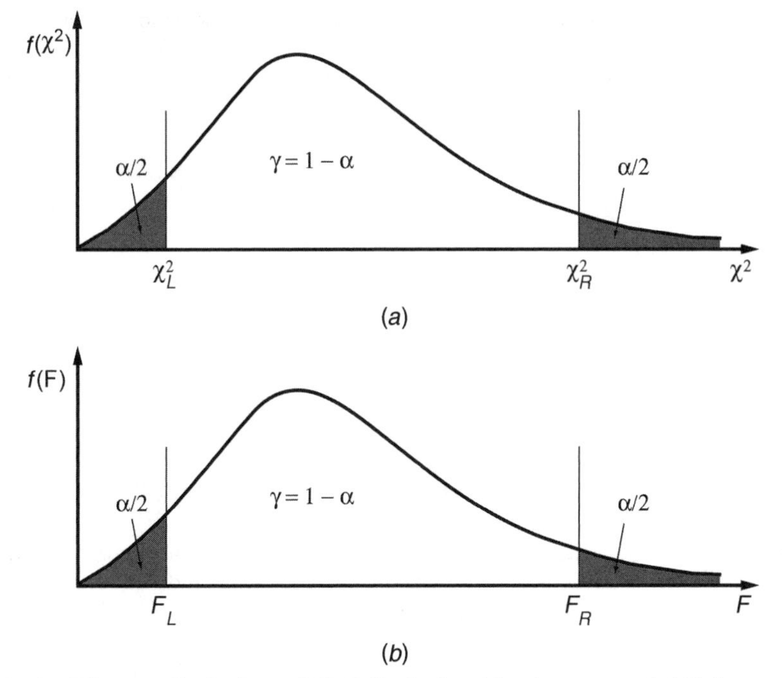

Figure 7 (*a*) Chi-square distribution and (*b*) *F*-distribution. Also shown are probabilistic statements of the distributions such as that shown in Eq. (52).

$$P[F_L(\nu_1,\nu_2) < \left(\frac{\hat{\sigma}_1}{\hat{\sigma}_2}\right)^2 < F_R(\nu_1,\nu_2)] = \gamma$$

$$P[F_{0.95}(15,11) < \left(\frac{\hat{\sigma}_1}{\hat{\sigma}_2}\right)2 < F_{0.05}(15,11)] = 0.90$$

$$F_{0.95}(15,11) = \frac{1}{F_{0.05}(11,\ 15)} = \frac{1}{2.51} = 0.398 \qquad P[0.398 < \left(\frac{\hat{\sigma}_1}{\hat{\sigma}_2}\right)^2 < 2.72] = 0.90$$

Since there was a probability of 90% of $\hat{\sigma}_1^2/\hat{\sigma}_2^2$ ranging between 0.398 and 2.72 and with the actual value of $\hat{\sigma}_1^2/\hat{\sigma}_2^2 = 2.0$, H_0 cannot be rejected at the 90% confidence level.

Comparison of Means

Industrial experimentation often compares two treatments of a set of parts to determine if a part characteristic such as strength, hardness, or lifetime has been improved. If it is assumed that the treatment does not change the variability of items tested (H_0), then the *t*-distribution determines if the treatment had a significant effect on the part characteristics (H_1). The *t*-statistic is

$$t = \frac{d - \mu_d}{\sigma_d} \tag{54}$$

where $d = \bar{x}_1 - \bar{x}_2$ and $\mu_d = \mu_1 - \mu_2$. From the propagation of variance, σ_d^2 becomes

$$\sigma_d^2 = \sigma_{\bar{x}_1}^2 + \sigma_{\bar{x}_2}^2 = \frac{\sigma_1^2}{n_1} + \frac{\sigma_2^2}{n_2} \tag{55}$$

where σ_1^2 and σ_2^2 are each estimates of the population variance σ^2. A better estimate of σ^2 is the combined variance σ_c^2, and it replaces both σ_1^2 and σ_2^2 in Eq. (55). The combined variance is determined by weighting the individual estimates of variance based on their degrees of freedom according to the relation

$$\sigma_c^2 = \frac{\hat{\sigma}_1^2 v_1 + \hat{\sigma}_2^2 v_2}{v_1 + v_2} \tag{56}$$

Then

$$\sigma_d^2 = \frac{\hat{\sigma}_c^2}{n_1} + \frac{\hat{\sigma}_c^2}{n_2} = \sigma_c^2\left(\frac{1}{n_1} + \frac{1}{n_2}\right) \tag{57}$$

Under the hypothesis H_0 that $\mu_1 = \mu_2$ (no effect due to treatment), the resulting probabilistic statement is

$$P\left[-t\sigma_c\sqrt{\frac{1}{n_1} + \frac{1}{n_2}} < \bar{x}_1 - \bar{x}_2 < t\sigma_c\sqrt{\frac{1}{n_1} + \frac{1}{n_2}}\right] = \gamma \tag{58}$$

If the variances of the items being compared are not equal (homogeneous), a modified t- (or d-) statistic is used,[9,15] where d depends on confidence level γ, degrees of freedom v, and a parameter θ that depends on the ratio of standard deviations according to

$$\tan\theta = \frac{\hat{\sigma}_1/\sqrt{n_1}}{\hat{\sigma}_2/\sqrt{n_2}} \tag{59}$$

The procedure for using the d-statistic is the same as described for the t-statistic.

Example 8 Testing for Homogeneous Means. A part manufacturer has the following data:

Sample Number	Number of Parts	Mean Lifetime (h)	Variance (h)
1	15	2530	490
2	11	2850	360

Determine if the lifetime of the parts is due to chance at a 10% significance level (90% confidence level).

Check variance first (H_0—homogeneous variances and H_1—nonhomogeneous variances):

$$\hat{\sigma}_1^2 = 490\left(\frac{15}{14}\right) = 525$$

$$\hat{\sigma}_2^2 = 360\left(\frac{11}{10}\right) = 396$$

$$\sigma_c^2 = \frac{(14)(525) + (10)(396)}{24} = 471$$

$$P[F_L < F_{\exp} < F_R] = \gamma \qquad F_{\exp} = \frac{\hat{\sigma}_1^2}{\hat{\sigma}_2^2} = \frac{525}{396} = 1.33$$

$$F_R = F_{0.05}(\nu_1, \nu_2) = F_{0.05}(14{,}10) = 2.8$$

$$F_L = F_{0.95}(\nu_1, \nu_2) = \frac{1}{F_{0.05}(10{,}14)} = \frac{1}{2.60} = 0.385$$

Therefore, accept H_0 (variances are homogeneous).

Check means next (H_0 is $\mu_1 = \mu_2$ and H_1 is $\mu_2 \neq \mu_1$):

$$P[-t < t_{\text{calc}} < t] = \gamma$$

$$t = t(\nu_1 + \nu_2, \gamma) = t(24{,}0.90) = 1.711$$

Therefore, $P[-1.71 < t_{\exp} < 1.71] = 0.90$:

$$t_{\exp} = \frac{|\bar{x}_1 - \bar{x}_2|}{\sqrt{\sigma_1^2/n_1 + \sigma_2^2/n_2}} = \frac{|\bar{x}_1 - \bar{x}_2|}{\sqrt{\sigma_c^2\,(1/n_1 + 1/n_2)}}$$

$$= \frac{|2530 - 2850|}{\sqrt{(471)(0.1576)}}$$

$$= \frac{|-320|}{\sqrt{73.6}} = \frac{320}{8.58}$$

$$= 32.3$$

Therefore, reject H_0 and accept H_1 of $\mu_1 \neq \mu_2$ and infer differences in samples are due not to chance alone but to some real cause.

Comparing Distributions

A chi-square distribution is also used for testing whether or not an observed phenomenon fits an expected or theoretical behavior.[15] The chi-square statistic is defined as

$$\sum \left[\frac{(O_j - E_j)^2}{E_j} \right]$$

where O_j is the observed frequency of occurrence in the *j*th class interval and E_j is the expected frequency of occurrence in the *j*th class interval. The expected class frequency is based on an assumed distribution or a hypothesis H_0. This statistical test is used to compare the performance of machines or other items. For example, lifetimes of certain manufactured parts, locations of hole center lines on rectangular plates, and locations of misses from targets in artillery and bombing missions follow chi-square distributions.

The probabilistic statements depend on whether a one-sided test or a two-sided test is being performed.[16] The typical probabilistic statements are

$$P[\chi^2_{\exp} > \chi^2(\alpha, \nu)] = \alpha \tag{60}$$

for the one-sided test and

$$P[\chi^2_L < \chi^2_{\exp} < \chi^2_R] = \gamma \tag{61}$$

for the two-sided test, where

$$\chi^2_{exp} = \frac{\Sigma(O_j - E_j)^2}{E_j} \tag{62}$$

When using Eq. (62) at least five items per class interval are normally used, and a continuity correction must be made if less than three class intervals are used.

Example 9 Use of the Chi-Square Distribution. In a test laboratory the following record shows the number of failures of a certain type of manufactured part.

	Number of Parts	
Number of Failures	Observed	Expected[a]
0	364	325
1	376	396
2	218	243
3	89	97
4	33	30
5	13	7
6	4	1
7	3	0
8	2	0
9	1	0
	$\Sigma = 1103$	

[a]From an assumed statistical model that the failures are random.

At a 95% confidence level, determine whether the failures are attributable to chance alone or to some real cause. For H_0, assume failures are random and not related to a cause so that the expected distribution of failures follows the Poisson distribution. For H_1, assume failures are not random and are cause related. The Poisson probability is $P_r = e^{-\mu}\mu^r/r!$, where $\mu = \sum x_i P_i$:

$$\mu = (0)\left(\frac{364}{1103}\right) + (1)\left(\frac{376}{1103}\right) + (2)\left(\frac{218}{1103}\right) + (3)\left(\frac{89}{1103}\right) + (4)\left(\frac{33}{1103}\right) + (5)\left(\frac{13}{1103}\right)$$

$$+ \frac{24}{1103} + \frac{21}{1103} + \frac{16}{1103} + \frac{9}{1103}$$

$$= \frac{1346}{1103} = 1.22$$

Then $P_0 = 0.295$ and $E_0 = P_0 n = 325$. Similarly $P_1 = 0.360$, $P_2 = 0.220$, $P_3 = 0.088, \ldots$ and, correspondingly, $E_1 = 396$, $E_2 = 243$, $E_3 = 97, \ldots$ are tabulated above for the expected number of parts to have the number of failures listed.

Goodness-of-fit test—H_0 represents random failures and H_1 represents real cause:

$$P[\chi^2_{exp} > \chi^2_{Table}] = \alpha = 0.05$$

$$\chi^2_{Table} = \chi^2(\nu, \alpha) = \chi^2(5{,}0.05) = 11.070$$

$$\chi^2_{exp} = \frac{\Sigma(0 - E)^2}{E}$$

O	E	$O - E$	$(O - E)^2$	$(O - E)^2/E$
364	325	39	1520	4.6800
376	396	−20	400	1.0100
218	243	−25	625	2.5700
89	97	−8	64	0.6598
33	30	3	9	0.3000
23	8	15	225	28.1000

$$\chi^2_{\text{exp}} = 37.320$$

Since there is only a 5% chance of obtaining a calculated chi-square statistic as large as 11.07 under the hypothesis chosen, the null hypothesis is rejected and it is inferred that some real (rather than random) cause exists for the failures with a 95% probability or confidence level.

REFERENCES

1. E. J. Minnar (ed.), *ISA Transducer Compendium,* Instrument Society of America/Plenum, New York, 1963.
2. "Electrical Transducer Nomenclature and Terminology," ANSI Standard MC 6.1-1975 (ISA S37.1), Instrument Society of America, Research Triangle Park, NC, 1975.
3. Anonymous (ed.), *Standards and Practices for Instrumentation,* Instrument Society of America, Research Triangle Park, NC, 1988.
4. E. O. Doebelin, *Measurement Systems: Application and Design,* 4th ed., McGraw-Hill, New York, 1990.
5. J. G. Webster, *The Measurement, Instrumentation, and Sensors Handbook,* CRC Press in cooperation with IEEE, Boca Raton, FL, 1999.
6. R. S. Figliola and D. E. Beasley, *Theory and Design for Mechanical Measurements,* 3rd ed., Wiley, New York, 2000.
7. C. L. Nachtigal (ed.), *Instrumentation and Control: Fundamentals and Applications,* Wiley, New York, 1990.
8. I. B. Gertsbakh, *Measurement Theory for Engineers,* Springer, New York, 2003.
9. J. B. Kennedy and A. M. Neville, *Basic Statistical Methods for Engineers and Scientists,* 3rd ed., Harper & Row, New York, 1986.
10. A. G. Worthing and J. Geffner, *Treatment of Experimental Data,* Wiley, New York, 1943.
11. C. R. Mischke, *Mathematical Model Building (an Introduction to Engineering)*, 2nd ed., Iowa State University Press, Ames, IA, 1980.
12. C. Lipson and N. J. Sheth, *Statistical Design and Analysis of Engineering Experiments,* McGraw-Hill, New York, 1973.
13. B. Ostle and R. W. Mensing, *Statistics in Research,* 3rd ed., Iowa State University Press, Ames, IA, 1975.
14. D. C. Montgomery and G. C. Runger, *Applied Statistics and Probability for Engineers,* Wiley, New York, 2002.
15. S. B. Vardeman, *Statistics for Engineering Problem Solving,* PWS Publishing, Boston, 1994.
16. R. E. Walpole, S. L. Myers, R. H. Myers, and K. Ye, *Probability and Statistics for Engineers and Scientists,* 7th ed., Prentice-Hall, Englewood Cliffs, NJ, 2002.

CHAPTER 2
INPUT AND OUTPUT CHARACTERISTICS

Adam C. Bell
Dartmouth, Nova Scotia

1 INTRODUCTION

Everyone is familiar with the interaction of devices connected to form a system, although they may not think of their observations in those terms. Familiar examples include the following:

Reprinted from *Instrumentation and Control,* Wiley, New York, 1990, by permission of the publisher.

1. Dimming of the headlights while starting a car
2. Slowdown of an electric mixer lowered into heavy batter
3. Freezing a showerer by starting the dishwasher
4. Speedup of a vacuum cleaner when the hose plugs
5. Two-minute wait for a fever thermometer to rise
6. Special connectors required for TV antennas
7. Speedup of a fan in the window with the wind against it
8. Shifting of an automatic transmission on a hill

These effects happen because one part of a system loads another. Most mechanical engineers would guess that weighing an automobile by placing a bathroom-type scale under its wheels one at a time and summing the four measurements will yield a higher result than would be obtained if the scale was flush with the floor. Most electrical engineers understand that loading a potentiometer's wiper with too low a resistance makes its dial nonlinear for voltage division. Instrumentation engineers know that a heavy accelerometer mounted on a thin panel will not measure the true natural frequencies of the panel. Audiophiles are aware that loudspeaker impedances must be matched to amplifier impedance. We have all seen the 75- and 300-Ω markings under the antenna connections on TV sets, and most cable subscribers have seen balun transformers for connecting a coaxial cable to the flat-lead terminals of an older TV.

Every one of these examples involves a desired or undesirable interaction between a source and a receiver of energy. In every case, there are properties of the source part and the load part of the system that determine the efficiency of the interaction. This chapter deals exclusively with interactions between static and dynamic subsystems intended to function together in a task and with how best to configure and characterize those subsystems.

Consider the analysis of dynamic systems. To create mathematical models of these systems requires that we idealize our view of the physical world. First, the system must be identified and separated from its environment. The environment of a system is the universe outside the free body, control volume, or isolated circuit. The combination of these, which is the system under study and the external sources, provides or removes energy from the system in a known way. Next, in the system itself, we must arrange a restricted set of ideal elements connected in a way that will correctly represent the energy storages and dissipations of the physical system while, at the same time, we need the mathematical handles that explore the system's behavior in its environment. The external environment of the system being modeled must then itself be modeled and connected and is usually represented by special ideal elements called sources.

We expect, as a result of these sources, that the system under study will not alter the important variables in its environment. The water rushing from a kitchen faucet will not normally alter the atmospheric pressure; our electric circuit will not measurably slow the turbines in the local power plant; the penstock will not draw down the level of the reservoir (in a time frame consistent with a study of penstock dynamics, anyway); the cantilever beam will not distort the wall it is built into; and so on. In this last instance, for example, the wall is a special source of zero displacement and zero rotation no matter what forces and moments are applied.

In this chapter, we consider, instead of the behavior of a single system in a known environment, the interaction between pairs of connected dynamic systems at their interface, often called the driving point. The fundamental currency is, as always, the energy or power exchanged through the interface. In an instrumentation or control system, the objective of the energy exchange might be information transmission, but this is not considered here (we would like information exchanges to take place at the lowest possible energy costs, but the second law of thermodynamics rules out a free transmission).

As always, energy factors into two variables, such as voltage and current in electrical systems, and we are concerned with the behavior of these in the energetic interaction. The major difference in this perspective is that the system supplying energy cannot do so at a fixed value. Neither the source nor the system receiving energy can fix its values for a changing demand without a change in the value of a supply variable. The two subsystems are in an equilibrium with each other and are forced by their connection to have the same value of both of the appropriate energy variables. We concern ourselves with determining and controlling the value of these energy variables at the interface where, obviously, only one is determined by each of the interacting systems.

2 FAMILIAR EXAMPLES* OF INPUT–OUTPUT INTERACTIONS

2.1 Power Exchange

In the real world, pure sources and sinks are difficult to find. They are idealized, convenient constructs or approximations that give our system analyses independent forcing functions. We commonly think of an automobile storage battery as a source of 12.6 V independent of the needed current, and yet we have all observed dimming headlights while starting an engine. Clearly, the voltage of this battery is a function of the current demanded by its load. Similarly, we cannot charge the battery unless our alternator provides more than 12.6 V, and the charging rate depends on the overvoltage supplied. Thus, when the current demanded or supplied to a battery approaches its limits, we must consider that the battery really looks like an ideal 12.6-V source in series with a small resistance. The voltage at the battery terminals is a function of the current demanded and is not independent of the system loading or charging it in the interaction. This small internal resistance is termed the output impedance (or input impedance or driving-point impedance) of the battery.

If we measure the voltage on this battery with a voltmeter, we should draw so little current that the voltage we see is truly the source voltage without any loss in the internal resistance. The power delivered from the battery to the voltmeter is negligible (but not zero) because the current is so small. Alternatively, if we do a short-circuit test of the battery, its terminal voltage should fall to zero while we measure the very large current that results. Again, the power delivered to the ammeter is negligible because, although the current is very large, the voltage is vanishingly small.

At these two extremes the power delivered is essentially zero, so clearly at some intermediate load the power delivered will be a maximum. We will show later that this occurs when the load resistance is equal to the internal resistance of the battery (a point at which batteries are usually not designed to operate). The discussion above illustrates a simple concept: Impedances should be matched to maximize power or energy transfer but should be maximally mismatched for making a measurement without loading the system in which the measurement is to be made. We will return to the details of this statement later.

---

*Many of the examples in this chapter are drawn from Chapter 6 of a manuscript of unpublished notes, "Dynamic Systems and Measurements," by C. L. Nachtigal, used in the School of Mechanical Engineering, Purdue University, 1978.

2.2 Energy Exchange

Interactions between systems are not restricted to resistive behavior, nor is the concept of impedance matching restricted to real, as opposed to reactive, impedances. Consider a pair of billiard balls on a frictionless table (to avoid the complexities of spin), and consider that their impact is governed by a coefficient of restitution, ϵ. Before impact, only one ball is moving, but afterward both may be. The initial and final energies are as follows:

$$\text{Initial energy} = \tfrac{1}{2}M_1 v_{1i}^2$$

$$\text{Final energy} = \tfrac{1}{2}M_1 v_{1f}^2 + \tfrac{1}{2}M_2 v_{2f}^2 \qquad (1)$$

where the subscript 1 refers to the striker and 2 to the struck ball, M is mass, v is velocity, and the subscripts i and f refer to initial and final conditions, respectively.

Since no external forces act on this system of two balls during their interaction, the total momentum of the system is conserved:

$$M_1 v_{1i} + M_2 v_{2i} = M_1 v_{1f} + M_2 v_{2f}$$

or

$$v_{1i} + \mathbf{m} v_{2i} = v_{1f} + \mathbf{m} v_{2f} \qquad (2)$$

where $\mathbf{m} = M_2/M_1$. The second equation, required to solve for the final velocities, derives from impulse and momentum considerations for the balls considered one at a time. Since no external forces act on either ball during their interaction except those exerted by the other ball, the impulses,* or integrals of the force acting over the time of interaction on the two, are equal. (See *impact* in virtually any dynamics text.) From this it can be shown that the initial and final velocities must be related:

$$\epsilon(v_{1i} - v_{2i}) = (v_{1f} - v_{2f}) \qquad (3)$$

where $v_{2i} = 0$ in this case and the coefficient of restitution ϵ is a number between 0 and 1. A 0 corresponds to a plastic impact while a 1 corresponds to a perfectly elastic impact. Equations (2) and (3) can be solved for the final velocities of the two balls:

$$v_{1f} = \frac{1 - \mathbf{m}\epsilon}{1 + \mathbf{m}} v_{1i} \qquad \text{and} \qquad v_{2f} = \frac{1 + \epsilon}{1 + \mathbf{m}} v_{1i} \qquad (4)$$

Now assume that one ball strikes the other squarely† and that the coefficient of restitution ϵ is unity (perfectly elastic impact). Consider three cases:

1. The two balls have equal mass, so $\mathbf{m} = 1$, and $\epsilon = 1$. Then the striking ball, M_1, will stop, and the struck ball, M_2, will move away from the impact with exactly the initial velocity of the striking ball. *All the initial energy is transferred.*
2. The struck ball is more massive than the striking ball, $\mathbf{m} > 1$, $\epsilon = 1$. Then the striker will rebound along its initial path, and the struck ball will move away with less than the initial velocity of the striker. *The initial energy is shared between the balls.*

---

*Impulse $= \int_{t=0}^{t}$ Force dt, where Force is the vector sum of all the forces acting over the period of interaction, t.

†Referred to in dynamics as *direct central impact.*

3. The striker is the more massive of the two, $\mathbf{m} < 1$, $\epsilon = 1$. Then the striker, M_1, will follow at reduced velocity behind the struck ball after their impact, and the struck ball will move away faster than the initial velocity of the striker (because it has less mass). Again, *the initial energy is shared between the balls.*

Thus, the initial energy is conserved in all of these transactions. But the energy can be transferred completely from one ball to the other *if and only if* the two balls have the same mass.

If these balls were made of clay so that the impact was perfectly plastic (no rebound whatsoever), then $\epsilon = 0$, so the striker and struck balls would move off together at the same velocity after impact no matter what the masses of the two balls. They would be effectively stuck together. The final momentum of the pair would equal the initial momentum of the striker because, on a frictionless surface, there are no external forces acting, but energy could not be conserved because of the losses in plastic deformation during the impact. The final velocities for the same three cases are

$$v_f = \frac{1}{1 + \mathbf{m}} v_i \tag{5}$$

Since the task at hand, however, is to transfer kinetic (KE) from the first ball to the second, we are interested in maximizing the energy in the second ball after impact with respect to the energy in the first ball before impact:

$$\frac{\mathrm{KE}_{(M_2,\text{after})}}{\mathrm{KE}_{(M_1,\text{before})}} = \frac{\frac{1}{2}M_2(v_{2f})^2}{\frac{1}{2}M_1(v_{1i})^2} = \frac{M_2(1/(1+\mathbf{m}))^2(v_{1i})^2}{M_1(v_{1i})^2} = \frac{\mathbf{m}}{(1+\mathbf{m})^2} \tag{6}$$

This takes on a maximum value of $\frac{1}{4}$ when $\mathbf{m} = 1$ and falls off rapidly as $\mathbf{m}$ departs from 1.

Thus, after the impact of two clay balls of equal mass, one-fourth of the initial energy remains in the striker, one-fourth is transferred to the struck ball, and one-half of the initial energy of the striker is lost in the impact. If the struck ball is either larger or smaller than the striker, however, then a greater fraction of the initial energy is dissipated in the impact and a smaller fraction is transferred to the second ball. The reader should reflect on how this influences the severity of automobile accidents between vehicles of different sizes.

2.3 A Human Example

Those in good health can try the following experiment. Run up a long flight of stairs one at a time and record the elapsed time. After a rest, try again, but run the stairs two at a time. Still later, try a third time, but run three steps at a time. Most runners will find that their best time is recorded for two steps at a time.

In the first test, the runner's legs are velocity limited: Too much work is expended simply moving legs and feet, and the forces required are too low to use the full power of the legs effectively. In the third test, although the runner's legs do not have to move very quickly, they are on the upper edge of their force capabilities for continued three-step jumps; the forces required are too high and the runner could, at lower forces, move his or her legs much faster. In the intermediate case there is a match between the task and the force–velocity characteristics of the runner's legs.

Bicycle riders assure this match with a variable-speed transmission that they adjust so they can crank the pedals at approximately 60 RPM. We will later look at other means of ensuring the match between source capabilities and load requirements when neither of them is changeable, but the answer is always a transformer or gyrator of some type (a gear ratio in this case).

3 ENERGY, POWER, IMPEDANCE

3.1 Definitions and Analogies

Energy is the fundamental currency in the interactions between elements of a physical system no matter how the elements are defined. In engineering systems, it is convenient to describe these transactions in terms of a complementary pair of variables whose product is the power or flow rate of the energy in the transaction. These product pairs are familiar to all engineers: voltage × current = power, force × displacement = energy, torque × angular velocity = power, pressure × flow = power, and pressure × time rate of change of volume exchanged = power. Some are less familiar: flux linkage × current = energy, charge × voltage = energy, and absolute temperature × entropy flux = thermal power. Henry M. Paynter's[1] *tetrahedron of state* shows how these are related (Fig. 1). Typically, one of these factors is *extensive,* a flux or *flow,* such as current, velocity, volume flow rate, or angular velocity. The other is *intensive,* a potential or *effort,** such as voltage, force, pressure, or torque. Thus $\mathcal{P}$ = *extensive* × *intensive* for any of these domains of physical activity.

This factoring is quite independent of the analogies between the factors of power in different domains, for which any arbitrary selection is acceptable. In essence, velocity is not like voltage or force like current, just as velocity is not like current or force like voltage. It is convenient, however, before defining impedance and working with it to choose an analogy so that generalizations can be made across the domains of engineering activity. There are two standard ways to do this: the *Firestone* analogy[2] and the *mobility* analogy. Electrical engineers are most familiar with the Firestone analogy, while mechanical engineers are probably more comfortable with the mobility analogy. The results derived in this chapter are independent of the analogy chosen. To avoid confusion, both will be introduced, but only the mobility analogy will be used in this chapter.

The Firestone analogy gives circuitlike properties to mechanical systems: All systems consist of nodes like a circuit and only of lumped elements considered to be two-terminal or four-terminal devices. For masses and tanks of liquid, one of the terminals must be understood to be ground, the inertial reference frame, or atmosphere. Then one of the energy

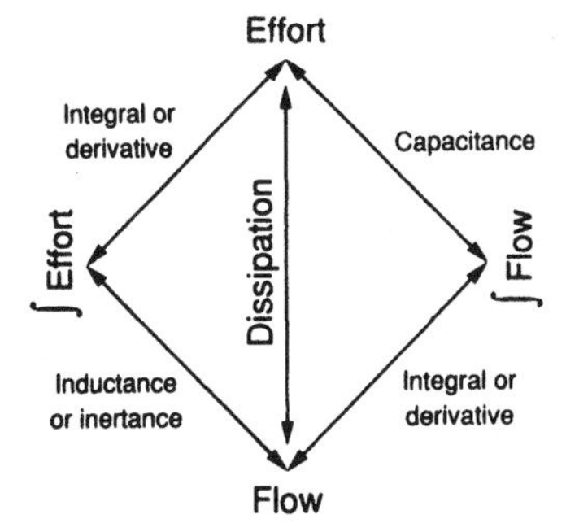

Figure 1 H. M. Paynter's *tetrahedron of state.*

*This is Paynter's terminology, used with reference to his "Bond Graphs."

variables is measured *across* the terminals of the element and the other passes *through* the element. In a circuit, voltage is across and current passes through. For a spring, however, velocity difference is *across* and the force passes through. Thus this analogy linked voltage to velocity, angular velocity, and pressure as across variables and linked current to force, torque, and flow rate as through variables. Clearly, across $\times$ through = power.

The mobility analogy, in contrast, considers the complementary power variables to consist of a potential and a flux, an intrinsic and extrinsic variable. The potentials, or *efforts,* are voltage, force, torque, and pressure, while the fluxes, or *flows,* are current, velocity, angular velocity, and fluid flow rate.

3.2 Impedance and Admittance

Impedance, in the most general sense, is the relationship between the factors of power. Because only the constitutive relationships for the dissipative elements are expressed directly in terms of the power variables, $\Delta V_R = R \cdot i_R$ for example, while the equations for the energy storage elements are expressed in terms of the derivative of one of the power variables* with respect to the other, $i_C = C(dV_C/dt)$ for example, these are most conveniently expressed in Laplace transform terms. Impedances are really self-transfer functions at a point in a system. Since the concept was probably defined first for electrical systems, that definition is most standardized: Electrical impedance $Z_{\text{electrical}}$ is defined as the rate of change of voltage with current:

$$Z_{\text{electrical}} = \frac{d(\text{voltage})}{d(\text{current})} = \frac{d(\text{effort})}{d(\text{flow})} \tag{7}$$

By analogy, impedance can be similarly defined for the other engineering domains:

$$Z_{\text{translation}} = \frac{d(\text{force})}{d(\text{velocity})} \tag{8}$$

$$Z_{\text{rotation}} = \frac{d(\text{torque})}{d(\text{angular velocity})} \tag{9}$$

$$Z_{\text{fluid}} = \frac{d(\text{pressure})}{d(\text{flow rate})} \tag{10}$$

Table 1 is an impedance table using these definitions of the fundamental lumped linear elements. Note that these are derived from the Laplace transforms of the constitutive equa-

Table 1 Impedances of Lumped Linear Elements

Domain	Kinetic Storage	Dissipation	Potential Storage
Translational	Mass: Ms	Damping: b	Spring: k/s
Rotational	Inertia: Js	Damping: B	Torsion spring: k_f/s
Electrical	Inductance: Ls	Resistance: R	Capacitance: $1/Cs$
Fluid	Inertance: Is	Fluid resistance: R	Fluid capacitance: $1/Cs$

*See Fig. 1 again. Capacitance is a relationship between the integral of the flow and the effort, which is the same as saying that capacitance relates the flow to the derivative of the effort.

tions for these elements; they are the transfer functions of the elements and are expressed in terms of the Laplace operator s. The familiar $F = M \cdot a$, for example, becomes, in power-variable terms, $F = M(dv/dt)$; it transforms as $F(s) = Msv(s)$, so

$$(Z_{\text{translation}})_{\text{mass}} = \frac{dF_{\text{mass}}}{dv_{\text{mass}}} = Ms \tag{11}$$

Because these involve the Laplace operator s, they can be manipulated algebraically to derive combined impedances. The reciprocal of the impedance, the *admittance,* is also useful. Formally, admittance is defined as

$$\text{Admittance: } Y = \frac{1}{Z} = \frac{d(\text{flow})}{d(\text{effort})} \tag{12}$$

3.3 Combining Impedances and/or Admittances

Elements in series are those for which the flow variable is common to both elements and the efforts sum. Elements in parallel are those for which the effort variable is common to both elements and the flows sum. By analogy to electrical resistors, we can deduce that the impedance sum for series elements and the admittance sum for parallel elements form the combined impedance or admittance of the elements:

Impedances *in series:*

$$Z_1 + Z_2 = Z_{\text{total}} \quad \text{(common flow)} \qquad \frac{1}{Y_1} + \frac{1}{Y_2} = \frac{1}{Y_{\text{total}}} \tag{13}$$

Impedances *in parallel:*

$$Y_1 + Y_2 = Y_{\text{total}} \quad \text{(common effort)} \qquad \frac{1}{Z_1} + \frac{1}{Z_2} = \frac{1}{Z_{\text{total}}} \tag{14}$$

When applying these relationships to electrical or fluid elements, there is rarely any confusion about what constitutes series and parallel. In the mobility analogy, however, a pair of springs connected end to end are in parallel *because they experience a common force,* regardless of the topological appearance, whereas springs connected side by side are in series *because they experience a common velocity difference.** For a pair of springs *end to end,* the total admittance is

$$\frac{s}{k_{\text{total}}} = \frac{s}{k_1} + \frac{s}{k_2}$$

so the impedance is

$$\frac{k_{\text{total}}}{s} = \frac{k_1 k_2}{s(k_1 + k_2)}$$

For the same springs *side by side,* the total impedance is

*For many, the appeal of the Firestone analogy is that springs are equivalent to inductors, and there can be no ambiguity about series and parallel connections. End to end is series.

$$\frac{k_{\text{total}}}{s} = \frac{k_1 + k_2}{s}$$

3.4 Computing Impedance or Admittance at an Input or Output

There are basically two ways in which an input or output admittance can be computed. The first, and most direct, is to compute the transfer function between the effort and the flow at the driving point and take the derivative with respect to the flow. For a mechanical rotational system, for example, torque as a function of angular velocity is expressed and differentiated with respect to angular velocity. This method must be used if the system being considered is nonlinear because the derivative must be taken at an operating point. If the system is linear, then the ratio of flow or effort will suffice; in the rotational system, the impedance is simply the ratio τ/ω (torque/speed).

The second method takes the impedances of the elements one at a time and combines them. This approach is particularly useful for linear (or linearized) systems. The question then arises of determining the impedance of any sources in the subsystem being considered. Flow sources, such as current sources, velocity sources, angular velocity sources, and fluid flow sources, all have the relationship flow = constant. Their impedance is therefore infinite ($Z_{\text{flow source}} = \infty$) because any change in effort results in zero change in flow. Effort sources, such as voltage sources, force sources, torque sources, and pressure sources, will provide any flow to maintain the effort required; the change in effort for a change in flow remains zero, so their impedance is $Z_{\text{effort source}} = 0$. An effort source therefore represents a short circuit from an impedance point of view; it connects together two nodes that were separate. A flow source represents a null element; since its impedance is infinite, it represents an open circuit. Flow sources are simply removed in impedance calculations.

An example will distinguish between these two approaches. Figure 2 shows, on the left, a simple circuit disconnected at a driving point from its load. The load is of no consequence in this calculation; we simply require the driving-point impedance of the circuit. In the first approach, we derive the voltage at the driving point as a function of the current leaving those terminals and the *source,* V_s and I_s to obtain the relationship

$$V_o = \frac{R_3}{R_1 + R_3} V_s + \left(\frac{R_1R_2 + R_2R_3 + R_1R_3}{R_1 + R_3}\right)(I_s - i_o) \tag{15}$$

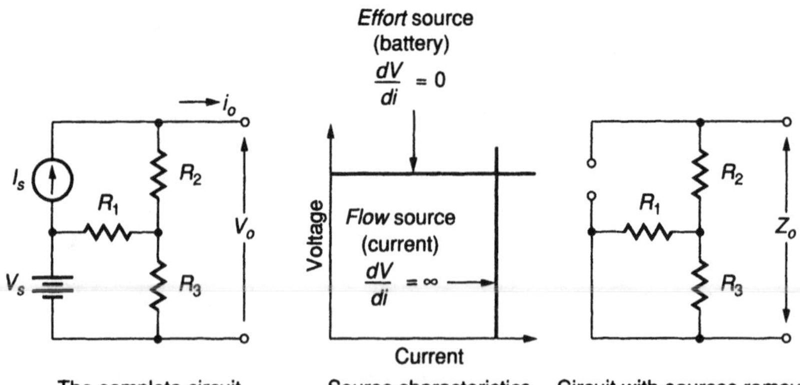

Figure 2 A simple circuit as a source.

Clearly, the negative derivative of the voltage (V_o) with respect to the current (i_o) is given by

$$-\frac{dV_o}{di_o} = Z_o = \frac{R_1R_2 + R_2R_3 + R_1R_3}{R_1 + R_3} \tag{16}$$

In the second method, the voltage source (V_s) can be set to zero, a short circuit, without affecting the impedances, and the current source (I_s) can be removed to yield the circuit on the right in Fig. 2. Then the impedances need only be combined as R_2 in series with the parallel pair R_1 and R_3. Thus,

$$Z_o = R_2 + \frac{R_1R_3}{R_1 + R_3} = \frac{R_1R_2 + R_2R_3 + R_1R_3}{R_1 + R_3} \quad \text{as before} \tag{17}$$

3.5 Transforming or Gyrating Impedances

Ideal transformers, transducers, and gyrators play an important part in dynamic systems and an equally important part in obtaining optimal performance from source–load combinations. They share several vital features: All are two-port (or four-terminal) devices, all are energetically conservative, and all are considered lumped elements. Each of the many types requires two equations for its description, always of the same form. Table 2 lists many of the more common linear two-ports, and Fig. 3 illustrates them.

All of these transducing devices alter the *effort–flow* relationships of elements connected at their far end. Consider Fig. 4, which shows a simple model of a front wheel of an automobile suspension. Because the spring and damper are mounted inboard of the wheel, their effectiveness is reduced by the mechanical disadvantage of the suspension arm. What is the impedance of the spring and shock absorber at $(\cdot)_1$ as viewed from the wheel at $(\cdot)_2$?

Since the spring and shock absorber share a common velocity (both ends share nodes or points of common velocity), their impedances add to give the impedance of the pair at their point of attachment, $(\cdot)_1$, to the lever:

$$(Z_1)_{\text{total}} = Z_{\text{spring}} + Z_{\text{damper}} = b + \frac{k}{s} = \frac{bs + k}{s} \tag{18}$$

At the $(\cdot)_2$ end of the lever, the force, from Table 2, is $F_2 = (L_1/L_2)F_1$, but from the definition of Z for linear elements, $F_1 = (Z_1)_{\text{total}}v_1$. So we get the following:

$$F_2 = \frac{L_1}{L_2}\left(\frac{bs + k}{s}\right)v_1 \tag{19}$$

The second equation for the lever is $v_1 = (L_1/L_2)v_2$, and substituting this into Eq. (19) yields

$$F_2 = \frac{L_1}{L_2}\left(\frac{bs + k}{s}\right)\left(\frac{L_1}{L_2}\right)v_2 = \left(\frac{L_1}{L_2}\right)^2\left(\frac{bs + k}{s}\right)v_2 \tag{20}$$

$$(Z_2)_{\text{total}} = \left(\frac{L_1}{L_2}\right)^2 (Z_1)_{\text{total}} \tag{21}$$

Thus, the impedance of the suspension, observed from the wheel end of the lever arm, is multiplied by the *square* of the lever ratio. This general result applies to all transduction elements.

Table 2 Ideal Linear Lumped Two-Ports

Domain to Domain	Name	Figure Number	Governing Equations (Left to Right in Figures)
Translation–translation	Lever, L = distance to function	3*a*	$F_1 = \frac{L_2}{L_1} F_2$
			$v_1 = \frac{L_1}{L_2} v_2$
Rotation–rotation	Gears, N = number of teeth	3*b*	$\tau_1 = \frac{N_1}{N_2} \tau_2$
			$\omega_1 = \frac{N_2}{N_1} \omega_2$
Rotation–translation	Crank ($\theta \ll 1$) or rack and pinion	3*c* 3*d*	$\begin{cases} \tau = RF \\ \omega = \frac{1}{R} v \end{cases}$
	•	•	•
	Screw (p = length/radius)	3*e*	$\begin{cases} \tau = pF \\ \omega = \frac{1}{p} v \end{cases}$
Electrical–electrical	Transformer	3*f*	$V_1 = \frac{N_1}{N_2} V_2$
			$i_1 = \frac{N_2}{N_1} i_2$
Electrical–rotation	Permanent magnet dc motor	3*g*	$V_a = K_b\omega$
			$i_a = \frac{1}{K_m} \tau$
Electrical–translation	Voice coil	3*h*	$V_a = K_c v$
			$i_a = \frac{1}{K_c} F$
Fluid–translation	Piston area = A	3*j*	$P = \frac{1}{A} F$
			$Q = Av$
Rotation–fluid	Fixed-displacement pump-motor, displacement/radius = D	3*k*	$\tau = DP$
			$\omega = \frac{1}{D} Q$

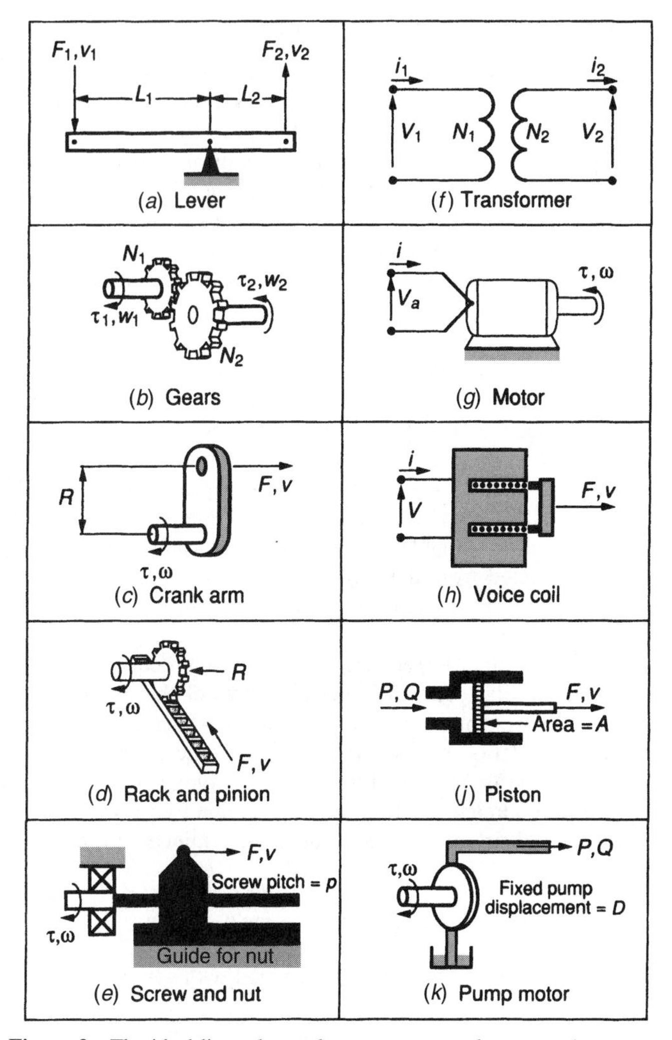

Figure 3 The ideal linear lumped two-ports: *transformers* and *gyrators.*

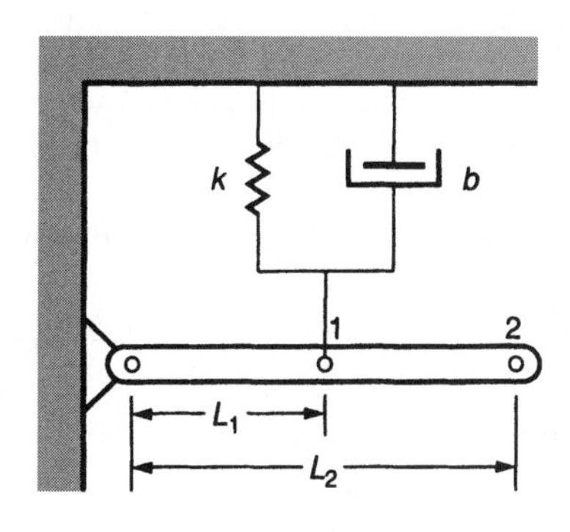

Figure 4 An abstraction of an automobile suspension linkage.

3.6 Source Equivalents: Thévenin and Norton

Thévenin's and Norton's theorems were both originally developed for electric circuits. They are duals; that is, while Thévenin uses a series voltage source, Norton uses a parallel current source to construct an equivalent to a real subsystem being considered as a source. Both of these theorems are completely generalizable to any system, provided that the appropriate sources are used. In this chapter, the mobility analogy has been adapted so that Thévenin equivalents are constructed using sources of voltage, force, pressure, or torque, and Norton equivalents are constructed using sources of current, velocity, flow, or angular velocity. In the development below, these two classes of sources will be referred to as effort and flow sources.

Thévenin Equivalent

Assume that a subsystem being considered as a source contains within its structure one or more ideal sources. Insofar as can be measured externally at the driving point (the point of connection between any two systems considered to be source and load), any active subsystem, no matter what its load, can be replaced by a new effort source added in series with the original subsystem; all the original internal sources are set to zero. The value of the new source effort variable is the output that would appear at the driving point if the load were disconnected from the original subsystem. Setting an effort source to zero is equivalent to connecting its nodes together; setting a flow source to zero is equivalent to removing it from the system.

The output impedance of this Thévenin equivalent is clearly the impedance looking in at the driving point; it is the derivative of the driving-point effort variable with respect to the driving-point flow variable with the load disconnected. Thévenin's theorem simplifies this calculation because it tells us that the internal sources can be set to zero before the calculation is made, and the system topology is often substantially simplified by this move. We have already seen an example of this [see Eqs. (15)–(17)]. The Thévenin equivalent of this circuit is simply a new source determined with the current $i_o = 0$ (see Fig. 2):

$$V_{\text{Thévenin}} = \frac{R_3}{R_1 + R_3} V_s + \left(\frac{R_1 R_2 + R_2 R_3 + R_1 R_3}{R_1 + R_3} \right) I_s \tag{22}$$

in series with a resistance determined with sources V_s and I_s zeroed:

$$R_{\text{Thévenin}} = \frac{R_1 R_2 + R_2 R_3 + R_1 R_3}{R_1 + R_3} \tag{23}$$

Norton Equivalent

Assume the same subsystem considered above. Insofar as can be measured externally at the driving point, any active subsystem, no matter what its load, can be replaced by a new flow source added in parallel with the original subsystem, all the original sources set to zero. The value of the new source flow variable is the flow that would pass through a short circuit substituted at the driving point for the original load.

Referring again to Eq. (15), with V_o set to zero, we obtain the following:

$$0 = \frac{R_3}{R_1 + R_3} V_s + \left(\frac{R_1 R_2 + R_2 R_3 + R_1 R_3}{R_1 + R_3} \right) (I_s - i_o) \tag{24}$$

we see that i_o for a short circuit would be

$$i_o = I_s + \left(\frac{R_3}{R_1R_2 + R_2R_3 + R_1R_3}\right)V_s \tag{25}$$

and, as for the Thévenin equivalent, the circuit impedance is

$$R_{\text{Norton}} = \frac{R_1R_2 + R_2R_3 + R_1R_3}{R_1 + R_3} \tag{26}$$

Note that $R_{\text{Thévenin}} = R_{\text{Norton}}$, *always.*

4 OPERATING POINT OF STATIC SYSTEMS

A static system is a system without energy storage, a system in which there are only sources, transducers, and dissipation elements. Such systems have no transient response: They respond instantly to their inputs algebraically. The relationships among any of their variables are proportionalities—simple static gains. If the inputs to a stable dynamic system are held constant for long enough, it will become stationary; its variables will not change with time provided only that there is sufficient dissipation in the system to damp out any oscillations. Such a system is not static; it is at steady state. There is no exchange of energy among its energy storage elements, and the dissipative elements completely determine the state of the system.

4.1 Exchange of Real Power

If one system is supplying real power to another system in steady-state operation, then for the purposes of a static analysis, energy storage elements can be ignored. Both the source and the load can be considered to be purely resistive. If the source is separated from the load at the point of interest (at least conceptually), then the characteristics of source and load can be measured or computed. The load will be a power absorber—an electrical fluid resistance or a mechanical damper—and its characteristics can be represented as a line in a power plane coordinate system: voltage versus current, force versus velocity, torque versus angular velocity, or pressure versus flow. There is no requirement that this line be straight, and except that the measurement might be more difficult, there is no necessity that this line be single valued or that it start at the origin. Figure 5 shows a selection of common dissipative load characteristics.

Similarly, when the source portion of the system is loaded with a variable dissipation, the line representing its output characteristics can be plotted on the same coordinates as the load. Such characteristics are often given for pumps, servomotors, transistors, hydraulic valves, electrical supplies, and fans. Again, there is no requirement that this line be straight or that it be single valued, but it is very unusual for it to pass through the origin. If the source or the load is not constant in time or can be controlled, then a family of these characteristics will be required for variations in the parameters of the source or load. Figure 6 shows a selection of these sources. With very few exceptions, real sources cannot operate at the maximum values of their power variables simultaneously: A battery cannot deliver its maximum voltage and current at the same time. In more general terms, this means that in spite of local variations, real source characteristics tend to droop from left to right in the power plane; their average slope is negative.

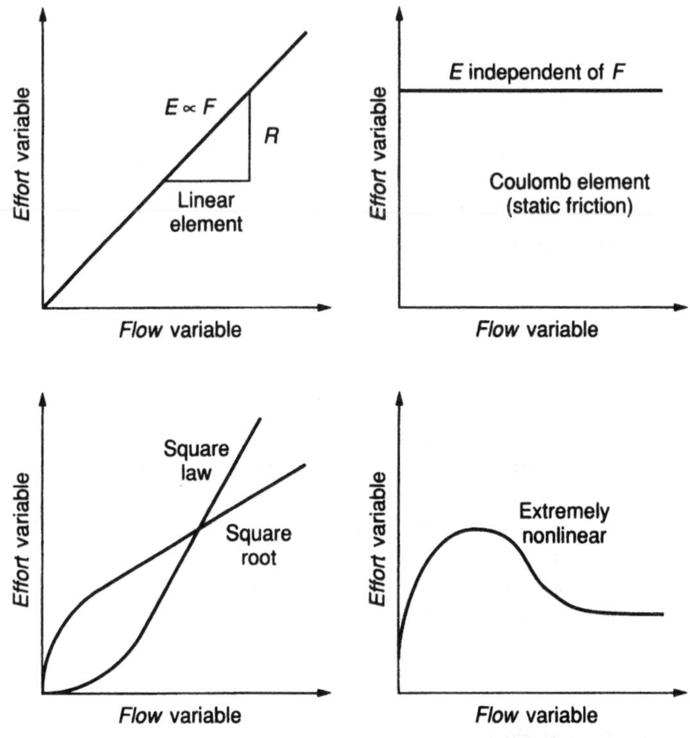

Figure 5 Common dissipative load characteristics.

4.2 Operating Points in an Exchange of Power or Energy

When resistive source and load characteristics are plotted on the same coordinates, they intersect at least once. The coordinate values at that point, or at those points if there are several, are the values of the power variables at which that combination of source and load must operate if they are connected. This is called an operating point. From a computational point of view, the source *causes* one of the power variables given the other and the load causes the other given the source. They must operate at the same point in the power plane to satisfy continuity (common flows) and compatibility (common efforts) conditions.

When there are multiple intersections, all are possible operating points, but not all will be stable operating points; for example, any disturbance from equilibrium might result in a transition to another operating point. The condition for a stable intersection is best seen graphically in Fig. 7. For a stable intersection, as shown on the left, it is required that a small perturbation of the load, which increases its demand for power, be countered by a shortage of power from the supply side of the system and a small perturbation of the load, which decreases its demand for power, be met with an excess from the source. In either case, the load will be driven back to the intersection by the excess or deficit in the source capability. A reversal of these conditions is an unstable operating point because disturbances will be driven further in the direction of their initial departure.

At the unstable intersection in Fig. 7 (2, on the right-hand side), a slight increase in the flow demand of the load will result in an overwhelming increase in the supply flow available to drive the load, which will then cause a traversal to point 1 in the figure. Similarly, if the

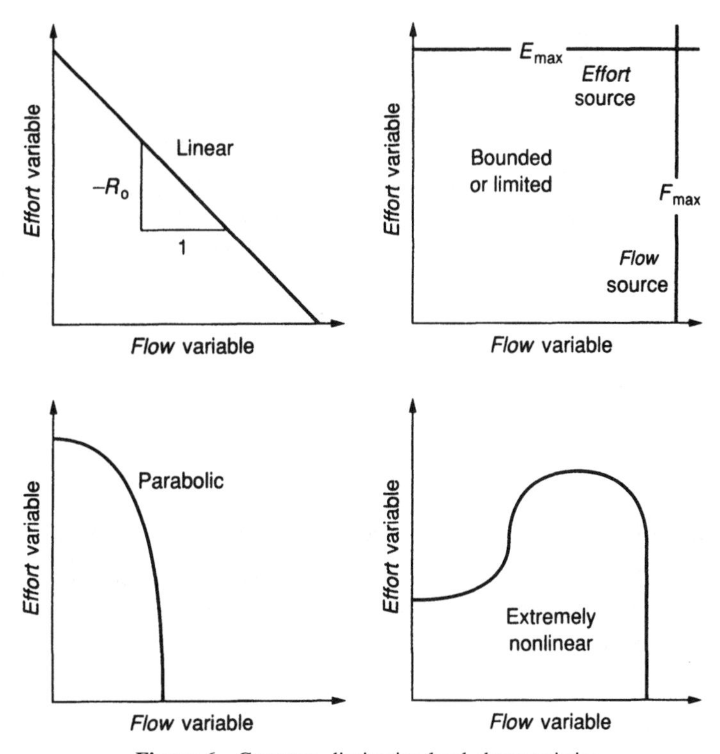

Figure 6 Common dissipative load characteristics.

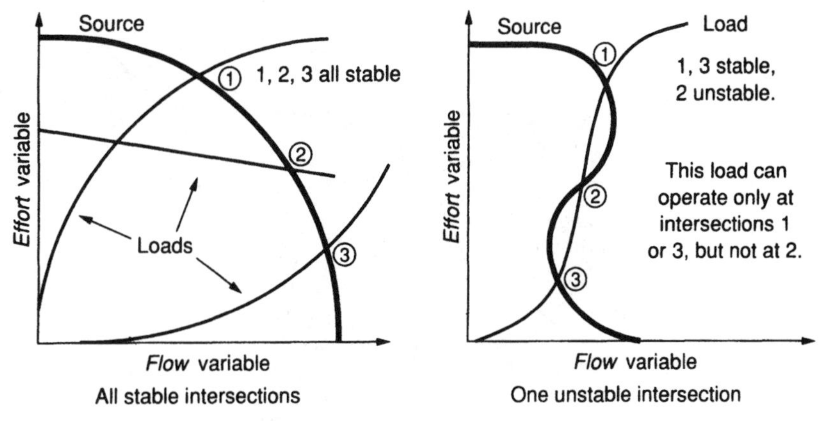

Figure 7 Stable and unstable operating point intersections.

effort decreases slightly from operating point 2, then the source will be starved compared to the demand of the load at that effort level, so the effort will fall until point 3 is reached.

4.3 Input and Output Impedance at the Operating Point

Lines of constant power are hyperbolas in the power plane, with increasing values of the power at increasing distance from the origin. The usual sign conventions imply that sources deliver power in the first and third quadrants while loads absorb power in those quadrants. Conversely, sources absorb power in the second and fourth quadrants and loads return it. The output impedance of a source is defined as minus the slope of the output characteristic. For nonlinear characteristics, the output impedance at any point is defined as minus the slope at that point. For loads, the input impedance is defined as the slope of the load line, but for nonlinear characteristics there are two possibilities of importance: the slope of the line at a point (the incremental or local input impedance) and the slope of the chord to the point from the origin (the chordal impedance). Figure 8 summarizes these features of the power plane.

4.4 Operating Point and Load for Maximum Transfer of Power

Consider a battery with the voltage–current characteristic shown in Fig. 9. The maximum unloaded terminal voltage (open circuit) of the battery is V_{oc} volts and the short-circuit current is i_{sc} amperes. The equation of the line shown is

$$V_t = V_{oc} - R_o i_b$$

or

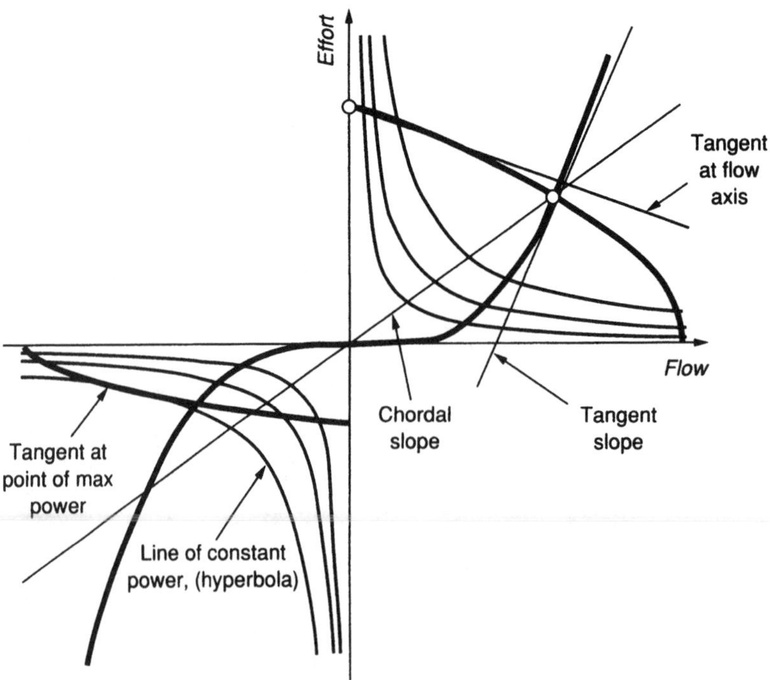

Figure 8 Definitions in the power plane.

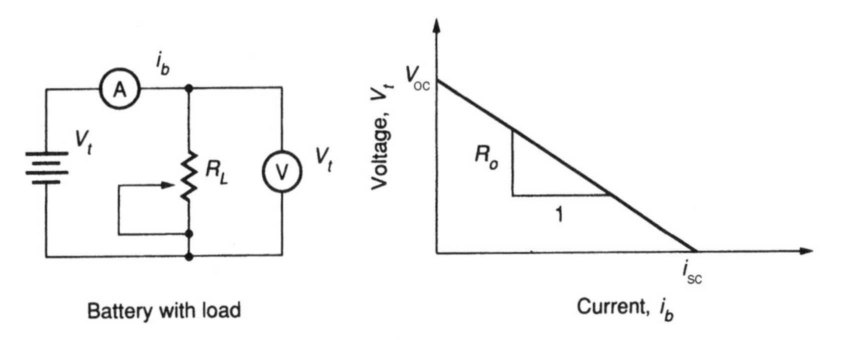

Figure 9 A battery with load.

$$i_b = i_{sc} - \frac{V_t}{R_o} \tag{27}$$

where V_t is the terminal voltage and i_b is the battery current.

The output impedance $Z_o = -dV_t/di_b = R_o = V_{oc}/i_{sc}$ (a pure resistance in this case). If the battery is loaded by a resistor across its terminals (R_L), the terminal voltage must be

$$V_t = R_L i_b \tag{28}$$

Solving Eqs. (27) and (28) simultaneously for V_t and i_b yields the following operating point coordinates:

$$V_t = \frac{V_{oc}}{1 + R_o/R_L} \qquad i_b = \frac{V_{oc}}{R_o + R_L} \tag{29}$$

The output power at the operating point is [from (29)]

$$\mathcal{P} = V_t i_b = \frac{(V_{oc})^2 R_L}{(R_o + R_L)^2} \tag{30}$$

Clearly, the output power depends on the load resistance. If R_L is zero or infinite, no power is drawn from the battery. To measure V_{oc}, we would want a voltmeter with an infinite input impedance, and to measure i_{sc}, we would want an ammeter with zero resistance. In practice, we would use a voltmeter with an input resistance very large compared to R_o and an ammeter with a resistance very small compared to R_o.

If our objective is to deliver power, then a best value of R_L is that for which the derivative $d\mathcal{P}/dR_L = 0$. This value is the point at which $R_L = R_o$. Alternatively, the maximum power output of the battery for any load occurs at the current (i_b) that maximizes $V_t i_b$. Equation (27) can be restated as

$$V_t = V_{oc}\left(1 - \frac{i_b}{i_{sc}}\right) \tag{31}$$

so that

$$\mathcal{P} = V_t i_b = V_{oc}\left(1 - \frac{i_b}{i_{sc}}\right) i_b \tag{32}$$

which is maximized at $\mathcal{P} = V_{oc} i_{sc}/4$ when $i_b = i_{sc}/2$.

For the battery characteristic, the operating point $i_b = i_{sc}/2$ yields $V_t = V_{oc}/2$ [substitution in Eq. (29)]. A loading resistor characteristic must pass through this point to draw maximum power from the battery, so that the load resistance must be $R_L = V_{oc}/i_{sc} = R_o$. At this operating point, the equivalent resistor within the battery (representing the internal losses in the battery) is dissipating exactly as much power as is being delivered to the load.

If we want maximum power delivery, impedance should match the load to the source, but if we want to minimize power delivery from a source, then impedance mismatching is the key. Impedance matching assures that the source and load will divide the power equally; all other impedances will result in less power transfer.

4.5 An Unstable Energy Exchange: Tension-Testing Machine

Although tensile studies of material properties require only a simple test apparatus, it is not simple to interpret the data from such tests. The problem is that the tensile test machine and the specimen can interact[3] in an unstable way. Almost any desired stress–strain curve can be obtained in a given material by a suitable choice of the test machine's elastic compliance compared to the specimen.

A tensile test involves the interaction between two springs, one that represents the specimen and the other combined with a velocity source that represents the testing machine. Figure 10 shows this simple model. While the testing machine is linearly elastic and does not yield, the test specimen is not elastic. It undergoes a large plastic deformation in a typical load–elongation test. Normally in such a test, the specimen is to be elongated at a constant cross-head velocity (v). The test machine, however, is not a velocity source as is commonly supposed; that source is really in series with a spring (K) representing the elastic deformations of the testing machine structure between the source of the motion and the jaws of the machine.

In the course of a test, the specimen undergoes an elongation (δ) comprised of both elastic and plastic displacements. The test machine undergoes only an elastic displacement (ξ) given by F/K, where F is the load applied to the specimen by the machine at a given cross-head displacement (y), which is really the sum of both the specimen (δ) and the machine (ξ) displacements. Figure 11 shows the components of this force–displacement situation.

The cross-head velocity of the test machine is made up of two components as well:

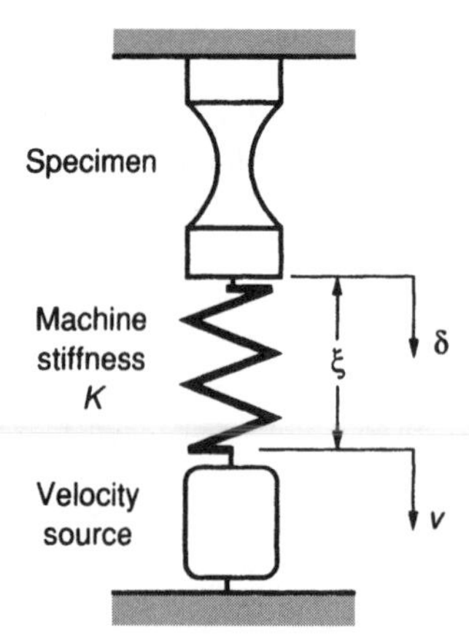

Figure 10 Simple model of a tensile-testing machine.

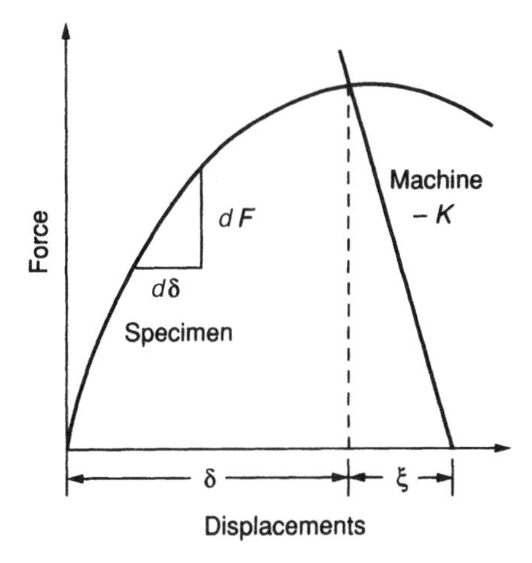

Figure 11 Components of the force–displacement curve.

$$v = \frac{dy}{dt} = \frac{d\delta}{dt} + \frac{d\xi}{dt} \tag{33}$$

The machine displacement (ξ), however, is a function of the force being applied, so it is therefore a function of the elongation of the specimen (δ). Equation (33) must be rewritten:

$$v = \frac{d\delta}{dt} + \frac{d\xi}{d\delta}\frac{d\delta}{dt} = \frac{d\delta}{dt}\left(1 + \frac{d\xi}{d\delta}\right) \tag{34}$$

Since $\xi = F/K$, presuming linearity for the machine, we get the following derivative:

$$\frac{d\xi}{d\delta} = \frac{1}{K}\frac{dF}{d\delta} \tag{35}$$

where the term $dF/d\delta$ is the slope of the force–elongation curve of the specimen. The slope, in other words, is the driving-point stiffness at any point along the test curve. If these last two equations are combined, the specimen's elongation rate is found in terms of the cross-head velocity (v), which is normally constant, the machine stiffness (K), and the driving-point stiffness of the machine ($dF/d\delta$):

$$\frac{d\delta}{dt} = \frac{v}{1 + \dfrac{1}{K}\dfrac{dF}{d\delta}} \tag{36}$$

As a test proceeds, however, the specimen starts to neck down, its stiffness begins to decrease, and then it becomes negative. When the driving-point stiffness ($dF/d\delta$) passes through zero, the specimen has reached the maximum force. The stiffness thereafter decreases until it equals $-K$ and at that point the specimen elongation rate ($d\delta/dt$) becomes infinite for any cross-head rate (v), and the system is mechanically unstable. This instability point in a tensile test is the point at which the machine load line is tangential to the load–elongation curve of the material being tested. At that point, the specimen breaks and the energy stored in the machine structure dumps into the specimen at this unstable intersection.

Clearly, percent elongation at failure specifications is not very meaningful because it depends on the test machine stiffness. Figure 12 shows that for accurate measurements the

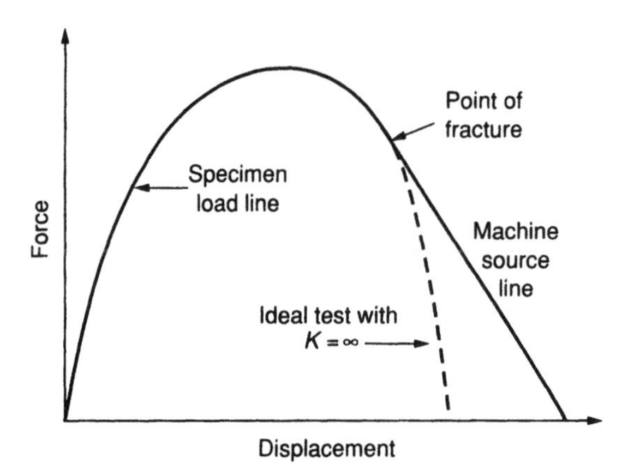

Figure 12 The point of instability in a tensile test.

stiffness of the test machine must be many times as large as the stiffness of the specimen near failure or the machine will dump its energy into the specimen and force a premature failure.

4.6 Fatigue in Bolted Assemblies

The mechanical engineering reader will perhaps recall that preloading a bolt stretches the bolt and compresses the part being bolted, but it is the relative stiffness of the bolt and part that determines what fraction of external loads applied to separate the part from the bolt will be felt by the bolt.* If the objective is to relieve the bolt of these loads, as it is in the head bolts of an automobile engine, then the designer tries to make the part much stiffer than the bolt; he or she mismatches the stiffness of the bolt and the part by specifying a hard head gasket. If the stiffnesses of the bolt and the part were the same, then they would share the external load equally: The bolt tension would increase by half the applied load while the part compression would decrease by half the applied load. If the gasket were very soft compared to the bolts, then the bolts would see virtually all of the applied load.

4.7 Operating Point for Nonlinear Characteristics

Let us continue to use the battery as an example of a linear source. It should be obvious that the maximum power point for the battery is independent of the load it must drive, but the load characteristic, however nonlinear, must pass through this point for maximum power transfer. Figure 13 illustrates this. It is not the slope of the load impedance that must match the source impedance, it is the *chordal slope,* the slope of a line from the origin of the power plane to the maximum power point of the source.

An orifice supplied from a constant upstream pressure and loaded at its output is a good example of a nonlinear source. For a short, sharp-edged orifice, the orifice equation† applies:

---

*Refer to any text on machine design under the indexed heading "fatigue in bolts."

†This is derived from Bernoulli's equation. The discharge coefficient (C_d) corrects for viscous effects not considered and for the existence of a vena contracta or convergence in the flow through the orifice, which makes the area of the jet smaller than the orifice itself. For most oils $C_d \approx 0.62$.

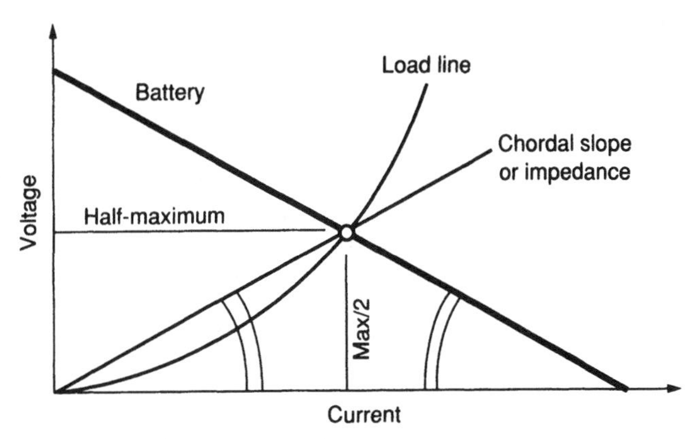

Figure 13 Chordal impedance matching.

$$Q_o = C_d A_o \sqrt{\frac{2}{\rho} \Delta P} = C_d A_o \sqrt{\frac{2}{\rho} (P_{\text{up}} - P_{\text{down}})} \tag{37}$$

where Q_o is the orifice flow, C_d is the discharge coefficient of the orifice, A_o is the orifice area, ρ is the density of the fluid, and $\Delta P = P_{\text{up}} - P_{\text{down}}$ is the pressure drop across the orifice. If the upstream pressure is kept constant, then P_{down} may be considered the output pressure and Q_o may be considered the output flow from this orifice, a characteristic typical of many hydraulic valves. Where is the maximum power point on this characteristic?

Hydraulic power $\mathcal{P}$ is the product $Q \cdot P$. It is a maximum when it is stationary with respect to either Q or P. If we nondimensionalize Eq. (37) with respect to the maximum flow through the orifice when $P_{\text{down}} = 0$, then the orifice equation becomes the following:*

$$\mathbf{Q}^* = \frac{Q_o}{Q_{\max}} = \sqrt{1 - \frac{P_{\text{down}}}{P_{\text{up}}}} = \sqrt{1 - \mathbf{P}^*} \tag{38}$$

$$\mathcal{P}^* = \frac{\mathcal{P}}{\mathcal{P}_{\max}} = \mathbf{P}^*\mathbf{Q}^* = \mathbf{P}^*\sqrt{1 - \mathbf{P}^*} \tag{39}$$

$$\frac{d\mathcal{P}^*}{d\mathbf{P}^*} = \sqrt{1 - \mathbf{P}^*} - \frac{\mathbf{P}^*}{2\sqrt{1 - \mathbf{P}^*}} = 0 \tag{40}$$

for which $\mathbf{P}^* = \frac{2}{3}$, and by substitution into (38), the flow at that point is $\mathbf{Q}^* = 1/\sqrt{3}$, and the maximum power delivered is $(\sqrt{2}/3)P_{\max}Q_{\max}$. The output admittance of the orifice is the slope of the curve at any operating point:

$$Z_o = \frac{d\mathbf{Q}^*}{d\mathbf{P}^*} = \frac{1}{2\sqrt{1 - \mathbf{P}^*}} = \left.\frac{\sqrt{3}}{2}\right|_{\mathbf{P}^*=2/3} \tag{41}$$

The load for which maximum power will be delivered, whether it has a linear or nonlinear characteristic, must pass through the maximum power point. Its *chordal admittance* must be

---

* Where bold letters will be used to indicate the nondimensional forms.

$$Z_{\text{chordal}} = \frac{\mathbf{Q}^*_{\text{op}}}{\mathbf{P}^*_{\text{op}}} = \frac{1/\sqrt{3}}{2/3} = \frac{\sqrt{3}}{2} \tag{42}$$

which is exactly the slope in Eq. (41). Figure 14 shows this graphically.

Note that for this power plane, the ordinate is *flow* and abscissa is *pressure* so that the slopes shown are admittances rather than impedances. While in other figures the abscissa has always been effort and the ordinate flow, it is conventional in hydraulic systems to show these figures the other way probably because pressure is usually the independent variable in hydraulic system characteristics: Pressure is controlled and flow is measured.

4.8 Graphical Determination of Output Impedance for Nonlinear Systems

A three-way hydraulic valve supplies or drains its load through a pair of variable orifices, one connecting the supply to the load and the other connecting the drain or tank to the load. These orifices are operated (typically on a single moving part of the valve) in a push–pull fashion; that is, as one opens, the other closes. If both orifices are partially open together when the valve is centered, the three-way valve is open centered. If one is wide open just as the other closes, the valve has no overlap.

Suppose it is required to find the P–Q characteristics of the load port for flows both into and out of the valve. The system is shown in Fig. 15. Thus for the upstream and downstream orifices respectively, Eq. (37) becomes

$$Q_u = C_d A_u \sqrt{\frac{2}{\rho}(P_S - P_L)} \qquad \text{and} \qquad Q_d = C_d A_d \sqrt{\frac{2}{\rho}(P_L - P_T)} \tag{43}$$

The load flow (Q_L) is the difference between Q_u, and Q_d. Also, the tank pressure (P_T) can be taken as zero. These equations combined become

$$Q_L = C_d A_u \sqrt{\frac{2}{\rho}(P_S - P_L)} - C_d A_d \sqrt{\frac{2}{\rho} P_L} \tag{44}$$

It is much more convenient to work with Eq. (44) in dimensionless form. If we assume that the maximum upstream and downstream areas are the same and that they are truly push–

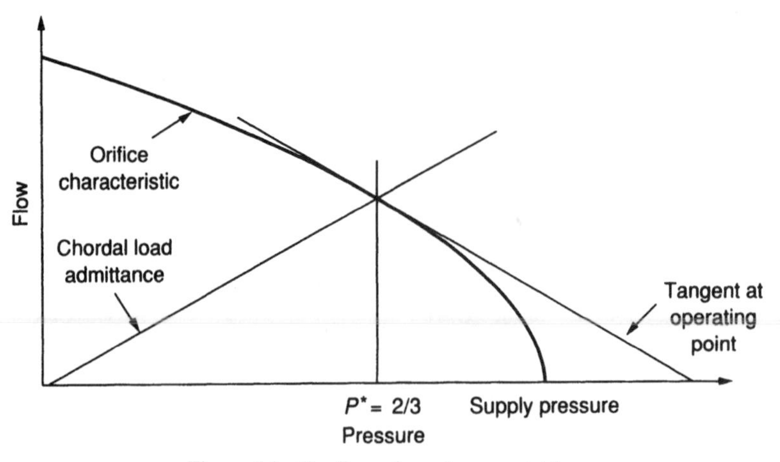

Figure 14 Nonlinear impedance matching.

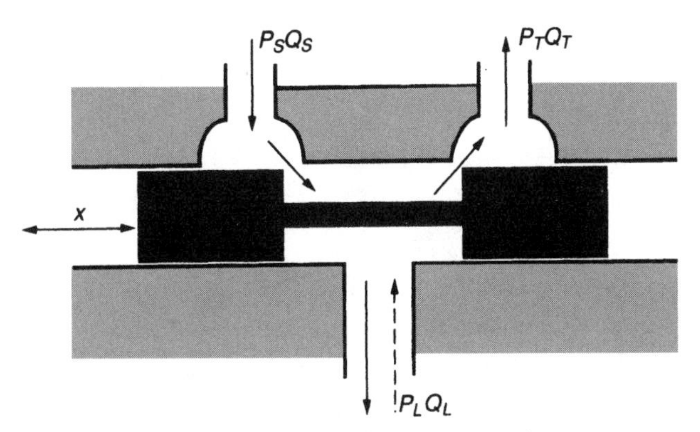

Figure 15 A three-way valve geometry and output.

pull, then we can nondimensionalize the area using $A_{\max}$. The discharge coefficient (C_d) is a constant for conventional valve geometries. The supply pressure (P_S) is a convenient term for the pressure nondimensionalization. Flows can be nondimensionalized with respect to a maximum flow that would pass through either orifice at full area $A_{\max}$ with P_S acting across it:

$$Q_{\max} = C_d A_{\max}\sqrt{\frac{2}{\rho} P_S} \tag{45}$$

If we set $P_L/P_S = \mathbf{P}$ and $Q_L/Q_S = \mathbf{Q}$ and express the upstream and downstream orifice sizes as push–pull fractions of $A_{\max}$, $0.5(1 + x)A_{\max}$ upstream and $0.5(1 - x)A_{\max}$ downstream, where $-1 \leq x \leq 1$ and the valve is centered for $x = 0$, then Eq. (44) combined with (45) yields

$$\mathbf{Q} = \mathbf{Q}_u - \mathbf{Q}_d = 0.5(1 + x)\sqrt{1 - \mathbf{P}} - 0.5(1 - x)\sqrt{\mathbf{P}} \tag{46}$$

which is one of those unfortunate equations in which the radical cannot be eliminated by squaring both sides. While Eq. (46) can be readily plotted, $\mathbf{Q}$ versus P with x (the valve stroke) as a parameter, it is instructive to construct it instead from its parts.

The term $0.5(1 + x)\sqrt{1 - \mathbf{P}}$ is the characteristic family for the upstream orifice, and the term $0.5(1 - x)\sqrt{\mathbf{P}}$ is the characteristic family for the downstream orifice. The first of these are parabolas to the left (on their sides because they are roots), starting at $\mathbf{Q}_u = 0.5(1 + x)$, $\mathbf{P} = 0$ and ending at $\mathbf{Q}_u = 0$, $\mathbf{P} = 1$ (there is no flow when the downstream pressure equals the upstream pressure). The second term starts at $\mathbf{Q}_d = 0$, $\mathbf{P} = 0$ and rises to the right to the points $\mathbf{Q}_d = 0.5(1 - x)$, $\mathbf{P} = 1$. All this is shown in Fig. 16. If there is no load *flow,* then the curves for the upstream orifice show its output characteristic while those for the downstream orifice represent the only load. The intersections predict the operating pressures for $\mathbf{Q} = 0$ as the valve is stroked, $-1 \leq x \leq 1$.

If there is a load flow ($\mathbf{Q}$), then continuity must be served. This requires that

$$\mathbf{Q} = \mathbf{Q}_u - \mathbf{Q}_d \rightarrow \mathbf{Q}_u = \mathbf{Q} + \mathbf{Q}_d \tag{47}$$

In Fig. 16, $\mathbf{Q}$ is simply a vertical bar between the curves for $\mathbf{Q}_u$ and $\mathbf{Q}_d$ whose length is the load flow. Thus for any load pressure value on the abscissa, the load flow is determined as the vertical distance between the input orifice curve and the output orifice curve. The load flow is positive if the upstream orifice curve is above the downstream orifice curve but

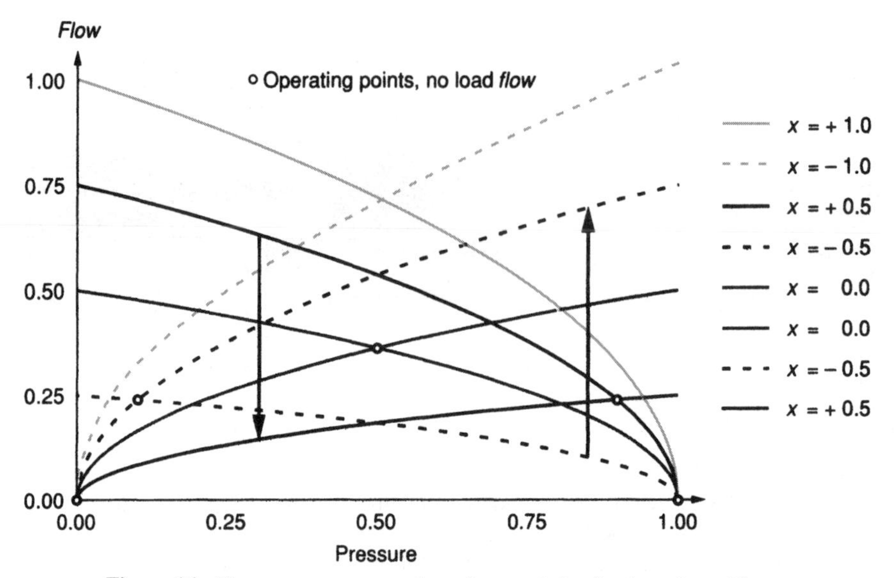

Figure 16 Flow versus pressure drop characteristics for the valve orifices.

negative otherwise. These data can be picked off and plotted as the **Q–P** characteristic of the valve, as shown in Fig. 17.

Figure 17 illustrates that an operating point on an input–output characteristic forces compatibility and continuity, but as long as those are preserved, a second energy exchange as at the $P_L - Q_L$ port of this valve can still be accommodated. This approach can be extremely useful in systems such as air-conditioning ducting where the fan characteristic is known only graphically, and then its output impedance at some point along the ducting must be determined. Exactly the same procedure is used.

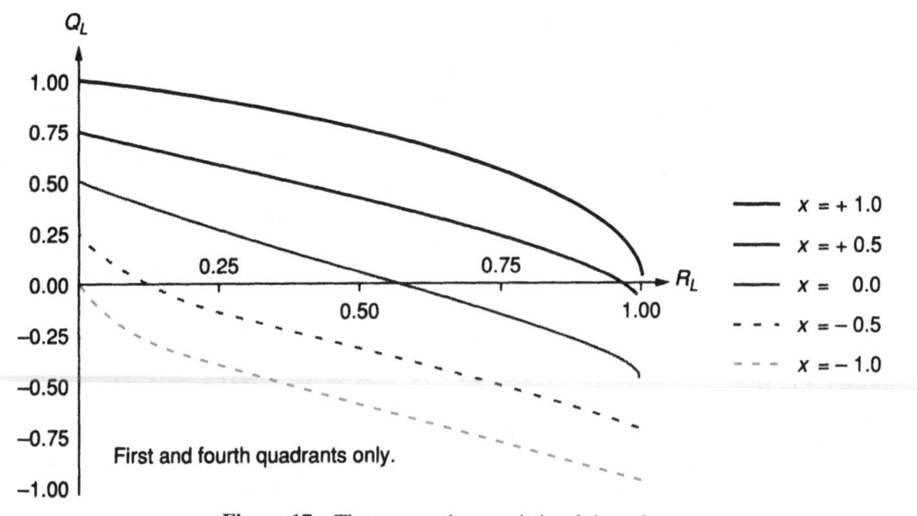

Figure 17 The output characteristic of the valve.

5 TRANSFORMING THE OPERATING POINT

It is often not among the system designer's options to choose either the output impedance of the power source in the system or the input impedance of the load that must drive. The only recourse at that point is to insert a transformer, transducer, or gyrator in the system if it does not already contain one or to vary the modulus of the two-port element if it does. In the old tube-type audio systems, the output impedance of the push–pull power tubes exceeded 1000 Ω while the input impedance of the speakers available then was 4, 8, or 16 Ω. To match the amplifier to the load, each channel had a large transformer for speaker connections with taps having turns ratios of $\sqrt{1000/4}$, $\sqrt{1000/8}$, and $\sqrt{1000/16}$. With this arrangement, maximum power transfer was assured down to the lowest frequencies for which the transformers were designed.

5.1 Transducer-Matched Impedances

Suppose a permanent direct current (dc) magnet servomotor is to drive a screw that, in turn, drives a mass, perhaps a machine-tool table. If the objective is to minimize the move time from stationary start to full stop, then the optimal trajectory for the servo, assuming equal and constant acceleration and deceleration, is well known: maximum acceleration to either half of the move distance or maximum velocity, whichever comes first, followed by maximum deceleration to the finish. These trajectories are illustrated in Fig. 18.

The motor operates in two modes: at maximum acceleration (maximum torque), which is set by the maximum short-term current permitted by the coercivity of the motor magnets and the capacity of the commutation, and at maximum speed, if that is reached, set by the maximum voltage available or by whatever the commutation allows. Often, servo designers accomplish these two modes by using an overvoltage (two to three times rating) during acceleration and deceleration, which runs the power amplifier as a current source, and then as maximum speed is attained, switching the amplifier voltage limit to the motor rating, which runs the amplifier as a voltage source. This achieves a constant acceleration and a constant top speed.*

During the acceleration or deceleration phase of the trajectory, the electrical torque available is accelerating two inertias: the motor itself and the load. Since the motor is acting

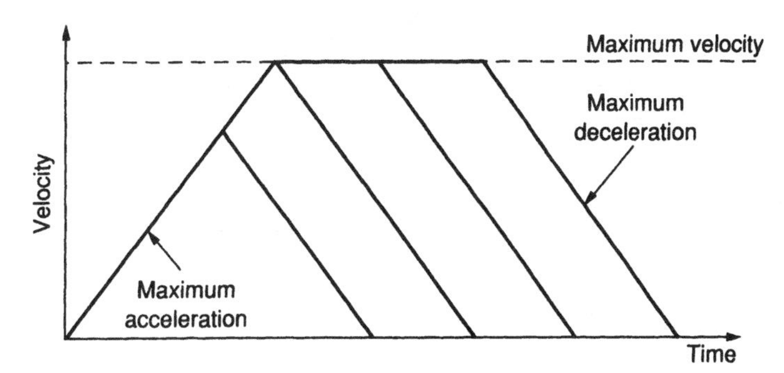

Figure 18 Optimal trajectories for point-to-point move.

*Actually, designers rarely use maximum acceleration or maximum velocity because such full-scale values leave no overhead for control. A servo running in saturation is an open-loop system.

as an electromagnetic torque source while constant current is supplied, $\tau_{\text{motor}} = K_m i_{\text{armature}}$, the mechanical output impedance viewed at the motor shaft is the rotor inertia between the output shaft and the electromagnetic torque source:

$$\mathbf{Z}_o\Big|_{\substack{\textit{motor shaft with}\\ \textit{constant armature current}}} = \boldsymbol{J}_{\text{armature}}\boldsymbol{s} \tag{48}$$

During acceleration and deceleration, the optimal load impedance (Z_{load}) will be equal to the motor inertia, and the available electromagnetic torque will be shared equally by the motor armature and the load.

The load given in this example is primarily massive, so with reflection through the screw, the load inertia is computed as the load inertia times the square of the transducer ratio (p for the screw; see Table 2):

$$Z_{\text{load}} = p^2 M_{\text{load}} s \tag{49}$$

To the extent possible, the pitch of the screw or the inertia of the motor should be chosen to achieve this match. Failing both of those options, a gear box should be placed between the screw and the motor to accomplish the match required.

5.2 Impedance Requirements for Mixed Systems

When a source characteristic is primarily real or static (so that the source impedance is resistive) and the load is reactive or dynamic (dominated by energy storage elements), then impedance matching in the strictest sense is impossible, and the concept of passing the load line through the source characteristic at the maximum power point does not make sense. How then does one match a static source to a dynamic load or the reverse?

Figure 19 shows an electrohydraulic position servo driving a mass load with negligible damping losses, and Fig. 20 shows the pressure flow characteristics of the servovalve: a family of parabolas in the power plane used for hydraulic systems. The transducer between this hydraulic power plane and the force–velocity power plane in which the load operates is a piston of area A. There are two equations:

$$P = \frac{1}{A} F \qquad \text{and} \qquad Q = Av \tag{50}$$

With these equations, a load in F–v coordinates can be transformed to a load in P–Q coordinates for superposition on the source characteristics.

The impedance of the mass load, however, is simply $Z_{\text{load}} = (M)(s)$, but this cannot be plotted in an F–v coordinate system because it is the slope of a straight line in energetic

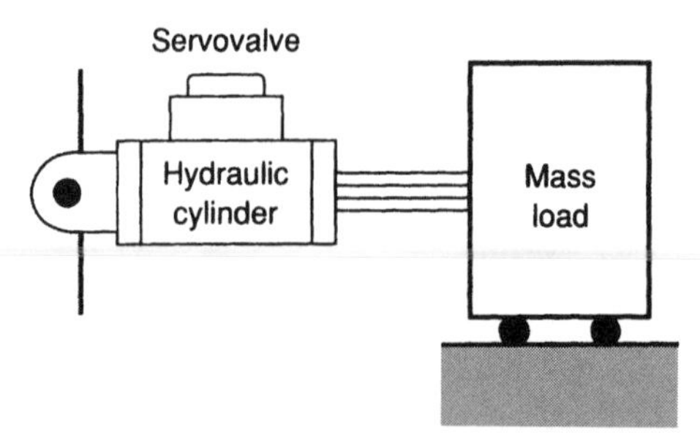

Figure 19 Electrohydraulic position servo.

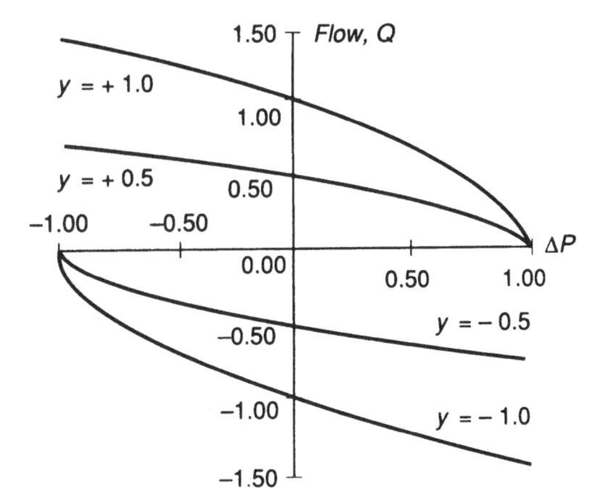

Figure 20 Servovalve characteristic.

coordinates: F versus dv/dt. To match load to source, we need the force–velocity relationship for the load. A servo designer normally has a frequency response in mind for his or her application or can determine one from a linearization of the system. Suppose, in this example, that the valve operating between a supply pressure (P_s) and a maximum load flow ($Q_{\max}$) must drive the mass load (M) through the piston of the area (A), with a frequency response flat in position to a frequency of ω (radians per second), at an amplitude of $D_{\max}$. With these conditions, the force–velocity relationship for the load can be found as follows.

If y is the displacement, $\dot{y}$ is the velocity, and $\ddot{y}$ is the acceleration, then the most taxing demand on the servo will be $y = D_{\max} \sin(\omega t)$, where t is time. Then $\dot{y} = D_{\max}\omega \cos(\omega t)$ and $\ddot{y} = -D_{\max}\omega^2 \sin(\omega t)$. Given that $F = M\ddot{y}$ is the load equation, however, the load can now be expressed parametrically as a pair of equations:

$$v = \dot{y} = D_{\max}\omega \cos(\omega t) \qquad \text{and} \qquad F = -MD_{\max}\omega^2 \sin(\omega t) \tag{51}$$

If these are cross-plotted in the force–velocity plane with time as a parameter, the plot traces out an ellipse as time goes from zero through multiples of $2\pi/\omega$. Transforming these to the P–Q plane requires application of the piston equations (50) to yield

$$Q = AD_{\max}\omega \cos(\omega t) \qquad \text{and} \qquad P = -\frac{MD_{\max}\omega^2}{A} \sin(\omega t) \tag{52}$$

If the valve is to drive the mass through the piston around the trajectory, $y = D_{\max} \sin(\omega t)$, then the valve output characteristic for the maximum valve stroke must *entirely enclose* the elliptical load characteristic derived in Eqs. 52. Furthermore, if the valve and load are to be perfectly matched, then the valve characteristic and the load ellipse must be *tangent* at the maximum power point for the valve: $P = \frac{2}{3}P_S$, $Q = (1/\sqrt{3})Q_{\max}$. This is shown in Fig. 21.

Note that this requirement means that neither maximum valve output pressure nor maximum valve flow will ever be reached while the load executes its maximal sinusoid. If $P_s = -MD_{\max}\omega^2/A$ and $Q_{\max} = AD_{\max}\omega$, the valve sizing would have provided inadequate power for reaching any but those points on the trajectory, which would therefore have followed the valve characteristic instead. The correct matching relationships are those which satisfy the following equations:

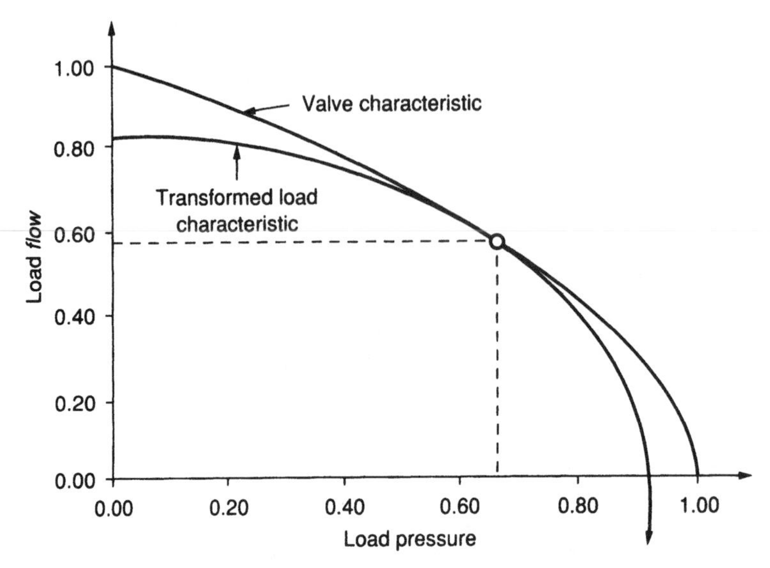

Figure 21 Matching power requirements for a dynamic load.

$$\frac{1}{\sqrt{3}} Q_{max} = AD_{max}\omega \cos(\omega t) \qquad \text{and} \qquad \frac{2}{3} P_s = \frac{MD_{max}\omega^2}{A} \sin(\omega t) \tag{53}$$

at the appropriate time. These equations relate D_{max}, ω, M, Q_{max}, P_s, and A, so any five of these determine t and the sixth. If the load mass, supply pressure, piston area, peak frequency, and maximum amplitude are all known, for example, Eqs. (53) size the valve by determining its maximum required flow. If the valve has been selected, these equations will size the piston.

In the example above, the load was purely massive. Most real loads are dissipative as well. The procedure outlined above, however, need only be modified by adding the damping term to the force equation:

$$F = M\ddot{y} + B\dot{y} = -MD_{max}\omega^2 \sin(\omega t) + BD_{max}\omega \cos(\omega t) \tag{54}$$

where B is the damping coefficient. This has the effect of tipping the axis of the elliptical load line up to the right, but the remainder of the development follows as before with the added term. Figure 22 shows the result. If the load mass were to be negligible, then the elliptical transformed load line in Fig. 22 would collapse to a line passing through the maximum power point of the valve characteristic, as discussed previously.

6 MEASUREMENT SYSTEMS

Measurement extracts information about the state of a measured system, usually in the form of one of the factors in the measurement domain whose product is power or energy flux, such as voltage or current, pressure or flow, and so on. A measurement usually extracts some energy from the system being measured, if only in a transient sense; each link in the chain from measurement to display involves a further exchange of energy or power. If we measure

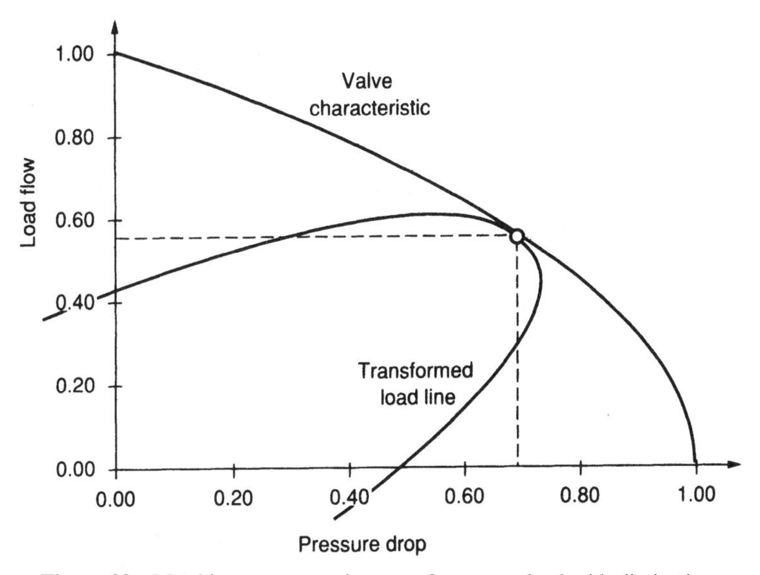

Figure 22 Matching power requirements for a mass load with dissipation.

voltage in a circuit, for example, we must draw some current, at least instantaneously, so that the power is dissipated by effective resistances, or energy is stored capacitively, inductively, or both.

At the measurement interface, we wish to disturb the measured system as little as possible by extracting as little energy or power as can be managed. It is usually our objective to pass power or energy along the chain of elements that form the measurement system, and there is a best combination of the energy variables to attain the optimum transfer. This chapter has dealt with these issues: the maximization of energy transfer within the system where we want it and the minimization of energy theft at the measurement interface where we do not.

6.1 Interaction in Instrument Systems

The generalized instrument consists of a number of interconnected parts with both abstract and physical embodiments. An orifice flow-metering system might consist of the following chain: An orifice plate converts the flow to a pressure drop; a diaphragm converts the pressure drop to a force; a spring (perhaps an unbonded strain-gage bridge) converts the force to a displacement; the strain gages convert the displacement to a resistance change; a bridge arrangement converts the resistance change to a differential voltage; and a galvanometer converts the voltage to a trace on paper. Figure 23 illustrates this chain.

In setting up this chain, the orifice is sized to minimize the *pressure drop* resulting from our flow measurement, and the diaphragm must be sized for minimum pumping volume during transients in pressure, or it will alter the flow reading in a transient sense. The diaphragm, however, has an output stiffness; the force it transmits decreases with increasing displacement, and it is driving the unbonded gages, converting force to displacement. The combined stiffness of the unbonded strain gages, which will be linear springs, must equal

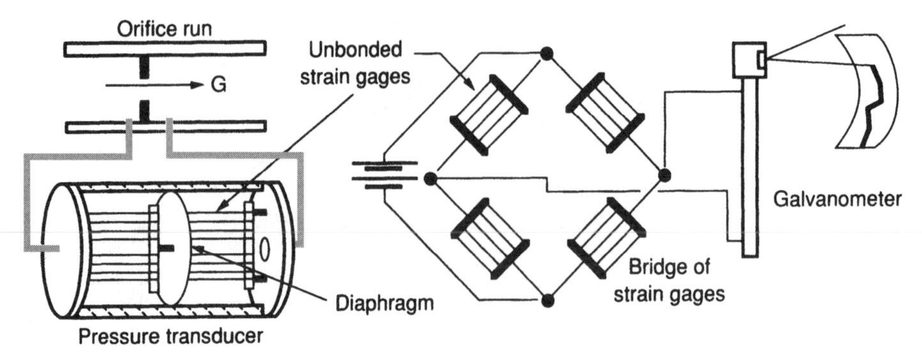

Figure 23 The chain of a flow measurement.

the average stiffness of the diaphragm* in order to ensure that the strain energy transfer is maximized for a given input energy from the fluid.

The displacement–strain relationship is a definition. Further along the chain, however, this strain is converted to a differential voltage whose magnitude depends on the input impedance of the galvanometer, which is loading the bridge. For bridges constructed with the same resistance in each arm, whether the resistors are active gages or not, the input and output impedances are the same: the resistance of one arm, usually the gage resistance. If we ignore dynamic considerations (such as galvanometer damping), the optimal galvanometer will therefore have the same resistance for maximum deflection per unit strain as one gage in the bridge.

At the measurement interface, therefore, the objective is to mismatch the impedances between the measurement system and the measured system as much as possible. There is more to this than the selections of voltage-measuring devices with higher impedance than the system in which voltage is being measured and current-measuring devices with lower impedance than the circuits in which they are placed. The measurement and control engineer must always be aware of dynamic loading as well.

The output impedance of any piezoelectric device, for example, is almost purely capacitive and is typically only a few picofarads for small devices. The input impedance of most oscilloscopes is a parallel combination of 1 MΩ and 100 pF. Either or both of those would load a piezoelectric device to near uselessness, even though for most other purposes they are *high* impedances The resistance would reduce the charge on the crystal much too quickly, and the capacitance would steal charge and reduce the voltage output drastically. An attempt to measure the pressure in a small volume with a transducer that has a large *swept* volume itself would meet with the same failure: The displacement of the transducer would alter the volume in which the pressure was measured. Holography has become popular in the study of the vibration of thin plates and shells because it does not *load* the structure by adding mass as an array of accelerometers would.

*If the diagram were rigidly supported in the center, then the force on the support would be the pressure times the effective area of the diaphragm. If the support were not rigid, this output force would depend nonlinearly on the support displacement. The plot's negative slope of force transmitted vs. displacement is the output stiffness of the diaphragm, and pressure × area is the force source in this model.

6.2 Dynamic Interactions in Instrument Systems

It is not only the steady-state *loading* of a measurement that is of concern; under many circumstances, the unsuspected dynamics of the instrument being used will lead to erroneous results. Consider the simple measurement interface shown in Fig. 24. The readout instrument, an oscilloscope for example, has both resistance and capacitance. The load impedance is therefore the parallel combination of these:

$$Z_{\text{inst}} = \frac{R_i(1/C_i s)}{R_i + 1/C_i s} = \frac{R_i}{R_i C_i s + 1} = \frac{R_i}{\tau_i s + 1} \quad \text{with} \quad \tau_i = R_i C_i \tag{55}$$

Thus the load depends on the frequency of the voltage being measured. This might not concern us in the sense that we can predict it and compensate for the known phase shift which this will induce, but when the interaction of the two systems is considered, the problem becomes more obvious. The total impedance loading the measurement of V_s includes the source impedance Z_o. Suppose this is purely resistive (R_o):

$$Z_{\text{total}} = R_o + Z_{\text{inst}} = \frac{R_o R_i C_i s + (R_o + R_i)}{R_i C_i s + 1} \tag{56}$$

The readout instrument is sensitive only to the voltage it sees, which has been reduced by the voltage drop in R_o. In general terms, the measured voltage, V_{meas}, is the voltage across the instrument's input resistor and capacitor, R_i and C_i:

$$\frac{V_{\text{meas}}}{V_s} = \frac{Z_{\text{inst}}}{Z_{\text{total}}} = \frac{R_i}{R_o R_i C_i s + (R_o + R_i)} = \frac{1}{R_o C_i s + (R_o/R_i + 1)} \tag{57}$$

Unless we know the output impedance at the point of measurement, in this case R_o, we do not even know the time constant or break frequency germane to the measurement. In the event that Z_o is also complex (has reactive terms), this situation is more complicated, and if $Z_o = L_o s + R_o$, for example, the system could even be oscillatory.

Sometimes, frequency-dependent impedances are intentionally introduced into a measurement system, most commonly in the form of passive filters. Figure 25* shows a first-

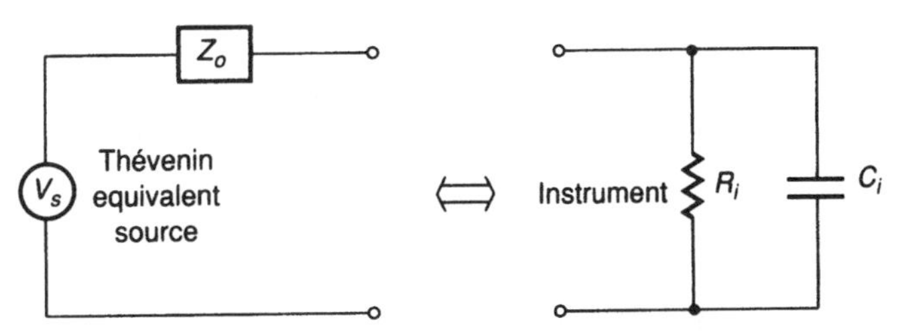

Figure 24 A measurement with a dynamic instrument.

---

*This example is drawn entirely from Ref. 4, with the permission of Dr. C. L. Nachtigal. In this and the previous example, Figs. 24 and 25, the source voltage is referred to as a Thévenin equivalent source. In single-loop circuits the source voltage is by definition the Thévenin voltage, since removing one element from the loop causes it to become open circuited. If the source is a sensor, for example, the magnitude of the voltage is governed by the value of the sensed variable and, of course, its own design parameters. The Thévenin, or open-circuit voltage, is specified on the sensor data sheet.

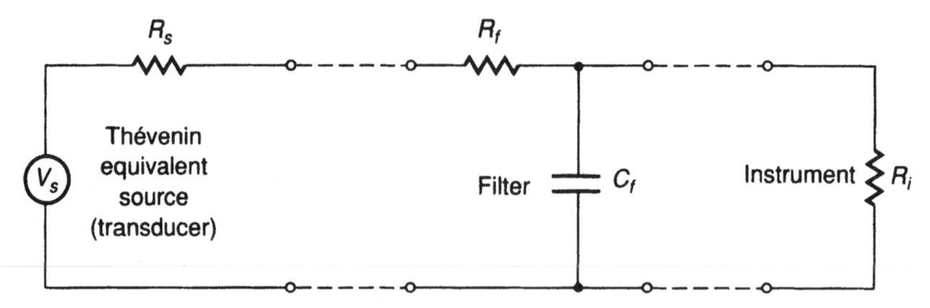

Figure 25 Schematic of cascaded instrument system.

order, low-pass filter being driven by a source with a nonzero output impedance (R_s) and being loaded by a readout instrument with finite input impedance (R_i).

The loading problem is potentially present at the interface of each stage indicated by the broken lines and terminals. What is not so obvious is that the dynamics of this filter are quite sensitive to its source and load. An unwary designer might choose the filter time constant ($\tau_f = R_f C_f$) with little regard for the values of R_f and C_f and essentially design the filter under the simple assumption that source and load were ideal.

To the instrument, however, the filter is part of the source impedance, so the true source impedance combines R_s, R_f, and C_f. and the Thévenin equivalent source voltage (V_T) is no longer simply V_s. The Thévenin voltage (V_T) and the new output impedance (Z_T) with the filter included become

$$V_T = \frac{V_s}{(R_s + R_f)C_f s + 1} \quad \text{and} \quad Z_T = \frac{R_s + R_f}{(R_s + R_f)C_f s + 1} \tag{58}$$

Figure 26 shows the new equivalent circuit for the cascaded instrument system.

The instrument is now loading this system, and the output (V_o) is equal to the voltage across R_i. Therefore, in terms of the original measured voltage (V_s)

$$\frac{V_o}{V_s} = \frac{1/(1 + (R_s + R_f)/R_i)}{[(1 + R_s/R_f)/(1 + (R_s + R_f)/R_i)]R_f C_f s + 1} = \frac{K_f}{\tau s + 1} \tag{59}$$

where

$$K_f = \frac{1}{1 + (R_s/R_i) + (R_f/R_i)} = \frac{1}{1 + \beta} \tag{60}$$

and

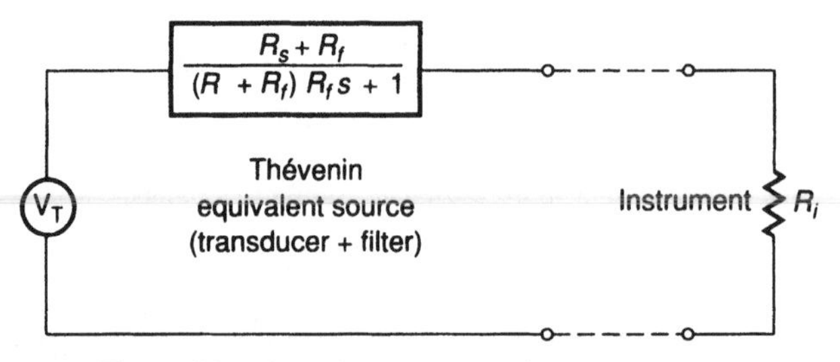

Figure 26 Thévenin equivalent of cascade system.

$$\tau = \left(\frac{1 + (R_s/R_f)}{1 + (R_s/R_i) + (R_f/R_i)}\right)R_fC_f = \left(\frac{1 + \alpha}{1 + \beta}\right)R_fC_f \tag{61}$$

Our filter has its expected characteristics, $\tau = \tau_f = R_fC_f$ and $K_f = 1$, only if both α and β approach zero, a circumstance that requires $R_s \to 0$ and $R_i \to \infty$ while R_f remains finite. As the instrument resistance (R_i) decreases, the static gain and time constant decrease as well, which means that the break frequency of the filter increases from the designed value because the instrument provides another path for the discharge of the capacitor in the filter. As the transducer resistance (R_s) increases from zero, the static sensitivity (K) again decreases, but this time the break frequency decreases as well.

This analysis shows us that source loading by a filter and filter loading by a readout instrument can cause significant changes in both the designed filter gain and break frequency. Only if α and β in Eqs. (60) and (61) remain equal does the filter break frequency remain unscathed by nonideal source and load impedances. This condition requires that

$$\frac{R_s}{R_f} = \frac{R_s + R_f}{R_i}$$

or

$$R_f^2 + R_sR_f - R_sR_i = 0 \tag{62}$$

Dividing the second of Eqs. (62) by R_s^2 and solving it for R_f/R_s yield

$$\frac{R_f}{R_s} = \frac{1}{2}\left(-1 \pm \sqrt{1 + 4\frac{R_i}{R_s}}\right) \cong \sqrt{\frac{R_i}{R_s}} \tag{63}$$

since any realistic measurement system has $R_i/R_s \gg 1$, and since the resistances must be positive, the negative solution is discarded. This approximation is equivalent to saying that R_f should be chosen to be the geometric mean of the estimated or known values of R_i and R_s; that is, $R_f = \sqrt{R_iR_s}$ presuming only that $R_i \gg R_s$. For this choice of filter resistance, the resulting gain is given by

$$K = \frac{1}{1 + \sqrt{R_s/R_i} + R_s/R_i} = \frac{1}{1 + (R_s/R_i)^{3/2}} \cong 1 \qquad \text{for} \qquad R_i \gg R_s \tag{64}$$

If the filter fails these conditions, it must either be carefully designed for the task or it should be an active* filter with a high input impedance and low output impedance.

6.3 Null Instruments

Many instruments are servos, active systems designed to oppose the variable they measure so as to decrease their demand on the complementary energy or power variable to zero. In a steady-state sense, these instruments draw no power or energy from the measurement interface because these energetic requirements are provided by the instrument's power supply. These instruments therefore have infinite or zero input impedance in steady state. Examples abound: Slidewire potentiometers for strain readout and thermocouple readout both measure voltage by servoing to zero current. Servo accelerometers avoid the problems with temperature inherent in the elasticity variations of metals and crystals by servoing the motion of

---

*Incorporating amplification, usually an operational amplifier.

the proof mass to zero and measuring the force required to do it. The *differential pressure-cells* used in the process control industry measure differential pressure across a diaphragm or *capsule* while preventing the diaphragm or capsule (separating the two pressures) from moving. They thus avoid having to worry about the nonlinear elastic characteristics of the capsule and seals in the system.* Their output is either the current in a voice coil or the pressure in a bellows necessary to oppose the motion. In all of these cases, however, there is a transient *displacement* until the servo zeros it out, and *some energy must be lost to transfer the information required by the servo in the instrument.* For this reason, null instruments are no better at measuring stored energy, the voltage on a small capacitor for example, than passive but high-impedance instruments, particularly if the energy consumed to reach null is not extremely low.

7 DISTRIBUTED SYSTEMS IN BRIEF

While a detailed study of the input–output relationships for distributed systems is beyond the scope of this chapter, a brief discussion can tie these in to the concepts already covered. All of the systems discussed to this point have been *lumped,* a label that implies physical dimensions are not of importance; the system parts can be considered as *point* objects. Considering a tank, for example, to be a point does not mean that the tank has no dimensions; it merely means that the internal pressure is considered to be the same everywhere within it; conditions in its interior are absolutely uniform. When studying lumped circuits, we are not concerned with the dimensions of the circuit elements or with their distances from each other on the circuit board. In reality, the nodes of the circuit have lengths, but they are ignored for the purposes of analyzing the lumped circuit.

In the mechanical domain, we consider that masses are rigid and behave as if the forces acting were applied at a point, and if we are interested in the distributed properties of an object, we consider only its moment of inertia. In each of these examples, changes in the physical variables of our model are assumed to propagate instantaneously, even though it is well known that all have finite propagation velocities. A system element must be considered to be distributed to have properties that vary with a physical dimension if that assumption is not true. This occurs whenever the dimensions of the object are large compared to the characteristic size of the *events* occurring.

Mechanical disturbances of all kinds propagate at the speed of sound in the medium involved, and electromagnetic disturbances propagate at speeds near the speed of light, depending again on the medium. A hammer blow to the end of a long, slender bar, for example, induces a strain pulse at the struck end which travels into the bar (*informing* the interior of the event) at the speed of sound in compression in the bar material ($c = \sqrt{E/\rho}$, where E is Young's modulus of the material, ρ is its mass density, and c is the propagation speed of compressive or tensile events). If the pulse duration is short, its physical length approaches that of the bar or may be shorter than the bar. Our simple lumped models of the bar's behavior [$F = M\ddot{y}$ and $\int F\,dt = Mv(t) - Mv(0)$], which treat it as a solid rigid object, are incorrect. Similarly, if a small explosion, a spark, for example, is initiated in a tank, the pressure in the tank will not remain uniform throughout. Instead, pressure waves will propagate within the tank at the speed of sound until damping takes its toll; we can no longer

---

*Dry friction would nonetheless be fatal to the instrument because it would be fatal to its servo and would lead to dynamic instabilities of the limit-cycle variety.

consider the tank to be a simple lumped capacitor, at least in the time scale of the spark event.

7.1 Impedance of a Distributed System

Imagine a long, slender tank of water, a trough, open at the top, and perform the following *thought* experiment. If one end wall were moved inward suddenly, the level of water at that end would rise higher up than was required by the change in tank volume because the remaining water further along the tank would not change level instantaneously. Then, this wave coming down off the tank wall would travel to the other end where it would slosh up the far wall and return. This wave would continue to slosh back and forth at decreasing amplitude as viscosity took its toll, until finally the surface would be calm again at a new, slightly higher level. Each time the wave reached an end, it would be returned in kind, that is, with the same sign.

Now suppose that the far end of the tank opened into a large lake, so large that no level changes would take place when an end wall was moved, and again move the remaining end wall inward quickly. Then, when the first wave reached the opening to the lake, it would leave the tank and be lost in the lake. But that would involve water leaving the tank in excess of the volume change in the tank, and a negative wave would return from the open end to signal the new, lower level required by the loss. The closed end would reflect this wave as a rarefaction, and when that returned to the open end, lake water would spill back in as an upward wave. Eventually these alternating processes would return the level in the tank to that of the lake.

A closed end returns a wave of like kind, and an open end returns a wave of opposite kind. If there are no losses as the waves hit the ends of the tank, a wave of strength $+1$ is reflected with strength $+1$ from a closed end and with a strength -1 from an open end. This implies that there is an end condition somewhere between *closed* and *open* from which a wave will not be reflected at all. A suitably constructed porous wall, in this example, would simply absorb the wave completely by accepting and dissipating all of its energy. The impedance of this wave-matched wall is the wave impedance of the channel and is a characteristic of it, depending on the inductive and capacitive properties of the medium.

The 75- and 300-Ω markings on the antenna connections of a television receiver imply two things: First the input impedances of those terminals are resistive at 75 and 300 Ω, respectively, and, second, the coaxial cable and flat-lead antenna wiring are really waveguides whose wave impedances are 75 and 300 Ω. By matching the impedance of the cable at the receiver terminals, we are assured that all the incoming *wave energy* will be absorbed by the receiver and none will be reflected back up the cable to the antenna and thus lost.

8 CONCLUDING REMARKS

This chapter has demonstrated an alternative viewpoint for the interaction of systems with each other. Control engineers are quite accustomed to transfer functions: relationships in the frequency (s or Laplace) domain between a variable at one point in a system and another at some other point, most often between inputs to and outputs from a controlled system. This chapter has dealt with a special class of these relationships between the complementary variables of power at a single point in a system. These special transfer functions are called driving-point impedances or admittances, and they determine how one subsystem will load or be loaded by another.

Admittances are the reciprocals of impedances, and both are unique properties of a system. The Laplace operator (s) expresses these properties as a polynomial ratio. The denominator polynomial (when set to zero, it becomes the characteristic equation of the system) is always the same, and the numerator polynomial is a function of the location of the point considered in the system. It was also shown that driving-point impedances are not a function of the controlled variables on any ideal sources the system contains. Instead, all effort sources may be replaced by solid connections, and all flow sources may be removed before the driving-point impedance is computed.

When two systems are connected together at a driving point, port, or pair of terminals, usually so that one can pass energy or information to the other, then there is a favorable relationship between the impedances of the two systems that depends on the objective of the connection. When it is desired to pass energy or power from one system to the other, then the output impedance of the driving system should match the input impedance of the driven system. If neither the driver nor the driven are adjustable, then a transducer, gyrator, or transformer is used to match them by selecting the modulus to achieve the match. Any impedance seen through a transformer, for example, appears to be increased or diminished by the square of the transformer ratio. In a chain of subsystems, it is not necessary to install the transformer at the driving point under consideration; the correct ratio can be determined no matter where it is placed within the chain because the square of the modulus will always appear in one of the driving-point impedances.

If the interconnection represents a measurement interface, then the most favorable relationship between the driving-point impedances is the largest possible mismatch consistent with obtaining the measurement. The ideal instruments for measuring efforts have infinite input impedance and the ideal instruments for measuring flows have infinite input admittances. Instruments that measure the integral of flows, such things as volume, charge, and displacements, should have very low compliance (should displace easily, have low volume themselves, or have small capacitances), while instruments that measure the integral of efforts, such things as flux linkage or momentum, must have low mass or inductance.

The operating point of a pair of coupled systems is at the intersection of their input and output characteristics in the power or energy plane. If one of these, for example the source or output characteristic, exists in the power plane, that is, is static, but the other is energetic (i.e., dynamic: massive, inductive, capacitive, etc.), then the source characteristic must enclose the trajectory of the load characteristic at the highest frequency of interest, and ideally, the source characteristic and load trajectory should be tangent at the maximum power point or should be made tangent there by suitable choice of system parameters.

The key issue in this chapter is this: Whenever two dynamic systems are connected, an interaction occurs. If the connection is to meet its objectives, the nature of this interaction must be explored and controlled.

REFERENCES

1. H. M. Paynter, *Analysis and Design of Engineering Systems,* MIT Press, Cambridge, MA, 1960.
2. F. A. Firestone, "A New Analogy between Mechanical and Electrical Systems," *Journal of the Acoustical Society of American,* **4,** 249–267, (1932/33).
3. A. C. Bell and S. Ramalingam, "Design and Application of a Tensile Testing Stage for the SEM," *Journal of Engineering Materials and Technology,* **96,** 157–162 (July 1974).
4. L. C. Nachtigal, "Dynamic Systems and Measurements." Unpublished notes, Purdue University, 1978.

CHAPTER 3

BRIDGE TRANSDUCERS

Patrick L. Walter
Department of Engineering
Texas Christian University
Fort Worth, Texas

1 TERMINOLOGY

A telemetry system responding to a measurand consists of four basic parts—the transducer, the transmitting system, the receiving system, and the data output or display system:

Telemetry. The transmission of information about a measurand.

Measurand. The object of a measurement. The process to be defined.

Transducer. A component in the telemetry system which provides information about a process and, as a by-product, transfers energy from the process. Typical bridge transducers convert physical quantities such as force, pressure, displacement, velocity, acceleration, temperature, and humidity into electrical quantities for input to the transmitting system.

Reprinted from *Instrumentation and Control,* Wiley, New York, 1990, by permission of the publisher.

Transmitting System. The transmitting system typically consists of some or all of the following devices: cable, amplifier, subcarrier oscillator, filter, analog-to-digital (A/D) converter, transmitter, and antenna.

Receiving System. The receiving system typically consists of some or all of the following devices: antenna, preamplifier, multicoupler, receiver, tape or disc recorder, discriminator, decommutator, digital-to-analog (D/A) converter, and output filter.

Data Output and Display System. The data output and display system typically consist of some or all of the following devices: oscilloscope, analog meter, digital meter, graphic display, and digital printer. Either these devices may be connected directly to the output of the receiving system or a computer may process the data from the receiving system before display.

2 FLEXURAL DEVICES IN MEASUREMENT SYSTEMS

Bridge transducers depend on a measurand to directly modify some electrical or magnetic property of a conductive element. For example, the thermal coefficient of impedance can result in a change in impedance of a conductive element proportional to temperature (e.g., resistance thermometer). Similarly, hygroscopic materials can have their impedance change in a deterministic fashion due to humidity (e.g., humidity sensor). Most bridge transducers, however, depend on the displacement of a flexure to vary the impedance of a conductive element, resulting in an electrical signal proportional to the measurand. Advantage is taken of either the strain pattern on the surface of the flexure or the motion of this surface. Among the gamut of flexure elements associated with bridge transducers are cantilever beams, Bourdon tubes, and clamped diaphragms.

2.1 Cantilever Beams

Cantilever beams are routinely designed into bridge transducers. Strain near the clamped end of the beam can be correlated to displacement of the free beam end, force or torque applied to the free beam end, dynamic pressure associated with fluid flow acting over the beam surface, and so on. The compliance of a cantilever beam is defined as

$$\frac{y}{F} = \frac{L^3}{3EI} \tag{1}$$

where y is deflection of the beam free end, F is the force applied to this end, L is the beam length, E is the modulus of elasticity of the beam material, and I is the beam area moment of inertia. A compliant flexure will result in a bridge transducer with a large electrical signal output. Equation (1) indicates a compliant flexure design can be achieved by a long, thin, narrow beam of low-modulus material. The penalty attached to such a design in application could be a transducer which is bulky, displays undesirable response to physical inputs orthogonal to its sensing direction, and has poor dynamic response.

2.2 Bourdon Tubes

Bourdon tubes are one of the most widely used flexures for sensing pressure. The original patent for this device was granted to Eugene Bourdon in 1852. Bourdon tubes are hollow tubes that are twisted or curved along their length. The application of pressure deforms the tube wall which, depending on tube shape, causes it to untwist or unwind. Motion of the tube is typically used to modify the alternating current (ac) impedance of bridge transducers. Bourdon tubes can be integrated into transducers to achieve extremely high accuracies and have been manufactured from perfectly elastic materials such as quartz.

Transducers employing Bourdon tubes tend to be physically large and easily damaged by environmental inputs such as acceleration. In addition, the tubes themselves afford poor frequency response to time-varying pressure.

2.3 Clamped Diaphragms

Clamped diaphragms are another flexure used to transform a measurand into a strain or displacement proportional to applied pressure. A small, flat, circular diaphragm can be made simply, and it can be placed flush against surfaces whose flow dynamics are being studied. This type of diaphragm is typically designed to deflect in accord with theory associated with clamped circular plates. Corrugated diaphragms provide extensibility over a greater linear operating range than do flat diaphragms. A catenary diaphragm consists of a flexurally weak seal diaphragm bearing against a thin cylinder whose motion is measured. The compliance of a flat, clamped circular diaphragm is defined as

$$\frac{y}{P} = \frac{3R_0^4(1 - \nu^2)}{16t^3E} \tag{2}$$

where y is the deflection of the center of the diaphragm, P is the applied pressure, R_0 is the diaphragm radius, ν is Poisson's ratio, t is the diaphragm thickness, and E is the modulus of elasticity of the diaphragm material. Somewhat analogous to the cantilever beam, a compliant diaphragm will have a large radius, be thin, and be made of a low-modulus material. Equation (2) holds for deflections no greater than t.

Figure 1 shows the radial and tangential strain distribution in a flat, clamped, circular diaphragm. The radial and tangential strains at the center of the diaphragm are identical. The tangential strain decreases to zero at the periphery while the radial strain becomes negative. Figure 2 describes a strain gage pattern designed to take advantage of this strain distribution. The central sensing elements measure the higher tangential strain while the radial sensing elements measure the high radial strains near the periphery. Resistance strain gages are discussed beginning in Section 3.

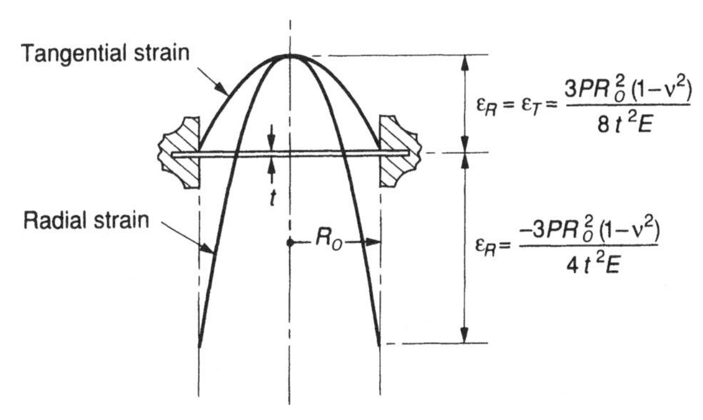

Figure 1 Radial and tangential strain distribution in a flat, clamped, circular diaphragm. (Courtesy of Measurements Group, Inc., Raleigh, NC.)

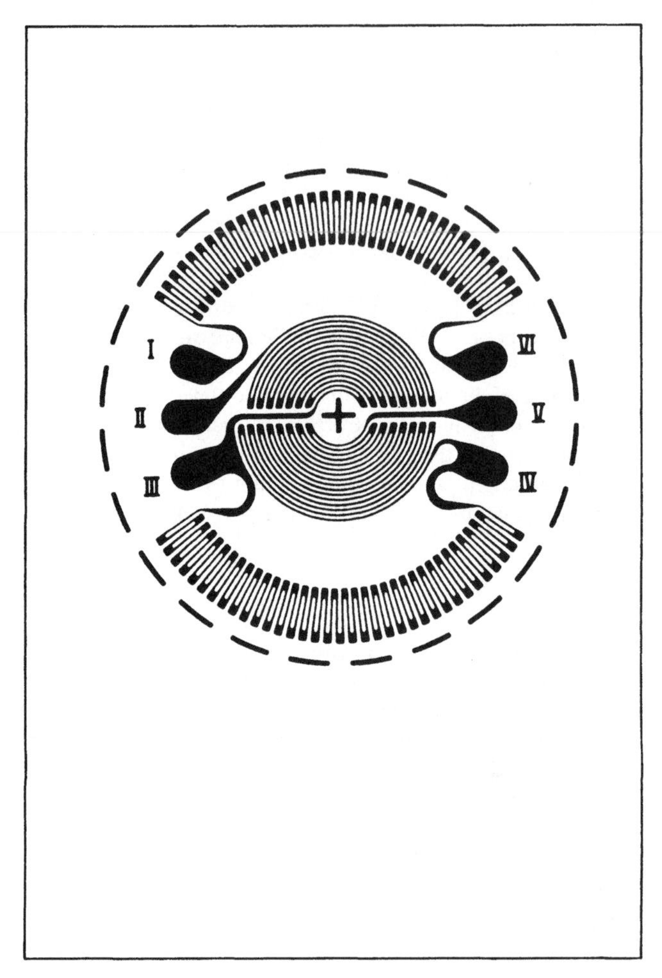

Figure 2 Micro-Measurements' "JB" pattern strain gage for circular diaphragm pressure transducers. (Courtesy of Measurements Group, Inc., Raleigh, NC.)

2.4 Error Contributions from the Flexure Properties

When flexures are designed for bridge transducers, the final transducer may have to possess an accuracy over its operating temperature range of from a few to a fractional percent. Knowledge of the inelasticities and metallurgical behaviors of flexural elements must be considered in transducer design. Metals under a constant load experience a minute deformation with time, called creep. Differences between the loading and unloading curve of a flexure, due to energy absorbed by the material as internal friction, introduce another effect, known as hysteresis. The modulus of elasticity of materials changes with temperature. Corrosion resistance, machinability, magnetics, fatigue effects, thermal conductivity, and thermal expansion are other properties of flexural materials to consider in design application. The 300 series stainless steels are useful flexural materials due to their corrosion resistance, desirable low-temperature properties, and good creep properties at elevated temperatures.

Inconel is a good flexural material in corrosive salt water environments. These materials and others are discussed in an extremely good article on transducer flexures in Chapter 11 of Ref. 1.

3 THE RESISTANCE STRAIN GAGE

Strain gages are used to measure the strain pattern on the surface of the flexure in bridge transducers. In 1938, Simmons, at the California Institute of Technology, and Ruge, at the Massachusetts Institute of Technology, discovered independently that fine wire bonded directly to a surface being studied would respond to surface strain. Ruge's original gage was made by unwinding a constantan wire-wound vitrified resistor, gluing a portion of this wire with Duco cement to a celluloid bar, attaching brass shim stock as terminals, and interfacing the completed assembly to a galvanometer. The first strain gage manufacturer established in the United States was the Baldwin Lima Hamilton (BLH) Corporation (now part of the Vishay Measurements Group). BLH gages operating on the principle discovered by Simmons and Ruge were designated the SR-4 gage to include the initials of both men. The evolution of the bonded resistance wire strain gage occurred during the early 1940s. The first practical bridge transducer load cell was built by Baldwin Lima Hamilton in 1941. By the mid-1950s, Baldwin Lima Hamilton remained the only major strain gage manufacturer and the foil strain gage was beginning to appear. Subsequent work by W. P. Mason and R. N. Thurston (reported in the *Journal of the Acoustical Society of America,* vol. 29, 1957) resulted in the introduction of the commercial semiconductor strain gage. Continued maturation of the bonded resistance strain gage has enabled a transducer industry centered around this technology to develop.

Other sources provide the derivation of all the equations dealing with ensuing topics in this chapter.[2–5]

3.1 Strain Gage Types and Fabrication

Paper-backed wire strain gages typically consist of a grid of resistance wire to which a paper backing has been attached with nitrocellulose cement. The wire is manufactured by drawing the selected alloy through progressive forming dies. To protect it during handling and assembly, the wire is usually sandwiched between two thin layers of paper. Typical grid wires are 0.02 mm in diameter.

Foil strain gages are essentially small printed circuits. Art work for a master gage pattern is first prepared. This pattern is then photographically reduced, and multiple images are placed on a photographic plate. A sheet of foil (typically 0.003–0.005 mm thick) of the appropriate alloy has a light-sensitive emulsion applied, is exposed to the photographic plate, and then undergoes a development process. Chemical etching removes all but the grid material. The resultant grid cross section is square as opposed to round for wire. Advantages inherent in foil gages include better strain transmission due to improved bonding of the grid to the backing, a better thermal path for dissipation of electrically generated heat, and a grid that can more readily be configured to minimize sensitivity to transverse strains.

The total combination of wire and foil gages span grid lengths from 0.2 mm to more than 250 mm. Foil gages satisfy the smaller of these requirements. Both wire and foil gages have associated with them a variety of ohmic values, such as 120, 175, 350, 1000, and so on. Standard values which have historically evolved are 120 and 350 Ω. These values are carry-overs from impedance-matching requirements for galvanometers which were formally used for strain recording. Figure 3 displays numerous configurations of wire and foil strain gages.

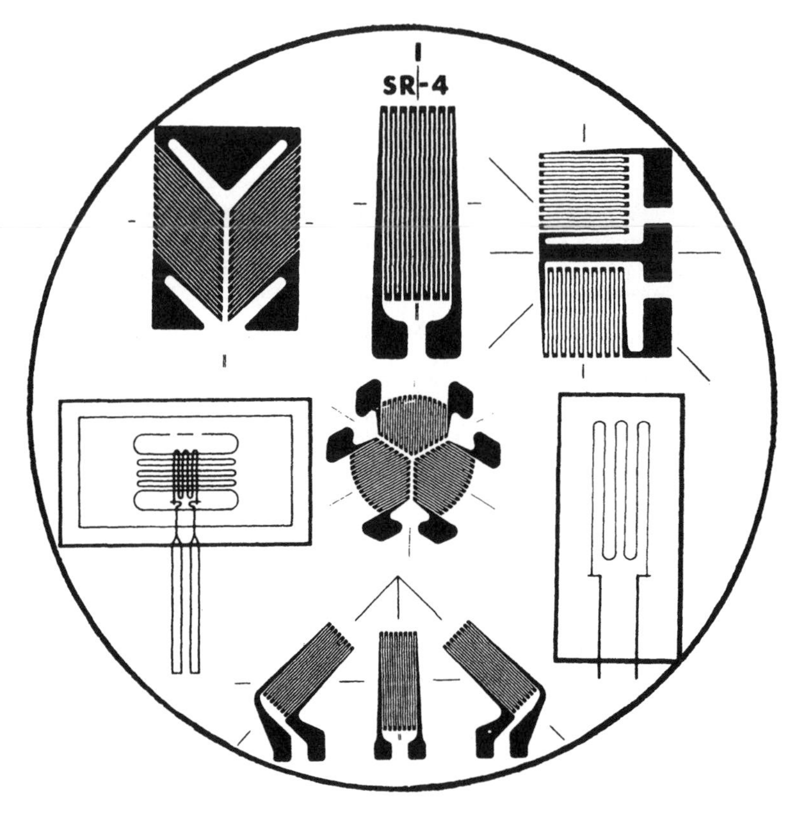

Figure 3 Numerous configurations of wire and foil strain gages—not to scale. (Courtesy of BLH Electronics, now Vishay Measurements Group.)

In the manufacture of bridge transducers using metallic strain gages, vacuum deposition of the gages is an alternate technique to bonding individual gages to the transducer flexure. The flexure is coated with aluminum oxide and then the metal gages are selectively deposited. This process yields the closest match of thermal and electrical characteristics for each bridge element.

The manufacture of semiconductor strain gages starts with a single high-purity silicon crystal. Atoms such as phosphorus (n type) or boron (p type) are doped into the material to lower its resistivity. The parent crystal is sliced into wafers before the dopant is added in a furnace at high temperature (>1000°C). The wafer is masked and etched to produce a suitable grid pattern (usually either a straight element or U-shape). This grid can remain unbacked or can be mounted on a suitable carrier.

Alternately, in diffused semiconductor transducers, the transducer flexure itself may be made of silicon. Its surface can be passivated, etched, and doped to form gage elements integral to the flexure. Similar to vacuum deposition of metal gages, diffused semiconductor transducers offer a more nearly optimum match of thermal and electrical properties for each bridge element. Problems with slippage associated with the bond of a gage carrier or backing are nonexistent.

Bridge transducers using semiconductor gages typically possess poorer thermal and linearity specifications than those using metal gages. However, the sensitivity of semiconductor gages to surface strain is much greater than metal gages. This allows them to be used in transducers providing more signal output (typically 100–500 mV versus 30 mV) and on stiffer flexures, resulting in smaller transducer size, higher frequency response, and increased ruggedness. Although not strictly correct, by convention it has become equivalent to refer to semiconductor-based transducers as either piezoresistive or solid-state transducers.

3.2 Gage Factor

The gage factor F for a strain gage is defined as

$$F = \frac{\Delta R}{R\epsilon} \tag{3}$$

where R is resistance and ϵ is strain equal to $\Delta l / l$ (Δl is the change in length of l_0). Equation (3) may be redefined as

$$F = 1 + 2\nu + \frac{d\rho}{\rho\epsilon} \tag{4}$$

where ν is Poisson's ratio and ρ is the resistivity of the grid material. Most metal gages have a gage factor between 2 and 4.5. For strain gages made from a semiconductor material, the change in resistivity with applied stress is the dominant factor and values as high as 170 are possible. Table 1 lists properties and gage factors for various grid materials.

3.3 Mechanical Aspects of Gage Operation

To build effective bridge transducers, one must be aware of the interaction between the gage and the surface of the transducer flexure to which it is mounted. Mechanical aspects of this interaction include the influences of temperature, backing material, size, orientation, trans-

Table 1 Grid Material Composition, Trade Name, Properties, and Gage Factor

Composition	Trade Name	Properties	Gage Factor
1. Copper–nickel (57%–43%)	Constantan	Strain sensitivity relatively independent of level and temperature; used to 200°C; high resistivity applicable to small grids; measures strains to 20% in annealed form	2.0
2. Nickel–chromium–iron–molybdenum (36%–8%–55.5%–0.5%)	Isoelastic	High gage factor; high fatigue life; used to 200°C; high temperature coefficient of resistance; nonlinear at strain levels above 5%	3.5
3. Nickel–chromium (80%–20%), Nickel–chromium (75%–20%) plus iron and aluminum	Nichrome V Karma	Good fatigue life; stable; high resistance applicable to small grids; used to 400°C	2.2

Note: Nickel alloy gages are susceptible to magnetic fields.

verse sensitivity, distance from the surface, bonding and installation, and gage frequency response.

Temperature

A qualitative discussion of temperature effects on bonded strain gages indicates the effects to be attributable to three principal causes: (1) The transducer flexure to which the gage is attached expands or contracts, (2) the strain gage resistance changes with temperature, and (3) the strain gage grid expands or contracts. With some gages (particularly semiconductors), the change of gage factor with temperature is also extremely significant. These temperature effects are accounted for by temperature–strain calibration; self-temperature-compensated gages where combined effects 1, 2, and 3 above are minimized over a given temperature range for a given combination of grid and flexure material; and a dummy gage integrated into a bridge circuit (discussed later) to electrically subtract temperature-induced strain.

Backing Material

The purpose of the backing material used in constructing strain gages is to provide support, dimensional stability, and mechanical protection for the grid element. The backing material of the gage element(s) acts as a spring in parallel with the flexure to which it is attached and can potentially modify flexure mechanical behavior. In addition, the temperature operating range of the gage can be constrained by its backing material. Most backings are epoxies or glass fiber–reinforced epoxies. Some gages are encapsulated for chemical and mechanical protection as well as extended fatigue life. For high-temperature applications, some gages have strippable backings for mounting with ceramic adhesives. Still other metal gages can be welded. The frequency response of welded gages, due to uncertainties in dynamic response, is a subject area still requiring investigation.

Size

The major factors to be considered in determining the size of strain gage to use are available space for gage mounting, strain gradient at the test location, and character of the material under test. The strain gage must be small enough to be compatible with mounting location and concentrated strain field. It must be large enough so that, on metals with large grain size, it measures average strain as opposed to local effects. Grid elements greater than 3 mm generally have greater fatigue resistance.

Transverse Sensitivity and Orientation

Strain gage transverse sensitivity and mounting orientation are concurrent considerations. Transverse sensitivity in strain gages is important due to the fact that part of the geometry of the gage grid is oriented in directions other than parallel to the principal gage sensing direction. Values of transverse sensitivities are provided with individual gages but typically vary between fractional and several percent. The position of the strain gage axis relative to the numerically larger principal strain on the surface to which it is mounted will have an influence on indicated strain.

Distance from the Surface

The grid element of a strain gage is separated from the transducer flexure by its backing material and cement. The grid then responds to strain at a location removed from the flexure surface. The strain on flexures such as thin plates in bending can vary considerably from that measured by the strain gage.

Bonding Adhesives

Resistance strain gage performance is entirely dependent on the bond attaching it to the transducer flexure. The grid element must have the strain transmitted to it undiminished by the bonding adhesive. The elimination of this bond is one of the principal advantages of vacuum-deposited metallic and diffused semiconductor bridge transducers. Typical adhesives are as follows:

Epoxy Adhesives. Epoxy adhesives are useful over a temperature range of −270 to +320°C. The two classes are either room-temperature curing or thermal setting type; both are available with various organic fillers to optimize performance for individual test requirements.

Phenolic Adhesives. Bakelite, or phenolic adhesive, requires high bonding pressure and long curing cycles. It is used in some transducer applications because of long-term stability under load. The maximum operating temperature for static loads is 180°C.

Polyimide Adhesives. Polyimide adhesives are used to install gages backed by polyimide carriers or high-temperature epoxies. They are a one-part thermal setting resin and are used from −200 to +400°C.

Ceramic cements (applicable from −270 to +550°C) and welding are other mounting techniques.

Frequency Response

The frequency response of bridge transducers cannot be addressed without considering the frequency response of the strain gage as well. It is assumed that the transducer is used in such a manner that mounting variables do not influence its frequency response.

Piping in front of pressure transducer diaphragms and mounting blocks under accelerometers are two examples of variables which can violate this assumption. Transducers, particularly those which measure force, pressure, and acceleration, typically are dynamically modeled as single-degree-of-freedom systems characterized by a linear second-order differential equation with constant mass, damping, and stiffness coefficients. In reality, transducers possess multiple resonant frequencies associated with their flexure and their case. Figure 4 presents the actual frequency response of a bridge-type accelerometer; The response indicates this single-degree-of-freedom model to be adequate through the first major transducer resonance. Such devices have a frequency response usable (constant within 4% referenced to their dc response) to one-fifth of the value of this major resonance. The strain gage itself acts as a spatial averaging device whose frequency response is a function of both its gage length and the sound velocity of the material on which it is mounted. Reference 6 discusses this relationship from which Fig. 5 is extracted. Figure 5 contains curves for three different length gages. Its abscissa must be multiplied by a specific sound velocity. For most bridge transducers, the structural resonance of the flexure constrains its frequency response.

3.4 Electrical Aspects of Gage Operation

The resistance strain gage, which manifests a change in resistance proportional to strain, must form part of an electrical circuit such that a current passed through the gage transforms this change in resistance into a current, voltage, or power change to be measured. The electrical aspects of gage operation to be considered include current in the gage, resistance to ground, and shielding.

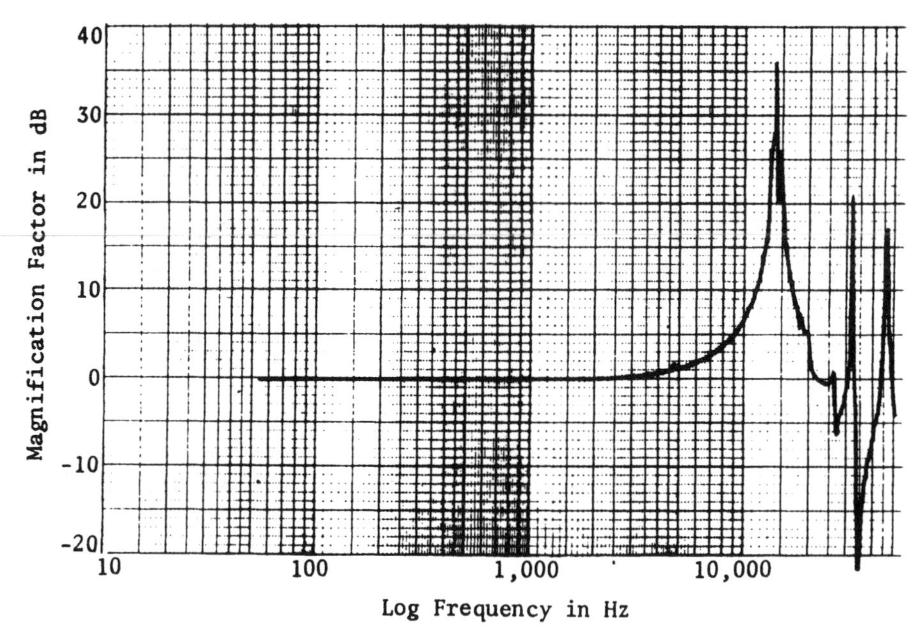

Figure 4 Magnitude of transfer function of piezoresistive accelerometer.

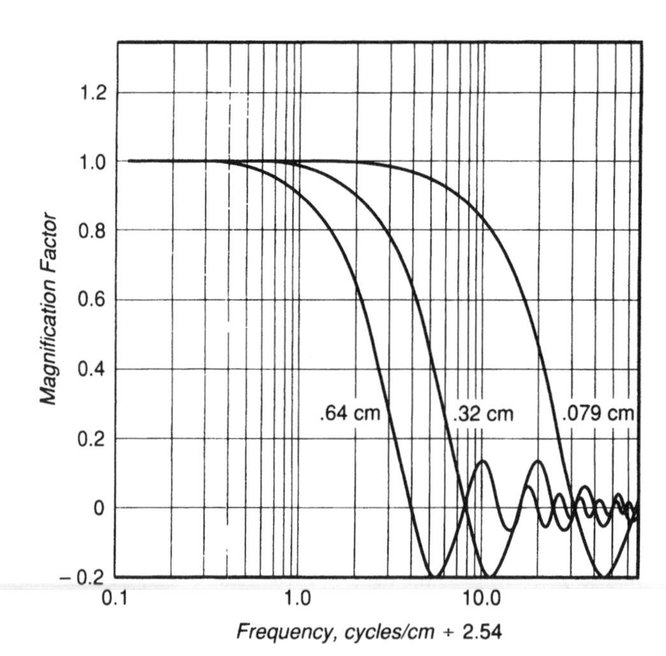

Figure 5 Transfer function for strain gages of varying lengths when analyzed as spatial averaging transducers. (Reprinted with permission from *ISA Transactions,* Vol. 19, Issue 3. Copyright 1980 by the Instrument Society of America.)

Strain gages are seldom damaged by excitation voltages in excess of proper values, but performance degrades. The voltage applied to a strain gage bridge creates a power loss in each arm which must be dissipated in the form of heat. By its basic design, all of the power input to the bridge is dissipated in the bridge with none available to the output circuit. The sensing grid of every strain gage then operates at a higher temperature than the transducer flexure to which it is bonded. The heat generated within the gage must be transferred by conduction to the flexure. Heat flow into the flexure causes a temperature rise which is a function of its heat sink capacity and gage power level. The optimum excitation level for strain gage applications is a function of the strain gage grid area, gage resistance, heat sink properties of the mounting surface, environmental operating temperature range of the gage installation, required operational specifications, and installation and wiring techniques. Rigid operating requirements for precision transducers require performance verification of the optimum excitation level. Zero shift versus load and stability under load at the maximum operating temperature are the performance tests most sensitive to excessive excitation voltage.

Table 2 and Figs. 6–8 allow a first approximation at optimizing bridge excitation levels. Table 2 defines the suitability of various structural materials for providing an adequate heat sink for gage mounting dependent on both accuracy requirements and static or dynamic measurements. Figures 6–8 define the recommended excitation voltage for specific gages as a function of the power density capability of the heat sink and gage grid area.

Resistance to ground is an important parameter in strain gage mounting since insulation leakage paths produce shunting of the gage resistance between the gage and metal structure to which it is bonded, producing false compressive strain readings. The ingress of fluids typically leads to this breakdown in resistance-to-ground value and can also change the

Table 2 Suitability of Various Materails as Heat Sink for Strain Gage Mounting

Accuracy Requirements	Excellent, Heavy Aluminum or Copper Specimen	Good, Thick Steel	Fair, Thin Stainless Steel or Titanium	Poor, Filled Plastic Such as Fiberglass/ Epoxy	Very Poor, Unfilled Plastic Such as Acrylic or Polystyrene
			Static		
High	2–5	1–2	0.5–1	0.1–0.2	0.01–0.02
	3.1–7.8	*1.6–3.1*	*0.78–1.6*	*0.16–0.31*	*0.016–0.031*
Moderate	5–10	2–5	1–2	0.2–0.5	0.02–0.05
	7.8–16	*3.1–7.8*	*1.6–3.1*	*0.31–0.78*	*0.031–0.078*
Low	10–20	5–10	2–5	0.5–1	0.05–0.1
	16–31	*7.8–16*	*3.1–7.8*	*0.78–1.6*	*0.078–0.16*
			Dynamic		
High	5–10	5–10	2–5	0.5–1	0.01–0.05
	7.8–16	*7.8–16*	*3.1–7.8*	*0.78–1.6*	*0.016–0.078*
Moderate	10–20	10–20	5–10	1–2	0.05–0.2
	16–31	*16–31*	*7.8–16*	*1.6–3.1*	*0.078–0.31*
Low	20–50	20–50	10–20	2–5	0.2–0.5
	31–78	*31–78*	*16–31*	*3.1–7.8*	*0.31–0.78*

Note: Units are W/in^2 on top, kW/m^2 in italics underneath. Courtesy of Measurement Group, Inc., Raleigh, NC.

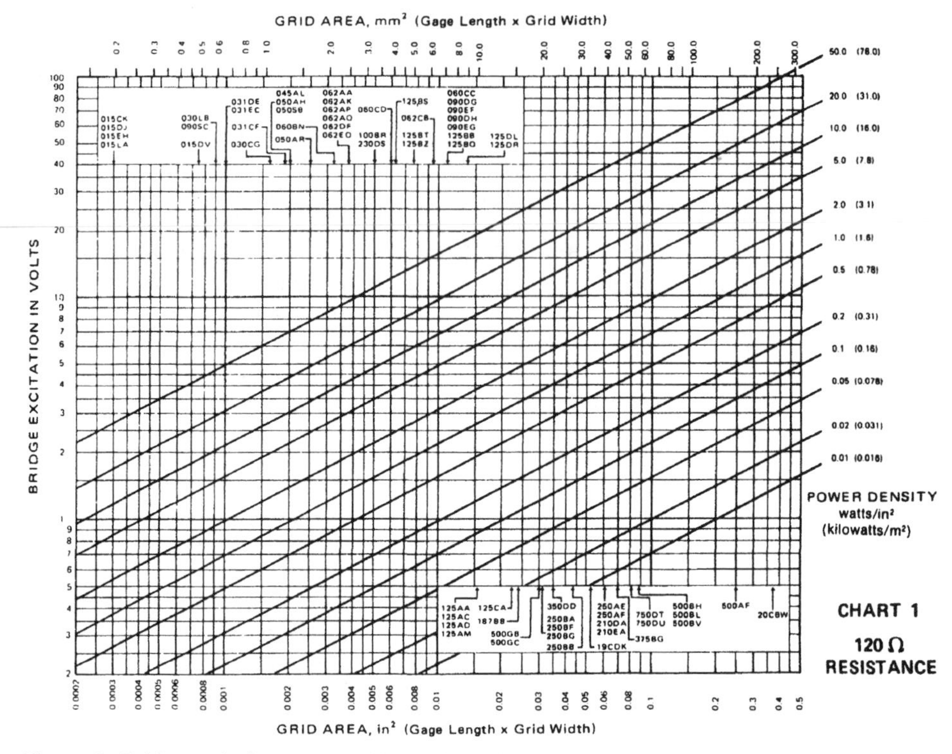

Figure 6 Bridge excitation versus grid area for various power densities and 120-Ω gages. (Courtesy of Measurements Group, Inc., Raleigh, NC.)

mechanical properties of the adhesive. A minimum gage-to-mounting-surface resistance-to-ground value of 50 MΩ is recommended.

Since signals of interest from strain gage bridges are typically on the order of a few millivolts, shielding of the bridge from stray pickup is important. Gage leads should also be shielded and proper grounding procedures followed. Stray pickup may be introduced by 60-Hz line voltage associated with other electronic equipment, electrical noise from motors, radio frequency interference, and so on. Note that shielding materials for electrical fields are different from those for magnetic fields. Nickel alloy strain gages are particularly susceptible to magnetic fields.

3.5 Technical Societies and Strain Gage Manufacturers

In concluding a discussion of the resistance strain gage, it is appropriate to identify some of the technical societies dealing with strain gages and some of the manufacturers of strain gages. In 1956, to accelerate the development of the resistance strain gage, BLH Electronics established a users' group to accomplish this purpose and to further the state of the art in strain gage technology in general. This users' group was formed primarily of various aircraft companies in the western United States and is entitled the Western Regional Strain Gage Committee (WRSGC). For 15 years, the WRSGC was an autonomous organization financed by BLH Electronics. Since 1971, WRSGC has operated under the auspices of the Technical

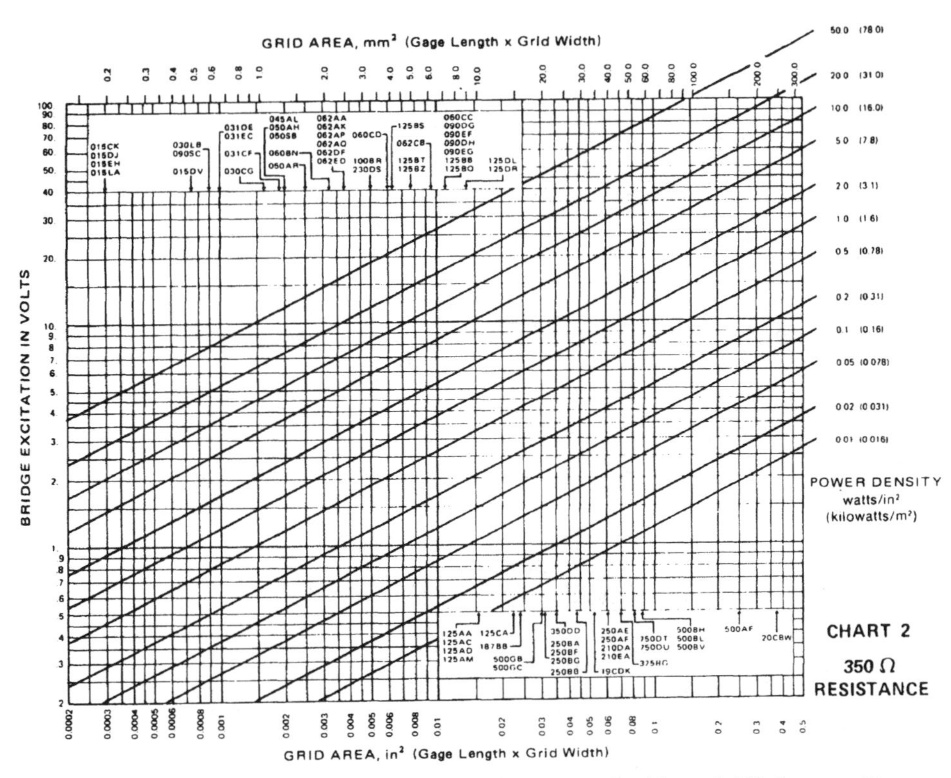

Figure 7 Bridge excitation versus grid area for various power densities and 350-Ω gages. (Courtesy of Measurements Group, Inc., Raleigh, NC.)

Committee on Strain Gages (TCSG) of the Society for Experimental Mechanics (SEM). The SEM is the premier organization in the United States involved with strain gages and experimental mechanics in general. The SEM (formerly Society for Experimental Stress Analysis) was founded by William Murray at the Massachusetts Institute of Technology. Publications of this society include *Experimental Mechanics* and *Experimental Techniques.* A similar European organization is the Joint British Committee for Stress Analysis, whose publication is *The Journal of Strain Analysis for Engineering Design.*

By 2004, essentially the entire strain gage manufacturing capability in the United States was consolidated by Vishay Measurements Group, corporate headquarters at Vishay Intertechnology, Inc., 63 Lincoln Highway, Malvern, Pennsylvania 19355-2120. Its associated website has an extensive array of in-depth articles on strain gage technology.

4 THE WHEATSTONE BRIDGE

Best transducer performance can be achieved by minimizing the strain level in the transducer flexure. Lower strains allow increased safety without mechanical overload protection. Effective overload stops are usually troublesome to design and an added expense to make. Reduced strain levels almost always produce an improvement in linearity accompanied by a reduction in the hysteresis originating in the transducer flexure material.

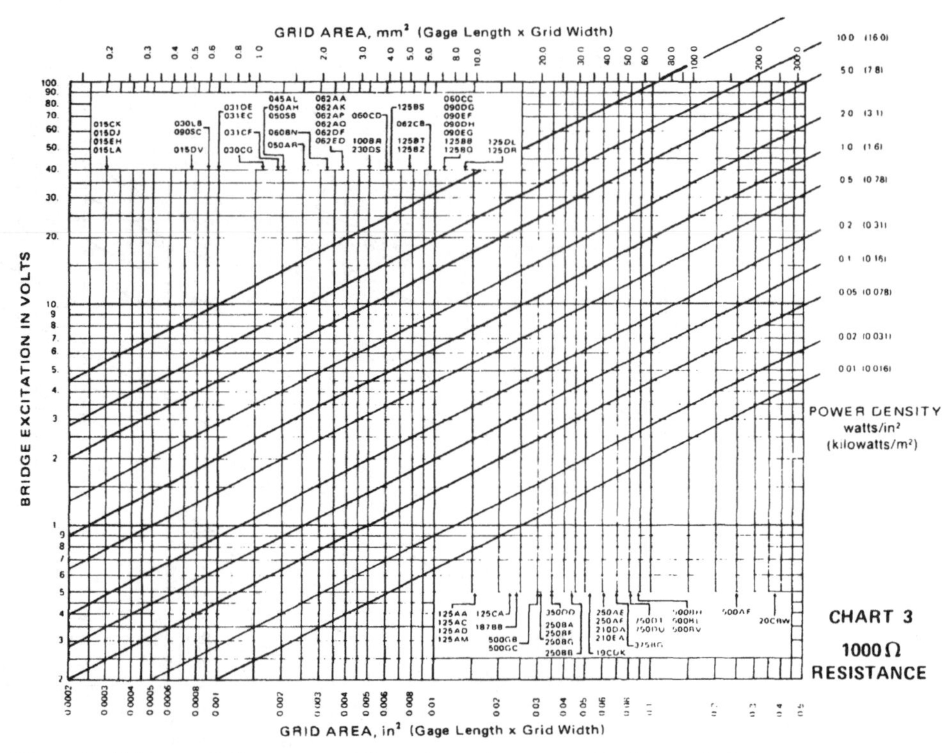

Figure 8 Bridge excitation versus grid area for various power densities and 1000-Ω gages. (Courtesy of Measurements Group, Inc., Raleigh, NC.)

Small strains result in small impedance changes in resistive strain gage elements. Electromechanical transducers use a Wheatstone bridge circuit to detect a small change in impedance to a high degree of accuracy.

4.1 Bridge Equations

The circuit most often used with strain gages is a four-arm bridge with a constant-voltage power supply. Figure 9 shows a basic bridge configuration. The supply voltage E_{ex} can be

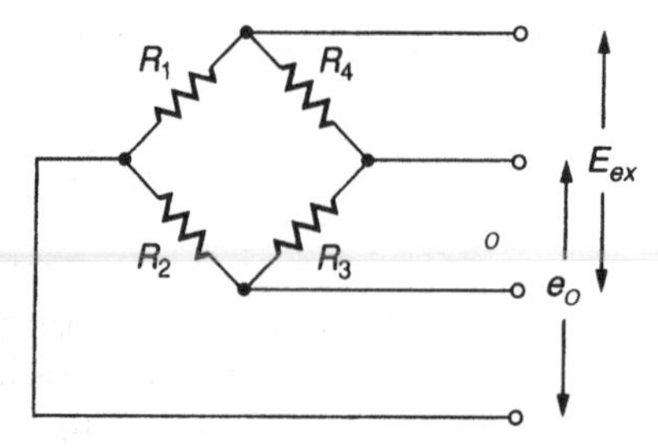

Figure 9 Four-arm bridge with constant-voltage (E_{ex}) power supply.

either ac or dc, but for now we assume it is dc so equations can be written in terms of resistance R rather than a complex impedance. The condition for a balanced bridge with e_o equal to zero is

$$\frac{R_1}{R_2} = \frac{R_4}{R_3} \tag{5}$$

Next, an expression is presented for e_o due to *small* changes in R_1, R_2, R_3, and R_4:

$$e_o = \left[-\frac{R_3\,dR_4}{(R_3 + R_4)^2} + \frac{R_4\,dR_3}{(R_3 + R_4)^2} - \frac{R_1\,dR_2}{(R_1 + R_2)^2} + \frac{R_2\,dR_1}{(R_1 + R_2)^2} \right] E_{\text{ex}} \tag{6}$$

In many cases, the bridge circuit is made up of equal resistances. Substituting for individual resistances, a strain gage resistance R, and using the definition of the gage factor from Eq. (3), Eq. (6) becomes

$$e_o = \frac{FE_{\text{ex}}}{4}(-\epsilon_4 + \epsilon_3 - \epsilon_2 + \epsilon_1) \tag{7}$$

The unbalance of the bridge is seen to be proportional to the sum of the strain (or resistance changes) in opposite arms and to the difference of strain (or resistance changes) in adjacent arms.

Equations (6) and (7) indicate one technique to compensate strain gage circuits to minimize the influence of temperature-induced strain. This was referred to in Section 3 as the dummy gage method.

Assume that we have a bridge circuit with one active arm and arbitrarily let this arm be number 4. Equation (7) becomes

$$e_o = \frac{FE_{\text{ex}}}{4}(-\epsilon_4) \tag{8}$$

Arm 4 responds to the total strain induced in it, which is comprised of both thermal (t) and mechanical (m) strain

$$\epsilon_4 = \epsilon_m + \epsilon_t \tag{9}$$

A problem arises if it is desired to isolate the mechanical strain component. One solution is to take another strain gage (the dummy gage) and mount it on a strain-isolated piece of the same material as that on which gage 4 is mounted. If placed in the same thermal environment as gage 4, the output from the dummy gage becomes simply ϵ_t. If the dummy gage is wired in an adjacent bridge arm to 4 (1 or 3), Eq. (7) becomes

$$e_o = \frac{FE_{\text{ex}}}{4}(-\epsilon_m - \epsilon_t + \epsilon_t) \tag{10}$$

Equation (10) indicates that thermal strain effects are canceled. Similarly, in Fig. 2, four gages were shown mounted on a transducer diaphragm. Equation (7) indicates that thermal strain effects from this circuit should be canceled.

In reality, perfect temperature compensation is not achieved since no two strain gages from a lot track one another identically. However, compensation adequate for many applications can be accomplished.

The biggest thermal problem with bridge transducers occurs in transient situations, such as explosive or combustion environments. Here, due to individual physical locations, gages in a bridge are not in the same time-varying temperature, and compensation cannot be achieved. The only technique which can be used in this situation is either to cool the transducer by circulating water or gas around it or to delay the thermal transient until the measurement is complete.

The alternating signs in Eq. (7) are useful in isolating various strain components when using bridge circuits containing strain gages. Figure 10 shows a beam flexure used in an accelerometer. Four gages are mounted on the beam—two on the top and two on the bottom. Notches are placed in the beam to intensify the strain field under the gages. Due to symmetry, the tension gages see the same strain as do the compression gages. If the tension gages occupy two adjacent arms of the bridge and the compression gages the other two, Eq. (7) indicates that the net bridge output will be zero. However, if the tension and compression gages are in opposite arms, Eq. (7) indicates that a bending strain signal four times that of an individual gage will be produced with temperature compensation also achieved.

Equation (6) presented the generalized form of the bridge equation for four active arms. If only one arm (e.g., arm 4) is active, this equation becomes

$$e_o = \left[\frac{-R_3 dR_4}{(R_3 + R_4)^2}\right] E_{\text{ex}} \tag{11}$$

This equation was specifically presented for *small* changes in resistance, such as those associated with metallic strain gages. If the change in resistance in arm 4 is large, Eq. (11) is better expressed as

$$e_o = \frac{(R_4 + \Delta R_4)E_{\text{ex}}}{(R_4 + \Delta R_4) + R_3} - \frac{R_4 E_{\text{ex}}}{R_4 + R_3} \tag{12}$$

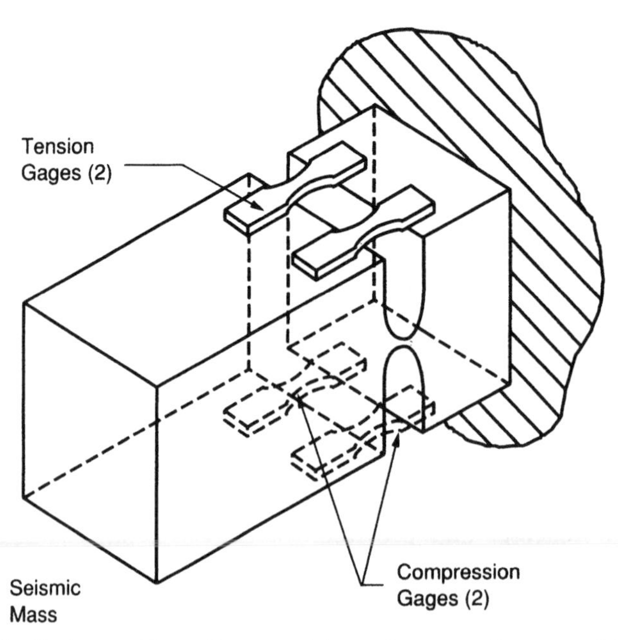

Figure 10 Strain gaged beam flexure used in accelerometer. (Courtesy of Endevco.)

For an equal-arm bridge, this becomes

$$e_o = \frac{\Delta R\, E_{ex}}{4R + 2\,\Delta R} = \frac{FE_{ex}\epsilon}{4 + 2F\epsilon} \tag{13}$$

For an equal-arm bridge, Eq. (11) becomes

$$e_o = \frac{dR\, E_{ex}}{4R} = \frac{FE_{ex}\epsilon}{4} \tag{14}$$

The difference between Eqs. (14) and Eq. (13) is that Eq. (14) describes a linear process while Eq. (13) describes a nonlinear one. Semiconductor gages, because of their large gage factor, require analysis using Eq. (13).

Semiconductor gages may be used in constant-voltage four-arm bridge circuits when two or four gages are used in adjacent arms and strained so that their outputs are additive. Analysis of the bridge equations for this situation will show that if gages in adjacent arms are subjected to equal but opposite values of ΔR, the output signal is doubled and the nonlinearity in the bridge output is eliminated. Another approach to eliminating this nonlinearity is to design a circuit where the current through the strain gage remains constant.

Table 3 provides generalized bridge equations for one, two, and four equal-active-arm bridges of various configurations. The dimensionless bridge output is presented in millivots per volts for a constant-voltage power supply. Strain is presented in microstrain. No small-strain assumption is built into these equations. For large strains with semiconductor gages, F may not be a constant and this correction also has to be built into the equations. In this table, the Poisson gage is one which measures the lateral compressive strain accompanying an axial tension strain. As noted earlier, only for two adjacent active gages with equal and opposite strains or for four active gages with pairs subjected to equal and opposite strains is the bridge output a linear function of strain.

4.2 Lead Wire Effects

There has been a historical lack of agreement between manufacturers of strain gages as to color codes and wiring designations. This is particularly true in bridge transducers. Figures 11 and 12 are suggested industry standards which have assisted in lessening this confusion. Figure 11 covers the situation where all bridge elements are remote from the power supply, while Fig. 12 covers the situation where only one bridge arm is remote from the power supply. The bridge balance network and shunt calibration are discussed in Sections 5 and 6, respectively. Table 4 presents guidelines for multiconductor strain gage cable.

The previous discussion has assumed that the only resistive elements in the circuits are the gages themselves. Resistance of circuit lead wires also is a consideration.

One possible need for remote recording occurs when the bridge power supply and the readout instrumentation are at one location and the bridge transducer is at a remote location. In this situation, the resistance R_L of each lead wire between the bridge and the power supply or readout must be accounted for. Most readout instruments have very high input impedances, so the effect of R_L in series with them can be ignored. The significant effect of lead-wire resistance is to modify the resistance in series with the power supply from R_{bridge} to $R_{bridge} + 2R_L$. For example, a lead-wire resistance of 3 Ω and a bridge resistance of 120 Ω will produce loading effects which, if not corrected, will result in a 5% error in bridge transducer output.

There are at least three simple techniques to eliminate this error source:

Table 3 Equations for One, Two, and Four Equal-Active-Arm Bridges

Bridge/Strain Arrangement	Description	Output Equation–e_o/E_{ex} (mV/V)
ϵ, e_o, E_{ex}	Single active gage in uniaxial tension or compression	$\frac{e_o}{E_{ex}} = \frac{F\epsilon \times 10^{-3}}{4 + 2F\epsilon \times 10^{-6}}$
ϵ, $-\nu\epsilon$, e_o, E_{ex}	Two active gages in uniaxial stress field—one aligned with maximum principal strain, one "Poisson" gage	$\frac{e_o}{E_{ex}} = \frac{F\epsilon(1 + \nu) \times 10^{-3}}{4 + 2F\epsilon(1 - \nu) \times 10^{-6}}$
ϵ, $-\epsilon$, e_o, E_{ex}	Two active gages with equal and opposite strains—typical of bending-beam arrangement	$\frac{e_o}{E_{ex}} = \frac{F\epsilon}{2} \times 10^{-3}$
ϵ, ϵ, e_o, E_{ex}	Two active gages with equal strains of same sign—used on opposite sides of column with low temperature gradient (e.g., bending cancellation)	$\frac{e_o}{E_{ex}} = \frac{F\epsilon \times 10^{-3}}{2 + F\epsilon \times 10^{-6}}$
ϵ, $-\nu\epsilon$, $-\nu\epsilon$, ϵ, e_o, E_{ex}	Four active gages in uniaxial stress field—two aligned with maximum principal strain, two "Poisson" gages (column)	$\frac{e_o}{E_{ex}} = \frac{F\epsilon(1 + \nu) \times 10^{-3}}{2 + F\epsilon(1 - \nu) \times 10^{-6}}$
ϵ, $-\nu\epsilon$, $-\epsilon$, $\nu\epsilon$, e_o, E_{ex}	Four active gages in uniaxial stress field—two aligned with max. principal strain, two "Poisson" gages (beam).	$\frac{e_o}{E_{ex}} = \frac{F\epsilon(1 + \nu) \times 10^{-3}}{2}$
ϵ, $-\epsilon$, $-\epsilon$, ϵ, e_o, E_{ex}	Four active gages with pairs subjected to equal and opposite strains (beam in bending or shaft in torsion)	$\frac{e_o}{E_{ex}} = F\epsilon \times 10^{-3}$

Courtesy of Measurements Group Inc., Raleigh, NC.

1. The bridge transducer can be calibrated with the long length of cable with which it will operate.
2. The excitation voltage E_{ex} can be measured at the bridge itself instead of at the power supply and appropriate values substituted in Eq. (6) or equivalent versions of it.
3. The bridge voltage E_{ex} can be determined by measuring the current to the bridge (I_{bridge}) and calculating E_{ex} as the product $I_{bridge} \times R_{bridge}$.

Another possible need for remote recording occurs when two gages (either both active or one active and one for temperature compensation) are at the test site. The other two bridge completion resistors are in parallel with the power supply and located adjacent to it. In Fig. 9, assume R_3 and R_4 are the two remote active arms. In Eq. (6), the last two terms would be zero since these arms are not active. In this equation R_3 and R_4 would become respectively $R_3 + R_L$ and $R_4 + R_L$. If the strain gages in both arms are identical, Eq. (6) reduces to

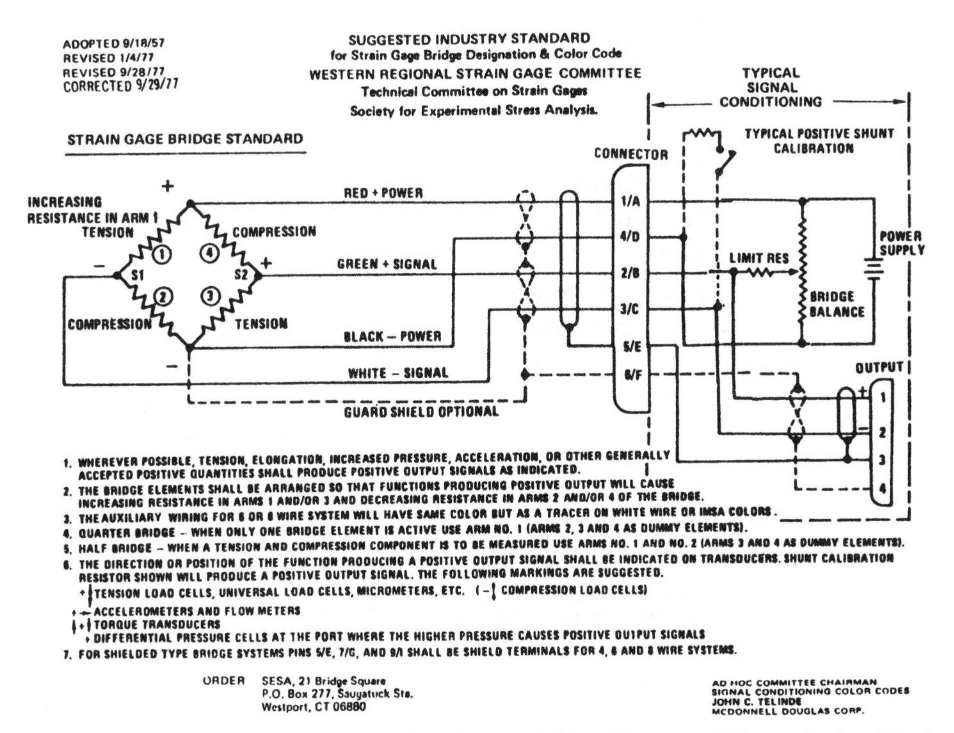

Figure 11 Color code and wiring designation, four-arm bridge. (Courtesy of Western Regional Strain Gage Committee.)

$$e_o = \frac{FE_{\text{ex}}R}{4(R + R_L)}(-\epsilon_4 + \epsilon_3) \tag{15}$$

Other situations can be investigated by substituting appropriate values for the resistance in each bridge arm (including lead-wire resistance) in the governing bridge equation. In addition, Section 6 will show that shunt calibration is one technique that can be used to compensate the system for the effects of lead-wire resistance.

4.3 Temperature Compensation

Before leaving the analysis of the Wheatstone bridge circuit, temperature compensation of bridge-type transducers should receive additional discussion. An ideal transducer would yield an output voltage which is a constant calibration factor times its mechanical input, independent of other environmental factors. Ambient temperature variations are one of the major error sources in precision transducers. Even when using self-temperature-compensated strain gages and taking advantage of the ability of the Wheatstone bridge circuit to subtract in the dummy gage method, some residual error remains. These remaining errors are of two types.

First, the transducer zero output can change with temperature. Unequal mechanical expansion of transducer members can cause this effect. Second, the calibration factor, span, or sensitivity also can change with temperature. This can be caused, for example, by a change in the stiffness of the transducer flexure with temperature.

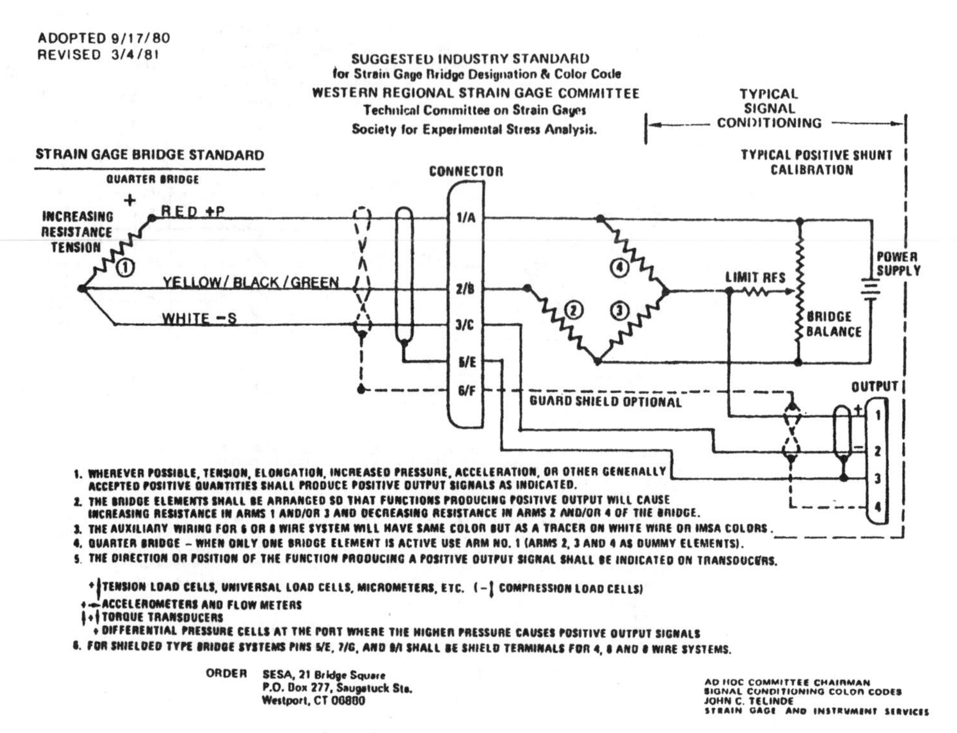

Figure 12 Color code and wiring designation, single-arm bridge. (Courtesy of Western Regional Strain Gage Committee.)

Table 4 Multiconductor Strain Gage Cable Guideline, Western Regional Strain Gage Committee

A need exists for low-millivolt signal levels to be transmitted by better quality multiple conductor cables of considerable length.

I. Conductors	Three through eight twisted, stranded conductors of tinned copper per ASTM-B-174, AWG 20-16/32, or AWG 18-16/30
II. Color code	Jacket: orange, grey, white, or black Conductors: Per ISA-S37.3, ANSI-MC6.2-1975, and WRSGC/SESA 5-6-1960
III. Insulation	Superior to the PVC materials currently in use. The dielectric material should be nonhygroscopic and approach zero water absorption and zero water permeability. Target jacket thickness of .016 in. or less and conductor insulation of .012 in. or less. Target resistance values should be constant as practical humid and wet environments and as high as possible (target value of 500 Ω per 1000 ft). The breakdown level of the dielectric materials shall be greater than 150 V dc.
IV. Construction	The cable shield shall be aluminized polyester tape with 100% coverage of all conductors. A 22-AWG drain wire shall be in intimate contact with the shield throughout the entire cable length. The cable shall have as a small a diameter as practical and be flexible enough to have a bend radius less than six cable diameters. Overall cable strength sufficiently high to be pulled through conduits.

Courtesy of Society for Experimental Mechanics.

The following discussion provides one compensation scheme for each type (metallic and semiconductor) of bridge transducer. References 7 and 8 are sources of more detailed information. An equal-arm bridge transducer operating with a constant-voltage supply is assumed. Metallic strain gages are discussed first.

Figure 13 shows one scheme for compensating for transducer zero shift. A corner of the bridge is brought out to terminals, and a temperature-sensitive resistor, r, is placed in one side of the bridge. Typically, a wire resistor such as Balco, nickel, or copper with a positive temperature coefficient is used.

The transducer must first be temperature calibrated and the change in zero reading for a given temperature range determined. This can be characterized in volts of output change per volt of input. Definitions are

b = output voltage change per degree per input volt
a = temperature coefficient of resistance of r
R = bridge arm resistance
T = temperature change from reference temperature

If the bridge supply voltage is E_{ex} and R is changed a small amount by the addition of r, the bridge output is

$$e_o = \frac{E_{ex} r}{4R} \tag{16}$$

Equation (16) can further be expressed as

$$e_o = \frac{E_{ex} r_0 (1 + aT)}{4R} \tag{17}$$

or

$$e_o = \text{const} + E_{ex} bT \tag{18}$$

where r_0 is the value of r at the reference temperature. The effect of the constant term is eliminated by a temperature-insensitive trim resistor in an adjacent arm. The above equations indicate that at the reference temperature r_0 should be selected equal to 4 Rb/a. If the transducer is properly designed, b is very small compared to a, keeping the compensating resistor small in value. The compensating resistor should be located in an arm causing a voltage change of opposite sign to the zero drift with increasing temperature.

After zero shift is compensated, the calibration or span factor remains to be compensated. Most metal strain gage transducers give larger outputs with increasing temperature, so the temperature coefficient of the calibration scale factor, K, is positive. The trick in span compensation is to hold the transducer supply voltage constant while automatically varying

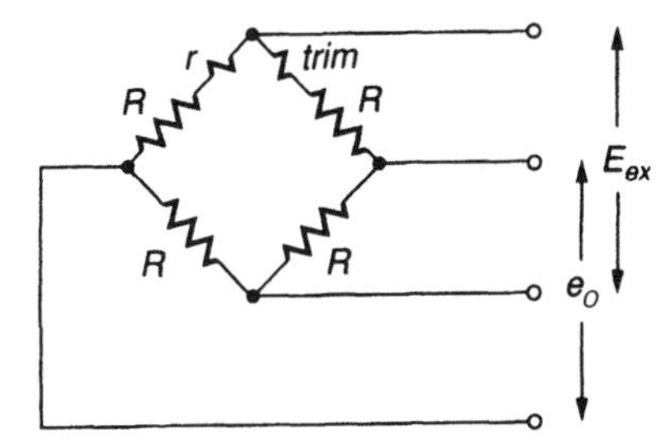

Figure 13 Transducer bridge compensation for zero shift, metal gages.

the bridge current, causing it to decrease with increasing temperature. In this discussion, r is identified to be a positive series resistor (Fig. 14). Definitions are

$r = r_0(1 + aT)$
a = positive temperature coefficient of r
T = temperature difference from reference temperature
c = temperature coefficient of the calibration factor K, so $K = K_0(1 + cT)$
E_{ex} = transducer supply voltage

The voltage on the transducer at the reference temperature is $RE_{ex}/(R + r_0)$ and at temperature T is $RE_{ex}/(R + r)$. The ratio by which it changes is $(R + r_0)/(R + r)$, which is used to correct for the variation in K. This variation is corrected for when $K_0(1 + cT)(R + r_0)/[R + r_0(1 + aT)]$ = constant. The value of r_0 which satisfies this requirement can be shown to be

$$r_0 = \frac{cR}{a - c} \tag{19}$$

Note that in span and zero compensation as discussed thus far, the compensating resistors must be at the same temperature as the transducer. Usually, this is accomplished by mounting the resistors inside the transducer.

Figure 15 shows one technique for correcting for zero shift due to the temperature in semiconductor bridges. Temperature compensation is performed by adding non-temperature-sensitive resistors in series and parallel to the gage having the highest resistance change with temperature. The objective of this method is to achieve both zero balance and temperature compensation together. Since the compensation resistors are non–temperature sensitive, they can be added wherever convenient in the circuit.

The bridge is first balanced using a series resistor at ambient room temperature. Next, the transducer is cycled over the temperature extremes. A parallel resistor is installed across the gage having the greatest resistance change. The bridge is then rebalanced and the procedure repeated until satisfactory performance is achieved.

Semiconductor bridge transducers are typically compensated for calibration or span factor with a circuit as in Fig. 14. However, r for this situation is a non-temperature-sensitive resistor. For p-type silicon gages, the strain sensitivity drops with temperature while the resistance rises. The increase in resistance occurs at a greater rate than does the decrease in sensitivity. Figure 14 shows that the effect of an increase in resistance R, with r constant, is to increase the voltage applied to the bridge, offsetting the decrease in strain sensitivity. Alternately, in Fig. 14, r can be replaced by a thermistor instead of a fixed dropping resistor. The thermistor is generally a more efficient method of compensation but must be in the same thermal environment as the bridge network.

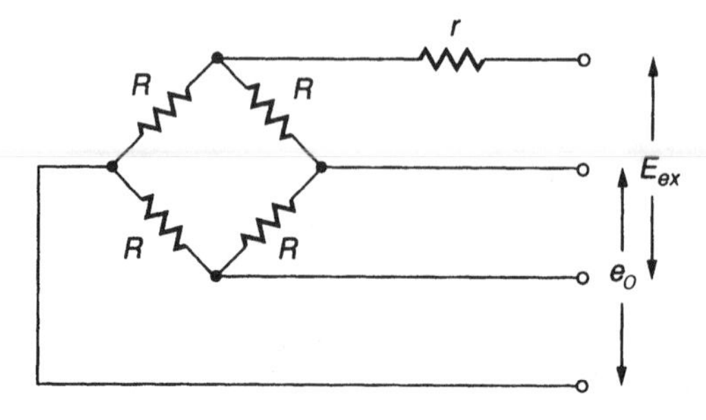

Figure 14 Transducer bridge compensation for span, metal gages.

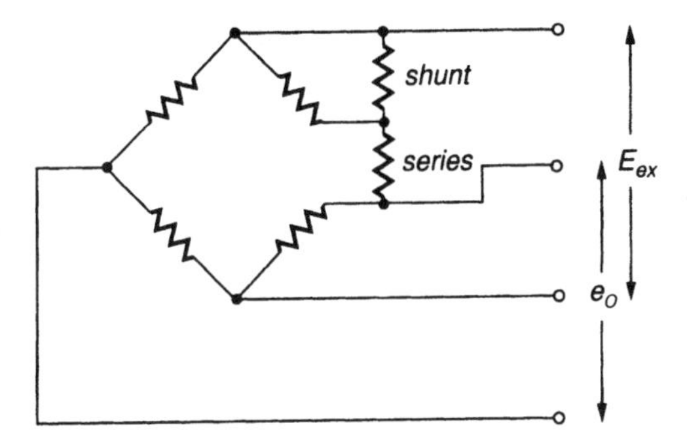

Figure 15 Transducer bridge compensation for zero shift, semiconductor gages.

When balancing Wheatstone bridges, it must be determined that the balancing circuit does not significantly alter the thermal compensation network. Balancing methods are discussed next.

5 RESISTANCE BRIDGE BALANCE METHODS

Even when a best attempt is made at matching resistors, the output from a bridge transducer with zero measurand applied is always something other than zero volts. With microprocessors and scanners, this is of little consequence. The initial bridge output can be acquired and stored in the memory of the microprocessor and then subtracted from all subsequent readings. Frequently, however, it is desired to initialize a bridge circuit such that a zero value of measurand corresponds to zero voltage. For example, assume it is desired to acquire a vibration measurement on a space vehicle using a bridge transducer. Assume the channel is to be calibrated for $\pm 20g$ and the accelerometer has a sensitivity of 1 mV/g (g = standard acceleration of gravity). If the data channel range were ± 20 mV, and the accelerometer acquiring the measurement had a zero offset of 5 mV, the channel could transmit only in the range of $+15g$ to $-25g$ as opposed to $\pm 20g$. Balancing the bridge would solve this problem.

Equation (5) presented the requirement for a balanced bridge. Basically, the resistance ratio of any two adjacent bridge arms must be equal to the resistance ratio of the other two arms. Any bridge-balancing network must then have as its objective the satisfying of this criterion. The two main types of zero balancing methods are those which manipulate one arm of a transducer bridge to bring its output to the desired condition and those which manipulate two adjacent arms of the transducer bridge.

Figure 16 presents the most common circuit for manipulating a single bridge arm. A variable resistor R_B is placed across one of the resistors (say R_4) whose value needs to be

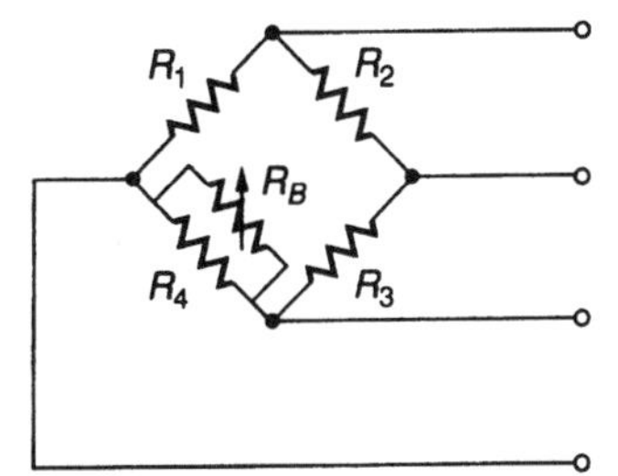

Figure 16 Circuit for manipulating a single bridge arm.

lessened such that $R_1/R_4 = R_2/R_3$. The effect of R_B in parallel with R_4 is to lessen the value of the bridge arm from R_4 to some new value R_T.

The overall combination of R_B in parallel with R_4 must be variable over a range at least equal to the maximum possible initial unbalance of the bridge. Selecting this range, other than by trial and error, requires knowledge of the strain gage resistance R, its tolerance in percentage m, and the number of active gages n in the bridge. The range of the balancing circuit should be

$$\frac{2Rmn}{100} \tag{20}$$

Note that the presence of the variable resistor R_B desensitizes the bridge network since $\Delta R_4/R_4$ is not equal to $\Delta R_T/R_T$. If the strain gages are initially closely matched, the influence of this effect is small since R_B will remain large and R_T will closely approximate R_4. For optimum precision, the best method to minimize the influence of the variable resistor is to calibrate the transducer once the bridge is balanced. Of course, if less than four arms of the bridge are active and balancing is performed across a dummy completion resistor, no desensitizing of the bridge occurs.

Two techniques are available to manipulate two adjacent arms in a bridge. Again, the rationale for this manipulation is to satisfy Eq. (5). The first technique is series manipulation, which assumes the bridge is open such that the variable resistance may be inserted in series with two arms of the bridge. This technique is not applicable to a closed bridge.

Figure 17 shows a variable series resistor R_S inserted in one corner of the bridge. The insertion of R_S, which is typically quite small, allows adjustment of the ratio of R_2 to R_3 to achieve balance. Reference 2 provides the best discussion of bridge balance networks and indicates that minimum bridge desensitization occurs when bridge power is applied across the vertical terminals of Fig. 17 as opposed to the horizontal.

The second technique discussed (and the one typically used) is parallel manipulation of two adjacent bridge arms. Figure 18 illustrates this technique. The parallel variable resistor R_B allows the ratio of R_2/R_3 to be adjusted. A pad resistor, R_{PAD}, serves simply to keep the individual bridge arms from being shorted out at the end of travel of R_B. Again, the secret is to keep the combination of R_B and R_{PAD} as high as possible to avoid bridge desensitization. If no other guidance is available, start out with a pad resistor about 100 times the bridge resistance and a variable resistor about 20 times the bridge resistance. Again, maximum accuracy is achieved when the bridge transducer is calibrated with the balance network with which it will be used.

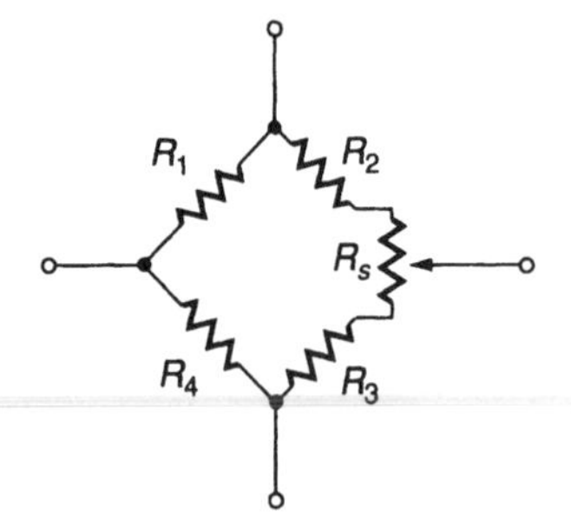

Figure 17 Circuit for series manipulation of two adjacent bridge arms.

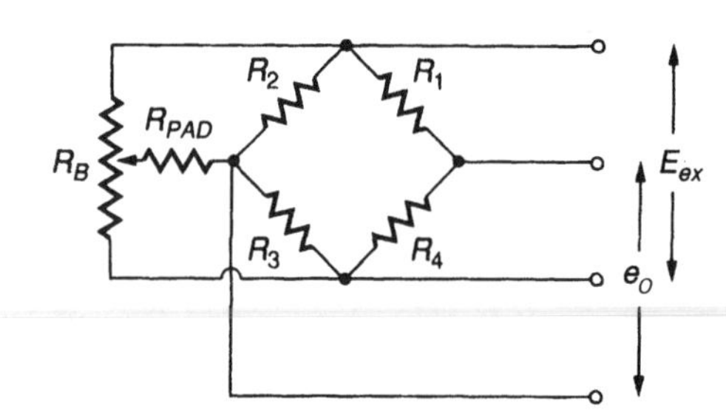

Figure 18 Circuit for parallel manipulation of two adjacent bridge arms.

As alluded to earlier, the addition of a balance network to a bridge transducer may react unfavorably with temperature compensation resistors placed in the transducer's circuitry by its manufacturer. Temperature compensation can be severely modified by the presence of this balancing network. The prerequisite to insertion of a balance network should be an exact knowledge of the circuit of the transducer. For this reason, and for reasons associated with desensitization of the transducer, balance networks should be avoided unless required.

6 RESISTANCE BRIDGE TRANSDUCER MEASUREMENT SYSTEM CALIBRATION

A basic component in any measurement system is the transducer. The measurement system can be as simple as a transducer, a cable from the transducer, and a recorder. Alternately, the measuring system can contain many more of the elements of the transmitting and receiving system defined in Section 1. Cables, amplifiers, filters, digitizers, tape recorders, and so on all have the capability, when inserted into a measurement system, to modify both the amplitude and spectral content of the signal from the transducer defining the measurand. The response of these components may also drift with time.

To obtain measurements of the highest possible quality, one must accurately and carefully calibrate the entire measurement system as near to the time of actual measurement as possible. The calibrations may be conducted prior to, immediately after, or even during the time of actual measurement. Such calibration of an entire measurement system is referred to as "end-to-end" calibration. This calibration ordinarily does not replace the evaluation of individual components of the measurement system.

One group concerned with "end-to-end" calibration of measurement systems is the Telemetry Group/Range Commanders Council whose Secretariat is headquartered at White Sands Missile Range, New Mexico. The Transducer Committee of the Telemetry Group coordinated the writing of Chapter 2 of Ref. 9, entitled "Test Methods for Transducer-Based System Calibrations." The following information is largely extracted from that chapter, which also deals with piezoelectric transducers, servo transducers, capacitive and inductive transducers, and thermoelectric transducers.

A preferred calibration procedure is one in which a known value of the measurand is applied directly to the transducer in the measurement system (transmitting, receiving, and display) in which it will be used. This procedure permits the output display to be read directly in terms of units of the measurand.

6.1 Static Calibration

The basic equipment for the static calibration of transducer systems consists of a measurand source supplying accurately known and precisely repeatable values of the measurand and an output-indicating or recording system. The combined errors or uncertainties of the calibration system should be sufficiently smaller than the permissible tolerance of the system performance characteristic under evaluation so as to result in meaningful calibration values. All calibration system components should be periodically checked against standards. Environmental conditions during calibrations should be constant and specified to permit corrections to the data, as required.

The procedures which will be specified are based on the assumption that the measuring system is linear. For systems that will ultimately measure dynamic data, linearity is a prerequisite.

The static calibration sequence consists of the following steps:

1. Zero-measurand output verification
2. Sensitivity verification
3. Linearity and hysteresis verification
4. Repeatability verification

Zero-measurand output verification starts with a measurement of system output with zero measurand applied to the transducer. Zero measurand is an important measurement for several reasons:

1. In many measurement systems, the transducers are exposed to zero measurand before the test begins.
2. For many transducers, zero measurand means that no external input to the transducer is required, thus greatly simplifying the test procedure.
3. Measurement system malfunctions, including drift, will frequently appear when the system output is monitored over some reasonable time period with zero measurand applied.
4. In some measurement systems with no external measurand input, the ambient environment will furnish an important reference point for the system calibration. For example, for an absolute-pressure-measuring system, the ambient atmospheric pressure provides this reference. Similarly, for certain accelerometer systems, the measurable attitude of the test vehicle prior to launch represents a known component of the earth's gravitational field as input to the accelerometer.

For a measurement system with a linear response, the slope of the line characterizing the input measurand versus system output represents the sensitivity of the system. There are a number of straight lines which may be chosen to provide this sensitivity verification. These include the following:

1. Endpoint line—The straight line between the outputs at the specified upper and lower limits of the range.
2. Best straight line—The line midway between the two parallel straight lines closest together that enclose all output versus measurand values on a calibration curve.
3. Terminal line—The straight line for which the endpoints are 0 and 100% of both measurand and output.
4. Theoretical slope line—The straight line connecting the specified points between which a specified theoretical curve has been established.
5. Least-squares line—The straight line for which the sum of the squares of the residuals is minimized for all calibration points.

Procedures used in the verification of sensitivity will depend on specific accuracy, calibration time, and expense trade-off choices for each system. For unidirectional transducers, it is typical to calibrate from zero to full scale and back again in 10% of full-scale increments (21 points). For bidirectional transducers, a 21-point calibration cycles the transducer from negative full scale to positive full scale and back again in 20% of full-scale increments.

Data extracted from these calibrations are typically linearity and hysteresis. Linearity is the closeness of a calibration curve to the specified straight line, expressed as the maximum deviation of any calibration point from that line during any one calibration cycle. Hysteresis is the maximum difference in output at any measured value within the specified range when the value is approached first with increasing and then with decreasing measurand.

The reference straight line selected is often the linear least-squares line. This line is based on the following principle: The most probable value of an observed quantity is such that the sum of the squares of the deviations of the observations from this value is a minimum. This is based on the fact that most measurements of physical quantities show a normal distribution with both positive and negative deviations from the mean probable and very large deviations less likely than small deviations. The line can be defined unequivocally in terms of the quantities measured. The line also is statistically significant, and standard deviations can be assigned to estimates of the slope, intercept, and other parameters derived from it.

An additional parameter describing the performance of the measurement system is obtained by repeating the static calibration of the system. A minimum of two, but preferably three, consecutive static calibrations yield data from which the "repeatability" of the system is verified. Repeatability is the ability of the measurement system to reproduce output readings when the same measurand value is applied to it consecutively under the same conditions and in the same direction. It is expressed as the maximum difference between corresponding values from at least two consecutive calibrations. Although there is no universal agreement as to the particular values selected, a value close to full-scale output is commonly used.

If the bridge transducer will be used to acquire time-varying measurements, the measuring system must be both dynamically and statically calibrated. The dynamic response of any system is described by a frequency response function which is a complex function of frequency. The frequency response function relates system output to system input in the frequency domain. For measurement systems, this frequency response function is typically represented by Bode plots which are log amplitude and phase versus log frequency.

6.2 Dynamic Calibration

Dynamic calibrations are inherently more difficult to perform than are static calibrations and usually require specialized equipment. Dynamic calibrations can be performed using several types of well-defined input signals, such as applications of sinusoids, transients, or broadband noise. The principal requirement that the input must satisfy is that it must contain significant energy at frequencies over the range of the frequency response function of interest.

The dynamic calibration sequence consists of the following steps:

1. Dynamic sensitivity determination
2. Dynamic amplitude linearity determination
3. Amplitude–frequency verification
4. Phase–frequency verification

If the measuring system does not have zero frequency response, its end-to-end calibration is of necessity made by dynamic methods. The simplest approach to dynamically determining system sensitivity is the application of a sinusoidally varying measurand to the transducer. The amplitude of the measurand should be equal to the range of the transducer. For unidirectional transducers, this test involves biasing the transducer to its half-range point. At the test frequency, it is possible to relate the peak amplitude of the system response to the amplitude of the measurand and determine system dynamic sensitivity.

It is further desired to acquire dynamic amplitude linearity by performing tests equivalent to dynamic sensitivity determination at several levels of the measurand (usually levels of 25, 50, 75, and 100% of full scale are adequate). This testing should be performed at several different frequencies over the range of the frequency response function of interest.

If the measurement system cannot be verified to be linear, it should not be used to acquire time-varying measurements.

The concept of a measuring system having a unique amplitude–frequency and phase–frequency response is only meaningful for systems which have been verified to be dynamically linear. Amplitude–frequency response tests consist essentially of a series of dynamic sensitivity determinations at a number of frequencies within the bandwidth of the system. Three is the *minimum* number of test frequencies. One test should be performed close to the upper limit of the frequency band where the response has not been affected by the high-frequency roll-off characteristic of the system. The second frequency should be sufficiently higher than the first to provide some indication of the roll-off rate of the system. The third frequency should be about halfway between zero frequency and the first test frequency to verify a flat response to the upper band edge. More improved definition obviously can be provided by increasing the number of test frequencies.

Phase–frequency response characteristics of a measuring system can often be acquired simultaneously with the amplitude–frequency response. An output recording device is required with two identically responding channels. The system output is recorded on one channel. The second channel records the measurand, which is typically acquired by a previously calibrated monitoring transducer whose amplitude–frequency and phase–frequency response characteristics are well established. A time correlation between the system output and this monitoring transducer can establish measuring system phase–frequency response. For systems measuring signals whose time history is important, a linear phase response with frequency is required. For those signals about which only statistical information is to be acquired (e.g., random vibration), phase response is not an important system characteristic.

With today's technology, frequency response functions can also be characterized by transient or random system excitation. Dual-channel spectrum analyzers can ratio input-to-output measuring system Fourier transforms in near-real time. Recall that the system input stimulus must contain significant signal content at all frequencies of interest.

6.3 Electrical Substitution Techniques

If actual values of the measurand cannot be used to calibrate resistance bridge transducers, electrical substitution techniques can be used. Test equipment required includes a precision voltage source, precision resistors or decade box, and a signal generator. The techniques include shunt calibration, series calibration, and bridge substitution. Shunt calibration techniques are discussed first.

Inserting a resistor of known value in parallel with one arm of a strain gage bridge is single-shunt calibration. The calibration resistor is inserted across the arm opposite the strain gage conditioning system. The conditioning system may contain a balance potentiometer, a limit or pad resistor, modulus resistors, and temperature compensation resistors. Standard practice is to insert the shunt resistor between the negative input (excitation) and the negative output (Fig. 19). This reduces errors caused by shunting some of the bridge-conditioning resistors.

The value of the shunt resistor R_c is determined by first applying a value of the measurand to the transducer and monitoring the voltage change at the transducer output terminals (Fig. 19). With the measurand removed, a decade box is substituted for R_c and its resistance adjusted until a voltage change results with a magnitude equal to that caused by the measurand. For subsequent calibrations, a fixed resistor R_c can be substituted for the decade box. When the switch in series with R_c is closed, it will produce a step voltage through the measuring system of amplitude equal to that produced by the measurand. When shunting one arm of the bridge, the resistance change produced in that arm is $-R^2/(R_c + R)$.

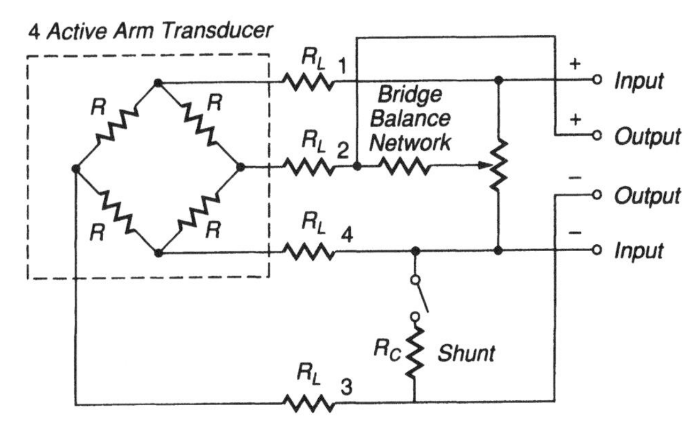

Figure 19 Single-shunt calibration of bridge transducer.

In the calibration laboratory, the small lead length associated with the transducer introduces no error in establishing R_c. The application of the bridge transducer in the field can require significant lengths of cable with significant transmission line resistance. Figure 20 illustrates the situation where R_c must be applied remotely. If R_c were applied directly at the bridge, loading errors introduced by transmission lines would be accounted for. If (as in Fig. 20) R_c is applied at a remote location, the effect of the transmission line resistance $2R_L$ in series with R_c must be considered.

Bipolar shunting is used when the physical loading creates both positive- and negative-going signals and it is desired to create positive and negative calibration outputs. The calibration resistor is alternately inserted across the two arms opposite the bridge-conditioning network. If line resistance is significant, it must be considered as in Fig. 21.

Series calibration of bridge transducers is considerably different from shunt calibration. Figure 22 describes this process. Series calibration consists of two distinct calibration phases.

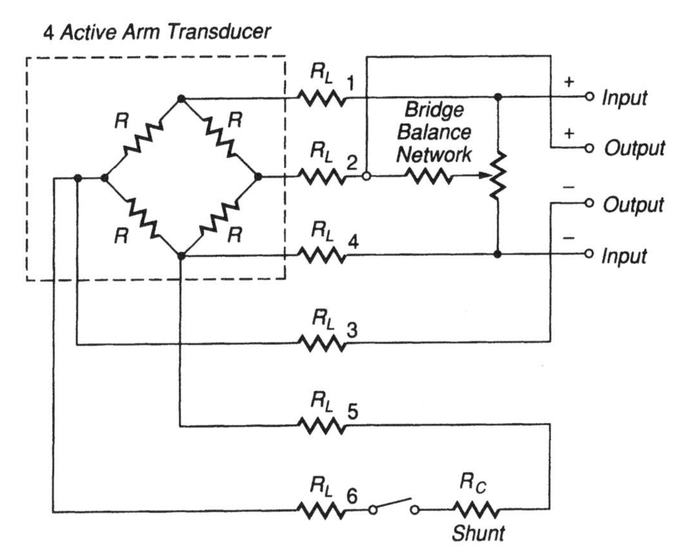

Figure 20 Remote-location single-shunt calibration of bridge transducer.

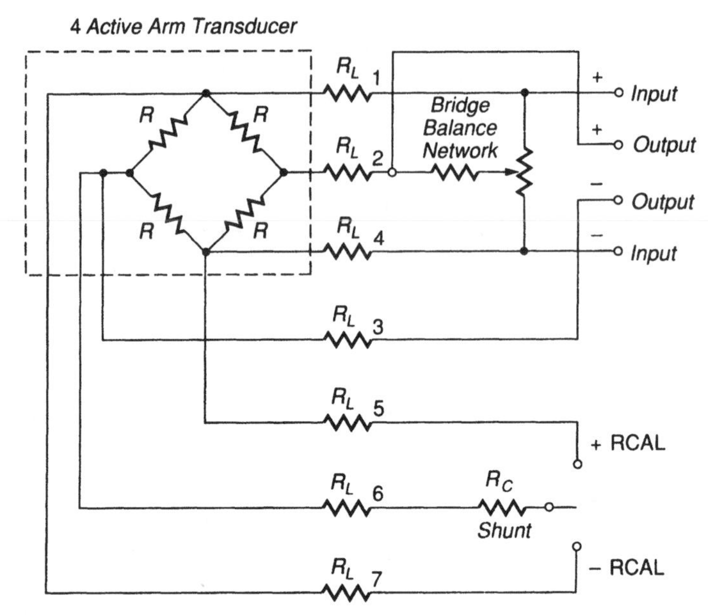

Figure 21 Remote-location double-shunt calibration of bridge transducer.

In the zero calibration phase the Zero Cal switch is moved downward so excitation is removed from the bridge. The sensitivity resistor (R_{SENS}) is concurrently placed across the bridge input terminals, simulating the power supply impedance to result in the same overall system impedance encountered in the data circuit. Zero bridge transducer output is recorded.

The next calibration phase is the series phase. In the series calibration mode the two switches in Fig. 22 are closed with the Zero Cal switch back in its original position, introducing R_{c1} and R_{c2} into the circuit. This removes power from one corner of the bridge and puts a calibration resistor R_{c1} in series with the sensitivity resistor and one side of the bridge output. The second calibration resistor becomes intermediate between the R_{SENS}-to-R_{c1} connection and excitation return. This second resistor, R_{c2}, is selected to maintain the approximate equivalent bridge impedance across the excitation. The calibration circuit then electrically simulates the bridge transducer. The value of R_{c1} is determined experimentally, corresponding to some measurand equivalent.

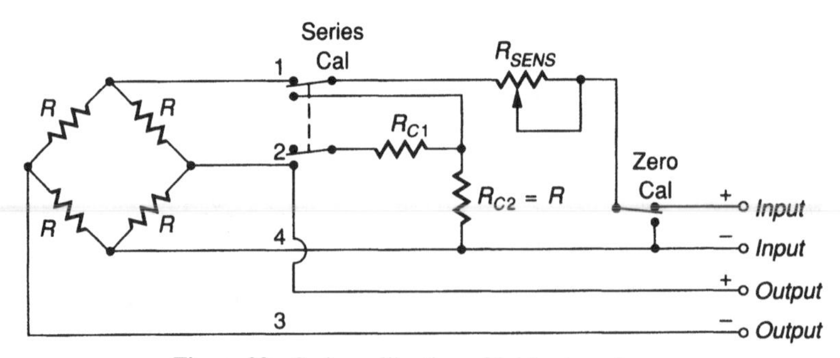

Figure 22 Series calibration of bridge transducer.

Series calibration overcomes a serious shortcoming of shunt calibration. During application of a shunt resistor, the transducer can still respond to mechanical input. The calibration step is superimposed upon any mechanically induced signal present. If the mechanical input is static and of sufficient magnitude, overranging will invalidate the calibration step. If the mechanical input is dynamic, it may be impossible to accurately measure the magnitude of the calibration step. The magnitude of the series calibration step is significantly more independent of this mechanical input. As in all calibration, transmission line resistance must be considered where significant. Similarly, a change in sensitivity resistance modifies the effect of the series calibration resistance. However, the typical error incurred is negligible.

The final electrical substitution technique discussed is bridge substitution. This technique involves substitution of a model for the bridge transducer itself. Figure 23 represents a typical low-level bridge system.

An accurate bridge transducer model has the same terminal impedance as the transducer and provides a fast and simple method of generating a static and dynamic output equivalent to that generated by the transducer for a given physical load. It also provides a convenient method for verifying the calibration resistor's measurand equivalency for shunt and series systems. The two types of bridge transducer models employed for system calibrations are the shunt resistor adapter and the shunt resistor bridge.

Figure 24 describes the shunt resistor adapter, which is simple, inexpensive to construct, and an exact model since it is used in conjunction with the actual transducer. The adapter is inserted between the transducer and the rest of the measurement system. It performs three primary functions.

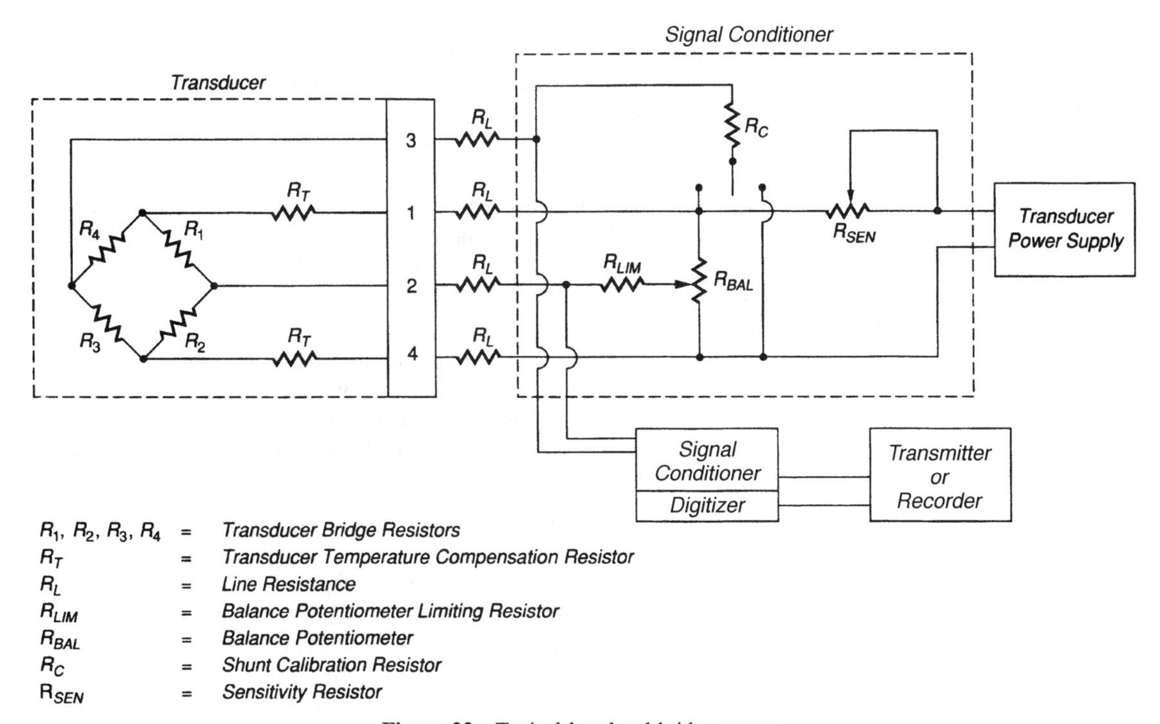

Figure 23 Typical low-level bridge system.

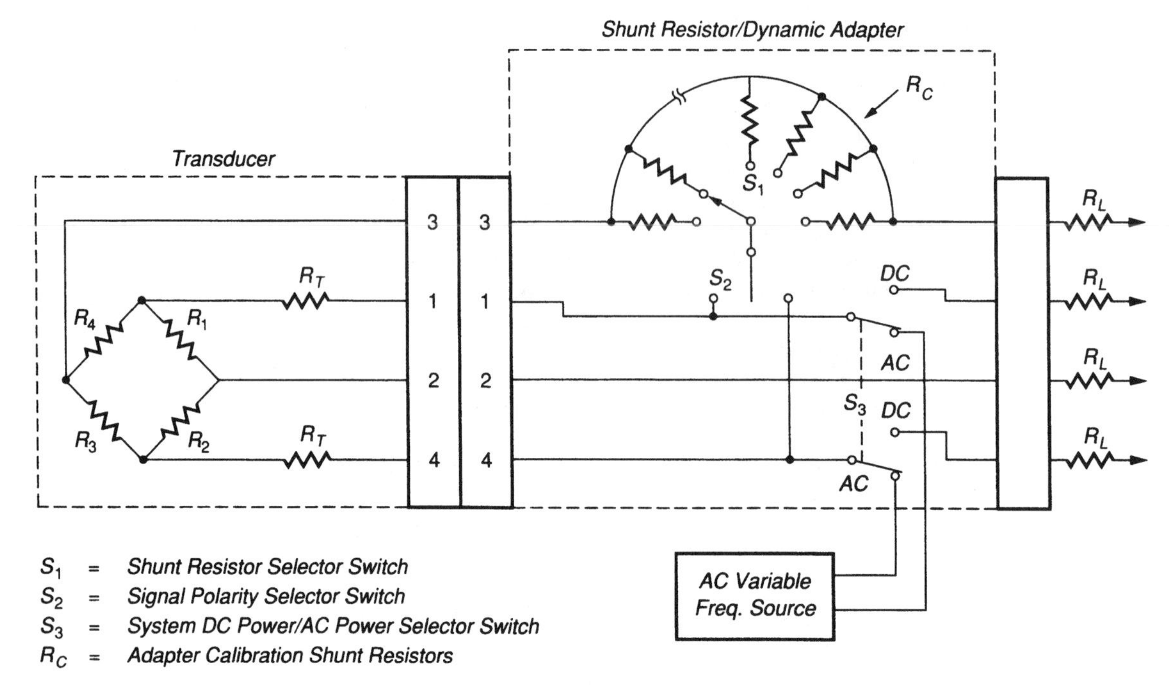

Figure 24 Shunt resistor adapter.

1. It supplies the stimulus for performance of system end-to-end calibrations. Shunting the arms of a transducer bridge with the appropriate resistors produces an unbalance in the bridge equivalent to that produced by a given measurand. The adapter provides a convenient method of applying these shunt resistors directly to the bridge with negligible line loss (S_1 and S_2).
2. It performs a system frequency response test. A convenient system frequency response can be performed by selecting the appropriate shunt resistor and sweeping the adapter's ac power supply over the desired range. Figure 25 shows a typical oscillograph display of the results.
3. It provides a convenient check of the system's calibration resistors (R_c) and equivalents. The system's R_c equivalent will differ from the values established by the laboratory calibration as a function of line resistance, calibration resistor tolerance, and so on. Since the adapter shunt resistors are precision resistors that are applied directly to the bridge, the equivalency of the adapter shunt resistors will not be affected by lead resistance and other variables.

Although the shunt resistor adapter model is a very powerful and simple calibration tool, it has two undesirable characteristics. The least desirable characteristic is that the system calibration and calibration resistor equivalents generated by the adapter are incremental values superimposed on the transducer output resulting from the physical stimulus acting at the time of the test. Also, the adapter does not provide a fixed independent reference since it is used in conjunction with the transducer.

The undesirable features of the shunt resistor adapter can be eliminated by replacing the actual transducer with a bridge model (Fig. 26). Since the shunt resistor bridge (bridge

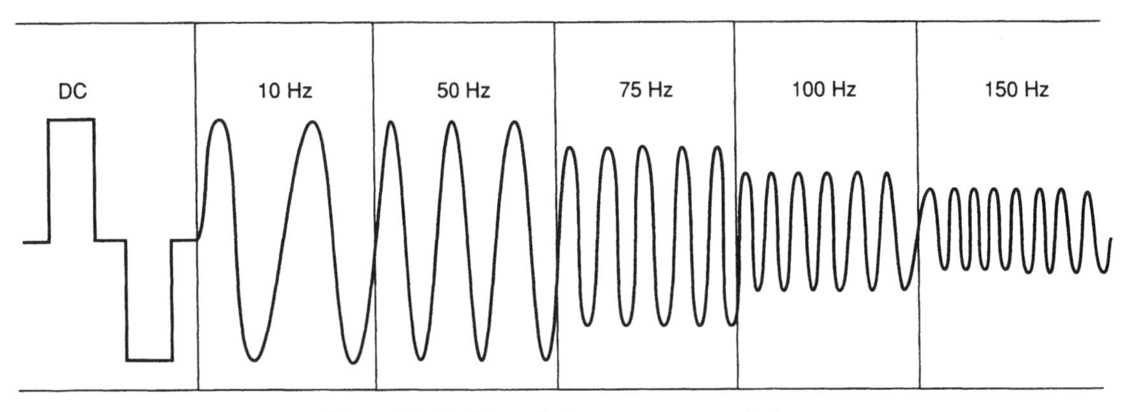

Figure 25 Oscillograph frequency response display.

model plus shunt resistor adapter) is a stable, complete model of the transducer, it can be used to perform an absolute end-to-end system calibration and can be a valuable tool in troubleshooting.

Several disadvantages are encountered when using the shunt resistor bridge as a calibration tool. Since some transducers are hard to model, it is difficult to ensure that the bridge is a representative model of the transducer under all conditions. Furthermore, a different bridge model is required for each major transducer design.

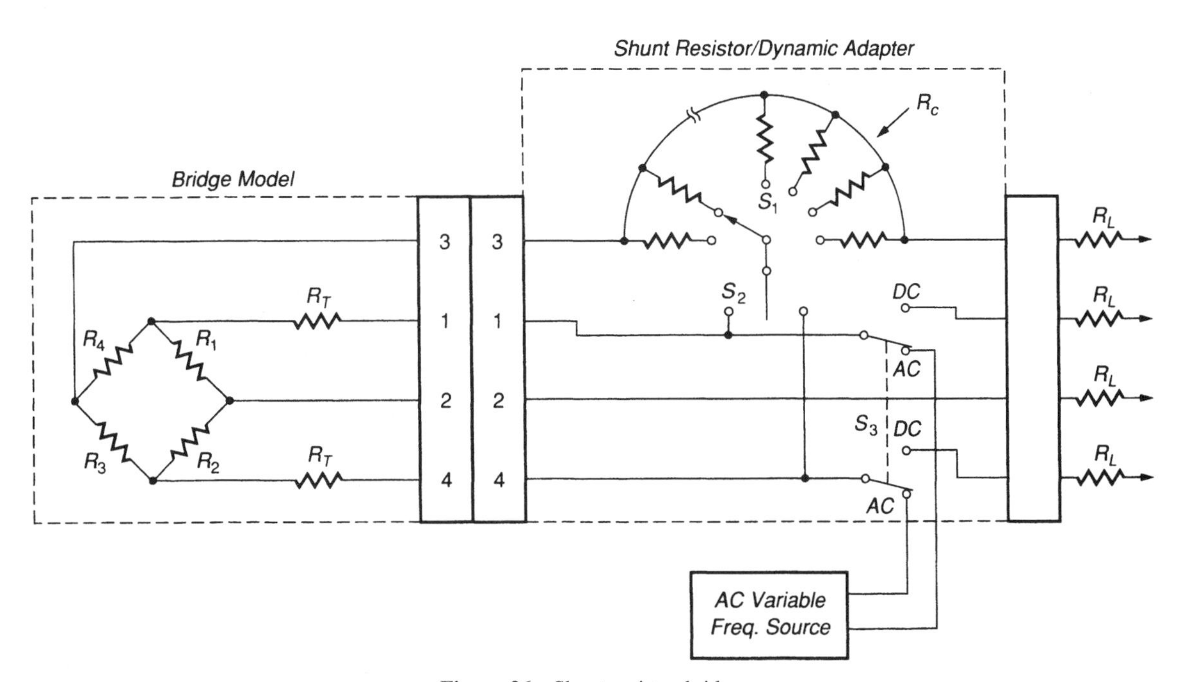

Figure 26 Shunt resistor bridge.

As a final note, remember that the resistance of semiconductor bridge transducers is strongly a function of temperature. When using shunt or series calibration techniques on semiconductor bridges, ambient-temperature changes should be taken into account.

7 RESISTANCE BRIDGE TRANSDUCER MEASUREMENT SYSTEM CONSIDERATIONS

7.1 Bridge Excitation

When amplifiers and power supplies were formally designed around vacuum tubes, component drift was a problem in bridge transducer measurement systems. Alternating-current power supplies in bridge circuits eliminated many of these problems by operating at frequencies above dc. Most bridge transducer power supplies today are dc. When comparing dc supplies with ac, the following advantages are associated with dc:

1. Simpler circuitry
2. Wider resultant instrumentation system frequency response
3. No cable capacitive or inductive effects due to the excitation
4. Simpler shunt calibration and bridge balance circuitry

Independent of type of supply, the power level selected has to take account of all variables which affect the measurement. These include gage resistance, gage grid area, thermal conductivity of flexure to which gage is mounted, flexure mass, ambient test temperature, whether used on a static or dynamic test, accuracy requirements, and long- or short-term measurement. These variables account for the fact that a strain gage is a resistance which has to dissipate heat when current passes through it. Most of the heat is conducted away from the gage grid to the transducer flexure. The result of inadequate heat conduction is gage drift.

For transient measurements, a steady transducer zero reference is not as important as for static measurements. Bridge power can be significantly elevated to increase measurement system signal-to-noise ratio.

The following specifications define key performance parameters of dc output instrumentation power supplies. Input supply can be either ac or dc.

1. *Warmup Time.* The time necessary for the power supply to deliver nominal output voltage at full-rated load. It is usually specified over the range of operating temperatures.
2. *Line Regulation.* The change in steady-state dc output voltage resulting from an input voltage change over the specified range.
3. *Load Regulation.* The change in steady-state dc output voltage resulting from a full-range load change.
4. *Efficiency.* The ratio of the output power to the input power.
5. *Load Transient Recovery.* The time required for the output dc voltage to recover and stay within a specified band following a step change in load.
6. *Periodic and Random Deviation.* The ac ripple and the noise of the dc output voltage over a specified bandwidth with all other parameters held constant.
7. *Stability (Drift).* The deviation in the dc output voltage from dc to an upper limit which coincides with the lower limit as specified above in 6.

8. *Temperature Coefficient.* The change in output voltage per degree change in ambient temperature.

Reference 10 defines test procedures for these specifications.

7.2 Signal Amplification

In addition to providing a precision power source to bridge transducers, the resultant millivolt signals from these transducers often require amplification. This amplification is usually performed by a differential dc amplifier. A differential dc amplifier is an electronic circuit whose input lines are conductively isolated from the output lines, power, and chassis ground and whose output voltage is proportional to the differential input signal voltage. Ideally, both input lines have equal impedance and transfer characteristics with respect to the amplifier ground structure. The amplifier has a frequency response from 0 Hz to a value determined by the bandwidth of the amplifier.

Selecting amplifiers can be difficult because specification terminology is not universally standardized. Amplifier specifications are either referred to input (RTI) or referred to output (RTO). Discussing these specifications can lead to an understanding of the amplifiers themselves.

1. *Input Impedance.* The minimum impedance the amplifier will present when operated within its specification. It is the impedance seen between the two ungrounded input lines of the amplifier.
2. *Source Current.* The bias current flowing through the circuit comprised of the amplifier input terminals closed through the source resistance. The amplifier input transistors act as constant-current generators in series with the input terminals. This current can result in both offset voltage and common-mode voltage.
3. *Common Mode Rejection.* The measure of the conversion of common-mode voltage to normal differential signal. The common-mode input voltage is the voltage common with both inputs to the amplifier. A common-mode rejection of 60 dB implies that a 10-V signal applied simultaneously to both inputs produces an error signal RTI of 10 mV.
4. *Linearity.* The maximum deviation from the least-squares straight line established through the output voltage versus differential input voltage characteristic. In evaluating linearity, it is usually sufficient to test at the highest and lowest gains, since linearity will be worst at these settings.
5. *Gain Range.* The slope of the least-squares straight line established through the output voltage versus the differential input voltage characteristic of the amplifier. The gain range is the maximum and minimum values of gain available from the amplifier without causing any degradation in performance beyond the limits of the specification.
6. *Gain Stability with Temperature.* The change in amplifier gain as a function of ambient temperature for any gain in the specified gain range.
7. *Zero Stability with Temperature.* The change in output voltage with temperature. It must be specified as RTI or RTO, and this test is typically performed with the amplifier input leads terminated with the maximum source impedance and no signal applied. A warm-up period is usually specified for both this test and gain stability with temperature.

8. *Frequency Response.* The minimum frequency range over which the amplifier gain is within ±3 dB of the dc level for all specified gains for any output signal amplitude within the linear output voltage range. In writing specifications, it is not uncommon for a user to also specify the desired phase characteristics over the frequency range of interest and the number of filter poles.
9. *Slew Rate.* The maximum rate at which the amplifier can change output voltage from the minimum to the maximum limit of linear output voltage range. It is expressed in volts per microsecond with a large-amplitude step voltage applied to the input of the amplifier and the amplifier driving a specified capacitive load. The usual source of slew rate difficulty is current limiting, and this specification (a nonlinear process) should not be confused with rise time (a linear process).
10. *Settling Time.* The time following the application of a step voltage input for the amplifier output voltage to settle to within a specified percentage of its final value.
11. *Overload Recovery.* The time required for the amplifier to recover from a specified differential input signal overload. It is specified as the number of microseconds from the end of the input overload to the time that the amplifier dc output voltage recovers to within the linear output voltage range. Amplifier gain must be specified.
12. *Noise.* Noise is divided into two components: RTI and RTO. RTI noise is that component of noise that varies directly with gain. It is measured with the amplifier input leads terminated in the maximum source impedance and no signal applied. The RTO noise is that component of noise which remains fixed with gain.
13. *Harmonic Distortion.* The maximum harmonic content for any amplifier frequency or output amplitude within the specified limits.
14. *Output Impedance.* The maximum impedance the amplifier will present when it is operated anywhere within its specification. This specification is important in resistive loading ratings or in determining the amount of capacitance which can be connected across the output without causing instability.

Reference 10 describes test procedures for these specifications and discusses them further. Figure 27 presents the basic dc amplifier circuit. Reference 11 provides additional discussion directed toward understanding dc instrumentation amplifiers.

7.3 Slip Rings

In many measurement applications, it is necessary to acquire data from rotating machinery. Turbines, rate tables, and centrifuges are examples of such machinery. If it is necessary to

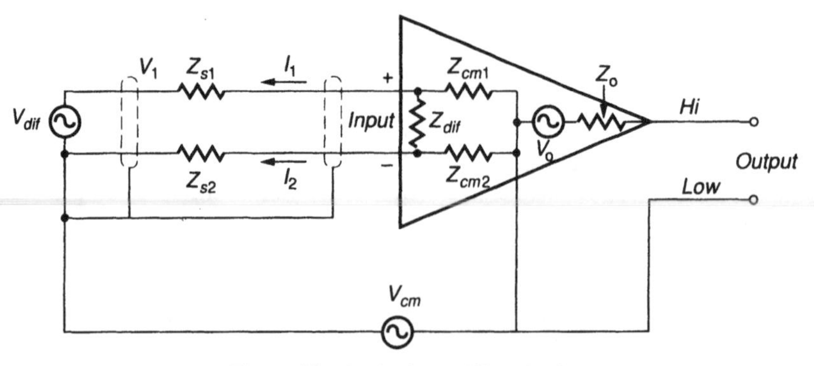

Figure 27 Basic dc amplifier circuit.

measure strain, pressure, torque, force, and so on on the rotating machine component, signals from bridge transducers must be coupled from this component to a stationary instrumentation system. Instrumentation slip rings accomplish this function.

In their simplest form, slip rings consist of a metal ring on the rotating machine component against which a brush attached to the stationary machine portion is spring loaded to make ohmic contact. Precious metals are generally used for mating surfaces to minimize contact resistance.

Slip rings came into existence in the 1940s with initial application in the aircraft industry. In the 1950s, mercury slip rings came into existence. These latter rings, which first found application at Rolls Royce in England, use mercury as the signal transfer medium. The mercury is entrapped between the rotor and the stator of the ring assembly. Today, slip rings are capable of operating from very low RPM to tens of thousands of RPM.

Noise induced in slip rings is of the ohmic contact type, that is, it is roughly proportional to current. A high brush pressure reduces noise at the expense of increased brush wear. Brush wear is a function of the brush pressure, material, finish (usually microinch), and flatness. One technique for lowering contact noise is to mount several brushes in parallel on the same ring.

Because ohmic changes in the slip rings can be of the same order of magnitude as resistance changes in the bridge transducer, full bridges are almost always used on the rotating part to avoid inserting slip rings within the bridge itself. Slip rings in the output circuits of bridge transducers using voltage monitoring do not create significant problems because any small resistance changes in the rings are in series with the large input impedance of the voltage-measuring device and are effectively ignored. Slip rings in the input circuits of bridge transducers operating from a constant-voltage source can create problems if they cause fluctuating voltage drops in series with the transducer. For this reason, constant-current sources are preferred when using slip rings.

Other techniques for extracting data from rotating machinery have evolved over the years. These include rotary transformers, light modulation, and radio frequency (RF) telemetry. Of these schemes, RF telemetry has displayed the most promise with commercially available low-power transmitters capable of operating up to 30,000*g*.

7.4 Noise Considerations

Many other sources besides slip rings can induce unwanted spurious signals in these transducers. Since the unamplified output from bridge transducers is typically ones or tens of millivolts and never more than a few hundred, they are easily influenced by noise sources. The following discussion defines noise, documents how to verify its existence (or hopefully nonexistence), and provides some hints as to how to suppress noise in bridge transducer measuring systems. Reference 12 provides a basis for this discussion.

The output of measuring system components represents combinations of responses to environments. These environments can be divided into two categories: desired and all others (undesired). For example, consider a bridge pressure transducer in a hostile explosive environment. Its desired environment is pressure. Other undesired environments it encounters are temperature, acceleration, ionized gas, and so on. Ideally, the transducer would respond to pressure alone. In practice, an additional response is elicited from the transducer due to the other environments; usually, but not always, the response is small compared to the pressure response.

Two response types exist for a bridge transducer: self-generating and non-self-generating. Non-self-generating responses are due to changes in the material properties or geometries within a transducer. Power has to be applied to the transducer to elicit a non-self-generating response. For example, pressure applied to the diaphragm of a pressure transducer with bridge electrical power supplied modifies the impedance of the strain gage circuit and results in a millivolt output (non-self-generating). Self-generating responses are those attributable to various measurands applied to bridge transducers without electrical power supplied. Examples of these responses include thermoelectric-, photoelectric-, pyroelectric-, and magnetoelectric-induced voltages within the bridge circuit. Thus, there exist four environment–response combinations in bridge transducers. The non-self-generating response to the desired environment is defined as signal. The non-self-generating response to the undesired environment, as well as the self-generating response to both the desired and undesired environment, is noise. Figure 28 illustrates the paths associated with these four combinations with path 4 being signal and paths 1, 2, and 3 being noise.

The quantifications of paths 1–4 can be accomplished by switching. If at some time during the test bridge power is switched off, Fig. 28 indicates that only paths 1 and 3 will exist. Since these paths are both noise, the bridge transducer response ideally should approach zero. Similarly, if the desired environment can be switched off for some time period, only paths 1 and 2 remain. If path 1 was verified as being noise free when bridge power was removed, path 2 also becomes quantified. If paths 1, 2, and 3 are all shown to provide negligible signal level, transducer output becomes attributable to path 4, which has been defined as signal. In summary:

Remove bridge power: Document paths 1 and 3.

Reapply power and remove desired environment: Document paths 1 and 2.

If the documented signal paths are of sufficiently small magnitude to be considered inconsequential during test, the non-self-generating response to the desired environment (signal) is recorded.

Some question may arise as to how to implement these procedures, particularly in transient measurement situations. For example, assume a bridge accelerometer is to be used to measure a transient acceleration event. Three accelerometers can be fielded in close physical

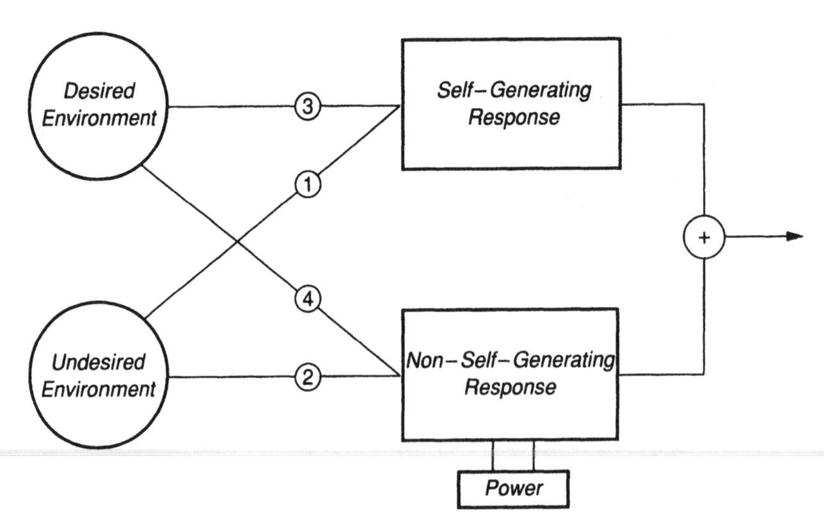

Figure 28 Bridge transducer model for noise hunting and documentation. (Adapted with permission from "Information as a 'Noise Suppression' Method," by Peter Stein, Stein Engineering Services, Inc., Phoenix, AZ, LR/MSE Publ. 66, 1975.)

proximity. The first can be mounted without power applied to document paths 1 and 3. The second can have power applied but be mounted in a piece of foam to isolate it from the acceleration environment, resulting in documentation of paths 1 and 2. The third can be powered and properly mounted to measure the acceleration environment. If the first two accelerometers produce no output, then the output from the third is the noise-free signal.

If noise is present in measuring systems containing bridge transducers, noise control efforts are dictated by the specific noise type. Electric and magnetic fields can be shielded, noise components at specific frequencies can be filtered, thermal transient effects can be absorbed or delayed, steady-state temperature effects can be compensated, and so on. The prerequisite to any noise control is documentation of its presence.

As noted earlier, most modern resistance bridge transducers use dc power supplies as opposed to ac power supplies. However, ac power supplies still have an important role to play with resistance bridge transducers. The ac power supplies accomplish noise suppression by separating the self-generating responses from the non-self-generating responses in a trans-

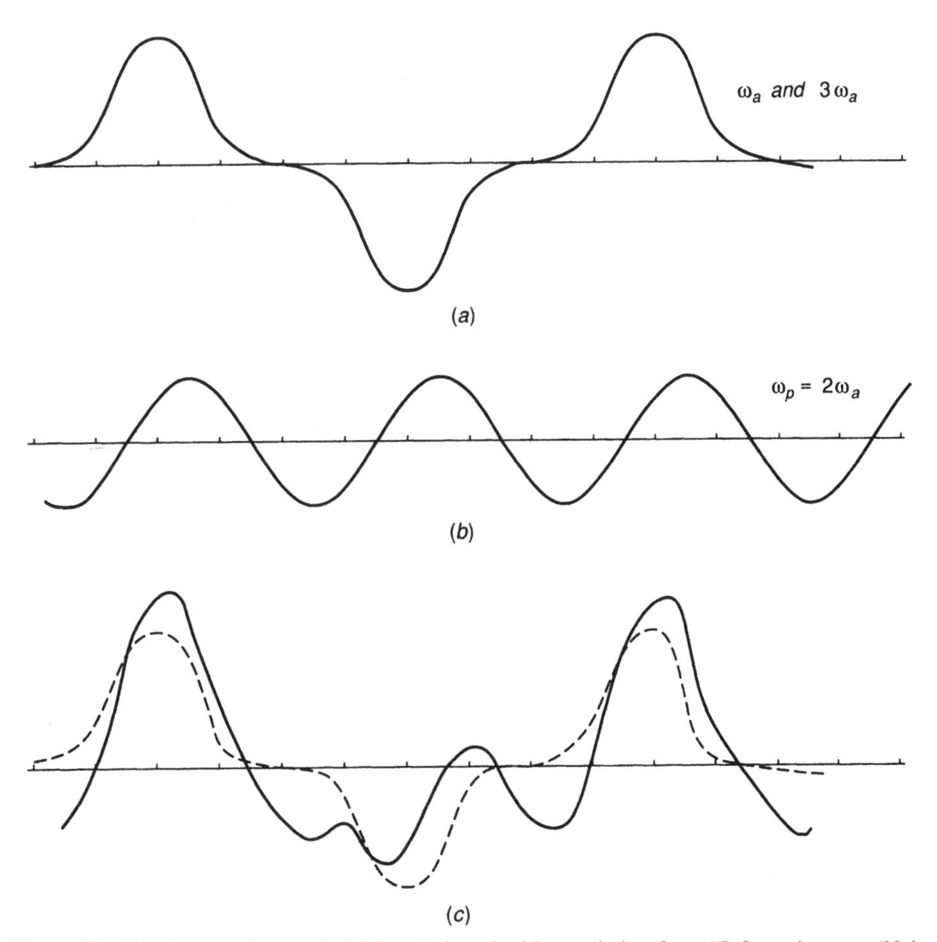

Figure 29 Signal wave shapes, dc bridge. (Adapted with permission from "Information as a 'Noise Suppression' Method," by Peter Stein, Stein Engineering Services, Inc., Phoenix, AZ, LR/MSE Publ. 66, 1975.)

ducer by moving the frequency content of a signal into some new range of the frequency spectrum. This procedure is known as amplitude modulation. Reference 13 forms the basis for the following discussion.

Referring to Fig. 28, amplitude modulation can eliminate paths 1 and 3 (noise) from the net output signal. Thus, self-generating electromotive force (emf), such as thermoelectric and electromagnetic ones, can be separated from emf attributable to resistance changes in a bridge transducer.

A design procedure is presented for an ac-powered bridge where the signal input to the self-generating response extends to frequency ω_1 and the input to the non-self-generating response extends to ω_2. Power supplied is at frequency ω_3. The design method developed requires a knowledge of these frequencies. A procedure to determine these frequencies involves the following:

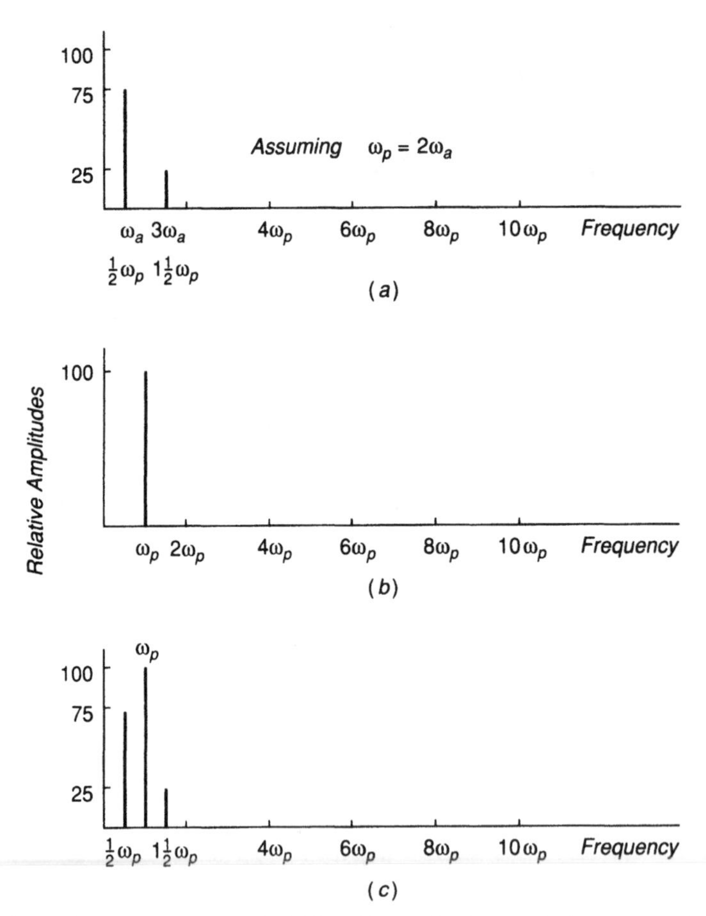

Figure 30 Signal frequency contents, dc bridge. (Adapted with permission from "Information as a 'Noise Suppression' Method," by Peter Stein, Stein Engineering Services, Inc., Phoeniz, AZ, LR/MSE Publ. 66, 1975.)

1. Performing a frequency analysis of the signal recorded with no power supplied to determine ω_1
2. Applying power under normal operating conditions and comparing signal frequency content to the results of 1 to determine ω_2

An example follows where noise is present as a self-generating response and the bridge is powered first by a dc supply and then by an ac supply.

In this example, the non-self-generating response (signal) occurs at frequency ω_p and the self-generating response (noise) occurs at two frequencies bracketing ω_p ($\omega_a = \omega_p/2$ and $3\omega_a$). Figure 29 illustrates the wave shapes assumed for the self-generating and non-self-generating responses with dc power supplied ($\omega_3 = 0$). Figure 29*a* represents both self-generating response input and output, Fig. 29*b* represents non-self-generating response input and output, and Fig. 29*c* represents the total transducer output (summation of 29*a* and 29*b*). It is seen that the two responses are hopelessly intermingled and that the signal cannot be separated from the noise. Figure 30 shows the frequency content of the wave shapes, with Fig. 30*a* corresponding to 29*a*, Fig. 30*b* corresponding to 29*b*, and Fig. 30*c* corresponding to 29*c*. The ac-powered bridge will be presented as a solution to measurement problems such as this. Frequency ω_3 is typically selected as 10 times ω_p.

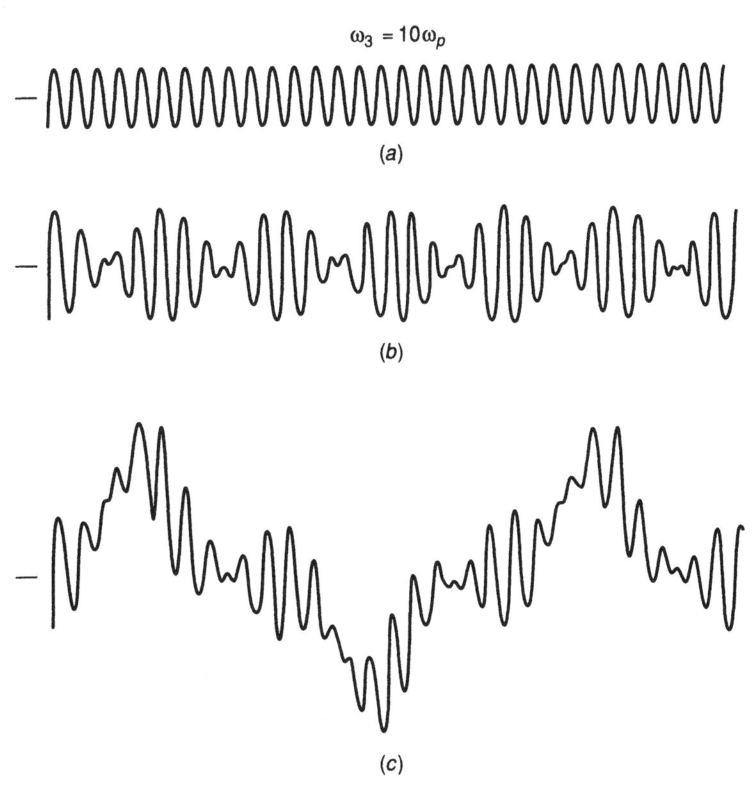

Figure 31 Signal wave shapes, ac bridge. (Adapted with permission from "Information as a 'Noise Suppression' Method," by Peter Stein, Stein Engineering Services, Inc., Phoeniz, AZ, LR/MSE Publ. 66, 1975.)

Figure 31 describes this situation for the ac bridge. Figure 31*a* describes bridge power. Figure 31*b* describes the output from the non-self-generating response which now contains frequencies at $\omega_c - \omega_p$ and $\omega_c + \omega_p$. Here ω_c is defined to be the carrier frequency, ω_3. Figure 31*c* describes the net transducer output, which is a summation of Fig. 29*a* and Fig. 3l*b*. Figure 32 describes the frequency content associated with Fig. 31 respectively. The frequency content in Fig. 32*c* associated with the time history of Fig. 3l*c* shows conclusively that the non-self-generating information has been moved from its original frequency, ω_p, to occupy a new frequency range, $\omega_c - \omega_p$ to $\omega_c + \omega_p$, while the self-generating response is left at ω_a and $3\omega_a$. The non-self-generating response is then in that part of the frequency spectrum where no appreciable noise exists and can be separated by bandpass filtering.

After bandpassing, a problem still remains in phase sensing. Figure 33 describes this problem. Figure 33 illustrates an amplitude-modulated signal after bandpassing to remove the effects of any self-generating response which may be present. This amplitude-modulated signal is ambiguous in that it could correspond to any of the lower four signal inputs to the non-self-generating response. The problem of phase sensing associated with a modulated signal is that of determining which portion of the modulated wave shape is positive and which is negative.

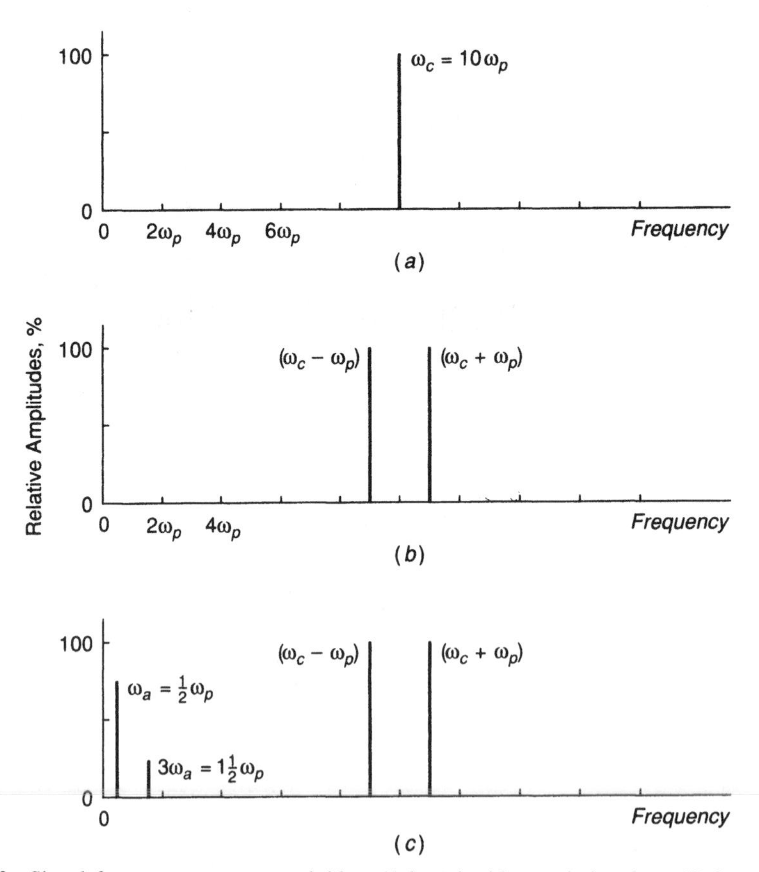

Figure 32 Signal frequency contents, ac bridge. (Adapted with permission from "Information as a 'Noise Suppression' Method," by Peter Stein, Stein Engineering Services, Inc., Phoeniz, AZ, LR/MSE Publ. 66, 1975.)

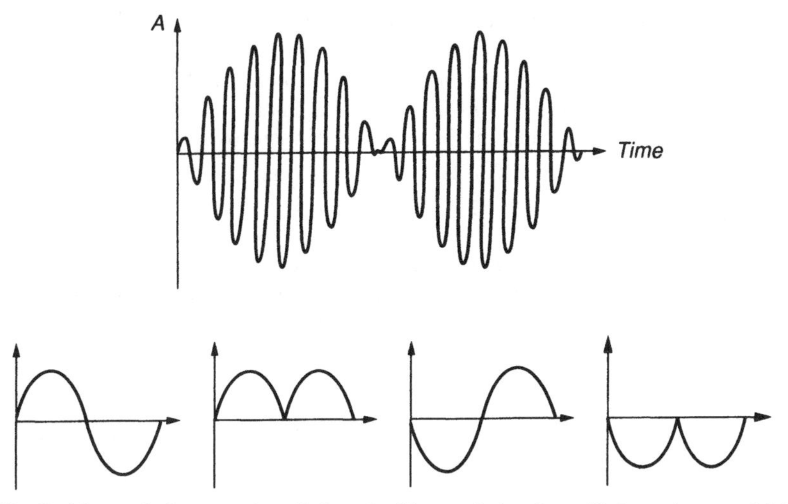

Figure 33 Problems of phase sensing. (Adapted with permission from "Information as a 'Noise Suppression' Method," by Peter Stein, Stein Engineering Services, Inc., Phoeniz, AZ, LR/MSE Publ. 66, 1975.)

If a modulated signal emerges from a measuring system which is initially balanced (zero output for zero input), phase sensing must be done in a phase-sensitive manner. The general principle for all phase sensors is as follows:

> If the system output is of the same sign at the same time as the time-varying supply power, the measurand must have been positive. If the signs are opposite, the measurand must have been negative.

Phase sensing is accomplished by a phase-sensitive demodulator. A reference signal is fed from the bridge supply power to the phase-sensitive demodulator. This signal is compared with the amplitude-modulated signal for phase determination. A half-wave rectifier with transformer coupled reference and amplitude-modulated signals forms one basis for a demodulator.

After phase sensing, a low-pass filter is required to separate the non-self-generating analog signal proportional to the measurand from the other frequencies which appear as sidebands around harmonics of power supply frequency. Final selection of a power supply frequency is a trade-off between the maximum frequency content in the measurand and the low-pass filter roll-off characteristics. While a 10:1 ratio is typical, the power supply frequency may vary between 3 and 20 times the maximum nonself-generating signal frequency.

Note that introduction of an ac power supply requires a bridge-balancing network incorporating complex impedance in the balance controls.

8 AC IMPEDANCE BRIDGE TRANSDUCERS

Having discussed bridge transducers that use resistive sensing elements and ac power supplies (amplitude modulation), a logical question is whether bridge transducer sensing elements can be capacitive or inductive. In practice, many bridge transducers do employ capacitive or inductive elements. While resistance bridge-type transducers typically possess resonant frequencies in the tens or hundreds of kilohertz range, ac impedance bridge transducers typically possess resonant frequencies of less than 10 kHz. The larger physical size of ac impedance bridge transducers makes them environmentally more fragile but improves their performance by increasing their sensitivity to low-level measurands.

8.1 Inductive Bridges

Figure 34 shows an example of how variable-reluctance sensing elements can be incorporated into an ac bridge transducer. A differential pressure transducer containing a magnetically permeable stainless steel diaphragm as the mechanical flexure is portrayed. This diaphragm

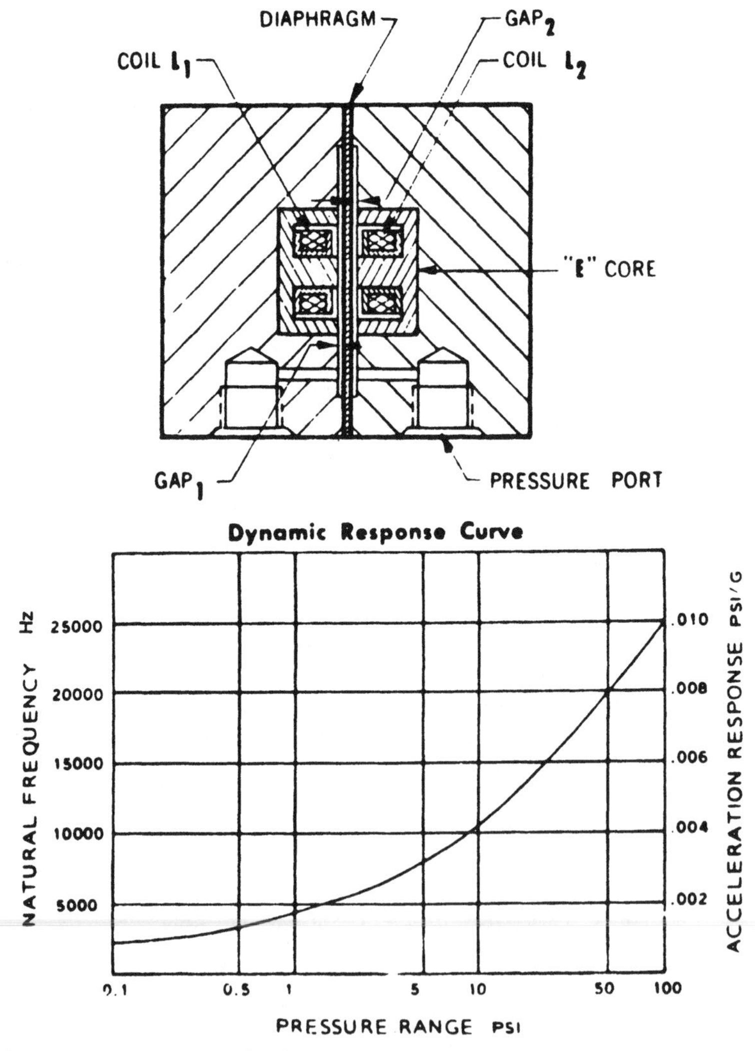

Figure 34 Variable reluctance sensing in a bridge transducer. (Courtesy of Validyne Engineering Corporation.)

is clamped between two blocks and deflects when a pressure difference is created across it through the two ports shown. An E-core and coil assembly is embedded in each block. A small gap exists in front of each E-core. When the diaphragm is undeflected, a condition of equal inductance exists in each coil. When the diaphragm does deflect, an increase of gap in the magnetic flux path of one core occurs, with a resultant decrease in the gap in the magnetic flux path of the other. Magnetic reluctance varies with gap, determining the inductance value. The diaphragm motion then changes the inductance of the two coils, one increasing and one decreasing. These two coils can be placed in adjacent arms of an ac-powered bridge. Resistive elements can be used to complete the bridge. Once the bridge is balanced, an amplitude-modulated signal results when a differential pressure is applied across the ports of the transducer. When the resultant signal is properly demodulated, the applied pressure can be quantified.

Eddy current inductive displacement-measuring systems are another example of the use of impedance as opposed to resistive bridges. Placing a coil with an ac flowing in it a nominal distance from a metal target induces a current flow on the surface and within the target. The induced current produces a secondary magnetic field that opposes and reduces the intensity of the first field. Changes in the impedance of the exciting coil provide information about the target. The target can be the diaphragm of a pressure transducer, the seismic mass of an accelerometer, the flexure of a load cell, and so on. The coil is one leg of a balanced bridge network. Unbalanced bridge conditions are sensed and converted into a signal directly proportional to the distance between coil and target. Figure 35 schematically illustrates this conversion.

The electrical parameters of resistivity and permeability in the target material influence performance of eddy current transducers. For a specific material, displacement sensitivity is influenced by coil geometry and operating frequency.

Target thickness is generally not a limiting factor. At one "skin depth," the eddy current density is only 36% of the maximum encountered on the target surface; at two "skin depths," it is 13%. Figure 36 defines skin depth as a function of target resistivity and permeability.

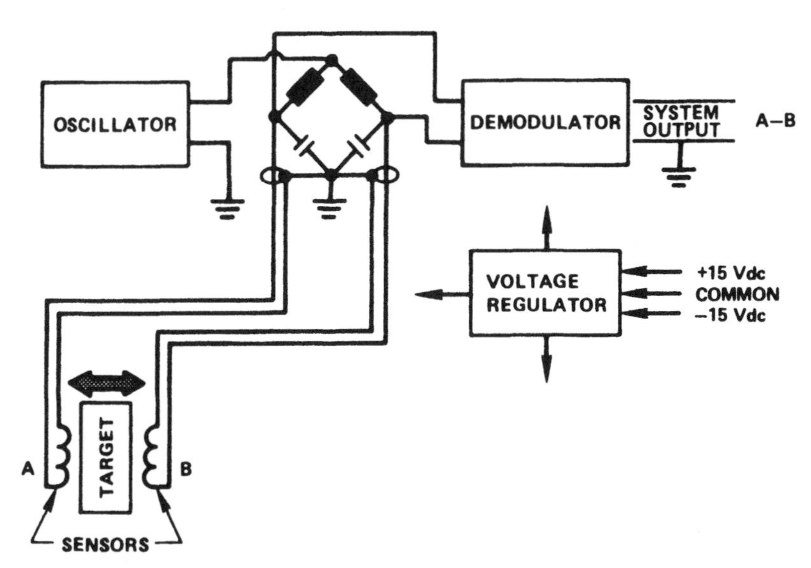

Figure 35 Eddy current inductive displacement measuring system. (Courtesy of Kaman Instrumentation Corporation, Measurement Systems Group, Colorado Springs, CO.)

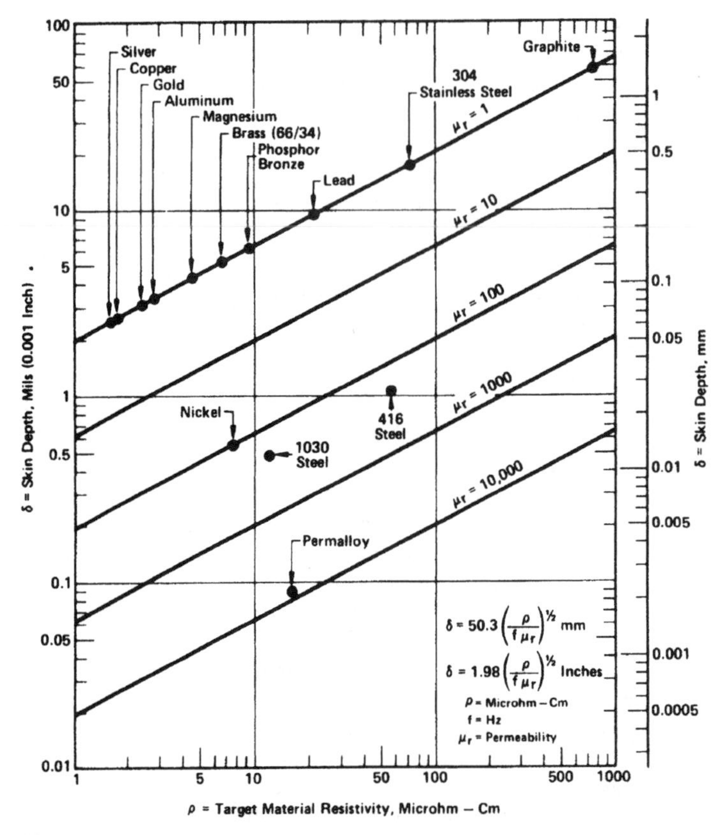

Figure 36 Skin depth versus target resistivity and permeability at 1 MHz. (Courtesy of Kaman Instrumentation Corporation, Measurement Systems Group, Colorado Springs, CO.)

Target shape and alignment also should be considered in application. A flat, circular target equal to the coil diameter appears as an infinite plane. Smaller target diameters produce smaller voltage unbalances in the impedance bridge. Since the transducer senses the average distance to the target, the nonparallelism effect is small up to 15°.

A differential transformer also is briefly mentioned here, although it does not operate in an impedance bridge. A linear variable differential transformer (LVDT) consists of three symmetrically spaced coils wound onto an insulated bobbin. A magnetic core moving through the bobbin provides a path for magnetic flux linkage between coils. The center coil is the primary and has an ac voltage applied. The two secondary coils are wired in a series-opposing circuit. When the core is centered between two secondary coils, the voltages in the two coils cancel. As the core is displaced, the phase-referenced and demodulated output signal provides a linear voltage output with displacement.

8.2 Capacitive Bridges

Capacitance sensors can be integrated into bridge transducers. The capacitance between two metal plates separated by an air gap is $C = kKA/h$, where C is capacitance, K is the dielectric constant for the material between the plates, A is the plate overlapping area, h is the gap thickness between the two plates, and k is a proportionality constant. The range of a capacitance sensor can be shown to remain linear with changes in area but become nonlinear when the change in gap displacement becomes a significant portion of the original gap.

Again, it should be pointed out that advantages of small size and enhanced dynamic response are to be found with resistance bridge transducers. Increased sensitivity will be displayed by impedance bridge transducers. Both impedance bridges and ac-powered resistance bridges offer noise suppression through separating non-self-generating responses from self-generating responses.

REFERENCES

1. P. K. Stein, *Measurement Engineering,* Stein Engineering Services, Phoenix, AZ, 1964.
2. W. M. Murray and P. K. Stein, *Strain Gage Techniques,* Engineering Extension UCLA and Society for Experimental Stress Analysis, 1953.
3. C. C. Perry and H. R. Lissner, *The Strain Gage Primer,* 2nd ed., McGraw-Hill, New York, 1962.
4. R. C. Dove and P. H. Adams, *Experimental Stress Analysis and Motion Measurements,* Charles E. Merrill Books, Columbus, OH, 1964.
5. A. L. Window and G. S. Hollister, *Strain Gauge Technology,* Applied Science Publishers, London and New York, 1982.
6. P. L. Walter, "Deriving the Transfer Function of Spatial Averaging Transducers," *ISA Transactions* **19**(3) (1980).
7. "Temperature Compensation of Bridge Type Transducers," Statham Instrument Notes Number 5, Gould, Oxnard, CA, August 1951.
8. TECHNICAL DATA TD4354-1, "When and How—Semiconductor Strain Gages," BLH Electronics, Waltham, MA, June 1975.
9. "End-to-End Test Methods for Telemetry Systems," Document 118-79, in *Test Methods for Telemetry Systems and Subsystems,* Vol. 1, Secretariat, Range Commanders Council, White Sands Missile Range, NM, 1979.
10. "Test Methods for Vehicle Telemetry Systems," Document 118-03, in *Test Methods for Telemetry Systems and Subsystems,"* Vol. 5, Secretariat, Range Commanders Council, White Sands Missile Range, NM, 2003.
11. J. W. Jaquay, "Understanding DC Instrumentation Amplifiers," *Instruments and Controls,* September 1972.
12. P. K. Stein, "A Unified Approach to Handling of Noise in Measuring Systems," AGARD LS-50, NATO, Nevilly-sur-Seine, France, September 1972.
13. P. K. Stein, "Information Conversion as a Noise Suppression Method," Lf/MSE Publication 66, Stein Engineering Services, Phoenix, AZ, January 1975.

BIBLIOGRAPHY

Dally, J. W., W. F. Riley, and K. McConnell, *Instrumentation for Engineering Measurements,* Wiley, New York, 1988.

Doebelin, Ernesto O., *Measurement Systems: Application and Design,* McGraw-Hill Science/Engineering/Math; 5th Pkg edition, ISBN: 007292201X (June 4, 2003).

CHAPTER 4
MEASUREMENTS

E. L. Hixson and E. A. Ripperger
University of Texas
Austin, Texas

1 STANDARDS AND ACCURACY

1.1 Standards

Measurement is the process by which a quantitative comparison is made between a standard and a measurand. The measurand is the particular quantity of interest—the thing that is to be quantified. The standard of comparison is of the same character as the measurand, and so far as mechanical engineering is concerned the standards are defined by law and maintained by the National Institute of Standards and Technology (NIST, formerly known as the National Bureau of Standards). The four independent standards which have been defined are length, time, mass, and temperature.[1] All other standards are derived from these four. Before 1960 the standard for length was the international prototype meter, kept at Sevres, France. In 1960 the meter was redefined as 1,650,763.73 wavelengths of krypton light. Then in 1983, at the Seventeenth General Conference on Weights and Measures, a new standard was adopted: A meter is the distance traveled in a vacuum by light in 1/299,792,458 seconds.[2] However, there is a copy of the international prototype meter, known as the national prototype meter, kept by NIST. Below that level there are several bars known as national reference standards and below that there are the working standards. Interlaboratory standards in factories and laboratories are sent to NIST for comparison with the working standards. These interlaboratory standards are the ones usually available to engineers.

Standards for the other three basic quantities have also been adopted by NIST, and accurate measuring devices for those quantities should be calibrated against those standards.

The standard mass is a cylinder of platinum–iridium, the international kilogram, also kept at Sevres, France. It is the only one of the basic standards that is still established by a prototype. In the United States the basic unit of mass is the basic prototype kilogram No. 20. Working copies of this standard are used to determine the accuracy of interlaboratory standards. Force is not one of the fundamental quantities, but in the United States the standard unit of force is the pound, defined as the gravitational attraction for a certain platinum mass at sea level and 45° latitude.

Absolute time, or the time when some event occurred in history, is not of much interest to engineers. Engineers are more likely to need to measure time intervals, that is, the time between two events. The basic unit for time measurements is the *second.* At one time the second was defined as 1/86,400 of the average period of rotation of the earth on its axis, but that is not a practical standard. The period varies and the earth is slowing up. Consequently a new standard based on the oscillations associated with a certain transition within the cesium atom was defined and adopted. That standard, the cesium clock, has now been superceded by the cesium fountain atomic clock as the primary time and frequency standard of the United States.[3] Although this cesium "clock" is the basic frequency standard, it is not generally usable by mechanical engineers. Secondary standards such as tuning forks, crystals, electronic oscillators, and so on are used, but from time to time access to time standards of a higher order of accuracy may be required. To help meet these requirements, NIST broadcasts 24 hours per day, 7 days per week time and frequency information from radio stations WWV, WWVB, and WWVL located in Fort Collins, Colorado, and WWVH located in Hawaii. Other nations also broadcast timing signals. For details on the time signal broadcasts, potential users should consult NIST.[4]

Temperature is one of four fundamental quantities in the international measuring system. Temperature is fundamentally different in nature from length, time, and mass. It is an intensive quantity, whereas the others are extensive. Join together two bodies that have the same temperature and you will have a larger body at that same temperature. If you join two bodies which have a certain mass, you will have one body of twice the mass of the original body. Two bodies are said to be at the same temperature if they are in thermal equilibrium. The international practical temperature scale, adopted in 1990 (ITS-90) by the International Committee on Weights and Measurement is the one now in effect and the one with which engineers are primarily concerned. In this system the kelvin (K) is the basic unit of temperature. It is 1/273.16 of the temperature at the triple point of water, the temperature at which the solid, liquid, and vapor phases of water exist in equilibrium.[5] Degrees Celsius (°C) is related to degrees kelvin by the equation

$$t = T - 273.15$$

where t = degrees Celsius
T = degrees kelvin

1.2 Accuracy and Precision

In measurement practice four terms are frequently used to describe an instrument. They are accuracy, precision, sensitivity, and linearity. Accuracy, as applied to an instrument, is the closeness with which a reading approaches the true value. Since there is some error in every reading, the "true value" is never known. In the discussion of error analysis which follows, methods of estimating the "closeness" with which the determination of a measured value approaches the true value will be presented. Precision is the degree to which readings agree among themselves. If the same value is measured many times and all the measurements agree very closely, the instrument is said to have a high degree of precision. It may not, however, be a very accurate instrument. Accurate calibration is necessary for accurate measurement. Measuring instruments must, for accuracy, be from time to time compared to a standard. These will usually be laboratory or company standards which are in turn compared from time to time with a working standard at NIST. This chain can be thought of as the pedigree of the instrument, and the calibration of the instrument is said to be traceable to NIST.

1.3 Sensitivity and Resolution

These two terms, as applied to a measuring instrument, refer to the smallest change in the measured quantity to which the instrument responds. Obviously the accuracy of an instrument will depend to some extent on the sensitivity. If, for example, the sensitivity of a pressure transducer is 1 kPa, any particular reading of the transducer has a potential error of at least 1 kPa. If the readings expected are in the range of 100 kPa and a possible error of 1% is acceptable, then the transducer with a sensitivity of 1 kPa may be acceptable, depending upon what other sources of error may be present in the measurement. A highly sensitive instrument is difficult to use. Therefore a sensitivity significantly greater than that necessary to obtain the desired accuracy is no more desirable than one with insufficient sensitivity.

Many instruments today have digital readouts. For such instruments the concepts of sensitivity and resolution are defined somewhat differently than they are for analog-type instruments. For example, the resolution of a digital voltmeter depends on the "bit" specification and the voltage range. The relationship between the two is expressed by the equation

$$R = \frac{V}{2^n}$$

where R = resolution in volts
V = voltage range
n = number of bits

Thus an 8-bit instrument on a 1-V scale would have a resolution of 1/256, or 0.004, volt. On a 10-V scale that would increase to 0.04 V. As in analog instruments, the higher the resolution, the more difficult it is to use the instrument, so if the choice is available, one should use the instrument which just gives the desired resolution and no more.

1.4 Linearity

The calibration curve for an instrument does not have to be a straight line. However, conversion from a scale reading to the corresponding measured value is most convenient if it can be done by multiplying by a constant rather than by referring to a nonlinear calibration curve or by computing from an equation. Consequently instrument manufacturers generally try to produce instruments with a linear readout, and the degree to which an instrument approaches this ideal is indicated by its *linearity*. Several definitions of linearity are used in instrument specification practice.[6] The so-called independent linearity is probably the most commonly used in specifications. For this definition the data for the instrument readout versus the input are plotted and then a "best straight line" fit is made using the method of least squares. Linearity is then a measure of the maximum deviation of any of the calibration points from this straight line. This deviation can be expressed as a percentage of the actual reading or a percentage of the full-scale reading. The latter is probably the most commonly used, but it may make an instrument appear to be much more linear than it actually is. A better specification is a combination of the two. Thus, linearity equals $+A$ percent of reading or $+B$ percent of full scale, whichever is greater. Sometimes the term independent linearity is used to describe linearity limits based on actual readings. Since both are given in terms of a fixed percentage, an instrument with A percent proportional linearity is much more accurate at low reading values than an instrument with A percent independent linearity.

It should be noted that although specifications may refer to an instrument as having A percent linearity, what is really meant is A percent nonlinearity. If the linearity is specified as independent linearity, the user of the instrument should try to minimize the error in readings by selecting a scale, if that option is available, such that the actual reading is close to full scale. A reading should never be taken near the low end of a scale if it can possibly be avoided.

For instruments that use digital processing, linearity is still an issue since the analog-to-digital converter used can be nonlinear. Thus linearity specifications are still essential.

2 IMPEDANCE CONCEPTS

Two basic questions which must be considered when any measurement is made are: How has the measured quantity been affected by the instrument used to measure it? Is the quantity the same as it would have been had the instrument not been there? If the answers to these questions are no, the effect of the instrument is called *loading*. To characterize the loading, the concepts of *stiffness* and *input impedance* are used.[7] At the input of each component in a measuring system there exists a variable q_{i1} which is the one we are primarily concerned with in the transmission of information. At the same point, however, there is associated with q_{i1} another variable q_{i2} such that the product $q_{i1}q_{i2}$ has the dimensions of power and represents the rate at which energy is being withdrawn from the system. When these two quantities are identified, the generalized input impedance Z_{gi} can be defined by

$$Z_{gi} = \frac{q_{i1}}{q_{i2}} \tag{1}$$

if q_{i1} is an *effort variable.* The effort variable is also sometimes called the *across variable.* The quantity q_{i2} is called the *flow variable* or *through variable.* In the dynamic case these variables can be represented in the frequency domain by their Fourier transform. Then the quantity Z is a complex number.The application of these concepts is illustrated by the example in Fig. 1. The output of the linear network in the blackbox (Fig. 1*a*) is the open-circuit voltage E_o until the load Z_L is attached across the terminals *A–B*. If Thévenin's theorem is applied after the load Z_L is attached, the system in Fig. 1*b* is obtained. For that system the current is given by

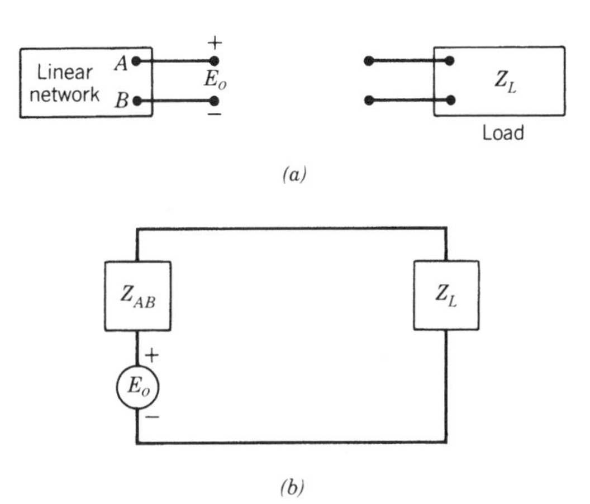

Figure 1 Application of Thévenin's theorem.

$$i_m = \frac{E_o}{Z_{AB} + Z_L} \tag{2}$$

and the voltage E_L across Z_L is

$$E_L = i_m Z_L = \frac{E_o Z_L}{Z_{AB} + Z_L}$$

or

$$E_L = \frac{E_o}{1 + Z_{AB}/Z_L} \tag{3}$$

Equations (2)–(7) are frequency-domain equations.

In a measurement situation E_L would be the voltage indicated by the voltmeter, Z_L would be the input impedance of the voltmeter, and Z_{AB} would be the output impedance of the linear network. The true output voltage, E_o, has been reduced by the voltmeter, but it can be computed from the voltmeter reading if Z_{AB} and Z_L are known. From Eq. (3) it is seen that the effect of the voltmeter on the reading is minimized by making Z_L as large as possible.

If the generalized input and output impedances Z_{gi} and Z_{go} are defined for nonelectrical systems as well as electrical systems, Eq (3) can be generalized to

$$q_{im} = \frac{q_{iu}}{1 + Z_{go}/Z_{gi}} \tag{4}$$

where q_{im} is the measured value of the effort variable and q_{iu} is the undisturbed value of the effort variable. The output impedance Z_{go} is not always defined or easy to determine; consequently Z_{gi} should be large. If it is large enough, knowing Z_{go} is unimportant.

If q_{i1} is a flow variable rather than an effort variable (current is a flow variable, voltage an effort variable), it is better to define an input admittance

$$Y_{gi} = \frac{q_{i1}}{q_{i2}} \tag{5}$$

rather than the generalized input impedance

$$Z_{gi} = \frac{\text{effort variable}}{\text{flow variable}}$$

The power drain of the instrument is

$$P = q_{i1} q_{i2} = \frac{q_{i2}^2}{Y_{gi}} \tag{6}$$

Hence, to minimize power drain, Y_{gi} must be large. For an electrical circuit

$$I_m = \frac{I_u}{1 + Y_o/Y_i} \tag{7}$$

where I_m = measured current
I_u = actual current
Y_o = output admittance of circuit
Y_i = input admittance of meter

When the power drain is zero and the deflection is zero, as in structures in equilibrium, for example when deflection is to be measured, the concepts of impedance and admittance are replaced with the concepts of *static stiffness* and *static compliance*. Consider the idealized structure in Fig. 2.

To measure the force in member K_2, an elastic link with a spring constant K_m is inserted in series with K_2. This link would undergo a deformation proportional to the force in K_2. If the link is very soft in comparison with K_1, no force can be transmitted to K_2. On the other hand, if the link is very stiff, it does not affect the force in K_2 but it will not provide a very good measure of the force. The measured variable is an effort variable, and in general, when it is measured, it is altered somewhat. To apply the impedance concept, a flow variable whose product with the effort variable gives power is selected. Thus,

$$\text{Flow variable} = \frac{\text{power}}{\text{effort variable}}$$

Mechanical impedance is then defined as force divided by velocity, or

$$Z = \frac{\text{force}}{\text{velocity}}$$

where force and velocity are dynamic quantities represented by their Fourier transform and Z is a complex number. This is the equivalent of electrical impedance. However, if the static mechanical impedance is calculated for the application of a constant force, the impossible result

$$Z = \frac{\text{force}}{0} = \infty$$

is obtained.

This difficulty is overcome if energy rather than power is used in defining the variable associated with the measured variable. In that case the static mechanical impedance becomes the *stiffness:*

$$\text{Stiffness} = S_g = \frac{\text{effort}}{\int \text{flow } dt}$$

In structures,

$$S_g = \frac{\text{effort variable}}{\text{displacement}}$$

When these changes are made, the same formulas used for calculating the error caused by the loading of an instrument in terms of impedances can be used for structures by inserting S for Z. Thus

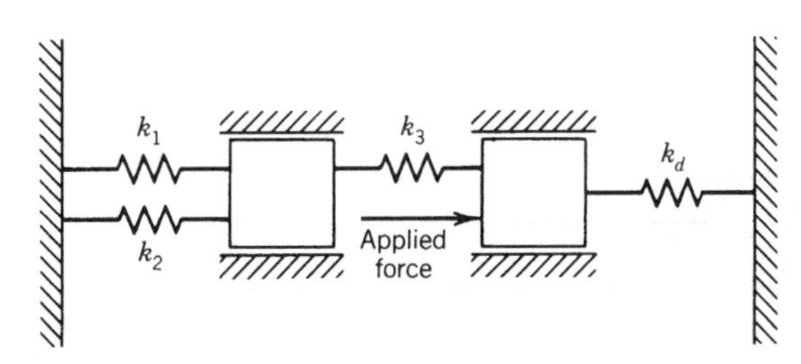

Figure 2 Idealized elastic structure.

$$q_{im} = \frac{q_{iu}}{1 + S_{go}/S_{gi}} \tag{8}$$

where q_{im} = measured value of effort variable
q_{iu} = undisturbed value of effort variable
S_{go} = static output stiffness of measured system
S_{gi} = static stiffness of measuring system

For an elastic force-measuring device such as a load cell S_{gi} is the spring constant K_m. As an example, consider the problem of measuring the reactive force at the end of a propped cantilever beam, as in Fig. 3.

According to Eq. (8), the force indicated by the load cell will be

$$F_m = \frac{F_u}{1 + S_{go}/S_{gi}}$$

$$S_{gi} = K_m \qquad \text{and} \qquad S_{go} = \frac{3EI}{L^3}$$

The latter is obtained by noting that the deflection at the tip of a tip-loaded cantilever is given by

$$\delta = \frac{PL^3}{3EI}$$

The stiffness is the quantity by which the deflection must be multiplied to obtain the force producing the deflection.

For the cantilever beam

$$F_m = \frac{F_u}{1 + 3EI/K_m L^3} \tag{9}$$

or

$$F_u = F_m\left(1 + \frac{3EI}{K_m L^3}\right) \tag{10}$$

Clearly, if $K_m >> 3EI/L^3$, the effect of the load cell on the measurement will be negligible.

To measure displacement rather than force, the concept of compliance is introduced and defined as

$$C_g = \frac{\text{flow variable}}{\int \text{effort variable } dt}$$

Then

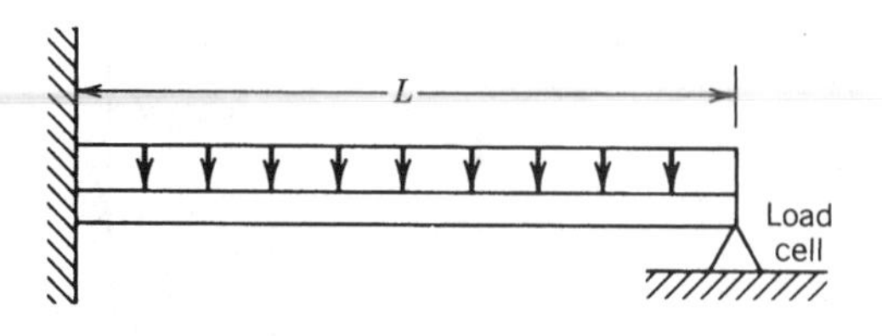

Figure 3 Measuring the reactive force at the tip.

$$q_m = \frac{q_u}{1 + C_{go}/C_{gi}} \tag{11}$$

If displacements in an elastic structure are considered, the compliance becomes the reciprocal of stiffness, or the quantity by which the force must be multiplied to obtain the displacement caused by the force. The cantilever beam in Fig. 4 again provides a simple illustrative example.

If the deflection at the tip of this cantilever is to be measured using a dial gage with a spring constant K_m,

$$C_{gi} = \frac{1}{K_m} \quad \text{and} \quad C_{go} = \frac{L^3}{3EI}$$

Thus

$$\delta_m = \delta_u\left(1 + \frac{K_m L^3}{3EI}\right) \tag{12}$$

Not all interactions between a system and a measuring device lend themselves to this type of analysis. A pitot tube, for example, inserted into a flow field distorts the flow field but does not extract energy from the field. Impedance concepts cannot be used to determine how the flow field will be affected.

There are also applications in which it is not desirable for a force-measuring system to have the highest possible stiffness. A subsoil pressure gage is an example. Such a gage, if it is much stiffer than the surrounding soil, will take a disproportionate share of the total load and will consequently indicate a higher pressure than would have existed in the soil if the gage had not been there.

3 ERROR ANALYSIS

It may be accepted as axiomatic that there will always be errors in measured values. Thus if a quantity X is measured, the correct value q, and X will differ by some amount e. Hence

$$\pm(q - X) = e$$

or

$$q = X \pm e \tag{13}$$

It is essential, therefore, in all measurement work that a realistic estimate of e be made. Without such an estimate the measurement of X is of no value. There are two ways of

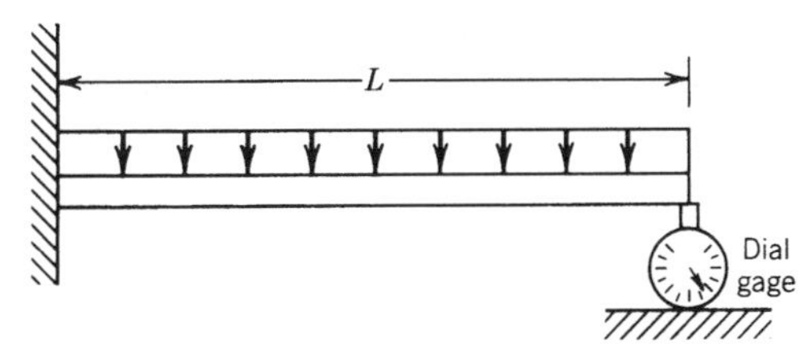

Figure 4 Measuring the tip deflection.

estimating the error in a measurement. The first is the external estimate, or €$_E$, where € = e/q. This estimate is based on knowledge of the experiment and measuring equipment and to some extent on the internal estimate €$_I$.

The internal estimate is based on an analysis of the data using statistical concepts.

3.1 Internal Estimates

If a measurement is repeated many times, the repeat values will not, in general, be the same. Engineers, it may be noted, do not usually have the luxury of repeating measurements many times. Nevertheless the standardized means for treating results of repeated measurements are useful, even in the error analysis for a single measurement.[8]

If some quantity is measured many times and it is assumed that the errors occur in a completely random manner, that small errors are more likely to occur than large errors, and that errors are just as likely to be positive as negative, the distribution of errors can be represented by the curve

$$F(X) = \frac{Y_o e^{-(X-U)}}{2\sigma^2} \tag{14}$$

where $F(X)$ = number of measurements for a given value of $(X - U)$
Y_o = maximum height of curve or number of measurements for which $X = U$
U = value of X at point where maximum height of curve occurs σ determines lateral spread of the curve

This curve is the normal, or Gaussian, frequency distribution. The area under the curve between X and δX represents the number of data points which fall between these limits and the total area under the curve denotes the total number of measurements made. If the normal distribution is defined so that the area between X and $X + \delta X$ is the probability that a data point will fall between those limits, the total area under the curve will be unity and

$$F(X) = \frac{\exp -(X - U)^2/2\sigma^2}{\sigma\sqrt{2\Pi}} \tag{15}$$

and

$$P_x = \int \frac{\exp -(X - U)^2/2\sigma^2}{\sigma\sqrt{2\Pi}}\, dx \tag{16}$$

Now if U is defined as the average of all the measurements and s as the standard deviation,

$$\sigma = \left[\frac{\Sigma(X - U)^2}{N}\right]^{1/2} \tag{17}$$

where N is the total number of measurements. Actually this definition is used as the best estimate for a universe standard deviation, that is, for a very large number of measurements. For smaller subsets of measurements the best estimate of σ is given by

$$\sigma = \left(\frac{\Sigma(X - U)^2}{n - 1}\right)^{1/2} \tag{18}$$

where n is the number of measurements in the subset. Obviously the difference between the two values of σ becomes negligible as n becomes very large (or as $n \rightarrow N$).

The probability curve based on these definitions is shown in Fig. 5.

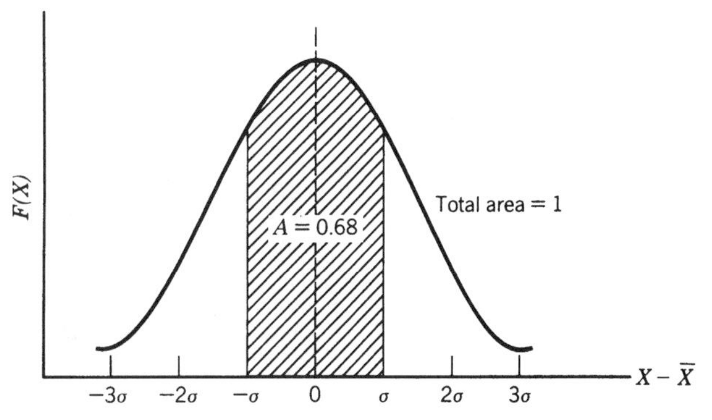

Figure 5 Probability curve.

The area under this curve between $-\sigma$ and $+\sigma$ is 0.68. Hence 68% of the measurements can be expected to have errors that fall in the range of $\pm\delta$. Thus the chances are 68/32, or better than 2 to 1, that the error in a measurement will fall in this range. For the range $\pm 2\sigma$ the area is 0.95. Hence 95% of all the measurement errors will fall in this range and the odds are about 20:1 that a reading will be within this range. The odds are about 384:1 that any given error will be in the range of $\pm 3\sigma$.

Some other definitions related to the normal distribution curve are as follows:

1. *Probable Error.* The error likely to be exceeded in half of all the measurements and not reached in the other half of the measurements. This error in Fig. 5 is about 0.67δ.
2. *Mean Error.* The arithmetic mean of all the errors regardless of sign. This is about 0.8σ.
3. *Limit of Error.* The error that is so large it is most unlikely ever to occur. It is usually taken as 4δ.

3.2 Use of Normal Distribution to Calculate Probable Error in X

The foregoing statements apply strictly only if the number of measurements is very large. Suppose that n measurements have been made. That is a sample of n data points out of an infinite number. From that sample U and σ are calculated as above. How good are these numbers? To determine that, we proceed as follows: Let

$$U = F(X_1, X_2, X_3, \ldots, X_n) = \frac{\Sigma X_i}{n} \tag{19}$$

$$e_u = \Sigma \frac{\partial F}{\partial X_i} e_{xi} \tag{20}$$

where e_u = error in U
e_{xi} = error in X_i

$$(e_u)^2 = \Sigma\left(\frac{\partial F}{\partial X_i} e_{xi}\right)^2 + \Sigma\left(\frac{\partial F}{\delta X_i} e_{xi}\right)\left(\frac{\delta F}{\delta X_j} e_{xj}\right) \tag{21}$$

where $I \neq j$. If the errors e_i to e_n are independent and symmetrical, the cross-product terms will tend to disappear and

$$(e_u)^2 = \Sigma\left(\frac{\partial F}{\partial X_i} e_{xi}\right)^2 \tag{22}$$

Since $\partial F/\partial X_i = 1/n$,

$$e_u = \left[\Sigma\left(\frac{1}{n}\right)^2 e_{xi}^{\;2}\right]^{1/2} \tag{23}$$

or

$$e_u = \left[\left(\frac{1}{n}\right)^2 \Sigma(e_{xi})^2\right]^{1/2} \tag{24}$$

from the definition of σ

$$\Sigma(e_{xi})^2 = n\sigma^2 \tag{25}$$

and

$$e_u = \frac{\sigma}{\sqrt{n}}$$

This equation must be corrected because the real errors in X are not known. If the number n were to approach infinity, the equation would be correct. Since n is a finite number, the corrected equation is written as

$$e_u = \frac{\sigma}{(n-1)^{1/2}} \tag{26}$$

and

$$q = U \pm \frac{\sigma}{(n-1)^{1/2}} \tag{27}$$

This says that if one reading is likely to differ from the true value by an amount σ, then the average of 10 readings will be in error by only $\sigma/3$ and the average of 100 readings will be in error by $\sigma/10$. To reduce the error by a factor of 2, the number of readings must be increased by a factor of 4.

3.3 External Estimates

In almost all experiments several steps are involved in making a measurement. It may be assumed that in each measurement there will be some error, and if the measuring devices are adequately calibrated, errors are as likely to be positive as negative. The worst condition insofar as accuracy of the experiment is concerned would be for all errors to have the same sign. In that case, assuming the errors are all much less than 1, the resultant error will be the sum of the individual errors, that is,

$$€_E = €_1 + €_2 + €_3 + \cdots \tag{28}$$

It would be very unusual for all errors to have the same sign. Likewise it would be very unusual for the errors to be distributed in such a way that

$$€_E = 0$$

A general method follows for treating problems that involve a combination of errors to determine what error is to be expected as a result of the combination.

Suppose that

$$V = F(a, b, c, d, e, \cdots, x, y, z) \tag{29}$$

where $a, b, c, \cdots x, y, z$ represent quantities which must be individually measured to determine V. Then

$$\delta V = \sum \left(\frac{\partial F}{\partial n}\right) \delta n$$

and

$$€_E = \sum \left(\frac{\partial F}{\partial n}\right) e_n \tag{30}$$

The sum of the squares of the error contributions is given by

$$e_E^2 = \left[\sum \left(\frac{\partial F}{\partial n}\right) e_n\right]^2 \tag{31}$$

Now, as in the discussion of internal errors, assume that errors e_n are independent and symmetrical. This justifies taking the sum of the cross products as zero:

$$\sum \left(\frac{\partial F}{\partial n}\right)\left(\frac{\partial F}{\partial m}\right) e^n e^m = 0 \qquad n \neq m \tag{32}$$

Hence

$$(€_E)^2 = \sum \left(\frac{\partial F}{\partial n}\right)^2 e_n^2$$

or

$$e_E = \left[\sum \left(\frac{\partial F}{\partial n}\right)^2 e_n^2\right]^{1/2} \tag{33}$$

This is the *most probable value* of e_E. It is much less than the worst case:

$$€_e = [|€_a| + |€_b| + |€_c| \cdots + |€_z|] \tag{34}$$

As an application, the determination of g, the local acceleration of gravity, by use of a simple pendulum will be considered:

$$g = \frac{4\Pi^2 L}{T^2} \tag{35}$$

where L = length of pendulum
T = period of pendulum

If an experiment is performed to determine g, the length L and the period T would be measured. To determine how the accuracy of g will be influenced by errors in measuring L and T write

$$\frac{\partial g}{\partial L} = \frac{4\Pi^2}{T^2} \qquad \text{and} \qquad \frac{\partial g}{\partial T} = \frac{-8\Pi^2 L}{T^3} \tag{36}$$

The error in g is the variation in g written as follows:

$$\delta g = \left(\frac{\partial g}{\partial L}\right) \Delta L + \left(\frac{\partial g}{\partial T}\right) \Delta T \quad (37)$$

or

$$\delta g = \left(\frac{4\Pi^2}{T^2}\right) \Delta L - \left(\frac{8\Pi^2 L}{T^3}\right) \Delta T \quad (38)$$

It is always better to write the errors in terms of percentages. Consequently Eq. (38) is rewritten as

$$\delta g = \frac{(4\Pi^2 L/T^2)\ \Delta \mathrm{L}}{L} - \frac{2(4\Pi^2 L/T^2)\ \Delta T}{T} \quad (39)$$

or

$$\frac{\delta g}{g} = \frac{\Delta L}{L} - \frac{2\ \Delta T}{T} \quad (40)$$

Then

$$e_g = [e_L^2 + (2e_T)^2]^{1/2} \quad (41)$$

where e_g is the *most probable error* in the measured value of g. That is,

$$g = \frac{4\Pi^2 L}{T^2} \pm e_g \quad (42)$$

where L and T are the measured values. Note that even though a positive error in T causes a negative error in the calculated value of g, the contribution of the error in T to the most probable error is taken as positive. Note also that an error in T contributes four times as much to the most probable error as an error in L contributes. It is fundamental in measurements of this type that those quantities which appear in the functional relationship raised to some power greater than unity contribute more heavily to the most probable error than other quantities and must, therefore, be measured with greater care.

The determination of the most probable error is simple and straightforward. The question is how are the errors, such as $\Delta L/L$ and $\Delta T/T$, determined. If the measurements could be repeated often enough, the statistical methods discussed in the internal error evaluation could be used to arrive at a value. Even in that case it would be necessary to choose some representative error such as the standard deviation or the mean error. Unfortunately, as was noted previously, in engineering experiments it usually is not possible to repeat measurements enough times to make statistical treatments meaningful. Engineers engaged in making measurements will have to use what knowledge they have of the measuring instruments and the conditions under which the measurements are made to make a reasonable estimate of the accuracy of each measurement. When all of this has been done and a most probable error has been calculated, it should be remembered that the result is not the actual error in the quantity being determined but is, rather, the engineer's best estimate of the magnitude of the uncertainty in the final result.[9,10]

Consider again the problem of determining g. Suppose that the length L of the pendulum has been determined by means of a meter stick with 1-mm calibration marks and the error in the calibration is considered negligible in comparison with other errors. Suppose the value of L is determined to be 91.7 cm. Since the calibration marks are 1 mm apart, it can be assumed that ΔL is no greater than 0.5 mm. Hence the maximum

$$\frac{\Delta L}{L} = 5.5 \times 10^{-4}$$

Suppose T is determined with the pendulum swinging in a vacuum with an arc of $\pm 5°$ using a stop watch that has an inherent accuracy of one part in 10,000. (If the arc is greater than $\pm 5°$, a nonisochronous swing error enters the picture.) This means that the error in the watch reading will be no more than 10^{-4} s. However, errors are introduced in the period determination by human error in starting and stopping the watch as the pendulum passes a selected point in the arc. This error can be minimized by selecting the highest point in the arc because the pendulum has zero velocity at that point and timing a large number of swings so as to spread the error out over that number of swings. Human reaction time may vary from as low as 0.2 s to as high as 0.7 s. A value of 0.5 s will be assumed. Thus the estimated maximum error in starting and stopping the watch will be 1 s (± 0.5 s at the start and ± 0.5 s at the stop). A total of 100 swings will be timed. Thus the estimated maximum error in the period will be 1/100 s. If the period is determined to be 1.92 s, the estimated maximum error will be $0.01/1.92 = 0.005$. Compared to this, the error in the period due to the inherent inaccuracy of the watch is negligible. The nominal value of g calculated from the measured values of L and T is 982.03 cm/s^2. The most probable error [Eq. (29)] is

$$[4(0.005)^2 + (5.5 \times 10^{-4})^2]^{1/2} = 0.01 \tag{43}$$

The uncertainty in the value of g is then ± 9.82 cm/s^2, or in other words the value of g will be somewhere between 972.21 and 991.85 cm/s^2.

Often it is necessary for the engineer to determine in advance how accurately the measurements must be made in order to achieve a given accuracy in the final calculated result. For example, in the pendulum problem it may be noted that the contribution of the error in T to the most probable error is more than 300 times the contribution of the error in the length measurement. This suggests, of course, that the uncertainty in the value of g could be greatly reduced if the error in T could be reduced. Two possibilities for doing this might be (1) find a way to do the timing that does not involve human reaction time or (2) if that is not possible, increase the number of cycles timed. If the latter alternative is selected and other factors remain the same, the error in T timed over 200 swings is 1/200 or 0.005, second. As a percentage the error is $0.005/1.92 = 0.0026$. The most probable error in g then becomes

$$e_g = [4 \times (2.6 \times 10^{-3})^2 + (5.5 \times 10^{-4})^2]^{1/2} = 0.005 \tag{44}$$

This is approximately half of the most probable error in the result obtained by timing just 100 swings. With this new value of e_g the uncertainty in the value of g becomes ± 4.91 cm/s^2 and g then can be said to be somewhere between 977.12 and 986.94 cm/s^2. The procedure for reducing this uncertainty still further is now self-evident.

Clearly the value of this type of error analysis depends upon the skill and objectivity of the engineer in estimating the errors in the individual measurements. Such skills are acquired only by practice and careful attention to all the details of the measurements.

REFERENCES

1. W. A. Wildhack, "NBS Source of American Standards," *ISA Journal* **8**(2) (February 1961).
2. P. Giacomo "News from the IBPM," *Metrologia* **20**(1) (April, 1984).
3. NIST-F1 Cesium Fountain Atomic Clock.

4. "NIST Time and Frequency Services," NIST Special Publication 432, 2002.

5. R. E. Bentley (ed.) *Handbook of Temperature Measurement,* CSIRO Springer.

6. E. A. Doebelin, *Measurement Systems—Application and Design,* 5th ed., McGraw Hill, New York, 2004, pp. 85–91.

7. C. M. Harris and A. G. Piersol, "Mechanical Impedance," in *Shock and Vibration Handbook,* 5th ed. McGraw-Hill, New York, 2002, Chapter 10, pp. 10.1–10.14.

8. N. H. Cook and E. Rabinowicz, *Physical Measurement and Analysis,* Addison Wesley, Readng, MA, 1963, pp. 29–68.

9. S. J. Kline and F. A. McClintock, "Describing Uncertainties in Single Sample Experiments," *Mechanical Engineering,* **75**(3) (January 1953).

10. B. N. Taylor and C. E. Kuyatt, "Guidelines for Evaluating and Expressing the Uncertainty of NIST Measurement Results," NIST Technical Note 1297, 1994.

CHAPTER 5

TEMPERATURE AND FLOW TRANSDUCERS

Robert J. Moffat
Department of Mechanical Engineering
Stanford University
Stanford, California

Reprinted from *Instrumentation and Control,* Wiley, New York, 1990, by permission of the publisher.

1 INTRODUCTION

There are hundreds of different transducers for temperature and flow measurements. The most common types will be discussed in this chapter; some others will be mentioned only briefly.

In flow measurements, only closed-channel flow measurement techniques are considered here, and only for "clean" fluids. Slurries and liquids carrying large objects are not treated.

In temperature measurements, current interests range from cryogenics (a few kelvins) to plasmas (upward of 10,000 K). Most applications, however, are in the range from room temperature to 2000 K, and that is where the bulk of this chapter will be concentrated.

The accuracy of a temperature or flow measurement depends not only on the sensor characteristics but also on the interaction between the sensor and the system being instrumented. There are two primary classes of interactions: *system disturbance errors* (i.e., changes in the behavior of the system caused by the presence of the sensor) and *system/ sensor interactions* (the sensor responding to more than one parameter of the system).

High-temperature measurements are subject to installation errors caused by heat transfer between the system and the transducer. The term *error* is defined as the difference between the observed value and the true value of the intended measurand. The output of a temperature sensor describes its own temperature, the *achieved temperature,* but the objective is usually to measure the temperature at a particular point in the solid, liquid, or gas into which the sensor is installed—the *available temperature.* There is often a significant difference between the available value and the achieved value because the sensor exchanges heat with its entire surroundings, not just with the immediate region around the sensor. It is not uncommon in high-temperature gas temperature measurements with unshielded sensors, for example, to have errors of several hundreds of degrees caused by radiation error, velocity error, or conduction error effects on the sensor. These errors cannot be accounted for by calibration of the sensor, nor can corrections be applied with any degree of certainty. The sensor must be protected by appropriate shielding. In most applications at high temperature, the installation errors are far larger than the calibration error of the sensor; hence, sensor accuracy does not mean the same as measurement accuracy.

Individual transducers are discussed in the following sections.

2 THERMOCOUPLES*

Thermocouples are the most commonly used electrical output transducers that measure temperature. They are inexpensive, small in size, and remarkably accurate when their peculiarities are understood.

2.1 Types and Ranges

Any pair of thermoelectrically dissimilar wires can be used as a thermocouple. The wires need only be joined together at one end (the measuring junction) and connected to a voltage-measuring instrument at the other end (the reference junction) to form a usable system. Whenever the measuring junction is at a different temperature than the reference junction, a voltage will be developed, which is related to the temperature difference between the two junctions. Several metallic materials are listed in Table 1 in order of thermoelectric polarity;

---

*Materials in this section is substantially derived from Ref. 2 with permission, except where otherwise referenced.

Table 1 Thermoelectric Polarity Order of Metallic Materials

100°C	500°C	900°C
Antimony	Chromel	Chromel
Chromel	Nichrome	Nichrome
Iron	Copper	Silver
Nichrome	Silver	Gold
Copper	Gold	Iron
Silver	Iron	$Pt_{90}Rh_{10}$
$Pt_{90}Rh_{10}$	$Pt_{90}Rh_{10}$	Platinum
Platinum	Platinum	Cobalt
Palladium	Cobalt	Alumel
Cobalt	Palladium	Nickel
Alumel	Alumel	Palladium
Nickel	Nickel	Constantan
Constantan	Constantan	
Copel	Copel	
Bismuth		

Source: Reference 5.

each material in the listing is positive with respect to all beneath it. In an iron–palladium thermocouple, for example, the cold end of the iron wire will be positive with respect to the cold end of the palladium.

In some instances the operating temperatures of machinery elements have been measured using the machine structure as part of the thermoelectric circuit (cutting-tool tip temperatures, cam shaft/rocker arm contact temperatures, etc.). In such cases each material in the circuit must be calibrated, and all intermediate temperatures must be measured in order to interpret the signal.

The alloys usually used for thermoelectric temperature measurement are listed in Table 2. These have been developed over the years for the linearity, stability, and reproducibility of their electromotive force (emf)–temperature characteristics and for their high-temperature capability. Tables of thermocouple emf versus temperature [referenced to the international practical temperature scale of 1990 (IPTS9O)][1] are available as DOS files on a disk from the National Institute of Standards and Technology (NIST). The standardized letter-designated thermocouple pairs are treated: B, F, J, K, N, R, S, and T. This is the primary source for reliable thermocouple data.

The noble metal and refractory metal thermocouples are used generally with extension wires of substitute materials, which are cheaper and easier to handle (more ductile). The extension wires used are described in Table 3. Except for the substitute alloys, thermocouple extension wire is of the same nominal composition as thermocouple wire and differs from it mainly in the accuracy of its calibration and the type of insulation used. Extension wire is not calibrated as accurately as thermocouple-grade wire.

Thermocouple material can be purchased as individual bare wires, as flexible, insulated pairs of wires, or as mineral-insulated pairs swaged into stainless steel tubes for high-temperature service. Prices range from a few cents to several dollars per foot, depending on the wire and the insulation. There are many suppliers.

2.2 Peripheral Equipment

Any instrument capable of reading low-dc voltages (on the order of millivolts) with 5–10 μV resolution will suffice for temperature measurements; the accuracy depends upon the

Table 2 Common Alloys Used in Thermoelectric Temperature Measurement

Maximum Temperature		Allowable Atmosphere (Hot)	Material Names	Type ANSI[a]	Color Code	Average Output, mV/100°F	Accuracy[a]	
°F	°C						Std.	Spec.
5072	2800	Inert, H_2, vac.	Tungsten/tungsten–26% rhenium	—	—	0.86	—	—
5000	2760	Inert, H_2, vac.	Tungsten–5% rhenium/tungsten–26% rhenium	—	—	0.76	—	—
4000	2210	Inert, H_2, vac.	Tungsten-3% rhenium/tungsten-25% rhenium	—	—	0.74	—	—
3270	1800	Oxidizing[b]	Platinum–30% rhodium/platinum–6% rhodium	B	—	0.43	1/2%	1/4%
2900	1600	Oxidizing[b]	Platinum–13% rhodium/platinum	R	—	0.64	1/4%	1/4%
2800	1540	Oxidizing[b]	Platinum–10% rhodium/platinum	S	—	0.57	1/4%	1/4%
2372	1300	Oxidizing[b]	Platinel II (5355)/platinel II (7674)[c]	—	—	2.20	5/8%	—
2300	1260	Oxidizing	Chromel/Alumel,[d] Tophel/Nial,[e] Advance T1/T2,[f] Therm. Kanthal P/N[g]	K	Yellow-red	2.20	4°F,3/4%	2°F,3/8%
1800	980	Reducing	Chromel[a]/constantan	E	Purple-red	4.20	1/2%	3/8%
1600	875	Reducing	Iron/constantan	J	White-red	3.00	4°F,3/4%	2°F,3/8%
750	400	Reducing	Copper/constantan	T	Blue-red	2.50	3/4%	3/8%

Source: Reference 4.

[a] Per American National Standards Institute C96.1 standard

[b] Avoid contact with carbon, hydrogen, metallic vapors, silica, reducing atmosphere.

[c] © Englehard Corp.

[d] © Hoskins Mfg. Co.

[e] Wilber B. Driver Co.

[f] Driver–Harris Co.

[g] The Kanthal Corp.

Table 3 Extension Wires for Thermocouples

Thermocouple Material	Type	Extension Wire, Type[a]	Color[a] (+)	Color[a] (−)	Color[a] Overall
Tungsten/tungsten–26% rhenium	—	Alloys 200/226[b]	—	—	—
Tungsten–5% rhenium/tungsten–26% rhenium	—	Alloys (405/426)[b]	White	Red	Red[b]
Tungsten–3% rhenium/tungsten–25% rhenium	—	Alloys (203/225)[b]	White/yellow	White/red	Yellow/red
Platinum/platinum–rhodium	S,R	SX, SR	Black	Red	Green
Platinel II-5355/platinel II-7674	—	P2X[c]	Yellow	Red	Black[c]
Chromel/alumel, tophel/nial, Advance, Thermokanthal[d]	E	KK	Yellow	Red	Yellow
Chromel/constantan	K	EX	Purple	Red	Purple
Iron/constantan	J	JX	White	Red	Black
Copper/constantan	T	TX	Blue	Red	Blue

Source: Reference 4

[a] ANSI, except where noted otherwise.

[b] Designations affixed by Hoskins Mfg. Co.

[c] Englehardt Mfg. Co.

[d] Registered trademark names, see Table 2 for identification of ownership.

voltmeter. The signal from a thermocouple depends upon the difference in temperature between the two ends of the loop; hence, the accuracy of the temperature measurement depends upon the accuracy with which the *reference junction* temperature is known as well as the accuracy with which the electrical signal is measured.

Galvanometric measuring instruments can be used, but since they draw current, the voltage available at the terminals of the instrument depends not only on the voltage output of the thermocouple loop but also on the resistances of the instrument and the loop. Such instruments are normally marked to indicate the external resistance for which they have been calibrated. Potentiometric instruments, either manually or automatically balanced, draw no current when in balance and therefore can be used with thermocouple loops of any resistance without error. High-input-impedance voltmeters draw only minute currents and, except for very high resistance circuits, pose no problems. When in doubt, check the input impedance of the instrument against the circuit resistance.

The input stages of many instruments have one side grounded. Ground loops can result from using a grounded-junction thermocouple with such an instrument. If the ground potential where the thermocouple is attached is different from the potential where the instrument is grounded, then a current may flow through the thermocouple wire. The voltage drop in the wire due to the ground-loop current will mix with the thermoelectric signal and may cause an error.

2.3 Thermoelectric Theory

The emf–temperature calibrations of the more common materials are shown qualitatively in Fig. 1. The emf is the electromotive force *mv* that would be derived from thermocouples made of material *X* used with platinum when the cold end is at 0°C and the hot end is at *T*. Those elements commonly used as first names for thermocouple pairs [i.e., Chromel (–Alumel), iron (–constantan), copper (–constantan), etc.] have positive slopes in Fig. 1.

It can be shown from either the free-electron theory of metals or thermodynamic arguments alone that the output of a thermocouple can be rigorously described as the sum of

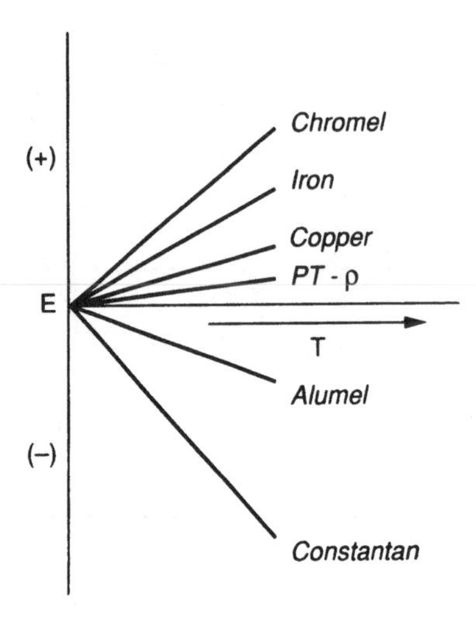

Figure 1 The emf vs. temperature calibrations for several materials. (Reproduced from Ref. 2, with permission.)

the contributions from each length of material in the circuit—the junctions are merely electrical connections between the wires. Formally,

$$E_{\text{net}} = \int_0^L \epsilon_1 \frac{dT}{dx}\, dx + \int_L^0 \epsilon_2 \frac{dT}{dx}\, dx \tag{1}$$

where ϵ = total thermoelectric power of material; equal to sum of Thomson coefficient and temperature derivative of Peltier coefficient
T = temperature
x = distance along the wire
L = length of the wire

When the wire is uniform in composition, so that ϵ is not a function of position,

$$E_{\text{net}} = \int_{T_0}^{T_L} \epsilon_1\, d\theta + \int_{T_L}^{T_0} \epsilon_2\, d\theta \tag{2}$$

Commercial wire is homogeneous within close limits, but used thermocouples may be far from uniform.

When only two wires are used in a circuit, it is customary to further simplify the problem. If both wires begin at one temperature (say T_0) and end at another (say T_L), the two integrals above can be collected:

$$E_{\text{net}} = \int_{T_0}^{T_L} (\epsilon_1 - \epsilon_2)\, d\theta \tag{3}$$

Three simplifications are built into this reduced equation:

1. ϵ is not a function of position (i.e., the wires are homogeneous).
2. There are only two wires.
3. Each wire begins at T_0 and ends at T_L.

These are the conditions for which the emf–temperature tables are intended. If any of the three conditions is not met, the tables cannot be used to interpret the output.

2.4 Graphical Analysis of Circuits

It is possible to graphically analyze a thermocouple circuit and describe its output in terms of the calibration of its wires. The simplest practical circuit consists of two wires joined together at one end and connected directly to a measuring instrument, as shown in Fig. 2. This is referred to as the *pattern circuit.* The pattern circuit is thermoelectrically ideal (providing that the materials of the instrument do not affect the reading) since it contains no switches, connectors, or lead wires. The output of this system can be analyzed graphically using the calibration data shown in Fig. 1. The method used here is described in detail by Moffat.[3]

The circuit in Fig. 2 requires frequent measurement of the temperature at the instrument terminals and is seldom used where accuracy of better than ± 2°F is required (chiefly because it is difficult to measure the temperatures of points 1 and 3 more accurately than ± 1°F).

A reference zone of controlled temperature eliminates the need for frequent measurements of the ambient temperature. The reference zone may be an ice point, a triple point, or an electrically controlled, high-temperature reference zone box. A circuit as in Fig. 3 assumes the reference temperature to be an ice-point bath.

The output of this circuit is the emf between points 1 and 5 (those connected to the instrument terminals). The ideal circuit would have had the output given by emf(2–4). The graphical construction shows that emf(2–4) is equal to emf(1–5) since the segments 1–2 and 4–5 each represent the same material (copper) over the same temperature interval ($T_{amb} - T_{ref}$), and the wires are connected so as to cancel these emfs. Thus, the actual circuit is thermoelectrically equivalent to the ideal circuit. Note that the copper lead wires (1–2 and 4–5) play no role in determining the output of the circuit, provided that (1) the calibrations

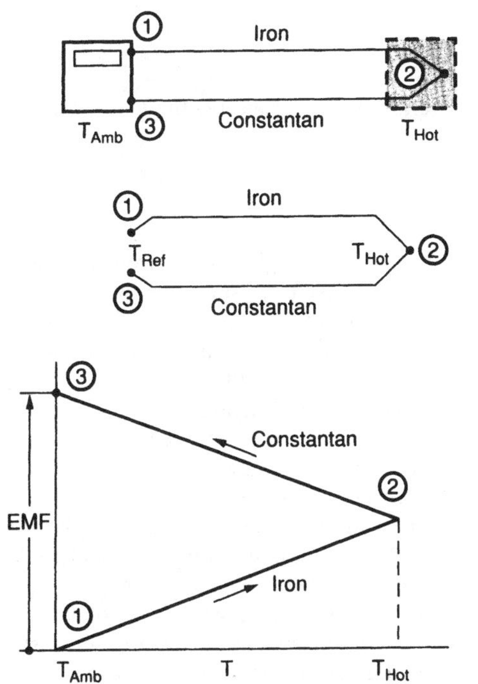

Figure 2 Temperature measurement using the ambient temperature as the reference. (Reproduced from Ref. 2, with permission.)

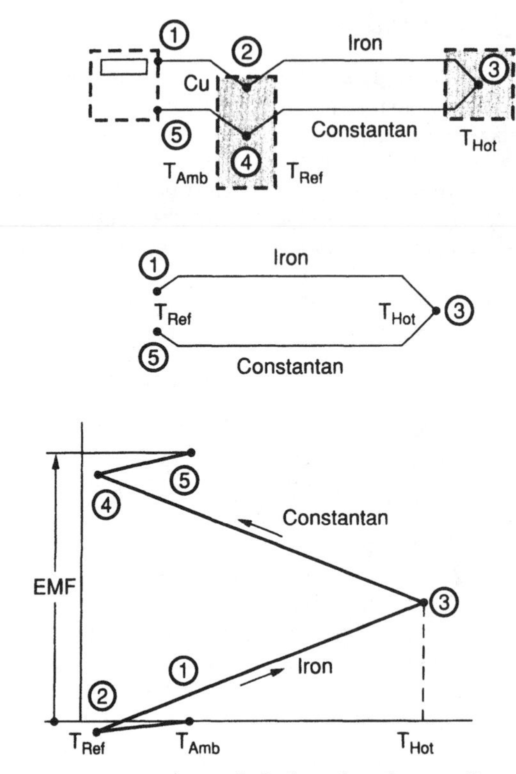

Figure 3 Temperature measurement using an icebath as the reference. (Reproduced from Ref. 2, with permission.)

of the two pieces of copper wire are the same and (2) the temperature intervals across the two copper wires are the same.

2.5 Zone-Box Circuits

When several thermocouples are used in a single test far from the measuring station, significant economies can sometimes be achieved by using a common zone box and substituting copper lead wires for much of the thermoelectric material. Such a circuit is shown in Fig. 4. The objective of the analysis is to determine the conditions under which the actual circuit is thermoelectrically equivalent to the pattern circuit.

The function of the zone box is to provide a region of uniform temperature within which connections can be made. The temperature of the zone box need not be constant, and it need not be known—it need only be uniform.

The circuit consists of a reference bath, a selector switch, a readout instrument, and a set of thermocouples extending from the zone box to their individual sensing points connected together with copper wires. Providing the selector switch and the lead wires introduce no spurious emf, the behavior of any one thermocouple for this circuit should be the same as that of the pattern circuit shown. In the *E–T* diagram, the copper lead wires are shown passing through the selector switch with no acknowledgment of its existence—the switch is assumed to have a uniform temperature at the ambient temperature.

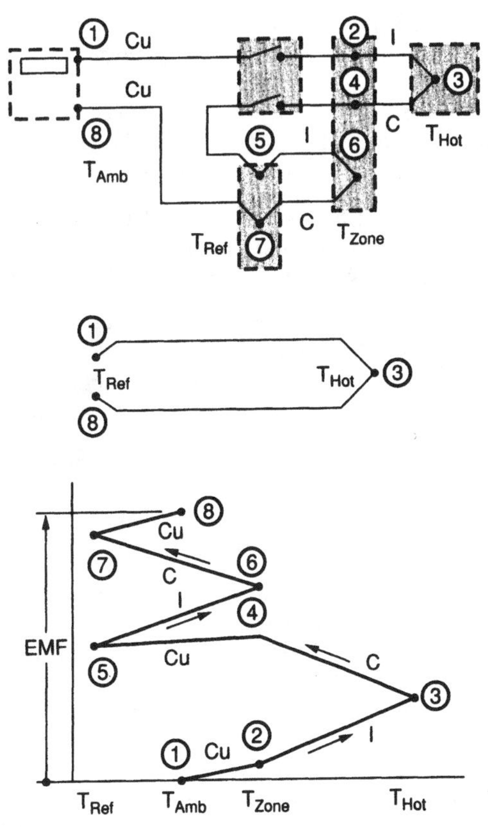

Figure 4 Multiple measurements using a zone box and selector switch. (Reproduced from Ref. 2, with permission.)

Commercially available zone-box and selector switch assemblies are sometimes made with the assumption that the junction 6 will be placed in the reference bath and points 5 and 7 in the zone box. This requires reversing the polarity of the reference junction (i.e., the wire between points 5 and 6 must then be constantan in this example, and the wire between 6 and 7 must then be iron).

If a wiring diagram is not available, a test for reference junction polarity should be made. With the system connected, at any arbitrary temperature, the instrument reading should go up if the temperature of the reference junction goes down, and conversely.

2.6 Laws of Thermoelectricity

Various authors have attempted to summarize the behavior of thermocouples through sets of laws ranging from three to six in number. One of the more detailed sets is given by Doebelin.[6] Each law can easily be proven by recourse to an emf–temperature sketch. The first three from Doebelin's list are used as examples:

1. The thermal emf of a thermocouple with junctions at T_{hot} and T_{ref} is totally unaffected by temperature elsewhere in the circuit if the two metals used are each homogeneous.

In Fig. 5, it is presumed that $T_{hot} < T_{candle}$. If the wire is uniform in calibration on both sides of the hot spot, the potential hysteresis loop closes and no net emf is generated because of the hot spot. The principal importance of this law is that it establishes the conditions under which a thermocouple is a point sensor. If and only if wires A and B are uniform in composition, the thermocouple output is determined only by the temperatures T_{ref} and T_{hot} and is independent of the temperature distribution along the wire.

2. If a third homogeneous metal C is inserted into either A or B, as long as the two new thermojunctions are at like temperature, the net emf of the circuit is unchanged irrespective of the temperature of C away from the junctions.

Figure 6 shows a third material inserted into the A leg and then heated locally. It is presumed that the temperatures at the two ends of C remain equal. If the material C is homogeneous, the emf induced by the excursion in temperature from point 2 to point 3 is canceled by that from point 3 to point 4, and no net signal is produced.

3. If metal C is inserted between A and B at one of the junctions, the temperature of C at any point away from the AC and BC junctions is immaterial. So long as the junctions AC and BC are both at the temperature T_1, the net emf is the same as if C were not there.

Figure 7 illustrates this case. An intermediate material, C, is inserted between A and B at the measuring junction. The diagram once more shows no change in the net emf if the inserted material is homogeneous and does not undergo a net temperature change.

The situation involving intermediate materials at the junction is of great practical importance because it addresses questions of manufacturing technique and how they affect thermocouple calibration. For example, the third material (C), used to connected the two materials A and B, might be the soft solder, silver solder, or braze material. The output of the thermocouple is independent of that third material, provided that the third material is homogeneous and begins and ends at the same temperature. In practice, these conditions are usually satisfied because the joining material is isothermal. If the third material is isothermal,

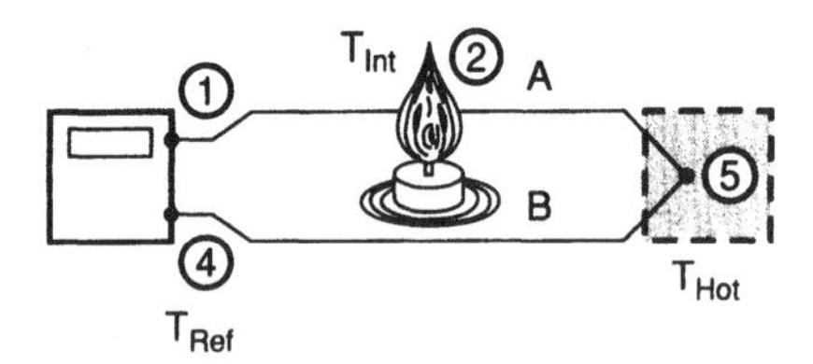

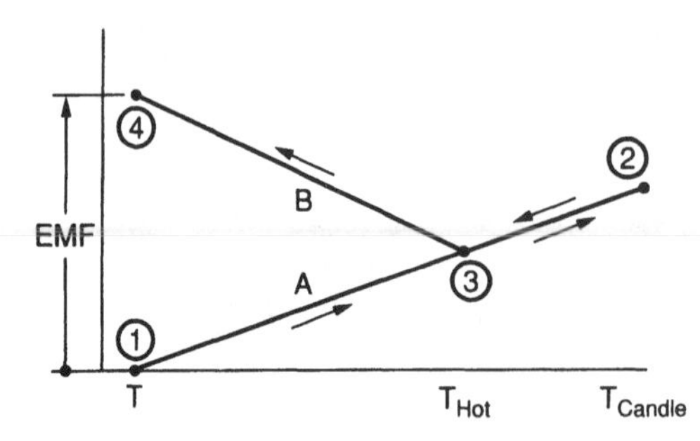

Figure 5 Illustration of the law of interior temperatures. (Reproduced from Ref. 2, with permission.)

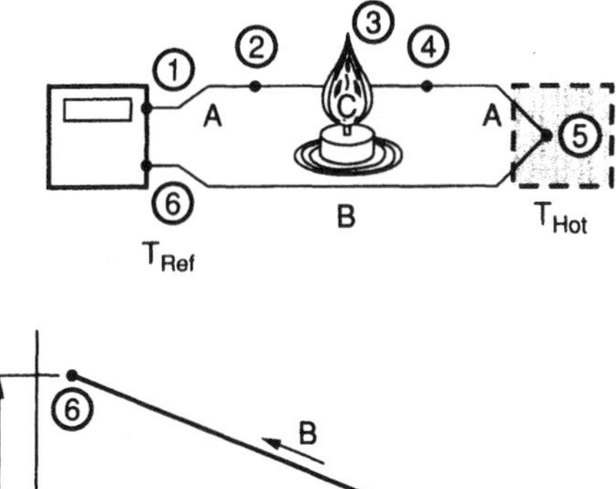

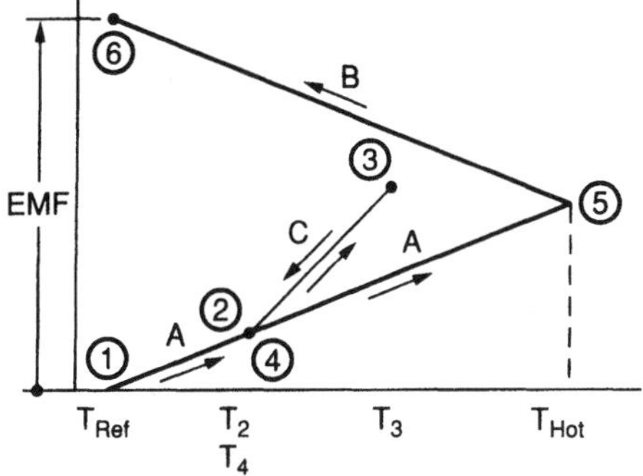

Figure 6 Illustration of the law of inserted materials. (Reproduced from Ref. 2, with permission.)

it makes no contribution to the output of the thermocouple. This proof should not be taken as a blanket license to connect thermocouple wires together without due care—some installations may result in temperature gradients near the junction, and those are sensitive to the presence of a joining material. However, a well-designed probe assures an isothermal zone around the junction, and those probes are insensitive to the material used to join the thermoelements together.

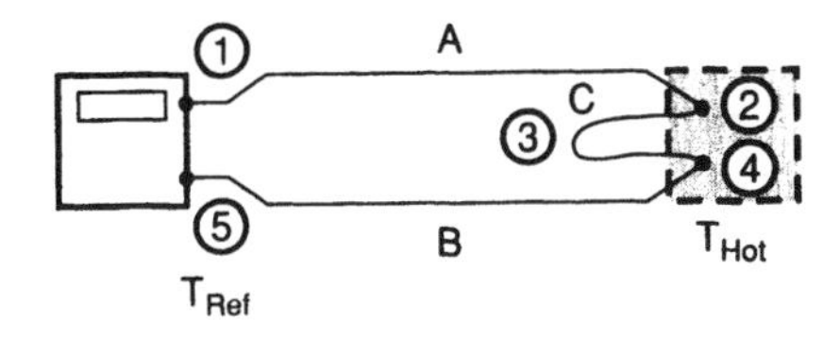

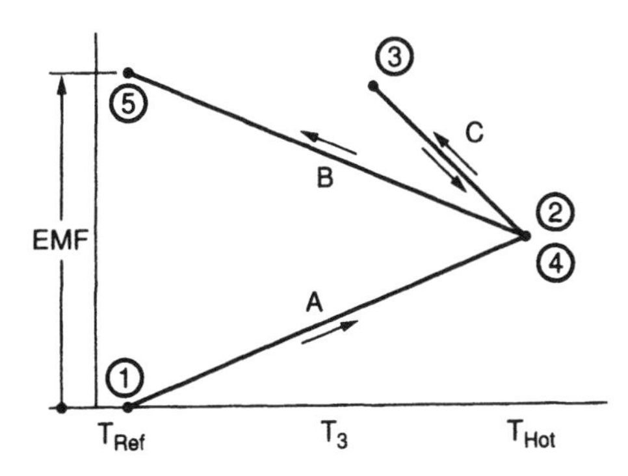

Figure 7 Illustration of the law of intermediate materials. (Reproduced from Ref. 2, with permission.)

2.7 Switches, Connectors, Zone Boxes, and Reference Baths

A thermocouple switch or connector must not produce emf that would contaminate the temperature signal. By their very nature, switches and connectors connect several materials together and are susceptible to generation of thermoelectric emf.

The principal defense against spurious emf is to ensure that the switch or connector is isothermal, not only on the whole but in detail. The mechanical energy dissipated as heat when the switch is moved appears first as a high-temperature spot on the oxide films of the two contacts. Substantial temperature gradients may persist for several milliseconds after a switch movement.

Connectors frequently are used to join a thermocouple to lead wires, often in a location near the test apparatus. Temperature gradients within connectors may generate spurious signals. It is important to insulate the outer shell and provide a good conduction path inside the connector.

Switches and connectors made of thermocouple grade alloys will minimize the troubles caused by poor thermal protection, but good thermal design is still necessary.

The errors that can be introduced by using a nonisothermal connector are illustrated in Fig. 8. An all-copper connector is presumed, with its connection points 2, 3, 5, and 6 as shown. Two cases are examined: one in which both the A and B wires enter at one temperature and leave at another ($T_2 = T_6$ and $T_3 = T_5$, but $T_2 \neq T_3$) and one in which the A wire

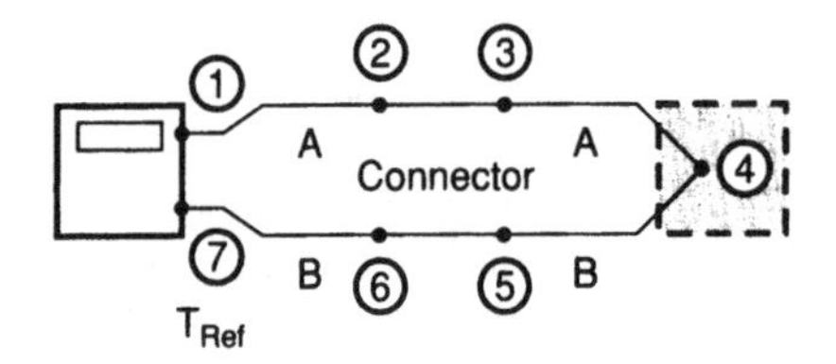

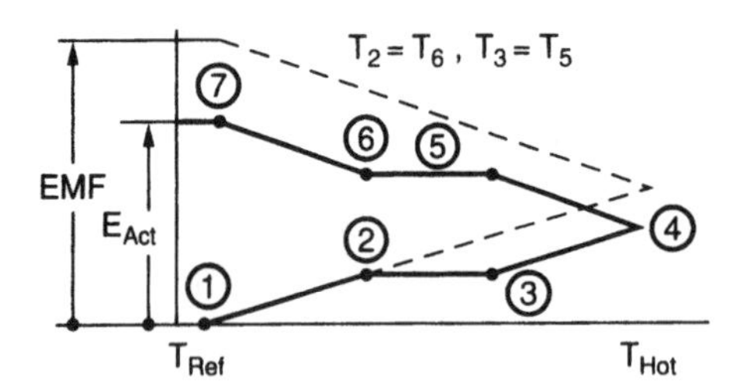

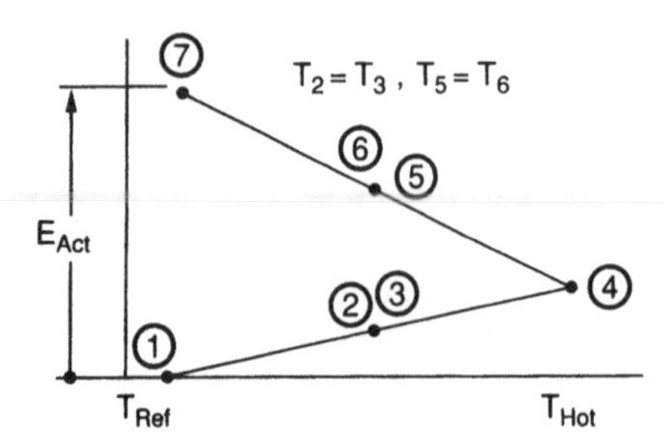

Figure 8 The effects of connectors. (Reproduced from Ref. 2, with permission.)

enters and leaves at one temperature while the *B* wire enters and leaves at some other temperature ($T_2 = T_3$ and $T_5 = T_6$, but $T_2 \neq T_5$). The *E–T* diagram shows the latter situation causes no error regardless of the type of material in the connector, since there is never a temperature gradient along the connector material. In the first case, the resultant signal is in error, having "lost" the entire amount of the temperature interval across the connector.

Reference baths which provide stable and uniform zones for termination of thermocouple systems are available. These generally consist of a heated or cooled zone box thermostatically controlled to remain at some specified temperature. Laboratory users may well use an ice bath or triple-point cell. For most engineering purposes, a bath made with ice and water sufficiently pure for human consumption will be within a few hundredths of a degree of 0°C if certain simple precautions are followed.

A Dewar flask or vacuum-insulated bottle of at least 0.5 liter capacity should be completely filled with ice crushed to particles about $\frac{1}{4}$ to $\frac{1}{2}$ cm in diameter. The flask is then flooded with water to fill the interstices between the ice particles. A glass tube resembling a small-diameter (0.5-cm), thin-walled (0.5-mm) test tube should be inserted 6 or 8 cm deep into the ice pack and supported there with a cork or float. A small amount (0.5 cm deep) of silicone oil should be placed in the bottom of the tube to improve the thermal contact between the junction and the ice-water mixture. The reference thermocouple junction is then inserted into the tube until it touches the bottom of the tube and is secured at the top of the tube with a gas-tight seal. The object of the seal is to prevent atmospheric moisture from condensing inside the tube and causing corrosion of the thermocouple.

The assembly is shown in Fig. 9, along with a proper connection diagram. Note that the relative polarity of the connection is different from that shown in Fig. 4. If a connection like Fig. 4 is desired, two glass tubes must be prepared, one each for the positive and negative

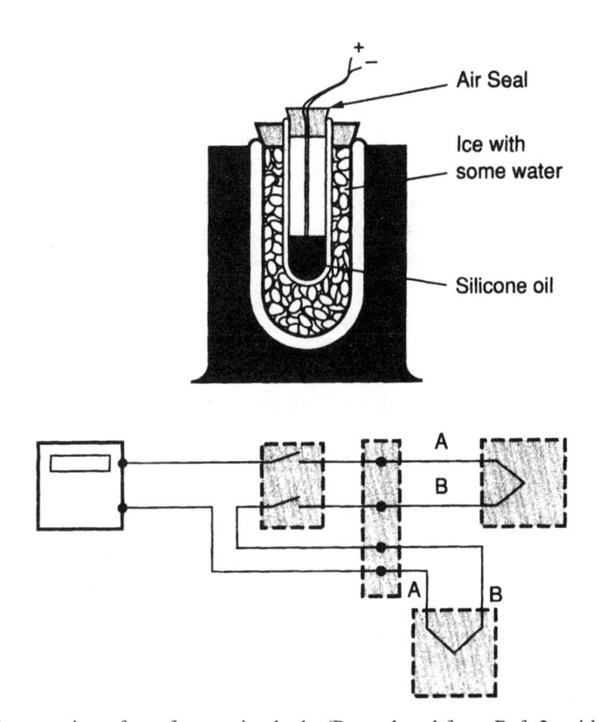

Figure 9 Construction of a reference ice bath. (Reproduced from Ref. 2, with permission.)

elements. These tubes must be mounted farther apart than the size of the ice particles to assure adequate cooling. One thermoelement and one copper lead wire are put into each tube.

The principal requirement of the reference bath is that its temperature be known accurately. Any region of known temperature can serve as a reference bath.

Many instruments that provide their output in temperature units contain local reference regions and compensating circuits that augment the thermoelectric signal to account for the local reference temperature. Such instruments can be used only with the type of thermocouple for which they were intended, since the compensating network is specific to the calibration of the thermocouple being used.

2.8 Obtaining High Accuracy with Thermocouples

The temperature emf tolerances quoted for thermocouples account for two types of deviations from the expectation values: (1) batch-to-batch differences in average calibration and (2) point-to-point differences in local calibration along an individual thermocouple.

Calibration of individual thermocouples can account for the batch-to-batch differences but not the point-to-point variations along the wire.

For highest precision, three precautions should be taken:

1. Calibrate the individual thermocouples.
2. Minimize the working temperature difference (i.e., use a reference temperature near the working temperature and physically close by).
3. Install the thermocouple so the working temperature difference is stretched over as long a length of wire as possible.

2.9 Service-Induced Inhomogeneity Errors

When thermocouples are used in unfavorable environments or for very long times, the output voltage may drift with time.

There are many possible causes for this drift, among which are selective oxidation, which changes the composition of the alloy; diffusion of one or more of the components from the thermocouple alloys to the sheath of a mineral-insulated, metal-sheathed assembly; and local annealing of previous cold work. Most of these are "high-temperature" effects, occurring mainly between 500 and 1500°C, as described by Campari and Garribba[7] and Bentley.[8] Many of these effects are attributable to the complex composition of the thermoelectric alloys, according to Schuh and Frost,[9] who pointed out that the alloys were developed in the early and mid-1900s to generate an emf that was linearly proportional to temperature. This constraint, imposed by the widespread use of simple analog instruments, led to the "tailoring" of the emf characteristic by adding trace amounts of several constituents. With a complex composition, even small changes could significantly change the emf at a given temperature. Schuh and Frost pointed out that linearity is no longer an important issue since digital processing does not require linearity. They recommended increased attention to the use of high-temperature structural alloys as thermoelements, relying on "smart" instruments.

In the low-temperature domain, such as electronics cooling, one of the sources of inhomogeneity error is cold working of the thermocouple wire by bending. Type K is significantly vulnerable to this effect. The calibration of a type K pair can be lowered by 1% simply by bending the wires by hand. The error caused by this cold work depends on the severity of the cold work, the length of the damaged region of wire, and the temperature difference across that length when the thermocouple is in service. The cold-worked region is often very short, which makes it difficult to detect.

A similar drop in calibration was reported by Bentley and Morgan[10] for PtRh versus Pt thermocouples in response to cold work introduced by handling. They reported a drop of about 0.4% in the cold-worked region. Annealing at 200°C did not entirely remove the drift in the PtRh leg. It is difficult to identify this problem by subsequent recalibration, since the region of partially degraded material may be placed in a uniform temperature zone during recalibration and hence play no part in generating the signal under calibration conditions. If a thermocouple is suspected of being inhomogeneous, it should be tested for homogeneity along its entire length. If the test shows that the wire is homogeneous, then no calibration is required, since the used portion of the wire is the same as the unused portion. If the wire is not homogeneous, no recalibration can be of value because it will be impossible to place the temperature gradient in the same location for calibration as it was for service. This situation is described in Figs. 10–12.

Assume that the thermocouple was exposed to the unfavorable environment only near the hot end, as shown in Fig. 10, and that both wires became less active as a result of the reaction. The output will drop, as shown. If this defective thermocouple were placed in a usual calibration facility, all of the affected material would be in the region of uniform temperature. The emf would be generated entirely by the material near the entrance of the furnace, which was never changed—the wires between points 1 and 2 and between 6 and 7 in Fig. 11. Under such conditions, a perfectly normal signal would be developed.

Figure 12 illustrates a test for homogeneity that can identify a defective thermocouple. To test a thermocouple for inhomogeneity, clamp the junction at a constant temperature and

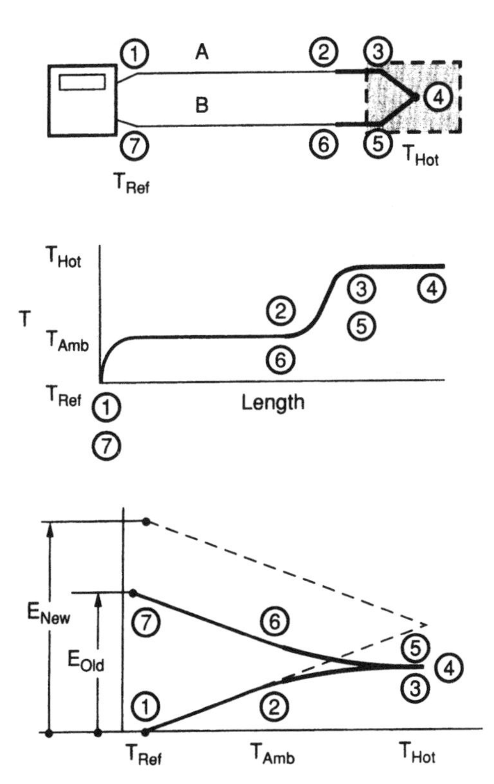

Figure 10 Temperature distribution along the thermocouple, in service, and the resulting output after deterioration of the wires. (Reproduced from Ref. 2, with permission.)

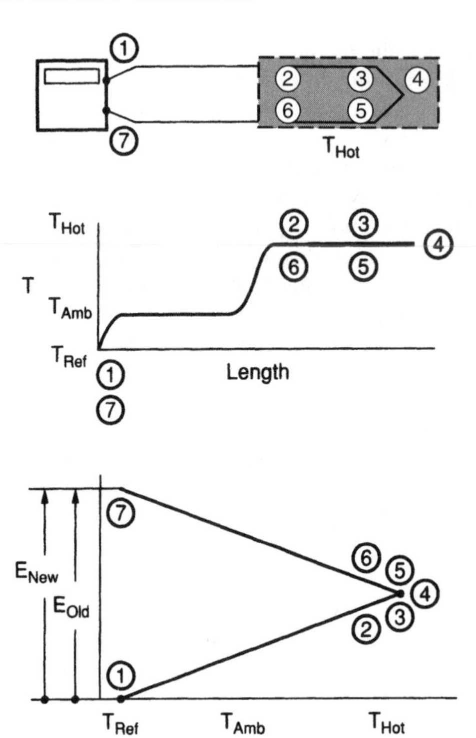

Figure 11 Temperature distribution along the thermocouple, during recalibration, and the resulting output. (Reproduced from Ref. 2, with permission.)

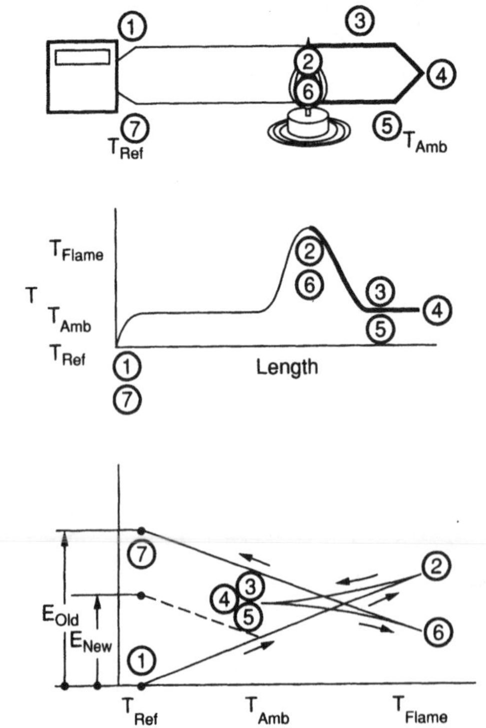

Figure 12 Homogeneity testing with a local hot spot and the resulting output. (Reproduced from Ref. 2, with permission.)

heat the wire in the suspected region. A homogeneous material will not produce any emf, as shown earlier in Fig. 5. An inhomogeneous material will have different calibrations on the upslope and downslope sides of the temperature excursion and will produce a net emf. This process is illustrated in Fig. 12.

The best technique for identifying a service-induced inhomogeneity is "side-by-side" running or comparison by replacement. Neither is very convenient, but on the other hand, there is no alternative. It is not within the present state of the art, regardless of how much effort is expended, to be able to interpret the readings from an inhomogeneous thermocouple in an arbitrary environment.

2.10 Thermoelectric Materials Connected in Parallel

Thermocouple materials are usually thought of as being connected in series with respect to the direction of the temperature gradient. There are situations, however, where two materials are in parallel electrical contact along their length. In such cases, the thermoelectric potential causes a distributed current to circulate in the materials. The net effect of such a configuration can be computed if the geometries and material properties are known.

There are three applications for which this effect is known to be of importance: (1) in the design of plated-junction thermopiles, (2) in attempts to precisely measure surface temperatures using thermocouples attached with solder, and (3) in the case of distributed failure of thermocouple insulation, usually at high temperatures.

The essential features of this parallel circuitry can be illustrated by discussion of the thermopile design and the shunted thermocouple situation.

Copper may be plated onto a constantan wire to form a cylindrical thermocouple pair in which the copper and constantan contact throughout their entire length. When the two ends of this plated material are held at different temperatures, an emf will be generated, which is a function of the relative electrical resistance of the two materials, as well as the usual thermoelectric parameters. The physical situation is shown in Fig. 13.

Gerashenko and Ionova[11] related the net emf to that of a conventional copper–constantan thermocouple by the following:

$$\frac{\mathrm{EMF}}{\mathrm{EMF}_{\mathrm{std}}} = \frac{R_{\mathrm{const}}}{R_{\mathrm{const}} + R_{\mathrm{copper}}} \tag{4}$$

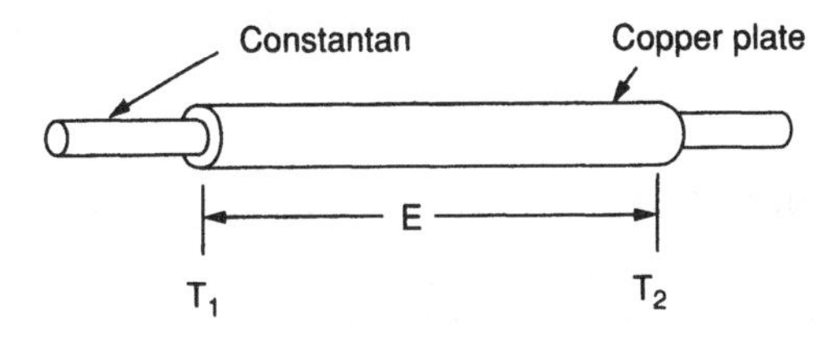

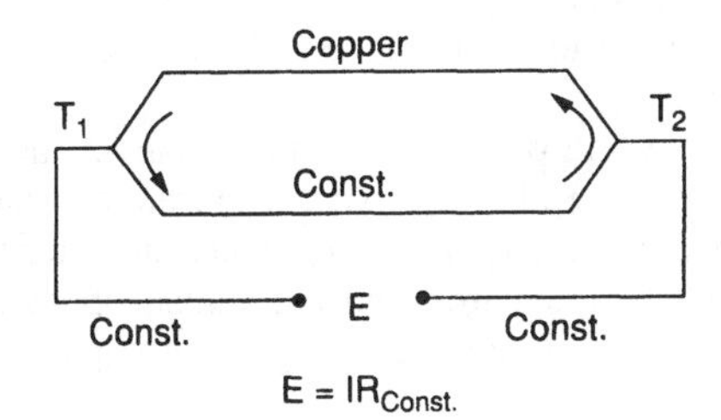

Figure 13 The plated junction thermopile element and its equivalent circuit. (Reproduced from Ref. 2, with permission.)

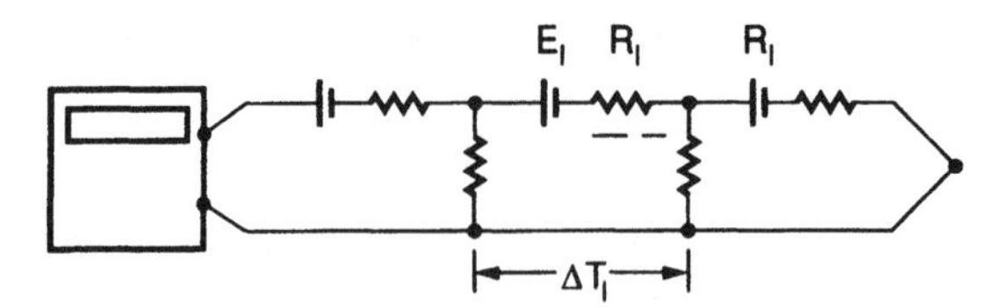

Figure 14 Electrical network model of the distributed shunt effect. (Reproduced from Ref. 2, with permission.)

where R is the electrical resistance for the component named in the length direction.

Thick copper platings will produce low resistance in the copper, and almost the entire output of the copper–constantan pair will be recovered.

There is another circumstance in which thermocouple elements become connected in parallel. When long thermocouples are used in hostile environments, there can be significant effects on the output due to partial shunting of the thermocouple by breakdown of the insulation between the wires. This effect was discussed in 1964 by Tallman,[12] who presented a model for numerical calculations to evaluate the output which is useful in explaining the behavior. The circuit shown in Fig. 14 describes the behavior when the insulators serve only as resistive connections between the wires (i.e., do not participate by generating emf). Only one node is shown in Fig. 14, although a real circuit would require many nodes for an accurate model:

$$E_1 = \epsilon_{AB}(\Delta T_i)$$
$$\epsilon_{AB} = \text{total thermoelectric power of } AB \text{ pair}$$
$$R_i = R_A + R_B \text{ for } i\text{th interval}$$
$$r_i = \text{resistance between } A \text{ and } B \text{ wires in } i\text{th interval}$$
$$\Delta T_i = \text{temperature difference across } i\text{th interval}$$

This network can be solved by several techniques and seems to accurately model the physical behavior of the distributed shunt effect.

The problem of resistive shunting was revisited in 1999 by Gill and Nakos,[13] who were interested in measuring the temperatures of large pools of liquid fuel (diesel or jet fuel). Long runs of mineral-insulated thermocouple material were exposed to high temperatures (up to 1200°C). The resistance of the mineral-insulated thermocouple stock went down rapidly at high temperature, shunting the thermocouple signals. As a consequence, the apparent temperature was affected by the temperature in the shunt region as well as by the temperature at the measuring junction.

2.11 Spurious emf due to Corrosion and Strain

Thermocouples produce dc signals in response to temperature differences. They can also become sensitive to ac electrical noise and ground loops, all of which are tractable and can be handled as noise problems would be handled in any system.

There are two problems peculiar to the thermocouple that bear mention: galvanic emf generation and strain-induced emf.

Thermocouples necessarily involve pairs of dissimilar materials and hence are liable to galvanic emf generation in the presence of electrolytes. The iron–constantan pair is the most vigorous, being capable of generating a 250-mV signal when immersed in an electrolyte.

The galvanic emf generated between iron and constantan opposes the thermoelectric signal; that is, the voltage drop of the current through the measuring loop opposes the thermo-

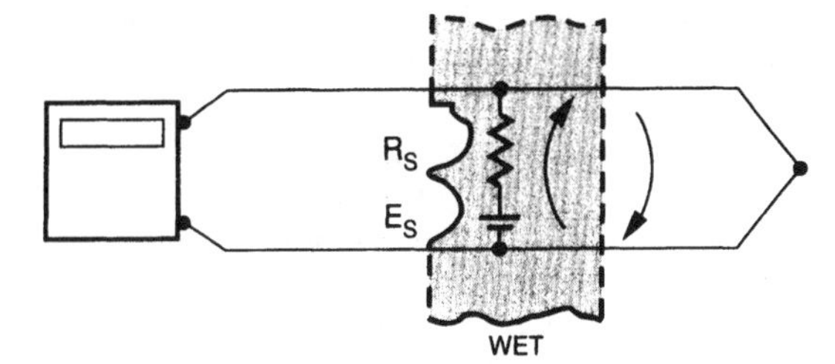

Figure 15 Electrical network model of the galvanic emf effect on a wet thermocouple. (Reproduced from Ref. 2, with permission.)

electric signal. The net effect, then, is that the thermocouple reads low. If the galvanic connection is near the junction, as it would be if certain soldering fluxes were not completely removed, the corrosive "necking down" of the iron wire increases the resistance to the galvanic current more and more as the wire is eaten away. This in turn means that the thermocouple readout circuit picks up a larger and larger fraction of the 250-mV galvanic signal. In bench tests of this mechanism,[14] it has been shown possible to have an iron–constantan thermocouple read negative (i.e., below the reference bath temperature).

Neither copper–constantan nor type K material shows significant galvanic effects.

The equivalent circuit is shown in Fig. 15. It is similar to the shunt problem, but the active emf source in the shunt dominates it.

The output of the system will be the thermoelectric emf corresponding to $T_{\text{hot}} - T_{\text{ref}}$ plus or minus the IR drop in the circulating current loop.

Typical values for E_{shunt} and R_{shunt} are 250 mV and 1 MΩ/cm of wetted length (24-gauge, duplex-fiberglass-insulated iron–constantan wires, wet with distilled water). The error induced by this signal depends on the resistance in the thermocouple loop and the location of the wet spot along the thermocouple. Wet iron–constantan thermocouples can produce large error signals.

Although type K materials do not display appreciable galvanic emf, they are strain sensitive (i.e., cold work causes a change in calibration), and during the act of straining, they generate appreciable emf. Temperatures measured on vibrating equipment may appear to fluctuate as a result of flexing of the thermocouple wires. Copper–constantan and iron–constantan are less active than type K, with copper–constantan least active. Spurious signals on the order of 50°C have been observed using type K materials on a vibrating apparatus (e.g., whole engine tests).

2.12 Self-Validating Thermocouples

Most thermocouple users blame the thermocouple for any "bad" data when, often, the fault lies elsewhere in the system—unrealistic expectations, poor installation of the sensor, and so on. As a consequence, there has been considerable interest in methods of "validating" a thermocouple's output while it is still in service.

One approach was to install the thermocouple junction in a capsule containing a metal or alloy whose melting temperature was known and was below the operating temperature of the thermocouple at steady state. On each "fire-up," as the indicated temperature rose, the thermocouple trace would display a plateau at the melting point, thus confirming that the output was correct. A recent example of this approach was reported by Sachenko et al.,[15] who outfitted a Chromel–Alumel thermocouple with a capsule containing lead.

Industrial thermocouples are frequently installed in thermowells to isolate the sensor from the system. Often, the thermowell is partly filled with a fluid of high thermal conductivity to assure good thermal contact between the thermocouple and the system through the wall of the thermowell. An empty thermowell is one of the main causes of "faulty thermocouple" reports. This type of failure can be detected by an electrical heating test known as the loop current step response proposed by Hashemian[16] and more recently incorporated into the Oxford University SEVA (Self Evaluation) protocol by Yang and Clarke.[17] This procedure involves passing a current through the thermocouple to heat it above its operating temperature and recording the cooling curve obtained when the heating current is shut off. This procedure is executed periodically and the traces compared with a reference trace obtained when the thermocouple was first installed. Any loss of thermal contact between the thermocouple and its thermowell results in a change in the shape of the response curve in the first few seconds of the trace.

2.13 Thermocouple Probe Designs for Gas Temperature Measurements

Thermocouples respond to their total environment, including all heat transfer modes: radiation, conduction, and convection. Radiation and conduction tend to pull the thermocouple temperature away from the gas temperature toward the temperature of the radiating source or the conducting support structure. The thermocouple is "connected" to the gas temperature by convection through the heat transfer coefficient. The equilibrium temperature is that which results from an energy balance between these competing tendencies. In cases where chemical reactions may be catalyzed on the surface of the thermocouple, those mechanisms must also be considered. Any difference between the thermocouple temperature and the true gas temperature is considered an error. Three types of error are recognized: radiation error, conduction error, and velocity error. Each describes the difference between the true gas temperature and the indicated temperature due to one mechanism: radiant heat transfer, conductive heat transfer, or high-velocity viscous dissipation effects. The heat transfer characteristics of a thermocouple (which governs its response to its environment) is relatively well known and cannot be changed.

Designing a thermocouple probe for accurate gas temperature measurement consists in designing a protective envelope within which the thermocouple will be acceptably accurate, that is, within which the sum of all three errors will be acceptably small. No thermocouple probe can respond instantaneously to changes in gas temperature. Thermocouples (the junctions themselves, not including any surrounding probe structure) act as first-order systems in response to changes in temperature and compensation methods for first-order systems are well understood. The general procedure for design of gas temperature probes for steady-state accuracy and transient behavior has been described by Moffat.[18]

The following citations provide examples of these principles applied to probe design. Elmore and Watkins[19] developed an analog compensation system that used signals from two first-order thermocouples of different diameters to determine transient gas temperatures up to 1 kHz. The fundamental approach could be implemented using high-speed digital processing as well. Moffat[20] proposed a transient method for estimating the (steady) temperature of a very hot gas stream (above the melting temperature of the thermocouple). His method is based on briefly exposing the thermocouple to the gas stream and correcting the transient behavior using a first-order time constant. The appropriate value of the time constant is found by minimizing the variance in the corrected signal over the time interval of exposure of the thermocouple to the hot gas stream.

Radiation error plays an important role at high temperatures, especially in low-velocity situations. One approach to this problem is to use an aspirated probe to increase the gas velocity past the thermocouple junction, increasing the convective heat transfer coefficient and reducing radiation error. Suction pyrometer probes were investigated at the National Bureau of Standards during the 1950s for use in combustion chamber and afterburner studies, as reported by Lalos.[21] Norton et al.[22] adapted this method for use in a high-temperature glass furnace. Another approach is to use two probes with different (but predictable) error sensitivities and calculate the radiation error from the difference in the readings of the two probes. This approach has been used by Moffat[23] and Brohez et al.[24] The main problem with this approach is that the correction can be very large, and if the uncertainty in the raw data exceeds a few percent, then the correction becomes highly uncertain. The behavior of thermocouple probe designs in fire environments was investigated by Blevins and Pitts,[25] who modeled the behavior of bare thermocouples, single-shielded aspirated probes, and double-shielded aspirated probes under fire conditions. Pitts et al.,[26] of NIST's Building and Fire Research Laboratory, investigated the uncertainties in the corrected readings of bare and shielded, aspirated probes. Their results confirmed that the largest uncertainties were found when the corrections were largest.

Measurements in high-velocity gases suffer mainly from problems with the recovery factor of the probes. High-recovery-factor probes read close to stagnation temperature, but their recovery factor is sensitive to manufacturing tolerances and, therefore, is somewhat uncertain. Low-recovery-factor probes require larger corrections for the residual velocity error, but their simpler geometry can reduce the uncertainty in the final reading. Moffat[27] recommended a probe using a spherical tip (a ball bearing) whose recovery factor is very well known and repeatable, while Vasquez and Sanchez[28] describe the design of a system using a Kiel-type shield for a high-recovery probe. Both references dealt with applications requiring high accuracy in the measurement of small temperature differences—the temperature rise across a single stage of a compressor or turbine.

2.14 Thermocouple Installations for Surface Temperature Measurement

Two problems dominate the accuracy of surface temperature measurement with thermocouples: ensuring good contact between the sensor and the surface and avoiding disturbing the surface temperature by the presence of the thermocouple.

Keltner and Beck[29] analyzed the steady-state and transient errors of thermocouples attached to thick walls using the unsteady surface element method. Sobolik, Keltner, and Beck[30] dealt with thin plates, with special emphasis on the effect of the measurement errors on the interpretation of data from thin-foil heat flux gages.

Analysis of the surface measurement problem shows that thin-film thermocouples should be ideal sensors. Han and Wei[31] describe a method of making thin-film thermocouples on nonmetallic surfaces that yields excellent steady-state and transient measurement accuracy. They applied this technique to a Zirconia coating on an engine piston, but the thin-film concept would also be well suited to the electronics industry, which uses nonmetallic packaging and requires accurate measurement.

Not all thermocouple probes are attached to the surface—spring-loaded contact probes are often used for measurements in situations where no permanent attachment would be permitted. The error in such applications is dominated by the contact resistance. Osman, Eilers, and Beck[32] used an inverse conduction solution to develop a method for correcting transient measurements based on a steady-state calibration.

3 RESISTANCE TEMPERATURE DETECTORS

The terms *resistance temperature detector* and *resistance thermometer* are used interchangeably to describe temperature sensors containing either a fine-wire or a thin-film metallic element, whose resistance increases with temperature. A small current (ac or dc) is passed through the element, and its resistance is measured. The temperature of the element is then deduced from the measured resistance using a calibration equation or a table.

3.1 Types and Ranges

Resistance temperature detectors can be designed for standards and calibration laboratories or for field service, although the probe designs are vastly different. Field service probes are generally encased in stainless steel protective tubes, with the wire or film elements bonded to sturdy support structures. They are made to handle considerable physical abuse. Laboratory standards probes are often enclosed in quartz tubes, with the resistance wire mounted in a strain-free manner on a delicate mandrel. High-precision temperature measurement requires that the element be strain free, so the electrical resistance is a function only of temperature, not of strain.

Resistance temperature detectors are well suited for single-point measurements in steady-state service at temperatures below 1000°C where long-time stability and traceable accuracy are required and where reasonably good heat transfer conditions exist between the probe and its environment.

These detectors are not recommended for use in still air or in low-conductivity environments unless the self-heating effect can be accounted for appropriately. They are not recommended for averaging service unless computer data acquisition and software averaging are contemplated. They are not recommended for transient service or dynamic temperature measurements unless specifically designed for such characteristics, since the usual probe is not amenable to simple time constant compensation.

Resistance temperature detector probes tend to be larger than some of the alternative sensors and to require more peripheral equipment (i.e., bridges and linearizers).

3.2 Physical Characteristics of Typical Probes

Probes intended for field service concentrate on ruggedness and repeatability. Their calibrations may not agree with the standard expectation values of laboratory-grade probes but should be repeatable. Drift tolerances on field service probes may be stated in terms of percent drift per 100 hours.

Probes for laboratory service are designed to ensure freedom from mechanical strain either due to fabrication or thermal expansion during service.

The bare sensing elements of resistance temperature detectors range in size from wafers $0.5 \times 1.0 \times 2.0$ mm, with pig-tail leads 0.25 mm in diameter and 2.5 cm in length, to wire-wrapped mandrels, again with pig-tail leads 4 mm in diameter and 2 cm in length. With protective tubes in place, typical probes range from 1.0 to 5.0 mm in diameter.

Some typical sensors and probes for field service are shown in Fig. 16. Stainless steel is used for the protection tubes on most such units. The sensing element can be either a wire wound on a mandrel or a thin film deposited on an insulating substrate. Wire resistance elements may also be bonded to their support structures and encapsulated in glass or ceramic, but strain-free, steel-jacketed probes are also available. Thin-film elements are usually formed directly on the substrate by sputtering or vapor deposition.

Some, but not all, laboratory-grade probes use quartz for the protection tubes. For highest accuracy and best long-term stability, the laboratory-grade probes use strain-free rigging

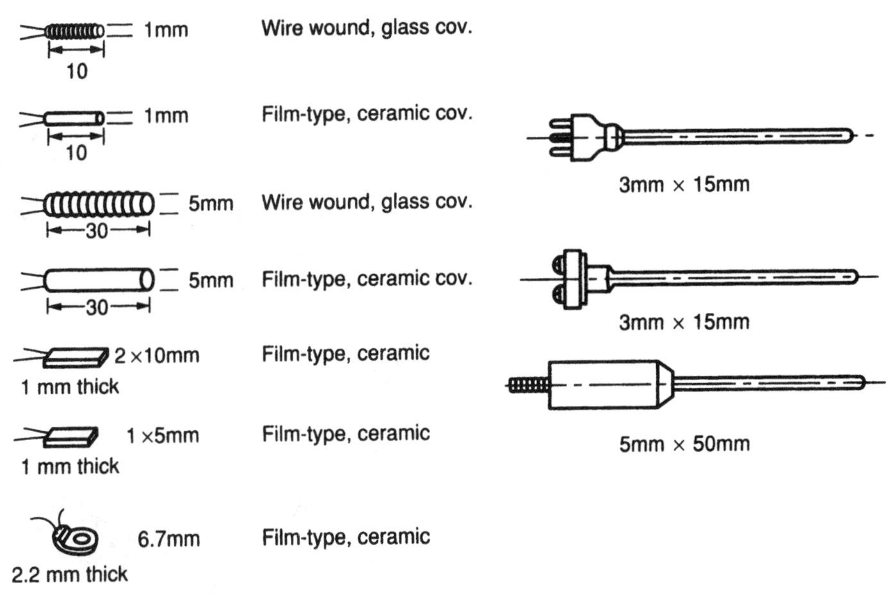

Figure 16 Physical characteristics of some sensors and probes.[33]

and a bifilar winding. This combination avoids the effects of mechanical strain on the resistance and also reduces pickup from electromagnetic or electrostatic fields. The windings are fragile, however, and the probes must be treated with great care. Figure 17 shows cutaway views of two laboratory-grade probes.

3.3 Electrical Characteristics of Typical Probes

Resistance–temperature detectors are available commercially with resistances from 20 to 2000 Ω, 100 Ω being common. Platinum, nickel, copper, and manganin have been used in commercial resistance thermometers.

Typically, the resistance of a platinum resistance thermometer will change by about 0.39%/°C. The resistance of a 100-Ω probe, according to the European calibration curve, would change by about 0.385 Ω/°C in the range 0–100°C. If the probe were in still air at 0°C and were being driven at a constant current of 3.16 mA (which would, on an average probe, produce a self-heating effect of about 0.25°C in still air), the voltage drop across the probe would be 316.304 mV, which would increase by 1.2166 mV for each degree Celsius increase in air temperature. By contrast, this compares to about 0.050 mV/°C for a typical base-metal thermocouple.

Twisted-pair lead wires are recommended. Care must be taken to avoid thermoelectric signal generation in resistance thermometer circuits using dc excitation (ac circuits convey the temperature information at carrier frequency, and they are not affected by thermoelectric signals).

3.4 Thermal Characteristics of Typical Probes

Interrogation of a resistance–temperature detector dissipates power in the element, which goes off as heat transfer to the surroundings. This self-heating causes the sensing element

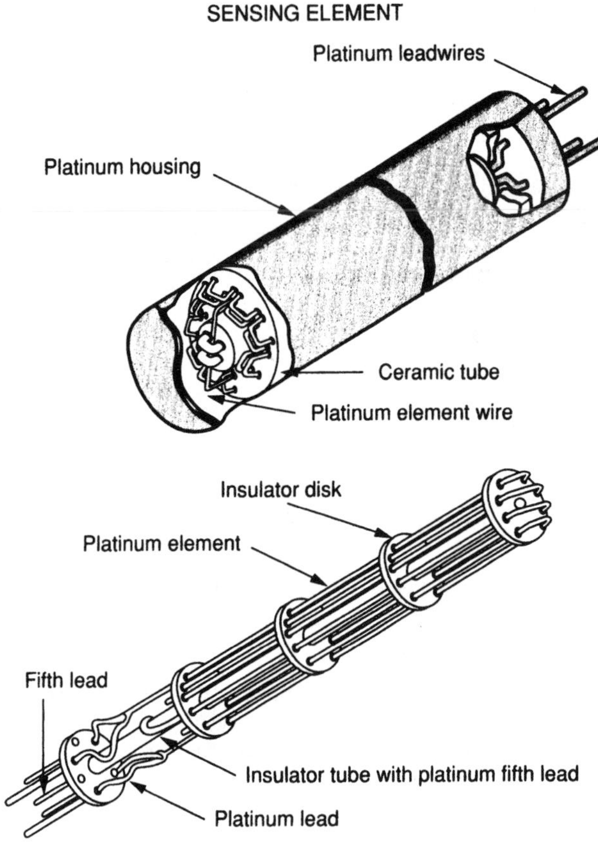

Figure 17 Cutaway views of two laboratory-grade probes.[34,35]

to stabilize at a temperature higher than its surroundings. The amount of self-heating depends on three factors: (1) the power dissipated in the element, (2) the probe's internal thermal resistance, and (3) the external thermal resistance between the surface of the probe and the surrounding material whose temperature is to be measured. The self-heating temperature rise is given by

$$T_{sens} - T_{spec} = W(R_{int} + R_{ext}) \tag{5}$$

A typical probe exposed to still air will display self-heating errors on the order of 0.1–1.0°C/mW (commercial probes of 1.5–5 mm in diameter). At 1 m/s air velocity, the self-heating error may only be 0.03–0.3°C, while in water (at 1 m/s velocity), the self-heating effect would be reduced by a factor of 4 or 5, depending on the relative importance of the internal and the external thermal resistances, compared to the values in moving air.

Self-heating error can be kept small by using probes with low internal thermal resistance, by locating the probes in regions of high fluid velocity (or placing the probes in tight-fitting holes in structures), or by operating at very low power dissipation. Pulse interrogation can also be used; it will reduce self-heating regardless of the internal and external resistances. The temperature rise of an element subjected to a pulse of current depends on the duration of the pulse and the thermal capacitance of the sensor rather than on the resistances between the sensor and the specimen.

Resistance–temperature detectors (RTDs) are subject to all of the installation errors and environmental errors of any immersion sensor and, because of the detectors' larger size, are usually affected more than thermocouples or thermistors. Since RTDs are usually selected by investigators who wish to claim high accuracy for their data, the higher susceptibility to environmental error may be a significant disadvantage.

The internal structure of most RTDs is sufficiently complex that their thermal response is not *first order.* As a consequence, it is very difficult to interpret transients. The term *time constant,* which applies only to first-order systems, should not be applied to RTDs in general. If a *time constant* is quoted for a RTD, it may only mean "the time required for 63.2% completion of the response to a step change," and it may not imply the other important consequences of first-order response. Quoted values, from one probe supplier, of "time to 90% completion" range from 9 to 140 s for probes of 1.0–4.5 mm in diameter exposed to air at 1 m/s.

3.5 Measuring Circuits

Resistance–temperature detectors require a source of power and a means of measuring resistance or voltage.

Resistance can be deduced by voltage drop measurements across the resistor when the current is known or by comparison with a known resistor in a bridge circuit. Six circuits frequently used for *resistance thermometry* are shown in Fig. 18.

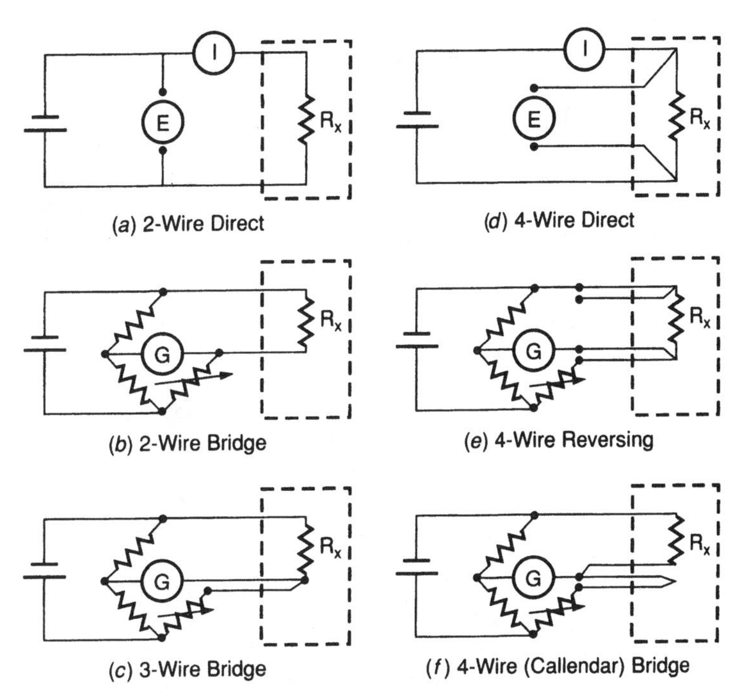

Figure 18 Six circuits for the measurement of resistance.

In the following paragraphs, these six circuits used for reading the resistance of the thermometer will be briefly discussed. The determination of probe temperature from probe resistance is discussed in a future section.

The two-wire, constant-current method is the simplest way to use a resistance thermometer. One simply provides it with a known current and measures the voltage drop across the probe and its lead wire (Fig. 18*a*). The resistance of the circuit is determined from Ohm's law in the cold and the hot condition, and the temperature is determined from the resistance ratio.

In this approach, the lead-wire resistance is ignored, which leads to underestimating the temperature change by approximately the ratio of the lead-wire resistance to the total circuit resistance:

$$\frac{R_{\mathrm{TOT},H}}{R_{\mathrm{TOT},C}} = 1 + \left(\frac{\Delta R_P}{R_{P,C} + R_L}\right) \tag{6}$$

Errors can be avoided by properly accounting for the lead-wire resistance, either by measuring it using a shorting plug to replace the probe or by calculating it from standard tabular data on wire resistance.

The two-wire bridge circuit (Fig. 18*b*) suffers from the same lead-wire error as the two-wire direct method but has one advantage: It can be used with an unstable power supply. The probe resistance can be measured in terms of the calibrated reference resistor, independent of the bridge voltage value.

The three-wire bridge circuit (Fig. 18*c*) compensates, to a considerable extent, for the effect of lead-wire resistance. The circuit uses a matched pair of lead wires to connect the bridge to the probe. One member of the pair is inserted into each of the two arms of the bridge, so changes in the lead-wire resistance affect both arms. For a bridge with equal resistances in both arms, this provides an approximate compensation for lead-wire resistance.

Many commercial probes are supplied with three-wire lead connection.

The most accurate technique for measuring the probe resistance is the four-wire direct method (Fig. 18*d*). Lewis[36] describes the advantages of the four-wire direct system in process industry measurements and points out that significant advantages in accuracy can be achieved at relatively low cost.

The probe is driven from a known source of constant current through one pair of leads, while the voltage drop across the sensing resistor is measured using the other pair. Since there is no current flow in the voltage-sensing leads, there is no error introduced by the lead-wire resistance. With current and voltage drop known, the resistance can be calculated directly using Ohm's law. Modern, regulated power supplies and high-impedance voltmeters make this an increasingly attractive option.

Two four-wire bridges are shown: the four-wire Callendar bridge and the reversing bridge. The Callendar bridge inserts equal lengths of lead wire in each of the two arms of the bridge, which provides good compensation for the effect of lead-wire resistance. The reversing bridge allows the operator to exchange lead wires, thus measuring the resistance in each arm with two different pairs of leads. The average of the two values is usually used as the best estimate, since the two measurements are equally believable.

Several manufacturers offer linearizing amplifiers as accessories for resistance thermometers. These devices produce signals that are linearly proportional to sensor temperature by correcting its nonlinear response.

The increasing use of computers in data interpretation has reduced the need for linearizing circuitry in laboratory and research work, but the commercial market still prefers linear systems for control and monitoring.

In some cases, the linearizing amplifier is provided as an integral part of the probe, which may also contain its own power supply. The output of such a probe is linearly proportional to the probe temperature.

3.6 The Standard Relationships for Temperature versus Resistance

Platinum resistance thermometers are sold in the United States under two different calibrations (U.S. and European) and are subject to tolerances, which can also be described by either of two standards.

The two calibrations differ in their expected values of α: the average temperature coefficient of resistivity ($\Omega/\Omega\cdot$C) over the interval 0–100°C. The European standard [Deutsches Institute für Normung (DIN) 43670] specifies α to be 0.003850 while the U.S. standard specifies 0.003925/°C. The IPTS-68 specified 0.003925 for probes acceptable as standards. The value of α increases with increasing purity, and values as high as 0.003927 have been observed, according to Norton.[37] Norton also mentions that thin films of platinum do not follow either of the two standard calibration curves just mentioned for bulk platinum but tend to approach the European standard value.

Table 4 illustrates the differences in resistance of U.S. and European calibrations.

From the user's standpoint, the important issue is, "How much different are the temperatures deduced from a measured resistance?" Figure 19 shows the temperature difference (European–U.S.) as a function of temperature level.

The equations proposed for data interpretation (i.e., the Callendar equation or the Callendar–Van Dusen equation) are attempts at fitting the tabular data—they are not the sources of the tables.

Just as there are different standards for the resistance–temperature relationship, so there are different standards for interchangeability among commercial probes. The hot resistance of a probe at a particular temperature depends on its cold resistance and its average value of α up to the operating temperature. Standards for interchangeability must acknowledge both sources of difference. Interchangeability is typically discussed in terms of "percent of reading" (in degree Celsius), with values from 0.1 to 0.6% being available in commercial probes for field service.

Table 4 Resistances of a Platinum Resistance Thermometer

Temperature, °C	U.S. Curve, $\alpha = 0.003925$	European Curve, $\alpha = 0.003850$
−200	17.14	18.53
−100	59.57	60.20
0	100.00	100.00
+100	139.16	138.50
+200	177.13	175.84
+300	213.93	212.03
+400	249.56	247.08
+500	284.02	280.93
+600	317.28	313.65

Source: Reference 37.

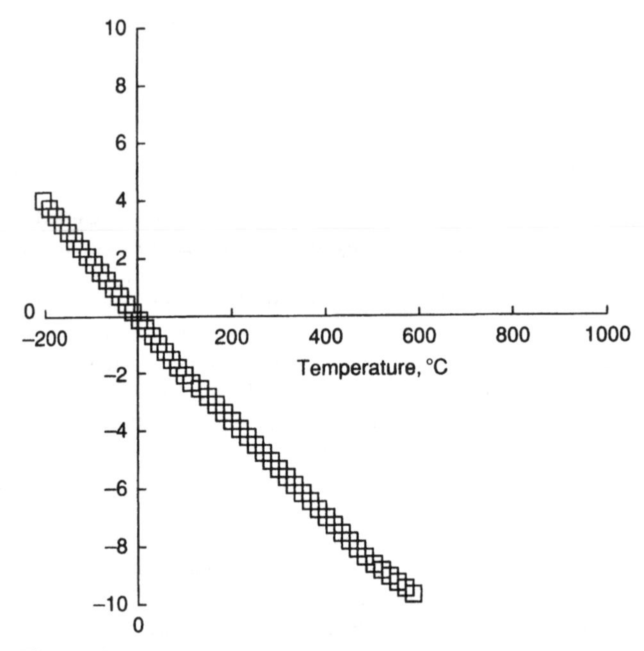

Figure 19 Comparison of the European and U.S. resistance curves.

The resistance–temperature characteristic of a probe may drift (i.e., change with time) while the probe is in service. Manufacturers of laboratory-grade probes will specify the expected drift rate, usually in terms of the expected temperature error over an interval of time. Two sample specifications are "0.01°C per 100 hours" for a low-resistance, high-temperature probe (0.22 Ω at 0°C, 1100°C maximum service temperature) and "0.01°C per year" for a moderate-resistance, moderate-temperature probe (25.5 Ω at 0°C, 250°C maximum service temperature). Drift of the resistance–temperature relationship appears related to changes in the grain structure of the platinum. Such changes take place more rapidly at high temperatures. Probes are annealed at high temperatures after fabrication in order to reduce the effects of manufacturing strains.

3.7 Interpreting Temperature from Resistance: Common Practice

The final step in determining the measured temperature is to interpret the probe resistance through a calibration relationship. There are several options depending on the accuracy required.

For rough work near the ambient, it is often good enough to simply take the nominal value of α as a constant and divide the observed change in resistance by the value of α to find the temperature change.

For work between 0 and 100°C, the *constant-*α method is quite accurate, since α was determined in this interval.

In general, manufacturers will provide resistance–temperature tables for their probes. These tables should be used only with the probes for which they were intended because they may not have any generality. Manufacturers' tables are slightly more accurate than the constant-α method.

For high-quality, laboratory-grade platinum resistance thermometers (used between 0 and 630°C), the Callendar equation describes the IPTS-68 temperature within ±0.045°C, according to Benedict.[38] The difference between IPTS-68 and the Callendar equation varies across the range of temperatures, as shown in Fig. 20.

The Callendar equation is

$$t = 100\left(\frac{R_t - R_0}{R_{100} - R_0}\right) + \delta\left(\frac{t}{100} - 1\right)\left(\frac{t}{100}\right) \tag{7}$$

where R_t = resistance at temperature t
R_0 = resistance at ice point
R_{100} = resistance at 100°C
δ = constant evaluated at zinc point (419.58)

As shown in Fig. 20, the tolerance of ±0.045°C represents an agreed-upon difference between the Callendar equation and the IPTS-68 equation.

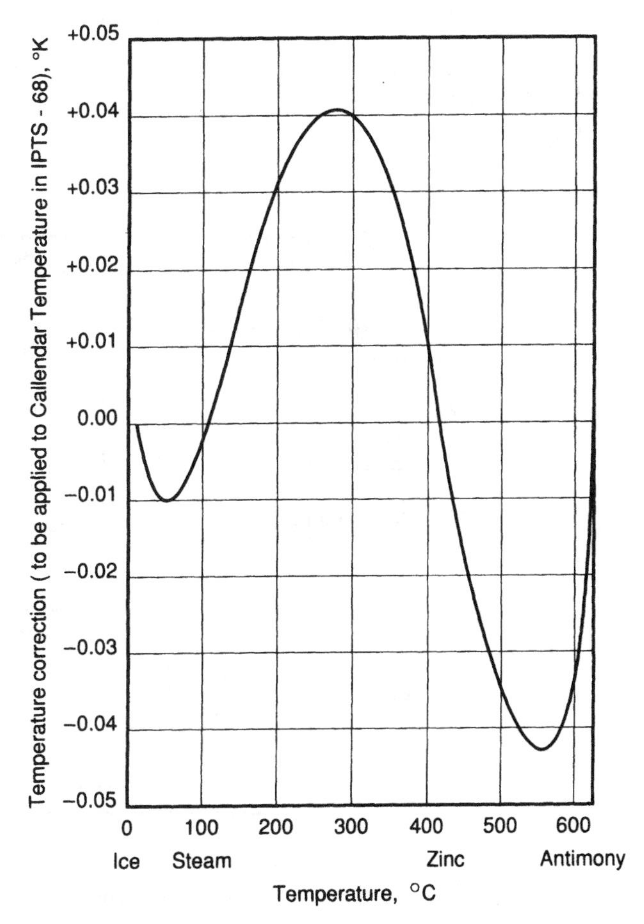

Figure 20 The difference between the Callendar equation and IPTS-68. (From Ref. 38, with permission.)

For most measurement purposes, the Callendar equation can be considered accurate within ±0.045°C over the range 0–630°C.

Below 0°C, temperature according to the Callendar equation deviates significantly from the IPTS-68 temperature. An additional term converts the Callendar equation into the Callendar–Van Dusen equation, which is much closer to the IPTS-68 temperature and is recommended for field service work between −190°C and 0°C.

The Callendar–Van Dusen equation is

$$t = 100\left(\frac{R_t - R_0}{R_{100} - R_0}\right) + \delta\left(\frac{t}{100} - 1\right)\left(\frac{t}{100}\right) + \beta\left(\frac{t}{100} - 1\right)\left(\frac{t}{100}\right)^3 \qquad (8)$$

where R_t = resistance at temperature t
R_0 = resistance at ice point
R_{100} = resistance at 100°C
δ = constant determined at zinc point (419.58)

and

$$\beta = \begin{cases} 0 & t > 0 \\ \text{constant determined at oxygen point } (-182.962°\text{C}) & t < 0 \end{cases}$$

According to Benedict,[38] the disagreement between the Callendar–Van Dusen equation and IPTS-68 does not exceed ±0.03°C from −190 to 0°C. The difference varies, as shown in Fig. 21.

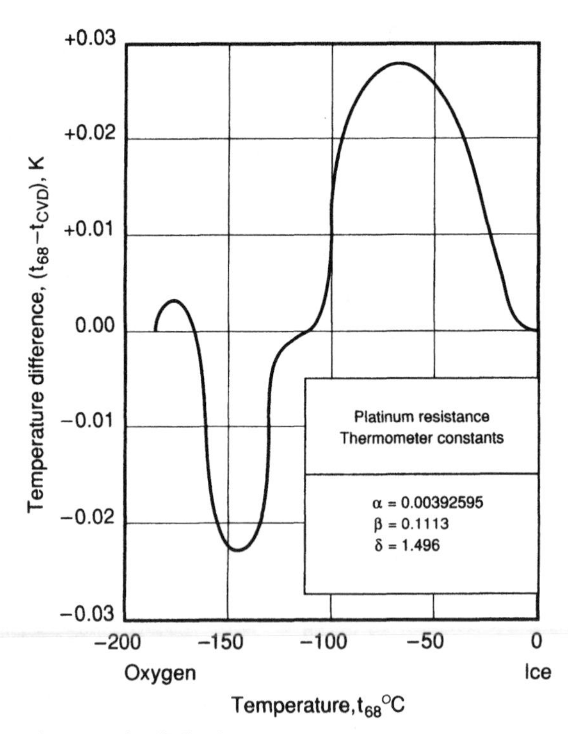

Figure 21 Difference between the Callendar–Van Dusen equation and IPTS-68. (From Ref. 38, with permission).

4 THERMISTORS

Thermistors are temperature-sensitive resistors whose resistance varies inversely with temperature. The resistance of a 5000-Ω thermistor temperature sensor may go down by 20 Ω for each degree Celsius increase in temperature (in the vicinity of the initial temperature). Driven by a 1.0-mA current source, this yields a signal of 200 mV/°C.

Thermistors are used frequently in systems where high sensitivity is required. It is not uncommon to find thermistor data logged to the nearest 0.001°C. This does not mean the data are accurate to 0.001°C, but the data are readable to that precision.

A thermistor probe is sensitive to the same environmental errors that afflict any immersion sensor: Its accuracy depends on the care with which it was designed for that particular environment.

4.1 Types and Ranges

Thermistor probes can be used between −183°C (the oxygen point) and +327°C (the lead point).[39] But most applications are between −80 and +150°C.[40] The sensitivity of a thermistor (i.e., the percent change in resistance per degree Celsius change in thermistor temperature) varies markedly with temperature, being highest at the lowest temperatures.

The long-time stability of thermistor probes is open to some question, although several months of accurate usage between calibrations seem attainable. The evidence on drift and its causes is not clear. It would be prudent, as with any temperature-measuring system, to make provision for periodic recertification of thermistor probes on a time scale established by experience within the system itself.

If accurate measurements are required, calibration facilities are needed, and this need poses some problems. Few instruments are capable of providing a transfer calibration for thermistors to their limit of readability, since few are that sensitive. Melting point baths and precision-grade resistance thermometry are needed to capitalize on the available precision.

Thermistors have strongly nonlinear output. Linearizing bridges are available, but these add to cost. Nonlinearity is not a significant issue when the data will be interpreted by computer.

4.2 Physical Characteristics of Typical Probes

Thermistor probes range in size from 0.25 mm spherical beads (glass covered) to 6-mm-diameter steel-jacketed cylinders. Lead wires are proportionately sized. Disks and pad-mounted sensors are available in a wide range of shapes, usually representing a custom design gone commercial. Aside from the unmounted spherical probes and the cylindrical probes, there is nothing standard about the probe shapes.

Figure 22 shows some representative shapes of commercially available probes.

For medical applications, thermistor probes are often encapsulated in sterilizable, flexible vinyl material. Such probes are frequently taped to a patient's skin and used as the control sensor for the temperature-regulating system.

The thermistor element itself is fabricated using the techniques of powder metallurgy. A mixture of metallic oxides is compressed into a disk and sintered. The mixture's composition, the sintering temperature, and the atmosphere in the furnace determine the resistance and the resistance–temperature coefficient of the thermistor. The faces of the disk are plated with silver, and the resistance of the thermistor is adjusted by removing material from the edges until the desired value has been obtained. The lead wires are attached, and the assembled thermistor is then potted in epoxy, rubber, or glass.

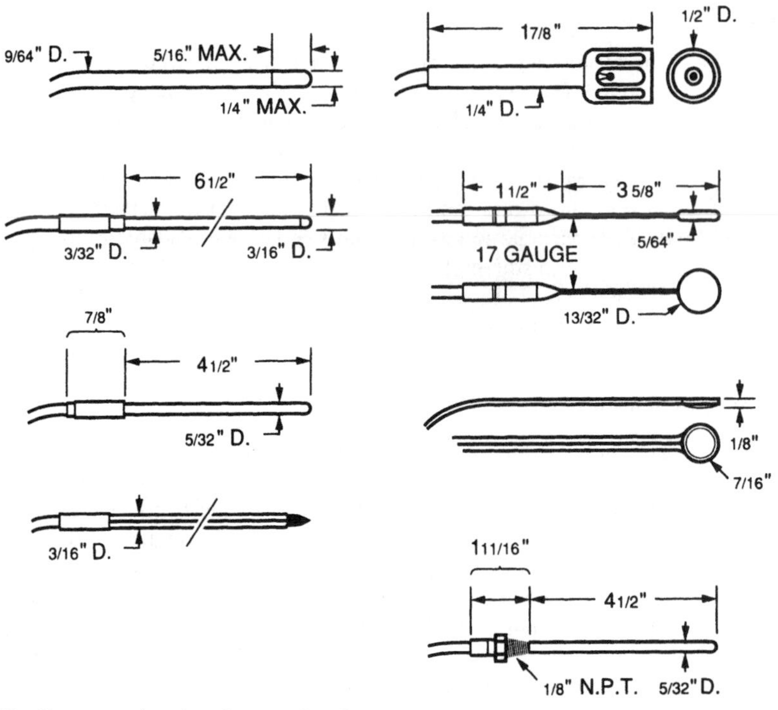

Figure 22 Representative thermistor probes for temperature measurement. (Courtesy of Omega Engineering, F4 of 1985 catalog.)

Resistance drift with time is a problem with some thermistor materials, especially at higher temperatures. Park[41] recommends the addition of SiO_2 to the mixture of oxides generally used for negative-coefficient thermistor material as a means of stabilizing the resistance–temperature characteristic.

The process is shown schematically in Fig. 23.

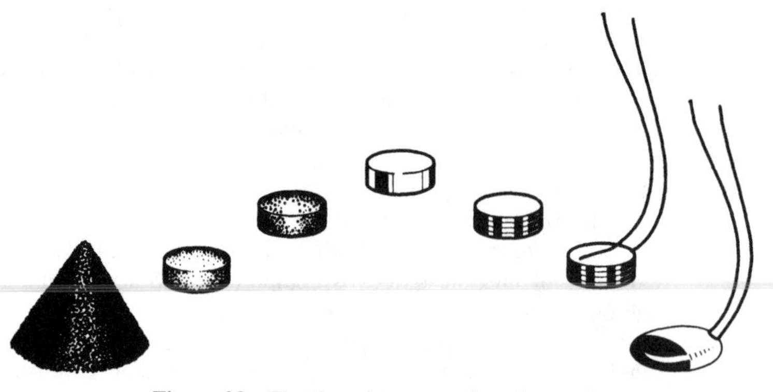

Figure 23 The thermistor manufacturing cycle.

High-temperature thermistors (1000°C) have been made from doped ceramic materials but with only moderate success.

4.3 Electrical Characteristics of Typical Probes

Thermistor probes vary in resistance from a few hundred ohms to megohms. Resistance is frequently quoted at 25°C with no power dissipation in the thermistor. The commercial range is from about 2000 to 30,000 Ω.

The resistance–temperature characteristic of a thermistor depends on the material of which it is made. Representative values of the sensitivity coefficient (percent change in resistance per degree Celsius) are given in Table 5.

To illustrate the range of resistances encountered in practice, Table 6 shows the resistance as a function of temperature for a typical probe whose resistance is 2252 Ω at 25°C.

Thermistor resistance data are frequently shown as logarithm of resistance ratio versus $1/T$. Such a presentation emphasizes the "almost linearity" of a single-term exponential description of the resistance–temperature characteristic. The data listed in Table 6 are shown in Fig. 24.

Proprietary probes are available that *linearize* thermistors by placing them in combination with other resistors to form a circuit whose overall resistance varies linearly with temperature over some range. These compound probes can be summed, differenced, and averaged like any linear sensor.

Modem manufacturing practices allow matched sets of thermistors to be made interchangeable within ±0.1°C.

4.4 Thermal Characteristics of Typical Probes

Thermistor probes are generally interrogated using a current of 1–10 μA, either ac or dc. With a probe resistance of 10 Ω, this results in power dissipation of 0.01 W that must be transferred from the probe into its surrounding material. At the first instant of application, this current has no significant effect on the measured resistance of the probe, but in steady state, it results in the probe running slightly above the temperature of the medium into which it is installed. This is referred to as the *self-heating* effect. Since thermistors are often used where very small changes in temperature are important, even small amounts of self-heating may be important.

Table 5 Variations of Thermistor Temperature Coefficient with Temperature

Temperature (°C)	Condition	%/°C
−183	Liquid oxygen	−61.8
−80	Dry ice	−13.4
−40	Frozen mercury	−9.2
0	Ice point	−6.7
25	Room temperature	−5.2
100	Boiling water	−3.6
327	Melting lead	−1.4

Source: Reference 39.

Table 6 Variation of Thermistor Resistance with Temperature

Temperature (°C)	Resistance (Ω)	Temperature (°C)	Resistance (Ω)
−80	1.66 M	0	7355.0
−40	75.79 K	25	2252.0
−30	39.86 K	100	152.8
−20	21.87 K	120	87.7
−10	12.46 K	150	41.9

Source: Reference 43.

The self-heating response is frequently discussed in terms of the *dissipation constant* of the probe, in milliwatts per degree Celsium. Dissipation constants can range from 0.005 mW/°C to several watts per degree Celsius and can be estimated with acceptable accuracy, given the geometry and thermal properties of the probe and its installation. The dissipation constant should not be used to correct a reading for the self-heating effect; the uncertainty in the correction is too high. If a calculation shows the self-heating effect to be significant, pulse interrogation should be used.

Dissipation constants for two representative probes are given in Table 7.

The self-heating effect must be considered in calibration as well as in use. Calibration baths frequently use stirred oil as the medium. Table 7 shows the self-heating effect in the oil bath to be very low compared to that in air at 5 m/s. A probe calibrated in an oil bath and used in air would be subject to different self-heating effects, and the user should be aware of that.

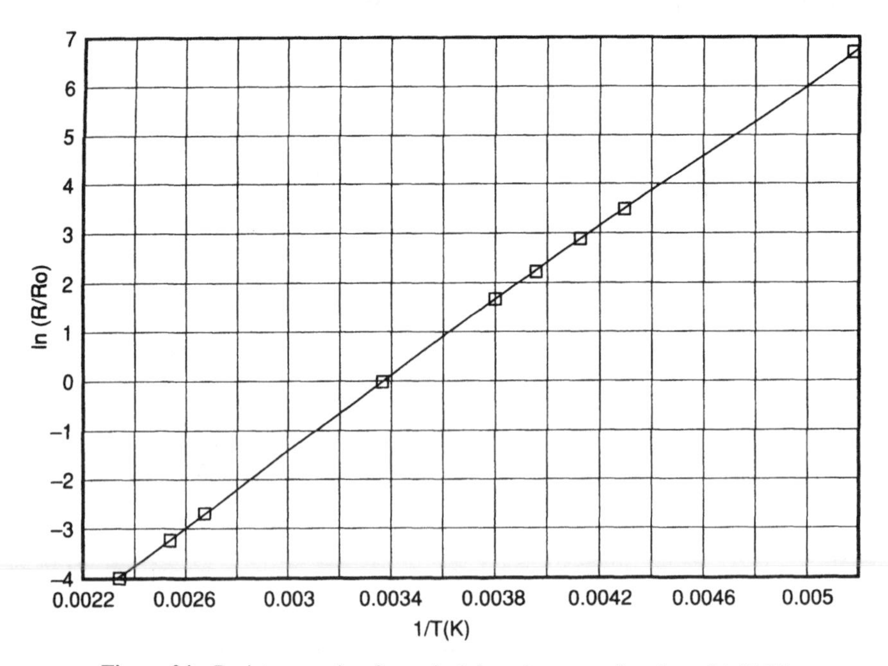

Figure 24 Resistance ratio of a typical thermistor as a function of $1/T$ (K).

Table 7 Representative Thermal Dissipation Constants for Two Different Thermistor Probe Designs

Environment	1.0-cm Disk	5.0-cm Glass Cylinder
Still air	8 (mW/°C)	1 (mW/°C)
Air at 5 m/s	35	—
Still oil	55	3.5
Still water	—	5
Oil at 1 m/s	250	—

Source: Reference 39.

The transient response of thermistors is more complex than that of thermocouples, and they are not as well suited to transient measurements. The difference arises from the thermal response of the thermistor material itself. Thermistor materials typically have lower thermal diffusivity than metals and hence do not equilibrate their internal temperatures as rapidly, all other factors being the same. Thus the highest frequency at which a thermistor will yield a first-order response is far lower than that for a thermocouple. During rapid changes, the temperature is not uniform inside the thermistor, and the average resistance no longer describes the average temperature. Since the thermistor resistance is not a linear function of temperature, the transient resistance cannot easily be converted to temperature.

Thermistor probes are susceptible to environmental errors (radiation, velocity, and conduction errors) as are all other immersion temperature sensors, and the same guidelines and design rules apply.

Thermistor sensors can be built into complex probe designs, as can any temperature sensor, and the structure of the probe will determine its steady-state and transient behavior. The thermistor has relatively low thermal conductivity and is frequently encapsulated in epoxy, glass, or vinyl, which also have low thermal conductivity. On this account, the transient response of a thermistor probe will be slower and more complex than that of a metallic sensor.

Tagawa et al.[42] analyzed the transient response of spherical and planar thermistors and showed that some configurations can be treated as first-order systems for slow changes in temperature, although they display second- or third-order characteristics at higher frequencies. They reported success using time constant compensation to reduce the compensated time constant by factors as large 50.

4.5 Measuring Circuits and Peripheral Equipment

Electrical resistance is easily measured within ordinary levels of accuracy by a number of techniques, and any method can be used with thermistors. The logarithmic variation of resistance with temperature means that over small ranges a constant percent error in resistance translates into a fixed error in temperature level. For example, with a thermistor whose sensitivity coefficient (see Table 5) is 5%/°C, a resistance measurement accurate to $\pm 0.1\%$ of reading will yield a temperature measurement accurate within ± 0.02°C everywhere in the range. Accuracy of this order can be achieved by two-wire and four-wire systems using off-the-shelf instruments and thermistors.

When thermistors are used to resolve very small differences in temperature (e.g., into the milliKelvin resolution range), special precautions must be taken to measure the resistance with the required accuracy. For the sensitivity mentioned above, 50 ppm corresponds to about 1 mK. Measurements in the 1–100-mK range are more typical of standards room practice than of field measurements, yet some applications require this precision.

4.6 Determining Temperature from Resistance

Experimental evidence shows that the resistance of a thermistor varies inversely and nearly exponentially with temperature. This suggests expressing the resistance–temperature relationship as a polynomial in ln(R):

$$\frac{1}{T} = A_0 + A_1 \ln(R) + A_2 \ln(R)^2 + A_3 \ln(R)^3 + \cdots + A_N \ln(R)^N \tag{9}$$

Note the R is dimensional in this equation; hence the values of the constants would depend on the units used for R.

A form retaining terms through the cubic, sufficiently accurate for small ranges in temperature, was used by Steinhart and Hart.[44] Their results indicated that the second-order terms had no effect, but later work has shown some advantage in retaining these terms. However, the Steinhart–Hart equation is still widely used:

$$\frac{1}{T} = A_0 + A_1 \ln(R) + A_3[\ln(R)]^3 \tag{10}$$

Arguments concerning the need for dimensional homogeneity led Bennet[45] to propose a dimensionless form which is independent of the units used:

$$\frac{1}{T} - \frac{1}{T_0} = A_1 \ln\left(\frac{R}{R_0}\right) + A_2 \ln\left(\frac{R}{R_0}\right)^2 + \cdots + A_N \ln\left(\frac{R}{R_0}\right)^N \tag{11}$$

Justification for this form can be seen in the near-linearity of Fig. 24.

This general form has given rise to several simplified forms. For a first approximation, only one term is needed:

$$R = R_0 \exp\left[A_1\left(\frac{1}{T_0} - \frac{1}{T}\right)\right] \tag{12}$$

This form reveals the exponential nature of the thermistor resistance response to changes in temperature.

Seren et al.[46] compared four different forms of the calibration equation for goodness of fit to a data set covering a range of 200 K, published by Bosson, Guttmann, and Simmons,[47] and accepted the data set as reliable. Their results are shown in Table 8. The following calculations were used to arrive at those results:

Table 8 Accuracy of Least-Squares Fits of Four Different Polynomial Expressions for Thermistor Calibration

Form of Equation for $1/T$	SRE	RMS Error	Mean Error
1. $A_0 + A_1 \ln(R)$	0.00601	1.999	0.1925
2. $A_0 + A_1 \ln(R) + A_2[\ln(R)]^2$	0.000428	0.1353	−0.0026
3. $A_0 + A_1 \ln(R) + A_2[\ln(R)]^2 + A_3[\ln(R)]^3$	0.000139	0.0392	0.00008
4. $A_0 + A_1 \ln(R) + A_3[\ln(R)]^3$	0.000138	0.0380	−0.00001

Source: Reference 47.

SRE = standard relative error, the root-mean-square of the relative errors at the test points:

$$\text{SRE} = \left[\sum_{i=1}^{n} \frac{(T_{c,i} - T_{0,i})^2 / T_{0,i}^2}{n} \right]^{1/2} \tag{13}$$

RMS error = root-mean-square error:

$$\text{RMS} = \left[\frac{1}{n} \sum_{i=1}^{n} (T_{c,i} - T_{0,i})^2 \right]^{1/2} \tag{14}$$

Mean error = algebraic mean error:

$$\text{Mean} = \frac{1}{n} \sum_{i=1}^{n} (T_{c,i} - T_{0,i}) \tag{15}$$

where T_c = calculated temperature (K) at measured resistance R_i
T_i = observed temperature at measured resistance R_i

The fourth form is the Steinhart–Hart equation. It seems, from this comparison, that there is little to choose from between the two cubic fits as far as goodness of fit is concerned, and since the Steinhart–Hart form requires only three constants, it would be preferred.

A calibration equation such as the Steinhart–Hart equation would be used with a commercial probe only when the highest accuracy is required. In most cases, the manufacturer's calibration is sufficiently accurate.

Thermistor probes are sold with resistance–temperature calibration tables accurate within ± 0.1 or 0.2 K, depending on the probe grade purchased. These tables are typically in 1-K increments. For computer interpretation, they should be fit to the Steinhart–Hart form and the coefficients determined for least error.

For the highest precision in measurement of temperature level, the purchased thermistors must be calibrated against laboratory-grade temperature standards at enough points to determine all of the constants in the model equation selected. Good practice would suggest one redundant calibration point to provide a closure check.

When an application requires more precision than provided by the manufacturer's tables but less than the highest achievable, a transfer standard instrument can be used. A strain-free platinum resistance thermometer or a quartz crystal thermometer can be used as such a transfer standard. Typically, the comparison is made in a well-stirred, temperature-regulated oil bath or in a comparison block of nickel or copper in an electrically heated oven.

Values of resistance and resistance divided by resistance at 0°C are listed in Table 9 for two typical commercial probes as functions of temperature (°C) and $1/T$ (K).

5 OPTICAL METHODS

Optical temperature measurements have become increasingly reliable and accurate as the electronic arts have advanced. This is especially true in the range below 600°C. Above 600°C, radiation methods have long enjoyed a position of trust in the measurements community. Low-temperature data interpretation and analysis that were once only conceptually possible can now be routinely handled with on-board computing.

There was a veritable explosion of instruments for optical temperature measurement in the period between 1980 and 2000. Only a few of the many types of optical temperature-sensing systems will be discussed here.

Bigelow et al.[48] describe the characteristics of commercial radiation temperature detectors for the range -173 to $+3750$ K and estimate the accuracy as between 0.5 and 1.5% of

Table 9 Resistance and Relative Resistance of Typical Probes

		Probe 1		Probe 2	
Temperature (°C)	$1/T$ (K)	Resistance	Relative Resistance	Resistance	Relative Resistance
−80	.00518	1660000	225.6968	2211000	225.7044
−70	.00493	702300	95.4861	935600	95.5084
−60	.00469	316500	43.0320	421500	43.0278
−50	.00448	151050	20.5370	201100	20.5288
−40	.00429	75790	10.3046	101000	10.3103
−30	.00412	39860	5.4194	53100	5.4206
−20	.00395	21870	2.9735	29130	2.9737
−10	.00380	12460	1.6941	16600	1.6946
0	.00366	7355	1.0000	9796	1.0000
10	.00353	4482	.6094	5971	.6095
20	.00341	2814	.3826	3748	.3826
30	.00330	1815	.2468	2417	.2467
40	.00319	1200	.1632	1598	.1631
50	.00310	811.30	.1103	1081	.1104
60	.00300	560.30	.0762	746.30	.0762
70	.00292	394.50	.0536	525.40	.0536
80	.00283	282.70	.0384	376.90	.0385
90	.00275	206.10	.0280	274.90	.0281
100	.00268	152.80	.0208	203.80	.0208
110	.00261	115	.0156	153.20	.0156
120	.00254	87.70	.0119	116.80	.0119
130	.00248	67.80	.0092	90.20	.0092
140	.00242	53	.0072	70.40	.0072
150	.00236	41.90	.0057	55.60	.0057

reading depending on the temperature level. The accuracy is better at high temperatures. These devices are vulnerable to errors from several sources: calibration, size of source, emissivity uncertainty, background radiation (especially in the lower temperature ranges), and gas path absorption of the infrared by water vapor or CO_2.

One way to avoid gas path absorption and source emissivity uncertainty is to use a blackbody cavity source and transmit the energy via a closed optical path. This is the principle behind the fiber-optic blackbody-sensing system described by Dils.[49] A fiber-optic blackbody radiation sensor measures the intensity of the radiation emitted from a blackbody cavity at the end of an optical fiber and deduces temperature from the intensity, using well-established laws of radiant emission.

The system has four main elements: the cavity, the high-temperature fiber, the low-temperature fiber, and the processor. Two of these elements are exposed to the high-temperature environment. A typical high-temperature fiber is a single-crystal sapphire fiber about 1.25 mm in diameter and usually between 100 and 250 mm long. The cavity can be formed by sputtering a thin metallic film onto the end of the sapphire fiber, typically about 5 μm. The metal layer can be covered by a film of alumina (also about 5 μm) to protect it from oxidation or erosion.

The general arrangement and the principal features of the data interpretation are shown in Fig. 25.

The cavity radiates as a "dark-gray body" (almost a blackbody) and the high-temperature fiber conducts the radiant energy from the cavity to the low-temperature fiber,

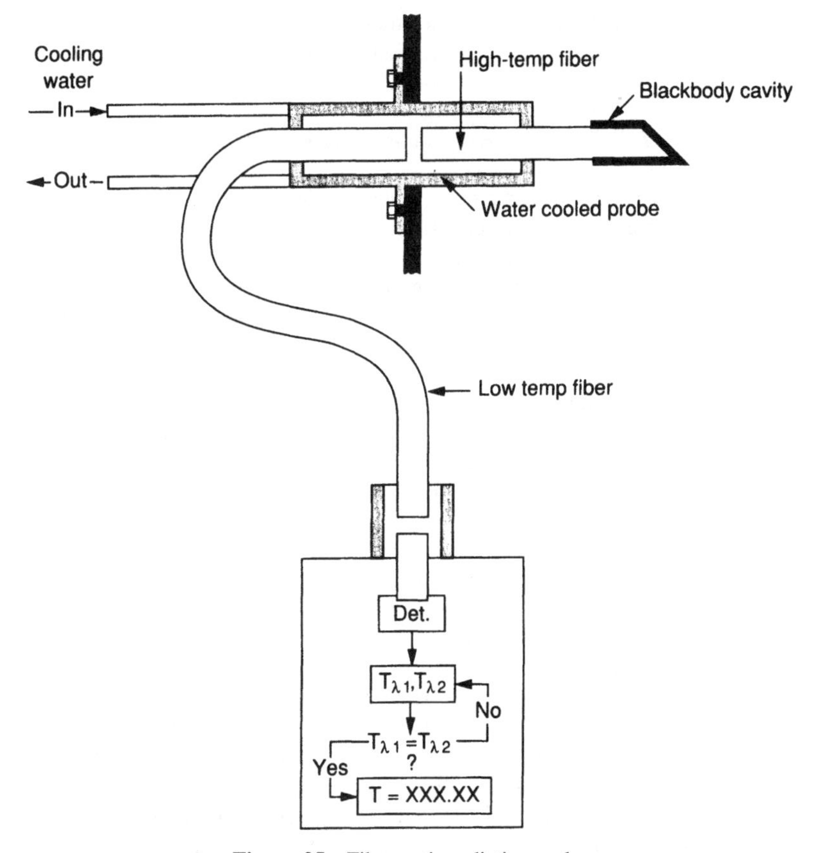

Figure 25 Fiber-optic radiation probe.

which in turn delivers it to the signal processor. The low-temperature fiber can be of great length, without degrading the signal, since loss coefficients for the low-temperature fiber material are quoted in percent per kilometer.

The processing of the radiant signal can be discussed in terms of the system pattern function: a set of equations that show how the system is supposed to work. The pattern function for the fiber-optic system is presented below. A final calibration adjusts the output to bring it into conformance with the calibration results.

$$I'(T_i) = \int_0^\infty \frac{C_1 C_d(\lambda, T_i)}{\lambda^5[\exp(C_2, \lambda T) - 1]} F(\lambda)\mathrm{BS}(\lambda, T_i)\mathrm{LTF}(T_i)\, d\lambda \tag{16}$$

and

$$I = I' K_{\text{sensor}} K_{\text{cable}} K_{\text{detector}} F(T_i) \tag{17}$$

where $I'(T_i)$ = detector current, pattern function
$I(T_i)$ = actual detector current, corrected for T_i
λ = wavelength
T = temperature, K
T_i = temperature of signal processor box
C_1 = radiation constant: 3.743×10^8, $\mathrm{W}\cdot\mu\mathrm{m}^4/\mathrm{m}^2$

C_2 = radiation constant: 1.4387×10^4, μm·K
$C_d(\lambda, T_i)$ = calibration constant for photodiode
$F(\lambda)$ = narrow-bandpass filter function
$BS(\lambda, T_i)$ = beamsplitter function
$LTF(T_i)$ = low-temperature fiber transmission function

The dependence on T_i reflects changes in the optical properties of the mirrors and prisms used in steering the beams to their respective sensors.

The temperature range is 600–2000°C. Accuracy begins to fail at lower temperatures, while the sapphire fiber cannot tolerate higher temperatures.

The calibration precision is limited only by the precision with which the electrical signals can be read, assuming the equation governing the radiant emission to be absolutely accurate. The measurement accuracy depends on how well the sensor is brought into equilibrium with the specimen temperature.

The fiber-optic probe is sensitive to all the usual installation errors, which could be calculated from its physical and thermal properties. But it introduces one new concern: radiation loss along the transparent fiber. This is not an error mode that has been recorded in the literature, but it amounts to only a small fraction of the sensor's total radiation error under usual conditions. In addition, there is a small (3–10-K) error introduced because of radiation lost through the sides of the hot fiber near the cavity, the Brewster loss. These new error modes are not viewed as deterrents to the use of the probe as a reference system for high temperatures; rather they are part of the correction which must be applied at final calibration.

The sensing capsule is 1.25 mm in diameter and can be from 1 to 15 mm long. The optimum length is calculated on the basis of minimizing the radiation error of the cavity.

Standard thin-film units will respond to 10-kHz fluctuations in temperature. Response is flatter than that of a thermocouple, rolling off at only 3 dB per octave instead of 6.

Wickersham[50] developed a temperature-measuring system based upon the temperature sensitivity of the response of certain phosphors: fluor-optic temperature sensing. He found a family of phosphors whose fluorescent emission characteristics were strongly temperature dependent. Pulsing these phosphors with ultraviolet light resulted in an emission spectrum with two properties that depended on temperature: the ratios of amplitudes of certain bands and the decay times of the principal energy-containing bands. Two classes of instruments were then developed: ones that used the ratios of wavelengths and others that used the decay time. Experience favored the decay time instrument, and those are now the dominant type in the marketplace. Phosphors exist whose emissions are useful at temperatures of only a few kelvins, while others emit best in the range of 2000°C. Fluor-optic temperature-measuring systems are under development for high-temperature measurements on aircraft engine components and also for low-temperature measurements in systems with high magnetic fields or electric fields. Considerable development has proceeded on phosphors.[51] Since the sensor is an optical fiber, the system is not vulnerable to electromagnetic interference (EMI) and does not present a shock hazard to the operator.[52] Recent developments have focused on the illumination scheme, searching for more efficient coupling of the source and the phosphor by using blue light-emitting diodes (LEDs) as the illuminant source.[53]

Thermo-chromic liquid crystals have been used for mapping temperature fields in low-temperature systems (room temperature up to 100°C). These coatings have a reversible reaction to temperature change. Colorless below their lowest transition temperature, they display red when heated to the bottom of their range, with their color moving through the spectrum to blue at their highest active temperature. The full range of colors can be compressed into a temperature band of 1°C (narrow-band material) or stretched out over 10°C (wide-band material). Narrow-band paints can be mixed; as long as the active bands do not overlap, the calibration of each band is unaffected by the presence of the other materials. A mixture of narrow-band materials painted on a surface will display a set of rainbow-colored lines representing isotherms centered around each active band. A wide-band paint displays a gradual shift in hue from one end of the range to the other. Digital image processing is required for interpretation of the wide-band images, especially when the incident light may change its spectrum over time, as discussed by Farina et al.[54] Liquid crystal material can be suspended in water and used to make visible the temperature distribution in the water.[55] If a liquid crystal is painted on a surface with known heat release per unit area, then the surface temperature distribution can be interpreted to learn the heat transfer coefficient distribution.[56]

Rajendran et al.[57] described a novel optical temperature measurement sensing system based on time-domain reflectometry. A long optical fiber was etched with 150 Bragg gratings regularly spaced along the fiber. The fiber was then embedded in the stator windings of an electric motor, along with a reference set of resistance–temperature detectors. The fiber was then pulsed from one end and time-domain reflectometry, using the back-scattered light from the gratings, was used to infer the temperature distribution along the fiber. The optical measurements agreed with the RTD measurements within $\pm 3°C$ over the length of the fiber. Many features of the distribution were evident in the optical measurements that were not visible in the RTD measurements since only a few RTDs were installed. An acoustic counterpart to this technique in described in Section 7.

6 ELECTRON NOISE THERMOMETERS

The electron noise method of temperature measurement uses for its signal the voltage developed by thermal agitation of the electrons in a resistor. The voltage is small (on the order of microvolts) and at high frequencies (up to 1 GHz). But, of greatest importance, the signal is linearly related to the absolute temperature (for frequencies less than kT/h, defined later) by a known physical law. Furthermore, the voltage is independent of the resistor material. The signal using multiple measurements can be made independent of the resistor value and insensitive to background noise (electrical noise). The signal is broadband with a zero mean and a high bandwidth. These characteristics allow rejection, by filtering, of environmental noise without loss of measurement accuracy.

The theory is described by Decreton et al.[58]

Electric noise due to thermal agitation was first described by Nyquist (1928)[59] and Johnson (1928)[60] An unloaded passive network always presents at its ends a voltage V_n fluctuating statistically around zero. This mean-squared noise voltage is given by

$$\overline{V_n^2} = 4kTR \frac{(hf)/(kT)}{\exp[(hf)/(kT)] - 1} \tag{18}$$

where h and k are Planck's and Boltzmann's constants respectively, T the absolute temperature, f the frequency, and R the resistance. For temperatures above 100 K and frequencies below 1 GHz, Eq. (18) can be accurately approximated by

$$\overline{V_n^2} = 4kTR \tag{19}$$

Equation (19) represents a frequency-independent white-noise signal. In a practical measurement, a given frequency bandwidth (df) is imposed, and the true rms voltage V_n is then given by

$$V_n = (4kTR\,df)^{(1/2)} \tag{20a}$$

When the current I_n is measured instead of the voltage, the equivalent model of the noise signal gives the rms current

$$I_n = \left(\frac{4kTdf}{R}\right)^{(1/2)} \tag{20b}$$

On multiplying Eqs. (20a) and (20b), the noise power P_n is obtained:

$$P_n = 4kT\,df \tag{21}$$

which is independent of the resistance value and thus is strictly linearly related to the absolute temperature.

Sophisticated signal extraction techniques are required. This technique is useful mainly at the temperature extremes (very low and very high) when other sensors fail. The technique is promising. Results have been reported up to 1000°C with errors less than 0.5% of reading over a period of several months.[61]

Brixy et al.[62] place a high degree of confidence on noise thermometry since "it is the only method of contact thermometry that is not affected by the unavoidable changes in the sensor at high temperatures." They describe the use of noise thermometry as the reference method in a study to determine the drift characteristics of WRe thermocouples at temperatures up to 2000°C.

7 ACOUSTIC VELOCITY PROBES

The velocity of stress-wave propagation through a gas, liquid, or solid can be used to measure temperature. In gases and liquids, the acoustic velocity is used (normal stress or pressure-wave propagation), whereas both normal and shear stress waves have been used in solids. Two signal-processing techniques are available: detection of resonance and measurement of the time of travel. These techniques have been used in both gases and solids, as described by Tasman and Richter.[63]

The acoustic velocity in a perfect gas is given by

$$a = \left[\left(\frac{C_p}{C_v}\right)RT\right]^{1/2} \tag{22}$$

Where C_p and C_v are the specific heats at constant pressure and at constant volume, respectively, R is the gas constant, and T is the temperature. Shear stress waves are not present in gases.

The normal stress propagation velocity in a solid (the extensional wave velocity) is

$$V_{\text{ex}} = \left(\frac{E}{D}\right)^{1/2} \tag{23}$$

where E and D are the modulus of elasticity and the density, respectively.

The shear stress propagation velocity in a solid is

$$V_{\text{sh}} = K\left(\frac{G}{D}\right)^{1/2} \tag{24}$$

where K is a shape factor and G and D are the shear modulus and density, respectively. Torsional waves sent down a cylindrical rod are one embodiment of shear waves. The

Hewlett-Packard quartz crystal thermometer uses shear-wave propagation in the resonance mode to achieve its very high precision (on the order of 0.0001 K).[64]

Tuning-fork resonance probes of pure sapphire have been run at 2000°C in a laboratory environment with an apparent stability of 1% or better by Bell et al.[65]

While capable of high precision under favorable conditions, the resonance mode is very sensitive to deposits or accidental contact between the resonator and its support. It is also inherently a line-averaging technique since it is the temperature of the oscillating stem, not that of the mass at the end, that determines the resonant frequency. For those reasons, most acoustic velocity systems proposed for high-temperature work use the time-of-flight technique known as the *pulse-echo* technique.

Tasman and Richter[63] offer the following description of the pulse-echo system:

> The pulse-echo technique measures the transit time of a single sound pulse through a thin wire. A magnetostrictive transducer transforms an electrical pulse into a sound pulse, which is injected into a transmission line. Discontinuities in the line, and the end of the line, reflect (part of) the sound pulse, which is then converted back into electrical signals by the transmitting transducer. The actual sensor is the wire joining the discontinuities. Almost any kind of discontinuity will do to produce an echo, even a kink in the wire. One sensor may contain several discontinuities, apart from its end, and thus measure the longitudinal temperature profile over several consecutive sections.

For good resolution, the sensor material should have a large change in propagation velocity over the temperature range to be encountered. Thoriated tungsten is one candidate material. Tasman and Richter quote the round-trip time difference between echos as increasing from about 4.5 to about 5.5 μs/cm when the temperature changes from 20 to 2700°C. Time differences can be measured to about ± 1 ns, which means that a 50-mm-long sensor could resolve temperature to ± 0.2 K at 2000°C (in theory). Practical considerations seem to limit the resolution to about ± 3 K for the 50-mm sensor. Tungsten is very sensitive to oxidation at high temperatures, and pure tungsten cannot be used as an acoustic velocity sensor because it continues to recrystallize at high temperatures, causing a continuous shift in calibration. Thoria blocks the grain growth, however, and the calibration of thoriated tungsten is stable (in the absence of oxygen).

8 TEMPERATURE-SENSITIVE COATINGS

Paints and crayons are available that are designed as temperature indicators,[66] up to 2500°F (1371°C) with a quoted accuracy of $\pm 1\%$. Two types are available: phase change and color change. When the phase change materials melt, they yield easily discernible evidence that their event temperature has been exceeded. Color change materials are subtler and less easy to interpret. There have been some complaints of calibration shifts of these paints when used on heavily oxidized materials which are believed due to alloying of the oxide with the paint.

These paints are nonmetallic and, therefore, have different radiation properties than metals. They should be used only over small areas of metallic surfaces (small compared with the metal thickness), or else their different emissivities will lead to a shift in the operating temperature of the parts. Since nonmetals tend to have higher emissivities than metals, the painted regions may have different radiation properties than the substrate material. Trial specimens should be checked if precise data are needed. If the specimens are to be heated by radiation from a high-temperature source, the different radiation properties can have a significant effect (2–3% in temperature).

Paints are useful in examining the distribution of temperature over a surface. Discrete spots should be used, rather than lines, since contact with molten material facilitates the melting of neighboring material even if it has not yet reached its own melting point.

The paints are cheap and easy to apply. They provide a good "first-look" capability.

The principal disadvantages of the paints and crayons are that they require visual interpretation and they are one-shot, irreversible indicators. They read peak temperature during the test cycle, yet they cannot record whether the peak was reached during steady state or during soak-back.

9 FLOW RATE

The objective of flow rate measurement is to determine the quantity of matter flowing. In some instances a flowmeter returns this information directly, but in most cases the flowmeter signal is derived from some property of the flow: volume, heat transfer rate, momentum flux, and so on. In most cases the flowmeter signal requires correction for pressure, temperature, or viscosity before the flow rate is known.

Interest in mass flow rate stems from a basic principle of engineering: The creation rate of mass is zero. If a flow is measured in mass flow units, then unless the pipe leaks or more fluid is added, the mass flow rate is the same everywhere along the pipe, regardless of changes in the density of the fluid flowing. This is not so for a volume-based measurement. The volume flow rate represented by a fixed mass flow rate of gas depends upon the molecular weight, the temperature, and the pressure of the gas. Hence, as either T or P change, the volume flow measure would change, even though the mass flow measure remained constant.

The term *SCFM* (*standard cubic feet per minute*) is frequently used in gas flow metering. The number of SCFM corresponding to a particular flow rate is a measure of the volume flow rate that would have been observed had the same mass flow rate of the same gas occurred at standard conditions of temperature and pressure. In other words, it is the volume flow rate that the actual mass flow rate would have produced had it been delivered at standard temperature and pressure.

9.1 Nomenclature

Mass. Quantity of matter. When weighed under standard gravitational conditions (i.e., on the average surface of the earth), one pound mass weighs one pound (exerts a force of one pound) and 1 kg exerts a force of 9.8 N. Symbol: lb_m or kg.

Mass Flow Rate. The rate at which matter passes the measuring location. Pounds mass (or kg) per unit time. Symbol: W or $\dot{m}$ depending on the author.

Volume Flow Rate. The volume occupied by the mass passing the measuring location. Cubic feet (or cubic meters) per unit time. Symbol: Q.

Standard Conditions. An arbitrarily chosen pair of values (different organizations may use different values) for temperature and pressure used to describe a standard state for measurement of the density of a gas. (e.g., 70°F and 14.7 psia or 20°C and 760 mm Hg).

Standard Density (. . . of a gas). The density (lb_m per cubic foot or kg/m^3) of the gas in question when its temperature and pressure are those of standard conditions.

The following symbols and terms are commonly used in flow-metering situations. In the rate statements that follow, the time base is taken as minutes or seconds.

W = mass flow rate, lb_m/min or kg/s
Q = volume flow rate, ft^3/min or m^3/s

Also frequently used:

GPM = gallons per minute
CFM = cubic feet per minute
ACFM = actual cubic feet per minute, ft^3/min
SCFM = standard cubic feet per minute, ft^3/min

Consider a flowmeter operating at 2 atm and 70°F. The density of a gas at those conditions is twice its density at standard conditions. Consider a flow through the meter such that 100 ft^3 of gas at 70°F and 2 atm passes through the meter each minute. This 100 ft^3 of gas would occupy 200 ft^3 if it were at standard conditions. In this case, the flow rate could be described either as 100 ACFM (at 70°F and 2 atm) or 200 SCFM. If ACFM is quoted, then the temperature and pressure must also be specified. The term *SCFM* describes the mass flow but expresses it in terms of the volume that that mass flow would occupy if it were at standard conditions.

9.2 Basic Principles Used in Flow Measurement

Measurement of flow rate can be accomplished using many different physical principles. The basic flow rate equation is

$$W = \rho A V \tag{25}$$

where W = flow rate, lb_m/s (kg/s)
A = area, ft^2 (m^2)
V = velocity, ft/s (m/s)
ρ = density, lb_m/ft^3 (kg/m^3)

To measure flow, any metering system must provide enough information to evaluate all three terms. Usually, two of the terms are fixed, and flow is evaluated by observing the change in the remaining term. An orifice used on water (presumed to be constant density) fixes A and ρ, leaving W proportional to V. A variable-area meter fixes ρ and V, leaving W proportional to A. Any combination of physical laws that permits evaluation of ρ, A, and V can form the basis for a flow-metering system.

Flowmeters can be divided into three generic groups depending on their sensing principle: conservation based, rate based, and dynamic.

The first group consists of flowmeters which depend upon a conservation principle for their output. There are three conservation laws which can be related to mass flow: conservation of mass, conservation of momentum, and conservation of energy. Conservation of momentum provides the basis for a large class of meters: orifices, nozzles, Venturi meters, drag disks, and obstruction meters. Conservation of energy (usually thermal energy) has been used in the construction of flowmeters for small flow rate. The conservation-of-mass law is used implicitly in all systems.

The second large group of flowmeters depend upon *rate equations*. There are many natural phenomena whose rate depends upon fluid velocity: viscous drag, heat transfer, mass transfer, displacement of a tracer particle, and so on. Any rate process sensitive to flow rate can be used as the basis for a flowmeter.

The third class of flowmeters, the dynamic meters, contains those whose signal depends on some dynamic aspect of a flow field. This class of meters depends upon the repeatability of certain unstable behaviors of flow fields, for example, vortex precession in a swirling flow in an adverse pressure gradient and vortex shedding from a bluff body.

In the remaining sections the more common types of flowmeters will be discussed in terms of the following items: physical appearance, output data and peripheral equipment, equations computed for off-design operation, and sources for more information.

9.3 Orifice, Nozzle, and Venturi Meters

Orifice, nozzle, and Venturi meters are momentum-based, fixed-area meters. The flow-related signal reveals the pressure difference between two points on the meter body, as shown in Fig. 26. These meters accelerate the fluid stream by imposing a contraction on the flow area and then decelerate the flow by expanding back to the initial pipe diameter. As the fluid accelerates, its static pressure goes down. As it decelerates, the pressure rises again but not without loss. Losses are related to irreversibilities in the flow field, such as eddy structures and turbulence. Mechanical energy dissipated by these mechanisms comes from flow work done by the main stream. As a result, there is a loss in total pressure.

A principal determinant of "Which meter to choose?" is the irrecoverable loss in pressure associated with the meter. As shown in Fig. 26, orifice meters have the largest loss for a given signal among these three candidates and Venturi meters have the smallest. As might be expected, Venturi meters are more expensive than orifice meters. *Fluid Meters—Their Theory and Application*[68] provides extensive data on the losses for different styles of orifice, nozzle, and Venturi meters and data on the precautions that must be observed in installing them.

Measurement of flow rate with these meters requires two pressure sensors and one temperature sensor. The accuracy of the flow rate measurement will depend upon the accuracy of these instruments.

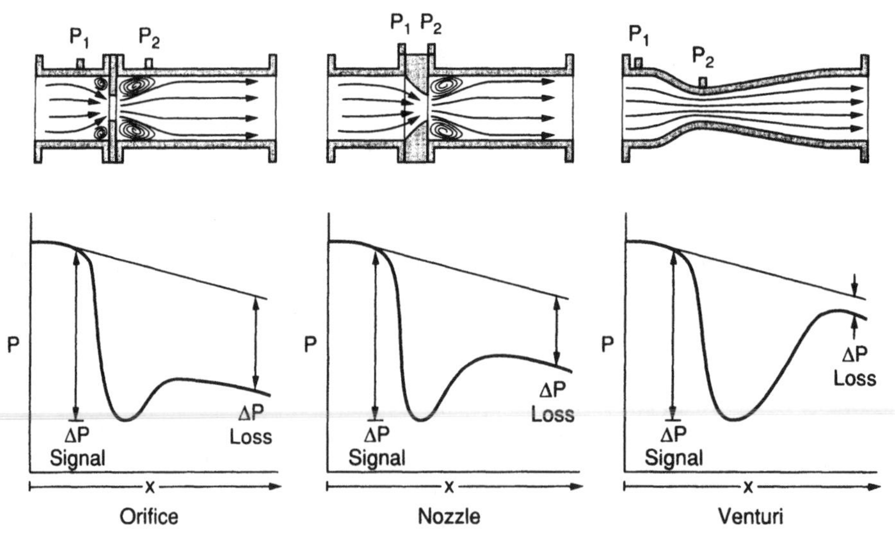

Figure 26 Orifice, nozzle, and Venturi meter pressure distributions. (From Ref. 67, with permission.)

The equation given below takes its nomenclature from the previous reference:

$$W_h = 359CFd^2F_aY\sqrt{h_w\gamma} \tag{26}$$

where W_h = flow rate, lb_m/h
C = coefficient of discharge, dimensionless
F = coefficient of approach, dimensionless
d = minimum diameter of the orifice, nozzle, or Venturi, in. (m)
F_a = coefficient of expansion, dimensionless
Y = compressibility factor, dimensionless
h_w = pressure difference $P_1 - P_2$, in. H_2O
γ = density of fluid flowing at location of measurement of P_1, lb_m/ft^3

Of the seven required terms, two are data or derived from data (h_w and γ), while five are calibration coefficients tabulated within the reference. The product of C and F is tabulated directly (K).

The coefficient K is a function of the flow rate; hence Eq. (26) cannot be solved explicitly—an iterative procedure is required. A first estimate of $K = C \times F = 0.60$ should be used for a first try. Using this K value, the calculated flow will generally be within a few percent and will serve to fix the exact value of K (within fractions of a percent).

It is apparent from Eq. (26) that the signal (ΔP) is proportional to W_n^2, the square of the flow rate. If the flow is fluctuating and an average ΔP is observed, then this corresponds to the time average of the square of the instantaneous flow, which is not equal to the square of the time-average flow. As a result, orifice, nozzle, and Venturi meters are not recommended for pulsating flows.

The most authoritative source of data regarding installation precautions for this class of meters is found in Reference 68. A good general reference is Benedict.[69]

In general, orifice flowmeters are accurate within $\pm 1/2\%$ of flow rate even without individual calibrations, when used properly. To achieve this accuracy, the orifice meter must be installed in strict compliance with the prescribed conditions for the approach and exit flow paths. The specifications in Ref. 68 cannot be abridged without loss of generality and should be treated as the final arbiter of an installation.

9.4 Variable-Area Meters

A variable-area meter consists of a tapered tube (usually of glass) with a float inside, as shown in Fig. 27. The clearance area between the outer diameter of the float and the inner diameter of the tube is a function of float position. When flow passes through the meter, it lifts the float under constant-pressure-drop conditions until the clearance area is sufficient to pass the existing flow. The output of the meter is *float position.* For gas flow measurements, pressure and temperature must be measured upstream of the float to establish the density.

Variable-area meters are usually purchased for use in a specific application and delivered calibrated for those conditions. The scale is calibrated directly in flow rate units. If the pressure and temperature of the flow remain at their design values, the float position gives flow rate directly.

A typical calibration legend for such a meter would read:

Full scale = 1.44 SCFM

Gas specific gravity of 1.38

When metered at 25 psig and 70°F

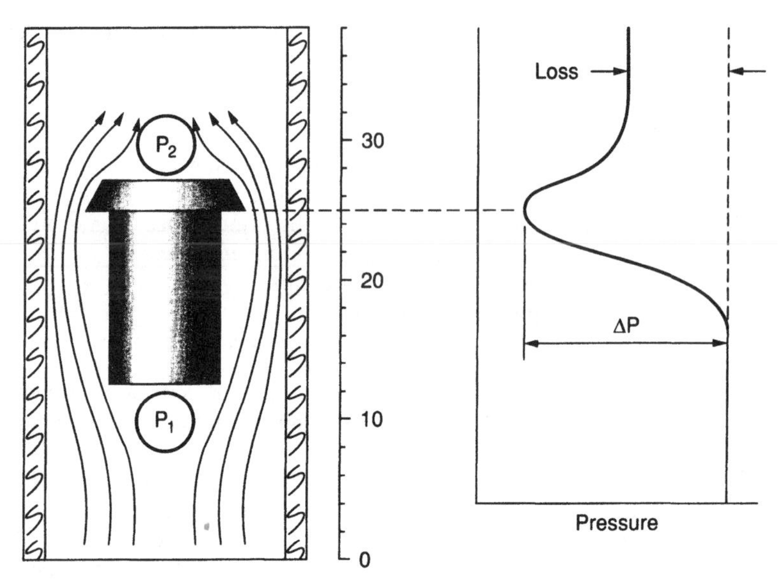

Figure 27 Variable area flowmeter pressure distribution. (From Ref. 67, with permission.)

The float position then directly reads in percentage of full-scale flow provided the inlet pressure is 25 psig, the inlet temperature is 70°F, and the specific gravity of the gas flowing is 1.38.

For off-design operation of the variable-area meter it is necessary to correct for changes in pressure, temperature, or specific gravity.

A sufficiently accurate equation for gas flow service can be derived from a force balance on the float in combination with Eq. (25), thus resulting in the fundamental equation for flow measurement

$$W_{\text{USE}} = W_{\text{calib}} \sqrt{\frac{\rho_{\text{USE}}}{\rho_{\text{calib}}}} \tag{27}$$

where W_{USE} = mass flow rate under actual conditions of use, $\text{lb}_\text{m}/\text{s}$ (kg/s)
W_{calib} = mass flow rate under calibration conditions for same float position, $\text{lb}_\text{m}/\text{s}$ (kg/s)
ρ_{USE} = density of fluid flowing, $\text{lb}_\text{m}/\text{ft}^3$ (kg/m^3)
ρ_{calib} = density of calibration fluid, $\text{lb}_\text{m}/\text{ft}^3$ (kg/m^3)

Example. Assume that the meter previously described is to be used in a new application where the fluid is air (Specific gravity = 1.00), the pressure is 100 psig, and the temperature is 70°F. What would be the full-scale flow rate in SCFM (air)?

$$W_{\text{USE}} = W_{\text{calib}} \sqrt{\frac{\rho_{\text{USE}}}{\rho_{\text{calib}}}}$$

$$\rho_{\text{USE}} = \frac{P_{\text{USE}}}{R_{\text{AIR}} T_{\text{USE}}} = \frac{(100 + 14.7)(144)}{53.3 \times (460 + 70)}$$

$$= 0.585\ \text{lb}_\text{m}/\text{ft}^3$$

$$\rho_{\text{calib}} = 1.38\left(\frac{P_{\text{calib}}}{R_{\text{AIR}}T_{\text{calib}}}\right)$$

$$= 1.38\,\frac{(25 + 14.7)(144)}{(53.3)(460 + 70^\circ\text{F})} = 0.279$$

$$W_{\text{calib}} = (1.44)(1.38)(.075) = 0.149\ \text{lb}_{\text{m}}/\text{s}$$

$$W_{\text{USE}} = (0.149)\sqrt{\frac{0.585}{0.279}} = 0.216\ \text{lb}_{\text{m}}/\text{s}$$

$$\text{SCFM}_{(\text{USE})} = \frac{W_{\text{USE}}}{\rho_{\text{STD,AIR}}} = \frac{W_{\text{USE}}}{0.075} = 2.877\ \text{SCFM}$$

This correction equation was derived under the assumption that the float was held up only by the pressure force (i.e., buoyancy and viscous drag were both negligible). When metering liquids of high density, the buoyant force must be accounted for, and if the viscosity of the fluid exceeds the *viscosity immunity ceiling* for a given meter, then the viscous drag must be taken into account. Precautions for dealing with these two cases are covered by the manufacturer's instructions. A meter purchased for gas service cannot be converted to liquid service by the preceding equation, or vice versa; a new calibration is required.

Variable-area meters are generally limited in accuracy to ± 1 or $\pm 2\%$ of full-scale reading.

It is important to measure the actual density of the fluid flowing in the meter to ensure that the calibration conditions are properly met. Pressure and temperature must be measured at the meter, just upstream of the float.

Irreversibilities involved in the mixing of the annular jet introduce losses in pressure that are nearly independent of flow rate and roughly equal to the pressure required to hold up the float. More accurate meters generally have higher losses (up to 3 or 4 psi used with air).

9.5 Laminar Flowmeters

Commercial laminar flowmeters consist of a *matrix* or *core* of small-diameter passages arranged so that the pressure drop across this core can be measured. The meter must be sized properly so the flow in these passages will remain laminar, even at the highest rated flow. If the flow remains laminar, then the pressure drop is linearly related to the volume flow rate through the core. General commercial practice is to provide flow-straightening sections upstream and downstream of the core and measure the pressures in the space between the flow straighteners and the core, as shown in Fig. 28.

Laminar flowmeters produce a pressure difference, usually 0–8 in. H_2O, which must be measured using an appropriate auxiliary instrument. The meter responds to volume flow; hence it is also necessary to measure the density of the fluid flowing. If the composition is known, density can be calculated knowing temperature and pressure at the inlet.

The pressure drop across a well-designed laminar flowmeter is linearly proportional to the volume flow rate (ACFM) multiplied by the viscosity of the fluid flowing. The pressure drop is independent of density.

Each meter is accompanied by a calibration curve, which typically reads:

Air Flow in Cubic Feet per Minute at 70°F and 29.92 Inches of Mercury Absolute vs. Pressure Drop, Inches of Water.

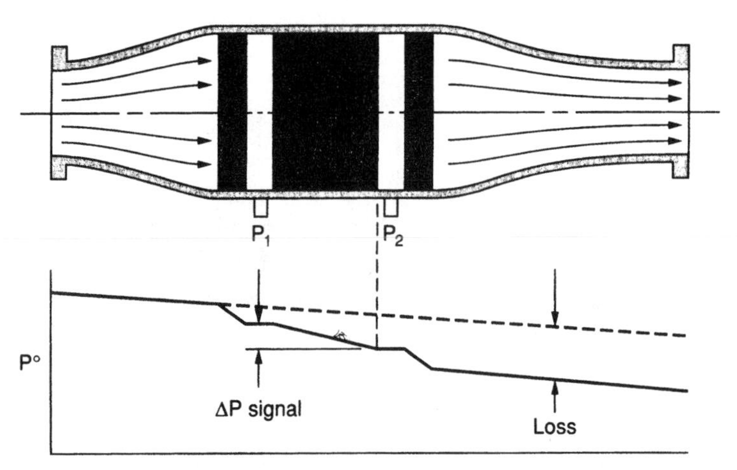

Figure 28 Laminar flowmeter pressure distribution. (From Ref. 67, with permission.)

The SCFM flow rate can be taken directly from the calibration curve if the meter is being used at 29.92 in. Hg and 70°F. If the pressure is 2 atm, however, then the density, the mass flow, and the number of SCFM corresponding to the same pressure drop signal would be doubled. Tables of pressure corrections, often provided with each meter, simplify this calculation.

If the temperature of the metered flow is higher than 70°F, the density is less than standard density and the viscosity of the flow is higher. Temperature correction factors, often provided in tabular form, account for both effects. The temperature correction factor for air flow is roughly proportional to absolute temperature raised to the 1.7 power, and it accounts for both the decrease in density and the increase in viscosity.

The equation for a laminar flowmeter in terms of the calibration chart is

$$\text{SCFM} = (\text{chart CFM})(CP)(CT)$$

or

$$\text{SCFM} = (\text{chart CFM})\left(\frac{P_{\text{act}}}{29.92}\right)\left(\frac{530}{460 + T_{\text{act}}}\right)^{1.7} \tag{28}$$

Laminar flowmeters respond linearly to flow; hence they can be used to directly measure the time average of a fluctuating flow. Commercial meters will usually average properly up to about 100 Hz.

9.6 Instability Meters

Two modern flow-metering systems derive their signal from the frequency of occurrence of an unstable fluid-dynamic phenomenon. One uses a swirling flow in a divergent passage to generate a precessing stall region. The other introduces a prismatic bluff body into a pipe and generates a periodic vortex trail. In each case, the frequency of the unstable event is related to the volume flow rate through the meter. Each claims 100:1 usable range, and each produces a pulse train whose repetition rate is linearly related to flow.

The Swirlmeter (registered trademark of the Fischer and Porter Co.) is shown schematically in Fig. 29. The meter has no moving parts. Stationary blades impart a swirling motion

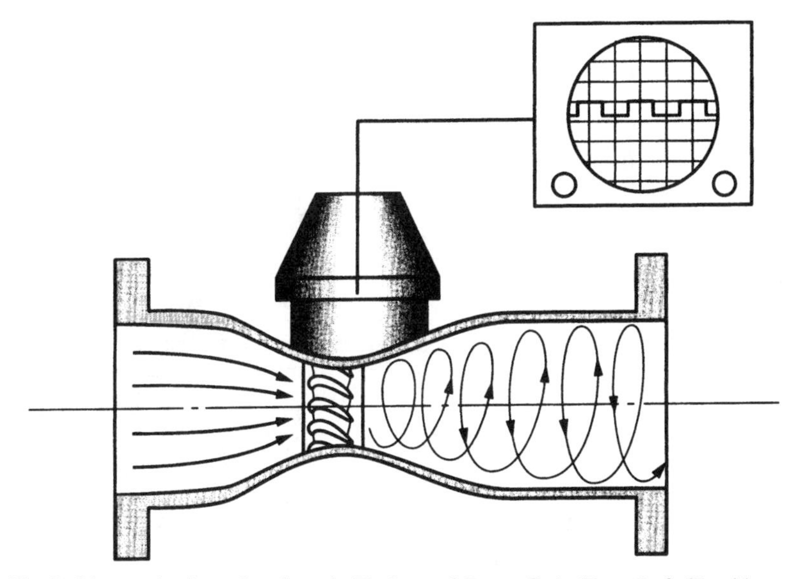

Figure 29 Swirlmeter (registered trademark, Fischer and Porter Co.). (From Ref. 67, with permission.)

to the flow, which then enters a diverging passage. The rotating flow field develops a stall (a region of relatively low axial velocity), which rotates in the diverging passage. An anemometer button in the wall senses the low-speed region each time it passes, generating a pulse train. The frequency of this pulse train is linearly related to the volume flow rate.

Mass flow rate calculation requires knowledge of the volume flow and the density. The equation is

$$\text{ACFM} = \frac{\text{total counts per minute}}{\text{calibration constant}} \tag{29}$$

$$W = \text{ACFM} \times \rho_{\text{act}} \tag{30}$$

$$\text{SCFM} = \text{ACFM} \times \frac{\rho_{\text{act}}}{\rho_{\text{STD}}} \tag{31}$$

Vortex-shedding flowmeters use a prismatic bluff body to generate the instability and one of several means to sense the oscillations of the stagnation streamline, which occur when the bluff body is shedding a vortex trail. A plan view schematic is shown in Fig. 30.

Again, these meters can operate over a turn-down ratio of 100:1 or better and claim $\pm 0.25\%$ accuracy.

The advantages of the instability meters are due to their pulse train signal, which is not easily obscured by random noise and can be transmitted over long distances.

9.7 Ultrasonic Flowmeters

An ultrasonic flowmeter is a device for estimating the flow rate in a duct based on measuring the propagation velocity of sound waves along one or more paths through the fluid, assuming the velocity profile in the duct to be known. There are two general types of ultrasonic flowmeters: transit time and Doppler. In both types, a transmitter unit injects an ultrasonic

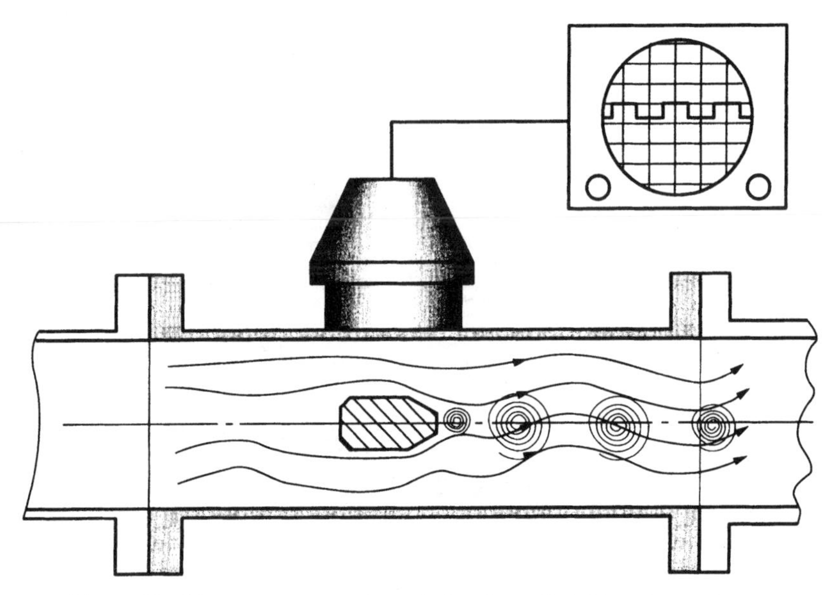

Figure 30 The vortex meter, schematically. (From Ref. 67, with permission.)

beam (an acoustic signal of at least 20 kHz or higher, more typically up to 1 MHz) into the fluid and a second unit receives the signal. The transmitter and receiver can be directly in contact with the fluid (a "wetted" system) or attached to the outside of the pipe wall (a "clamp-on" or "nonwetted" system).

Doppler Meter Principles

In a Doppler meter, the injected beam is reflected off particles, bubbles, or turbulence structures (very high frequency required) in the flow and picked up by the receiver. The frequency of the reflected beam is compared with the frequency of the injected beam and the flow rate inferred from the frequency shift and the system geometry. If all of the particles were moving at the same velocity V, that velocity could be determined from the frequency shift using the following equation[70,71]:

$$V = a\frac{\Delta f}{2f\cos\theta} \tag{32}$$

where V = velocity of particles
a = acoustic velocity in fluid
Δf = frequency shift
f = injected frequency
θ = angle between injected beam and pipe centerline

The velocity is not uniform across a pipe; hence a calibration factor must always be used. This factor will depend on the piping layout as well as the installation geometry of the flowmeter. Elbows, tees, and valves upstream of the measuring location will affect the velocity distribution and, therefore, the calibration factor. Doppler systems using very high frequency (around 1 MHz) can be used with clean, turbulent flows, but, for the most part, Doppler systems are used on flows with low to moderate (100-ppm) particle or bubble loading.

Yamanaka et al.[72] describe a velocity profile measuring technique using a time-domain correlation method which they claim has two advantages over the conventional pulse Doppler: higher time resolution and avoidance of the Nyquist limit on maximum velocity.

Transit Time Meter Principles

Transit time ultrasonic flowmeters are discussed by Lynnworth,[73] who traces the history of such meters back to the 1930s, and by Yoder[74] as part of a survey article. A transit time ultrasonic flowmeter compares the transit times of pulses transmitted upstream and downstream and infers the average fluid velocity along the beam path from the difference of the two transit times. Typically this is accomplished using pairs of transmitter–receiver units. Each unit serves alternately as a transmitter and a receiver, sending out a burst and then receiving a burst from the other unit. Commercial units are available.[75]

The time required for the pulse to travel from the upstream transmitter to the downstream receiver is shorter than the time required for the pulse originating at the downstream unit to move upstream, against the flow velocity. This time difference, Δt, is the basis for the flow measurement, as described in Efunda[76]:

$$\Delta t \triangleq (t_u - t_d) = \frac{L}{a - V\cos\theta} - \frac{L}{a + V\cos\theta} \tag{33}$$

$$= L\frac{2V\cos\theta}{a^2 - V^2\cos^2\theta} = \frac{\dfrac{2VX}{a^2}}{1 - (V^2/a^2)\cos^2\theta} \tag{34}$$

If $V/a \ll 1$ (flow Mach number $\ll 1$), then

$$\Delta t \approx \frac{2VX}{a^2} \tag{35}$$

$$V \approx \frac{a^2\,\Delta t}{2X} \tag{36}$$

where a = acoustic velocity in fluid
t_u = upwind transit time
t_d = downwind transit time
$\Delta t = t_u - t_d$
L = line-of-sight distance between transponders
θ = angle between line of sight and pipe centerline
V = fluid velocity (assumed uniform along line of sight)
X = axial distance between transponders, $L\cos\theta$

Flood[71] introduces an additional term to account for the transmission time of the pulse through the body of the transducer and the wall of the pipe, so that the calibration equation can be based on the overall time difference. Eq. (36) deals only with the transit times within the fluid and does not include the wall delay.

Both types of ultrasonic flowmeters use information gathered along the propagation path (or paths) to infer the average flow over the entire flow area, as though the velocity were uniform. Considerable effort has been put into the development of methods for dealing with nonuniform flow.

Nonuniform Velocity

The simplest nonuniformities to deal with are those found in isothermal, fully developed laminar or turbulent flows. These distributions are symmetric about the centerline and, if the propagation path intersects the centerline, a simple Reynolds-number-dependent calibration factor could be used. Such flows are rare, however, in industry. More realistic flows present more complex nonuniformities. Pipe bends, elbows, and tees lying in one plane generate centrifugal forces that move the highest velocity fluid from the centerline of the pipe toward the outside of each turn. This movement, in turn, causes secondary flows that roll up into pairs of vortices near the inside walls of the turns. If the elbows and bends are "out of plane," the resulting velocity distributions may have several local maxima, not just one. Such effects have been discussed by Ruppel and Peters,[77] who experimentally investigated the effects of several upstream piping configurations on the accuracy of an ultrasonic flowmeter. One approach to this problem has been to use several acoustic paths disbursed across the cross section of the pipe. Flowmeters with up to five paths have been tested. There are also different ways of "weighting" the data from the individual paths in forming the average.

A complex set of velocity distributions has been studied by Moore et al.[78] who used closed-form descriptions of representative multimodal velocity distributions (constant density) to challenge a family of multipath transit time flowmeter designs. These distributions were used as inputs to a program that calculated the "reading" of the flowmeter. These readings were then compared to the "real" flow deduced by integrating the velocity distribution over the flow area. Three different mathematical schemes were compared for weighting the readings from different paths as well as different numbers of acoustic paths from 2 to 5. For each flowmeter configuration, 90 trials were made, with the velocity profile being rotated 2 degrees each time. The range of calibration factors (maximum to minimum) for these 90 trials was recorded as a measure of the orientation sensitivity of the configuration. The authors offer this approach as a good tool for optimizing the design of multipath flowmeters.

Yeh et al.[79] combined a four-path transit time ultrasonic flowmeter with pattern recognition software and a "learning algorithm." The system was "educated" on a set of experimental and computational flow fields and "learned" to recognize and accommodate flow fields and adjust its calibration according to the flow field it encountered.

Temperature Gradients

If there is heat transfer to or from the fluid, the resulting density gradient is capable of distorting the flow calibration even in a straight run of pipe with fully developed flow. In situations with bends, the distortions may be worse with heat transfer than the isothermal cases.

Willatzen[80] looked at the effect of the radial density gradient resulting from heat transfer out of a water flow with a fully developed parabolic velocity distribution. The author considered a system in which idealized (permeable) transducers of different diameters centered within the pipe generated ultrasonic waves. The smallest diameter transducers generated waves containing up to eight modes; the largest ($R/R_0 = 1$) produced only the fundamental mode. In the absence of density gradients, all modes except the fundamental overestimated the mean flow while excitation of the fundamental mode alone gave error-free measurements. The predicted error was zero when the transducer disks occupied the entire pipe cross section (fundamental mode only). With a radial temperature gradient in the water (heat transfer out of the fluid into the surroundings) the indicated flow was low by about 2% even with only the fundamental mode excited.

Pulsing Flow

Pulsations in the flow can cause errors if the pulsation frequency aliases with the meter pulse frequency. This topic has been treated by Berrebi et al.[81] and by Carlander and Delsing,[82] who reported experimentally measured errors of more than 2–4% induced by unsteady pressure pulsations and velocity profile disturbances downstream of compound, out-of-plane elbows and bends. Carlander and Delsing proposed using the variance in a set of measurements as a diagnostic clue to the existence of an error in the mean value, since an increased variance in indicated flow accompanied the onset of errors due to pulsations.

Aliasing with Process Noise

Vermeulen et al.[83] examined the error introduced by aliasing of the noise from a throttling control valve with the signal of an ultrasonic flowmeter. They report that, under certain predictable conditions, the ultrasonic component of noise from the control valve can be larger than the signal generated by a commercial ultrasonic flowmeter and, therefore, can swamp out the meter.

Non-Newtonian Fluids

Non-Newtonian fluids display velocity distributions that are significantly different from Newtonian fluids and, hence, would be expected to cause trouble with flowmeters that require an assumption of the profile shape. Fyrippi, Owen, and Escudier[84] tested a single-beam transit time ultrasonic flowmeter in a non-Newtonian fluid and found significant errors.

Turbulence

A fundamental assertion used in developing the equations for transit time flowmeters is that the flow field is the same during the up-wind and the down-wind pulse transits. This may not be the case in a turbulent flow. This possibility was investigated by Weber et al.[85]

Fouling

Lansing[86] Investigated the effect of fouling on the performance of a transit time ultrasonic flowmeter.

REFERENCES

1. NIST ITS90 Thermocouple Data base; www.nist.gov.
2. R. J. Moffat, *Experimental Methods in the Thermosciences,* Department of Mechanical Engineering, Stanford University, Stanford, CA, 1978, pp. 1–26.
3. R. J. Moffat, "The Gradient Approach to Thermocouple Circuitry," in *Temperature, Its Measurement and Control in Science and Industry,* Charles M. Herzfeld (ed.), Vol. 3, Part 2, Reinhold New York, 1962, pp. 33–38.
4. R. J. Moffat, *Experimental Methods in the Thermosciences,* Department of Mechanical Engineering, Stanford University, Stanford, CA, 1980, pp. 2–3.
5. R. P. Benedict, *Fundamentals of Temperature, Pressure, and Flow Measurement,* 2nd ed., Wiley, New York, 1977.
6. E. O. Doebelin, *Measurement Systems: Application and Design,* McGraw-Hill, New York, 1966.
7. M. Campari and S. Garribba, "The Behavior of Type K Thermocouples in Temperature Measurement: The Chromel-P-Alumel Thermocouple," *Review of Scientific Instruments* **42**(5), 644–653 (May 1971).
8. R. E. Bentley, "Thermoelectric Hysteresis in Nickel-Based Thermocouple Alloys," *Journal of Physics D: Applied Physics* **22,** 1902–1907 (1989).
9. W. Schuh and N. Frost, "Improving Industrial Thermocouples Temperature Measurement," *NEWS 2002,* www.watlow.com.
10. R. E. Bentley and T. L. Morgan, "Thermoelectric Effects of Cold Work in Pt/10%Rh and Pt/13%Rh versus Pt Thermocouples," *Metrologia* **20**(2), 61–66 (June 1984).
11. O. A. Geraschenko and N. N. Ionova, "Thermal EMF of Plated Thermocouples" UDC 536,532, Translated from *Ismeritel'naya Tekhnika,* No. 1, January 1966, pp. 65–66.

12. C. Tallman, personal communication, 1964.
13. W. Gill and J. T. Nakos, "Temperature/Heat Flux Errors Caused by High Temperature Resistive Shunting along Mineral-Insulated, Metal Sheathed Thermocouples," *ASME HTD* **364-4,** 43–56 (1999).
14. R. J. Moffat, unpublished research results, 2002.
15. A. A. Sachenko, V. V. Kochlan, and V. Y. Mil'chenko, "Check of Thermoelectric Transducers with Built-in Temperature Calibrators," *Measurement Techniques* (English translation of *Izmeritel'naya Tekhnika*) **31**(7), 679–682 (December 1988).
16. H. M. Hashemian, "In-Situ Response Time Testing of Thermocouples," in *Proc. 1989 International Instrumentation Symposium,* 35th, Orlando, FL, May 1–4, 1989. ISA, Research Triangle Park, NC, pp. 587–593.
17. C. Y. Yang and D. W. Clarke, "A Self-Validating Thermocouple," *IEEE Transactions on Control Systems Technology* **5**(2), 239–253 (March 1997).
18. R. J. Moffat, "Gas Temperature Measurement," in *Temperature, Its Measurement and Control in Science and Industry,* Vol. 3, Part 2, Reinhold, New York, 1962, p. 553.
19. D. Elmore and W. B. Watkins, "Dynamic Gas Temperature Measuring System," *ISA Transactions* **24**(2), 73–82 (1985).
20. R. J. Moffat, "A Transient Method for Measuring Very High Gas Temperature," *ISA Proceedings of the International Instrumentation Symposium,* **47,** 413–422 (2001).
21. G. T. Lalos, "A Sonic Flow Pyrometer for Measuring Gas Temperatures," *Journal of Research National Bureau of Standards* **47,** 179–190 (1951), RP2242.
22. O. P. Norton, R. Arunkumar, and R. L. Cook, "Suction Thermocouple Measurements in High Temperature Gas Streams," *ISA Proceedings of the International Instrumentation Symposium,* **47,** 451–459 (2001).
23. R. J. Moffat, "Gas Temperature Measurement: A Two-Sensor Method for Canceling Radiation Error," *ISA Proceedings of the International Instrumentation Symposium,* **47,** 405–412 (2001).
24. S. Brohez, C. Delvosalle, and G. Marlair, "A Two-Thermocouple Probe for Radiation Corrections of Measured Temperatures in Compartment Fires," *Fire Safety Journal* **39**(5), 399–411 (July 2004).
25. L. G. Blevins and W. M. Pitts, "Modeling of Bare and Aspirated Thermocouples in Compartment Fires," *Fire Safety Journal* **33**(4), 239–259 (1999).
26. W. M. Pitts, E. Braun, R. D. Peacock, H. E. Mitler, E. L. Johnsson, P. A. Reneke, and L. G. Blevins, "Temperature Uncertainties for Bare-Beaded and Aspirated Thermocouple Measurements in Fire Environments" Special Technical Publication No. 1427, American Society for Testing and Materials, West Conshohocken, PA, pp. 3–15.
27. R. J. Moffat, "A Proposed Temperature Probe for Inter-Stage Compressor Measurements," *ISA Proceedings of the International Instrumentation Symposium,* **47,** 423–430 (2001).
28. R. Vasquez and J. M. Sanchez, "Temperature Measurement System for Low Pressure Ratio Turbine Testing," in *ASME Turbo Expo 2003: Power for Land, Sea, and Air,* Vol. 1, ASME International Gas Turbine Institute, Atlanta, GA, June 16–19, 2003, pp. 527–539.
29. N. R. Keltner and J. V. Beck, "Surface Temperature Measurement Errors," *ASME Transactions, Journal of Heat Transfer,* **105,** 312–318 (May 1983).
30. K. B. Sobolik, N. R. Keltner, and J. V. Beck, "Measurement Errors for Thermocouples Attached to Thin Plates. Application to Heat Flux Measurement Devices," *ASME HTD* **112,** 15–22 (1989).
31. P. C. Han and X. X. Wei, "Film Thermocouple Technique for Measurement of Nonmetal Surface Temperature," in *Proceedings of the International Centre for Heat and Mass Transfer,* 1991, pp. 137–144.
32. A. M. Osman, L. H. Eilers, and J. V. Beck, "Correction of Surface Temperature Measurements using an Inverse Method," in *Proceedings of the International Conference on Inverse Problems in Engineering,* 1998. pp. 599–606.
33. Omega Engineering, *Temperature Measurement Handbook and Encyclopedia,* Stamford, CT, 1984, pp. El–E18.
34. Product Bulletin 2086, Rosemount Engineering, Chanbassan, MI, rev. October 1978.
35. Product Bulletin 2108, Rosemount Engineering, Chanbassan, MI, rev. June 1974.
36. C. W. Lewis, "Four-wire Techniques Minimize Temperature Measurement Errors," *InTech* **42**(5), 40–43 (1995).

37. H. N. Norton, *Sensor and Analyser Handbook,* Prentice-Hall, Englewood Cliff, NJ, 1982, p. 337.

38. R. P. Benedict, *Fundamentals of Temperature, Pressure, and Flow Measurement,* 2nd ed., Wiley, New York, 1977, pp. 38–9.

39. Technical Bulletin MCT-18l, Victory Engineering Corporation, Springfield, NJ, 1968.

40. "YSI Precision Thermistor Products," Yellow Springs Instrument Co., Yellow Springs, OH, 1980.

41. K. Park, "Improvement in Electrical Stability by Addition of SiO 2 in $(Mn_{1.2}Ni_{0.78}Co_{0.87}Cu_{0.15}Si_x)O_4$ Negative Temperature Coefficient Thermistors," *Scripta Materialia* **50,** 551–554 (2004).

42. M. Tagawa, K. Kato, and Y. Ohta, "Response Compensation of Thermistors: Frequency Response and Identification of Thermal Time Constant," *Review of Scientific Instruments* **74**(3), 1350–1358 (2003).

43. *Temperature Measurement Handbook and Encyclopedia,* Omega Engineering, Stamford, CT, 1985, pp. T75–T76.

44. J. S. Steinhart and S. R. Hart, "Calibration Curves for Thermistors," *Deep Sea Research* **15,** 497 (1968).

45. A. S. Bennet, "The Calibration of Thermistors over the Temperature Range of 0–30°C," Deep Sea Research, **19,** 157 (1972).

46. L. Seren, C. B. Panchal, and D. M. Rote, "Temperature Sensors for OTEC Applications," ANL/OTEC-PS-12, Argonne National Laboratory, U.S. Department of Energy, May 1984.

47. G. Bosson, F. Guttmann, and L. M. Simmons, "A Relationship between Resistance and Temperature of Thermistors," *Journal of Applied Physics* **21,** 1267–1268 (1950).

48. R. N. Bigelow, P. S. Carlson, A. M. Hunter, and S. R. King, "Radiation Thermometry," in *Measurement Science Conference Proceedings,* 1996.

49. R. R. Dils, "High Temperature Optical Fiber Thermometer," *Journal of Applied Physics* **54,** 1198–1201 (1983).

50. K. A. Wickersham, "New Thermometry Technique Measures Component Temperature," in *Electronic Packaging and Production,* Cahners Publishing, September 1981.

51. D. M. Cunningham, S. W. Allison, and D. B. Smith, "Thermographic Properties of Eight Blue-Emitting Phosphors," ISA 1993, Paper No. 93-111, in *Proc. 39th Int. Inst. Symp.* Albuquerque, NM, May 1993.

52. S. C. Jensen, S. D. Tilstra, G. A. Barnabo, D. C. Thomas, and R. W. Phillips, "Fiber Optic Temperature Sensor for Aerospace Applications," in *Proceedings of SPIE,* Vol. 1369, International Society for Optical Engineering, 1991, pp. 87–95.

53. R. Raja and G. T. Cunningham, "Experimental Investigation of Phosphor Thermography Using a Blue LED optical excitation Source," in *ISA TECH/EXPO Technology Update Conference Proceedings,* Vols. 424–425, 2002, pp. 1085–1092.

54. D. Farina, J. Hacker, J. K. Eaton, and R. J. Moffat, "Illuminant Invariant Calibration of Thermochromic Liquid Crystals," paper presented at the ASME National Heat transfer Conference, August 8–11, 1993.

55. H. S. Rhee, J. R. Koseff, and R. L. Street, "Flow Visualization of a Recirculating Flow by Rheoscopic Liquid and Liquid Crystal Techniques," *Experiments in Fluids* **2,** 57–64 (1984).

56. K. Hollingsworth, A. L. Boehman, E. G. Smith, and R. J. Moffat, "Measurement of Temperature and Heat Transfer Coefficient Distributions in a Complex Flow Using Liquid Crystal Thermography and True-Color Image Processing," Collected Papers in Heat Transfer, *ASME HTD* **123,** 35–42 (1989).

57. V. Rajendran, M. Deblock, T. Wetzel, M. Lusted, C. Kaminski, and B. Childers, "Use of Fiber Optic Based Distributed Temperature Measurement System for Electrical Machines," *Proceedings of SPIE—The International Society for Optical Engineering;* **5191,** 214–225 (2001).

58. M. Decreton, L. Binard, C. Delrez, W. Hebel, and W. Schubert, "High-Temperature Measurements by Noise Thermometer," *High Temperature—High Pressure,* **12,** 395–402 (1980).

59. H. Nyquist, *Physics,* **32,** 110–113 (1928, rev. ed.).

60. J. B. Johnson, Physics **32,** 97–110 (1928, rev. ed.)

61. H. Brixy, R. Hecker, K. F. Rittinghaus, and H. Howener, "Applications of Noise Thermometry in Industry under Plant Conditions," in *Temperature, Its Measurement and Control in Science and Industry,* J. F. Schooley (ed.), Reinhold, New York, 1982, pp. 1225–1237.

62. H. Brixy, R. Hecker, J. Oehmen, and E. Zimmermann, "Temperature Measurements in the High-Temperature Range (1000–2000C) by Means of Noise Thermometry," *High temperatures–High Pressures* **23**(6) 625–631 (1991).

63. H. A. Tasman and J. Richter, "Unconventional Methods of Temperature Measurement," *High Temperatures—High Pressure* **11,** 87–101 (1974).

64. A. Benjaminson and F. Rowland, *Temperature, Its Measurement and Control in Science and Industry,* Vol. 4, Instrument Society of America, Pittsburgh, PA, 1972, p. 701.

65. J. F. W. Bell, A. A. Fathamani, and T. N. Seth, 1975, cited in Von der Hardt et al., vol. 2, 1975, pp. 649–680.

66. *The Temperature Measurement Handbook and Encyclopedia,* Omega Engineering, Stamford, CT, 1985, p. Q4.

67. R. J. Moffat, "Introduction to Flow Rate Measurement," in *Experimental Methods in the Thermosciences,* Mechanical Engineering Department, Stanford University, Stanford, CA, 1978.

68. H. S. Bean, *Fluid Meters—Their Theory and Application,* 6th ed., (American Society of Mechanical Engineers, New York, 1971).

69. R. P. Benedict, Fundamentals of Temperature, Pressure, and Flow Measurement, 2nd ed., Wiley, New York, 1977.

70. Ultrasonic Doppler and time-of-flight flowmeters, www.enginering toolbox.com

71. J. Flood, "Ultrasonic Flowmeter Basics," *Sensors Magazine,* October 1997.

72. G. Yamanaka, H. Kikura, and M. Aritomi, "Study on Velocity Profile Measurement Using Ultrasound Time-Domain Correlation Method," *ASME HTD* **369**(3), 259–264 (2001).

73. L. Lynnworth, "Clamp-On Flowmeters for Fluids," *Sensors,* August 2001.

74. J. Yoder, "Flowmeters and Their Apps: An Overview," *Sensor Technology and Design,* October 2003.

75. Omega.com, technical reference section: ultrasonic flowmeters.

76. http://www.efunda.com/DesignStandards/sensors/flowmeters/flowmeter_ustt.cfm.

77. C. Ruppel and F. Peters, "Effects of Upstream Installations on the Reading of an Ultrasonic Flowmeter," *Flow Measurement and Instrumentation* **15**(3), 167–177 (June 2004).

78. P. I. Moore, G. J. Brown, and B. P. Stimson, "Ultrasonic Transit-Time Flowmeters Modeled with Theoretical Velocity Profiles: Methodology," *Measurement Science and Technology,* **11,** pp. 1802–1811 (2000).

79. T. T. Yeh, P. I. Espina, and S. A. Osella, "An Intelligent Ultrasonic Flow Meter for Improved Flow Measurement and Flow Calibration Facility," in *Conference Record—IEEE Instrumentation and Measurement Technology Conference,* Vol. 3, 2001, pp. 1741–1746.

80. M. Willatzen, "Ultrasonic Flowmeters: Temperature Gradients and Transducer Geometry Effects," *Ultrasonics* **41,** 105–114 (2003).

81. J. Berrebi, P.-E. Martinsson, M. Willatzen, and J. Delsing, "Ultrasonic Flow Metering Errors Due to Pulsating Flow," *Flow Measurement and Instrumentation* **15**(3), 179–185 (June 2004).

82. C. Carlander, and J. Delsing, "Installation Effects on an Ultrasonic Flow Meter with Implications for Self Diagnosis," *Flow Measurement and Instrumentation* **11**(2), 109–122 (2000).

83. M. J. Vermeulen, G. De Boer, and J. Bowen, "A Model for Estimation of the Ultrasonic Noise Level Emitted by Pressure Regulating Valves and Its Influence on Ultrasonic Flowmeters," in *Proceedings of the Annual ACM-SIAM Symposium on Discrete Algorithms,* 2001, pp. 435–444.

84. I. Fyrippi, I. Owen, and M. P. Escudier, "Flowmetering of Non-Newtonian Liquids," *Flow Measurement and Instrumentation* **15**(3), 131–138 (June 2004).

85. F. J. Weber, H. Johari, and W. W. Durgin, "Ultrasonic Beam Propagation in Turbulent Flow," in *Proceedings of the ASME Fluids Engineering Division Summer Meeting,* Vol. 1, 2003, pp. 85–91.

86. J. Lansing, "Dirty vs Clean Ultrasonic Gas Flow Meter Performance," in *Proceedings of the Annual ACM-SIAM Symposium on Discrete Algorithms,* 2002, pp. 989–999.

CHAPTER 6
SIGNAL PROCESSING

John Turnbull
Case Western Reserve University
Cleveland, Ohio

1 FREQUENCY-DOMAIN ANALYSIS OF LINEAR SYSTEMS

Signals are any carriers of information. Our objective in signal processing involves the encoding of information for the purpose of transmission of information or decoding the information at the receiving end of the transmission. Unfortunately, the signal is often corrupted by noise during our transmission, and hence it is our objective to extract the information from the noise. The standard method most commonly used for this involves filters that exploit some separation of the signal and noise in the frequency domain. To this end, it is useful to use frequency-domain tools such as the Fourier transform and the Laplace transform in designing and analyzing various filters. The Fourier transform of a function of a time is

$$\mathcal{F}\{f(t)\} = F(\omega) = \frac{1}{\sqrt{2\pi}} \int_{-\infty}^{\infty} f(t) e^{j\omega t}\, dt \qquad j^2 = -1 \tag{1}$$

For continuous systems, the transfer characteristics of a filter system is a function that gives information of the gain versus frequency. The Laplace transform for a given time-domain function is

$$\mathcal{L}\{f(t)\} = F(s) = \int_{0}^{\infty} f(t) e^{-st}\, dt \tag{2}$$

The steady-state Laplace transform (i.e., neglecting transients) for the derivative and integral of a given function is

$$\mathcal{L}\left\{\frac{df(t)}{dt}\right\} = sF(s) \qquad \mathcal{L}\left\{\int f(t)\,dt\right\} = \frac{F(s)}{s} \tag{3}$$

By convention, functions in the time domain use t as the independent variable and functions in the Laplace domain use s as the independent variable. For this reason, the Laplace domain is commonly called the *S-domain.* The Fourier transform and the Laplace transform are similar but different in two respects: (1) The Fourier transform integrates the signal over all time while the Laplace transform integrates for positive times only and (2) the exponent of the kernel in the Laplace transform is complex with both real and imaginary components while the exponent in the Fourier transform has imaginary component and no real component. By using Euler's identity, $e^{j\omega} = \cos(\omega) + j\sin(\omega)$, we see that

$$\begin{aligned}\mathcal{F}\{f(t)\} = F(\omega) &= \int_{-\infty}^{\infty} f(t)e^{j\omega}\,dt \\ &= \int_{-\infty}^{\infty} f(t)\cos(\omega t)\,dt + j\int_{-\infty}^{\infty} f(t)\sin(\omega t)\,dt \\ &= \langle\cos(\omega t)\rangle + j\langle\sin(\omega t)\rangle. \end{aligned} \tag{4}$$

We approximate the Fourier transform from discrete samples $f(k) \leftrightarrow F(k)$, where $0 \leq k \leq N$ and $F(k) = \alpha_k + j\beta_k$:

$$\alpha_k = \sum_{i=0}^{N-1} f(i)\cos\left(2\pi\frac{i\cdot k}{N}\right) \qquad \beta_k = \sum_{i=0}^{N-1} f(i)\sin\left(2\pi\frac{i\cdot k}{N}\right) \tag{5}$$

However, if the number of points we transform is a composite number and not a prime number, we can restructure our calculations to eliminate some of the calculations. Furthermore, of the factors that are themselves composite, we can further factor and eliminate more calculations. For this reason, the most efficient vector sizes are those that are highly composite. As an example, consider the simple case of transforming four points. We can express Eq. (5) for this case in matrix form as

$$\begin{bmatrix} c_0 \\ c_1 \\ c_2 \\ c_3 \end{bmatrix} = \begin{bmatrix} 1 & 1 & 1 & 1 \\ 1 & \rho & \rho^2 & \rho^3 \\ 1 & \rho^2 & \rho^4 & \rho^6 \\ 1 & \rho^3 & \rho^6 & \rho^9 \end{bmatrix} \begin{bmatrix} f_0 \\ f_1 \\ f_2 \\ f_3 \end{bmatrix}$$

where ρ is the principal root of 1. The nth principal root of 1 is $\cos(2\pi/n) + j\sin(2\pi/n)$. We can then factor the matrix into two sparse matrices:

$$\begin{bmatrix} c_0 \\ c_1 \\ c_2 \\ c_3 \end{bmatrix} = \begin{bmatrix} 1 & 1 & 0 & 0 \\ 1 & \rho^2 & 0 & 0 \\ 0 & 0 & 1 & \rho \\ 0 & 0 & 1 & \rho^3 \end{bmatrix} \begin{bmatrix} 1 & 0 & 1 & 0 \\ 0 & 1 & 0 & 1 \\ 1 & 0 & \rho^2 & 0 \\ 0 & 1 & 0 & \rho^2 \end{bmatrix} \begin{bmatrix} f_0 \\ f_1 \\ f_2 \\ f_3 \end{bmatrix}$$

Although it is possible to implement this efficient algorithm—known as the fast Fourier transform (FFT)—for any vector size that is a composite number, it is most commonly implemented for vector sizes in which all factors are 2. The effort in evaluating Eq. (5) increases with the square of the number of points to transform, or $O(n^2)$. In contrast, the effort for the FFT of order 2^p increases proportionately to $O(n\log(n))$. Finally, we can use the FFT to estimate the power spectral density (PSD) of a given discrete signal by computing the square of the magnitude of the FFT. The following algorithm describes the method for computing the Fast Fourier Transform.[1]

FAST FOURIER TRANSFORM 1 *Given a vector* $\{x_1, \ldots, x_n\}$ *of complex numbers, where n is some integer and there exists some integer p such that* $n = 2^p$. *This algorithm outputs the Fourier transform overwriting the input vector* **x**.

```
k ← 1
For i = 1 To n
    If k > i
        swap(x_i, x_k)
    End If
    m = n/2
    While m ≥ 1 And k > m
        k ← k − m
        m ← m/2
    End While
    k ← k + m
Next i
M_max ← 1
While n > M_max
    i_c ← 2·M_max
    θ ← π / M_max
    wp ← −2 sin² (θ/2) − j·sin(θ)
    w ← 1
    For m = 1 To M_max
        For i = m To n By i_c
            k ← i + M_max
            xtemp ← x_i − w·x_k
            x_i ← x_i + w·x_k
            x_k ← xtemp
        Next i
        w ← w·(wp + 1)
    Next m
    M_max = i_c
End While
```

2 BASIC ANALOG FILTERS

Linear filters apply frequency-specific gains to a signal. This is often done to enhance desired portions of the spectrum while attenuating or eliminating other portions. Four common filters are low pass, high pass, bandpass, and band reject. The objective of an ideal low-pass filter is to eliminate a range of undesired high frequencies from a signal and leave the remaining portion undistorted. To this end, an ideal low-pass filter will have a gain of 1 for all frequencies less than some desired cutoff frequency f_c and a gain of 0 for all frequencies greater than f_c, as seen in Fig. 1. There are various rational functions that approximate this ideal. But because of the discontinuity in the ideal low-pass response, all realizations of this ideal with be an approximation. The various approximation functions generally trade off between three characteristics: passband ripple, stop-band ripple, and the transition width, shown in Fig. 2. Four common rational function approximations for low-pass filters are the Butterworth, the Tchebyshev Types I and II, and the elliptical filter. By convention, we use $H(s)$ as the transfer function from which we determine the frequency response, where

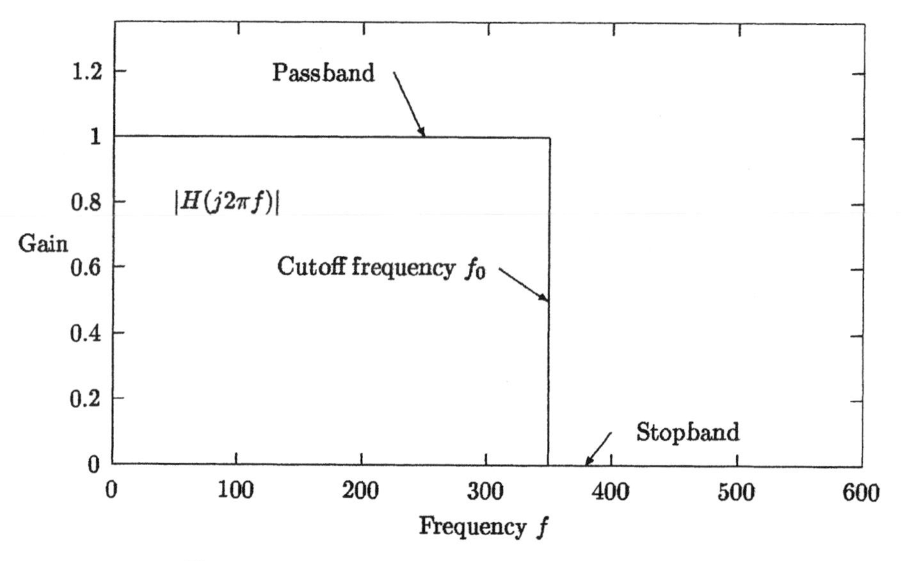

Figure 1 Frequency response for an ideal low-pass filter.

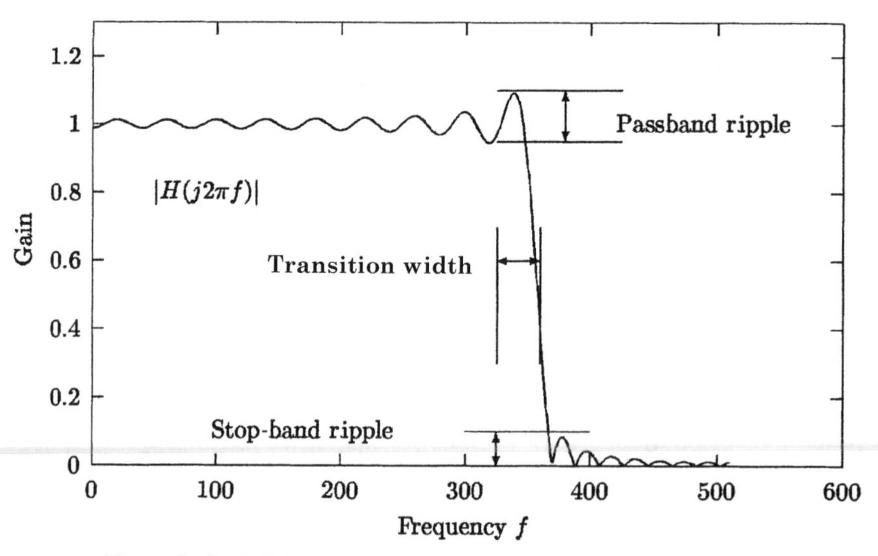

Figure 2 Typical frequency response for an approximate low-pass filter.

$$H(s) = \frac{V_{\text{output}}(s)}{V_{\text{input}}(s)} \tag{6}$$

is the output-to-input gain of a standard two-port system.

2.1 Butterworth

The Butterworth filter has a smooth passband region (frequencies less than f_c hertz) and a smooth stop band (frequencies greater than f_c) and a comparatively wide transition region as shown in Figs. 3 and 4. Let S_c be the cutoff frequency; then a low-pass rational function approximation is as follows:

$$|H(s)|^2 = \frac{1}{1 + (s/s_c)^{2N}} \tag{7}$$

where $2\pi f_c = s_c$. In factored form,

$$H(s) = \frac{1}{(s - p_0)(s - p_1) \cdots (s - p_{N-1})} \tag{8}$$

where $p_i = \alpha_i + j\beta_i$ and

$$\begin{aligned} \alpha_i &= 2\pi f_c \cos\left(\frac{\pi}{2N}(N + 2i + 1)\right) \qquad i = 0, \ldots, N-1 \\ \beta_i &= 2\pi f_c \sin\left(\frac{\pi}{2N}(N + 2i + 1)\right) \qquad i = 0, \ldots, N-1 \end{aligned} \tag{9}$$

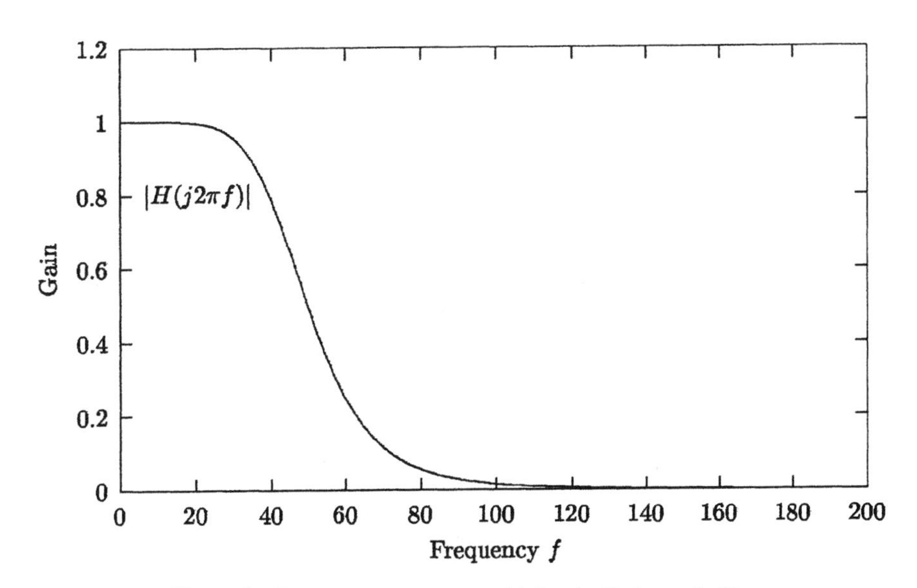

Figure 3 Frequency response to a third-order Butterworth filter.

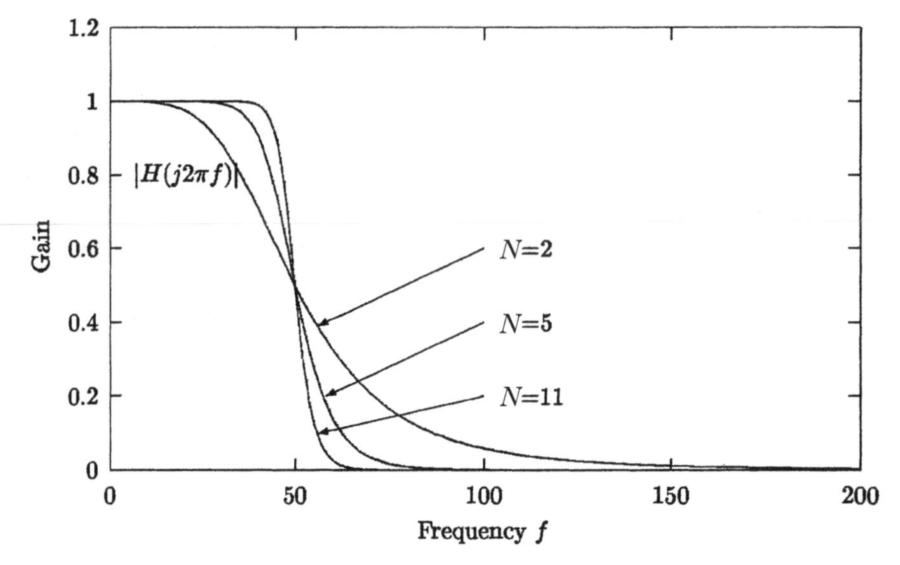

Figure 4 Frequency response to Butterworth filters of order 2, 5, and 11.

2.2 Tchebyshev

Unlike the Butterworth rational function, the Tchebyshev (Type I) rational function permits some ripple to occur in the passband (frequencies less than f_c in the low-pass filter), in exchange for a sharper transition region compared to a Butterworth filter of equal order N, as seen in Fig. 5:

$$|H(s)|^2 = \frac{1}{1 + \epsilon^2 T_N^2 (s/s_c)} \tag{10}$$

Where T_N is the Tchebyshev polynomial defined as

$$T_N(s) = \cos[n \cos^{-1}(s)] \tag{11}$$

The pole placements $p_i = \alpha_i + j\beta_i$ for the Tchebyshev Type I filter are

$$\begin{aligned} \alpha_i &= 2\pi f_c \sinh(v_0) \cos\left(\frac{\pi}{2N}(N + 2i + 1)\right) \qquad i = 0, \ldots, N-1 \\ \beta_i &= 2\pi f_c \cosh(v_0) \sin\left(\frac{\pi}{2N}(N + 2i + 1)\right) \qquad i = 0, \ldots, N-1 \end{aligned} \tag{12}$$

where

$$v_0 = \frac{1}{N} \sinh^{-1}\left(\frac{1}{\epsilon}\right) \tag{13}$$

$$\epsilon = \sqrt{\frac{1}{(1 - r)^2} - 1} \qquad 0 < r < 1 \tag{14}$$

where r is this amplitude of the ripple in proportion to the gain of the passband.

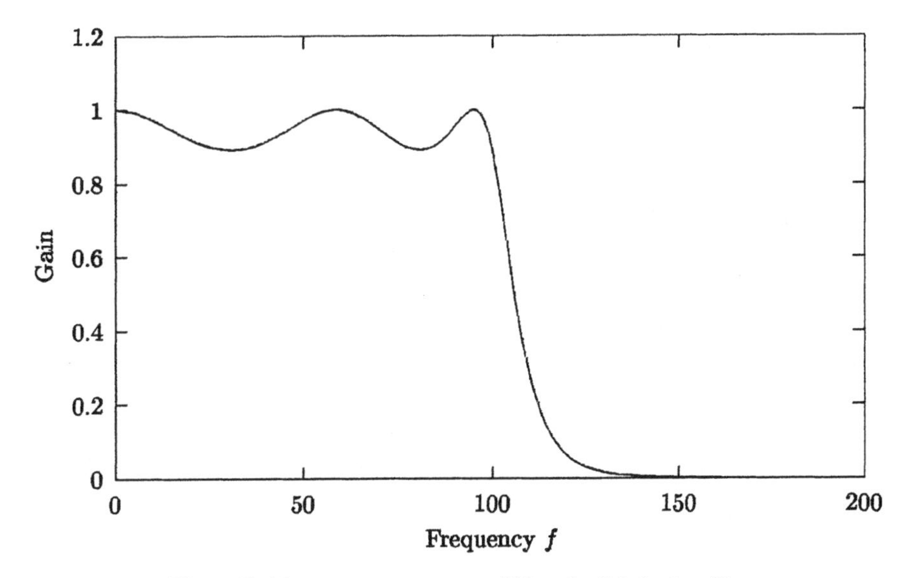

Figure 5 Frequency response to a fifth-order Tchebyshev filter.

2.3 Inverse Tchebyshev

The inverse Tchebyshev (or Tchebyshev Type II) filter has a smooth passband, ripple in the stop band (frequencies less than f_c for the low-pass filter), and a sharper transition region compared to the Butterworth function of equal order N, as seen in Fig. 6:

$$|H(s)|^2 = \frac{\epsilon^2 T_N^2\,(s_c/s)}{1 + \epsilon^2 T_N^2\,(s_c/s)} \tag{15}$$

The zero placements $\zeta_i = \alpha_i + \beta_i$ for the Tchebyshev type II filter are

$$\zeta_i = \frac{1}{\sin(i\pi/2N)} \qquad 0 \le i \le N-1 \tag{16}$$

The pole placements $p_i = \alpha_i + j\beta_i$ for the Tchebyshev Type II filter are simply the reciprocals of the pole placements computed for the Tchebyshev Type I filter.

2.4 Elliptical

The Elliptical filter has ripple in both the passband and the stop band but, in exchange, has the narrowest transition region for equal filter order N, as seen in Fig. 7. The derivation and implementation for the determination of the poles and zeros involve the Jacobian elliptic function:

$$f(t, k) = \int_0^t \frac{dx}{\sqrt{1 - k^2 \sin^2(x)}} \tag{17}$$

The method is beyond the scope of this chapter. The interested reader is referred to Refs. 2 and 3.

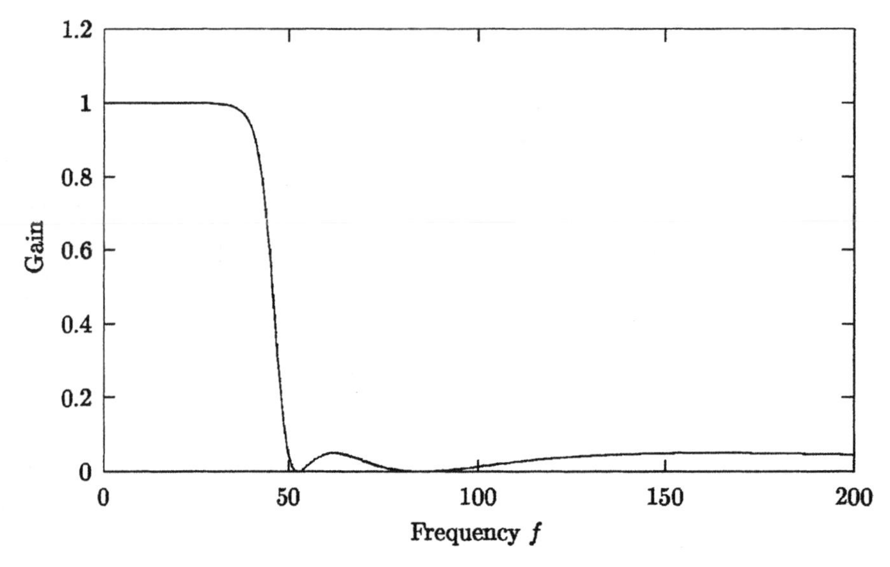

Figure 6 Frequency response to a fifth-order inverse Tchebyshev filter.

2.5 Arbitrary Frequency Response Curve Fitting by Method of Least Squares

It is possible to design a filter to approximate a desired frequency response $F(\omega)$ by the method of least squares. Consider a transfer function in factored form as

$$H(s) = G \frac{\prod_{i=i}^{M} s - \zeta_i}{\prod_{k=1}^{N} s - p_k} \tag{18}$$

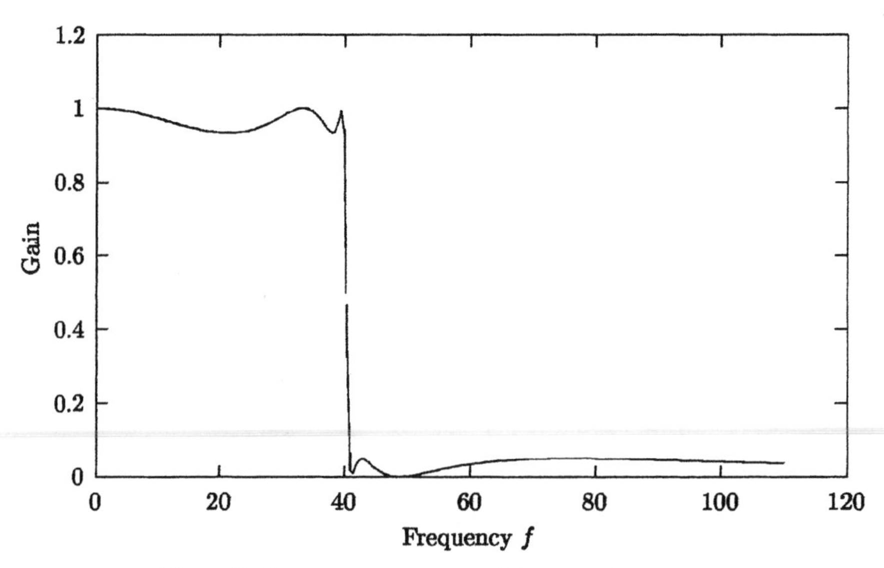

Figure 7 Frequency response to a fifth-order elliptical filter.

Minimize

$$\chi^2(G;\boldsymbol{\zeta};\mathbf{p}) = \int_D [|H(j\omega)| - F(\omega)]^2 \, d\omega \tag{19}$$

subject to

$$\Re\text{e}(p_k) < 0 \qquad \text{for all } k = 1, \ldots, N \tag{20}$$

Unfortunately, this results in a nonlinear system of equations. Furthermore, the topography of this objective function is generally complicated with many local minima, making standard gradient-descent methods unfeasible. It is usually best to use finite-impulse-response (FIR) filtering or filtering in the frequency domain for these types of problems. If an infinite-impulse-response (IIR) filter is desired, the reader is referred to Prony's method, which linearizes this system and finds an approximate optimal solution.[4]

2.6 Circuit Prototypes for Pole and Zero Placement for Realization of Filters Designed from Rational Functions

The voltage–current relation for a resistor (R), inductor (L), and capacitor (C) is

$$v_r = iR \qquad v_l = L\frac{di}{dt} \qquad v_c = \frac{1}{C}\int_{-\infty}^{t} i \, dr \tag{21}$$

These relationships are represented in the S-domain as

$$V_r(s) = RI(s) \qquad V_l(s) = sLI(s) \qquad V_c = \frac{I(s)}{sC} \tag{22}$$

Thus, the general transfer function for any linear circuit involving standard passive, active, and reactive devices is a rational function, that is, a ratio of polynomials:

$$\begin{aligned} H(s) &= \frac{\sum_{i=0}^{M-1} a_i s^i}{\sum_{k=0}^{N-1} b_k s^k} \\ &= \frac{V_0}{V_i} = -\frac{Z_2}{Z_1} \end{aligned} \tag{23}$$

Thus one can construct any arbitrary transfer function through a serial placement of this building block circuit prototype shown in Fig. 8. Table 1 gives circuit elements for Z_1 and Z_2 to construct the basic prototype circuits.

3 BASIC DIGITAL FILTER

Basic linear digital filters are of two types: those that have a finite response to an impulse, or FIR, and those that have an infinite response to an impulse (IIR). The general form of a linear digital filter is

$$y_k = b_1 y_{k-1} + b_2 y_{k-2} + \cdots + b_{k-m} y_{k-m} + a_0 x_k + a_1 x_{k-1} + \cdots + a_n x_{k-n} \tag{24}$$

where $k - i$ represents the ith delay. That is, the kth output from a linear digital filter is some linear combination of previous inputs and outputs. The filter will have a finite response to an impulse if $b_1 = b_2 = \cdots = b_m = 0$; otherwise, the filter is of type IIR.

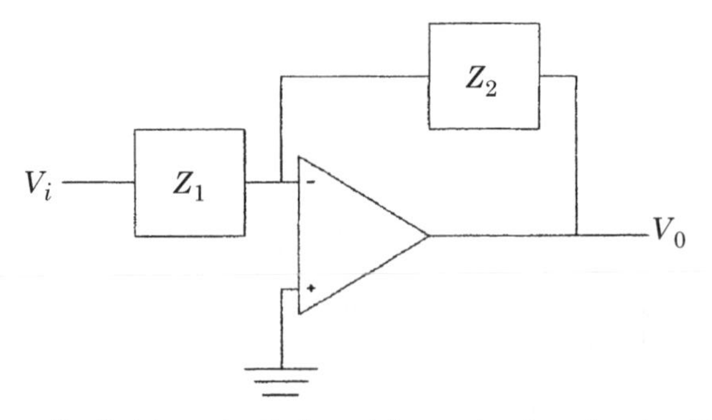

Figure 8 Prototype circuit element for construction of analog filter.

3.1 *z*-Transforms

The *z*-transform is used to analyze the frequency response and stability of a system of difference equations in much the same way that the Laplace transform is used to analyze the frequency response and stability of a system of differential equations. The *z*-transform of Eq. (24) is

$$H(z) = \frac{a_0 + a_1 z^{-1} + \cdots + a_M z^{-M}}{1 - b_1 z^{-1} - \cdots b_N z^{-N}} = \frac{\sum_0^M a_j z^{-j}}{1 - \sum_{j=1}^N b_j z^{-j}} \tag{25}$$

We determine the frequency response by $|H(e^{j\omega})|$, where $0 \le \omega \le 2\pi$ and corresponds to the scaled frequencies of our sampled system from 0 Hz up to the sampled frequency. The function $e^{j\omega}$ is a periodic signal. For $\pi < \omega \le 2\pi$, $e^{j\omega} = e^{j(\pi-\omega)}$, and for $2\pi < \omega \le 4\pi$, $e^{j\omega} = e^{j(\omega-2\pi)}$. Thus frequencies greater than π, corresponding to the Nyquist frequency, or one-half of the sampling frequency assume an identical characteristic to an analogous frequency less than π. This phenomenon is called *aliasing* and is illustrated in Fig. 9.

3.2 Design of FIR Filters

It is possible to use the Fourier transform to determine the coefficients to an FIR filter. However, the Fourier coefficients are generally complex numbers, and when working with real signals, it is desirable to have a real coefficients in our filter. To do this, we apply Euler's identity and observe from Eq. (4) that the coefficients will be real if the inner products with

Table 1 Circuit Elements for the Construction of Basic Transfer Function Prototypes

Single pole	$Z_1 \leftarrow$ resistor
	$Z_2 \leftarrow$ *RC* in parallel
Single zero	$Z_1 \leftarrow$ *RC* in parallel
	$Z_2 \leftarrow$ resistor
Complex-conjugate pole pair	$Z_1 \leftarrow$ *LRC* in series
	$Z_2 \leftarrow$ capacitor
Complex-conjugate zero pair	$Z_1 \leftarrow$ capacitor
	$Z_2 \leftarrow$ *LRC* series

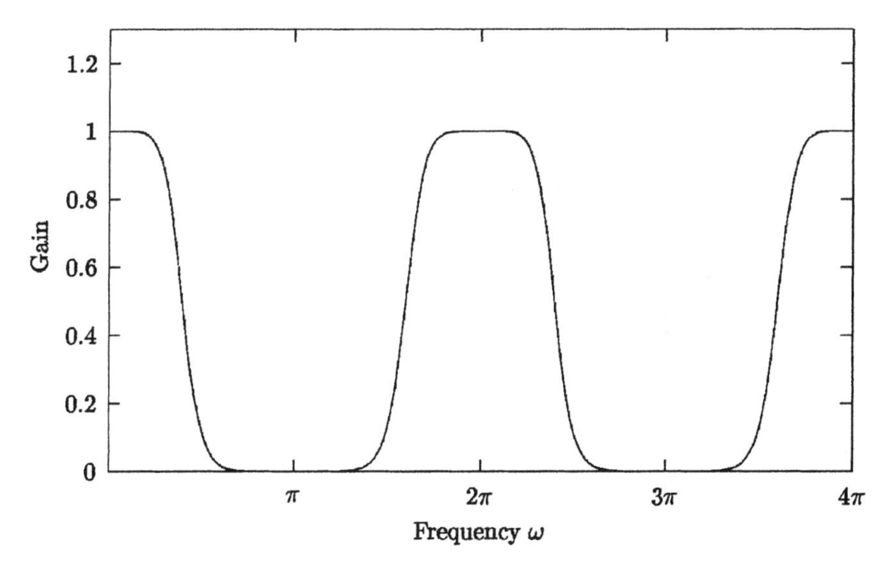

Figure 9 Demonstration of aliasing from digital filter response.

the sine function are all zero. We can artificially construct our desired frequency response so that this will be so. To do this, suppose $F(\omega)$ is the desired frequency response where $0 \leq \omega \leq \pi$:

Step 1. Augment the domain of the function over $0 \leq \omega \leq 2\pi$.

Step 2. Augment the function values from $\pi \leq \omega \leq 2\pi$ as $F(\omega) = F(2\pi - \omega)$.

Step 3. Compute the discrete cosine Transform [α_i coefficients from Eq. (5)] over the range $0 \leq \omega \leq 2\pi$.

For a discrete system with a desired FIR filter of length N:

Step 1. Augment the domain of the function over $1 \leq i \leq 2N$.

Step 2. Augment the function values from $N + 1 \leq i \leq 2N$ as $F(i) = F(2N - i)$.

Step 3. Construct the FIR filter with $2N$ points from the discrete cosine transform of the desired frequency response.

Step 4. Keep the first N coefficients and truncate the remaining coefficients.

The discrete cosine Transform of an ideal low-pass filter is a sinc function, defined as

$$\text{sinc}(x) = \begin{cases} \dfrac{\sin(\pi x)}{\pi x} & \text{if } x \neq 0 \\ 1 & \text{if } x = 0 \end{cases} \tag{26}$$

Therefore, the coefficients to a low-pass FIR filter with cutoff frequency f_c and length $2N - 1$ are determined using Eq. (27), where ω_c is the desired cutoff frequency divided by the Nyquist frequency:

$$h(i) = \omega_c \operatorname{sinc}\left[\omega_c\left(i - \left\lfloor \frac{N}{2} \right\rfloor\right)\right] \qquad 1 \le i \le N \tag{27}$$

Where $\lfloor x \rfloor$ is the greatest integer less than or equal to x and ω_c is the desired cutoff frequency divided by the Nyquist frequency.

Windowing

This FIR filter will have ripple in the passband and in the stop band. It is possible to suppress these ripples and smooth the frequency response, but the trade-off will be an increased transition width. The method for suppressing these ripples is with the application of a windowing function. There is a large class of windowing functions that allow the designer to determine how he or she wishes to trade off the transition width and how much ripple is to be tolerated. The design of an FIR filter with windowing involves the use of Eq. (26) for the determination of the FIR coefficients followed by the component-by-component product of the coefficients with the windowing values, that is, $h'(i) = h(i)\,\mathrm{win}(i)$. Below, is a list of common windows[5]:

Rectangular:

$$\mathrm{win}(i) = 1 \qquad 0 \le i \le N - 1 \tag{28}$$

Bartlett:

$$\mathrm{win}(i) = \begin{cases} \dfrac{2i}{N-1} & 0 \le i \le \dfrac{N-1}{2} \\ 2 - \dfrac{2i}{N-1} & \dfrac{N-1}{2} \le i \le N-1 \end{cases} \tag{29}$$

Hanning:

$$\mathrm{win}(i) = \tfrac{1}{2}\left[1 - \cos\left(\frac{2\pi i}{N-1}\right)\right] \qquad 0 \le i \le N - 1 \tag{30}$$

Hamming:

$$\mathrm{win}(i) = 0.54 - 0.46 \cos\left(\frac{2\pi i}{N-1}\right) \qquad 0 \le i \le N - 1 \tag{31}$$

Blackman:

$$\mathrm{win}(i) = 0.42 - 0.5 \cos\left(\frac{2\pi i}{N-1}\right) + 0.08 \cos\left(\frac{4\pi i}{N-1}\right) \qquad 0 \le i \le N - 1 \tag{32}$$

Table 2 gives a list of several common windowing functions together with their characteristics.[5]

Figure 10 demonstrates the effect of a Hanning window.

FIR High-Pass and Bandpass Design

The design of a high-pass filter is simply $1 - H(z)$. In the time domain, this is

$$h(i) = \omega_c \operatorname{sinc}\left[\omega_c\left(i - \left\lfloor \frac{N}{2} \right\rfloor\right)\right] - \omega_c \delta\left(\left\lfloor \frac{N}{2} \right\rfloor\right) \qquad 1 \le i \le N \tag{33}$$

where

Table 2 Comparison of Characteristics for Commonly Used Windowing Functions

Window Name	Minimum Stop-Band Attenuation (dB)
Rectangular	−21
Bartlett	−25
Hanning	−44
Hamming	−53
Blackman	−74

$$\delta(i) = \begin{cases} 1 & i = 0 \\ 0 & i \neq 0 \end{cases} \tag{34}$$

We construct a bandpass by filtering the data with a high pass-filter, then filtering the output with a low-pass filter. Or we can combine the two filters together into a single filter by convolving the coefficients.

3.3 Design of IIR Filters

The common strategy in designing IIR filters is as follows:

Step 1. Design a rational function in the S-domain in factored form that best approximates the desired frequency response characteristics (using Butterworth, Tchebyshev, elliptical, etc.).

Step 2. Transform the poles and zeros into the z-domain

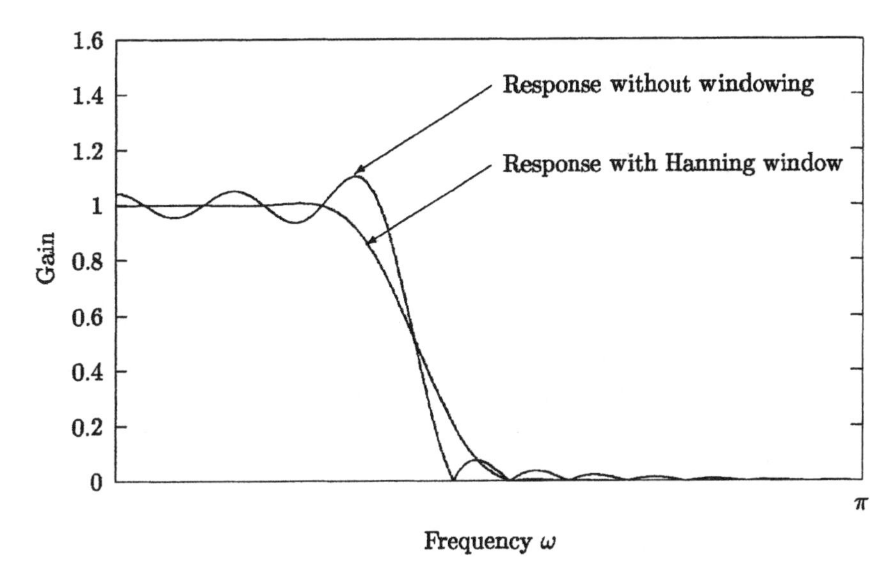

Figure 10 Comparison of FIR filter with Hanning window and with windowing.

Step 3. Reconstruct the rational function.

Step 4. Realize the difference equation by inverse z-transform of the rational polynomials.

There are different S- to z-transforms. The two most common are the impulse-invariant and the bilinear transformation. The impulse-invariant transformation is

$$z = e^{sT} \tag{35}$$

where T is the sampling period. This method is usually not used because it can cause aliasing. The bilinear transformation avoids this but distorts the mapping in other ways for which we need to compensate. The bilinear transformation is

$$z = \frac{2/T + s}{2/T - s} \tag{36}$$

The S- to z-transformation is not exact and always involves various trade-offs. Because of this, the actual placement of the cutoff in a designed digital filter is misplaced and the error increases for cutoff frequencies closer to the Nyquist frequency. To compensate for this effect, we apply Eq. (37) in the design of our filter:

$$f_c' = \frac{1}{\pi T} \tan(\pi f_c T) \tag{37}$$

This process is called "prewarping."

Example IIR Design

For a digital system with a sampling rate of 100 Hz, a third-order low-pass Tchebyshev filter with a cutoff frequency at 15 Hz and with 20% ripple in the passband is designed as follows:

Step 1. Compute the Nyquist frequency $f_N = 100\ \text{Hz}/2 = 50\ \text{Hz}$.

Step 2. Compute f' using Eq. (37) to prewarp the cutoff frequency:

$$f' = 16.22\ \text{Hz}$$

Step 3. Compute v_0 using Eqs. (13) and (14):

$$\epsilon = 0.75 \qquad v_0 = 0.3662$$

Step 4. Determine the poles and zeros of the analog system using Eq. (12):

$$p_1 = -19.0789 + j94.2364$$

$$p_2 = -38.1578$$

$$p_3 = -19.0789 - j94.2364$$

Step 5. Map the poles from the S-domain into the z-domain using the bilinear transformation equation (36) and $T = (1/\text{sampling rate}) = 0.01$ s:

$$pz_1 = 0.5407 + j0.6627$$

$$pz_2 = 0.6796$$

$$pz_3 = 0.5407 - j0.6627$$

Step 6. Map the zeros from the S-domain into the z-domain also using Eq. (36). Although we may be tempted to conclude that there are no zeros in the S-domain, since our numerator is constant, we note that $H(s) \to 0$ as $s \to \infty$. Thus mapping the bilinear transformation for this case yields

$$\lim_{s \to \infty} \frac{2/T + s}{2/T - s} = -1$$

And since this is a third-order system,

$$z_1 = -1 \qquad z_2 = -1 \qquad z_3 = -1$$

Step 7. Expand the numerator and denominator polynomials:

$$\frac{(z - \zeta_1)(z - \zeta_2)(z - \zeta_3)}{(z - pz_1)(z - pz_2)(z - pz_3)} = \frac{1 + 3z^{-1} + 3z^{-2} + z^{-3}}{-1 + 1.7611z^{-1} - 1.466z^{-2} + 0.4972z^{-3}}$$

Step 8. Normalize the filter so that the gain in the passband will be 1. In this case, we know that the gain should be 1 at $\omega = 0$. Hence, we evaluate $|H(e^{j\omega})|$ for $\omega = 0$: $|H(e^0)| = 38.4$. The normalized transfer function is then

$$H(z) = \frac{0.026 + 0.078z^{-1} + 0.078z^{-2} + 0.026z^{-3}}{-1 + 1.7611z^{-1} - 1.466z^{-2} + 0.4972z^{-3}}$$

Step 9. Realize the difference equation from inverse z-transformation of the derived transfer function:

$$y_n = 1.7611y_{n-1} - 1.466y_{n-2} + 0.4972y_{n-3} + 0.026x_n + 0.078x_{n-1} + 0.078x_{n-2} + 0.026x_{n-3}$$

3.4 Design of Various Filters from Low-Pass Prototypes

The procedure for designing a high-pass, bandpass, or band-reject IIR filter is as follows: First, design, by pole–zero placement, a low-pass filter (Butterworth, Tchebyshev, etc.) with an arbitrary cutoff frequency (though for practical considerations, it is best to choose a value midway between 0 Hz and Nyquist), transform the poles and zeros according to the following formulas, reconstitute a new transfer function from the transformed poles and zeros, then realize the digital filter by taking the inverse z-transform of the new transfer function. The formulas with an example are given below, where ω_L and ω_H are the low- and high-frequency cutoffs, respectively, normalized between 0 and π, that is, $\omega_L = 2\pi f_L/\text{sample rate}$, where f_L is the cutoff frequency in hertz; ϕ_L is the normalized cutoff frequency of the low-pass prototype.[5]

High Pass

$$z' = -\frac{1 + AZ}{Z + A} \tag{38}$$

where Z is the pole or zero to be transformed and z' is the transformed pole or zero and

$$A = -\frac{\cos[(\omega_H + \phi_L)/2]}{\cos[(\omega_H - \phi_L)/2]} \tag{39}$$

Bandpass

The bandpass filter has two transitions: a rising edge and a falling edge. For this reason, we need twice as many coefficients for the same approximate transition width as the prototype filter. Thus the order of these polynomials will be twice the order of polynomials in the prototype. Each pole from the prototype will transform into a pair of poles (z_1' and z_2'). Likewise, each zero will transform into a pair of zeros:

$$z_1' = \frac{(A + AZ) + \sqrt{(A + AZ)^2 - 4(Z + B)(BZ + 1)}}{2(Z + B)} \tag{40}$$

$$z_2' = \frac{(A + AZ) - \sqrt{(A + AZ)^2 - 4(Z + B)(BZ + 1)}}{2(Z + B)} \tag{41}$$

$$A = \frac{2CD}{D + 1} \tag{42}$$

$$B = \frac{D - 1}{D + 1} \tag{43}$$

$$C = \frac{\cos[(\omega_H + \omega_L)/2]}{\cos[(\omega_H - \omega_L)/2]} \tag{44}$$

$$D = \cot\left(\frac{\omega_H - \omega_L}{2}\right) \tan \frac{\phi_L}{2} \tag{45}$$

Band Reject

$$z_1' = \frac{(AZ - A) + \sqrt{(AZ - A)^2 - 4(Z - B)(BZ - 1)}}{2(Z - B)} \tag{46}$$

$$z_2' = \frac{(AZ - A) - \sqrt{(AZ - A)^2 - 4(Z - B)(BZ - 1)}}{2(Z - B)} \tag{47}$$

$$C = \frac{\cos[(\omega_H + \omega_L)/2]}{\cos[(\omega_H - \omega_L)/2]} \tag{48}$$

$$D = \tan\left(\frac{\omega_H - \omega_L}{2}\right) \tan \frac{\phi_L}{2} \tag{49}$$

As an example, design a Tchebyshev (type I) bandpass filter for a system sampled at 100 Hz with a low cutoff frequency of 20 Hz and a high cutoff frequency of 35 Hz.

Step 1. We can use the poles and zeros designed in Section 3.3 as the low-pass prototype.

Step 2. Convert the low and high cutoff frequencies to values normalized between 0 and π, where π corresponds to the Nyquist frequency:

$$\omega_L = 2\pi \cdot \tfrac{20}{100} = 1.2566 \qquad \omega_H = 2\pi \cdot \tfrac{35}{100} = 2.1991 \tag{50}$$

Step 3. Map the poles from the prototype using Eqs. (40) and (41):

$$p_1 \rightarrow \begin{cases} -0.6095 - j0.9105 \\ 0.2940 + j1.0722 \end{cases} \tag{51}$$

$$p_2 \rightarrow \begin{cases} -0.2302 - j1.2527 \\ -0.2302 + j1.2527 \end{cases} \tag{52}$$

$$p_3 \rightarrow \begin{cases} -0.6095 + j0.9105 \\ 0.2940 - j1.0722 \end{cases} \tag{53}$$

Then map the zeros from the prototype using Eqs. (40) and (41):

$$\zeta_1 \rightarrow \begin{cases} -1 \\ 1 \end{cases} \tag{54}$$

$$\zeta_2 \rightarrow \begin{cases} -1 \\ 1 \end{cases} \tag{55}$$

$$\zeta_3 \rightarrow \begin{cases} -1 \\ 1 \end{cases} \tag{56}$$

Step 4. Expand the numerator and denominator polynomials:

$$\frac{(z - \zeta_1) \cdots (z - \zeta_6)}{(z - pz_1) \cdots (z - pz_6)}$$

$$= \frac{1 - 3z^{-2} + 3z^{-4} - z^{-6}}{(1 + 1.091z^{-1} + 3.632z^{-2} + 2.616z^{-3} + 4.642z^{-4} + 1.982z^{-5} + 2.407z^{-6}} \tag{57}$$

Step 5. Normalize the transfer function so that it will have unity gain in the passband. For this, we estimate

$$M = \max_{0 \leq \omega \leq \pi} |H(e^{j\omega})] \tag{58}$$

Then compute

$$H_{\text{normalized}}(z) = \frac{1}{M} H(z) \tag{59}$$

For this example, $M = 14.45$.

Step 6. Realize the difference equation from inverse z-transformation of the derived transfer function:

$$\begin{aligned} y_n = {} & -1.091y_{n-1} - 3.632y_{n-2} - 2.616y_{n-3} - 4.642y_{n-4} - 1.982y_{n-5} \\ & -2.407y_{n-6} + 0.0692 * x_n - 0.2076x_{n-2} + 0.2076x_{n-4} - 0.0692x_{n-6} \end{aligned} \tag{60}$$

3.5 Frequency-Domain Filtering

It is possible to filter the data in the frequency domain. The method involves the use of the Fourier transform. We Fourier transform the data, multiply by the desired frequency response, then inverse Fourier transform the data. This is similar to the FIR filters discussed earlier. Deriving the FIR coefficients by performing a discrete cosine transform (DCT) of the desired frequency response and then convolving the coefficients with the data is equivalent to filtering the data in the frequency domain. One difference, however, is that the frequency domain filtering is generally done on blocks of data and not on streaming data, as is done in the time domain, which can be of concern when processing highly nonstationary data with abrupt transients. The inverse Fourier transform is

$$\mathcal{F}^{-1}\{F(\omega)\} = \frac{1}{\sqrt{2\pi}} \int_{-\infty}^{\infty} f(\omega)e^{-j\omega t}\, d\omega \tag{61}$$

4 STABILITY AND PHASE ANALYSIS

4.1 Stability Analysis

Consider a transfer function

$$H(s) = \frac{1}{s - p} \tag{62}$$

where the pole p is a complex number $\alpha + j\beta$. The inverse Laplace transform of this function is $e^{\alpha t}[\cos(\beta t) + j\ \sin(\beta t)]$. This function is bounded as $t \to \infty$ if and only if $\alpha < 0$. From this, we can determine the stability of a function by inspecting the real components of all poles of a given transfer function. The procedure for a rational function (a ratio of polynomials) would be to factor the polynomials in the denominator and inspect to ensure that the real components to all of the poles are less than zero. Suppose

$$\begin{aligned} H(s) &= \frac{a_0 + a_1 s + \cdots + a_m s^m}{b_0 + b_1 s + \cdots + b_n s^2} \\ &= \frac{(s - \zeta_0)(s - \zeta_1) \cdots (s - \zeta_m)}{(s - p_0)(s - p_1) \cdots (s - p_n)} \end{aligned} \tag{63}$$

In a similar way, by inspection of the S-to-z transformation $z = e^s$, we see that the entire left half of the plane in the S-domain maps inside the unit circle in the z-domain. For this reason, we analyze the stability of systems in the z-domain by inspecting the poles of the transfer function. The system is stable if the norm of all poles is less than 1.

4.2 Phase Analysis

While processing the data in real time, our filters must act on the signal history. For this reason, there will always be some delay in the output of our process. Worse, certain filters will delay some frequency components by more or less than other frequency components. This results in a phase distortion of the filter. For a certain class of FIR filters, it is possible to design filters that shift each frequency component by a time delay in proportion to the frequency. In this way, all frequency components are shifted by an equal time delay. Though it is possible to design certain non-real-time, noncausal IIR filters that are phase shift distortionless, in general, IIR filters will produce some phase shift distortion. We can determine the actual phase shift for each frequency component by computing

Table 3 Comparison of FIR and IIR Characteristics

	FIR	IIR
Run time efficiency	Less efficient; requires high-order filter.	Higher efficiency; usually possible to achieve a desired design specification in fewer computations
Stability	Always stable	Stable if all poles are inside the unit circle
Phase shift distortion	Can be designed to be phase shift distortionless	Generally distorts phase
Ease of design	Simpler design process, usually involving Fourier transforms or solving linear systems	Design is more complex, involving special functions or solving nonlinear systems

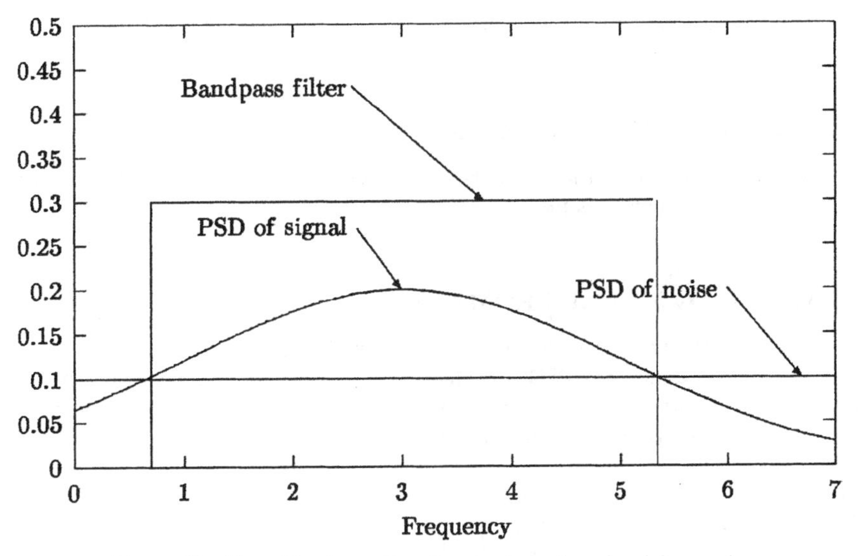

Figure 11 Use of bandpass filter for discriminating signal from noise.

$$\arg H(j\omega) = \angle H_{\mathrm{real}} + jH_{\mathrm{imag}} = \tan^{-1}\frac{H_{\mathrm{imag}}}{H_{\mathrm{real}}}$$

The arctan will produce the principal value of the phase shift, not necessarily the cumulative phase shift, since the arctan function produces principal value $-\pi \leq \tan^{-1}(\phi) \leq \pi$. It is possible to recover the accumulated phase shift by factoring the rational function into its binomial parts, then expressing this in exponential form as a summation. One can then

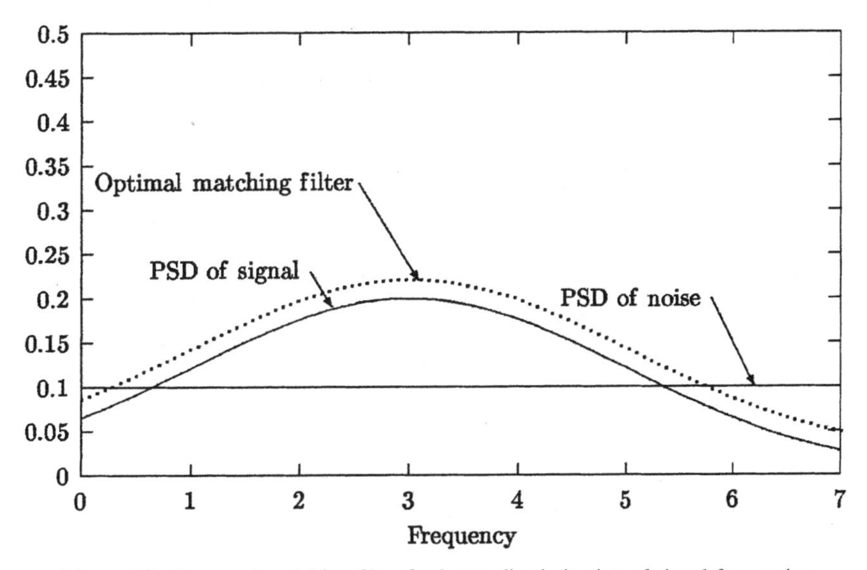

Figure 12 Improved matching filter for better discrimination of signal from noise.

determine the principal angle on each part and the accumulated phase shift by summing the parts.

4.3 Comparison of FIR and IIR Filters

There are various factors when deciding on a particular filter for a given application. Table 3 summarizes these.

5 EXTRACTING SIGNAL FROM NOISE

The PSD of white noise is uniformly distributed over all frequencies. Therefore, it is possible to detect the PSD signature of a signal corrupted by white noise by inspecting spectral components that rise above some baseline. From this we can design a matching filter to optimally extract the signal from the noise. Figures 11 and 12 illustrate this procedure.

REFERENCES

1. W. Press, B. Flannery, A. Teukolsky, and W. Vetterline, *Numerical Recipes in C,* Cambridge University Press, Cambridge, UK, 1988.
2. T. Parks and C. Burrus, *Digital Filter Design,* Wiley, New York, 1987.
3. M. Vlček and R. Unbehauen, "Analytical Solutions for Design of IIR Equiripple Filters," *IEEE Transactions on Acoustics, Speech, and Signal Processing* **37**(10) (October 1989).
4. S. Marple, *Digital Spectral Analysis with Applications,* Prentice-Hall, Englewood Cliffs, NJ, 1987.
5. A. Oppenheim and R. Schafer, *Digital Signal Processing,* Prentice-Hall, Englewood Cliffs, NJ, 1975.

CHAPTER 7

DATA ACQUISITION AND DISPLAY SYSTEMS

Philip C. Milliman
Weyerhaeuser Company
Federal Way, Washington

1 INTRODUCTION

The industry has changed significantly since this chapter was first written in the months before 1990. The personal computer has become part and parcel of everyday life. Control systems have become increasingly based on standard systems and interfaces; sensors themselves are often based on just smaller versions of the same operating system as large manufacturing systems. This has tended to change the focus from the technology of data acquisition to the software and systems to support data acquisition.

The trend has been away from requiring the engineer to understand the science of how sensors work and the lowest levels of data acquisition and more toward the engineer understanding the collection, coordination, storage, access, and manipulation of data. With that in mind, this chapter has been updated to focus more on the latter and less on the former. Other chapters in this book cover details of the electronics, transducers, sampling, and calibration.

To control any process or understand what occurs during the life cycle of a process, the system (a human or machine) must have information about what is occurring. In the simplest of control loops, the measured variable must be converted to usable units, comparison in some form to a target occurs, and a response is determined. At the plant level, improvement of plant operation relies upon understanding the relationships between processes within the plant (not only current, but historical), which in turn requires collecting data throughout the plant, characterizing the relationship of the data with other data, storing the data in such a way as to be retrievable in a useful, timely way, and manipulating the data for presentation and hopefully providing an aid to understanding the relationships between processes. In today's competitive environment, focusing on local control and ignoring the interaction between processes, both internal to the plant and external, can be disastrous. If one is not focused on improvement, one can bet the competitor is. Larger corporations, especially, can bring analytical tools to bear to improve local processes, plantwide processes, and their relationships to external influences, such as the supply chain. On the other hand, today's computing tools bring very powerful data acquisition and analysis capability within the reach of the average technician with a little bit of motivation.

Data acquisition and display systems have changed dramatically. Twenty years ago, terms referring to specialized systems such as SCADA (supervisory control and data acquisition) and data loggers were common terms. Now, with the proliferation and broadening role of computer systems and their intrusion into every aspect of manufacturing, many of the features that used to be in specialized instruments and systems are now part of the everyday tools available to anyone with a computer. This chapter attempts to cover aspects of data acquisition and manipulation that may help the engineer better understand issues and give a foundation for using and even constructing tools. The organization is as follows:

- The initial sections cover the nature of data and the acquisition and conversion of data to usable units and includes some discussion of useful display techniques. The discussion attempts to identify issues of which the engineer should be aware and give guidelines on how to manage data.
- The latter sections cover the coordination, storage, access, and manipulation of data. A discussion of pros and cons of different strategies should help the reader understand the trade-offs in system selection and construction. It is difficult to do this without describing specific technologies and brands, but the author has endeavored to level the discussion in such a way that changes in technology will not change the value of the discussion. Time will tell if the approach is effective.

2 DATA ACQUISITION

Data acquisition includes the following: (1) acquiring raw data from the process being measured and (2) converting data to usable units. Included in this section are also some topics of data display closely related to the nature of the data being acquired. Other aspects of data display will be covered in later sections.

In process industries, much of the data are analog in nature, such as pressure, temperature, and flow rate. The values acquired are sampled representations of process data that have a scale and a range, with various issues around effective range and whether values over a range are linear or more complex. When acquired in a data acquisition system there are a number of issues that must be addressed related to how data is sampled; how it is represented in the computer as a digital value but still able to be manipulated as an analog number; and how a continuously changing value can be stored without exceeding the capacity of storage or computation of the acquisition system. Discrete manufacturing still has a number of analog data sources, but a larger proportion involves discrete data, such as motor stops, starts, and pulses. These have their own issues of acquisition and storage and are often related to attributes of the process. The next section deals primarily with process data, additionally covering some discrete data and issues around data collection, representation, and storage.

As businesses begin to broaden the scope of optimization and understand their global processes, the context of the data in terms of product, plant conditions, market conditions, and other environmental aspects has increasingly added discrete data to the set of data to be obtained. The interaction of the process with factors such as which crew is managing the process, which customer's needs are highest priority, legal controls such as environmental limits impacting allowable process rates, operator decisions, which product is being manufactured, grade achieved, and a large number of other factors become important when a company is competing with other companies that have already achieved excellent local control of processes. These data involve less understanding how to deal with continuous data and more with the coordination of data within and between processes. These can be termed manufacturing attributes to emphasize their importance in providing an environment around process data. Later sections deal with manufacturing attribute data and issues around their collection and coordination with process data.

3 PROCESS DATA ACQUISITION

Most modern data acquisition is via digital systems that may have a lower level analog collection mechanism but is now so removed from the engineer that the engineer is only concerned with the digital portion of the system. The ability to use digital microprocessors as building blocks for data collection, the prevalence of computer tools, and the creation of widely available operating systems that operate on small footprints have virtually eliminated the need for analog instrumentation. While the data may be analog in nature, the technology has been developed to such a degree that the engineer decreasingly needs to pay attention to the analog aspects of the data.

A digital-to-analog (D/A) and analog-to-digital (A/D) converter performs the actual processing required to bring analog information from or to the process. While the resulting signal may be digital, it is a representation of a continuous number that has characteristics that, if not understood, can result in erroneous conclusions from data, including missing data, misinterpreting trends, or improperly weighting certain values.

The engineer should be aware of several features of analog data to ensure that the data are used properly. An understanding of sampling interval, scaling, and linearization will facilitate the use of data once collected.

3.1 Sampling Interval

One of the important steps with any data collection process includes the proper choice of sampling interval. As an example of the impact of selection of sampling interval, or frequency, a sine wave with a period of 1-s (Fig. 1*a*) is measured with several sampling intervals. 0.5-s (Fig. 1*b*) and 0.1-s (Fig. 1*c*) intervals both provide different impressions of what is actually happening. The 1-s sampling rate being in phase with the sine wave yields the impression that we are measuring a nonvarying level. The 0.5-s sampling rate yields several

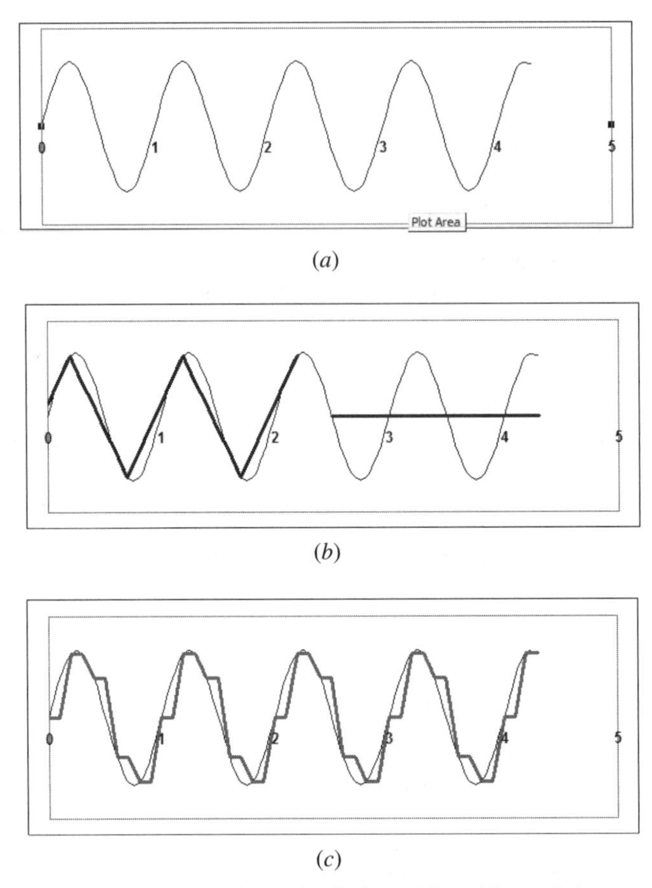

Figure 1 Sampling: (*a*) sine wave form; (*b*) aliasing of data; (*c*) sampled every tenth second.

different results depending on what phase shift is encountered. This is known as aliasing (see Ref. 1, pp. 122–125). If a 0.1-s period is used, we finally begin to obtain a realistic idea of what the waveform truly looks like.

The sampling interval has a different impact when collecting manufacturing attribute data. With manufacturing attribute data, every change in value has importance. They provide a context to the process that assists with the tieback to business interactions. The values are often coded. The sampling must be close enough to the time of occurrence to allow determination of state in relation to other events. The sampling interval must be fast enough to capture any change in state. Sampling at slower than the change rate of the data will mean lost events—potentially critical in a situation where the count of items processed is important. Sampling at a rate slightly faster than the maximum change rate of the manufacturing attribute data assures that no change will be missed. Another important consideration is to know at what time an event occurred. For instance, if a value changes only infrequently but other related manufacturing attribute values are changing at a faster rate, then the scan rate has to be fast enough to match the fastest change rate of all the related manufacturing attribute data. This is sometimes called the master scan rate, meaning that the frequency of scanning must be fast enough to capture faster events and determine the state of other variables relative to those events. Similarly, if there are related analog data, the scan rate may have to be fast enough to even characterize the curve of the analog data (remember Fig. 1).

The capacity of the target system must also be taken into account. Storing large volumes of data is becoming more feasible as systems increase in speed and power, but the retrieval and organization of those data may become a time-consuming, overly complex task with too much data or poorly organized data. Consequently, even though storage itself is less of an issue, other factors impact how much data are retained and how organized for later retrieval. Later sections examine approaches for organizing and retrieving data.

3.2 Accuracy and Precision of Data

Accuracy and precision are dependent on the sampling interval as well as the resolution of the system (Ref. 2; Ref. 3, pp. 78–80; and Chapter 1 in this volume). When dealing with the A/D conversion process, the step size or number of bits used is critical when determining the system precision and accuracy (Ref. 1, pp. 78–81). Figure 2 illustrates the difference between accuracy and precision of data. Table 1 illustrates the effect the number of bits has on the precision.

This also interacts with range, which will be discussed later, since having an accurate number over a small percent of the desired range would not allow the ability to fully characterize the process. For example, highly accurate readings with 1% moisture resolution over a range from 10 to 20% moisture content would be inadequate if one were attempting to measure moisture over a 5–40% range. When selecting transducers, it is necessary that they have both the accuracy and range needed for the process being observed. When selecting converters, one should be aware of the settling time (governs how often readings can be obtained), resolution of the converter (affects range and detail of measurements), and accuracy of sensor.

Chapter 1 includes some characteristics of transducers, including calibration, and the sampling of data.

One should be aware that an event that has been stabilized in a data collection system may be offset in time, resulting in a potential discrepancy between events or values when values are compared from different sources or from multiplexed data. It should be verified

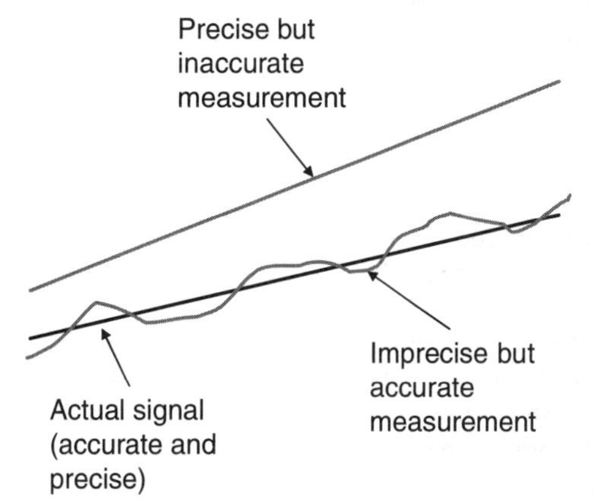

Figure 2 Difference between accuracy and precision.

Table 1 Relationship between Number of Bits and Precision

Number of Bits	Steps	Resolution on 5-V Measurement	Percent Resolution
8	256	0.01950	0.3900
10	1024	0.00488	0.0980
12	4096	0.00122	0.0240
16	65,536	0.00008	0.0015

that the transducer is collecting the data fast enough to allow one to have relevant times of collection in the data acquisition system. Also one should ensure that the potential relationship between events from different sources and their intended use is understood when considering the speed and accuracy of transducers.

3.3 Time-Based versus Event-Driven Collection

There are two major approaches when collecting data with a general-purpose data acquisition system. In one approach data are collected on a regular frequency based on time, such as once per second. This is easy to institute and it is relatively easy to analyze the data and their relationships after the fact. This approach tends to require more data storage and can make it difficult to identify events or the interactions with manufacturing attribute data. The other approach is event-based acquisition. An event is identified, such as when a package is dropped onto a platform, the time of that event is recorded, and the values of related variables are collected for that time. The sampling rate of the transducers to acquire the other variables may be important, as their values may become irrelevant if too long a time interval has passed after the event has occurred when the related variables are sampled. Batch processes, such as mixing a tankful of chemicals, often have some data collected only at the start and end of the process. Other data may be recorded at fixed time intervals during the batch process. Depending on the needs, the data during the actual reaction may be of great or of little use. The time between sampled events may be several minutes, hours, or even days in length, but the time of the event may be critical, as well as detailed data at the time of the event, resulting in a common tactic of using high-speed scanning to detect the occurrence of an infrequent event. Approaches for combining and analyzing data will be covered in a later section.

4 DATA CONDITIONING

Often the data obtained from a process are not in the form or units desired. This section describes several methods of transforming data to produce proper units, reduce storage quantity, and reduce noise.

There are many reasons why process measurements might need to be transformed in order to be useful. Usually the signals obtained will be values whose units (e.g., voltage, current) are other than the desired units (e.g., temperature, pressure). For example, the measurement from a pressure transducer may be in the range 4–20 mA. To use this as a pressure measurement in pounds per square inch (PSI), one would need to convert it using some equation. As environmental conditions change, the performance characteristics of many sensors change. A parametric model (equation) can be used to convert between types of units or to correct for changes in the parameters of the model. The parameters for this equation may be derived through a process known as calibration (Chapter 1 covers much of the process of calibration and sampling). This involves determining the parameters of some equation by placing the sensor in known environmental conditions (such as freezing or boiling water) and recording the voltage or other measurable quantity it produces. Some (normally simple) calculations will then produce the parameters desired. (See the following discussion of simple linear fit for the procedure for a simple, two-parameter equation.) The complexity of the model increases when the measured value is not directly proportional to the desired units (nonlinear). Additionally, as sensors get dirty or age, the parameters might need adjusting. There are a variety of control techniques to assist with compensating for changes in the environment around the sensor, including adaptive control.

4.1 Simple Linear Fit

The simplest formula for converting a measured value to the desired units is a simple linear equation. The form of the equation is $y = ax + b$, where x is the measured value, y represents the value in the units desired, and a and b are parameters to adjust the slope and offset, respectively. The procedure for finding a and b is as follows:

1. Create a known state for the sensor in the low range. An example would be to put a temperature sensor in ice water.
2. Determine the value obtained from the sensor.
3. Create a known state for the sensor in the high range. An example would be to immerse the sensor in boiling water.
4. Determine the value obtained from the sensor
5. Calculate the values of a and b from these values using the equations

$$a = \frac{\text{actual high value} - \text{actual low value}}{\text{measured high value} - \text{measured low value}} \tag{1}$$

$$b = \text{actual low value} - (a \times \text{measured low value}) \tag{2}$$

Figure 3 demonstrates example relationships between measured values and engineering units.

4.2 Nonlinear Relationships

Often, there is not a simple linear relationship between the engineering units and the measured units (Fig. 4). Instead, for a constantly rising pressure or temperature, the measured value would form some curve. If possible, we use a portion of the sensor's range where it is linear, and we can use Eqs. (1) and (2). When this is not possible, we have to characterize the sensor by a different equation, which could be a polynomial, a transcendental, or a combination of a series of functions.

One can imagine several sensors which are linear in different ranges to be used in conjunction to create a larger range of operational data. This variety of formulas should make one point clear: Without an understanding of the basic model of the sensor, one cannot know what type of conversion to use. Many sensors have known differences in output depending on the range of sensed data. Be aware of the effect environmental conditions have on the sensor readings. If the characteristics of a sensor are unknown, then the sensor must be measured under a variety of conditions to determine the basic relationship between the

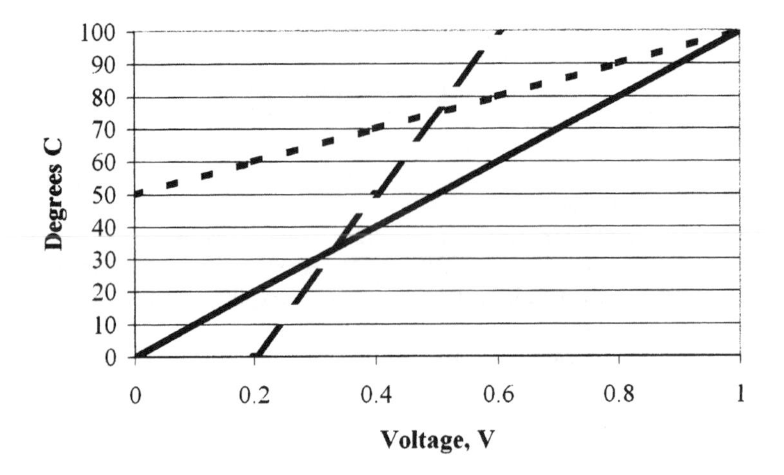

Legend						
Line	Voltage for 0°C	Voltage for 100°C	Equation for *a*	Equation for *b*	Value for *b*	Equation for °C
- - - -	0	1	$\frac{100-0}{1-0}$	0 – 100	0	C=100V
— — —	–1	1	$\frac{100-0}{1-(-1)}$	0 – 50(–1)	50	C=50V + 50
———	.2	.6	$\frac{100-0}{.6-.2}$	0 – 250(.2)	-50	C=250V – 50

Figure 3 Relationship between measured values and engineering units.

measured values and engineering units. Some knowledge of the theory of the sensor's mechanism will help to give an idea of which model to use. The development and evaluation of a model is beyond the scope of this chapter, but other chapters in this volume provide assistance.

4.3 Filtering

Even after data are converted to the appropriate units, the data may have characteristics that inhibit understanding the important relationships for which one is looking. For instance, the data may have occasional fluctuations caused by factors other than the process or the process may have short-term perturbations, which are not really an indication of the major process factors.

Filtering is a technique that allows one to retain the essence of the data while minimizing the effects of fluctuations. The data may then appear to be "smoothed." In fact, the terms "filtering" and "smoothing" are often interchanged. Filtering can occur when the data are still in an analog state (Ref. 1, p. 54) or can occur after the data are converted to digital data (digital filtering). Measurement variability comes from a variety of sources. The process itself may undergo fluctuations that result in variation in measurement but that are only temporary and should be ignored. For instance, if the level of an open tank of water were to be measured but waves cause fluctuations in the height of a float, then the exact value at any given time would not be an accurate reflection of the level of the tank. The sensor itself may have fluctuations due to variability in its method for acquiring data. For instance, the

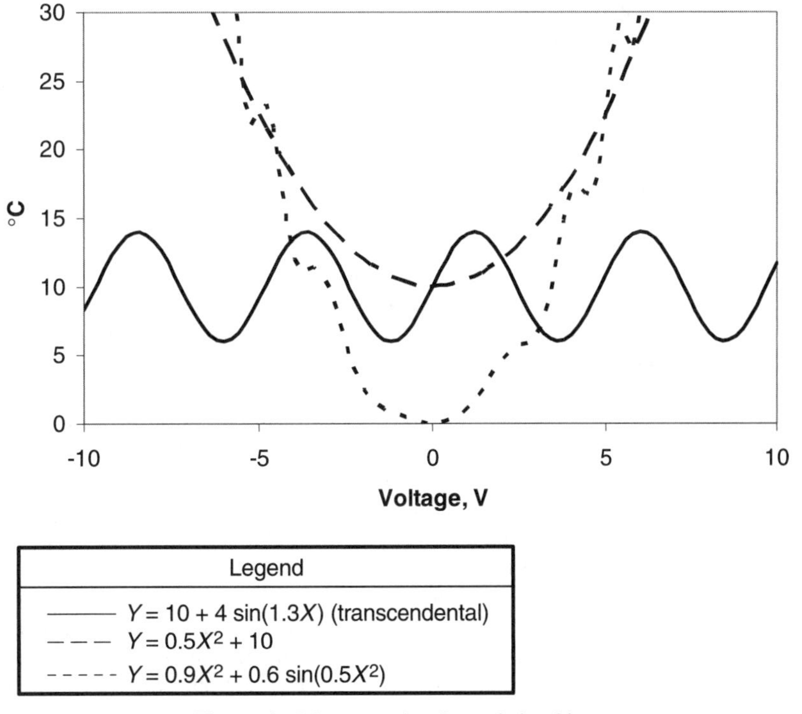

Figure 4 More complex data relationships.

proximity of a 60-Hz line may induce a 60-Hz sinusoidal variation in the signal (measured value). Examples of filtering approaches are as follows and are also given in other chapters in this volume (Ref. 12, p. 538):

(a) Repeated sample average: Take a number N of samples at once and average them:

$$\frac{1}{N}\sum \text{Value}(i)$$

(b) Finite-length average (moving): Take the average of the last N measurements, averaging them to obtain a current calculated value.

(c) Digital filters:

$$y = (1 - \alpha)y_{i-1} + \alpha x_{i-1}$$

The simple average is useful when repeated samples are taken at approximately the same point in time. The more samples, the more random noise is removed. Chapter 1 addresses some of the issues with sampling and the concept of population distribution. The formula for an average is shown in (a) above.

However, if the noise appeared for all the samples (as when all the samples are taken at just the time that a wave ripples through a tank), then this average would still have the noise value. A moving average can be taken over time [see (b) above] with the same formula, but each value would be from the same sensor, only displaced in time. A disadvantage of this approach is that one has to keep a list of previous values at least as long as the time span one wishes to average.

A simpler approach is the first-order digital filter, where a portion of the current sample is combined with a portion of previous samples. This composite value, since it contains more than one measured value, will tend to discard transitory information and retain information that has existed over more than one scan. The formula for a first-order digital filter is described in (c) above, where α is a factor selected by the user. The more one wants the data filtered, the smaller the choice of α; the less one wants filtered, the larger the choice of α. Alpha must be between 0 and 1 inclusive.

The moving average or digital filter can tend to make the appearance of important events to be later than the event occurred in the real world. This can be mitigated somewhat for moving averages by including data centered on the point of time of interest in the moving-average calculation. These filters can be cascaded, that is, the output of a filter can be used as the input to another filter.

A danger with any filter is that valuable information might be lost. This is related to the concept of compression, which is covered in the next section. When data are not continuous, with peaks or exceptions being important elements to record, simple filters such as moving-average or digital filters are not adequate. Some laboratory instruments such as a gas chromatograph may have profiles that correspond to certain types of data (a peak may correspond to the existence of an element). The data acquisition system can be trained to look for these profiles through a preexisting set of instructions or the human operator could indicate which profiles correspond to an element and the data acquisition system would build a set of rules. An example is to record the average of a sample of a set of data but also record the minimum and maximum. In situations where moisture or other physical attributes are measured, this is a common practice. Voice recognition systems often operate on a similar set of procedures. The operator speaks some words on request into the computer and it builds an internal profile of how the operator speaks to use later on new words. One area where pattern matching is used is in error-correcting serial data transmission. When serial data are being transmitted, a common practice to reduce errors is to insert known patterns into the data stream before and after the data. What if there is noise on the line? One can then look for a start-of-message pattern and an end-of-message pattern. Any data coming over the line that are not bracketed by these characters would be ignored or flagged as extraneous transmission.

4.4 Compression Techniques

For high-speed or long-duration data collection sessions there may be massive amounts of data collected. It is a difficult decision to determine how much detail to retain. The trade-offs are not just in space but also in the time required to store and later retrieve the data. Sampling techniques, also covered in other chapters, provide a way of retaining much of the important features of the data while eliminating the less important noise or redundant data. As an example, 1000 points of data collected each second for 1 year in a database could easily approach 0.5 terabyte when index files and other overhead are taken into account. Even if one has a large disk farm, the time required to get to a specific data element can be prohibitive. Often, systems are indexed by time of collection of the data point. This speeds up retrieval when a specific time frame is desired but is notoriously slow when relationships between data are explored or when events are searched for based on value and not on time. Approaches to reducing data volume include the following:

- Reduce the volume of data stored by various compression tools, such as discovering repeating data, storing one copy, and then indicating how many times the data are repeated. There are many techniques for compressing data, covered elsewhere. Zip files are a common instance of compression to make the data consume less storage and to take less time in transmission from one computer to another. Compression tends to increase the storage and retrieval time slightly. Increasingly, file systems associated with common operating systems include compression as a standard option or feature of mass storage. These systems are quite good at compressing repeated data but are less effective when data vary but have a mathematical relationship, such as a straight line between two points.

- Normalize the data. When the developer knows relationships between data, redundancy can be avoiding by normalizing the data—following some basic principles to organize the data in such a way that redundancy is avoided.[5] C. J. Date describes the levels of normalization of data in a relational database.[5] For instance, if a person has several addresses, then one could store the person's name once, store each address, and store the links from the person to the address. While very similar to compression, it relies on the developer identifying and taking advantage of the relationships between data to eliminate redundancy and reduce space. This creates significant effort in planning for acquisition and storage of data. It pays off in reduced storage and significantly improved retrieval and analysis times.
- Eliminate nonessential data. If one is not interested in the shape of a sinusoidal signal, for instance, but only interested in how many cycles occurred during a given time frame, then sampling techniques can be used to characterize the data without having to store significant data.

The engineer or researcher has to make assumptions about how the data will be used and factor those into the acquisition and storage system. A project attempting to discover the relationships between waveforms would require high-frequency sampling and probably time-based storage. A project attempting to record the number of times a boiler went over a certain temperature level might have a high-speed scanning capability but only store those values that were above the temperature limit. An inventory tracking system may have triggers that cause scanning only when some event occurs.

Often, a batch or pallet of product may contain a large number of items. The items can be sampled for some process attribute. The customer may want to know summary statistics about the pallet, but the storage of all the data may not be feasible. In this case, statistical results can normally be derived from summary data. Average, standard deviation, total, correlation, maximum, and minimum are easily calculated from summary, accumulated data*:

Averages. Keep a running sum of data and count of readings:

$$\text{Average} = \text{Sum Of Values/Count Of Values}$$

Totals. Keep a running sum of data.

Standard Deviation

$$\sqrt{\frac{n \sum_{i=1}^{n} x_i^2 - (\sum_{i=1}^{n} x_i)^2}{n(n-1)}}$$

*Standard deviation and correlation from Ref. 6, pp. 473, 477.

Correlation

$$r = \frac{n \sum x_i y_i - (\sum x_i)(\sum y_i)}{\sqrt{[n\sum x_i^2 - (\sum x_i)^2][n\sum y_i^2 - (\sum y_i)^2]}} \quad (3)$$

Range. Save largest and smallest values.

Median. Find the middle value of a distribution, which requires keeping all values.

The median (the true center of the data) requires the raw data to be calculated. A compromise for depicting the distribution of data without having to store the full details is to store a distribution of the data. For instance, the range of possible important data can be broken into a series of totals, reflecting the count of items that fit into the particular total. A histogram representing the distribution of data can be created from the totals without requiring the full set of original data. In addition, the median can be approximated using this technique. The distribution can also be used to supply data for statistics based on distribution of data, such as the Taguchi loss function (Ref. 7, pp. 397–400).

4.5 More on Sampling and Compression

Rather than just sample the data, why not save all the changed values of the data, discarding values which are the same or within some limits of the previous reading? This really applies best to continuous processes. Quite significant space reduction can be maintained in processes that are slowly changing and have only occasional large upsets. Variations of this technique can provide additional improvements. For instance, rather than just checking to see if the current value is the same as or within some limits from the previous reading, see if it is on the same line or curve as the previous value. This can result in a great reduction of storage requirements at the loss of a slight amount of accuracy in reconstruction. The more flexible the compression technique, the more work must be done to reconstruct the data later for examination. For instance, if the user of the data acquisition system wants to retrieve a data point within data that has been reduced to a line segment, the user or the system must determine which line segment is wanted using the time stamp for the beginning and ending of the line segment interval and then recalculate the point from the equation. This is referred to as a boxcar algorithm. Values that are close to the line segment can be treated as on the line segment if one can afford to lose some accuracy.[8] The formula for a boxcar has to take into account the length of the interval (maximum), the height of the box (how much noise is allowed), and how peak or exception values are treated. For instance, Table 2 presents a set of data with several types of compression applied for data sampled at a constant interval.

In Table 2, the simple repeating-value compression will not lose data but will result in little or no compression if the data are changing value frequently, including having any noise. The boxcar compression technique results in much higher compression for slowly changing data with only the loss of fine detail data (depending on the height of the window). For data that are nonlinear or changing frequently, the boxcar compression method results in little compression. Process information systems often use the boxcar compression method. If data are slow moving with occasional bursts of activity, the boxcar and repeating-value methods can result in dramatic reductions in space required. If data changes tend to be linear, then the boxcar algorithm tends to be superior to the repeated-value approach. For an extreme example, see Table 3.

The raw data would have resulted in 631 data points being stored. The boxcar method would result in 5 data points being stored, less than 1% of the storage required. In the

Table 2 Examples of Compression Techniques

Time Stamp and Raw Value		Simple Repeating Value			Boxcar Compression		
Time	Value	Time	Value	Count	Start Time	Start Value	Slope
12:00	1	12:00	1.0	3	12:00	1	0
12:01	1	12:03	2.0	1	12:03	2	1
12:02	1	12:04	3.0	1	12:08	7	−2
12:03	2	12:05	4.0	1	12:11	1	4
12:04	3	12:06	5.0	1	12:12	5	0
12:05	4	12:07	6.0	1	12:16	5	−4
12:06	5	12:08	7.0	1			
12:07	6	12:09	5.0	1			
12:08	7	12:10	3.0	1			
12:09	5	12:11	1.0	1			
12:10	3	12:12	5.0	5			
12:11	1	12:17	1.0	1			
12:12	5						
12:13	5						
12:14	5						
12:15	5						
12:16	5						
12:17	1						

example above, the repeated-value method would have resulted in almost exactly the same compression as the boxcar. However, if there had been a 0.1% slope in data throughout the period, the boxcar would remain the same but the repeated-value method would have resulted in no compression. A deadband (the height of the boxcar) around the repeated value (meaning two values within some small deviation from each other would be counted as the same value) would result in very high compression in the above example. A long ramp-up of the value during that time frame would have further differentiated the two compression methods.

Table 3 Data Compression: Raw Data versus Boxcar

Raw Data[a]		Boxcar Method		
Start Time	Start Value	Start Time	Start Value	Slope
12:00	1	12:00	1	0
12:01	1	15:00	1	1
...		15:01	2	3
14:59	1	15:02	5	−4
15:00	1	15:03	1	0
15:01	2			
15:02	5			
15:03	1			
15:04	1			
...				
18:59	1			
19:00	1			

[a] Gaps represent no change in data.

There are many techniques for compression of data. As mentioned above, many rely on assumptions about the underlying nature of the data, such as being continuous data. Where data are directly related to events, more traditional compression techniques which look for repeating patterns in the data may be used. These are typically performed by operating systems and database systems and can therefore be taken advantage of with little or no work on the part of the engineer.

5 DATA STORAGE

In whatever ways data are sampled, collected, filtered, smoothed, and/or compressed, at some point the data must be stored on some media if long-term data are to be analyzed (covered later in this chapter). There are several approaches to data storage that will be discussed in brief here.

5.1 In-Memory Storage

There are normally limitations on how much data can be stored, particularly when low-frequency events have high-frequency data surrounding them that are of interest. For example, if scientists are monitoring Mt. St. Helens for seismic data, it would be prohibitive to capture millisecond data for years while waiting for an eruption. It would be of interest to capture data at high density just before, during, and after each eruption but not in the quiet times in the intervening years. Collecting the millisecond data on many sensors would overflow the storage capability of most systems. There are techniques to store subsets of the data that allow high-density data from constrained time intervals to be stored.

An approach for collecting and later reporting high-density data around an event of interest is to collect the data continuously using the triggered snapshot method. High-speed data are temporarily retained for a fixed time interval or memory capacity, with the start and end time of the data moving forward with time. Older data are discarded as the time range moves past it. This moving window is useful for creating trends and summary data for that interval. The user can be shown dynamic displays that update over time and reflect characteristics of the moving window of the process. Periodically, a set of the data can be extracted to mass storage, especially triggered by some event of interest. An event is recognized by some means (automatic or user generated) that the engineer has preconfigured to cause the transfer of the current instance of the moving window to permanent storage.

The relationship between the trigger and the moving window can be configured several ways, as depicted in Fig. 5. The handling of the data involves moving values through a data array, adding more recent values at the end and pushing the rest toward the beginning—a queue.

The triggered snapshot is particularly useful when knowledge about the sequence of events just before the event of interest can help discover problems. As an example in manufacturing, in sawmills there are often very high speed sequences of events, such as where a board may come out of one conveyor and is transferred to another conveyor and some event such as the board leaving the conveyor occurs. High-speed video can be always in progress, and the detection of the board leaving the system can be used to trigger the transfer to permanent storage of the video. Events of concern can be safety issues and the triggered snapshot method can be used to help eliminate potential life-threatening situations. The triggered snapshot method is particularly useful for discovering the causes of unusual events.

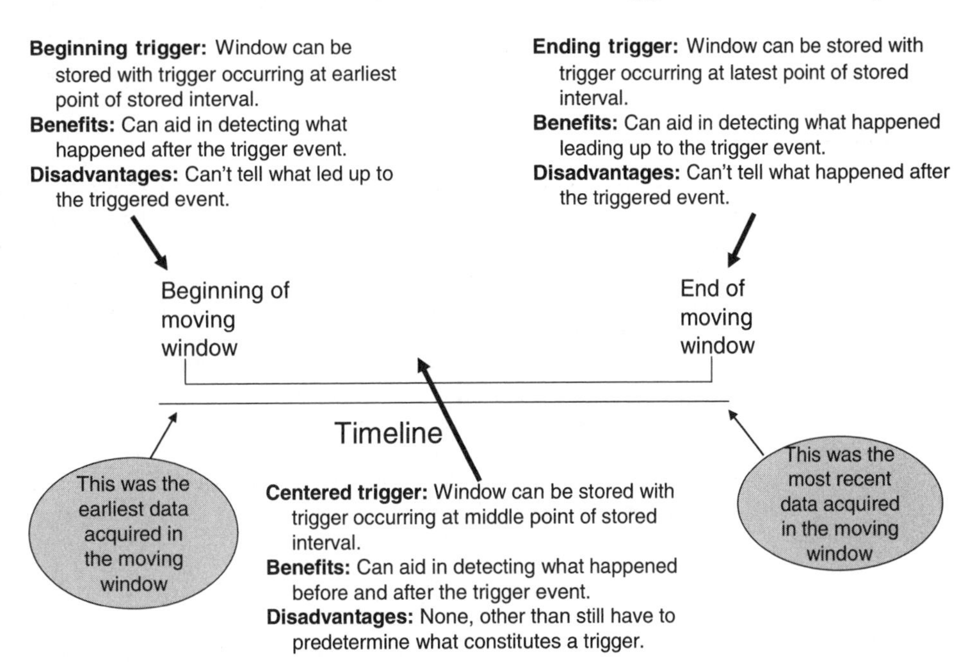

Figure 5 Relationship between trigger and moving window.

This has the advantage of allowing monitoring and analysis of high-speed events and still capturing some data to enable determining some data relationships. This is most useful if the engineer has some idea of what events may yield valuable relationships. It is much less useful when events, triggers, or relationships are unknown or unexpected. Sampling data at slower intervals may serve to allow accidental capture and identification of useful relationships, but the work required to find those relationships is much higher and of questionable probability of success.

As an example, in the sawmill many variables are changing state at high speed. For diagnostic purposes, it is valuable to see a high-speed snapshot of states of photo eyes compared to saw drops, gate openings, and grade decisions. However, the volume of data is normally too great for storage and later analysis. There are some events that are of more importance than others, such as when a gate is opening early or a saw is failing to drop consistently. These can often be recognized and the data captured in the window of time before and after the event can be stored, allowing later analysis of what led to the event and what happened shortly afterward. Some characteristic data can be summarized for each time window, stored, and used later for analysis, such as the number of photo eye changes, number of saw drops, and number of gate openings. More complex relationships can be tallied to aid in diagnostics, such as number of gate openings for grade 2. The more complex the relationship, the more difficult the programming task to ensure capturing the incidence to storage. A typical pattern is to collect process variables that may be of interest, often from a programmable logic controller. As a given problem begins to be identified, additional logic can be added to examine relationships between process inputs and sequences of events, creating a new variable that reflects some attribute of that relationship. The data collection system can store the results of that variable, such as the sum of the number of times it occurred in some larger time interval, allowing it to be low enough in volume to be mass stored for later analysis.

5.2 File Storage

An easy way to store data is in a file, often a comma-delimited file. This is easy to program and can easily be imported into analysis tools such as spreadsheets. It is not well suited to the compression techniques mentioned above because of the complexity of storing and interpreting data. However, for storing records of multiple variables collected at a time interval this can be a very useful technique. An example is shown in Table 4.

In the above example, the filename was chosen so that it would be unique. The date and time were concatenated together, eliminating every invalid character with an underscore. This helps to prevent files from being overwritten accidentally and facilitates store-and-forward techniques described below. Files can be sorted by date or name. Often, they are stored in a directory structure so the number of files in any one directory does not get too great. This speeds up file search in a given directory but can make programs more complex that search for files across directories. An example directory structure is the following:

```
C:
 DataDirectory
  2003
   11—files created during November 2003 are stored here
   12
  2004
   01
   02
   03
```

Archiving files to backup media is easy in this file organization because one only needs to reference the particular directory for the time frame desired. Files can fill up mass media and so either a manual process to check for file limits, back up old files, and delete them must be instituted or a program to provide the same functions would need to be created.

A major deficiency with a file-based system is that when the time range of a search is larger than one file, the analysis can become very difficult. One may be searching for events,

Table 4 Example of Storing Multiple Variables in Files

Filename	ProcessData20040312142305
Data in file:	Timestamp,TemperatureA,TemperatureB,TemperatureC,Rate1,Rate2,Rate3
	2004-03-20 12:34,15,14,15,12,13,13
	2004-03-20 12:35,15,14,14,12,13,13
	2004-03-20 12:36,16,14,13,12,13,13
	2004-03-20 12:37,15,14,12,12,13,13
	2004-03-20 12:38,15,14,11,12,13,13
	2004-03-20 12:39,14,14,11,12,13,13
	2004-03-20 12:40,15,14,10,12,13,13
	2004-03-20 12:41,15,14,10,12,13,13
	2004-03-20 12:42,16,14,10,12,13,13
	2004-03-20 12:43,15,14,09,12,13,13

specific time frames that go across file boundaries, or relationships between variables that may not be effectively evaluated within the time frame of one file. The analysis task usually consists of importing a number of files into some analysis tool and then using the analysis tool to look for relationships. This means that the importation process has to include the organizing of data and identifying relationships between events, a difficult task at best. A common tactic is to import data via a script or macro for a given time range, so the user only has to specify a beginning and ending time.

5.3 Database Storage

Database technology has been improving for many years, resulting in database management systems being increasingly the storage tool of choice for data acquisition systems. Database management systems provide organization tools, compression of data, access aids in the form of indexes, and easy access for analysis tools. A special benefit of database management systems is that they allow the combination of discrete data and time-based data collected on different time intervals. Relational databases are now the dominant database management system type. Data are organized in tables. Each table is composed of a set of rows, each row having a fixed set of columns. Indexes are provided to speed access to data. In data acquisition systems, the designer often adds a time stamp column to each row to facilitate retrieval and analysis of data. An example of a simple database is given in Table 5.

The TimeData table contains data that are sampled every second, while the BatchEvent table contains a record for each batch that has occurred. The SQL language is a common language used to examine and extract data in the tables. An example of its power is that a query can be constructed to use the BeginBatchTime and EndBatchTime to extract data from the TimeData table and combine it with related batch events in the BatchEvent table. A sample query to combine event- and time-based data is as follows:

```
Select BatchEvent.Timestamp, Batchevent.BatchNumber,
Batchevent.MixPercent,
Min(TimeData.TemperatureA),Max(TimeData.TemperatureA)
From BatchEvent, TimeData
Where BatchEvent.OutputProductType='BENCH' and TimeData.Timestamp
between BatchEvent.BeginBatchTime and BatchEvent.EndBatchTime
Group by BatchEvent.BatchNumber
Order by BatchEvent.BatchNumber
```

The above query searches for batches that created a certain output product type (BENCH) and reports the maximum and minimum temperatures from those batches. This

Table 5 Example Database Structure

Time Data Table		Batch Event Table	
Timestamp	datetime	Timestamp	datetime
TemperatureA	float	BatchNumber	integer
TemperatureB	float	MixPercent	float
TemperatureC	float	InputMaterialAQty	integer
Rate1	integer	InputMaterialBQty	integer
Rate2	integer	OutputProductType	varchar(50)
Rate3	integer	OutputProductQty	integer
		BeginBatchTime	datetime
		EndBatchTime	datetime

can facilitate research, for example, on what conditions lead to the best yield of a particular product.

The ease of performing this operation is a particular advantage of relational databases. There are some disadvantages, including overhead due to the access methods, extra space requirements due to the creation of indexes, and costs and complexity associated with the database management system. Indexes may add as much or more than 100% to the size of a database. Old data must be managed and removed as with any other storage system. This typically is via an automated program, since the structure is not as simple as just looking for the file creation date of a file.

5.4 Using Third-Party Data Acquisition Systems

When data storage is fairly simple, it is not hard to store data in the above-mentioned methods, but when one is using sophisticated methods of data compression, it is recommended that systems that have robust implementations of those methods be used rather than attempting to reinvent the wheel. They can be quite complex to implement reliably. Transfer of data to those systems can occur through a variety of methods, including creation of files that are captured by the other systems, insertion of data into standard interfaces such as OPC or message buses, or insertion into database tables which are monitored by the other systems (often ODBC links). Third-party systems often have software development kit (SDK) interfaces that allow the engineer with some programming skills to store data directly into the system.

Process historians are optimized for storing time-based data. A technique used to provide some relational capability to the data is the following:

- Store related data at exactly the same time stamp (time stored in the database for when the data elements were collected).
- Treat data stored with the same time stamp as being part of the same record.
- Select a set of these "records" based on a time range.
- Search a variable for some attribute value (e.g., having some value or range of values).
- Provide to the display system the values of some related variable in the same record having the same time stamp as the desired attribute variables.

This is functionally the same as performing a relational database query on a set of records in a table, with criteria based on values in some columns.

6 DATA DISPLAY AND REPORTING

There are a variety of ways to reference and display data acquired from a sensor and stored in suitable media. The current value can be inspected, values can be stored for inspection later, values can be trended, alarm conditions can be detected and reported, or some output back to the process can be performed.

6.1 Current-Value Inspection

Often, one wants to see the data as they are being collected. This can be of critical importance in experiments which are hard or costly to repeat, allowing the researcher to react to situations as they occur. As it is collected, each data item will be called the current value for that sensor. Current data are usually stored in high-speed storage (the computer main memory). As new values are obtained, they replace the value from the last reading. The collection rate can vary widely (Table 6).

For instance, detecting the profile at 10-mm intervals for a log moving at 100 m/min requires values to be obtained for each sensor 167 times per second. In continuous processes, data may only need to be acquired once per minute, as in monitoring the level of a large vat. It is useful to remember that human reaction time is in terms of tenths of a second, so displaying data at a faster rate would only be useful if it were easier to program the data. Do not waste time and energy attempting to record data at a high frequency if the only reason is for display to an operator, even if the operator must immediately react to an alert. Human–machine interfaces often show data changes at the time the new value arrives from a sensor. They have display elements that are tied to sensing points. Process historians (data acquisition systems architected for the long-term storage of process data) provide tools to extract data and present the data to the analyst. Their display systems normally provide update tools that automatically refresh the user's display at some display refresh rate (often in terms of seconds, such as 10 s). The data being collected by the historian may be changing faster, and the data may be stored at a faster rate, but the display normally is still refreshed at the standard refresh rate. It is useful to have the time displayed when the value was collected, as there is often a time delay between the collection and the display of data values. This is especially true where the data collection system may be disconnected temporarily from the display system. The data may come back in a rush when the link is reconnected and is useful for the user (and systems performing analysis) to provide a context for what time the data represent.

6.2 Display of Individual Data Points

Display of the data is normally in text or some simple bar graph representation. Other techniques include button or light indicators where color may represent the state of some value or range of values of the current value. Coded values may take the current value and translate it into some form that provides more value to the user such as zero being translated to the string "FALSE" on the display.

Often, one can better understand the data being obtained by using an analog representation (Ref. 9, pp. 243–254, and Ref. 10). This involves representing the measured quantity by some other continuously variable quantity such as position, intensity, or rotation. A common example is the traditional wristwatch. The hours and minutes are determined by the position of a line indicator on a circle. A common analog representation for data acquisition

Table 6 Data Collection Rates: Examples

Type of Operation	Time per Event
Discrete manufacturing operations	
Assembly line manufacturing/assembly	0.01 to multiple seconds
Video image processing	0.001 s
Parts machining	0.002–0.02 s
Continuous processes	
Paper machine	1–60 s
Boiler	Several seconds to several minutes
Refinery	Seconds to minutes
Dissolving operations	1 s–20 min

is the faceplate. This is a bar graph where the height of the bar corresponds to the value being measured. Often, lines or symbols may be overlaid on the bar to indicate high or low ranges. A frequent indicator is the meter. A needle rotates in a circle with the degrees of movement corresponding to the value obtained from a sensor. Many voltmeters use this technique (Fig. 6). Increasingly sophisticated calculations can be established to translate a flow rate, for example, into a cost number, providing the user with immediate feedback on the costs being incurred by the current process rate.

A common technique for representing trends of current value is to create a simple array and plot it as a trend line on the display. As new values are gathered, the array values are shifted through the array, with old values shifted out at one end of the array while the new values are shifted in at the other. This is a simple technique that provides some of the benefits of data storage without requiring the complexity of actually storing data in mass storage and managing it.

6.3 Display of Historical Data

There are two main issues with display of historical data:

- Selection of the data
- The representation the data will have

Selection of Historical Data

Selection of historical data involves several factors, including time frame and attributes of the data. Identifying a time frame is probably the most common activity in selecting historical data. How the data are updated can be an important consideration when comparing third-party historians.

The time frame is often referenced by a span (such as a number of hours) and a starting point which can be absolute time (e.g., 2004-12-14 16:22:03) or relative time (e.g., −4H for starting 4 h in the past). Another option is to provide an absolute start time and an absolute end time. It is common to have the time frame updating (moving forward with time) if the start point is relative (but check your particular vendor's software for their practice) and to be fixed if the start point is absolute. For example, if the span is 2 h and the start point is

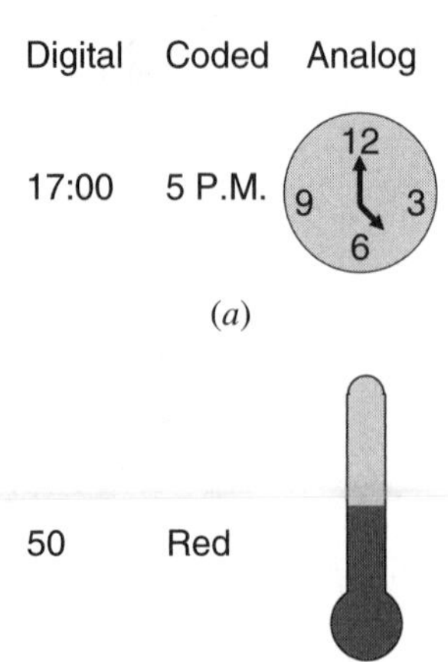

Figure 6 Comparison of digital, coded, and analog data representation: (*a*) time; (*b*) temperature, °F.

−2 h at 15:00, the starting point on a trend line would be 13:00 and the ending point would be 15:00. Ten minutes later the starting point on the trend line would be 13:10 and the ending point would be 15:10.

For relational data, there is often a desired time frame as described above (to restrict the size of the data to be searched) and some selection criteria based on characteristics of the data itself or of related data, including events. For example, one may wish to find those manufactured units for the past month that were for product X and see how many had quality defects. As described above under third-party data acquisition systems, there are techniques for selecting data from process history databases that approach (but do not equal) relational capability.

Representation of Historical Data

The primary difference between current data and historical data is that there are multiple data points, normally with an order defined by the time they were acquired (for process history data) and/or by their relationship to other variables and events (for relational data). Once the time frame and relationships are selected, the data must have some method of representation on a display.

Historical data have a number of potential representation techniques. Multiple data values can be combined into a single number such as average, standard deviation, mode, maximum, minimum, range, variance, and so on. These can then be represented by techniques for individual data elements as described above. The equations in Section 4.4 are examples of summary statistics formulas.

The simplest form to represent historical data is the list. Create a column for each variable of interest. If they are all collected at the same time (the "records" described above), then each time data were collected can be used as the first value on the left in each row. The data for each variable of interest can be placed on the row corresponding to its time stamp, similar to that for files shown in Table 4.

Where data were collected with different time stamps, a new row can be created whenever a new value is obtained and values only placed in the column–row combinations where there is a corresponding time stamp between the row's time stamp and the time stamp of the variable in question. While useful, the problem with this is that there are now holes in the data. This list is very useful when viewing a trend line or graphical tool to validate numbers, to verify time stamps, and diagnose problems with collection.

Often, one desires to view the relationships between data and time or other variables. Trend plots typically are used to show the relationship between variables and time. The X axis is typically represents the time range and the Y axis represents the value of the desired variable. The range of data on the Y axis can vary depending on how one wishes to view the data. The maximum and minimum of the data can be used to set the top and bottom of the range of the Y axis, respectively. This can have two undesired effects. It may make small movements in data appear to be very large when the range is small. In cases where there are spikes in the data where a value is disproportionately high or low, representing the Y axis based on the maximum and minimum could make it difficult to view normal variation in the data. One has to understand the potential use of the data to choose the Y-axis scale appropriately.

When more than one variable is shown on a trend chart, the selection of scale of the Y axis becomes more complex. If all the variables are representative of the same domain, such as all temperatures, then perhaps the same Y axis can be used for all of them. Often, however, the viewer is attempting to compare relative variations in data, sort of a poor man's correlation analysis. In this case it may be useful to have multiple Y axes and select the range of each of them such that they represent the range of one or more of the variables being viewed (Fig. 7).

Another approach to compare variation between two variables is to use one variable for the X value and the other variable for the Y value (an X–Y chart). This is useful when two variables are related by sample time or some other selection technique that results in a paired relationship between the two variables. The correlation function in Section 4.4 represents the mathematical correlation between two variables and can be used to determine the strength of that relationship. Chapter 1 discusses correlation and the calculation of the line through a distribution of data.

7 DATA ANALYSIS

7.1 Distributed Systems

Distributed systems are a powerful approach to data acquisition systems because they combine some of the best of both stand-alone and host-based systems. The data acquisition portion is located on a small processor that has communication capability to a host computer system. The small system collects the data, possibly reducing some to a more compact form, and then sends the data to the host systems for analysis. The host system can analyze the data when it has the available time to do so. Only the data acquisition portion needs to be very responsive to the process. If the data acquisition task gets too big for the small system, the cost of expansion is limited to moving the data acquisition software to a new computer or splitting it up over several computers and changes to the host computer portion are not required. The major disadvantage of distributed systems is that they suffer from a more complex overall architecture even though the individual parts are simple. This leads to problems with understanding error sources and increases the potential errors because of more parts. Unless the communications are designed carefully, the messages sent between the small systems and the host system may be inflexible, causing increased effort when one wants to change the type of data being collected. Distributed systems may be expensive because of the number of individual components and the complexity required but often fit well with environments where one already has a host computer.

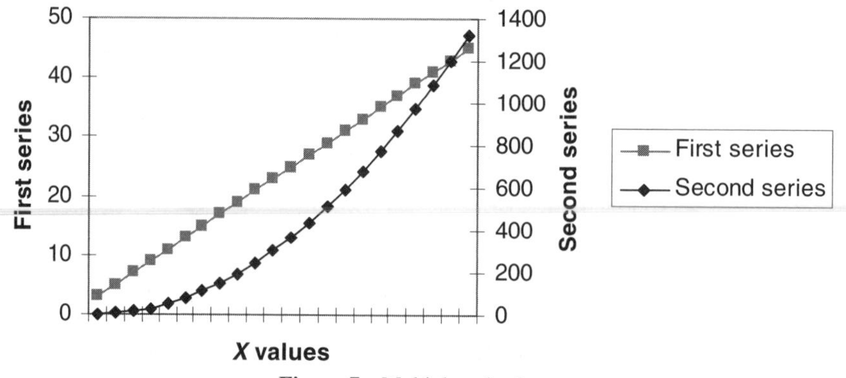

Figure 7 Multiple-axis chart.

7.2 System Error Analysis

The errors that can occur at different stages in the data acquisition process must be analyzed, as they can add up to make the data meaningless. For instance, one may have very accurate sensors, but by the time the data reach the host computer they might have been converted to integer data or from real to integer and back to real again. This can lead to dangerous assumptions about the accuracy of the received numbers, because each conversion can cause rounding or other errors. It is the responsibility of the person setting up the acquisition system and the analyst to examine each source of potential error, discover its magnitude, and reduce it to the point where it will not have a significant impact on the conclusions to be derived from the data. Use of the filtering techniques described above under data collection can be of use to eliminate random error. It is not within the scope of this chapter to cover system error analysis, but Chapter 1 gives some foundation.

8 DATA COMMUNICATIONS

Data communications are involved in many aspects of data acquisition systems. The communications between the sensing and control elements and data acquisition devices, as well as the communications between the data acquisition system and other computer systems, can be carried out in many ways. This section will cover some aspects of communications, especially as they pertain to computer systems.

8.1 Serial Communications

A serial communication link means that data sent over a communications line is spread out over time on one physical data path. For instance, if a character is sent from a sensor to a computer, each bit making up the character (normally eight bits) will be sent one after the other (Table 7). This is often useful for low-cost, low-speed (usually less than 10,000 cps) rates of data transfer.

Table 7 Time Sequence of Bits Sent over Serial Communication Line

Character "A" is (in bit form)
Bit number 6 5 4 3 2 1 0
Bit value 1 0 0 0 0 0 1
The communications are using the RS232C communications standard and sending the ASCII character A.

Bit Value	Time
Start bit	0 s after start
1 (bit 0 of A)	1/9600 s after start
0 (bit 1 of A)	2/9600 s after start
0 (bit 2 of A)	3/9600 s after start
0 (bit 3 of A)	4/9600 s after start
0 (bit 4 of A)	5/9600 s after start
0 (bit 5 of A)	6/9600 s after start
1 (bit 6 of A)	7/9600 s after start
parity bit	8/9600 s after start
stop bit	9/9600 s after start

8.2 Parallel Communications

A serial communication link may not require very many wires, but the time spent to transfer data can add up. A way to improve the speed of communications is to use parallel communication links. This is done by having a number of wires to carry data. For instance, sending the same "A" over a nine-wire bus would only require one transfer (Table 8).

8.3 Networks

Ethernet with Transmission Control Protocol/Internet Protocol (TCP/IP) has become the dominant communications network protocol for data collection. There are proprietary process control and data acquisition networks that serve special purposes, but Ethernet has proven to be versatile for everything from office communications to data collection from smart sensors. Many computers can be connected to the same network segments. The use of switches and routers provides ways to isolate and limit communications to improve performance and security. Firewalls provide filters and protection for entire classes of messages and sources.

While Ethernet cannot guarantee delivery (being based on a collision detection and retransmit strategy), it has been shown to provide excellent response to moderate network activity. Communications speeds are regularly being improved to provide an even greater range of applicability.

8.4 OSI Standard

The International Standards Organization has developed a set of standards for discussing communications between cooperating systems called the Open Systems Interconnect (OSI) model (see Table 9). This defines communications protocols in terms of seven layers.[11] While

Table 8 Time Sequence of Bits Sent over Parallel Communications Interface

Character "A" is (in bit form)
Bit number 6 5 4 3 2 1 0
Bit value 1 0 0 0 0 0 1
The communications are using a hypothetical nine-line parallel communications bus sending the ASCII character A.

Bit Value	Time
Start bit	0 s after start
1 (bit 0 of A)	0/9600 s after start
0 (bit 1 of A)	0/9600 s after start
0 (bit 2 of A)	0/9600 s after start
0 (bit 3 of A)	0/9600 s after start
0 (bit 4 of A)	0/9600 s after start
0 (bit 5 of A)	0/9600 s after start
1 (bit 6 of A)	0/9600 s after start
parity bit	0/9600 s after start

If the bus could handle the same rate of change of bits as the serial interface, then the next character could be sent 1/9600 s after the first character (the A).

Table 9 Open Systems Interconnect Model

Layer	Principle	Example
7. Application	Application	Millwide reporting
6. Presentation	Display, format, edit	Convert ASCII to EBCDIC
5. Session	Establish communications	Log onto remote computer
4. Transport	Virtual circuits	Make sure all message parts got there in order
3. Network	Route to other networks	Talk to Internet
2. Data link	Correct errors	Send character downline
	Synchronize communications	Send Ack-Nak
1. Physical	Electrical interface	Wire and voltages

not providing for specific interface protocols, the OSI model has had a significant impact on communications because it has provided a framework for compartmentalizing aspects of communications to allow the handoff of information from one device to another in a standard way. For instance, the transmission of data from one media type to another (such as copper wire to fiber to satellite to copper wire and then to wireless) is a result of standards enabling the seamless transfer of messages in a way that is transparent to the user.

8.5 OPC Standard

A recent standard of use in manufacturing is the OPC (OLE for process control) standard, which provides for a standard way of communicating with process equipment. It is sponsored by the OPC Foundation and originated as an extension for process control from the Microsoft OLE functionality.[4]

Functions provided by OPC include ability to browse the variable database of a device and monitor data on demand or when events occur. The capability of OPC has been expanded to work with Web communication methods such as XML and cover complex data such as record structures. The power of OPC is that data from an instrument can be available via a standard network interface so that any data acquisition program that uses the OPC interface can gain access to any OPC device. The need to know the protocol of each device or adhere to the wiring of specialized communications or use custom database access methods is eliminated through the use of a standard protocol. Multiple programs can be simultaneously monitoring the same piece of data, performing different functions at different time intervals, as events occur. The last point is particularly significant, because much of the work of data acquisition systems is spent in polling for changes in data or otherwise attempting to determine when an event has taken place. A program can subscribe to an OPC item and it will be notified when the item changes value, reducing the complexity of monitoring data dramatically.

As an example of the power of this approach, consider the following example (Fig. 8). A device can collect the identification from a unit of material such as unit number, color, and manufacturing date and make it available via OPC. As the unit is processed in a manufacturing center, another device collects defect counts and makes the unit number available via OPC. A human–machine interface program can monitor both sources with the same interface protocol and software and display it live for an operator to see. Simultaneously, another program can collect the defect counts and summarize them into totals. Yet another program can monitor the totals and wait for the unit number to change, triggering a transaction to a database or an email if there was a problem. The power of the OPC interface is that it provides real-time access to the data from each of the sources and multiple programs

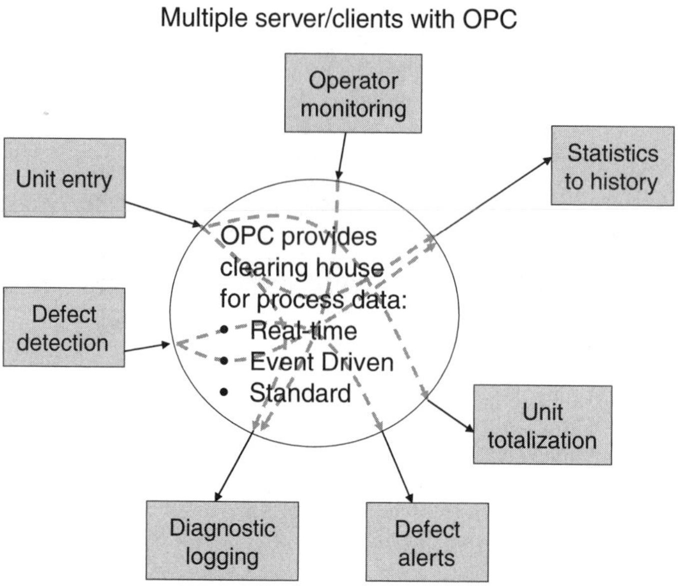

Figure 8 Example use of OPC communications.

can monitor the same OPC sources to perform work. Diagnostics can monitor the same data to evaluate system processing, downtime, or quality issues. Other programs can sample and store data to log files or diagnostic files for further analysis. All can be operating in parallel with no need to understand the internals of the other programs, via a standard interface and standard protocols that support asynchronous delivery of data.

8.6 Benefits of Standard Communications

When implementing data acquisition and display systems, the ability to communicate in a standard fashion can play a large role in the cost of the system. This is realized in a variety of ways:

1. Different sensors can be connected to a system without having to buy a whole new system.
2. Data can be sent to other systems as needed for further processing.
3. As technology or need changes, portions of a system can be mixed and matched.
4. Increased competition from vendors tends to bring prices down.
5. A standard will have many people using products based on the standard, resulting in more vendors and greater availability of parts with a greater variety of options.

9 OTHER DATA ACQUISITION AND DISPLAY TOPICS

9.1 Data Chain

As materials and parts move through a manufacturing operation, the data collected are separated by time and type of data. Combining that data together presents a number of challenges. One can consider the first piece of data collected about some object to be the beginning of a chain of data and each successive data acquisition point in the manufacturing process to be a link in that chain until the end of the chain, where the last piece of data about a manufactured item is collected. Often, the steps of the manufacturing process may proceed from raw material to some intermediate work-in-process unit, to some other step that may be time based, then to some other step that may be finished-unit based. Many varieties of the above exist. Each link can represent one of the following:

- Data collected from a start time to an end time
- A set of attributes about a particular manufacturing unit with an associated time of processing

Often referred to as the genealogy, the steps can be linked together through:

- Some assumptions based on time stamp relationship of one link back to the previous one
- Recording of units that entered or left the time-based portion of the process and the beginning and ending of the entry time
- Some assumptions about the mixing of elements of the manufactured item

Using the techniques described above for combination of time-based and event data, a set of data for the whole life cycle of a manufactured item can be created. Where mixing occurs, the data will be less accurate but may provide clues as to the factors that went into the final product, such as proportions of additives. As an example, Table 10 shows various queries that can be combined to provide one picture of all data sources for a manufactured unit.

Starting at the finished-good item, the batch ID of the previous step acts as a link into the range of time data in the previous step. If the list of items broken down is retained from the prior step, then those can be used to link back to the previous time frame.

Depending on the amount of mixing, the results will be more or less indicative of what actually happened. The smaller the batch sizes, the easier the tracking back to source data will be. Time lags between process steps can dramatically impact the ability to assume when the raw materials for a particular item were processed.

9.2 Web Programs and Interfaces

Web interfaces have improved to the point that user interfaces can be written in Web browser screens. This eliminates the effort and organization required to deploy code across a company. The application is written for a Web interface. When the user uses a Web browser to access the page, functionality is downloaded to the user's page or is executed in such a fashion as to obtain the results of a query and transmit a data page to the user. The developer does not have to get involved in the process of manually installing software on the user's machine. This reduces the demand on the desktop computer and allows the developer to make a change and have it proliferated to all users when they next reference the Web.

9.3 Configuration versus Implementation

As a general rule of thumb, third-party data acquisition and storage systems provide configurable tools for acquisition and display. For simple applications which do not require great flexibility in program functions, such as generating alarms, unusual graphics types, control of the process, or integration into larger systems, it is appropriate to use these, often simple,

Table 10 Data Chain Sample Queries

State in Process	Important Data (Typical)	Example Data	Simplified Query to Combine Data with Previous Step
Raw material inventory	Identifiers: raw material batch ID Data: supplier, quality, time in inventory	Lot 1: 5% rejects; lot 2: 7% rejects	
Intermediate goods processing	Identifiers: raw material lots consumed; raw material batch ID; start and end time of entry to process; time delay through process Data: piece count, rejects, downgrades, process characteristics such as rate, temperature, modifications to materials	3:00–4:00 raw materials from lot 1; 1000 pieces; 10 rejects; 200 degrees 4:00–5:00 raw materials from lot 2; 1045 pieces; 14 rejects; 205 degrees; :10 residence time in process	Use raw material batch ID to match to raw materials characteristics; material run at 4:00–5:00 had 7% rejects when delivered to plant
Intermediate package creation	Identifiers: raw package ID, start/end time of creation Data: piece count, package dimensions	Package PR1 created: 3:30–4:10; 100 pieces; package PR2 created: 4:10–5:10; 150 pieces	Use time of manufacture, time lag through process to identify characteristics from intermediate process and raw materials lots; package PR1 was processed at 200 degrees, was created from lot 1 and was from a lot that had 5% rejects detected when delivered to the plant
Intermediate inventory	Identifier: intermediate package ID Data: location in inventory	Package PR1 location warehouse 1	Use ID of intermediate package to match to package at intermediate package creation; package PR1 had 100 pieces
Finished goods processing	Identifer: finishing batch ID, intermediate package consumed, start/end time of consumption; time delay through process Data: piece count, rejects, downgrades, process characteristics such as rate, temperature, modification to materials	Intermediate packages consumed at breakdown: 5:30–6:00 PR1; 95 pieces; 400 degrees 6:00–6:30 PR2; 147 pieces; 390 degrees 6:30–6:40 PR3; 25 pieces; 395 degrees 6:40–7:00 PR4; 40 pieces; 410 degrees Typical residence time :15	Use ID of intermediate package to match to intermediate inventory; in this example, package PR1 came from warehouse 1, lost 5 pieces in consumption
Finished goods package creation	Identifiers: finished package ID; start/end time of creation Data: piece count, grade, package dimensions, customer order	Finished package PF1 created from 6:00 to 6:30; 40 pieces; prime grade; order PO5670	Use time of manufacture, time lag through process to identify characteristics from breakdown, and which raw packages sourced this finished package. There may be significant mixing. In this example, PR1 and PR2 would be sources for PF1. PF1 was possibly created at 400 degrees, stored in warehouse 1, lost 5 pieces when loading into finished goods process, was probably processed at 200 degrees, and was probably from lot 1

question–answer or menu-type systems. When the system must be very flexible or customized, it may be more appropriate to write a custom program. When considering this approach, be cautious, for the cost of implementing a program is often much higher than expected. For instance, if one wanted to perform simple data acquisition and storage from a sensor, the cost to write a program would probably be higher than buying a small off-the-shelf system and entering the parameters for data collection. Writing a program involves analysis, design, development, debugging the program, and testing of results. The cost of documenting a program is often a large unplanned cost. If the results are intended to be used to make economic or process-related decisions, then the program must be tested carefully. Additionally, maintenance of the program can be quite expensive. Someone must be trained in the technologies used to build the program, the logic of the program, and the installation of the program. Another factor to consider is that costs of improvement of third-party software are borne by many customers and driven by many customers. The net result to the user of this software is that it is normally constantly improving, constantly tested, and maintained by a group of developers whose primary job is software development. One reason to build and maintain software internally is that a company can keep special knowledge within the company and thus maintain competitive advantage.

9.4 Store and Forward

When data acquisition and data storage are on two separate machines, it is important to provide methods to retain data in case the link between systems is broken. Message buses provide automated methods of maintaining a link between data acquisition systems. The developer inserts data into the message bus. If the link between the two systems is broken, the message bus queues up the data messages on the collection machine. When the data storage machine connection is reestablished, the message bus passes on the data to the data storage machine.

When a message bus is not available or feasible, a simplified mechanism can be created where a file representing each sample of data is created. If the data collection and data storage system are linked, then the data storage system monitors the directory of the collector for a new data file. If the data storage system detects one or more files on the data collection computer, then it will process them into storage. If the link is broken, then the files build up until the link is reestablished. A related technique is to store data in a database or similar mechanism on the data collection computer and scan it periodically for missing data from the storage computer. This is particularly useful when connection to the data collection computer is unreliable.

9.5 Additional Communications Topics

When considering transmission media, some points may provide value to the engineer. Fiber-optic cabling is less sensitive to noise than other transmission media. Wireless access points provide increased flexibility in positioning of sensors and greatly reduce wiring costs. Particularly, if one wishes to collect data from sites that may move, such as environmental sampling sites, the costs of wiring and rewiring can be quite significant. Using wireless transmission technology eliminates much of the wiring costs and facilitates moving the sensors from one location to another. Wireless transmission has a set of concerns that must be taken into account by the engineer, including security, since other units can monitor signals (still evolving) and ability to be jammed.

10 SUMMARY

The tremendous change in technology for data acquisition and display systems since this chapter was first written has driven us to take a different approach than with the first edition. The technologies for data acquisition and display have become more standardized. Engineers are increasingly reliant upon and versed in computing technologies. The combination of data from various sources into an integrated view of the process has facilitated process improvement and leads to competitive advantage.

This chapter has attempted to provide tools and techniques to aid in the acquisition, storage, and manipulation of process data, expanding from the previous edition into techniques to aid in the manipulation of data for integration and analysis.

REFERENCES

1. C. D. Johnson, *Microprocessor-Based Process Control,* Prentice-Hall, Englewood Cliffs, NJ, 1984.
2. P. W. Murrill, *Fundamentals of Process Control Theory,* Instrument Society of America, Research Triangle Park, NC, 1981.
3. B. G. Liptak, "System Accuracy," in B. G. Liptak ed in chief, *Instrument Engineer's Handbook,* 4th ed., Vol. 1: *Process Measurement and Analysis,* CRC, Boca Raton, FL, 2003.
4. http://www.opcfoundation.org.
5. C. J. Date, *An Introduction to Database Systems,* 5th ed., Addison-Wesley, 1990.
6. W. H. Beyer (ed.), *CRC Standard Mathematical Tables,* 24th ed., CRC, Boca Raton, FL, 1976.
7. M. L. Crossley, *The Desk Reference of Statistical Quality Methods,* ASQ Quality Press, Milwaukee, WI, 2000.
8. http://www.aspentech.com/publication_files/White_Paper_for_IP_21.pdf
9. R. W. Bailey, *Human Performance Engineering: A Guide for Systems Designers,* Prentice-Hall, Englewood Cliffs, NJ, 1982.
10. E. R. Tufte, *The Visual Display of Quantitative Information,* Graphics, Cheshire, England, 1983.
11. "Open Systems Interconnection—Basic Reference Model," Draft Proposal 7498, 97/16 N719, American National Standards Institute, New York, 1981.
12. J. D. Wright and T. F. Edgar, "Digital Computer Control and Signal Processing Algorithms," in *Real-Time Computing,* D. A. Mellichamp (ed.), Van Nostrand Reinhold, New York, 1983.

MAGAZINES THAT CARRY RELEVANT INFORMATION

Control Engineering International: http://www.controleng.com/.
Design Engineering: http://www.designengineering.co.uk/.
IEEE Control Systems Magazine: http://www.ieee.org/organizations/pubs/magazines/cs.htm.
Industrial Technology: http://www.industrialtechnology.co.uk/.
Instrumentation and Automation News: http://www.ianmag.com/.
Pollution Engineering Online: http://www.pollutionengineering.com/.
Scientific Computing and Instrumentation: http://www.scamag.com/.
Sensors Magazine: www.sensorsmag.com.

CHAPTER 8

DIGITAL INTEGRATED CIRCUITS: A PRACTICAL APPLICATION

Todd Rhoad
Austin, Texas

Keith Folken
Peoria, Illinois

1 REDUCTION OF PROCESSOR LOAD

Today's processors can make many calculations in a very short amount of time. Embedded controllers make many repeated calculations several times a second. In many embedded controllers, such as an engine control or plant process control, the processor takes several inputs from outside sources, does some computation with those values, and then takes some action. This may be in the form of an output to the system, an alarm, or just a log of the data. The processor's mathematical calculations can be very time intensive. If accuracy is required, these calculations must done using several digits, sometimes with floating-point numbers. To take some of this load off of the processor, a field programmable gate array (FPGA) may be used.

An FPGA is a device of logic gates. These logic gates can be used to take the place of the computationally intensive mathematical operations that the processor would normally perform. For example, a process controller is asked to measure 10 analog inputs every 200 ms. The controller must compute a running average of the last three samples of each input. Using this average, several multiplication and division computations are made. To offload the processor, an FPGA could be used to make these repetitive calculations and feed the results back to the processor.

2 FIELD-PROGRAMMABLE GATE ARRAYS

Field programmable gate array technology permits the design of many different complex digital circuits using a single off-the-shelf device. The general structure for FPGAs is shown in Fig. 1. *Field programmable* refers to the ability of the FPGA function to be designed by the user in the field, as opposed to the manufacturer. Depending on the type of device, the function either is "burned" in permanently or semipermanently as part of the board assembly process or is loaded in from external memory every time the FPGA is powered on. FPGAs comprise an array of uncommitted circuit elements, referred to as logic blocks (LBs), programmable input/output (I/O) blocks, and programmable interconnects. Configuration of an FPGA is performed by the end user through programming. The I/O blocks function as the interface between the external device pins, the central processing unit (CPU) in this example, and the internal logic. The array of logic blocks provides the functional elements from which the end user's logic is constructed. Each logic block can independently take on any one of a limited set of personalities. These personalities are typically implemented through both Boolean logic and latching data. The individual logic and I/O blocks are interconnected by a matrix of wires and programmable switches.

FPGA architectures can vary in the size, structure, and number of logic and I/O blocks as well as the amount of connectivity of the interconnects. The secret to density and performance in these devices lies in the logic contained in their logic blocks and the performance and efficiency of their routing architecture. There are two primary classes of FPGA architectures: fine grained and coarse grained. Coarse-grained architectures consist of fairly large logic blocks, often containing two or more lookup tables and two or more flip-flops (FFs). In a majority of these architectures, a four-input lookup table implements the actual logic. The larger logic block usually corresponds to improved performance. The other architecture type is called fine grained. In these devices, there are a large number of relatively simple logic blocks. The logic block usually contains either a two-input logic function or a 4-to-1 multiplexer and a FF. These devices are good at systolic functions and have some benefits for designs created by logic synthesis. Another difference in architectures is the underlying process technology used to manufacture the device. Currently, the highest density FPGAs

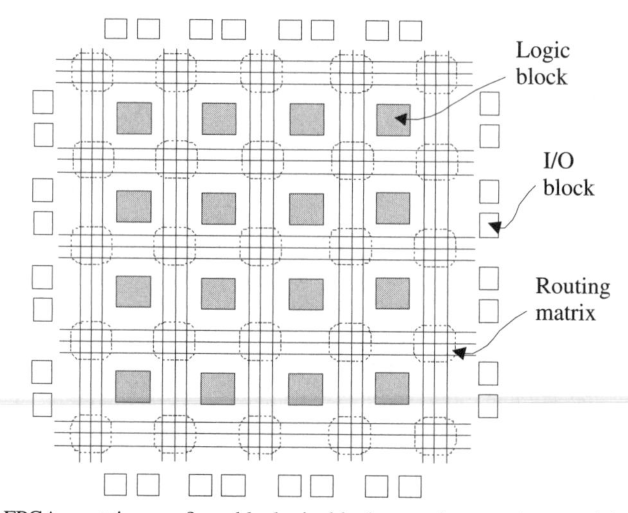

Figure 1 The FPGA contains configurable logic blocks, routing matrices, and input/output (I/O) blocks that can configure each I/O pin for a different function.

are built using static random-access memory (SRAM) technology, similar to microprocessors. The other common process technology is called antifuse, which has benefits for more plentiful programmable interconnect.

Reprogrammable FPGA technologies tend to be SRAM based. The routing matrices are usually implemented using pass gates driven by the value of the SRAM memory cell assigned to it. Lookup table (LUT) function generators are used to program the function of the logic block. These LUTs are small SRAM cells that have four or five inputs, which are the address lines of the SRAM cell. The output of the SRAM cell is the output of the function generator. The programmable FF after the logic block is programmed using pass gates. Then, to program the function of the logic block, the contents of the SRAM cell are loaded and configure the block to be either registered or combinatorial. Without any sort of external memory, SRAM-based FPGAs will lose their configuration if the device is turned off. Therefore, erasable programmable read-only memory (EPROM), or flash memory, connected serial or parallel, is used to provide the configuration to the FPGA during startup. That is, this configuration memory holds the program that defines how each of the logic blocks functions, which I/O blocks are inputs and outputs, and how the blocks are interconnected together.

There are a few advantages and disadvantages in using FPGA-assisted designs. Pipelining is a hardware technique of segmenting a complex operation into stages so that multiple computations can be performed simultaneously. Pipelining is a natural choice due to the discrete and regular quantities of FPGA logic blocks. Memory located at the end of the logic blocks completes the design and uses the storage as pipeline stage registers. The biggest feature of FPGA pipelining is the ability to combine numerous complex computations into a single pipeline. FPGAs have the ability to chain together any sequence of operations as required by the algorithm. Such deep pipelines are difficult to abstract for general-purpose CPUs; therefore, FPGAs offer computational potential found elsewhere only in ASICs (application-specific integrated circuits). Another major advantage is the ability to implement a large degree of computational parallelism. The parallelism most concerned here is data and algorithmic parallelism. Data parallelism is the ability to regularly process a large data set concurrently. Algorithmic parallelism refers to the ability to operate multiple independent algorithms at the same time. A combination of these two types can result in high-throughput devices with low-speed parts, which is simply streaming high-speed data through an array of low-speed computational units. FPGAs can also minimize computational logic and delay by utilizing partial evaluation techniques to reduce multivariable functions to less complicated functions based on information provided at compilation.

FPGAs do suffer physical constraints in that they are slower and are degraded in density, when compared to ASICs, due to the use of configurable SRAM-based devices to implement gates and interconnections. Gate density is extremely important since the size of the gate array determines how much logic can be implemented. Aside from physical constraints, FPGAs require relatively long delays for reprogramming the device. Current chips require configurations to be downloaded one at a time, resulting in significant dead time on the order of a few milliseconds. For applications where the process time is of the same order, significant issues can arise. Other weaknesses for FPGA-based designs include poor floating-point performance, poor data proximity, and overall design complexity.

3 LOGIC BLOCKS

The logic block of an FPGA is the core building block and is used to implement combinational and sequential logic. This block implements the actual logic functions for configurable computing. A general representation of a logic block is shown in Fig. 2. This model

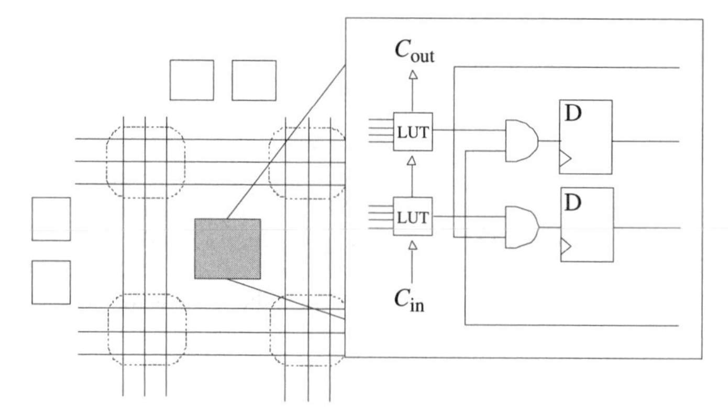

Figure 2 General logic block with two-input logic and D-type FFs.

uses three stages. The first stage is function generation, which utilizes an array of programmable LUTs. These LUTs typically have between three and five inputs, with four inputs being cited as the best compromise between LUT access time and the average desired function delay.[1] The second stage, the programmable routing stage, follows the function generators and allows function results to be supplied to the logic block's final stage with minimal delays. Finally, in the memory stage, the logic block uses D FFs to implement bit storage within the logic block. The FFs are beneficial for distributed storage in configured designs, including their use as registers between pipeline stages.

Logic blocks have enough flexibility to operate in three distinct modes: combinational logic, arithmetic or ripple modes, and dedicated memory storage. In combinational logic mode, the LUTs are loaded at configuration time with truth tables to implement a logic function that is independent of LUT inputs. The arithmetic or ripple mode uses carry gates to rapidly propagate signals from a logic block to its neighbors. Lastly, the memory stage allows the LUT to be configured as RAM or ROM units. This mode increases the amount of internal memory within the device and is useful for embedding memory elements throughout a design.

4 LOOKUP TABLES

A LUT of a regular, single-context FPGA consists of a bank of SRAM cells and a multiplexer. The depth of the SRAM cell is one, and it should be as wide as the data input of the multiplexer. The output bits of the SRAM are directly connected to the data input pins of the multiplexer. The inputs to the LUT control the select signals of the multiplexer. Based on the LUT inputs, an SRAM output is selected by the multiplexer and sent to the LUT output.

A K-input lookup table (K-LUT) is a digital memory with K address lines and a one-bit output. This memory contains $2K$ bits and is capable of implementing any Boolean function of K variables. A LUT of K inputs can implement $22K$ different Boolean functions. A simple example of a two-input LUT implementing a function $f = a + \overline{b}$ is shown in Fig. 3, where the LUT is described in terms of a multiplexer. In the LUT, there are two address lines (a and b) and four memory cells. The inputs to the LUT are the select lines of the multiplexer, and the memory cells serve as the inputs to the multiplexer. If $a = 0$ and $b = 1$, for example, then the memory cell corresponding to select line 01 will be connected to

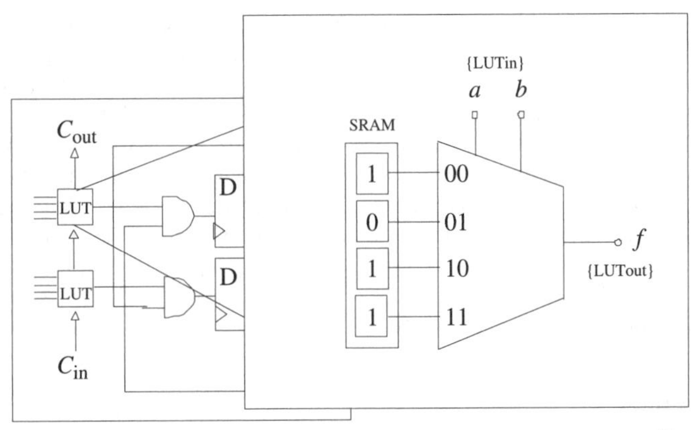

Figure 3 A two-input LUT implementing the function $f = a + \bar{b}$.

the output f. The contents of the memory cells are calculated from the evaluation of $f = a + \bar{b}$ for all the combinations of logic values of a and b.

5 FLIP-FLOPS

Flip-flop devices are used in the digital field for a variety of purposes such as storing data temporarily, multiplication or division, counting, or receiving and transferring information. Flip-flops are bistable multivibrators. Multivibrators are classified into three types: bistable, monostable, and astable. Their behavior is illustrated by the simple mechanical analogies in Fig. 4. The bistable monovibrator, or FF, is analogous to the seesaw or teeter-totter of a children's playground. If an input in the form of a sufficient downward force is applied to the up end, then it will assume the down position and the other end will move into the up

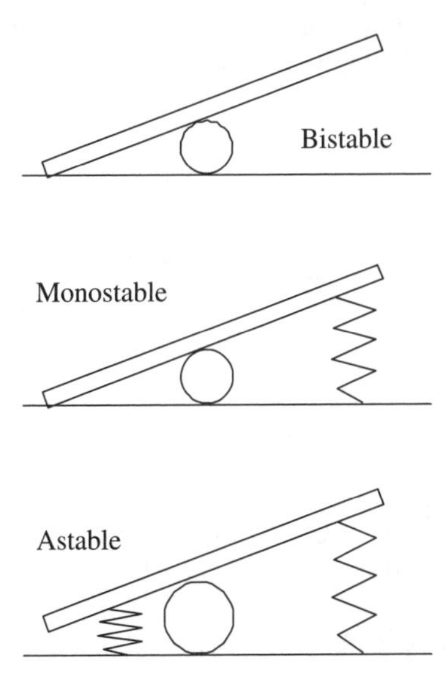

Figure 4 Mechanical analogies for bistable, monostable, and astable multivibrators.

position and remain there after the input is removed. Further inputs corresponding to downward forces on this end will have no effect. Also, the state of the system can be described by the position of either end, and if logical variables are introduced to designate these states, the variables are complementary. The system exhibits the characteristics of memory in that it remains in the state it has been set to with no further input.

The types of FFs used in digital equipment are identified by the inputs. They may have from two up to five inputs depending on the type. They are all common in one respect in that they have only two distinct output states. The outputs are normally labeled Q and $\overline{Q}$ and should always be complementary. If $Q = 1$, then $\overline{Q} = 0$, and vice versa. In this section, the four types of FFs that are most common to digital equipment will be discussed and are the D, R–S, T, and J–K FFs.

5.1 D Flip-Flops

The D FF gets its name from its ability to hold data in its internal storage. It is sometimes referred to as a gated D latch and is shown in Fig. 5. This is because the presence of the clock pulse at the enable pin will transfer the input D to the output of the FF. The D FF has two inputs: D and an Enable or Clock. The state of its output, Q, follows that of input D at the time of the clock signal's transition from low to high and is delayed by the device propagation time. In other words, the D-type latch changes its output state when the clock input is in the high state and retains that value when the clock input is in the low state. Thus regardless of the initial state condition the final state assumes the value of the data input. This value will be held until the next clock pulse, at which time it may be updated. The influence of the clock exists for a finite albeit short time around the edge. For proper operation it is necessary that the logical level D be established for an appropriate time preceding the clock edge. This time is referred to as the setup time. The level must also be maintained for a time following the edge. This is referred to as the hold time. Clearly the time interval between clock pulses, which controls the ultimate speed with which a digital system performs, must always exceed the sum of the setup and hold times.

5.2 R–S Flip-Flops

The R–S FF is used to temporarily hold or store information until it is needed. A single R–S FF will store one binary digit, either a 1 or a 0. Storing a four-digit binary number would

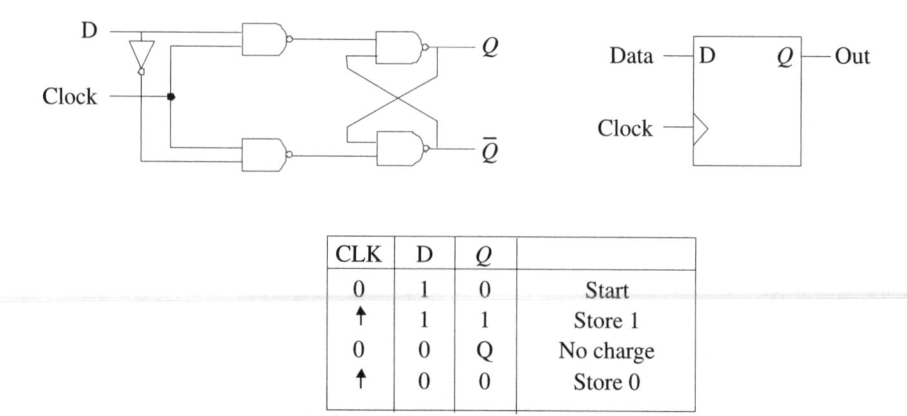

CLK	D	Q	
0	1	0	Start
↑	1	1	Store 1
0	0	Q	No charge
↑	0	0	Store 0

Figure 5 Clocked D FF configuration (upper left), symbol (upper right), and truth table (bottom).

require four R–S FFs. The name for the R–S FF is derived from the inputs, R for reset and S for set, and is often referred to as an R–S latch. The R–S FF has two output conditions. When the Q output is high and $\overline{Q}$ is low, the FF is set. When Q is low and $\overline{Q}$ is high, the FF is reset. When the R and S inputs are low, the Q and $\overline{Q}$ outputs will both be HIGH. When this condition exists, the FF is considered to be jammed and the outputs cannot be used. The jammed condition is corrected when either S or R goes high. To set the FF requires a high on the S input and a low on the R input. To reset, the opposite is required; S input low and R input high. When both R and S are high, the FF will hold, or "latch," the condition that existed before both inputs went high.

In a complex digital system the signal levels are dynamic, and it is important to define precisely the time at which transitions take place, thereby defining the time at which the correct state is assumed. This is accomplished by issuing a clock pulse generated by an astable multivibrator to all FFs in the system. Such a configuration is shown in Fig. 6 where the AND gates ensure that the set and reset logic levels are only applied at clock time. Here the AND gates play the role of clock-controlled switches. In the absence of the clock signal, shown as the triangle on the FF, the system is in the hold mode with S and R at a value of 0. The clock pulse may be of significant duration. However, these devices are typically fairly sensitive to changes on the input lines during the period when the clock pulse is present. Therefore, the clock pulse is normally shortened to a narrow spike before being applied to the two AND gates. This can be accomplished by differentiating the clock pulse. The assembly of differentiator and clocked FF constitutes an edge-triggered FF since transitions are restricted to a narrow time at the edge of the clock pulse. These circuits are triggered by either the rising edge or the falling edge of the clock pulse. This presents two device states of concern: the state prior to the edge of the clock pulse (Q) and the new state after the edge of the clock pulse (Q'). Figure 6 shows the possible states for this configuration.

5.3 T Flip-Flops

The T, or toggle, FF is a bistable device that changes state on command from a common input terminal. The T-type FF is often used in counting applications. As with the other types of FFs, the T FF is a derivative of the R–S FF. Comparatively speaking, the S and R inputs are internally connected, R to Q and S to its complement, so that the only external input is the clock. If an inverter precedes the T input, it indicates that the FF will toggle on a high-to-low transition of the input pulse. The absence of an inverter indicates the FF will toggle on a low-to-high transition of the pulse.

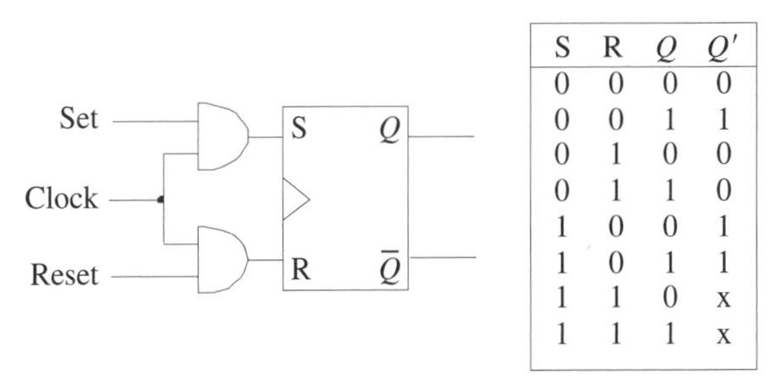

S	R	Q	Q'
0	0	0	0
0	0	1	1
0	1	0	0
0	1	1	0
1	0	0	1
1	0	1	1
1	1	0	x
1	1	1	x

Figure 6 R–S FF configuration (left) and truth table (right).

5.4 J–K Flip-Flops

The J–K FF is very similar to the R–S FF except that there are no indeterminate states. From Fig. 7, it can be observed that for the first three combinations of the truth table, the J–K FF behavior is identical to the R–S FF with J = S and K = R. The observable difference is in the combination J = 1, K = 1. For in this state, the device becomes a toggle FF. A convenient aspect of this arrangement is that the indeterminate state of the RS FF cannot occur and all four combinations of the external inputs are allowed.

6 BOOLEAN LOGIC NOTATION

George Boole created Boolean logic, sometimes referred to as Boolean algebra, in the nineteenth century. He based his concepts on the assumption that most quantities have two possible conditions: true and false. A Boolean expression is a description of the input conditions required to get the desired output, which are based on Boole's laws and theorems. Boolean algebra is primarily used by design engineers to arrange logic gates to accomplish specific tasks while enabling the designers to achieve the desired output by using the fewest number of logic gates. Naturally, space, weight, and cost are important factors in the design of equipment; one would usually want to use as few parts as possible.

A good starting point is the existence of only two binary values—true (T or 1) and false (F or 0)—and three operators—AND (Boolean multiply), OR (Boolean addition), and NOT (Boolean inversion)—with behavior described by their truth tables. Boolean algebraic formulas follow certain conventions. Assume our function is $f = A + B$. This basically states that "f is TRUE when A or B is TRUE." Boolean logic also follows commutation, association, and distribution laws. AND and OR operators commute, such as $A + B \equiv B + A$. Distributive property is such that $A + (B + C) \equiv (A + B) + C$. Lastly, the distributive property is evident from $A \cdot (B + C) \equiv (A \cdot B) + (A \cdot C)$. The conventional hierarchy of operator action in Boolean expressions is NOT, then AND, last OR. Another important law useful in Boolean algebra to convert AND to OR and OR to AND is De Morgan's law, which states that $\overline{A \cdot B} \equiv \bar{A} + \bar{B}$ and $\overline{A + B} \equiv \bar{A} \cdot \bar{B}$.

7 LOGIC GATES

Given the laws of commutation and association, many logic gates can be implemented with more than two inputs, and for reasons of space in circuits, usually multiple-input, complex gates are made. The gates discussed in this section are the NOT, AND, OR, NAND, and NOR. The logical NOT is denoted by the overscore and is defined by variables listed in the truth tables, which show all possible combinations of inputs and outputs. For example, if we have input A and its current value is 0, then NOT A is equal to 1. Figure 8 represents the truth table for a single input variable and its NOT function values.

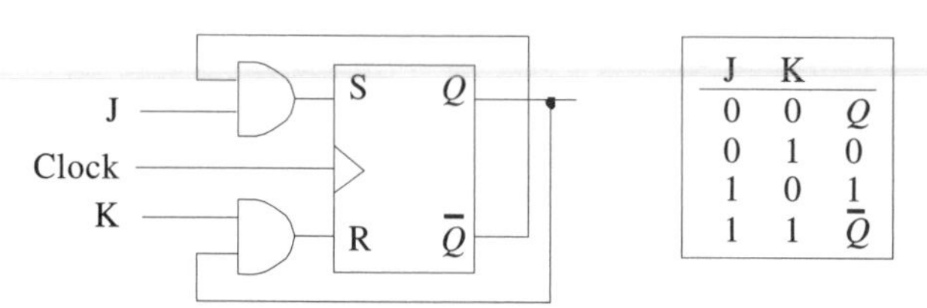

J	K	
0	0	Q
0	1	0
1	0	1
1	1	$\bar{Q}$

Figure 7 J–K FF configuration (left) and truth table (right).

A	$\overline{A}$
0	1
1	0

Figure 8 Truth table for input A and NOT A.

The AND gate performs logical multiplication, commonly known as the AND function. The AND gate has two or more inputs and single output. The output of the AND gate, denoted as f in Fig. 9, is high only when all its inputs are high. The Boolean expression for this operation is $f = A \cdot B$. It may also be written without the dot, or simply $f = AB$. The expression is spoken, "$f = A$ AND B." The dot, or lack of, indicates the AND function. A simple way to remember the operation of the AND gate: In order to produce a 1 output, all the inputs must be 1. If any or all of the inputs are 0, then the output will be 0. By placing an inverter on input A of the AND gate, we get the NAND gate. The function that would describe this operation is $f = \overline{A} \cdot B$ and may be spoken as "$f =$ Not A AND B." By inverting input A, we are specifying that the only condition that will cause a high output (or 1) is for input A to be a 0 and input B to be a 1. Every other combination will result in a low (or 0) output.

The logical OR gate differs from the AND gate in that only one input signal needs to be high to get a high (or 1) output. As with the AND gate, there can be numerous input signals. However, for simplicity, we will only consider two inputs. For an example, we will use the gates shown in Fig. 10 to produce a high output: variable A, variable B, or both must be high. The Boolean expression for this operation is $f = A + B$ and is spoken "$f = A$ OR B." The plus sign indicates the OR function and should not be confused with addition. By inverting inputs A and B (shown as small circles on the inputs of the gate), the output changes from any high generating a high output as in the OR to any low (or 0) input generating a high output.

8 COUNTERS

A counter is simply a device that counts. Counters may be used to count operations, quantities, or periods of time. They may also be used for dividing frequencies, for addressing information in storage, or for temporary storage. Counters are a series of FFs wired together to perform the type of counting desired. They will count up or down by 1s, 2s, or more. A counter's modulus refers to the total number of counts or stable states a counter can indicate. For example, the modulus of a four-stage counter would be 16 (base 10), since it is capable

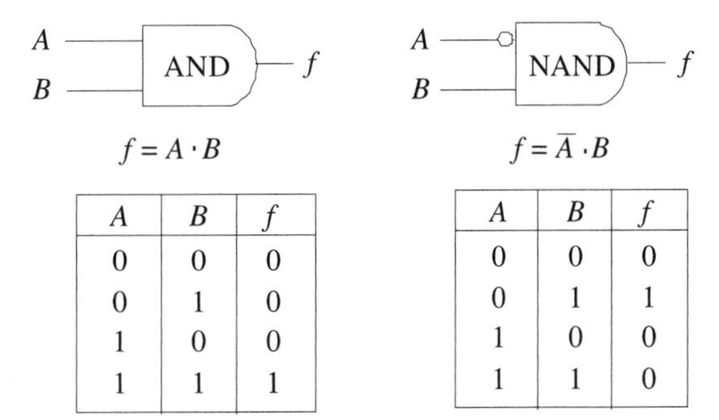

A	B	f
0	0	0
0	1	0
1	0	0
1	1	1

A	B	f
0	0	0
0	1	1
1	0	0
1	1	0

Figure 9 Gate logic symbol for AND and NAND gates with their respective truth table.

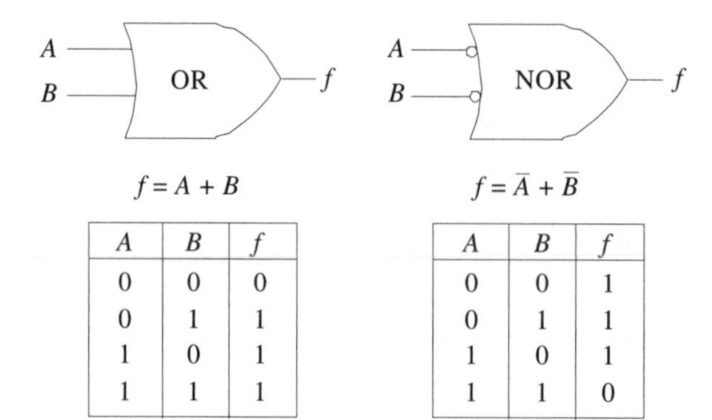

$f = A + B$

A	B	f
0	0	0
0	1	1
1	0	1
1	1	1

$f = \bar{A} + \bar{B}$

A	B	f
0	0	1
0	1	1
1	0	1
1	1	0

Figure 10 Gate logic symbol for OR and NOR gates with their respective truth table.

of indicating 0000 to 1111 (base 2). The term *modulo* is used to describe the count capability of counters; that is, modulo-16 for a four-stage binary counter, modulo-11 for a decade counter, modulo-8 for a three-stage binary counter, and so on. Ripple counters, as shown in Fig. 11, are so named because the count is like a chain reaction that ripples through the counter because of the time involved. The ripple counter is also called an asynchronous counter. Asynchronous means that the events (setting and resetting of FFs) occur one after the other rather than all at once. Because the ripple count is asynchronous, it can produce erroneous indications when the clock speed is high. A high-speed clock can cause the lower

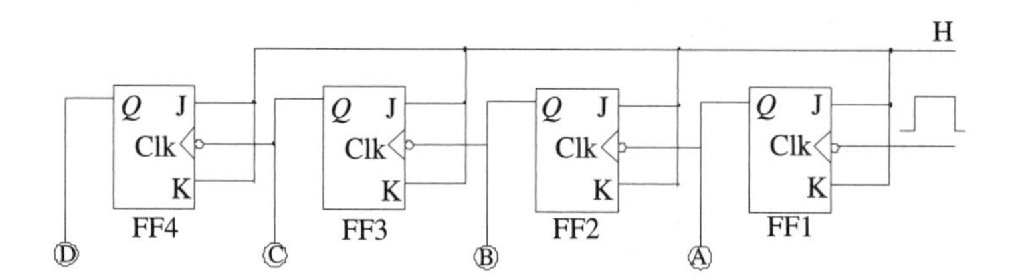

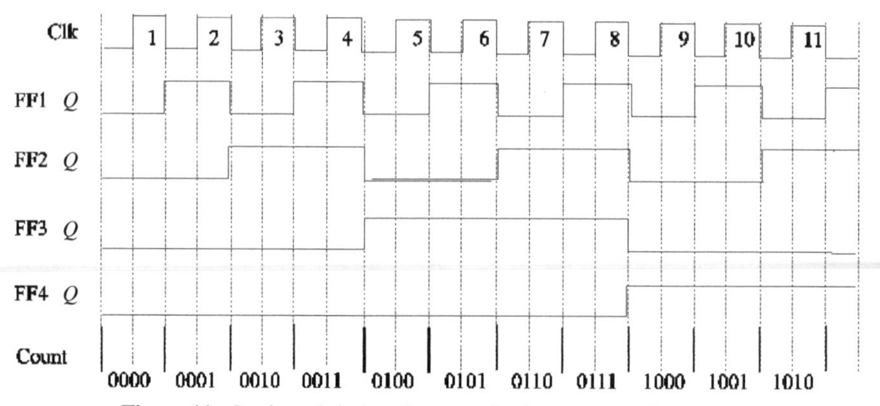

Figure 11 Logic and timing diagram of a four-stage ripple counter.

stage FFs to change state before the upper stages have reacted to the previous clock pulse. The errors are produced by the FFs' inability to keep up with the clock. High-frequency operations require that all the FFs of a counter be triggered at the same time to prevent errors. The synchronous counter is similar to a ripple counter with two exceptions: The clock pulses are applied to each FF, and additional gates are added to ensure that the FFs toggle in the proper sequence.

9 ADDERS, REGISTERS, AND MULTIPLEXERS

Development of most digital systems is a fairly complicated task that usually requires designers to divide the work into manageable building blocks and standardize subunits. For digital designers, these building blocks are typically adders, registers, and multiplexers. These circuits are either combinational or sequential. Combinational circuits have outputs that depend only on the present value or state of the input. Sequential circuits take into consideration the state of the past outputs. A multiplexer is a device for selecting one of several possible input signals and making it available to a specified output. A simple mechanical analogy is that of a manual selector switch such as the dial on a radio that can select AM, FM, cassette, CD, and so on. A register is an ordered set of FFs and is often used for temporary storage of a set of bits for a particular operation. One of the most elegant of registers is that of the D register because of the ability to connect the system clock directly to the device's clock input. Other registers such as the shift register perform lateral movements of data from one bit position to an adjacent position. Shift registers are commonly configured as serial in, parallel out; parallel in, serial out; parallel in, parallel out; and serial in, serial out.

In the previously mentioned FPGA example, we discussed using the FPGA to offload mathematical calculations to reduce the workload on the main processor. To do this, the logic blocks must contain circuits to do these calculations. For example, multiplication is easily accomplished with repeated applications of an adder to find the product of two numbers. Consider, for example, the case where we wanted to use decimal arithmetic to sum two numbers and determine the hundreds digit. This requires not only the addition of the hundreds digit of each number but also the carryover of the tens digits, assuming one exists. This requirement is also true in binary arithmetic. It can be computed by first adding the two bits corresponding to the 2^n digit and then adding the resultant to the carry from the 2^{n-1} digit. With only two inputs, a half adder can be constructed to perform this operation. If three inputs are provided, a full adder can be used.

10 MEMORIES

Read-only memory is programmed with transistors to supply the desired values. These values are usually set by programming them when the memory device is offline or inactive. Programming of ROM is typically done before inserting it into an embedded system. ROM has many purposes, such as the storing of a software program for use by a general-purpose processor, converting binary code to another type of code, and implementing a combinational circuit. ROM can be used to implement any combinational function of k variables by using a $2^k \times 1$ ROM and can implement n functions of the same k variables by using a $2^k \times n$ ROM. For this application, the ROM is simply programmed to implement a truth table for these functions.

Most other systems use user-programmable ROM devices, or PROM, which can be programmed by the chip's user well after the chip has been manufactured. These devices are best suited to prototyping and to low-volume applications. Programming a PROM is accomplished by the user who provides a file indicating the desired ROM contents. A tool called a ROM programmer then configures each programmable connection according to the user-specified file. A basic PROM uses a fuse for each programmable connection. The ROM programmer blows the selected fuses in the array by passing a large current at high voltage through them. However, once a fuse is blown, the connection can never be reestablished. This type of programming cannot be undone. For this reason, basic PROM is often referred to as one-time-programmable device, or OTP. As a result of this ability, OTP devices are frequently used in military and space applications for their robustness in the presence of radiation.[3]

Programmable ROMs do come in erasable versions. One type of erasable PROM, called EPROM, utilizes a floating-gate complementary metal–oxide–semiconductor (CMOS) device as the programmable component, as shown in Fig. 12.[4] Typically, a higher voltage (12–25 V) is applied to the control gate and the drain. This causes avalanche injection whereby electrons tunnel into the floating gate. Once the voltage is removed, the charges are trapped in the floating gate and programming is complete. To erase the programming, the electrons must be excited enough to escape from the gate. One source of energy used is ultraviolet light, or UV. The exposure time to UV is typically 5–30 minutes through a small quartz window on top of the chip. In an effort to eliminate the time spent erasing the programming, the development of an electronically erasable device, or EEPROM, was developed. These devices are more expensive but are also more convenient to use. One type of EEPROM is FLASH memory, which has the advantage of being able to reprogram certain regions of the memory rather than the whole memory at once.

Random-access memory, or RAM, has the advantage of being able to read from and write to. In contrast to ROM, RAM is not programmed. It is empty when placed in the embedded system and is written to or read from during its execution. There are two basic types of RAM: static and dynamic. Static RAM, or SRAM, is the faster version and uses a memory cell consisting of a FF to store a bit. Six transistors are required to hold this bit. The term static refers to the fact that the RAM will hold its data only as long as power is applied. SRAM is typically used for high-end applications such as cache memory.

Dynamic RAM (DRAM) is more compact than SRAM because it has one MOS transistor and a capacitor to store each bit. The disadvantage is that the capacitor cannot store the information indefinitely because it will leak charge and eventually lose the data. This forces the requirement to recharge or refresh this memory. Refresh rates are typically around 15.625 μs. Refreshing can be done by reading the contents of the cell, which forces the data to be stored in a buffer and then rewritten to the memory cell. Refreshing the memory cells tends to make DRAM slower than SRAM.

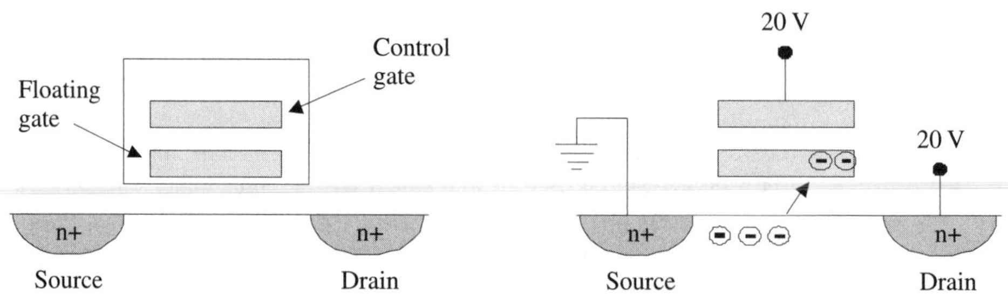

Figure 12 Floating gate transistor programming for erasable PROMs. EPROMs, EEPROMS, and Flash use different methods for controlling the charge of the floating gate.

11 PROGRAMMABLE I/O BLOCKS

An I/O interface block is peripheral logic built to interface the chip to external devices through its package pins. The term *programmable* indicates that the pins can be programmed for a wide variety of I/O standards such as an input, output, or bidirectional. In state-of-the-art FPGA devices, these I/O resources are typically grouped into banks. The compatibility of these standards is based on the standard's supply and reference voltage requirements. In complicated designs that use more than one standard, complications arise that lead to constrained I/O pad placement issues, such as not being able to place certain I/O pads in the same bank. This is a real problem if there is only one bank in the design. Current FPGAs support several I/O standards. For example, some standards require use of a differential amplifier input. With this standard, the external reference voltage must be applied to the amplifier. This is an advantage in that it reduces the I/O voltages swings, which allows for faster switching. Another standard may require a power supply for the I/O blocks. Figure 13 illustrates some standard requirements and an I/O bank layout.

Multiple functionality of I/O resources is important because the number of pins available is limited by the size of the device package. Input/output block (IOB) paths will be placed close to the device I/O pins to overcome issues at input and output. At output, long delays from internal logic block outputs to the IOBs make it difficult to connect at very high clock rates to external synchronous systems. External system setup and hold time issues occur if there are long delays from the I/O pins to the logic block inputs if the external inputs are clocked directly into the logic block FFs without being captured initially by a FF at the IOB pin. A simplified I/O block configuration is shown in Fig. 14.

12 PROGRAMMABLE INTERCONNECTS

The logic blocks in an FPGA are arranged in rows and columns. The number of rows and columns must be equal. Communication among the logic blocks and the I/O cells is typically done through a programmable interconnect network consisting of many wires organized into channels in both the horizontal and vertical directions. Connections between the intersections of the wires (horizontal and vertical) are made at programmable interconnect points (PIPs), sometimes referred to as switch blocks. These points are controlled by writing the configuration RAM, same as with the logic blocks. Wire segments in the programmable interconnect network are bounded by these PIPs and are considered to be either global or local routing

I/O Standard	Direction	V_{ref} Requirement	V_{cco} Requirement
Peripheral component interface (PCI)	Input	Not req.	3.3 V
	Output	Not req.	3.3 V
	Bidirectional	Not req.	3.3 V
Gunning transceiver logic (GTL)	Input	0.8 V	Not req.
	Output	Not req.	Not req.
	Bidirectional	0.8 V	Not req.
High-speed transceiver logic class I (HSTL I)	Input	0.75 V	Not req.
	Output	Not req.	1.5 V
	Bidirectional	0.75 V	1.5 V

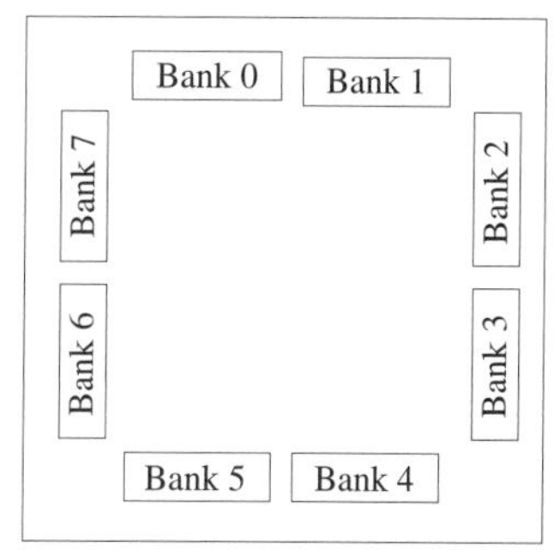

Figure 13 Some I/O standard voltage requirements (left) and organization of Virtex-E FPGA (right).[2]

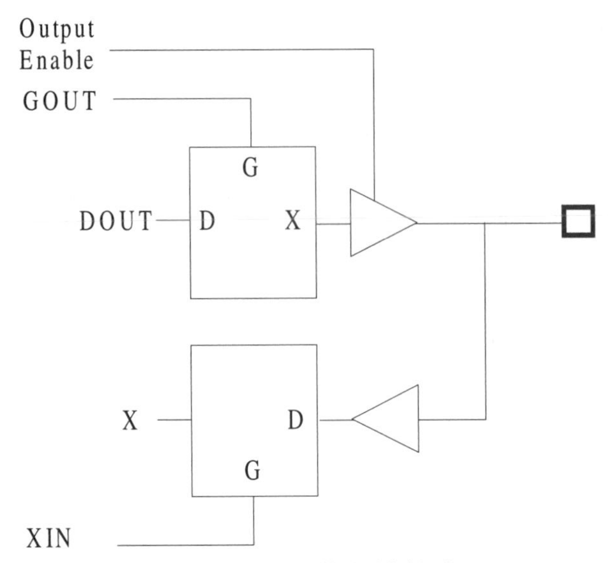

Figure 14 Simplified I/O block.

resources. Global routing resources connect nonadjacent logic blocks and are used to define the combinations of channels. Local routing resources connect a logic block to global routing resources or to adjacent logic blocks. In short, local routing refers to which wire is in each channel. Programming technologies used include Antifuse, SRAM, and EEPROM.

The Antifuse switch, shown in Fig. 15, is a two-terminal device. It is designed so that it offers high resistance when not programmed and once programmed the resistance drops close to 0 Ω. This method offers only one-time programming and any error made during programming cannot be corrected. The Antifuse array is formed with horizontal and vertical wires and with an Antifuse switch at every intersection. Based on the required configuration of the FPGA, high voltage is applied to particular switches. The main disadvantage of this technology is that it requires a separate on-chip circuitry to deliver the high voltage for programming.[5]

In SRAM-based technology, pass gates are used as switches controlled by RAM cells. These RAM cells are usually implemented by LUTs and sometimes multiplexers. Some disadvantages are the huge space requirements needed and the large *RC* delay to the signal, which makes SRAM-based technology slower than Antifuse.[6] Another major disadvantage is that it is volatile, or it loses its memory every time it is powered up. The major advantage is the reprogrammability, which allows for full testing after fabrication.

Lastly, the EEPROM is a viable option. As mentioned previously, it has a control gate and a floating gate. During normal operation, these transistors look like open circuits, and

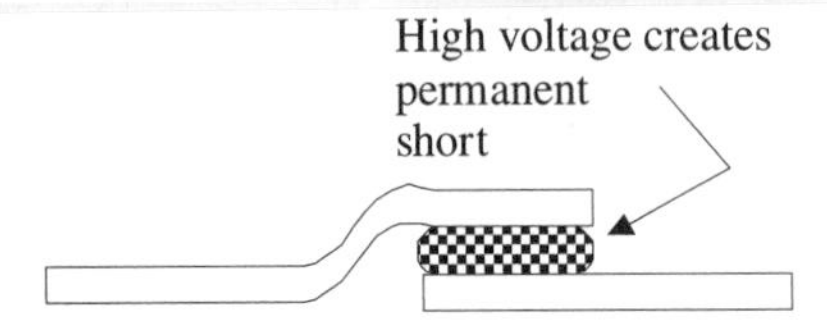

Figure 15 Antifuse technology.

during power-down, the data are held due to the floating gate. This technology has the nonvolatility of the Antifuse with the programmability of the SRAM. Unfortunately, the on-state resistance is higher than that of the Antifuse (before high voltage is applied) and the programming time required is larger than the SRAM technology.

REFERENCES

1. V. Betz and J. Rose, "How Much Logic Should Go in an FPGA Block?" *IEEE Design and Test of Computers,* January–March 1998, pp. 10–15.
2. J. Anderson, J. Saunders, S. Nag, C. Madabhushi, and R. Jayaraman, "A Placement Algorithm for FPGA Designs with Multiple I/O Standards," paper presented at the International Conference on Field-Programmable Logic and Applications (FPL), Villach, Austria, 2000, in *Lecture Notes in Computer Science,* Vol. 1896, Springer-Verlag, pp. 211–220.
3. I. Cyliax, "Learning the Ropes: The FPGA Tour." *Circuit Cellular Online,* Nov. 1999, www.circuitcellular.com/online.
4. R. Patel, M. Wong, J. Costello, D. Reese, V. Bocchino, M. Chu, and J. Turner, "A 90.7 MHz-2.5 Million Transistors CMOS PLD with JTAG Boundary Scan and In-System Programmability," in *Proc. IEEE Custom Integrated Circuits Conference,* May 1–4, 1995, Santa Clara, California, pp. 507–510.
5. M. Ahrens, A. El Gamal, D. Galbraith, J. Greene, and S. Kaptanoglu, K. R. Dharmarajan, L. Hutchings, S. Ku, P. McGibney, J. McGowan, A. Sanie, K. Shaw, N. Stiawalt, T. Whitney, T. Wong, W. Wong, and B. Wu, "An FPGA Family Optimized for High Densities and Reduced Routing Delay," in *Proc. IEEE Custom Integrated Circuits Conference,* May 13–16, 1990, San Jose, California, pp. 31.5.1–31.5.4.
6. H. Hsieh, W. Carter, J. Y. Ja, E. Cheung, S. Schreifels, C. Erickson, P. Freidin, and L. Tinkey, "Third-Generation Architecture Boosts Speed and Density of Field-Programmable Gate Arrays," in *Proc. Custom Integrated Circuits Conf.,* 1990, pp. 31.2.1–31.2.7.

PART 2

SYSTEMS, CONTROLS, AND MEMS

CHAPTER 9
SYSTEMS ENGINEERING: ANALYSIS, DESIGN, AND INFORMATION PROCESSING FOR ANALYSIS AND DESIGN

Andrew P. Sage
School of Information Technology and Engineering
George Mason University
Fairfax, Virginia

1 INTRODUCTION

Systems engineering is a management technology. Technology involves the organization and delivery of science for the (presumed) betterment of humankind. Management involves the interaction of the organization, and the humans in the organization, with the environment. Here, we interpret environment in a very general sense to include the complete external milieu surrounding individuals and organizations. Hence, systems engineering as a management technology involves three ingredients: science, organizations, and their environments. Information, and knowledge, is ubiquitous throughout systems engineering and management efforts and is, in reality, a fourth ingredient. Systems engineering is thus seen to involve science, organizations and humans, environments, technologies, and information and knowledge.

The process of systems engineering involves working with clients in order to assist them in the organization of information and knowledge to aid in judgment and choice of activities that lead to the engineering of trustworthy systems. These activities result in the making of decisions and associated resource allocations through enhanced efficiency, effectiveness, equity, and explicability as a result of systems engineering efforts.

This set of action alternatives is selected from a larger set, in accordance with a value system, in order to influence future conditions. Development of a set of rational policy or action alternatives must be based on formation and identification of candidate alternative policies and objectives against which to evaluate the impacts of these proposed activities, such as to enable selection of efficient, effective, and equitable alternatives for implementation.

In this chapter, we are concerned with the engineering of large-scale systems, or *systems engineering.*[1] We are especially concerned with strategic-level systems engineering, or *systems management.*[2] We begin by first discussing the need for systems engineering and then providing some definitions of systems engineering. We next present a structure describing the systems engineering process. The result of this is a *life-cycle model* for systems engineering processes. This is used to motivate discussion of the functional levels, or considerations, involved in systems engineering efforts: *measurements, systems engineering methods and tools, systems methodology or processes,* and *systems management.* Considerably more details are presented in Refs. 1 and 2, which are the sources from which most of this chapter is derived.

Systems engineering is an appropriate combination of mathematical, behavioral, and management theories in a useful setting appropriate for the resolution of complex real-world issues of large scale and scope. As such, systems engineering consists of the use of management, behavioral, and mathematical constructs to identify, structure, analyze, evaluate, and interpret generally incomplete, uncertain, imprecise, and otherwise imperfect information. When associated with a value system, this information leads to knowledge to permit decisions that have been evolved with maximum possible understanding of their impacts. A central need, but by no means the only need, in systems engineering is to select an appropriate life cycle, or process, that is explicit, rational, and compatible with the implementation framework extant and the perspectives and knowledge bases of those responsible for decision activities. When this is accomplished, an appropriate choice of systems engineering methods and tools may be made to enable full implementation of the life-cycle process.

Information is a very important quantity that is assumed to be present in the management technology that is systems engineering. This strongly couples notions of systems engineering with those of technical direction or systems management of technological development, rather than exclusively with one or more of the methods of systems engineering, important as they may be for the ultimate success of a systems engineering effort. It suggests that *systems engineering is the management technology that controls a total system life-cycle process, which involves and which results in the definition, development, and deployment of a system that is of high quality, trustworthy, and cost-effective in meeting user needs.* This process-oriented notion of systems engineering and systems management will be emphasized here.

Among the appropriate conditions for use of systems engineering are the following:

- There are many considerations and interrelations.
- There are far-reaching and controversial value judgments.
- There are multidisciplinary and interdisciplinary considerations.
- The available information is uncertain, imprecise, incomplete, or otherwise flawed.
- Future events are uncertain and difficult to predict.

- Institutional and organizational considerations play an important role.
- There is a need for explicit and explicable consideration of the efficiency, effectiveness, and equity of alternative courses of action.

There are a number of results potentially attainable from use of systems engineering approaches. These include:

- Identification of perceived needs in terms of identified objectives and values of a client group
- Identification or definition of a set of user or client requirements for the product system or service system that will ultimately be fielded
- Enhanced identification of a wide range of proposed alternatives or policies that might satisfy these needs, achieve the objectives of the clients in a high-quality and trustworthy fashion, and fulfill the requirements definition
- Increased understanding of issues that led to the effort and the impacts of alternative actions upon these issues
- Ranking of these identified alternative courses of action in terms of the utility (benefits and costs) in achieving objectives, satisfying needs, and fulfilling requirements
- A set of alternatives that is selected for implementation, generally by a group of content specialists responsible for detailed design and implementation, and an appropriate plan for action to achieve this implementation

Ultimately these action plans result in a working product or service, each of which is maintained over time in subsequent phases of the postdeployment efforts that also involve systems engineering.

To develop professionals capable of coping satisfactorily with diverse factors involved in widescope problem solving is a primary goal of systems engineering and systems engineering education. This does not imply that a single individual or even a small group can, despite its strong motivation, solve all of the problems involved in a systems study. Such a requirement would demand total and absolute intellectual maturity on the part of the systems engineer and such is surely not realistic. It is also unrealistic to believe that issues can be resolved without very close association with a number of people who have stakes, and who thereby become stakeholders, in problem solution efforts. Consequently, systems engineers must be capable of facilitation and communication of knowledge between the diverse groups of professionals, and their publics, that are involved in wide-scope problem solving. This requires that systems engineers be knowledgeable and able to use not only the technical methods-based tools that are needed for issue and problem resolution, but also the behavioral constructs and management abilities that are needed for resolution of complex, large-scale problems. Intelligence, imagination, and creativity are necessary but not sufficient for proper use of the procedures of systems engineering. Facility in human relations and effectiveness as a broker of information among parties at interest in a systems engineering program are very much needed as well.

It is this blending of the technical, managerial, and behavioral that is a normative goal of success for systems engineering education and for systems engineering professional practice. Thus, systems engineering involves:

- The sciences and the various methods, analysis, and measurement perspectives associated with the sciences
- Life-cycle process models for definition, development, and deployment of systems

- The systems management issues associated with choice of an appropriate process
- Organizations and humans and the understanding of organizational and human behavior
- Environments and understanding of the diverse interactions of organizations of people, technologies, and institutions with their environments
- Information and the way in which it can and should be processed to facilitate all aspects of systems engineering efforts

Successful systems engineering must be practiced at three levels: systems methods and measurements, systems processes and methodology, and systems management. Systems engineers must be aware of a wide variety of methods that assist in the formulation, analysis, and interpretation of contemporary issues. They must be familiar with systems engineering process life cycles (or methodology, as an open set of problem-solving procedures) in order to be able to select eclectic approaches that are best suited to the task at hand. Finally, a knowledge of systems management is necessary in order to be able to select life-cycle processes that are best matched to behavioral and organizational concerns and realities.

All three of these levels, suggested in Fig. 1, are important. To neglect any of them in the practice of systems engineering is to invite failure. It is generally not fully meaningful to talk only of a method or algorithm as a useful system-fielding or life-cycle process. It is ultimately meaningful to talk of a particular process as being useful. A process or product line that is truly useful for the fielding of a system will depend on the methods that are available, the operational environment, and leadership facets associated with use of the system and the system-fielding process. Thus systems management, systems engineering processes, and systems engineering methods and measurements do, separately and collectively, play a fundamental role in systems engineering.

2 THE SYSTEM LIFE CYCLE AND FUNCTIONAL ELEMENTS OF SYSTEMS ENGINEERING

We have provided one definition of systems engineering thus far. It is primarily a structural and process-oriented definition. A related definition, in terms of purpose, is that "systems

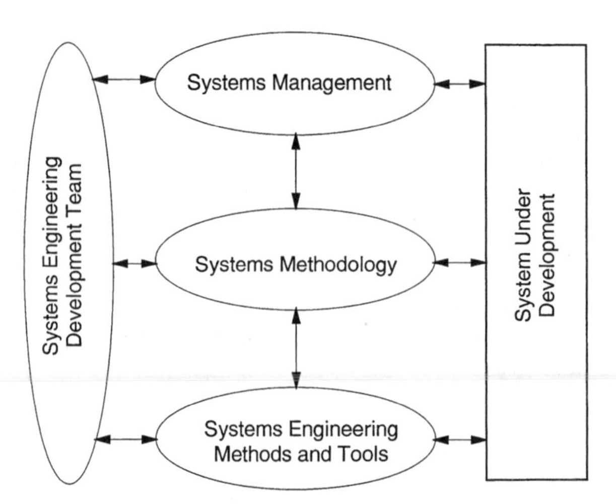

Figure 1 Conceptual illustration of the three levels for systems engineering.

engineering is management technology to assist and support policy-making, planning, decision-making, and associated resource allocation or action deployment for the purpose of acquiring a product desired by customers or clients. Systems engineers accomplish this by quantitative and qualitative formulation, analysis, and interpretation of the impacts of action alternatives upon the needs perspectives, the institutional perspectives, and the value perspectives of their clients or customers." Each of these three steps is generally needed in solving systems engineering problems. Issue *formulation* is an effort to identify the needs to be fulfilled and the requirements associated with these in terms of objectives to be satisfied, constraints and alterables that affect issue resolution, and generation of potential alternative courses of action. Issue *analysis* enables us to determine the impacts of the identified alternative courses of action, including possible refinement of these alternatives. Issue *interpretation* enables us to rank in order the alternatives in terms of need satisfaction and to select one for implementation or additional study. This particular listing of three systems engineering steps and their descriptions is rather formal. Often, issues are resolved this way. The steps of formulation, analysis, and interpretation may also be accomplished on as "as-if" basis by application of a variety of often useful heuristic approaches. These may well be quite appropriate in situations where the problem solver is experientially familiar with the task at hand and the environment into which the task is imbedded.[1]

The key words in this definition are "formulation," "analysis," and "interpretation." In fact, all of systems engineering can be thought of as consisting of formulation, analysis or assessment, and interpretation efforts, together with the systems management and technical direction efforts necessary to bring this about. We may exercise these in a formal sense throughout each of the several phases of a systems engineering life cycle or in an "as-if" or experientially based intuitive sense. These formulation, analysis, and interpretation efforts are the stepwise or microlevel components that comprise a part of the structural framework for systems methodology. They are needed for each phase in a systems engineering effort, although the specific formulation methods, analysis methods, and interpretation methods may differ considerably across the phases.

We can also think of a functional definition of systems engineering: "Systems engineering is the art and science of producing a product, based on phased efforts, that satisfies user needs. The system is functional, reliable, of high quality, and trustworthy, and has been developed within cost and time constraints through use of an appropriate set of methods and tools."

Systems engineers are very concerned with the appropriate *definition, development,* and *deployment* of product systems and service systems. These comprise a set of phases for a systems engineering life cycle. There are many ways to describe the life-cycle phases of the systems engineering process, and we have described a number of them in Refs. 1 and 2. Each of these basic life-cycle models, and those that are outgrowths of them, is comprised of these three phases of definition, development, and deployment. For pragmatic reasons, a typical life cycle will almost always contain more than three phases. Often, it takes on the "waterfall" pattern illustrated in Fig. 2, although there are a number of modifications of the basic waterfall, or "grand-design," life cycles that allow for incremental and evolutionary development of systems life-cycle processes.[2]

A successful approach to systems engineering as an intellectual and action-based approach for increased innovation and productivity and other contemporary challenges must be capable of issue formulation, analysis, and interpretation at the level of institutions and values as well as at the level of symptoms. Systems engineering approaches must allow for the incorporation of need and value perspectives as well as technology perspectives into models and postulates used to evolve and evaluate policies or activities that may result in technological and other innovations.

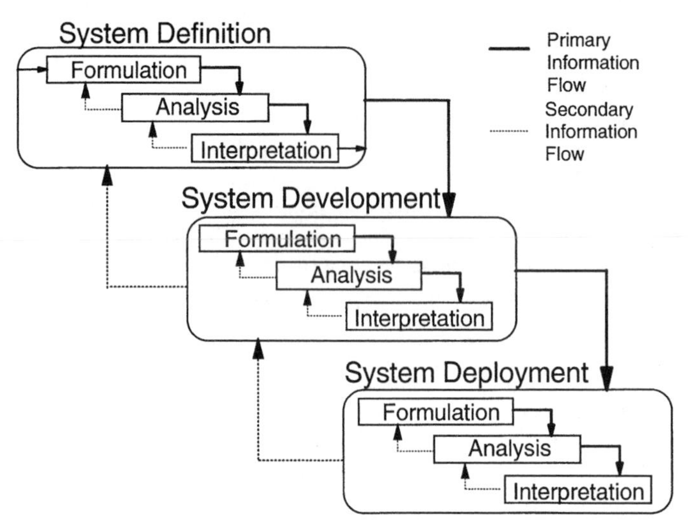

Figure 2 One representation of three systems engineering steps within each of three life-cycle phases.

In actual practice, the steps of the systems process (formulation, analysis, and interpretation) are applied iteratively, across each of the phases of a systems engineering effort, and there is much feedback from one step (and one phase) to the other. This occurs because of the learning that is accomplished in the process of problem solution. Underlying all of this is the need for a general understanding of the diversity of the many systems engineering methods and algorithms that are available and their role in a systems engineering process. The knowledge taxonomy for systems engineering, which consists of the major intellectual categories into which systems efforts may be categorized, is of considerable importance. The categories include systems methods and measurements, systems engineering processes or systems methodology, and systems management. These are used, as suggested in Fig. 3, to produce a *system,* which is a generic term that we use to describe a product or a service.

The methods and metrics associated with systems engineering involve the development and application of concepts that form the basis for problem formulation and solution in systems engineering. Numerous tools for mathematical systems theory have been developed, including operations research (linear programming, nonlinear programming, dynamic programming, graph theory, etc.), decision and control theory, statistical analysis, economic systems analysis, and modeling and simulation. Systems science is also concerned with psychology and human factors concepts, social interaction and human judgment research, nominal group processes, and other behavioral science efforts. Of very special significance for systems engineering is the interaction of the behavioral and the algorithmic components of systems science in the choice-making process. The combination of a set of systems science and operations research methods and a set of relations among these methods and activities constitutes what is known as a *methodology.* References 3 and 4 discuss a number of systems engineering methods and associated methodologies for systems engineering.

As we use it here, a methodology is an open set of procedures that provides the means for solving problems. The tools or the content of systems engineering consists of a variety of algorithms and concepts that use words, mathematics, and graphics. These are structured in ways that enable various problem-solving activities within systems engineering. Particular sets of relations among tools and activities, which constitute the framework for systems

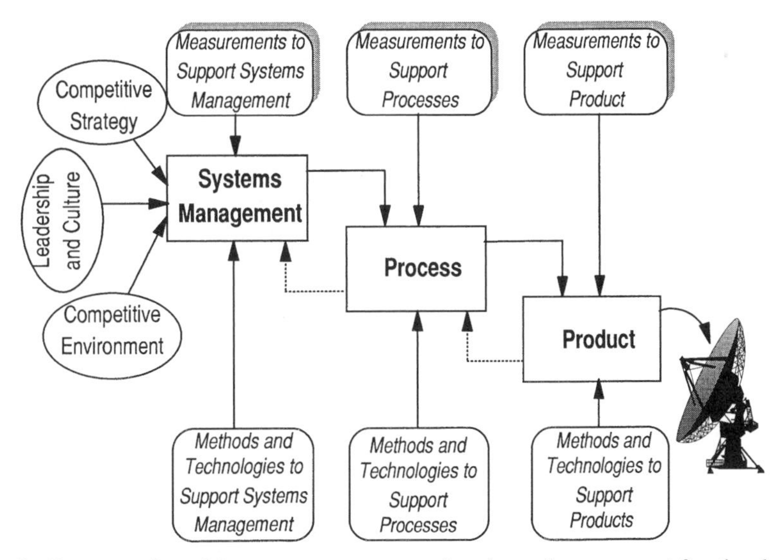

Figure 3 Representation of the structure systems engineering and management functional efforts.

engineering, are of special importance here. Existence and use of an appropriate systems engineering process are of considerable utility in dealing with the many considerations, interrelations, and controversial value judgments associated with contemporary problems.

Systems engineering can be and has been described in many ways. Of particular importance is a morphological description, that is, in terms of form. This description leads to a specific methodology that results in a process* that is useful for fielding a system and/or issue resolution. We can discuss the knowledge dimension of systems engineering. This would include the various disciplines and professions that may be needed in a systems team to allow it to accomplish intended purposes of the team, such as provision of the knowledge base. Alternatively, we may speak of the phases or time dimension of a systems effort. These include system definition, development, and deployment. The deployment phase includes system operation, maintenance, and finally modification or reengineering or ultimate retirement and phase-out of the system. Of special interest are the steps of the logic structure or logic dimension of systems engineering:

- Formulation of issues, or identification of problems or issues in terms of needs and constraints, objectives, or values associated with issue resolution, and alternative policies, controls, hypotheses, or complete systems that might resolve or ameliorate issues
- Analysis of impacts of alternative policies, courses of action, or complete systems
- Interpretation or evaluation of the utility of alternatives and their impacts upon the affected stakeholder group and selection of a set of action alternatives for implementation

* As noted in Refs. 1 and 2, there are life cycles for systems engineering efforts in research, development, test, and evaluation (RDT&E); systems acquisition, production, or manufacturing; and systems planning and marketing. Here, we restrict ourselves to discussions of the life cycle associated with acquisition or production of a system.

We could also associate feedback and learning steps to interconnect these steps one to another. The systems process is typically very iterative. We shall not explicitly show feedback and learning in our conceptual models of the systems process, although it is ideally always there.

Here we have described a three-dimensional morphology of systems engineering. There are a number of systems engineering morphologies or frameworks. In many of these, the logic dimension is divided into a larger number of steps that are iterative in nature. A particular seven-step framework, due to Hall,[5,6] involves:

1. *Problem definition,* in which a descriptive and normative scenario of needs, constraints, and alterables associated with an issue is developed. Problem definition clarifies the issues under consideration to allow other steps of a systems engineering effort to be carried out.
2. *Value system design,* in which objectives and objectives measures or attributes with which to determine success in achieving objectives are determined. Also, the interrelationship between objectives and objectives measures and the interaction between objectives and the elements in the problem definition step are determined. This establishes a measurement framework, which is needed to establish the extent to which the impacts of proposed policies or decisions will achieve objectives.
3. *System synthesis,* in which candidate or alternative decisions, hypotheses, options, policies, or systems that might result in needs satisfaction and objective attainment are postulated.
4. *Systems analysis and modeling,* in which models are constructed to allow determination of the consequences of pursuing policies. Systems analysis and modeling determine the behavior or subsequent conditions resulting from alternative policies and systems. Forecasting and impact analysis are, therefore, the most important objectives of systems analysis and modeling.
5. *Optimization or refinement* of each alternative, in which the individual policies and/or systems are tuned, often by means of parameter adjustment methods, so that each individual policy or system is refined in some "best" fashion in accordance with the value system that has been identified earlier.
6. *Evaluation and decision making,* in which systems and/or policies and/or alternatives are evaluated in terms of the extent to which the impacts of the alternatives achieve objectives and satisfy needs. Needed to accomplish evaluation are the attributes of the impacts of proposed policies and associated objective and/or subjective measurement of attribute satisfaction for each proposed alternative. Often this results in a prioritization of alternatives, with one or more being selected for further planning and resource allocation.
7. *Planning for action,* in which implementation efforts, resource and management allocations, or plans for the next phase of a systems engineering effort are delineated.

More often than not, the information required to accomplish these seven steps is not perfect due to information uncertainty, imprecision, or incompleteness effects. This presents a major challenge to the design of processes and for systems engineering practice as well.

Figure 4 illustrates a not-untypical 49-element morphological box for systems engineering. This is obtained by expanding our initial three systems engineering steps of formulation, analysis, and interpretation to the seven just discussed. The three basic phases of definition, development, and deployment are expanded to a total of seven phases. These seven steps, and the seven phases that we associate with them, are essentially those identified

			Steps of Systems Engineering						
			Formulation			Analysis		Interpretation	
			Problem Definition	Value System Design	System Synthesis	Systems Analysis	Alternative Refinement	Decision Making	Planning for Action
Phases of Systems Engineering	Definition	Program Planning	1	2	3	4	5	6	7
		Project Planning	8						
	Develop-ment	System Development		16					
		Production							
	Deployment	Distribution							
		Operations			38				
		Reengineering or Retirement						48	49

Figure 4 The phases and steps in one 49-element two-dimensional systems engineering framework with activities shown sequentially for waterfall implementation of effort.

by Hall in his pioneering efforts in systems engineering.[5,6] The specific methods we need to use in each of these seven steps are clearly dependent upon the phase of activity that is being completed, and there are a plethora of systems engineering methods available.[3,4] Using a seven-phase, seven-step framework raises the number of activity cells to 49 for a single life cycle. A very large number of systems engineering methods may be needed to fill in this matrix, especially since more than one method will almost invariably be associated with many of the entries.

The requirements and specification phase of the systems engineering life cycle has as its goal the identification of client or stakeholder needs, activities, and objectives for the functionally operational system. This phase should result in the identification and description of preliminary conceptual design considerations for the next phase. It is necessary to translate operational deployment needs into requirement specifications so that these needs may be addressed by the system design efforts. As a result of the requirement specification phase, there should exist a clear definition of development issues such that it becomes possible to make a decision concerning whether to undertake preliminary conceptual design. If the requirement specification effort indicates that client needs can be satisfied in a functionally satisfactory manner, then documentation is typically prepared concerning system-level specifications for the preliminary conceptual design phase. Initial specifications for the following three phases of effort are typically also prepared, and a concept design team is selected to implement the next phase of the life-cycle effort. This effort is sometimes called *system-level architecting*.[7,8] Many[9,10] have discussed technical-level architectures. It is only recently that the need for major attention to architectures at the systems level has also been identified.

Preliminary conceptual system design typically includes, or results in, an effort to specify the content and associated architecture and general algorithms for the system product in question. The desired product of this phase of activity is a set of detailed design and architectural specifications that should result in a useful system product. There should exist a high degree of user confidence that a useful product will result from detailed design or the entire design effort should be redone or possibly abandoned. Another product of this phase is a refined set of specifications for the evaluation and operational deployment phases of the life cycle. In the third phase, these are translated into detailed representations in logical form so that system development may occur. A product, process, or system is produced in the fourth phase of the life cycle. This is not the final system design, but rather the result of implementation of the design that resulted from the conceptual design effort.

Evaluation of the detailed design and the resulting product, process, or system is achieved in the sixth phase of the systems engineering life cycle. Depending upon the specific application being considered, an entire systems engineering life-cycle process could be called *design,* or *manufacturing,* or some other appropriate designator. *System acquisition* is an often-used term to describe the entire systems engineering process that results in an operational systems engineering product. Generally, an acquisition life cycle primarily involves knowledge practices or standard procedures to produce or manufacture a product based on established practices. An RDT&E life cycle is generally associated with an emerging technology and involves knowledge principles. A marketing life cycle is concerned with product planning and other efforts to determine market potential for a product or service and generally involves knowledge perspectives.

The intensity of effort needed for the steps of systems engineering varies greatly with the type of problem being considered. Problems of large scale and scope will generally involve a number of perspectives. These interact and the intensity of their interaction and involvement with the issue under consideration determines the scope and type of effort needed in the various steps of the systems process. Selection of appropriate algorithms or approaches to enable completion of these steps and satisfactory transition to the next step, and ultimately to completion of each phase of the systems engineering effort, are major systems engineering tasks.

Each of these phases of a systems engineering life cycle is very important for sound development of physical systems or products and such service systems as information systems. Relatively less attention appears to have been paid to the requirement specification phase than to the other phases of the systems engineering life-cycle process. In many ways, the requirement specification phase of a systems engineering design effort is the most important. It is this phase that has as its goal the detailed definition of the needs, activities, and objectives to be fulfilled or achieved by the process to be ultimately developed. Thus, this phase strongly influences all the phases that follow. It is this phase that describes preliminary design considerations that are needed to achieve successfully the fundamental goals underlying a systems engineering study. It is in this phase that the information requirements and the method of judgment and choice used for selection of alternatives are determined. Effective systems engineering, which inherently involves design efforts, must also include an operational evaluation component that will consider the extent to which the product or service is useful in fulfilling the requirements that it is intended to satisfy.

3 SYSTEMS ENGINEERING OBJECTIVES

Ten performance objectives appear to be of primary importance to those who desire to evolve quality plans, forecasts, decisions, or alternatives for action implementation:

1. Identify needs, constraints, and alterables associated with the problem, issue, or requirement to be resolved (problem definition).
2. Identify a planning horizon or time interval for alternative action implementation, information flow, and objective satisfaction (planning horizon, identification).
3. Identify all significant objectives to be fulfilled, values implied by the choice of objectives, and objectives measures or attributes associated with various outcome states, with which to measure objective attainment (value system design).

4. Identify decisions, events, and event outcomes and the relations among them such that a structure of the possible paths among options, alternatives, or decisions and the possible outcomes of these emerge (impact assessment).
5. Identify uncertainties and risks associated with the environmental influences affecting alternative decision outcomes (probability identification).
6. Identify measures associated with the costs and benefits or attributes of the various outcomes or impacts that result from judgment and choice (worth, value, or utility measurement).
7. Search for and evaluate new information, and the cost-effectiveness of obtaining this information, relevant to improved knowledge of the time-varying nature of event outcomes that follow decisions or choice of alternatives (information acquisition and evaluation).
8. Enable selection of a best course of action in accordance with a rational procedure (decision assessment and choice making).
9. Reexamine the expected effectiveness of all feasible alternative courses of action, including those initially regarded as unacceptable, prior to making a final alternative selection (sensitivity analysis).
10. Make detailed and explicit provisions for implementation of the selected action alternative, including contingency plans, as needed (planning for implementation of action).

These objectives are, of course, very closely related to the aforementioned steps of the framework for systems engineering. To accomplish them requires attention to and knowledge of the methods of systems engineering such that we are able to design product systems and service systems and also enabling systems to support products and services. We also need to select an appropriate process, or product line, to use for management of the many activities associated with fielding a system. Also required is much effort at the level of systems management so that the resulting process is efficient, effective, equitable, and explicable. Thus, it is necessary to ensure that those involved in systems engineering efforts be concerned with technical knowledge of the issue under consideration, able to cope effectively with administrative concerns relative to the human elements of the issue, interested in and able to communicate across those actors involved in the issue, and capable of innovation and outscoping of relevant elements of the issue under consideration. These attributes (technical knowledge, human understanding and administrative ability, communicability, and innovativeness) are, of course, primary attributes of effective management.

4 SYSTEMS ENGINEERING METHODOLOGY AND METHODS

A variety of methods are suitable to accomplish the various steps of systems engineering. We shall briefly describe some of them here.

4.1 Issue Formulation

As indicated above, issue formulation is the step in the systems engineering effort in which the problem or issue is defined (problem definition) in terms of the objectives of a client group (value system design) and where potential alternatives that might resolve needs are identified (system synthesis). Many studies have shown that the way in which an issue is resolved is critically dependent on the way in which the issue is formulated or framed. The issue formulation effort is concerned primarily with identification and description of the elements of the issue under consideration, with, perhaps, some initial effort at structuring these in order to enhance understanding of the relations among these elements. Structural concerns are also of importance in the analysis effort. The systems process is iterative and interactive, and the results of preliminary analysis are used to refine the issue formulation effort. Thus, the primary intent of issue formulation is to identify relevant elements that represent and are associated with issue definition, the objectives that should be achieved in order to satisfy needs, and potential action alternatives.

There are at least four ways to accomplish issue formulation, or to identify requirements for a system, or to accomplish the initial part of the definition phase of systems engineering:

1. Asking stakeholders in the issue under consideration for the requirements
2. Descriptive identification of the requirements from a study of presently existing systems
3. Normative synthesis of the requirements from a study of documents describing what "should be," such as planning documents
4. Experimental discovery of requirements, based on experimentation with an evolving system

These approaches are neither mutually exclusive nor exhaustive. Generally, the most appropriate efforts will use a combination of these approaches.

There are conflicting concerns with respect to which blend of these requirement identification approaches is most appropriate for a specific task. The asking approach seems very appropriate when there is little uncertainty and imprecision associated with the issue under consideration, so that the issue is relatively well understood and may be easily structured, and where members of the client group possess much relevant expertise concerning the issue and the environment in which the issue is embedded. When these characteristics of the issue—lack of imprecision and presence of expert experiential knowledge—are present, then a direct declarative approach based on direct "asking" of "experts" is a simple and efficient approach. When there is considerable imprecision or a lack of experiential familiarity with the issue under concern, the other approaches take on greater significance. The asking approach is also prone to a number of human information-processing biases, as will be discussed in Section 4.5. This is not as much of a problem in the other approaches.

Unfortunately, however, there are other difficulties with each of the other three approaches. Descriptive identification, from a study of existing systems of issue formulation elements, will very likely result in a new system that is based or anchored on an existing system and tuned, adjusted, or perturbed from this existing system to yield incremental improvements. Thus, it is likely to result in incremental improvements to existing systems but not to result in major innovations or totally new systems and concepts.

Normative synthesis from a study of planning documents will result in an issue formulation or requirement identification effort that is based on what have been identified as desirable objectives and needs of a client group. A plan at any given phase may well not exist or it may be flawed in any of several ways. Thus, the information base may well not be present or may be flawed. When these circumstances exist, it will not be a simple task to accomplish effective normative synthesis of issue formulation elements for the next phase of activity from a study of planning documents relative to the previous phase.

Often it is not easily possible to determine an appropriate set of issue formulation elements or requirements. Often it will not be possible to define an appropriate set of issue formulation efforts prior to actual implementation of a preliminary system design. There are many important issues where there is an insufficient experiential basis to judge the effectiveness and completeness of a set of issue formulation efforts or requirements. Often, for example, clients will have difficulty in coping with very abstract formulation requirements and in visualizing the system that may ultimately evolve. Thus, it may be useful to identify an initial set of issue formulation elements and accomplish subsequent analysis and interpretation based on these, without extraordinary concern for completeness of the issue formulation efforts. A system designed with ease of adaptation and change as a primary requirement is implemented on a trial basis. As users become familiar with this new system or process, additions and modifications to the initially identified issue formulation elements result. Such a system is generally known as a *prototype*. One very useful support for the identification of requirements is to build a prototype and allow the users of the system to be fielded to experiment with the prototype and, through this experimentation, to identify system requirements.[11] This heuristic approach allows users to identify the requirements for a system by experimenting with an easily changeable set of system design requirements and to improve their identification of these issue formulation elements as their experiential familiarity with the evolving prototype system grows.

The key parts of the problem definition step of issue formulation involve identification of needs, constraints, and alterables and determination of the interactions among these elements and the group that they impact. Need is a condition requiring supply or relief or is a lack of something required, desired, or useful. In order to define a problem satisfactorily, we must determine the alterables or those items pertaining to the needs that can be changed. Alterables can be separated into those over which control is or is not possible. The controllable alterables are of special concern in systems engineering since they can be changed or modified to assist in achieving particular outcomes. To define a problem adequately, we must also determine the limitations or constraints under which the needs can or must be satisfied and the range over which it is permissible to vary the controllable alterables. Finally, we must determine relevant groups of people who are affected by a given problem.

Value system design is concerned with defining objectives, determining their interactions, and ordering these into a hierarchical structure. Objectives and their attainment are, of course, related to the needs, alterables, and constraints associated with problem definition. Thus, the objectives can and should be related to these problem definition elements. Finally, a set of measures is needed whereby to measure objective attainment. Generally, these are called *attributes of objectives* or *objectives measures*. It is necessary to ensure that all needs are satisfied by attainment of at least one objective.

The first step in system synthesis is to identify activities and alternatives for attaining each of the objectives or the postulation of complete systems to this end. It is then desirable to determine interactions among the proposed activities and to illustrate relationships between the activities and the needs and objectives. Activities measures are needed to gauge the degree of accomplishment of proposed activities. Systemic methods useful for problem definition are generally useful for value system design and system synthesis as well. This is another reason that suggests the efficacy of aggregating these three steps under a single heading: *issue formulation*.

Complex issues will have a structure associated with them. In some problem areas, structure is well understood and well articulated. In other areas, it is not possible to articulate structure in such a clear fashion. There exists considerable motivation to develop techniques with which to enhance structure determination, as a system structure must always be dealt with by individuals or groups, regardless of whether the structure is articulated or not. Furthermore, an individual or a group can deal much more effectively with systems and make better decisions when the structure of the underlying system is well defined and exposed and communicated clearly. One of the fundamental objectives of systems engineering is to structure knowledge elements such that they are capable of being better understood and communicated.

We now discuss several formal methods appropriate for "asking" as a method of issue formulation. Most of these, and other, approaches are described in Refs. 1, 3, and 4. Then we shall very briefly contrast and compare some of these approaches. The methods associated with the other three generic approaches to issue formulation also involve approaches to analysis that will be discussed in the next section.

Several of the formal methods that are particularly helpful in the identification, through asking, of issue formulation elements are based on principles of collective inquiry, in which interested and motivated people are brought together to stimulate each other's creativity in generating issue formulation elements. We may distinguish two groups of collective-inquiry methods:

1. *Brainwriting, Brainstorming, Synectics, Nominal Group Technique,* and *Charette.* These approaches typically require a few hours of time, a group of knowledgeable people gathered in one place, and a group leader or facilitator. Brainwriting is typically better than brainstorming in reducing the influence of dominant individuals. Both methods can be very productive: 50–150 ideas or elements might be generated in less than an hour. Synectics, based on problem analogies, might be appropriate if there is a need for truly unconventional, innovative ideas. Considerable experience with the method is a requirement, however, particularly for the group leader. The nominal group technique is based on a sequence of idea generation, discussion, and prioritization. It can be very useful when an initial screening of a large number of ideas or elements is needed. Charette offers a conference or workshop-type format for generation and discussion of ideas and/or elements.
2. *Questionnaires, Surveys,* and *Delphi.* These three methods of collective-inquiry modeling do not require the group of participants to gather at one place and time, but they typically take more time to achieve results than the first group of methods. In questionnaires and surveys, a usually large number of participants are asked, on an individual basis, for ideas or opinions, which are then processed to achieve an overall result. There is no interaction among participants. Delphi usually provides for written interaction among participants in several rounds. Results of previous rounds are fed back to participants, who are asked to comment, revise their views as desired, and so on. A Delphi exercise can be very instructive but usually takes several weeks or months to complete.

Use of most structuring methods, in addition to leading to greater clarity of the problem formulation elements, will also typically lead to identification of new elements and revision of element definitions. As we have indicated, most structuring methods contain an analytical component; they may, therefore, be more properly labeled analysis methods. The following element-structuring aids are among the many modeling aids available:

- *Interaction matrices* may be used to identify clusters of closely related elements in a large set, in which case we have a self-interaction matrix, or to structure and identify the couplings between elements of different sets, such as objectives and alternatives. In this case, we produce cross-interaction matrices, such as shown in Fig. 5. Interaction matrices are useful for initial, comprehensive exploration of sets of elements. Learning about problem interrelationships during the process of constructing an interaction matrix is a major result of use of these matrices.
- *Trees* are graphical aids particularly useful in portraying hierarchical or branching-type structures. They are excellent for communication, illustration, and clarification. Trees may be useful in all steps and phases of a systems effort. Figure 6 presents an

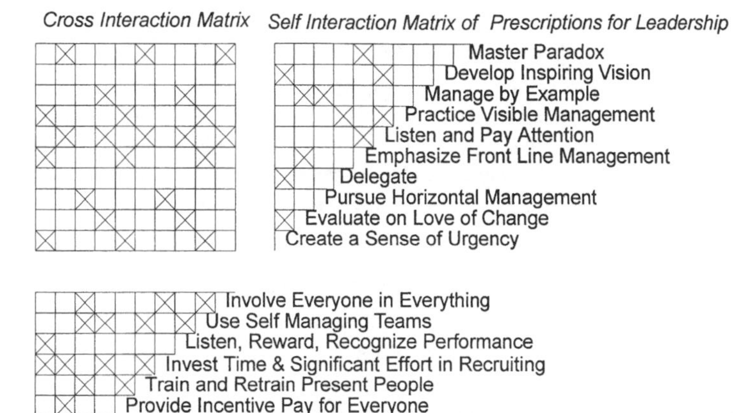

Figure 5 Hypothetical self- and cross-interaction matrices for prescriptions for leadership and for empowering people at all levels.

1. Understanding of Problem
   - 1.1 Navy Cost Credentialing Process
   - 1.2 NAVELEX Cost Analysis Methodology
   - 1.3 DoD Procurement Procedures
2. Technical Approach
   - 2.1 Establishment of a Standard Methodology
   - 2.2 Compatibility with Navy Acquisition Process
3. Staff Experience
   - 3.1 Directly Related Experience in Cost Credentialing, Cost Analysis, and Procurement Procedures
   - 3.2 Direct Experience with Navy R&D Programs
4. Corporate Qualification

   Relevant Experience in Cost Analysis for Navy R&D Programs
5. Management Approach
   - 5.1 Quality/Relevance
   - 5.2 Organization and Control Effectiveness
6. Cost
   - 6.1 Manner in which Elements of Cost Contribute Directly to Project Success
   - 6.2 Appropriate of Cost Mix to the Technical Effort

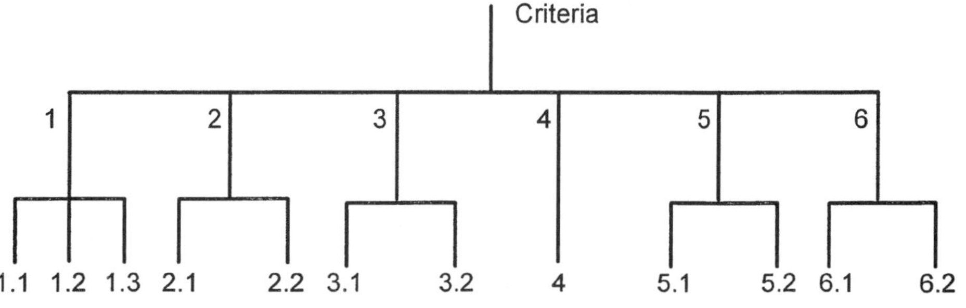

Figure 6 Possible attribute tree for evaluation of proposals concerning cost credentialing.

attribute tree that represents those aspects that will be formally considered in the evaluation and prioritization of a set of proposals.

- *Causal loop diagrams,* or influence diagrams, represent graphical pictures of causal interactions between sets of variables. They are particularly helpful in making explicit one's perception of the causes of change in a system and can serve very well as communication aids. A causal loop diagram is also useful as the initial part of a detailed simulation model. Figure 7 represents a causal loop diagram of a belief structure.

Two other descriptive methods are potentially useful for issue formulation:

- The *system definition matrix,* options profile, decision balance sheet, or checklist provides a framework for specification of the essential aspects, options, or characteristics of an issue, a plan, a policy, or a proposed or existing system. It can be helpful for the design and specification of alternative policies, designs, or other options or alternatives. The system definition matrix is just a table that shows important aspects of the options that are important for judgment relative to selection of approaches to issue formulation or requirement determination.
- *Scenario writing* is based on narrative and creative descriptions of existing or possible situations or developments. Scenario descriptions can be helpful for clarification and communication of ideas and obtaining feedback on those ideas. Scenarios may also be helpful in conjunction with various analysis and forecasting methods, where they may represent alternative or opposing views.

Clearly, successful formulation of issues through "asking" requires creativity. Creativity may be much enhanced through use of a structured systems engineering framework. For example, group meetings for issue formulation involve idea formulation, idea analysis, and idea interpretation. The structure of a group meeting may be conceptualized within a systems engineering framework. This framework is especially useful for visualizing the trade-offs that

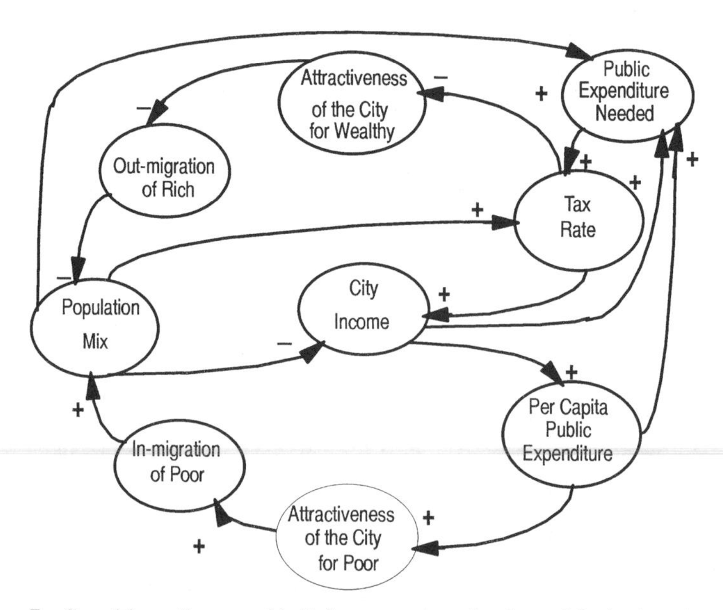

Figure 7 Causal loop diagram of belief structure in a simple model of urban dynamics.

must be made among allocation of resources for formulation, analysis, and interpretation of ideas in the issue formulation step itself. If there is an emphasis on idea formulation, we shall likely generate too many ideas to cope with easily. This will lead to a lack of attention to detail. On the other hand, if idea formulation is deemphasized, we shall typically encourage defensive avoidance through undue efforts to support the present situation or a rapid unconflicted change to a new situation. An overemphasis on analysis of ideas is usually time consuming and results in a meeting that seems to drown in details. There is inherent merit in encouraging a group to reach consensus, but the effort may also be inappropriate, since it may encourage arguments over concerns that are ineffective in influencing judgments.

Deemphasizing the analysis of identified ideas will usually result in disorganized meetings in which hasty, poorly thought-out ideas are accepted. Postmeeting disagreements concerning the results of the meeting are another common disadvantage. An emphasis on interpretation of ideas will produce a meeting that is emotional and people centered. Misunderstandings will be frequent as issues become entrenched in an adversarial, personality-centered process. On the other hand, deemphasizing the interpretation of ideas results in meetings in which important information is not elicited. Consequently, the meeting is awkward and empty, and routine acceptance of ideas is a likely outcome.

4.2 Issue Analysis

In systems engineering, issue analysis involves forecasting and assessing of the impacts of proposed alternative courses of action. In turn, this suggests construction, testing, and validation of models. Impact assessment in systems engineering includes system analysis and modeling and optimization and ranking or refinement of alternatives. First, the options or alternatives defined in issue formulation are structured, often as part of the issue formulation effort, and then analyzed in order to assess the anticipated impacts that may result from their implementation. Second, a refinement or optimization effort is often desirable. This is directed toward refinement or fine-tuning a viable alternative and parameters within an alternative, so as to obtain maximum needs satisfaction, within given constraints, from a proposed policy.

To determine the structure of systems in the most effective manner requires the use of quantitative analysis to direct the structuring effort along the most important and productive paths. This is especially needed when time available to construct structural models is limited. Formally, there are at least four types of self-interaction matrices: nondirected graphs, directed graphs (or digraphs), signed digraphs, and weighted digraphs. The theory of digraphs and structural modeling is authoritatively presented in Ref. 12 and a number of applications to what is called *interpretative structural modeling* are described in Refs. 3, 13, 14, and 15. Cognitive map structural models are considered in Ref. 16. A development of structural modeling concepts based on signed digraphs is discussed in Ref. 17. Geoffrion has been especially concerned with the development of a structured modeling methodology[18,19] and environment. He has noted[20] that a modeling environment needs five quality- and productivity-related properties. A modeling environment should:

1. Nurture the entire modeling life cycle, not just a part of it
2. Be hospitable to decision- and policy-makers as well as to modeling professionals
3. Facilitate the maintenance and ongoing evolution of those models and systems that are contained therein
4. Encourage and support those who use it to speak the same paradigm-neutral language in order to best support the development of modeling applications
5. Facilitate management of all the resources contained therein

Structured modeling is a general conceptual framework for modeling. Cognitive maps, interaction matrices, intent structures, Delta charts, objective and attribute trees, causal loop diagrams, decision outcome trees, signal flow graphs, and so on are all structural models that are very useful graphic aids to communications. The following are requirements for the processes of structural modeling:

1. An object system, which is typically a poorly defined, initially unstructured set of elements to be described by a model
2. A representation system, which is a presumably well-defined set of relations
3. An embedding of perceptions of some relevant features of the object system into the representation system

Structural modeling, which has been of fundamental concern for some time, refers to the systemic iterative application of typically graph-theoretic notions such that an easily communicable directed-graph representation of complex patterns of a particular contextual relationship among a set of elements results. There are a number of computer software realizations of various structural modeling constructs, such as cognitive policy evaluation (COPE), interpretive structural modeling (ISM), and various multiattribute utility theory-based representations, as typically found in most decision-aiding software.

Transformation of a number of identified issue formulation elements, which typically represent unclear, poorly articulated mental models of a system, into visible, well-defined models useful for many purposes is the object of systems analysis and modeling. The principal objective of systems analysis and modeling is to create a process with which to produce information concerning consequences of proposed actions or policies. From the issue formulation steps of problem definition, value system design, and system synthesis, we have various descriptive and normative scenarios available for use. Ultimately, as a part of the issue interpretation step, we wish to evaluate and compare alternative courses of action with respect to the value system through use of a systems model. A model is always a substitute for reality but is, one hopes, descriptive enough of the system elements under consideration to be useful. By posing a variety of questions using the model, we can, from the results obtained, learn how to cope with that subset of the real world being modeled.

A model must depend on much more than the particular problem definition elements being modeled; it must also depend strongly on the value system and the purpose behind construction and utilization of the model. These influence, generally strongly, the structure of the situation and the elements that comprise this structure. Which elements a client believes important enough to include in a model depend on the client's value system.

We wish to be able to determine correctness of predictions and forecasts that are based on model usage. Given the definition of a problem, a value system, and a set of proposed policies, we wish to be able to design a simulation model consisting of relevant elements of these three steps and to determine the results or impacts of implementing proposed policies. Following this, we wish to be able to validate a simulation model to determine the extent to which it represents reality sufficiently to be useful. Validation must, if we are to have confidence in what we are doing with a model, precede actual use of model-based results.

There are three essential steps in constructing a simulation model:

1. Determination of those problem definitions, value systems, and system synthesis elements that are most relevant to a particular problem
2. Determination of the structural relationships among these elements
3. Determination of parametric coefficients within the structure

There are three uses to which models may normally be put. Model categories corresponding to these three uses are descriptive models, predictive or forecasting models, and policy or planning models. Representation and replication of important features of a given problem are the objects of a descriptive model. Good descriptive models are of considerable value in that they reveal much about the substance of complex issues and how, typically in a retrospective sense, change over time has occurred. One of the primary purposes behind constructing a descriptive model is to learn about the past. Often the past will be a good guide to the future.

In building a predictive or forecasting model, we must be especially concerned with determination of proper cause-and-effect relationships. If the future is to be predicted with integrity, we must have a method with which to determine exogenous variables, or input variables that result from external causes, accurately. Also, the model structure and parameters within the structure must be valid for the model to be valid. Often, it will not be possible to predict accurately all exogenous variables; in that case, conditional predictions can be made from particular assumed values of unknown exogenous variables.

The future is inherently uncertain. Consequently, predictive or forecasting models are often used to generate a variety of future scenarios, each a conditional prediction of the future based on some conditioning assumptions. In other words, we develop an "if–then" model.

Policy or planning models are much more than predictive or forecasting models, although any policy or planning model is also a predictive or forecasting model. The outcome from a policy or planning model must be evaluated in terms of a value system. Policy or planning efforts must not only predict outcomes from implementing alternative policies but also present these outcomes in terms of the value system that is in a form useful and suitable for alternative ranking, evaluation, and decision making. Thus, a policy model must contain some provision for impact interpretation.

Model usefulness cannot be determined by objective truth criteria alone. Well-defined and well-stated functions and purposes for the simulation model are needed to determine simulation model usefulness. Fully objective criteria for model validity do not typically exist. Development of a general-purpose, context-free simulation model appears unlikely; the task is simply far too complicated. We must build models for specific purposes, and thus the question of model validity is context dependent.

Model credibility depends on the interaction between the model and model user. One of the major potential difficulties is that of building a model that reflects the outlook of the modeler. This activity is proscribed in effective systems engineering practice, since the purpose of a model is to describe systematically the "view" of a situation held by the client, not that held by the analyst.

A great variety of approaches have been designed and used for the forecasting and assessment that are the primary goals of systems analysis. There are basically two classes of methods that we describe here: expert-opinion methods and modeling and/or simulation methods.

Expert-opinion methods are based on the assumption that knowledgeable people will be capable of saying sensible things about the impacts of alternative policies on the system, as a result of their experience with or insight into the issue or problem area. These methods are generally useful, particularly when there are no established theories or data concerning system operation, precluding the use of more precise analytical tools. Among the most prominent expert-opinion-based forecasting methods are surveys and Delphi. There are, of course, many other methods of asking experts for their opinion—for example, hearings, meetings, and conferences. A particular problem with such methods is that cognitive bias and value incoherence are widespread, often resulting in inconsistent and self-contradictory results.

There exists a strong need in the forecasting and assessment community to recognize and ameliorate, by appropriate procedures, the effects of cognitive bias and value incoherence in expert-opinion-modeling efforts. Expert-opinion methods are often appropriate for the "asking" approach to issue formulation. They may be of considerably less value, especially when used as stand-alone approaches, for impact assessment and forecasting.

Simulation and modeling methods are based on the conceptualization and use of an abstraction or model of the real world intended to behave in a similar way to the real system. Impacts of policy alternatives are studied in the model, which will, it is hoped, lead to increased insight with respect to the actual situation.

Most simulation and modeling methods use the power of mathematical formulations and computers to keep track of many pieces of information at the same time. Two methods in which the power of the computer is combined with subjective expert judgments are cross-impact analysis and workshop dynamic models. Typically, experts provide subjective estimates of event probabilities and event interactions. These are processed by a computer to explore their consequences and fed back to the analysts and thereafter to the experts for further study. The computer derives the resulting behavior of various model elements over time, giving rise to renewed discussion and revision of assumptions.

Expert judgment is virtually always included in all modeling methods. Scenario writing can be an expert-opinion-modeling method, but typically this is done in a less direct and explicit way than in Delphi, survey, ISM, cross-impact, or workshop dynamic models. As a result, internal inconsistency problems are reduced with those methods based on mathematical modeling. The following list describes six additional forecasting methods based on mathematical modeling and simulation. In these methods, a structural model is generally formed on the basis of expert opinion and physical or social laws. Available data are then processed to determine parameters within the structure. Unfortunately, these methods are sometimes very data intensive and, therefore, expensive and time consuming to implement.

- Trend extrapolation/time-series forecasting is particularly useful when sufficient data about past and present developments are available, but there is little theory about underlying mechanisms causing change. The method is based on the identification of a mathematical description or structure that will be capable of reproducing the data into the future, typically over the short to medium term.
- Continuous-time dynamic simulation is based on postulation and qualification of a causal structure underlying change over time. A computer is used to explore long-range behavior as it follows from the postulated causal structure. The method can be very useful as a learning and qualitative forecasting device, but its application may be rather costly and time consuming.
- Discrete-event digital simulation models are based on applications of queuing theory to determine the conditions under which system outputs or states will switch from one condition to another.
- Input–output analysis has been specially designed for study of equilibrium situations and requirements in economic systems in which many industries are interdependent. Many economic data fit in directly to the method, which mathematically is relatively simple and can handle many details.
- Econometrics is a method mainly applied to economic description and forecasting problems. It is based on both theory and data, with, usually, the main emphasis on specification of structural relations based on macroeconomic theory and the derivation of unknown parameters in behavioral equations from available economic data.

- Microeconomic models represent an application of economic theories of firms and consumers who desire to maximize the profit and utility of their production and consumption alternatives.

Parameter estimation is a very important subject with respect to model construction and validation. Observation of basic data and estimation or identification of parameters within an assumed structure, often denoted as system identification, are essential steps in the construction and validation of system models. The simplest estimation procedure, in both concept and implementation, appears to be the least-squares error estimator. Many estimation algorithms to accomplish this are available and are in actual use. The subjects of parameter estimation and system identification are being actively explored in both economics and systems engineering. There are numerous contemporary results, including algorithms for system identification and parameter estimation in very-large-scale systems representative of actual physical processes and organizations.

Verification of a model is necessary to ensure that the model behaves in a fashion intended by the model builder. If we can determine that the structure of the model corresponds to the structure of the elements obtained in the problem definition, value system design, and system synthesis steps, then the model is verified with respect to behaving in a gross, or structural, fashion, as the model builder intends.

Even if a model is verified in a structural as well as parametric sense, there is still no assurance that the model is valid in the sense that predictions made from the model will occur. We can determine validity only with respect to the past. That is all that we can possibly have available at the present. Forecasts and predictions inherently involve the future. Since there may be structural and parametric changes as the future evolves, and since knowledge concerning results of policies not implemented may never be available, there is usually no way to validate a model completely. Nevertheless, there are several steps that can be used to validate a model. These include a reasonableness test in which we determine that the overall model, as well as model subsystems, responds to inputs in a reasonable way, as determined by "knowledgeable" people. The model should also be valid according to statistical time series used to determine parameters within the model. Finally, the model should be epistemologically valid, in that the policy interpretations of the various model parameters, structure, and recommendations are consistent with ethical, professional, and moral standards of the group affected by the model.

Once a model has been constructed, it is often desirable to determine, in some best fashion, various policy parameters or controls that are subject to negotiation. The optimization or refinement-of-alternatives step is concerned with choosing parameters or controls to maximize or minimize a given performance index or criterion. Invariably, there are constraints that must be respected in seeking this extremum. As previously noted, the analysis step of systems engineering consists of systems analysis and modeling and optimization or refinement of alternatives and related methods that are appropriate in aiding effective judgment and choice.

There exist a number of methods for fine tuning, refinement, or optimization of individual specific alternative policies or systems. These are useful in determining the best (in terms of needs satisfaction) control settings or rules of operation in a well-defined, quantitatively describable system. A single scalar indicator of performance or desirability is typically needed. There are, however, approaches to multiple objective optimization that are based on welfare-type optimization concepts. It is these individually optimized policies or systems that are an input to the evaluation and decision-making effort in the interpretation step of systems engineering.

Among the many methods for optimization and refinement of alternatives are:

- *Mathematical programming,* which is used extensively for operations research and analysis practice, resource allocation under constraints, resolution of planning or scheduling problems, and similar applications. It is particularly useful when the best equilibrium or one-time setting has to be determined for a given policy or system.
- *Optimum systems control,* which addresses the problem of determining the best controls or actions when the system, the controls or actions, the constraints, and the performance index may change over time. A mathematical description of system change is necessary to use this approach. Optimum systems control is particularly suitable for refining controls or parameters in systems in which changes over time play an important part.

Application of the various refinement or optimization methods, like those described here, typically requires significant training and experience on the part of the systems analyst. Some of the many characteristics of analysis that are of importance for systemic efforts include the following:

1. Analysis methods are invaluable for understanding the impacts of proposed policy.
2. Analysis methods lead to consistent results if cognitive bias issues associated with expert forecasting and assessment methods are resolved.
3. Analysis methods may not necessarily lead to correct results since "formulation" may be flawed, perhaps by cognitive bias and value incoherence.

Unfortunately, however, large models and large optimization efforts are often expensive and difficult to understand and interpret. There are a number of possibilities for "paralysis through analysis" in the unwise use of systems analysis. On the other hand, models and associated analysis can help provide a framework for debate. It is important to note that small "back-of-the-envelope" models can be very useful. They have advantages that large models often lack, such as cost, simplicity, and ease of understanding and, therefore, explicability.

It is important to distinguish between analysis and interpretation in systems engineering efforts. Analysis cannot substitute, or will generally be a foolish substitute for, judgment, evaluation, and interpretation as exercised by a well-informed decision-maker. In some cases, refinement of individual alternative policies is not needed in the analysis step. But evaluation of alternatives is always needed, since, if there is but a single policy alternative, there really is no alternative at all. The option to do nothing at all must always be considered as a policy alternative. It is especially important to avoid a large number of cognitive biases, poor judgment heuristics, and value incoherence in the activities of evaluation and decision making. The efforts involved in evaluation and choice making interact strongly with the efforts in the other steps of the systems process, and these are also influenced by cognitive bias, judgment heuristics, and value incoherence. One of the fundamental tenets of the systems process is that making the complete issue resolution process as explicit as possible makes it easier to detect and connect these deficiencies than it is in holistic intuitive processes.

4.3 Information Processing by Humans and Organizations

After completion of the analysis step, we begin the evaluation and decision-making effort of interpretation. Decisions must typically be made and policies formulated, evaluated, and applied in an atmosphere of uncertainty. The outcome of any proposed policy is seldom known with certainty. One of the purposes of analysis is to reduce, to the extent possible, uncertainties associated with the outcomes of proposed policies. Most planning, design, and resource allocation issues will involve a large number of decision-makers who act according to their varied preferences. Often, these decision-makers will have diverse and conflicting data available to them and the decision situation will be quite fragmented. Furthermore, outcomes resulting from actions can often only be adequately characterized by a large number of incommensurable attributes. Explicit informed comparison of alternatives across these attributes by many stakeholders in an evaluation and choice-making process is typically most difficult.

As a consequence of this, people will often search for and use some form of a dominance structure to enable rejection of alternatives that are perceived to be dominated by one or more other alternatives. An alternative is said to be "dominated" by another alternative when the other alternative has attribute scores at least as large as those associated with the dominated alternative and at least one attribute score that is larger. However, biases have been shown to be systematic and prevalent in most unaided cognitive activities. Decisions and judgments are influenced by differential weights of information and by a variety of human information-processing deficiencies, such as base rates, representativeness, availability, adjustment, and anchoring. Often it is very difficult to disaggregate values of policy outcomes from causal relations determining these outcomes. Often correlation is used to infer causality. Wishful thinking and other forms of selective perception encourage us not to obtain potentially disconfirming information. The resulting confounding of values with facts can lead to great difficulties in discourse and related decision making.

It is especially important to avoid the large number of potential cognitive biases and flaws in the process of formulation, analysis, and interpretation for judgment and choice. These may well occur due to flaws in human information processing associated with the identification of problem elements, structuring of decision situations, and the probabilistic and utility assessment portions of the judgmental tasks of evaluation and decision making.

Among the cognitive biases and information-processing flaws that have been identified are several that affect information formulation or acquisition, information analysis, and interpretation. These and related material are described in Ref. 21 and the references contained therein. Among these biases, which are not independent, are the following:

1. *Adjustment and Anchoring*. Often a person finds that difficulty in problem solving is due not to the lack of data and information but rather to an excess of data and information. In such situations, the person often resorts to heuristics, which may reduce the mental efforts required to arrive at a solution. In using the anchoring and adjustment heuristic when confronted with a large number of data, the person selects a particular datum, such as the mean, as an initial or starting point or anchor and then adjusts that value improperly in order to incorporate the rest of these data, resulting in flawed information analysis.
2. *Availability*. The decision-maker uses only easily available information and ignores sources of significant but not easily available information. An event is believed to occur frequently, that is, with high probability, if it is easy to recall similar events.
3. *Base Rate*. The likelihood of occurrence of two events is often compared by contrasting the number of times the two events occur and ignoring the rate of occurrence of each event. This bias often arises when the decision-maker has concrete experience with one event but only statistical or abstract information on the other. Generally, abstract information will be ignored at the expense of concrete information. A base rate determined primarily from concrete information may be called a *causal base rate,* whereas that determined from abstract information is an *inci-*

dental base rate. When information updates occur, this individuating information is often given much more weight than it deserves. It is much easier for the impact of individuating information to override incidental base rates than causal base rates.

4. *Conservatism*. The failure to revise estimates as much as they should be revised, based on receipt of new significant information, is known as *conservatism*. This is related to data saturation and regression effects biases.
5. *Data Presentation Context*. The impact of summarized data, for example, may be much greater than that of the same data presented in detailed, nonsummarized form. Also, different scales may be used to change the impact of the same data considerably.
6. *Data Saturation*. People often reach premature conclusions on the basis of too small a sample of information while ignoring the rest of the data, which is received later, or stopping acquisition of data prematurely.
7. *Desire for Self-Fulfilling Prophecies*. The decision-maker values a certain outcome, interpretation, or conclusion and acquires and analyzes only information that supports this conclusion. This is another form of selective perception.
8. *Ease of Recall*. Data that can easily be recalled or assessed will affect perception of the likelihood of similar events reoccurring. People typically weigh easily recalled data more in decision making than those data that cannot easily be recalled.
9. *Expectations*. People often remember and attach higher validity to information that confirms their previously held beliefs and expectations than they do to disconfirming information. Thus, the presence of large amounts of information makes it easier for one to selectively ignore disconfirming information such as to reach any conclusion and thereby prove anything that one desires to prove.
10. *Fact–Value Confusion*. Strongly held values may often be regarded and presented as facts. That type of information is sought that confirms or lends credibility to one's views and values. Information that contradicts one's views or values is ignored. This is related to wishful thinking in that both are forms of selective perception.
11. *Fundamental Attribution Error* (success/failure error). The decision-maker associates success with personal inherent ability and associates failure with poor luck in chance events. This is related to availability and representativeness.
12. *Habit*. Familiarity with a particular rule for solving a problem may result in reuse of the same procedure and selection of the same alternative when confronted with a similar type of problem and similar information. We choose an alternative because it has previously been acceptable for a perceived similar purpose or because of superstition.
13. *Hindsight*. People are often unable to think objectively if they receive information that an outcome has occurred and they are told to ignore this information. With hindsight, outcomes that have occurred seem to have been inevitable. We see relationships much more easily in hindsight than in foresight and find it easy to change our predictions after the fact to correspond to what we know has occurred.
14. *Illusion of Control*. A good outcome in a chance situation may well have resulted from a poor decision. The decision-maker may assume an unreasonable feeling of control over events.
15. *Illusion of Correlation*. This is a mistaken belief that two events covary when they do not covary.

16. *Law of Small Numbers*. People are insufficiently sensitive to quality of evidence. They often express greater confidence in predictions based on small samples of data with nondisconfirming evidence than in much larger samples with minor disconfirming evidence. Sample size and reliability often have little influence on confidence.
17. *Order Effects*. The order in which information is presented affects information retention in memory. Typically, the first piece of information presented (primacy effect) and the last presented (recency effect) assume undue importance in the mind of the decision-maker.
18. *Outcome-Irrelevant Learning System*. Use of an inferior processing or decision rule can lead to poor results that the decision-maker can believe are good because of inability to evaluate the impacts of the choices not selected and the hypotheses not tested.
19. *Representativeness*. When making inference from data, too much weight is given to results of small samples. As sample size is increased, the results of small samples are taken to be representative of the larger population. The "laws" of representativeness differ considerably from the laws of probability and violations of the conjunction rule $P(A \cap B) < P(A)$ are often observed.
20. *Selective Perceptions*. People often seek only information that confirms their views and values and disregard or ignore disconfirming evidence. Issues are structured on the basis of personal experience and wishful thinking. There are many illustrations of selective perception. One is "reading between the lines"—for example, to deny antecedent statements and, as a consequence, accept "if you don't promote me, I won't perform well" as following inferentially from "I will perform well if you promote me."

Of particular interest are circumstances under which these biases occur and their effects on activities such as the identification of requirements for a system or for planning and design. Through this, it may be possible to develop approaches that might result in debiasing or amelioration of the effects of cognitive bias. A number of studies have compared unaided expert performance with simple quantitative models for judgment and decision making. While there is controversy, most studies have shown that simple quantitative models perform better in human judgment and decision-making tasks, including information processing, than holistic expert performance in similar tasks. There are a number of prescriptions that might be given to encourage avoidance of possible cognitive biases and to debias those that do occur:

1. Sample information from a broad database and be especially careful to include databases that might contain disconfirming information.
2. Include sample size, confidence intervals, and other measures of information validity in addition to mean values.
3. Encourage use of models and quantitative aids to improve upon information analysis through proper aggregation of acquired information.
4. Avoid the hindsight bias by providing access to information at critical past times.
5. Encourage people to distinguish good and bad decisions from good and bad outcomes.
6. Encourage effective learning from experience. Encourage understanding of the decision situation and methods and rules used in practice to process information and make decisions so as to avoid outcome-irrelevant learning systems.

A definitive discussion of debiasing methods for hindsight and overconfidence is presented by Fischhoff, a definitive chapter in an excellent edited work.[22] He suggests identifying faulty judges, faulty tasks, and mismatches between judges and tasks. Strategies for each of these situations are given.

Not everyone agrees with the conclusions just reached about cognitive human information processing and inferential behavior. Several arguments have been advanced for a decidedly less pessimistic view of human inference and decision. Jonathan Cohen,[23,24] for example, argues that all of this research is based upon a conventional model for probabilistic reasoning, which Cohen calls the "Pascalian" probability calculus. He expresses the view that human behavior does not appear "biased" at all when it is viewed in terms of other equally appropriate schemes for probabilistic reasoning, such as his own "inductive probability" system. Cohen states that human irrationality can never be demonstrated in laboratory experiments, especially experiments based upon the use of what he calls "probabilistic conundrums."

There are a number of other contrasting viewpoints as well. In their definitive study of behavioral and normative decision analysis, von Winterfeld and Edwards[25] refer to these information-processing biases as "cognitive illusions." They indicate that there are four fundamental elements to every cognitive illusion:

1. A *formal operational* rule that determines *the* correct solution to an intellectual question
2. An intellectual question that almost invariably includes all of the information required to obtain the correct answer through use of the formal rule
3. A human judgment, generally made without the use of these analytical tools, that is intended to answer the posed question
4. A systematic and generally large and unforgivable discrepancy between the correct answer and the human judgment

They also, as does Phillips,[26] describe some of the ways in which subjects might have been put at a disadvantage in this research on cognitive heuristics and information-processing biases. Much of this centers around the fact that the subjects have little experiential familiarity with the tasks that they are asked to perform. It is suggested that as inference tasks are decomposed and better structured, it is very likely that a large number of information-processing biases will disappear. Thus, concern should be expressed about the structuring of inference and decision problems and the learning that is reflected by revisions of problem structure in the light of new knowledge. In any case, there is strong evidence that humans are very strongly motivated to understand, to cope with, and to improve themselves and the environment in which they function. One of the purposes of systems engineering is to aid in this effort.

4.4 Interpretation

While there are a number of fundamental limitations to systems engineering efforts to assist in bettering the quality of human judgment, choice, decisions, and designs, there are also a number of desirable activities. These have resulted in several important holistic approaches that provide formal assistance in the evaluation and interpretation of the impacts of alternatives, including the following:

- *Decision analysis,* which is a very general approach to option evaluation and selection, involves identification of action alternatives and possible consequence identification of the probabilities of these consequences, identification of the valuation placed by the decision-maker on these consequences, computation of the expected utilities of the consequences, and aggregating or summarizing these values for all consequences of each action. In doing this, we obtain an expected utility evaluation of each alternative act. The one with the highest value is the most preferred action or option. Figure 7 presents some of the salient features involved in the decision analysis of a simplified problem.

- *Multiattribute utility theory* (MAUT) has been designed to facilitate comparison and ranking of alternatives with many attributes or characteristics. The relevant attributes are identified and structured and a weight or relative utility is assigned by the decision-maker to each basic attribute. The attribute measurements for each alternative are used to compute an overall worth or utility for each attribute. Multiattribute utility theory allows for explicit recognition and incorporation of the decision-maker's attitude toward risk in the utility computations. There are a number of variants of MAUT; many of them are simpler, more straightforward processes in which risk and uncertainty considerations are not taken into account. The method is very helpful to the decision-maker in making values and preferences explicit and in making decisions consistent with those values. The tree structure of Fig. 6 also indicates some salient features of the MAUT approach for the particular case where there are no risks or uncertainties involved in the decision situation. We simply need to associate importance weights with the attributes and then provide scores for each alternative on each of the lowest level attributes.
- *Policy capture* (or social judgment theory) has also been designed to assist decision-makers in making their values explicit and their decisions consistent with their values. In policy capture, the decision-maker is asked to rank order a set of alternatives in a gestalt or holistic fashion. Alternative attributes and associated attribute measures are then determined by elicitation from the decision-maker. A mathematical procedure involving regression analysis is used to determine that relative importance weight of each attribute that will lead to a ranking as specified by the decision-maker. The result is fed back to the decision-maker, who, typically, will express the view that his or her values are different. In an iterative learning process, preference weights and/or overall rankings are modified until the decision-maker is satisfied with both the weights and the overall alternative ranking.

There are many advantages to formal interpretation efforts in systems engineering, including the following:

1. Developing decision situation models to aid in making the choice-making effort explicit helps one both to identify and to overcome the inadequacies of implicit mental models.
2. The decision situation model elements, especially the attributes of the outcomes of alternative actions, remind us of information we need to obtain about alternatives and their outcomes.
3. We avoid such poor information-processing heuristics as evaluating one alternative on attribute A and another on attribute B and then comparing them without any basis for compensatory trade-offs across the different attributes.
4. We improve our ability to process information and, consequently, reduce the possibilities for cognitive bias.
5. We can aggregate facts and values in a prescribed systemic fashion rather than by adopting an agenda-dependent or intellect-limited approach.
6. We enhance brokerage, facilitation, and communication abilities among stakeholders to complex technological and social issues.

There is a plethora of literature describing the decision assessment or decision-making part of the interpretation step of systems engineering. In addition to the discussions in Refs. 1, 2, and 3, excellent discussions are to be found in Refs. 27, 28, and 29.

4.5 The Central Role of Information in Systems Engineering

Information is certainly a key ingredient supporting quality decisions; all of systems engineering efforts are based on appropriate acquisition and use of information. There are three basic types of information, which are fundamentally related to the three-step framework of systems engineering:

1. Formulation information
   - **a.** Information concerning the problem and associated needs, constraints, and alterables
   - **b.** Information concerning the value system
   - **c.** Information concerning possible option alternatives
   - **d.** Information concerning possible future alternative outcomes, states, and scenarios
2. Analysis information
   - **a.** Information concerning probabilities of future scenarios
   - **b.** Information concerning impacts of alternative options
   - **c.** Information concerning the importance of various criteria or attributes
3. Interpretation information
   - **a.** Information concerning evaluation and aggregation of facts and values
   - **b.** Information concerning implementation

We see that useful and appropriate formulation, analysis, and interpretation of information is one of the most important and vital tasks in systems engineering efforts, since it is the efficient processing of information by the decision-maker that produces effective decisions. A useful definition of information for our purposes is that it is data of value for decision making. The decision-making process is influenced by many contingency and environmental influences. A purpose of the management technology that is systems engineering is to provide systemic support processes to further enhance efficient decision-making activities.

After completion of evaluation and decision-making efforts, it is generally necessary to become involved in planning for action to implement the chosen alternative option or the next phase of a systems engineering effort. More often than not, it will be necessary to iterate the steps of systems engineering several times to obtain satisfactory closure upon one or more appropriate action alternatives. Planning for action also leads to questions concerning resource allocation, schedules, and management plans. There are, of course, a number of methods from systems science and operations research that support determination of schedules and implementation plans. Each of the steps is needed, with different focus and emphasis, at each phase of a systems effort. These phases depend on the particular effort under consideration but will typically include such phases as policy and program planning, project planning, and system development.

There are a number of complexities affecting "rational" planning, design, and decision making. We must cope with these in the design of effective systemic processes. The majority of these complexities involve systems management considerations. Many have indicated that the capacity of the human mind for formulating, analyzing, and interpreting complex large-scale issues is very small compared with the size and scope of the issues whose resolution is required for objective, substantive, and procedurally rational behavior. Among the limits to rationality are the fact that we can formulate, analyze, and interpret only a restricted amount of information; can devote only a limited amount of time to decision making; and can become involved in many more activities than we can effectively consider and cope with simultaneously. We must therefore necessarily focus attention only on a portion of the major competing concerns. The direct effect of these is the presence of cognitive bias in information acquisition and processing and the use of cognitive heuristics for evaluation of alternatives.

Although in many cases these cognitive heuristics will be flawed, this is not necessarily so. One of the hoped-for results of the use of systems engineering approaches is the development of effective and efficient heuristics for enhanced judgment and choice through effective decision support systems.[30]

There are many cognitive biases prevalent in most information acquisition activities. The use of cognitive heuristics and decision rules is also prevalent and necessary to enable us to cope with the many demands on our time. One such heuristic is satisfying or searching for a solution that is "good enough." This may be quite appropriate if the stakes are small. In general, the quality of cognitive heuristics will be task dependent, and often the use of heuristics for evaluation will be both reasonable and appropriate. Rational decision making requires time, skill, wisdom, and other resources. It must, therefore, be reserved for the more important decisions. A goal of systems engineering is to enhance information acquisition, processing, and evaluation so that efficient and effective use of information is made in a process that is appropriate to the cognitive styles and time constraints of management.

5 SYSTEM DESIGN

This section discusses several topics relevant to the design and evaluation of systems. In order to develop our design methodology, we first discuss the purpose and objectives of systems engineering and systems design. Development of performance objectives for quality systems is important, since evaluation of the logical soundness and performance of a system can be determined by measuring achievement of these objectives with and without the system. A discussion of general objectives for quality system design is followed by a presentation of a five-phase design methodology for system design. The section continues with leadership and training requirements for use of the resulting system and the impact of these requirements upon design considerations. While it is doubtless true that not every design process should, could, or would precisely follow each component in the detailed phases outlined here, we feel that this approach to systems design is sufficiently robust and generic that it can be used as a normative model of the design process and as a guide to the structuring and implementation of appropriate systems evaluation practices.

5.1 The Purposes of Systems Design

Contemporary issues that may result in the need for systems design are invariably complex. They typically involve a number of competing concerns, contain much uncertainty, and require expertise from a number of disparate disciplines for resolution. Thus, it is not surprising that intuitive and affective judgments, often based on incomplete data, form the usual basis used for contemporary design and associated choice making. At the other extreme of the cognitive inquiry scale are the highly analytical, theoretical, and experimental approaches of the mathematical, physical, and engineering sciences. When intuitive judgment is appropriately skill based, it is generally effective and appropriate. One of the major challenges in system design engineering is to develop processes that are appropriate for a variety of process users, some of whom may approach the design issue from a skill-based perspective, some from a rule-based perspective, and some from a knowledge-based perspective.

A central purpose of systems engineering and management is to incorporate appropriate methods and metrics into a methodology for problem solving, or a systems engineering process or life cycle, such that, when it is associated with human judgment through systems management, it results in a high-quality systems design procedure. By high-quality design, we mean one that will, with high probability, produce a system that is effective and efficient and trustworthy.

A systems design procedure must be specifically related to the operational environment for which the final system is intended. Control group testing and evaluation may serve many useful purposes with respect to determination of many aspects of algorithmic and behavioral efficacy of a system. Ultimate effectiveness involves user acceptability of the resulting system, and evaluation of this process effectiveness will often involve testing and evaluation in the environment, or at least a closely simulated model of the environment, in which the system would be potentially deployed.

The potential benefits of systems engineering approaches to design can be interpreted as attributes or criteria for evaluation of the design approach itself. Achievement of many of these attributes may often not be experimentally measured except by inference, anecdotal, or testimonial and case study evidence taken in the operational environment for which the system is designed. Explicit evaluation of attribute achievement is a very important part of the overall systemic design process. This section describes the following:

1. A methodological framework for the design of systems, such as planning and decision support systems
2. An evaluation methodology that may be incorporated with or used independently of the design framework

A number of characteristics of effective systems efforts can be identified. These form the basis for determining the attributes of systems and systemic design procedures. Some of these attributes will be more important for a given environment than others. Effective design must typically include an operational evaluation component that will consider the strong interaction between the system and the situational issues that led to the systems design requirement. This operational evaluation is needed in order to determine whether a product system or a service consisting of humans and machines:

1. Is logically sound
2. Is matched to the operational and organizational situation and environment extant
3. Supports a variety of cognitive skills, styles, and knowledge of the humans who must use the system
4. Assists users of the system to develop and use their own cognitive skills, styles, and knowledge
5. Is sufficiently flexible to allow use and adaptation by users with differing cognitive skills, styles, and knowledge
6. Encourages more effective solution of unstructured and unfamiliar issues, allowing the application of job-specific experiences in a way compatible with various acceptability constraints
7. Promotes effective long-term management

It is certainly possible that the product, or system, that results from a systems engineering effort may be used as a process or life cycle in some other application. Thus, what we have to say here refers both to the design of products and to the design of processes.

5.2 Operational Environments and Decision Situation Models

In order to develop robust scenarios of planning and design situations in various operational environments and specific instruments for evaluation, we first identify a mathematical and situational taxonomy:

- Algorithmic constructs used in systemic design
- Performance objectives for quality design
- Operational environments for design

One of the initial goals in systems design engineering is to obtain the conceptual specifications for a product such that development of the system will be based on customer or client information, objectives, and existing situations and needs. An aid to the process of design should assist in or support the evaluation of alternatives relative to some criteria. It is generally necessary that design information be described in ways that lead to effective structuring of the design problem. Of equal importance is the need to be aware of the role of the affective in design tasks such as to support different cognitive styles and needs, which vary from formal knowledge-based to rule-based to skill-based behavior.[31] We desire to design efficient and effective physical systems, problem-solving service systems, and interfaces between the two. This section is concerned with each of these.

Not all of the performance objectives for quality systems engineering will be, or need be, fully attained in all design instances, but it is generally true that the quality of a system or of a systems design process necessarily improves as more and more of these objectives are attained. Measures of quality or effectiveness of the resulting system, and therefore systems design process quality or effectiveness, may be obtained by assessing the degree of achievement of these performance criteria by the resulting system, generally in an operational environment. In this way, an evaluation of the effectiveness of a design decision support system may be conducted.

A taxonomy based on operational environments is necessary to describe particular situation models through which design decision support may be achieved. We are able to describe a large number of situations using elements or features of the three-component taxonomy described earlier. With these, we are able to evolve test instruments to establish quantitative and qualitative evaluations of a design support system within an operational environment. The structural and functional properties of such a system, or of the design process itself, must be described in order that a purposeful evaluation can be accomplished. This purposeful evaluation of a systemic process is obtained by embedding the process into specific operational planning, design, or decision situations. Thus, an evaluation effort also allows iteration and feedback to ultimately improve the overall systems design process. The evaluation methodology to be described is useful, therefore, as a part or phase of the design process. Also, it is useful, in and of itself, to evaluate and prioritize a set of systemic aids for planning, design, and decision support. It is also useful for evaluation of resulting system designs and operational systems providing a methodological framework both for the design and evaluation of physical systems and for systems that assist in the planning and design of systems.

5.3 The Development of Aids for the Systems Design Process

This section describes five important phases in the development of systems and systemic aids for the design process. These phases serve as a guide not only for the sound design and development of systems and systemic aids for design decision support but also for their evaluation and ultimate operational deployment:

- Requirements specification
- Preliminary conceptual design and architecting
- Detailed design, integration, testing, and implementation
- Evaluation (and potential modification)
- Operational deployment

These five phases are applicable to design in general. Although the five phases will be described as if they are to be sequenced in a chronological fashion, sound design practice will generally necessitate iteration and feedback from a given phase to earlier phases.

Requirements Specification Phase

The requirements specification phase has as its goal the detailed definition of those needs, activities, and objectives to be fulfilled or achieved by the system or process that is to result from the system design effort. Furthermore, the effort in this phase should result in a description of preliminary conceptual design considerations appropriate for the next phase. This must be accomplished in order to translate operational deployment needs, activities, and objectives into requirements specifications if, for example, that is the phase of the systems engineering design effort under consideration.

Among the many objectives of the requirements specifications phase of systems engineering are the following:

1. To define the problem to be solved, or range of problems to be solved, or issue to be resolved or ameliorated, including identification of needs, constraints, alterables, and stakeholder groups associated with operational deployment of the system or the systemic process
2. To determine objectives for operational system or the operational aids for planning, design, and decision support
3. To obtain commitment for prototype design of a system or systemic process aid from user group and management
4. To search the literature and seek other expert opinions concerning the approach that is most appropriate for the particular situation extant
5. To determine the estimated frequency and extent of need for the system or the systemic process
6. To determine the possible need to modify the system or the systemic process to meet changed requirements
7. To determine the degree and type of accuracy expected from the system or systemic process
8. To estimate expected effectiveness improvement or benefits due to the use of the system or systemic process
9. To estimate the expected costs of using the system or systemic process, including design and development costs, operational costs, and maintenance costs

10. To determine typical planning horizons and periods to which the system or systemic process must be responsive
11. To determine the extent of tolerable operational environment alteration due to use of the system or systemic process
12. To determine what particular planning, design, or decision process appears best
13. To determine the most appropriate roles for the system or systemic process to perform within the context of the planning, design, or decision situation and operational environment under consideration
14. To estimate potential leadership requirements for use of the final system itself
15. To estimate user group training requirements
16. To estimate the qualifications required of the design team
17. To determine preliminary operational evaluation plans and criteria
18. To determine political acceptability and institutional constraints affecting use of an aided support process and those of the system itself
19. To document analytical and behavioral specifications to be satisfied by the support process and the system itself
20. To determine the extent to which the user group can require changes during and after system development
21. To determine potential requirements for contractor availability after completion of development and operational tests for additional needs determined by the user group, perhaps as a result of the evaluation effort
22. To develop requirements specifications for prototype design of a support process and the operational system itself

As a result of this phase, to which the four issue requirements identification approaches of Section 4.1 are fully applicable, there should exist a clear definition of typical planning, design, and decision issues, or problems requiring support, and other requirements specifications, so that it is possible to make a decision whether to undertake preliminary conceptual design. If the result of this phase indicates that the user group or client needs can potentially be satisfied in a cost-effective manner, by a systemic process aid, for example, then documentation should be prepared concerning detailed specifications for the next phase, preliminary conceptual design, and initial specifications for the last three phases of effort. A design team is then selected to implement the next phase of the system life cycle. This discussion emphasizes the inherently coupled nature of these phases of the system life cycle and illustrates why it is not reasonable to consider the phases as if they are uncoupled.

Preliminary Conceptual Design and Architecting Phase

The preliminary conceptual design and architecting phase includes specification of the mathematical and behavioral content and associated algorithms for the system or process that should ultimately result from the effort as well as the possible need for computer support to implement these. The primary goal of this phase is to develop conceptualization of a prototype system or process in response to the requirements specifications developed in the previous phase. Preliminary design according to the requirements specifications should be achieved. Objectives for preliminary conceptual design include the following:

1. To search the literature and seek other expert opinion concerning the particular approach to design and implementation that is likely to be most responsive to requirements specifications

2. To determine the specific analytic algorithms to be implemented by the system or process
3. To determine the specific behavioral situation and operational environment in which the system or process is to operate
4. To determine the specific leadership requirements for use of the system in the operational environment extant
5. To determine specific hardware and software implementation requirements, including type of computer programming language and input devices
6. To determine specific information input requirements for the system or process
7. To determine the specific type of output and interpretation of the output to be obtained from the system or process that will result from the design procedure
8. To reevaluate objectives obtained in the previous phase, to provide documentation of minor changes, and to conduct an extensive reexamination of the effort if major changes are detected that could result in major modification and iteration through requirements specification or even termination of effort
9. To develop a preliminary conceptual design of, or architecture for, prototype aid that is responsive to the requirements specifications

The expected product of this phase is a set of detailed design and testing specifications that, if followed, should result in a usable prototype system or process. User group confidence that an ultimately useful product should result from detailed design should be above some threshold or the entire design effort should be redone. Another product of this phase is a refined set of specifications for the evaluation and operational deployment phases.

If the result of this phase is successful, the detailed design, testing, and implementation phase is begun. This phase is based on the products of the preliminary conceptual design phase, which should result in a common understanding among all interested parties about the planning and decision support design effort concerning the following:

1. Who the user group or responsive stakeholder is
2. The structure of the operational environment in which plans, designs, and decisions are made
3. What constitutes a plan, a design, or a decision
4. How plans, designs, and decisions are made without the process or system and how they will be made with it
5. What implementation, political acceptability, and institutional constraints affect the use of the system or process
6. What specific analysis algorithms will be used in the system or process and how these algorithms will be interconnected to form the methodological construction of the system or process

Detailed Design, Integration, Testing, and Implementation Phase
In the third phase of design, a system or process that is presumably useful in the operational environment is produced. Among the objectives to be attained in this phase are the following:

1. To obtain and design appropriate physical facilities (physical hardware, computer hardware, output device, room, etc.)
2. To prepare computer software

3. To document computer software
4. To integrate these effectively
5. To prepare a user's guide to the system and the process in which the system is embedded
6. To prepare a leader's guide for the system and the associated process
7. To conduct control group or operational (simulated operational) tests of the system and make minor changes in the aid as a result of the tests
8. To complete detailed design and associated testing of a prototype system based on the results of the previous phase
9. To implement the prototype system in the operational environment as a process

The products of this phase are detailed guides to use of the system as well as, of course, the prototype system itself. It is very important that the user's guide and the leader's guide address, at levels appropriate for the parties interested in the effort, the way in which the performance objectives identified in Section 5.3 are satisfied. The description of system usage and leadership topics should be addressed in terms of the analytic and behavioral constructs of the system and the resulting process as well as in terms of operational environment situation concerns. These concerns include the following:

1. Frequency of occurrence of need for the system or process
2. Time available from recognition of need for a plan, design, or decision to identification of an appropriate plan, design, or decision
3. Time available from determination of an appropriate plan, design, or decision to implementation of the plan, design, or decision
4. Value of time
5. Possible interactions with the plans, designs, or decisions of others
6. Information base characteristics
7. Organizational structure
8. Top-management support for the resulting system or process

It is especially important that the portion of this phase that concerns implementation of the prototype system specifically address important questions concerning cognitive style and organizational differences among parties at interest and institutions associated with the design effort. Stakeholder understanding of environmental changes and side effects that will result from use of the system is critical for ultimate success. This need must be addressed. Evaluation specification and operational deployment specifications will be further refined as a result of this phase.

Evaluation Phase

Evaluation of the system in accordance with evaluation criteria, determined in the requirements specification phase and modified in the subsequent two design phases, is accomplished in the fourth phase of systems development. This evaluation should always be assisted to the extent possible by all parties at interest to the systems design effort and the resultant systemic process. The evaluation effort must be adapted to other phases of the design effort so that it becomes an integral functional part of the overall design process. As noted, evaluation may well be an effort distinct from design that is used to determine usefulness or appropriateness for specified purposes of one or more previously designed systems. Among the objectives of system or process evaluation are the following:

1. To identify a methodology for evaluation
2. To identify criteria on which the success of the system or process may be judged
3. To determine effectiveness of the system in terms of success criteria
4. To determine an appropriate balance between the operational environment evaluation and the control group evaluation
5. To determine performance objective achievement of the system
6. To determine behavioral or human factor effectiveness of the system
7. To determine the most useful strategy for employment of the existing system
8. To determine user group acceptance of the system
9. To suggest refinements in existing systems for greater effectiveness of the process in which the new system has been embedded
10. To evaluate the effectiveness of the system or process

These objectives are obtained from a critical evaluation issue specification or evaluation need specification, which is the first, or problem definition, step of the evaluation methodology. Generally, the critical issues for evaluation are minor adaptations of the elements that are present in the requirements specifications step of the design process outlined in the previous section. A set of specific evaluation test requirements and tests is evolved from these objectives and needs. These must be such that each objective measure and critical evaluation issue component can be determined from at least one evaluation test instrument.

If it is determined that the system and the resulting process support cannot meet user needs, the systems design process iterates to an earlier phase and development continues. An important by-product of evaluation is the determination of ultimate performance limitations and the establishment of a protocol and procedure for use of the system that results in maximum user group satisfaction. A report is written concerning results of the evaluation process, especially those factors relating to user group satisfaction with the designed system. The evaluation process should result in suggestions for improvement in design and in better methodologies for future evaluations.

Section 5.6 will present additional details of the methodologies framework for evaluation. These have applicability to cases where evaluation is a separate and independent effort as well as cases where it is one of the phases of the design process.

Operational Deployment Phase

The last phase of design concerns operational deployment and final implementation. This must be accomplished in such a way that all user groups obtain adequate instructions in use of the system and complete operating and maintenance documentation and instructions. Specific objectives for the operational deployment phase of the system design effort are as follows:

1. To enhance operational deployment
2. To accomplish final design of the system
3. To provide for continuous monitoring of post implementation effectiveness of the system and the process into which the system is embedded
4. To provide for redesign of the system as indicated by effectiveness monitoring
5. To provide proper training and leadership for successful continued operational use of the system

6. To identify barriers to successful implementation of the final design product
7. To provide for "maintenance" of the system

5.4 Leadership Requirements for Design

The actual use, as contrasted with potential usefulness, of a system is directly dependent on the value that the user group of stakeholders associates with use of the system and the resulting process in an operational environment. This in turn is dependent, in part, on how well the system satisfies performance objectives and on how well it is able to cope with one or more of the pathologies or pitfalls of planning, design, and/or decision making under potentially stressful operational environment conditions.

Quality planning, design, and decision support are dependent on the ability to obtain relatively complete identification of pertinent factors that influence plans, designs, and decisions. The careful, comprehensive formulation of issues and associated requirements for issue resolution will lead to identification of pertinent critical factors for system design. These factors are ideally illuminated in a relatively easy-to-understand fashion that facilitates the interpretation necessary to evaluate and subsequently select plans, designs, and decisions for implementation. Success in this is, however, strongly dependent on adroitness in use of the system. It is generally not fully meaningful to talk only of an algorithm or even a complete system—which is, typically, a piece of hardware and software but which may well be a carefully written set of protocols and procedures—as useful by itself. It is meaningful to talk of a particular systemic process as being useful. This process involves the interaction of a methodology with systems management at the cognitive process or human judgment level. A systemic process depends on the system, the operational environment, and leadership associated with use of the system. A process involves design integration of a methodology with the behavioral concerns of human cognitive judgment in an operational environment.

Operational evaluation of a systemic process that involves human interaction, such as an integrated manufacturing complex, appears the only realistic way to extract truly meaningful information concerning process effectiveness of a given system design. This must necessarily include leadership and training requirements to use the system. There are necessary trade-offs associated with leadership and training for using a system and these are addressed in operational evaluation.

5.5 System Evaluation

Previous sections have described a framework for a general system design procedure. They have indicated the role of evaluation in this process. Successful evaluation, especially operational evaluation, is strongly dependent on explicit development of a plan for evaluation developed prior to, and perhaps modified and improved during the course of, an actual evaluation. This section will concern itself with development of a methodological framework for system evaluation, especially for operational evaluation of systemic processes for planning, design, and decision support.

Evaluation Methodology and Evaluation Criteria

Objectives for evaluation of a system concern the following:

1. Identification of a methodology for operational evaluation
2. Establishing criteria on which the success of the system may be judged

3. Determining the effectiveness of the support in terms of these criteria
4. Determining the most useful strategy for employment of an existing system and potential improvements such that effectiveness of the newly implemented system and the overall process might be improved

Figure 8 illustrates a partial intent structure or objectives tree, which contributes to system evaluation. The lowest level objectives contribute to satisfaction of the 10 performance objectives for systems engineering and systems design outlined in Section 3. These lowest level elements form pertinent criteria for the operational system evaluation. They concern the algorithmic effectiveness or performance objective achievement of the system, the behavioral or human factor effectiveness of the system in the operational environment, and the system efficacy. Each of these three elements become top-level criteria or attributes and each should be evaluated to determine evaluation of the system itself.

Subcriteria that support the three lowest level criteria of Fig. 8 may be identified. These are dependent on the requirements identified for the specific system that has been designed. Attainment of each of these criteria by the system may be measured by observation of the system within the operational environment and by test instruments and surveys of user groups involved with the operational system and process.

Algorithmic Effectiveness of Performance Objectives Achievement Evaluation
A number of performance objectives can be cited that, if achieved, should lead to a quality system. Achievement of these objectives is measured by logical soundness of the operational system and process; improved system quality as a result of using the system; and improvements in the way an overall process functions, compared to the way it typically functions without the system or with an alternative system.

Behavioral or Human Factors Evaluation
A system may be well structured algorithmically in the sense of achieving a high degree of satisfaction of the performance objectives, yet the process incorporating the system may seriously violate behavioral and human factor sensibilities. This will typically result in misuse or underuse. There are many cases where technically innovative systems have failed to achieve broad scope objectives because of human factor failures. Strongly influencing the acceptability of system implementation in operational settings are such factors as organizational slack; natural human resistance to change; and the present sophistication, attitude, and past experience of the user group and its management with similar systems and processes.

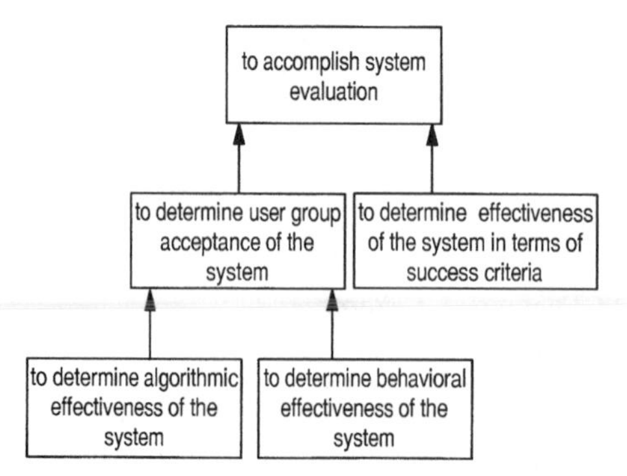

Figure 8 Objectives tree for evaluation of decision support system.

Behavioral or human factor evaluation criteria used to evaluate performance include political acceptability, institutional constraint satisfaction, implementability evaluation, human workload evaluation, management procedural change evaluation, and side-effect evaluation.

Efficacy Evaluation

Two of the three first-level evaluation criteria concern algorithmic effectiveness or performance objective achievement and behavioral or human factors effectiveness. It is necessary for a system to be effective in each of these for it to be potentially capable of truly aiding in terms of improving process quality and being acceptable for implementation in the operational environment for which it was designed. There are a number of criteria or attributes related to usefulness, service support, or efficacy to which a system must be responsive. Thus, evaluation of the efficacy of a system and the associated process is important in determining the service support value of the process. There are seven attributes of efficacy:

1. *Time Requirements*. The time requirements to use a system form an important service support criterion. If a system is potentially capable of excellent results but the results can only be obtained after critical deadlines have passed, the overall process must be given a low rating with respect to a time responsiveness criterion.
2. *Leadership and Training*. Leadership and training requirements for use of a system are important design considerations. It is important that there be an evaluation component directed at assessing leadership and training needs and trade-offs associated with the use of a system.
3. *Communication Accomplishments*. Effective communication is important for two reasons. (1) Implementation action is often accomplished at a different hierarchical level, and therefore by a different set of actors, than the hierarchical level at which selection of alternative plans, designs, or decisions was made. Implementation action agents often behave poorly when an action alternative is selected that they regard as threatening or arbitrary, either personally or professionally, on an individual or a group basis. Widened perspectives of a situation are made possible by effective communication. Enhanced understanding will often lead to commitment to successful action implementation as contrasted with unconscious or conscious efforts to subvert implementation action. (2) Recordkeeping and retrospective improvements to systems and processes are enhanced by the availability of well-documented constructions of planning and decision situations and communicable explanations of the rationale for the results of using the system.
4. *Educational Accomplishments*. There may exist values to a system other than those directly associated with improvement in process quality. The participating group may, for example, learn a considerable amount about the issues for which a system was constructed. The possibility of enhanced ability and learning with respect to the issues for which the system was constructed should be evaluated.
5. *Documentation*. The value of the service support provided by a system will be dependent on the quality of the user's guide and its usefulness to potential users of the system.
6. *Reliability and Maintainability*. To be operationally useful, a planning-and-decision-support system must be, and be perceived by potential users to be, reliable and maintainable.
7. *Convenience of Access*. A system should be readily available and convenient to access or usage will potentially suffer. While these last three service support measures are

not of special significance with respect to justification of the need for a system, they may be important in determining operational usage and, therefore, operational effectiveness of a system and the associated process.

5.6 Evaluation Test Instruments

Several special evaluation test instruments to satisfy test requirements and measure achievement of the evaluation criteria will generally need to be developed. These include investigations of effectiveness in terms of performance objective attainment; selection of appropriate scenarios that affect use of the system, use of the system subject to these scenarios by a test group, and completion of evaluation questionnaires; and questionnaires and interviews with operational users of the system.

Every effort must be made to ensure, to the extent possible, that evaluation test results will be credible and valid. Intentional redundancy should be provided to allow correlation of results obtained from the test instruments to ensure maximum supportability and reliability of the facts and opinions to be obtained from test procedures.

The evaluation team should take advantage of every opportunity to observe use of the system within the operational environment. Evaluation of personnel reactions to the aid should be based on observations, designed to be responsive to critical evaluation issues, and the response of operational environment personnel to test questionnaires. When any of a number of constraints make it difficult to obtain real-time operational environment observation, experiential and anecdotal information becomes of increased value. Also, retrospective evaluation of use of a system is definitely possible and desirable if sufficiently documented records of past usage of an aided process are available.

Many other effectiveness questions will likely arise as an evaluation proceeds. Questions specific to a given evaluation are determined after study of the particular situation and the system being evaluated. It is, however, important to have an initial set of questions to guide the evaluation investigation and a purpose of this section to provide a framework for accomplishing this.

One of the important concerns in evaluation is that of those parts of the efficacy evaluation that deal with various "abilities" of a system. These include producibility, reliability, maintainability, and marketability. Figure 9 presents a listing of attributes that may be used to "score" the performance of systems on relevant effectiveness criteria.

6 CONCLUSIONS

In this chapter, we have discussed salient aspects concerning the systems engineering of large and complex systems. We have been concerned especially with systems design engineering and associated information-processing and analysis efforts. To this end, we suggested a process for the design and evaluation of systems and how we might go about fielding a design decision support system. There are a number of effectiveness attributes or aspects of effective systems. Design of an effective large-scale system necessarily involves integration of operational environment concerns involving human behavior and judgment with mechanistic and physical science concerns. An effective systemic design process should:

1. Allow a thorough and carefully conducted requirements specification effort to determine and specify needs of stakeholders prior to conceptual design of a system process to accomplish the desired task
2. Be capable of dealing with both quantitative and qualitative criteria representing costs and effectiveness from their economic, social, environmental, and other perspectives

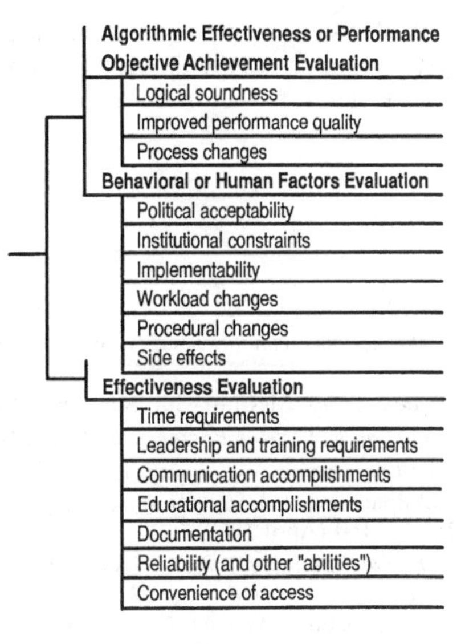

Figure 9 Attribute tree and criteria for evaluation of decision support system for design and other end uses.

3. Be capable of minimizing opportunities for cognitive bias and provide debiasing procedures for those biases that occur
4. Allow separation of opinions and facts from values, and separation of ends from means, or values from alternative acts
5. Provide an objective communicable framework that allows identification, formulation, and display of the structure of the issue under consideration as well as the rationale of the choice process
6. Allow for considerations of trade-offs among conflicting and incommensurate criteria
7. Provide flexibility and monitoring support to allow design process evaluation rule selection with due consideration to the task structure and operational environment constraints on the decision-maker
8. Provide an open process to allow consideration of new criteria and alternatives as values change and broad-scope awareness of issues grows

There are a number of potential benefits of the systems approach that should follow high achievement of each of the criteria for effective systems design processes. An appropriate systems design process will:

1. Provide structure to relatively unstructured issues
2. Facilitate conceptual formulation of issues
3. Provide cognitive cues to search and discovery
4. Encourage parsimonious collection, organization, and utilization of relevant data
5. Extend and debias information-processing abilities
6. Encourage vigilant cognitive style
7. Provide brokerage between parties at interest

There are many imperfections and limits to processes designed using the methodologies from what we know as systems engineering and systems analysis, design, and integration.[32] Some of these have been documented in this chapter. Others are documented in the references provided here and in a recent handbook of systems engineering and management.[33] But what are the alternatives to appropriate systemic processes for the resolution of issues associated with the design of complex large-scale systems and are not the fundamental limitations to these alternatives even greater?

REFERENCES

1. A. P. Sage, *Systems Engineering,* Wiley, New York, 1992.
2. A. P. Sage, *Systems Management for Information Technology and Software Engineering,* Wiley, New York, 1995.
3. A. P. Sage, *Methodology for Large Scale Systems,* McGraw-Hill, New York, 1977.
4. J. E. Armstrong and A. P. Sage, *Introduction to Systems Engineering,* Wiley, New York, 2000.
5. A. D. Hall, *A Methodology for Systems Engineering,* Van Nostrand, New York, 1962.
6. A. D. Hall, "A Three Dimensional Morphology of Systems Engineering," *IEEE Transactions on System Science and Cybernetics* **5**(2), 156–160 (April 1969).
7. E. Rechtin, *Systems Architecting: Creating and Building Complex Systems,* Prentice-Hall, Englewood Cliffs, NJ, 1991.
8. E. Rechtin, "Foundations of Systems Architecting," *Systems Engineering: The Journal of the National Council on Systems Engineering* **1**(1), 35–42 (July/September 1994).
9. W. R. Beam, *Systems Engineering: Architecture and Design,* McGraw-Hill, New York, 1990.
10. D. N. Chorfas, *Systems Architecture and Systems Design,* McGraw-Hill, New York, 1989.
11. A. P. Sage and J. D. Palmer, *Software Systems Engineering,* Wiley, New York, 1990.
12. F. Harary, R. Z. Norman, and D. Cartwright, *Structural Models: An Introduction to the Theory of Directed Graphs,* Wiley, New York, 1965.
13. J. N. Warfield, *Societal Systems: Planning, Policy, and Complexity,* Wiley, New York, 1976.
14. D. V. Steward, *Systems Analysis and Management: Structure, Strategy, and Design,* Petrocelli, New York, 1981.
15. G. G. Lendaris, "Structural Modeling: A Tutorial Guide," *IEEE Transactions on Systems, Man, and Cybernetics* **SMC 10,** 807–840 (1980).
16. C. Eden, S. Jones, and D. Sims, *Messing about in Problems,* Pergamon, Oxford, 1983.
17. F. M. Roberts, *Discrete Mathematical Models,* Prentice-Hall, Englewood Cliffs, NJ, 1976.
18. A. M. Geoffrion, "An Introduction to Structured Modeling," *Management Science* **33,** 547–588 (1987).
19. A. M. Geoffrion, "The Formal Aspects of Structured Modeling," *Operations Research* **37**(1), 30–51 (January 1989).
20. A. M. Geoffrion, "Computer Based Modeling Environments," *European Journal of Operations Research* **41**(1), 33–43 (July 1989).
21. A. P. Sage (ed.), *Concise Encyclopedia of Information Processing in Systems and Organizations,* Pergamon, Oxford, 1990.
22. D. Kahneman, P. Slovic, and A. Tversky (eds.), *Judgment Under Uncertainty: Heuristics and Biases,* Cambridge University Press, New York, 1981.
23. L. J. Cohen, "On the Psychology of Prediction: Whose Is the Fallacy," *Cognition* **7,** 385–407 (1979).
24. L. J. Cohen, "Can Human Irrationality Be Experimentally Demonstrated?" *The Behavioral and Brain Sciences* **4,** 317–370 (1981).
25. D. von Winterfeldt and W. Edwards, *Decision Analysis and Behavioral Research,* Cambridge University Press, Cambridge, 1986.
26. L. Phillips, "Theoretical Perspectives on Heuristics and Biases in Probabilistic Thinking," in *Analyzing and Aiding Decision Problems,* P. C. Humphries, O. Svenson, and O. Vari (eds.), North Holland, Amsterdam, 1984.
27. R. T. Clemen, *Making Hard Decisions: An Introduction to Decision Analysis,* Duxbury, Belmont, CA, 1986.

28. R. L. Keeney, *Value Focused Thinking: A Path to Creative Decision Making,* Harvard University Press, Cambridge, MA, 1992.

29. C. W. Kirkwood, *Strategic Decision Making: Multiobjective Decision Analysis with Spreadsheets,* Duxbury, Belmont, CA, 1997.

30. A. P. Sage, *Decision Support Systems Engineering,* Wiley, New York, 1991.

31. J. Rasmussen, *Information Processing and Human–Machine Interaction,* North Holland, Amsterdam, 1986.

32. D. M. Buede, *The Engineering Design of Systems: Models and Methods,* Wiley, New York, 2000.

33. A. P. Sage and W. B. Rouse (eds.), *Handbook of Systems Engineering and Management,* Wiley, New York, 1999.

CHAPTER 10

MATHEMATICAL MODELS OF DYNAMIC PHYSICAL SYSTEMS

K. Preston White, Jr.
Department of Systems and Information Engineering
University of Virginia
Charlottesville, Virginia

1 RATIONALE

The design of modern control systems relies on the formulation and analysis of mathematical models of dynamic physical systems. This is simply because a model is more accessible to study than the physical system the model represents. Models typically are less costly and less time consuming to construct and test. Changes in the structure of a model are easier to implement, and changes in the behavior of a model are easier to isolate and understand. A model often can be used to achieve insight when the corresponding physical system cannot, because experimentation with the actual system is too dangerous or too demanding. Indeed, a model can be used to answer "what if" questions about a system that has not yet been realized or actually cannot be realized with current technologies.

The type of model used by the control engineer depends upon the nature of the system the model represents, the objectives of the engineer in developing the model, and the tools which the engineer has at his or her disposal for developing and analyzing the model. A mathematical model is a description of a system in terms of equations. Because the physical systems of primary interest to the control engineer are dynamic in nature, the mathematical models used to represent these systems most often incorporate difference or differential equations. Such equations, based on physical laws and observations, are statements of the fundamental relationships among the important variables that describe the system. Difference and differential equation models are expressions of the way in which the current values assumed by the variables combine to determine the future values of these variables.

Mathematical models are particularly useful because of the large body of mathematical and computational theory that exists for the study and solution of equations. Based on this theory, a wide range of techniques has been developed specifically for the study of control systems. In recent years, computer programs have been written that implement virtually all of these techniques. Computer software packages are now widely available for both simulation and computational assistance in the analysis and design of control systems.

It is important to understand that a variety of models can be realized for any given physical system. The choice of a particular model always represents a trade-off between the fidelity of the model and the effort required in model formulation and analysis. This trade-off is reflected in the nature and extent of simplifying assumptions used to derive the model. In general, the more faithful the model is as a description of the physical system modeled, the more difficult it is to obtain general solutions. In the final analysis, the best engineering model is not necessarily the most accurate or precise. It is, instead, the simplest model that yields the information needed to support a decision. A classification of various types of models commonly encountered by control engineers is given in Section 8.

A large and complicated model is justified if the underlying physical system is itself complex, if the individual relationships among the system variables are well understood, if it is important to understand the system with a great deal of accuracy and precision, and if time and budget exist to support an extensive study. In this case, the assumptions necessary to formulate the model can be minimized. Such complex models cannot be solved analytically, however. The model itself must be studied experimentally, using the techniques of computer simulation. This approach to model analysis is treated in Section 7.

Simpler models frequently can be justified, particularly during the initial stages of a control system study. In particular, systems that can be described by linear difference or differential equations permit the use of powerful analysis and design techniques. These include the transform methods of classical control theory and the state-variable methods of modern control theory. Descriptions of these standard forms for linear systems analysis are presented in Sections 4, 5, and 6.

During the past several decades, a unified approach for developing lumped-parameter models of physical systems has emerged. This approach is based on the idea of idealized system elements, which store, dissipate, or transform energy. Ideal elements apply equally well to the many kinds of physical systems encountered by control engineers. Indeed, because control engineers most frequently deal with systems that are part mechanical, part electrical, part fluid, and/or part thermal, a unified approach to these various physical systems is especially useful and economic. The modeling of physical systems using ideal elements is discussed further in Sections 2, 3, and 4.

Frequently, more than one model is used in the course of a control system study. Simple models that can be solved analytically are used to gain insight into the behavior of the system and to suggest candidate designs for controllers. These designs are then verified and refined in more complex models, using computer simulation. If physical components are developed during the course of a study, it is often practical to incorporate these components directly into the simulation, replacing the corresponding model components. An iterative, evolutionary approach to control systems analysis and design is depicted in Fig. 1.

2 IDEAL ELEMENTS

Differential equations describing the dynamic behavior of a physical system are derived by applying the appropriate physical laws. These laws reflect the ways in which energy can be stored and transferred within the system. Because of the common physical basis provided by the concept of energy, a general approach to deriving differential equation models is possible. This approach applies equally well to mechanical, electrical, fluid, and thermal systems and is particularly useful for systems that are combinations of these physical types.

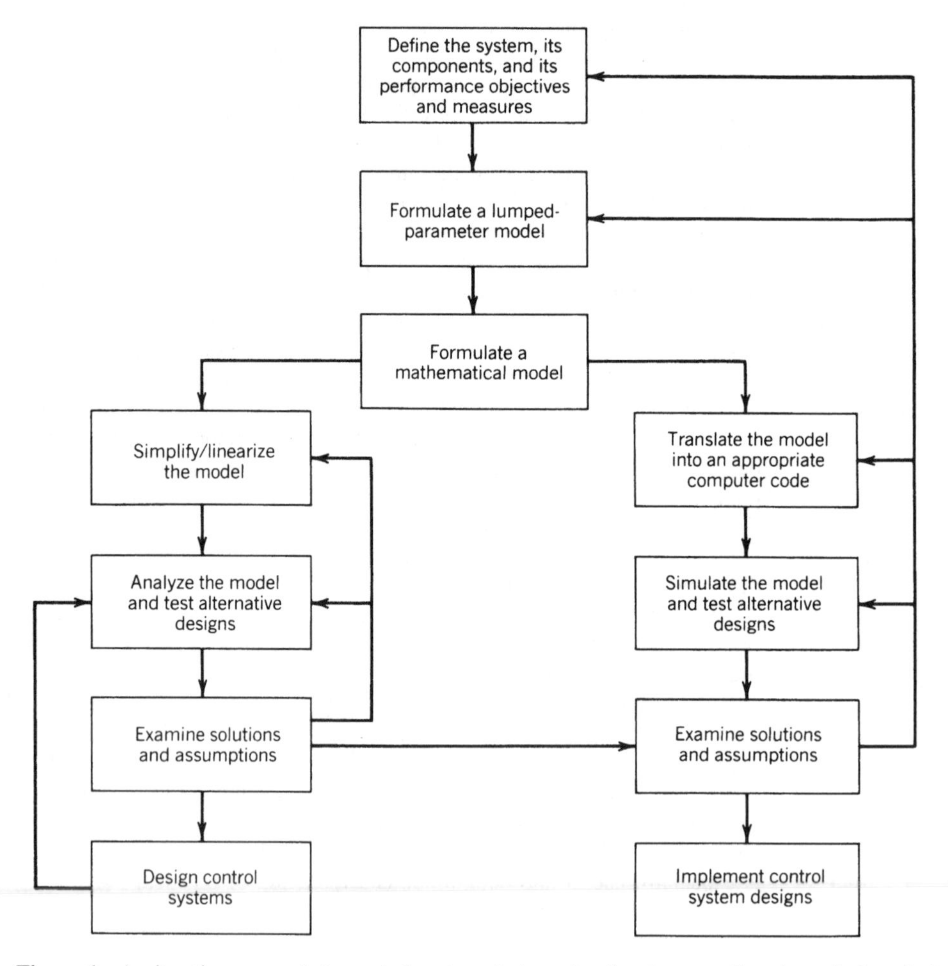

Figure 1 An iterative approach to control system design, showing the use of mathematical analysis and computer simulation.

2.1 Physical Variables

An idealized *two-terminal* or *one-port* element is shown in Fig. 2. Two *primary physical variables* are associated with the element: a through variable $f(t)$ and an across variable $v(t)$. *Through variables* represent quantities that are transmitted through the element, such as the force transmitted through a spring, the current transmitted through a resistor, or the flow of fluid through a pipe. Through variables have the same value at both ends or terminals of the element. *Across variables* represent the difference in state between the terminals of the element, such as the velocity difference across the ends of a spring, the voltage drop across a resistor, or the pressure drop across the ends of a pipe. *Secondary physical variables* are the integrated through variable $h(t)$ and the integrated across variable $x(t)$. These represent the accumulation of quantities within an element as a result of the integration of the associated through and across variables. For example, the momentum of a mass is an integrated through variable, representing the effect of forces on the mass integrated or accumulated over time. Table 1 defines the primary and secondary physical variables for various physical systems.

2.2 Power and Energy

The flow of *power* $P(t)$ into an element through the terminals 1 and 2 is the product of the through variable $f(t)$ and the difference between the across variables $v_2(t)$ and $v_1(t)$. Suppressing the notation for time dependence, this may be written as

$$P = f(v_2 - v_1) = fv_{21}$$

A negative value of power indicates that power flows out of the element. The *energy* $E(t_a, t_b)$ transferred to the element during the time interval from t_a to t_b is the integral of power, that is,

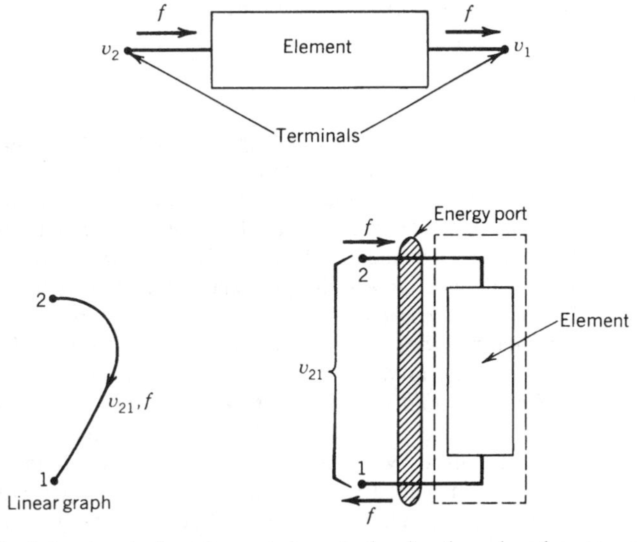

Figure 2 A two-terminal or one-port element, showing through and across variables.[1]

Table 1 Primary and Secondary Physical Variables for Various Systems[1]

System	Through Variable f	Integrated through Variable h	Across Variable v	Integrated across Variable x
Mechanical–translational	Force F	Translational momentum p	Velocity difference v_{21}	Displacement difference x_{21}
Mechanical–rotational	Torque T	Angular momentum h	Angular velocity difference Ω_{21}	Angular displacement difference Θ_{21}
Electrical	Current i	Charge q	Voltage difference v_{21}	Flux linkage λ_{21}
Fluid	Fluid flow Q	Volume V	Pressure difference P_{21}	Pressure–momentum Γ_{21}
Thermal	Heat flow q	Heat energy $\mathcal{H}$	Temperature difference θ_{21}	Not used in general

$$E = \int_{t_a}^{t_b} P\, dt = \int_{t_a}^{t_b} f v_{21}\, dt$$

A negative value of energy indicates a net transfer of energy out of the element during the corresponding time interval.

Thermal systems are an exception to these generalized energy relationships. For a thermal system, power is identically the through variable $q(t)$, heat flow. Energy is the integrated through variable $\mathcal{H}(t_a, t_b)$, the amount of heat transferred.

By the *first law of thermodynamics,* the net energy stored within a system at any given instant must equal the difference between all energy supplied to the system and all energy dissipated by the system. The generalized classification of elements given in the following sections is based on whether the element stores or dissipates energy within the system, supplies energy to the system, or transforms energy between parts of the system.

2.3 One-Port Element Laws

Physical devices are represented by idealized system elements, or by combinations of these elements. A physical device that exchanges energy with its environment through one pair of across and through variables is called a *one-port* or *two-terminal* element. The behavior of a one-port element expresses the relationship between the physical variables for that element. This behavior is defined mathematically by a *constitutive relationship.* Constitutive relationships are derived empirically, by experimentation, rather than from any more fundamental principles. The *element law,* derived from the corresponding constitutive relationship, describes the behavior of an element in terms of across and through variables and is the form most commonly used to derive mathematical models.

Table 2 summarizes the element laws and constitutive relationships for the one-port elements. Passive elements are classified into three types. *T-type* or *inductive storage* elements are defined by a single-valued constitutive relationship between the through variable $f(t)$ and the integrated across-variable difference $x_{21}(t)$. Differentiating the constitutive relationship yields the element law. For a linear (or ideal) T-type element, the element law states that the across-variable difference is proportional to the rate of change of the through variable. Pure translational and rotational compliance (springs), pure electrical inductance, and

Table 2 Element Laws and Constitutive Relationships for Various One-Port Elements[1]

Type of element	Physical element	Linear graph	Diagram	Constitutive relationship	Energy or power function	Ideal elemental equation	Ideal energy or power
T-type energy storage $\mathcal{E} \geq 0$	Translational spring	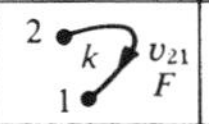 2, k, 1, v_{21}, F	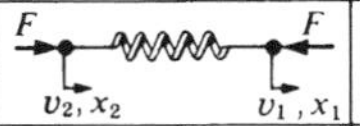 F, v_2, x_2, F, v_1, x_1	$x_{21} = f(F)$	$\mathcal{E} = \int_0^F F\, dx_{21}$	$v_{21} = \frac{1}{k}\frac{dF}{dt}$	$\mathcal{E} = \frac{1}{2}\frac{F^2}{k}$
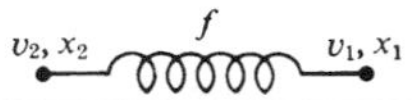 v_2, x_2 f v_1, x_1	Rotational spring	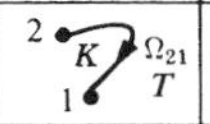 2, K, 1, Ω_{21}, T	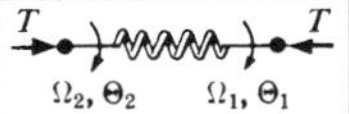 T, Ω_2, Θ_2, T, Ω_1, Θ_1	$\Theta_{21} = f(T)$	$\mathcal{E} = \int_0^T T\, d\Theta_{21}$	$\Omega_{21} = \frac{1}{K}\frac{dT}{dt}$	$\mathcal{E} = \frac{1}{2}\frac{T^2}{K}$
Pure: $x_{21} = f(f)$ Ideal: $x_{21} = Lf$	Inductance	 2, L, 1, v_{21}, i	 i, v_2, λ_2, i, v_1, λ_1	$\lambda_{21} = f(i)$	$\mathcal{E} = \int_0^i i\, d\lambda_{21}$	$v_{21} = L\frac{di}{dt}$	$\mathcal{E} = \frac{1}{2}Li^2$
Pure: $\mathcal{E} = \int_0^f f\, dx_{21}$ Ideal: $\mathcal{E} = \frac{1}{2}Lf^2$	Fluid inertance	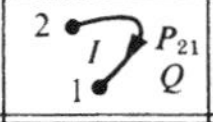 2, I, 1, P_{21}, Q	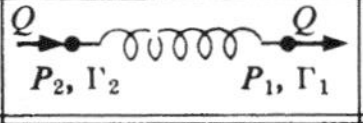 Q, P_2, Γ_2, Q, P_1, Γ_1	$\Gamma_{21} = f(Q)$	$\mathcal{E} = \int_0^Q Q\, d\Gamma_{21}$	$P_{21} = I\frac{dQ}{dt}$	$\mathcal{E} = \frac{1}{2}IQ^2$
A-type energy storage	Translational mass	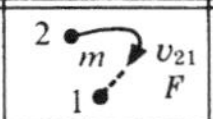 2, m, 1, v_{21}, F	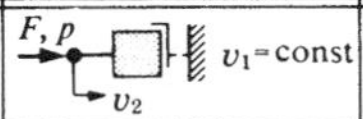 F, p, v_2, v_1 = const	$p = f(v_2)$	$\mathcal{E} = \int_0^{v_2} v_2\, dp$	$F = m\frac{dv_2}{dt}$	$\mathcal{E} = \frac{1}{2}mv_2^2$
$\mathcal{E} \geq 0$	Inertia	 2, J, 1, Ω_{21}, T	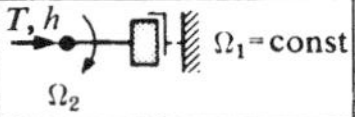 T, h, Ω_2, Ω_1 = const	$h = f(\Omega_2)$	$\mathcal{E} = \int_0^{\Omega_2} \Omega_2\, dh$	$T = J\frac{d\Omega_2}{dt}$	$\mathcal{E} = \frac{1}{2}J\Omega_2^2$
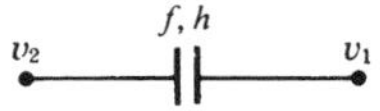 v_2 f, h v_1	Electrical capacitance	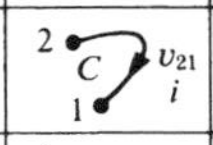 2, C, 1, v_{21}, i	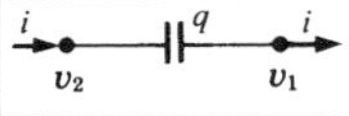 i, v_2, q, i, v_1	$q = f(v_{21})$	$\mathcal{E} = \int_0^{v_{21}} v_{21}\, dq$	$i = C\frac{dv_{21}}{dt}$	$\mathcal{E} = \frac{1}{2}Cv_{21}^2$
Pure: $h = f(v_{21})$ Ideal: $h = Cv_{21}$	Fluid capacitance	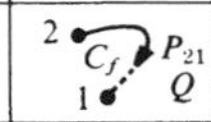 2, C_f, 1, P_{21}, Q	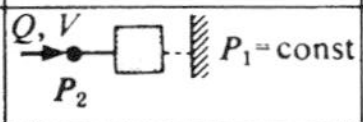 Q, V, P_2, P_1 = const	$V = f(P_2)$	$\mathcal{E} = \int_0^{P_2} P_2\, dV$	$Q = C_f\frac{dP_2}{dt}$	$\mathcal{E} = \frac{1}{2}C_f P_2^2$
Pure: $\mathcal{E} = \int_0^{v_{21}} v_{21}\, dh$ Ideal: $\mathcal{E} = \frac{1}{2}Cv_{21}^2$	Thermal capacitance	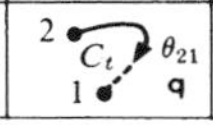 2, C_t, 1, θ_{21}, $\mathfrak{q}$	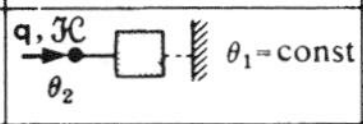 $\mathfrak{q}, \mathcal{H}$, θ_2, θ_1 = const	$\mathcal{H} = f(\theta_2)$	$\mathcal{E} = \int_0^{\theta_2} \mathfrak{q}\, dt = \mathcal{H}$	$\mathfrak{q} = C_t\frac{d\theta_2}{dt}$	$\mathcal{E} = C_t\theta_2$

Table 2 *(Continued)*

D-type energy dissipators $\mathcal{P} \geqslant 0$ (v_2, f, v_1)	Translational damper	2, b, 1, v_{21}, F	F, v_2, F, v_1	$F=f(v_{21})$	$\mathcal{P}=Fv_{21}$	$F=bv_{21}$	$\mathcal{P}=bv_{21}^2$
	Rotational damper	2, B, 1, Ω_{21}, T	T, Ω_2, Ω_1, T	$T=f(\Omega_{21})$	$\mathcal{P}=T\Omega_{21}$	$T=B\Omega_{21}$	$\mathcal{P}=B\Omega_{21}^2$
Pure / Ideal	Electrical resistance	2, R, 1, v_{21}, i	i, v_2, i, v_1	$i=f(v_{21})$	$\mathcal{P}=iv_{21}$	$i=\frac{1}{R}v_{21}$	$\mathcal{P}=\frac{1}{R}v_{21}^2$
Pure: $f=f(v_{21})$, $\mathcal{P}=v_{21}f(v_{21})$; Ideal: $f=\frac{1}{R}v_{21}$, $\mathcal{P}=\frac{1}{R}v_{21}^2=Rf^2$	Fluid resistance	2, R_f, 1, P_{21}, Q	Q, P_2, Q, P_1	$Q=f(P_{21})$	$\mathcal{P}=QP_{21}$	$Q=\frac{1}{R_f}P_{21}$	$\mathcal{P}=\frac{1}{R_f}P_{21}^2$
	Thermal resistance	2, R_t, 1, θ_{21}, q	q, θ_2, q, θ_1	$q=f(\theta_{21})$	$\mathcal{P}=q$	$q=\frac{1}{R_t}\theta_{21}$	$\mathcal{P}=\frac{1}{R_t}\theta_{21}$
Energy sources $\mathcal{P} \gtrless 0$	A-type across-variable source	2, v, 1	+, v, −, v_2, v_1	$v_{21}=f(t)$	$\mathcal{P}=fv_{21}$		
$\mathcal{E} \gtrless 0$	T-type through-variable source	2, f, 1	f, v_2, v_1	$f=f(t)$	$\mathcal{P}=fv_{21}$		

Nomenclature

λ = energy, $\mathcal{P}$ = power

f = generalized through variable, F = force, T = torque, i = current, Q = fluid flow rate, q = heat flow rate

h = generalized integrated through variable, p = translational momentum, h = angular momentum, q = charge, l' = fluid volume displaced, $\mathcal{H}$ = heat

v = generalized across variable, v = translational velocity, Ω = angular velocity, v = voltage, P = pressure, θ = temperature

x = generalized integrated across variable, x = translational displacement, Θ = angular displacement, λ = flux linkage, Γ = pressure-momentum

L = generalized ideal inductance, $1/k$ = reciprocal translational stiffness, $1/K$ = reciprocal rotational stiffness, L = inductance, I = fluid inertance

C = generalized ideal capacitance, m = mass, J = moment of insertia, C = capacitance, C_j = fluid capacitance, C_t = thermal capacitance

R = generalized ideal resistance, $1/b$ = reciprocal translational damping, $1/B$ = reciprocal rotational damping, R = electrical resistance, R_j = fluid resistance, R_t = thermal resistance

pure fluid inertance are examples of T-type storage elements. There is no corresponding thermal element.

A-type or *capacitive storage elements* are defined by a single-valued constitutive relationship between the across-variable difference $v_{21}(t)$ and the integrated through variable $h(t)$. These elements store energy by virtue of the across variable. Differentiating the constitutive relationship yields the element law. For a linear A-type element, the element law states that the through variable is proportional to the derivative of the across-variable difference. Pure translational and rotational inertia (masses) and pure electrical, fluid, and thermal capacitance are examples.

It is important to note that when a nonelectrical capacitance is represented by an A-type element, one terminal of the element must have a constant (reference) across variable, usually assumed to be zero. In a mechanical system, for example, this requirement expresses the fact that the velocity of a mass must be measured relative to a noninertial (nonaccelerating) reference frame. The constant-velocity terminal of a pure mass may be thought of as being attached in this sense to the reference frame.

D-type or *resistive elements* are defined by a single-valued constitutive relationship between the across and the through variables. These elements dissipate energy, generally by converting energy into heat. For this reason, power always flows into a D-type element. The element law for a D-type energy dissipator is the same as the constitutive relationship. For a linear dissipator, the through variable is proportional to the across-variable difference. Pure translational and rotational friction (dampers or dashpots) and pure electrical, fluid, and thermal resistance are examples.

Energy storage and energy-dissipating elements are called *passive* elements, because such elements do not supply outside energy to the system. The fourth set of one-port elements are *source elements,* which are examples of *active* or power-supplying elements. Ideal sources describe interactions between the system and its environment. A pure *A-type source* imposes an across-variable difference between its terminals, which is a prescribed function of time, regardless of the values assumed by the through variable. Similarly, a pure *T-type source* imposes a through-variable flow through the source element, which is a prescribed function of time, regardless of the corresponding across variable.

Pure system elements are used to represent physical devices. Such models are called *lumped-element models.* The derivation of lumped-element models typically requires some degree of approximation, since (1) there rarely is a one-to-one correspondence between a physical device and a set of pure elements and (2) there always is a desire to express an element law as simply as possible. For example, a coil spring has both mass and compliance. Depending on the context, the physical spring might be represented by a pure translational mass, or by a pure translational spring, or by some combination of pure springs and masses. In addition, the physical spring undoubtedly will have a nonlinear constitutive relationship over its full range of extension and compression. The compliance of the coil spring may well be represented by an ideal translational spring, however, if the physical spring is approximately linear over the range of extension and compression of concern.

2.4 Multiport Elements

A physical device that exchanges energy with its environment through two or more pairs of through and across variables is called a *multiport element.* The simplest of these, the idealized *four-terminal* or *two-port* element, is shown in Fig. 3. Two-port elements provide for transformations between the physical variables at different energy ports, while maintaining instantaneous continuity of power. In other words, net power flow into a two-port element is always identically zero:

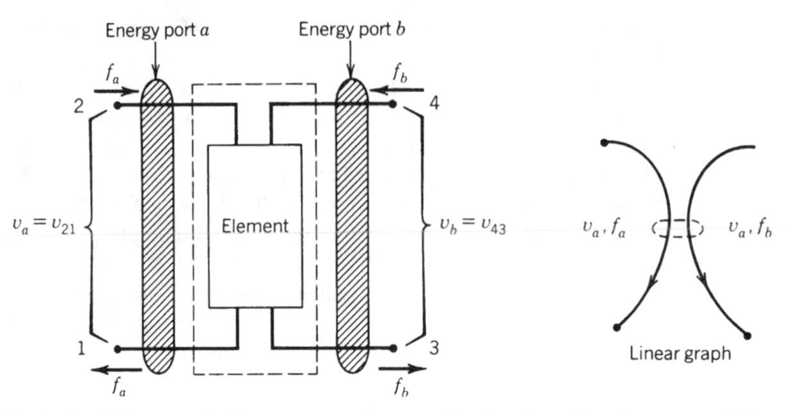

Figure 3 A four-terminal or two-port element, showing through and across variables.

$$P = f_a v_a + f_b v_b = 0$$

The particulars of the transformation between the variables define different categories of two-port elements.

A *pure transformer* is defined by a single-valued constitutive relationship between the integrated across variables or between the integrated through variables at each port:

$$x_b = f(x_a) \qquad \text{or} \qquad h_b = f(h_a)$$

For a linear (or ideal) transformer, the relationship is proportional, implying the following relationships between the primary variables:

$$v_b = nv_a \qquad f_b = -\frac{1}{n} f_a$$

where the constant of proportionality n is called the *transformation ratio.* Levers, mechanical linkages, pulleys, gear trains, electrical transformers, and differential-area fluid pistons are examples of physical devices that typically can be approximated by pure or ideal transformers. Figure 4 depicts some examples. *Pure transmitters,* which serve to transmit energy over a distance, frequently can be thought of as transformers with $n = 1$.

A *pure gyrator* is defined by a single-valued constitutive relationship between the across variable at one energy port and the through variable at the other energy port. For a linear gyrator, the following relations apply:

$$v_b = rf_a \qquad f_b = \frac{-1}{r} v_a$$

where the constant of proportionality is called the *gyration ratio* or *gyrational resistance.* Physical devices that perform pure gyration are not as common as those performing pure transformation. A mechanical gyroscope is one example of a system that might be modeled as a gyrator.

In the preceding discussion of two-port elements, it has been assumed that the type of energy is the same at both energy ports. A *pure transducer,* on the other hand, changes energy from one physical medium to another. This change may be accomplished as either a transformation or a gyration. Examples of *transforming transducers* are gears with racks (mechanical rotation to mechanical translation) and electric motors and electric generators

System	Symbol	Pure transformer	Ideal transformer	Transformation ratio
Mechanical translation (lever)	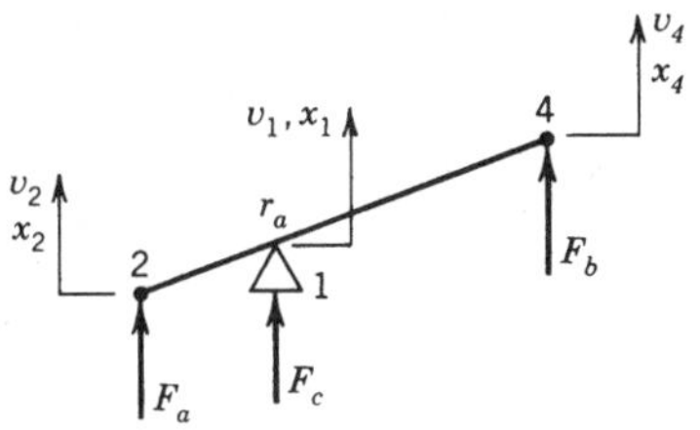 v_4, x_4, v_1, x_1, 4, v_2, x_2, r_a, 2, 1, F_b, F_a, F_c	$x_{41} = f(x_{21})$	$v_{41} = nv_{21}$ $F_b = -\frac{1}{n} F_a$	$n = -\frac{r_b}{r_a}$ Lever ratio
Mechanical rotational (gears)	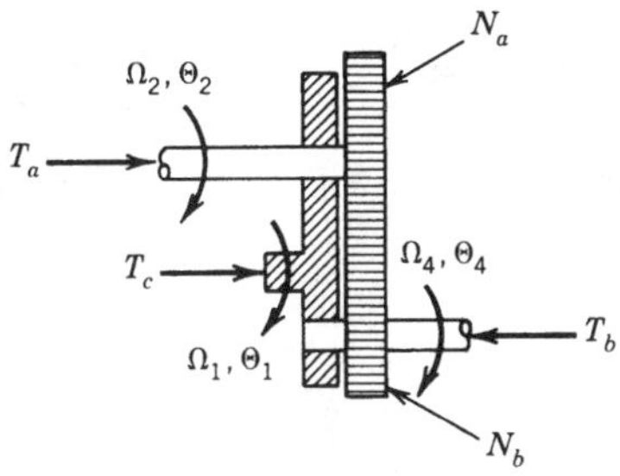 N_a, Ω_2, Θ_2, T_a, T_c, Ω_4, Θ_4, Ω_1, Θ_1, T_b, N_b	$\Theta_{41} = f(\Theta_2)$	$\Omega_{41} = n\Omega_{21}$ $T_b = -\frac{1}{n} T_a$	$n = -\frac{N_a}{N_b}$ Gear ratio
Electrical (magnetic)	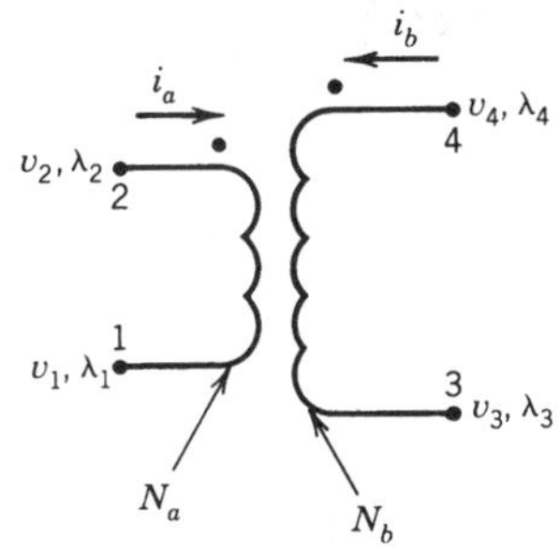 i_b, i_a, v_4, λ_4, 4, v_2, λ_2, 2, 1, v_1, λ_1, 3, v_3, λ_3, N_a, N_b	$\lambda_{43} = f(\lambda_{21})$	$v_{43} = nv_{21}$ $i_b = -\frac{1}{n} i_a$	$n = \frac{N_b}{N_a}$ Turns ratio
Fluid (differential piston)	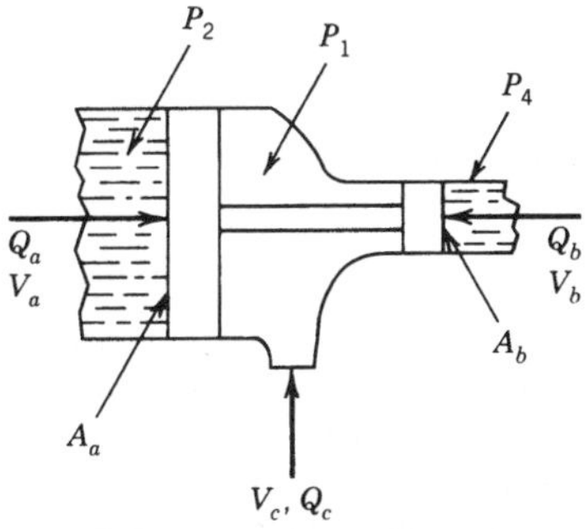 P_2, P_1, P_4, Q_a, V_a, Q_b, V_b, A_b, A_a, V_c, Q_c	$V_b = f(V_a)$	$P_{41} = nP_{21}$ $Q_b = -\frac{1}{n} Q_a$	$n = \frac{A_a}{A_b}$ Area ratio

Figure 4*a* Examples of transforms and transducers: pure transformers.[1]

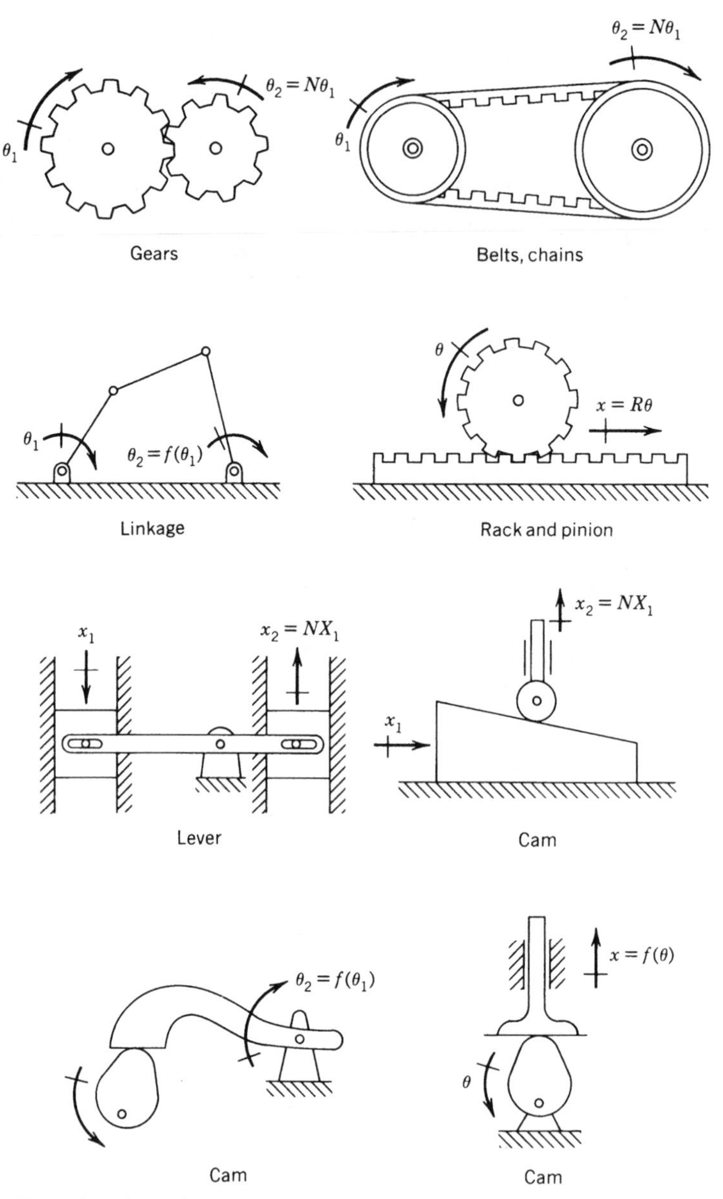

Figure 4*b* Examples of transformers and transducers: pure mechanical transformers and transforming transducers.[2]

(electrical to mechanical rotation and vice versa). Examples of *gyrating transducers* are the piston-and-cylinder (fluid to mechanical) and piezoelectric crystals (mechanical to electrical).

More complex systems may have a large number of energy ports. A common *six-terminal* or *three-port element* called a *modulator* is depicted in Fig. 5. The flow of energy between ports *a* and *b* is controlled by the energy input at the modulating port *c*. Such devices inherently dissipate energy, since

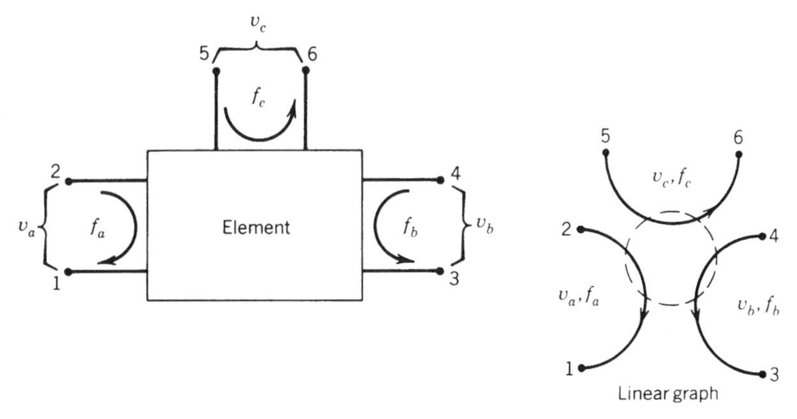

Figure 5 A six-terminal or three-port element, showing through and across variables.

$$P_a + P_c \geq P_b$$

although most often the modulating power P_c is much smaller than the power input P_a or the power output P_b. When port a is connected to a pure source element, the combination of source and modulator is called a *pure dependent source.* When the modulating power P_c is considered the input and the modulated power P_b is considered the output, the modulator is called an *amplifier.* Physical devices that often can be modeled as modulators include clutches, fluid valves and couplings, switches, relays, transistors, and variable resistors.

3 SYSTEM STRUCTURE AND INTERCONNECTION LAWS

3.1 A Simple Example

Physical systems are represented by connecting the terminals of pure elements in patterns that approximate the relationships among the properties of component devices. As an example, consider the mechanical–translational system depicted in Fig. 6*a*, which might represent an idealized automobile suspension system. The inertial properties associated with the masses of the chassis, passenger compartment, engine, and so on, all have been lumped together as the pure mass m_1. The inertial properties of the unsprung components (wheels, axles, etc.) have been lumped into the pure mass m_2. The compliance of the suspension is modeled as a pure spring with stiffness k_1 and the frictional effects (principally from the shock absorbers) as a pure damper with damping coefficient b. The road is represented as an input or source of vertical velocity, which is transmitted to the system through a spring of stiffness k_2, representing the compliance of the tires.

3.2 Structure and Graphs

The *pattern of interconnections* among elements is called the *structure* of the system. For a one-dimensional system, structure is conveniently represented by a *system graph.* The system graph for the idealized automobile suspension system of Fig. 6*a* is shown in Fig. 6*b*. Note that each distinct across variable (velocity) becomes a distinct *node* in the graph. Each distinct through variable (force) becomes a *branch* in the graph. Nodes coincide with the

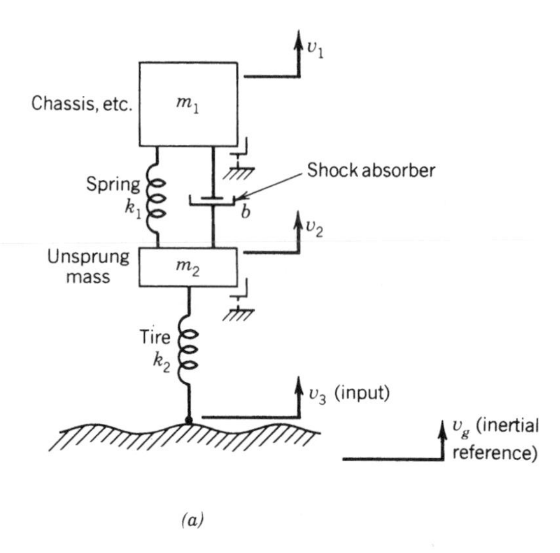

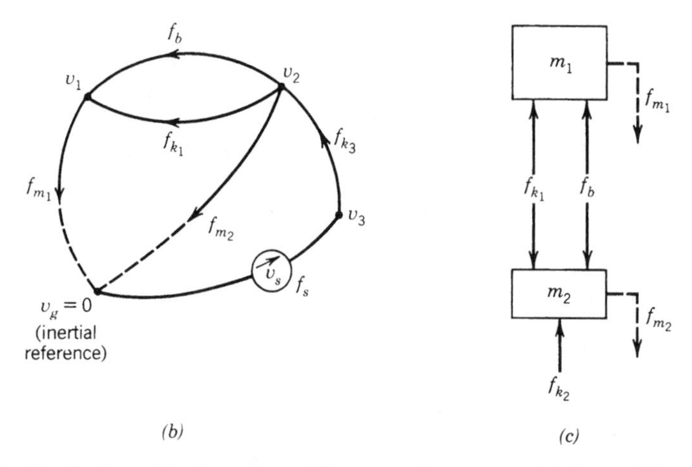

Figure 6 An idealized model of an automobile suspension system: (*a*) lumped-element model, (*b*) system graph, (*c*) free-body diagram.

terminals of elements and branches coincide with the elements themselves. One node always represents *ground* (the constant velocity of the inertial reference frame v_g), and this is usually assumed to be zero for convenience. For nonelectrical systems, all the A-type elements (masses) have one terminal connection to the reference node. Because the masses are not physically connected to ground, however, the convention is to represent the corresponding branches in the graph by dashed lines.

System graphs are oriented by placing arrows on the branches. The orientation is arbitrary and serves to assign reference directions for both the through-variable and the across-variable difference. For example, the branch representing the damper in Fig. 6*b* is directed from node 2 (tail) to node 1 (head). This assigns $v_b = v_{21} = v_2 - v_1$ as the across-variable difference to be used in writing the damper elemental equation

$$f_b = bv_b = bv_{21}$$

The reference direction for the through variable is determined by the convention that power flow $P_b = f_b v_b$ into an element is positive. Referring to Fig. 6*a*, when v_{21} is positive, the damper is in compression. Therefore, f_b must be positive for compressive forces in order to obey the sign convention for power. By similar reasoning, tensile forces will be negative.

3.3 System Relations

The structure of a system gives rise to two sets of *interconnection laws* or *system relations.* Continuity relations apply to through variables and compatibility relations apply to across variables. The interpretation of system relations for various physical systems is given in Table 3.

Continuity is a general expression of dynamic equilibrium. In terms of the system graph, continuity states that the algebraic sum of all through variables entering a given node must be zero. Continuity applies at each node in the graph. For a graph with n nodes, continuity gives rise to n continuity equations, $n - 1$ of which are independent. For node i, the continuity equation is

$$\sum_j f_{ij} = 0$$

where the sum is taken over all branches (i, j) incident on i.

For the system graph depicted in Fig. 6*b*, the four continuity equations are

$$\begin{aligned} &\text{node 1:} & f_{k_1} + f_b - f_{m_1} &= 0 \\ &\text{node 2:} & f_{k_2} - f_{k_1} - f_b - f_{m_2} &= 0 \\ &\text{node 3:} & f_s - f_{k_2} &= 0 \\ &\text{node } g\text{:} & f_{m_1} + f_{m_2} - f_s &= 0 \end{aligned}$$

Only three of these four equations are independent. Note, also, that the equations for nodes 1–3 could have been obtained from the conventional *free-body diagrams* shown in Fig. 6*c*, where f_{m_1} and f_{m_2} are the *D'Alembert forces* associated with the pure masses. Continuity relations are also known as *vertex, node, flow,* and *equilibrium relations.*

Compatibility expresses the fact that the magnitudes of all across variables are scalar quantities. In terms of the system graph, compatibility states that the algebraic sum of the across-variable differences around any closed path in the graph must be zero. Compatibility

Table 3 System Relations for Various Systems

System	Continuity	Compatibility
Mechanical	Newton's first and third laws (conservation of momentum)	Geometrical constraints (distance is a scalar)
Electrical	Kirchhoff's current law (conservation of charge)	Kirchhoff's voltage law (potential is a scalar)
Fluid	Conservation of matter	Pressure is a scalar
Thermal	Conservation of energy	Temperature is a scalar

applies to any closed path in the system. For convenience and to ensure the independence of the resulting equations, continuity is usually applied to the *meshes* or "windows" of the graph. A one-part graph with n nodes and b branches will have $b - n + 1$ meshes, each mesh yielding one independent compatibility equation. A planar graph with p separate parts (resulting from multiport elements) will have $b - n + p$ independent compatibility equations. For a closed path q, the compatibility equation is

$$\sum_{q} v_{ij} = 0$$

where the summation is taken over all branches (i, j) on the path.

For the system graph depicted in Fig. 6*b*, the three compatibility equations based on the meshes are

$$\begin{aligned} &\text{path } 1 \rightarrow 2 \rightarrow g \rightarrow 1: && -v_b + v_{m_2} - v_{m_1} = 0 \\ &\text{path } 1 \rightarrow 2 \rightarrow 1: && -v_{k_1} + v_b = 0 \\ &\text{path } 2 \rightarrow 3 \rightarrow g \rightarrow 2: && -v_{k_2} - v_s - v_{m_2} = 0 \end{aligned}$$

These equations are all mutually independent and express apparent geometric identities. The first equation, for example, states that the velocity difference between the ends of the damper is identically the difference between the velocities of the masses it connects. Compatibility relations are also known as *path, loop,* and *connectedness* relations.

3.4 Analogs and Duals

Taken together, the element laws and system relations are a complete mathematical model of a system. When expressed in terms of generalized through and across variables, the model applies not only to the physical system for which it was derived, but also to any physical system with the same generalized system graph. Different physical systems with the same generalized model are called *analogs.* The mechanical rotational, electrical, and fluid analogs of the mechanical translational system of Fig. 6*a* are shown in Fig. 7. Note that because the original system contains an inductive storage element, there is no thermal analog.

Systems of the same physical type but in which the roles of the through variables and the across variables have been interchanged are called *duals.* The analog of a dual—or, equivalently, the dual of an analog—is sometimes called a *dualog.* The concepts of analogy and duality can be exploited in many different ways.

4 STANDARD FORMS FOR LINEAR MODELS

The element laws and system relations together constitute a complete mathematical description of a physical system. For a system graph with n nodes, b branches, and s sources, there will be $b - s$ element laws, $n - 1$ continuity equations, and $b - n + 1$ compatibility equations. This is a total of $2b - s$ differential and algebraic equations. For systems composed entirely of linear elements, it is always possible to reduce these $2b - s$ equations to either of two standard forms. The *input/output,* or *I/O, form* is the basis for *transform* or so-called *classical linear systems analysis.* The *state-variable form* is the basis for *state-variable* or so-called *modern linear systems analysis.*

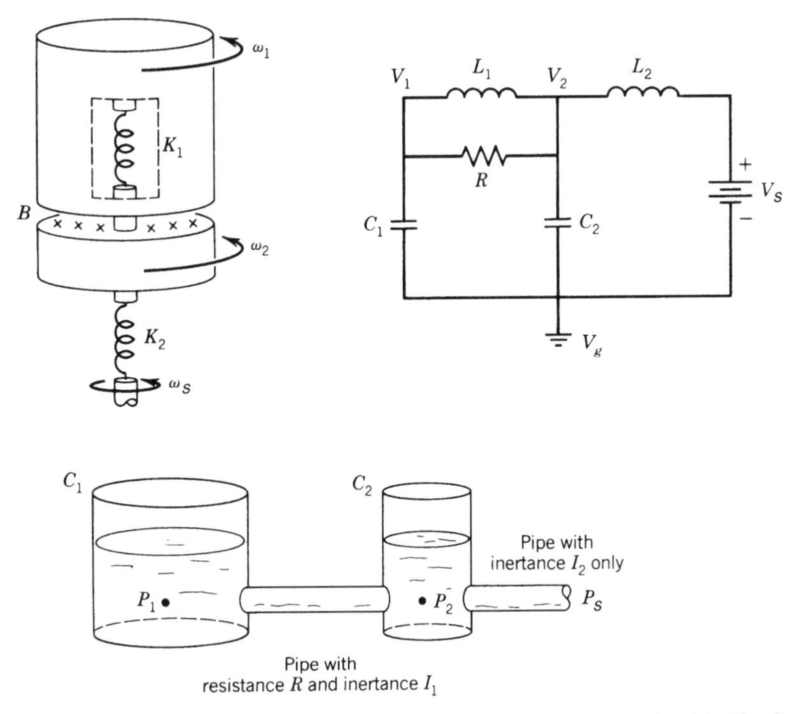

Figure 7 Analogs of the idealized automobile suspension system depicted in Fig. 6.

4.1 I/O Form

The classical representation of a system is the "blackbox," depicted in Fig. 8. The system has a set of p *inputs* (also called *excitations* or *forcing functions*), $u_j(t)$, $j = 1, 2, \ldots, p$. The system also has a set of q *outputs* (also called *response variables*), $y_k(t)$, $k = 1, 2, \ldots, q$. Inputs correspond to sources and are assumed to be known functions of time. Outputs correspond to physical variables that are to be measured or calculated.

Linear systems represented in I/O form can be modeled mathematically by *I/O differential equations*. Denoting as $y_{kj}(t)$ that part of the kth output $y_k(t)$ that is attributable to the jth input $u_j(t)$, there are $(p \times q)$ I/O equations of the form

$$\frac{d^n y_{kj}}{dt^n} + a_{n-1}\frac{d^{n-1} y_{kj}}{dt^{n-1}} + \cdots + a_1 \frac{dy_{kj}}{dt} + a_0 y_{kj}(t)$$
$$= b_m \frac{d^m u_j}{dt^m} + b_{m-1}\frac{d^{m-1} u_j}{dt^{m-1}} + \cdots + b_1 \frac{du_j}{dt} + b_0 u_j(t)$$

where $j = 1, 2, \ldots, p$ and $k = 1, 2, \ldots, q$. Each equation represents the dependence of one output and its derivatives on one input and its derivatives. By the *principle of superposition,* the kth output in response to all of the inputs acting simultaneously is

$$y_k(t) = \sum_{j=1}^{p} y_{kj}(t)$$

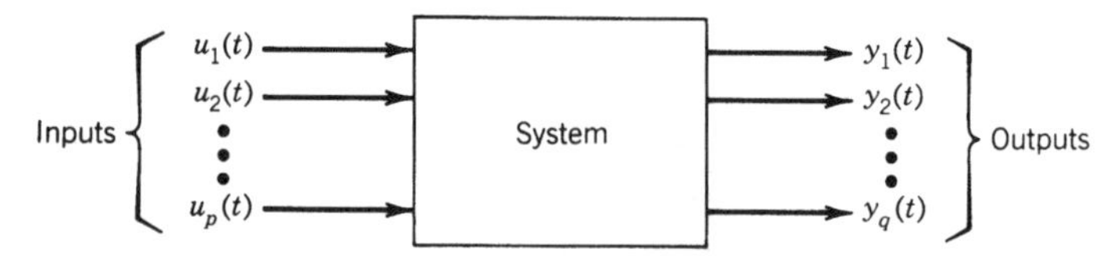

Figure 8 Input/output (I/O), or blackbox, representation of a dynamic system.

A system represented by nth-order I/O equations is called an n*th-order system.* In general, the order of a system is determined by the number of *independent* energy storage elements within the system, that is, by the combined number of T-type and A-type elements for which the initial energy stored can be independently specified.

The coefficients $a_0, a_1, \ldots, a_{n-1}$ and $b_0, b_1, \ldots, b_m$ are parameter groups made up of algebraic combinations of the system physical parameters. For a system with constant parameters, therefore, these coefficients are also constant. Systems with constant parameters are called *time-invariant* systems and are the basis for classical analysis.

4.2 Deriving the I/O Form—An Example

I/O differential equations are obtained by combining element laws and continuity and compatibility equations in order to eliminate all variables except the input and the output. As an example, consider the mechanical system depicted in Fig. 9*a*, which might represent an idealized milling machine. A rotational motor is used to position the table of the machine tool through a rack and pinion. The motor is represented as a torque source T with inertia J and internal friction B. A flexible shaft, represented as a torsional spring K, is connected to a pinion gear of radius R. The pinion meshes with a rack, which is rigidly attached to the table of mass m. Damper b represents the friction opposing the motion of the table. The problem is to determine the I/O equation that expresses the relationship between the input torque T and the position of the table x.

The corresponding system graph is depicted in Fig. 9*b*. Applying continuity at nodes 1, 2, and 3 yields

$$\text{node 1:} \quad T - T_J - T_B - T_K = 0$$

$$\text{node 2:} \quad T_K - T_p = 0$$

$$\text{node 3:} \quad -f_r - f_m - f_b = 0$$

Substituting the elemental equation for each of the one-port elements into the continuity equations and assuming zero ground velocities yield

$$\text{node 1:} \quad T - J\dot{\omega}_1 - B\omega_1 - K \int (\omega_1 - \omega_2)\, dt = 0$$

$$\text{node 2:} \quad K \int (\omega_1 - \omega_2)\, dt - T_p = 0$$

$$\text{node 3:} \quad -f_r - m\dot{v} - bv = 0$$

Note that the definition of the across variables for each element in terms of the node variables, as above, guarantees that the compatibility equations are satisfied. With the addition of the constitutive relationships for the rack and pinion

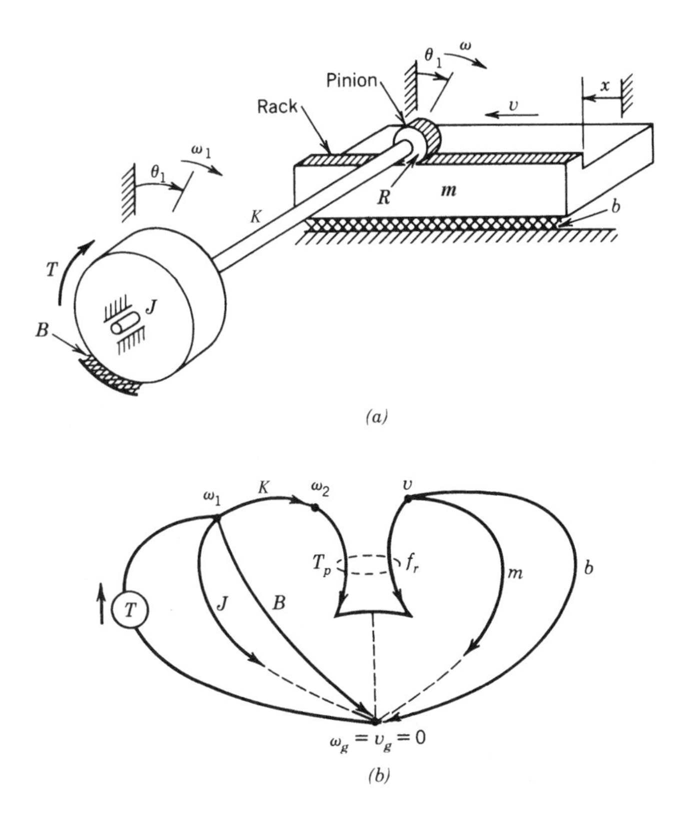

Figure 9 An idealized model of a milling machine: (*a*) lumped-element model,[3] (*b*) system graph.

$$\omega_2 = \frac{1}{R} v \qquad \text{and} \qquad T_p = -Rf_r$$

there are now five equations in the five unknowns ω_1, ω_2, v, T_p, and f_r. Combining these equations to eliminate all of the unknowns except v yields, after some manipulation,

$$a_3 \frac{d^3v}{dt^3} + a_2 \frac{d^2v}{dt^2} + a_1 \frac{dv}{dt} + a_0 v = b_1 T$$

where

$$a_3 = Jm \qquad a_1 = \frac{JK}{R^2} + Bb + mK \qquad b_1 = \frac{K}{R}$$

$$a_2 = Jb + mB \qquad a_0 = \frac{BK}{R^2} + Kb$$

Differentiating yields the desired I/O equation

$$a_3 \frac{d^3x}{dt^3} + a_2 \frac{d^2x}{dt^2} + a_1 \frac{dx}{dt} + a_0 x = b_1 \frac{dT}{dt}$$

where the coefficients are unchanged.

For many systems, combining element laws and system relations can best be achieved by ad hoc procedures. For more complicated systems, formal methods are available for the orderly combination and reduction of equations. These are the so-called *loop method* and *node method* and correspond to procedures of the same names originally developed in connection with electrical networks. The interested reader should consult Ref. 1.

4.3 State-Variable Form

For systems with multiple inputs and outputs, the I/O model form can become unwieldy. In addition, important aspects of system behavior can be suppressed in deriving I/O equations. The "modern" representation of dynamic systems, called the *state-variable form,* largely eliminates these problems. A state-variable model is the maximum reduction of the original element laws and system relations that can be achieved without the loss of any information concerning the behavior of a system. State-variable models also provide a convenient representation for systems with multiple inputs and outputs and for systems analysis using computer simulation.

State variables are a set of variables $x_1(t)$, $x_2(t)$, . . . , $x_n(t)$ internal to the system from which any set of outputs can be derived, as depicted schematically in Fig. 10. A set of state variables is the minimum number of independent variables such that by knowing the values of these variables at any time t_0 and by knowing the values of the inputs for all time $t \geq t_0$, the values of the state variables for all future time $t \geq t_0$ can be calculated. For a given system, the number n of state variables is unique and is equal to the order of the system. The definition of the state variables is not unique, however, and various combinations of one set of state variables can be used to generate alternative sets of state variables. For a physical system, the state variables summarize the *energy state* of the system at any given time.

A complete state-variable model consists of two sets of equations, the *state* or *plant equations* and the *output equations.* For the most general case, the state equations have the form

$$\dot{x}_1(t) = f_1[x_1(t),x_2(t), \ldots , x_n(t),u_1(t),u_2(t), \ldots , u_p(t)]$$

$$\dot{x}_2(t) = f_2[x_1(t),x_2(t), \ldots , x_n(t),u_1(t),u_2(t), \ldots , u_p(t)]$$

$$\vdots$$

$$\dot{x}_n(t) = f_n[x_1(t),x_2(t), \ldots , x_n(t),u_1(t),u_2(t), \ldots , u_p(t)]$$

and the output equations have the form

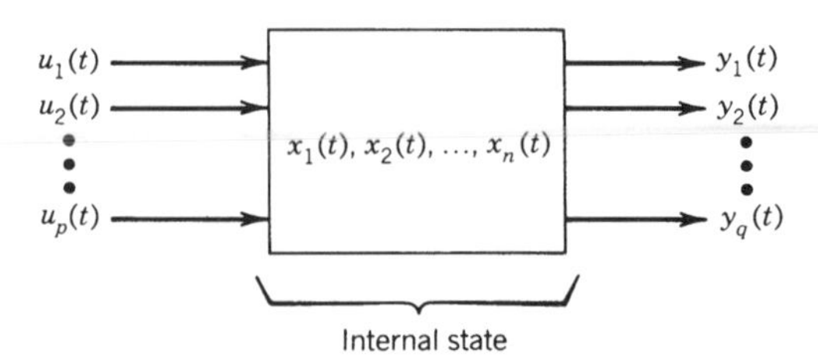

Figure 10 State-variable representation of a dynamic system.

$$y_1(t) = g_1[x_1(t),x_2(t), \ldots, x_n(t),u_1(t),u_2(t), \ldots, u_p(t)]$$

$$y_2(t) = g_2[x_1(t),x_2(t), \ldots, x_n(t),u_1(t),u_2(t), \ldots, u_p(t)]$$

$$\vdots$$

$$y_q(t) = g_q[x_1(t),x_2(t), \ldots, x_n(t),u_1(t),u_2(t), \ldots, u_p(t)]$$

These equations are expressed more compactly as the two vector equations

$$\dot{x}(t) = f[x(t),u(t)]$$

$$y(t) = g[x(t),u(t)]$$

where

$\dot{x}(t) = n \times 1$ *state vector*

$u(t) = p \times 1$ *input* or *control vector*

$y(t) = q \times 1$ *output* or *response vector*

and f and g are vector-valued functions.

For linear systems, the state equations have the form

$$\dot{x}_1(t) = a_{11}(t)x_1(t) + \cdots + a_{1n}(t)x_n(t) + b_{11}(t)u_1(t) + \cdots + b_{1p}(t)u_p(t)$$

$$\dot{x}_2(t) = a_{21}(t)x_1(t) + \cdots + a_{2n}(t)x_n(t) + b_{21}(t)u_1(t) + \cdots + b_{2p}(t)u_p(t)$$

$$\vdots$$

$$\dot{x}_n(t) = a_{n1}(t)x_1(t) + \cdots + a_{nn}(t)x_n(t) + b_{n1}(t)u_1(t) + \cdots + b_{np}(t)u_p(t)$$

and the output equations have the form

$$y_1(t) = c_{11}(t)x_1(t) + \cdots + c_{1n}(t)x_n(t) + d_{11}(t)u_1(t) + \cdots + d_{1p}(t)u_p(t)$$

$$y_2(t) = c_{21}(t)x_1(t) + \cdots + c_{2n}(t)x_n(t) + d_{21}(t)u_1(t) + \cdots + d_{2p}(t)u_p(t)$$

$$\vdots$$

$$y_n(t) = c_{q1}(t)x_1(t) + \cdots + c_{qn}(t)x_n(t) + d_{q1}(t)u_1(t) + \cdots + d_{qp}(t)u_p(t)$$

where the coefficients are groups of parameters. The linear model is expressed more compactly as the two linear vector equations

$$\dot{x}(t) = A(t)x(t) + B(t)u(t)$$

$$y(t) = C(t)x(t) + D(t)u(t)$$

where the vectors x, u, and y are the same as the general case and the matrices are defined as

$A = [a_{ij}]$ is the $n \times n$ *system matrix*

$B = [b_{jk}]$ is the $n \times p$ *control*, *input*, or *distribution matrix*

$C = [c_{lj}]$ is the $q \times n$ *output matrix*

$D = [d_{lk}]$ is the $q \times p$ *output distribution matrix*

For a time-invariant linear system, all of these matrices are constant.

4.4 Deriving the "Natural" State Variables—A Procedure

Because the state variables for a system are not unique, there are an unlimited number of alternative (but equivalent) state-variable models for the system. Since energy is stored only in generalized system storage elements, however, a natural choice for the state variables is the set of through and across variables corresponding to the independent T-type and A-type elements, respectively. This definition is sometimes called the set of *natural state variables* for the system.

For linear systems, the following procedure can be used to reduce the set of element laws and system relations to the natural state-variable model.

Step 1. For each independent T-type storage, write the element law with the derivative of the through variable isolated on the left-hand side, that is, $\dot{f} = L^{-1}v$.

Step 2. For each independent A-type storage, write the element law with the derivative of the across variable isolated on the left-hand side, that is, $\dot{v} = C^{-1}f$.

Step 3. Solve the compatibility equations, together with the element laws for the appropriate D-type and multiport elements, to obtain each of the across variables of the independent T-type elements in terms of the natural state variables and specified sources.

Step 4. Solve the continuity equations, together with the element laws for the appropriate D-type and multiport elements, to obtain the through variables of the A-type elements in terms of the natural state variables and specified sources.

Step 5. Substitute the results of step 3 into the results of step 1; substitute the results of step 4 into the results of step 2.

Step 6. Collect terms on the right-hand side and write in vector form.

4.5 Deriving the "Natural" State Variables—An Example

The six-step process for deriving a natural state-variable representation, outlined in the preceding section, is demonstrated for the idealized automobile suspension depicted in Fig. 6:

Step 1

$$\dot{f}_{k_1} = k_1 v_{k_1} \qquad \dot{f}_{k_2} = k_2 v_{k_2}$$

Step 2

$$\dot{v}_{m_1} = m_1^{-1} f_{m_1} \qquad \dot{v}_{m_2} = m_2^{-1} f_{m_2}$$

Step 3

$$v_{k_1} = v_b = v_{m_2} - v_{m_1} \qquad v_{k_2} = -v_{m_2} - v_s$$

Step 4

$$f_{m_1} = f_{k_1} + f_b = f_{k_1} + b^{-1}(v_{m_2} - v_{m_1})$$

$$f_{m_2} = f_{k_2} - f_{k_1} - f_b = f_{k_2} - f_{k_1} - b^{-1}(v_{m_2} - v_{m_1})$$

Step 5

$$\dot{f}_{k_1} = k_1(v_{m_2} - v_{m_1}) \qquad \dot{v}_{m_1} = m_1^{-1}[f_{k_1} + b^{-1}(v_{m_2} - v_{m_1})]$$

$$\dot{f}_{k_2} = k_2(-v_{m_2} - v_s) \qquad \dot{v}_{m_2} = m_2^{-1}[f_{k_2} - f_{k_1} - b^{-1}(v_{m_2} - v_{m_1})]$$

Step 6

$$\frac{d}{dt}\begin{bmatrix} f_{k1} \\ f_{k2} \\ v_{m_1} \\ v_{m_2} \end{bmatrix} = \begin{bmatrix} 0 & 0 & -k_1 & k_1 \\ 0 & 0 & 0 & -k_2 \\ 1/m_1 & 0 & -1/m_1 b & 1/m_1 b \\ -1/m_2 & 1/m_2 & 1/m_2 b & -1/m_2 b \end{bmatrix}\begin{bmatrix} f_{k1} \\ f_{k2} \\ v_{m_1} \\ v_{m_2} \end{bmatrix} + \begin{bmatrix} 0 \\ -k_2 \\ 0 \\ 0 \end{bmatrix} v_s$$

4.6 Converting from I/O to "Phase-Variable" Form

Frequently, it is desired to determine a state-variable model for a dynamic system for which the I/O equation is already known. Although an unlimited number of such models is possible, the easiest to determine uses a special set of state variables called the *phase variables.* The phase variables are defined in terms of the output and its derivatives as follows:

$$x_1(t) = y(t)$$

$$x_2(t) = \dot{x}_1(t) = \frac{d}{dt}\,y(t)$$

$$x_3(t) = \dot{x}_2(t) = \frac{d^2}{dt^2}\,y(t)$$

$$\vdots$$

$$x_n(t) = \dot{x}_{n-1}(t) = \frac{d^{n-1}}{dt^{n-1}}\,y(t)$$

This definition of the phase variables, together with the I/O equation of Section 4.1, can be shown to result in a state equation of the form

$$\frac{d}{dt}\begin{bmatrix} x_1(t) \\ x_2(t) \\ \vdots \\ x_{n-1}(t) \\ x_n(t) \end{bmatrix} = \begin{bmatrix} 0 & 1 & 0 & \cdots & 0 \\ 0 & 0 & 1 & \cdots & 0 \\ \vdots & \vdots & \vdots & \ddots & \vdots \\ 0 & 0 & 0 & \cdots & 1 \\ -a_0 & -a_1 & -a_2 & \cdots & -a_{n-1} \end{bmatrix}\begin{bmatrix} x_1(t) \\ x_2(t) \\ \vdots \\ x_{n-1}(t) \\ x_n(t) \end{bmatrix} + \begin{bmatrix} 0 \\ 0 \\ \vdots \\ 0 \\ 1 \end{bmatrix} u(t)$$

and an output equation of the form

$$y(t) = [b_0 \quad b_1 \cdots b_m]\begin{bmatrix} x_1(t) \\ x_2(t) \\ \vdots \\ x_n(t) \end{bmatrix}$$

This special form of the system matrix, with 1s along the upper off-diagonal and 0s elsewhere except for the bottom row, is called a *companion matrix.*

5 APPROACHES TO LINEAR SYSTEMS ANALYSIS

There are two fundamental approaches to the analysis of linear, time-invariant systems. *Transform methods* use rational functions obtained from the Laplace transformation of the system I/O equations. Transform methods provide a particularly convenient algebra for combining the component submodels of a system and form the basis of so-called *classical control theory. State-variable methods* use the vector state and output equations directly. State-variable methods permit the adaptation of important ideas from linear algebra and form the basis for so-called *modern control theory.* Despite the deceiving names of "classical" and "modern," the two approaches are complementary. Both approaches are widely used in current practice and the control engineer must be conversant with both.

5.1 Transform Methods

A *transformation* converts a given mathematical problem into an equivalent problem, according to some well-defined rule called a *transform.* Prudent selection of a transform frequently results in an equivalent problem that is easier to solve than the original. If the solution to the original problem can be recovered by an inverse transformation, the three-step process of (1) transformation, (2) solution in the *transform domain,* and (3) inverse transformation may prove more attractive than direct solution of the problem in the original problem domain. This is true for fixed linear dynamic systems under the *Laplace transform,* which converts differential equations into equivalent algebraic equations.

Laplace Transforms: Definition

The one-sided Laplace transform is defined as

$$F(s) = \mathcal{L}[f(t)] = \int_0^\infty f(t)e^{-st}\,dt$$

and the inverse transform as

$$f(t) = \mathcal{L}^{-1}[F(s)] = \frac{1}{2\pi j}\int_{\sigma-j\omega}^{\sigma+j\omega} F(s)e^{-st}\,ds$$

The Laplace transform converts the function $f(t)$ into the transformed function $F(s)$; the inverse transform recovers $f(t)$ from $F(s)$. The symbol $\mathcal{L}$ stands for the "Laplace transform of"; the symbol $\mathcal{L}^{-1}$ stands for "the inverse Laplace transform of."

The Laplace transform takes a problem given in the *time domain,* where all physical variables are functions of the *real variable t,* into the *complex-frequency domain,* where all physical variables are functions of the complex frequency $s = \sigma + j\omega$, where $j = \sqrt{-1}$ is the imaginary operator. Laplace transform pairs consist of the function $f(t)$ and its transform $F(s)$. Transform pairs can be calculated by substituting $f(t)$ into the defining equation and then evaluating the integral with s held constant. For a transform pair to exist, the corresponding integral must converge, that is,

$$\int_0^\infty |f(t)|e^{-\sigma^* t}\,dt < \infty$$

for some real $\sigma^* > 0$. Signals that are physically realizable always have a Laplace transform.

Tables of Transform Pairs and Transform Properties

Transform pairs for functions commonly encountered in the analysis of dynamic systems rarely need to be calculated. Instead, pairs are determined by reference to a *table of transforms* such as that given in Table 4. In addition, the Laplace transform has a number of properties that are useful in determining the transforms and inverse transforms of functions

Table 4 Laplace Transform Pairs

$F(s)$	$f(t),\ t \geq 0$
1. 1	$\delta(t)$, the unit impulse at $t = 0$
2. $\frac{1}{s}$	1, the unit step
3. $\frac{n!}{s^{n+1}}$	t^n
4. $\frac{1}{s+a}$	e^{-at}
5. $\frac{1}{(s+a)^n}$	$\frac{1}{(n-1)!} t^{n-1} e^{-at}$
6. $\frac{a}{s(s+a)}$	$1 - e^{-at}$
7. $\frac{1}{(s+a)(s+b)}$	$\frac{1}{b-a}(e^{-at} - e^{-bt})$
8. $\frac{s+p}{(s+a)(s+b)}$	$\frac{1}{b-a}[(p-a)e^{-at} - (p-b)e^{-bt}]$
9. $\frac{1}{(s+a)(s+b)(s+c)}$	$\frac{e^{-at}}{(b-a)(c-a)} + \frac{e^{-bt}}{(c-b)(a-b)} + \frac{e^{-ct}}{(a-c)(b-c)}$
10. $\frac{s+p}{(s+a)(s+b)(s+c)}$	$\frac{(p-a)e^{-at}}{(b-a)(c-a)} + \frac{(p-b)e^{-bt}}{(c-b)(a-b)} + \frac{(p-c)e^{-ct}}{(a-c)(b-c)}$
11. $\frac{b}{s^2+b^2}$	$\sin bt$
12. $\frac{s}{s^2+b^2}$	$\cos bt$
13. $\frac{b}{(s+a)^2+b^2}$	$e^{-at} \sin bt$
14. $\frac{s+a}{(s+a)^2+b^2}$	$e^{-at} \cos bt$
15. $\frac{\omega_n^2}{s^2 + 2\zeta\omega_n s + \omega_n^2}$	$\frac{\omega_n}{\sqrt{1-\zeta^2}} e^{-\zeta\omega_n t} \sin \omega_n \sqrt{1-\zeta^2}\, t, \quad \zeta < 1$
16. $\frac{\omega_n^2}{s(s^2 + 2\zeta\omega_n s + \omega_n^2)}$	$1 + \frac{1}{\sqrt{1-\zeta^2}} e^{-\zeta\omega_n t} \sin(\omega_n \sqrt{1-\zeta^2}\, t + \phi)$ $\phi = \tan^{-1} \frac{\sqrt{1-\zeta^2}}{\zeta} + \pi$ (third quadrant)

in terms of the tabulated pairs. The most important of these are given in a *table of transform properties* such as that given in Table 5.

Poles and Zeros

The response of a dynamic system most often assumes the following form in the complex-frequency domain:

Table 5 Laplace Transform Properties

$f(t)$	$F(s) = \int_0^\infty f(t)e^{-st}\,dt$
1. $af_1(t) + bf_2(t)$	$aF_1(s) + bF_2(s)$
2. $\dfrac{df}{dt}$	$sF(s) - f(0)$
3. $\dfrac{d^2f}{dt^2}$	$s^2F(s) - sf(0) - \left.\dfrac{df}{dt}\right\rvert_{t=0}$
4. $\dfrac{d^nf}{dt^n}$	$s^nF(s) - \sum_{k=1}^{n} s^{n-k}g_{k-1}$ $g_{k-1} = \left.\dfrac{d^{k-1}f}{dt^{k-1}}\right\rvert_{t=0}$
5. $\int_0^t f(t)\,dt$	$\dfrac{F(s)}{s} + \dfrac{h(0)}{s}$ $h(0) = \left.\int f(t)\,dt\right\rvert_{t=0}$
6. $\begin{Bmatrix} 0, & t < D \\ f(t-D), & t \geqslant D \end{Bmatrix}$	$e^{-sD}F(s)$
7. $e^{-at}f(t)$	$F(s+a)$
8. $f\left(\dfrac{t}{a}\right)$	$aF(as)$
9. $f(t) = \int_0^t x(t-\tau)y(\tau)\,d\tau$ $= \int_0^t y(t-\tau)x(\tau)\,d\tau$	$F(s) = X(s)Y(s)$
10. $f(\infty) = \lim_{s\to 0} sF(s)$	
11. $f(0+) = \lim_{s\to\infty} sF(s)$	

$$F(s) = \frac{N(s)}{D(s)} = \frac{b_m s^m + b_{m-1}s^{m-1} + \cdots + b_1 s + b_0}{s^n + a_{n-1}s^{n-1} + \cdots + a_1 s + a_0} \tag{1}$$

Functions of this form are called *rational functions,* because these are the ratio of two polynomials $N(s)$ and $D(s)$. If $n \geq m$, then $F(s)$ is a *proper rational function;* if $n > m$, then $F(s)$ is a *strictly proper rational function.*

In factored form, the rational function $F(s)$ can be written as

$$F(s) = \frac{N(s)}{D(s)} = \frac{b_m(s - z_1)(s - z_2)\cdots(s - z_m)}{(s - p_1)(s - p_2)\cdots(s - p_n)} \tag{2}$$

The roots of the numerator polynomial $N(s)$ are denoted by z_j, $j = 1, 2, \ldots, m$. These numbers are called the *zeros* of $F(s)$, since $F(z_j) = 0$. The roots of the denominator polynomial are denoted by p_i, 1, 2, . . . , n. These numbers are called the *poles* of $F(s)$, since $\lim_{s\to p_i} F(s) = \pm\infty$.

Inversion by Partial-Fraction Expansion

The *partial-fraction expansion theorem* states that a strictly proper rational function $F(s)$ with *distinct* (*nonrepeated*) *poles* p_i, $i = 1, 2, \ldots, n$, can be written as the sum

$$F(s) = \frac{A_1}{s - p_1} + \frac{A_2}{s - p_2} + \cdots + \frac{A_n}{s - p_n} = \sum_{i=1}^{n} A_i \left(\frac{1}{s - p_i}\right) \tag{3}$$

where the A_i, $i = 1, 2, \ldots, n$, are constants called *residues*. The inverse transform of $F(s)$ has the simple form

$$f(t) = A_1 e^{p_1 t} + A_2 e^{p_2 t} + \cdots + A_n e^{p_n t} = \sum_{i=1}^{n} A_i e^{p_i t}$$

The *Heaviside expansion theorem* gives the following expression for calculating the residue at the pole p_i,

$$A_i = (s - p_i)F(s)|_{s=p_i} \qquad \text{for } i = 1, 2, \ldots, n$$

These values can be checked by substituting into Eq. (3), combining the terms on the right-hand side of Eq. (3), and showing that the result yields the values for all the coefficients b_j, $j = 1, 2, \ldots, m$, originally specified in the form of Eq. (3).

Repeated Poles

When two or more poles of a strictly proper rational function are identical, the poles are said to be *repeated* or *nondistinct.* If a pole is repeated q times, that is, if $p_i = p_{i+1} = \cdots = p_{i+q-1}$, then the pole is said to be of *multiplicity* q. A strictly proper rational function with a pole of multiplicity q will contain q terms of the form

$$\frac{A_{i1}}{(s - p_i)^q} + \frac{A_{i2}}{(s - p_i)^{q-1}} + \cdots + \frac{A_{iq}}{s - p_i}$$

in addition to the terms associated with the distinct poles. The corresponding terms in the inverse transform are

$$\left(\frac{1}{(q-1)!} A_{i1} t^{(q-1)} + \frac{1}{(q-2)!} A_{i2} t^{(q-2)} + \cdots + A_{iq}\right) e^{p_i t}$$

The corresponding residues are

$$A_{i1} = (s - p_i)^q F(s)|_{s=p_i}$$

$$A_{i2} = \left(\frac{d}{ds}\left[(s - p_i)^q F(s)\right]\right)\bigg|_{s=p_i}$$

$$\vdots$$

$$A_{iq} = \frac{1}{(q-1)!}\left(\frac{d^{(q-1)}}{ds^{(q-1)}}\left[(s - p_i)^q F(s)\right]\right)\bigg|_{s=p_i}$$

Complex Poles

A strictly proper rational function with complex-conjugate poles can be inverted using partial-fraction expansion. Using a method called *completing the square,* however, is almost always easier. Consider the function

$$F(s) = \frac{B_1 s + B_2}{(s + \sigma - j\omega)(s + \sigma + j\omega)}$$

$$= \frac{B_1 s + B_2}{s^2 + 2\sigma s + \sigma^2 + \omega^2}$$

$$= \frac{B_1 s + B_2}{(s + \sigma)^2 + \omega_2}$$

From the transform tables the Laplace inverse is

$$f(t) = e^{-\sigma t}[B_1 \cos \omega t + B_3 \sin \omega t]$$

$$= Ke^{-\sigma t} \cos(\omega t + \phi)$$

where $B_3 = (1/\omega)(B_2 - aB_1)$
$K = \sqrt{B_1^2 + B_3^2}$
$\phi = -\tan^{-1}(B_3/B_1)$

Proper and Improper Rational Functions

If $F(s)$ is not a strictly proper rational function, then $N(s)$ must be divided by $D(s)$ using *synthetic division.* The result is

$$F(s) = \frac{N(s)}{D(s)} = P(s) + \frac{N^*(s)}{D(s)}$$

where $P(s)$ is a polynomial of degree $m - n$ and $N^*(s)$ is a polynomial of degree $n - 1$. Each term of $P(s)$ may be inverted directly using the transform tables. The strictly proper rational function $N^*(s)/D(s)$ may be inverted using partial-fraction expansion.

Initial-Value and Final-Value Theorems

The limits of $f(t)$ as time approaches zero or infinity frequently can be determined directly from the transform $F(s)$ without inverting. The *initial-value theorem* states that

$$f(0_+) = \lim_{s \to \infty} sF(s)$$

where the limit exists. If the limit does not exist (i.e., is infinite), the value of $f(0_+)$ is undefined. The *final-value theorem* states that

$$f(\infty) = \lim_{s \to 0} sF(s)$$

provided that (with the possible exception of a single pole at $s = 0$) $F(s)$ has no poles with nonnegative real parts.

Transfer Functions

The Laplace transform of the system I/O equation may be written in terms of the transform $Y(s)$ of the system response $y(t)$ as

$$Y(s) = \frac{G(s)N(s) + F(s)D(s)}{P(s)D(s)}$$

$$= \left(\frac{G(s)}{P(s)}\right)\left(\frac{N(s)}{D(s)}\right) + \frac{F(s)}{P(s)}$$

where (a) $P(s) = a_n s^n + a_{n-1} + \cdots + a_1 s + a_0$ is the *characteristic polynomial* of the system

(b) $G(s) = b_m s^m + b_{m-1} s^{m-1} + \cdots + b_1 s + b_0$ represents the *numerator dynamics* of the system

(c) $U(s) = N(s)/D(s)$ is the transform of the input to the system, $u(t)$, assumed to be a rational function

(d) $$F(s) = a_n y(0)s^{n-1} + \left(a_n \frac{dy}{dt}(0) + a_{n-1}y(0)\right) s^{n-2} + \cdots + \left(a_n \frac{d^{n-1}y}{dt^{n-1}}(0) + a_{n-1}\frac{d^{n-2}y}{dt}(0) + \cdots + a_1 y(0)\right)$$

reflects the initial system state [i.e., the initial conditions on $y(t)$ and its first $n - 1$ derivatives]

The transformed response can be thought of as the sum of two components,

$$Y(s) = Y_{zs}(s) + Y_{zi}(s)$$

where (e) $Y_{zs}(s) = [G(s)/P(s)][N(s)/D(s)] = H(s)U(s)$ is the transform of the *zero-state response,* that is, the response of the system to the input alone

(f) $Y_{zi}(s) = F(s)/P(s)$ is the transform of the *zero-input response,* that is, the response of the system to the initial state alone

The rational function

(g) $H(s) = Y_{zs}(s)/U(s) = G(s)/P(s)$ is the *transfer function* of the system, defined as the Laplace transform of the ratio of the system response to the system input, assuming zero initial conditions

The transfer function plays a crucial role in the analysis of fixed linear systems using transforms and can be written directly from knowledge of the system I/O equation as

$$H(s) = \frac{b_m s^m + \cdots + b_0}{a_n s^n + a_{n-1}s^{n-1} + \cdots + a_1 s + a_0}$$

Impulse Response

Since $U(s) = 1$ for a unit impulse function, the transform of the zero-state response to a unit impulse input is given by the relation (g) as

$$Y_{zs}(s) = H(s)$$

that is, the system transfer function. In the time domain, therefore, the unit *impulse response* is

$$h(t) = \begin{cases} 0 & \text{for } t \leq 0 \\ \mathcal{L}^{-1}[H(s)] & \text{for } t > 0 \end{cases}$$

This simple relationship is profound for several reasons. First, this provides for a direct characterization of time-domain response $h(t)$ in terms of the properties (poles and zeros) of the rational function $H(s)$ in the complex-frequency domain. Second, applying the convolution transform pair (Table 5) to relation (e) above yields

$$Y_{zs}(t) = \int_0^t h(\tau)u(t - \tau)\, d\tau$$

In words, the zero-state output corresponding to an arbitrary input $u(t)$ can be determined by convolution with the impulse response $h(t)$. In other words, the impulse response completely characterizes the system. The impulse response is also called the system *weighing function.*

Block Diagrams

Block diagrams are an important conceptual tool for the analysis and design of dynamic systems, because block diagrams provide a graphic means for depicting the relationships among system variables and components. A block diagram consists of unidirectional blocks representing specified system components or subsystems interconnected by arrows representing system variables. Causality follows in the direction of the arrows, as in Fig. 11, indicating that the output is caused by the input acting on the system defined in the block.

Combining transform variables, transfer functions, and block diagrams provides a powerful graphical means for determining the overall transfer function of a system when the transfer functions of its component subsystems are known. The basic blocks in such diagrams are given in Fig. 12. A block diagram comprising many blocks and summers can be reduced to a single transfer function block by using the diagram transformations given in Fig. 13.

5.2 Transient Analysis Using Transform Methods

Basic to the study of dynamic systems are the concepts and terminology used to characterize system behavior or performance. These ideas are aids in *defining* behavior in order to consider for a given context those features of behavior which are desirable and undesirable; in *describing* behavior in order to communicate concisely and unambiguously various behavioral attributes of a given system; and in *specifying* behavior in order to formulate desired behavioral norms for system design. Characterization of dynamic behavior in terms of standard concepts also leads in many cases to analytical shortcuts, since key features of the system response frequently can be determined without actually solving the system model.

Parts of the Complete Response

A variety of names are used to identify terms in the response of a fixed linear system. The complete response of a system may be thought of alternatively as the sum of the following:

1. The *free response* (or complementary or homogeneous solution) and the *forced response* (or particular solution). The free response represents the natural response of a system when inputs are removed and the system responds to some initial stored energy. The forced response of the system depends on the form of the input only.
2. The *transient response* and the *steady-state response.* The transient response is that part of the output that decays to zero as time progresses. The steady-state response is that part of the output that remains after all the transients disappear.

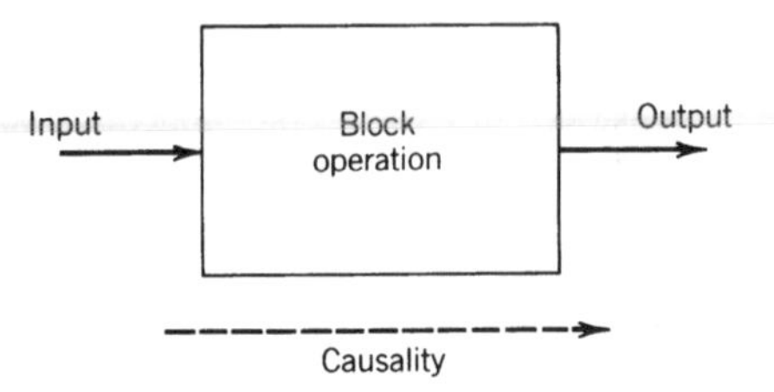

Figure 11 Basic block diagram, showing assumed direction of causality or loading.

Type	Input–Output Relations: Time Domain	Input–Output Relations: Transform Domain	Symbol
(*a*) Multiplier	$y(t) = Kv(t)$	$Y(s) = KV(s)$	$V(s)$ → K → $Y(s)$
(*b*) General transfer function	$y(t) = \mathcal{L}^{-1}[T(s)V(s)]$	$Y(s) = T(s)V(s)$	$V(s)$ → $T(s)$ → $Y(s)$
(*c*) Summer	$y(t) = v_1(t) + v_2(t)$	$Y(s) = V_1(s) + V_2(s)$	$V_1(s)$ +, $V_2(s)$ +, $Y(s)$
(*d*) Comparator	$y(t) = v_1(t) - v_2(t)$	$Y(s) = V_1(s) - V_2(s)$	$V_1(s)$ +, $V_2(s)$ −, $Y(s)$
(*e*) Takeoff point	$y(t) = v(t)$	$Y(s) = V(s)$	$V(s)$, $Y(s)$, $Y(s)$

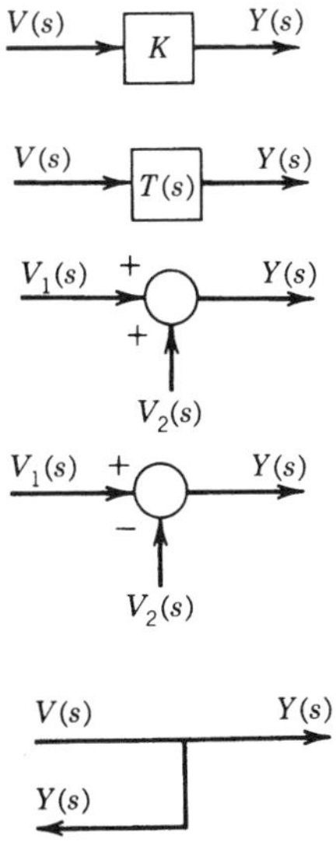

Figure 12 Basic block diagram elements.[4]

3. The *zero-state response* and the *zero-input response.* The zero-state response is the complete response (both free and forced responses) to the input when the initial state is zero. The zero-input response is the complete response of the system to the initial state when the input is zero.

Test Inputs or Singularity Functions

For a stable system, the response to a specific input signal will provide several measures of system performance. Since the actual inputs to a system are not usually known a priori, characterization of the system behavior is generally given in terms of the response to one of a standard set of *test input signals.* This approach provides a common basis for the comparison of different systems. In addition, many inputs actually encountered can be approximated by some combination of standard inputs. The most commonly used test inputs are members of the family of *singularity functions,* depicted in Fig. 14.

First-Order Transient Response

The standard form of the I/O equation for a first-order system is

$$\frac{dy}{dt} + \frac{1}{\tau}y(t) = \frac{1}{\tau}u(t)$$

where the parameter τ is called the system *time constant.* The response of this standard first-order system to three test inputs is depicted in Fig. 15, assuming zero initial conditions on the output $y(t)$. For all inputs, it is clear that the response approaches its steady state mon-

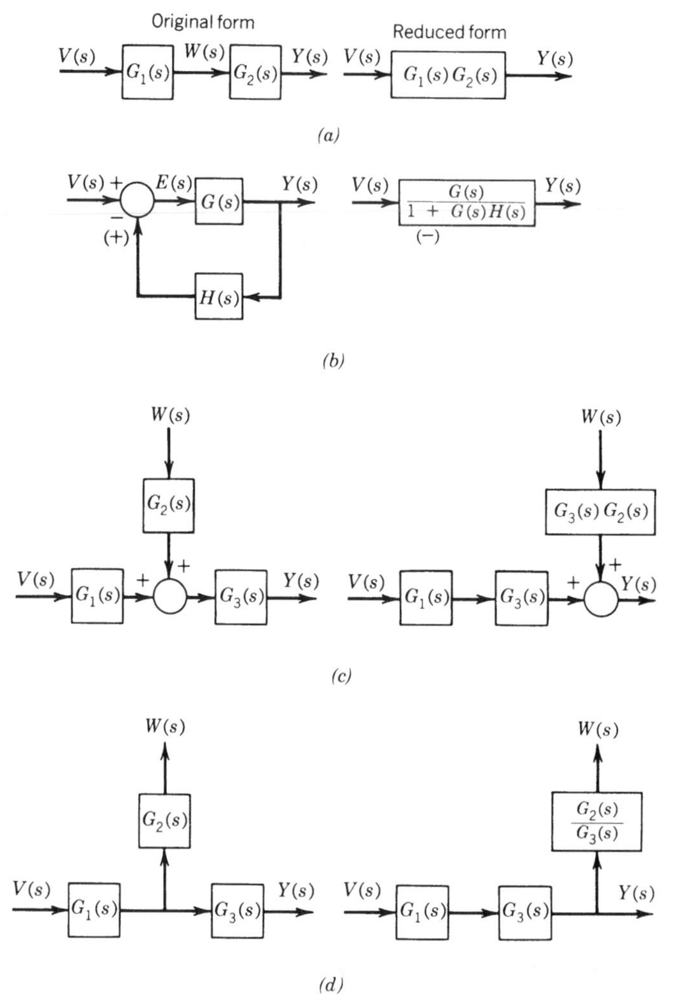

Figure 13 Representative block diagram transformations: (*a*) series or cascaded elements, (*b*) feedback loop, (*c*) relocated summer, (*d*) relocated takeoff point.[4]

otonically (i.e., without oscillations) and that the *speed of response* is completely characterized by the time constant τ. The transfer function of the system is

$$H(s) = \frac{Y(s)}{U(s)} = \frac{1/\tau}{s + 1/\tau}$$

and therefore $\tau = -p^{-1}$, where p is the system pole. As the absolute value of p increases, τ decreases and the response becomes faster.

The response of the standard first-order system to a step input of magnitude u for arbitrary initial condition $y(0) = y_0$ is

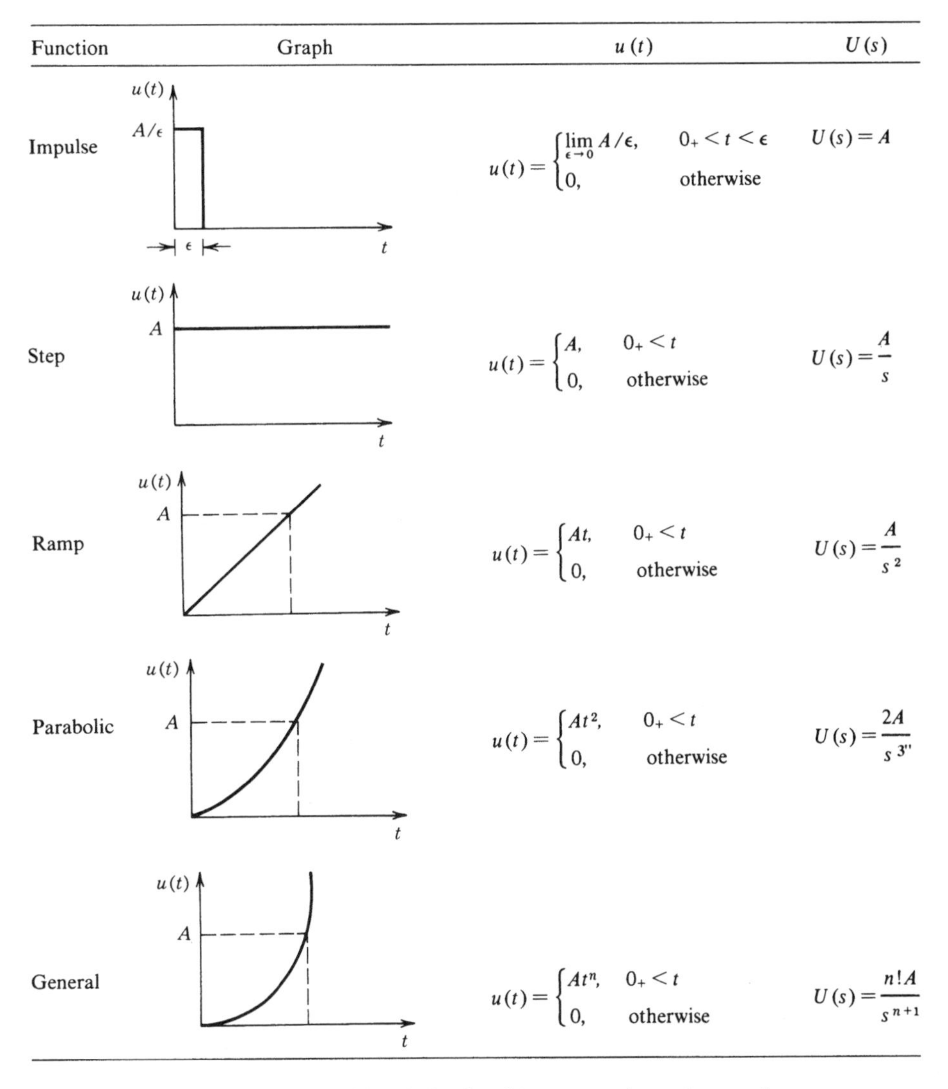

Figure 14 Family of singularity functions commonly used as test inputs.

$$y(t) = y_{ss} - [y_{ss} - y_0]e^{-t/r}$$

where $y_{ss} = u$ is the steady-state response. Table 6 and Fig. 16 record the values of $y(t)$ and $\dot{y}(t)$ for $t = k\tau$, $k = 0, 1, \ldots, 6$. Note that over any time interval of duration τ, the response increases approximately 63% of the difference between the steady-state value and the value at the beginning of the time interval, that is,

$$y(t + \tau) - y(t) \approx 0.63212[y_{ss} - y(t)]$$

Note also that the slope of the response at the beginning of any time interval of duration τ intersects the steady-state value y_{ss} at the end of the interval, that is,

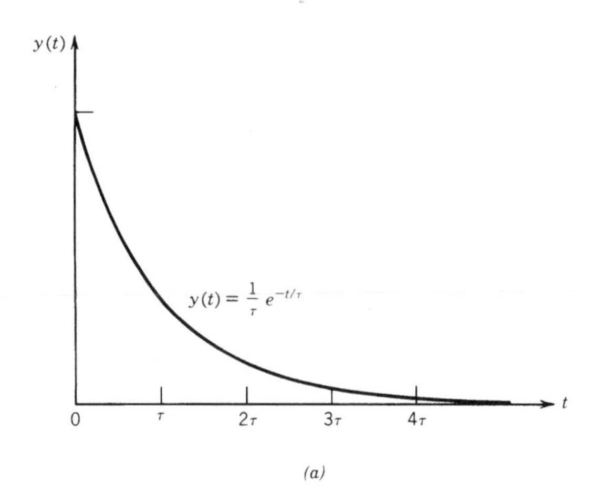

(a)

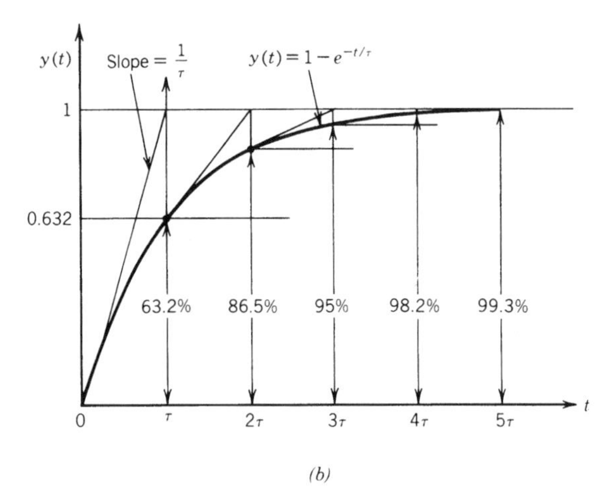

(b)

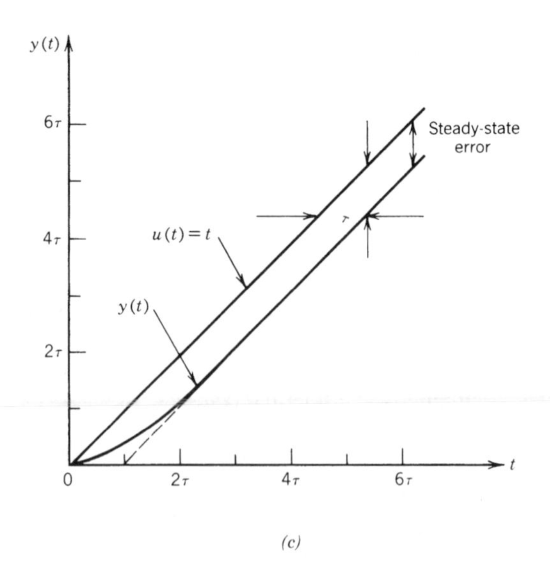

(c)

Figure 15 Response of a first-order system to (*a*) unit impulse, (*b*) unit step, and (*c*) unit ramp inputs.

Table 6 Tabulated Values of the Response of a First-Order System to a Unit Step Input

t	$y(t)$	$\dot{y}(t)$
0	0	τ^{-1}
τ	0.632	$0.368\tau^{-1}$
2τ	0.865	$0.135\tau^{-1}$
3τ	0.950	$0.050\tau^{-1}$
4τ	0.982	$0.018\tau^{-1}$
5τ	0.993	$0.007\tau^{-1}$
6τ	0.998	$0.002\tau^{-1}$

$$\frac{dy}{dt}(t) = \frac{y_{ss} - y(t)}{\tau}$$

Finally, note that after an interval of four time constants, the response is within 98% of the steady-state value, that is,

$$y(4\tau) \approx 0.98168(y_{ss} - y_0)$$

For this reason, $T_s = 4\tau$ is called the (2%) *setting time.*

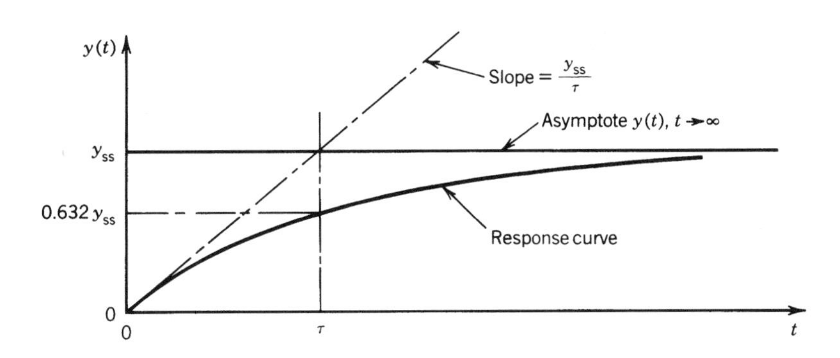

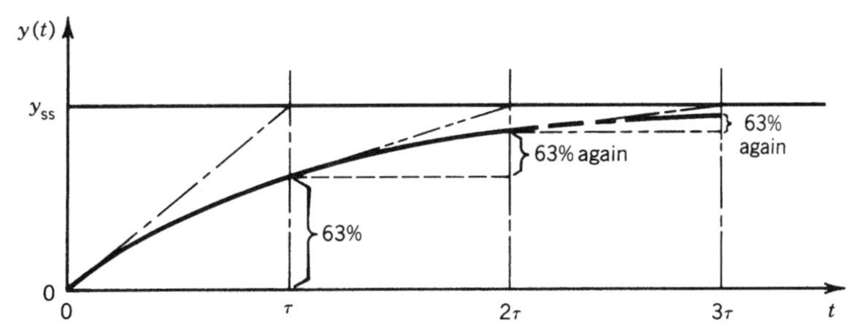

Figure 16 Response of a first-order system to a unit step input, showing the relationship to the time constant.

Second-Order Transient Response

The standard form of the I/O equation for a second-order system is

$$\frac{d^2y}{dt^2} + 2\zeta\omega_n \frac{dy}{dt} + \omega_n^2 y(t) = \omega_n^2 u(t)$$

with transfer function

$$H(s) = \frac{Y(s)}{U(s)} = \frac{\omega_n^2}{s^2 + 2\zeta\omega_n s + \omega_n^2}$$

The system poles are obtained by applying the quadratic formula to the characteristic equation as

$$p_{1.2} = -\zeta\omega_n \pm j\omega_n\sqrt{1 - \zeta^2}$$

where the following parameters are defined: ζ is the *damping ratio,* ω_n is the *natural frequency,* and $\omega_d = \omega_n\sqrt{1 - \zeta^2}$ is the *damped natural frequency.*

The nature of the response of the standard second-order system to a step input depends on the value of the damping ratio, as depicted in Fig. 17. For a stable system, four classes of response are defined.

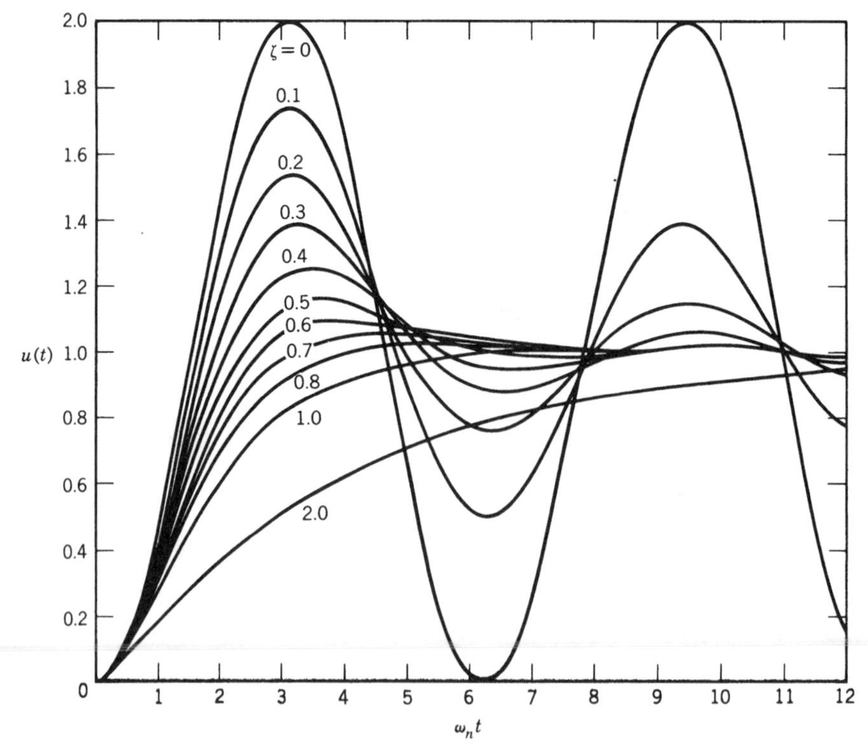

Figure 17 Response of a second-order system to a unit step input for selected values of the damping ratio.

1. *Overdamped Response* ($\zeta > 1$). The system poles are real and distinct. The response of the second-order system can be decomposed into the response of two cascaded first-order systems, as shown in Fig. 18.
2. *Critically Damped Response* ($\zeta = 1$). The system poles are real and repeated. This is the limiting case of overdamped response, where the response is as fast as possible without overshoot.
3. *Underdamped Response* ($1 > \zeta > 0$). The system poles are complex conjugates. The response oscillates at the damped frequency ω_d. The magnitude of the oscillations and the speed with which the oscillations decay depend on the damping ratio ζ.
4. *Harmonic Oscillation* ($\zeta = 0$). The system poles are pure imaginary numbers. The response oscillates at the natural frequency ω_n and the oscillations are undamped (i.e., the oscillations are sustained and do not decay).

The Complex s-Plane

The location of the system poles (roots of the characteristic equation) in the *complex s-plane* reveals the nature of the system response to test inputs. Figure 19 shows the relationship between the location of the poles in the complex plane and the parameters of the standard second-order system. Figure 20 shows the unit impulse response of a second-order system corresponding to various pole locations in the complex plane.

Transient Response of Higher Order Systems

The response of third- and higher order systems to test inputs is simply the sum of terms representing component first- and second-order responses. This is because the system poles must either be real, resulting in first-order terms, or complex, resulting in second-order underdamped terms. Furthermore, because the transients associated with those system poles having the largest real part decay the most slowly, these transients tend to dominate the output. The response of higher order systems therefore tends to have the same form as the response to the *dominant poles,* with the response to the *subdominant poles* superimposed over it. Note that the larger the relative difference between the real parts of the dominant and subdominant poles, the more the output tends to resemble the dominant mode of response.

For example, consider a fixed linear third-order system. The system has three poles. Either the poles may all be real or one may be real while the other pair is complex conjugate. This leads to the three forms of step response shown in Fig. 21, depending on the relative locations of the poles in the complex plane.

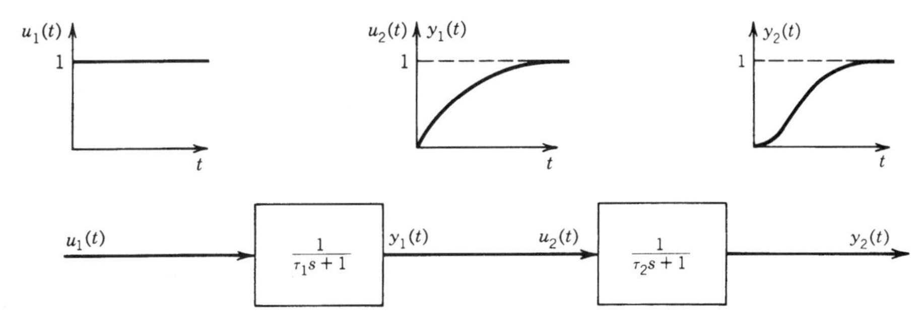

Figure 18 Overdamped response of a second-order system decomposed into the responses of two first-order systems.

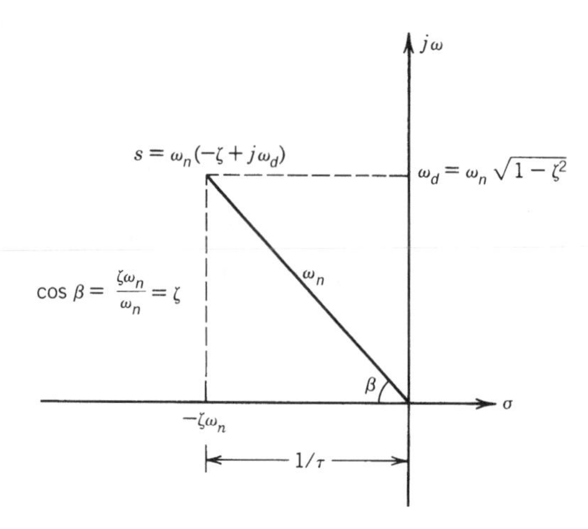

Figure 19 Location of the upper complex pole in the s-plane in terms of the parameters of the standard second-order system.

Transient Performance Measures

The transient response of a system is commonly described in terms of the measures defined in Table 7 and shown in Fig. 22. While these measures apply to any output, for a second-order system these can be calculated exactly in terms of the damping ratio and natural frequency, as shown in column 3 of the table. A common practice in control system design is to determine an initial design with dominant second-order poles that satisfy the performance specifications. Such a design can easily be calculated and then modified as necessary to achieve the desired performance.

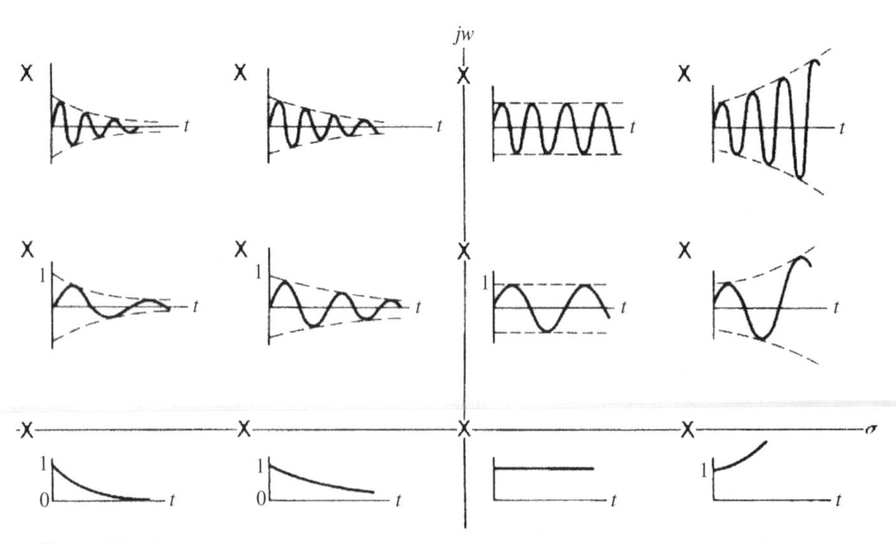

Figure 20 Unit impulse response for selected upper complex pole locations in the s-plane.

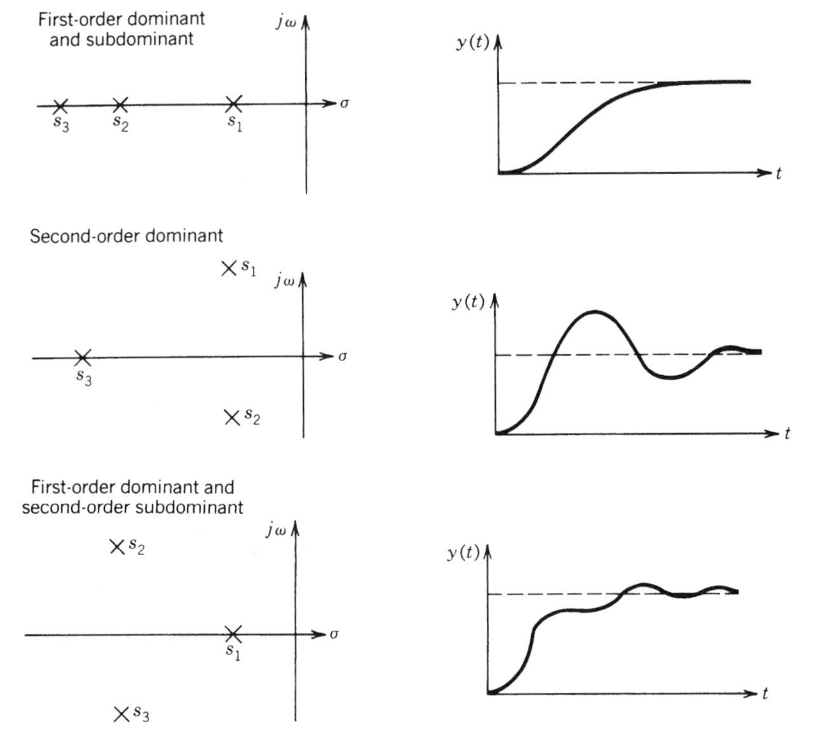

Figure 21 Step response of a third-order system for alternative upper complex pole locations in the s-plane.

The Effect of Zeros on the Transient Response

Zeros arise in a system transfer function through the inclusion of one or more derivatives of $u(t)$ among the inputs to the system. By sensing the rate(s) of change of $u(t)$, the system in effect *anticipates* the future values of $u(t)$. This tends to increase the speed of response of the system relative to the input $u(t)$.

The effect of a zero is greatest on the modes of response associated with neighboring poles. For example, consider the second-order system represented by the transfer function

$$H(s) = K\frac{s - z}{(s - p_1)(s - p_2)}$$

If $z = p_1$, then the system responds as a first-order system with $\tau = -p_2^{-1}$; whereas if $z = p_2$, then the system responds as a first-order system with $\tau = -p_1^{-1}$. Such *pole–zero cancellation* can only be achieved mathematically, but it can be approximated in physical systems. Note that by diminishing the residue associated with the response mode having the larger time constant, the system responds more quickly to changes in the input, confirming our earlier observation.

5.3 Response to Periodic Inputs Using Transform Methods

The response of a dynamic system to periodic inputs can be a critical concern to the control engineer. An input $u(t)$ is *periodic* if $u(t + T) = u(t)$ for all time t, where T is a constant

Table 7 Transient Performance Measures Based on Step Response

Performance Measure	Definition	Formula for a Second-Order System
Delay time, t_d	Time required for the response to reach half the final value for the first time	
10–90% rise time, t_r	Time required for the response to rise from 10 to 90% of the final response (used for overdamped responses)	
0–100% rise time, t_r	Time required for the response to rise from 0 to 100% of the final response (used for underdamped responses)	$t_r = \dfrac{\pi - \beta}{\omega_d}$ where $\beta = \cos^{-1} \zeta$
Peak time, t_p	Time required for the response to reach the first peak of the overshoot	$t_p = \dfrac{\pi}{\omega_d}$
Maximum overshoot, M_p	The difference in the response between the first peak of the overshoot and the final response	$M_p = e^{-\zeta\pi/\sqrt{1 - \zeta^2}}$
Percent overshoot, PO	The ratio of maximum overshoot to the final response expressed as a percentage	$\text{PO} = 100e^{-\zeta\pi/\sqrt{1 - \zeta^2}}$
Setting time, t_s	The time required for the response to reach and stay within a specified band centered on the final response (usually 2% or 5% of final response band)	$t_s = \dfrac{4}{\zeta\omega_n}$ (2% band) $t_s = \dfrac{3}{\zeta\omega_n}$ (5% band)

called the period. Periodic inputs are important because these are ubiquitous: rotating unbalanced machinery, reciprocating pumps and engines, ac electrical power, and a legion of noise and disturbance inputs can be approximated by periodic inputs. Sinusoids are the most important category of periodic inputs, because these are frequently occurring and easily analyzed and form the basis for analysis of general periodic inputs.

Frequency Response

The *frequency response* of a system is the steady-state response of the system to a sinusoidal input. For a linear system, the frequency response has the unique property that the response is a sinusoid of the same frequency as the input sinusoid, differing only in amplitude and phase. In addition, it is easy to show that the amplitude and phase of the response are functions of the input frequency, which are readily obtained from the system transfer function.

Consider a system defined by the transfer function $H(s)$. For an input

$$u(t) = A \sin \omega t$$

the corresponding steady-state output is

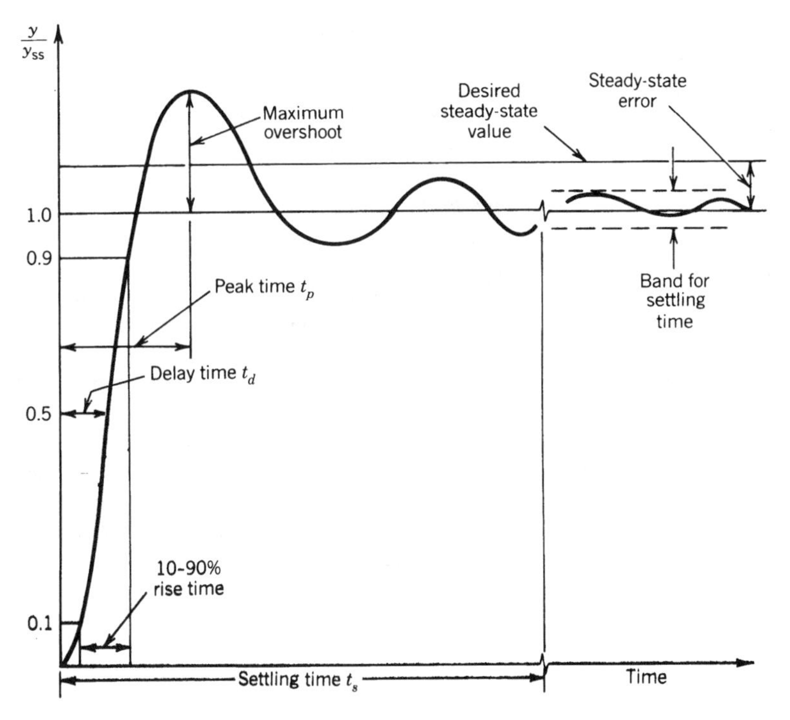

Figure 22 Transient performance measures based on step response.

$$y_{ss}(t) = AM(\omega) \sin[\omega t + \phi(\omega)]$$

where $M(\omega) = |H(j\omega)|$ is called the *magnitude ratio*
$\phi(\omega) = \angle H(j\omega)$ is called the *phase angle*
$H(j\omega) = H(s)|_{s=j\omega}$ is called the *frequency transfer function*

The frequency transfer function is obtained by substituting $j\omega$ for s in the transfer function $H(s)$. If the complex quantity $H(j\omega)$ is written in terms of its real and imaginary parts as $H(j\omega) = \text{Re}(\omega) + j\,\text{Im}(\omega)$, then

$$M(\omega) = [\text{Re}(\omega)^2 + \text{Im}(\omega)^2]^{1/2}$$

$$\phi(\omega) = \tan^{-1}\left[\frac{\text{Im}(\omega)}{\text{Re}(\omega)}\right]$$

and in polar form

$$H(j\omega) = M(\omega)e^{j\phi(\omega)}$$

Frequency Response Plots

The frequency response of a fixed linear system is typically represented graphically using one of three types of frequency response plots. A *polar plot* is simply a plot of the vector $H(j\omega)$ in the complex plane, where $\text{Re}(\omega)$ is the abscissa and $\text{Im}(\omega)$ is the ordinate. A *logarithmic plot* or *Bode diagram* consists of two displays: (1) the magnitude ratio in decibels $M_{dB}(\omega)$ [where $M_{dB}(\omega) = 20 \log M(\omega)$] versus $\log \omega$ and (2) the phase angle in degrees

$\phi(\omega)$ versus log ω. Bode diagrams for normalized first- and second-order systems are given in Fig. 23. Bode diagrams for higher order systems are obtained by adding these first- and second-order terms, appropriately scaled. A *Nichols diagram* can be obtained by cross plotting the Bode magnitude and phase diagrams, eliminating log ω. Polar plots and Bode and Nichols diagrams for common transfer functions are given in Table 8.

Frequency Response Performance Measures

Frequency response plots show that dynamic systems tend to behave like *filters,* "passing" or even amplifying certain ranges of input frequencies while blocking or attenuating other frequency ranges. The range of frequencies for which the amplitude ratio is no less than 3 dB of its maximum value is called the *bandwidth* of the system. The bandwidth is defined by upper and lower *cutoff frequencies* ω_c, or by $\omega = 0$ and an upper cutoff frequency if $M(0)$ is the maximum amplitude ratio. Although the choice of "down 3 dB" used to define the cutoff frequencies is somewhat arbitrary, the bandwidth is usually taken to be a measure of the range of frequencies for which a significant portion of the input is felt in the system output. The bandwidth is also taken to be a measure of the system speed of response, since attenuation of inputs in the higher frequency ranges generally results from the inability of the system to "follow" rapid changes in amplitude. Thus, a narrow bandwidth generally indicates a sluggish system response.

Response to General Periodic Inputs

The *Fourier series* provides a means for representing a general periodic input as the sum of a constant and terms containing sine and cosine. For this reason the *Fourier series,* together with the superposition principle for linear systems, extends the results of frequency response analysis to the general case of arbitrary periodic inputs. The Fourier series representation of a periodic function $f(t)$ with period $2T$ on the interval $t^* + 2T \geq t \geq t^*$ is

$$f(t) = \frac{a_0}{2} + \sum_{n=1}^{\infty} \left(a_n \cos \frac{n\pi t}{T} + b_n \sin \frac{n\pi t}{T} \right)$$

where

$$a_n = \frac{1}{T} \int_{t^*}^{t^*+2T} f(t) \cos \frac{n\pi t}{T}\, dt$$

$$b_n = \frac{1}{T} \int_{t^*}^{t^*+2T} f(t) \sin \frac{n\pi t}{T}\, dt$$

If $f(t)$ is defined outside the specified interval by a periodic extension of period $2T$ and if $f(t)$ and its first derivative are piecewise continuous, then the series converges to $f(t)$ if t is a point of continuity or to $\frac{1}{2}\,[f(t_+) + f(t_-)]$ if t is a point of discontinuity. Note that while the Fourier series in general is infinite, the notion of bandwidth can be used to reduce the number of terms required for a reasonable approximation.

6 STATE-VARIABLE METHODS

State-variable methods use the vector state and output equations introduced in Section 4 for analysis of dynamic systems directly in the time domain. These methods have several advantages over transform methods. First, state-variable methods are particularly advantageous for the study of multivariable (multiple-input/multiple-output) systems. Second, state-variable methods are more naturally extended for the study of linear time-varying and non-

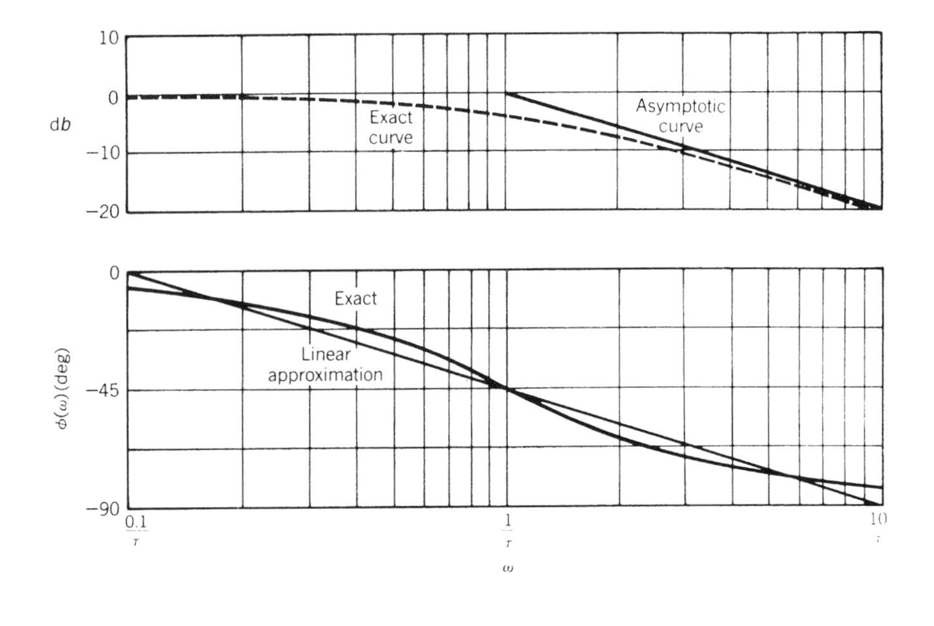

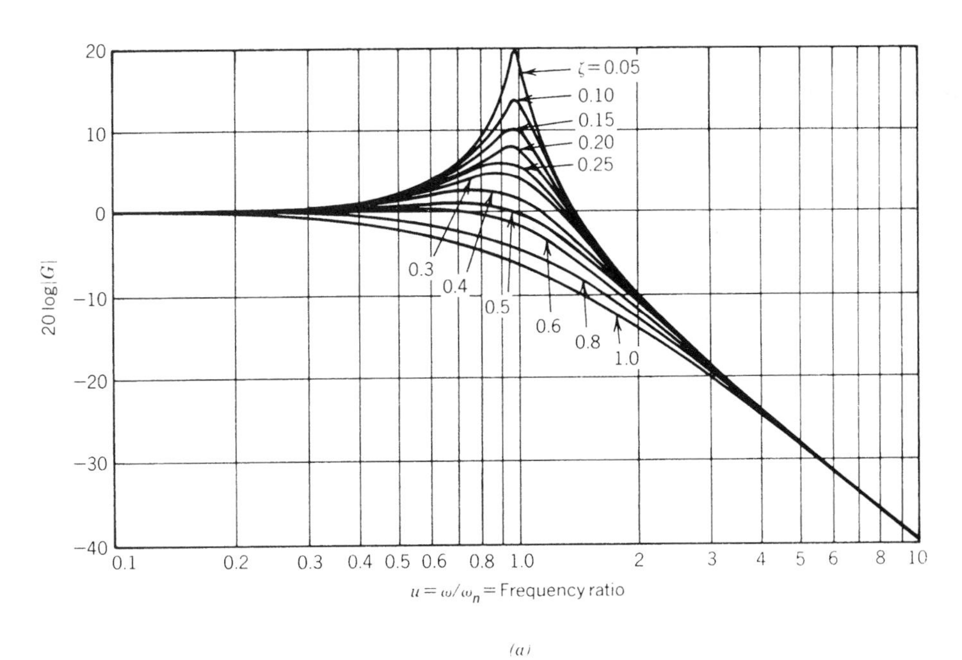

(*a*)

Figure 23 Bode diagrams for normalized (*a*) first-order and (*b*) second-order systems.

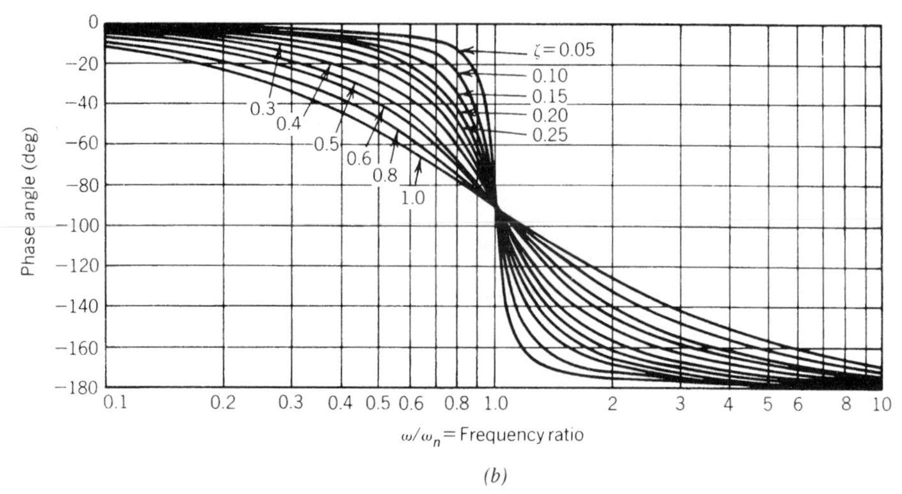

(b)

Figure 23 (*Continued*)

linear systems. Finally, state-variable methods are readily adapted to computer simulation studies.

6.1 Solution of the State Equation

Consider the vector equation of state for a fixed linear system:

$$\dot{x}(t) = Ax(t) + Bu(t)$$

The solution to this system is

$$x(t) = \Phi(t)x(0) + \int_0^t \Phi(t - \tau)Bu(\tau)\, d\tau$$

where the matrix $\Phi(t)$ is called the *state transition matrix.* The state transition matrix represents the free response of the system and is defined by the matrix exponential series

$$\Phi(t) = e^{At} = I + At + \frac{1}{2!}A^2t^2 + \cdots = \sum_{k=0}^{\infty} \frac{1}{k!} A^k t^k$$

where I is the identity matrix. The state transition matrix has the following useful properties:

$$\Phi(0) = I$$

$$\Phi^{-1}(t) = \Phi(-t)$$

$$\Phi^k(t) = \Phi(kt)$$

$$\Phi(t_1 + t_2) = \Phi(t_1)\Phi(t_2)$$

$$\Phi(t_2 - t_1)\Phi(t_1 - t_0) = \Phi(t_2 - t_0)$$

$$\dot{\Phi}(t) = A\Phi(t)$$

The Laplace transform of the state equation is

Table 8 Transfer Function Plots for Representative Transfer Functions[5]

$G(s)$	Polar Plot	Bode Diagram
1. $\dfrac{K}{s\tau_1 + 1}$	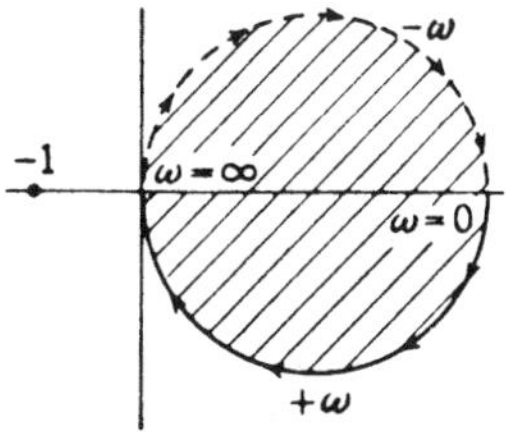	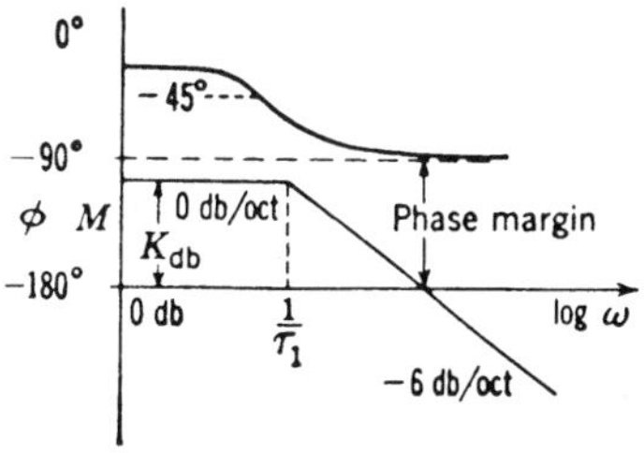
2. $\dfrac{K}{(s\tau_1 + 1)\ (s\tau_2 + 1)}$	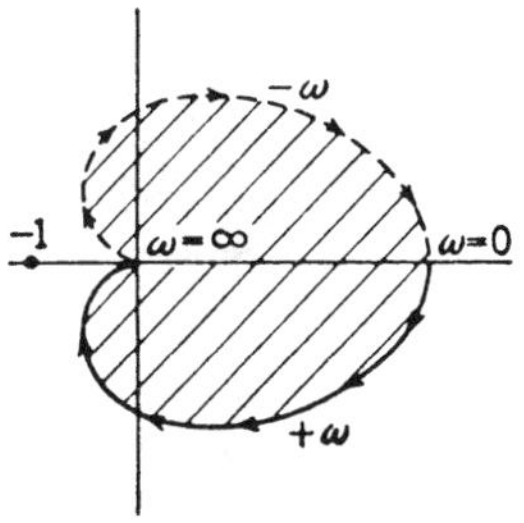	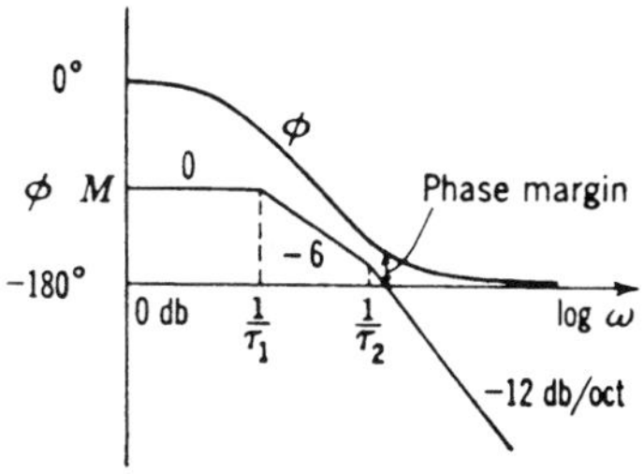
3. $\dfrac{K}{(s\tau_1 + 1)\ (s\tau_2 + 1)\ (s\tau_3 + 1)}$	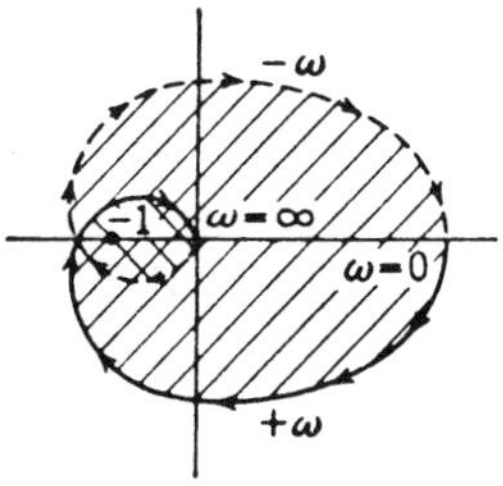	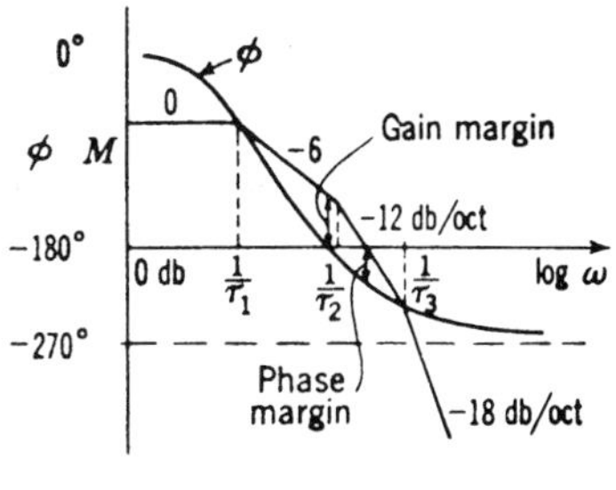
4. $\dfrac{K}{s}$	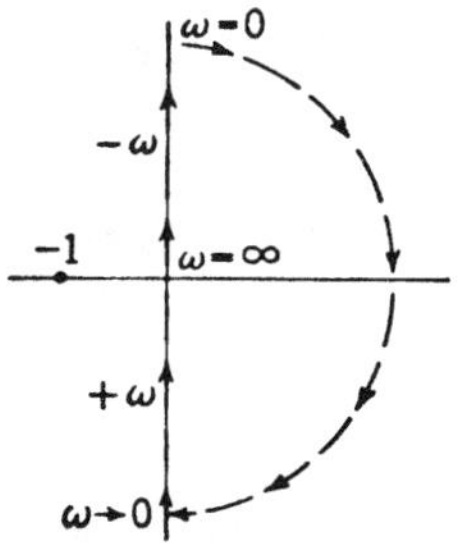	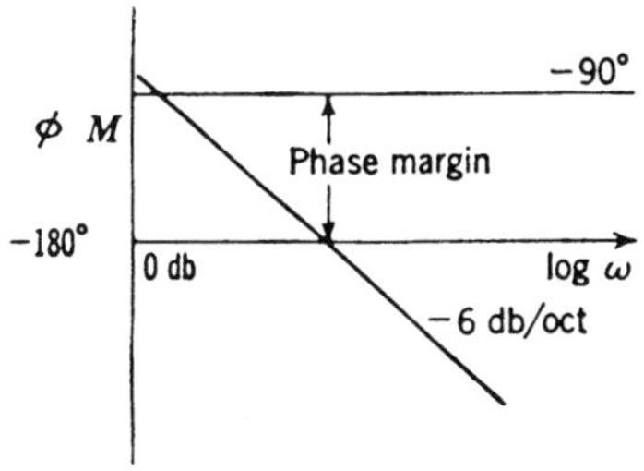

Table 8 (*Continued*)

Nichols Diagram	Root Locus	Comments
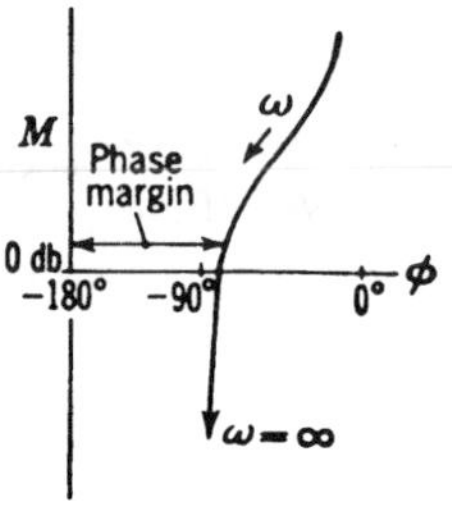 M; Phase margin; ω; 0 db; $-180°$; $-90°$; $0°$; ϕ; $\omega = \infty$	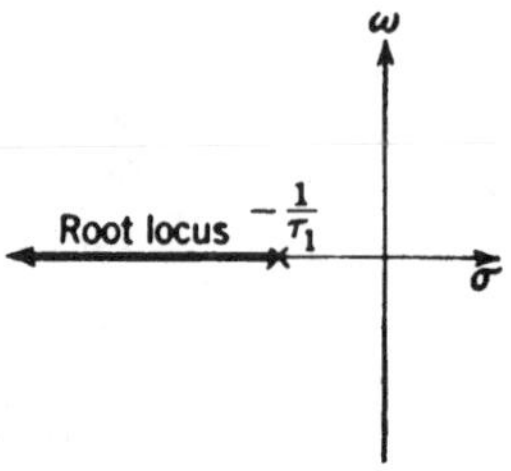 ω; Root locus; $-\frac{1}{\tau_1}$; σ	Stable; gain margin $=\infty$
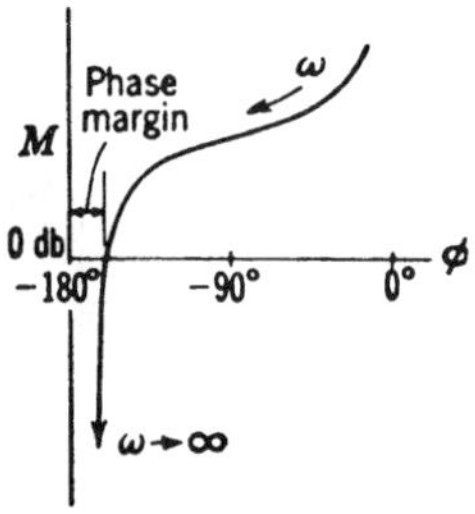 M; Phase margin; ω; 0 db; $-180°$; $-90°$; $0°$; ϕ; $\omega \to \infty$	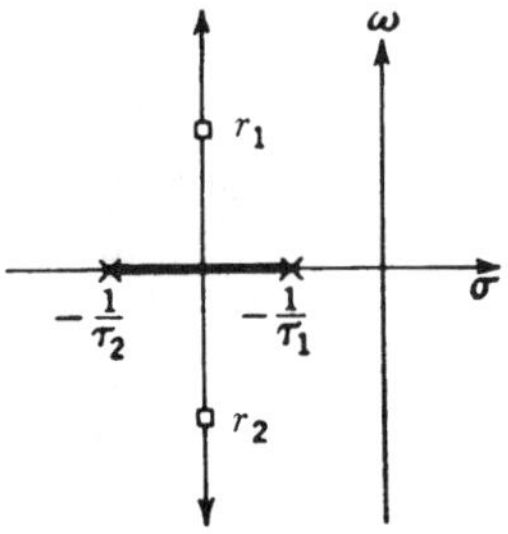 ω; r_1; r_2; $-\frac{1}{\tau_2}$; $-\frac{1}{\tau_1}$; σ	Elementary regulator; stable; gain margin $=\infty$
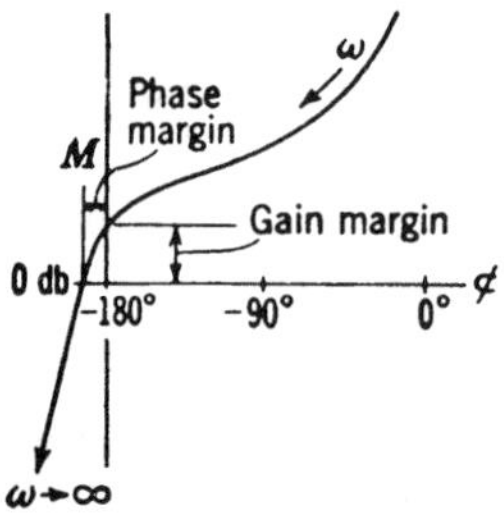 M; Phase margin; Gain margin; ω; 0 db; $-180°$; $-90°$; $0°$; ϕ; $\omega \to \infty$	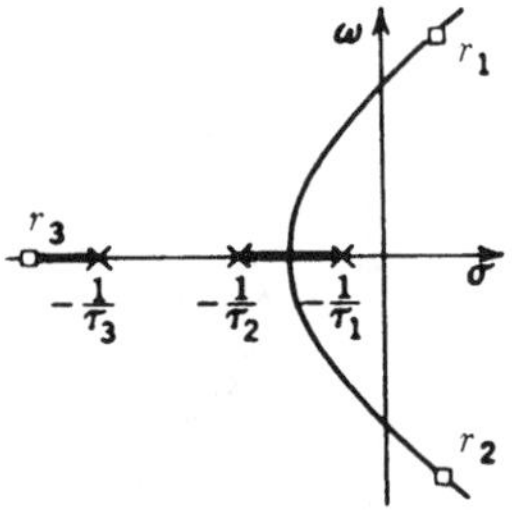 ω; r_1; r_2; r_3; $-\frac{1}{\tau_3}$; $-\frac{1}{\tau_2}$; $-\frac{1}{\tau_1}$; σ	Regulator with additional energy storage component; unstable, but can be made stable by reducing gain
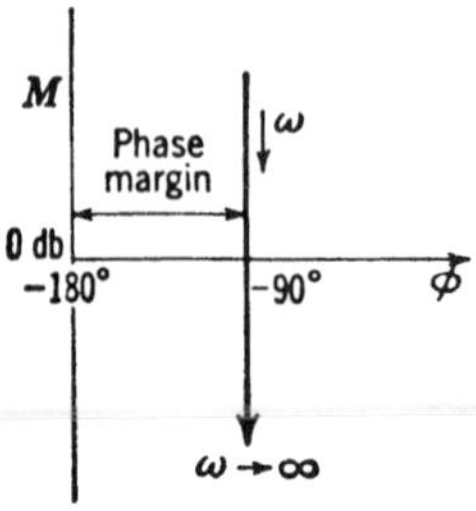 M; Phase margin; ω; 0 db; $-180°$; $-90°$; ϕ; $\omega \to \infty$	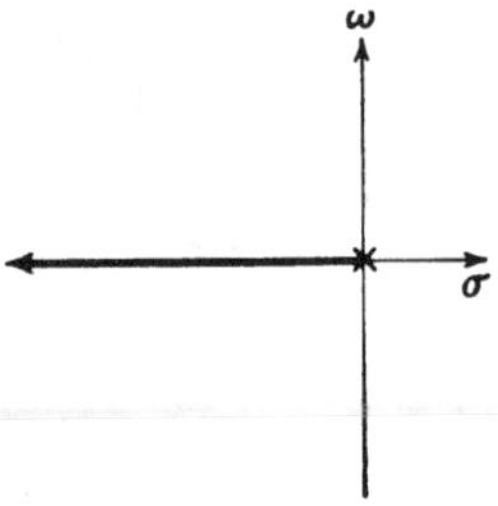 ω; σ	Ideal integrator; stable

Table 8 (*Continued*)

	$G(s)$	Polar Plot	Bode Diagram
5.	$\dfrac{K}{s(s\tau_1 + 1)}$	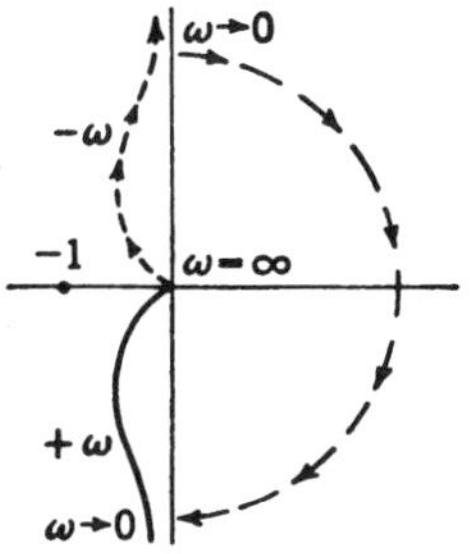	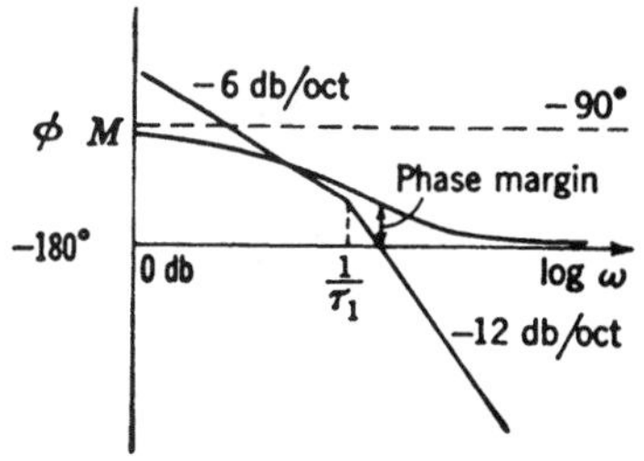
6.	$\dfrac{K}{s(s\tau_1 + 1)(s\tau_2 + 1)}$	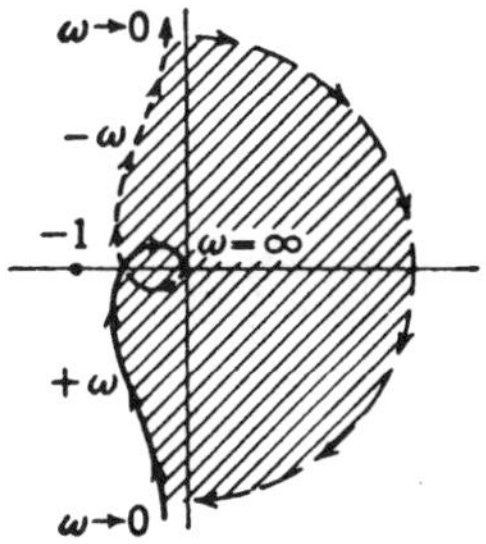	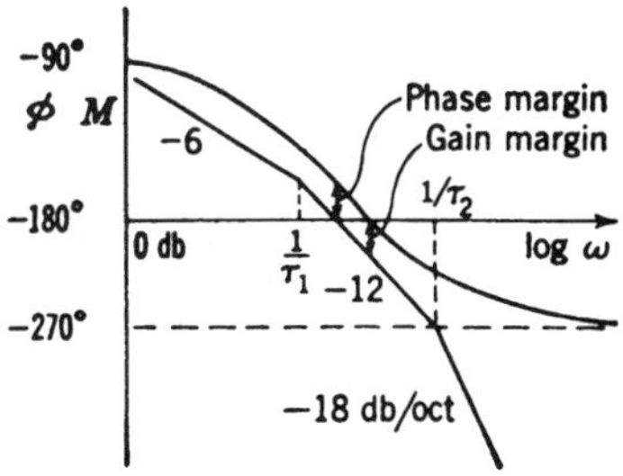
7.	$\dfrac{K(s\tau_a + 1)}{s(s\tau_1 + 1)(s\tau_2 + 1)}$	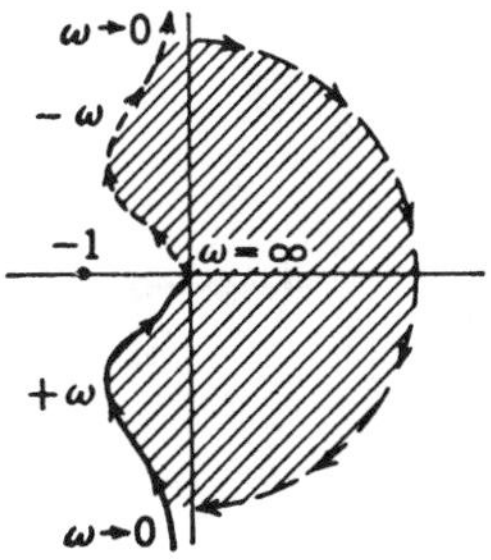	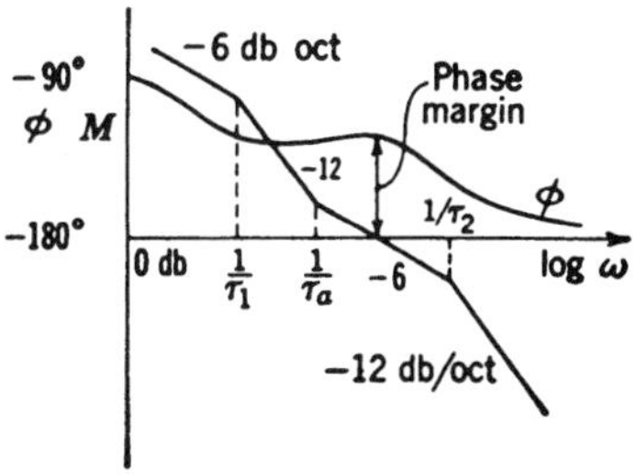
8.	$\dfrac{K}{s^2}$		

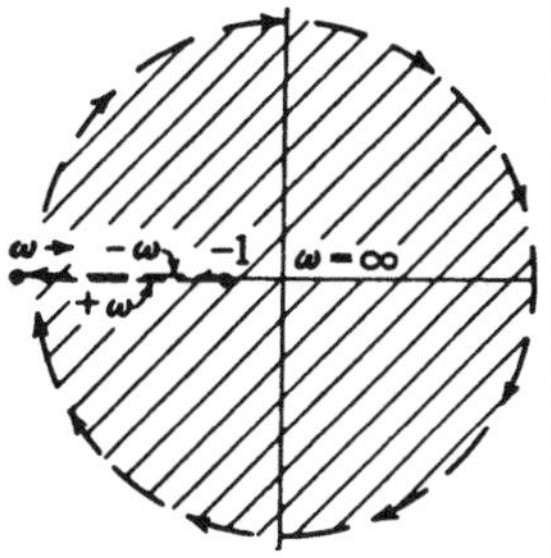

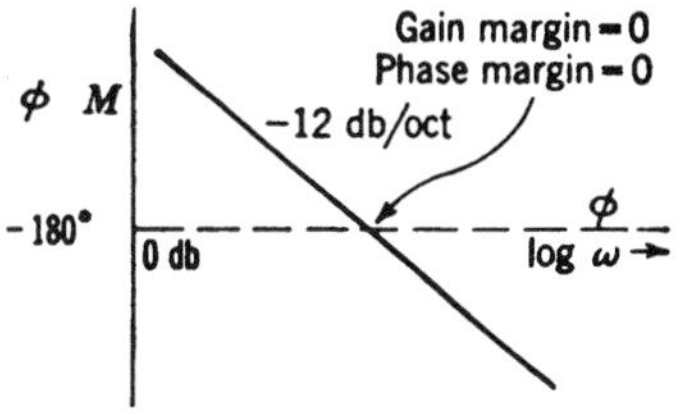

Table 8 (*Continued*)

Nichols Diagram	Root Locus	Comments
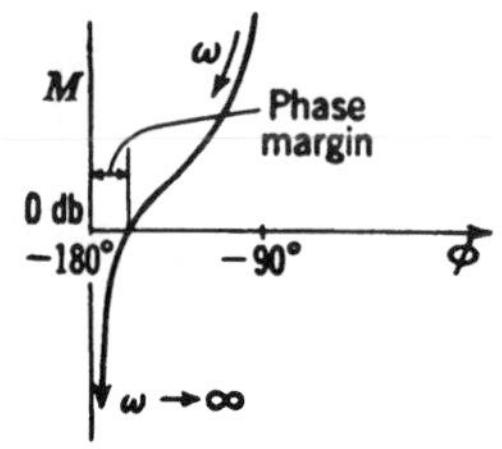	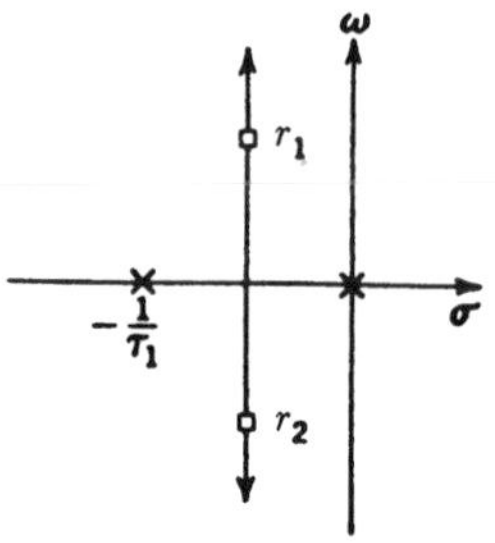	Elementary instrument servo; inherently stable; gain margin = ∞
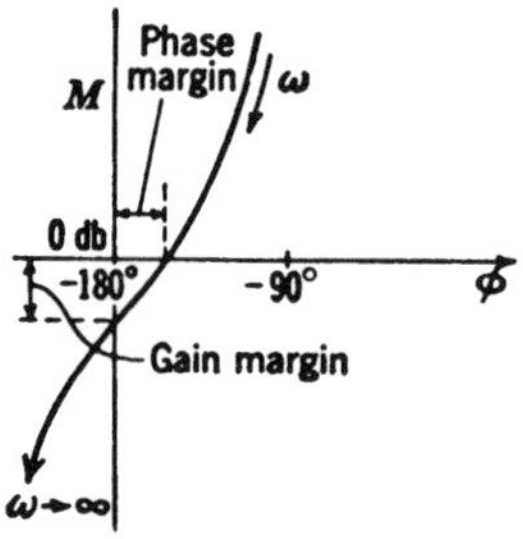	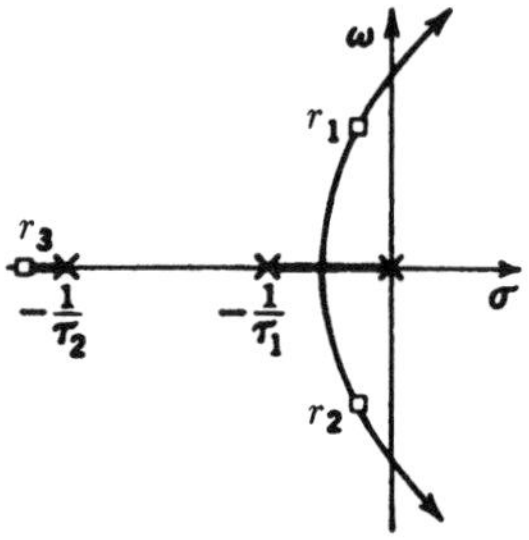	Instrument servo with field control motor or power servo with elementary Ward-Leonard drive; stable as shown but may become unstable with increased gain
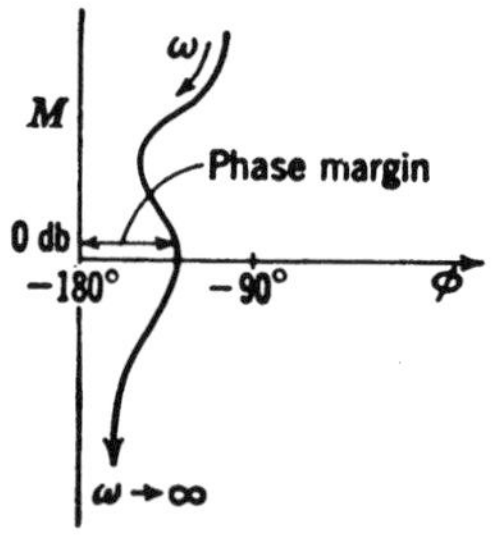	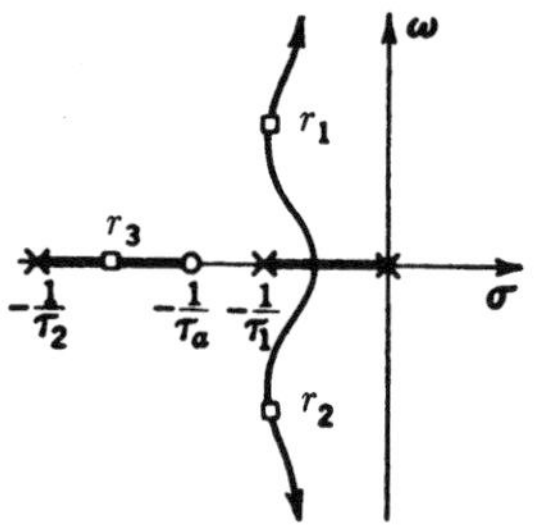	Elementary instrument servo with phase lead (derivative) compensator; stable
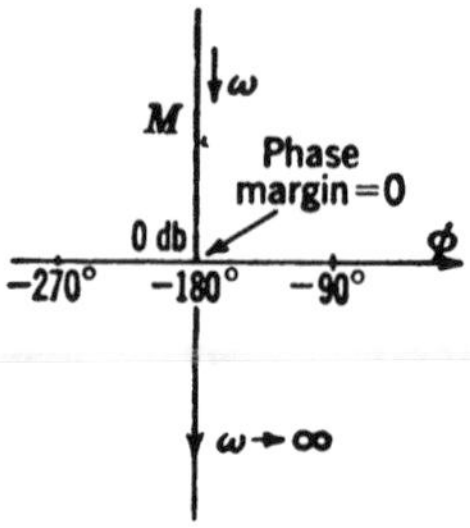	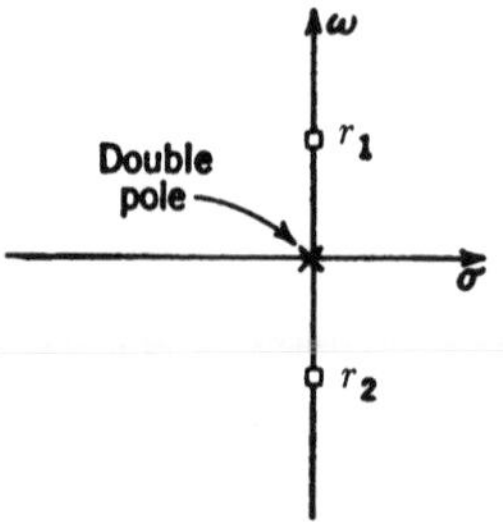	Inherently unstable; must be compensated

Table 8 *(Continued)*

$G(s)$	Polar Plot	Bode Diagram
9. $\dfrac{K}{s^2(s\tau_1 + 1)}$	$+\omega$; -1; $\omega = \infty$; $-\omega$	ϕ M; -12 db/oct; Phase margin; $-180°$; 0 db; $\frac{1}{\tau_1}$; log ω; $-270°$; ϕ; -18 db/oct
10. $\dfrac{K(s\tau_a + 1)}{s^2(s\tau_1 + 1)}$ $\tau_a > \tau_1$	$-\omega$; -1; $\omega = \infty$; $+\omega$	ϕ M; -12 db/oct; Phase margin; ϕ; $1/\tau_1$; $-180°$; 0 db; $\frac{1}{\tau_a}$ -6 db/oct; log ω; -12 db/oct
11. $\dfrac{K}{s^3}$	$+\omega$; -1; $\omega = \infty$; $-\omega$	ϕ M; -18 db/oct; $-180°$; 0 db; Phase margin; log ω; $-270°$; ϕ
12. $\dfrac{K(s\tau_a + 1)}{s^3}$	$+\omega$; -1; $\omega = \infty$; $-\omega$	ϕ M; -18 db/oct; Phase margin; log ω; $-180°$; 0 db; $\frac{1}{\tau_a}$; -12 db/oct; $-270°$; ϕ

Table 8 (*Continued*)

Nichols Diagram	Root Locus	Comments
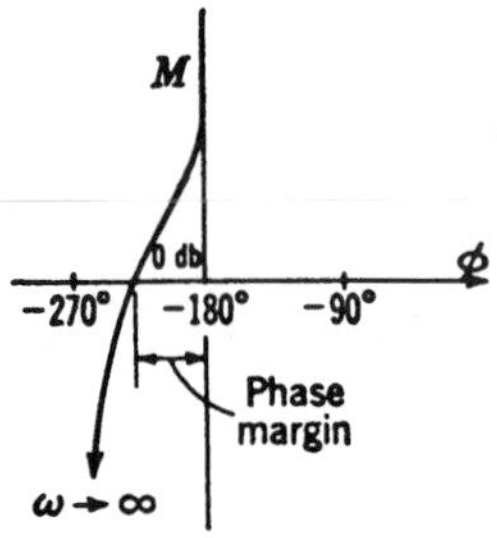 M 0 db ϕ −270° −180° −90° Phase margin $\omega \to \infty$	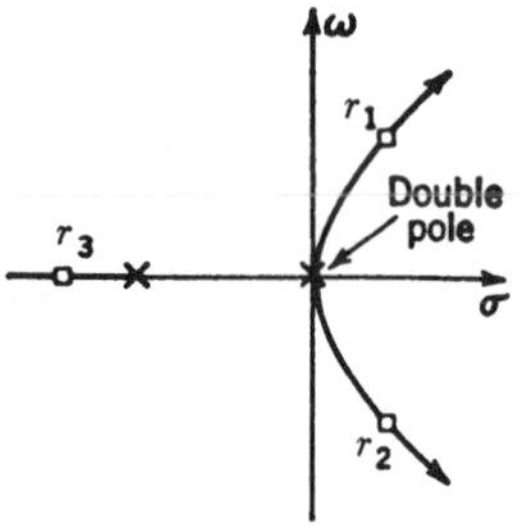 ω r_1 Double pole r_3 σ r_2	Inherently unstable; must be compensated
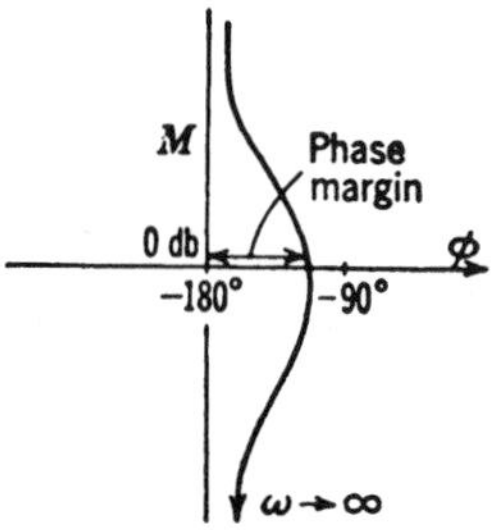 M Phase margin 0 db ϕ −180° −90° $\omega \to \infty$	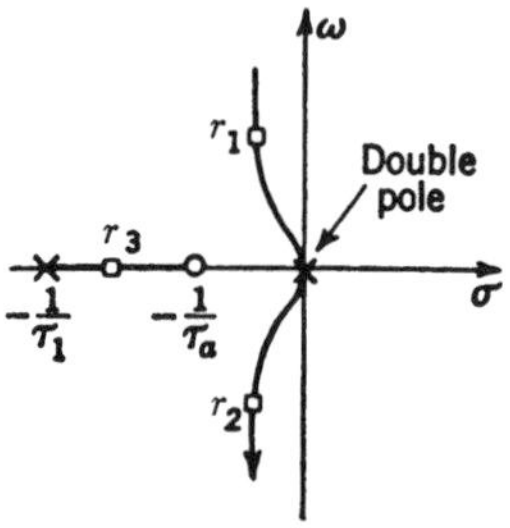 ω r_1 Double pole r_3 σ $-\frac{1}{\tau_1}$ $-\frac{1}{\tau_a}$ r_2	Stable for all gains
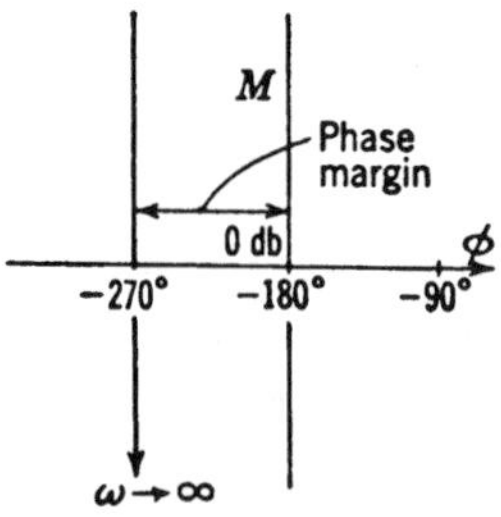 M Phase margin 0 db ϕ −270° −180° −90° $\omega \to \infty$	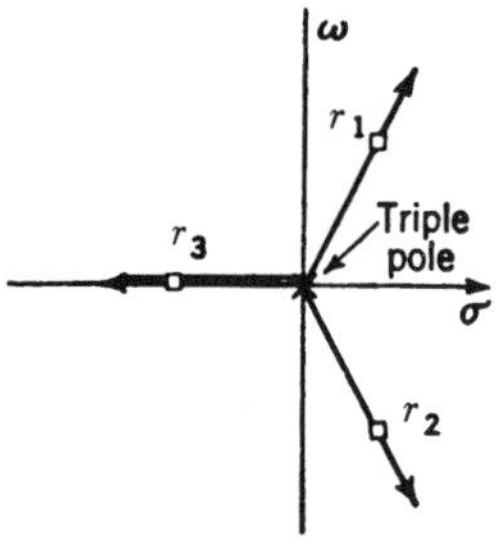 ω r_1 Triple pole r_3 σ r_2	Inherently unstable
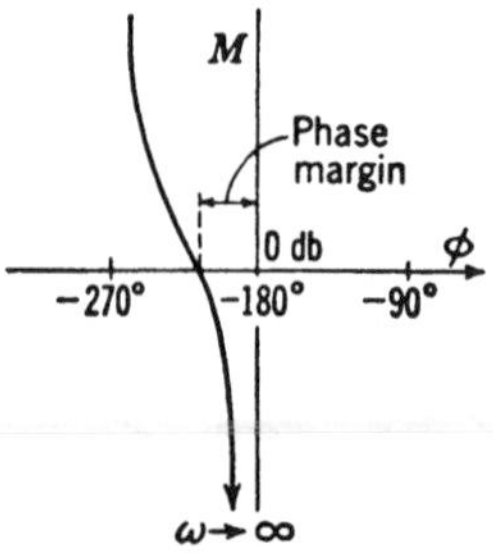 M Phase margin 0 db ϕ −270° −180° −90° $\omega \to \infty$	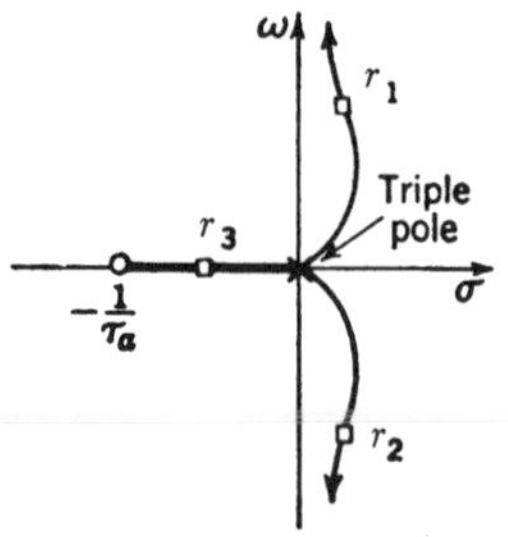 ω r_1 Triple pole r_3 σ $-\frac{1}{\tau_a}$ r_2	Inherently unstable

Table 8 (*Continued*)

$G(s)$	Polar Plot	Bode Diagram
13. $\dfrac{K(s\tau_a+1)(s\tau_b+1)}{s^3}$	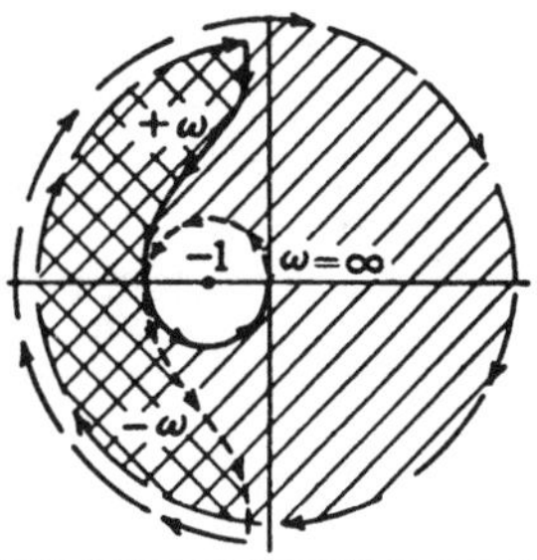 $+\omega$; -1; $\omega=\infty$; $-\omega$	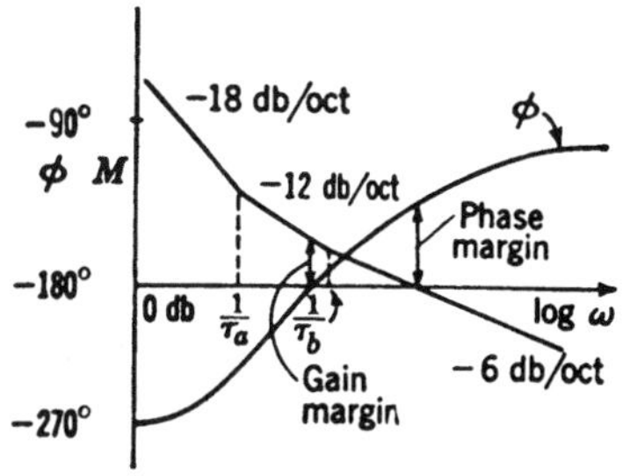 -90°; ϕ M; -180°; -270°; -18 db/oct; -12 db/oct; Phase margin; 0 db; $\frac{1}{\tau_a}$; $\frac{1}{\tau_b}$; log ω; Gain margin; -6 db/oct; ϕ
14. $\dfrac{K(s\tau_a+1)(s\tau_b+1)}{\tau_1+1)(s\tau_2+1)(s\tau_3+1)(s\tau_4+1)}$	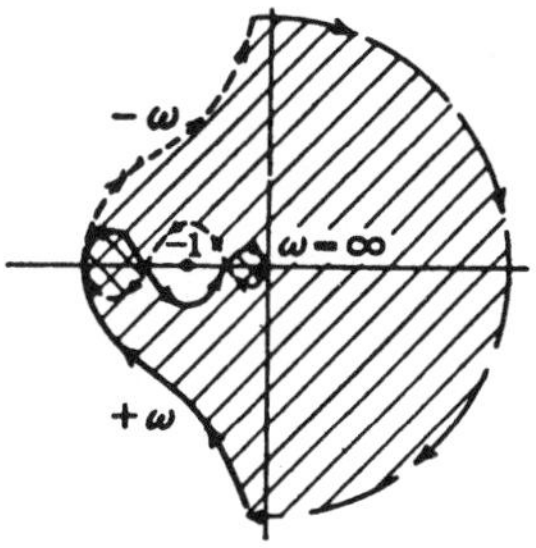 $-\omega$; -1; $\omega=\infty$; $+\omega$	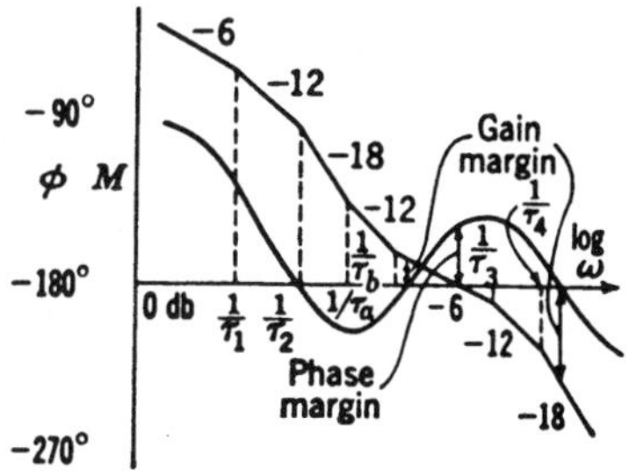 -90°; ϕ M; -180°; -270°; -6; -12; -18; -12; Gain margin; $\frac{1}{\tau_4}$; log ω; $\frac{1}{\tau_b}$; $\frac{1}{\tau_3}$; 0 db; $\frac{1}{\tau_1}$; $\frac{1}{\tau_2}$; $1/\tau_a$; -6; -12; Phase margin; -18
15. $\dfrac{K(s\tau_a+1)}{s^2(s\tau_1+1)(s\tau_2+1)}$	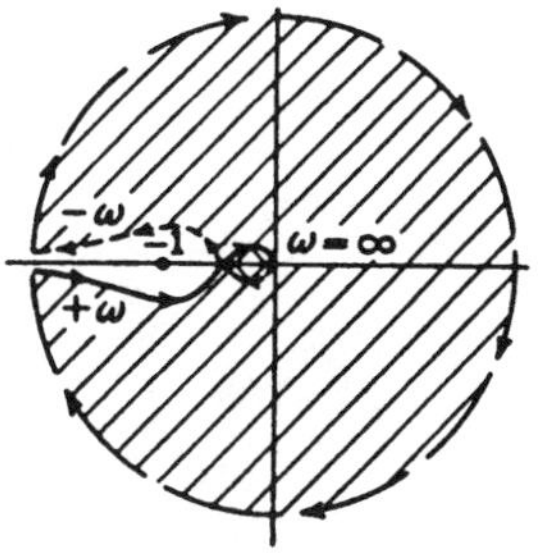 $-\omega$; -1; $\omega=\infty$; $+\omega$	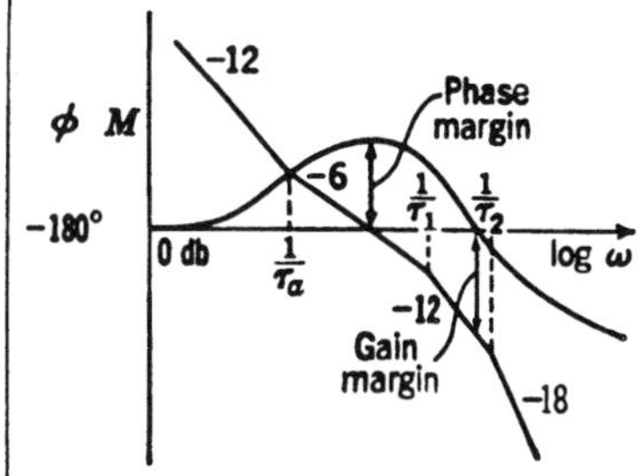 ϕ M; -180°; -12; Phase margin; -6; $\frac{1}{\tau_1}$; $\frac{1}{\tau_2}$; 0 db; $\frac{1}{\tau_a}$; log ω; -12; Gain margin; -18

$$sX(s) - x(0) = AX(s) + BU(s)$$

The solution to the fixed linear system therefore can be written as

$$\begin{aligned} x(t) &= \mathcal{L}^{-1}[X(s)] \\ &= \mathcal{L}^{-1}[\Phi(s)]x(0) + \mathcal{L}^{-1}[\Phi(s)BU(s)] \end{aligned}$$

where $\Phi(s)$ is called the *resolvent matrix* and

$$\Phi(t) = \mathcal{L}^{-1}[\Phi(s)] = \mathcal{L}^{-1}[sI - A]^{-1}$$

Table 8 (*Continued*)

Nichols Diagram	Root Locus	Comments
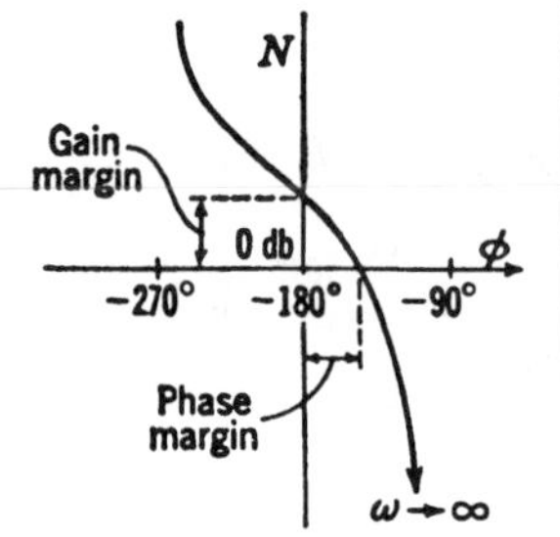	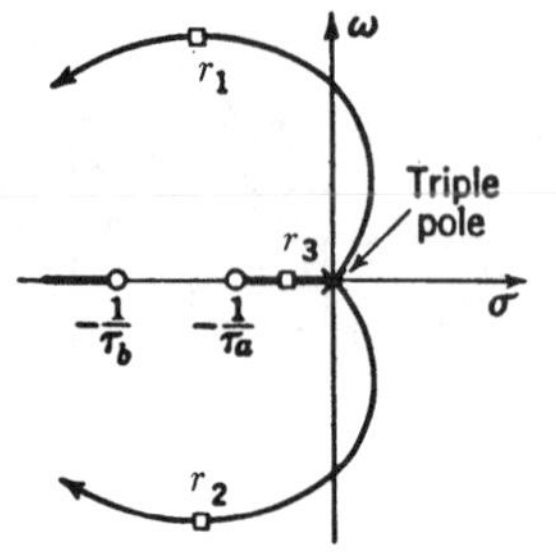	Conditionally stable; becomes unstable if gain is too low
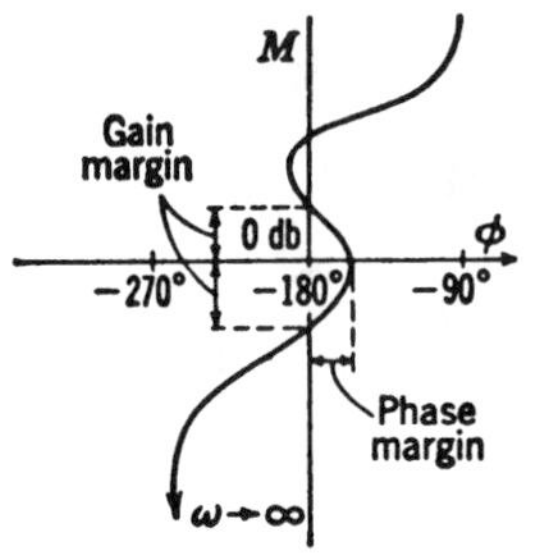	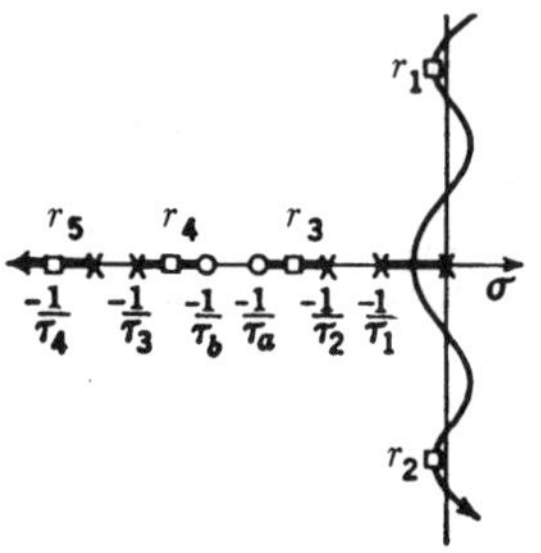	Conditionally stable; stable at low gain, becomes unstable as gain is raised, again becomes stable as gain is further increased, and becomes unstable for very high gains
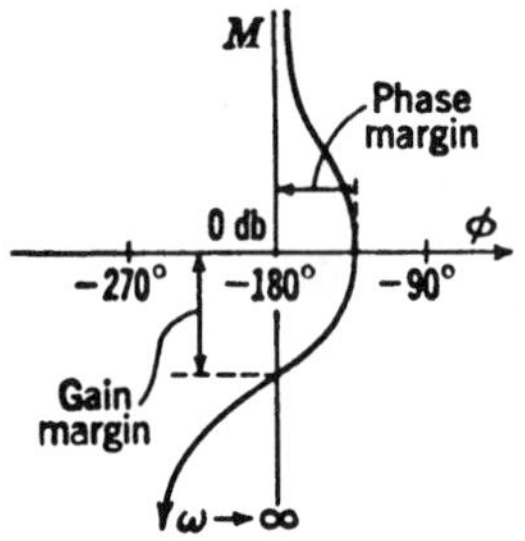	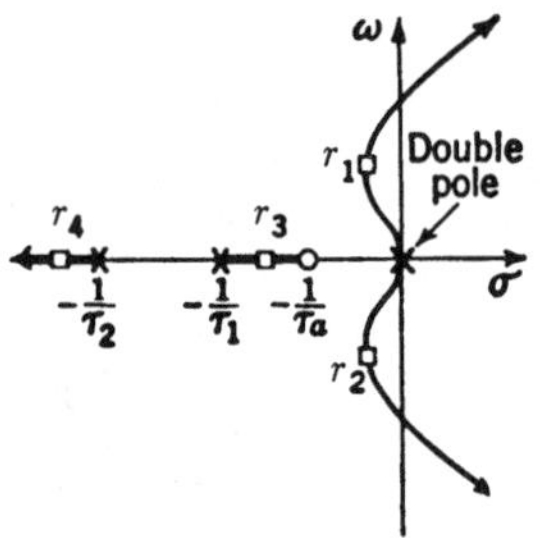	Conditionally stable; becomes unstable at high gain

6.2 Eigenstructure

The internal structure of a system (and therefore its free response) is defined entirely by the system matrix A. The concept of matrix *eigenstructure,* as defined by the eigenvalues and eigenvectors of the system matrix, can provide a great deal of insight into the fundamental behavior of a system. In particular, the system eigenvectors can be shown to define a special set of first-order subsystems embedded within the system. These subsystems behave independently of one another, a fact that greatly simplifies analysis.

System Eigenvalues and Eigenvectors

For a system with system matrix A, the system *eigenvectors* v_i and associated *eigenvalues* λ_i are defined by the equation

$$Av_i = \lambda_i v_i$$

Note that the eigenvectors represent a set of special directions in the state space. If the state vector is aligned in one of these directions, then the homogeneous state equation becomes $\dot{v}_i = A\dot{v}_i = \lambda v_i$, implying that each of the state variables changes at the *same* rate determined by the eigenvalue λ_i. This further implies that, in the absence of inputs to the system, a state vector that becomes aligned with an eigenvector will remain aligned with that eigenvector.

The system eigenvalues are calculated by solving the nth-order polynomial equation

$$|\lambda I - A| = \lambda^n + a_{n-1}\lambda^{n-1} + \cdots + a_1\lambda + a_0 = 0$$

This equation is called the *characteristic equation.* Thus the system eigenvalues are the roots of the characteristic equation, that is, the system eigenvalues are identically the system poles defined in transform analysis.

Each system eigenvector is determined by substituting the corresponding eigenvalue into the defining equation and then solving the resulting set of simultaneous linear equations. Only $n - 1$ of the n components of any eigenvector are independently defined, however. In other words, the magnitude of an eigenvector is arbitrary, and the eigenvector describes a direction in the state space.

Diagonalized Canonical Form

There will be one linearly independent eigenvector for each distinct (nonrepeated) eigenvalue. If all of the eigenvalues of an nth-order system are distinct, then the n independent eigenvectors form a new basis for the state space. This basis represents new coordinate axes defining a set of state variables $z_i(t)$, $i = 1, 2, \ldots, n$, called the *diagonalized canonical variables.* In terms of the diagonalized variables, the homogeneous state equation is

$$\dot{z}(t) = \Lambda z$$

where Λ is a diagonal system matrix of the eigenvectors, that is,

$$\Lambda = \begin{bmatrix} \lambda_1 & 0 & \cdots & 0 \\ 0 & \lambda_2 & \cdots & 0 \\ \vdots & \vdots & \ddots & \vdots \\ 0 & 0 & \vdots & \lambda_n \end{bmatrix}$$

The solution to the diagonalized homogeneous system is

$$z(t) = e^{\Lambda t} z(0)$$

where $e^{\Lambda t}$ is the diagonal state transition matrix

$$e^{\Lambda t} = \begin{bmatrix} e^{\lambda_1 t} & 0 & \cdots & 0 \\ 0 & e^{\lambda_2 t} & \cdots & 0 \\ \vdots & \vdots & \ddots & \vdots \\ 0 & 0 & \cdots & e^{\lambda_n t} \end{bmatrix}$$

Modal Matrix

Consider the state equation of the nth-order system

$$\dot{x}(t) = Ax(t) + Bu(t)$$

which has real, distinct eigenvalues. Since the system has a full set of eigenvectors, the state vector $x(t)$ can be expressed in terms of the canonical state variables as

$$x(t) = v_1 z_1(t) + v_2 z_2(t) + \cdots + v_n z_n(t) = Mz(t)$$

where M is the $n \times n$ matrix whose columns are the eigenvectors of A, called the *modal matrix.* Using the modal matrix, the state transition matrix for the original system can be written as

$$\Phi(t) = e^{\Lambda t} = Me^{\Lambda t}M^{-1}$$

where $e^{\Lambda t}$ is the diagonal state transition matrix. This frequently proves to be an attractive method for determining the state transition matrix of a system with real, distinct eigenvalues.

Jordan Canonical Form

For a system with one or more repeated eigenvalues, there is not in general a full set of eigenvectors. In this case, it is not possible to determine a diagonal representation for the system. Instead, the simplest representation that can be achieved is block diagonal. Let $L_k(\lambda)$ be the $k \times k$ matrix

$$L_k(\lambda) = \begin{bmatrix} \lambda & 1 & 0 & \cdots & 0 \\ 0 & \lambda & 1 & \cdots & 0 \\ \vdots & \vdots & \lambda & \ddots & 0 \\ \vdots & \vdots & \vdots & \ddots & 1 \\ 0 & 0 & 0 & 0 & \lambda \end{bmatrix}$$

Then for any $n \times n$ system matrix A there is certain to exist a nonsingular matrix T such that

$$T^{-1}AT = \begin{bmatrix} L_{k_1}(\lambda_1) & & & \\ & L_{k_2}(\lambda_2) & & \\ & & \ddots & \\ & & & L_{k_r}(\lambda_r) \end{bmatrix}$$

where $k_1 + k_2 + \cdots + k_r = n$ and λ_i, $i = 1, 2, \ldots, r$, are the (not necessarily distinct) eigenvalues of A. The matrix $T^{-1}AT$ is called the *Jordan canonical form.*

7 SIMULATION

7.1 Simulation—Experimental Analysis of Model Behavior

Closed-form solutions for nonlinear or time-varying systems are rarely available. In addition, while explicit solutions for time-invariant linear systems can always be found, for high-order systems this is often impractical. In such cases it may be convenient to study the dynamic behavior of the system using *simulation.*

Simulation is the *experimental* analysis of model behavior. A *simulation run* is a controlled experiment in which a specific realization of the model is manipulated in order to determine the response associated with that realization. A *simulation study* comprises *multiple runs,* each run for a different combination of model parameter values and/or initial conditions. The generalized solution of the model must then be inferred from a finite number of simulated data points.

Simulation is almost always carried out with the assistance of computing equipment. *Digital simulation* involves the *numerical solution* of model equations using a digital computer. *Analog simulation* involves solving model equations by analogy with the behavior of a physical system using an analog computer. *Hybrid simulation* employs digital and analog simulation together using a hybrid (part digital and part analog) computer.

7.2 Digital Simulation

Digital continuous-system simulation involves the approximate solution of a state-variable model over successive time steps. Consider the general state-variable equation

$$\dot{x}(t) = f[x(t), u(t)]$$

to be simulated over the time interval $t_0 \le t \le t_K$. The solution to this problem is based on the repeated solution of the single-variable, single-step subproblem depicted in Fig. 24. The subproblem may be stated formally as follows:

Given:

1. $\Delta t(k) = t_k - t_{k-1}$, the length of the kth *time step.*
2. $\dot{x}_i(t) = f_i[x(t), u(t)]$ for $t_{k-1} \le t \le t_k$, the ith equation of state defined for the state variable $x_i(t)$ over the kth time step.
3. $u(t)$ for $t_{k-1} \le t \le t_k$, the input vector defined for the kth time step.
4. $\tilde{x}(k-1) \simeq x(t_{k-1})$, an initial approximation for the state vector at the beginning of the time step.

Find:

5. $\tilde{x}_i(k) \simeq x_i(t_k)$, a final approximation for the state variable $x_i(t)$ at the end of the kth time step.

Solving this single-variable, single-step subproblem for each of the state variables $x_i(t)$, $i = 1, 2, \ldots, n$, yields a final approximation for the state vector $\tilde{x}(k) \simeq x(t_k)$ at the end of the kth time step. Solving the complete single-step problem K times over K time steps, beginning with the initial condition $\tilde{x}(0) = x(t_0)$ and using the final value of $\tilde{x}(t_k)$ from the

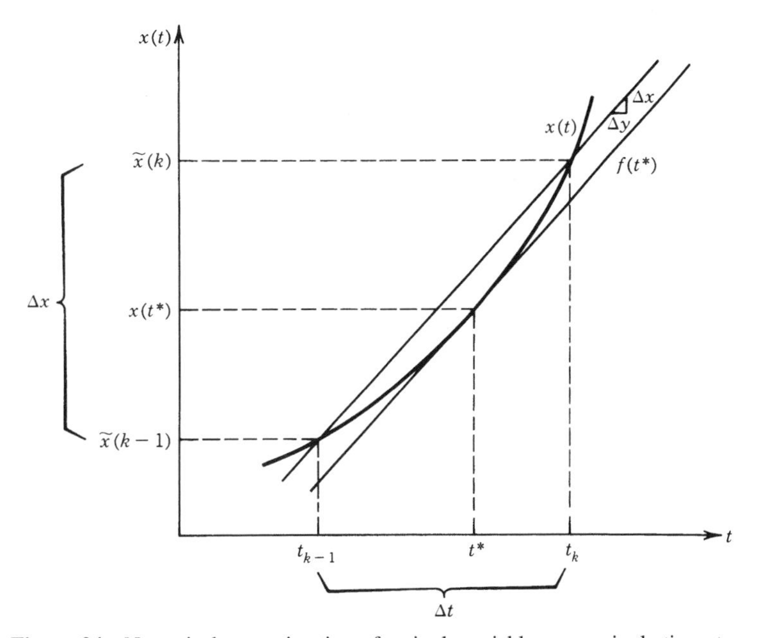

Figure 24 Numerical approximation of a single variable over a single time step.

kth time step as the initial value of the state for the $(k + 1)$st time step, yields a discrete succession of approximations $\tilde{x}(1) \simeq x(t_1)$, $\tilde{x}(2) \simeq x(t_2)$, . . . , $\tilde{x}(K) \simeq x(t_k)$ spanning the solution time interval.

The basic procedure for completing the single-variable, single-step problem is the same regardless of the particular integration method chosen. It consists of two parts: (1) calculation of the average value of the ith derivative over the time step as

$$\dot{x}_i(t^*) = f_i[x(t^*), u(t^*)] = \frac{\Delta x_i(k)}{\Delta t(k)} \simeq \tilde{f}_i(k)$$

and (2) calculation of the final value of the simulated variable at the end of the time step as

$$\tilde{x}_i(k) = \tilde{x}_i(k - 1) + \Delta x_i(k)$$
$$\simeq \tilde{x}_i(k - 1) + \Delta t(k)\tilde{f}_i(k)$$

If the function $f_i[x(t), u(t)]$ is continuous, then t^* is guaranteed to be on the time step, that is, $t_{k-1} \leq t^* \leq t_k$. Since the value of t^* is otherwise unknown, however, the value of $x(t^*)$ can only be approximated as $\tilde{f}(k)$.

Different *numerical integration* methods are distinguished by the means used to calculate the approximation $f_i(k)$. A wide variety of such methods is available for digital simulation of dynamic systems. The choice of a particular method depends on the nature of the model being simulated, the accuracy required in the simulated data, and the computing effort available for the simulation study. Several popular classes of integration methods are outlined in the following sections.

Euler Method

The simplest procedure for numerical integration is the Euler method. The standard Euler method approximates the average value of the ith derivative over the kth time step using the derivative evaluated at the beginning of the time step, that is,

$$\tilde{f}_i(k) = f_i[\tilde{x}(k - 1), u(t_{k-1})] \simeq f_i(t_{k-1})$$

$i = 1, 2, \ldots, n$ and $k = 1, 2, \ldots, K$. This is shown geometrically in Fig. 25 for the scalar single-step case. A modification of this method uses the newly calculated state variables in the derivative calculation as these new values become available. Assuming the state variables are computed in numerical order according to the subscripts, this implies

$$\tilde{f}_i(k) = f_i[\tilde{x}_1(k), \ldots, \tilde{x}_{i-1}(k), \tilde{x}_i(k - 1), \ldots, \tilde{x}_n(k - 1), u(t_{k-1})]$$

The modified Euler method is modestly more efficient than the standard procedure and, frequently, is more accurate. In addition, since the input vector $u(t)$ is usually known for the entire time step, using an average value of the input, such as

$$u(k) = \frac{1}{\Delta t(k)} \int_{t_{k-1}}^{t_k} u(\tau)\, d\tau$$

frequently leads to a superior approximation of $\tilde{f}_i(k)$.

The Euler method requires the least amount of computational effort per time step of any numerical integration scheme. Local truncation error is proportional to Δt^2, however, which means that the error within each time step is highly sensitive to step size. Because the accuracy of the method demands very small time steps, the number of time steps required to implement the method successfully can be large relative to other methods. This can imply

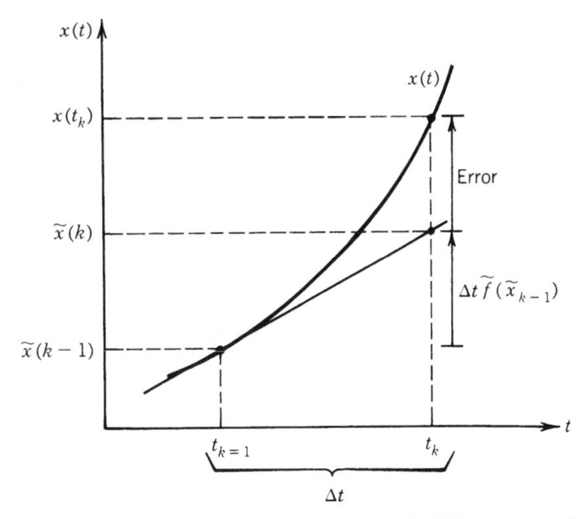

Figure 25 Geometric interpretation of the Euler method for numerical integration.

a large computational overhead and can lead to inaccuracies through the accumulation of roundoff error at each step.

Runge–Kutta Methods

Runge–Kutta methods precompute two or more values of $f_i[x(t), u(t)]$ in the time step $t_{k-1} \leq t \leq t_k$ and use some weighted average of these values to calculate $\tilde{f}_i(k)$. The *order* of a Runge–Kutta method refers to the number of derivative terms (or *derivative calls*) used in the scalar single-step calculation. A Runge–Kutta routine of order N therefore uses the approximation

$$\tilde{f}_i(k) = \sum_{j=1}^{N} w_j f_{ij}(k)$$

where the N approximations to the derivative are

$$f_{i1}(k) = f_i[\tilde{x}(k-1), u(t_{k-1})]$$

(the Euler approximation) and

$$f_{ij} = f_i\left[\tilde{x}(k-1) + \Delta t \sum_{t=1}^{j-1} I b_{jt} f_{il},\ u\left(t_{k-1} + \Delta t \sum_{t=1}^{j-1} b_{jl}\right)\right]$$

where I is the identity matrix. The weighting coefficients w_j and b_{jl} are not unique, but are selected such that the error in the approximation is zero when $x_i(t)$ is some specified Nth-degree polynomial in t. Coefficients commonly used for Runge–Kutta integration are given in Table 9.

Among the most popular of the Runge–Kutta methods is fourth-order Runge–Kutta. Using the defining equations for $N = 4$ and the weighting coefficients from Table 9 yields the derivative approximation

$$\tilde{f}_i(k) = \tfrac{1}{6}[f_{i1}(k) + 2f_{i2}(k) + 2f_{i3}(k) + f_{i4}(k)]$$

based on the four derivative calls

Table 9 Coefficients Commonly Used for Runge–Kutta Numerical Integration[6]

Common Name	N	b_{jl}	w_j
Open or explicit Euler	1	All zero	$w_1 = 1$
Improved polygon	2	$b_{21} = 1/2$	$w_1 = 0$ $w_2 = 1$
Modified Euler or Heun's method	2	$b_{21} = 1$	$w_1 = 1/2$ $w_2 = 1/2$
Third-order Runge–Kutta	3	$b_{21} = 1/2$ $b_{31} = -1$ $b_{32} = 2$	$w_1 = 1/6$ $w_2 = 2/3$ $w_3 = 1/6$
Fourth-order Runge–Kutta	4	$b_{21} = 1/2$ $b_{31} = 0$ $b_{32} = 1/2$ $b_{43} = 1$	$w_1 = 1/6$ $w_2 = 1/3$ $w_3 = 1/3$ $w_4 = 1/6$

$$f_{i1}(k) = f_i[\tilde{x}(k-1), u(t_{k-1})]$$

$$f_{i2}(k) = f_i\left[\tilde{x}(k-1) + \frac{\Delta t}{2} If_{i1}, u\left(t_{k-1} + \frac{\Delta t}{2}\right)\right]$$

$$f_{i3}(k) = f_i\left[\tilde{x}(k-1) + \frac{\Delta t}{2} If_{i2}, u\left(t_{k-1} + \frac{\Delta t}{2}\right)\right]$$

$$f_{i4}(k) = f_i[\tilde{x}(k-1) + \Delta t\, If_{i3}, u(t_k)]$$

where I is the identity matrix.

Because Runge–Kutta formulas are designed to be exact for a polynomial of order N, local truncation error is of the order Δt^{N+1}. This considerable improvement over the Euler method means that comparable accuracy can be achieved for larger step sizes. The penalty is that N derivative calls are required for each scalar evaluation within each time step.

Euler and Runge–Kutta methods are examples of *single-step methods* for numerical integration, so-called because the state $x(k)$ is calculated from knowledge of the state $x(k-1)$, without requiring knowledge of the state at any time prior to the beginning of the current time step. These methods are also referred to as *self-starting methods,* since calculations may proceed from any known state.

Multistep Methods

Multistep methods differ from the single-step methods previously described in that multistep methods use the stored values of two or more previously computed states and/or derivatives in order to compute the derivative approximation $\tilde{f}_i(k)$ for the current time step. The advantage of multistep methods over Runge–Kutta methods is that these require only one derivative call for each state variable at each time step for comparable accuracy. The disadvantage is that multistep methods are not self-starting, since calculations cannot proceed from the initial state alone. Multistep methods must be started, or restarted in the case of discontinuous derivatives, using a single-step method to calculate the first several steps.

The most popular of the multistep methods are the *Adams–Bashforth predictor methods* and the *Adams–Moulton corrector methods.* These methods use the derivative approximation

$$\tilde{f}_i(k) = \sum_{j=0}^{N} b_j f_i[\tilde{x}(k-j), u(k-j)]$$

where the b_j are weighting coefficients. These coefficients are selected such that the error in the approximation is zero when $x_i(t)$ is a specified polynomial. Table 10 gives the values of the weighting coefficients for several Adams–Bashforth–Moulton rules. Note that the predictor methods employ an *open* or *explicit rule,* since for these methods $b_0 = 0$ and a prior estimate of $x_i(k)$ is not required. The corrector methods use a *closed* or *implicit rule,* since for these methods $b_i \neq 0$ and a prior estimate of $x_i(k)$ is required. Note also that for all of these methods $\Sigma_{j=0}^{N} b_j = 1$, ensuring unity gain for the integration of a constant.

Predictor–Corrector Methods

Predictor–corrector methods use one of the multistep predictor equations to provide an initial estimate (or "prediction") of $x(k)$. This initial estimate is then used with one of the multistep corrector equations to provide a second and improved (or "corrected") estimate of $x(k)$ before proceeding to the next step. A popular choice is the four-point Adams–Bashforth predictor together with the four-point Adams–Moulton corrector, resulting in a prediction of

$$\tilde{x}_i(k) = \tilde{x}_i(k-1) + \frac{\Delta t}{24}[55\tilde{f}_i(k-1) - 59\tilde{f}_i(k-2) + 37\tilde{f}_i(k-3) - 9\tilde{f}_i(k-4)]$$

for $i = 1, 2, \ldots, n$ and a correction of

$$\tilde{x}_i(k) = \tilde{x}_i(k-1) + \frac{\Delta t}{24}\{9f_i[\tilde{x}(k), u(k)] + 19\tilde{f}_i(k-1) - 5\tilde{f}_i(k-2) + \tilde{f}_i(k-3)\}$$

Predictor–corrector methods generally incorporate a strategy for increasing or decreasing the size of the time step depending on the difference between the predicted and corrected $x(k)$ values. Such *variable time step methods* are particularly useful if the simulated system possesses local time constants that differ by several orders of magnitude or if there is little a priori knowledge about the system response.

Numerical Integration Errors

An inherent characteristic of digital simulation is that the discrete data points generated by the simulation $x(k)$ are only approximations to the exact solution $x(t_k)$ at the corresponding

Table 10 Coefficients Commonly Used for Adams–Bashforth–Moulton Numerical Integration[6]

Common Name	Predictor or Corrector	Points	b_{-1}	b_0	b_1	b_2	b_3
Open or explicit Euler	Predictor	1	0	1	0	0	0
Open trapezoidal	Predictor	2	0	$3/2$	$-1/2$	0	0
Adams three-point predictor	Predictor	3	0	$23/12$	$-16/12$	$5/12$	0
Adams four-point predictor	Predictor	4	0	$55/24$	$-59/24$	$37/24$	$-9/24$
Closed or implicit Euler	Corrector	1	1	0	0	0	0
Closed trapezoidal	Corrector	2	$1/2$	$1/2$	0	0	0
Adams three-point corrector	Corrector	3	$5/12$	$8/12$	$-1/12$	0	0
Adams four-point corrector	Corrector	4	$9/24$	$19/24$	$-5/24$	$1/24$	0

point in time. This results from two types of errors that are unavoidable in the numerical solutions. *Round-off errors* occur because numbers stored in a digital computer have finite word length (i.e., a finite number of bits per word) and therefore limited precision. Because the results of calculations cannot be stored exactly, round-off error tends to increase with the number of calculations performed. For a given total solution interval $t_0 \le t \le t_K$, therefore, round-off error tends to increase (1) with increasing integration-rule order (since more calculations must be performed at each time step) and (2) with decreasing step size Δt (since more time steps are required).

Truncation errors or *numerical approximation errors* occur because of the inherent limitations in the numerical integration methods themselves. Such errors would arise even if the digital computer had infinite precision. *Local* or *per-step truncation error* is defined as

$$e(k) = x(k) - x(t_k)$$

given that $x(k-1) = x(t_{k-1})$ and that the calculation at the kth time step is infinitely precise. For many integration methods, local truncation errors can be approximated at each step. *Global* or *total truncation error* is defined as

$$e(K) = x(K) - x(t_K)$$

given that $x(0) = x(t_0)$ and the calculations for all K time steps are infinitely precise. Global truncation error usually cannot be estimated, neither can efforts to reduce local truncation errors be guaranteed to yield acceptable global errors. In general, however, truncation errors can be decreased by using more sophisticated integration methods and by decreasing the step size Δt.

Time Constants and Time Steps

As a general rule, the step size Δt for simulation must be less than the smallest local time constant of the model simulated. This can be illustrated by considering the simple first-order system

$$\dot{x}(t) = \lambda x(t)$$

and the difference equation defining the corresponding Euler integration

$$x(k) = x(k-1) + \Delta t\ \lambda\ x(k-1)$$

The continuous system is stable for $\lambda < 0$, while the discrete approximation is stable for $|1 + \lambda\ \Delta t| < 1$. If the original system is stable, therefore, the simulated response will be stable for

$$\Delta t \le 2\left|\frac{1}{\lambda}\right|$$

where the equality defines the *critical step size*. For larger step sizes, the simulation will exhibit *numerical instability*. In general, while higher order integration methods will provide greater per-step accuracy, the critical step size itself will not be greatly reduced.

A major problem arises when the simulated model has one or more time constants $|1/\lambda_i|$ that are small when compared to the total solution time interval $t_0 \le t \le t_K$. Numerical stability will then require very small Δt, even though the transient response associated with the higher frequency (larger λ_i) subsystems may contribute little to the particular solution.

Such problems can be addressed either by neglecting the higher frequency components where appropriate or by adopting special numerical integration methods for *stiff systems.*

Selecting an Integration Method

The best numerical integration method for a specific simulation is the method that yields an acceptable global approximation error with the minimum amount of round-off error and computing effort. No single method is best for all applications. The selection of an integration method depends on the model simulated, the purpose of the simulation study, and the availability of computing hardware and software.

In general, for well-behaved problems with continuous derivatives and no stiffness, a lower order Adams predictor is often a good choice. Multistep methods also facilitate estimating local truncation error. Multistep methods should be avoided for systems with discontinuities, however, because of the need for frequent restarts. Runge–Kutta methods have the advantage that these are self-starting and provide fair stability. For stiff systems where high-frequency modes have little influence on the global response, special stiff-system methods enable the use of economically large step sizes. Variable-step rules are useful when little is known a priori about solutions. Variable-step rules often make a good choice as general-purpose integration methods.

Round-off error usually is not a major concern in the selection of an integration method, since the goal of minimizing computing effort typically obviates such problems. Double-precision simulation can be used where round-off is a potential concern. An upper bound on step size often exists because of discontinuities in derivative functions or because of the need for response output at closely spaced time intervals.

Continuous-System Simulation Languages

Digital simulation can be implemented for a specific model in any high-level language such as FORTRAN or C. The general process for implementing a simulation is shown in Fig. 26. In addition, many special-purpose continuous-system simulation languages are commonly available across a wide range of platforms. Such languages greatly simplify programming tasks and typically provide for good graphical output.

8 MODEL CLASSIFICATIONS

Mathematical models of dynamic systems are distinguished by several criteria which describe fundamental properties of model variables and equations. These criteria in turn prescribe the theory and mathematical techniques that can be used to study different models. Table 11 summarizes these distinguishing criteria. In the following sections, the approaches adopted for the analysis of important classes of systems are briefly outlined.

8.1 Stochastic Systems

Systems in which some of the dependent variables (input, state, output) contain random components are called *stochastic systems.* Randomness may result from environmental factors, such as wind gusts or electrical noise, or simply from a lack of precise knowledge of the system model, such as when a human operator is included within a control system. If the randomness in the system can be described by some rule, then it is often possible to derive a model in terms of probability distributions involving, for example, the means and variances of model variables or parameters.

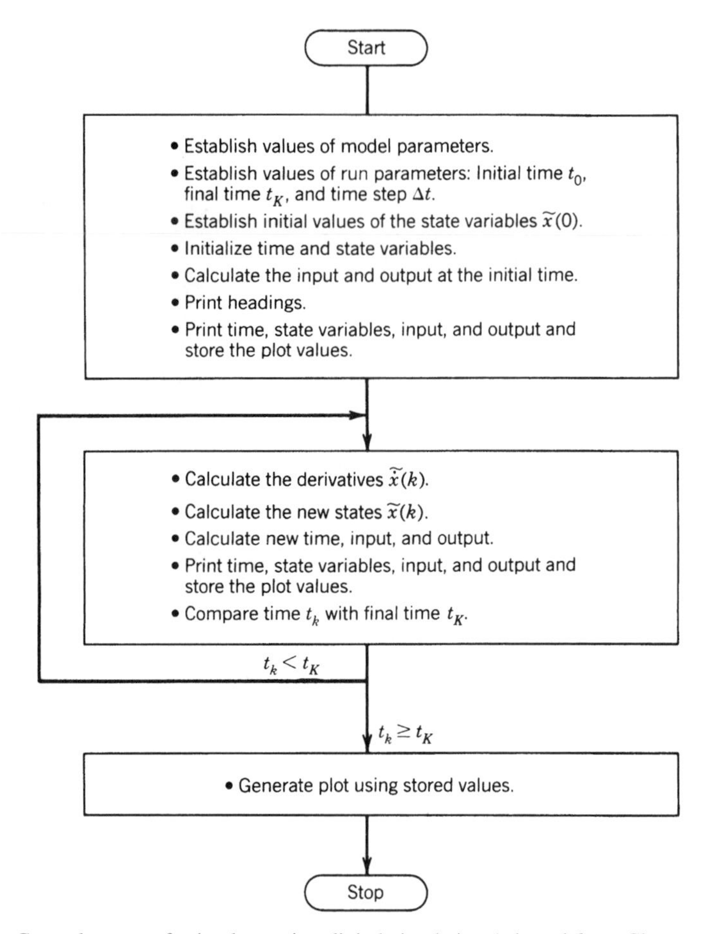

Figure 26 General process for implementing digital simulation (adapted from Close and Frederick[3]).

State-Variable Formulation

A common formulation is the fixed, linear model with additive noise

$$\dot{x}(t) = Ax(t) + Bu(t) + w(t)$$

$$y(t) = Cx(t) + v(t)$$

where $w(t)$ is a zero-mean Gaussian disturbance and $v(t)$ is a zero-mean Gaussian measurement noise. This formulation is the basis for many *estimation problems,* including the problem of *optimal filtering.* Estimation essentially involves the development of a rule or algorithm for determining the best estimate of the past, current, or future values of measured variables in the presence of disturbances or noise.

Random Variables

In the following, important concepts for characterizing random signals are developed. A *random variable* x is a variable that assumes values that cannot be precisely predicted a priori. The likelihood that a random variable will assume a particular value is measured as

Table 11 Classification of Mathematical Models of Dynamic Systems

Criterion	Classification	Description
Certainty	Deterministic	Model parameters and variables can be known with certainty. Common approximation when uncertainties are small.
	Stochastic	Uncertainty exists in the values of some parameters and/or variables. Model parameters and variables are expressed as random numbers or processes and are characterized by the parameters of probability distributions.
Spatial characteristics	Lumped	State of the system can be described by a finite set of state variables. Model is expressed as a discrete set of point functions described by ordinary differential or difference equations.
	Distributed	State depends on both time and spatial location. Model is usually described by variables that are continuous in time and space, resulting in partial differential equations. Frequently approximated by lumped elements. Typical in the study of structures and mass and heat transport.
Parameter variation	Fixed or time invariant	Model parameters are constant. Model described by differential or difference equations with constant coefficients. Model with same initial conditions and input delayed by t_d has the same response delayed by t_d.
	Time varying	Model parameters are time dependent.
Superposition property	Linear	Superposition applies. Model can be expressed as a system of linear difference or differential equations.
	Nonlinear	Superposition does not apply. Model is expressed as a system of nonlinear difference or differential equations. Frequently approximated by linear systems for analytical ease.
Continuity of independent variable (time)	Continuous	Dependent variables (input, output, state) are defined over a continuous range of the independent variable (time), even though the dependence is not necessarily described by a mathematically continuous function. Model is expressed as differential equations. Typical of physical systems.
	Discrete	Dependent variables are defined only at distinct instants of time. Model is expressed as difference equations. Typical of digital and nonphysical systems.
	Hybrid	System with continuous and discrete subsystems, most common in computer control and communication systems. Sampling and quantization typical in A/D (analog-to-digital) conversion; signal reconstruction for D/A conversion. Model frequently approximated as entirely continuous or entirely discrete.
Quantization of dependent variables	Nonquantized	Dependent variables are continuously variable over a range of values. Typical of physical systems at macroscopic resolution.
	Quantized	Dependent variables assume only a countable number of different values. Typical of computer control and communication systems (sample data systems).

the *probability* of that value. The probability *distribution function* $F(x)$ of a continuous random variable x is defined as the probability that x assumes a value no greater than x, that is,

$$F(x) = \Pr(X \leq x) = \int_{-\infty}^{x} f(x)\, dx$$

The probability *density function* $f(x)$ is defined as the derivative of $F(x)$.

The *mean* or *expected value* of a probability distribution is defined as

$$E(X) = \int_{-\infty}^{\infty} xf(x)\, dx = \overline{X}$$

The mean is the first moment of the distribution. The *nth moment* of the distribution is defined as

$$E(X^n) = \int_{-\infty}^{\infty} x^n f(x)\, dx$$

The mean square of the difference between the random variable and its mean is the *variance* or *second central moment* of the distribution,

$$\sigma^2(X) = E(X - \overline{X})^2 = \int_{-\infty}^{\infty} (x - \overline{X})^2 f(x)\, dx = E(X^2) - [E(X)]^2$$

The square root of the variance is the *standard deviation* of the distribution:

$$\sigma(X) = \sqrt{E(X^2) - [E(X)]^2}$$

The mean of the distribution therefore is a measure of the average magnitude of the random variable, while the variance and standard deviation are measures of the variability or dispersion of this magnitude.

The concepts of probability can be extended to more than one random variable. The *joint distribution* function of two random variables x and y is defined as

$$F(x,y) = \Pr(X < x \text{ and } Y < y) = \int_{-\infty}^{x} \int_{-\infty}^{y} f(x,y)\, dy\, dx$$

where $f(x,y)$ is the joint distribution. The ijth moment of the joint distribution is

$$E(X^i Y^j) = \int_{-\infty}^{\infty} x^i \int_{-\infty}^{\infty} y^j f(x,y)\, dy\, dx$$

The *covariance* of x and y is defined to be

$$E[(X - \overline{X})(Y - \overline{Y})]$$

and the normalized covariance or *correlation coefficient* as

$$\rho = \frac{E[(X - \overline{X})(Y - \overline{Y})]}{\sqrt{\sigma^2(X)\sigma^2(Y)}}$$

Although many distribution functions have proven useful in control engineering, far and away the most useful is the *Gaussian* or *normal distribution*

$$F(x) = \frac{1}{\sigma\sqrt{2\pi}} \exp\left[\frac{(-x - \mu)^2}{2\sigma^2}\right]$$

where μ is the mean of the distribution and σ is the standard deviation. The Gaussian distribution has a number of important properties. First, if the input to a linear system is Gaussian, the output also will be Gaussian. Second, if the input to a linear system is only approximately Gaussian, the output will tend to approximate a Gaussian distribution even more closely. Finally, a Gaussian distribution can be completely specified by two parameters, μ and σ, and therefore a zero-mean Gaussian variable is completely specified by its variance.

Random Processes

A *random process* is a set of random variables with time-dependent elements. If the statistical parameters of the process (such as σ for the zero-mean Gaussian process) do not vary with time, the process is *stationary.* The *autocorrelation function* of a stationary random variable $x(t)$ is defined by

$$\phi_{xx}(\tau) = \lim_{T\to\infty} \frac{1}{2T} \int_{-T}^{T} x(t)x(t + \tau)\, dt$$

a function of the fixed time interval τ. The autocorrelation function is a quantitative measure of the sequential dependence or time correlation of the random variable, that is, the relative effect of prior values of the variable on the present or future values of the variable. The autocorrelation function also gives information regarding how rapidly the variable is changing and about whether the signal is in part deterministic (specifically, periodic). The autocorrelation function of a zero-mean variable has the properties

$$\sigma^2 = \phi_{xx}(0) \geq \phi_{xx}(\tau) \qquad \phi_{xx}(\tau) = \phi_{xx}(-\tau)$$

In other words, the autocorrelation function for $\tau = 0$ is identically the variance and the variance is the maximum value of the autocorrelation function. From the definition of the function, it is clear that (1) for a purely random variable with zero mean, $\phi_{xx}(\tau) = 0$ for $\tau \neq 0$, and (2) for a deterministic variable, which is periodic with period T, $\phi_{xx}(k2\pi T) = \sigma^2$ for k integer. The concept of time correlation is readily extended to more than one random variable. The *cross-correlation function* between the random variables $x(t)$ and $y(t)$ is

$$\phi_{xy}(\tau) = \lim_{T\to\infty} \int_{-\infty}^{\infty} x(t)y(t + \tau)\, dt$$

For $\tau = 0$, the cross-correlation between two zero-mean variables is identically the covariance. A final characterization of a random variable is its *power spectrum,* defined as

$$G(\omega, x) = \lim_{T\to\infty} \frac{1}{2\pi T} \left| \int_{-T}^{T} x(t)e^{-j\omega t}\, dt \right|^2$$

For a stationary random process, the power spectrum function is identically the Fourier transform of the autocorrelation function

$$G(\omega, x) = \frac{1}{\pi} \int_{-\infty}^{\infty} \phi_{xx}(\tau)e^{-j\omega t}\, dt$$

with

$$\phi_{xx}(0) = \int_{-\infty}^{\infty} G(\omega, x)\, d\omega$$

8.2 Distributed-Parameter Models

There are many important applications in which the state of a system cannot be defined at a finite number of points in space. Instead, the system state is a continuously varying function of both time and location. When continuous spatial dependence is explicitly accounted for in a model, the independent variables must include spatial coordinates as well as time. The resulting *distributed-parameter model* is described in terms of *partial differential equations,* containing partial derivatives with respect to each of the independent variables.

Distributed-parameter models commonly arise in the study of mass and heat transport, the mechanics of structures and structural components, and electrical transmission. Consider as a simple example the unidirectional flow of heat through a wall, as depicted in Fig. 27. The temperature of the wall is not in general uniform but depends on both the time t and position within the wall x, that is, $\theta = \theta(x, t)$. A distributed-parameter model for this case might be the first-order partial differential equation

$$\frac{d}{dt}\theta(x,t) = \frac{1}{C_t}\frac{\partial}{\partial x}\left[\frac{1}{R_t}\frac{\partial}{\partial x}\theta(x,t)\right]$$

where C_t is the thermal capacitance and R_t is the thermal resistance of the wall (assumed uniform).

The complexity of distributed-parameter models is typically such that these models are avoided in the analysis and design of control systems. Instead, distributed-parameter systems are approximated by a finite number of spatial "lumps," each lump being characterized by some average value of the state. By eliminating the independent spatial variables, the result is a *lumped-parameter* (*or lumped-element*) *model* described by coupled ordinary differential equations. If a sufficiently fine-grained representation of the lumped microstructure can be achieved, a lumped model can be derived that will approximate the distributed model to any desired degree of accuracy. Consider, for example, the three temperature lumps shown in Fig. 28, used to approximate the wall of Fig. 27. The corresponding third-order lumped approximation is

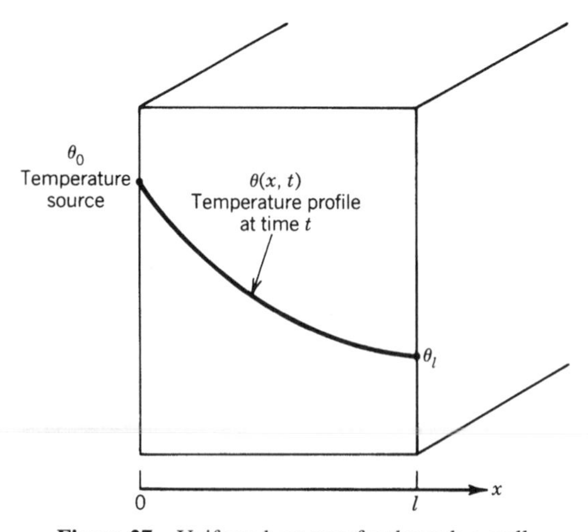

Figure 27 Uniform heat transfer through a wall.

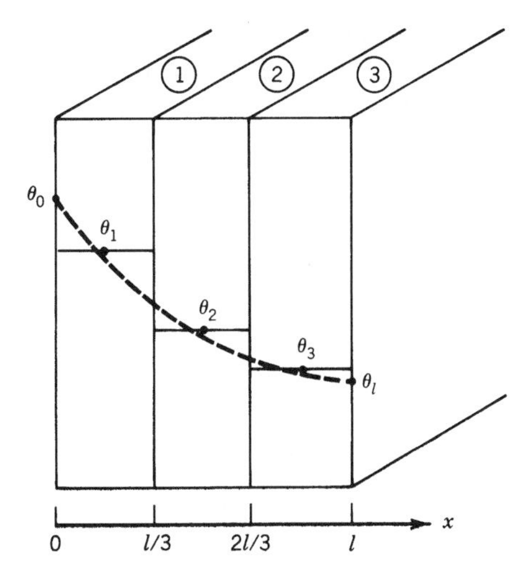

Figure 28 Lumped-parameter model for uniform heat transfer through a wall.

$$\frac{d}{dt}\begin{bmatrix}\theta_1(t)\\ \theta_2(t)\\ \theta_3(t)\end{bmatrix} = \begin{bmatrix}-\dfrac{9}{C_tR_t} & \dfrac{3}{C_tR_t} & 0\\ \dfrac{3}{C_tR_t} & -\dfrac{6}{C_tR_t} & \dfrac{3}{C_tR_t}\\ 0 & \dfrac{3}{C_tR_t} & -\dfrac{6}{C_tR_t}\end{bmatrix}\begin{bmatrix}\theta_1(t)\\ \theta_2(t)\\ \theta_3(t)\end{bmatrix} + \begin{bmatrix}\dfrac{6}{C_tR_t}\\ 0\\ 0\end{bmatrix}\theta_0(t)$$

If a more detailed approximation is required, this can always be achieved at the expense of adding additional, smaller lumps.

8.3 Time-Varying Systems

Time-varying systems are those with characteristics that change as a function of time. Such variation may result from environmental factors, such as temperature or radiation, or from factors related to the operation of the system, such as fuel consumption. While in general a model with variable parameters can be either linear or nonlinear, the name time varying is most frequently associated with linear systems described by the following state equation:

$$\dot{x}(t) = A(t)x(t) + B(t)u(t)$$

For this linear time-varying model, the superposition principle still applies. Superposition is a great aid in model formulation but unfortunately does not prove to be much help in determining the model solution.

Paradoxically, the form of the solution to the linear time-varying equation is well known[7]:

$$x(t) = \Phi(t,t_0)x(t_0) + \int_{t_0}^{t} \Phi(t,\tau)B(\tau)u(\tau)\,dt$$

where $\Phi(t,t_0)$ is the time-varying state transition matrix. This knowledge is typically of little value, however, since it is not usually possible to determine the state transition matrix by any straightforward method. By analogy with the first-order case, the relationship

$$\Phi(t,t_0) = \exp\left(\int_{t_0}^{t} A(\tau)\, d\tau\right)$$

can be proven valid *if and only if*

$$A(t) \int_{t_0}^{t} A(\tau)\, d\tau = \int_{t_0}^{t} A(\tau)\, d\tau\, A(t)$$

that is, if and only if $A(t)$ and its integral commute. This is a very stringent condition for all but a first-order system and, as a rule, it is usually easiest to obtain the solution using simulation.

Most of the properties of the fixed transition matrix extend to the time-varying case:

$$\Phi(t,t_0) = I$$

$$\Phi^{-1}(t,t_0) = \Phi(t_0,t)$$

$$\Phi(t_2,t_1)\Phi(t_1,t_0) = \Phi(t_2,t_0)$$

$$\Phi(t,t_0) = A(t)\Phi(t,t_0)$$

8.4 Nonlinear Systems

The theory of fixed, linear, lumped-parameter systems is highly developed and provides a powerful set of techniques for control system analysis and design. In practice, however, all physical systems are nonlinear to some greater or lesser degree. The linearity of a physical system is usually only a convenient approximation, restricted to a certain range of operation. In addition, nonlinearities such as dead zones, saturation, or on–off action are sometimes introduced into control systems intentionally, either to obtain some advantageous performance characteristic or to compensate for the effects of other (undesirable) nonlinearities.

Unfortunately, while nonlinear systems are important, ubiquitous, and potentially useful, the theory of nonlinear differential equations is comparatively meager. Except for specific cases, closed-form solutions to nonlinear systems are generally unavailable. The only universally applicable method for the study of nonlinear systems is *simulation*. As described in Section 7, however, simulation is an experimental approach, embodying all of the attending limitations of experimentation.

A number of special techniques are available for the analysis of nonlinear systems. All of these techniques are in some sense approximate, assuming, for example, either a restricted range of operation over which nonlinearities are mild or the relative isolation of lower order subsystems. When used in conjunction with more complex simulation models, however, these techniques often provide insights and design concepts that would be difficult to discover through the use of simulation alone.[8]

Linear versus Nonlinear Behaviors

There are several fundamental differences between the behavior of linear and nonlinear systems that are especially important. These differences not only account for the increased difficulty encountered in the analysis and design of nonlinear systems, but also imply entirely new types of behavior for nonlinear systems that are not possible for linear systems.

The fundamental property of linear systems is *superposition.* This property states that if $y_1(t)$ is the response of the system to $u_1(t)$ and $y_2(t)$ is the response of the system to $u_2(t)$, then the response of the system to the linear combination $a_1u_1(t) + a_2u_2(t)$ is the linear combination $a_1y_1(t) + a_2y_2(t)$. An immediate consequence of superposition is that the responses of a linear system to inputs differing only in amplitude is qualitatively the same. Since superposition does not apply to nonlinear systems, the responses of a nonlinear system to large and small changes may be fundamentally different.

This fundamental difference in linear and nonlinear behaviors has a second consequence. For a linear system, interchanging two elements connected in series does not affect the overall system behavior. Clearly, this cannot be true in general for nonlinear systems.

A third property peculiar to nonlinear systems is the potential existence of *limit cycles.* A linear oscillator oscillates at an amplitude that depends on its initial state. A limit cycle is an oscillation of fixed amplitude and period, independent of the initial state, that is unique to the nonlinear system.

A fourth property concerns the response of nonlinear systems to sinusoidal inputs. For a linear system, the response to sinusoidal input is a sinusoid of the same frequency, potentially differing only in magnitude and phase. For a nonlinear system, the output will in general contain other frequency components, including possibly harmonics, subharmonics, and aperiodic terms. Indeed, the response need not contain the input frequency at all.

Linearizing Approximations

Perhaps the most useful technique for analyzing nonlinear systems is to approximate these with linear systems. While many linearizing approximations are possible, linearization can frequently be achieved by considering small excursions of the system state about a reference trajectory. Consider the nonlinear state equation

$$\dot{x}(t) = f[x(t),u(t)]$$

together with a reference trajectory $x^0(t)$ and reference input $u^0(t)$ that together satisfy the state equation

$$\dot{x}^0(t) = f[x^0(t),u^0(t)]$$

Note that the simplest case is to choose a static equilibrium or *operating point* $\bar{x}$ as the reference "trajectory" such that $0 = t(\bar{x},0)$. The actual trajectory is then related to the reference trajectory by the relationships

$$x(t) = x^0(t) + \delta x(t)$$
$$u(t) = u^0(t) + \delta u(t)$$

where $\delta x(t)$ is some small perturbation about the reference state and $\delta u(t)$ is some small perturbation about the reference input. If these perturbations are indeed small, then applying Taylor's series expansion about the reference trajectory yields the linearized approximation

$$\delta\dot{x}(t) = A(t)\,\delta x(t) + B(t)\,\delta u(t)$$

where the state and distribution matrices are the *Jacobian matrices*

$$A(t) = \begin{bmatrix} \frac{\partial f_i}{\partial x_1} & \frac{\partial f_1}{\partial x_2} & \cdots & \frac{\partial f_1}{\partial x_n} \\ \frac{\partial f_2}{\partial x_1} & \frac{\partial f_2}{\partial x_2} & \cdots & \frac{\partial f_2}{\partial x_n} \\ \vdots & \vdots & & \vdots \\ \frac{\partial f_n}{\partial x_1} & \frac{\partial f_n}{\partial x_2} & \cdots & \frac{\partial f_n}{\partial x_n} \end{bmatrix}_{x(t)=x^0(t);\ u(t)=u^0(t)}$$

$$B(t) = \begin{bmatrix} \frac{\partial f_1}{\partial u_1} & \frac{\partial f_1}{\partial u_2} & \cdots & \frac{\partial f_1}{\partial u_m} \\ \frac{\partial f_2}{\partial u_1} & \frac{\partial f_2}{\partial u_2} & \cdots & \frac{\partial f_2}{\partial u_m} \\ \vdots & \vdots & & \vdots \\ \frac{\partial f_n}{\partial u_1} & \frac{\partial f_n}{\partial u_2} & \cdots & \frac{\partial f_n}{\partial u_m} \end{bmatrix}_{x(t)=x^0(t);\ u(t)=u^0(t)}$$

If the reference trajectory is a fixed operating point $\bar{x}$, then the resulting linearized system is time invariant and can be solved analytically. If the reference trajectory is a function of time, however, then the resulting system is linear but time varying.

Describing Functions

The describing function method is an extension of the frequency transfer function approach of linear systems, most often used to determine the stability of limit cycles of systems containing nonlinearities. The approach is approximate and its usefulness depends on two major assumptions:

1. All the nonlinearities within the system can be aggregated mathematically into a single block, denoted as $N(M)$ in Fig. 29, such that the equivalent gain and phase associated with this block depend only on the amplitude M_d of the sinusoidal input $m(\omega t) = M \sin(\omega t)$ and are independent of the input frequency ω.

2. All the harmonics, subharmonics, and any dc component of the output of the nonlinear block are filtered out by the linear portion of the system such that the effective output of the nonlinear block is well approximated by a periodic response having the same fundamental period as the input.

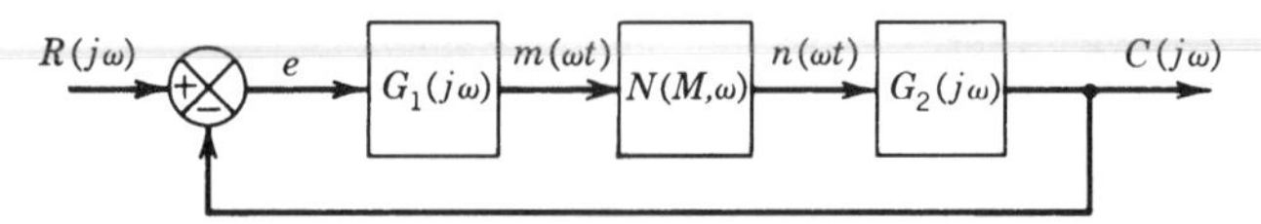

Figure 29 General nonlinear system for describing function analysis.

Although these assumptions appear to be rather limiting, the technique gives reasonable results for a large class of control systems. In particular, the second assumption is generally satisfied by higher order control systems with symmetric nonlinearities, since (a) symmetric nonlinearities do not generate dc terms, (b) the amplitudes of harmonics are generally small when compared with the fundamental term and subharmonics are uncommon, and (c) feedback within a control system typically provides low-pass filtering to further attenuate harmonics, especially for higher order systems. Because the method is relatively simple and can be used for systems of any order, describing functions have enjoyed wide practical application.

The describing function of a nonlinear block is defined as the ratio of the fundamental component of the output to the amplitude of a sinusoidal input. In general, the response of the nonlinearity to the input

$$m(\omega t) = M \sin \omega t$$

is the output

$$n(\omega t) = N_1 \sin(\omega t + \phi_1) + N_2 \sin(2\omega t + \phi_2) + N_3 \sin(3\omega t + \phi_3) + \cdots$$

and, hence, the describing function for the nonlinearity is defined as the complex quantity

$$N(M) = \frac{N_1}{M} e^{j\phi_1}$$

Derivation of the approximating function typically proceeds by representing the fundamental frequency by the Fourier series coefficients

$$A_1(M) = \frac{2}{T} \int_{-T/2}^{T/2} n(\omega t) \cos \omega t \, d(\omega t)$$

$$B_1(M) = \frac{2}{T} \int_{-T/2}^{T/2} n(\omega t) \sin \omega t \, d(\omega t)$$

The describing function is then written in terms of these coefficients as

$$N(M) = \frac{B_1(M)}{M} + j\frac{A_1(M)}{M} = \left[\left(\frac{B_1(M)}{M}\right)^2 + \left(\frac{A_1(M)}{M}\right)^2\right]^{1/2} \exp\left[j \tan^{-1}\left(\frac{A_1(M)}{B_1(M)}\right)\right]$$

Note that if $n(\omega t) = -n(-\omega t)$, then the describing function is odd, $A_1(M) = 0$, and there is no phase shift between the input and output. If $n(\omega t) = n(-\omega t)$, then the function is even, $B_1(M) = 0$, and the phase shift is $\pi/2$.

The describing functions for a number of typical nonlinearities are given in Fig. 30. Reference 9 contains an extensive catalog. The following derivation for a dead-zone nonlinearity demonstrates the general procedure for deriving a describing function. For the saturation element depicted in Fig. 30*a*, the relationship between the input $m(\omega t)$ and output $n(\omega t)$ can be written as

$$n(\omega t) = \begin{cases} 0 & \text{for} \quad -D < m < D \\ K_1 M(\sin \omega t - \sin \omega_1 t) & \text{for} \quad m > D \\ K_1 M(\sin \omega t + \sin \omega_1 t) & \text{for} \quad m < -D \end{cases}$$

Since the function is odd, $A_1 = 0$. By the symmetry over the four quarters of the response period,

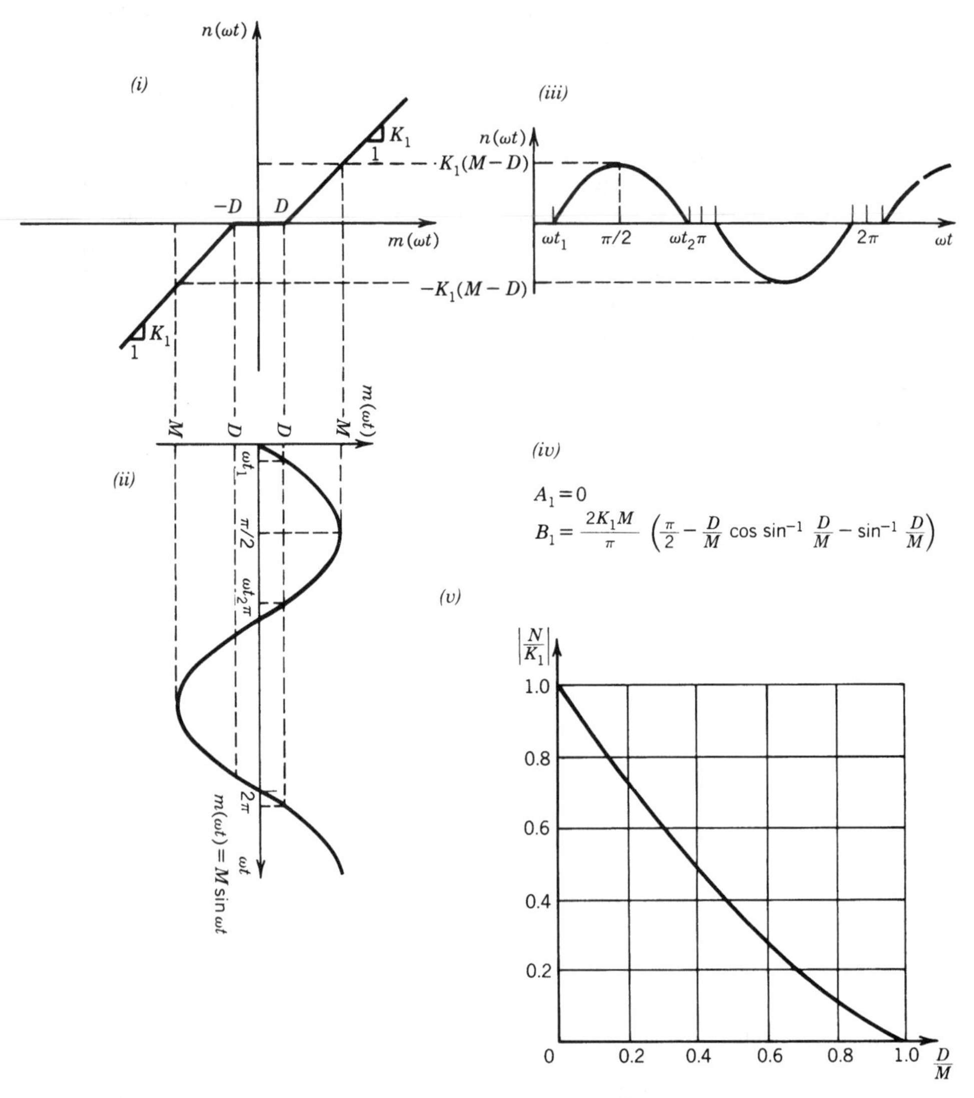

Figure 30*a* Describing functions for typical nonlinearities (after Refs. 9 and 10). Dead-zone nonlinearity: (*i*) nonlinear characteristic; (*ii*) sinusoidal input wave shape; (*iii*) output wave shape; (*iv*) describing-function coefficients; (*v*) normalized describing function.

$$
\begin{aligned}
B_1 &= 4\left[\frac{2}{\pi/2}\int_0^{\pi/2} n(\omega t)\sin\omega t\; d(\omega t)\right] \\
&= \frac{4}{\pi}\left[\int_0^{\omega t_1} (0)\sin\omega t\; d(\omega t) + \int_{\omega t_1}^{\pi/2} K_1 M(\sin\omega t - \sin\omega_1 t)\sin\omega t\; d(\omega t)\right]
\end{aligned}
$$

where $\omega t_1 = \sin^{-1}(D/M)$. Evaluating the integrals and dividing by M yields the describing function listed in Fig. 30.

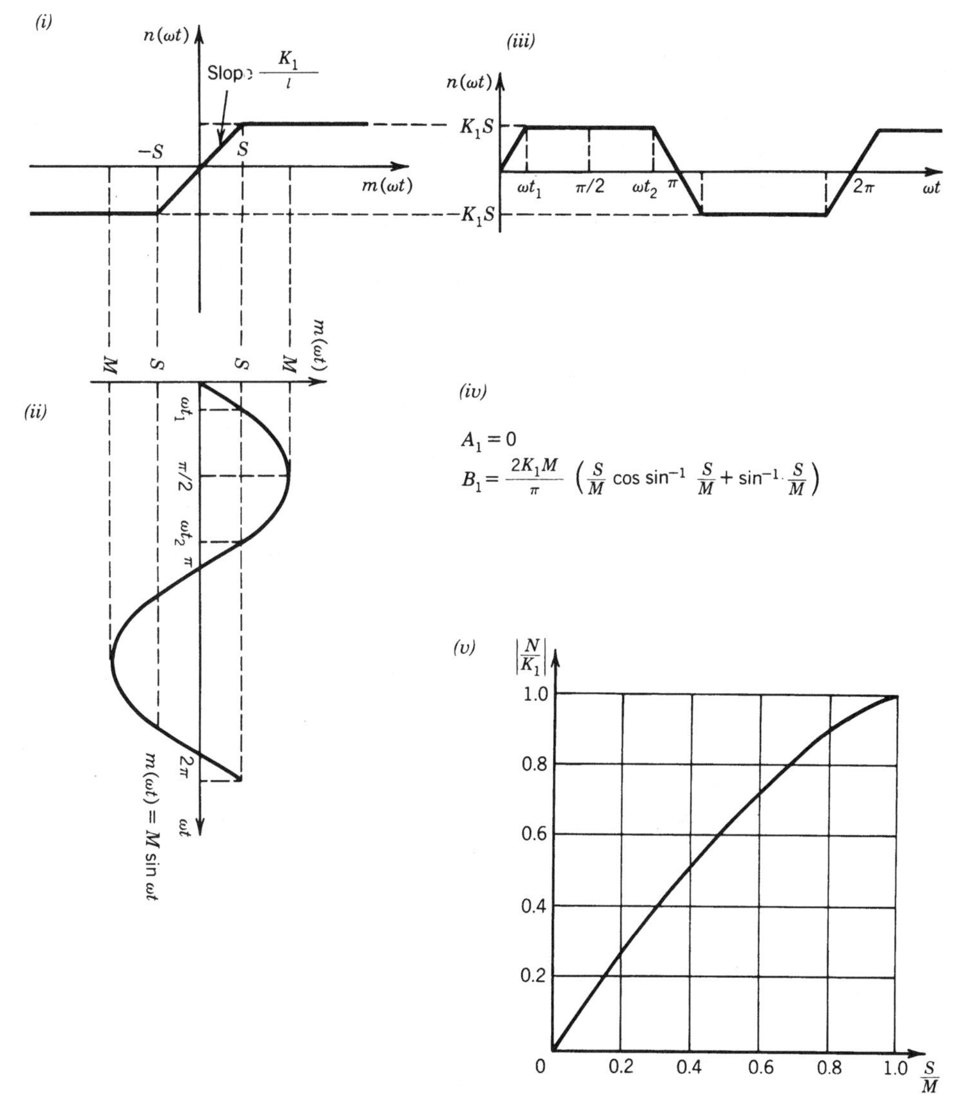

Figure 30*b* Saturation nonlinearity: (*i*) nonlinear characteristic; (*ii*) sinusoidal input wave shape; (*iii*) output wave shape; (*iv*) describing-function coefficients; (*v*) normalized describing function.

Phase-Plane Method

The *phase-plane method* is a graphical application of the state-space approach used to characterize the free response of second-order nonlinear systems. While any convenient pair of state variables can be used, the *phase variables* originally were taken to be the displacement and velocity of the mass of a second-order mechanical system. Using the two state variables as the coordinate axis, the transient response of a system is captured on the *phase plane* as the plot of one variable against the other, with time implicit on the resulting curve. The curve for a specific initial condition is called a *trajectory* in the phase plane; a representative sample of trajectories is called the *phase portrait* of the system. The phase portrait is a

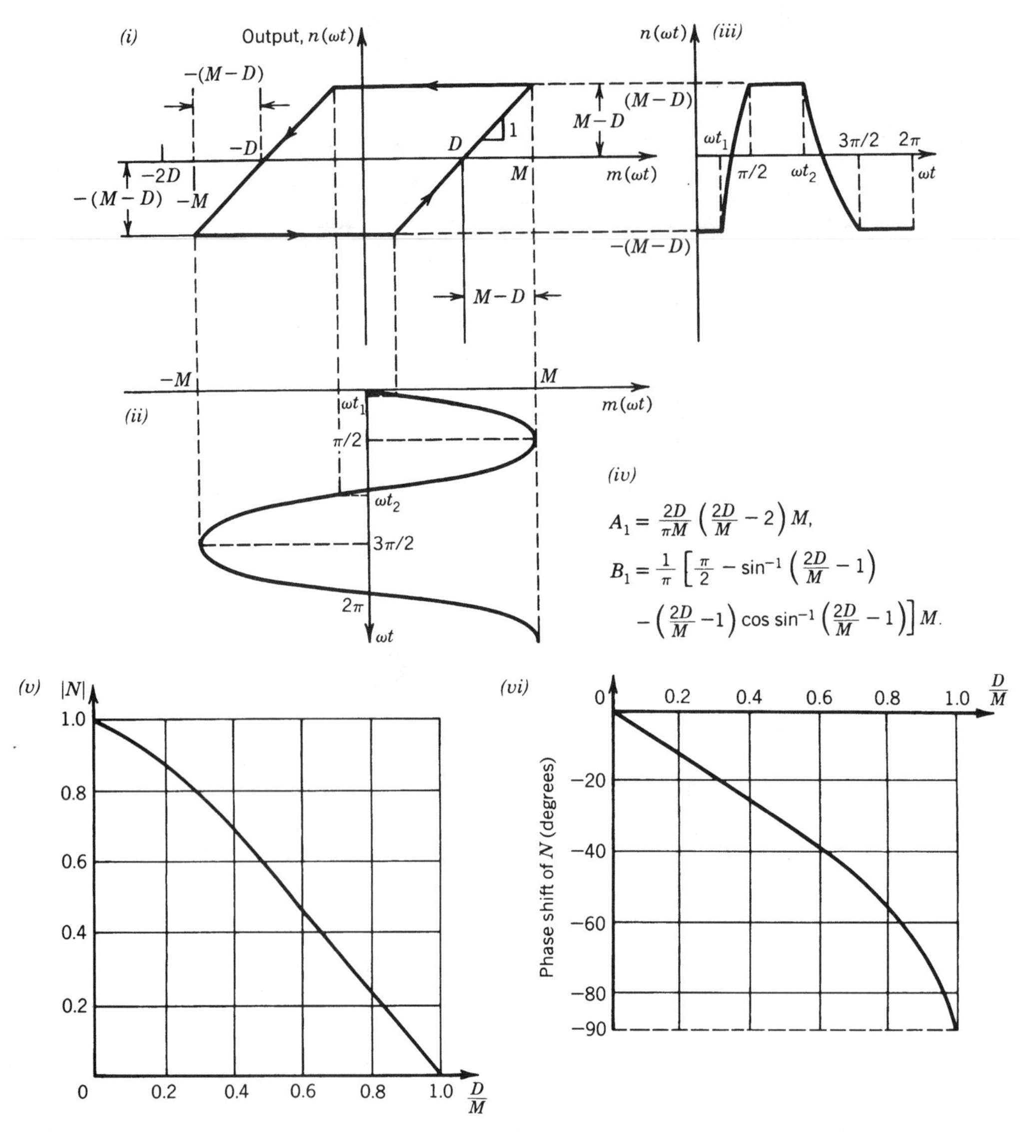

Figure 30c Backlash nonlinearity: (*i*) nonlinear characteristic; (*ii*) sinusoidal input wave shape; (*iii*) output wave shape; (*iv*) describing-function coefficients; (*v*) normalized amplitude characteristics for the describing function; (*vi*) normalized phase characteristics for the describing function.

compact and readily interpreted summary of the system response. Phase portraits for a sample of typical nonlinearities are shown in Fig. 31.

Four methods can be used to construct a phase portrait: (1) direct solution of the differential equation, (2) the graphical *method of isoclines,* (3) transformation of the second-order system (with time as the independent variable) into an equivalent first-order system (with one of the phase variables as the independent variable), and (4) numerical solution using simulation. The first and second methods are usually impractical; the third and fourth methods are frequently used in combination. For example, consider the second-order model

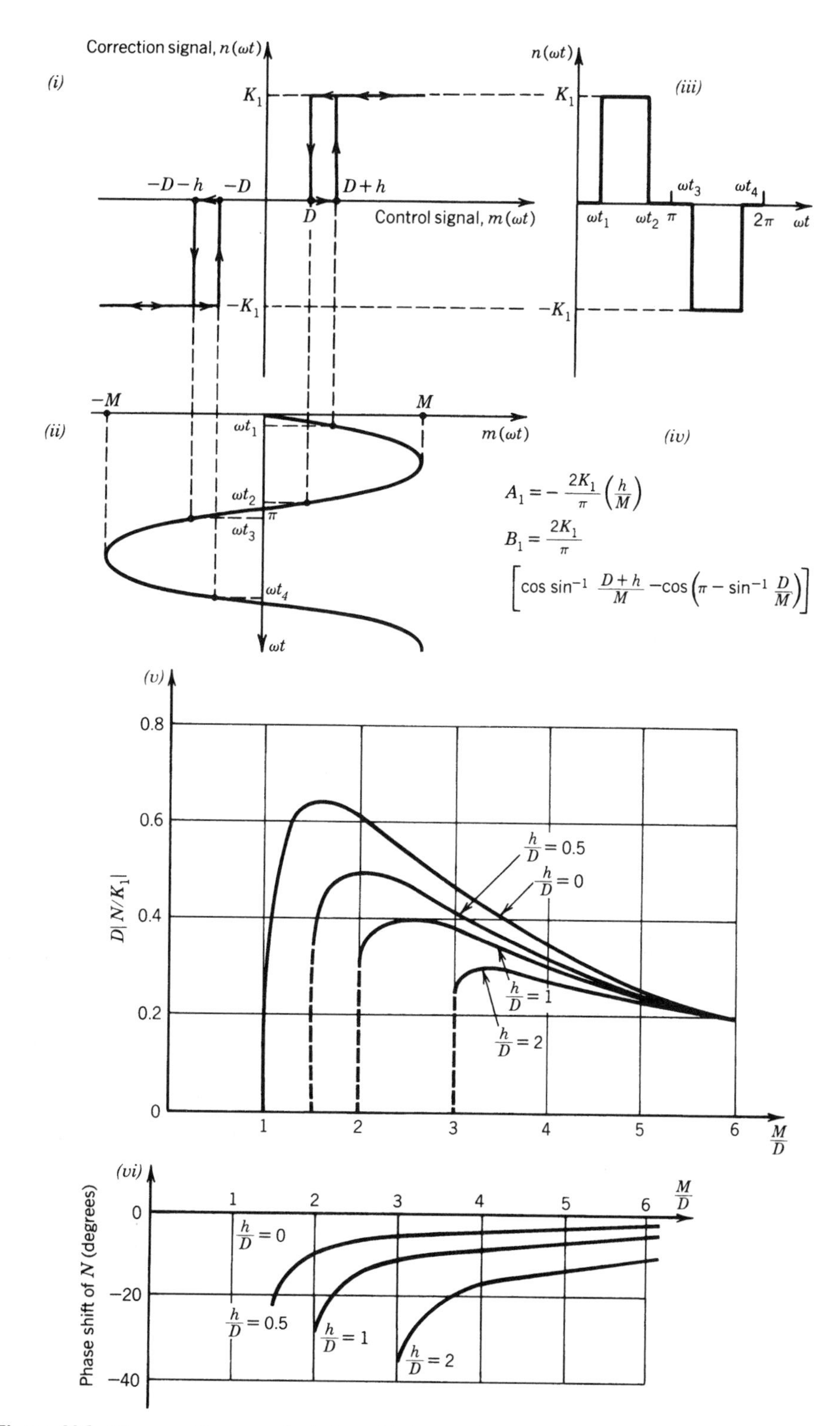

Figure 30*d* Three-position on–off device with hysteresis: (*i*) nonlinear characteristic; (*ii*) sinusoidal input wave shape; (*ii*) output wave shape; (*iv*) describing-function coefficients; (*v*) normalized amplitude characteristics for the describing function; (*vi*) normalized phase characteristics for the describing function.

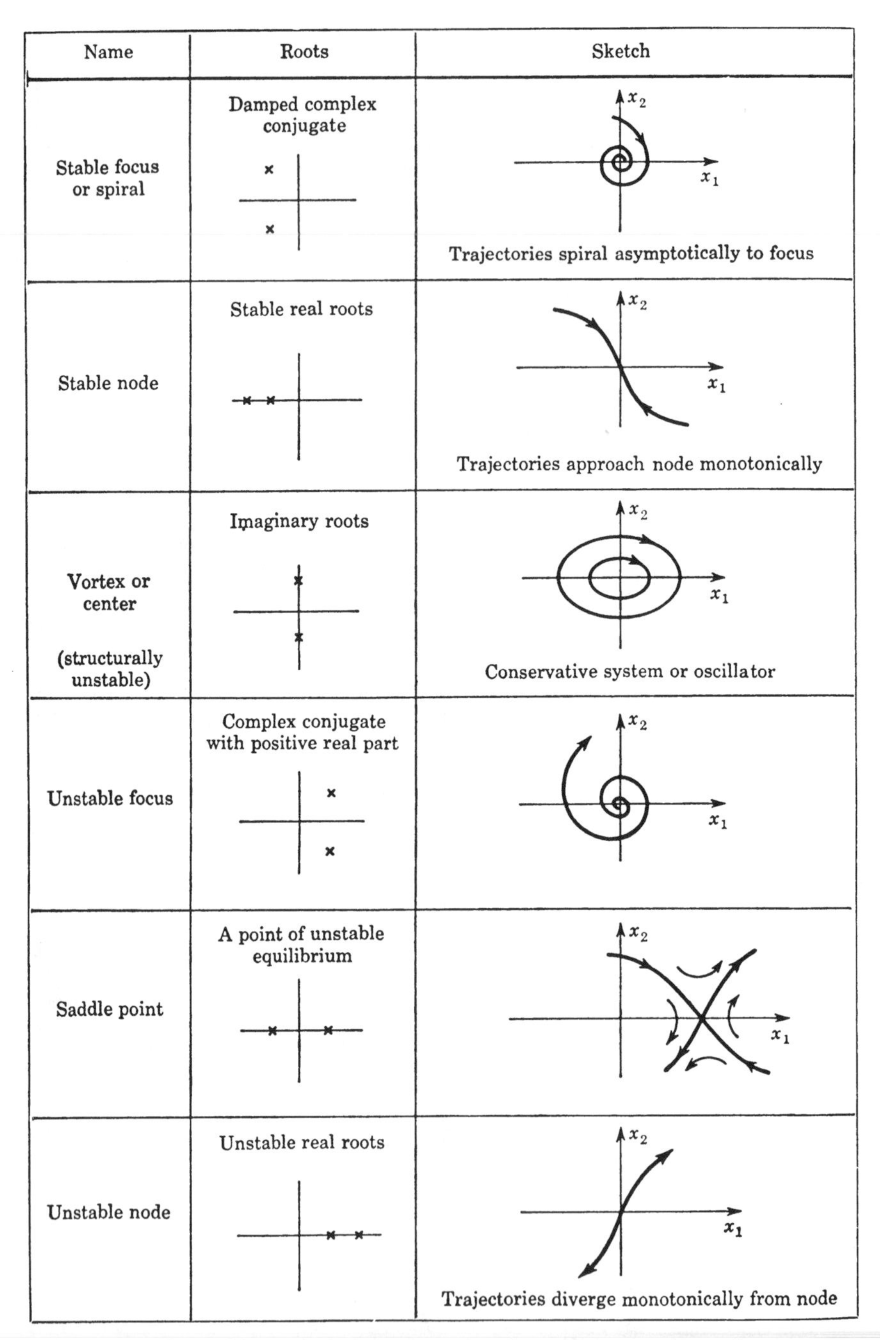

Name	Roots	Sketch
Stable focus or spiral	Damped complex conjugate	x_2 x_1 Trajectories spiral asymptotically to focus
Stable node	Stable real roots	x_2 x_1 Trajectories approach node monotonically
Vortex or center (structurally unstable)	Imaginary roots	x_2 x_1 Conservative system or oscillator
Unstable focus	Complex conjugate with positive real part	x_2 x_1
Saddle point	A point of unstable equilibrium	x_2 x_1
Unstable node	Unstable real roots	x_2 x_1 Trajectories diverge monotonically from node

Figure 31 Typical phase-plane plots for second-order systems.[9]

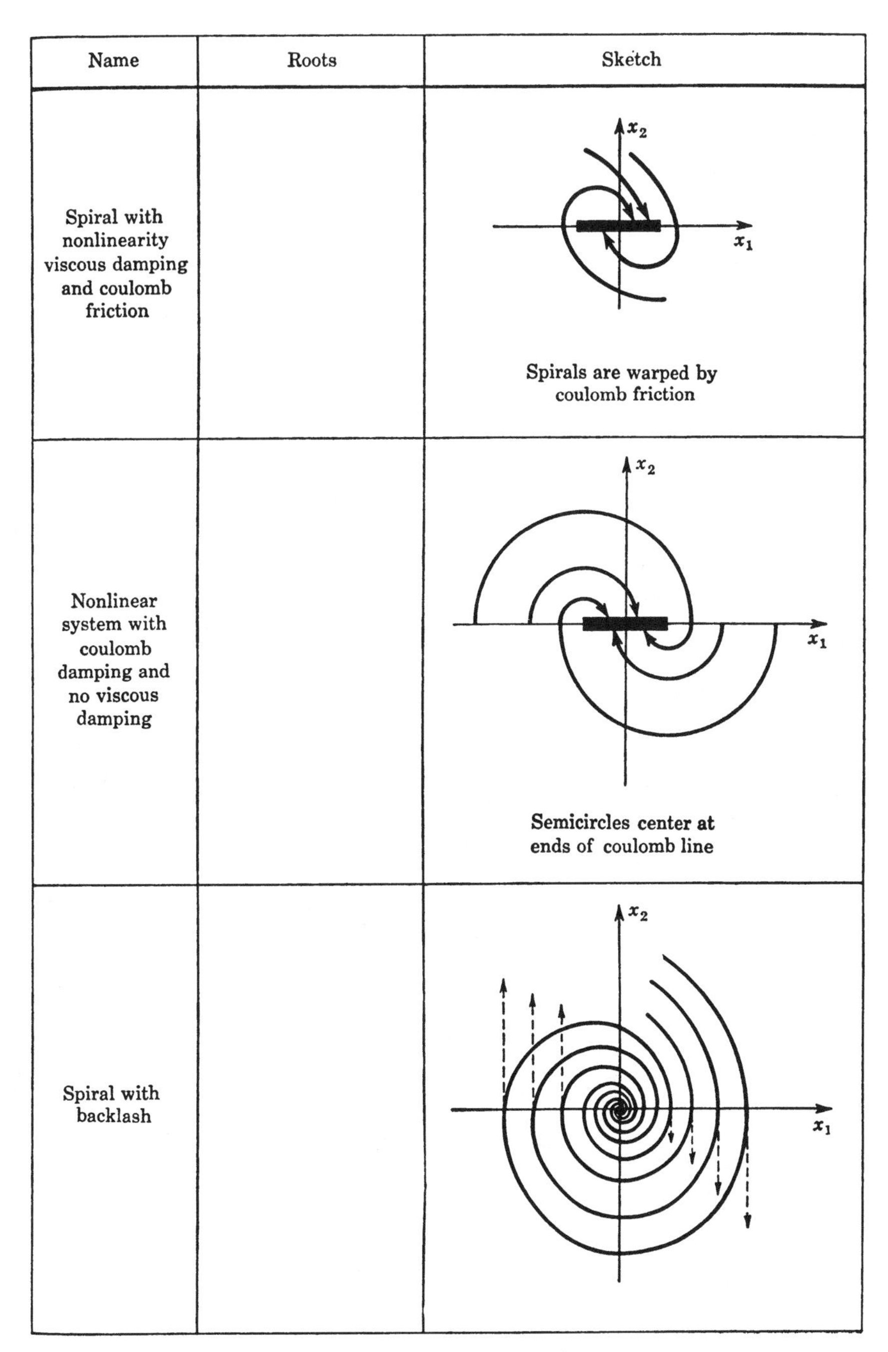

Figure 31 (*Continued*)

$$\frac{dx_1}{dt} = f_1(x_1,x_2) \qquad \frac{dx_2}{dt} = f_2(x_1,x_2)$$

Dividing the second equation by the first and eliminating the dt terms yield

$$\frac{dx_2}{dx_1} = \frac{f_2(x_1,x_2)}{f_1(x_1,x_2)}$$

This first-order equation describes the phase-plane trajectories. In many cases it can be solved analytically. If not, it always can be simulated.

The phase-plane method complements the describing-function approach. A describing function is an approximate representation of the sinusoidal response for systems of any order, while the phase plane is an exact representation of the (free) transient response for first- and second-order systems. Of course, the phase-plane method theoretically can be extended for higher order systems, but the difficulty of visualizing the nth-order state space typically makes such a direct extension impractical. An approximate extension of the method has been used with some considerable success,[8] however, in order to explore and validate the relationships among pairs of variables in complex simulation models. The approximation is based on the assumptions that the paired variables define a second-order subsystem which, for the purposes of analysis, is weakly coupled to the remainder of the system.

8.5 Discrete and Hybrid Systems

A *discrete-time system* is one for which the dependent variables are defined only at distinct instants of time. Discrete-time models occur in the representation of systems that are inherently discrete, in the analysis and design of digital measurement and control systems, and in the numerical solution of differential equations (see Section 7). Because most control systems are now implemented using digital computers (especially microprocessors), discrete-time models are extremely important in dynamic systems analysis. The discrete-time nature of a computer's sampling of continuous physical signals also leads to the occurrence of *hybrid systems,* that is, systems that are in part discrete and in part continuous. Discrete-time models of hybrid systems are called *sampled-data systems.*

Difference Equations

Dynamic models of discrete-time systems most naturally take the form of *difference equations.* The I/O form of an nth-order difference equation model is

$$f[y(k+n), y(k+n-1), \ldots, y(k), u(k+n-1), \ldots, u(k)] = 0$$

which expresses the dependence of the $(k+n)$th value of the output, $y(k+n)$, on the n preceding values of the output y and input u. For a linear system, the I/O form can be written as

$$y(k+n) + a_{n-1}(k)y(k+n-1) + \cdots + a_1(k)y(k+1) + a_0(k)y(k)$$
$$= b_{n-1}(k)u(k+n-1) + \cdots + b_0(k)u(k)$$

In state-variable form, the discrete-time model is the vector difference equation

$$x(k+1) = f[x(k), u(k)]$$
$$y(k) = g[x(k), u(k)]$$

where x is the state vector, u is the vector of inputs, and y is the vector of outputs. For a linear system, the discrete state-variable form can be written as

$$x(k + 1) = A(k)x(k) + B(k)u(k)$$

$$y(k) = C(k)x(k) + D(k)u(k)$$

The mathematics of difference equations parallels that of differential equations in many important respects. In general, the concepts applied to differential equations have direct analogies for difference equations, although the mechanics of their implementation may vary (see Ref. 11 for a development of dynamic modeling based on difference equations). One important difference is that the general solution of nonlinear and time-varying difference equations can usually be obtained through *recursion.* For example, consider the discrete nonlinear model

$$y(k + 1) = \frac{y(k)}{1 + y(k)}$$

Recursive evaluation of the equation beginning with the initial condition $y(0)$ yields

$$y(1) = \frac{y(0)}{1 + y(0)}$$

$$y(2) = \frac{y(1)}{1 + y(1)} = \left[\frac{y(0)}{1 + y(0)}\right] \Big/ \left[1 + \frac{y(0)}{1 + y(0)}\right] = \frac{y(0)}{1 + 2y(0)}$$

$$y(3) = \frac{y(2)}{1 + y(2)} = \frac{y(0)}{1 + 3y(0)}$$

$$\vdots$$

the pattern of which reveals, by induction,

$$y(k) = \frac{y(0)}{1 + ky(0)}$$

as the general solution.

Uniform Sampling

Uniform sampling is the most common mathematical approach to *A/D conversion,* that is, to extracting the discrete-time approximation $y^*(k)$ of the form

$$y^*(k) = y(t = kT)$$

from the continuous-time signal $y(t)$, where T is a constant interval of time called the *sampling period.* If the sampling period is too large, however, it may not be possible to represent the continuous signal accurately. The *sampling theorem* guarantees that $y(t)$ can be reconstructed from the uniformly sampled values $y^*(k)$ if the sampling period satisfies the inequality

$$T \leq \frac{\pi}{\omega_u}$$

where ω_u is the highest frequency contained in the Fourier transform $Y(\omega)$ of $y(t)$, that is, if

$$Y(\omega) = 0 \qquad \text{for all} \quad \omega > \omega_u$$

The Fourier transform of a signal is defined to be

$$\mathcal{F}[y(t)] = Y(\omega) = \int_{-\infty}^{\infty} y(t)e^{-j\omega t}\,dt$$

Note that if $y(t) = 0$ for $t \geq 0$, and if the region of convergence for the Laplace transform includes the imaginary axis, then the Fourier transform can be obtained from the Laplace transform as

$$Y(\omega) = [Y(s)]_{s=j\omega}$$

For cases where it is impossible to determine the Fourier transform analytically, such as when the signal is described graphically or by a table, numerical solution based on the *fast Fourier transform (FFT) algorithm* is usually satisfactory.

In general, the condition $T \leq \pi/\omega_u$ cannot be satisfied exactly, since most physical signals have no finite upper frequency ω_u. A useful approximation is to define the upper frequency as the frequency for which 99% of the signal "energy" lies in the frequency spectrum $0 \leq \omega \leq \omega_u$. This approximation is found from the relation

$$\int_{0}^{\omega_u} |Y(\omega)|^2\,d\omega = 0.99 \int_{0}^{\infty} |Y(\omega)|^2\,d\omega$$

where the square of the amplitude of the Fourier transform $|Y(\omega)|^2$ is said to be the *power spectrum* and its integral over the entire frequency spectrum is referred to as the "energy" of the signal. Using a sampling frequency 2–10 times this approximate upper frequency (depending on the required factor of safety) and inserting a low-pass filter (called a *guard filter*) before the sampler to eliminate frequencies above the *Nyquist frequency* π/T usually lead to satisfactory results.[4]

The z-Transform

The *z-transform* permits the development and application of transfer functions for discrete-time systems, in a manner analogous to continuous-time transfer functions based on the Laplace transform. A discrete signal may be represented as a series of impulses

$$\begin{aligned} y^*(t) &= y(0)\delta(t) + y(1)\delta(t - T) + y(2)\delta(t - 2T) + \cdots \\ &= \sum_{k=0}^{N} y(k)\delta(t - kT) \end{aligned}$$

where $y(k) = y^*(t = kT)$ are the values of the discrete signal, $\delta(t)$ is the unit impulse function, and N is the number of samples of the discrete signal. The Laplace transform of the series is

$$Y^*(s) = \sum_{k=0}^{N} y(k)e^{-ksT}$$

where the shifting property of the Laplace transform has been applied to the pulses. Defining the *shift* or *advance operator* as $z = e^{st}$, $Y^*(s)$ may now be written as a function of z:

$$Y^*(z) = \sum_{k=0}^{N} \frac{y(k)}{z^k} = \mathcal{Z}[y(t)]$$

where the transformed variable $Y^*(z)$ is called the z-transform of the function $y^*(t)$. The inverse of the shift operator $1/z$ is called the *delay operator* and corresponds to a time delay of T.

The z-transforms for many sampled functions can be expressed in closed form. A listing of the transforms of several commonly encountered functions is given in Table 12. Properties of the z-transform are listed in Table 13.

Pulse Transfer Functions

The transfer function concept developed for continuous systems has a direct analog for sampled-data systems. For a continuous system with sampled output $u(t)$ and sampled input $y(t)$, the *pulse* or *discrete transfer function* $G(z)$ is defined as the ratio of the z-transformed output $Y(z)$ to the z-transformed input $U(z)$, assuming zero initial conditions. In general, the pulse transfer function has the form

$$G(z) = \frac{Y(z)}{U(z)} = \frac{b_0 + b_1 z^{-1} + b_2 z^{-2} + \cdots + b_m z^{-m}}{1 + a_1 z^{-1} + a_2 z^{-1} + \cdots + a_n z^{-n}}$$

Zero-Order Hold

The *zero-order data hold* is the most common mathematical approach to *D/A conversion,* that is, to creating a piecewise continuous approximation $u(t)$ of the form

Table 12 z-Transform Pairs

	$X(s)$	$x(t)$ or $x(k)$	$X(z)$
1	1	$\delta(t)$	1
2	e^{-kTs}	$\delta(t - kT)$	z^{-k}
3	$\frac{1}{s}$	$1(t)$	$\frac{z}{z-1}$
4	$\frac{1}{s^2}$	t	$\frac{Tz}{(z-1)^2}$
5	$\frac{1}{s+a}$	e^{-at}	$\frac{z}{z - e^{-aT}}$
6	$\frac{a}{s(s+a)}$	$1 - e^{-at}$	$\frac{(1 - e^{-aT})z}{(z-1)(z - e^{-aT})}$
7	$\frac{\omega}{s^2 + \omega^2}$	$\sin \omega t$	$\frac{z \sin \omega T}{z^2 - 2z \cos \omega T + 1}$
8	$\frac{s}{s^2 + \omega^2}$	$\cos \omega t$	$\frac{z(z - \cos \omega T)}{z^2 - 2z \cos \omega T + 1}$
9	$\frac{1}{(s+a)^2}$	te^{-at}	$\frac{Tze^{-aT}}{(z - e^{-aT})^2}$
10	$\frac{\omega}{(s+a)^2 + \omega^2}$	$e^{-at} \sin \omega t$	$\frac{ze^{-aT} \sin \omega T}{z^2 - 2ze^{-aT} \cos \omega T + e^{-2aT}}$
11	$\frac{s+a}{(s+a)^2 + \omega^2}$	$e^{-at} \cos \omega t$	$\frac{z^2 - ze^{-aT} \cos \omega T}{z^2 - 2ze - ^{aT} \cos \omega T + e^{-2aT}}$
12	$\frac{2}{s^3}$	t^2	$\frac{T^2 z(z+1)}{(z-1)^3}$
13		a	$\frac{z}{z-a}$
14		$a^k \cos k\pi$	$\frac{z}{z+a}$

Table 13 z-Transform Properties

	$x(t)$ or $x(k)$	$\mathcal{Z}[x(t)]$ or $\mathcal{Z}[x(k)]$
1	$ax(t)$	$aX(z)$
2	$x_1(t) + x_2(t)$	$X_1(z) + X_2(z)$
3	$x(t + T)$ or $x(k + 1)$	$zX(z) - zx(0)$
4	$x(t + 2T)$	$z^2X(z) - z^2x(0) - zx(T)$
5	$x(k + 2)$	$z^2X(z) - z^2x(0) - zx(1)$
6	$x(t + kT)$	$z^kX(z) - z^kx(0) - z^{k-1}x(T) - \cdots - zx(kT - T)$
7	$x(k + m)$	$z^mX(z) - z^mx(0) - z^{m-1}x(1) - \cdots - zx(m - 1)$
8	$tx(t)$	$-Tz \dfrac{d}{dz}[X(z)]$
9	$kx(k)$	$-z \dfrac{d}{dz}[X(z)]$
10	$e^{-at}x(t)$	$X(ze^{aT})$
11	$e^{-ak}x(k)$	$X(ze^{a})$
12	$a^kx(k)$	$X\left(\dfrac{z}{a}\right)$
13	$ka^kx(k)$	$-z \dfrac{d}{dz}\left[X\left(\dfrac{z}{a}\right)\right]$
14	$x(0)$	$\lim_{z \to \infty} X(z)$ if the limit exists
15	$x(\infty)$	$\lim_{z \to 1} [(z - 1)X(z)]$ if $\dfrac{z-1}{z} X(z)$ is analytic on and outside the unit circle
16	$\sum_{k=0}^{\infty} x(k)$	$X(1)$
17	$\sum_{k=0}^{n} x(kT)y(nT - kT)$	$X(z)Y(z)$

$$u(t) = u^*(k) \qquad \text{for} \quad kT \le t < (k + 1)T$$

from the discrete-time signal $u^*(k)$, where T is the period of the hold. The effect of the zero-order hold is to convert a sequence of discrete impulses into a staircase pattern, as shown in Fig. 32. The transfer function of the zero-order hold is

$$G(s) = \frac{1}{s}(1 - e^{-Ts}) = \frac{1 - z^{-1}}{s}$$

Using this relationship, the pulse transfer function of the sampled-data system shown in Fig. 33 can be derived as

$$G(z) = (1 - z^{-1})\mathcal{Z}\left[\mathcal{L}^{-1}\frac{G(s)}{s}\right]$$

The continuous system with transfer function $G(s)$ has a sampler and a zero-order hold at its input and a sampler at its output. This is a common configuration in many computer control applications.

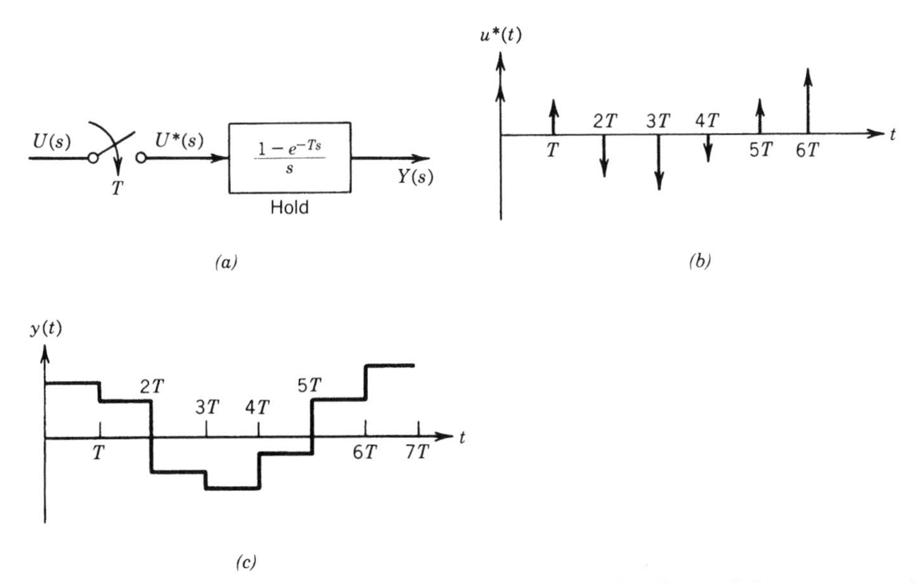

Figure 32 Zero-order hold: (*a*) block diagram of hold with a sampler, (*b*) sampled input sequence, (*c*) analog output for the corresponding input sequence.[4]

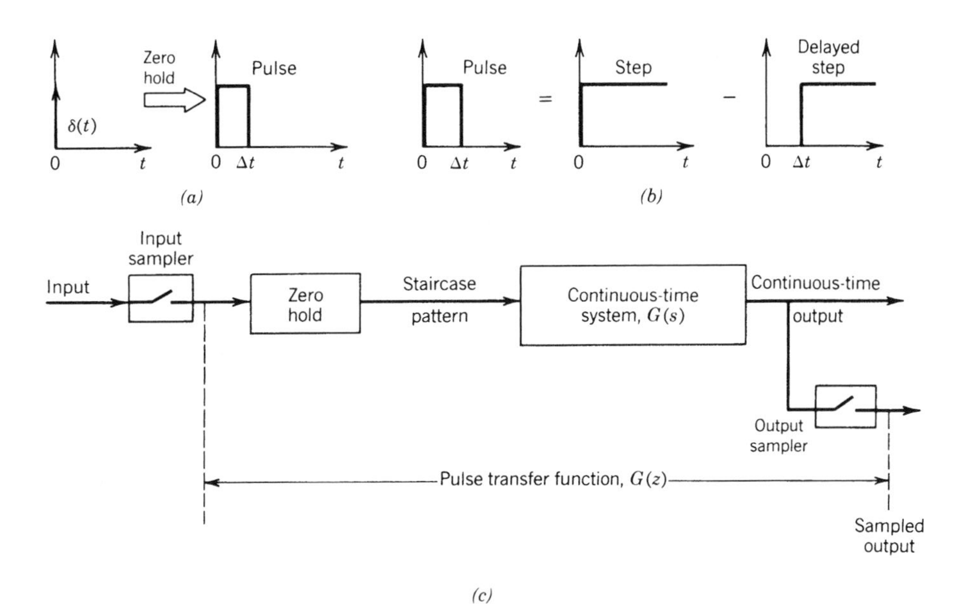

Figure 33 Pulse transfer function of a continuous system with sampler and zero hold.[12]

REFERENCES

1. J. L. Shearer, A. T. Murphy, and H. H. Richardson, *Introduction to System Dynamics,* Addison-Wesley, Reading, MA, 1971.
2. E. O. Doebelin, *System Dynamics: Modeling, Analysis, Simulation, and Design,* Merrill, Columbus, OH, 1998.
3. C. M. Close, D. K. Frederick, and J. C. Newell, *Modeling and Analysis of Dynamic Systems,* 3rd ed., Houghton Mifflin, Boston, 2001.
4. W. J. Palm III, *Modeling, Analysis, and Control of Dynamic Systems,* 2nd ed., Wiley, New York, 2000.
5. G. J. Thaler and R. G. Brown, *Analysis and Design of Feedback Control Systems,* 2nd ed., McGraw-Hill, New York, 1960.
6. G. A. Korn and J. V. Wait, *Digital Continuous System Simulation,* Prentice-Hall, Englewood Cliffs, NJ, 1975.
7. B. C. Kuo and F. Golnaraghi, *Automatic Control Systems,* 8th ed., Prentice-Hall, Englewood Cliffs, NJ, 2002.
8. W. Thissen, "Investigation into the World3 Model: Lessons for Understanding Complicated Models," *IEEE Transactions on Systems, Man, and Cybernetics,* **SMC-8**(3) (1978) pp. 183–193.
9. J. E. Gibson, *Nonlinear Automatic Control,* McGraw-Hill, New York, 1963.
10. S. M. Shinners, *Modern Control System Theory and Design,* 2nd ed., Wiley, New York, 1998.
11. D. G. Luenberger, *Introduction to Dynamic Systems: Theory, Models, and Applications,* Wiley, New York, 1979.
12. D. M. Auslander, Y. Takahashi, and M. J. Rabins, *Introducing Systems and Control,* McGraw-Hill, New York, 1974.

BIBLIOGRAPHY

Bateson, R. N., *Introduction to Control System Technology,* Prentice-Hall, Englewood Cliffs, NJ, 2001.

Bishop, R. H., and R. C. Dorf, *Modern Control Systems,* 10th ed., Prentice-Hall, 2004.

Brogan, W. L., *Modern Control Theory,* 3rd ed., Prentice-Hall, Englewood Cliffs, NJ, 1991.

Cannon, Jr., R. H., *Dynamics of Physical Systems,* McGraw-Hill, New York, 1967.

DeSilva, C. W., *Control Sensors and Actuators,* Prentice-Hall, Englewood Cliffs, NJ, 1989.

Doebelin, E. O., *Measurement Systems,* 5th ed., McGraw-Hill, New York, 2004.

Franklin, G. F., J. D. Powell, and A. Emami-Naeini, *Feedback Control of Dynamic Systems,* 4th ed., Addison-Wesley, Reading, MA, 2001.

Grace, A., A. J. Laub, J. N. Little, and C. Thompson, *Control System Toolbox Users Guide,* The Mathworks, Natick, MA, 1990.

Hartley, T. T., G. O. Beale, and S. P. Chicatelli, *Digital Simulation of Dynamic Systems: A Control Theory Approach,* Prentice-Hall, Englewood Cliffs, NJ, 1994.

Kheir, N. A., *Systems Modeling and Computer Simulation,* 2nd ed., Marcel Dekker, New York, 1996.

Kelton, W. D., R. P. Sadowski, and D. T. Sturrock, *Simulation with Arena,* 3rd ed., McGraw-Hill, New York, 2004.

Lay, D. C., *Linear Algebra and Its Applications,* 3rd ed., Addison-Wesley, Reading, MA, 2006.

Ljung, L., and T. Glad, *Modeling Simulation of Dynamic Systems,* Prentice-Hall, Englewood Cliffs, NJ, 1994.

The MathWorks on-line at ⟨www.mathworks.com⟩ NJ, 1995.

Phillips, C. L., and R. D. Harbor, *Feedback Control Systems,* 4th ed., Prentice-Hall, Englewood Cliffs, NJ, 1999.

Phillips, C. L., and H. T. Nagle, *Digital Control System Analysis and Design,* 3rd ed., Prentice-Hall, Englewood Cliffs, NJ, 1995.

Van Loan, C., *Computational Frameworks for the Fast Fourier Transform,* SIAM, Philadelphia, PA, 1992.

Wolfram, S., *Mathematica Book,* 5th ed., Wolfram Media, Inc., 2003.

CHAPTER 11

BASIC CONTROL SYSTEMS DESIGN

William J. Palm III
Department of Mechanical Engineering
University of Rhode Island
Kingston, Rhode Island

Revised from William J. Palm III, *Modeling, Analysis and Control of Dynamic Systems,* 2nd ed., Wiley, 2000, by permission of the publisher.

1 INTRODUCTION

The purpose of a *control system* is to produce a desired *output.* This output is usually specified by the command *input* and is often a function of time. For simple applications in well-structured situations, *sequencing* devices like timers can be used as the control system. But most systems are not that easy to control, and the controller must have the capability of reacting to disturbances, changes in its environment, and new input commands. The key element that allows a control system to do this is *feedback,* which is the process by which a system's output is used to influence its behavior. Feedback in the form of the room-temperature measurement is used to control the furnace in a thermostatically controlled heating system. Figure 1 shows the *feedback loop* in the system's *block diagram,* which is a graphical representation of the system's control structure and logic. Another commonly found control system is the pressure regulator shown in Fig. 2.

Feedback has several useful properties. A system whose individual elements are nonlinear can often be modeled as a linear one over a wider range of its variables with the proper use of feedback. This is because feedback tends to keep the system near its reference operation condition. Systems that can maintain the output near its desired value despite changes in the environment are said to have good *disturbance rejection.* Often we do not have accurate values for some system parameter or these values might change with age. Feedback can be used to minimize the effects of parameter changes and uncertainties. A system that has both good disturbance rejection and low sensitivity to parameter variation is *robust.* The application that resulted in the general understanding of the properties of feedback is shown in Fig. 3. The electronic amplifier gain A is large, but we are uncertain of its exact value. We use the resistors R_1 and R_2 to create a feedback loop around the amplifier and pick R_1 and R_2 to create a feedback loop around the amplifier and R_1 and R_2 so that $AR_2/R_1 >> 1$. Then the input–output relation becomes $e_o \approx R_1 e_i/R_2$, which is independent

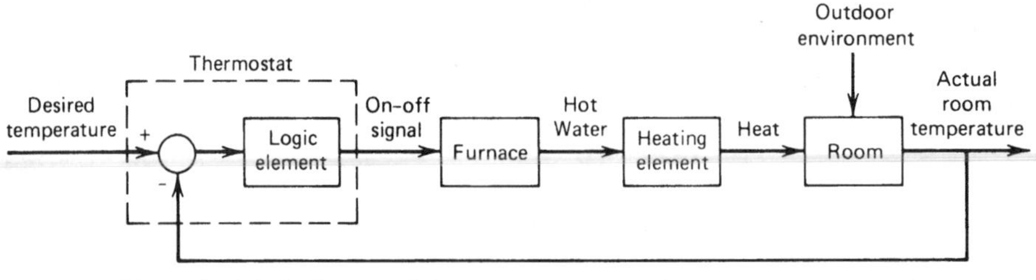

Figure 1 Block diagram of the thermostat system for temperature control.[1]

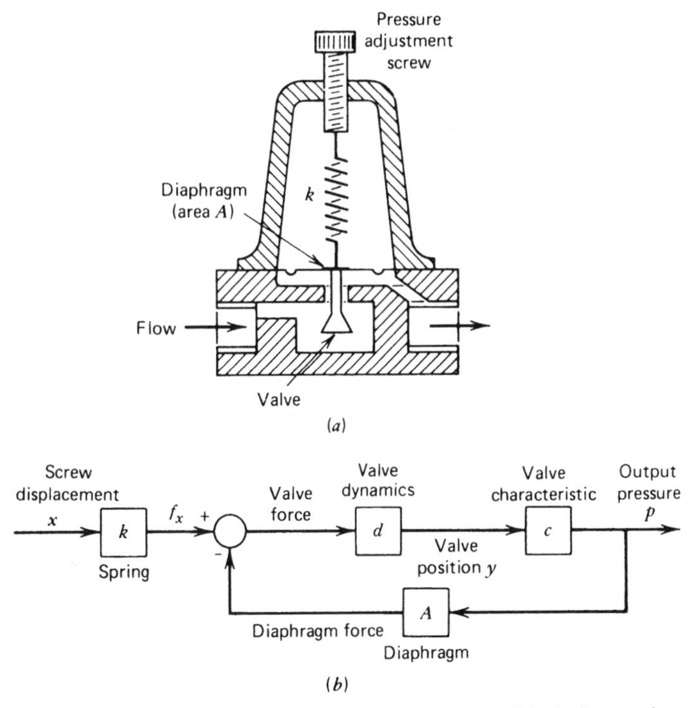

Figure 2 Pressure regulator: (*a*) cutaway view; (*b*) block diagram.[1]

of A as long as A remains large. If R_1 and R_2 are known accurately, then the system gain is now reliable.

Figure 4 shows the block diagram of a *closed-loop* system, which is a system with feedback. An *open-loop* system, such as a timer, has no feedback. Figure 4 serves as a focus for outlining the prerequisites for this chapter. The reader should be familiar with the *transfer function* concept based on the Laplace transform, the *pulse transfer* function based on the z-transform, for digital control, and the differential equation modeling techniques needed to obtain them. It is also necessary to understand block diagram algebra, characteristic roots, the final-value theorem, and their use in evaluating system response for common inputs like the step function. Also required are stability analysis techniques such as the Routh criterion

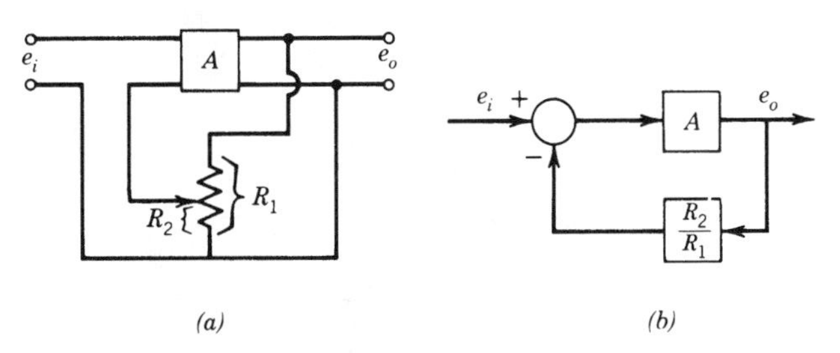

Figure 3 A closed-loop system.

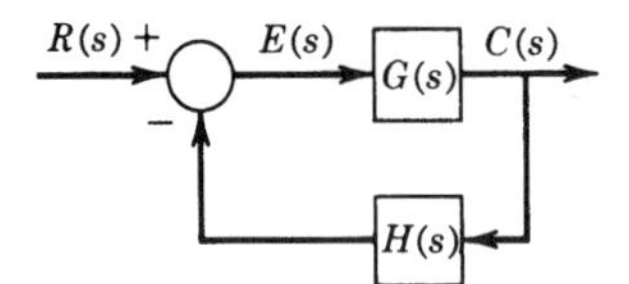

Figure 4 Feedback compensation of an amplifier.

and transient performance specifications such as the damping ratio ζ, natural frequency ω_n, dominant time constant τ, maximum overshoot, settling time, and bandwidth. The above material is reviewed in the previous chapter. Treatment in depth is given in Refs. 1–4.

2 CONTROL SYSTEM STRUCTURE

The electromechanical position control system shown in Fig. 5 illustrates the structure of a typical control system. A load with an inertia I is to be positioned at some desired angle θ_r. A dc motor is provided for this purpose. The system contains viscous damping, and a disturbance torque T_d acts on the load, in addition to the motor torque T. Because of the disturbance, the angular position θ of the load will not necessarily equal the desired value θ_r. For this reason, a potentiometer, or some other sensor such as an encoder, is used to measure the displacement θ. The potentiometer voltage representing the controlled position θ is compared to the voltage generated by the command potentiometer. This device enables the operator to dial in the desired angle θ_r. The amplifier sees the difference e between the two potentiometer voltages. The basic function of the amplifier is to increase the small error voltage e up to the voltage level required by the motor and to supply enough current required by the motor to drive the load. In addition, the amplifier may shape the voltage signal in certain ways to improve the performance of the system.

The control system is seen to provide two basic functions: (1) to respond to a command input that specifies a new desired value for the controlled variable and (2) to keep the controlled variable near the desired value in spite of disturbances. The presence of the feedback loop is vital to both functions. A block diagram of this system is shown in Fig. 6. The power supplies required for the potentiometers and the amplifier are not shown in block diagrams of control system logic because they do not contribute to the control logic.

2.1 A Standard Diagram

The electromechanical positioning system fits the general structure of a control system (Fig. 7). This figure also gives some standard terminology. Not all systems can be forced into this format, but it serves as a reference for discussion.

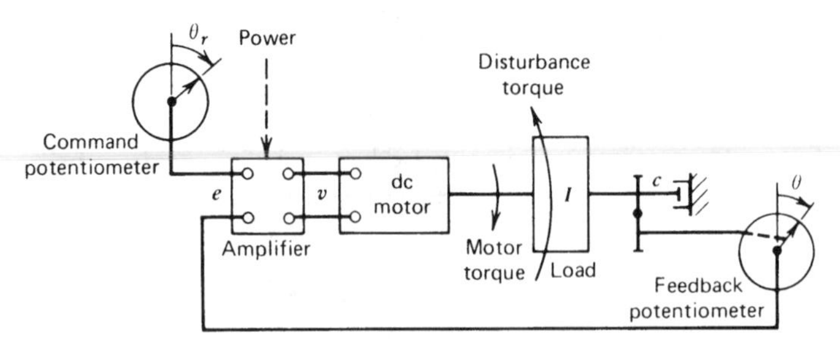

Figure 5 Position control system using a dc motor.[1]

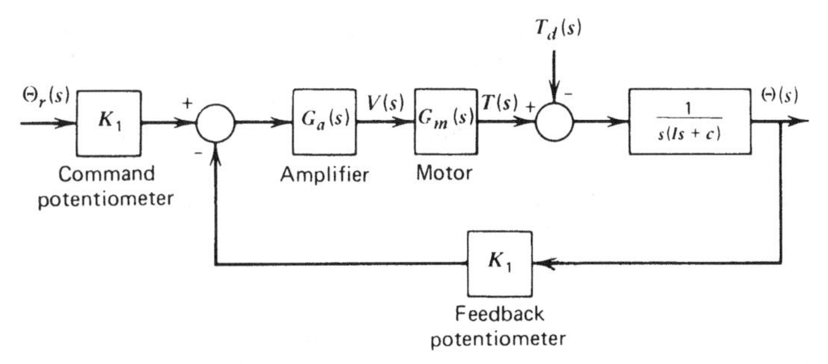

Figure 6 Block diagram of the position control system shown in Fig. 5.[1]

The controller is generally thought of as a logic element that compares the command with the measurement of the output and decides what should be done. The input and feedback elements are transducers for converting one type of signal into another type. This allows the error detector directly to compare two signals of the same type (e.g., two voltages). Not all functions show up as separate physical elements. The error detector in Fig. 5 is simply the input terminals of the amplifier.

The control logic elements produce the control signal, which is sent to the *final control elements*. These are the devices that develop enough torque, pressure, heat, and so on to influence the elements under control. Thus, the final control elements are the "muscle" of the system, while the control logic elements are the "brain." Here we are primarily concerned with the design of the logic to be used by this brain.

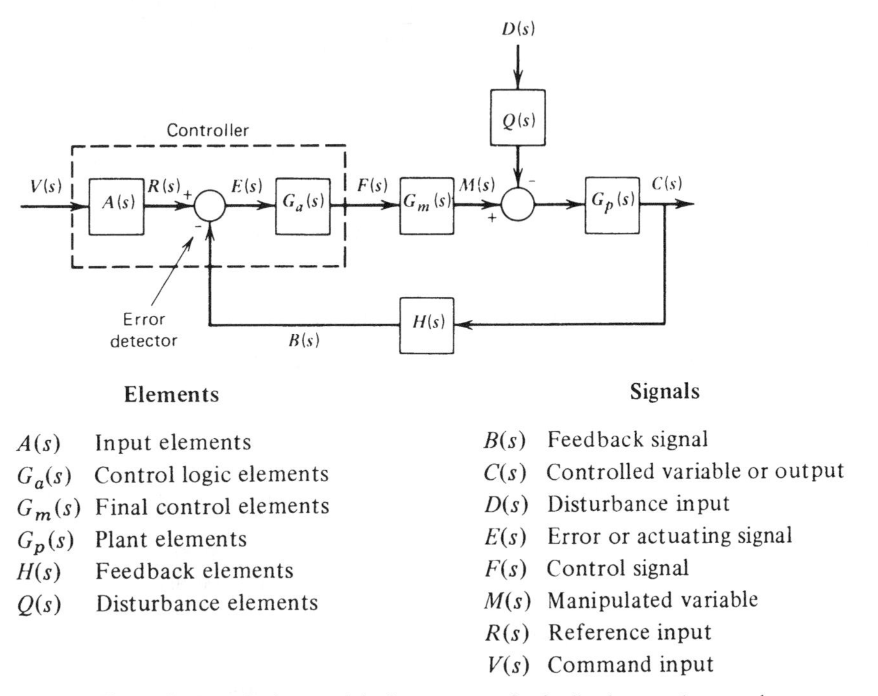

Figure 7 Terminology and basic structure of a feedback control system.[1]

The object to be controlled is the *plant.* The *manipulated* variable is generated by the final control elements for this purpose. The disturbance input also acts on the plant. This is an input over which the designer has no influence and perhaps for which little information is available as to the magnitude, functional form, or time of occurrence. The disturbance can be a random input, such as wind gust on a radar antenna, or deterministic, such as Coulomb friction effects. In the latter case, we can include the friction force in the system model by using a nominal value for the coefficient of friction. The disturbance input would then be the deviation of the friction force from this estimated value and would represent the uncertainty in our estimate.

Several control system classifications can be made with reference to Fig. 7. A *regulator* is a control system in which the controlled variable is to be kept constant in spite of disturbances. The command input for a regulator is its *set point.* A *follow-up system* is supposed to keep the control variable near a command value that is changing with time. An example of a follow-up system is a machine tool in which a cutting head must trace a specific path in order to shape the product properly. This is also an example of a *servomechanism,* which is a control system whose controlled variable is a mechanical position, velocity, or acceleration. A thermostat system is not a servomechanism, but a *process control system,* where the controlled variable describes a thermodynamic process. Typically, such variables are temperature, pressure, flow rate, liquid level, chemical concentration, and so on.

2.2 Transfer Functions

A transfer function is defined for each input–output pair of the system. A specific transfer function is found by setting all other inputs to zero and reducing the block diagram. The *primary* or *command* transfer function for Fig. 7 is

$$\frac{C(s)}{V(s)} = \frac{A(s)G_a(s)G_m(s)G_p(s)}{1 + G_a(s)G_m(s)G_p(s)H(s)} \tag{1}$$

The *disturbance* transfer function is

$$\frac{C(s)}{D(s)} = \frac{-Q(s)G_p(s)}{1 + G_a(s)G_m(s)G_p(s)H(s)} \tag{2}$$

The transfer functions of a given system all have the same denominator.

2.3 System-Type Number and Error Coefficients

The error signal in Fig. 4 is related to the input as

$$E(s) = \frac{1}{1 + G(s)H(s)} R(s) \tag{3}$$

If the final-value theorem can be applied, the steady-state error is

$$e_{ss} = \lim_{s \to 0} \frac{sR(s)}{1 + G(s)H(s)} \tag{4}$$

The *static error coefficient* c_i is defined as

$$c_i = \lim_{s \to 0} s^i G(s)H(s) \tag{5}$$

A system is of *type n* if $G(s)H(s)$ can be written as $s^n F(s)$. Table 1 relates the steady-state error to the system type for three common inputs and can be used to design systems for minimum error. The higher the system type, the better the system is able to follow a rapidly changing input. But higher type systems are more difficult to stabilize, so a compromise must be made in the design. The coefficients c_0, c_1, and c_2 are called the *position, velocity,* and *acceleration error coefficients.*

3 TRANSDUCERS AND ERROR DETECTORS

The control system structure shown in Fig. 7 indicates a need for physical devices to perform several types of functions. Here we present a brief overview of some available transducers and error detectors. Actuators and devices used to implement the control logic are discussed in Sections 4 and 5.

3.1 Displacement and Velocity Transducers

A *transducer* is a device that converts one type of signal into another type. An example is the potentiometer, which converts displacement into voltage, as in Fig. 8. In addition to this conversion, the transducer can be used to make measurements. In such applications, the term *sensor* is more appropriate. Displacement can also be measured electrically with a *linear variable differential transformer* (LVDT) or a *synchro.* An LVDT measures the linear displacement of a movable magnetic core through a primary winding and two secondary windings (Fig. 9). An ac voltage is applied to the primary. The secondaries are connected together and also to a detector that measures the voltage and phase difference. A phase difference of 0° corresponds to a positive core displacement, while 180° indicates a negative displacement. The amount of displacement is indicated by the amplitude of the ac voltage in the secondary. The detector converts this information into a dc voltage e_o, such that $e_o = Kx$. The LVDT is sensitive to small displacements. Two of them can be wired together to form an error detector.

A synchro is a rotary differential transformer, with angular displacement as either the input or output. They are often used in pairs (a *transmitter* and a *receiver*) where a remote indication of angular displacement is needed. When a transmitter is used with a synchro *control transformer,* two angular displacements can be measured and compared (Fig. 10). The output voltage e_o is approximately linear with angular difference within ±70°, so that $e_o = K(\theta_1 - \theta_2)$.

Table 1 Steady-State Error e_{ss} for Different System-Type Numbers

	System-Type Number n			
$R(s)$	0	1	2	3
Step $1/s$	$\frac{1}{1 + C_0}$	0	0	0
Ramp $1/s^2$	∞	$\frac{1}{C_1}$	0	0
Parabola $1/s^3$	∞	∞	$\frac{1}{C_2}$	0

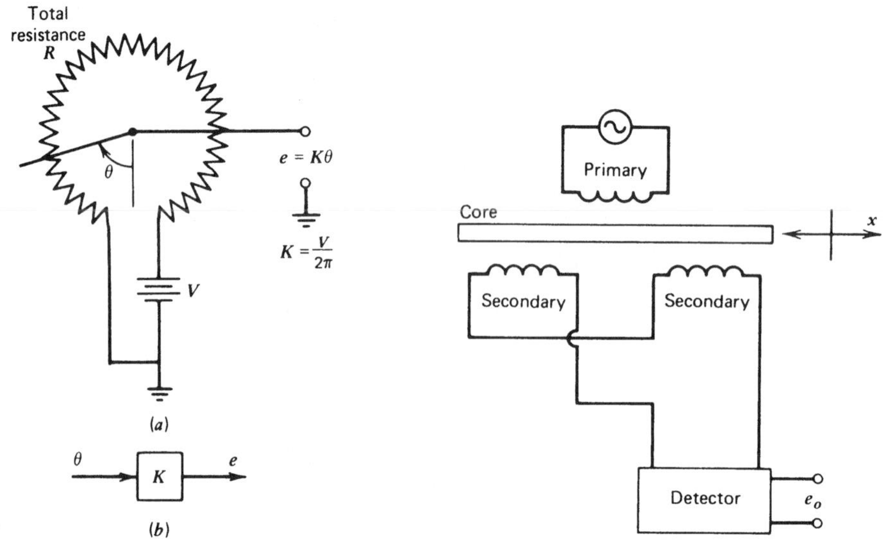

Figure 8 Rotary potentiometer.[1]

Figure 9 Linear variable differential transformer.[1]

Displacement measurements can be used to obtain forces and accelerations. For example, the displacement of a calibrated spring indicates the applied force. The accelerometer is another example. Still another is the strain gage used for force measurement. It is based on the fact that the resistance of a fine wire changes as it is stretched. The change in resistance is detected by a circuit that can be calibrated to indicate the applied force. Sensors utilizing piezoelectric elements are also available.

Velocity measurements in control systems are most commonly obtained with a *tachometer.* This is essentially a dc generator (the reverse of a dc motor). The input is mechanical (a velocity). The output is a generated voltage proportional to the velocity. Translational

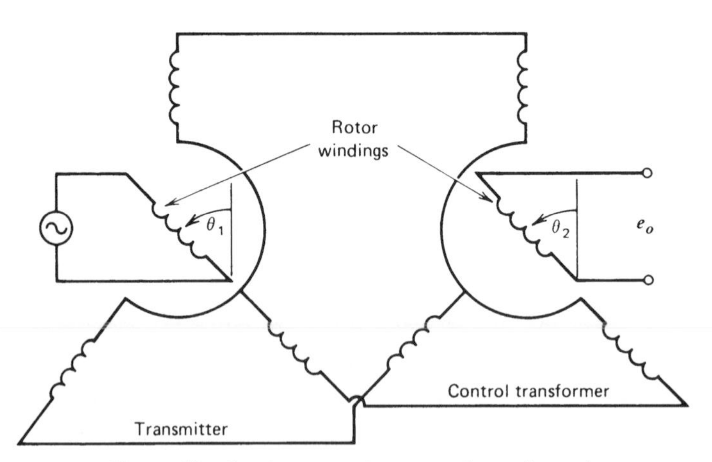

Figure 10 Synchro transmitter control transformer.[1]

velocity can be measured by converting it to angular velocity with gears, for example. Tachometers using ac signals are also available.

Other velocity transducers include a magnetic pickup that generates a pulse every time a gear tooth passes. If the number of gear teeth is known, a pulse counter and timer can be used to compute the angular velocity. This principle is also employed in turbine flowmeters.

A similar principle is employed by *optical encoders,* which are especially suitable for digital control purposes. These devices use a rotating disk with alternating transparent and opaque elements whose passage is sensed by light beams and a photosensor array, which generates a binary (on–off) train of pulses. There are two basic types: the absolute encoder and the incremental encoder. By counting the number of pulses in a given time interval, the incremental encoder can measure the rotational speed of the disk. By using multiple tracks of elements, the absolute encoder can produce a binary digit that indicates the amount of rotation. Hence, it can be used as a position sensor.

Most encoders generate a train of transistor–transistor logic (TTL) voltage level pulses for each channel. The incremental encoder output contains two channels that each produce N pulses every revolution. The encoder is mechanically constructed so that pulses from one channel are shifted relative to the other channel by a quarter of a pulse width. Thus, each pulse pair can be divided into four segments called *quadratures.* The encoder output consists of $4N$ *quadrature counts per revolution.* The pulse shift also allows the direction of rotation to be determined by detecting which channel leads the other. The encoder might contain a third channel, known as the zero, index, or marker channel, that produces a pulse once per revolution. This is used for initialization.

The gain of such an incremental encoder is $4N/2\pi$. Thus, an encoder with 1000 pulses per channel per revolution has a gain of 636 counts per radian. If an absolute encoder produces a binary signal with n bits, the maximum number of positions it can represent is $2n$, and its gain is $2^n/2\pi$. Thus, a 16-bit absolute encoder has a gain of $2^{16}/2\pi = 10{,}435$ counts per radian.

3.2 Temperature Transducers

When two wires of dissimilar metals are joined together, a voltage is generated if the junctions are at different temperatures. If the reference junction is kept at a fixed, known temperature, the thermocouple can be calibrated to indicate the temperature at the other junction in terms of the voltage v. Electrical resistance changes with temperature. Platinum gives a linear relation between resistance and temperature, while nickel is less expensive and gives a large resistance change for a given temperature change. Seminconductors designed with this property are called *thermistors.* Different metals expand at different rates when the temperature is increased. This fact is used in the bimetallic strip transducer found in most home thermostats. Two dissimilar metals are bonded together to form the strip. As the temperature rises, the strip curls, breaking contact and shutting off the furnace. The temperature gap can be adjusted by changing the distance between the contacts. The motion also moves a pointer on the temperature scale of the thermostat. Finally, the pressure of a fluid inside a bulb will change as its temperature changes. If the bulb fluid is air, the device is suitable for use in pneumatic temperature controllers.

3.3 Flow Transducers

A flow rate q can be measured by introducing a flow restriction, such as an orifice plate, and measuring the pressure drop Δp across the restriction. The relation is $\Delta p = Rq^2$, where

R can be found from calibration of the device. The pressure drop can be sensed by converting it into the motion of a diaphragm. Figure 11 illustrates a related technique. The Venturi-type flowmeter measures the static pressures in the constricted and unconstricted flow regions. Bernoulli's principle relates the pressure difference to the flow rate. This pressure difference produces the diaphragm displacement. Other types of flowmeters are available, such as turbine meters.

3.4 Error Detectors

The error detector is simply a device for finding the difference between two signals. This function is sometimes an integral feature of sensors, such as with the synchro transmitter–transformer combination. This concept is used with the diaphragm element shown in Fig. 11. A detector for voltage difference can be obtained, as with the position control system shown in Fig. 5. An amplifier intended for this purpose is a *differential amplifier.* Its output is proportional to the difference between the two inputs. In order to detect differences in other types of signals, such as temperature, they are usually converted to a displacement or pressure. One of the detectors mentioned previously can then be used.

3.5 Dynamic Response of Sensors

The usual transducer and detector models are static models and as such imply that the components respond instantaneously to the variable being sensed. Of course, any real component has a dynamic response of some sort, and this response time must be considered in relation to the controlled process when a sensor is selected. If the controlled process has a time constant at least 10 times greater than that of the sensor, we often would be justified in using a static sensor model.

4 ACTUATORS

An *actuator* is the final control element that operates on the low-level control signal to produce a signal containing enough power to drive the plant for the intended purpose. The armature-controlled dc motor, the hydraulic servomotor, and the pneumatic diaphragm and piston are common examples of actuators.

4.1 Electromechanical Actuators

Figure 12 shows an electromechanical system consisting of an armature-controlled dc motor driving a load inertia. The rotating armature consists of a wire conductor wrapped around

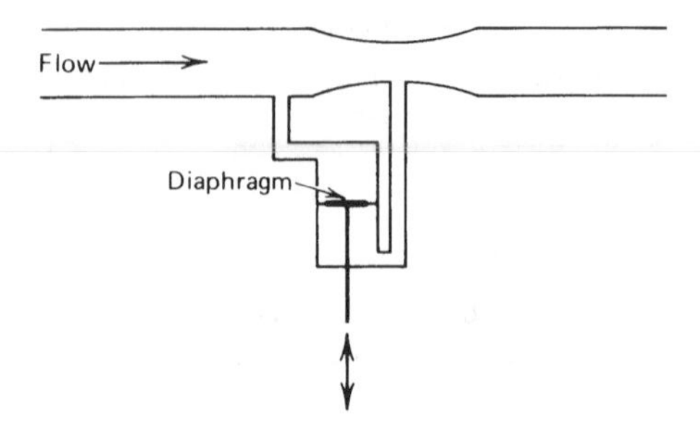

Figure 11 Venturi-type flowmeter. The diaphragm displacement indicates the flow rate.[1]

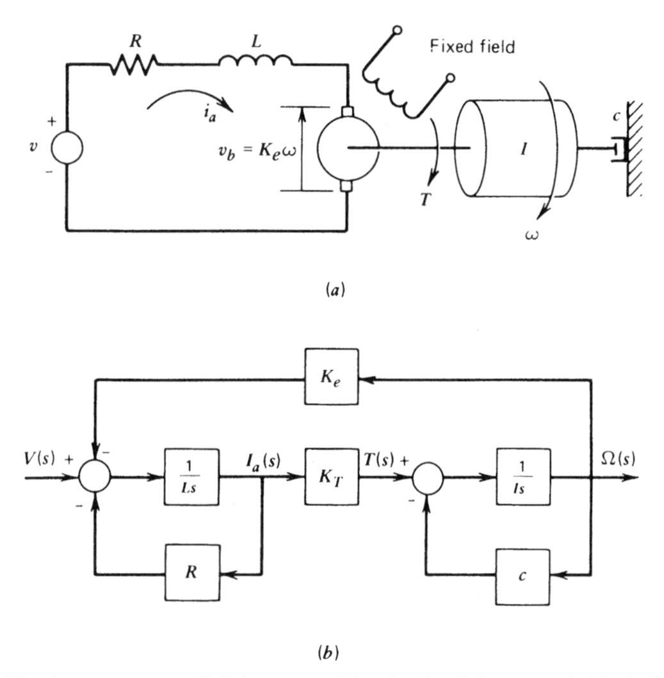

Figure 12 Armature-controlled dc motor with a load and the system's block diagram.[1]

an iron core. This winding has an inductance L. The resistance R represents the lumped value of the armature resistance and any external resistance deliberately introduced to change the motor's behavior. The armature is surrounded by a magnetic field. The reaction of this field with the armature current produces a torque that causes the armature to rotate. If the armature voltage v is used to control the motor, the motor is said to be *armature controlled.* In this case, the field is produced by an electromagnet supplied with a constant voltage or by a permanent magnet. This motor type produces a torque T that is proportional to the armature current i_a:

$$T = K_T i_a \tag{6}$$

The torque constant K_T depends on the strength of the field and other details of the motor's construction. The motion of a current-carrying conductor in a field produces a voltage in the conductor that opposes the current. This voltage is called the *back emf* (electromotive force). Its magnitude is proportional to the speed and is given by

$$e_b = K_e \omega \tag{7}$$

The transfer function for the armature-controlled dc motor is

$$\frac{\Omega(s)}{V(s)} = \frac{K_T}{LIs^2 + (RI + cL)s + cR + K_e K_T} \tag{8}$$

Another motor configuration is the *field-controlled* dc motor. In this case, the armature current is kept constant and the field voltage v is used to control the motor. The transfer function is

$$\frac{\Omega(s)}{V(s)} = \frac{K_T}{(Ls + R)(Is + c)} \tag{9}$$

where R and L are the resistance and inductance of the field circuit and K_T is the torque constant. No back emf exists in this motor to act as a self-braking mechanism.

Two-phase ac motors can be used to provide a low-power, variable-speed actuator. This motor type can accept the ac signals directly from LVDTs and synchros without demodulation. However, it is difficult to design ac amplifier circuitry to do other than proportional action. For this reason, the ac motor is not found in control systems as often as dc motors. The transfer function for this type is of the form of Eq. (9).

An actuator especially suitable for digital systems is the *stepper motor,* a special dc motor that takes a train of electrical input pulses and converts each pulse into an angular displacement of a fixed amount. Motors are available with resolutions ranging from about 4 steps per revolution to more than 800 steps per revolution. For 36 steps per revolution, the motor will rotate by 10° for each pulse received. When not being pulsed, the motors lock in place. Thus, they are excellent for precise positioning applications, such as required with printers and computer tape drives. A disadvantage is that they are low-torque devices. If the input pulse frequency is not near the resonant frequency of the motor, we can take the output rotation to be directly related to the number of input pulses and use that description as the motor model.

4.2 Hydraulic Actuators

Machine tools are one application of the hydraulic system shown in Fig. 13. The applied force f is supplied by the servomotor. The mass m represents that of a cutting tool and the power piston, while k represents the combined effects of the elasticity naturally present in the structure and that introduced by the designer to achieve proper performance. A similar statement applies to the damping c. The valve displacement z is generated by another control system in order to move the tool through its prescribed motion. The spool valve shown in

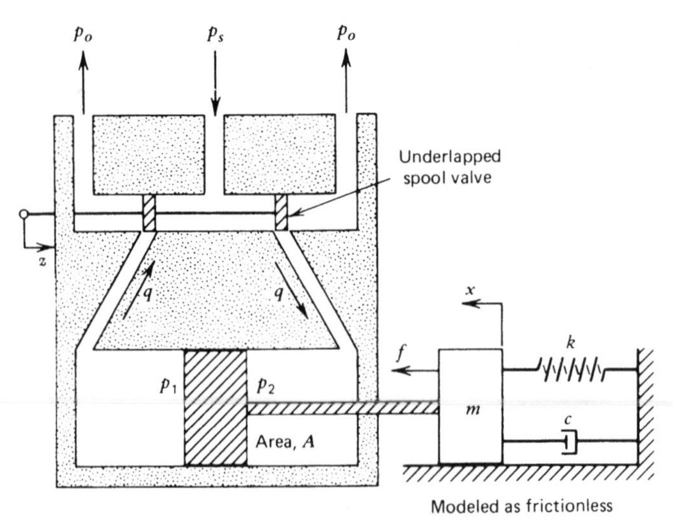

Figure 13 Hydraulic servomotor with a load.[1]

Fig. 13 had two *lands*. If the width of the land is greater than the port width, the valve is said to be *overlapped*. In this case, a dead zone exists in which a slight change in the displacement z produces no power piston motion. Such dead zones create control difficulties and are avoided by designing the valve to be *underlapped* (the land width is less the port width). For such valves there will be a small flow opening even when the valve is in the neutral position at $z = 0$. This gives it a higher sensitivity than an overlapped valve.

The variables z and $\Delta p = p_2 - p_1$ determine the volume flow rate, as

$$q = f(z, \Delta p)$$

For the reference equilibrium condition ($z = 0$, $\Delta p = 0$, $q = 0$), a linearization gives

$$q = C_1 z - C_2\, \Delta p \tag{10}$$

The linearization constants are available from theoretical and experimental results.[5] The transfer function for the system is[1,2]

$$T(s) = \frac{X(s)}{Z(s)} = \frac{C_1}{(C_2 m/A)\, s^2 + (cC_2/A + A)\, s + C_2 k/A} \tag{11}$$

The development of the steam engine led to the requirement for a speed control device to maintain constant speed in the presence of changes in load torque or steam pressure. In 1788, James Watt of Glasgow developed his now-famous flyball governor for this purpose (Fig. 14). Watt took the principle of sensing speed with the centrifugal pendulum of Thomas Mead and used it in a feedback loop on a steam engine. As the motor speed increases, the flyballs move outward and pull the slider upward. The upward motion of the slider closes the steam valve, thus causing the engine to slow down. If the engine speed is too slow, the spring force overcomes that due to the flyballs, and the slider moves down to open the steam valve. The desired speed can be set by moving the plate to change the compression in the spring. The principle of the flyball governor is still used for speed control applications. Typically, the pilot valve of a hydraulic servomotor is connected to the slider to provide the high forces required to move large supply valves.

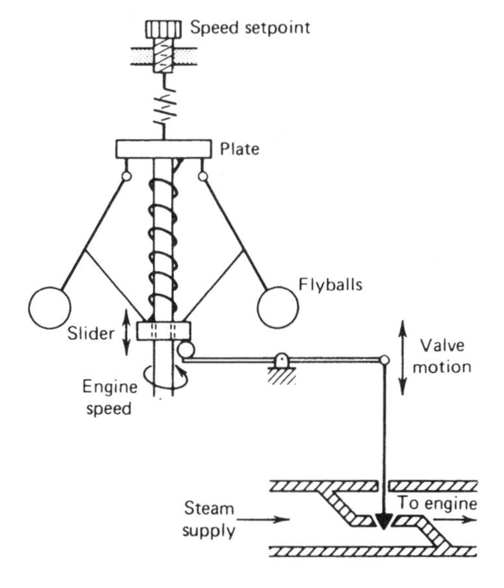

Figure 14 James Watt's flyball governor for speed control of a steam engine.[1]

Many hydraulic servomotors use multistage valves to obtain finer control and higher forces. A *two-stage valve* has a *slave valve,* similar to the pilot valve but situated between the pilot valve and the power piston.

Rotational motion can be obtained with a *hydraulic motor,* which is, in principle, a pump acting in reverse (fluid input and mechanical rotation output). Such motors can achieve higher torque levels than electric motors. A hydraulic pump driving a hydraulic motor constitutes a *hydraulic transmission.*

A popular actuator choice is the *electrohydraulic* system, which uses an electric actuator to control a hydraulic servomotor or transmission by moving the pilot valve or the swash-plate angle of the pump. Such systems combine the power of hydraulics with the advantages of electrical systems. Figure 15 shows a hydraulic motor whose pilot valve motion is caused by an armature-controlled dc motor. The transfer function between the motor voltage and the piston displacement is

$$\frac{X(s)}{V(s)} = \frac{K_1 K_2 C_1}{As^2(\tau s + 1)} \tag{12}$$

If the rotational inertia of the electric motor is small, then $\tau \approx 0$.

4.3 Pneumatic Actuators

Pneumatic actuators are commonly used because they are simple to maintain and use a readily available working medium. Compressed air supplies with the pressures required are commonly available in factories and laboratories. No flammable fluids or electrical sparks are present, so these devices are considered the safest to use with chemical processes. Their power output is less than that of hydraulic systems but greater than that of electric motors.

A device for converting pneumatic pressure into displacement is the bellows shown in Fig. 16. The transfer function for a linearized model of the bellows is of the form

$$\frac{X(s)}{P(s)} = \frac{K}{\tau s + 1} \tag{13}$$

where x and p are deviations of the bellows displacement and input pressure from nominal values.

In many control applications, a device is needed to convert small displacements into relatively large pressure changes. The nozzle–flapper serves this purpose (Fig. 17*a*). The input displacement y moves the flapper, with little effort required. This changes the opening at the nozzle orifice. For a large enough opening, the nozzle back pressure is approximately the same as atmospheric pressure p_a. At the other extreme position with the flapper completely blocking the orifice, the back pressure equals the supply pressure p_s. This variation is shown in Fig. 17*b*. Typical supply pressures are between 30 and 100 psia. The orifice

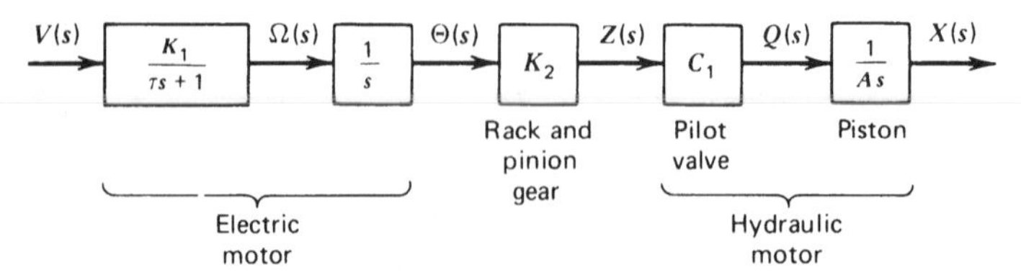

Figure 15 Electrohydraulic system for translation.[1]

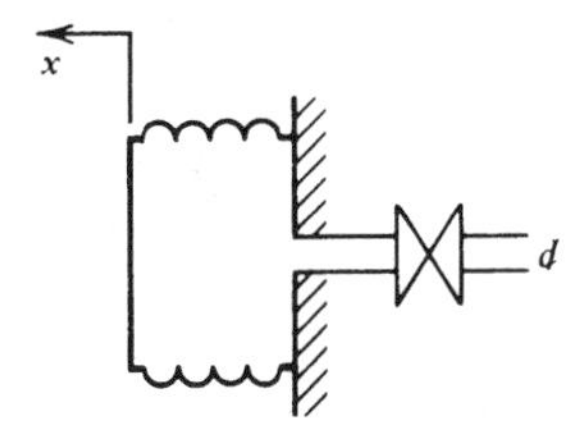

Figure 16 Pneumatic bellows.[1]

diameter is approximately 0.01 in. Flapper displacement is usually less than one orifice diameter.

The nozzle–flapper is operated in the linear portion of the back-pressure curve. The linearized back pressure relation is

$$p = -K_f x \tag{14}$$

where $-K_f$ is the slope of the curve and is a very large number. From the geometry of similar triangles, we have

$$p = -\frac{aK_f}{a + b} y \tag{15}$$

In its operating region, the nozzle–flapper's back pressure is well below the supply pressure.

The output pressure from a pneumatic device can be used to drive a final control element like the pneumatic actuating valve shown in Fig. 18. The pneumatic pressure acts on the upper side of the diaphragm and is opposed by the return spring.

Formerly, many control systems utilized pneumatic devices to implement the control law in analog form. Although the overall, or higher level, control algorithm is now usually implemented in digital form, pneumatic devices are still frequently used for final control corrections at the actuator level, where the control action must eventually be supplied by a mechanical device. An example of this is the electropneumatic valve positioner used in Valtek valves and illustrated in Fig. 19. The heart of the unit is a pilot valve capsule that moves up and down according to the pressure difference across its two supporting diaphragms. The capsule has a plunger at its top and at its bottom. Each plunger has an exhaust seat at one

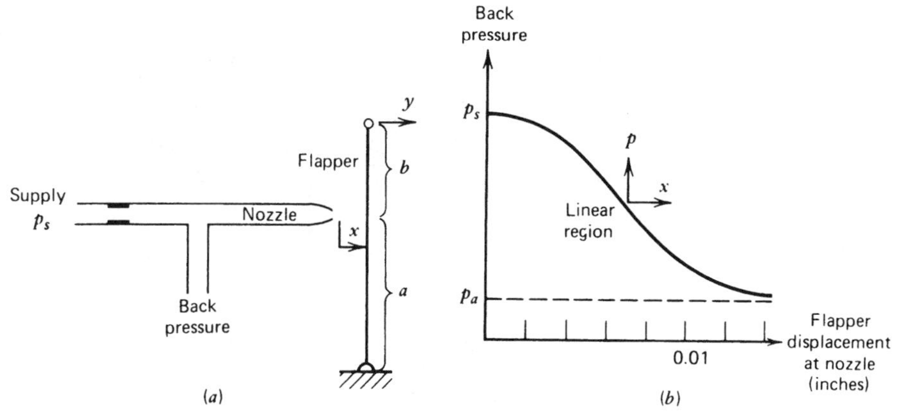

Figure 17 Pneumatic nozzle–flapper amplifier and its characteristic curve.[1]

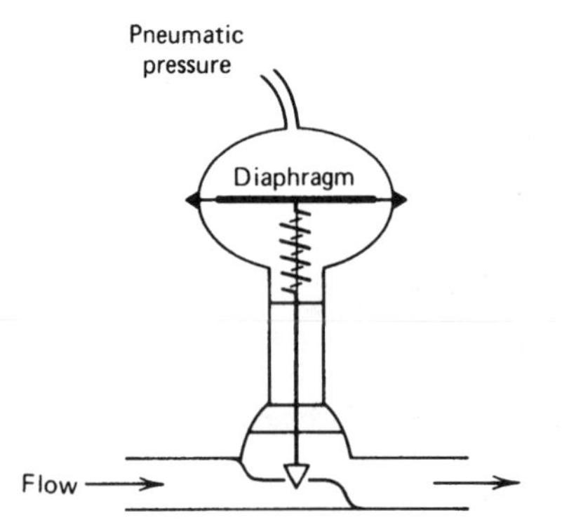

Figure 18 Pneumatic flow control valve.[1]

end and a supply seat at the other. When the capsule is in its equilibrium position, no air is supplied to or exhausted from the valve cylinder, so the valve does not move.

The process controller commands a change in the valve stem position by sending the 4–20-mA dc input signal to the positioner. Increasing this signal causes the electromagnetic actuator to rotate the lever counterclockwise about the pivot. This increases the air gap between the nozzle and flapper. This decreases the back pressure on top of the upper diaphragm and causes the capsule to move up. This motion lifts the upper plunger from its supply seat and allows the supply air to flow to the bottom of the valve cylinder. The lower plunger's exhaust seat is uncovered, thus decreasing the air pressure on top of the valve piston, and the valve stem moves upward. This motion causes the lever arm to rotate, increasing the tension in the feedback spring and decreasing the nozzle–flapper gap. The valve

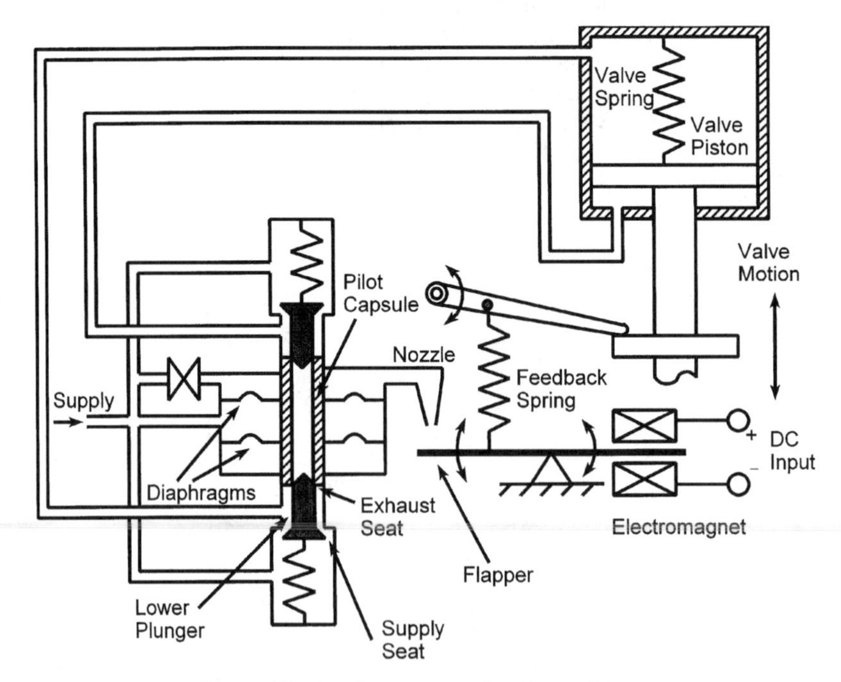

Figure 19 An electropneumatic valve positioner.

continues to move upward until the tension in the feedback spring counteracts the force produced by the electromagnetic actuator, thus returning the capsule to its equilibrium position.

A decrease in the dc input signal causes the opposite actions to occur, and the valve moves downward.

5 CONTROL LAWS

The control logic elements are designed to act on the error signal to produce the control signal. The algorithm that is used for this purpose is called the *control law,* the *control action,* or the *control algorithm.* A nonzero error signal results from either a change in command or a disturbance. The general function of the controller is to keep the controlled variable near its desired value when these occur. More specifically, the control objectives might be stated as follows:

1. Minimize the steady-state error.
2. Minimize the settling time.
3. Achieve other transient specifications, such as minimizing the overshoot.

In practice, the design specifications for a controller are more detailed. For example, the bandwidth might also be specified along with a safety margin for stability. We never know the numerical values of the system's parameters with true certainty, and some controller designs can be more sensitive to such parameter uncertainties than other designs. So a parameter sensitivity specification might also be included.

The following control laws form the basis of most control systems.

5.1 Proportional Control

Two-position control is the most familiar type, perhaps because of its use in home thermostats. The control output takes on one of two values. With the *on–off controller,* the controller output is either on or off (e.g., fully open or fully closed). Two-position control is acceptable for many applications in which the requirements are not too severe. However, many situations require finer control.

Consider a liquid-level system in which the input flow rate is controlled by a valve. We might try setting the control valve manually to achieve a flow rate that balances the system at the desired level. We might then add a controller that adjusts this setting in proportion to the deviation of the level from the desired value. This is *proportional control,* the algorithm in which the change in the control signal is proportional to the error. Block diagrams for controllers are often drawn in terms of the deviations from a zero-error equilibrium condition. Applying this convention to the general terminology of Fig. 6, we see that proportional control is described by

$$F(s) = K_P E(s)$$

where $F(s)$ is the deviation in the control signal and K_P is the *proportional gain.* If the total valve displacement is $y(t)$ and the manually created displacement is x, then

$$y(t) = K_P e(t) + x$$

The percent change in error needed to move the valve full scale is the *proportional band.* It is related to the gain as

$$K_P = \frac{100}{\text{band \%}}$$

The zero-error valve displacement x is the *manual reset.*

Proportional Control of a First-Order System

To investigate the behavior of proportional control, consider the speed control system shown in Fig. 20; it is identical to the position controller shown in Fig. 6, except that a tachometer replaces the feedback potentiometer. We can combine the amplifier gains into one, denoted K_P. The system is thus seen to have proportional control. We assume the motor is field controlled and has a negligible electrical time constant. The disturbance is a torque T_d, for example, resulting from friction. Choose the reference equilibrium condition to be $T_d = T = 0$ and $\omega_r = w = 0$. The block diagram is shown in Fig. 21. For a meaningful error signal to be generated, K_1 and K_2 should be chosen to be equal. With this simplification the diagram becomes that shown in Fig. 22, where $G(s) = K = K_1 K_P K_T / R$. A change in desired speed can be simulated by a unit step input for ω_r. For $\Omega_r(s) = 1/s$, the velocity approaches the steady-state value $\omega_{ss} = K/(c + K) < 1$. Thus, the final value is less than the desired value of 1, but it might be close enough if the damping c is small. The time required to reach this value is approximately four time constants, or $4\tau = 4I/(c + K)$. A sudden change in load torque can also be modeled by a unit step function $T_d(s) = 1/s$. The steady-state response due solely to the disturbance is $-1/(c + K)$. If $c + K$ is large, this error will be small.

The performance of the proportional control law thus far can be summarized as follows. For a first-order plant with step function inputs:

1. The output never reaches its desired value if damping is present ($c \neq 0$), although it can be made arbitrarily close by choosing the gain K large enough. This is called *offset error.*
2. The output approaches its final value without oscillation. The time to reach this value is inversely proportional to K.
3. The output deviation due to the disturbance at steady state is inversely proportional to the gain K. This error is present even in the absence of damping ($c = 0$).

As the gain K is increased, the time constant becomes smaller and the response faster. Thus, the chief disadvantage of proportional control is that it results in steady-state errors and can only be used when the gain can be selected large enough to reduce the effect of the largest expected disturbance. Since proportional control gives zero error only for one load condition (the reference equilibrium), the operator must change the manual reset by hand

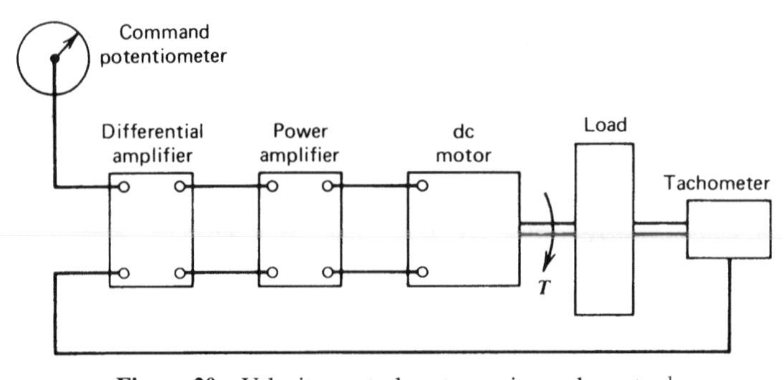

Figure 20 Velocity control system using a dc motor.[1]

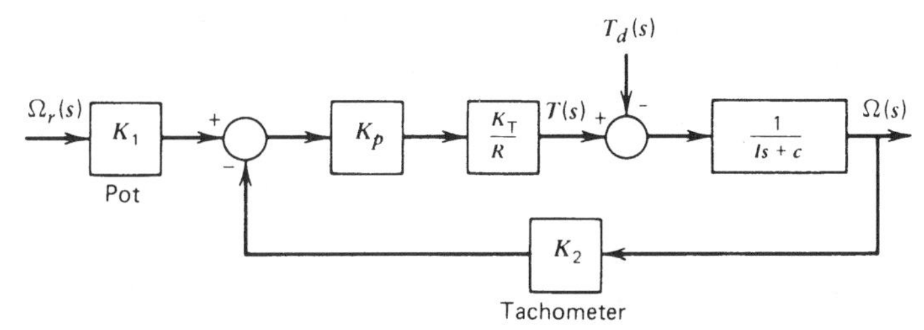

Figure 21 Block diagram of the velocity control system of Fig. 20.[1]

(hence the name). An advantage to proportional control is that the control signal responds to the error instantaneously (in theory at least). It is used in applications requiring rapid action. Processes with time constants too small for the use of two-position control are likely candidates for proportional control. The results of this analysis can be applied to any type of first-order system (e.g., liquid level, thermal, etc.) having the form in Fig. 22.

Proportional Control of a Second-Order System

Proportional control of a neutrally stable second-order plant is represented by the position controller of Fig. 6 if the amplifier transfer function is a constant $G_a(s) = K_a$. Let the motor transfer function be $G_m(s) = K_T/R$, as before. The modified block diagram is given in Fig. 23 with $G(s) = K = K_1 K_a K_T/R$. The closed-loop system is stable if I, c, and K are positive. For no damping ($c = 0$), the closed-loop system is neutrally stable. With no disturbance and a unit step command, $\Theta_r(s) = 1/s$, the steady-state output is $\omega_{ss} = 1$. The offset error is thus zero if the system is stable ($c > 0$, $K > 0$). The steady-state output deviation due to a unit step disturbance is $-1/K$. This deviation can be reduced by choosing K large. The transient behavior is indicated by the damping ratio, $\zeta = c/2\sqrt{IK}$.

For slight damping, the response to a step input will be very oscillatory and the overshoot large. The situation is aggravated if the gain K is made large to reduce the deviation due to the disturbance. We conclude, therefore, that proportional control of this type of second-order plant is not a good choice unless the damping constant c is large. We will see shortly how to improve the design.

5.2 Integral Control

The offset error that occurs with proportional control is a result of the system reaching an equilibrium in which the control signal no longer changes. This allows a constant error to exist. If the controller is modified to produce an increasing signal as long as the error is nonzero, the offset might be eliminated. This is the principle of *integral control.* In this mode

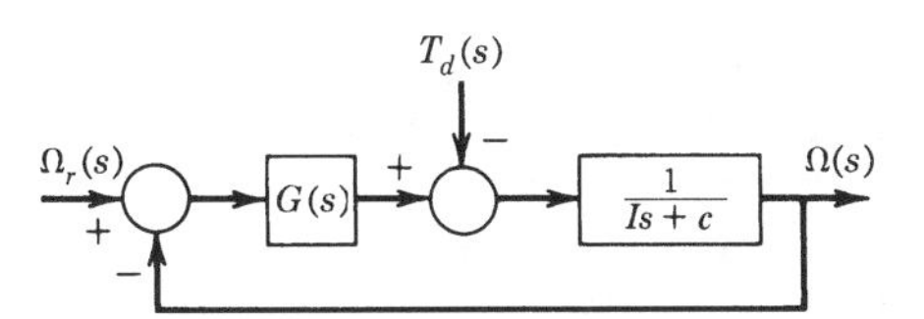

Figure 22 Simplified form of Fig. 21 for the case $K_1 = K_2$.

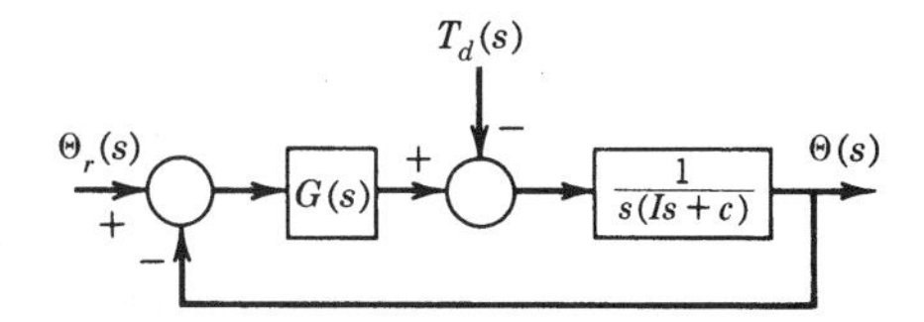

Figure 23 Position servo.

the change in the control signal is proportional to the *integral* of the error. In the terminology of Fig. 7, this gives

$$F(s) = \frac{K_I}{s} E(s) \tag{16}$$

where $F(s)$ is the deviation in the control signal and K_I is the *integral gain.* In the time domain, the relation is

$$f(t) = K_I \int_0^t e(t)\, dt \tag{17}$$

if $f(0) = 0$. In this form, it can be seen that the integration cannot continue indefinitely because it would theoretically produce an infinite value of $f(t)$ if $e(t)$ does not change sign. This implies that special care must be taken to reinitialize a controller that uses integral action.

Integral Control of a First-Order System

Integral control of the velocity in the system of Fig. 20 has the block diagram shown in Fig. 22, where $G(s) = K/s$, $K = K_1 K_I K_T/R$. The integrating action of the amplifier is physically obtained by the techniques to be presented in Section 6 or by the digital methods presented in Section 10. The control system is stable if I, c, and K are positive. For a unit step command input, $\omega_{ss} = 1$; so the offset error is zero. For a unit step disturbance, the steady-state deviation is zero if the system is stable. Thus, the steady-state performance using integral control is excellent for this plant with step inputs. The damping ratio is $\zeta = c/2\sqrt{IK}$. For slight damping, the response will be oscillatory rather than exponential as with proportional control. Improved steady-state performance has thus been obtained at the expense of degraded transient performance. The conflict between steady-state and transient specifications is a common theme in control system design. As long as the system is underdamped, the time constant is $\tau = 2I/c$ and is not affected by the gain K, which only influences the oscillation frequency in this case. It might by physically possible to make K small enough so that $\zeta >> 1$, and the nonoscillatory feature of proportional control recovered, but the response would tend to be sluggish. Transient specifications for fast response generally require that $\zeta < 1$. The difficulty with using $\zeta < 1$ is that τ is fixed by c and I. If c and I are such that $\zeta < 1$, then τ is large if $I >> c$.

Integral Control of a Second-Order System

Proportional control of the position servomechanism in Fig. 23 gives a nonzero steady-state deviation due to the disturbance. Integral control [$G(s) = K/s$] applied to this system results in the command transfer function

$$\frac{\Theta(s)}{\Theta_r(s)} = \frac{K}{Is^3 + cs^2 + K} \tag{18}$$

With the Routh criterion, we immediately see that the system is not stable because of the missing s term. Integral control is useful in improving steady-state performance, but in general it does not improve and may even degrade transient performance. Improperly applied, it can produce an unstable control system. It is best used in conjunction with other control modes.

5.3 Proportional-plus-Integral Control

Integral control raised the order of the system by 1 in the preceding examples but did not give a characteristic equation with enough flexibility to achieve acceptable transient behavior. The instantaneous response of proportional control action might introduce enough variability into the coefficients of the characteristic equation to allow both steady-state and transient specifications to be satisfied. This is the basis for using *proportional-plus-integral control* (PI control). The algorithm for this two-mode control is

$$F(s) = K_P E(s) + \frac{K_I}{s} E(s) \tag{19}$$

The integral action provides an automatic, not manual, reset of the controller in the presence of a disturbance. For this reason, it is often called *reset action.*

The algorithm is sometimes expressed as

$$F(s) = K_P \left(1 + \frac{1}{T_I s}\right) E(s) \tag{20}$$

where T_I is the *reset time.* The reset time is the time required for the integral action signal to equal that of the proportional term if a constant error exists (a hypothetical situation). The reciprocal of reset time is expressed as repeats per minute and is the frequency with which the integral action repeats the proportional correction signal.

The proportional control gain must be reduced when used with integral action. The integral term does not react instantaneously to a zero-error signal but continues to correct, which tends to cause oscillations if the designer does not take this effect into account.

PI Control of a First-Order System

PI action applied to the speed controller of Fig. 20 gives the diagram shown in Fig. 21 with $G(s) = K_P + K_I/s$. The gains K_P and K_I are related to the component gains, as before. The system is stable for positive values of K_P and K_I. For $\Omega_r(s) = 1/s$, $\omega_{ss} = 1$, and the offset error is zero, as with integral action only. Similarly, the deviation due to a unit step disturbance is zero at steady state. The damping ratio is $\zeta = (c + K_P)/2\sqrt{IK_I}$. The presence of K_P allows the damping ratio to be selected without fixing the value of the dominant time constant. For example, if the system is underdamped ($\zeta < 1$), the time constant is $\tau = 2I/(c + K_P)$. The gain K_P can be picked to obtain the desired time constant, while K_I is used to set the damping ratio. A similar flexibility exists if $\zeta = 1$. Complete description of the transient response requires that the numerator dynamics present in the transfer functions be accounted for.[1,2]

PI Control of a Second-Order System

Integral control for the position servomechanism of Fig. 23 resulted in a third-order system that is unstable. With proportional action, the diagram becomes that of Fig. 22, with $G(s) = K_P + K_I/s$. The steady-state performance is acceptable, as before, if the system is assumed to be stable. This is true if the Routh criterion is satisfied, that is, if I, c, K_P, and K_I are positive and $cK_P - IK_I > 0$. The difficulty here occurs when the damping is slight. For small c, the gain K_P must be large in order to satisfy the last condition, and this can be difficult to implement physically. Such a condition can also result in an unsatisfactory time constant. The root-locus method of Section 9 provides the tools for analyzing this design further.

5.4 Derivative Control

Integral action tends to produce a control signal even after the error has vanished, which suggests that the controller be made aware that the error is approaching zero. One way to accomplish this is to design the controller to react to the derivative of the error with *derivative control* action, which is

$$F(s) = K_D s E(s) \tag{21}$$

where K_D is the *derivative gain.* This algorithm is also called *rate action.* It is used to damp out oscillations. Since it depends only on the error rate, derivative control should never be used alone. When used with proportional action, the following PD control algorithm results:

$$F(s) = (K_P + K_D s)E(s) = K_P(1 + T_D s)E(s) \tag{22}$$

where T_D is the *rate time* or *derivative time.* With integral action included, the *proportional-plus-integral-plus-derivative (PID) control law* is obtained:

$$F(s) = \left(K_P + \frac{K_I}{s} + K_D s\right) E(s) \tag{23}$$

This is called a *three-mode controller.*

PD Control of a Second-Order System

The presence of integral action reduces steady-state error but tends to make the system less stable. There are applications of the position servomechanism in which a nonzero derivation resulting from the disturbance can be tolerated but an improvement in transient response over the proportional control result is desired. Integral action would not be required, but rate action can be added to improve the transient response. Application of PD control to this system gives the block diagram of Fig. 23 with $G(s) = K_P + K_D s$.

The system is stable for positive values of K_D and K_P. The presence of rate action does not affect the steady-state response, and the steady-state results are identical to those with proportional control; namely, zero offset error and a deviation of $-1/K_P$, due to the disturbance. The damping ratio is $\zeta = (c + K_D)/2\sqrt{IK_P}$. For proportional control, $\zeta = c/2\sqrt{IK_P}$. Introduction of rate action allows the proportional gain K_P to be selected large to reduce the steady-state deviation, while K_D can be used to achieve an acceptable damping ratio. The rate action also helps to stabilize the system by adding damping (if $c = 0$, the system with proportional control is not stable).

The equivalent of derivative action can be obtained by using a tachometer to measure the angular velocity of the load. The block diagram is shown in Fig. 24. The gain of the amplifier–motor–potentiometer combination is K_1, and K_2 is the tachometer gain. The advantage of this system is that it does not require signal differentiation, which is difficult to

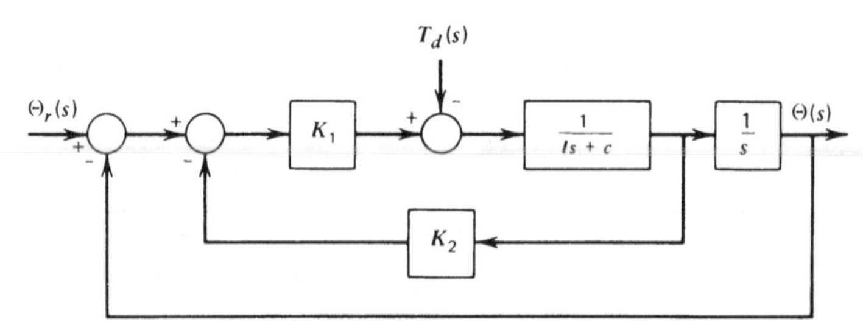

Figure 24 Tachometer feedback arrangement to replace PD control for the position servo.[1]

implement if signal noise is present. The gains K_1 and K_2 can be chosen to yield the desired damping ratio and steady-state deviation, as was done with K_P and K_I.

5.5 PID Control

The position servomechanism design with PI control is not completely satisfactory because of the difficulties encountered when the damping c is small. This problem can be solved by the use of the full PID control law, as shown in Fig. 23 with $G(s) = K_P + K_D s + K_I/s$.

A stable system results if all gains are positive and if $(c + K_D)K_P - IK_I > 0$. The presence of K_D relaxes somewhat the requirement that K_P be large to achieve stability. The steady-state errors are zero, and the transient response can be improved because three of the coefficients of the characteristic equation can be selected. To make further statements requires the root-locus technique presented in Section 9.

Proportional, integral, and derivative actions and their various combinations are not the only control laws possible, but they are the most common. PID controllers will remain for some time the standard against which any new designs must compete.

The conclusions reached concerning the performance of the various control laws are strictly true only for the plant model forms considered. These are the first-order model without numerator dynamics and the second-order model with a root at $s = 0$ and no numerator zeros. The analysis of a control law for any other linear system follows the preceding pattern. The overall system transfer functions are obtained, and all of the linear system analysis techniques can be applied to predict the system's performance. If the performance is unsatisfactory, a new control law is tried and the process repeated. When this process fails to achieve an acceptable design, more systematic methods of altering the system's structure are needed; they are discussed in later sections. We have used step functions as the test signals because they are the most common and perhaps represent the severest test of system performance. Impulse, ramp, and sinusoidal test signals are also employed. The type to use should be made clear in the design specifications.

6 CONTROLLER HARDWARE

The control law must be implemented by a physical device before the control engineer's task is complete. The earliest devices were purely kinematic and were mechanical elements such as gears, levers, and diaphragms that usually obtained their power from the controlled variable. Most controllers now are analog electronic, hydraulic, pneumatic, or digital electronic devices. We now consider the analog type. Digital controllers are covered starting in Section 10.

6.1 Feedback Compensation and Controller Design

Most controllers that implement versions of the PID algorithm are based on the following feedback principle. Consider the single-loop system shown in Fig. 1. If the open-loop transfer function is large enough that $|G(s)H(s)| >> 1$, the closed-loop transfer function is approximately given by

$$T(s) = \frac{G(s)}{1 + G(s)H(s)} \approx \frac{G(s)}{G(s)H(s)} = \frac{1}{H(s)} \tag{24}$$

The principle states that a power unit $G(s)$ can be used with a feedback element $H(s)$ to create a desired transfer function $T(s)$. The power unit must have a gain high enough that

$|G(s)H(s)| >> 1$, and the feedback elements must be selected so that $H(s) = 1/T(s)$. This principle was used in Section 1 to explain the design of a feedback amplifier.

6.2 Electronic Controllers

The *operational amplifier* (*op amp*) is a high-gain amplifier with a high input impedance. A diagram of an op amp with feedback and input elements with impedances $T_f(s)$ and $T_i(s)$ is shown in Fig. 25. An approximate relation is

$$\frac{E_o(s)}{E_i(s)} = -\frac{T_f(s)}{T_i(s)}$$

The various control modes can be obtained by proper selection of the impedances. A proportional controller can be constructed with a *multiplier,* which uses two resistors, as shown in Fig. 26. An *inverter* is a multiplier circuit with $R_f = R_i$. It is sometimes needed because of the sign reversal property of the op amp. The multiplier circuit can be modified to act as an adder (Fig. 27).

PI control can be implemented with the circuit of Fig. 28. Figure 29 shows a complete system using op amps for PI control. The inverter is needed to create an error detector. Many industrial controllers provide the operator with a choice of control modes, and the operator can switch from one mode to another when the process characteristics or control objectives change. When a switch occurs, it is necessary to provide any integrators with the proper initial voltages or else undesirable transients will occur when the integrator is switched into the system. Commercially available controllers usually have built-in circuits for this purpose.

In theory, a differentiator can be created by interchanging the resistance and capacitance in the integrating op amp. The difficulty with this design is that no electrical signal is "pure." Contamination always exists as a result of voltage spikes, ripple, and other transients generally categorized as "noise." These high-frequency signals have large slopes compared with the more slowly varying primary signal, and thus they will dominate the output of the differentiator. In practice, this problem is solved by filtering out high-frequency signals, either with a low-pass filter inserted in cascade with the differentiator or by using a redesigned differentiator such as the one shown in Fig. 30. For the ideal PD controller, $R_1 = 0$. The attenuation curve for the ideal controller breaks upward at $\omega = 1/R_2C$ with a slope of 20 dB/decade. The curve for the practical controller does the same but then becomes flat for $\omega > (R_1 + R_2)/R_1R_2C$. This provides the required limiting effect at high frequencies.

PID control can be implemented by joining the PI and PD controllers in parallel, but this is expensive because of the number of op amps and power supplies required. Instead, the usual implementation is that shown in Fig. 31. The circuit limits the effect of frequencies above $\omega = 1/\beta R_1C_1$. When $R_1 = 0$, ideal PID control results. This is sometimes called the *noninteractive* algorithm because the effect of each of the three modes is additive, and they do not interfere with one another. The form given for $R_1 \neq 0$ is the *real* or *interactive*

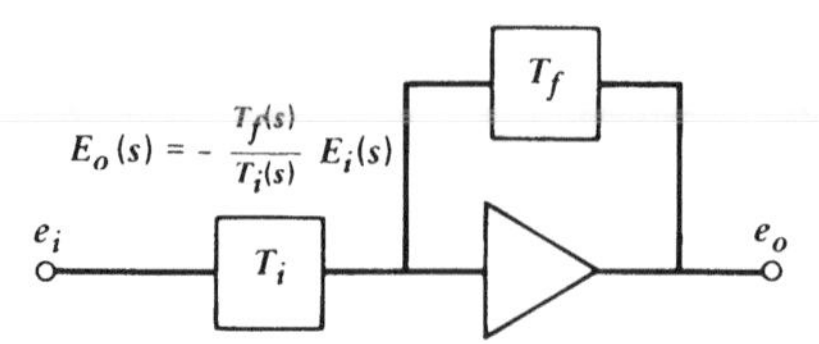

Figure 25 Operational amplifier (op amp).[1]

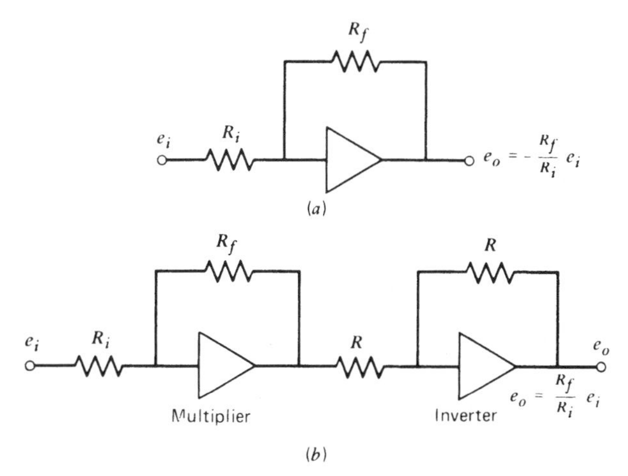

Figure 26 Op-amp implementation of proportional control.[1]

algorithm. This name results from the fact that historically it was difficult to implement noninteractive PID control with mechanical or pneumatic devices.

6.3 Pneumatic Controllers

The nozzle–flapper introduced in Section 4 is a high-gain device that is difficult to use without modification. The gain K_f is known only imprecisely and is sensitive to changes induced by temperature and other environmental factors. Also, the linear region over which Eq. (14) applies is very small. However, the device can be made useful by compensating it with feedback elements, as was illustrated with the electropneumatic valve positioner shown in Fig. 19.

6.4 Hydraulic Controllers

The basic unit for synthesis of hydraulic controllers is the hydraulic servomotor. The nozzle–flapper concept is also used in hydraulic controllers.[5] A PI controller is shown in Fig. 32. It can be modified for proportional action. Derivative action has not seen much use in hydraulic controllers. This action supplies damping to the system, but hydraulic systems are usually

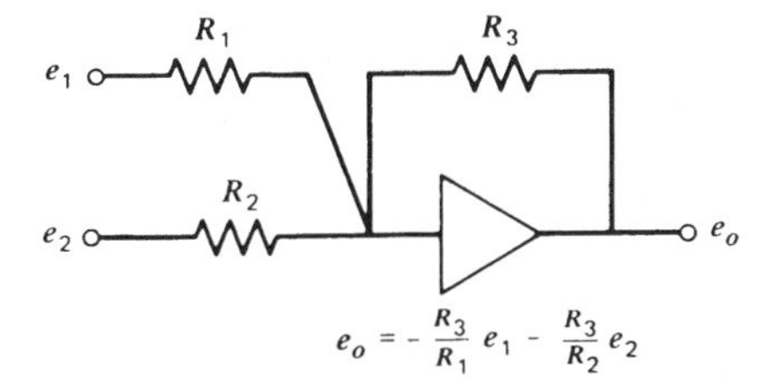

Figure 27 Op-amp adder circuit.[1]

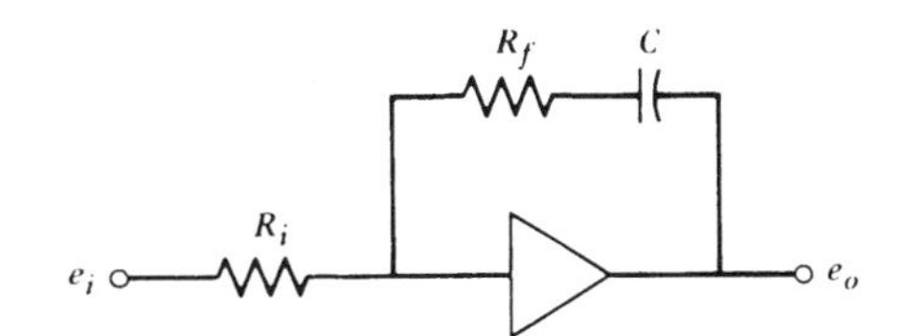

Figure 28 Op-amp implementation of PI control.[1]

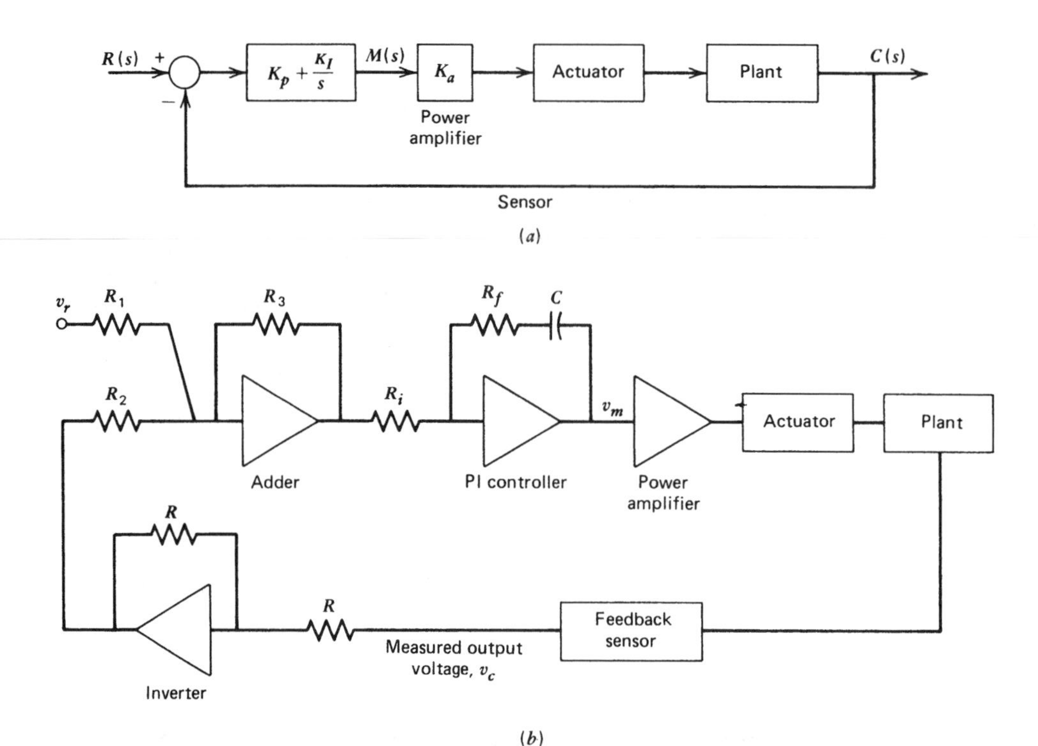

Figure 29 Implementation of a PI controller using op amps. (*a*) Diagram of the system. (*b*) Diagram showing how the op amps are connected.[2]

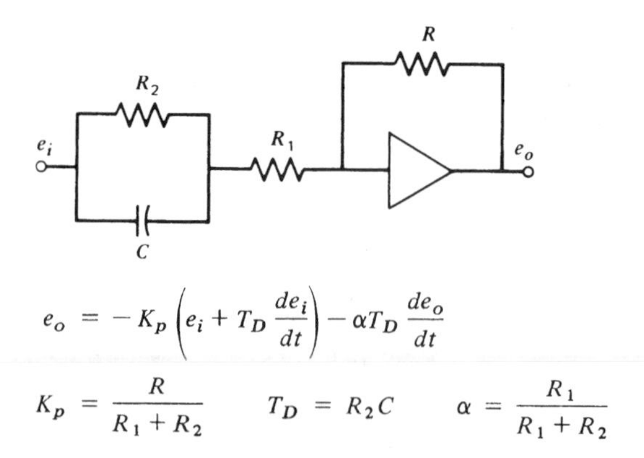

$$e_o = -K_p\left(e_i + T_D\frac{de_i}{dt}\right) - \alpha T_D\frac{de_o}{dt}$$

$$K_p = \frac{R}{R_1 + R_2} \qquad T_D = R_2 C \qquad \alpha = \frac{R_1}{R_1 + R_2}$$

Figure 30 Practical op-amp implementation of PD control.[1]

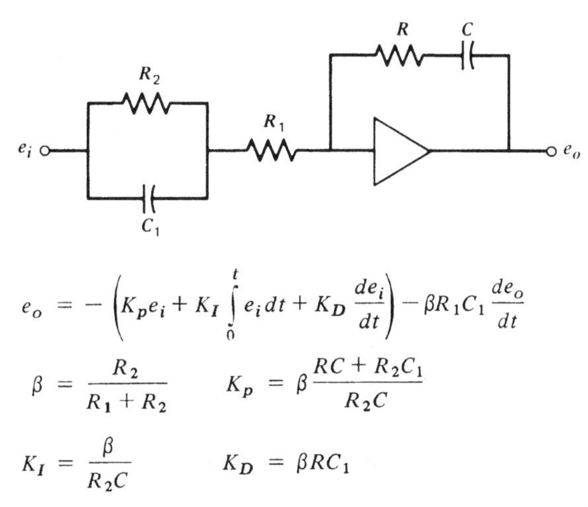

$$e_o = -\left(K_p e_i + K_I \int_0^t e_i\,dt + K_D \frac{de_i}{dt}\right) - \beta R_1 C_1 \frac{de_o}{dt}$$

$$\beta = \frac{R_2}{R_1 + R_2} \qquad K_p = \beta \frac{RC + R_2 C_1}{R_2 C}$$

$$K_I = \frac{\beta}{R_2 C} \qquad K_D = \beta R C_1$$

Figure 31 Practical op-amp implementation of PID control.[1]

highly damped intrinsically because of the viscous working fluid. PI control is the algorithm most commonly implemented with hydraulics.

7 FURTHER CRITERIA FOR GAIN SELECTION

Once the form of the control law has been selected, the gains must be computed in light of the performance specifications. In the examples of the PID family of control laws in Section 5, the damping ratio, dominant time constant, and steady-state error were taken to be the primary indicators of system performance in the interest of simplicity. In practice, the criteria are usually more detailed. For example, the rise time and maximum overshoot, as well as the other transient response specifications of the previous chapter, may be encountered. Requirements can also be stated in terms of frequency response characteristics, such as bandwidth, resonant frequency, and peak amplitude. Whatever specific form they take, a complete

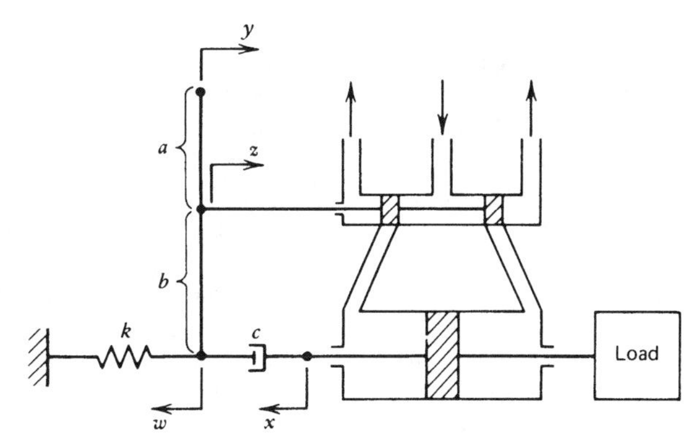

Figure 32 Hydraulic implementation of PI control.[1]

set of specifications for control system performance generally should include the following considerations for given forms of the command and disturbance inputs:

1. **Equilibrium specifications**
   - **(a)** Stability
   - **(b)** Steady-state error
2. **Transient specifications**
   - **(a)** Speed of response
   - **(b)** Form of response
3. **Sensitivity specifications**
   - **(a)** Sensitivity to parameter variations
   - **(b)** Sensitivity to model inaccuracies
   - **(c)** Noise rejection (bandwidth, etc.)

In addition to these performance stipulations, the usual engineering considerations of initial cost, weight, maintainability, and so on must be taken into account. The considerations are highly specific to the chosen hardware, and it is difficult to deal with such issues in a general way.

Two approaches exist for designing the controller. The proper one depends on the quality of the analytical description of the plant to be controlled. If an accurate model of the plant is easily developed, we can design a specialized controller for the particular application. The range of adjustment of controller gains in this case can usually be made small because the accurate plant model allows the gains to be precomputed with confidence. This technique reduces the cost of the controller and can often be applied to electromechanical systems.

The second approach is used when the plant is relatively difficult to model, which is often the case in process control. A standard controller with several control modes and wide ranges of gains is used, and the proper mode and gain settings are obtained by testing the controller on the process in the field. This approach should be considered when the cost of developing an accurate plant model might exceed the cost of controller tuning in the field. Of course, the plant must be available for testing for this approach to be feasible.

7.1 Performance Indices

The performance criteria encountered thus far require a set of conditions to be specified—for example, one for steady-state error, one for damping ratio, and one for the dominant time constant. If there are many such conditions, and if the system is of high order with several gains to be selected, the design process can get quite complicated because transient and steady-state criteria tend to drive the design in different directions. An alternative approach is to specify the system's desired performance by means of one analytical expression called a *performance index*. Powerful analytical and numerical methods are available that allow the gains to be systematically computed by minimizing (or maximizing) this index.

To be useful, a performance index must be selective. The index must have a sharply defined extremum in the vicinity of the gain values that give the desired performance. If the numerical value of the index does not change very much for large changes in the gains from their optimal values, the index will not be selective.

Any practical choice of a performance index must be easily computed, either analytically, numerically, or experimentally. Four common choices for an index are the following:

$$J = \int_0^\infty |e(t)|\, dt \qquad \text{(IAE Index)} \tag{25}$$

$$J = \int_0^\infty t|e(t)|\, dt \qquad \text{(ITAE Index)} \tag{26}$$

$$J = \int_0^\infty [e(t)]^2\, dt \qquad \text{(ISE Index)} \tag{27}$$

$$J = \int_0^\infty t[e(t)]^2\, dt \qquad \text{(ITSE Index)} \tag{28}$$

where $e(t)$ is the system error. This error usually is the difference between the desired and the actual values of the output. However, if $e(t)$ does not approach zero as $t \to \infty$, the preceding indices will not have finite values. In this case, $e(t)$ can be defined as $e(t) = c(\infty) - c(t)$, where $c(t)$ is the output variable. If the index is to be computed numerically or experimentally, the infinite upper limit can be replaced by a time t_f large enough that $e(t)$ is negligible for $t > t_f$.

The *integral absolute-error* (IAE) criterion (25) expresses mathematically that the designer is not concerned with the sign of the error, only its magnitude. In some applications, the IAE criterion describes the fuel consumption of the system. The index says nothing about the relative importance of an error occurring late in the response versus an error occurring early. Because of this, the index is not as selective as the *integral-of-time-multiplied absolute-error* (ITAE) criterion (26). Since the multiplier t is small in the early stages of the response, this index weights early errors less heavily than later errors. This makes sense physically. No system can respond instantaneously, and the index is lenient accordingly, while penalizing any design that allows a nonzero error to remain for a long time. Neither criterion allows highly underdamped or highly overdamped systems to be optimum. The ITAE criterion usually results in a system whose step response has a slight overshoot and well-damped oscillations.

The *integral squared-error* (ISE) and *integral-of-time-multiplied squared-error* (ITSE) criteria are analogous to the IAE and ITAE criteria, except that the square of the error is employed for three reasons: (1) in some applications, the squared error represents the system's power consumption; (2) squaring the error weights large errors much more heavily than small errors; (3) the squared error is much easier to handle analytically. The derivative of a squared term is easier to compute than that of an absolute value and does not have a discontinuity at $e = 0$. These differences are important when the system is of high order with multiple error terms.

The closed-form solution for the response is not required to evaluate a performance index. For a given set of parameter values, the response and the resulting index value can be computed numerically. The optimum solution can be obtained using systematic computer search procedures; this makes this approach suitable for use with nonlinear systems.

7.2 Optimal-Control Methods

Optimal-control theory includes a number of algorithms for systematic design of a control law to minimize a performance index, such as the following generalization of the ISE index, called the *quadratic* index:

$$J = \int_0^\infty (\mathbf{x}^T\mathbf{Q}\mathbf{x} + \mathbf{u}^T\mathbf{R}\mathbf{u})\, dt \tag{29}$$

where **x** and **u** are the deviations of the state and control vectors from the desired reference values. For example, in a servomechanism, the state vector might consist of the position and velocity, and the control vector might be a scalar—the force or torque produced by the actuator. The matrices **Q** and **R** are chosen by the designer to provide relative weighting for the elements of **x** and **u**. If the plant can be described by the linear state-variable model

$$\dot{\mathbf{x}} = \mathbf{A}\mathbf{x} + \mathbf{B}\mathbf{u} \tag{30}$$

$$\mathbf{y} = \mathbf{C}\mathbf{x} + \mathbf{D}\mathbf{u} \tag{31}$$

where **y** is the vector of outputs—for example, position and velocity—then the solution of this *linear-quadratic* control problem is the linear control law:

$$\mathbf{u} = \mathbf{K}\mathbf{y} \tag{32}$$

where **K** is a matrix of gains that can be found by several algorithms.[1,6,7] A valid solution is guaranteed to yield a stable closed-loop system, a major benefit of this method.

Even if it is possible to formulate the control problem in this way, several practical difficulties arise. Some of the terms in (29) might be beyond the influence of the control vector **u**; the system is then *uncontrollable.* Also, there might not be enough information in the output equation (31) to achieve control, and the system is then *unobservable.* Several tests are available to check controllability and observability. Not all of the necessary state variables might be available for feedback or the feedback measurements might be noisy or biased. Algorithms known as *observers, state reconstructors, estimators,* and *digital filters* are available to compensate for the missing information. Another source of error is the uncertainty in the values of the coefficient matrices **A**, **B**, **C**, and **D**. Identification schemes can be used to compare the predicted and the actual system performance and to adjust the coefficient values "on-line."

7.3 The Ziegler–Nichols Rules

The difficulty of obtaining accurate transfer function models for some processes has led to the development of empirically based rules of thumb for computing the optimum gain values for a controller. Commonly used guidelines are the *Ziegler–Nichols rules,* which have proved so helpful that they are still in use 50 years after their development. The rules actually consist of two separate methods. The first method requires the open-loop step response of the plant, while the second uses the results of experiments performed with the controller already installed. While primarily intended for use with systems for which no analytical model is available, the rules are also helpful even when a model can be developed.

Ziegler and Nichols developed their rules from experiments and analysis of various industrial processes. Using the IAE criterion with a unit step response, they found that controllers adjusted according to the following rules usually had a step response that was oscillatory but with enough damping so that the second overshoot was less than 25% of the first (peak) overshoot. This is the *quarter-decay* criterion and is sometimes used as a specification.

The first method is the *process reaction* method and relies on the fact that many processes have an open-loop step response like that shown in Fig. 33. This is the *process signature* and is characterized by two parameters, R and L, where R is the slope of a line tangent to the steepest part of the response curve and L is the time at which this line intersects the time axis. First- and second-order linear systems do not yield positive values for L, and so the method cannot be applied to such systems. However, third- and higher order linear systems with sufficient damping do yield such a response. If so, the Ziegler–Nichols rules recommend the controller settings given in Table 2.

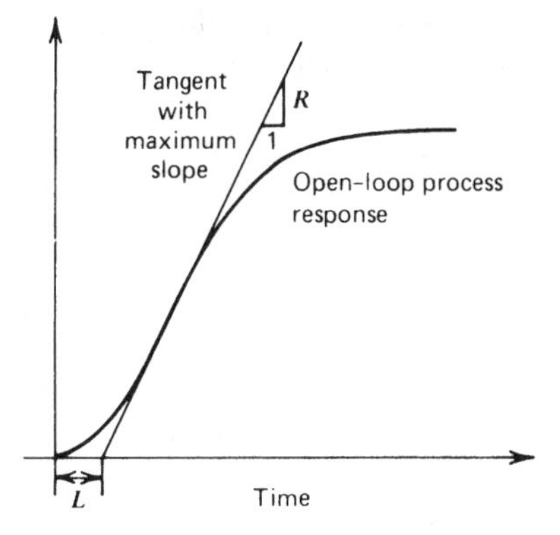

Figure 33 Process signature for a unit step input.[1]

The *ultimate-cycle* method uses experiments with the controller in place. All control modes except proportional are turned off, and the process is started with the proportional gain K_P set at a low value. The gain is slowly increased until the process begins to exhibit sustained oscillations. Denote the period of this oscillation by P_u and the corresponding *ultimate gain* by K_{Pu}. The Ziegler–Nichols recommendations are given in Table 2 in terms of these parameters. The proportional gain is lower for PI control than for proportional control and is higher for PID control because integral action increases the order of the system and thus tends to destabilize it; thus, a lower gain is needed. On the other hand, derivative action tends to stabilize the system; hence, the proportional gain can be increased without degrading the stability characteristics. Because the rules were developed for a typical case out of many types of processes, final tuning of the gains in the field is usually necessary.

7.4 Nonlinearities and Controller Performance

All physical systems have nonlinear characteristics of some sort, although they can often be modeled as linear systems provided the deviations from the linearization reference condition are not too great. Under certain conditions, however, the nonlinearities have significant effects

Table 2 The Ziegler–Nichols Rules

Controller transfer function $G(s) = K_p\left(1 + \frac{1}{T_I s} + T_D s\right)$

Control Mode	Process Reaction Method	Ultimate-Cycle Method
P control	$K_p = \frac{1}{RL}$	$K_p = 0.5K_{pu}$
PI control	$K_p = \frac{0.9}{RL}$	$K_p = 0.45K_{pu}$
	$T_I = 3.3L$	$T_I = 0.83P_u$
PID control	$K_p = \frac{1.2}{RL}$	$K_p = 0.6K_{pu}$
	$T_I = 2L$ $T_D = 0.5L$	$T_I = 0.5P_u$ $T_D = 0.125P_u$

on the system's performance. One such situation can occur during the start-up of a controller if the initial conditions are much different from the reference condition for linearization. The linearized model is then not accurate, and nonlinearities govern the behavior. If the nonlinearities are mild, there might not be much of a problem. Where the nonlinearities are severe, such as in process control, special consideration must be given to start-up. Usually, in such cases, the control signal sent to the final control elements is manually adjusted until the system variables are within the linear range of the controller. Then the system is switched into automatic mode. Digital computers are often used to replace the manual adjustment process because they can be readily coded to produce complicated functions for the start-up signals. Care must also be taken when switching from manual to automatic. For example, the integrators in electronic controllers must be provided with the proper initial conditions.

7.5 Reset Windup

In practice, all actuators and final control elements have a limited operating range. For example, a motor–amplifier combination can produce a torque proportional to the input voltage over only a limited range. No amplifier can supply an infinite current; there is a maximum current and thus a maximum torque that the system can produce. The final control elements are said to be *overdriven* when they are commanded by the controller to do something they cannot do. Since the limitations of the final control elements are ultimately due to the limited rate at which they can supply energy, it is important that all system performance specifications and controller designs be consistent with the energy delivery capabilities of the elements to be used.

Controllers using integral action can exhibit the phenomenon called *reset windup* or *integrator buildup* when overdriven, if they are not properly designed. For a step change in set point, the proportional term responds instantly and saturates immediately if the set-point change is large enough. On the other hand, the integral term does not respond as fast. It integrates the error signal and saturates some time later if the error remains large for a long enough time. As the error decreases, the proportional term no longer causes saturation. However, the integral term continues to increase as long as the error has not changed sign, and thus the manipulated variable remains saturated. Even though the output is very near its desired value, the manipulated variable remains saturated until after the error has reversed sign. The result can be an undesirable overshoot in the response of the controlled variable.

Limits on the controller prevent the voltages from exceeding the value required to saturate the actuator and thus protect the actuator, but they do not prevent the integral buildup that causes the overshoot. One way to prevent integrator buildup is to select the gains so that saturation will never occur. This requires knowledge of the maximum input magnitude that the system will encounter. General algorithms for doing this are not available; some methods for low-order systems are presented in Ref. 1, Chapter 7; Ref. 2, Chapter 7, and Ref. 4, Chapter 11. Integrator buildup is easier to prevent when using digital control; this is discussed in Section 10.

8 COMPENSATION AND ALTERNATIVE CONTROL STRUCTURES

A common design technique is to insert a *compensator* into the system when the PID control algorithm can be made to satisfy most but not all of the design specifications. A compensator is a device that alters the response of the controller so that the overall system will have satisfactory performance. The three categories of compensation techniques generally recognized are *series compensation, parallel* (or *feedback*) *compensation,* and *feedforward compensation.* The three structures are loosely illustrated in Fig. 34, where we assume the final control elements have a unity transfer function. The transfer function of the controller is $G_1(s)$. The feedback elements are represented by $H(s)$, and the compensator by $G_c(s)$. We assume that the plant is unalterable, as is usually the case in control system design. The choice of compensation structure depends on what type of specifications must be satisfied. The physical devices used as compensators are similar to the pneumatic, hydraulic, and electrical devices treated previously. Compensators can be implemented in software for digital control applications.

8.1 Series Compensation

The most commonly used series compensators are the *lead,* the *lag,* and the *lead–lag* compensators. Electrical implementations of these are shown in Fig. 35. Other physical implementations are available. Generally, the lead compensator improves the speed of response; the lag compensator decreases the steady-state error; and the lead–lag affects both. Graphical aids, such as the root-locus and frequency response plots, are usually needed to design these compensators (Ref. 1, Chapter 8; Ref. 2, Chapter 9; and Ref. 4, Chapter 11).

8.2 Feedback Compensation and Cascade Control

The use of a tachometer to obtain velocity feedback, as in Fig. 24, is a case of feedback compensation. The feedback compensation principle of Fig. 3 is another. Another form is *cascade control,* in which another controller is inserted within the loop of the original control system (Fig. 36). The new controller can be used to achieve better control of variables within the forward path of the system. Its set point is manipulated by the first controller.

Cascade control is frequently used when the plant cannot be satisfactorily approximated with a model of second order or lower. This is because the difficulty of analysis and control increases rapidly with system order. The characteristic roots of a second-order system can easily be expressed in analytical form. This is not so for third order or higher, and few general design rules are available. When faced with the problem of controlling a high-order system, the designer should first see if the performance requirements can be relaxed so that the system can be approximated with a low-order model. If this is not possible, the designer should attempt to divide the plant into subsystems, each of which is second order or lower. A controller is then designed for each subsystem. An application using cascade control is given in Section 11.

8.3 Feedforward Compensation

The control algorithms considered thus far have counteracted disturbances by using measurements of the output. One difficulty with this approach is that the effects of the disturbance must show up in the output of the plant before the controller can begin to take action. On the other hand, if we can measure the disturbance, the response of the controller can be improved by using the measurement to augment the control signal sent from the controller to the final control elements. This is the essence of feedforward compensation of the disturbance, as shown in Fig. 34*c*.

Feedforward compensation modified the output of the main controller. Instead of doing this by measuring the disturbance, another form of feedforward compensation utilizes the

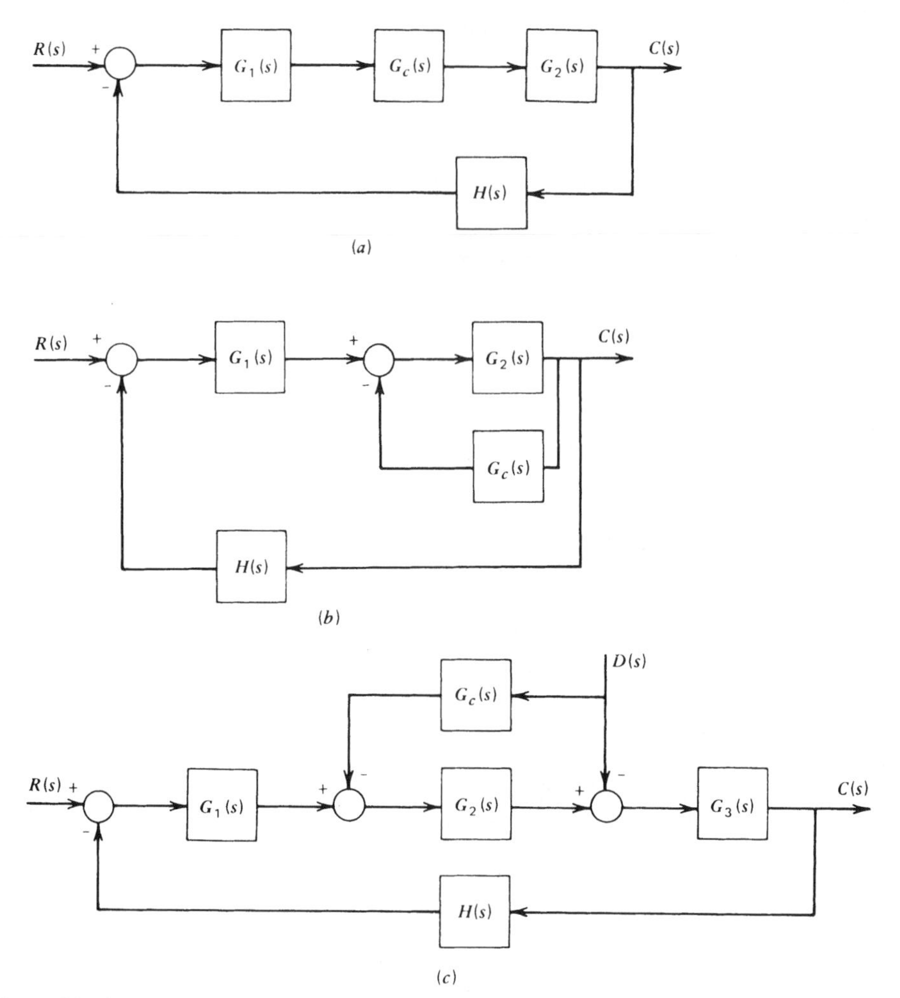

Figure 34 General structures of the three compensation types: (*a*) series; (*b*) parallel (or feedback); (*c*) feedforward. The compensator transfer function is $G_c(s)$.[1]

command input. Figure 37 is an example of this approach. The closed-loop transfer function is

$$\frac{\Omega(s)}{\Omega_r(s)} = \frac{K_f + K}{Is + c + K} \tag{33}$$

For a unit step input, the steady-state output is $\omega_{ss} = (K_f + K)/(c + K)$. Thus, if we choose the feedforward gain K_f to be $K_f = c$, then $\omega_{ss} = 1$ as desired, and the error is zero. Note that this form of feedforward compensation does not affect the disturbance response. Its effectiveness depends on how accurately we know the value of c. A digital application of feedforward compensation is presented in Section 11.

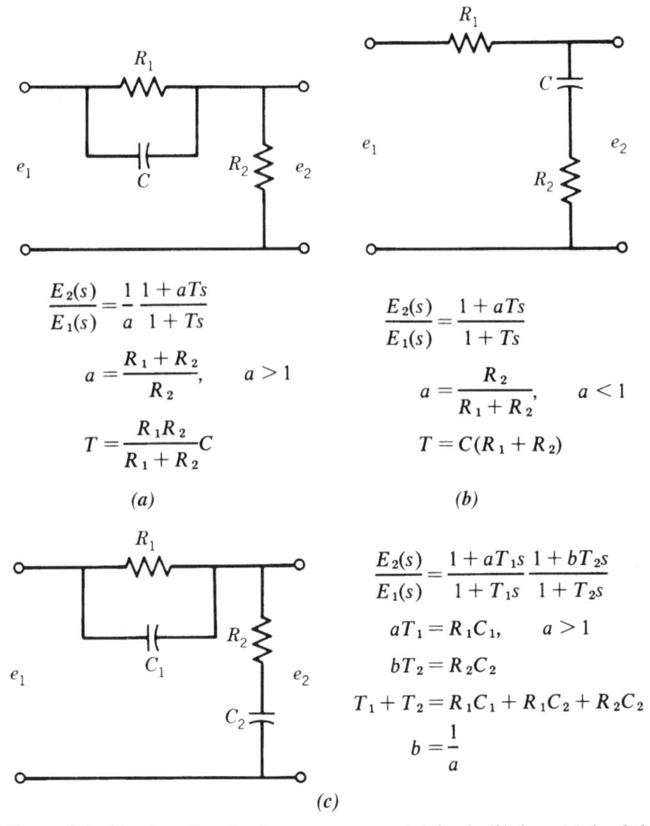

Figure 35 Passive electrical compensators: (*a*) lead; (*b*) lag; (*c*) lead–lag.

8.4 State-Variable Feedback

There are techniques for improving system performance that do not fall entirely into one of the three compensation categories considered previously. In some forms these techniques can be viewed as a type of feedback compensation, while in other forms they constitute a modification of the control law. *State-variable feedback* (SVFB) is a technique that uses information about all the system's state variables to modify either the control signal or the

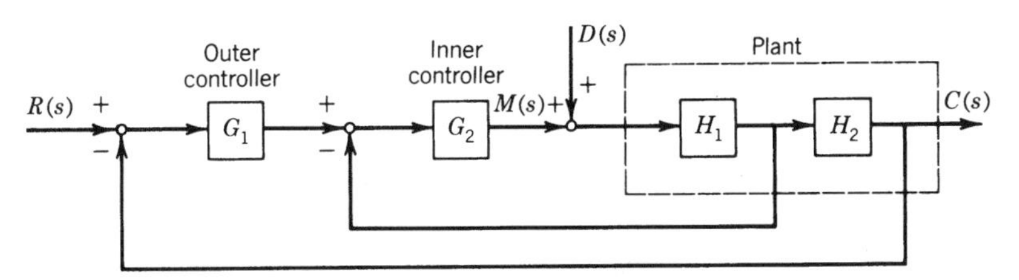

Figure 36 Cascade control structure.

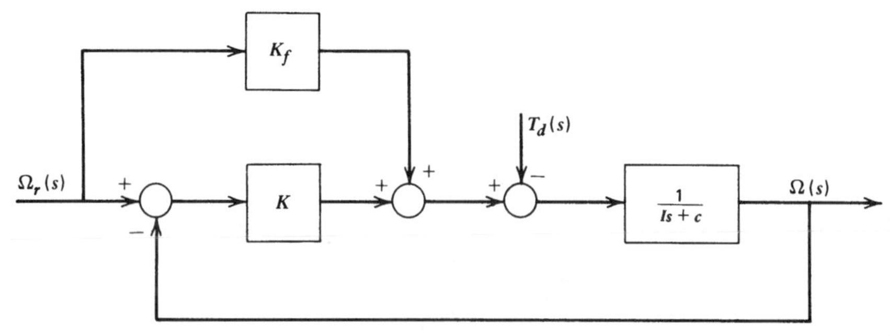

Figure 37 Feedforward compensation of the command input to augment proportional control.[2]

actuating signal. These two forms are illustrated in Fig. 38. Both forms require that the state vector $\boldsymbol{x}$ be measurable or at least derivable from other information. Devices or algorithms used to obtain state-variable information other than directly from measurements are variously termed *state reconstructors, estimators, observers,* or *filters* in the literature.

8.5 Pseudoderivative Feedback

Pseudoderivative feedback (PDF) is an extension of the velocity feedback compensation concept of Fig. 24.[1,2] It uses integral action in the forward path plus an internal feedback loop whose operator $H(s)$ depends on the plant (Fig. 39). For $G(s) = 1/(Is + c)$, $H(s) = K_1$. For $G(s) = 1/Is^2$, $H(s) = K_1 + K_2 s$. The primary advantage of PDF is that it does not need derivative action in the forward path to achieve the desired stability and damping characteristics.

9 GRAPHICAL DESIGN METHODS

Higher order models commonly arise in control systems design. For example, integral action is often used with a second-order plant, and this produces a third-order system to be designed. Although algebraic solutions are available for third- and fourth-order polynomials, these solutions are cumbersome for design purposes. Fortunately, there exist graphical techniques to aid the designer. Frequency response plots of both the open- and closed-loop transfer

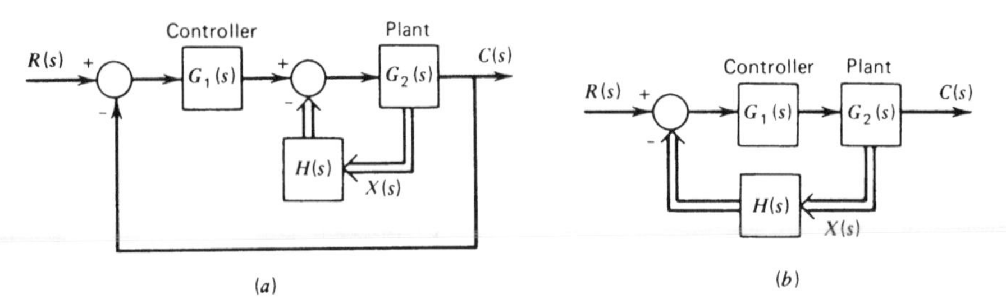

Figure 38 Two forms of state-variable feedback: (*a*) internal compensation of the control signal; (*b*) modification of the actuating signal.[1]

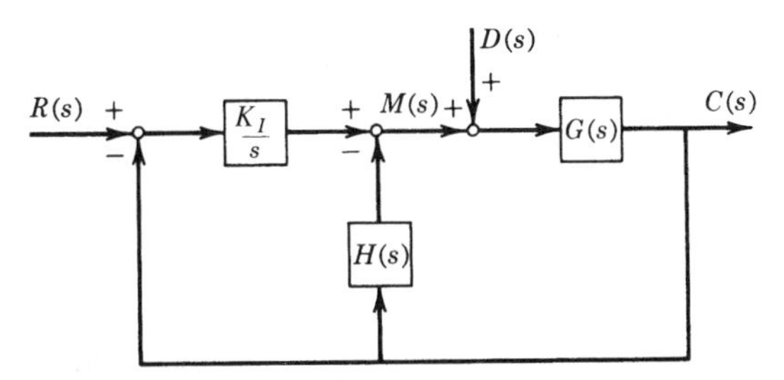

Figure 39 Structure of pseudoderivative feedback.

functions are useful. The *Bode plot* and the *Nyquist plot* present the frequency response information in different forms. Each form has its own advantages. The root-locus plot shows the location of the characteristic roots for a range of values of some parameters, such as a controller gain. A tabulation of these plots for typical transfer functions is given in the previous chapter (Fig. 27.8). The design of two-position and other nonlinear control systems is facilitated by the *describing function,* which is a linearized approximation based on the frequency response of the controller (see Section 27.8.4). Graphical design methods are discussed in more detail in Refs. 1–4.

9.1 The Nyquist Stability Theorem

The Nyquist stability theorem is a powerful tool for linear system analysis. If the open-loop system has no poles with positive real parts, we can concentrate our attention on the region around the point $-1 + i0$ on the polar plot of the open-loop transfer function. Figure 40 shows the polar plot of the open-loop transfer function of an arbitrary system that is assumed to be open-loop stable. The Nyquist stability theorem is stated as follows:

> A system is closed-loop stable if and only if the point $-1 + i0$ lies to the left of the open-loop Nyquist plot relative to an observer traveling along the plot in the direction of increasing frequency ω.

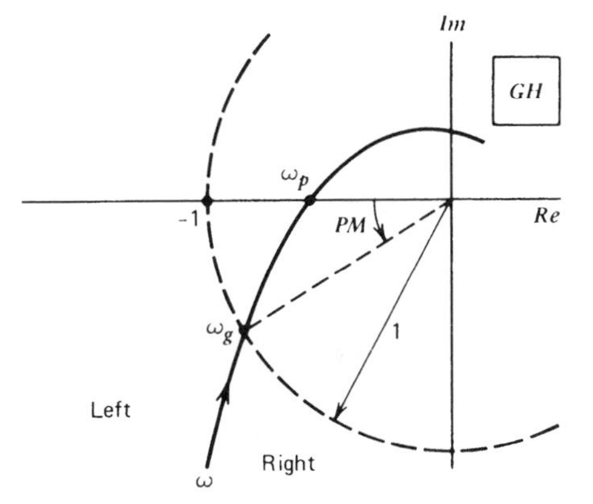

Figure 40 Nyquist plot for a stable system.[1]

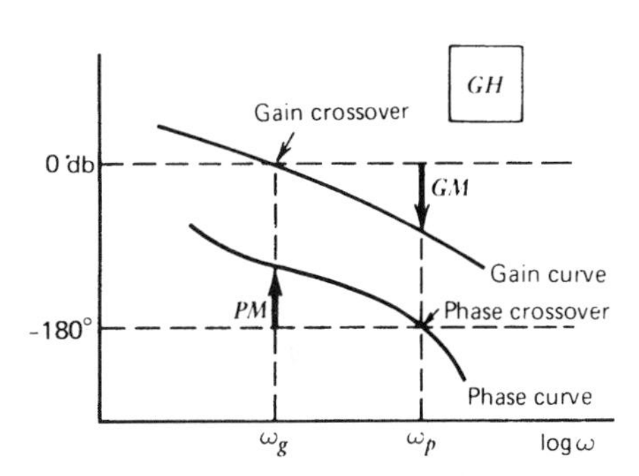

Figure 41 Bode plot showing definitions of phase and gain margin.[1]

Therefore, the system described by Fig. 39 is closed-loop stable.

The Nyquist theorem provides a convenient measure of the relative stability of a system. A measure of the proximity of the plot to the $-1 + i0$ point is given by the angle between the negative real axis and a line from the origin to the point where the plot crosses the unit circle (see Fig. 39). The frequency corresponding to this intersection is denoted ω_g. This angle is the *phase margin* (PM) and is positive when measured down from the negative real axis. The phase margin is the phase at the frequency ω_g where the magnitude ratio or "gain" of $G(i\omega)H(i\omega)$ is unity, or 0 decibels (dB). The frequency ω_p, the *phase crossover frequency,* is the frequency at which the phase angle is $-180°$. The *gain margin* (GM) is the difference in decibels between the unity gain condition (0 dB) and the value of $|G(\omega_p)H(\omega_p)|$ decibels at the phase crossover frequency ω_p. Thus,

$$\text{Gain margin} = -|G(\omega_p)H(\omega_p)| \quad \text{(dB)} \tag{34}$$

A system is stable only if the phase and gain margins are both positive.

The phase and gain margins can be illustrated on the Bode plots shown in Fig. 41. The phase and gain margins can be stated as safety margins in the design specifications. A typical set of such specifications is as follows:

$$\text{Gain margin} \geq 8 \text{ dB} \qquad \text{and} \qquad \text{Phase margin} \geq 30° \tag{35}$$

In common design situations, only one of these equalities can be met, and the other margin is allowed to be greater than its minimum value. It is not desirable to make the margins too large, because this results in a low gain, which might produce sluggish response and a large steady-state error. Another commonly used set of specifications is

$$\text{Gain margin} \geq 6 \text{ dB} \qquad \text{and} \qquad \text{Phase margin} \geq 40° \tag{36}$$

The 6-dB limit corresponds to the quarter amplitude decay response obtained with the gain settings given by the Ziegler–Nichols ultimate-cycle method (Table 2).

9.2 Systems with Dead-Time Elements

The Nyquist theorem is particularly useful for systems with dead-time elements, especially when the plant is of an order high enough to make the root-locus method cumbersome. A delay D in either the manipulated variable or the measurement will result in an open-loop transfer function of the form

$$G(s)H(s) = e^{-Ds}P(s) \tag{37}$$

Its magnitude and phase angle are

$$|G(i\omega)H(i\omega)| = |P(i\omega)||e^{-i\omega D}| = |P(i\omega)| \tag{38}$$

$$\angle G(i\omega)H(i\omega)) = \angle P(i\omega) + \angle e^{-i\omega D} = \angle P(i\omega) - \omega D \tag{39}$$

Thus, the dead time decreases the phase angle proportionally to the frequency ω, but it does not change the gain curve. This makes the analysis of its effects easier to accomplish with the open-loop frequency response plot.

9.3 Open-Loop Design for PID Control

Some general comments can be made about the effects of proportional, integral, and derivative control actions on the phase and gain margins. P action does not affect the phase curve at all and thus can be used to raise or lower the open-loop gain curve until the specifications for the gain and phase margins are satisfied. If I action or D action is included, the proportional gain is selected last. Therefore, when using this approach to the design, it is best to write the PID algorithm with the proportional gain factored out, as

$$F(s) = K_P \left(1 + \frac{1}{T_I s} + T_D s\right) E(s) \tag{40}$$

D action affects both the phase and gain curves. Therefore, the selection of the derivative gain is more difficult than the proportional gain. The increase in phase margin due to the positive phase angle introduced by D action is partly negated by the derivative gain, which reduces the gain margin. Increasing the derivative gain increases the speed of response, makes the system more stable, and allows a larger proportional gain to be used to improve the system's accuracy. However, if the phase curve is too steep near $-180°$, it is difficult to use D action to improve the performance. I action also affects both the gain and phase curves. It can be used to increase the open-loop gain at low frequencies. However, it lowers the phase crossover frequency ω_p and thus reduces some of the benefits provided by D action. If required, the D-action term is usually designed first, followed by I action and P action, respectively.

The classical design methods based on the Bode plots obviously have a large component of trial and error because usually both the phase and gain curves must be manipulated to achieve an acceptable design. Given the same set of specifications, two designers can use these methods and arrive at substantially different designs. Many rules of thumb and ad hoc procedures have been developed, but a general foolproof procedure does not exist. However, an experienced designer can often obtain a good design quickly with these techniques. The use of a computer plotting routine greatly speeds up the design process.

9.4 Design with the Root Locus

The effect of D action as a series compensator can be seen with the root locus. The term $1 + T_D s$ in Fig. 32 can be considered as a series compensator to the proportional controller. The D action adds an open-loop zero at $s = -1/T_D$. For example, a plant with the transfer function $1/s(s+1)(s+2)$, when subjected to proportional control, has the root locus shown in Fig. 42*a*. If the proportional gain is too high, the system will be unstable. The smallest achievable time constant corresponds to the root $s = -0.42$ and is $\tau = 1/0.42 = 2.4$. If D action is used to put an open-loop zero at $s = -1.5$, the resulting root locus is given by Fig. 42*b*. The D action prevents the system from becoming unstable and allows a smaller time constant to be achieved (τ can be made close to $1/0.75 = 1.3$ by using a high proportional gain).

The integral action in PI control can be considered to add an open-loop pole at $s = 0$ and a zero at $s = -1/T_I$. Proportional control of the plant $1/(s+1)(s+2)$ gives a root locus like that shown in Fig. 43, with $a = 1$ and $b = 2$. A steady-state error will exist for a step input. With the PI compensator applied to this plant, the root locus is given by Fig. 42*b*, with $T_I = 2/3$. The steady-state error is eliminated, but the response of the system has been slowed because the dominant paths of the root locus of the compensated system lie closer to the imaginary axis than those of the uncompensated system.

As another example, let the plant transfer function be

$$G_P(s) = \frac{1}{s^2 + a_2 s + a_1} \tag{41}$$

where $a_1 > 0$ and $a_2 > 0$. PI control applied to this plant gives the closed-loop command transfer function

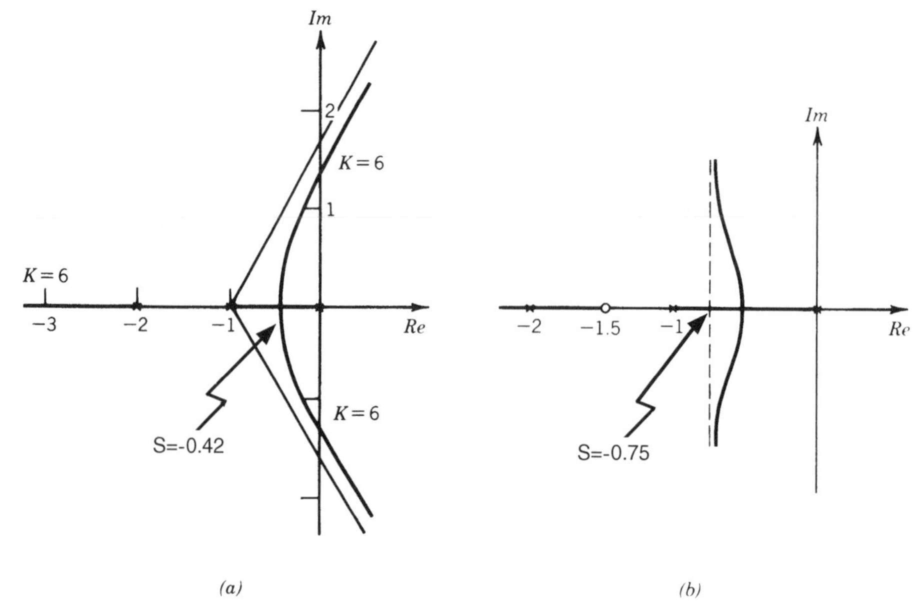

Figure 42 (*a*) Root-locus plot for $s(s + 1)(s + 2) + K = 0$, for $K \geq 0$. (*b*) The effect of PD control with $T_D = 2/3$.

$$T_1(s) = \frac{K_P s + K_I}{s^3 + a_2 s^2 + (a_1 + K_P)s + K_I} \tag{42}$$

Note that the Ziegler–Nichols rules cannot be used to set the gains K_P and K_I. The second-order plant, Eq. (41), does not have the S-shaped signature of Fig. 33, so the process reaction method does not apply. The ultimate-cycle method requires K_I to be set to zero and the ultimate gain K_{Pu} determined. With $K_I = 0$ in Eq. (42) the resulting system is stable for all $K_P > 0$, and thus a positive ultimate gain does not exist.

Take the form of the PI control law given by Eq. (42) with $T_D = 0$, and assume that the characteristic roots of the plant (Fig. 44) are real values $-r_1$ and $-r_2$ such that $-r_2 < -r_1$. In this case the open-loop transfer function of the control system is

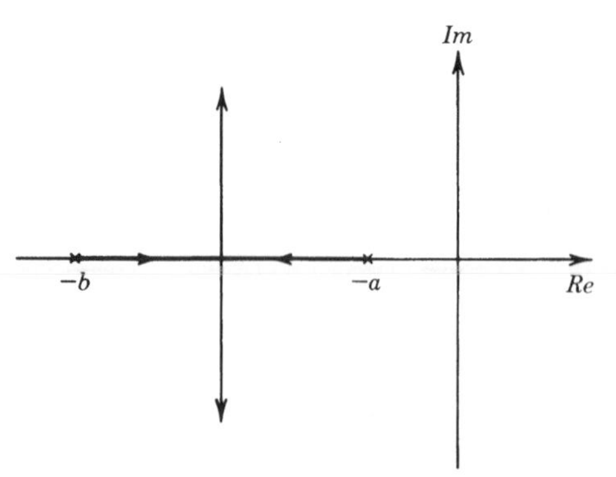

Figure 43 Root-locus plot for $(s + a)(s + b) + K = 0$.

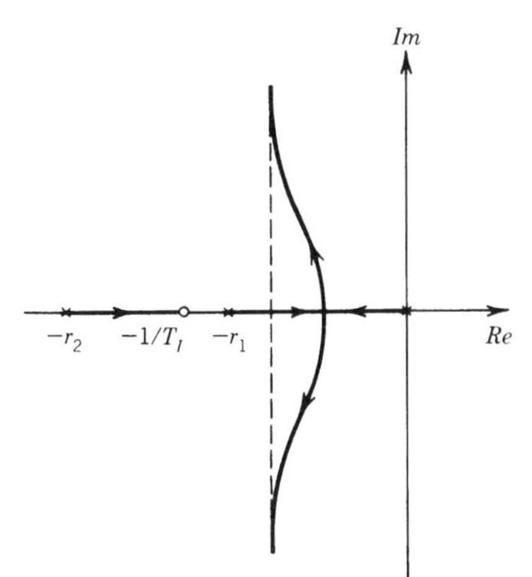

Figure 44 Root-locus plot for PI control of a second-order plant.

$$G(s)H(s) = \frac{K_P(s + 1/T_I)}{s(s + r_1)(s + r_2)} \tag{43}$$

One design approach is to select T_I and plot the locus with K_P as the parameter. If the zero at $s = -1/T_I$ is located to the right of $s = -r_1$, the dominant time constant cannot be made as small as is possible with the zero located between the poles at $s = -r_1$ and $s = -r_2$ (Fig. 44). A large integral gain (small T_I and/or large K_P) is desirable for reducing the overshoot due to a disturbance, but the zero should not be placed to the left of $s = -r_2$ because the dominant time constant will be larger than that obtainable with the placement shown in Fig. 44 for large values of K_P. Sketch the root-locus plots to see this. A similar situation exists if the poles of the plant are complex.

The effects of the lead compensator in terms of time-domain specifications (characteristic roots) can be shown with the root-locus plot. Consider the second-order plant with the real distinct roots $s = -\alpha$, $s = -\beta$. The root locus for this system with proportional control is shown in Fig. 45*a*. The smallest dominant time constant obtainable is τ_1, marked in the figure. A lead compensator introduces a pole at $s = -1/T$ and a zero at $s = -1/aT$, and the root locus becomes that shown in Fig. 45*b*. The pole and zero introduced by the compensator reshape the locus so that a smaller dominant time constant can be obtained. This is done by choosing the proportional gain high enough to place the roots close to the asymptotes.

With reference to the proportional control system whose root locus is shown in Fig. 45*a*, suppose that the desired damping ratio ζ_1 and desired time constant τ_1 are obtainable with a proportional gain of K_{P1}, but the resulting steady-state error $\alpha\beta/(\alpha\beta + K_{P1})$ due to a step input is too large. We need to increase the gain while preserving the desired damping ratio and time constant. With the lag compensator, the root locus is as shown in Fig. 45*c*. By considering specific numerical values, one can show that for the compensated system, roots with a damping ratio ζ_1 correspond to a high value of the proportional gain. Call this value K_{P2}. Thus $K_{P2} > K_{P1}$, and the steady-state error will be reduced. If the value of T is chosen large enough, the pole at $s = -1/T$ is approximately canceled by the zero at $s = -1/aT$, and the open-loop transfer function is given approximately by

$$G(s)H(s) = \frac{aK_P}{(s + \alpha)(s + \beta)} \tag{44}$$

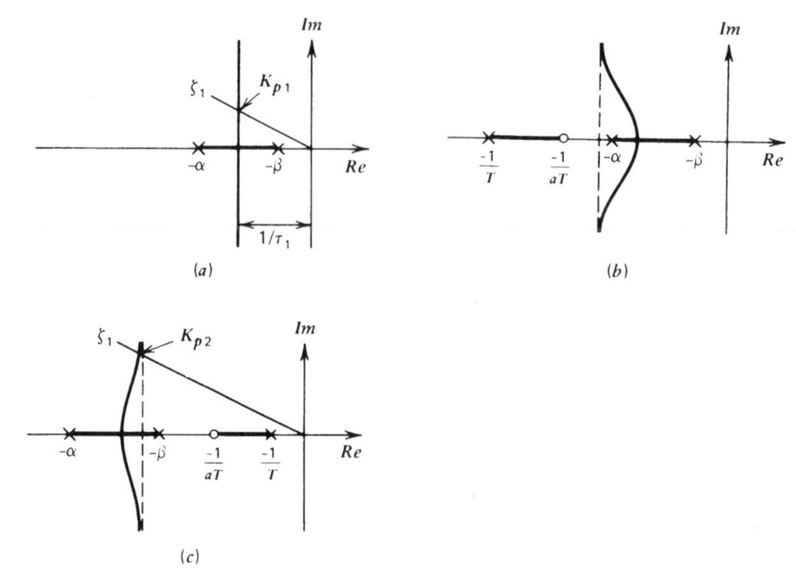

Figure 45 Effects of series lead and lag compensators: (*a*) uncompensated system's root locus; (*b*) root locus with lead compensation; (*c*) root locus with lag compensation.[1]

Thus, the system's response is governed approximately by the complex roots corresponding to the gain value K_{P2}. By comparing Fig. 45*a* with 45*c*, we see that the compensation leaves the time constant relatively unchanged. From Eq. (44) it can be seen that since $a < 1$, K_P can be selected as the larger value K_{P2}. The ratio of K_{P1} to K_{P2} is approximately given by the parameter a.

Design by pole–zero cancellation can be difficult to accomplish because a response pattern of the system is essentially ignored. The pattern corresponds to the behavior generated by the canceled pole and zero, and this response can be shown to be beyond the influence of the controller. In this example, the canceled pole gives a stable response because it lies in the left-hand plane. However, another input not modeled here, such as a disturbance, might excite the response and cause unexpected behavior. The designer should therefore proceed with caution. None of the physical parameters of the system are known exactly, so exact pole–zero cancellation is not possible. A root-locus study of the effects of parameter uncertainty and a simulation study of the response are often advised before the design is accepted as final.

10 PRINCIPLES OF DIGITAL CONTROL

Digital control has several advantages over analog devices. A greater variety of control algorithms is possible, including nonlinear algorithms and ones with time-varying coefficients. Also, greater accuracy is possible with digital systems. However, their additional hardware complexity can result in lower reliability, and their application is limited to signals whose time variation is slow enough to be handled by the samplers and the logic circuitry. This is now less of a problem because of the large increase in the speed of digital systems.

10.1 Digital Controller Structure

Sampling, discrete-time models, the z-transform, and pulse transfer functions were outlined in the previous chapter. The basic structure of a single-loop controller is shown in Fig. 46. The computer with its internal clock drives the *digital-to-analog* (D/A) and *analog-to-digital* (A/D) converters. It compares the command signals with the feedback signals and generates the control signals to be sent to the final control elements. These control signals are computed from the control algorithm stored in the memory. Slightly different structures exist, but Fig. 46 shows the important aspects. For example, the comparison between the command and feedback signals can be done with analog elements, and the A/D conversion made on the resulting error signal. The software must also provide for *interrupts,* which are conditions that call for the computer's attention to do something other than computing the control algorithm.

The time required for the control system to complete one loop of the algorithm is the time T, the *sampling time* of the control system. It depends on the time required for the computer to calculate the control algorithm and on the time required for the interfaces to convert data. Modern systems are capable of very high rates, with sample times under 1 μs.

In most digital control applications, the plant is an analog system, but the controller is a discrete-time system. Thus, to design a digital control system, we must either model the controller as an analog system or model the plant as a discrete-time system. Each approach has its own merits, and we will examine both.

If we model the controller as an analog system, we use methods based on *differential* equations to compute the gains. However, a digital control system requires *difference* equations to describe its behavior. Thus, from a strictly mathematical point of view, the gain values we will compute will not give the predicted response exactly. However, if the sampling time is small compared to the smallest time constant in the system, then the digital system will act like an analog system, and our designs will work properly. Because most physical systems of interest have time constants greater than 1 ms and controllers can now achieve sampling times less than 1 μs, controllers designed with analog methods will often be adequate.

10.2 Digital Forms of PID Control

There are a number of ways that PID control can be implemented in software in a digital control system, because the integral and derivative terms must be approximated with formulas chosen from a variety of available algorithms. The simplest integral approximation is to replace the integral with a sum of rectangular areas. With this rectangular approximation, the error integral is calculated as

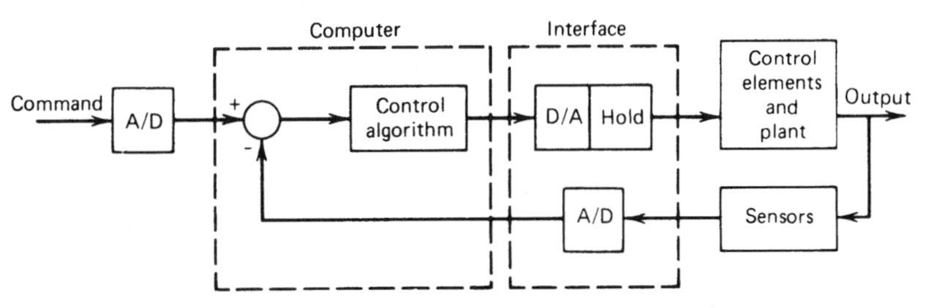

Figure 46 Structure of a digital control system.[1]

$$\int_0^{(k+1)T} e(t)\,dt \approx Te(0) + Te(t_1) + Te(t_2) + \cdots + Te(t_k) = T\sum_{i=0}^{k} e(t_i) \tag{45}$$

where $t_k = kT$ and the width of each rectangle is the sampling time $T = t_{i+1} - t_i$. The times t_i are the times at which the computer updates its calculation of the control algorithm after receiving an updated command signal and an updated measurement from the sensor through the A/D interfaces. If the time T is small, then the value of the sum in (45) is close to the value of the integral. After the control algorithm calculation is made, the calculated value of the control signal $f(t_k)$ is sent to the actuator via the output interface. This interface includes a D/A converter and a *hold* circuit that "holds" or keeps the analog voltage corresponding to the control signal applied to the actuator until the next updated value is passed along from the computer. The simplest digital form of PI control uses (45) for the integral term. It is

$$f(t_k) = K_P e(t_k) + K_I T\sum_{i=0}^{k} e(t_i) \tag{46}$$

This can be written in a more efficient form by noting that

$$f(t_{k-1}) = K_P e(t_{k-1}) + K_I T\sum_{i=0}^{k-1} e(t_i)$$

and subtracting this from (46) to obtain

$$f(t_k) = f(t_{k-1}) + K_P[e(t_k) - e(t_{k-1})] + K_I T e(t_k) \tag{47}$$

This form—called the *incremental* or *velocity* algorithm—is well suited for incremental output devices such as stepper motors. Its use also avoids the problem of integrator buildup, the condition in which the actuator saturates but the control algorithm continues to integrate the error.

The simplest approximation to the derivative is the first-order difference approximation

$$\frac{de}{dt} \approx \frac{e(t_k) - e(t_{k-1})}{T} \tag{48}$$

The corresponding PID approximation using the rectangular integral approximation is

$$f(t_k) = K_P e(t_k) + K_I T\sum_{i=0}^{k} e(t_i) + \frac{K_D}{T}[e(t_k) - e(t_{k-1})] \tag{49}$$

The accuracy of the integral approximation can be improved by substituting a more sophisticated algorithm, such as the following trapezoidal rule:

$$\int_0^{(k+1)T} e(t)\,dt \approx T\sum_{i=0}^{k} \frac{1}{2}[e(t_{i+1}) + e(t_i)] \tag{50}$$

The accuracy of the derivative approximation can be improved by using values of the sampled error signal at more instants. Using the four-point central-difference method (Refs. 1 and 2), the derivative term is approximated by

$$\frac{de}{dt} \approx \frac{1}{6T}[e(t_k) + 3e(t_{k-1}) - 3e(t_{k-2}) - e(t_{k-3})]$$

The derivative action is sensitive to the resulting rapid change in the error samples that follows a step input. This effect can be eliminated by reformulating the control algorithm as follows (Refs. 1 and 2):

$$\begin{aligned} f(t_k) &= f(t_{k-1}) + K_P[c(t_{k-1}) - c(t_k)] \\ &\quad + K_I T[r(t_k) - c(t_k)] \\ &\quad + \frac{K_D}{T}[-c(t_k) + 2c(t_{k-1}) - c(t_{k-2})] \end{aligned} \tag{51}$$

where $r(t_k)$ is the command input and $c(t_k)$ is the variable being controlled. Because the command input $r(t_k)$ appears in this algorithm only in the integral term, we cannot apply this algorithm to PD control; that is, the integral gain K_I must be nonzero.

11 UNIQUELY DIGITAL ALGORITHMS

Development of analog control algorithms was constrained by the need to design physical devices that could implement the algorithm. However, digital control algorithms simply need to be programmable and are thus less constrained than analog algorithms.

11.1 Digital Feedforward Compensation

Classical control system design methods depend on linear models of the plant. With linearization we can obtain an approximately linear model, which is valid only over a limited operating range. Digital control now allows us to deal with nonlinear models more directly using the concepts of feedforward compensation discussed in Section 8.

Computed Torque Method

Figure 47 illustrates a variation of feedforward compensation of the disturbance called the *computed torque method.* It is used to control the motion of robots. A simple model of a robot arm is the following nonlinear equation:

$$I\ddot{\theta} = T - mgL \sin \theta \tag{52}$$

where θ is the arm angle, I is its inertia, mg is its weight, and L is the distance from its mass center to the arm joint where the motor acts. The motor supplies the torque T. To position the arm at some desired angle θ_r, we can use PID control on the angle error $\theta_r - \theta$. This works well if the arm angle θ is never far from the desired angle θ_r so that we can linearize the plant model about θ_r. However, the controller will work for large-angle excursions if we compute the nonlinear gravity torque term $mgL \sin \theta$ and add it to the PID output. That is, part of the motor torque will be computed specifically to cancel the gravity torque, in effect producing a linear system for the PID algorithm to handle. The nonlinear torque calculations required to control multi-degree-of-freedom robots are very complicated and can be done only with a digital controller.

Feedforward Command Compensation

Computers can store lookup tables, which can be used to control systems that are difficult to model entirely with differential equations and analytical functions. Figure 48 shows a speed control system for an internal combustion engine. The fuel flow rate required to achieve

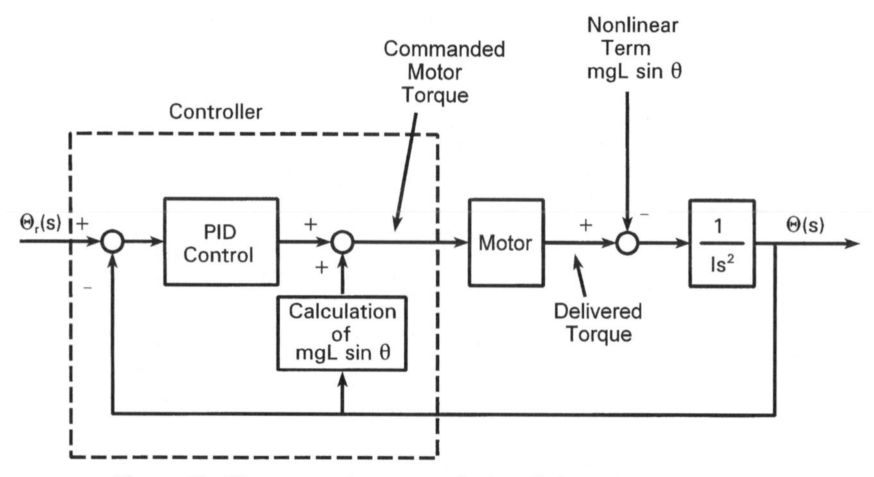

Figure 47 The computed torque method applied to robot arm control.

a desired speed depends in a complicated way on many variables not shown in the figure, such as temperature, humidity, and so on. This dependence can be summarized in tables stored in the control computer and can be used to estimate the required fuel flow rate. A PID algorithm can be used to adjust the estimate based on the speed error. This application is an example of feedforward compensation of the command input, and it requires a digital computer.

11.2 Control Design in the z-Plane

There are two common approaches to designing a digital controller:

1. The performance is specified in terms of the desired continuous-time response, and the controller design is done entirely in the s-plane, as with an analog controller. The

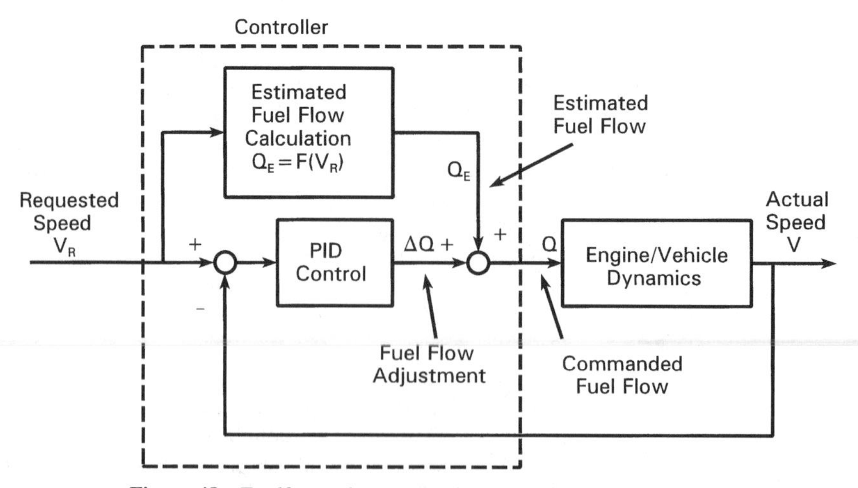

Figure 48 Feedforward compensation applied to engine control.

resulting control law is then converted to discrete-time form, using approximations for the integral and derivative terms. This method can be successfully applied if the sampling time is small. The technique is widely used for two reasons. When existing analog controllers are converted to digital control, the form of the control law and the values of its associated gains are known to have been satisfactory. Therefore, the digital version can use the same control law and gain values. Second, because analog design methods are well established, many engineers prefer to take this route and then convert the design into a discrete-time equivalent.

2. The performance specifications are given in terms of the desired continuous-time response and/or desired root locations in the s-plane. From these the corresponding root locations in the z-plane are found and a discrete control law is designed. This method avoids the derivative and integral approximation errors that are inherent in the first method and is the preferred method when the sampling time T is large. However, the algebraic manipulations are more cumbersome.

The second approach uses the z-transform and pulse transfer functions, which were outlined in the previous chapter. If we have an analog model of the plant, with its transfer function $G(s)$, we can obtain its pulse transfer function $G(z)$ by finding the z-transform of the impulse response $g(t) = \mathcal{L}^{-1}[G(s)]$; that is, $G(z) = \mathcal{Z}[g(t)]$. A table of transforms facilitates this process; see Refs. 1 and 2. Figure 49a shows the basic elements of a digital control system. Figure 49b is an equivalent diagram with the analog transfer functions inserted. Figure 49c represents the same system in terms of pulse transfer functions. From the diagram we can find the closed-loop pulse transfer function. It is

$$\frac{C(z)}{R(z)} = \frac{G(z)P(z)}{1 + G(z)P(z)} \tag{53}$$

The variable z is related to the Laplace variable s by

$$z = e^{sT} \tag{54}$$

If we know the desired root locations and the sampling time T, we can compute the z roots from this equation.

Digital PI Control Design

For example, the first-order plant $1/(2s + 1)$ with a zero-order hold has the following pulse transfer function (Refs. 1 and 2):

$$P(z) = \frac{1 - e^{-0.5T}}{z - e^{-0.5T}} \tag{55}$$

Suppose we use a control algorithm described by the following pulse transfer function:

$$G(z) = \frac{F(z)}{E(z)} = \frac{K_1 z + K_2}{z - 1} = \frac{K_1 + K_2 z^{-1}}{1 - z^{-1}} \tag{56}$$

The corresponding difference equation that the control computer must implement is

$$f(t_k) = f(t_{k-1}) + K_1 e(t_k) + K_2 e(t_{k-1}) \tag{57}$$

where $e(t_k) = r(t_k) - c(t_k)$. By comparing (57) with (47), it can be seen that this is the digital equivalent of PI control, where $K_P = -K_2$ and $K_I = (K_1 + K_2)/T$. Using the form of $G(z)$ given by (56), the closed-loop transfer function is

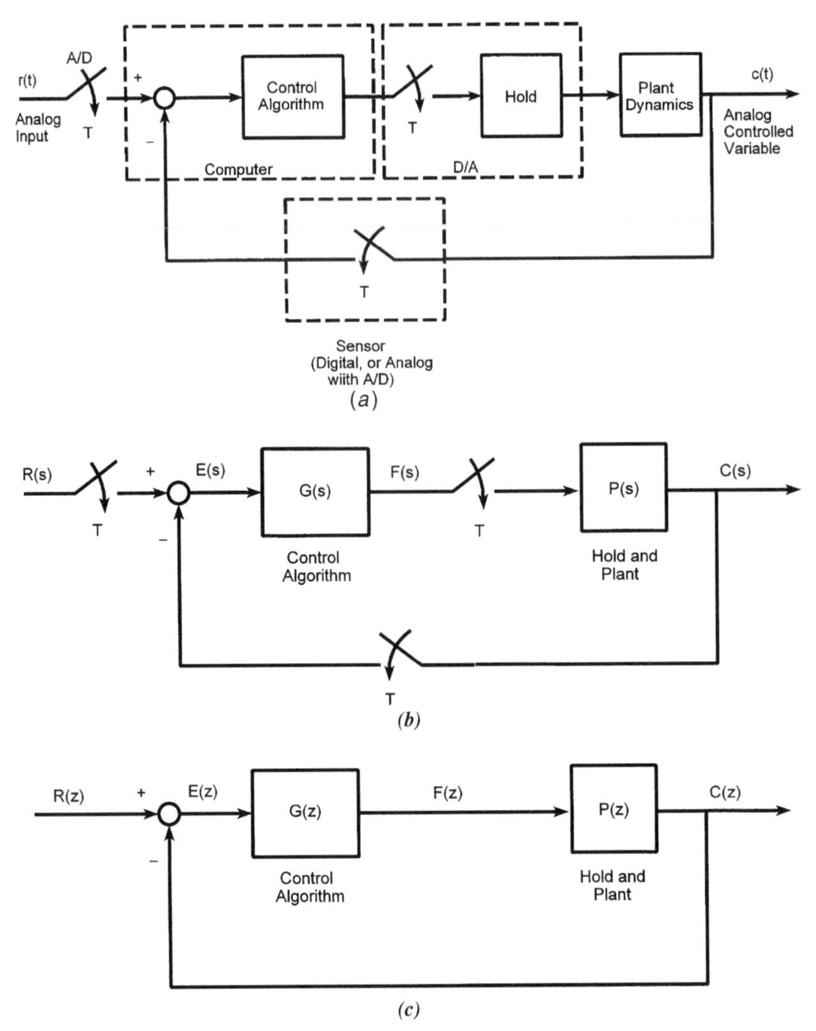

Figure 49 Block diagrams of a typical digital controller. (*a*) Diagram showing the components. (*b*) Diagram of the *s*-plane relations. (*c*) Diagram of the *z*-plane relations.

$$\frac{C(z)}{R(z)} = \frac{(1 - b)(K_1 z + K_2)}{z^2 + (K_1 - 1 - b - bK_1)z + b + K_2 - bK_2} \tag{58}$$

where $b = e^{-05T}$.

If the design specifications call for $\tau = 1$ and $\zeta = 1$, then the desired s roots are $s = -1, -1$, and the analog PI gains required to achieve these roots are $K_P = 3$ and $K_I = 2$. Using a sampling time of $T = 0.1$, the z roots must be $z = e^{-0.1}, e^{-0.1}$. To achieve these roots, the denominator of the transfer function (58) must be $z^2 - 2e^{-0.1}z + e^{-0.2}$. Thus the control gains must be $K_1 = 2.903$ and $K_2 = -2.717$. These values of K_1 and K_2 correspond to $K_P = 2.72$ and $K_I = 1.86$, which are close to the PI gains computed for an analog controller. If we had used a sampling time smaller than 0.1, say $T = 0.01$, the values of K_P and K_I computed from K_1 and K_2 would be $K_P = 2.97$ and $K_I = 1.98$, which are even closer to the analog gain values. This illustrates the earlier claim that analog design methods can be used when the sampling time is small enough.

Digital Series Compensation

Series compensation can be implemented digitally by applying suitable discrete-time approximations for the derivative and integral to the model represented by the compensator's transfer function $G_c(s)$. For example, the form of a lead or a lag compensator's transfer function is

$$G_c(s) = \frac{M(s)}{F(s)} = K\frac{s + c}{s + d} \tag{59}$$

where $m(t)$ is the actuator command and $f(t)$ is the control signal produced by the main (PID) controller. The differential equation corresponding to (59) is

$$\dot{m} + dm = K(\dot{f} + cf) \tag{60}$$

Using the simplest approximation for the derivative, Eq. (48), we obtain the following difference equation that the digital compensator must implement:

$$\frac{m(t_k) - m(t_{k-1})}{T} + dm(t_k) = K\left[\frac{f(t_k) - f(t_{k-1})}{T} + cf(t_k)\right]$$

In the z-plane, the equation becomes

$$\frac{1 - z^{-1}}{T}M(z) + dM(z) = K\left[\frac{1 - z^{-1}}{T}F(z) + cF(z)\right] \tag{61}$$

The compensator's pulse transfer function is thus seen to be

$$G_c(z) = \frac{M(z)}{F(z)} = \frac{K(1 - z^{-1}) + cT}{1 - z^{-1} + dT}$$

which has the form

$$G_c(z) = K_c\frac{z + a}{z + b} \tag{62}$$

where K_c, a, and b can be expressed in terms of K, c, d, and T if we wish to use analog design methods to design the compensator. When using commercial controllers, the user might be required to enter the values of the gain, the pole, and the zero of the compensator. The user must ascertain whether these values should be entered as s-plane values (i.e., K, c, and d) or as z-plane values (K_c, a, and b).

Note that the digital compensator has the same number of poles and zeros as the analog compensator. This is a result of the simple approximation used for the derivative. Note that Eq. (61) shows that when we use this approximation, we can simply replace s in the analog transfer function with $1 - z^{-1}$. Because the integration operation is the inverse of differentiation, we can replace $1/s$ with $1/(1 - z^{-1})$ when integration is used. [This is equivalent to using the rectangular approximation for the integral, and can be verified by finding the pulse transfer function of the incremental algorithm (47) with $K_P = 0$.]

Some commercial controllers treat the PID algorithm as a series compensator, and the user is expected to enter the controller's values, not as PID gains, but as pole and zero locations in the z-plane. The PID transfer function is

$$\frac{F(s)}{E(s)} = K_P + \frac{K_I}{s} + K_D s \tag{63}$$

Making the indicated replacements for the s terms, we obtain

$$\frac{F(z)}{E(z)} = K_P + \frac{K_I}{1 - z^{-1}} + K_D(1 - z^{-1})$$

which has the form

$$\frac{F(z)}{E(z)} = K_c \frac{z^2 - az + b}{z - 1} \tag{64}$$

where K_c, a, and b can be expressed in terms of K_P, K_I, K_D, and T. Note that the algorithm has two zeros and one pole, which is fixed at $z = 1$. Sometimes the algorithm is expressed in the more general form

$$\frac{F(z)}{E(z)} = K_c \frac{z^2 - az + b}{z - c} \tag{65}$$

to allow the user to select the pole as well.

Digital compensator design can be done with frequency response methods or with the root-locus plot applied to the z-plane rather than the s-plane. However, when better approximations are used for the derivative and integral, the digital series compensator will have more poles and zeros than its analog counterpart. This means that the root-locus plot will have more root paths, and the analysis will be more difficult. This topic is discussed in more detail in Refs. 1–3 and 8.

11.3 Direct Design of Digital Algorithms

Because almost any algorithm can be implemented digitally, we can specify the desired response and work backward to find the required control algorithm. This is the *direct-design* method. If we let $D(z)$ be the desired form of the closed-loop transfer function $C(z)/R(z)$ and solve (53 in Chapter 14) for the controller transfer function $G(z)$, we obtain

$$G(z) = \frac{D(z)}{P(z)[1 - D(z)]} \tag{66}$$

We can pick $D(z)$ directly or obtain it from the specified input transform $R(z)$ and the desired output transform $C(z)$, because $D(z) = C(z)/R(z)$.

Finite-Settling-Time Algorithm

This method can be used to design a controller to compensate for the effects of process dead time. A plant having such a response can often be approximately described by a first-order model with a dead-time element; that is,

$$G_P(s) = K \frac{e^{-Ds}}{\tau s + 1} \tag{67}$$

where D is the dead time. This model also approximately describes the S-shaped response curve used with the Ziegler–Nichols method (Fig. 33). When combined with a zero-order hold, this plant has the following pulse transfer function:

$$P(z) = Kz^{-n} \frac{1 - a}{z - a} \tag{68}$$

where $a = \exp(-T/\tau)$ and $n = D/T$. If we choose $D(z) = z^{-(n+1)}$, then with a step command input, the output $c(k)$ will reach its desired value in $n + 1$ sample times, one more than is in the dead time D. This is the fastest response possible. From (66) the required controller transfer function is

$$G(z) = \frac{1}{K(1 - a)} \frac{1 - az^{-1}}{1 - z^{-(n+1)}} \tag{69}$$

The corresponding difference equation that the control computer must implement is

$$f(t_k) = f(t_{k-n-1}) + \frac{1}{K(1 - a)} [e(t_k) - ae(t_{k-1})] \tag{70}$$

This algorithm is called a *finite-settling-time* algorithm because the response reaches its desired value in a finite, prescribed time. The maximum value of the manipulated variable required by this algorithm occurs at $t = 0$ and is $1/K(1 - a)$. If this value saturates the actuator, this method will not work as predicted. Its success depends also on the accuracy of the plant model.

Dahlin's Algorithm

This sensitivity to plant modeling errors can be reduced by relaxing the minimum-response-time requirement. For example, choosing $D(z)$ to have the same form as $P(z)$, namely,

$$D(z) = K_d z^{-n} \frac{1 - a_d}{z - a_d} \tag{71}$$

we obtain from (66) the following controller transfer function:

$$G(z) = \frac{K_d(1 - a_d)}{K(1 - a)} \frac{1 - az^{-1}}{1 - a_d z^{-1} - K_d(1 - a_d)z^{-(n+1)}} \tag{72}$$

This is *Dahlin's algorithm.*[3] The corresponding difference equation that the control computer must implement is

$$f(t_k) = a_d f(t_{k-1}) + K_d(1 - a_d)f(t_{k-n-1}) + \frac{K_d(1 - a_d)}{K(1 - a)} [e(t_k) - ae(t_{k-1})] \tag{73}$$

Normally we would first try setting $K_d = K$ and $a_d = a$, but since we might not have good estimates of K and a, we can use K_d and a_d as tuning parameters to adjust the controller's performance. The constant a_d is related to the time constant τ_d of the desired response: $a_d = \exp(-T/\tau_d)$. Choosing τ_d smaller gives faster response.

Algorithms such as these are often used for system startup, after which the control mode is switched to PID, which is more capable of handling disturbances.

12 HARDWARE AND SOFTWARE FOR DIGITAL CONTROL

This section provides an overview of the general categories of digital controllers that are commercially available. This is followed by a summary of the software currently available for digital control and for control system design.

12.1 Digital Control Hardware

Commercially available controllers have different capabilities, such as different speeds and operator interfaces, depending on their targeted application.

Programmable Logic Controllers (PLCs)

These are controllers that are programmed with relay ladder logic, which is based on Boolean algebra. Now designed around microprocessors, they are the successors to the large relay panels, mechanical counters, and drum programmers used up to the 1960s for sequencing control and control applications requiring only a finite set of output values (for example, opening and closing of valves). Some models now have the ability to perform advanced mathematical calculations required for PID control, thus allowing them to be used for modulated control as well as finite-state control. There are numerous manufacturers of PLCs.

Digital Signal Processors (DSPs)

A modern development is the *digital signal processor* (DSP), which has proved useful for feedback control as well as signal processing.[9] This special type of processor chip has separate buses for moving data and instructions and is constructed to perform rapidly the kind of mathematical operations required for digital filtering and signal processing. The separate buses allow the data and the instructions to move in parallel rather than sequentially. Because the PID control algorithm can be written in the form of a digital filter, DSPs can also be used as controllers.

The DSP architecture was developed to handle the types of calculations required for digital filters and discrete Fourier transforms, which form the basis of most signal-processing operations. DSPs usually lack the extensive memory management capabilities of general-purpose computers because they need not store large programs or large amounts of data. Some DSPs contain A/D and D/A converters, serial ports, timers, and other features. They are programmed with specialized software that runs on popular personal computers. Low-cost DSPs are now widely used in consumer electronics and automotive applications, with Texas Instruments being a major supplier.

Motion Controllers

Motion controllers are specialized control systems that provide feedback control for one or more motors. They also provide a convenient operator interface for generating the commanded trajectories. Motion controllers are particularly well suited for applications requiring coordinated motion of two or more axes and for applications where the commanded trajectory is complicated. A higher level host computer might transmit required distance, speed, and acceleration rates to the motion controller, which then constructs and implements the continuous position profile required for each motor. For example, the host computer would supply the required total displacement, the acceleration and deceleration times, and the desired slew speed (the speed during the zero acceleration phase). The motion controller would generate the commanded position versus time for each motor. The motion controller also has the task of providing feedback control for each motor to ensure that the system follows the required position profile.

Figure 50 shows the functional elements of a typical motion controller, such as those built by Galil Motion Control, Inc. Provision for both analog and digital input signals allows these controllers to perform other control tasks besides motion control. Compared to DSPs, such controllers generally have greater capabilities for motion control and have operator interfaces that are better suited for such applications. Motion controllers are available as plug-in cards for most computer bus types. Some are available as stand-alone units.

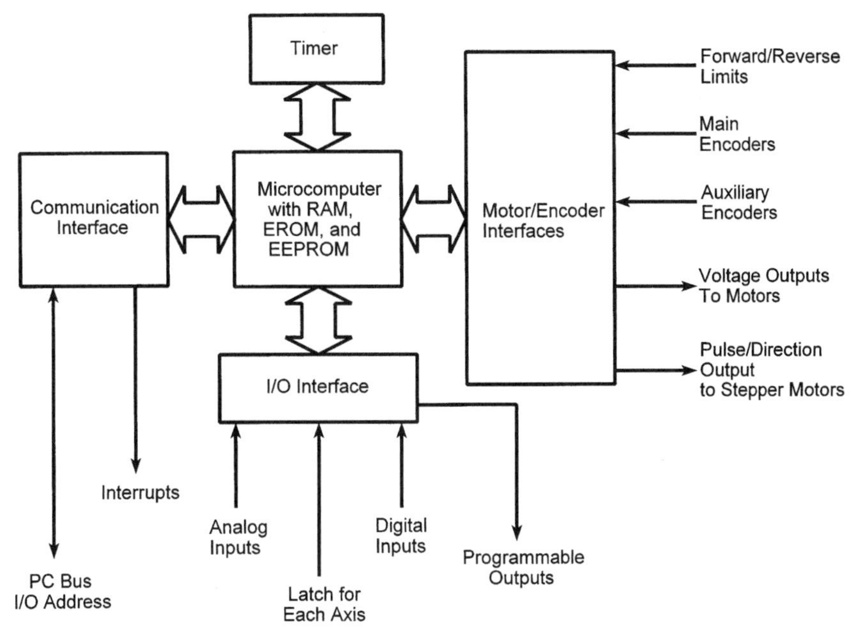

Figure 50 Functional diagram of a motion controller.

Motion controllers use a PID control algorithm to provide feedback control for each motor (some manufacturers call this algorithm a "filter"). The user enters the values of the PID gains (some manufacturers provide preset gain values, which can be changed; others provide tuning software that assists in selecting the proper gain values). Such controllers also have their own language for programming a variety of motion profiles and other applications. For example, they provide for linear and circular interpolation for two-dimensional coordinated motion, motion smoothing (to eliminate jerk), contouring, helical motion, and electronic gearing. The latter is a control mode that emulates mechanical gearing in software, in which one motor (the slave) is driven in proportion to the position of another motor (the master) or an encoder.

Process Controllers

Process controllers are designed to handle inputs from sensors, such as thermocouples, and outputs to actuators, such as valve positioners, that are commonly found in process control applications. Figure 51 illustrates the input–output capabilities of a typical process controller such as those manufactured by Honeywell, which is a major supplier of such devices. This device is a stand-alone unit designed to be mounted in an instrumentation panel. The voltage and current ranges of the analog inputs are those normally found with thermocouple-based temperature sensors. The current outputs are designed for devices like valve positioners, which usually require 4–20-mA signals.

The controller contains a microcomputer with built-in math functions normally required for process control, such as thermocouple linearization, weighted averaging, square roots, ratio/bias calculations, and the PID control algorithm. These controllers do not have the same software and memory capabilities as desktop computers, but they are less expensive. Their operator interface consists of a small keypad with typically fewer than 10 keys, a small

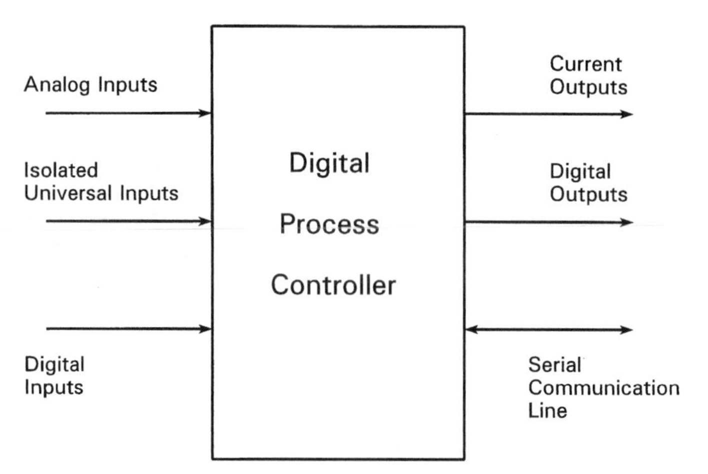

Figure 51 Functional diagram of a digital process controller.

graphical display for displaying bargraphs of the set points and the process variables, indicator lights, and an alphanumeric display for programming the controller.

The PID gains are entered by the user. Some units allow multiple sets of gains to be stored; the unit can be programmed to switch between gain settings when certain conditions occur. Some controllers have an adaptive tuning feature that is supposed to adjust the gains to prevent overshoot in startup mode, to adapt to changing process dynamics, and to adapt to disturbances. However, at this time, adaptive tuning cannot claim a 100% success rate, and further research and development in adaptive control is needed.

Some process controllers have more than one PID control loop for controlling several variables. Figure 52 illustrates a boiler feedwater control application for a controller with two PID loops arranged in a cascade control structure. Loop 1 is the main or outer loop controller for maintaining the desired water volume in the boiler. It uses sensing of the steam flow rate to implement feedforward compensation. Loop 2 is the inner loop controller that directly controls the feedwater control valve.

12.2 Software for Digital Control

The software available to the modern control engineer is quite varied and powerful and can be categorized according to the following tasks:

1. Control algorithm design, gain selection, and simulation
2. Tuning
3. Motion programming
4. Instrumentation configuration
5. Real-time control functions

Many analysis and simulation packages now contain algorithms of specific interest to control system designers. *MATLAB®* is one such package that is widely used. It contains built-in functions for generating root-locus and frequency response plots, system simulation,

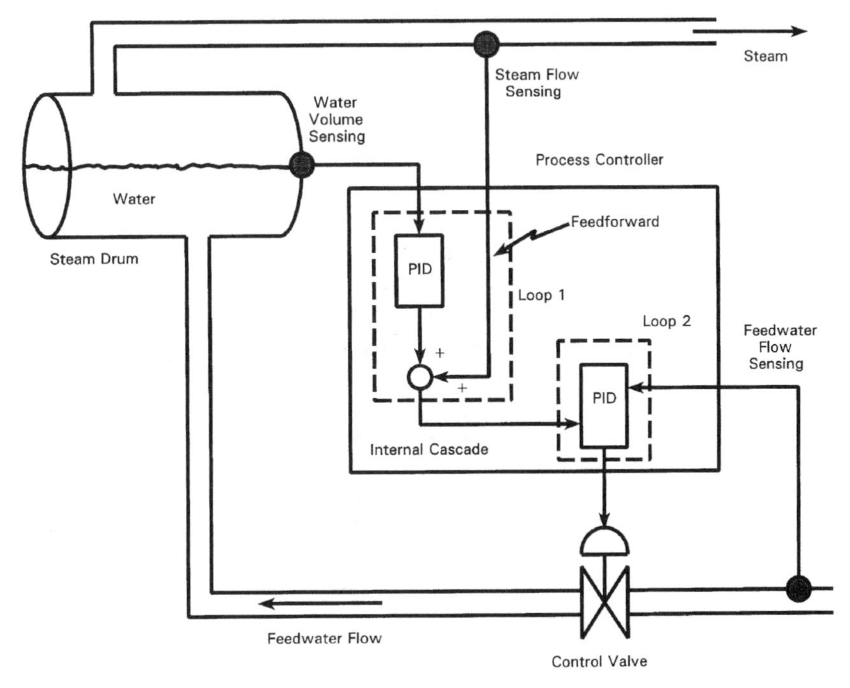

Figure 52 Application of a two-loop process controller for feedwater control.

digital filtering, calculation of control gains, and data analysis. It can accept model descriptions in the form of transfer functions or as state-variable equations.[1,4,10]

Some manufacturers provide software to assist the engineer in sizing and selecting components. An example is the *Motion Component Selector* (MCS) sold by Galil Motion Control, Inc. It assists the engineer in computing the load inertia, including the effects of the mechanical drive, and then selects the proper motor and amplifier based on the user's description of the desired motion profile.

Some hardware manufacturers supply software to assist the engineer in selecting control gains and modifying (*tuning*) them to achieve good response. This might require that the system to be controlled be available for experiments prior to installation. Some controllers, such as some Honeywell process controllers, have an autotuning feature that adjusts the gains in real time to improve performance.

Motion programming software supplied with motion controllers was mentioned previously. Some packages, such as Galil's, allow the user to simulate a multiaxis system having more than one motor and to display the resulting trajectory.

Instrumentation configuration software, such as *LabView*®, provides specialized programming languages for interacting with instruments and for creating graphical real-time displays of instrument outputs.

Until recently, development of real-time digital control software involved tedious programming, often in assembly language. Even when implemented in a higher level language, such as Fortran or C, programming real-time control algorithms can be very challenging, partly because of the need to provide adequately for interrupts. Software packages are now available that provide real-time control capability, usually a form of the PID algorithm, that can be programmed through user-friendly graphical interfaces. Examples include the Galil motion controllers and the add-on modules for Labview and MATLAB.

12.3 Embedded Control Systems and Hardware-in-the Loop Testing

An *embedded control system* is a microprocessor and sensor suite designed to be an integral part of a product. The aerospace and automotive industries have used embedded controllers for some time, but the decreased cost of components now makes embedded controllers feasible for more consumer and biomedical applications.

For example, embedded controllers can greatly increase the performance of orthopedic devices. One model of an artificial leg now uses sensors to measure in real time the walking speed, the knee joint angle, and the loading due to the foot and ankle. These measurements are used by the controller to adjust the hydraulic resistance of a piston to produce a stable, natural, and efficient gait. The controller algorithms are adaptive in that they can be tuned to an individual's characteristics and their settings changed to accommodate different physical activities.

Engines incorporate embedded controllers to improve efficiency. Embedded controllers in new active suspensions use actuators to improve on the performance of traditional passive systems consisting only of springs and dampers. One design phase of such systems is *hardware-in-the-loop testing,* in which the controlled object (the engine or vehicle suspension) is replaced with a real-time simulation of its behavior. This enables the embedded system hardware and software to be tested faster and less expensively than with the physical prototype and perhaps even before the prototype is available.

Simulink®, which is built on top of MATLAB and requires MATLAB to run, is often used to create the simulation model for hardware-in-the-loop testing. Some of the *toolboxes* available for MATLAB, such as the control systems toolbox, the signal-processing toolbox, and the DSP and fixed-point blocksets, are also useful for such applications.

13 SOFTWARE SUPPORT FOR CONTROL SYSTEM DESIGN

Software packages are available for graphical control system design methods and control system simulation. These greatly reduce the tedious manual computation, plotting, and programming formerly required for control system design and simulation.

13.1 Software for Graphical Design Methods

Several software packages are available to support graphical control system design methods. The most popular of these is MATLAB, which has extensive capabilities for generation and interactive analysis of root-locus plots and frequency response plots. Some of these capabilities are discussed in Refs. 1 and 4.

13.2 Software for Control Systems Simulation

It is difficult to obtain closed-form expressions for system response when the model contains *dead time* or nonlinear elements that represent realistic control system behavior. Dead time (also called *transport delay*), rate limiters, and actuator saturation are effects that often occur in real control systems, and simulation is often the only way to analyze their response. Several software packages are available to support system simulation. One of the most popular is Simulink.

Systems having dead-time elements are easily simulated in Simulink. Figure 53 shows a Simulink model for PID control of the plant $53/(3.44s^2 + 2.61s + 1)$, with a dead time between the output of the controller and the plant. The block implementing the dead-time transfer function e^{-Ds} is called the *transport delay* block. When you run this model, you will see the response in the scope block.

In addition to being limited by saturation, some actuators have limits on how fast they can react. This limitation is independent of the time constant of the actuator and might be due to deliberate restrictions placed on the unit by its manufacturer. An example is a flow control valve whose rate of opening and closing is controlled by a *rate limiter*. Simulink has such a block, and it can be used in series with the saturation block to model the valve behavior. Consider the model of the height h of liquid in a tank whose input is a flow rate q_i. For specific parameter values, such a model has the form $H(s)/Q_i(s) = 2/(5s + 1)$. A Simulink model is shown in Figure 54 for a specific PI controller whose gains are $K_P = 4$ and $K_I = 5/4$. The saturation block models the fact that the valve opening must be between 0 and 100%. The model enables us to experiment with the lower and upper limits of the rate limiter block to see its effect on the system performance.

An introduction to Simulink is given in Refs. 4 and 10. Applications of Simulink to control system simulation are given in Ref. 4.

14 FUTURE TRENDS IN CONTROL SYSTEMS

Microprocessors have rejuvenated the development of controllers for mechanical systems. Currently, there are several applications areas in which new control systems are indispensable to the product's success:

1. Active vibration control
2. Noise cancellation
3. Adaptive optics
4. Robotics
5. Micromachines
6. Precision engineering

Most of the design techniques presented here comprise "classical" control methods. These methods are widely used because when they are combined with some testing and computer simulation, an experienced engineer can rapidly achieve an acceptable design. Modern control algorithms, such as state-variable feedback and the linear–quadratic optimal controller, have had some significant mechanical engineering applications—for example, in the control of aerospace vehicles. The current approach to multivariable systems like the one shown in Fig. 55 is to use classical methods to design a controller for each subsystem because

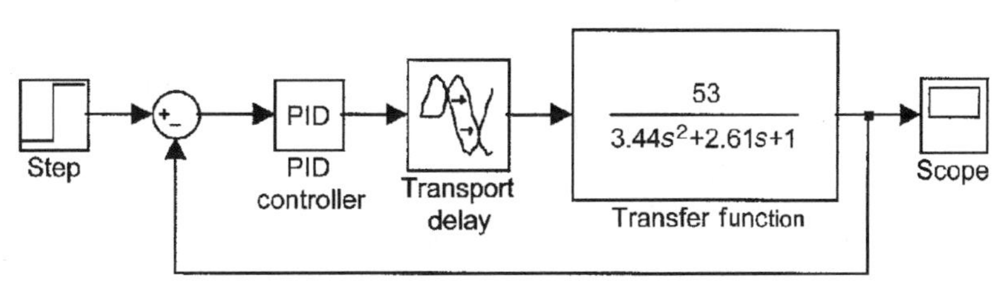

Figure 53 Simulink model of a system with transport delay.

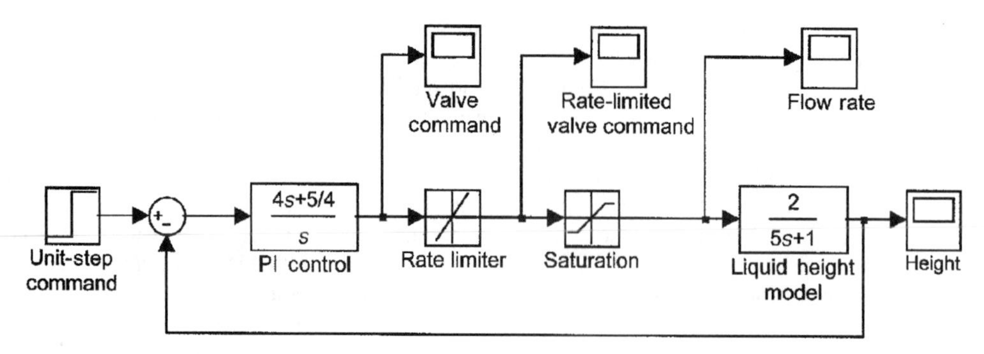

Figure 54 Simulink model of a system with actuator saturation and a rate limiter.

they can often be modeled with low-order linearized models. The coordination of the various low-level controllers is a nonlinear problem. High-order, nonlinear, multivariable systems that cannot be controlled with classical methods cannot yet be handled by modern control theory in a general way, and further research is needed.

In addition to the improvements, such as lower cost, brought on by digital hardware, microprocessors have allowed designers to incorporate algorithms of much greater complexity into control systems. The following is a summary of the areas currently receiving much attention in the control systems community.

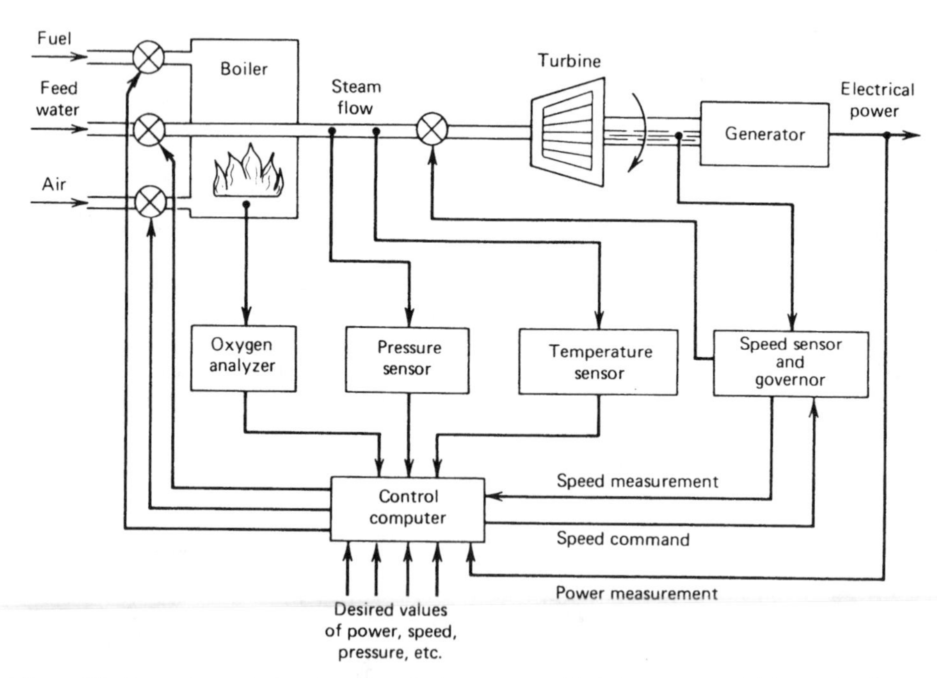

Figure 55 Computer control system for a boiler-generator. Each important variable requires its own controller. The interaction between variables calls for coordinated control of all loops.[1]

14.1 Fuzzy Logic Control

In classical set theory, an object's membership in a set is clearly defined and unambiguous. *Fuzzy logic control* is based on a generalization of classical set theory to allow objects to belong to several sets with various degrees of membership. Fuzzy logic can be used to describe processes that defy precise definition or precise measurement, and thus it can be used to model the inexact and subjective aspects of human reasoning. For example, room temperature can be described as cold, cool, just right, warm, or hot. Development of a fuzzy logic temperature controller would require the designer to specify the membership functions that describe "warm" as a function of temperature, and so on. The control logic would then be developed as a linguistic algorithm that models a human operator's decision process (for example, if the room temperature is "cold," then "greatly" increase the heater output; if the temperature is "cool," then increase the heater output "slightly").

Fuzzy logic controllers have been implemented in a number of applications. Proponents of fuzzy logic control point to its ability to convert a human operator's reasoning process into computer code. Its critics argue that because all the controller's fuzzy calculations must eventually reduce to a specific output that must be given to the actuator (e.g., a specific voltage value or a specific valve position), why not be unambiguous from the start, and define a "cool" temperature to be the range between 65° and 68°, for example? Perhaps the proper role of fuzzy logic is at the human operator interface. Research is active in this area, and the issue is not yet settled.[11,12]

14.2 Nonlinear Control

Most real systems are nonlinear, which means that they must be described by nonlinear differential equations. Control systems designed with the linear control theory described in this chapter depend on a linearized approximation to the original nonlinear model. This linearization can be explicitly performed, or implicitly made, as when we use the small-angle approximation: $\sin \theta \approx \theta$. This approach has been enormously successful because a well-designed controller will keep the system in the operating range where the linearization was done, thus preserving the accuracy of the linear model. However, it is difficult to control some systems accurately in this way because their operating range is too large. Robot arms are a good example.[13,14] Their equations of motion are very nonlinear, due primarily to the fact that their inertia varies greatly as their configuration changes.

Nonlinear systems encompass everything that is "not linear," and thus there is no general theory for nonlinear systems. There have been many nonlinear control methods proposed—too many to summarize here.[15] *Lyapunov's stability theory* and Popov's method play a central role in many such schemes. Adaptive control is a subcase of nonlinear control (see below).

The high speeds of modern digital computers now allow us to implement nonlinear control algorithms not possible with earlier hardware. An example is the computed-torque method for controlling robot arms, which was discussed in Section 11 (see Fig. 47).

14.3 Adaptive Control

The term *adaptive control,* which unfortunately has been loosely used, describes control systems that can change the form of the control algorithm or the values of the control gains in real time, as the controller improves its internal model of the process dynamics or in response to unmodeled disturbances.[16] Constant control gains do not provide adequate response for some systems that exhibit large changes in their dynamics over their entire operating range, and some adaptive controllers use several models of the process, each of which is accurate within a certain operating range. The adaptive controller switches between gain settings that are appropriate for each operating range. Adaptive controllers are difficult to design and are prone to instability. Most existing adaptive controllers change only the gain values, not the form of the control algorithm. Many problems remain to be solved before adaptive control theory becomes widely implemented.

14.4 Optimal Control

A rocket might be required to reach orbit using minimum fuel or it might need to reach a given intercept point in minimum time. These are examples of potential applications of *optimal-control theory.* Optimal-control problems often consist of two subproblems. For the rocket example, these subproblems are (1) the determination of the minimum-fuel (or minimum-time) trajectory and the open-loop control outputs (e.g., rocket thrust as a function of time) required to achieve the trajectory and (2) the design of a feedback controller to keep the system near the optimal trajectory.

Many optimal-control problems are nonlinear, and thus no general theory is available. Two classes of problems that have achieved some practical successes are the *bang-bang control* problem, in which the control variable switches between two fixed values (e.g., on and off or open and closed),[6] and the *linear-quadratic-regulator* (LQG), discussed in Section 7, which has proven useful for high-order systems.[1,6]

Closely related to optimal-control theory are methods based on stochastic process theory, including *stochastic control theory,*[17] *estimators, Kalman filters,* and *observers.*[1,6,17]

REFERENCES

1. W. J. Palm III, *Modeling, Analysis, and Control of Dynamic Systems,* 2nd ed., Wiley, New York, 2000.
2. W. J. Palm III, *Control Systems Engineering,* Wiley, New York, 1986.
3. D. E. Seborg, T. F. Edgar, and D. A. Mellichamp, *Process Dynamics and Control,* Wiley, New York, 1989.
4. W. J. Palm III, *System Dynamics,* McGraw-Hill, New York, 2005.
5. D. McCloy and H. Martin, *The Control of Fluid Power,* 2nd ed., Halsted, London, 1980.
6. A. E. Bryson and Y. C. Ho, *Applied Optimal Control,* Blaisdell, Waltham, MA, 1969.
7. F. Lewis, *Optimal Control,* Wiley, New York, 1986.
8. K. J. Astrom and B. Wittenmark, *Computer Controlled Systems,* Prentice-Hall, Englewood Cliffs, NJ, 1984.
9. Y. Dote, *Servo Motor and Motion Control Using Digital Signal Processors,* Prentice-Hall, Englewood Cliffs, NJ, 1990.
10. W. J. Palm III, *Introduction to MATLAB 7 for Engineers,* McGraw-Hill, New York, 2005.
11. G. Klir and B. Yuan, *Fuzzy Sets and Fuzzy Logic,* Prentice-Hall, Englewood Cliffs, NJ, 1995.
12. B. Kosko, *Neural Networks and Fuzzy Systems,* Prentice-Hall, Englewood Cliffs, NJ, 1992.
13. J. Craig, *Introduction to Robotics,* 3rd ed., Addison-Wesley, Reading, MA, 2005.
14. M. W. Spong and M. Vidyasagar, *Robot Dynamics and Control,* Wiley, New York, 1989.
15. J. Slotine and W. Li, *Applied Nonlinear Control,* Prentice-Hall, Englewood Cliffs, NJ, 1991.
16. K. J. Astrom, *Adaptive Control,* Addison-Wesley, Reading, MA, 1989.
17. R. Stengel, *Stochastic Optimal Control,* Wiley, New York, 1986.

CHAPTER 12

CLOSED-LOOP CONTROL SYSTEM ANALYSIS

Suhada Jayasuriya
Department of Mechanical Engineering
Texas A&M University
College Station, Texas

Reprinted from *Instrumentation and Control,* Wiley, New York, 1990, by permission of the publisher.

1 INTRODUCTION

The field of control has a rich heritage of intellectual depth and practical achievement. From the water clock of Ctesibius in ancient Alexandria, where feedback control was used to regulate the flow of water, to the space exploration and the automated manufacturing plants of today, control systems have played a very significant role in technological and scientific development. James Watt's flyball governor (1769) was essential for the operation of the steam engine, which was, in turn, a technology fueling the Industrial Revolution. The fundamental study of feedback begins with James Clerk Maxwell's analysis of system stability of steam engines with governors (1868). Giant strides in our understanding of feedback and its use in design resulted from the pioneering work of Black, Nyquist, and Bode at Bell Labs in the 1920s. Minorsky's work on ship steering was of exceptional practical and theoretical importance. Tremendous advances occurred during World War II in response to the pressing problems of that period. The technology developed during the war years led, over the next 20 years, to practical applications in many fields.

Since the 1960s, there have been many challenges and spectacular achievements in space. The guidance of the Apollo spacecraft on an optimized trajectory from the earth to the moon and the soft landing on the moon depended heavily on control engineering. Today, the shuttle relies on automatic control in all phases of its flight. In aeronautics, the aircraft autopilot, the control of high-performance jet engines, and ascent/descent trajectory optimization to conserve fuel are typical examples of control applications. Currently, feedback control makes it possible to design aircraft that are aerodynamically unstable (such as the X-29) so as to achieve high performance. The National Aerospace Plane will rely on advanced control algorithms to fly its demanding missions.

Control systems are providing dramatic new opportunities in the automotive industry. Feedback controls for engines permit federal emission levels to be met, while antiskid braking control systems provide enhanced levels of passenger safety. In consumer products, control systems are often a critical factor in performance and thus economic success. From simple thermostats that regulate temperature in buildings to the control of the optics for compact disk systems, from garage door openers to the head servos for computer hard disk drives, and from artificial hearts to remote manipulators, control applications have permeated every aspect of life in industrialized societies.

In process control, where systems may contain hundreds of control loops, adaptive controllers have been available commercially since 1983. Typically, even a small improvement in yield can be quite significant economically. Multivariable control algorithms are now being implemented by several large companies. Moreover, improved control algorithms also permit inventories to be reduced, a particularly important consideration in processing dangerous material. In nuclear reactor control, improved control algorithms can have significant safety and economic consequences. In power systems, coordinated computer control of a large number of variables is becoming common. Over 30,000 computer control systems have been installed in the United States alone. Again, the economic impact of control is vast.

Accomplishments in the defense area are legion. The accomplishments range from the antiaircraft gunsights and the bombsights of World War II to the missile autopilots of today and to the identification and estimation techniques used to track and designate multiple targets.

A large body of knowledge has come into existence as a result of these developments and continues to grow at a very rapid rate. A cross section of this body of knowledge is collected in this chapter. It includes analysis tools and design methodologies based on classical techniques. In presenting this material, some familiarity with general control system design principles is assumed. As a result, detailed derivations have been kept to a minimum with appropriate references.

1.1 Closed-Loop versus Open-Loop Control

In a closed-loop control system the output is measured and used to alter the control inputs applied to the plant under control. Figure 1 shows such a closed-loop system.

If output measurements are not utilized in determining the plant input, then the plant is said to be under open-loop control. An open-loop control system is shown in Fig. 2.

An open-loop system is very sensitive to any plant parameter perturbations and any disturbances that enter the plant. So an open-loop control system is effective only when the plant and disturbances are exactly known. In real applications, however, neither plants nor disturbances are exactly known a priori. In such situations open-loop control does not provide satisfactory performance. When the loop is closed as in Fig. 1 any plant perturbations or disturbances can be sensed by measuring the outputs, thus allowing plant input to be altered appropriately.

The main reason for closed-loop control is the need for systems to perform well in the presence of uncertainties. It can reduce the sensitivity of the system to plant parameter variations and help reject or mitigate external disturbances. Among other attributes of closed-loop control is the ability to alter the overall dynamics to provide adequate stability and good tracking characteristics. Consequently, a closed-loop system is more complex than an open-loop system due to the required sensing of the output. Sensor noise, however, tends to degrade the performance of a feedback system, thus making it necessary to have accurate sensing. Therefore sensors are usually the most expensive devices in a feedback control system.

1.2 Supervisory Control Computers[1]

With the advent of digital computer technology computers have found their way into control applications. The earliest applications of digital computers in industrial process control were in a so-called supervisory control mode, and the computers were generally large-scale mainframes (minis and micros had not yet been developed). In supervisory control the individual feedback loops are locally controlled by typical analog devices used prior to the installation of the computer. The main function of the computer is to gather information on how the entire process is operating, then feed this into an overall process model located in the computer memory, and then periodically send signals to the set points of individual local analog controllers so that the overall performance of the system is enhanced. A conceptual block diagram of such supervisory control is shown in Fig. 3.

1.3 Hierarchical Control Computers[1]

To continuously achieve optimum overall performance, the function of the supervisory control computer can be broken up into several levels and operated in a hierarchical manner.

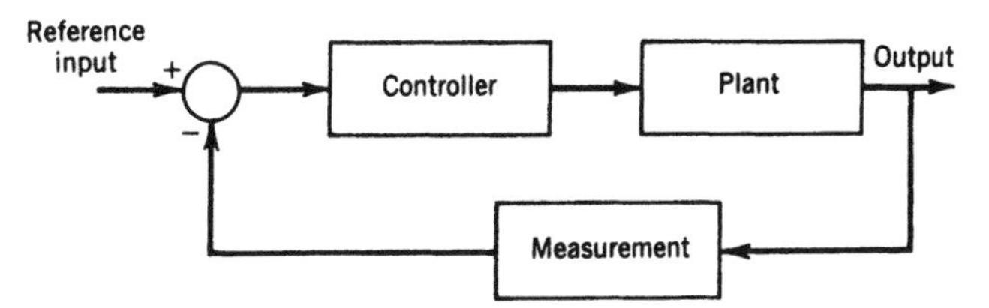

Figure 1 Closed-loop system configuration.

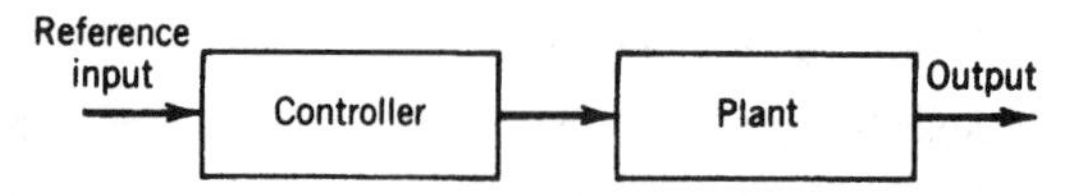

Figure 2 Open-loop system configuration.

Such a multilevel approach becomes useful in controlling very complex processes, with large numbers of control loops with many inputs and controlled variables. In the hierarchical approach the system is subdivided into a hierarchy of simpler control design problems rather than attempting direct synthesis of a single comprehensive controller for the entire process. Thus the controller on a given "level of control" can be less complex due to the existence of lower level controllers that remove frequently occurring disturbances. At each higher level in the hierarchy, the complexity of the control algorithm increases, but the required frequency of execution decreases. Such a system is shown in Fig. 4.

1.4 Direct Digital Control (DDC)

In direct digital control all the analog controllers at the lowest level or intermediate levels are replaced by a central computer serving as a single time-shared controller for all the individual feedback loops. Conventional control modes such as PI or PID (proportional integral differential) are still used for each loop, but the digital versions of the control laws for each loop reside in software in the central computer. In DDC the computer input and output are multiplexed sequentially through the list of control loops, updating each loop's control action in turn and holding this value constant until the next cycle. A typical DDC configuration is shown in Fig. 5.

1.5 Hybrid Control

Combinations of analog and digital methods based on individual loop controllers is known as hybrid control. It is clear that any real control system would have both analog and digital features. For example, in Fig. 5, the plant outputs and inputs are continuous (or analog) quantities. The inputs and outputs of the digital computer are nevertheless digital. The com-

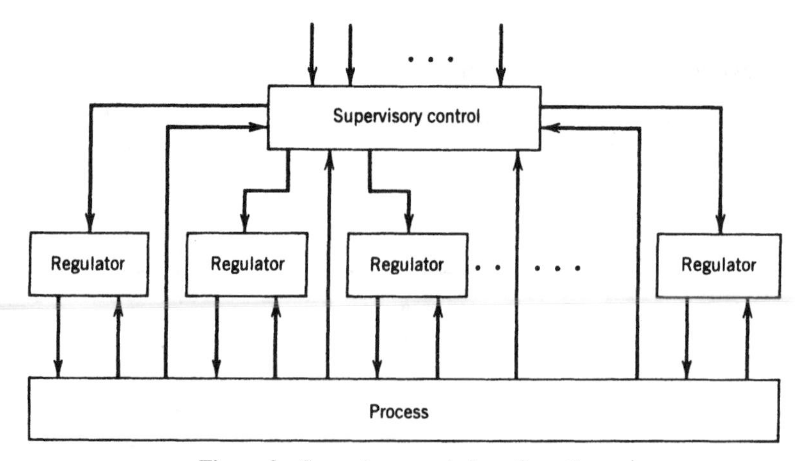

Figure 3 Supervisory control configuration.

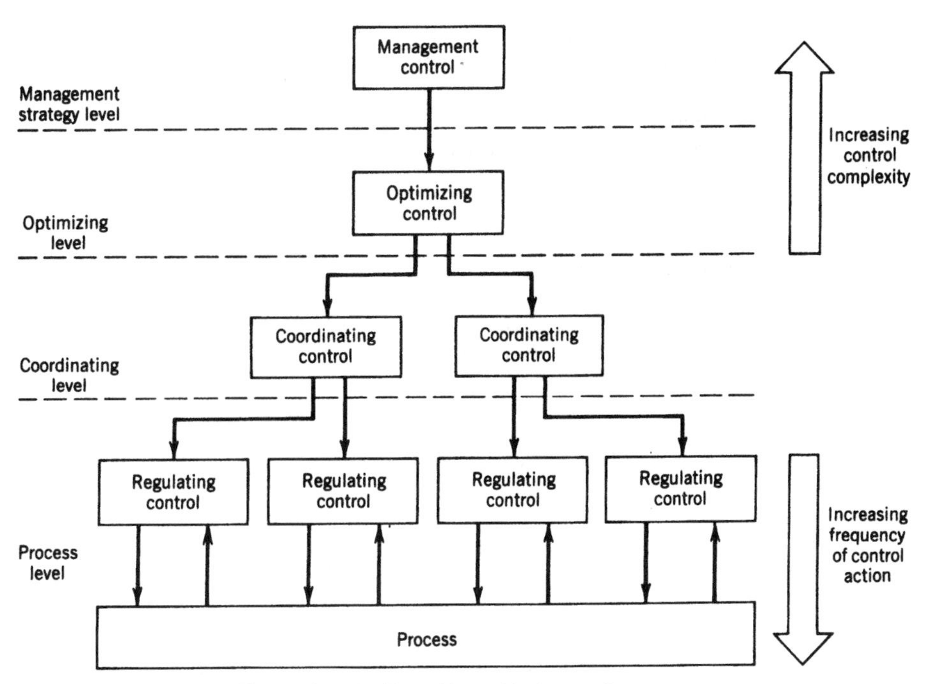

Figure 4 Multilevel hierarchical control structure.

puter processes only numbers. All control algorithms, however, need not be implemented digitally. As an example, for most feedback control loops, the PID algorithm implemented with analog circuitry is superior to the PID algorithm implemented in digital software. The derivative mode, for instance, requires very high resolution to approach the quality of analog performance. Thus it is sometimes advantageous to use hybrid systems for controller implementation.

1.6 Real Systems with Digital Control

Real systems are typically mixtures of analog and digital signals. Typical plant inputs and outputs are analog signals; that is, actuators and sensors are analog systems. If a digital

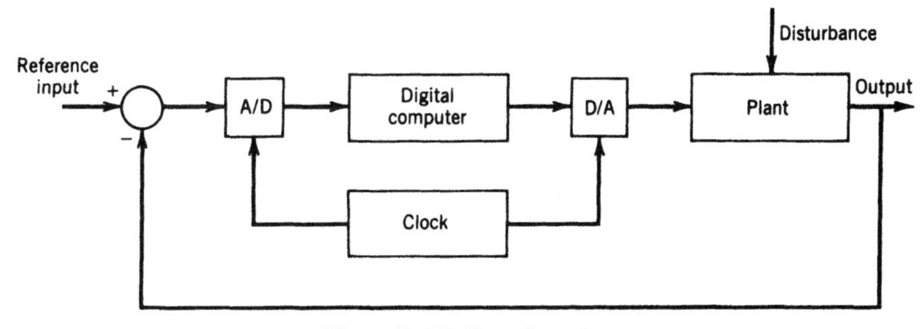

Figure 5 DDC configuration.

computer is included in the control loop, then digitized quantities are needed for their processing. These are usually accomplished by using analog-to-digital (A/D) converters. To use real-world actuators, computations done as numbers need to be converted to analog signals by employing digital-to-analog (D/A) converters. One of the main advantages of digital control is the ease with which an algorithm can be changed online by altering the software rather than hardware.

2 LAPLACE TRANSFORMS

Often in designing control systems, linear time-invariant (LTI) differential equation representations of physical systems to be controlled are sought. These are typically arrived at by appropriately linearizing the nonlinear equations about some operating point. The general form of such an LTI ordinary differential equation representation is

$$a_n \frac{d^n y(t)}{dt^n} + a_{n-1} \frac{d^{n-1} y(t)}{dt^{n-1}} + \cdots + a_1 \frac{dy(t)}{dt} + a_0 y(t)$$

$$= b_m \frac{d^m}{dt^m} u(t) + b_{m-1} \frac{d^{m-1}}{dt^{m-1}} u(t) + \cdots + b_1 \frac{du(t)}{dt} + b_0 u(t) \tag{1}$$

where $y(t)$ = output of the system
$u(t)$ = input to the system
t = time
a_j, b_j = physical parameters of the system

and $n \geq m$ for physical systems.

The ability to transform systems of the form given by Eq. (1) to algebraic equations relating the input to the output is the primary reason for employing Laplace transform techniques.

2.1 Single-Sided Laplace Transform

The Laplace transform $\mathcal{L}[f(t)]$ of the time function $f(t)$ defined as

$$f(t) = \begin{cases} 0 & t < 0 \\ f(t) & t \geq 0 \end{cases}$$

is given by

$$\mathcal{L}[f(t)] = F(s) = \int_0^\infty e^{-st} f(t)\, dt \qquad t > 0 \tag{2}$$

where s is a complex variable ($\sigma + j\omega$). The integral of Eq. (2) cannot be evaluated in a closed form for all $f(t)$, but when it can, it establishes a unique pair of functions, $f(t)$ in the time domain and its companion $F(s)$ in the s domain. It is conventional to use uppercase letters for s functions and lowercase for t functions.

Example 1 Determine the Laplace transform of the unit step function $u_s(t)$:

$$u_s(t) = \begin{cases} 0 & t < 0 \\ 1 & t \geq 0 \end{cases}$$

By definition

$$\mathcal{L}[u_s(t)] = U_s = \int_0^\infty e^{-ts}1\,dt = \frac{1}{s}$$

Example 2 Determine the Laplace transform of $f(t)$:

$$f(t) = \begin{cases} 0 & t < 0 \\ e^{-\alpha t} & t \geq 0 \end{cases}$$

$$\mathcal{L}[f(t)] = F(s) = \int_0^\infty e^{-ts}e^{-\alpha t}\,dt = \frac{1}{s + \alpha}$$

Example 3 Determine the Laplace transform of the function $f(t)$ given by

$$f(t) = \begin{cases} 0 & t < 0 \\ t & 0 \leq t \leq T \\ T & T \leq t \end{cases}$$

By definition

$$\begin{aligned} F(s) &= \int_0^\infty e^{-ts}f(t)\,dt \\ &= \int_0^T e^{-ts}t\,dt + \int_T^\infty e^{-ts}T\,dt \\ &= \frac{1}{s^2} - \frac{e^{-Ts}}{s^2} = \frac{1}{s^2}(1 - e^{-Ts}) \end{aligned}$$

In transforming differential equations, entire equations need to be transformed. Several theorems useful in such transformations are given next without proof.

T1. *Linearity Theorem*

$$\mathcal{L}[\alpha f(t) + \beta g(t)] = \alpha\mathcal{L}[f(t)] + \beta\mathcal{L}[g(t)] \tag{3}$$

T2. *Differentiation Theorem*

$$\mathcal{L}\left[\frac{df}{dt}\right] = sF(s) - f(0) \tag{4}$$

$$\mathcal{L}\left[\frac{d^2f}{dt^2}\right] = s^2F(s) - sf(0) = \frac{df}{dt}(0) \tag{5}$$

$$\mathcal{L}\left[\frac{d^nf}{dt^n}\right] = s^nF(s) - s^{n-1}f(0) - s^{n-2}\frac{df}{dt}(0) - \cdots - \frac{d^{n-1}}{dt^{n-1}}(0) \tag{6}$$

T3. *Translated Function* (Fig. 6)

$$\mathcal{L}[f(t - \alpha)u_s(t - \alpha)] = e^{-\alpha s}F(s + \alpha) \tag{7}$$

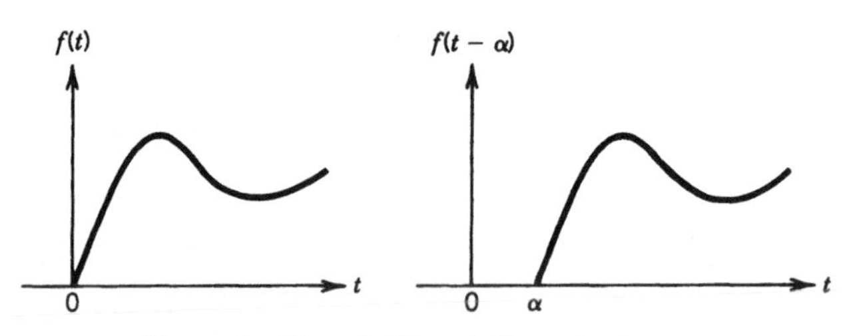

Figure 6 Plots of $f(t)$ and $f(t - \alpha)u_s(t - \alpha)$.

T4. *Multiplication of $f(t)$ by $e^{-\alpha t}$*

$$\mathcal{L}[e^{-\alpha t}f(t)] = F(s + \alpha) \tag{8}$$

T5. *Integration Theorem*

$$\mathcal{L}[\int f(t)\, dt] = \frac{F(s)}{s} + \frac{f^{-1}(0)}{s} \tag{9}$$

where $f^{-1}(0) = \int f(t)\, dt$ evaluated at $t = 0$.

T6. *Final-Value Theorem.* If $f(t)$ and $df(t)/dt$ are Laplace transformable, if $\lim_{t\to\infty} f(t)$ exists, and if $F(s)$ is analytic in the right-half s-plane including the $j\omega$ axis, except for a single pole at the origin, then

$$\lim_{t\to\infty} f(t) = \lim_{s\to 0} sF(s) \tag{10}$$

T7. *Initial-Value Theorem.* If $f(t)$ and $df(t)/dt$ are both Laplace transformable, and if $\lim_{s\to\infty} sF(s)$ exists, then

$$f(0) = \lim_{s\to 0} sF(s) \tag{11}$$

Example 4 The time function of Example 3 can be written as

$$f(t) = tu_s(t) - (t - T)u_s(t - T) \tag{12}$$

and

$$\mathcal{L}[f(t)] = \mathcal{L}[tu_s(t)] - \mathcal{L}[(t - T)u_s(t - T)] \tag{13}$$

But

$$\mathcal{L}[tu_s(t)] = \frac{1}{s^2} \tag{14}$$

By using Eqs. (14) and (7) in Eq. (13), we get

$$F(s) = \frac{1}{s^2} - e^{-Ts}\frac{1}{s^2} = \frac{1}{s^2}(1 - e^{-Ts})$$

2.2 Transforming LTI Ordinary Differential Equations

The Laplace transform method yields the complete solution (the particular solution plus the complementary solution) of linear differential equations. Classical methods of finding the complete solution of a differential equation require the evaluation of the integration constants by use of the initial conditions. In the case of the Laplace transform method, initial conditions are automatically included in the Laplace transform of the differential equation. If all initial conditions are zero, then the Laplace transform of the differential equation is obtained simply by replacing d/dt with s, d^2/dt^2 with s^2, and so on.

Consider the differential equation

$$5\frac{d^2y}{dt^2} + 3y = f(t) \tag{15}$$

with $\dot{y}(0) = 1$, $y(0) = 0$. By taking the Laplace transform of Eq. (15), we get

$$\mathcal{L}\left[5\frac{d^2y}{dt^2} + 3y\right] = \mathcal{L}[f(t)]$$

Now using Theorems T1 and T2 (page 449) yields

$$5[s^2Y(s) - sy(0) - \dot{y}(0)] + 3Y(s) = F(s)$$

Thus

$$Y(s) = \frac{F(s) + 5}{5s^2 + 3} \tag{16}$$

For a given $f(t)$, say $f(t) = u_s(t)$, the unit step function is given as

$$F(s) = \frac{1}{s} \tag{17}$$

Substituting Eq. (17) in Eq. (16) gives

$$Y(s) = \frac{5s + 1}{s(5s^2 + 3)}$$

Then $y(t)$ can be found by computing the inverse Laplace transform of $Y(s)$. That is,

$$y(t) = \mathcal{L}^{-1}[Y(s)] = \mathcal{L}^{-1}\left\{\frac{5s + 1}{s(5s^2 + 3)}\right\}$$

This will be discussed in Section 2.4.

2.3 Transfer Function

The transfer function of a LTI system is defined to be the ratio of the Laplace transform of the output to the Laplace transform of the input under the assumption that all initial conditions are zero.

For the system described by the LTI differential equation (1),

$$\frac{\mathcal{L}[\text{output}]}{\mathcal{L}[\text{input}]} = \frac{\mathcal{L}[y(t)]}{\mathcal{L}[u(t)]} = \frac{Y(s)}{U(s)}$$

$$= \frac{b_m s^m + b_{m-1}s^{m-1} + \cdots + b_1 s + b_0}{a_n s^n + a_{n-1}s^{n-1} + \cdots + a_1 s + a_0} \tag{18}$$

It should be noted that the transfer function depends only on the system parameters characterized by a_i's and b_i's in Eq. (18). The highest power of s in the denominator of the transfer function defines the order n of the system. The order n of a system is typically greater than or equal to the order of the numerator polynomial m.

Equation (18) can be further written in the form

$$\frac{Y(s)}{U(s)} = G(s) = \frac{b_m \prod_{j=1}^{m} (s + z_i)}{a_n \prod_{j=1}^{n} (s + p_j)} \tag{19}$$

The values of s making Eq. (19) equal to zero are called the system zeros, and the values of s making Eq. (19) go to ∞ are called poles of the system. Hence $s = -z_i$, $i = 1, \ldots, m$, are the system zeros and $s = -p_j$, $j = 1, \ldots, n$, are the system poles.

2.4 Partial-Fraction Expansion and Inverse Transform

The mathematical process of passing from the complex variable expression $F(s)$ to the time-domain expression $f(t)$ is called an inverse transformation. The notation for the inverse Laplace transformation is $\mathcal{L}^{-1}$, so that

$$\mathcal{L}^{-1}[F(s)] = f(t)$$

Example 5 The inverse Laplace transform of $1/s$ is the unit step function:

$$\mathcal{L}[u_s(t)] = \frac{1}{s}$$

Hence

$$\mathcal{L}^{-1}\left(\frac{1}{s}\right) = u_s(t)$$

Time functions for which Laplace transforms are found in a closed form can be readily inverse Laplace transformed by writing them in pairs. Such pairs are listed in Table 1. When transform pairs are not found in tables, other methods have to be used to find the inverse Laplace transform. One such method is the partial-fraction expansion of a given Laplace transform.

If $F(s)$, the Laplace transform of $f(t)$, is broken up into components

$$F(s) = F_1(s) + F_2(s) + \cdots + F_n(s)$$

and if the inverse Laplace transforms of $F_1(s)$, $F_2(s)$, . . . , $F_n(s)$ are readily available, then

$$\begin{aligned}\mathcal{L}^{-1}[F(s)] &= \mathcal{L}^{-1}[F_1(s)] + \mathcal{L}^{-1}[F_2(s)] + \cdots + \mathcal{L}^{-1}[F_n(s)] \\ &= f_1(t) + f_2(t) + \cdots + f_n(t)\end{aligned} \tag{20}$$

where $f_1(t)$, $f_2(t)$, . . . , $f_n(t)$ are the inverse Laplace transforms of $F_1(s)$, $F_2(s)$, . . . , $F_n(s)$, respectively.

For problems in control systems, $F(s)$ is frequently in the form

$$F(s) = \frac{B(s)}{A(s)}$$

where $A(s)$ and $B(s)$ are polynomials in s and the degree of $B(s)$ is equal to or higher than that of $A(s)$.

Table 1 Laplace Transform Pairs

No.	$f(t)$	$F(s)$
1	Unit impulse $\delta(t)$	1
2	Unit step $u_s(t)$	$\dfrac{1}{s}$
3	t	$\dfrac{1}{s^2}$
4	e^{-at}	$\dfrac{1}{s+a}$
5	te^{-at}	$\dfrac{1}{(s+a)^2}$
6	$\sin \omega t$	$\dfrac{\omega}{s^2+\omega^2}$
7	$\cos \omega t$	$\dfrac{s}{s^2+\omega^2}$
8	$t^n \ (n = 1, 2, 3, \ldots)$	$\dfrac{n!}{s^{n+1}}$
9	$t^n e^{-at} \ (n = 1, 2, 3, \ldots)$	$\dfrac{n!}{(s+a)^{n+1}}$
10	$\dfrac{1}{b-a}(e^{-at} - e^{-bt})$	$\dfrac{1}{(s+a)(s+b)}$
11	$\dfrac{1}{b-a}(be^{-bt} - ae^{-at})$	$\dfrac{s}{(s+a)(s+b)}$
12	$\dfrac{1}{ab}\left[1 + \dfrac{1}{a-b}(be^{-at} - ae^{-bt})\right]$	$\dfrac{1}{s(s+a)(s+b)}$
13	$e^{-at}\sin \omega t$	$\dfrac{\omega}{(s+a)^2+\omega^2}$
14	$e^{-at}\cos \omega t$	$\dfrac{s+a}{(s+a)^2+\omega^2}$
15	$\dfrac{1}{a^2}(at - 1 + e^{-at})$	$\dfrac{1}{s^2(s+a)}$
16	$\dfrac{\omega_n}{\sqrt{1-\zeta^2}} e^{-\zeta\omega_n t} \sin \omega_n \sqrt{1-\zeta^2}t$	$\dfrac{\omega_n^2}{s^2 + 2\zeta\omega_n s + \omega_n^2}$
17	$\dfrac{-1}{\sqrt{1-\zeta^2}} e^{-\zeta\omega_n t} \sin(\omega_n \sqrt{1-\zeta^2}t - \phi)$ $\phi = \tan^{-1}\dfrac{\sqrt{1-\zeta^{*2}}}{\zeta}$	$\dfrac{s}{s^2 + 2\zeta\omega_n s + \omega_n^2}$
18	$1 - \dfrac{1}{\sqrt{1-\zeta^2}} e^{-\zeta\omega_n t} \sin(\omega_n \sqrt{1-\zeta^2}t + \phi)$ $\phi = \tan^{-1}\dfrac{\sqrt{1-\zeta^2}}{\zeta}$	$\dfrac{\omega_n^2}{s(s^2 + 2\zeta\omega_n s + \omega_n^2)}$

If $F(s)$ is written as in Eq. (19) and if the poles of $F(s)$ are distinct, then $F(s)$ can always be expressed in terms of simple partial fractions as follows:

$$F(s) = \frac{B(s)}{A(s)} = \frac{\alpha_1}{s + p_1} + \frac{\alpha_2}{s + p_2} + \cdots + \frac{\alpha_n}{s + p_n} \tag{21}$$

where the α_j's are constant. Here α_j is called the residue at the pole $s = -p_j$. The value of α_j is found by multiplying both sides of Eq. (21) by $s + p_j$ and setting $s = -p_j$, giving

$$\alpha_j = \left[(s + p_j)\frac{B(s)}{A(s)}\right]_{s=p_j} \tag{22}$$

Noting that

$$\mathcal{L}^{-1}\left[\frac{1}{s + p_j}\right] = e^{-p_j t}$$

the inverse Laplace transform of Eq. (21) can be written as

$$f(t) = \mathcal{L}^{-1}[F(s)] = \sum_{j=1}^{n} \alpha_j e^{-p_j t} \tag{23}$$

Example 6 Find the inverse Laplace transform of

$$F(s) = \frac{s + 1}{(s + 2)(s + 3)}$$

The partial-fraction expansion of $F(s)$ is

$$F(s) = \frac{s + 1}{(s + 2)(s + 3)} = \frac{\alpha_1}{s + 2} + \frac{\alpha_2}{s + 3}$$

where α_1 and α_2 are found by using Eq. (22) as follows:

$$\alpha_1 = \left[(s + 2)\frac{s + 1}{(s + 2)(s + 3)}\right]_{s=-2} = -1$$

$$\alpha_2 = \left[(s + 3)\frac{s + 1}{(s + 2)(s + 3)}\right]_{s=-3} = 2$$

Thus

$$\begin{aligned} f(t) &= \mathcal{L}^{-1}\left[\frac{-1}{s + 2}\right] + \mathcal{L}^{-1}\left[\frac{2}{s + 3}\right] \\ &= -e^{-2t} + 2e^{-3t} \qquad t \geq 0 \end{aligned}$$

For partial-fraction expansion when $F(s)$ involves multiple poles, consider

$$F(s) = \frac{K\prod_{i=1}^{m}(s + z_i)}{(s + p_1)^r \prod_{j=r+1}^{n}(s + p_j)} = \frac{B(s)}{A(s)}$$

where the pole at $s = -p_1$ has multiplicity r. The partial-fraction expansion of $F(s)$ may then be written as

$$F(s) = \frac{B(s)}{A(s)} = \frac{\beta_1}{s + p_1} + \frac{\beta_2}{(s + p_1)^2} + \cdots + \frac{\beta_r}{(s + p_1)^r}$$

$$+ \sum_{j=r+1}^{n} \frac{\alpha_j}{s + p_j} \tag{24}$$

The α_j's can be evaluated as before using Eq. (22). The β_i's are given by

$$\beta_r = \left[\frac{B(s)}{A(s)}(s + p_1)^r\right]_{s=-p_1}$$

$$\beta_{r-1} = \left\{\frac{d}{ds}\left[\frac{B(s)}{A(s)}(s + p_1)^r\right]\right\}_{s=-p_1}$$

$$\vdots$$

$$\beta_{r-j} = \frac{1}{j!}\left\{\frac{d^j}{ds^j}\left[\frac{B(s)}{A(s)}(s + p_1)^r\right]\right\}_{s=-p_1}$$

$$\vdots$$

$$\beta_1 = \frac{1}{(r-1)!}\left\{\frac{d^{r-1}}{ds^{r-1}}\left[\frac{B(s)}{A(s)}(s + p_1)^r\right]\right\}_{s=-p_1}$$

The inverse Laplace transform of Eq. (24) can then be obtained as follows:

$$f(t) = \left[\frac{\beta_r}{(r-1)!}t^{r-1} + \frac{\beta_{r-1}}{(r-2)!}t^{r-2} + \cdots + \beta_2 t + \beta_1\right]e^{-p_1 t}$$

$$+ \sum_{j=r+1}^{n} \alpha_j e^{-p_1 t} \qquad t \geq 0 \tag{25}$$

Example 7 Find the inverse Laplace transform of the function

$$F(s) = \frac{s^2 + 2s + 3}{(s + 1)^3}$$

Expanding $F(s)$ into partial fractions, we obtain

$$F(s) = \frac{B(s)}{A(s)} = \frac{\beta_1}{s + 1} = \frac{\beta_2}{(s + 1)^2} + \frac{\beta_3}{(s + 1)^3}$$

where β_1, β_2, and β_3 are determined as already shown:

$$\beta_3 = \left[\frac{B(s)}{A(s)}(s + 1)^3\right]_{s=-1} = 2$$

similarly $\beta_2 = 0$ and $\beta_1 = 1$. Thus we get

$$f(t) = \mathcal{L}^{-1}\left[\frac{1}{s + 1}\right] + \mathcal{L}^{-1}\left[\frac{2}{(s + 1)^3}\right]$$

$$= e^{-t} + t^2 e^{-t} \qquad t \geq 0$$

2.5 Inverse Transform by a General Formula

Given the Laplace transform $F(s)$ of a time function $f(t)$, the following expression holds true:

$$f(t) = \frac{1}{2\pi j} \int_{c-j\infty}^{c+j\infty} F(s)e^{ts}\, ds \tag{26}$$

where c, the abscissa of convergence, is a real constant and is chosen larger than the real parts of all singular points of $F(s)$. Thus the path of integration is parallel to the $j\omega$ axis and is displaced by the amount c from it (see Fig. 7). This path of integration is to the right of all singular points. Equation (26) need be used only when other simpler methods cannot provide the inverse Laplace transformation.

3 BLOCK DIAGRAMS

A control system consists of a number of components. To show the functions performed by each component, we commonly use a diagram called a *block diagram.*

In a block diagram, all system variables are linked to each other through blocks. The block is a symbol for the mathematical operation on the input signal to the block that produces the output. The transfer functions of the components are usually entered in the blocks, with blocks connected by arrows to indicate the direction of the flow of signals. A basic assumption in block diagrams is that there is no loading between blocks. Figure 8 shows a simple block diagram with two blocks in cascade. The arrowheads pointing toward blocks indicate inputs and those pointing away indicate outputs. These arrows are referred to as signals. The following basic components allow the generation of many complex block diagrams.

Addition. Signal addition is represented by a summing point, as shown in Fig. 9.

It should be noted that there can be only one signal leaving a summing point though any number of signals can enter a summing point. For the summing point shown in Fig. 9, we have

$$-x_1 + x_2 + x_3 = y$$

The sign placed near an arrow indicates whether the signal is to be added or subtracted.

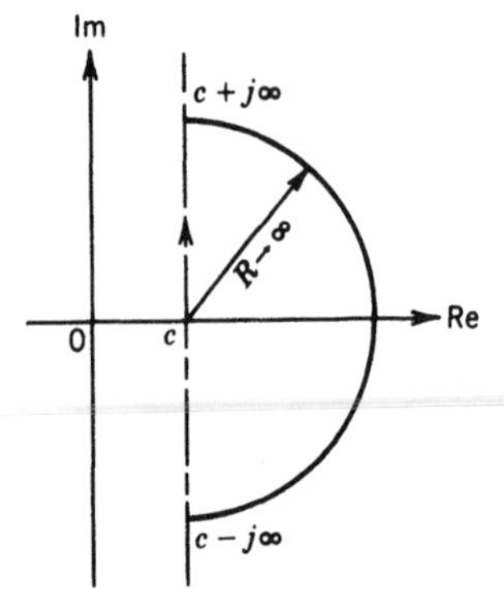

Figure 7 Path of integration.

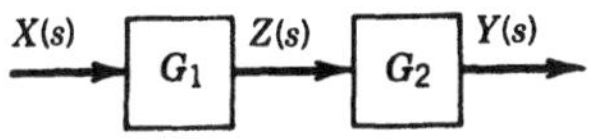

Figure 8 Two blocks in cascade.

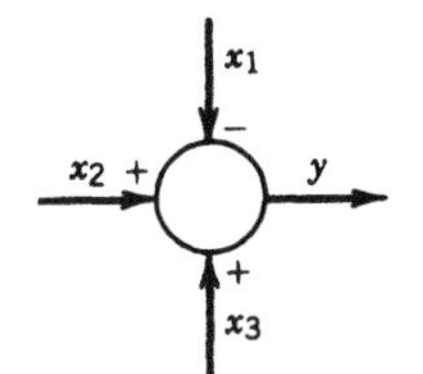

Figure 9 Summing point.

Multiplication. Multiplication is denoted by a symbol as shown in Fig. 10. Here the input X and the output Y are related by the expression

$$Y = GX$$

Takeoff Point. If a signal becomes an input to more than one element, then a takeoff point as shown in Fig. 11 is employed.

A typical block diagram using these elements is shown in Fig. 12.

In the block diagram of Fig. 12 it is assumed that G_2 will have no back reaction (or loading) on G_1; G_1 and G_2 usually represent two physical devices. If there are any loading effects between the devices, it is necessary to combine these components into a single block. Such a situation is given in Section 3.2.

3.1 Block Diagram Reduction

Any number of cascaded blocks representing nonloading components can be replaced by a single block, the transfer function of which is simply the product of the individual transfer functions. For example, the two elements in cascade shown in Fig. 12 can be replaced by a single element $G_0 = G_1G_2$. Some fundamental block diagram reduction rules are shown in Fig. 13.

In simplifying a block diagram a general rule is to first move takeoff points and summing points to form internal feedback loops as shown in Fig. 13*d*. Then remove the internal feedback loops and proceed in the same manner.

Example 8 Reduce the block diagram shown in Fig. 14.

First, to eliminate the loop $G_3G_4H_1$, we move H_2 behind block G_4 and therefore obtain Fig. 15*a*. Eliminating the loop $G_3G_4H_1$, we obtain Fig. 15*b*. Then eliminating the inner loop containing H_2/G_4, we obtain Fig. 15*c*. Finally, by reducing the loop containing H_3, we obtain the closed-loop system transfer function as shown in Fig. 15*d*.

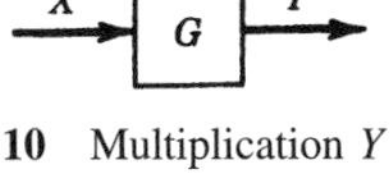

Figure 10 Multiplication $Y = GX$.

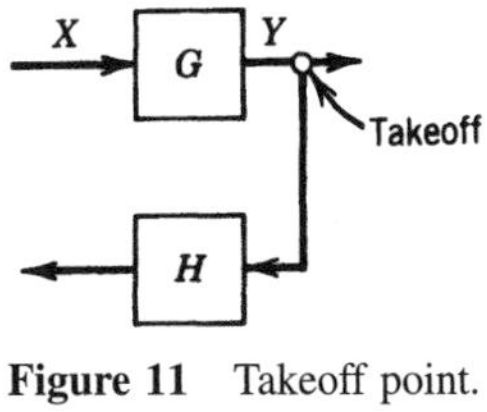

Figure 11 Takeoff point.

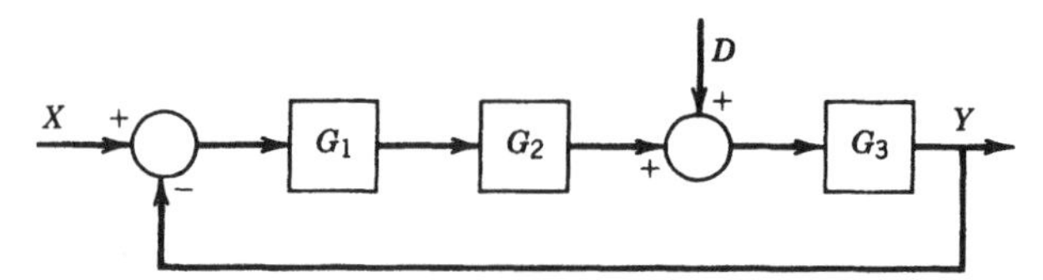

Figure 12 Typical block diagram.

3.2 Transfer Functions of Cascaded Elements[2]

Many feedback systems have components that load each other. Consider the system of Fig. 16. Assume that e_i is the input and e_o is the output. In this system the second stage of the circuit (R_2C_2 portion) produces a loading effect on the first stage (R_1C_1 portion). The equations for this system are

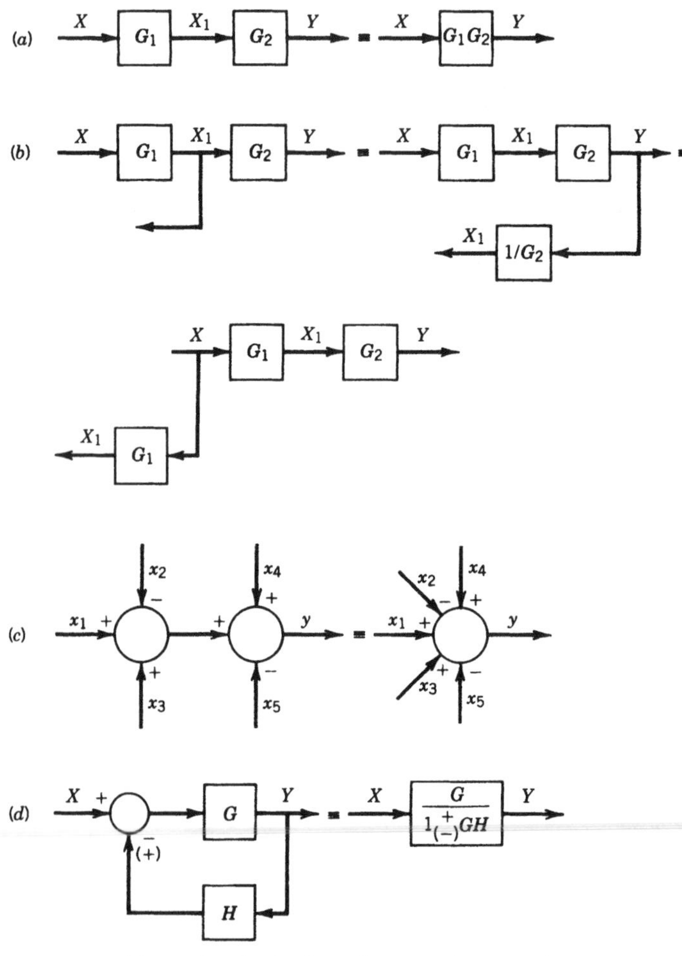

Figure 13 Basic block diagram reduction rules.

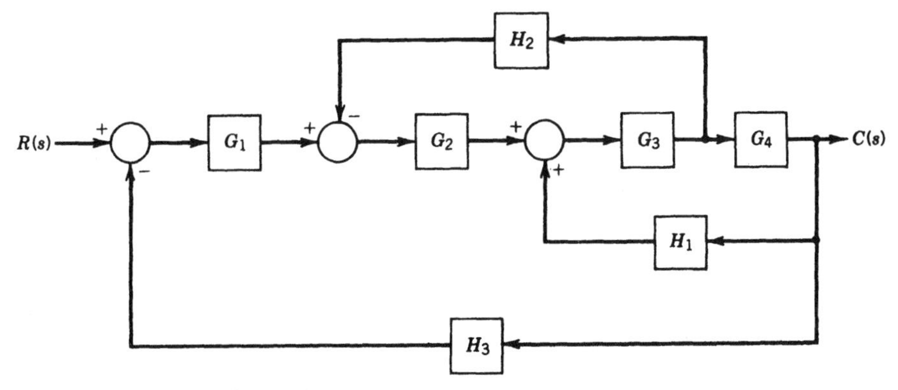

Figure 14 Multiple-loop feedback control system.

$$\frac{1}{C_1}\int (i_1 - i_2)\,dt + R_1 i_1 = e_i \tag{27}$$

and

$$\frac{1}{C_1}\int (i_1 - i_2)\,dt + R_2 i_2 = -\frac{1}{C_2}\int i_2\,dt = -e_o \tag{28}$$

Taking the Laplace transform of Eqs. (27) and (28), assuming zero initial conditions and simplifying yield

$$\frac{E_o(s)}{E_i(s)} = \frac{1}{R_1C_1R_2C_2s^2 + (R_1C_1 + R_2C_2 + R_1C_2)s + 1} \tag{29}$$

The term R_1C_2s in the denominator of the transfer function represents the interaction of two simple *RC* circuits.

This analysis shows that if two *RC* circuits are connected in cascade so that the output from the first circuit is the input to the second, the overall transfer function is not the product of $1/(R_1C_1s + 1)$ and $1/(R_2C_2s + 1)$. The reason for this is that when we derive the transfer function for an isolated circuit, we implicitly assume that the output is unloaded. In other words, the load impedance is assumed to be infinite, which means that no power is being withdrawn at the output. When the second circuit is connected to the output of the first, however, a certain amount of power is withdrawn and then the assumption of no loading is violated. Therefore, if the transfer function of this system is obtained under the assumption of no loading, then it is not valid. Chapter 2 deals with this type of problem in greater detail.

4 *z*-TRANSFORMS

One of the mathematical tools commonly used to deal with discrete-time systems is the *z*-transform. The role of the *z*-transform in discrete-time systems is similar to that of the Laplace transform in continuous-time systems. Laplace transforms allow the conversion of linear ordinary differential equations with constant coefficients into algebraic equations in *s*. The *z*-transformation transforms linear difference equations with constant coefficients into algebraic equations in *z*.

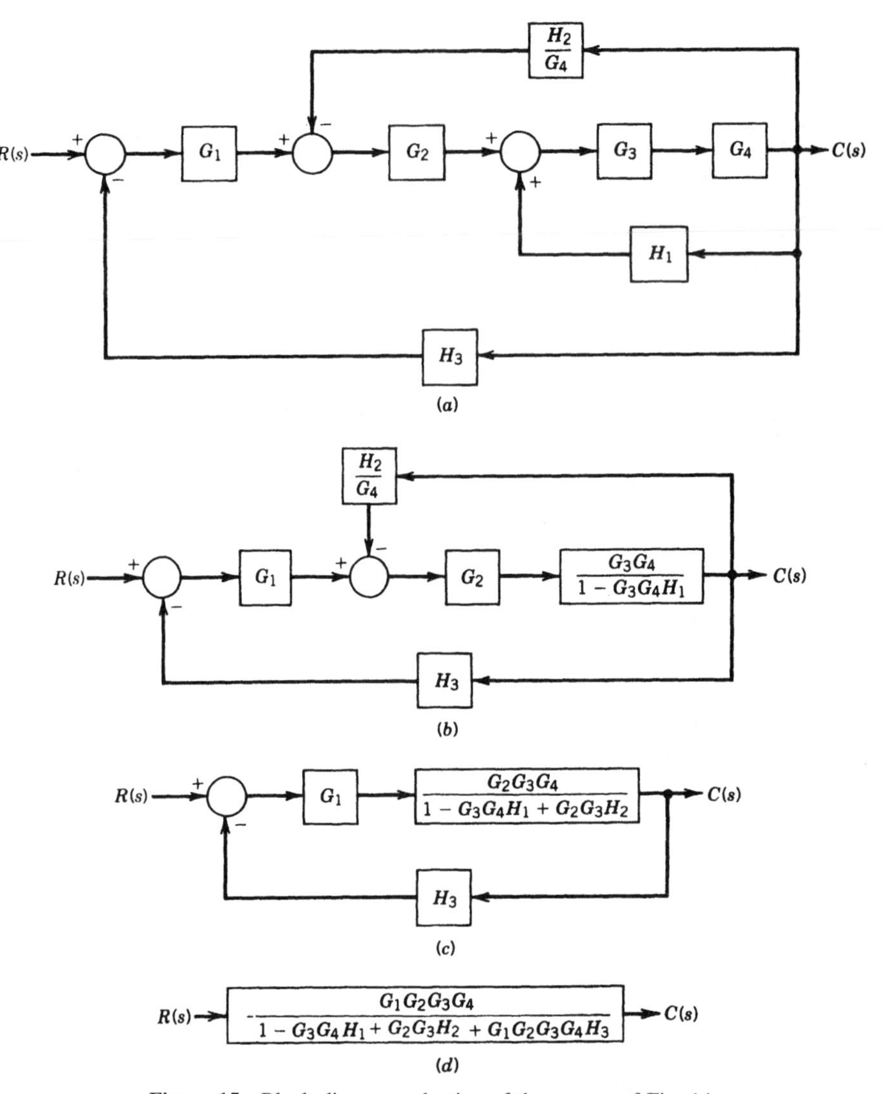

Figure 15 Block diagram reduction of the system of Fig. 14.

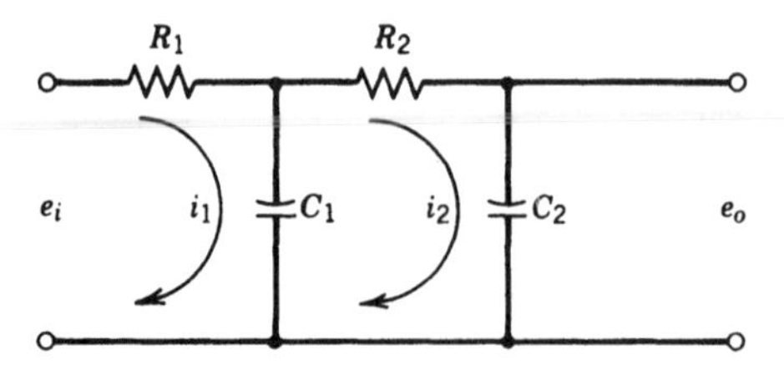

Figure 16 Electrical system.

4.1 Single-Sided z-Transform

If a signal has discrete values $f_0, f_1, \ldots, f_k, \ldots$, we define the z-transform of the signal as the function

$$F(z) = \mathcal{Z}\{f(k)\}$$

$$= \sum_{k=0}^{\infty} f(k)z^{-k} \qquad r_0 \le |z| \le R_0 \tag{30}$$

and is assumed that one can find a range of values of the magnitude of the complex variable z for which the series of Eq. (30) converges. This z-transform is referred to as the one-sided z-transform. The symbol $\mathcal{Z}$ denotes "the z-transform of." In the one-sided z-transform it is assumed $f(k) = 0$ for $k < 0$, which in the continuous-time case corresponds to $f(t) = 0$ for $t < 0$.

Expansion of the right-hand side of Eq. (30) gives

$$F(z) = f(0) + f(1)z^{-1} + f(2)z^{-2} + \cdots + f(k)z^{-k} + \cdots \tag{31}$$

The last equation implies that the z-transform of any continuous-time function $f(t)$ may be written in the series form by inspection. The z^{-k} in this series indicates the instant in time at which the amplitude $f(k)$ occurs. Conversely, if $F(z)$ is given in the series form of Eq. (31), the inverse z-transform can be obtained by inspection as a sequence of the function $f(k)$ that corresponds to the values of $f(t)$ at the respective instants of time. If the signal is sampled at a fixed sampling period T, then the sampled version of the signal $f(t)$ given by $f(0), f(1), \ldots, f(k)$ correspond to the signal values at the instants $0, T, 2T, \ldots, kT$.

4.2 Poles and Zeros in the z-Plane

If the z-transform of a function takes the form

$$F(z) = \frac{b_0 z^m + b_1 z^{m-1} + \cdots + b_m}{z^n + a_1 z^{n-1} + \cdots + a_n} \tag{32}$$

or

$$F(z) = \frac{b_0(z - z_1)(z - z_2)\cdots(z - z_m)}{(z - p_1)(z - p_2)\cdots(z - p_n)}$$

then p_i's are the poles of $F(z)$ and z_i's are the zeros of $F(z)$.

In control studies, $F(z)$ is frequently expressed as a ratio of polynomials in z^{-1} as follows:

$$F(z) = \frac{b_0 z^{-(n-m)} + b_1 z^{-(n-m+1)} + \cdots + b_m z^{-n}}{1 + a_1 z^{-1} + a_2 z^{-2} + \cdots + a_n z^{-n}} \tag{33}$$

where z^{-1} is interpreted as the unit delay operator.

4.3 z-Transforms of Some Elementary Functions

Unit Step Function

Consider the unit step function

$$u_s(t) = \begin{cases} 1 & t \ge 0 \\ 0 & t < 0 \end{cases}$$

whose discrete representation is

$$u_s(k) = \begin{cases} 1 & k \geq 0 \\ 0 & k < 0 \end{cases}$$

From Eq. (30)

$$\mathcal{Z}\{u_s(k)\} = \sum_{k=0}^{\infty} 1 \cdot z^{-k} = \sum_{k=0}^{\infty} z^{-k} = 1 + z^{-1} + z^{-2} + \cdots$$

$$= \frac{1}{1 - z^{-1}} = \frac{z}{z - 1} \tag{34}$$

Note that the series converges for $|z| > 1$. In finding the z-transform, the variable z acts as a dummy operator. It is not necessary to specify the region of z over which $\mathcal{Z}\{u_s(k)\}$ is convergent. The z-transform of a function obtained in this manner is valid throughout the z-plane except at the poles of the transformed function. The $u_s(k)$ is usually referred to as the unit step sequence.

Exponential Function

Let

$$f(t) = \begin{cases} e^{-at} & t \geq 0 \\ 0 & t < 0 \end{cases}$$

The sampled form of the function with sampling period T is

$$f(kT) = e^{-akT} \qquad k = 0, 1, 2, \ldots$$

By definition

$$F(z) = \mathcal{Z}\{f(k)\}$$

$$= \sum_{k=0}^{\infty} f(k) z^{-k}$$

$$= \sum_{k=0}^{\infty} e^{-akt} z^{-k}$$

$$= 1 + e^{-at} z^{-1} + e^{-2at} z^{-2} + \cdots$$

$$= \frac{1}{1 - e^{-at} z^{-1}}$$

$$= \frac{z}{z - e^{-at}}$$

4.4 Some Important Properties and Theorems of the z-Transform

In this section some useful properties and theorems are stated without proof. It is assumed that the time function $f(t)$ is z-transformable and that $f(t)$ is zero for $t < 0$.

P1. If

$$F(z) = \mathcal{Z}\{f(k)\}$$

then

$$\mathcal{Z}\{af(k)\} = aF(z) \tag{35}$$

P2. If $f_1(k)$ and $g_1(k)$ are z-transformable and α and β are scalars, then $f(k) = \alpha f_1(k) + \beta g_1(k)$ has the z-transform

$$F(z) = \alpha F_1(z) + \beta G_1(z) \tag{36}$$

where $F_1(z)$ and $G_1(z)$ are the z-transforms of $f_1(k)$ and $g_1(k)$, respectively.

P3. If

$$F(z) = \mathcal{Z}\{f(k)\}$$

then

$$\mathcal{Z}\{a^k f(k)\} = F\left(\frac{z}{a}\right) \tag{37}$$

T1. *Shifting Theorem.* If $f(t) \equiv 0$ for $t < 0$ and $f(t)$ has the z-transform $F(z)$, then

$$\mathcal{Z}\{f(t + nT)\} = z^{-n}F(z)$$

and

$$\mathcal{Z}\{f(t + nT)\} = z^n\left[F(z) - \sum_{k=0}^{n-1} f(kT)z^{-k}\right] \tag{38}$$

T2. *Complex Translation Theorem.* If

$$F(z) = \mathcal{Z}\{f(t)\}$$

then

$$\mathcal{Z}\{e^{-at}f(t)\} = F(ze^{at}) \tag{39}$$

T3. *Initial-Value Theorem.* If $F(z) = \mathcal{Z}\{f(t)\}$ and if $\lim_{z\to\infty} F(z)$ exists, then the initial value $f(0)$ of $f(t)$ or $f(k)$ is given by

$$f(0) = \lim_{z\to\infty} F(z) \tag{40}$$

T4. *Final-Value Theorem.* Suppose that $f(k)$, where $f(k) \equiv 0$ for $k < 0$, has the z-transform $F(z)$ and that all the poles of $F(z)$ lie inside the unit circle, with the possible exception of a simple pole at $z = 1$. Then the final value of $f(k)$ is given by

$$\lim_{k\to\infty} f(k) = \lim_{z\to 1} [(1 - z^{-1})F(z)] \tag{41}$$

4.5 Pulse Transfer Function

Consider the LTI discrete-time system characterized by the following linear difference equation:

$$y(k) + a_1 y(k-1) + \cdots + a_n y(k-n) = b_0 u(k) + b_1 u(k-1) + \cdots + b_m u(k-m) \tag{42}$$

where $u(k)$ and $y(k)$ are the system's input and output, respectively, at the kth sampling or at the real time kT; T is the sampling period. To convert the difference equation (42) to an algebraic equation, take the z-transform of both sides of Eq. (42) by definition:

$$\mathcal{Z}\{y(k)\} = Y(z) \tag{43a}$$

$$\mathcal{Z}\{u(k)\} = U(z) \tag{43b}$$

By referring to Table 2, the z-transform of Eq. (42) becomes

$$Y(z) + a_1 z^{-1}(Y(z) + \cdots + a_n z^{-n} y(z) = b_0 U(z) + b_1 z^{-1} U(z) + \cdots + b_m z^{-m} U(z)$$

or

$$[1 + a_1 z^{-1} + \cdots + a_n z^{-1}]Y(z) = [b_0 + b_1 z^{-1} + \cdots + b_m z^{-m}]U(z)$$

which can be written as

$$\frac{Y(z)}{U(z)} = \frac{b_0 + b_1 z^{-1} + \cdots + b_m z^{-m}}{1 + a_1 z^{-1} + \cdots + a_n z^{-n}} \tag{44}$$

Consider the response of the linear discrete-time system given by Eq. (44), initially at rest when the input $u(t)$ is the delta "function" $\delta(kT)$,

$$\delta(kT) = \begin{cases} 1 & k = 0 \\ 0 & k \neq 0 \end{cases}$$

since

$$\mathcal{Z}\{\delta(kT)\} = \sum_{k=0}^{\infty} \delta(kT) z^{-k} = 1$$

$$U(z) = \mathcal{Z}\{\delta(kT)\} = 1$$

and

$$Y(z) = \frac{b_0 + b_1 z^{-1} + \cdots + b_m z^{-m}}{1 + a_1 z^{-1} + \cdots + a_n z^{-n}} = G(z) \tag{45}$$

Thus $G(z)$ is the response of the system to the delta input (or unit impulse) and plays the same role as the transfer function in linear continuous-time systems. The function $G(z)$ is called the pulse transfer function.

4.6 Zero- and First-Order Hold

Discrete-time control systems may operate partly in discrete time and partly in continuous time. Replacing a continuous-time controller with a digital controller necessitates the conversion of numbers to continuous-time signals to be used as true actuating signals. The process by which a discrete-time sequence is converted to a continuous-time signal is called data hold.

In a conventional sampler, a switch closes to admit an input signal every sample period T. In practice, the sampling duration is very small compared with the most significant time constant of the plant. Suppose the discrete-time sequence is $f(kT)$; then the function of the data hold is to specify the values for a continuous equivalent $h(t)$ where $kT \le t < (k + 1)T$. In general, the signal $h(t)$ during the time interval $kT < t < (k + 1)T$ may be approximated by a polynomial in τ as follows:

$$h(kT + \tau) = a_n \tau^n + a_{n-1} \tau^{n-1} + \cdots + a_1 \tau + a_0 \tag{46}$$

where $0 \le \tau < T$. Since the value of the continuous equivalent must match at the sampling instants, one requires

Table 2 Table of z-Transforms[a]

No.	$\mathcal{F}(s)$	$f(nT)$	$F(z)$
1	—	$1,\ n = 0;\ 0,\ n \neq 0$	1
2	—	$1,\ n = k;\ 0,\ n \neq k$	z^{-k}
3	$\frac{1}{s}$	$1(nT)$	$\frac{z}{z-1}$
4	$\frac{1}{s^2}$	nT	$\frac{Tz}{(z-1)^2}$
5	$\frac{1}{s^3}$	$\frac{1}{2!}(nT)^2$	$\frac{T^2}{2}\frac{z(z+1)}{(z-1)^3}$
6	$\frac{1}{s^4}$	$\frac{1}{3!}(nT)^3$	$\frac{T^3}{6}\frac{z(z^2+4z+1)}{(z-1)^4}$
7	$\frac{1}{s^m}$	$\lim_{a\to 0}\frac{(-1)^{m-1}}{(m-1)!}\frac{\delta^{m-1}}{\delta a^{m-1}}e^{-unT}$	$\lim_{a\to 0}\frac{(-1)^{m-1}}{(m-1)!}\frac{\delta^{m-1}}{\delta a^{m-1}}\frac{z}{z-e^{-aT}}$
8	$\frac{1}{s+a}$	e^{-anT}	$\frac{z}{z-e^{-aT}}$
9	$\frac{1}{(s+a)^2}$	nTe^{-anT}	$\frac{Tze^{-aT}}{(z-e^{-aT})^2}$
10	$\frac{1}{(s+a)^3}$	$\frac{1}{2}(nT)^2e^{-anT}$	$\frac{T^2}{2}(e^{-aT})\frac{z(z+e^{-aT})}{(z-e^{-aT})^3}$
11	$\frac{1}{(s+a)^m}$	$\frac{(-1)^{m-1}}{(m-1)!}\frac{\delta^{m-1}}{\delta a^{m-1}}(e^{-anT})$	$\frac{(-1)^{m-1}}{(m-1)!}\frac{\delta^{m-1}}{\delta a^{m-1}}\frac{z}{z-e^{-aT}}$
12	$\frac{a}{s(s+a)}$	$1-e^{-anT}$	$\frac{z(1-e^{-aT})}{(z-1)(z-e^{-aT})}$
13	$\frac{a}{s^2(s+a)}$	$\frac{1}{a}(anT-1+e^{-anT})$	$\frac{z[(aT-1+e^{-aT})z+(1-e^{-aT}-aTe^{-aT})]}{a(z-1)^2(z-e^{-aT})}$
14	$\frac{b-a}{(s+a)(s+b)}$	$(e^{-anT}-e^{-bnT})$	$\frac{(e^{-aT}-e^{-bT})z}{(x-e^{-aT})(z-e^{-bT})}$
15	$\frac{s}{(s+a)^2}$	$(1-anT)e^{-anT}$	$\frac{z[z-e^{-aT}(1+aT)]}{(z-e^{-aT})^2}$
16	$\frac{a^2}{s(s+a)^2}$	$1-e^{-anT}(1+anT)$	$\frac{z[z(1-e^{-aT}-aTe^{-aT})+e^{-2aT}-e^{-aT}+aTe^{-aT}]}{(z-1)(z-e^{-aT})^2}$
17	$\frac{(b-a)s}{(s+a)(s+b)}$	$be^{-bnT}-ae^{-anT}$	$\frac{z[z(b-a)-(be^{-aT}-ae^{-bT})]}{(z-e^{-aT})(z-e^{-bT})}$
18	$\frac{a}{s^2+a^2}$	$\sin anT$	$\frac{z\sin aT}{z^2-(2\cos aT)z+1}$
19	$\frac{s}{s^2+a^2}$	$\cos anT$	$\frac{z(z-\cos aT)}{z^2-(2\cos aT)z+1}$
20	$\frac{s+a}{(s+a)^2+b^2}$	$e^{-anT}\cos bnT$	$\frac{z(z-e^{-aT}\cos bT)}{z^2-2e^{-aT}(\cos bT)z+e^{-2eT}}$
21	$\frac{b}{(s+a)^2+b^2}$	$e^{-anT}\sin bnT$	$\frac{ze^{-aT}\sin bT}{z^2-2e^{-aT}(\cos bT)z+e^{-2aT}}$
22	$\frac{a^2+b^2}{s((s+a)^2+b^2)}$	$1-e^{-anT}\left(\cos bnT+\frac{a}{b}\sin bnT\right)$	$\frac{z(Az+B)}{(z-1)(z^2-2e^{-aT}(\cos bT)z+e^{-2aT})}$ $A = 1-e^{-aT}\cos bT-\frac{a}{b}e^{-aT}\sin bT$ $B = e^{-2aT}+\frac{a}{b}e^{-aT}\sin bT-e^{-aT}\cos bT$

[a] $\mathcal{F}(s)$ is the Laplace transform of $f(t)$ and $F(z)$ is the transform of $f(nT)$. Unless otherwise noted, $f(t) = 0,\ t < 0$, and the region of convergence of $F(z)$ is outside a circle $r < |z|$ such that all poles of $F(z)$ are inside r.

$$h(kT) = f(kT)$$

Hence Eq. (46) can be written as

$$h(kT + \tau) = a_n\tau^n + a_{n-1}\tau^{n-1} + \cdots + a_1\tau + f(kT) \tag{47}$$

If an nth-order polynomial as in Eq. (47) is used to extrapolate the data, then the hold circuit is called an nth-order hold. If $n = 1$, it is called a first-order hold (the nth-order hold uses the past $n + 1$ discrete data $f[(k - n)T]$). The simplest data hold is obtained when $n = 0$ in Eq. (47), that is, when

$$h(kT + \tau) = f(kT) \tag{48}$$

where $0 \leq \tau < T$ and $k = 0, 1, 2, \ldots$. Equation (48) implies that the circuit holds the amplitude of the sample from one sampling instant to the next. Such a data hold is called a zero-order hold. The output of the zero-order hold is shown in Fig. 17.

Zero-Order Hold

Assuming that the sampled signal is 0 for $k < 0$, the output $h(t)$ can be related to $f(t)$ as follows:

$$\begin{aligned} h(t) &= f(0)[u_s(t) - u_s(t - T)] + f(T)[u_s(t - T) - u_s(t - 2T)] \\ &\quad + f(2T)[u_s(t - 2T)] - u_s(t - 3T)] + \cdots \\ &= \sum_{k=0}^{\infty} f(kT)\{u_s(t - kT) - u_s[t - (k + 1)T]\} \end{aligned} \tag{49}$$

Since $\mathcal{L}|u_s(t - kT)| = e^{-kTs}/s$, the Laplace transform of Eq. (49) becomes

$$\begin{aligned} \mathcal{L}[h(t)] = H(s) &= \sum_{k=0}^{\infty} f(kT)\frac{e^{-kTs} - e^{-(k+1)Ts}}{s} \\ &= \frac{1 - e^{-Ts}}{s}\sum_{k=0}^{\infty} f(kT)e^{-kTs} \end{aligned} \tag{50}$$

The right-hand side of Eq. (50) may be written as the product of two terms:

$$H(s) = G_{h0}(s)F^*(s) \tag{51}$$

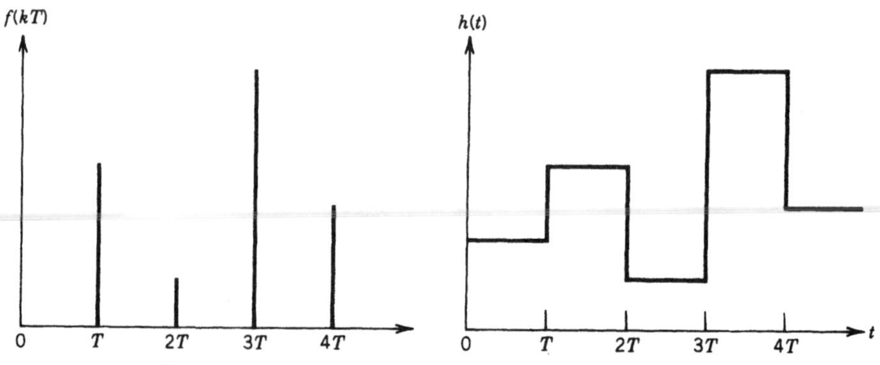

Figure 17 Input $f(kt)$ and output $h(t)$ of the zero-order hold.

where

$$G_{h0}(s) = \frac{1 - e^{-Ts}}{s}$$

$$F^*(s) = \sum_{k=0}^{\infty} f(kT)e^{-kTs} \tag{52}$$

In Eq. (51), $G_{h0}(s)$ may be considered the transfer function between the output $H(s)$ and the input $F^*(s)$ (see Fig. 18). Thus the transfer function of the zero-order hold device is

$$G_{h0}(s) = \frac{1 - e^{-Ts}}{s} \tag{53}$$

First-Order Hold

The transfer function of a first-order hold is given by

$$G_{h1}(s) = \frac{1 - e^{-Ts}}{s} \frac{Ts + 1}{T} \tag{54}$$

From Eq. (47) for $n = 1$,

$$h(kT + \tau) = a_1\tau + f(kT) \tag{55}$$

where $0 \leq \tau < T$ and $k = 0, 1, 2, \ldots$. By using the condition $h[(k-1)T] = f[(k-1)T]$ the constant a_1 can be determined as follows:

$$h[(k-1)T] = -a_1T + f(kT) = f[(k-1)T]$$

or

$$a_1 = \frac{f(kT)_f(k-1)T]}{T}$$

Hence Eq. (55) becomes

$$h(kT + \tau) = f(kT) + \frac{f(kT) - f[(k-1)T]}{T}\tau \tag{56}$$

where $0 \leq \tau < T$. The extrapolation process of the first-order hold is based on Eq. (56) and is a piecewise-linear signal as shown in Fig. 19.

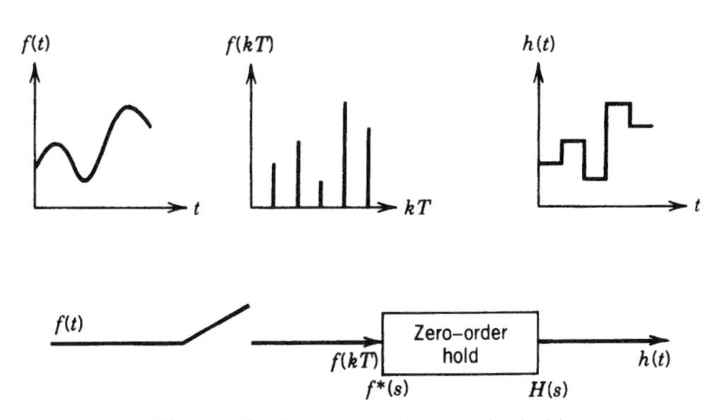

Figure 18 Sampler and zero-order hold.

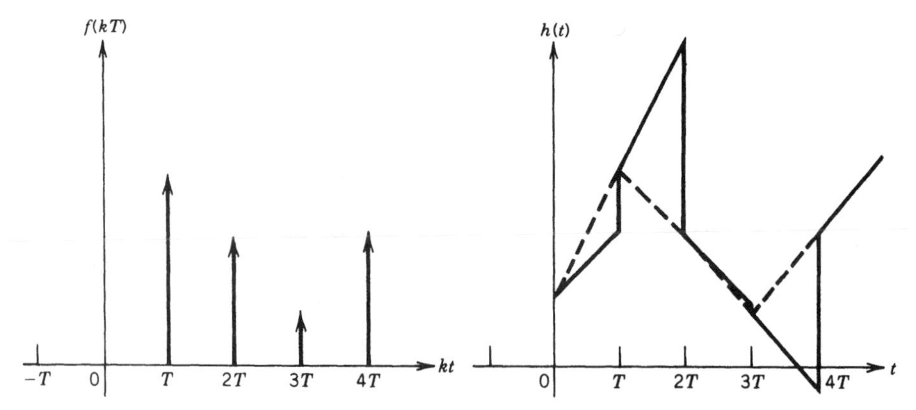

Figure 19 Input $f(kt)$ and output $h(t)$ of a first-order hold.

5 CLOSED-LOOP REPRESENTATION

The typical feedback control system has the feature that some output quantity is measured and then compared with a desired value, and the resulting error is used to correct the system output. A block diagram representation of a closed-loop or feedback system is shown in Fig. 20.

In this figure, r is the reference input, w is a disturbance, and y is the output. Transfer functions G_p, H, and G_c denote, respectively, the plant dynamics, sensor dynamics, and controller. The influence of r and w on the output y can be determined using elementary block diagram algebra as

$$Y(s) = \frac{G_c G_p}{1 + G_c G_p H} R(s) + \frac{G_p}{1 + G_c G_p H} W(s) \tag{57}$$

where $R(s) = \mathcal{L}[r(t)]$, $W(s) = \mathcal{L}[w(t)]$, and $Y(s) = \mathcal{L}[y(t)]$.

5.1 Closed-Loop Transfer Function

In Eq. (57), if $W(s) = 0$, then

$$Y(s) = \frac{G_c G_p}{1 + G_c G_p H} R(s)$$

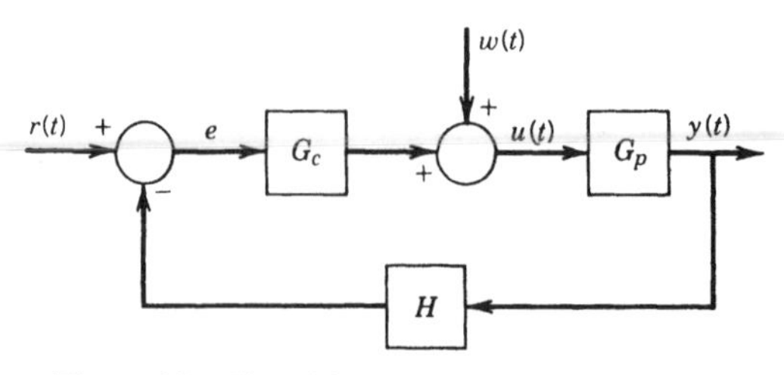

Figure 20 Closed-loop system with disturbance.

or alternatively a transfer function called the closed-loop transfer function between the reference input and the output is defined:

$$\frac{Y(s)}{R(s)} = G_{\mathrm{cl}}(s) = \frac{G_c G_p}{1 + G_c G_p H} \tag{58}$$

5.2 Open-Loop Transfer Function

The product of transfer functions within the loop, namely $G_c G_p H$, is referred to as the open-loop transfer function or simply the loop transfer function:

$$G_{\mathrm{ol}} = G_c G_p H \tag{59}$$

5.3 Characteristic Equation

The overall system dynamics given by Eq. (57) is primarily governed by the poles of the closed-loop system or the roots of the closed-loop characteristic equation (CLCE):

$$1 + G_c G_p H = 0 \tag{60}$$

It is important to note that the CLCE is simply

$$1 + G_{\mathrm{ol}} = 0 \tag{61}$$

This latter form is the basis of root-locus and frequency-domain design techniques discussed in Chapter 7.

The roots of the characteristic equation are referred to as poles. Specifically, the roots of the open-loop characteristic equation (OLCE) are referred to as open-loop poles and those of the closed loop are called closed-loop poles.

Example 9 Consider the block diagram shown in Fig. 21. The open-loop transfer function is

$$G_{\mathrm{ol}} = \frac{K_1(s+1)(s+4)}{s(s+2)(s+3)}$$

The closed-loop transfer function of the system is

$$G_{\mathrm{cl}} = \frac{K_1(s+1)(s+4)}{s(s+2)(s+3) + K_1(s+1)(s+4)}$$

The OLCE is the denominator polynomial of G_{ol} set equal to zero. Hence

$$\text{OLCE} \equiv s(s+2)(s+3) = 0$$

and the open-loop poles are 0, -2, and -3.

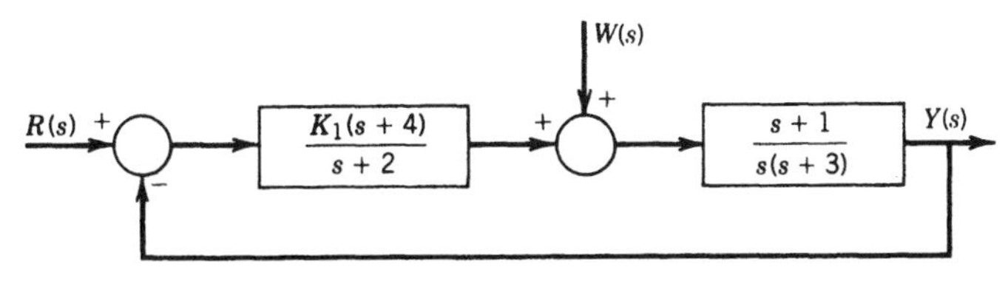

Figure 21 Closed-loop system.

The CLCE is

$$s(s + 2)(s + 3) + K_1(s + 1)(s + 4) = 0$$

and its roots are the closed-loop poles appropriate for the specific gain value K_1.

If the transfer functions are rational (i.e., are ratios of polynomials), then they can be written as

$$G(s) = \frac{K \prod_{i-1}^{m} (s + z_i)}{\prod_{j-1}^{n} (s + p_j)} \tag{62}$$

When the poles of the transfer function $G(s)$ are distinct, $G(s)$ may be written in partial-fraction form as

$$G(s) = \sum_{j=1}^{n} \frac{A_j}{S + p_j} \tag{63}$$

Hence

$$g(t) = \mathcal{L}^{-1}[G(s)] = \mathcal{L}^{-1}\left(\sum_{j=1}^{n} \frac{A_j}{s + p_j}\right)$$

$$= \sum_{j=1}^{n} A_j e^{-p_j t} \tag{64}$$

Since a transfer function $G(s) = \mathcal{L}[\text{output}]/\mathcal{L}[\text{input}]$, $g(t)$ of of Eq. (64) is the response of the system depicted by $G(s)$ for a unit impulse $\delta(t)$, since $\mathcal{L}[\delta(t)] = 1$.

The impulse response of a given system is key to its internal stability. The term *system* here is applicable to any part of (or the whole) closed-loop system.

It should be noted that the zeros of the transfer function only affect the residues A_j. In other words, the contribution from the corresponding transient term $e^{-p_j t}$ may or may not be significant depending on the relative size of A_j. If, for instance, a zero $-z_k$ is very close to a pole $-p_l$, then the transient term $Ae^{-p_j t}$ would have a value close to zero for its residue A_l. As an example, consider the unit impulse response of the two systems:

$$G_1(s) = \frac{1}{(s + 1)(s + 2)} = \frac{1}{s + 1} \frac{1}{s + 2} \tag{65}$$

$$G_2(s) = \frac{(s + 1.05)}{(s + 1)(s + 2)} = \frac{0.05}{s + 1} + \frac{0.95}{s + 2} \tag{66}$$

From Eq. (65), $g_1(t) = e^{-t} - e^{-2t}$, and from Eq. (66), $g_2(t) = 0.05e^{-t} + 0.95e^{-2t}$.

Note that the effect of the term e^{-t} has been modified from a residue of 1 in G_1 to a residue of 0.05 in G_2. This observation helps to reduce the order of a system when there are poles and zeros close together. In $G_2(s)$, for example, little error would be introduced if the zero at -1.05 and the pole at -1 are neglected and the transfer function is approximated by

$$G_2(s) = \frac{1}{s + 2}$$

From Eq. (64) it can be observed that the shape of the impulse response is determined primarily by the pole locations. A sketch of several pole locations and corresponding impulse responses is given in Fig. 22.

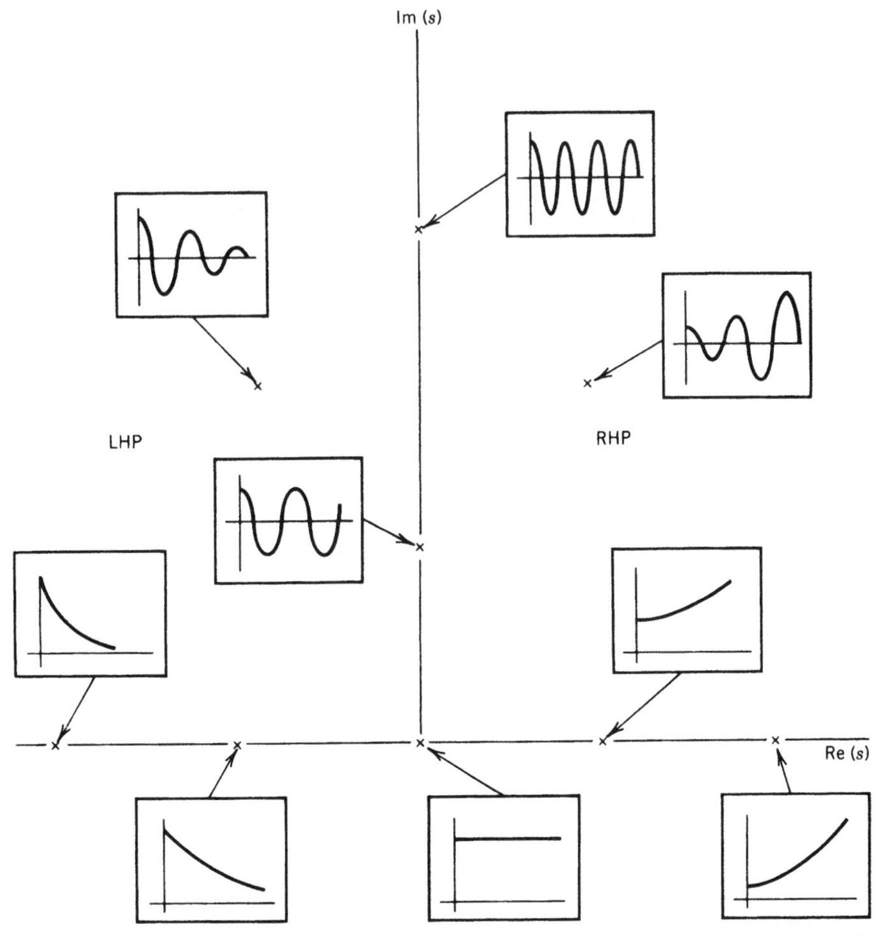

Figure 22 Impulse responses associated with pole locations. LHP, RHP: left- and right-half plane.

The fundamental responses that can be obtained are of the form $e^{-\alpha t}$ and $e^{\lambda t}\sin\beta t$ with $\beta > 0$. It should be noted that for a real pole its location completely characterizes the resulting impulse response. When a transfer function has complex-conjugate poles, the impulse response is more complicated.

5.4 Standard Second-Order Transfer Function

A standard second-order transfer function takes the form

$$G(s) = \frac{\omega_n^2}{s^2 + 2\zeta\omega_n s + \omega_n^2} = \frac{1}{s^2/\omega_n^2 + 2\zeta(s/\omega_n) + 1} \tag{67}$$

Parameter ζ is called the damping ratio, and ω_n is called the undamped natural frequency. The poles of the transfer function given by Eq. (67) can be determined by solving its characteristic equation:

$$s^2 + 2\zeta\omega_n s + \omega_n^2 = 0 \tag{68}$$

giving the poles

$$s_{1.2} = -\zeta\omega_n \pm j\sqrt{1 - \zeta^2}\omega_n = -\sigma \pm j\omega_d \tag{69}$$

where $\sigma = \zeta\omega_n$ and $\omega_d = \omega_n \sqrt{1 - \zeta^2}$, $\zeta < 1$.

The two complex-conjugate poles given by Eq. (69) are located in the complex s-plane, as shown in Fig. 23.

It can be easily seen that

$$|s_{1.2}| = \sqrt{\zeta^2\omega_n^2 + \omega_d^2} = \omega_n \tag{70}$$

and

$$\cos \beta = \zeta \qquad 0 \leq \beta \leq \tfrac{1}{2}\pi \tag{71}$$

When the system has no damping, $\zeta = 0$, the impulse response is a pure sinusoidal oscillation. In this case the undamped natural frequency ω_n is equal to the damped natural frequency ω_d.

5.5 Step Input Response of a Standard Second-Order System

When the transfer function has complex-conjugate poles, the step input response rather than the impulse response is used to characterize its transients. Moreover, these transients are almost always used as time-domain design specifications. The unit step input response of the second-order system given by Eq. (67) can be easily shown to be

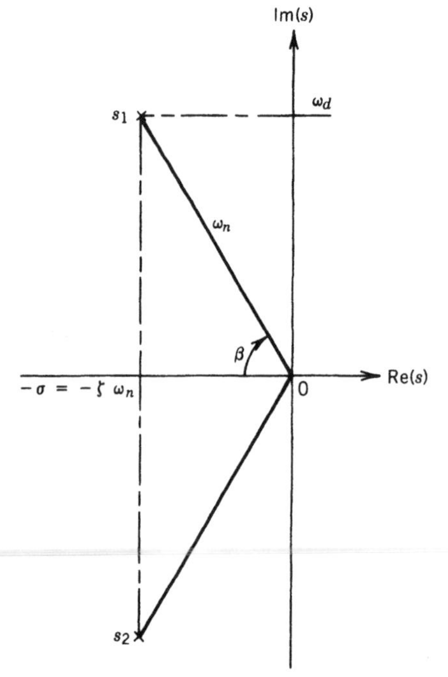

Figure 23 Compelx-conjugate poles in the s-plane.

$$y(t) = 1 - e^{\sigma T}\left(\cos \omega_d t + \frac{\sigma}{\omega_d} \sin \omega_d t\right) \tag{72}$$

where $y(t)$ is the output.

A typical unit step input response is shown in Fig. 24.

From the time response of Eq. (72) several key parameters characterizing the shape of the curve are usually used as time-domain design specifications. They are the rise time t_r, the settling time t_s, the peak overshoot M_p, and the peak time t_p.

Rise Time t_r

The time taken for the output to change from 10% of the steady-state value to 90% of the steady-state value is usually called the rise time. There are other definitions of rise time.[2] The basic idea, however, is that t_r is a characterization of how rapid the system responds to an input.

A rule of thumb for t_r is

$$t_r \simeq \frac{1.8}{\omega_n} \tag{73}$$

Settling Time t_s

This is the time required for the transients to decay to a small value so that $y(t)$ is almost at the steady-state level. Various measures of "smallness" are possible: 1, 2, and 5% are typical.

$$\begin{aligned} &\text{For 1\% settling:} \quad t_s \simeq \frac{4.6}{\zeta\omega_n} \\ &\text{For 2\% settling:} \quad t_s \simeq \frac{4}{\zeta\omega_n} \\ &\text{For 5\% settling:} \quad t_s \simeq \frac{3}{\zeta\omega_n} \end{aligned} \tag{74}$$

Peak Overshoot M_p

The peak overshoot is the maximum amount by which the output exceeds its steady-state value during the transients. It can be easily shown that

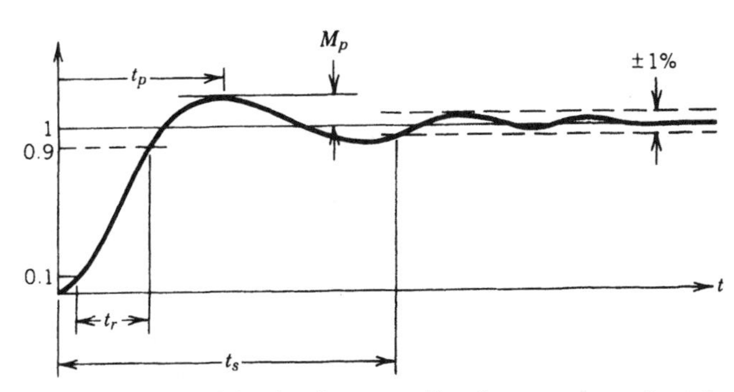

Figure 24 Definition of the rise time t_r, settling time t_s, and overshoot time M_p.

$$M_p = e^{-\pi\zeta/\sqrt{1-\zeta^2}} \tag{75}$$

Peak Time t_p
This is the time at which the peak occurs. It can be readily shown that

$$t_p = \frac{\pi}{\omega_d} \tag{76}$$

It is important to note that given a set of time-domain design specifications, they can be converted to an equivalent location of complex-conjugate poles. Figure 25 shows the allowable regions for complex-conjugate poles for different time-domain specifications.

For design purposes the following synthesis forms are useful:

$$\omega_n \geq \frac{1.8}{t_r} \tag{77a}$$

$$\zeta \geq 0.6(1 - M_p) \qquad \text{for} \qquad 0 \leq \zeta \leq 0.6 \tag{77b}$$

$$\sigma \geq \frac{4.6}{t_s} \tag{77c}$$

5.6 Effects of an Additional Zero and an Additional Pole[2]

If the standard second-order transfer function is modified due to a zero as

$$G_1(s) = \frac{(s/\alpha\zeta\omega_n + 1)\omega_n^2}{s^2 + 2\zeta\omega_n + \omega_n^2} \tag{78}$$

or due to an additional pole as

$$G_2(s) = \frac{\omega_n^2}{(s/\alpha\zeta\omega_n + 1)(s^2 + 2\zeta\omega_n s + \omega_n^2)} \tag{79}$$

it is important to know how close the resulting step input response is to the standard second-order step response. Following are several features of these additions:

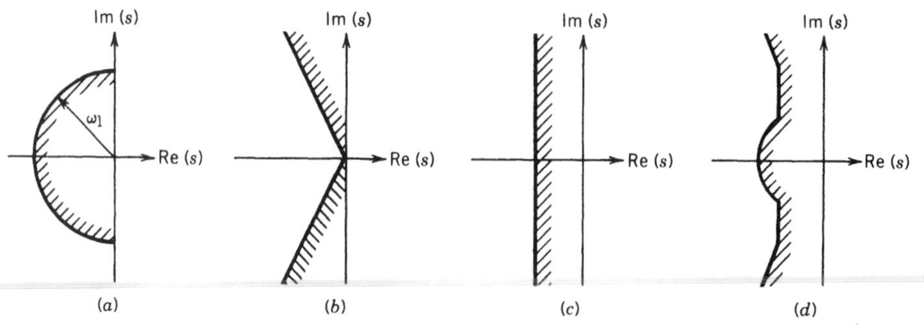

Figure 25 Graphs of regions in the s-plane for certain transient requirements to be met: (*a*) rise time requirements, (ω_n); (*b*) overshoot requirements, (ζ); (*c*) settling time requirements, (ζ); (*d*) composite of (*a*), (*b*), and (*c*).

1. For a second-order system with no finite zeros, the transient parameters can be approximated by Eq. (77).
2. An additional zero as in Eq. (78) in the left-half plane will increase the overshoot if the zero is within a factor of 4 of the real part of the complex poles. A plot is given in Fig. 26.
3. An additional zero in the right-half plane will depress the overshoot (and may cause the step response to undershoot). This is referred to as a non-minimum-phase system.
4. An additional pole in the left-half plane will increase the rise time significantly if the extra pole is within a factor of 4 of the real part of the complex poles. A plot is given in Fig. 27.

6 STABILITY

We shall distinguish between two types of stabilities: external and internal stability. The notion of external stability is concerned with whether or not a bounded input gives a bounded output. In this type of stability we notice that no reference is made to the internal variables of the system. The implication here is that it is possible for an internal variable to grow without bound while the output remains bounded. Whether or not the internal variables are well behaved is typically addressed by the notion of internal stability. Internal stability requires that in the absence of an external input the internal variables stay bounded for any perturbations of these variables. In other words internal stability is concerned with the response of the system due to nonzero initial conditions. It is reasonable to expect that a well-designed system should be both externally and internally stable.

The notion of asymptotic stability is usually discussed within the context of internal stability. Specifically, if the response due to nonzero initial conditions decays to zero asymptotically, then the system is said to be asymptotically stable. A LTI system is asymptotically stable if and only if all the system poles lie in the open left-half-plane (i.e., the left-half s-plane excluding the imaginary axis). This condition also guarantees external stability for LTI systems. So in the case of LTI systems the notions of internal and external stability may be considered equivalent.

For LTI systems, knowing the locations of the poles or the roots of the characteristic equation would suffice to predict stability. The Routh–Hurwitz stability criterion is frequently used to obtain stability information without explicitly computing the poles for LTI. This criterion will be discussed in Section 6.1.

For nonlinear systems, stability cannot be characterized that easily. As a matter of fact, there are many definitions and theorems for assessing stability of such systems. A discussion of these topics is beyond the scope of this handbook. Interested reader may refer to Ref. 3.

6.1 Routh–Hurwitz Stability Criterion

This criterion allows one to predict the status of stability of a system by knowing the coefficients of its characteristic polynomial. Consider the characteristic polynomial of an nth-order system:

$$P(s) = a_n s^n + a_{n-1} s^{n-1} + a_{n-2} s^{n-2} + \cdots + a_1 s + a_0$$

A necessary condition for asymptotic stability is that all the coefficients $\{a_i\}$'s be positive. If any of the coefficients are missing (i.e., are zero) or negative, then the system will have

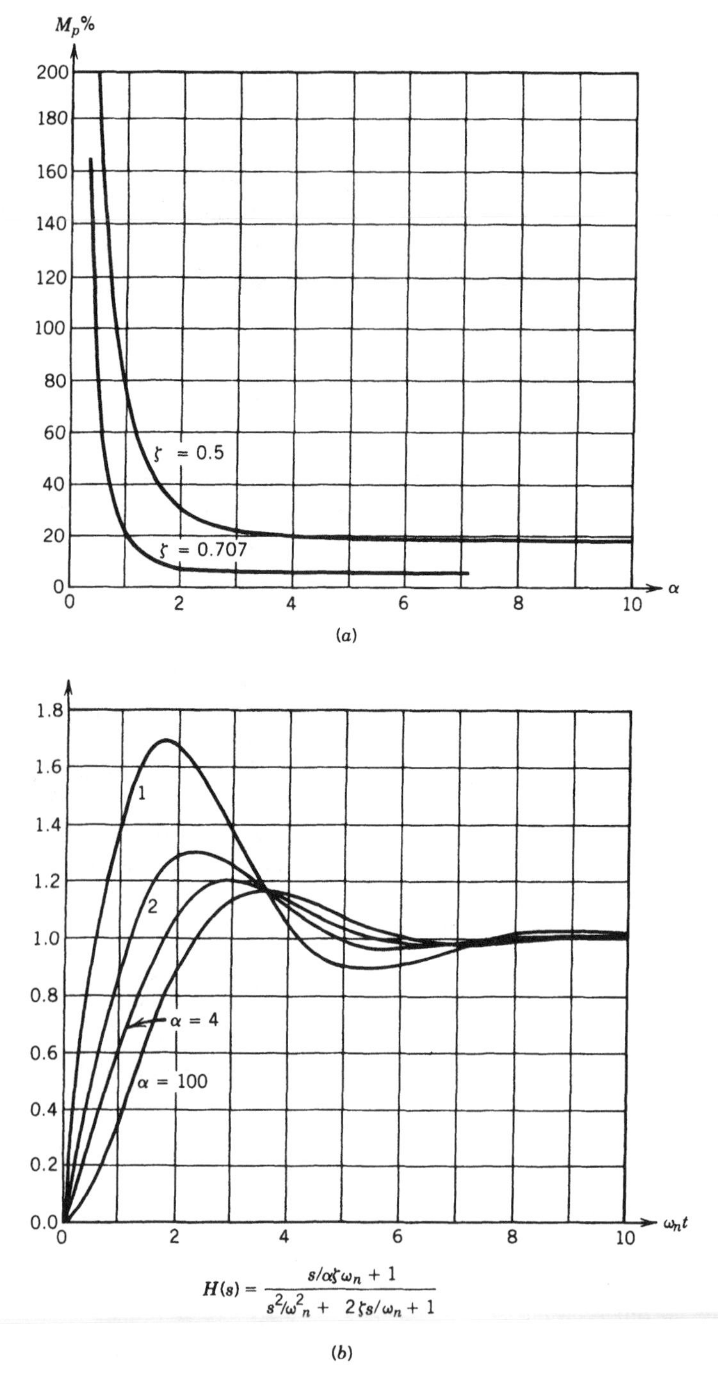

Figure 26 Effect of an extra zero on a standard second-order system: (*a*) normalized rise time versus α; (*b*) step response versus α, for $\zeta = 0.5$.

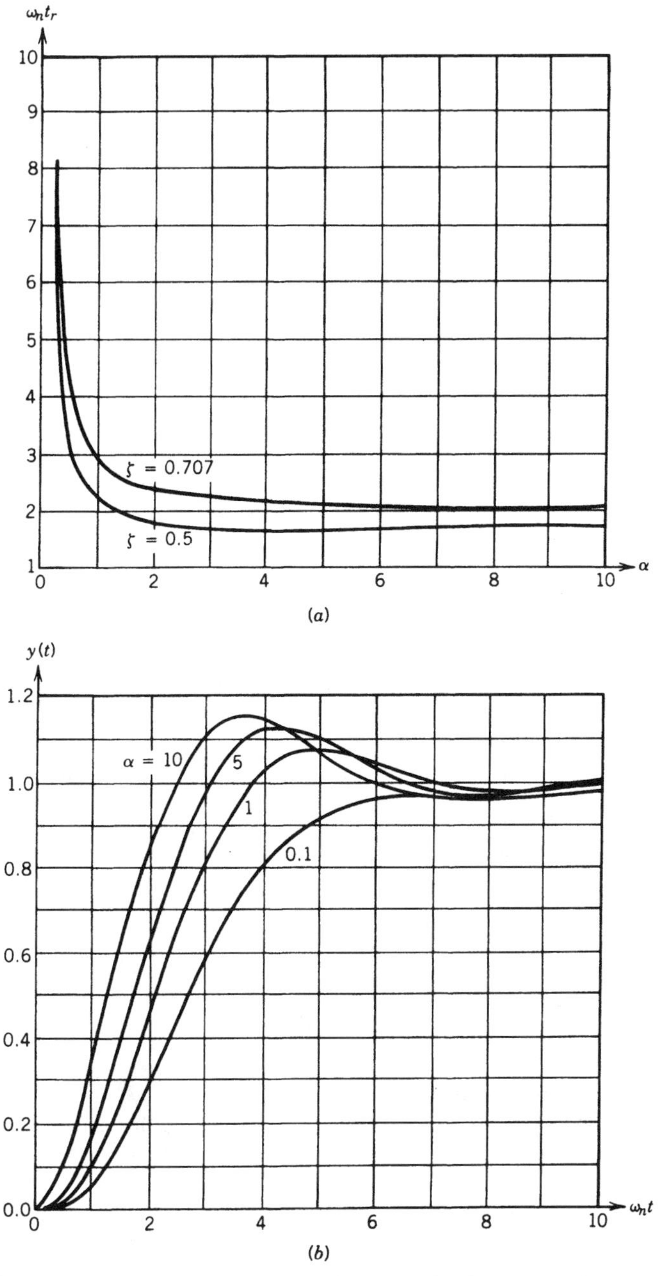

Figure 27 Effect of an additional pole on a standard second-order system: (*a*) normalized rise time versus α; (*b*) step response versus α, for $\zeta = 0.5$.

poles located in the closed right-half plane. If the necessary condition is satisfied, then the so-called Routh array needs to be formed to make conclusions about the stability. A necessary and sufficient condition for stability is that all the elements in the first column of the Routh array be positive.

To determine the Routh array, the coefficients of the characteristic polynomial are arranged in two rows, each beginning with the first and second coefficients and followed by even-numbered and odd-numbered coefficients as follows:

$$\begin{array}{lllll} s^n & a_n & a_{n-2} & a_{n-4} & \cdots \\ s^{n-1} & a_{n-1} & a_{n-3} & a_{n-5} & \cdots \end{array}$$

The following rows are subsequently added to complete the Routh array:

$$\begin{array}{lllll} s^n & a_n & a_{n-2} & a_{n-4} & \cdots \\ s^{n-1} & a_{n-1} & a_{n-3} & a_{n-5} & \cdots \\ s^{n-2} & b_1 & b_2 & b_3 & \cdots \\ s^{n-3} & c_1 & c_2 & c_3 & \cdots \\ \vdots & \vdots & \vdots & \vdots & \cdots \\ s^2 & * & * & & \\ s & * & & & \\ s^0 & & & & \end{array}$$

where the elements from the third row on are computed as follows:

$$b_1 = \frac{a_{n-1}a_{n-2} - a_n a_{n-3}}{a_{n-1}} \qquad c_1 = \frac{b_1 a_{n-3} - a_{n-1} b_2}{b_1}$$

$$b_2 = \frac{a_{n-1}a_{n-4} - a_n a_{n-5}}{a_{n-1}} \qquad c_2 = \frac{b_1 a_{n-5} - a_{n-1} b_3}{b_1}$$

$$b_3 = \frac{a_{n-1}a_{n-6} - a_n a_{n-7}}{a_{n-1}} \qquad c_3 = \frac{b_1 a_{n-7} - a_{n-1} b_4}{b_1}$$

Note that the elements of the third row and of rows thereafter are formed from the two previous rows using the two elements in the first column and other elements for successive columns. Normally there will be $n + 1$ elements in the first column when the array is completed.

The Routh–Hurwitz criterion states that the number of roots of $P(s)$ with positive real parts is equal to the number of changes of sign in the first column of the array. This criterion requires that there be no changes of sign in the first column for a stable system. A pattern of $+$, $-$, $+$ is counted as two sign changes, one going from $+$ to $-$ and another from $-$ to $+$.

If the first term in one of the rows is zero or if an entire row is zero, then the standard Routh array cannot be formed and the use of special techniques, described in the following discussion, become necessary.

Special Cases

1. Zero in the first column, while some other elements of the row containing a zero in the first column are nonzero
2. Zero in the first column, and the other elements of the row containing the zero are also zero

CASE 1: In this case the zero is replaced with a small positive constant $\epsilon > 0$, and the array is completed as before. The stability criterion is then applied by taking the limits of entries of the first column as $\epsilon \to 0$. For example, consider the following characteristic equation:

$$s^5 + 2s^4 + 2s^3 + 4s^2 + 11s + 10 = 0$$

The Routh array is then

$$\begin{array}{lll} 1 & 2 & 11 \\ 2 & 4 & 10 \\ 0 & 6 & 0 \\ \epsilon & 6 & 0 \quad \text{first column zero replaced by } \epsilon \\ c_1 & 10 & \\ d_1 & 0 & \end{array}$$

where

$$c_1 = \frac{4\epsilon - 12}{\epsilon} \qquad \text{and} \qquad d_1 = \frac{6c_1 - 10\epsilon}{c_1}$$

As $\epsilon \to 0$, we get $c_1 \simeq -12/\epsilon$, and $d_1 \simeq 6$. There are two sign changes due to the large negative number in the first column. Therefore the system is unstable, and two roots lie in the right-half plane. As a final example consider the characteristic polynomial

$$P(s) = s^4 + 5s^3 + 7s^2 + 5s + 6$$

The Routh array is

$$\begin{array}{lll} 1 & 7 & 6 \\ 5 & 5 & \\ 6 & 6 & \\ \epsilon & \leftarrow & \text{Zero replaced by } \epsilon > 0 \\ 6 & & \end{array}$$

If $\epsilon > 0$, there are no sign changes. If $\epsilon < 0$, there are two sign changes. Thus if $\epsilon = 0$, it indicates that there are two roots on the imaginary axis, and a slight perturbation would drive the roots into the right-half plane or the left-half plane. An alternative procedure is to define the auxiliary variable

$$z = \frac{1}{s}$$

and convert the characteristic polynomial so that it is in terms of z. This usually produces a Routh array with nonzero elements in the first column. The stability properties can then be deduced from this array.

CASE 2: This case corresponds to a situation where the characteristic equation has equal and opposite roots. In this case if the ith row is the vanishing row, an auxiliary equation is formed from the previous i_{-1} row as follows:

$$P_1(s) = \beta_1 s^{i+1} + \beta_2 s^{i-1} + \beta_3 s^{i-3} + \cdots$$

where the $\{\beta_i\}$'s are the coefficients of the $(i - 1)$th row of the array. The ith row is then replaced by the coefficients of the derivative of the auxiliary polynomial, and the array is completed. Moreover, the roots of the auxiliary polynomial are also roots of the characteristic equation. As an example, consider

$$s^5 + 2s^4 + 5s^3 + 10s^2 + 4s + 8 = 0$$

for which the Routh array is

$$\begin{array}{llll} s^5 & 1 & 5 & 4 \\ s^4 & 2 & 10 & 8 \\ s^3 & 0 & 0 & \\ s^3 & 8 & 20 & \\ s^2 & 5 & 8 & \\ s^1 & 7.2 & & \\ s^0 & 8 & & \end{array} \qquad \text{Auxiliary equation: } 2s^4 + 10s^2 + 8 = 0$$

There are no sign changes in the first column. Hence all the roots have nonpositive real parts with two pairs of roots on the imaginary axis, which are the roots of

$$2s^4 + 10s^2 + 8 = 0$$

$$= 2(s^2 + 4)(s^2 + 1)$$

Thus equal and opposite roots indicated by the vanishing row are $\pm j$ and $\pm 2j$.

The Routh–Hurwitz criterion may also be used in determining the range of parameters for which a feedback system remains stable. Consider, for example, the system described by the CLCE

$$s^4 + 3s^3 + 3s^2 + 2s + K = 0$$

The corresponding Routh array is

$$\begin{array}{llll} s^4 & 1 & 3 & K \\ s^3 & 3 & 2 & \\ s^2 & \frac{7}{3} & K & \\ s^1 & s - (9/7)K & & \\ s^0 & K & & \end{array}$$

If the system is to remain asymptotically stable, we must have

$$0 < K < \tfrac{14}{9}$$

The Routh–Hurwitz stability criterion can also be used to obtain additional insights. For instance information regarding the speed of response may be obtained by a coordinate transformation of the form $s + a$, where $-a$ characterizes rate. For this additional detail the reader is referred to Ref. 4.

6.2 Polar Plots

The frequency response of a system described by a transfer function $G(s)$ is given by

$$y_{ss}(t) = X|G(j\omega)| \sin[\omega t + \underline{/G(j\omega_1)}]$$

where $X \sin \omega t$ is the forcing input and $y_{ss}(t)$ is the steady-state output. A great deal of information about the dynamic response of the system can be obtained from a knowledge of $G(j\omega)$. The complex quantity $G(j\omega)$ is typically characterized by its frequency-dependent magnitude $|G(j\omega)|$ and the phase $\underline{/G(j\omega)}$. There are many different ways in which this information is handled. The manner in which the phasor $G(j\omega)$ traverses the $G(j\omega)$-plane as a function of ω is represented by a polar plot. This is represented in Fig. 28. The phasor OA corresponds to a magnitude $|G(j\omega_1)|$ and the phase $\underline{/G(j\omega_1)}$. The locus of point A as a function of the frequency is the polar plot. This representation is useful in stating the Nyquist stability criterion.

6.3 Nyquist Stability Criterion

This is a method by which the closed-loop stability of a LTI system can be predicted by knowing the frequency response of the loop transfer function. A typical CLCE may be written as

$$1 + GH = 0$$

where GH is referred to as the loop transfer function.

The Nyquist stability criterion[2] is based on a mapping of the so-called Nyquist contour by the loop transfer function. Figure 29*a* shows the Nyquist contour, and Fig. 29*b* shows its typical mapped form in the GH plane. Let C_1 be the contour in the s-plane to be mapped. Utilizing this information and the poles of GH, the Nyquist stability criterion can be stated as follows: A closed-loop system is stable if and only if the mapping of C_1 in the GH plane encircles the $(-1,0)$ point N number of times in the counterclockwise direction, where N is the number of poles of GH with positive real parts.

Clearly, if GH does not have any poles in the right-half plane, then C_1 should not encircle the $(-1,0)$ point if the closed-loop system is to be stable.

If there are poles of GH on the imaginary axis, then the Nyquist contour should be modified with appropriate indentations, as shown in Fig. 30. An example is given to map such an indented contour with the corresponding loop transfer function. Consider

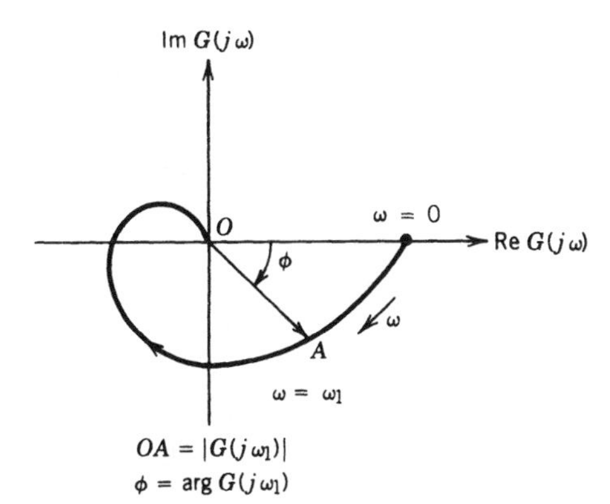

Figure 28 Polar plot.

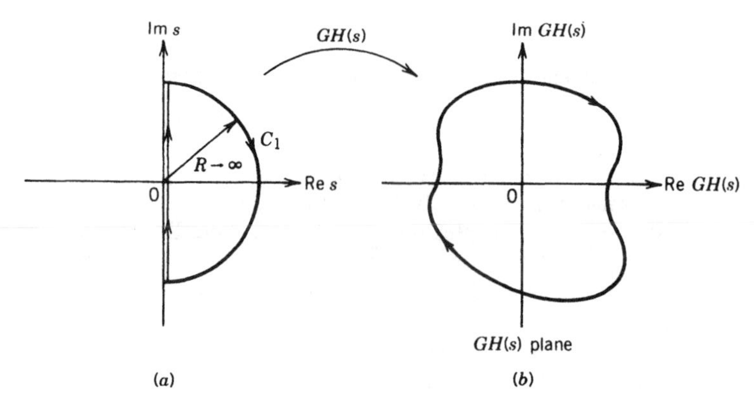

Figure 29 Nyquist contour and its mapping by $GH(s)$.

$$GH(s) = \frac{1}{s(s+1)}$$

The contour Γ in the s-plane is shown in Fig. 31*a*, where an indentation is affected by a small semicircle of radius ϵ, where $\epsilon > 0$. When mapped by $GH(s)$, the contour of Fig. 31*b* is obtained. To effect the mapping, the points on the semicircles are represented as $s = \epsilon e^{j\phi}$ on the small indentation with $\phi \epsilon [-\pi/2, \pi/2]$ and $\epsilon \to 0$ and $s = Re^{j\theta}$ with $\theta \epsilon [-\pi/2, \pi/2]$ and $R \to \infty$ on the infinite semicircle. We observe that the $(-1,0)$ point is not encircled. Since $GH(s)$ does not have any poles in the right-half s-plane from the Nyquist stability criterion, it follows that the closed-loop system is stable. In Table 3 several loop transfer functions with the appropriate Nyquist contour and mapped contours are given.

7 STEADY-STATE PERFORMANCE AND SYSTEM TYPE

In the design of feedback control systems the steady-state performance is also of importance in many instances. This is in addition to the stability and transient performance requirements. Consider the closed-loop system shown in Fig. 32.

Here $G_p(s)$ is the plant and $G_c(s)$ is the controller. When steady-state performance is important, it is advantageous to consider the system error due to an input. From the block diagram

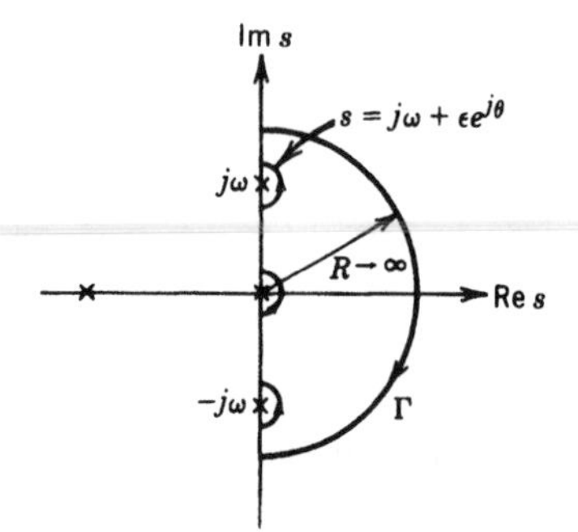

Figure 30 Nyquist contour with indentations.

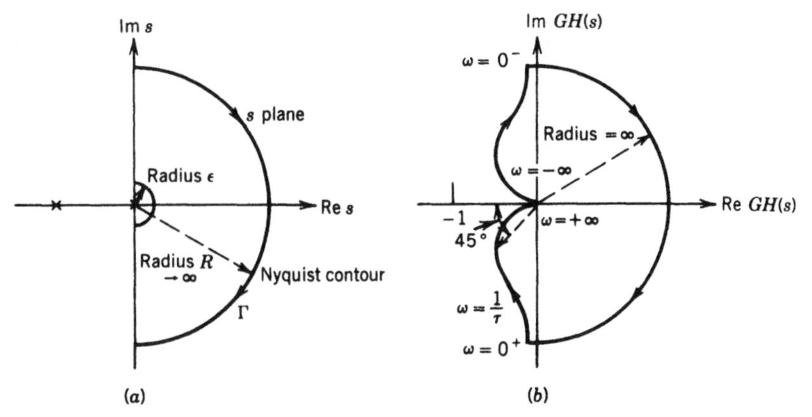

Figure 31 Nyquist contour and mapping for $GH(s) = K/s(\tau s + 1)$.

$$E(s) = R(s) - Y(s) \tag{80}$$

$$Y(s) = G_c(s)G_p(s)E(s) \tag{81}$$

Substituting Eq. (81) in (80) yields

$$\frac{E(s)}{R(s)} = \frac{1}{1 + G_c(s)G_p(s)} \tag{82}$$

called the error transfer function. It is important to note that the error dynamics are described by the same poles as those of the closed-loop transfer function. Namely, the roots of the characteristic equation $1 + G_c(s)G_p(s) = 0$.

Given any input $r(t)$ with $\mathcal{L}[r(t)] = R(s)$, the error can be analyzed by considering the inverse Laplace transform of $E(s)$ given by

$$e(t) = \mathcal{L}^{-1}\left[\frac{1}{1 + G_c(s)G_p(s)} R(s)\right] \tag{83}$$

For the system's error to be bounded, it is important to first assure that the closed-loop system is asymptotically stable. Once the closed-loop stability is assured, the steady-state error can be computed by using the final-value theorem. Hence

$$\lim_{t\to\infty} e(t) = e_{ss} = \lim_{s\to 0} sE(s) \tag{84}$$

By substituting for $E(s)$ in Eq. (84)

$$e_{ss} = \lim_{s\to 0} s \frac{1}{1 + G_c(s)G_p(s)} R(s) \tag{85}$$

7.1 Step Input

If the reference input is a step of magnitude c, then from Eq. (85)

$$e_{ss} = \lim_{s\to 0} s \frac{1}{1 + G_c(s)G_p(s)} \frac{c}{s} = \frac{c}{1 + G_cG_p(0)} \tag{86}$$

Equation (86) suggests that to have small steady-state error, the low-frequency gain of the open-loop transfer function $G_cG_p(0)$ must be very large. It is typical to define

Table 3 Loop Transfer Functions

No.	$G(s)$	Nyquist Contour	Polar Plot
1.	$\dfrac{K}{s\tau_1 + 1}$		$-\omega$, -1, $\omega = \infty$, $\omega = 0$, $+\omega$
2.	$\dfrac{K}{(s\tau_1 + 1)(s\tau_2 + 1)}$		$-\omega$, -1, $\omega = \infty$, $\omega = 0$, $+\omega$
3.	$\dfrac{K}{(s\tau_1 + 1)(s\tau_2 + 1)(s\tau_3 + 1)}$		$-\omega$, -1, $\omega = \infty$, $\omega = 0$, $+\omega$
4.	$\dfrac{K}{s}$		$\omega = 0$, $-\omega$, -1, $\omega = \infty$, $+\omega$, $\omega \to 0$
5.	$\dfrac{K}{(s\tau_1 + 1)}$		$\omega \to 0$, $-\omega$, -1, $\omega = \infty$, $+\omega$, $\omega \to 0$

Table 3 *(Continued)*

No.	$G(s)$	Nyquist Contour	Polar Plot
6.	$\dfrac{K}{s(s\tau_1 + 1)(s\tau_2 + 1)}$	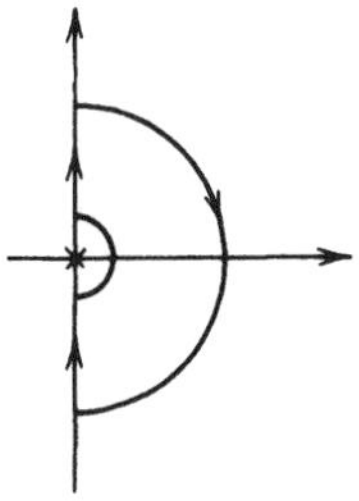	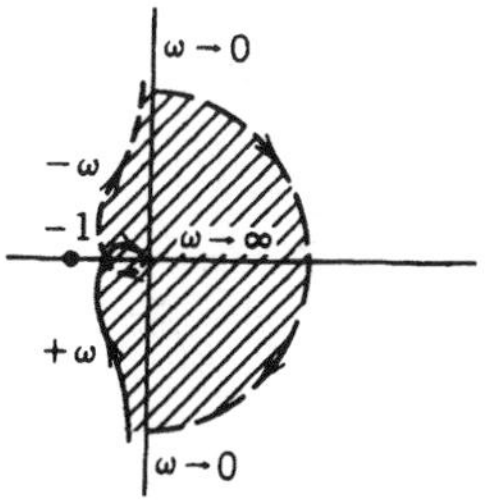
7.	$\dfrac{K(s\tau_a + 1)}{s(s\tau_1 + 1)(s\tau_2 + 1)}$ $\tau_a < \dfrac{\tau_1\tau_2}{\tau_1 + \tau_2}$ $\tau_a < \dfrac{\tau_1\tau_2}{\tau_1 + \tau_2}$	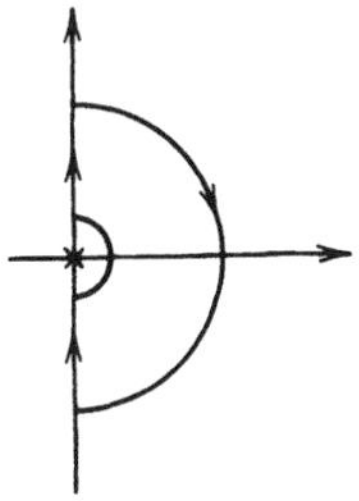	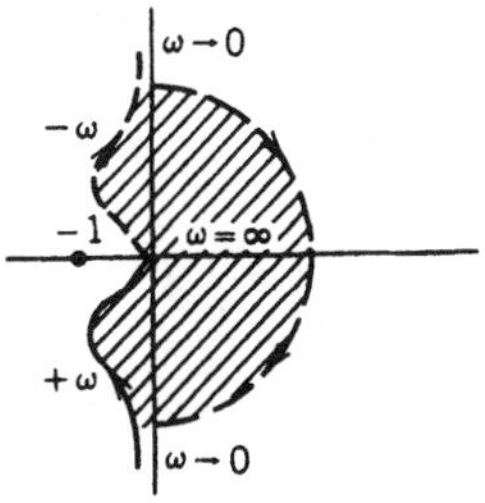
8.	$\dfrac{K}{s^2}$	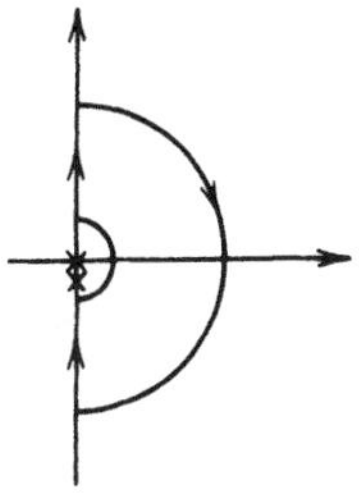	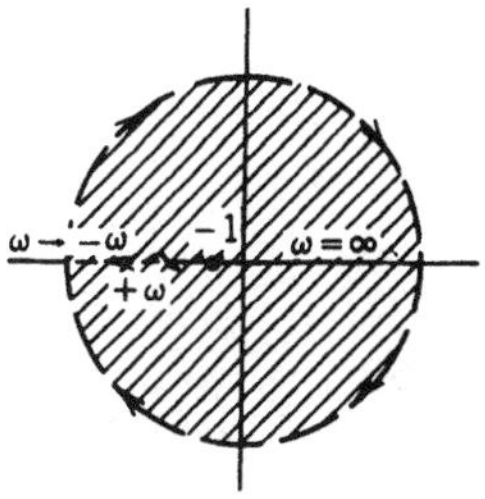
9.	$\dfrac{K}{s^2(s\tau_1 + 1)}$	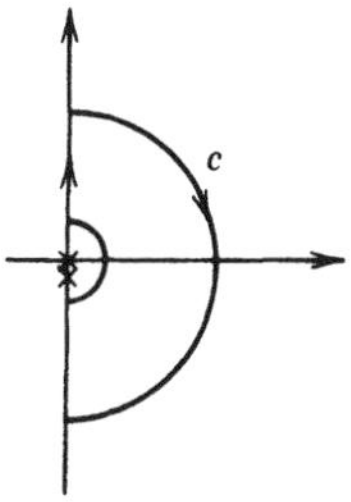	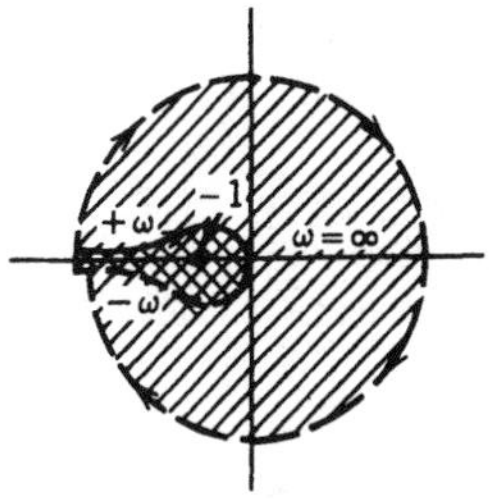
10.	$\dfrac{K(s\tau_1 + 1)}{s^2(s\tau_1 + 1)}$ $\tau_2 > \tau_1$ $\tau_a > \tau_1$	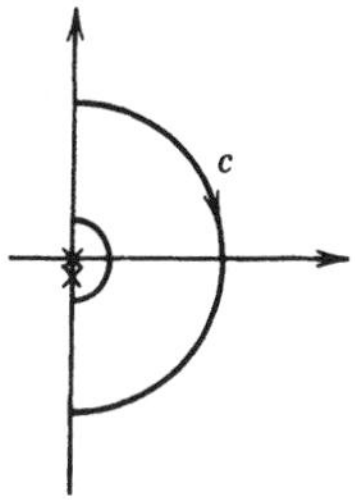	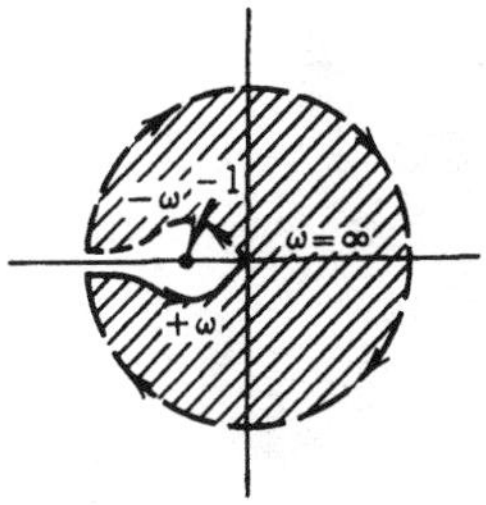

Table 3 (*Continued*)

No.	$G(s)$	Nyquist Contour	Polar Plot
11.	$\dfrac{K}{s^3}$	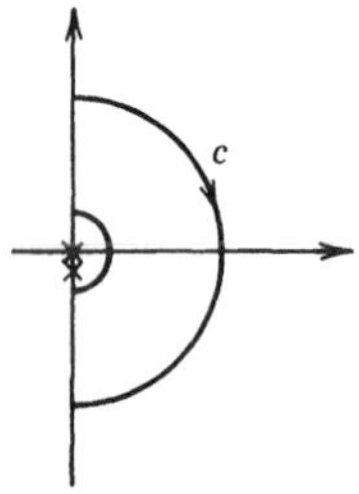	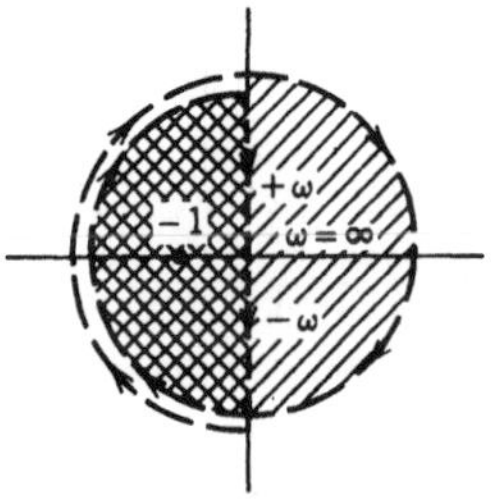
12.	$\dfrac{K(s\tau_a + 1)}{s^3}$	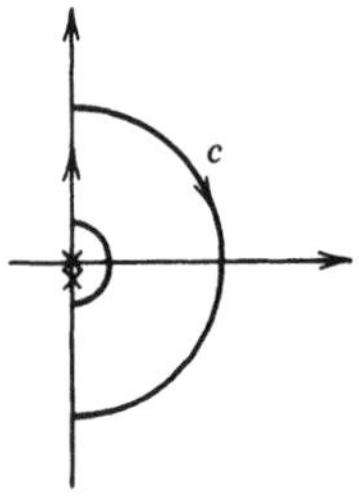	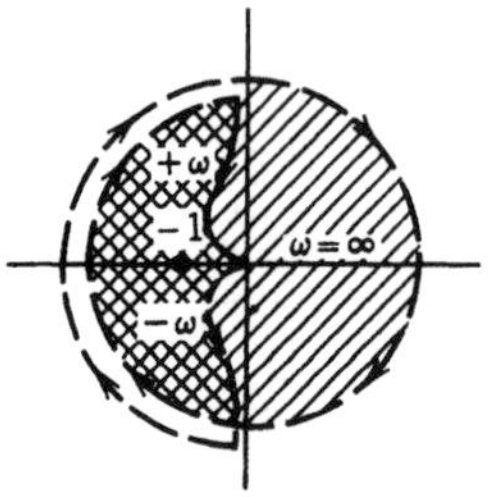
13.	$\dfrac{K(s\tau_a + 1)(s\tau_b + 1)}{s^3}$	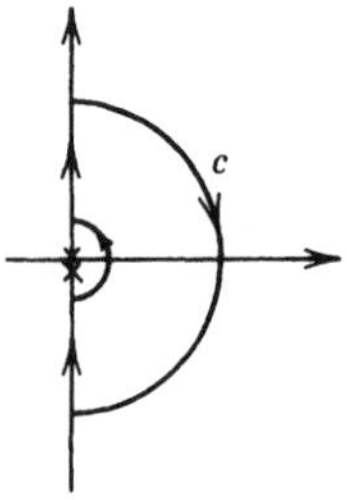	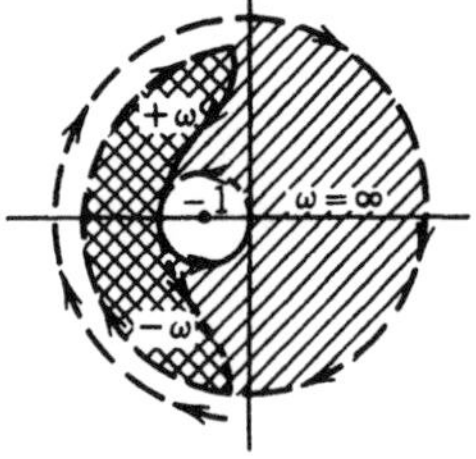
14.	$\dfrac{K(s\tau_a + 1)(s\tau_b + 1)}{s(s\tau_1 + 1)(s\tau_2 + 1)(s\tau_3 + 1)(s\tau_4 + 1)}$	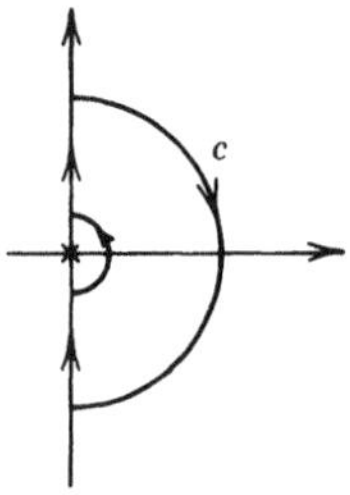	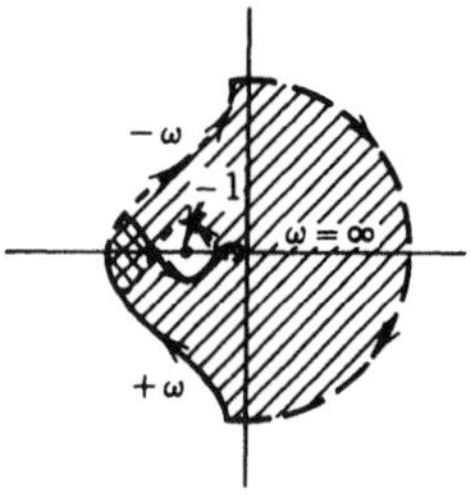
15.	$\dfrac{K(s\tau_a + 1)}{s^2(s\tau_1 + 1)(s\tau_2 + 1)}$	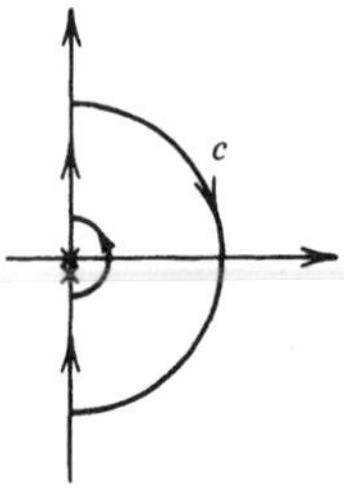	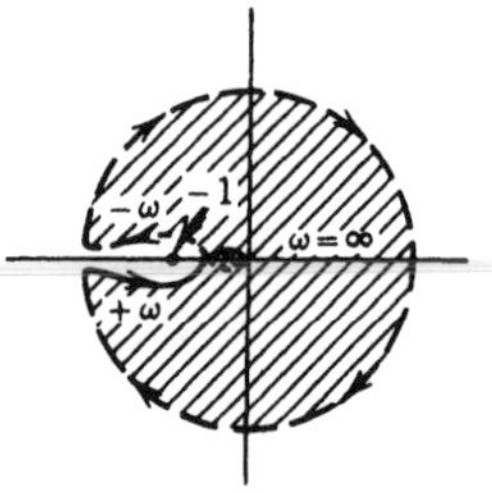

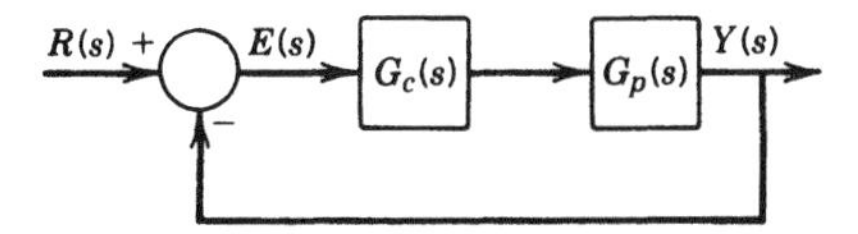

Figure 32 Closed-loop configuration.

$$K_p = G_c G_p(0) \tag{87}$$

as the position error constant. With this definition the steady-state error due to a step input of magnitude c can be written as

$$e_{ss} = \frac{c}{1 + K_p} \tag{88}$$

Thus a high value of K_p, corresponds to a low steady-state error. If the steady-state error is to be zero, then $K_p = \infty$. The only way that $K_p = \infty$ is if the open-loop transfer function has at least one pole at the origin, that is, $G_c G_p(s)$ must be of the form

$$G_c G_p(s) = \frac{1}{s^N} \frac{\prod_{i-1}^{m} (s + z_i)}{\prod_{j=1}^{n} (s + p_j)} \tag{89a}$$

where $N \geq 1$. When $N > 1$,

$$G_c G_p(0) = \frac{1}{0} \frac{\prod_{i-1}^{m} z_i}{\prod_{j=1}^{n} p_j} \longrightarrow \infty \tag{89b}$$

Hence, it can be concluded that for the steady-state error due to a step input to be zero, the open-loop transfer function must have at least one free integrator. The value of N specifies the type of system. If $N = 1$ it is called a type I, when $N = 2$ it is called a type II system, and so on. So to get zero steady-state error for a step input, the system loop transfer function must be at least type I.

7.2 Ramp Input

If the reference input is a ramp $ctu_s(t)$, where $u_s(t)$ is the unit step, then from Eq. (85)

$$e_{ss} = \lim_{s \to 0} s \frac{1}{1 + G_c G_p(s)} \frac{c}{s^2} = \lim_{s \to 0} \frac{c}{sG_c G_p(s)} \tag{90}$$

From Eq. (90) for small steady-state errors $\lim_{s \to 0} sG_c G_p(s) = K_v$ must be large, where K_v is the velocity error constant and

$$e_{ss} = \frac{c}{K_v} \tag{91}$$

As in the case with the step input, for e_{ss} to be small, K_v must be very large. For zero steady-state error with a ramp input, $K_v = \infty$. From Eq. (90) it is clear that for $K_v = \infty$, $G_c G_p(s)$ must be at least type II. Thus

$$K_v = \lim_{s \to 0} sG_c G_p(s) = \lim_{s \to 0} s \frac{1}{s^2} = \frac{\prod_{i=1}^{m} (s + z_i)}{\prod_{j=1}^{n} (s + p_j)} = \infty$$

7.3 Parabolic Input

If the reference input is a parabolic input of the form

$$r(t) = c\frac{t^2}{2}u_s(t)$$

then the steady-state error becomes

$$e_{ss} = \lim_{s \to 0} \frac{c}{s^2 G_c G_p(s)} \tag{92}$$

From Eq. (92) for small steady-state errors

$$K_a = \lim_{s \to 0} s^2 G_c G_p(s) \tag{93}$$

must be very large, where K_a is the parabolic error constant or the acceleration error constant. For zero steady-state error due to a parabolic input, $K_a = \infty$. Therefore the system open-loop transfer function must be at least type III.

Table 4 shows the steady-state errors in terms of the error constants K_p, K_v, and K_a.

In steady-state performance considerations it is important to guarantee both closed-loop stability and the system type. It should also be noticed that having a very high loop gain as given by Eq. (89b) is very similar to having a free integrator in the open-loop transfer function. This notion is useful in regulation-type problems. In tracking systems the problem is compounded by the fact that the system type must be increased. Increasing the system type is usually accompanied by instabilities. To illustrate this, consider the following example.

Example 10 Synthesize a controller for the system shown in Fig. 33*a* so that the steady-state error due to a ramp input is zero.

The open-loop system is unstable. For tracking purposes transform the system to a closed-loop one as shown in Fig. 33*b*. To have zero steady-state error for a ramp, the system's open-loop transfer function must be at least type II and must be of a form that guarantees closed-loop stability. Therefore let

$$G_c(s) = \frac{G_1(s)}{s^2} \tag{94}$$

Table 4 Steady-State Error in Terms of Error Constants

Type	$cu_s(t)$	$ctu_s(t)$	$\frac{t^2}{2}u_s(t)$
0	$\frac{c}{1+K_p}$	∞	∞
1	0	$\frac{c}{K_v}$	∞
2	0	0	$\frac{c}{K_a}$
3	0	0	0

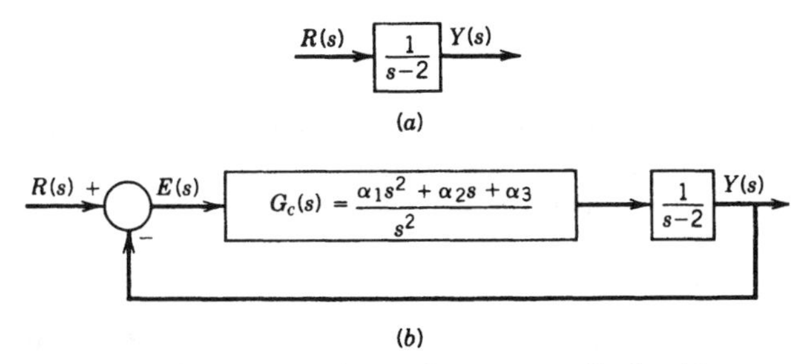

Figure 33 Tracking system: (*a*) open-loop system; (*b*) closed-loop system.

The CLCE is

$$1 + \frac{G_1(s)}{s^2}\frac{1}{s - 2} = 0 \tag{95}$$

That is, $s^3 - 2s^2 + G_1(s) = 0$.

Now for closed-loop stability, $G_1(s)$ must be of the form

$$G_1(s) = \alpha_1 s^2 + \alpha_2 s + \alpha_3$$

yielding the CLCE

$$s^3 + (\alpha_1 - 2)s^2 + \alpha_2 s + \alpha_3 = 0 \tag{96}$$

From the Routh–Hurwitz stability criterion it can be shown that for asymptotic stability $\alpha_1 > 2$, $\alpha_3 > 0$, and $(\alpha_1 - 2)\alpha_2 - \alpha_3 > 0$. So pick $\alpha_1 = 4$, $\alpha_2 = 1$, $\alpha_3 = 1$, yielding the controller

$$G_c(s) = \frac{4s^2 + s + 1}{s^2} \tag{97}$$

In an actual design situation the choice of α_1, α_2, and α_3 will be dictated by the transient design specifications.

7.4 Indices of Performance

Sometimes the performance of control systems is given in terms of a performance index. A performance index is a number that indicates the "goodness" of system performance. Such performance indices are usually employed to get an optimal design in the sense of minimizing or maximizing the index of performance. The optimal parameter values depend directly upon the performance index chosen. Typical performance indices are $\int_0^\infty e^2(t)\,dt$, $\int_0^\infty te^2(t)\,dt$, $\int_0^\infty |e(t)|\,dt$, $\int_0^\infty t|e(t)|\,dt$. In control system design the task is to select the controller parameters to minimize the chosen index.

7.5 Integral-Square-Error (ISE) Criterion

According to the ISE criterion, the quality of system performance is evaluated by minimizing the integral

$$J = \int_0^\infty e^2(t)\, dt \tag{98}$$

A system designed by this criterion tends to show a rapid decrease in a large initial error. Hence the response is fast and oscillatory. Thus the system has poor relative stability. The ISE is of practical significance because $\int e^2(t)\, dt$ resembles power consumption for some systems.

7.6 Integral of Time-Multiplied Absolute-Error (ITAE) Criterion

According to the ITAE criterion, the optimum system is the one that minimizes the performance index:

$$J = \int_0^\infty t|e(t)|\, dt \tag{99}$$

This criterion weighs large initial errors lightly, and errors occurring late in the transient response are penalized heavily. A system designed by use of the ITAE has a characteristic that the overshoot in the transient response is small and oscillations are well damped.

7.7 Comparison of Various Error Criteria[2]

Figure 34 shows several error performance curves. The system considered is

$$\frac{C(s)}{R(s)} = \frac{1}{s^2 + 2\zeta s + 1} \tag{100}$$

The curves of Fig. 34 indicate that $\zeta = 0.7$ corresponds to a near-optimal value with respect to each of the performance indices used. At $\zeta = 0.7$ the system given by Eq. (100) results in rapid response to a step input with approximately 5% overshoot.

Table 5 summarizes the coefficients that will minimize the ITAE performance criterion for a step input to the closed-loop transfer function[5]

$$\frac{C(s)}{R(s)} = \frac{a_0}{s^n + a_{n-1}s^{n-1} + \cdots + a_1 s + a_0} \tag{101}$$

Table 6[2] summarizes the coefficients that will minimize the ITAE performance criterion for a ramp input applied to the closed-loop transfer function:

$$\frac{C(s)}{R(s)} = \frac{b_1 s + b_0}{s^n + b_{n-1}s^{n-1} + \cdots + b_1 s + b_0} \tag{102}$$

Figure 35 shows the response resulting from optimum coefficients for a step input applied to the normalized closed-loop transfer function given in Eq. (101), for ISE, IAE (integral absolute error), and ITAE.

8 SIMULATION FOR CONTROL SYSTEM ANALYSIS

Often in control system design the governing differential equations are reduced to a form that facilitates the shaping of a controller. The model order reductions involve things such as linearization, neglecting fast dynamics, parasitics, time delays, and so on. Once a controller is synthesized for a system, after making simplifying assumptions, it becomes nec-

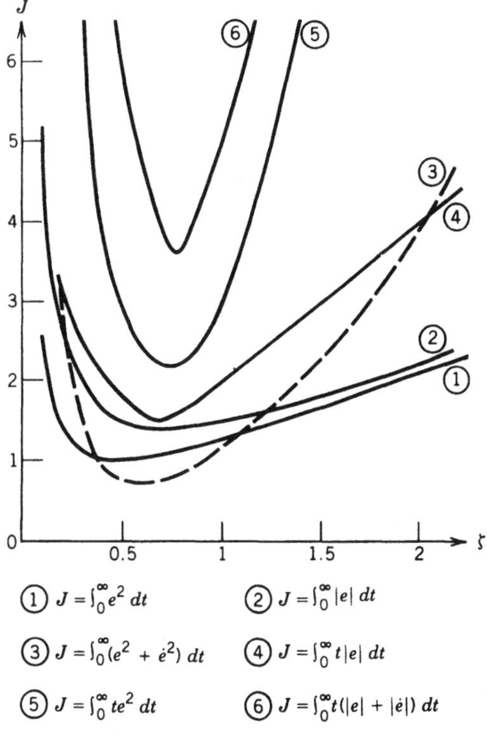

Figure 34 Error performance curves.

essary to try out the controller on the actual system. This involves experimentation. Computer simulation may also be viewed as experimentation where the synthesized controller is inserted into a very sophisticated model of the actual physical system and the system response computed. Extensive simulation studies can reduce the cost of experimentation and may even at times serve as the final controller to be implemented on the actual physical plant. So in simulation the burden of experiment is transformed into generating accurate models for all the components in a system without oversimplifying assumptions. Simulations can be performed on analog, digital, or hybrid computers.[7]

Table 5 Optimal Form of the Closed-Loop Transfer Function Based on the ITAE Criterion for Step Inputs

$\dfrac{C(s)}{R(s)} = \dfrac{a_0}{s^n + a_{n-1}s^{n-1} + \cdots + a_1 s + a_0},\ a_0 = \omega_n^n$
$s + \omega_n$
$s^2 + 1.4\omega_n s + \omega_n^2$
$s^3 + 1.75\omega_n s^2 + 2.15\omega_n^2 s + \omega_n^3$
$s^4 + 2.1\omega_n s^3 + 3.4\omega_n^2 s^2 + 2.7\omega_n^3 s + \omega_n^4$
$s^5 + 2.8\omega_n s^4 + 5.0\omega_n^2 s^3 + 5.5\omega_n^3 s^2 + 3.4\omega_n^4 s + \omega_n^5$
$s^6 + 3.25\omega_n s^5 + 6.60\omega_n^2 s^4 + 8.60\omega_n^3 s^3 + 7.45\omega_n^4 s^2 + 3.95\omega_n^5 s + \omega_n^6$

Table 6 Optimum Coefficients of $T(s)$ Based on the ITAE Criterion for a Ramp Input

$s^2 + 3.2\omega_n s + \omega_n^2$
$s^3 + 1.75\omega_n s^2 + 3.25\omega_n^2 s + \omega_n^3$
$s^4 + 2.41\omega_n s^3 + 4.93\omega_n^2 s^2 + 5.14\omega_n^3 s + \omega_n^4$
$s^5 + 2.19\omega_n s^4 + 6.50\omega_n^2 s^3 + 6.30\omega_n^3 s^2 + 5.24\omega_n^4 s + \omega_n^5$

8.1 Analog Computation[8]

An analog computer is a machine in which several physical components can be selected and interconnected in such a way that the equations describing the operation of the computer are analogous to that of the actual physical system to be studied. It is a continuous-time device operating in a real-time parallel mode, making it particularly suitable for the solution of differential equations and hence for the simulation of dynamic systems. The most commonly used analog computer is the electronic analog computer in which voltages at various places within the computer are proportional to the variable terms in the actual system. The ease of use and the direct interactive control over the running of such a computer allow full scope for engineering intuition and make it a valuable tool for the analysis of dynamic systems and the synthesis of any associated controllers. A facility frequently useful is the ability to slow down or speed up the problem solution. The accuracy of solution, since it is dependent on analog devices, is generally of the order of a few percent, but for the purposes of system analysis and design, higher accuracy is seldom necessary; also, this accuracy often matches the quality of the available input data.

The basic building block of the analog computer is the high-gain dc amplifier, represented schematically by Fig. 36. When the input voltage is $e_i(t)$, the output voltage is given by

$$e_o(t) = -Ae_i(t) \tag{103}$$

where A, the amplifier voltage gain, is a large constant value.

If the voltage to the input of an amplifier, commonly referred to as the summing junction, exceeds a few microvolts, then the amplifier becomes saturated or overloaded because the power supply cannot force the output voltage high enough to give the correct gain. Therefore, if an amplifier is to be operated correctly, its summing junction must be very close to ground potential and is usually treated as such in programming.

When this is used in conjunction with a resistance network as shown in Fig. 37, the resulting circuit can be used to add a number of voltages.

If

$$R_1 = R_2 = R_3$$

then

$$V_0 = -(V_1 + V_2 + V_3) \tag{104}$$

If R_1, R_2, R_3 are arbitrary, then

$$V_0 = -\left(\frac{R_f}{R_1} V_1 + \frac{R_f}{R_2} V_2 + \frac{R_f}{R_3} V_3\right) \tag{105}$$

If there is only one voltage input, then

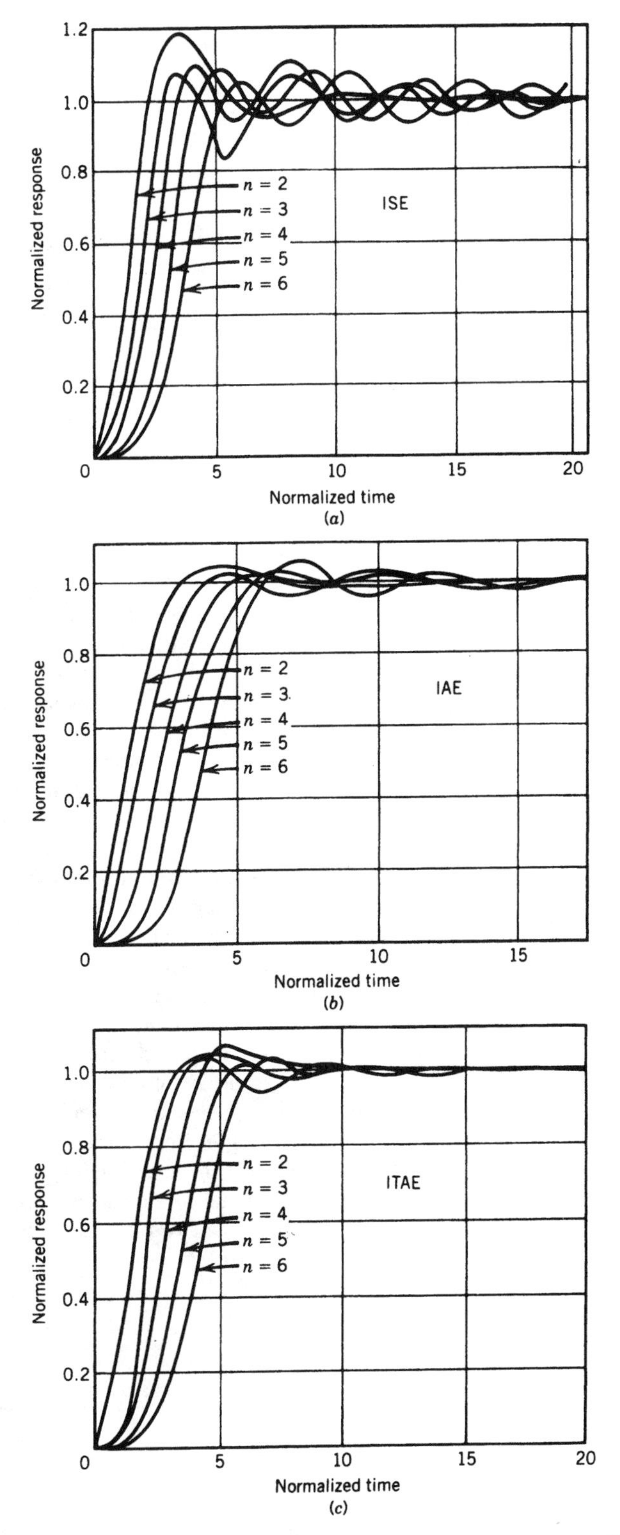

Figure 35 (*a*) Step response of a normalized transfer function using optimum coefficients of ISE, (*b*) IAE, and (*c*) ITAE.

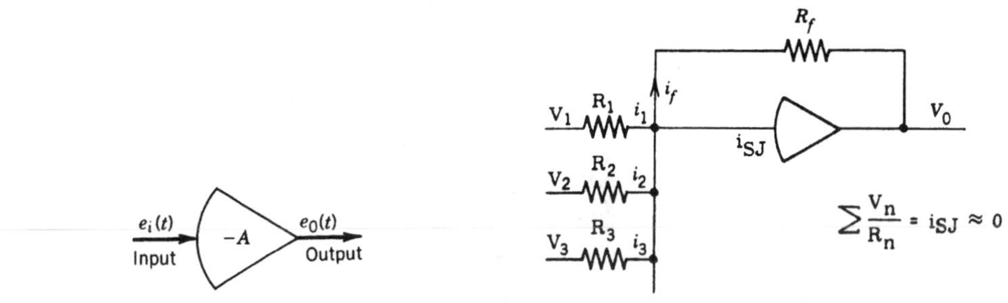

Figure 36 High-gain dc amplifier.

Figure 37 Current balance.

$$V_0 = \frac{R_f}{R_1} V_1 \cdots \text{multiplication by a constant} \tag{106}$$

It should be noted that in all cases there is a sign inversion. Usually the available ratios R_f/R_1 and so on are standardized to 1 and 10, the appropriate gain being selectable as required. The complete circuit comprising the high-gain amplifier, input resistors, and the feedback element is termed an *operational amplifier.* It is given symbolically in Fig. 38.

To multiply a voltage by a constant other than 10, use is made of a grounded potentiometer (usually a 10-turn helical potentiometer), as shown in Fig. 39. This permits multiplication by a constant in the range 0–1.

Since electrical circuits are very nearly linear, the generation of even the simplest nonlinearity, like a product, requires special consideration. One way this can be done is to use the feedback control concept shown in Fig. 40. Two or more potentiometers are mounted rigidly with a common shaft turning the sliders exactly together. A negative unit voltage, which commonly is 10–100 V, is applied across one of the potentiometers. The slider voltage is added to the voltage representing one of the factors. If they differ, then the motor turns the shaft in the direction that eliminates the error. The second factor voltage is applied across the other potentiometer. Its slider then reads the product. This device is called a servomultiplier. It is quite slow in operation because of the motor. Faster multiplication requires the use of special nonlinear networks using diodes. A diode allows current to flow in one direction only if polarities are in the indicated directions in the symbolic representations of Fig. 41. Combinations of diodes and resistors can be made such that the current–voltage curve is a series of straight-line segments. The circuit of Fig. 42 gives the current indicated there. Almost any shape of curve can be approximated by straight-line segments in this way; in particular a network can be made in which the current into the summing junction of an amplifier is approximately proportional to the square of the voltage applied.

The quarter-square multiplier uses two of these circuits, based on the identity

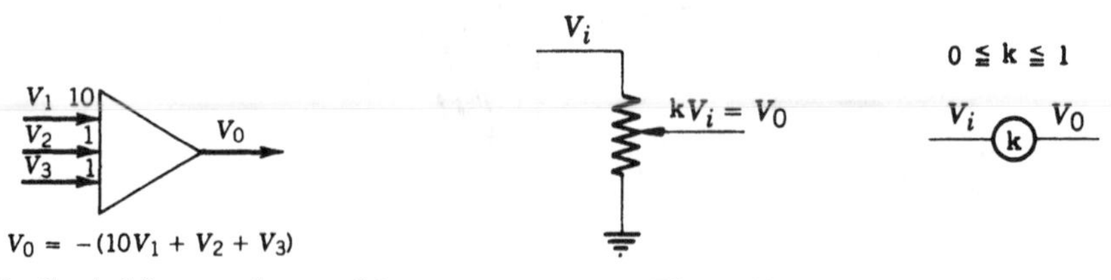

Figure 38 Symbol for summing amplifier.

Figure 39 Potentiometer.

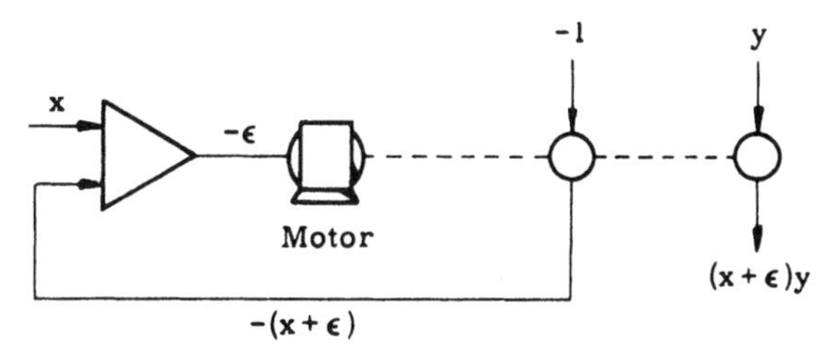

Figure 40 Servomultiplier.

$$xy = \frac{1}{4}\left[(x + y)^2 - (x - y)^2\right] \tag{107}$$

Its symbol is shown in Fig. 43. In most modern computers both polarities of input voltage must be provided. This has been indicated by two amplifiers on the inputs.

Division of voltages is accomplished by using a multiplier in the feedback path of an amplifier, as indicated in Fig. 44.

With the quarter-square multiplier, multiplication and division may be performed accurately at high frequencies. These complete the description of how the ordinary algebraic operations are performed on an analog computer.

Next, we turn to the operations of calculus, which enable differential equations to be solved. Consider the circuit of Fig. 45. The charge on a capacitor is the integral of the current; therefore the derivative of the charge is the current. The charge is the potential times the capacitance, which is fixed; thus current balance around the summing junction gives

$$\frac{V_1}{R} + C\dot{V}_2 = 0 \tag{108}$$

or

$$V_2(t) = V_0(0) - \frac{1}{RC}\int_0^t V_1(\tau)\, d\tau \tag{109}$$

Thus this circuit is an accurate integrator with a proportionality factor or time constant of *RC*. A common symbol is shown in Fig. 46. The initial condition can be set in with a special network furnished with most computers.

Differentiation can be performed with an input capacitor and a feedback resistor; however, it is not generally recommended because input signals are often contaminated by high-frequency noise, and the differentiator circuit amplifies this noise in direct proportion to its frequency. Since differential equations can always be written in terms of integrals, this causes no inconvenience.

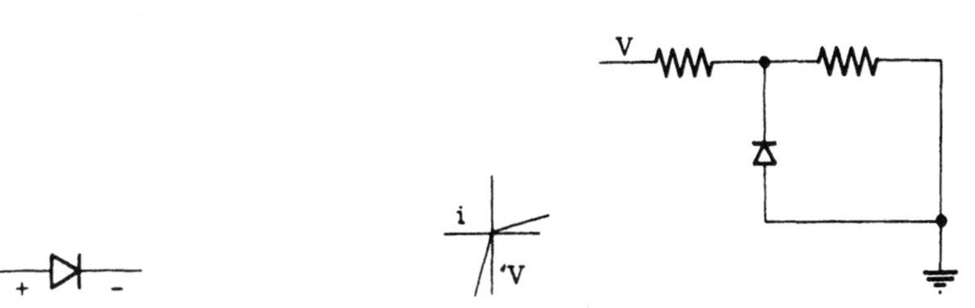

Figure 41 Diode.

Figure 42 Diode network.

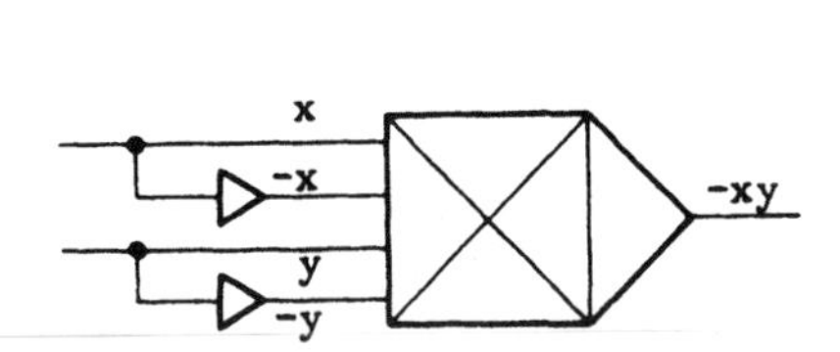

Figure 43 Quarter-square multiplier.

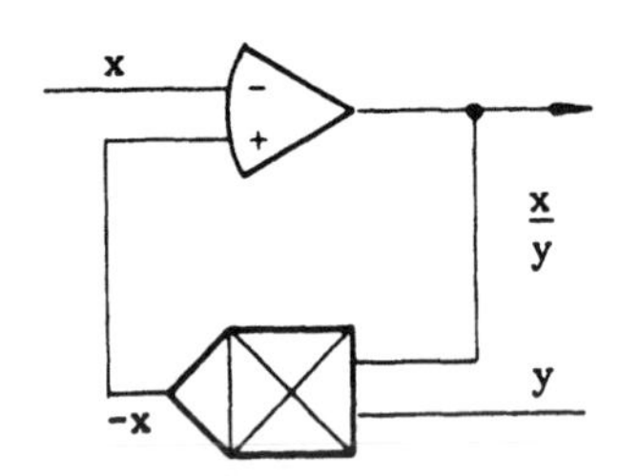

Figure 44 Division.

The circuit of Fig. 47 generates a simple time lag, as can be derived from its current balance:

$$\frac{V_1}{R_1} + \dot{V}_2 C + \frac{V_2}{R_2} = 0 \tag{110}$$

or

$$\dot{V}_2 + \frac{1}{R_2 C} V_2 = -\frac{V_1}{R_1 C} \tag{111}$$

The transfer function of this circuit is

$$G(s) = \frac{V_2}{V_1}(s) = \frac{-R_2/R_1}{R_2 Cs + 1} = \frac{-R_2/R_1}{\tau s + 1}$$

Its time response to a pulse input is a simple exponential decay, with its time constant equal to $R_2 C$.

The circuit corresponding to a damped sinusoid is shown in Fig. 48. Its differential equation is

$$\ddot{y} + 2\zeta\omega_n\dot{y} + \omega_n^2 y = \omega_n^2 f(t) \tag{112}$$

The transfer function is

$$G(s) = \frac{\omega_n^2}{s^2 + 2\omega_n\zeta s + \omega_n^2} = \frac{1}{s^2/\omega_n^2 + (2\zeta/\omega_n)\, s + 1} \tag{113}$$

A particular precaution that must be taken with analog computers is in scaling voltages to represent other physical variables. This must be done consistently for all variables and must be done in such a way that no amplifier is ever overloaded. Since amplifier outputs can never be fully predicted, some problems may require rescaling because of a bad initial

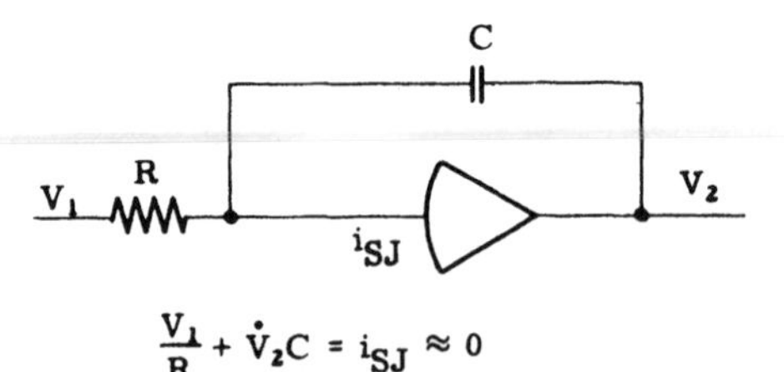

Figure 45 Current balance.

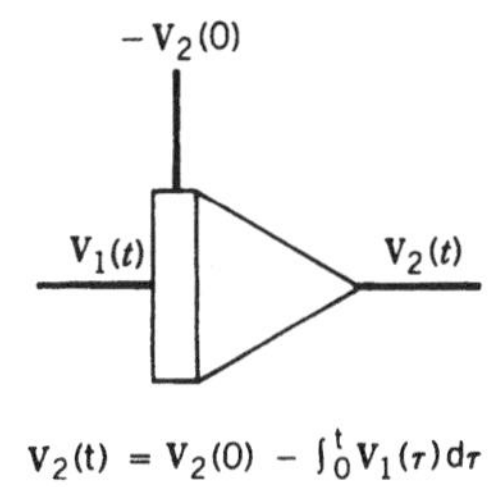

Figure 46 Integration.

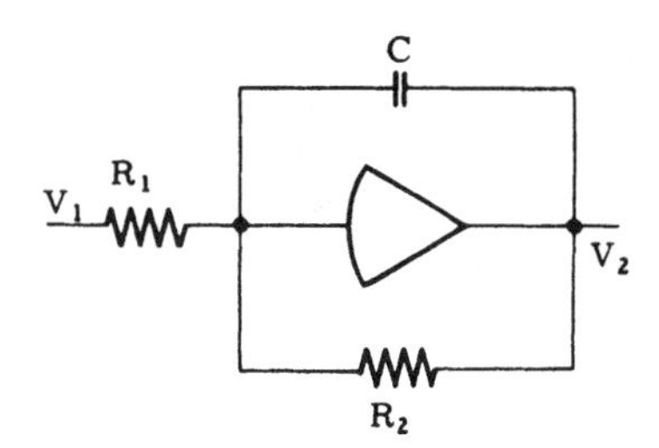

Figure 47 Time lag.

guess about the range of a dependent variable. This is particularly troublesome if nonlinear equipment is involved in the simulation. The necessity for proper prior scaling is one of the major disadvantages of analog computation.

With these basic circuit elements, the control engineer can proceed directly from a block diagram of a physical system to a wired program board without writing down differential equations because the circuits corresponding to the common transfer functions are combinations of those shown in Figs. 46–48. Special nonlinearities can be simulated by special networks. Simple ones such as multiplication, squaring, exponentiation, and generation of logarithms are often prewired directly to the program board.

Therefore, by suitable interconnection of components, even the most complicated system can be simulated in full dynamic character. Frequency response, stability limits, sensitivity to parameter changes, and effects of noise can all be tested and studied. Adjustments of the simulated controls for optimum rejection of noise can be done. Different types of control can be tried experimentally.

If care has been taken in formulating the system dynamics, then the analog computer can be made to behave precisely like the device it simulates.

Figures 49–53 show some special diode networks for simulating some of the special discontinuous nonlinearities that have been discussed previously.

Modem analog computers are generally equipped with additional types of special equipment, some of which are now listed and described below:

1. Comparator: Equivalent to the on–off control of Fig. 50.
2. Resolver: A circuit for rapidly and continuously converting from polar to Cartesian coordinates or the reverse.
3. Relays and switches: Operated by voltages generated in the simulation.
4. Memory amplifiers: An integrator into which an initial condition is set but to which no integrating input is patched will "remember" the initial condition.

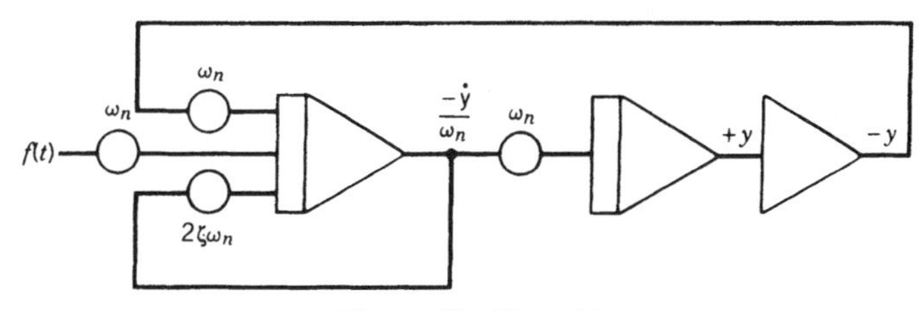

Figure 48 Sinusoid.

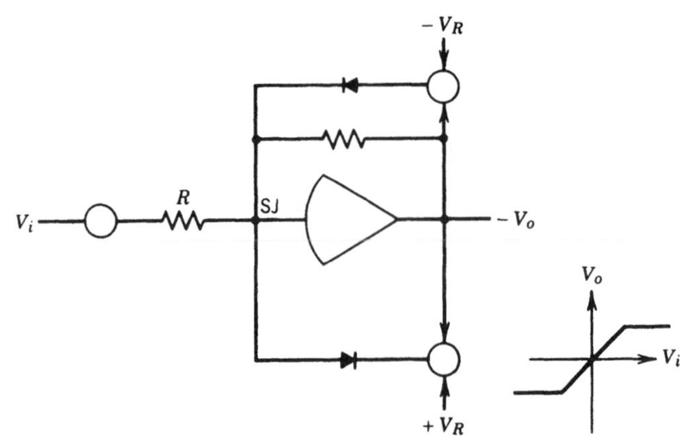

Figure 49 Saturation.

5. Repetitive operation at high speed: Solutions can be generated to repeat hundreds of times per second.
6. Outputs can be displayed on an oscilloscope or plotted on coordinate paper or on continuous recording equipment.

All these features make an analog computer a highly useful tool for design of control systems.

8.2 Digital Computation

An analog computer simultaneously solves all of the differential equations that model a physical system, and continuous voltage signals represent the variables of interest. This enables the machine to operate in real time. Significant disadvantages of analog simulation are the high cost of the computer due to the multiplicity of elements with demanding performance specifications, difficulties with scaling to avoid overloading of amplifiers, and rel-

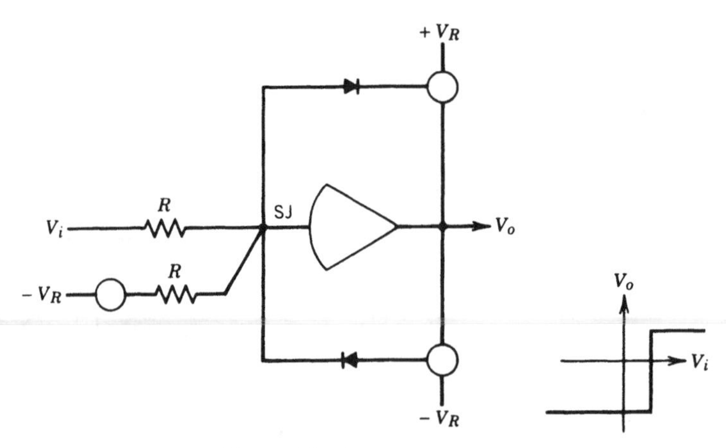

Figure 50 On–off controller.

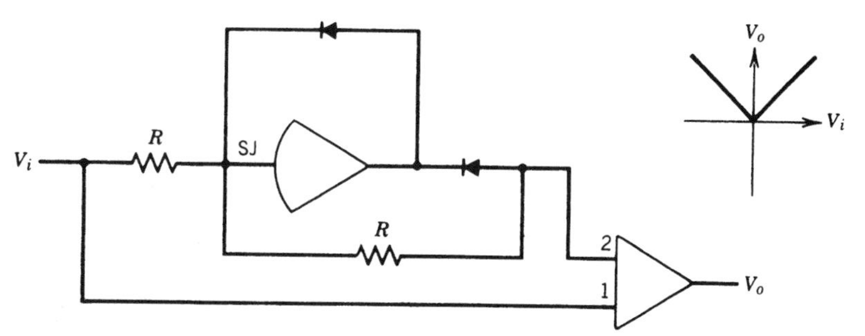

Figure 51 Rectifier.

atively limited accuracy and repeatability due in part to amplifier drift. As a consequence of the very rapid development of digital computer hardware and software giving ever greater capability and flexibility at reducing cost, system simulation is inevitably being carried out more and more on the digital computer. There is effectively no problem of overloading, enabling wide ranges of parameter variation to be accommodated.

The solution of a differential equation involves the process of integration, and for the digital computer analytical integration must be replaced by some numerical method that yields an approximation to the true solution. A number of special programming languages referred to as continuous system simulation languages (CSSLs) or simply as simulation languages are available as analytical tools to study the dynamic behavior of a wide range of systems without the need for a detailed knowledge of computing procedures. The languages are designed to be simple to understand and use, and they minimize programming difficulty by allowing the program to be written as a sequence of relatively self-descriptive statements. Numerous different languages with acronyms such as ACSL, CSMP, CSSL, DYNAMO, DARE, MIMIC, TELSIM, ENPORT, SCEPTRE, and SIMNON have been developed by computer manufacturers, software companies, universities, and others, some for specific families of machines and others for wider applications. Due to standardization, a number of the languages tend to be broadly similar. Symbolic names are used for the system variables, and the main body of the program is written as a series of simple statements based on the system state equations, block diagrams, or bond graphs. To these are added statements specifying initial parameter values and values of system constants and simple command statements controlling the running of the program and specifying the form in which the output is

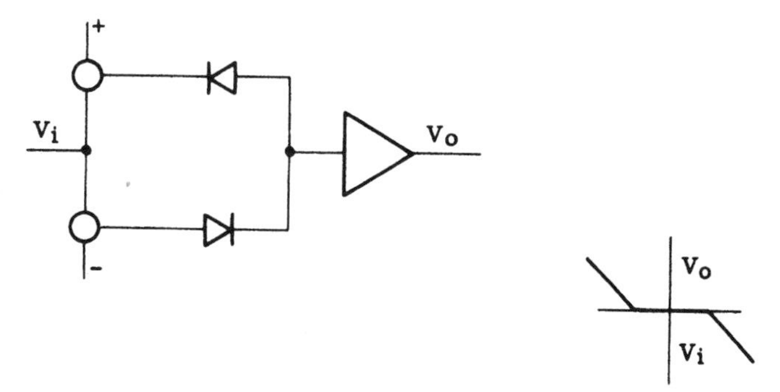

Figure 52 Dead zone.

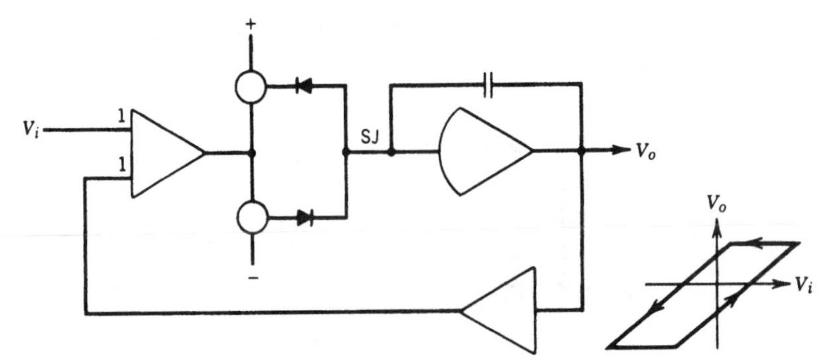

Figure 53 Hysteresis or backlash.

required. The short user-written program is then automatically translated into a FORTRAN (or other high-level language) program that is then compiled, loaded, and executed to produce a time history of the variables of interest in print or plot form. System constants and initial conditions can be altered and the program rerun without the need to retranslate and recompile. Many of the languages are designed to be run interactively from a graphics terminal and have the facility of displaying on the screen whichever output is of interest and, if the solution is not developing as desired, the capability of interrupting the program, changing parameters, and rerunning immediately.

The typical steps to be taken by a CSSL are shown in Fig. 54. The major process blocks include:

1. Reading the initial problem description, in which the circuit, bond graph, schematic, or block diagram information is communicated to the computer
2. Formulating the system and state equations
3. Performing the numerical integration of the state equations by a suitable method
4. Storing and displaying results

Most simulation languages differ with regard to use by the engineer in steps 1 and 2; for example, SCEPTRE for circuit descriptions, ENPORT for bond graph descriptions,

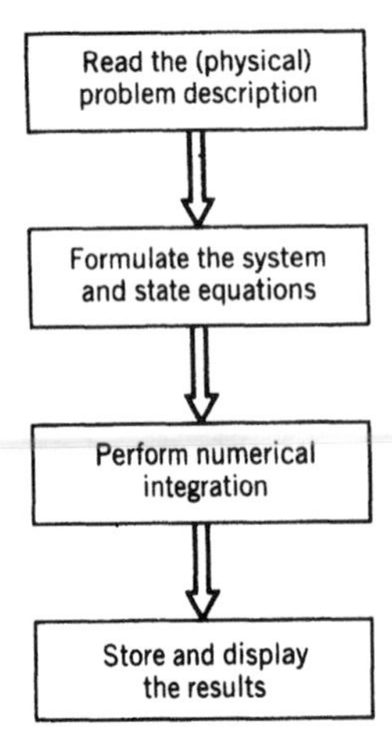

Figure 54 Typical simulation flowchart of the CSSL.

CSMP for block diagrams, and DARE for state equations. The best guide to a program's characteristics is its user's manual.

In summary, the advantages of digital simulation are (i) simple program writing, (ii) accuracy and reproducibility, (iii) cost-effectiveness, (iv) where interactive facilities exist, keyboard entry of program and running, (v) inspection of plots on graphics display screen, and (vi) online modification and rerunning.

8.3 Hybrid Computation

Hybrid computation is a combination of analog and digital computations. In hybrid computations analog and digital machines are used in such a way that a problem can be programmed by exploiting the most efficient features of each. Much of the cost of such a machine arises from the rather complicated interface equipment required to make the necessary conversions between analog and digital signals and vice versa. The cost of an analog or hybrid computer is now justified only for certain large simulations where fast computing speed attained by parallel operation is important. Even this advantage may disappear as parallel computation is introduced into digital systems.

REFERENCES

1. E. O. Doebelin, *Control System Principles and Design,* Wiley, New York, 1985.
2. K. Ogata, *Modern Control Engineering,* Prentice-Hall, Englewood Cliffs, NJ, 1970.
3. M. Vidyasagar, *Nonlinear Systems Analysis,* Prentice-Hall, Englewood Cliffs, NJ, 1978.
4. Y. Takahashi, M. J. Rabins, and D. M. Auslander, *Control and Dynamic Systems,* Addison-Wesley, Reading, MA, 1972.
5. D. Graham and R. C. Lathrop, "The Synthesis of Optimum Response: Criteria and Standard Forms," AIEE *Transactions,* **72**, (Pt. II), 273–288 (1953).
6. R. C. Dorf, *Modern Control Systems,* Addison-Wesley, Reading, MA, 1986.
7. G. A. Korn and J. V. Wait, *Digital Continuous System Simulation,* Prentice-Hall, Englewood Cliffs, NJ, 1978.
8. O. W. Eshbach and M. Souders (eds.), *Handbook of Engineering Fundamentals,* 3rd ed., Wiley, New York, 1975.

BIBLIOGRAPHY

Bode, H. W., *Network Analysis and Feedback Amplifier Design,* Van Nostrand, New York, 1945.

Chestnut, H., and R. W. Mayer, *Servomechanisms and Regulating Systems Design,* 2nd ed., Vol. 1, Wiley, New York, 1959.

D'Azzo, J. J., and C. H. Houpis, *Linear Control System Analysis and Design,* McGraw-Hill, New York, 1988.

Distefano, J. J. III, A. R. Stubberud, and I. J. Williams, *Feedback and Control Systems* (Schaum's Outline Series), Shaum Publishing, New York, 1967.

Dransfield, P., *Engineering Systems and Automatic Control,* Prentice-Hall, Englewood Cliffs, NJ, 1968.

Elgerd, O. I., *Control Systems Theory,* McGraw-Hill, New York, 1967.

Evans, W. R., *Control-System Dynamics,* McGraw-Hill, New York, 1954.

Eveleigh, V. W., *Introduction to Control Systems Design,* McGraw-Hill, New York, 1972.

Franklin, G. F., J. D. Powell, and A. Emami-Naeini, *Feedback Control of Dynamic Systems,* Addison-Wesley, Reading, MA, 1986.

Franklin, G. F., and J. D. Powell, *Digital Control of Dynamic Systems,* Addison-Wesley, Reading, MA, 1980.

Horowitz, I. M., *Synthesis of Feedback Systems,* Academic, New York, 1963.

Houpis, C. H., and G. B. Lamont, *Digital Control Systems Theory, Hardware, Software,* McGraw-Hill, New York, 1985.

Kuo, B. C., *Digital Control Systems,* Holt, Rinehart and Winston, New York, 1980.

Kuo, B. C., *Automatic Control Systems,* Prentice-Hall, Englewood Cliffs, NJ, 1982.

Melsa, J. L., and D. G. Schultz, *Linear Control Systems,* McGraw-Hill, New York, 1969.

Nyquist, H., "Regeneration Theory," Bell *System Technical Journal* **II,** 126–147 (1932).

Palm, W. J. III, *Modeling, Analysis and Control of Dynamic Systems,* Wiley, New York, 1983.

Phillips, C. L., and H. T. Nagle, Jr., *Digital Control System Analysis and Design,* Prentice-Hall, Englewood Cliffs, NJ, 1984.

Ragazzini, J. R., and G. F. Franklin, *Sampled Data Control Systems,* McGraw-Hill, New York, 1958.

Raven, F. H., *Automatic Control Engineering,* 4th ed., McGraw-Hill, New York, 1987.

Rosenberg, R. C., and D. C. Karnopp, *Introduction to Physical System Dynamics,* McGraw-Hill, New York, 1983.

Shinners, S. M., *Modern Control Systems Theory and Application,* Addison-Wesley, Reading, MA, 1972.

Truxal, J. G., *Automatic Feedback Control Synthesis,* McGraw-Hill, New York, 1955.

CHAPTER 13

CONTROL SYSTEM PERFORMANCE MODIFICATION

Suhada Jayasuriya
Department of Mechanical Engineering
Texas A&M University
College Station, Texas

1 INTRODUCTION

Chapter 12 presents a wide variety of tools for analyzing closed-loop control systems. This chapter focuses on the development of additional analytical tools for closed-loop control systems, tools aimed at performance modification and improvement. Each successive tool presented in this chapter is useful in its ability to predict system performance and to pinpoint the appropriate modifications so that the closed-loop system will meet its required performance objectives.

Chapter 14 also addresses the problem of system modifications, but there the emphasis is to work with the adjustable parameters of a device called the servocontroller to achieve the necessary result.

Reprinted from *Instrumentation and Control,* Wiley, New York, 1990, by permission of the publisher.

2 GAIN AND PHASE MARGIN

The Nyquist stability criterion may be conveniently used to define certain measures of relative stability or robustness. We note that the $(-1,0)$ point in the GH plane plays a crucial role in determining the closed-loop stability of a system. If a system's stability status is known, one might be interested in knowing how stable the system is due to changes in parameters. For example, if the system remains stable despite large changes in parameters, then it is said to possess a high degree of relative stability or robustness. Gain margin and phase margin are two measures typically employed to characterize this robustness. They characterize how close the Nyquist plot is to encircling the $(-1,0)$ point in the GH plane.

2.1 Gain Margin

This is a measure of how much the loop gain can be raised before closed-loop instability results. The basic definition of gain margin (GM) is apparent from Fig. 1.

2.2 Phase Margin

The phase margin (PM) is the difference between the phase of $GH(j\omega)$ and 180° when the $GH(j\omega)$ crosses the circle with unit magnitude. A positive PM corresponds to a case where the Nyquist locus does not encircle the $(-1,0)$ point. This is shown in Fig. 2.

A stable system corresponds to gain and phase margins that are positive. In some cases, however, the PM and GM notions break down. For first- and second-order systems, the phase never crosses the 180° line; hence the GM is always ∞. For higher order systems, it is possible to have more than one crossing of the unit amplitude circle and more than one crossing of the 180° line. In such situations the GM and PM are somewhat misleading. Furthermore, non-minimum-phase systems exhibit stability criteria that are opposite to those previously defined.

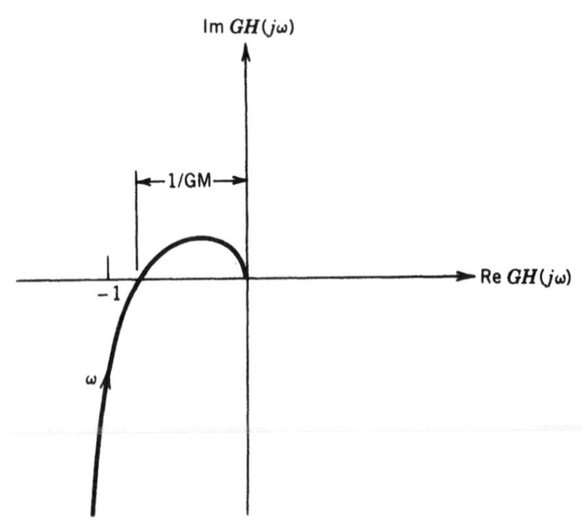

Figure 1 Gain margin.

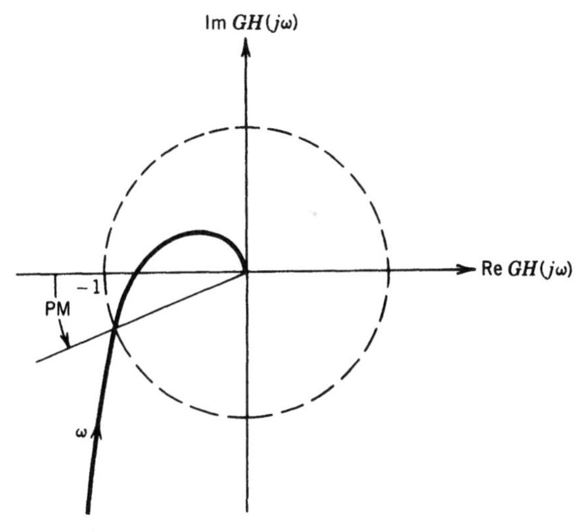

Figure 2 Phase margin.

2.3 Gain-Phase Plots

The graphical representation of the frequency response of the system $G(s)$ using either

$$G(j\omega) = G(s)|_{s=j\omega} = \text{Re } G(j\omega) + j \text{ Im } G(j\omega)$$

or

$$G(j\omega) = |G(j\omega)|e^{j\phi(\omega)}$$

where

$$\phi(\omega) = \underline{/G(j\omega)}$$

is known as the polar plot. The coordinates of the polar plot are the real and imaginary parts of $G(j\omega)$, as shown in Fig. 3.

Example 1 Obtain the polar plot of the transfer function

$$G(s) = \frac{K}{s(\tau s + 1)}$$

The frequency response is given by

$$G(j\omega) = \frac{K}{j\omega(j\tau\omega + 1)}$$

Then the magnitude and the phase can be written as

$$|G(j\omega)| = \frac{K}{\omega\sqrt{\omega^2\tau^2 + 1}}$$

and

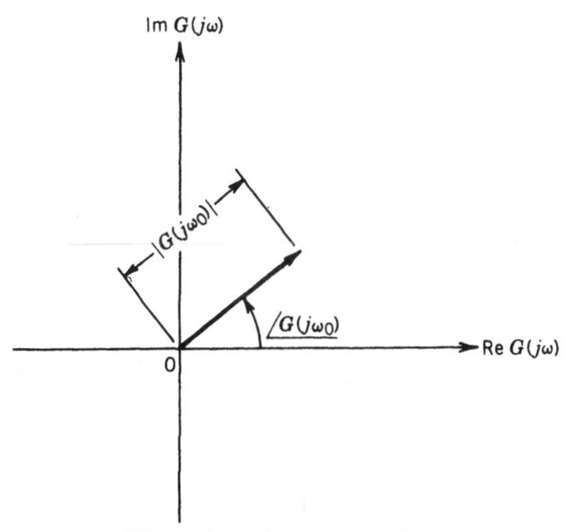

Figure 3 Polar representation.

$$G(j\omega) = -\frac{\pi}{2} - \tan^{-1} \omega\tau$$

If $|G(j\omega)|$ and $\underline{/G(j\omega)}$ are computed for different frequencies, an accurate plot can be obtained. A quick idea can, however, be gained by simply doing a limiting analysis at $\omega = 0$, $\omega = \infty$, and the corner frequency $\omega = 1/\tau$.

We note that

$$|G(j\omega)| \to \infty \qquad \underline{/G(j\omega)} = \frac{-\pi}{2} \qquad \text{for } \omega = 0$$

$$|G(j\omega)| \to 0 \qquad \underline{/G(j\omega)} = -\pi \qquad \text{for } \omega = \infty$$

$$|G(j\omega)| = \frac{K\tau}{\sqrt{2}} \qquad \underline{/G(j\omega)} = -\frac{3\pi}{4} \qquad \text{for } \omega = \frac{1}{\tau}$$

The polar plot is shown in Fig. 4.

A gain-phase plot is where the frequency response information is given with respect to a Cartesian frame with vertical axis for gain and horizontal axis for phase.

2.4 Polar Plot as a Design Tool in the Frequency Domain

As a design tool its best use is in determining relative stability with respect to GM and PM. If the uncertainty in the transfer function can be characterized by bounds on the gain-phase plot, then it allows one to determine what type of compensation needs to be provided for the system to perform in the presence of such uncertainties. As an example consider the closed-loop system shown in Fig. 5.

Suppose the gain-phase plot is known to lie in the shaded region in the $G(j\omega)$ plane, as shown in Fig. 6.

Since the shaded region includes the $(-1,0)$ point, and the true gain-phase plot for the plant can lie anywhere in the shaded region, the system can potentially be unstable. If

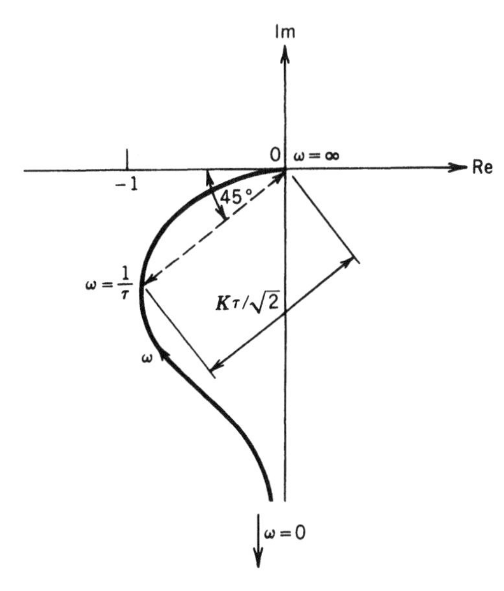

Figure 4 Polar plot for $G(s) = K/s(\tau s + 1)$.

stabilization in the presence of uncertainty is the primary design issue, then we would require a $G_c(s)$ so that it would reshape the high-frequency part of the polar plot with a reduced band of uncertainty. The reduction in the band of uncertainty is a required feature of any sound feedback system design. Qualitatively one would expect to reshape the polar plot to something that looks like what is shown in Fig. 6*b*. Moreover, knowing the important frequency ranges will allow one to be more concerned with relevant portions of the gain-phase plot for reshaping. To further illustrate the basic philosophy of a typical design in the frequency domain, consider the plant transfer function

$$G_p(s) = \frac{K}{s(1 + s)(1 + 0.0125s)}$$

in the feedback configuration shown in Fig. 7.

It is required that when a ramp input is applied to the closed-loop system, the steady-state error of the system does not exceed 1% of the amplitude of the input ramp. Using steady-state error computations we find that the minimum K should be such that

$$\text{Steady-state error} = e_{\text{ss}} = \lim_{s \to 0} \frac{1}{sG_p(s)} = \frac{1}{K} \leq 0.01$$

that is, $K \geq 100$.

It can be easily verified that with $G_c(s) = 1$ the system is unstable for $K > 81$, implying that a controller $G_c(s)$ must be designed to satisfy the steady-state performance and relative stability requirements. Putting it another way, the controller must be able to keep the zero-

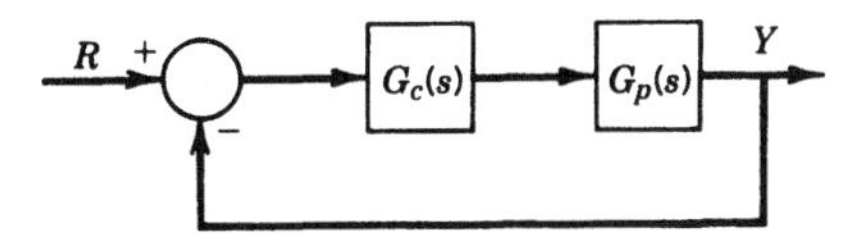

Figure 5 Closed-loop system.

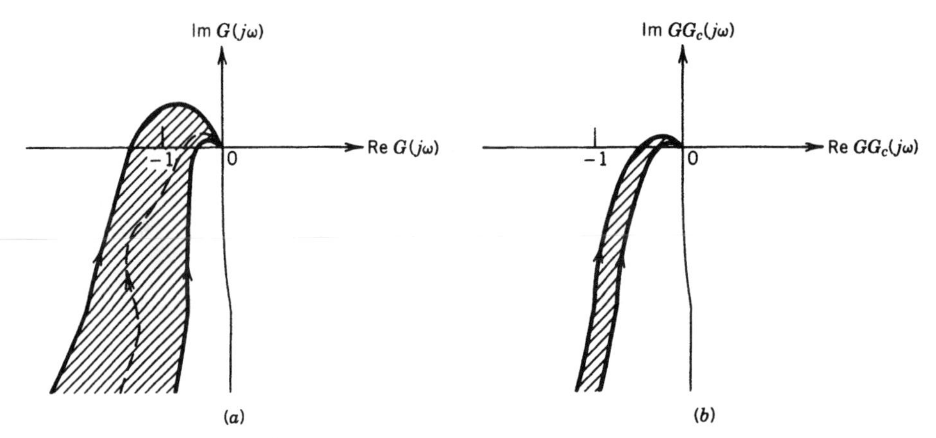

Figure 6 Polar plots for an uncertain system.

frequency gain of $sG_pG_c(s)$ effectively at 100 while maintaining a prescribed degree of relative stability. The principle of the design in the frequency domain is best illustrated by the polar plot of $G_p(s)$ shown in Fig. 8. In practice, the Bode diagram is preferred for design purposes because it is simpler to construct. The polar plot is used mainly for analysis and added insight.

As shown in Fig. 8, when $K = 100$, the polar plot of $G_p(s)$ encloses the $(-1,0)$ point, and the closed-loop system is unstable. Let us assume that we wish to realize that PM = 30°. This means that the polar plot must pass through point A (with magnitude 1 and phase −150°). If K is the only adjustable parameter to achieve this PM, the desired value $K \simeq 3.4$, as shown in Fig. 8. But, K cannot be set to 3.4 since the ramp error constant would only be 3.4 s^{-1}, and the steady-state error requirement will not be satisfied.

Since the steady-state performance of the system is governed by the characteristics of the transfer function at low frequency, and the damping or the transient behavior of the system is governed by the relatively high-frequency characteristics, as Fig. 8 shows, to simultaneously satisfy the transient and the steady-state requirements, the frequency locus of $G_p(s)$ has to be reshaped so that the high-frequency portion of the plot follows the $K = 3.4$ trajectory and the low-frequency portion follows the $K = 100$ trajectory. The significance of this reshaping of the frequency locus is that the compensated locus shown in Fig. 8 will be coincident with the high-frequency portion yielding PM = 30°, while the zero-frequency gain is maintained at 100 to satisfy the steady state requirement.

When we inspect the loci of Fig. 8, we see that there are at least two alternatives in arriving at the compensated locus:

1. Starting from the $K = 100$ locus and reshaping the locus in the region near the gain crossover frequency ω_g while keeping the low-frequency region of $G_p(s)$ relatively unaltered

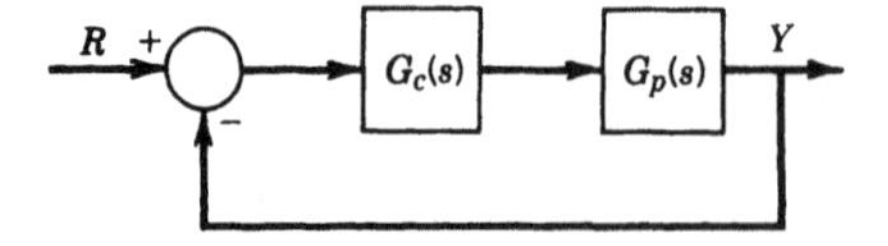

Figure 7 Closed-loop system with $G(s) = K/[s(s + 1)(0.0125s + 1)]$.

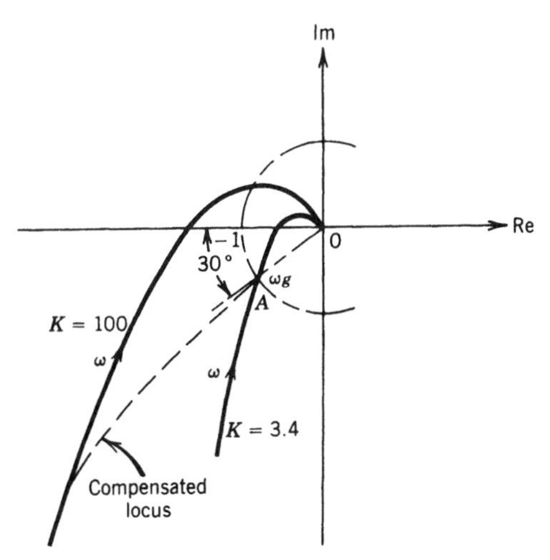

Figure 8 Polar plot for open-loop system transfer function of Fig. 7.

2. Starting from the $K = 3.4$ locus and reshaping the low-frequency portion of $G_p(s)$ to obtain an error constant of 100 while keeping the locus near $\omega = \omega_g$ relatively unchanged

In the first approach, the high-frequency portion of $G_p(s)$ is pushed in the counterclockwise (CCW) direction, which means that more phase is added to the system in the positive direction in the proper frequency range. This scheme is basically phase-lead compensation and controllers used for this purpose are often of the high-pass filter type. The second approach apparently involves the shifting of the low-frequency part of the $K = 3.4$ trajectory in the clockwise (CW) direction, or alternatively reducing the magnitude of $G_p(s)$ with $K = 100$ at the high-frequency range. This scheme is often referred to as phase-lag compensation since more phase lag is introduced to the system in the low-frequency range. The controllers used for this purpose are often referred to as low-pass filters.

3 HALL CHART

In typical frequency response design only the open-loop transfer function is plotted. Therefore it is useful to know how the closed-loop performance is related to the open loop. Hall charts provide a convenient way of carrying out a frequency response design with closed-loop performance specifications. One important consideration is the maximum closed-loop gain. Another is the closed-loop phase. A Hall chart primarily consists of constant closed-loop gain loci and constant closed-loop phase loci. A design would then proceed by drawing the open-loop polar plot on the Hall chart.

For a unity negative-feedback system as shown in Fig. 9 the closed-loop transfer function is

$$\frac{C(s)}{R(s)} = \frac{G(s)}{1 + G(s)} \tag{1}$$

In the following discussion we assume that the polar plot of $G(j\omega)$ is known.

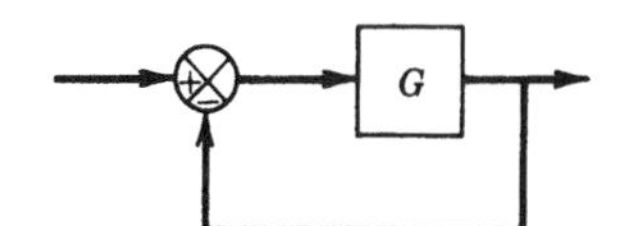

Figure 9 Unity negative-feedback system.

3.1 Constant-Magnitude Circles

The loci on which the closed-loop magnitude

$$\left|\frac{C(s)}{R(s)}\right| = \left|\frac{G(s)}{1 + G(s)}\right| = M = \text{const}$$

are referred to as constant-magnitude loci. In fact these loci are circles in the $G(j\omega)$-plane. This can be established by noting a typical point on the $G(j\omega)$ plot as $X + jY$.

Then

$$M = \frac{|X + jY|}{|1 + X + jY|}$$

and

$$M^2 = \frac{X^2 + Y^2}{(1 + X)^2 + Y^2}$$

Hence

$$X^2 + \frac{2M^2}{M^2 - 1}X + \frac{M^2}{M^2 - 1} + Y^2 = 0$$

which can be written as

$$\left(X + \frac{M^2}{M^2 - 1}\right)^2 + Y^2 = \frac{M^2}{(M^2 - 1)^2} \tag{2}$$

Equation (2) is the equation of a circle with center at $X = -M^2/(M^2 - 1)$, $Y = 0$ and with radius $|M/(M^2 - 1)|$. A family of constant-M circles is shown in Fig. 10. Given a point $P \equiv (X_1, Y_1)$ on an open-loop polar plot $G(j\omega)$, the corresponding closed-loop magnitude can be determined by locating the M circle passing through that point.

Graphically the intersection of the $G(j\omega)$ plot and the constant-M locus gives the value of M at the frequency denoted on the $G(j\omega)$ curve. If it is desired to keep the value of the maximum closed-loop gain M_r less than a certain value, the $G(j\omega)$ curve must not intersect the corresponding M circle at any point and at the same time must not enclose the $(-1,j0)$ point. The constant-M circle with the smallest radius that is tangent to the $G(j\omega)$ curve gives the value of M_r, and the resonant frequency ω_r is read off at the tangent point on the $G(j\omega)$ curve.

3.2 Constant-Phase Circles

The loci of constant phase of the closed-loop system can also be determined in the $G(j\omega)$-plane by a method similar to that used for constant-M loci. With reference to Eq. (1) the phase of the closed-loop system corresponding to the point $P = X + jY$ is written as

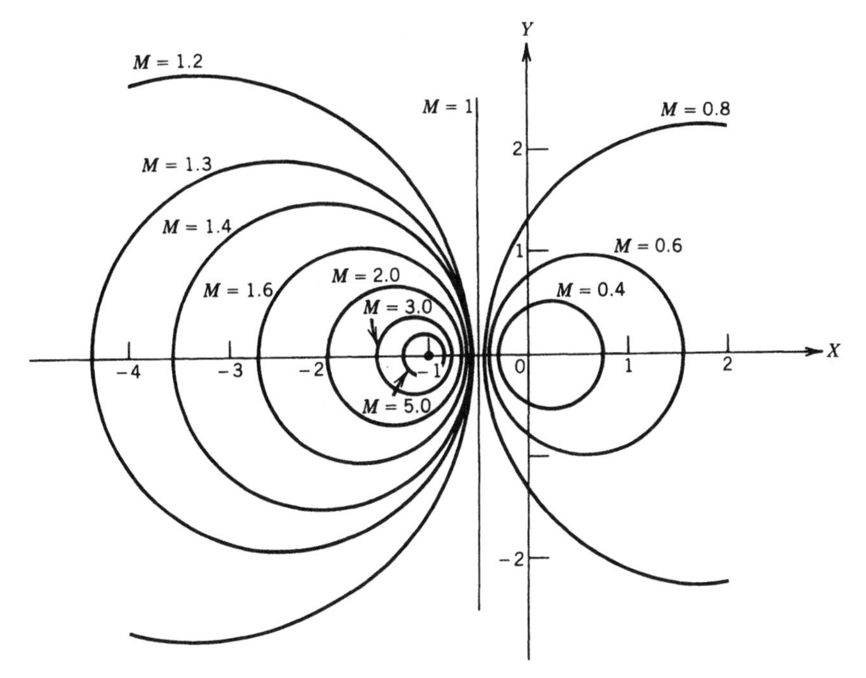

Figure 10 Family of constant-M circles.

$$\phi = \tan^{1}\left(\frac{Y}{X}\right) - \tan^{1}\left(\frac{Y}{1+X}\right) \tag{3}$$

Taking the tangent on both sides of Eq. (3) and rearranging yields

$$\left(X + \frac{1}{2}\right)^2 + \left(Y - \frac{1}{2N}\right)^2 = \frac{1}{4} + \left(\frac{1}{2N}\right)^2 \tag{4}$$

where $N = \tan \phi$.

Equation (4) represents a family of circles with center at $(-1/2, 1/2N)$ and with radius $\sqrt{1/4 + 1/(2N)^2}$. The constant-phase loci are shown in Fig. 11.

The use of constant-magnitude and constant-phase circles enables one to find the entire closed-loop frequency response from the open-loop frequency response $G(j\omega)$ without calculating the magnitude and phase of the closed-loop transfer function at each frequency. The intersections of the $G(j\omega)$ locus and the M circles and N circles give the values of M and N at frequency points on the $G(j\omega)$ locus.

3.3 Closed-Loop Frequency Response for Nonunity Feedback Systems

The constant-M and constant-N circles are limited to closed-loop systems with unity negative feedback, whose transfer function is given by Eq. (1). When a system has nonunity feedback, the closed-loop transfer function is

$$\frac{C(s)}{R(s)} = \frac{G(s)}{1 + G(s)H(s)} \tag{5}$$

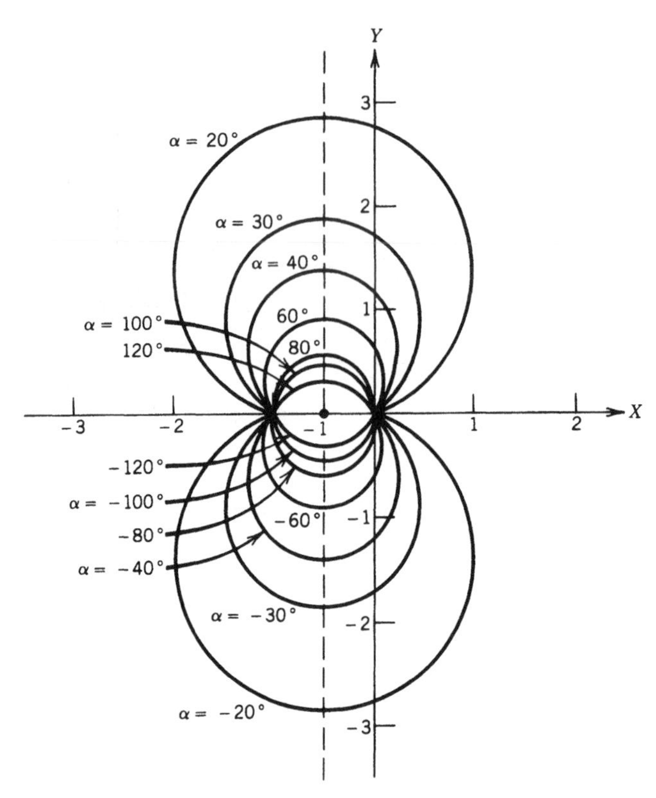

Figure 11 Family of constant-N circles.

and constant-M loci derived earlier cannot be directly applied. However, with a slight modification constant-M and constant-N loci can still be applied to systems with nonunity feedback. We modify Eq. (5) as

$$\frac{C(s)}{R(s)} = \frac{1}{H(s)} \frac{G(s)H(s)}{1 + G(s)H(s)}$$

The magnitude and phase angle of $G_1(s)/[1 + G_1(s)]$, where $G_1(s) = G(s)H(s)$, may be obtained easily by plotting the $G_1(j\omega)$ locus and reading the values of M and N at various frequency points. The closed-loop frequency response $C(j\omega)/R(j\omega)$ may then be obtained by multiplying $G_1(j\omega)/[1 + G_1(j\omega)]$ by $1/H(j\omega)$.

3.4 Closed-Loop Amplitude Ratio

In obtaining suitable performance, the adjustment of gain is usually the first consideration. The adjustment of gain is usually based on the maximum closed-loop gain or the resonant peak. That is the gain K which must be chosen so that over the entire frequency range the closed-loop amplitude ratio M_r is not exceeded.

Consider first isolating the circle corresponding to M_r as shown in Fig. 12. Then a tangent line to the M_r circle is drawn from the origin, which makes an angle ψ with the real line.

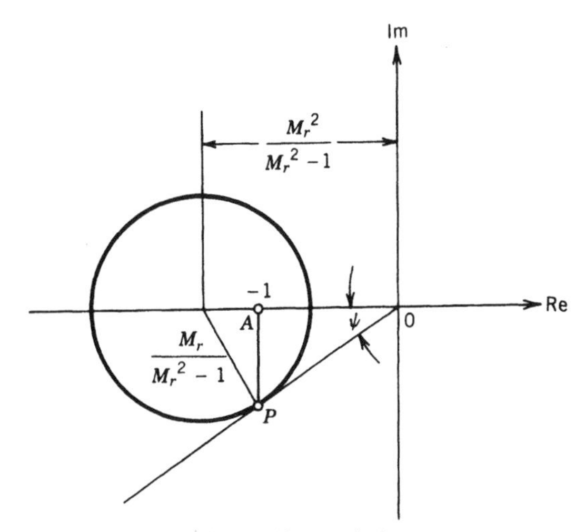

Figure 12 *M* circle.

If $M_r > 1$, then

$$\sin \psi = \left| \frac{M_r/(M_r^2 - 1)}{M_r^2/(M_r^2 - 1)} \right| = \frac{1}{M_r}$$

It can be shown that the line drawn from P perpendicular to the negative real axis intersects this axis at the $(-1,0)$ point. These two facts, namely $\sin \psi = 1/M_r$ and that the normal from P passes through $(-1,0)$, can be used to determine the appropriate gain K.

Example 2 Consider the system shown in Fig. 13*a*: Determine K so that $M_r = 1.4$. First sketch the polar plot of

$$\frac{G(j\omega)}{K} = \frac{1}{j\omega(1 + j\omega)}$$

as shown in Fig. 13*b*. The value of ψ corresponding to $M_r = 1.4$ is obtained from

$$\psi = \sin^{-1} \frac{1}{M_r} = \sin^{-1} \frac{1}{1.4} = 45.6°$$

The next step is to draw a line OP that makes an angle $\psi = 45.6°$ with the negative real axis. Then draw the circle that is tangent to both the $G(j\omega)/K$ locus and the line OP. The perpendicular line drawn from the point P intersects the negative real axis at $(-0.63,0)$. Then the gain K of the system is determined as follows:

$$K = \frac{1}{0.63} = 1.58$$

4 NICHOLS CHART

Both the gain and phase plots are generally required to analyze the performance of a closed-loop system. A major disadvantage in working with polar plots is that the curve no longer

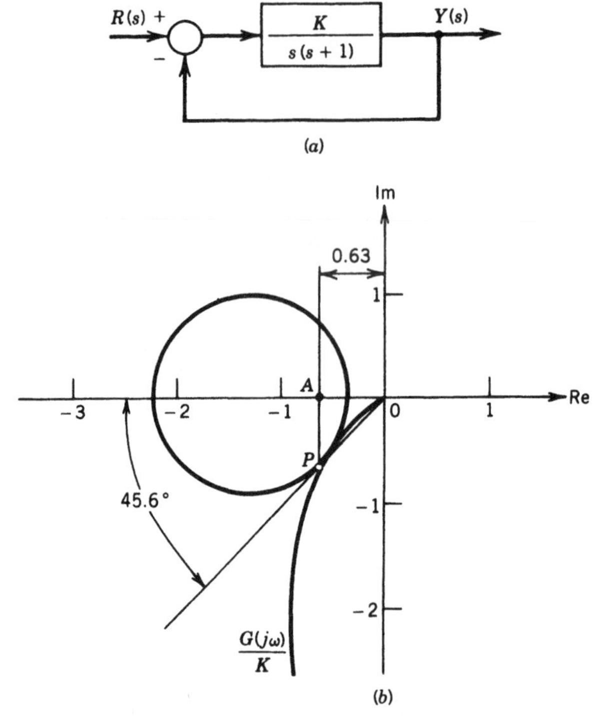

Figure 13 (*a*) Closed-loop system; (*b*) determination of the gain K using an M circle.

retains its original shape when a simple modification such as the change of the loop gain is made to the system. In design, however, not only the loop gain must be altered but often series or feedback controllers are to be added to the original system that require the complete reconstruction of the resulting open-loop transfer function. For design purposes it is more convenient to work with Bode diagrams or gain-versus-phase plots. The latter representation with corresponding M and N circles superimposed on it is referred to as the *Nichols chart.* In a gain-versus-phase plot the entire $G(j\omega)$ is shifted up or down vertically when the gain is altered. A Nichols chart is shown in Fig. 14.

This chart is symmetric about the $-180°$ axis. The M and N loci repeat for every 360°, and there is symmetry at every 180° interval. The M loci are centered about the critical point (0 dB, $-180°$).

4.1 Closed-Loop Frequency Response from That of Open Loop

It is quite easy to determine the closed-loop frequency response from that of the open loop by using the Nichols chart. If the open-loop frequency response curve is superimposed on the Nichols chart, the intersections of the open-loop frequency response curve $G(j\omega)$ and the M and N loci give the magnitude M and phase angle ϕ of the closed-loop frequency response at each frequency point. If the $G(j\omega)$ locus does not intersect the $M = M_r$ locus but is tangent to it, then the resonant peak value of the closed-loop frequency response is given by M_r. The resonant frequency is given by the frequency at the point of tangency.

As an example consider the unity negative-feedback system with the following open-loop transfer function:

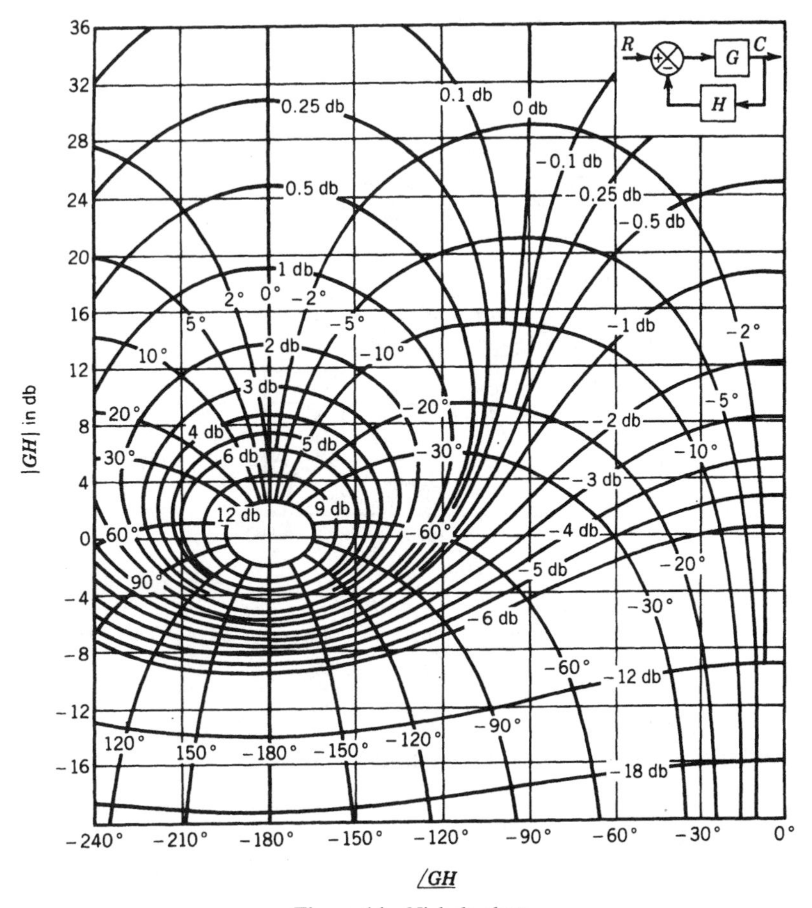

Figure 14 Nichols chart.

$$G(s) = \frac{K}{s(s+1)(0.5s+1)} \qquad K = 1$$

To find the closed-loop frequency response by use of the Nichols chart, the $G(j\omega)$ locus is first constructed. (It is easy to first construct the Bode diagram and then transfer values to the Nichols chart.) The closed-loop frequency response curves (gain and phase) may be constructed by reading the magnitude and phase angles at various frequency points on the $G(j\omega)$ locus from the M and N loci as shown in Fig. 15. Since the $G(j\omega)$ locus is tangent to the M = 5-dB locus, the peak value of the closed-loop frequency response is M_r = 5 dB, and the resonant frequency is 0.8 rad/s.

The bandwidth of the closed-loop system can easily be found from the $G(j\omega)$ locus in the Nichols chart. The frequency at the intersection of the $G(j\omega)$ locus and the $M = -3$-dB locus gives the bandwidth. The gain and phase margins can be read directly from the Nichols chart.

If the open-loop gain K is varied, the shape of the $G(j\omega)$ locus in the Nichols chart remains the same but is shifted up (for increasing K) or down (for decreasing K) along the vertical axis. Therefore the modified $G(j\omega)$ locus intersects the M and N loci differently, resulting in a different closed-loop frequency response curve.

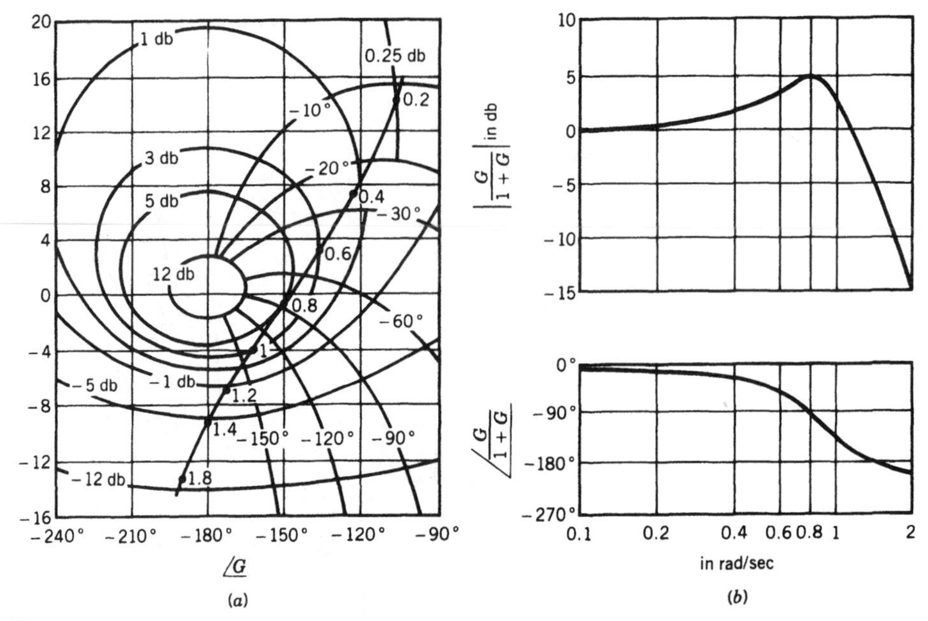

Figure 15 (*a*) Plot of $G(j\omega)$ superimposed on Nichols chart; (*b*) closed-loop frequency response curves.

4.2 Sensitivity Analysis Using the Nichols Chart[1]

Consider a unity feedback system with the transfer function

$$\frac{C(s)}{R(s)} = \frac{G(s)}{1 + G(s)} = G_{\text{cl}}(s)$$

The sensitivity of $G_{\text{cl}}(s)$ with respect to $G(s)$ is defined as

$$S_G^{G_{\text{cl}}}(s) = \frac{dG_{\text{cl}}(s)/G_{\text{cl}}(s)}{dG(s)/G(s)}$$

which yields

$$S_G^{G_{\text{cl}}}(s) = \frac{1}{1 + G(s)} \tag{6}$$

Clearly the sensitivity function is a function of the complex variable s.

To design a system with a prescribed sensitivity, the Nichols chart is quite convenient. Equation (6) is written as

$$S_G^{G_{\text{cl}}}(j\omega) = \frac{G^{-1}(j\omega)}{1 + G^{-1}(j\omega)}$$

which clearly indicates that the magnitude and phase of $S_G^{G_{\text{cl}}}(j\omega)$ can be obtained by plotting $G^{-1}(j\omega)$ on the Nichols chart and making use of the constant-M loci for a constant sensitivity function. Since the vertical coordinate of the Nichols chart is in decibels, the $G^{-1}(j\omega)$ curve on the Nichols chart can be easily obtained if $G(j\omega)$ is already available since

$$|G^{-1}(j\omega)|_{\text{dB}} = -|G(j\omega)|_{\text{dB}}$$

$$\underline{/G^{-1}(j\omega)} = -\underline{/G(j\omega)}$$

As an example consider the unity feedback system with the open-loop transfer function

$$G(s) = \frac{400{,}000K}{s(s + 49)(s + 991)}$$

the function $G^{-1}(j\omega)$ is plotted on the Nichols chart, as shown in Fig. 16, for $K = 2.94$. The intersections of the $G^{-1}(j\omega)$ curve with the M loci give the magnitude of $S_G^{G_{\text{cl}}}(j\omega)$ at the corresponding frequencies. Figure 16 indicates several interesting points with regard to the sensitivity function of the feedback system. The sensitivity function approaches 0 dB or unity as $\omega \rightarrow \infty$: $S_G^{G_{\text{cl}}} \rightarrow 0$ as $\omega \rightarrow 0$. A peak value of 1.1 dB is reached at $\omega = 25$ rad/s. This means that the closed-loop system is most sensitive to a change of $G(j\omega)$ at this frequency and more generally in this frequency range.

5 ROOT LOCUS

Poles and zero locations of a dynamic system characterize the system performance in a significant way. The root-locus method allows one to investigate the closed-loop pole patterns of a dynamic system with respect to a single parameter.

A typical closed-loop characteristic equation (CLCE) of a feedback system can be written as

$$1 + G(s)H(s) = 0 \tag{7}$$

where $G(s)H(s)$ is the open-loop transfer function.

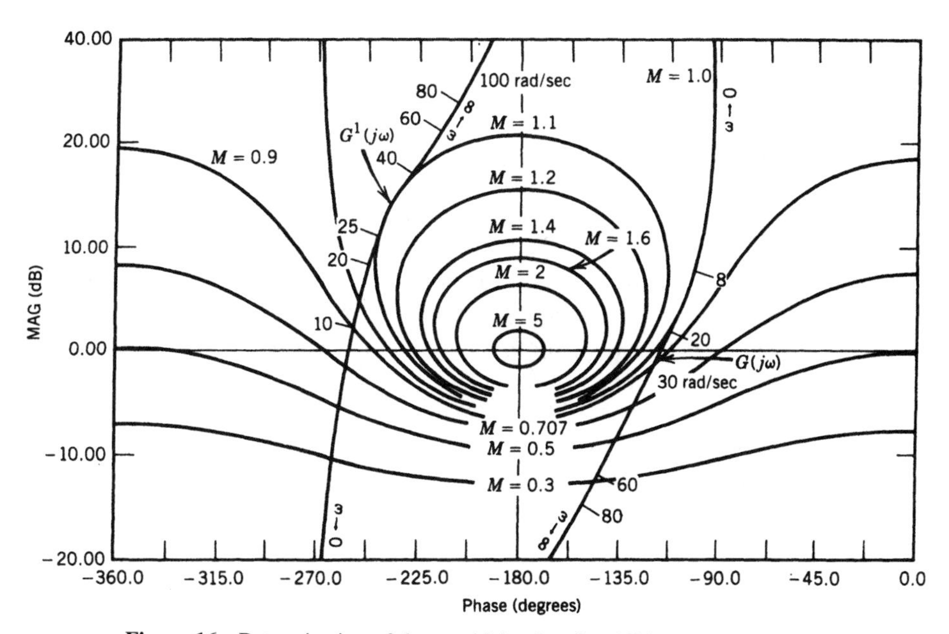

Figure 16 Determination of the sensitivity function S_G^M in the Nichols chart.

If $G(s)H(s)$ has a single parameter K as a variable, then by rewriting as

$$1 + KG(s)H(s) = 0 \tag{8}$$

a standard procedure for obtaining the closed-loop poles corresponding to any K is the Evans root-locus method.

5.1 Angle and Magnitude Conditions

The CLCE [Eq. (8)] can be written as

$$KG(s)H(s) = -1 = e^{j(2\pi l \pm \pi)} \tag{9}$$

Thus, any point s_0 satisfying the condition

$$\underline{/KG(s_0)H(s_0)} = (2l \pm 1)\pi \tag{10}$$

satisfies Eq. (8). If $K > 0$, then Eq. (10) reduces to

$$\underline{/G(s_0)H(s_0)} = (2l \pm 1)\pi \tag{11}$$

and is commonly called the angle condition. All points s_0 in the complex plane satisfying this angle condition satisfy the closed-loop characteristic equation and hence are said to lie on the root locus. If s_0 is a point on the root loci, then the corresponding value of K may be computed by noting that

$$|K||G(s_0)||H(s_0)| = 1 \tag{12}$$

which is called the amplitude condition.

By studying the angle condition in detail of the CLCE,

$$\underline{/1 + KG(s)H(s)} = 1 + K\frac{\prod_{i=1}^{m}(s + z_i)}{\prod_{j=1}^{n}(s + p_j)}$$

a set of rules can be developed for constructing the root locus easily. These rules are given next without proof.[2]

Rule 1. The system root loci have n branches originating at the n open-loop poles $-p_j$, $j = 1, \ldots, n$, with the value $K = 0$.

Rule 2. Out of the n branches m number of branches will terminate on m finite zeros $-z_i$, $i = 1, 2, \ldots, m$, of the open-loop transfer function at $K = \infty$.

Rule 3. The remaining $n - m$ branches will go to ∞ along asymptotes as $K \to \infty$. The asymptotes are straight lines meeting at a point on the real line called the hub with specific orientation as given in rule 4.

Rule 4. *a.* The asymptotes meet at the hub

$$\sigma = \frac{\sum_{j=1}^{n} \text{poles} - \sum_{i=1}^{m} \text{zeros}}{n - m}$$

$$= \frac{\sum_{j=1}^{n}(-p_j) - \sum_{i=1}^{m}(-z_i)}{n - m}$$

b. The $n - m$ asymptote angles are given by

$$\theta_N = \pm\frac{180°N}{n - m}$$

where N takes on values 1, 3, 5, 7, ... For each N, two angles are computed and the procedure is repeated until $n - m$ angles are obtained.

Rule 5. If to the right of a point on the real axis there lies an odd number of open-loop poles and zeros, then it is a point on the root loci.

Rule 6. If two open-loop poles or two open-loop zeros are connected, then there must be a break point between the two (Fig. 17).

If an open-loop pole $-p_l$ and an open-loop zero $-z_q$ are connected, in most cases it may be considered as a full branch of the root loci, that is, that the closed-loop pole corresponding to the open-loop pole $-p_l$ starts at $-p_l$ for $K = 0$ and reaches the closed-loop pole signified by the open-loop zero $-z_q$ as $K = \infty$.

Note: Exceptions to this rule exist. Some typical situations are depicted in Fig. 18*a*.

To determine the occurrence of such multiple break points, the next rule may be used.

Rule 7. The break points may be computed by determining points for which $dK/ds = 0$:

$$1 + KGH = 0$$

$$-K = \frac{1}{GH} = \frac{B(s)}{A(s)}$$

$$\frac{dK}{ds} = B(s)\frac{dA(s)}{ds} - A(s)\frac{dB(s)}{ds} = 0$$

Break points coupled with information from rule 6 make it rather easy to pin down the branches.

Rule 8. The points at which the branches cross the imaginary axis can be determined by letting $s = j\omega$ in the characteristic equation.

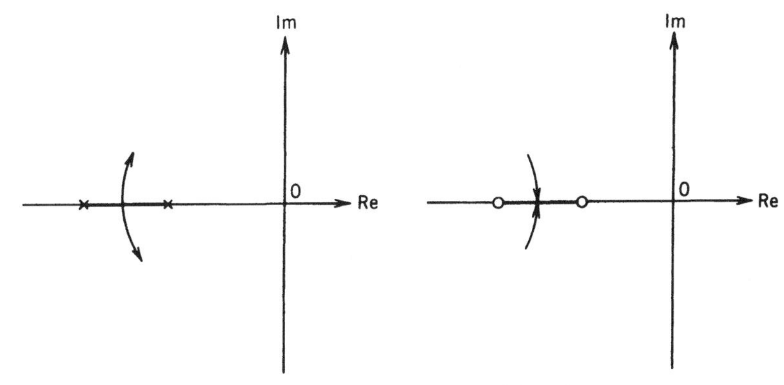

Figure 17 Breakaway and break-in points.

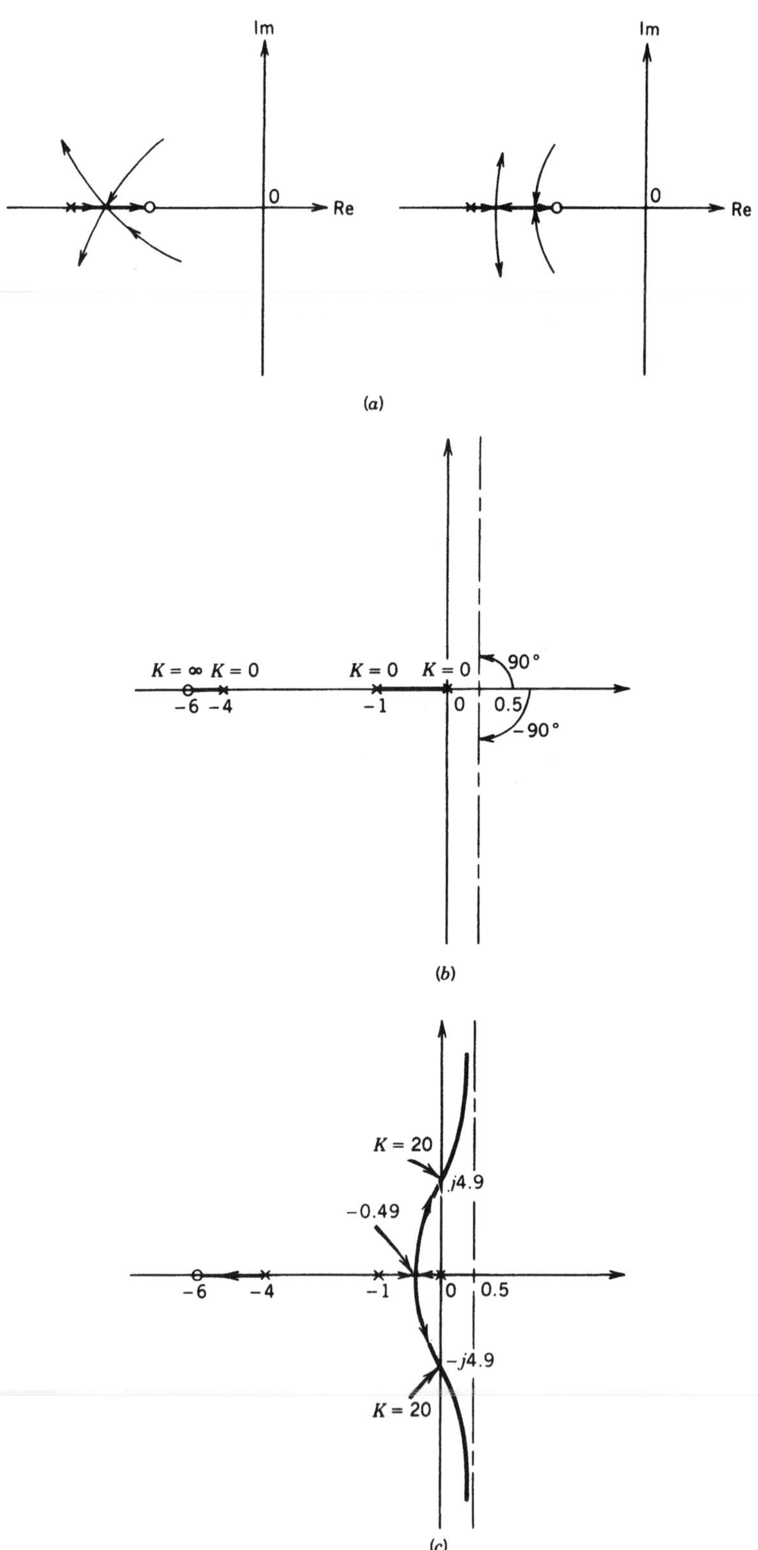

Figure 18 (*a*) Breakaway and break-in possibilities; (*b*) sketch of root loci of Example 3 resulting from rules R1–R4; (*c*) root loci for Example 3 where $G(s) = K(s + 6)/s(s + 1)(s + 4)$.

Rule 9. The angle condition is made use of to determine the angle by which a branch would depart from a pole or would arrive at a zero as $K \to \infty$.

A point s_0 is considered very near the pole (zero) and the angle $G(s_0)H(s_0)$ is computed. The fact that s_0 is very near the pole (zero) makes all but one angle fixed. Thus by employing the angle condition, the unknown angle of departure (arrival) can be computed.

An example is given next to illustrate the various rules for constructing a root locus.

Example 3 Consider CLCE

$$1 + KG(s)H(s) = 1 + \frac{K(s+6)}{s(s+1)(s+4)}$$

R1: $n = 3 \Rightarrow 3$ branches originating at $0, -1, -4$ at $K = 0$.

R2: $m = 1 \Rightarrow 1$ branch terminates at -6, at $K\ \infty$.

R3: $n - m = 2$ branches approach ∞ along asymptotes.

R4: Hub $\sigma = (0 - 1 - 4 - (-6))/2 = 0.5$.

$$\text{Asymptote angles:} \qquad \pm\frac{180^\circ N}{n-m} = \pm\frac{180^\circ N}{2}$$

$$\text{Set } N = 1 \Rightarrow \pm\ 90^\circ$$

R1–R4: Yield the sketch of Fig. 18*b*.

R5: Sections on the real line are 0 to -1 and -4 to -6.

R6: There must be a breakaway point between 0 and -1.

R7: Break points $dK/ds = 0$.

$$\Rightarrow (s+6)(3s^2 + 10s + 4) - (s^3 + 5s^2 + 4s) = 0$$

$$s^3 + 11.5s^2 + 30s + 12 = 0$$

$$(s + 0.49)(s + 7.89)(s + 3.12) = 0$$

Values for s of -7.89 and -3.12 are unacceptable from R5. Therefore the only breakaway point is at -0.49. A sketch of the root loci is given in Fig. 18*c*.

R8: Imaginary axis crossings:

$$\text{CLCE} = s^3 + 5s^2 + (4 + K)s + 6K = 0$$

Now lets $s = j\omega$:

$$(j\omega)^3 + 5(j\omega)^2 + (4 + K)j\omega + 6K = 0$$

$$(6k - 5\omega^2) + j\omega[(4 + K) - \omega^2] = 0$$

Therefore

$$6K = 5\omega^2$$

$$\omega[4 + K - \omega^2] = 0$$

yielding

$$K = 0 \qquad \omega = 0$$

and

$$K = 20 \qquad \omega = \pm 4.9$$

Example 4 Consider the unity negative-feedback system shown in Fig. 19*a*. Obtain the loci of the closed-loop poles as α is varied from 0 to ∞, that is, obtain the root loci for $0 \leq \alpha \leq \infty$.

Solution: The root loci are the points s satisfying the CLCE:

$$1 + \frac{750}{(s+5)(s+10)(s+\alpha)} = 0$$

To utilize the rules for constructing the root loci, the CLCE is rearranged in the form of Eq. (8).

By expanding and rearranging, we get

$$\text{CLCE} = s(s+5)(s+10) + \alpha(s+5)(s+10) + 750 = 0$$

or

$$1 + \frac{\alpha(s+5)(s+10)}{(s+15)(s^2+50)} = 0 \tag{13}$$

Equation (13) has three poles at -15, $+j\sqrt{50}$, $-j\sqrt{50}$ and two finite zeros at -5, -10.

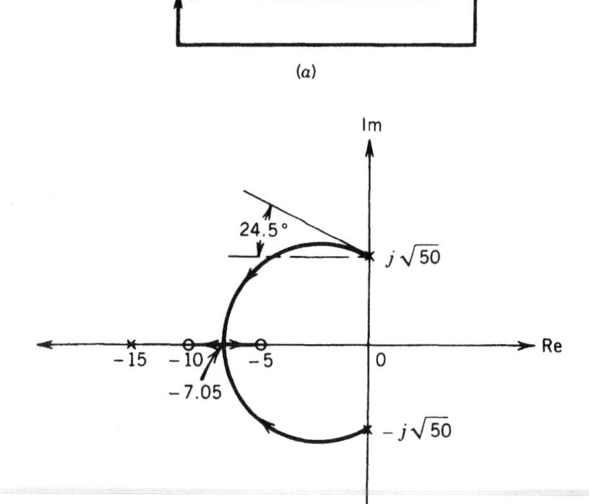

Figure 19 (*a*) Closed-loop system of Example 4; (*b*) root loci of system of Example 4 where $G(s) = \alpha(s+5)(s+10)/(s+15)(s+50)$.

R1: $n = 3$ implies there are three branches originating at -15, $+j\sqrt{50}$, $+j\sqrt{50}$ at $K = 0$.

R2: $m = 2$ implies two of the three branches terminate on -5, -10 at $K = \infty$.

R3: $n - m = 1$ implies that there is one asymptote.

R4: For a single asymptote a hub does not exist,

$$\text{Asymptote angle} = \pm\frac{180^\circ\ \text{N}}{n - m} = \pm\frac{180^\circ\ \text{N}}{1}$$

that is, $\theta_1 = 180^\circ$

R5: There is a section of the root loci between $-\infty$ and -15 and between -10 and -5.

R6: Since the two zeros -5 and -10 are connected, there must be a break-in point between -5 and -10. The section -15 to $-\infty$ forms a full branch.

R7: Break points: Since

$$G(s)H(s) = \frac{A(s)}{B(s)} = \frac{s^2 + 15s + 50}{s^3 + 15s^2 + 50s + 750}$$

$$\frac{d\alpha}{ds} = (s^2 + 15s + 50)(3s^2 + 30s + 50) - (s^3 + 15s^2 + 50s + 750)(2s + 15) = 0$$

that is,

$$s^4 + 30s^3 - 85s^2 - 8750 = 0 \tag{14}$$

Thus $s = -7.05$ is a break point.

Remark

To obtain the break points, the fourth-order polynomial in s given in (14) must be factored. However, knowing that there must be a break point in the range -5 and -10 (rule 6), it is quite easy to find the breakaway point. If all roots of (14) are found anyway, then only those points yielding $\alpha \leq 0$ are admissible as break points.

For this example R1–R6 give all the essential information to sketch the root loci of Fig. 19*b*. If the angles of departure are needed, we can employ rule 9.

Consider the point s_0 closer to $+j\sqrt{50}$ and write down the angle condition:

$$\angle(s_0 + 5) + \angle(s_0 + 10) - \angle(s_0 + 15) - \angle(s_0 + j\sqrt{50}) - \angle(s_0 - j\sqrt{50}) = \pi$$

Now let $s_0 \to j\sqrt{50}$ to yield

$$\angle(5 + j\sqrt{50}) + \angle(10 + j\sqrt{50}) - \angle(15 + j\sqrt{50} - j2\sqrt{50}) - \theta = \pi$$

$$\tan^{-1}\frac{\sqrt{50}}{5} + \tan^{-1}\frac{\sqrt{50}}{10} - \tan^{-1}\frac{\sqrt{50}}{15} - \frac{\pi}{2} - \theta = \pi$$

$$\theta = -3.57\ \text{rad} = -204.5^\circ$$

Some typical root-loci plots are shown in Table 1.

5.2 Time-Domain Design Using the Root Locus

Time-domain performance specifications can often be related in an approximate sense to closed-loop pole locations. If suitable pole locations for a certain time-domain performance can be effectively identified, then the root loci can be used to locate the closed-loop poles

Table 1 Typical Root-Loci Plots

No.	$G_0(s)$	Root Loci	No.	$G_0(s)$	Root Loci
1	$\frac{1}{s - p_1}$	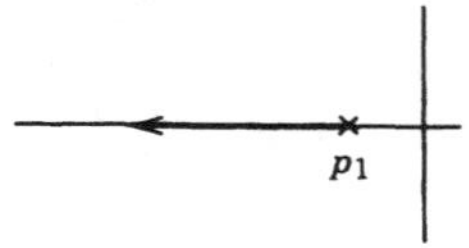 p_1	6	Same as 5 $p_2 < z_1 < p_1$	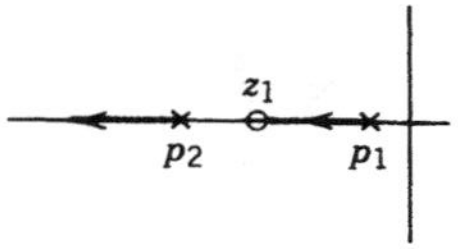 z_1 p_2 p_1
2	$\frac{s - z_1}{s - p_1}$	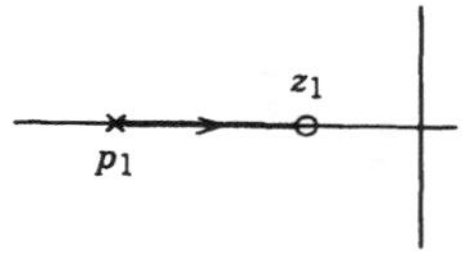 z_1 p_1	7	Same as 5 $p_1, p_2 < z_1$	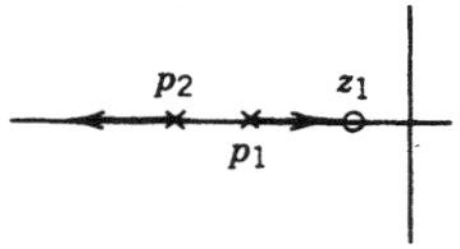 p_2 z_1 p_1
3	$\frac{1}{(s - p_1)(s - p_2)}$	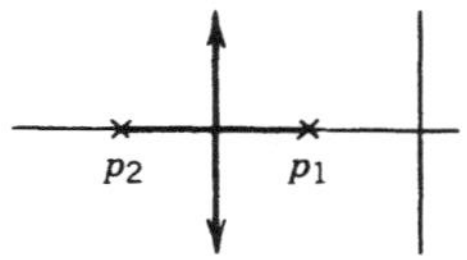 p_2 p_1	8	Same as 5 p_1, p_2 complex	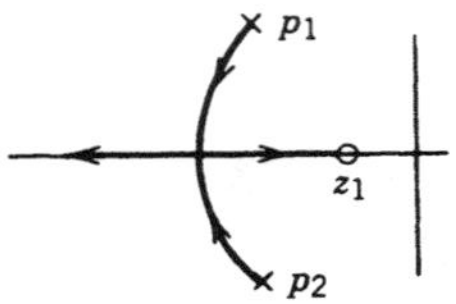 p_1 z_1 p_2
4	$\frac{1}{(s - p_1)(s - p_2)}$ p_1, p_2 complex	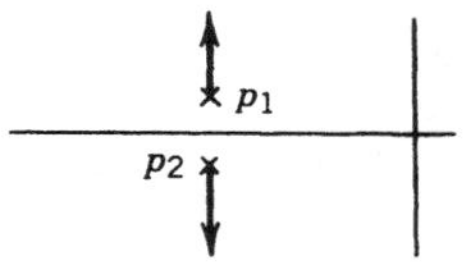 p_1 p_2	9	$\frac{1}{(s - p_1)(s - p_2)(s - p_3)}$	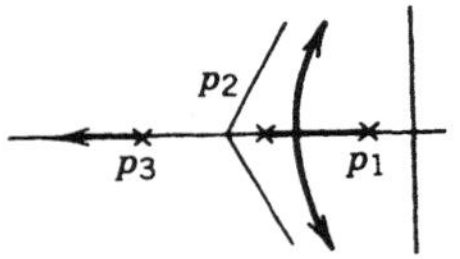 p_2 p_3 p_1
5	$\frac{(s - z_1)}{(s - p_1)(s - p_2)}$ $z_1 < p_1, p_2$	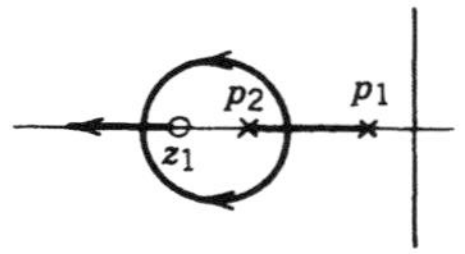 p_2 p_1 z_1	10	$\frac{1}{s(s - p_2)(s - p_3)}$	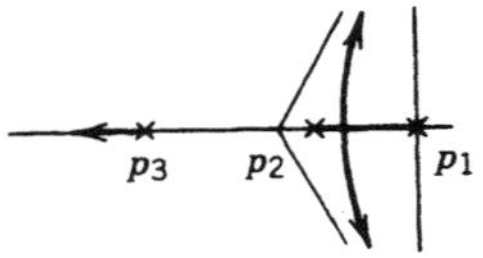 p_3 p_2 p_1

11	$\dfrac{1}{(s-p)^3)}$	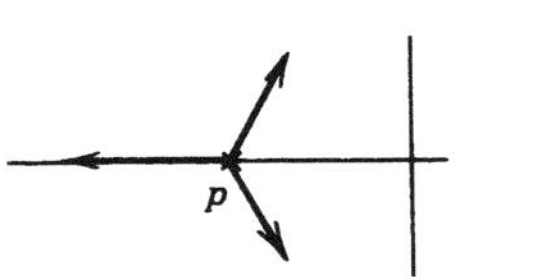
12	$\dfrac{1}{(s-p_1)(s-p_2)(s-p_3)}$ p_2, p_3 complex	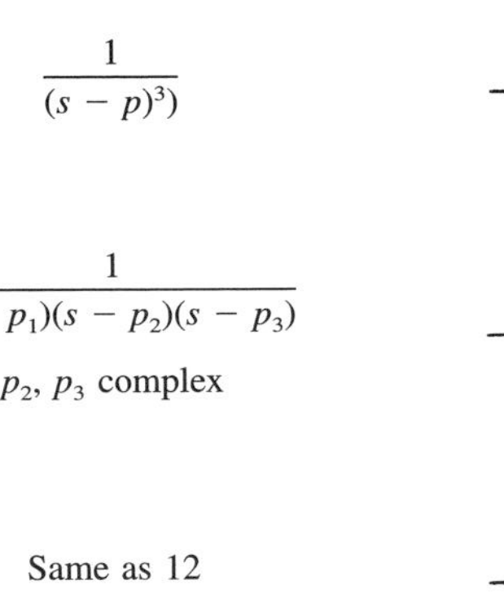 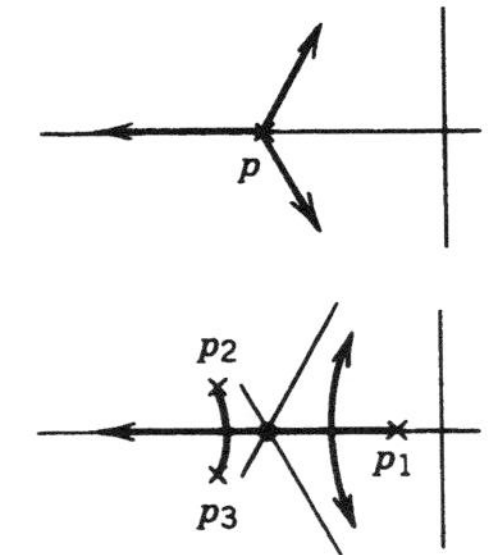
13	Same as 12	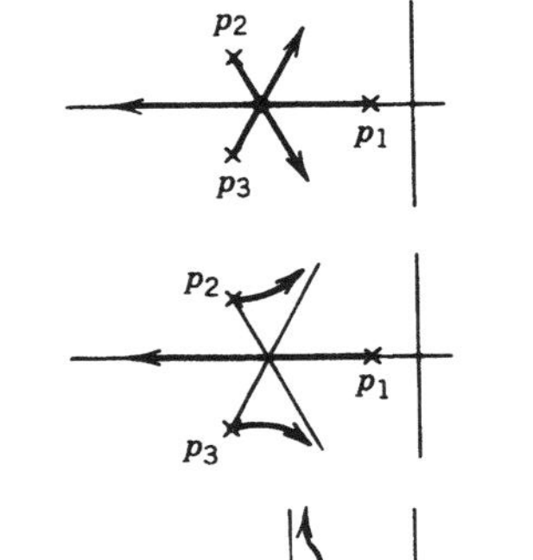
14	Same as 12	
15	$\dfrac{(s-z_1)}{(s-p_1)(s-p_2)(s-p_3)}$	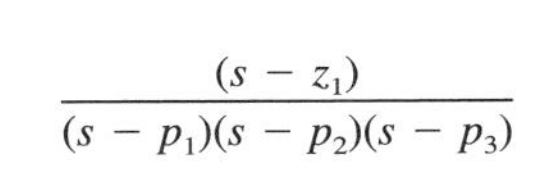
16	Same as 15	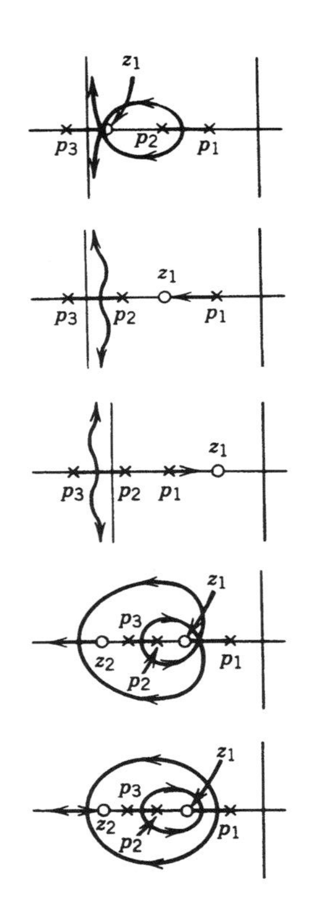
17	Same as 15	
18	Sams as 15	
19	$\dfrac{(s-z_1)(s-z_2)}{(s-p_1)(s-p_2)(s-p_3)}$	
20	Same as 19	

Table 1 (*Continued*)

No.	$G_0(s)$	Root Loci	No.	$G_0(s)$	Root Loci
21	$\frac{(s - z_1)(s - z_2)}{(s - p)^3}$ $z_2 < z_1 < p$	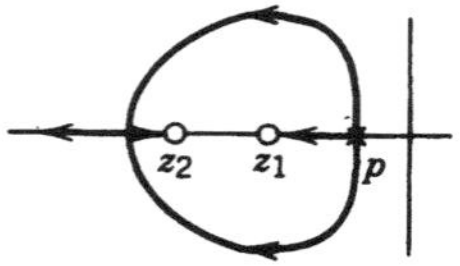 z_2 z_1 p	25	$\frac{1}{(s - p_1)(s - p_2)(s - p_3)(s - p_4)}$ All poles real	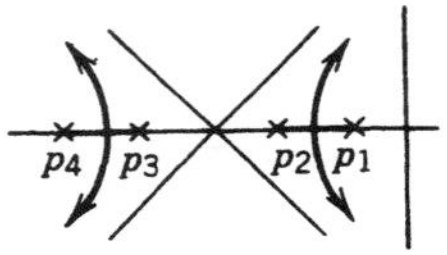 p_4 p_3 p_2 p_1
22	$\frac{(s - z_1)(s - z_2)}{s(s - p_2)(s - p_3)}$ $z_2 < p_3 < z_1 < p_2 < 0$	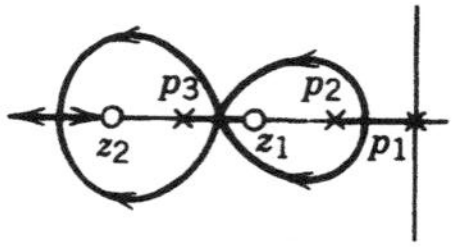 p_3 p_2 z_2 z_1 p_1	26	Same as 25 Two poles real, two poles complex	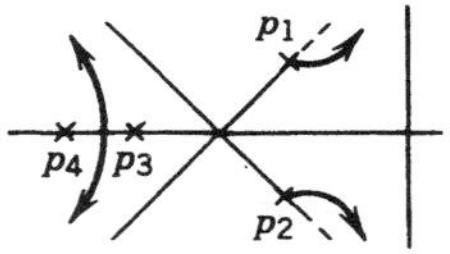 p_1 p_4 p_3 p_2
23	Same as 22	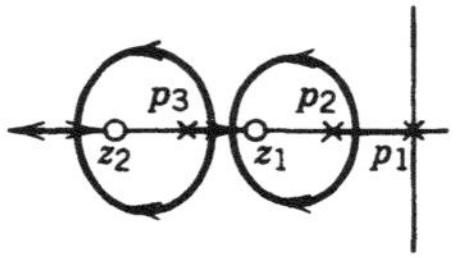 p_3 p_2 z_2 z_1 p_1	27	Same as 26	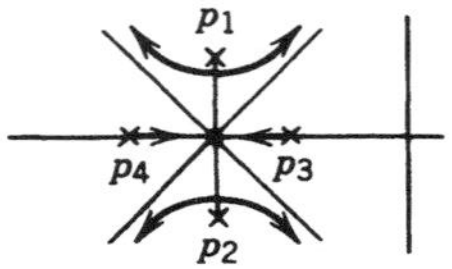 p_1 p_4 p_3 p_2
24	Same as 22	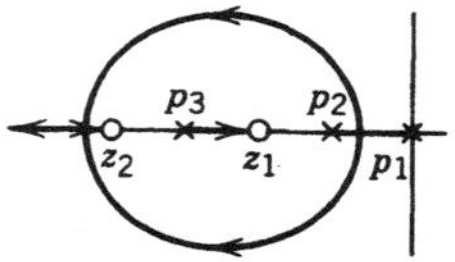 p_3 p_2 z_2 z_1 p_1	28	Same as 25 All poles complex	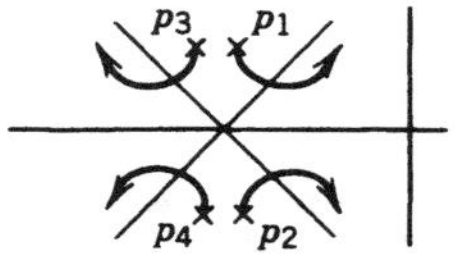 p_3 p_1 p_4 p_2

at those locations by appropriate compensation. Compensation can be provided by introducing additional dynamics into the feedback system in the form of increased poles and zeros [proportional-integral-derivative (PID) control, lead, lag, lead–lag, etc.]. We shall now consider some examples to illustrate this time-domain design philosophy.

Example 5 Obtain the root loci for a system with an open-loop transfer function

$$G(s) = \frac{K}{s(s+2)}$$

(a) Indicate the location of closed-loop poles when $K = 4$ and determine the damping ratio ζ and the natural frequency ω_n corresponding to $K = 4$.

(b) It is now required to double the natural frequency while keeping the same damping ratio. Design a compensator for satisfying the new design specifications.

Solution: (a) Suppose the closed-loop system is as shown in Fig. 20*a*. Then its root loci are as shown in Fig. 20*b*. To find the poles at $K = 4$, solve the CLCE

$$1 + \frac{4}{s(s+2)} = 0 \qquad \text{or} \qquad s^2 + 2s + 4 = 0$$

The roots are $s_{1,2} = -1 \pm j\sqrt{3}$.

$$\zeta = \cos\beta = 0.5$$

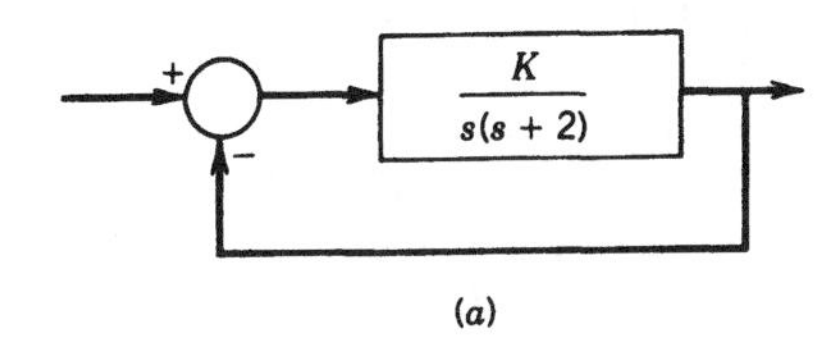

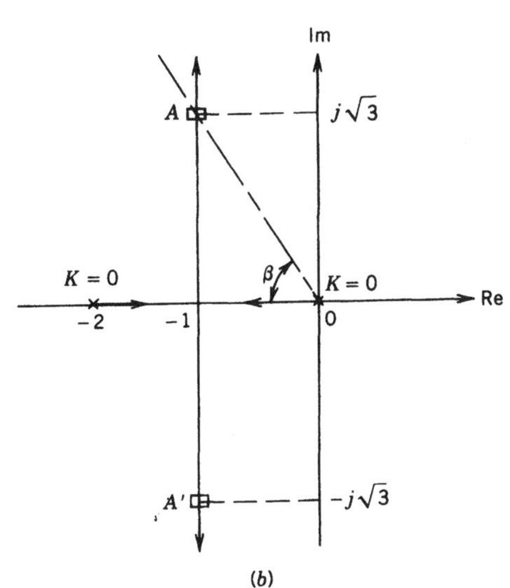

Figure 20 (*a*) Time-domain design example; (*b*) sketch of root loci.

(b) Since the natural frequency is to be doubled keeping the same damping ratio, we need to move the closed-loop poles at A and A' in Fig. 21 so that $\overline{OB} = 2\overline{OA}$.

To satisfy the design specifications, the modified root loci must be made to pass through B and B'. To reshape the original root loci, additional poles and zeros are required. We give below a simple way to appropriately modify the root loci. Conceptually, the modification takes place as shown in Fig. 22.

If $G_c(s)$ is chosen to cancel the pole at -2 by selecting

$$G_c(s) = \frac{s+2}{s+p}$$

then we only need to find p so that the modified root locus passes through B. This is quite easy to do by noting that the pole–zero cancellation at -2 leaves us with a second-order system with the two open-loop poles at 0 and $-p$.

By selecting $p = +4$, it is easy to verify that the modified root loci are as shown in Fig. 22. So the compensator

$$G_c(s) = \frac{s+2}{s+4}$$

will work.

Remark

Pole–zero cancellation as was done here must be avoided if it lies in the right-half plane. Since any real system model has parameter uncertainty, exact cancellation is almost impossible to achieve. When this is the case, such an attempted cancellation will leave an uncompensated unstable mode in the closed-loop system. Even in the case of a stable approximate cancellation the dynamics can change. To see this, consider in Example 5 the pole at -2 to be uncertain (say $-2 \pm \epsilon$, $\epsilon > 0$) and that a zero is exactly located at -2. Let us consider the two cases with the pole at $-2 + \epsilon$ and $-2 - \epsilon$ (Fig. 23).

We note that the modified root loci do not pass through B, B' when $\epsilon \neq 0$, implying that the time-domain performance will be affected.

Example 6 Consider the system shown in Fig. 24*a* where $K \geq 0$ and α and β are unknown constants. To identify K, α, and β, the following information about the system is provided:

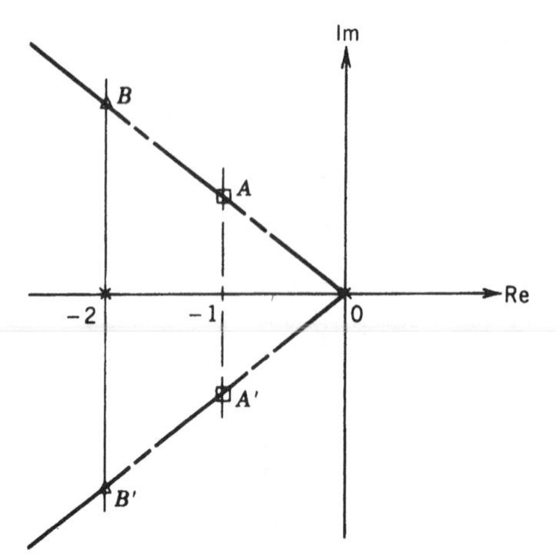

Figure 21 Desired pole locations.

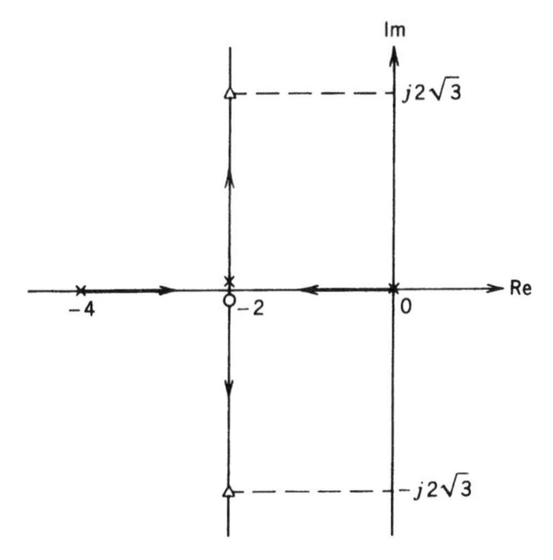

Figure 22 Root loci with pole–zero cancellation.

1. All the poles of the open-loop transfer function are in the closed left-half s-plane.
2. When the closed-loop system is excited by the input $r(t) = tu_s(t)$, the trace of Fig. 24b is obtained ($\infty > e_1 > 0$).
3. When the gain K is doubled, the impulse response of Fig. 24c is observed.

Determine K, α, and β.

Solution: Since the closed-loop system has a finite steady-state error $e_1 < \infty$ for a ramp input, the system should be type I. Thus we require either α or β to be zero. Let $\alpha = 0$. So it only remains to determine β and K.

Now the root loci for the system can be sketched in the following manner.

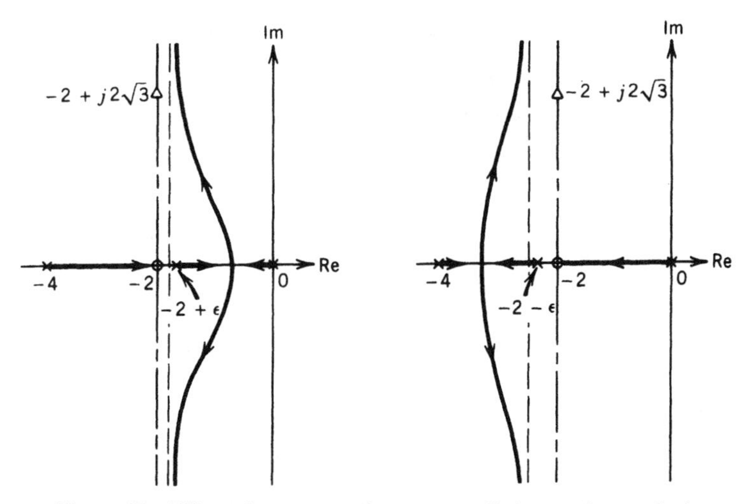

Figure 23 Effect of nonexact pole–zero cancellation on the root loci.

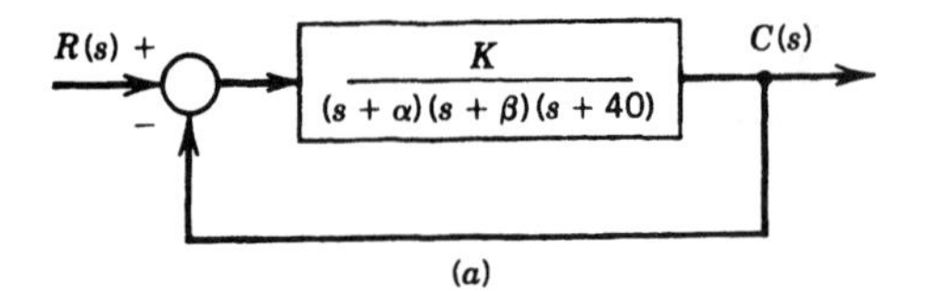

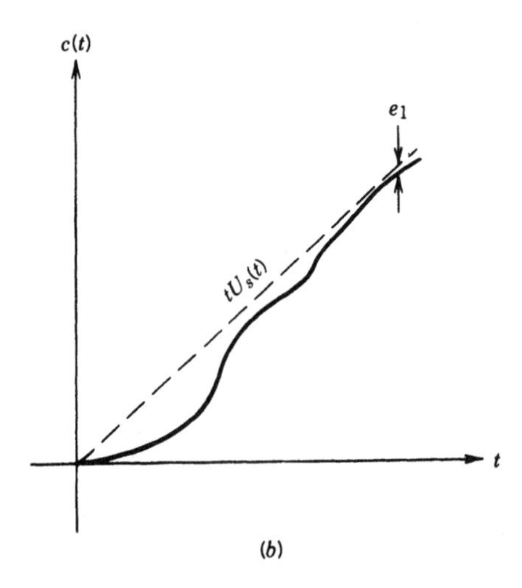

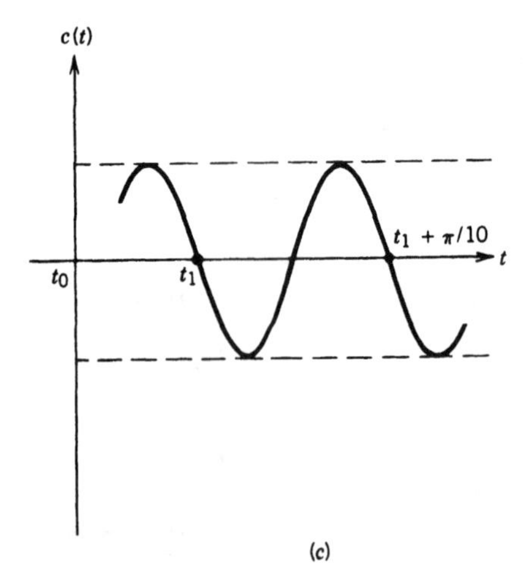

Figure 24 (*a*) System of Example 5; (*b*) response due to a ramp; (*c*) response due to an impulse.

From the root locus it is clear that at a certain gain value the system goes unstable. From Fig. 24*c* we know that when the gain is doubled, the system has two closed-loop eigenvalues at A and A'. From the impulse response trace the frequency of oscillation is

$$\omega = \frac{2\pi}{\pi/10}$$

or

$$\omega = 20 \text{ rad/s}$$

Thus we know that when the gain is doubled, there are two closed-loop poles at $\pm j20$. The corresponding CLCE therefore is

$$P(s) = (s + a)(s + j20)(s - j20) = 0$$

or

$$s^3 + as^2 + 400s + 400a = 0$$

We also know that

$$P(s) = 1 + \frac{2K}{s(s + \beta)(s + 40)} = 0$$

or

$$s^3 + (40 + \beta)s^2 + 40\beta s + 2K = 0$$

By matching coefficients

$$40 + \beta = \alpha$$
$$40\beta = 400$$
$$2K = 400a$$

Therefore $\beta = 10$, $a = 50$, and $K = 10{,}000$. Hence the open-loop transfer function is

$$G(s) = \frac{10{,}000}{s(s + 10)(s + 40)}$$

5.3 Time-Domain Response versus *s*-Domain Pole Locations

Given a transfer function $G(s)$ the pole locations can be found. These pole locations essentially describe the type of time response to be expected. The basic response can be effectively characterized by the impulse response $g(t)$ given by

$$g(t) = \mathcal{L}^{-1}[G(s)]$$

If

$$G(s) = \frac{K\prod_{i=1}^{m}(s + z_i)}{\prod_{j=1}^{n}(s + p_j)} \qquad n \geq m$$

then

$$g(t) = \sum_{j=1}^{n} a_j e^{-p_j t}$$

We note that any real pole contributes an exponential behavior into the time response and a complex-conjugate pair contributes an exponential oscillation. A pure imaginary pair of poles leads to a sustained oscillation. Various components to be expected are shown in Fig. 25.

The role of zeros of the transfer function is to affect the relative weights a_j in the impulse response. For example, if a pole and a zero are close together, the net contribution to the overall response from such a pair will be negligible. If they cancel each other (say $-p_k$ by $-z_j$), then the coefficient associated with the term $e^{-p_k t}$ is zero. This idea can often be used to reduce the order of a dynamic system, that is, remove all pole–zero pairs close to one another. However, care should be exercised not to remove right-half-plane poles and zeros. (See Example 5 of Section 5.2.)

To note the effect of zero locations on the time response consider a second-order oscillatory system with a single zero, that is, consider the transfer function written in the normalized form

$$G(s) = \frac{(s/\alpha\zeta\omega_n) + 1}{(s/\omega_n)^2 + 2\zeta s/\omega_n + 1} = \left(\frac{s}{\alpha\zeta\omega_n} + 1\right) G_0(s)$$

The zero is located at $s = -\alpha\zeta\omega_n$, so if α is large, the zero is far removed from the poles and will have little effect on the response of $G_0(s)$. If $\alpha = 1$, the zero is at the value of the real part of the poles and could be expected to have a substantial influence on the response of $G_0(s)$. The step response curves for $\zeta = 0.5$ and for several values of α are plotted in Fig. 26. We see that the major effect of the zeros is to increase the overshoot M_p with very little influence on the settling time. A plot of M_p versus α is given in Fig. 27. If α is negative, then the zero is in the right-half s-plane. In this case an undershooting phenomenon as shown in Fig. 28 occurs.

In addition, it is useful to know the effect of an extra pole on the standard second-order response $G_0(s)$. In this case consider the transfer function

$$G(s) = \frac{1}{(s/\alpha\zeta\omega_n + 1)} G_0(s)$$

Plots of the step response for this case are shown in Fig. 29 for $\zeta = 0.5$ and for several values of α. In this case the major effect is to increase the rise time, shown in Fig. 30. For a detailed discussion of the effect of a zero and a pole location on a standard second-order response the reader may refer to Ref. 3.

6 POLE LOCATIONS IN THE z-DOMAIN

For discrete-time systems the input–output relation is given by the pulse transfer function. A typical pulse transfer function $G(z)$ is of the form

$$G(z) = \frac{K\prod_{i=1}^{m}(z - \beta_i)}{\prod_{j=1}^{n}(z - p_j)} \qquad n \geq m$$

The poles of the pulse transfer function are p_j, $j = 1, \ldots, n$. As in the case of continuous time, the pole locations determine the stability properties of the system represented by its pulse transfer function. In the z-domain poles have to lie inside the unit circle $|z| = 1$ for

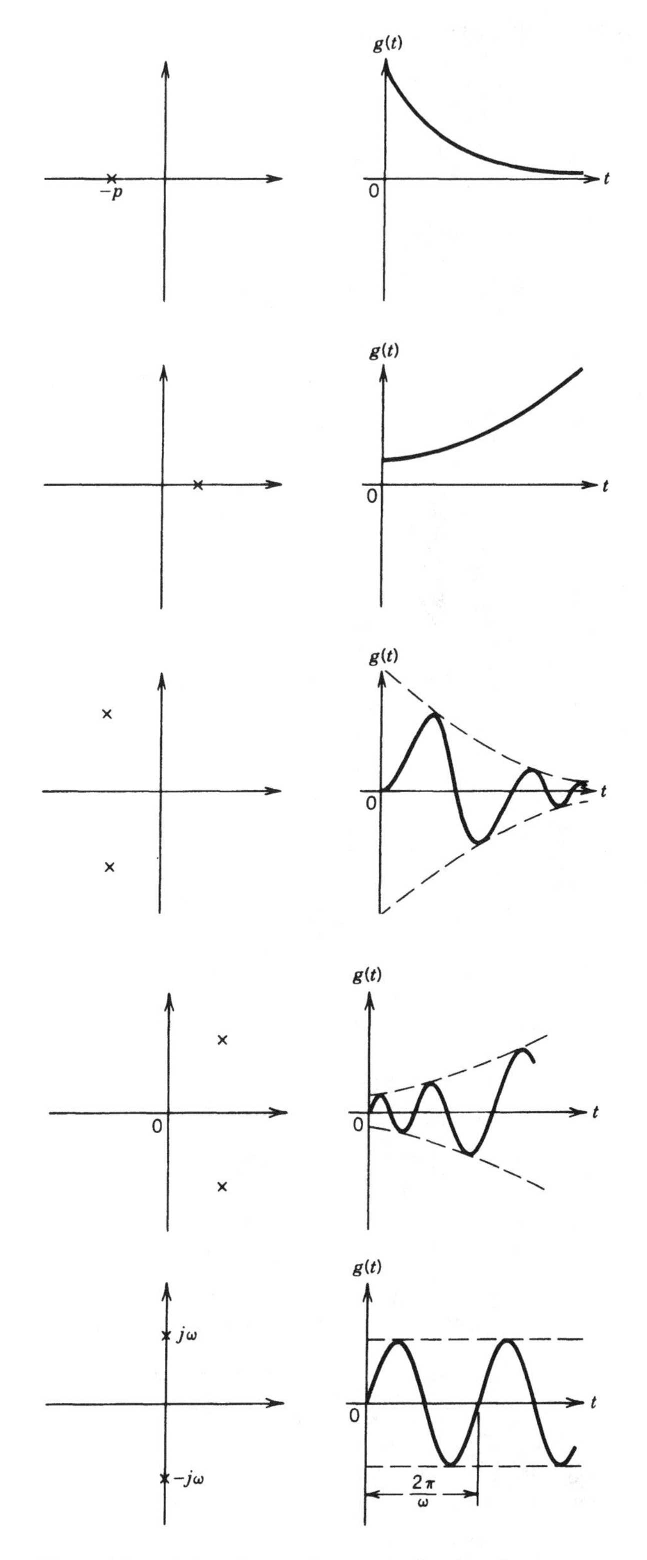

Figure 25 Pole locations and corresponding impulse responses.

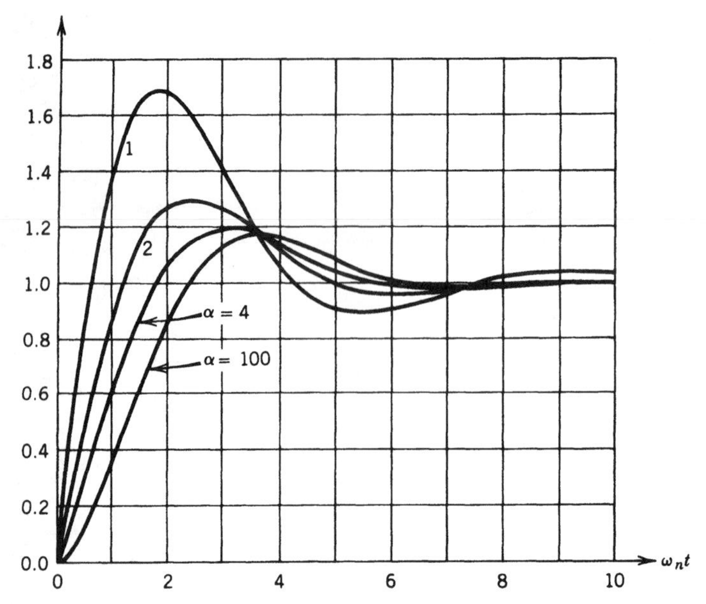

Figure 26 Plots of the step response of a second-order system with an extra zero ($\zeta = 0.5$).

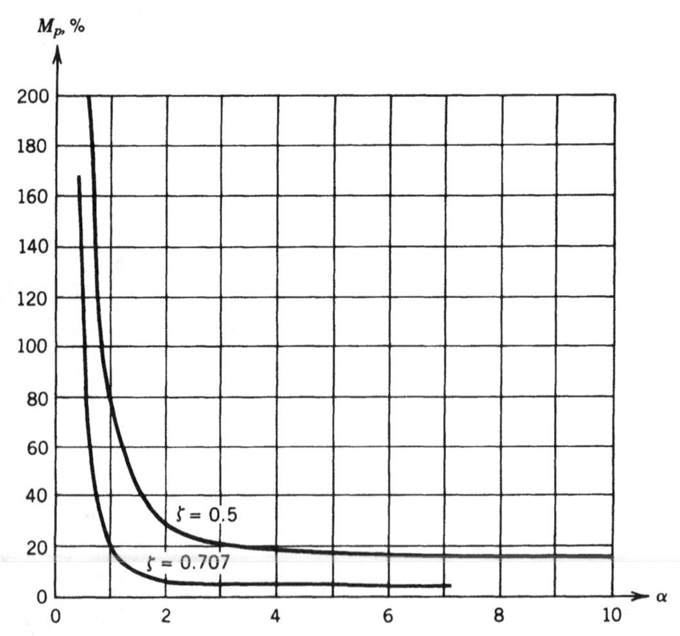

Figure 27 Plot of overshoot M_p as a function of normalized zero location α. At $\alpha = 1$, the real part of the zero equals the real part of the pole.

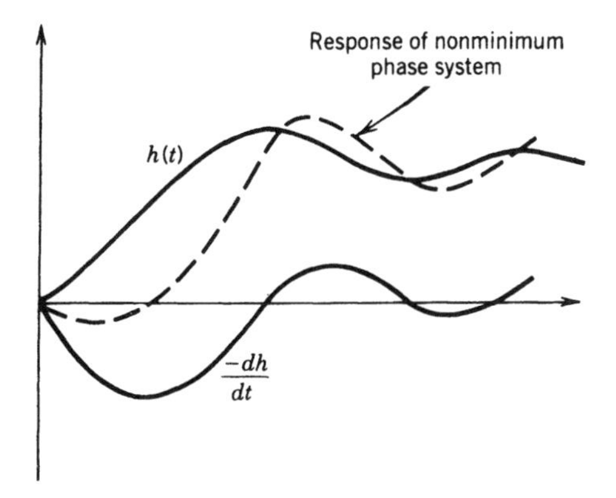

Figure 28 Plot of the response of a second-order system with a right-half-plane zero: a non-minimum-phase system.

asymptotic stability. Thus the open left-half s-plane is equivalent to the interior of the unit circle in the z-domain. The exterior of the unit circle (i.e., $|z| > 1$) represents the unstable region in the z-domain.

6.1 Stability Analysis of Closed-Loop Systems in the z-Domain

Consider a unity negative-feedback system with the closed-loop pulse transfer function

$$\frac{C(z)}{R(z)} = \frac{G(z)}{1 + G(z)} \tag{15}$$

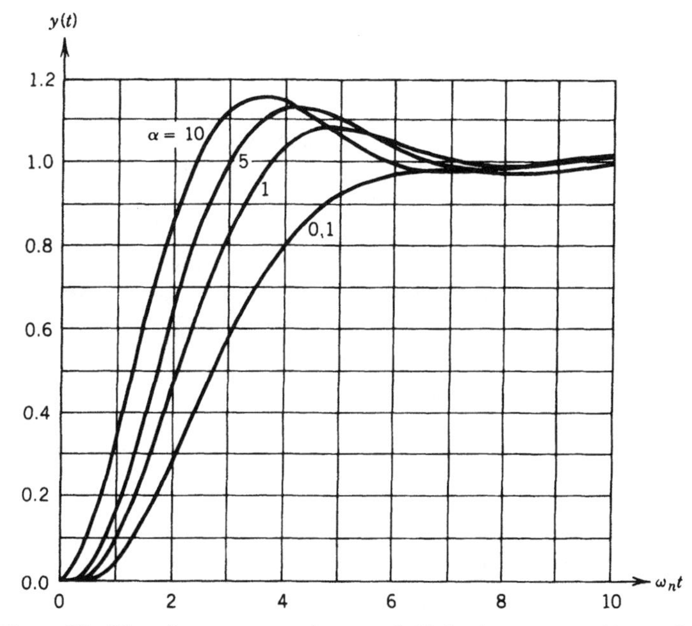

Figure 29 Plot of step response for several third-order systems with $\zeta = 0.5$.

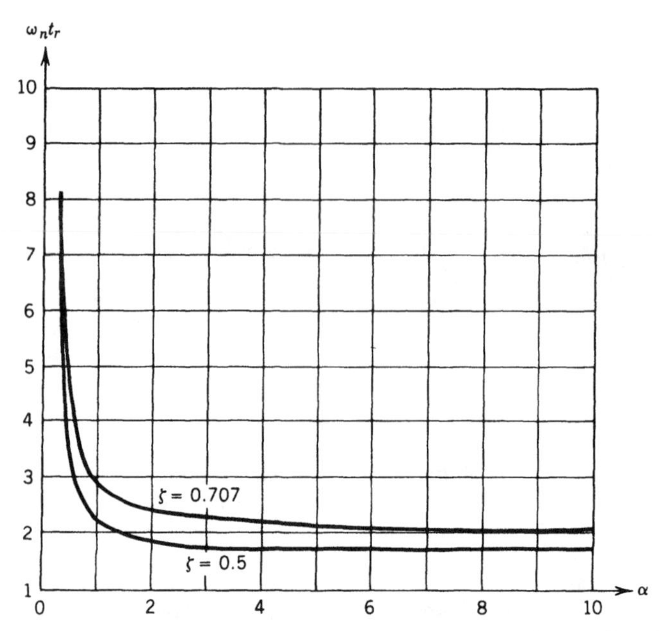

Figure 30 Plot of normalized rise time for several locations of an additional pole.

The stability of the system defined by Eq. (15), as well as of other types of discrete-time control systems, may be determined from the locations of the closed-loop poles in the z-plane or the roots of the closed-loop characteristic equation

$$P(z) = 1 + G(z) = 0$$

as follows:

1. For the system to be stable, the closed-loop poles or the roots of the characteristic equation must lie within the unit circle in the z-domain. Any closed-loop pole outside the unit circle makes the system unstable.
2. If a simple pole lies at $z = 1$ or $z = -1$, then the system becomes marginally stable. Also, the system becomes marginally stable if a single pair of complex-conjugate poles lie on the unit circle in the z-domain. Any multiple closed-loop pole on the unit circle makes the system unstable.
3. Closed-loop zeros do not affect the absolute stability and therefore may be located anywhere in the z-plane. Thus, a linear time-invariant single-input–single-output discrete-time closed-loop system becomes unstable if any closed-loop poles lies outside the unit circle or any multiple closed-loop pole lies on the unit circle in the z-domain.

6.2 Performance Related to Proximity of Closed-Loop Poles to the Unit Circle

In the continuous-time case or in the s-domain the transient performance of a system can be characterized by the s-plane pole locations. Recall that the overshoot is related to the damping ratio ζ.

Damping Ratio ζ

In the s-plane a constant damping ratio may be represented by a radial line from the origin. A constant damping ratio locus (for $0 \le \zeta \le 1$) in the z-plane is a logarithmic spiral. Figure 31 shows constant ζ loci in both the s-plane and the z-plane.

If all the poles in the s-plane are specified as having a damping ratio not less than a specified value ζ_1, then the poles must lie to the left of the constant-damping-ratio line in the s-plane (shaded region). In the z-plane, the poles must lie in the region bounded by logarithmic spirals corresponding to $\zeta = \zeta_1$ (shaded region).

Damped Natural Frequency ω_d

The rise time or the speed of response depends on the damped natural frequency ω_d and the damping ratio ζ of the dominant complex-conjugate closed-loop poles. In the s-plane the constant ω_d loci are horizontal lines, while in the z-plane they are radial lines emanating from the origin.

Settling Time t_s

The settling time is determined by the value of attenuation σ of the dominant closed-loop poles $-\sigma \pm j\omega_d$. If the settling time is specified, it is possible to draw a line $\alpha = -\sigma_1$, in the s-plane corresponding to a given settling time. The region to the left of the line $\sigma = -\sigma_1$ in the s-plane corresponds to the interior of a circle with radius $e^{-\sigma_1 T}$ in the z-plane, as shown in Fig. 32.

Remark

To transform s-plane pole locations to the z-domain, the transformation $z = e^{sT}$, where T is the sampling time, is employed.

6.3 Root Locus in the z-Domain

The root locus method for continuous-time systems can be extended to discrete-time systems without modifications, except that the stability boundary is changed from the $j\omega$ axis in the s-plane to the unit circle in the z-plane. The reason for being able to extend the root-locus method is that the characteristic equation for the discrete-time system is of the same form as that for the root loci in the s-plane. For the discrete-time case the CLCE is

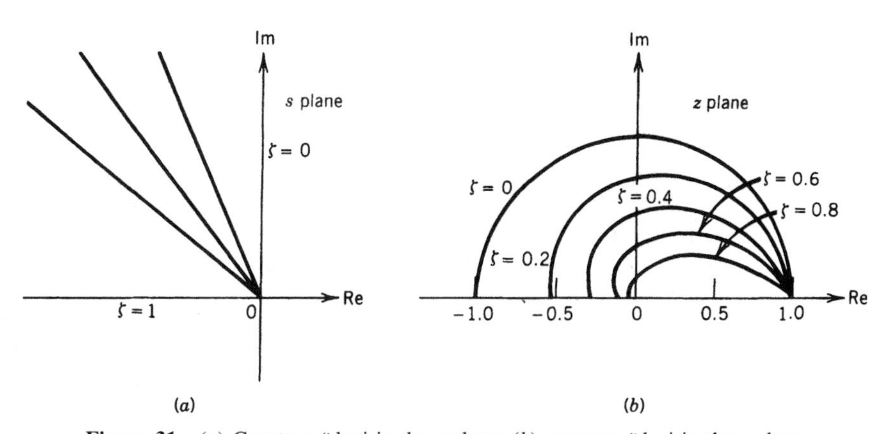

Figure 31 (*a*) Constant ζ loci in the s-plane; (*b*) constant ζ loci in the z-plane.

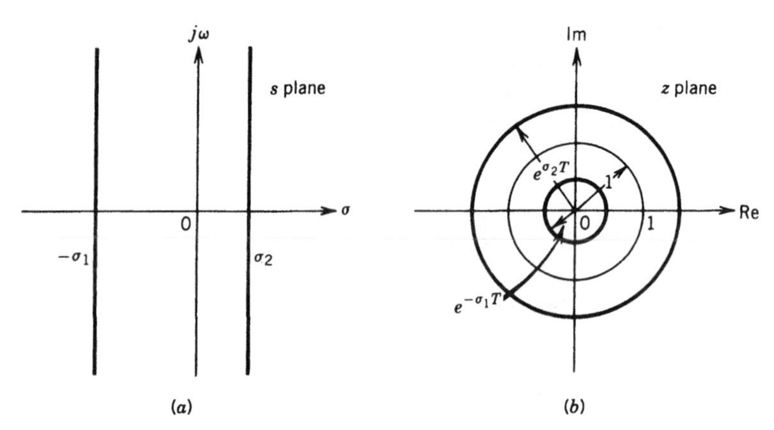

Figure 32 (*a*) Constant attenuation lines in the *s*-plane; (*b*) the corresponding loci in the *z*-plane.

$$1 + G(z) = 0$$

Exactly the same rules as used for the continuous-time case in the *s*-plane can be used for the discrete case too. (See Section 5 for root-loci construction in the *s*-plane.)

Example 7 Consider the closed-loop characteristic equation

$$1 + K\frac{(z + 1)(z - 0.5)}{(z - 1)(z - 0.9)(z + 0.6)} = 0$$

R1: The system has $n = 3$ poles indicating three branches. They start at 1, 0.9, and -0.6 with $K = 0$.

R2: There are two finite zeros ($m = 2$). Hence two of the branches terminate on the zeros -0.9 and 0.5 at $K = \infty$.

R3: One branch ($n - m = 1$) will go to ∞ along an asymptote.

R4: Sections of the root loci on the real line are between 0.9 and 1.0, -1.0 and $-\infty$, and -0.6 and 0.5; with this information the sketch shown in Fig. 33 can be easily obtained. If additional features are needed, the rest of the root-loci construction rules can be applied without change.

7 CONTROLLER DESIGN

In Sections 1–6 some useful tools for designing single-input–single-output systems were given. Once the open-loop system is described, either by its set of poles and zeros or by its frequency response, the root-locus method or frequency response method will indicate whether the feedback system can be given an acceptable transient response by adjustment of the loop gain. The steady-state accuracy can then be determined.

Often good transient performance and good steady-state performance cannot both be achieved simply by adjusting a single parameter such as a loop gain. When this is the case, it is necessary to modify the system dynamics. Either the dynamic properties of some components in the loop need to be altered or additional components need to be inserted into the loop. The process of modifying the system dynamics so as to allow the performance specifications to be met by subsequent loop gain adjustment is known as compensation.

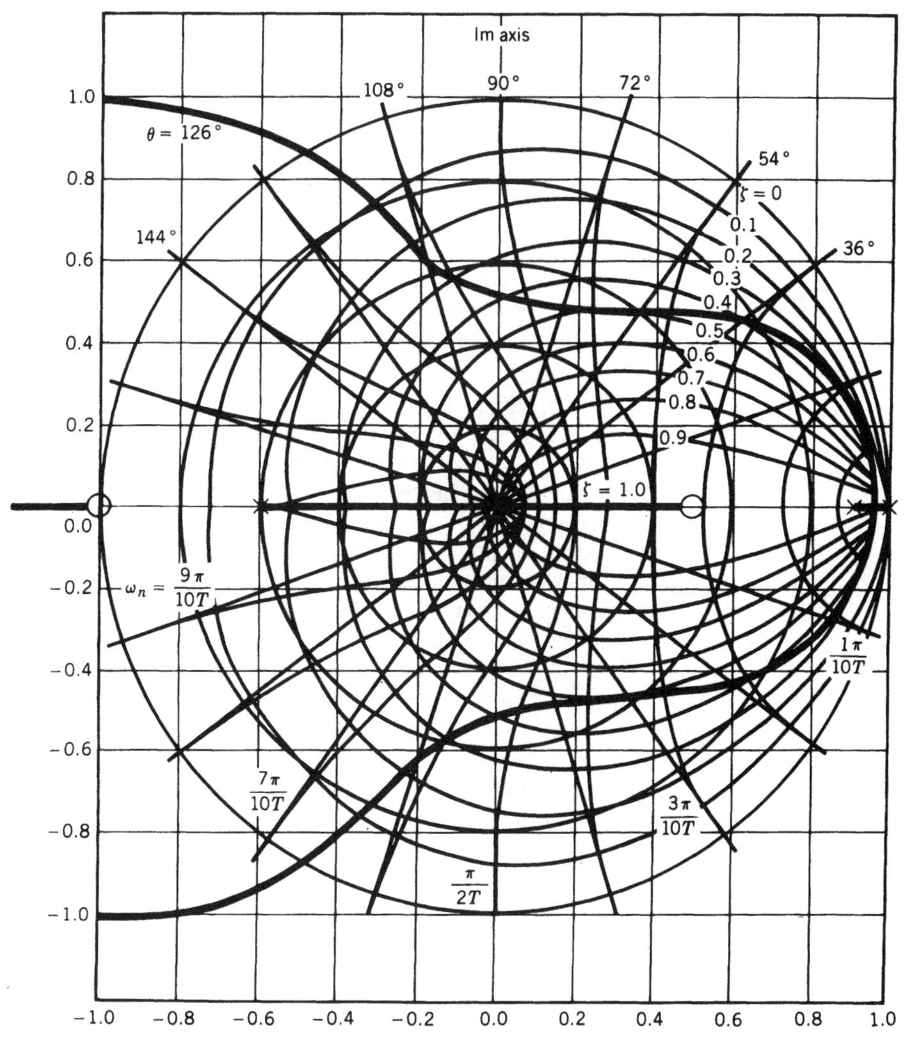

Figure 33 Root loci for Example 7.

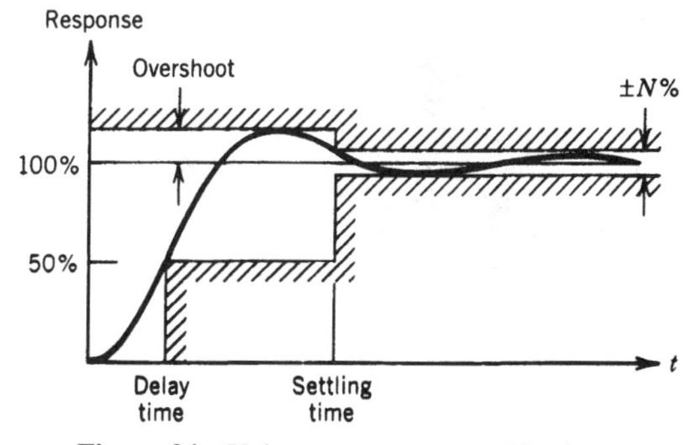

Figure 34 Unit step response specifications.

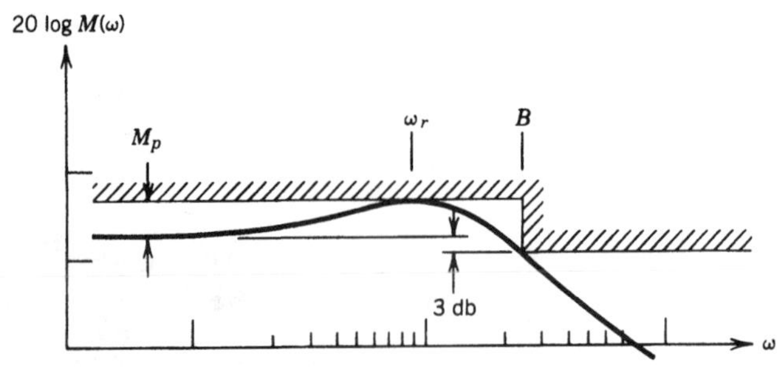

Figure 35 Closed-loop frequency response specifications.

Transient specifications are typically based on a step input response. By specifying the rise time, overshoot, and settling time, the response is confined to within the shaded region of Fig. 34. It is then assumed that a system whose step response satisfies these constraints will have an acceptable transient response to any kind of input.

In the frequency domain the bandwidth and the resonant peak of the closed-loop frequency response are measures roughly corresponding to rise time and overshoot, respectively. Specification of these parameters constrains the magnitude of the closed-loop frequency response to the region shown in Fig. 35. An alternative way of constraining the transient response by frequency-domain criteria is to stipulate the smallest acceptable gain and phase margins.

Often used compensators are the so-called three-term controllers (PID) and lag and lead compensators. These controller or compensator designs are discussed later. They are often done on a trial-and-error basis and can be designed in the s-domain or the z-domain depending on the type of application. Continuous-time or s-domain compensators can often be converted to equivalent z-domain compensators by techniques such as pole–zero maps, hold equivalence, and Butterworth pole configurations.[4]

REFERENCES

1. B. C. Kuo, *Automatic Control Systems,* Prentice-Hall, Englewood Cliffs, NJ, 1982.
2. K. Ogata, *Modern Control Engineering,* Prentice-Hall, Englewood Cliffs, NJ, 1970.
3. G. F. Franklin, J. D. Powell, and A. Emami-Naeini, *Feedback Control of Dynamic Systems,* Addison-Wesley, Reading, MA, 1986.
4. G. F. Franklin and J. D. Powell, *Digital Control of Dynamic Systems,* Addison-Wesley, Reading, MA, 1980.

BIBLIOGRAPHY

Bode, H. W., *Network Analysis and Feedback Amplifier Design,* Van Nostrand, New York, 1945.

Chestnut, H., and R. W. Mayer, *Servomechanisms and Regulating Systems Design,* 2nd ed., Vol. 1, Wiley, New York, 1959.

D'Azzo, J. J., and C. H. Houpis, *Linear Control System Analysis and Design,* McGraw-Hill, New York, 1988.

Distefano, J. J., III, A. R. Stubberud, and I. J. Williams, *Feedback and Control Systems* (Schaum's Outline Series), Schaum Publishing, New York, 1967.

Doebelin, E. O., *Control System Principles and Design,* Wiley, New York, 1985.

Dorf, R. C., *Modern Control Systems,* Addison-Wesley, Reading, MA, 1986.

Dransfield, P., *Engineering Systems and Automatic Control,* Prentice-Hall, Englewood Cliffs, NJ, 1968.

Elgerd, O. I., *Control Systems Theory,* McGraw-Hill, New York, 1967.

Eshbach, O. W., and M. Souders (eds.), *Handbook of Engineering Fundamentals,* 3rd ed., Wiley, New York, 1975.

Evans, W. R., *Control-System Dynamics,* McGraw-Hill, New York, 1954.

Eveleigh, V. W., *Introduction to Control Systems Design,* McGraw-Hill, New York, 1972.

Graham, D., and R. C. Lathrop, "The Synthesis of Optimum Response: Criteria and Standard Forms," *AIEE Transactions,* **72** (Pt. II), 273–288, (1953).

Horowitz, I. M., *Synthesis of Feedback Systems,* Academic, New York, 1963.

Houpis, C. H., and G. B. Lamont, *Digital Control Systems: Theory, Hardware, Software,* McGraw-Hill, New York, 1985.

Korn, G. A., and J. V. Wait, *Digital Continuous System Simulation,* Prentice-Hall, Englewood Cliffs, NJ, 1978.

Kuo, B. C., *Digital Control Systems,* Holt, Rinehart and Winston, New York, 1980.

Melsa, J. L., and D. G. Schultz, *Linear Control Systems,* McGraw-Hill, New York, 1969.

Nyquist, H., "Regeneration Theory," *Bell System Technical Journal,* **II**, 126–147, (1932).

Palm, N. J., III, *Modeling, Analysis and Control of Dynamic Systems,* Wiley, New York, 1983.

Phillips, C. L., and H. T. Nagle, Jr., *Digital Control System Analysis and Design,* Prentice-Hall, Englewood Cliffs, NJ, 1984.

Ragazzini, J. R., and G. F. Franklin, *Sampled Data Control Systems,* McGraw-Hill, New York, 1958.

Raven, F. H., *Automatic Control Engineering,* 4th ed., McGraw-Hill, New York, 1987.

Rosenberg, R. C., and D. C. Karnopp, *Introduction to Physical System Dynamics,* McGraw-Hill, New York, 1983.

Shinners, S. M., *Modern Control Systems Theory and Application,* Addison-Wesley, Reading, MA, 1972.

Takahashi, Y., M. J. Rabins, and D. M. Auslander, *Control and Dynamic Systems,* Addison-Wesley, Reading, MA, 1972.

Truxal, J. G., *Automatic Feedback Control Synthesis,* McGraw-Hill, New York, 1955.

Vidyasagar, M., *Nonlinear Systems Analysis,* Prentice Hall, Englewood Cliffs, NJ, 1978.

CHAPTER 14

SERVOACTUATORS FOR CLOSED-LOOP CONTROL

Karl N. Reid
College of Engineering, Architecture and Technology
Oklahoma State University
Stillwater, Oklahoma

Syed Hamid
Halliburton Services
Duncan, Oklahoma

Reprinted from *Instrumentation and Control,* Wiley, New York, 1990 by permission of the publisher.

1 INTRODUCTION

1.1 Definitions

A servoactuator is an open-loop system that controls the linear or rotary motion of a load in response to an input command (Fig. 1*a*). Feedback may be used with a servoactuator to produce a closed-loop system referred to as a servosystem (Fig. 1*b*). Servoactuators are normally "rate-type" systems, in that an input command results in an output velocity for steady-state operation. Position feedback must be used with the rate-type system to produce a servosystem for position control. If high-accuracy velocity control is required, velocity feedback may be used with the servoactuator. Or, if high-accuracy force (or torque) control is required, force (or torque) feedback may be used.

The term "servomotor" designates the various types of higher level energy converters such as electrical and hydraulic motors. The servomotor provides the muscle function of the servoactuator. The "modulator" provides a conversion of the low-power input command (for the servoactuator) or the error signal (for the servosystem) to a high-power output that operates the servomotor. The "transducer" provides the feedback in the case of the servosystem. The input to the servoactuator or servosystem can be electronic, mechanical, hydraulic, or pneumatic. And depending on the energy conversion medium, servoactuators can be of the electromechanical, electrohydraulic, electropneumatic, or hydromechanical types.

1.2 Applications

Early development of servoactuators and servosystems was predominantly in electropneumatics (in the process control industry).[1] With the advent of microprocessors and the development of high-coercive-strength magnetic materials (such as samarium cobalt and neodium). electromechanical servosystems find the largest applications in modem industry.[2] Table 1 describes the servocomponents (modulator, servomotor, and transducer) for the various implementations. Applications range from fairly simple open-loop systems such as the hydraulic

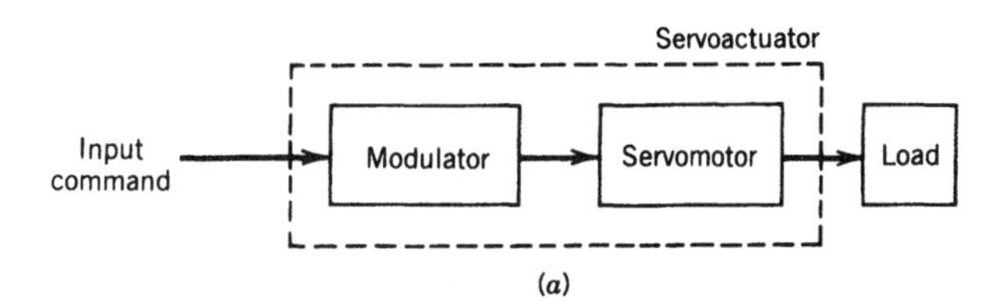

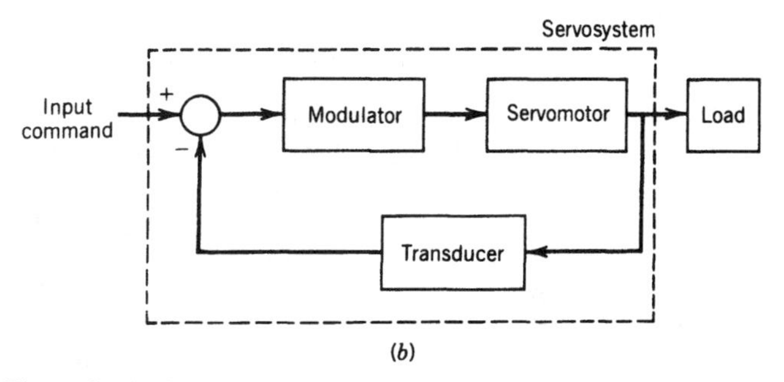

Figure 1 (*a*) Servoactuator (open loop); (*b*) servosystem (closed loop).

Table 1 Servosystem Components

System Type	Modulator	Servomotor	Transducer
Electromechanical	Amplifier	Servomotor ac, dc Linear/Rotary	Position, velocity, torque
	Driver	Brushless servomotor	
	Translator/driver	Stepper motor	
Electrohydraulic	Servovalve	Hydraulic motor cylinder	Position, velocity force, pressure torque
Electropneumatic	Servovalve	Airmotor, cylinder	Position, velocity, force, pressure torque,

controls on a backhoe to complex feedback systems in robotics and aerospace vehicles. Figure 2 shows a typical linear output servosystem designed for use in a wide variety of motion control applications. A threephase brushless motor is modulated by a pulse-width-modulated controller (not shown in Fig. 2; see Ref. 3). A ballscrew is used to convert rotary motion to linear motion. Feedback is provided by a tachometer.

1.3 Mathematical Models

Mathematical models of the various components of a servoactuator are needed for component selection to meet a given set of performance specifications. These specifications may consist of moving a given load through a given displacement or velocity profile in a specified time or, equivalently, following displacement or velocity commands generated by other subsystems or by an operator. The mathematical model of a component describes the steady-state and/or dynamic performance characteristics of that component. Mathematical models are presented in this chapter for the components typically used in high-performance servoactuators and servosystems. Examples are presented to illustrate the use of mathematical models in the prediction of steady-state and dynamic performance of servoactuators and servosystems.

2 ELECTRICAL SERVOMOTORS

Electrical servomotors may be classified by the following characteristics:

1. Type of power [direct current (dc) or alternating current (ac)]
2. Type of motion executed (continuous or discrete, rotary or translatory)
3. Type of commutation (mechanical or electronic)
4. Method of magnetic field generation (permanent magnet or electromagnetic)

Accordingly there are dc and ac servomotors of both the permanent-magnet and field-wound types. Stepper motors belong to the discrete motion type. The rather uncommon linear motor executes translatory motion. Brushless dc motors are of the electronic commutation type. For the sake of simplicity, electrical servomotors are broadly classified here into four categories: dc and ac servomotors, stepper motors, and linear servomotors.

Electrical servomotors offer several advantages over their hydraulic and pneumatic counterparts. These advantages include (a) compactness (facilitated by availability of high-coercive-strength magnetic materials such as samarium–cobalt or neodium), (b) low cost, (c)

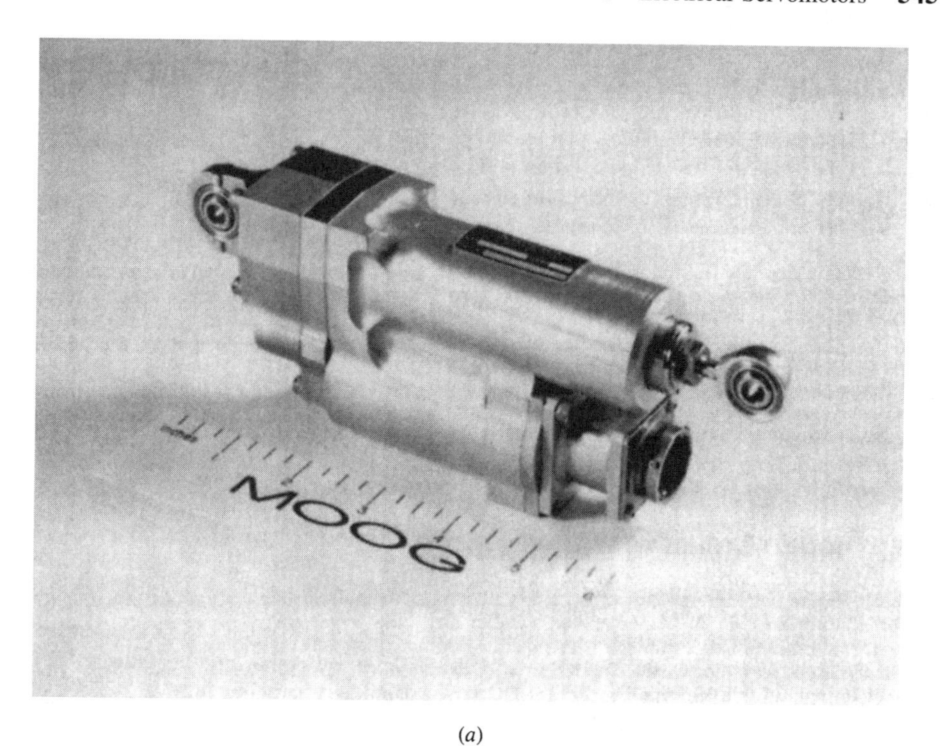

(a)

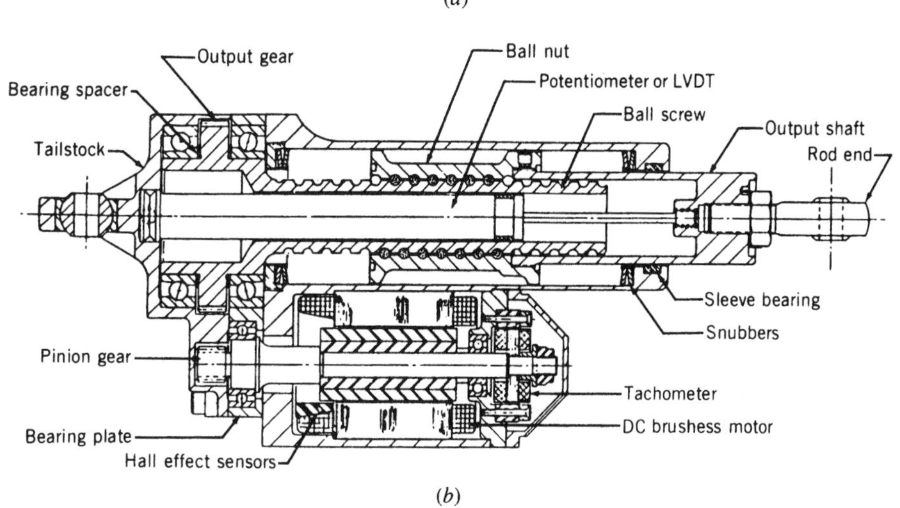

(b)

Figure 2 (a) Electromechanical servosystem; (b) cross section of an electromechanical servosystem. (Courtesy of Moog, Inc., East Aurora, NY.)

high reliability, (d) cleanliness, (e) ease of control function implementation, (f) portability due to operation at low dc voltage levels, and (g) large bandwidth due to high torque/inertia ratios.

3 DIRECT-CURRENT SERVOMOTORS

Direct-current servomotors offer certain advantages over ac servomotors. These advantages are higher reliability, smaller size, and lower cost. Use of epoxy resins and improved brush designs combined with superior magnetic materials contribute to these advantages. Direct-current servomotors are compatible with thyristors (silicon controlled rectifiers, SCR) and transistor amplifiers, which facilitates control implementation. Typical dc servomotors range in power from fractional horsepower to several thousand horsepower. Conventional brushed dc motors theoretically can be used as servomotors. However, in lower horsepower levels (10 hp or less) they are not preferred.

3.1 Brushed dc Servomotors

In the dc servomotor, the interaction of two magnetic fields (either one or both generated electrically) results in mechanical motion of an armature. A typical permanent-magnet dc motor is illustrated in Fig. 3. The permanent magnet is sometimes replaced by a field winding to generate the magnetic field. The field winding may be connected in three different ways to the armature winding: series, shunt, or compound. Table 2 summarizes the basic features of the various configurations along with the resultant performance characteristics. Table 3 shows typical upper limits of dc servomotor performance.[4]

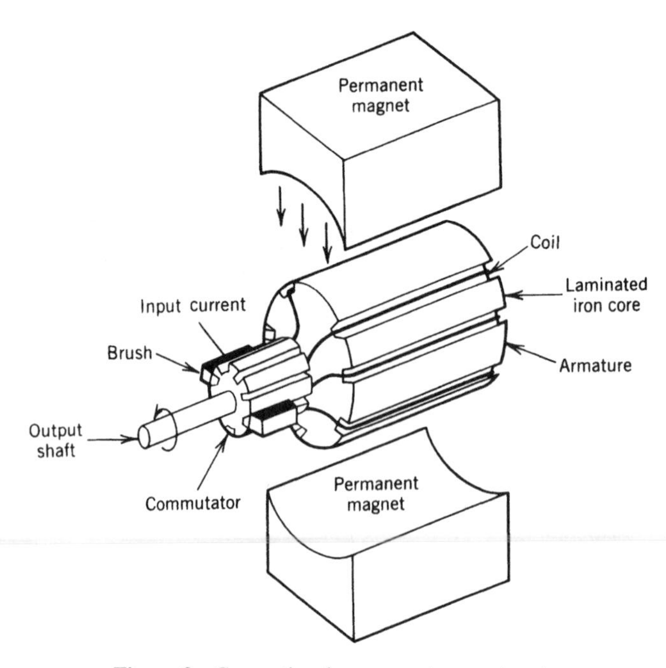

Figure 3 Conventional permanent-magnet motor.

Table 2 dc Servomotor Classification

Motor Type	Configuration	Typical Steady-State Characteristics	Salient Features
Permanent magnet	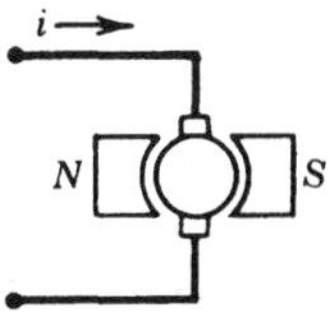	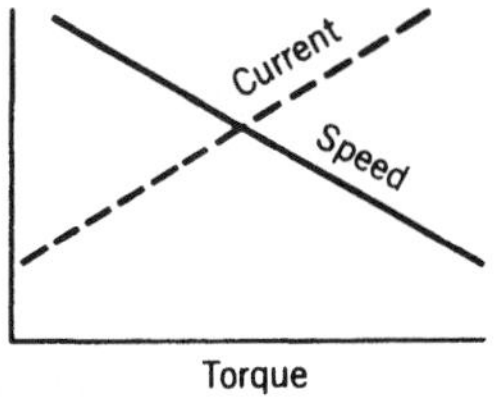	No power required for field generation Runs Cooler Torque-speed characteristics is linear Compactness
Straight series	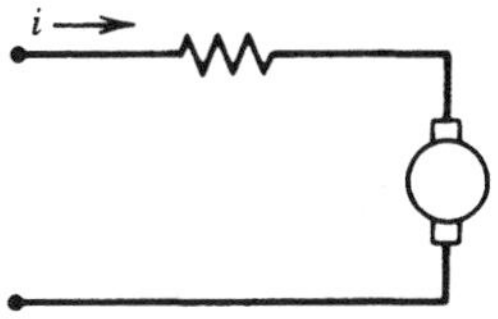	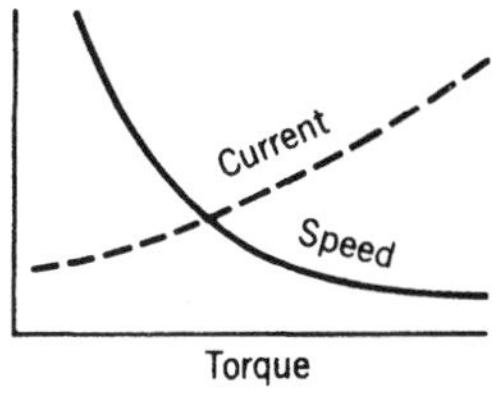	Large starting torque
Split series	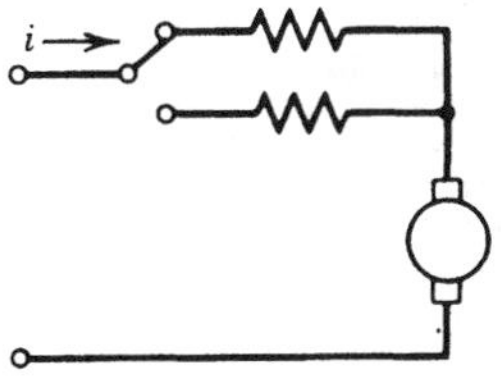	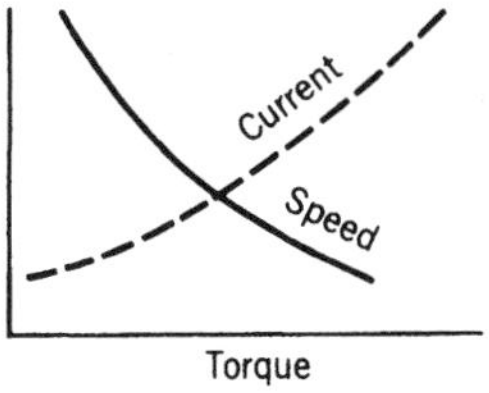	Allows quick reversing
Shunt	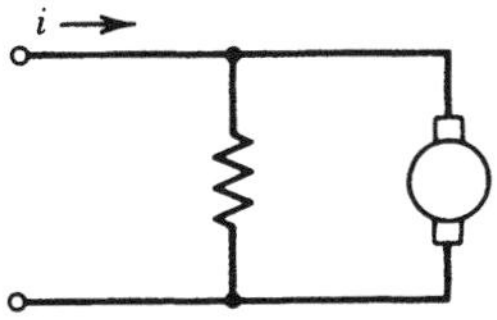	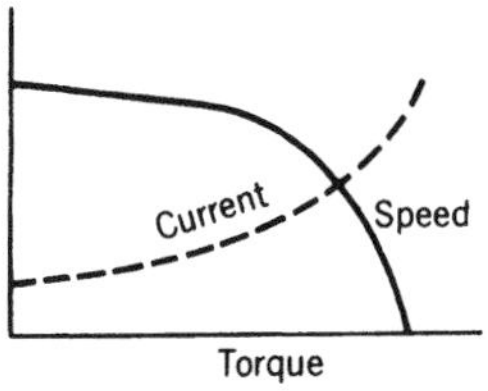	Low starting torque Finite speed at zero torque
Compound	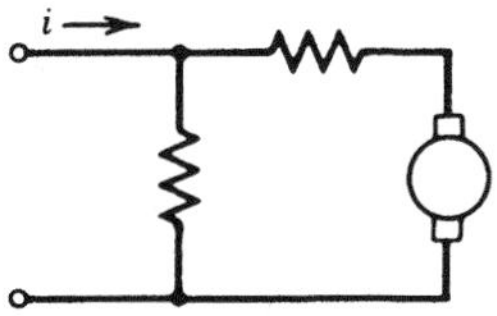	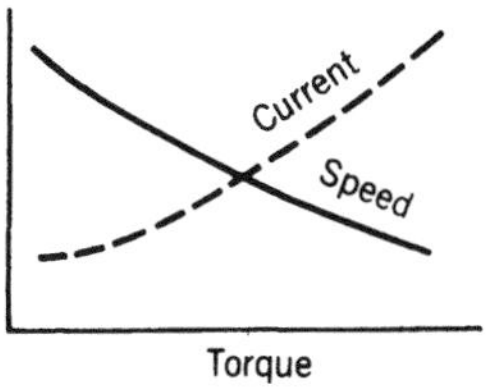	High starting torque Complex circuitry required for reversing

Table 3 Upper Limits of dc Servomotor Performance

Motor Type	Maximum Power (hp)	Maximum Speed (rmp)	Torque/Inertia Ratio (rad/s^2)	Maximum Bandwidth (rad/s)
Moving coil	0.5–1	4500–5500	200–250	1500
Printed circuit	7	3000–4000	130–220	1000
Permanent magnet	10–15	850–3000	15–30	100

Source: From Ref. 4.

The split-series field-wound motor has two windings, one for each direction of rotation. A manual switch is usually employed to activate the appropriate winding. The two windings of the compound motor are always excited and result in a high starting torque with good linearity. All of the fieldwound motors are self-excited with the residual magnetism.

Permanent-Magnet Motors

Permanent-magnet (PM) motors are the most extensively used for servomotors because they generate less heat and have higher efficiency and more compactness than field-excited motors. There are three types of PM motors with mechanical commutation: (1) iron core, (2) surface wound, and (3) moving coil. Figure 4 shows the construction of the three types. Details of the advantages and disadvantages of each type may be found in Ref. 5.

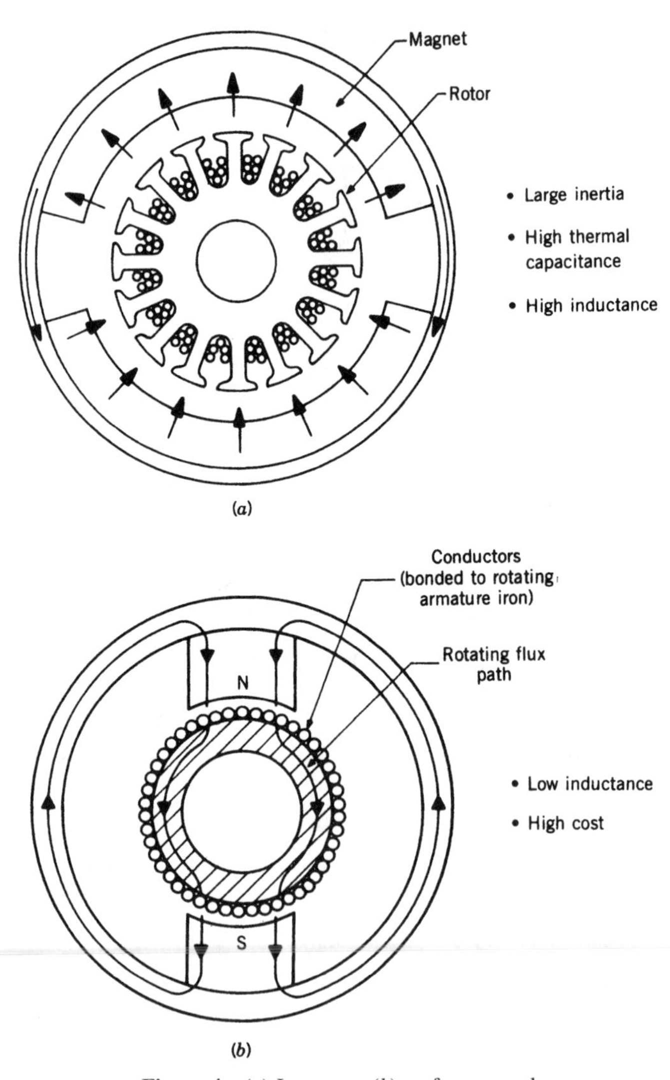

Figure 4 (*a*) Iron core; (*b*) surface wound.

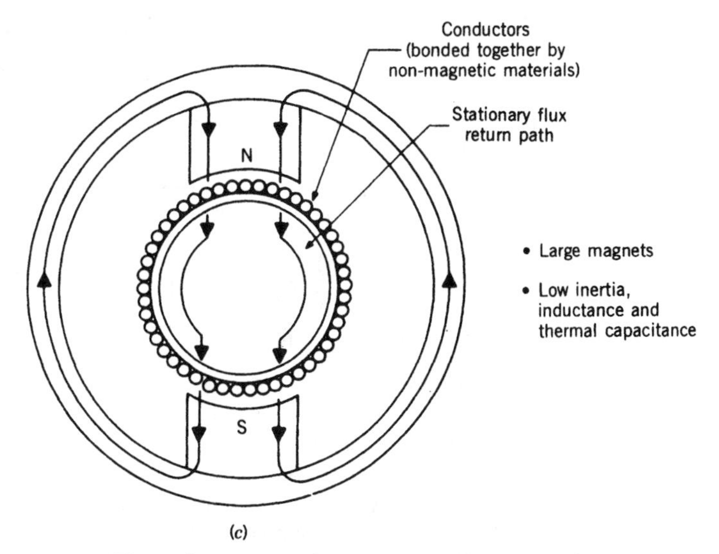

Figure 4 (*Continued*) (*c*) Moving coil. (From Ref. 5.)

Mathematical Model of a Permanent-Magnet Servomotor

Comprehensive presentations on mathematical modeling of dc servomotors are given in Refs. 6–9. A simplified dynamic model is presented here.

The mathematical model of a permanent-magnet dc motor is obtained by lumping the inductance and resistance of the armature winding as shown in Fig. 5. The resulting equations are given:

Voltage equations:

$$v_a = L_a \frac{di}{dt} + R_a i + e_b \tag{1}$$

$$e_b = K_E \Omega_m \tag{2}$$

Torque balance equation:

$$K_T i = J_m \frac{d\Omega_m}{dt} + B_m \Omega_m + T_{fm} + T_L \tag{3}$$

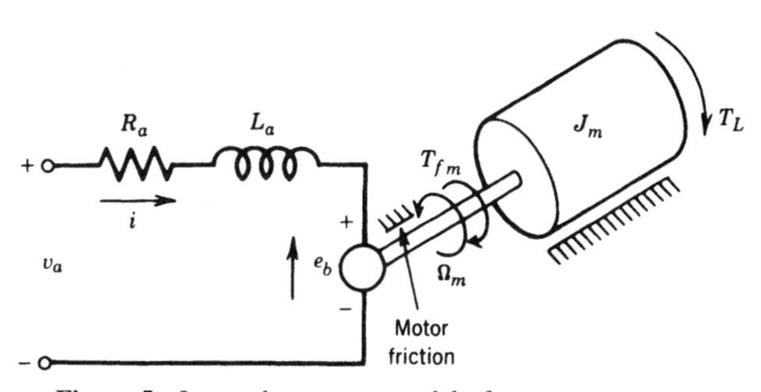

Figure 5 Lumped-parameter model of a permanent motor.

Taking Laplace transforms of Eqs. (1)–(3) gives, after algebraic manipulation,

$$\Omega_m(s) = G_1(s)V_a(s) - G_2(s)\,[T_{fm}(s) + T_L(s)] \tag{4}$$

where the transfer functions G_1 and G_2 are given by

$$G_1(s) = \frac{K_T}{R_aB_m(\tau_e s + 1)(\tau_m s + 1) + K_TK_E} \tag{5}$$

$$G_2(s) = \frac{R_a(\tau_e s + 1)}{R_aB_m(\tau_e s + 1)(\tau_m s + 1) + K_TK_E} \tag{6}$$

and the parameters are defined as follows:

B_m = viscous damping in motor (N·m·s/rad)
e_b = back electromotive force (emf) (V)
i = current through armature (A)
J_m = polar moment of inertia of armature (N·m·s²/rad)
$K_E = z'\phi P/60$ = motor voltage constant or back emf constant (V·s/rad)
$K_T = z'\phi P/2\pi$ = motor torque constant (N·m/A)
L_a = armature inductance (H)
P = number of poles
R_a = armature resistance (Ω)
s = Laplace operator
t = time (s)
T_{fm} = Coulomb friction torque in motor (N·m)
T_L = external load torque
v_a = voltage applied to armature (V)
$V_a(s)$ = Laplace transform of armature voltage $v_a(t)$
z' = number of conductors per parallel path in armature
θ_m = angular position of motor shaft (rad)
ϕ = magnetic flux per pole (Wb)
$\tau_e = L_a/R_a$ = electrical time constant (s)
$\tau_m = J_m/B_m$ = mechanical time constant (s)†
Ω_m = angular velocity of motor (rad/s)
$\Omega_m(s)$ = laplace transform of motor angular velocity

Equation (4) can be simplified if the armature inductance is small (making the electrical time constant τ_e negligible) and the Coulomb friction and load torque are assumed zero. The result is

$$\frac{\Omega_m(s)}{V_a(s)} = \frac{K_m}{\tau s + 1} \tag{7}$$

where

$$K_m = \frac{K_T}{R_aB_m + K_TK_E} \quad \text{(motor constant)} \tag{8}$$

and

---

*Some servomotor manufacturers define τ_m differently. For example, Electro-Craft[9] defines the mechanical time constant as $\tau_m = (R_aJ_m)/(K_TK_E)$.

$$\tau = \frac{R_a J_m}{R_a B_m + K_T K_E} \tag{9}$$

Reference 9 discusses cases where the electrical time constant cannot be neglected.

The preceding mathematical models assume a voltage input. For applications where a current amplifier is used, the following approximate model should be used:

$$\frac{\Omega_m(s)}{I(s)} = \frac{K'_m}{\tau_m s + 1} \tag{10}$$

where $I(s)$ is the Laplace transform of the current input i. The motor constant in this case is

$$K'_m = \frac{K_T}{B_m} \tag{11}$$

In principle, the models developed can be applied to all of the dc motors of the various types with the appropriate input conditions. These models describe the open-loop response. For closed-loop systems with velocity or position feedback, an appropriate closed-loop transfer function can be derived easily by making use of the motor dynamic model. An example of a closed-loop system is given in Section 10.2.

Numerical Example

For Motomatic PM servomotor model number E350-MG,[10] the following specifications are given:

K_T = 3.4 in.·oz/A (0.024 N·m/A)
K_E = 2.5 V/krpm (0.024 V·s/rad)
R_a = 12.4 Ω
J_m = 2.5 × 10^{-4} in.·oz·s²/rad (1.8 × 10^{-6} N·m·s²/rad)
B_m = 0.015 in.·oz/krpm (1.01 × 10^{-6} N·m·s/rad)
T_{fm} = 0.5 in.·oz (3.5 × 10^{-3} N·m)
T_{max} = 2.5 in.·oz (1.8 × 10^{-2} N·m)
I_{max} = 0.75 A
Ω_{max} = 10,500 rpm at no load (1099 rad/s)
L_a = 3.1 mH
τ_e = 0.25 × 10^{-3} s
R_{th} = thermal resistance = 13°C/W

The mechanical time constant can be computed as

$$\tau_m = \frac{J_m}{B_m} = 1.75 \text{ s} \tag{12}$$

Since $\tau_e \ll \tau_m$, Eq. (7) can be used to determine the dynamic response if the Coulomb friction and load torque are neglected. In this case, the time constant is

$$\tau = \frac{R_a J_m}{R_a B_m + K_T K_E} = 0.037 \text{ s} \tag{13}$$

and the motor constant is

$$K_m = \frac{K_T}{R_a B_m + K_T K_E} = 0.39 \text{ krpm/V} \quad (40.8 \text{ rad/V·s}) \tag{14}$$

The transfer function of Eq. 7 becomes

$$\frac{\Omega_m(s)}{V_a(s)} = \frac{40.8}{0.037s + 1} \tag{15}$$

For a step input of 1 V, the motor speed is given by

$$\Omega_m(s) = \frac{1}{s}\left(\frac{40.8}{0.037s + 1}\right) \tag{16}$$

The inverse Laplace transform gives the step response as

$$\Omega_m(t) = 40.8(1 - e^{-t/0.037}) \tag{17}$$

3.2 Brushless dc Servomotors

The development of brushless dc servomotors was an outgrowth of semiconductor devices even though the first patent was obtained with vacuum tube technology.[11] The basic construction of a brushless dc motor eliminates mechanical commutation. Instead, the commutation process is accomplished electronically with no moving contacts. Hence, the problems associated with mechanical commutation such as brush wear particles, electromagnetic interference (EMI), or arcing are eliminated. Elimination of arcing makes dc servomotors excellent candidates for applications requiring explosion-proof safety classification.

Construction

Typically, brushless motors have an inner rotor and outer stator and a configuration such as the one shown in Fig. 6*a*. However, the other configuration (i.e., inner stator and outer rotor) is also possible (see Fig. 6*b*). The former configuration with the outer stator carrying electrical windings provides excellent thermal dissipation characteristics, since both the iron and copper losses occur in the stator and the stator is better exposed to the ambient for convective heat transfer. This feature allows brushless motors to be operated at higher speeds and hence provides higher power-to-weight ratio.

Brushless dc motors range from 1 to 40 in. (0.025 to 1.02 m) in diameter with 6 in·oz (4.24×10^{-2} N·m) to 1650 ft·lb (2237 N·m) of torque capability (see Table 4). Typical applications include memory disk drives, videotape recorders, and position servos in cryogenic compressors and fuel pumps.

The rotors are permanent magnets made from one of three primary materials: ceramic, AlNiCo, and rare earth (such as samarium cobalt). Ceramic rotors are used in applications where cost consideration is important. Rare-earth magnets are the most expensive but provide exceptional performance. AlNiCo magnets are of medium cost and provide medium magnetic strengths.

Operation

The brushless motor is operated by generating a rotating magnetic field that is 90° (electrical) out of phase with the rotor. Position sensors are used to determine the rotor position. These position sensors are of three types: phototransistor, electromagnetic, and Hall effect generators.

Commutation

An electronic module consisting of logic circuits and power amplification circuits is used to drive the motor.[3,4,12,13] This module receives rotor position information from the position sensors. The angle through which the rotor turns during the firing of a winding is called the "conduction angle." Figure 7 shows schematically a two-phase brushless motor with the driver electronics.

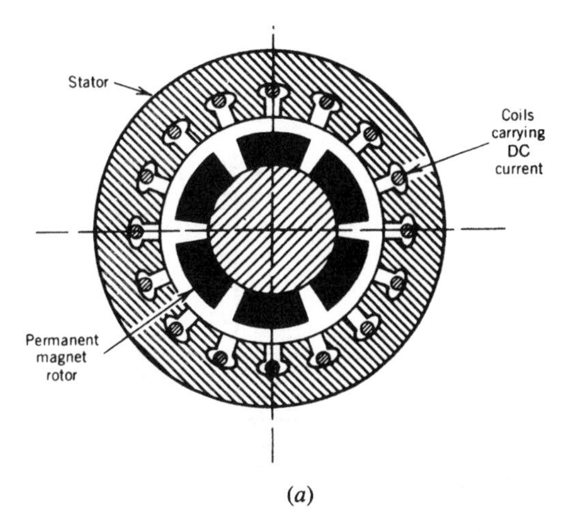

(*a*)

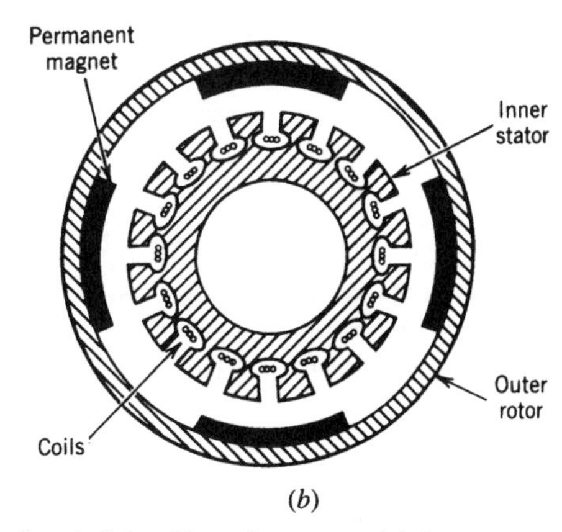

(*b*)

Figure 6 Cross section of typical brushless dc motors: (*a*) inner rotor–outer stator type; (*b*) inner stator–outer rotor type. (From Ref. 5.)

Table 4 Brushless dc Motor Performance Data

Magnet Type	Power (W)	Peak Torque (in.·oz)	Electrical Time Constant (s)	Mechanical Time Constant (s)	Torque/Inertia ratio (rad/s^2)
Ceramic	25–900	10–600	0.0002–0.0016	0.0221–0.7400	413–11,400
AlNiCo	20–280	6–5,000	0.0001–0.0030	0.0065–0.1330	465–57,500
Rare earth	25–6000	10–316,000	0.0001–0.0140	0.0024–0.0291	137–100,000

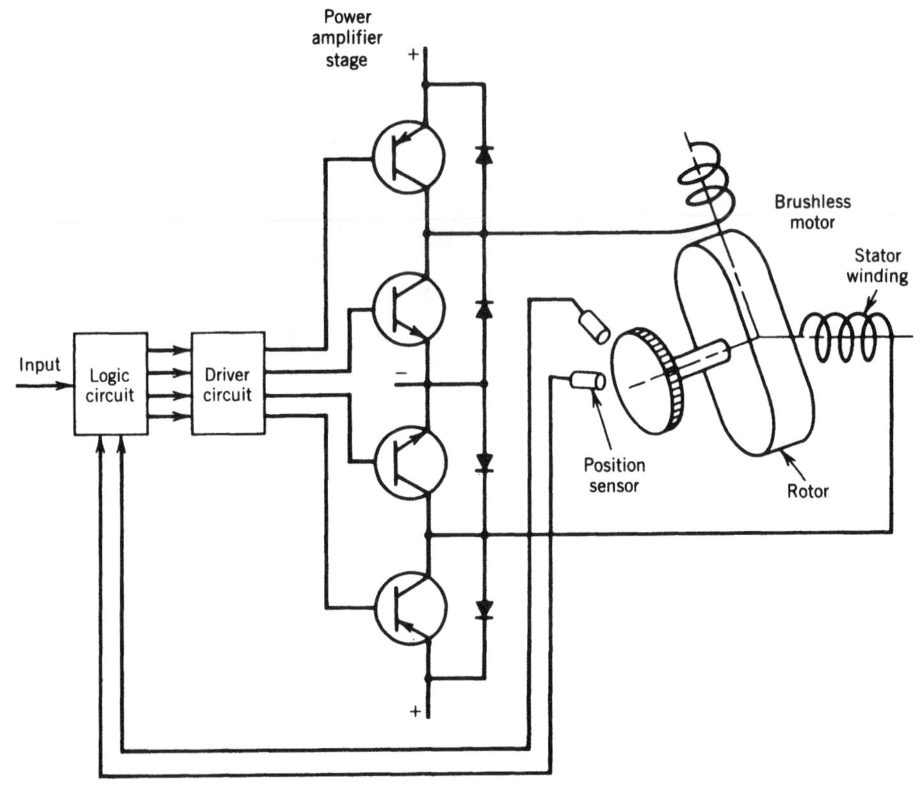

Figure 7 Two-phase brushless motor with driver electronics.

Figure 8 shows the controller circuit for a three-phase brushless motor. Each phase requires a pair of switches for commutation. Since the cost of the motor is dependent on the number of switches, there is a tendency to keep the number of phases to a minimum. Typically, three-phase motors with six switches are used. The current through the windings may be varied in a sinusoidal or a square-wave manner. The latter excitation results in a small torque ripple (17% average to peak for a two-phase motor and 7% for a three-phase motor).

Ideally, a sinusoidal torque function results in a constant torque. But sinusoidal torque function generation is technically difficult and uneconomical. An alternate approach is to design the spatial variation of the magnetic field (possible by means of high-coercive-strength magnets) to obtain a trapezoidal torque function while the input current has a square waveform (easily generated by simple transistor control circuitry, such as shown in Fig. 8). The motor torque is then approximately constant and is proportional to the maximum value of current during each cycle. The trapezoidal torque generation scheme also results in higher efficiency.

The locations of the position sensors relative to the rotor are aligned to result in appropriate timing for proper commutation. When properly commutated, a brushless motor duplicates the torque–speed characteristics of a brush-type dc motor.

The power output of the brushless motor is effectively controlled by pulse-width modulation (PWM) or pulse-frequency modulation (PFM) methods. A linear (i.e., class A) power amplifier can also be used for power control. However, use of this type of amplifier produces

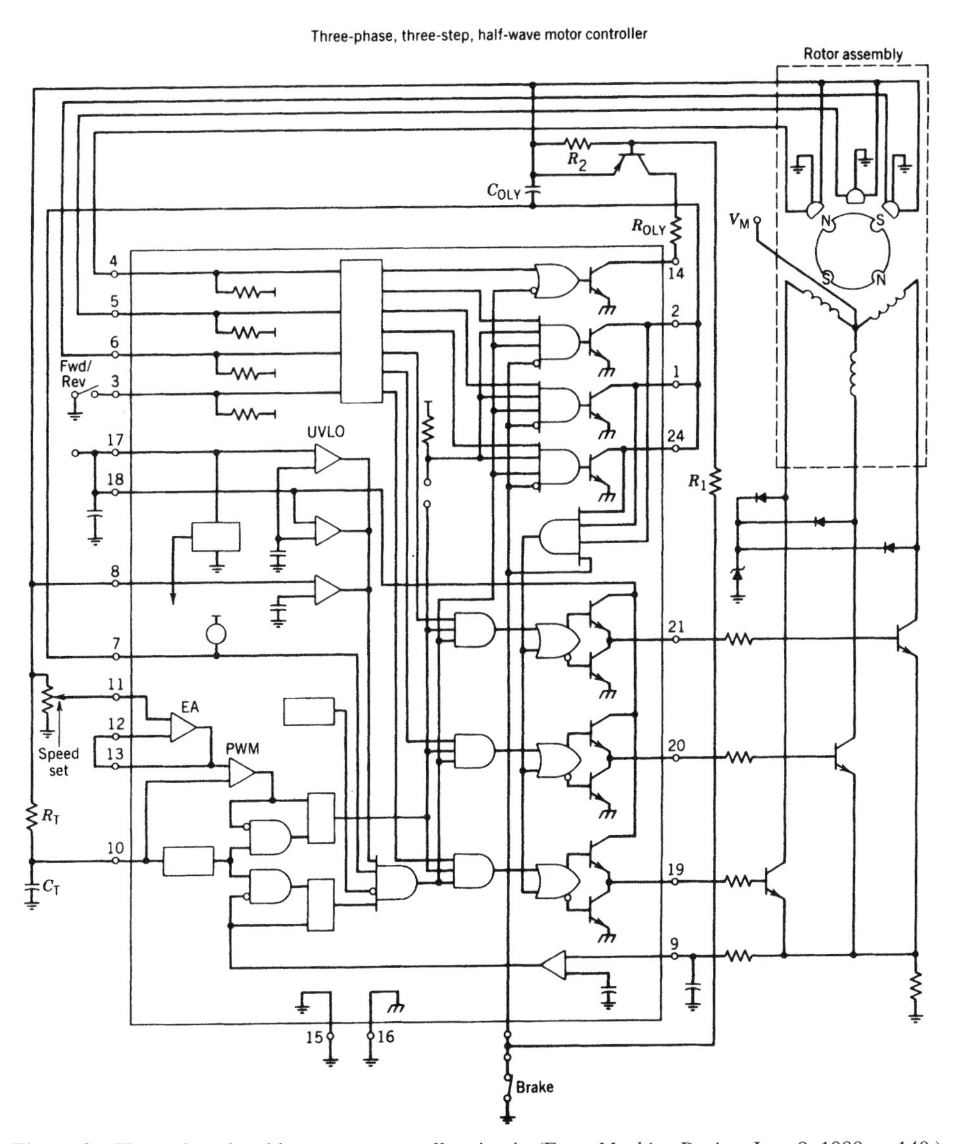

Figure 8 Three-phase brushless motor controller circuit. (From *Machine Design,* June 9, 1988, p. 140.)

back-emf conduction during the zero-voltage portions of the voltage modulation and thereby increases viscous damping. This effect can be eliminated by using a current amplifier rather than a voltage amplifier.[9]

Figure 9 shows a cross section of a brushless motor developed for use as a fin actuator. Hall effect sensors are used for position measurement.

Mathematical Model

The mathematical model required to represent a brushless dc motor is identical to that of a brush-type dc motor. Therefore the equations given in Section 3.1 are applicable.

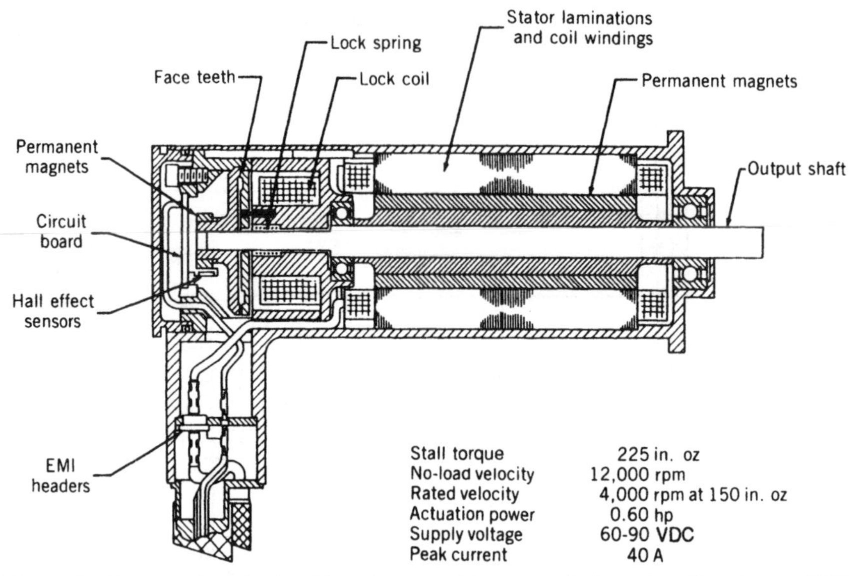

Figure 9 Cross section of a brushless dc motor. (Courtesy of Moog, Inc., East Aurora, NY.)

Numerical Example

Table 5 shows the specifications for a ceramic magnet, inside rotation-type dc brushless motor manufactured by Magnetic Technology.[14,15]

If this motor is operating at peak torque, the steady-state motor speed is given by [from Eq. (4), dropping the dynamic terms]

Table 5 Performance Data for Magnetic Technology Model 2800-153-084 Brushless dc Motor

Peak torque	40 in.·oz
Power at peak torque	175 W
Electrical time constant	0.0005 s
Mechanical time constant	0.054 s
Damping factor	0.064 in.·oz·s/rad
Moment of inertia	0.0035 in.·oz·s^2/rad
Total breakaway torque	1.5 in.·oz
Temperature rise	2°C/W
Maximum allowable winding temperature	155°C
Weight	19 oz
Number of poles	8
Number of phases	2
Resistance	8.4 Ω
Inductance	4.2 mH
Voltage at peak torque	38.2 V
Current at peak torque	4.57 A
Torque constant	8.7 in.·oz/A
Voltage constant	0.0617 V·s/rad

$$\Omega_m = \frac{K_T v_a}{R_a B_m + K_T K_E} = 309 \text{ rad/s} = 2953 \text{ rpm} \tag{18}$$

and the temperature rise is

$$\Delta T = (2°\text{C/W})\ (175\ \text{W}) = 350°\text{C} \tag{19}$$

This temperature rise is greater than the 155°C maximum allowable winding temperature. Thus the motor cannot be operated at peak torque indefinitely. The ambient temperature should be added to the temperature rise to arrive at the operating temperature of the winding.

4 ALTERNATING-CURRENT SERVOMOTORS

Alternating-current servomotors are used in applications requiring smooth speed control. These motors find widespread use in stationary industrial applications due to the ready availability of ac power.[16,17] In a majority of these applications, two-phase ac servomotors are used because of simplicity of the associated controls. Figure 10 shows a schematic of a two-phase ac servomotor. The operation of the two-phase ac servomotor is similar to an induction motor except the voltages applied to the two windings (fixed phase and control phase) are generally unequal and out of phase. The ac voltage applied to the fixed phase is held constant, and the one applied to the control phase is varied to control the motor speed. The two phases are generally 90° out of phase. Changing the phase angle from +90° to −90° reverses the direction of rotation of the motor. The appropriate phase angle is achieved through capacitors or other phase-shift circuits.

4.1 Types of ac Servomotors

Depending on the rotor construction, ac servomotors are classified into three types: (1) squirrel cage, (2) solid iron, and (3) drag cup (see Fig. 11). The squirrel-cage construction of the rotor is exactly the same as that of a standard induction motor. The inherent disadvantage of the squirrel-cage construction is cogging, or nonuniform armature rotation, which is minimized by skewing the bars of the cage relative to the rotor axis. The solid-iron rotor eliminates cogging. However, this configuration has a low torque-to-inertia ratio. The torque-

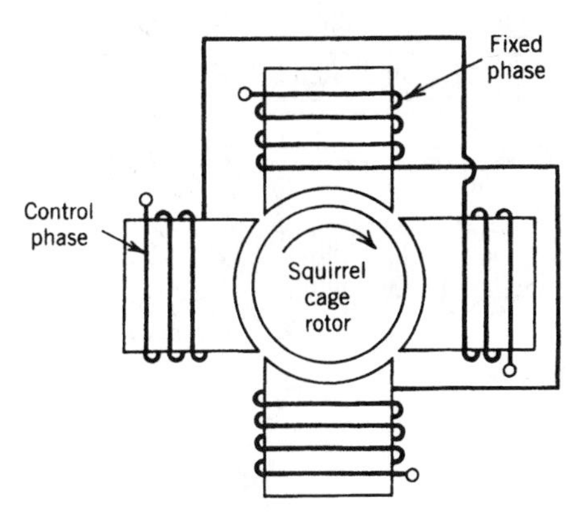

Figure 10 Two-phase ac servomotor.

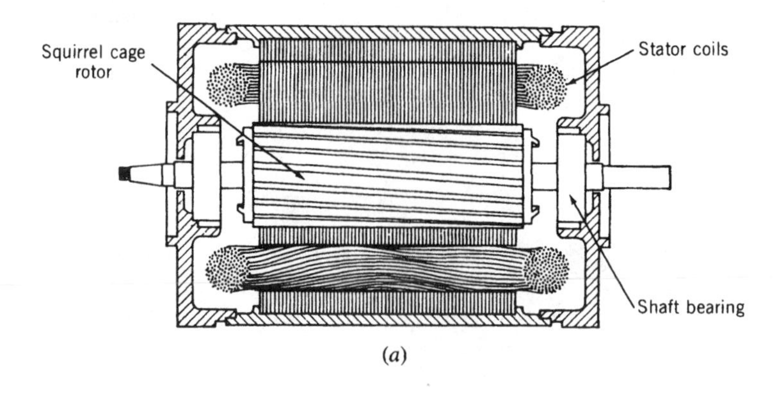

(*a*)

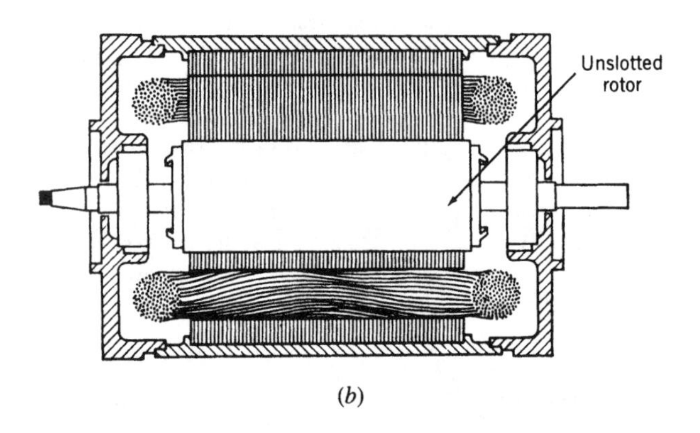

(*b*)

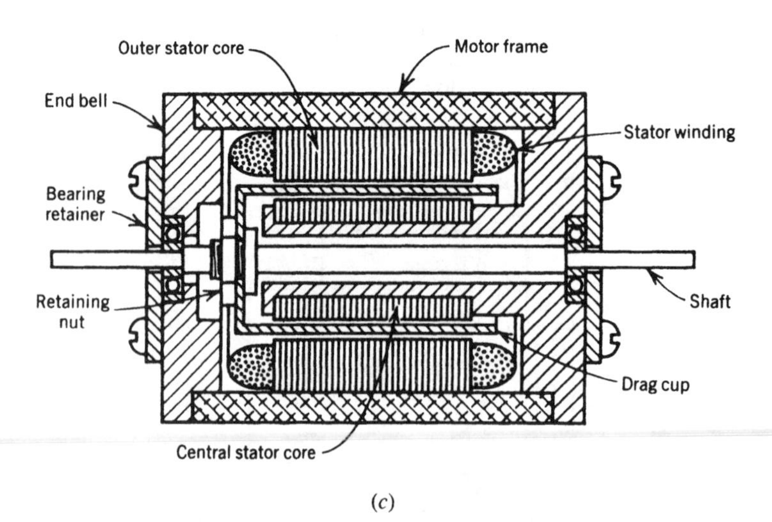

(*c*)

Figure 11 Types of ac servomotors; (*a*) squirrel-case rotor; (*b*) solid-iron rotor; (*c*) drag-cup rotor. (From J. E. Gibson and F. B. Tuteur, *Control System Components,* McGraw-Hill, New York, pp. 279–280.)

to-inertia ratio is improved by the use of drag-cup construction (see Fig. 11c). The efficiency of ac servomotors is fairly low (5–20%), which necessitates external cooling.

4.2 Mathematical Model

Steady-State Model

Assuming linearity (i.e., operation without magnetic saturation) and using the method of symmetrical components,[18] mathematical models can be developed for the ac servomotor. Figure 12 shows an equivalent circuit of one phase of the ac servomotor. The torque developed by a two-phase servomotor in which the control phase lags the reference phase by an angle θ is given by

$$T_m = \tfrac{1}{4}\left[F_1(1 + 2k \sin\theta + k^2) - F_2(1 - 2k \sin\theta + k^2)\right]|v_m|^2 \tag{20}$$

The functions F_1 and F_2 are defined as follows:

$$F_1 = \frac{2}{\Omega_s}\left|\frac{Z_m}{Z_1(Z_2 + Z_m) + Z_2 Z_m}\right|^2 \frac{R_2'}{S_R} \tag{21}$$

$$F_2 = \frac{2}{\Omega_s}\left|\frac{Z_m}{Z_1(Z_2' + Z_m) + Z_2' Z_m}\right|^2 \frac{R_2'}{2 - S_R} \tag{22}$$

where f = frequency of ac voltages (Hz)
$k = v_c / v_m$
n_1 = number of turns in winding of one pole of the stator
n_2 = number of turns of rotor winding ($n_2 = 1$ for squirrel-cage rotor)
R_1 = resistance of stator (Ω)
R_2 = resistance of rotor (Ω)
$R_2' = R_2(n_1/n_2)^2$
R_m = resistance due to magnetic field (Ω)
S_R = slip ratio = $(\Omega_s - \Omega_m)/\Omega_s$
v_c = voltage applied to control phase (V)
v_m = voltage applied to fixed phase (V)
X_1 = inductive reactance of stator
X_2 = inductive reactance of rotor

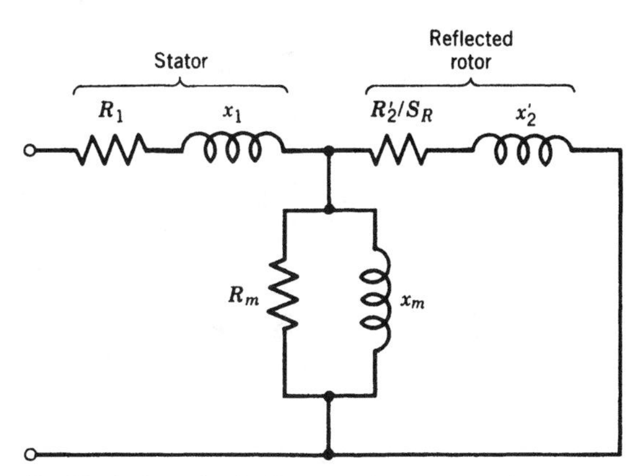

Figure 12 Equivalent circuit diagram of one phase of an ac servomotor.

$X_2' = X_2(n_1/n_2)^2$
X_m = inductive reactance due to magnetic field
Z_1 = stator impedance = $R_1 + jX_1$
Z_2 = reflected impedance of rotor at $S_R = (R_2'/S_R) + jX_2'$
Z_2' = reflected impedance of rotor at $(2 - S_R) = R_2'/(2 - S_R) + jX_2'$
Z_m = impedance due to magnetic field generation in the stator = $[(1/R_m) + (1/jX_m)]^{-1}$
θ = phase angle between fixed and control voltages
Ω_m = motor speed (rad/s)
Ω_s = synchronous speed (rad/s)

Figure 13 shows typical torque–speed characteristics for a two-phase ac motor. The characteristics are clearly nonlinear. However, about the origin they may be treated as linear.

Dynamic Model

Combining the equation of motion, the electrical lag (due to stator inductance and resistance), and the torque–speed characteristic of Eq. (20) yields a nonlinear dynamic model. For an approximate static analysis, Eq. (20) may be linearized about an operating point. For many servosystem applications the most interesting and useful operating point is $k = 0$ and $S_R = 1.0$. This linearization yields

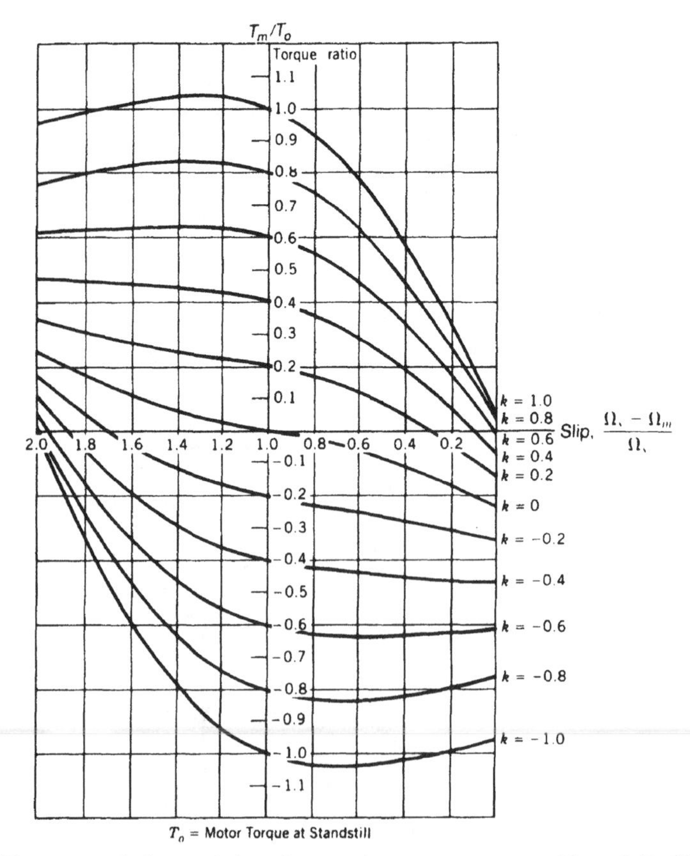

Figure 13 Torque–speed characteristics of a two-phase ac servomotor. (From J. E. Gibson and F. B. Tueter, *Control System Components,* McGraw-Hill, New York, 1958, p. 288.)

$$\Delta\Omega_m = A\,\Delta v_c - \frac{1}{B}\,\Delta T_m \tag{23}$$

where

$$A = \frac{\partial\Omega_m}{\partial v_c} \tag{24}$$

$$\frac{1}{B} = -\frac{\partial\Omega_m}{\partial T_m} \tag{25}$$

The symbol Δ indicates small variations from the steady-state operating point. Including the electrical and mechanical dynamics gives the transfer function

$$\Delta\Omega_m(s) = \frac{A\,\Delta V_c(s) - (1/B)\,\Delta T_m(s)}{(\tau_e s + 1)(\tau_m s + 1)} \tag{26}$$

where B_m = damping coefficient of motor (N·m·s)
J_m = polar moment of inertia of rotor (N·m·s²/rad)
L_1 = inductance of stator (H)
R_1 = resistance of stator (Ω)
s = Laplace variable
ΔT_m = change in load torque reflected to motor shaft
$\Delta T_m(s)$ = Laplace transform of change in load torque reflected to motor shaft
$\tau_e = L_1/R_1$ = electrical time constant (s)
$\tau_m = J_m/B_m$ = mechanical time constant (s)
$\Delta\Omega_m(s)$ = Laplace transform of change in motor speed

Computation of the constants A and B of Eq. (26) from the torque expression of Eq. (20) is rather tedious and requires measurements of the rotor impedances over a speed range of $-\Omega_s$ to $+\Omega_s$. As an alternative, an approximate expression for the torque has been developed[18] as follows:

$$T_m = \frac{S_R}{C_1 + C_2 S_R^2} \tag{27}$$

where C_1 and C_2 are constants which are determined from two points on an experimentally measured torque–speed characteristic. The constants A and B in Eq. (23) are related to C_1 and C_2 as follows:

$$A = \frac{2\Omega_s}{v_m}\,\frac{(C_1 + C_2)}{(C_1 - C_2)}\sin\theta \tag{28}$$

$$B = \frac{1}{2\Omega_s}\,\frac{(C_1 - C_2)}{(C_1 + C_2)^2} \tag{29}$$

Numerical Example

Specifications for a typical two-phase ac servomotor are given as follows:

Number of poles = $P = 4$

Stator resistance = $R_1 = 10\ \Omega$

Stator inductance = $L_1 = 3$ mH

Moment of inertia of rotor = $J_m = 5.4 \times 10^{-5}$ in.·oz·s²/rad (3.8×10^{-7} N·m·s²/rad)

Locked torque = 9.5 in.·oz (6.71×10^{-2} N·m)

Torque at 1000 rpm = 5.4 in.·oz (3.8×10^{-2} N·m)
Voltages: v_c = 110 V at 90°; v_m = 110 V at 0°
Frequency = 60 Hz
Synchronous speed $\Omega_s = 120f/P$ = 1800 rpm (188.3 rad/s)
Slip ratio at standstill = $S_{R1} = 1$
Slip ratio at 1000 rpm = $S_{R2} = (\Omega_s - \Omega_m)/\Omega_s = 0.44$

Substituting $T = 9.5$ in.·oz at $S_R = 1$ in Eq. (27) gives

$$\frac{1}{C_1 + C_2} = 9.5 \tag{30}$$

Similarly, substituting $T = 5.4$ in.·oz at $S_R = 0.44$ gives

$$\frac{0.44}{C_1 + C_2(0.44^2)} = 5.4 \tag{31}$$

Solving Eqs. (30) and (31) gives

$$C_1 = 0.076 \text{ 1/in.·oz (11 1/N·m)} \qquad C_2 = 0.029 \text{ 1/in.·oz (4.1 1/N·m)}$$

Substituting numerical values in Eqs. (28) and (29) gives

$$A = 7.79 \text{ rad/V·s}$$

$$B = 1.59 \times 10^{-2} \text{ in.·oz·s/rad } (1.12 \times 10^{-4} \text{ N·m·s/rad})$$

From the definitions for the motor time constants

$$\tau_e = 3 \times 10^{-4} \text{ s} \qquad \tau_m = 4.6 \times 10^{-2} \text{ s}$$

From Eq. (26), the motor transfer function is given as

$$\Delta\Omega_m(s) = \frac{7.79\ \Delta V_c(s) - 62.9\ \Delta T_m(s)}{(3 \times 10^{-4}\ s + 1)(4.6 \times 10^{-2}\ s + 1)} \tag{32}$$

Generally the mechanical time constant is much greater than the electrical time constant, as is the case for this example. The term $\tau_e s + 1$ in the transfer function of Eq. (26) often may be neglected without introducing significant error.

5 STEPPER MOTORS

Stepper motors (also called step motors) represent a significant breakthrough in the area of electromechanical actuation. These are incremental motors that by their very nature are compatible with digital systems. A stepper motor converts an electrical pulse into an equivalent rotary displacement. Since their introduction in the early 1930s, stepper motors have evolved into sophisticated designs.[19–22]

The stepper motor possesses some inherent advantages over a conventional servomotor: (1) it is compatible with digital processors; (2) open-loop control is possible, which eliminates stability problems associated with closed-loop servos; (3) the step error is noncumulative; and (4) the brushless design provides easy maintenance and ruggedness. As a result of these advantages, stepper motors find widespread usage in industrial applications such as drives for TV antennas, numerical control (NC) machines, computer drives, hydraulic valve positioning, and other high-performance feedback control systems. The primary disadvantage of stepper motors is their low efficiency, which restricts their use to fractional horsepower applications.

5.1 Operation

The principle of operation of a stepper motor is illustrated by the single-stack, variable-reluctance, three-phase motor shown in Fig. 14. The stator has six poles that are wound in a three-phase configuration. The soft-iron rotor has four poles. When phase 1 is powered by a dc voltage, one pair of rotor poles will line up with the phase 1 stator poles. When phase 1 is switched off and phase 2 is turned on, the rotor will turn clockwise through 30° until one pair of teeth align with the phase 2 poles at C. Similarly, if phase 2 is switched off and phase 3 is turned on, the rotor will rotate another 30° in the clockwise direction. Thus, with each switching the rotor advances through one step of 30°. So the "step angle" is 30°. (The most commonly used step angle is 1.8°.) The direction of rotation may be reversed by switching in a 1.3.2.1.3.2 . . . sequence.

In the preceding scheme only one phase winding is switched on at a time. This scheme is termed "one-phase-on" switching. An alternate switching scheme is called "two-phase-on" switching. In two-phase-on switching, two windings are turned on simultaneously. Referring to Fig. 14, if the switching sequence is 12.23.31.12 . . . , the rotor will rotate in a clockwise direction one step of 30° at a time. So, the two-phase-on scheme does not alter the step size for a three-phase stepper motor such as shown in Fig. 14. However, for a four-phase variable-reluctance (VR) motor a switching sequence such as 1.12.2.23.3.34.4.41.1 . . . results in reducing the step size in half. This sequence is called "half stepping." Half stepping results in about 41% more torque (for a four phase motor) compared to the single-stepping scheme. By varying the relative magnitude of the voltages applied to the two windings, the rotor can be made to rotate in fractional increments of a step. This scheme is termed "microstepping."

When the motor is energized, the holding torque of a stepper motor theoretically varies with the rotor position as a sinusoidal. Figure 15 shows a typical torque-speed characteristic for a stepper motor. The torque above which the stepper motor definitely loses steps is termed

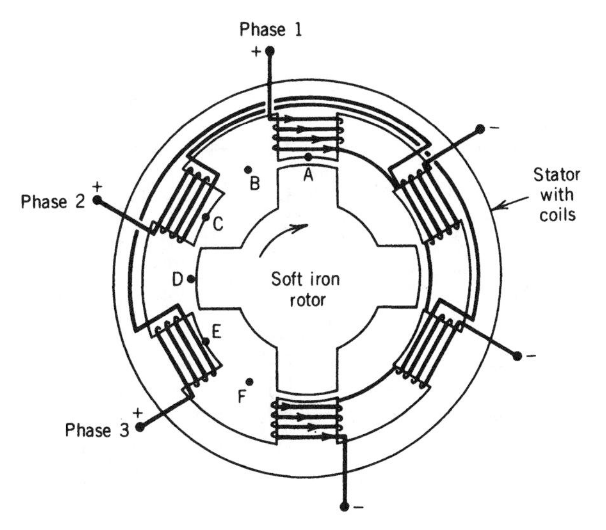

Figure 14 Single-stack variable-reluctance three-phase stepper motor.

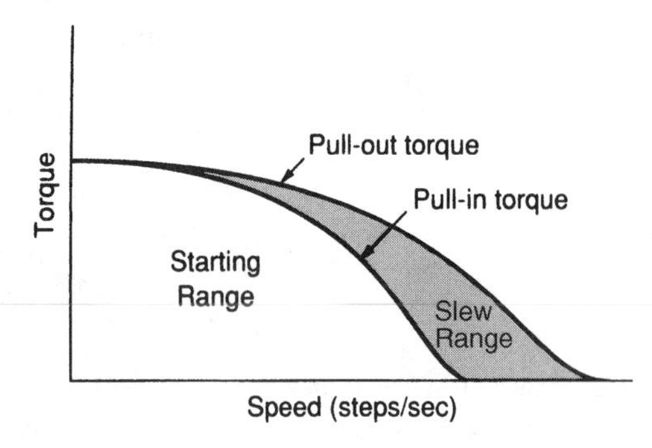

Figure 15 Typical relation between pull-in torque and pull-out torque of a stepper motor. (From B.C. Kuo, *Incremental Motion Control-Step Motors and Control Systems,* Vol. II, SRL Publishing, Champaign, IL, 1979, p. 101).

the "pull-out torque." The torque below which a stepper does not lose any steps (in an open-loop mode) is termed the "pull-in torque." The speed range between the pull-in and pull-out torques is called the "slew range." The slew range represents an unstable region of operation.

5.2 Types of Stepper Motors

In the early years of stepper motor development, stepper design included mechanical detenting and solenoid controls.[19] However, these designs have been replaced by more rugged and efficient designs. The latter designs may be broadly classified as (1) permanent-magnet steppers, (2) variable-reluctance steppers, (3) hybrid steppers, (4) electromechanical steppers, and (5) electrohydraulic steppers. Figure 16 shows configurations of each of these various types of stepper motors. The most commonly used are the permanent-magnet and the three-phase and four-phase variable-reluctance types of stepper motors.

The stator windings of variable-reluctance stepper motors sometimes are wound with two windings of opposite polarity per pole. This approach is termed a "bipolar winding." Figure 17 shows a bipolar-wound stepper motor. This arrangement provides more torque and improves damping, but the disadvantage is more complex circuitry.

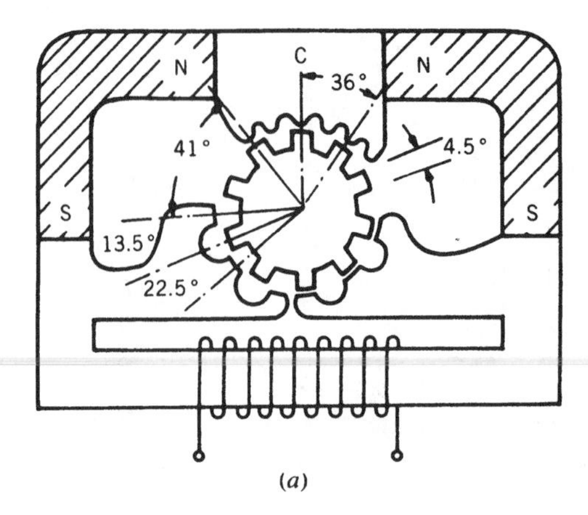

Figure 16 Stepper motor configurations. (From B. C. Kuo, *Incremental Motion Control-Step Motors and Control Systems,* Vol. II, SRL Publishing Co., Champaign, IL, 1979, pp. 12–14.): (*a*) permanent stator stepper motor.

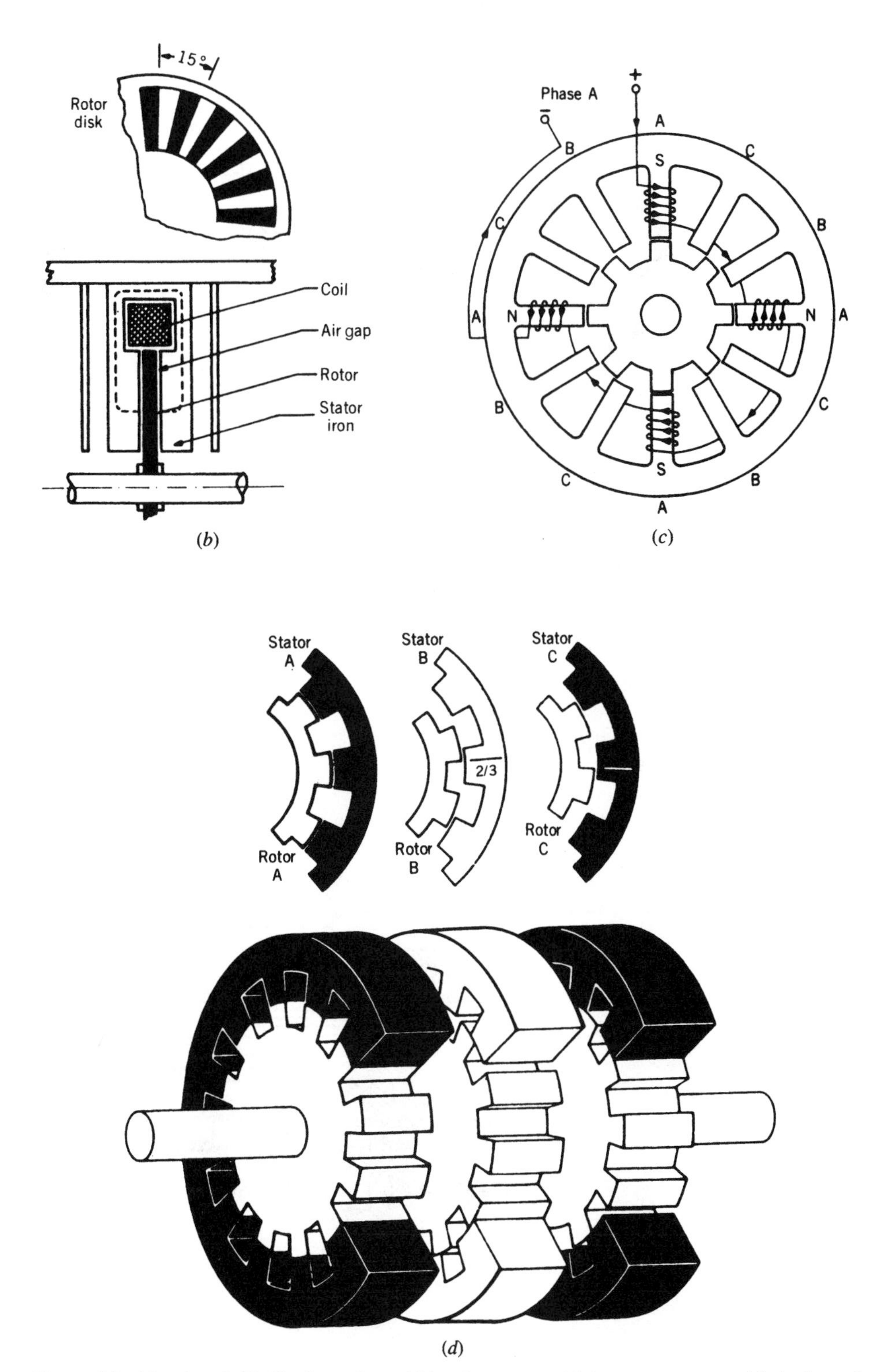

Figure 16 (*Continued*) (*b*) Single-stack, variable-reluctance, axial-gap stepper motor; (*c*) single-stack, variable-reluctance, radial-gap stepper motor; (*d*) multi stack, variable-reluctance, radial-gap stepper motor.

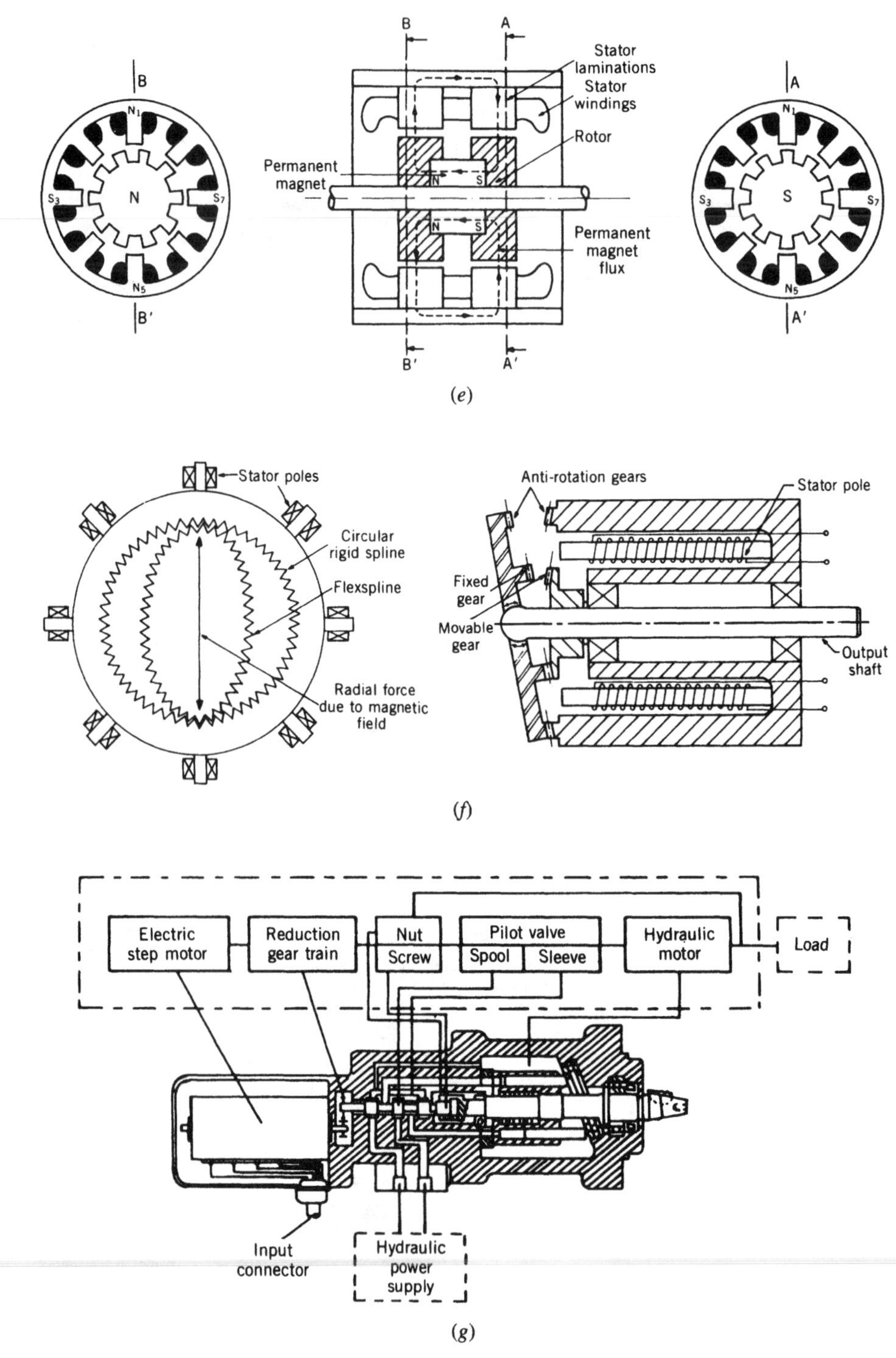

Figure 16 (*Continued*) (*e*) Hybrid stepper motor; (*f*) electromechanical stepper motor; (*g*) electrohydraulic stepper motor.

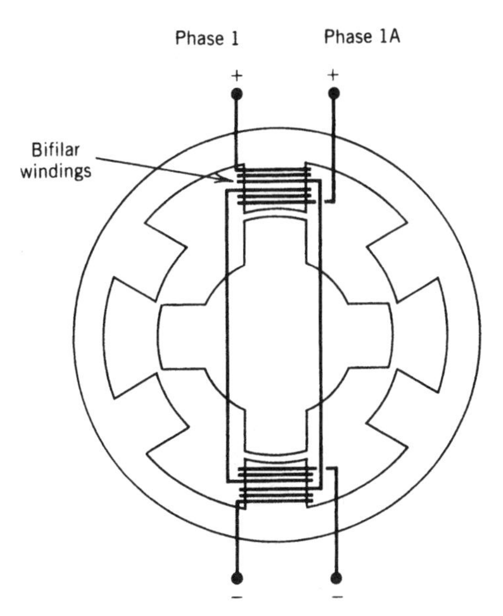

Figure 17 Bipolar-wound stepper motor.

5.3 Mathematical Model of a Permanent-Magnet Stepper Motor

The mathematical model of a stepper is generally much more complex than a conventional dc motor since the voltages applied to the various phases change in a discontinuous fashion. These discontinuities in the applied voltages result directly in corresponding discontinuities in the phase currents. This effect is further complicated by the spatial variation of the magnetic reluctance. Reference 20 gives detailed mathematical models for permanent-magnet and variable-reluctance stepper motors. Computer codes (in FORTRAN IV) are available in Ref. 20 for these stepper motors.

The mathematical models of stepper motors are inherently nonlinear due to discontinuities in input voltages and due to the transcendental spatial variation of the self and mutual inductances. Hence these models do not lend themselves to a frequency-domain analysis.

5.4 Numerical Example

Table 6 gives the specifications of a Crouzet model no. 82 940.0 stepper motor. The motor is to be used to drive a rotary viscometer that has a rotary inertia of 3.88×10^{-3} in.·oz·s^2/rad (2.74×10^{-5} N·m·s^2/rad), a constant frictional torque of 1.3 in.·oz (9.18×10^{-3} N·m). and a viscous damping coefficient of 0.96 in.·oz·s/rad (6.8×10^{-3} N·m·s/rad). The motor is required to accelerate the viscometer from 5.2 to 13.1 rad/s in a maximum of 0.1 s.

The maximum torque developed by the motor may be estimated as follows:

$$T_m = (J_m + J_L)\frac{d\Omega_m}{dt} + B_L\,\Omega_m + T_f \qquad (33)$$

where B_L = rotary damping coefficient of viscometer cup
J_L = polar moment of inertia of viscometer cup
J_m = polar moment of inertia of motor

Table 6 Stepper Motor Specifications (Crouzet Model 82940.0)

Step angle	7.5°
Number of phases	2 (bipolar)
Resistance per phase	9 Ω
Inductance per phase	24 mH
Current per phase	0.55 A
Maximum input power	10 W
Maximum voltage	6.7 V (continuous duty)
Holding torque at 6 V	21.9 in.·oz
Detent torque	1.67 in.·oz
Rotor inertia	1.19×10^{-3} in.·oz·s²/rad
Maximum coil temperature	248°F

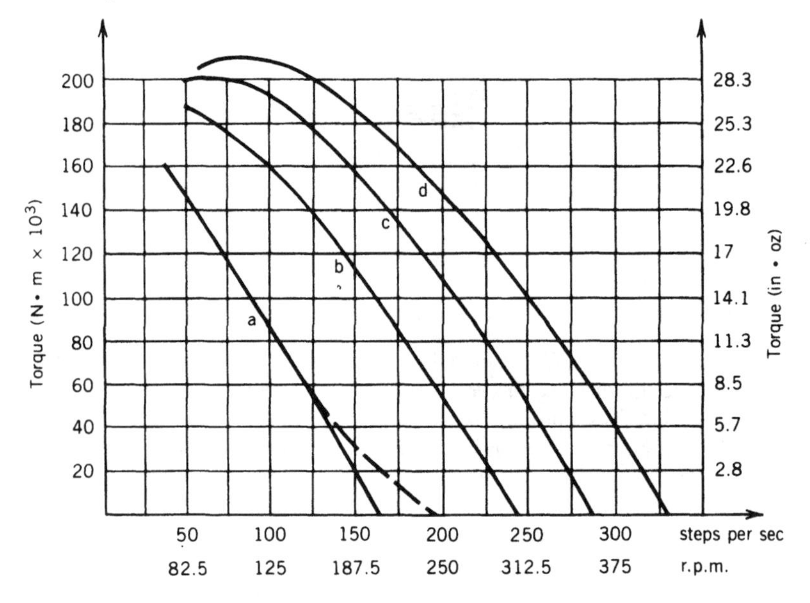

a = Using constant voltage drive with *Rs* (series resistance) = 0 (*L*/*R*)
b = Using constant voltage drive with *Rs* (series resistance) = *R Motor* (*L*/2*R*)
c = Using constant voltage drive with *Rs* (series resistance) = 2*R* Motor (*L*/3*R*)
d = Using constant voltage drive with *Rs* (series resistance) = 3*R* Motor (*L*/4*R*)

T_f = friction torque on viscometer cup
Ω_m = motor shaft speed

From the specifications and Eq. (33): $d\Omega_m/dt = 78.5$ rad/s² and $T_m = 14.2$ in.·oz (0.10 N·m). From the torque–speed characteristics of Table 6 it can be seen that the characteristic labeled *b* can be used. So a series resistance of 9 Ω should be used to meet the torque requirements.

The electrical and mechanical time constants may be estimated as follows:

$$\tau_e \approx \frac{L}{R} = 2.7 \times 10^{-3} \text{ s}$$

$$\tau_m \approx \frac{J_m + J_L}{B_L} = 5.3 \times 10^{-3} \text{ s}$$

Since the time allowed for the acceleration of the load is 0. 1 s, which is more than an order of magnitude greater than either time constant, the motor should have adequate dynamic response.

For a more exact dynamic response determination, the motor parameters may be determined experimentally (as described in Chapter 6 of Ref. 20) and used in the dynamic models.

6 ELECTRICAL MODULATORS

This section describes electrical modulators used for the various types of servomotors described in Sections 3–5. The term "modulator," as defined earlier, designates components employed for conversion of the command signal to appropriate means to modulate the power flow to the servomotor (see Fig. 1). Modulators for the various types of electrical servomotors differ significantly.

6.1 Direct-Current Motor Modulators

Modulators used in servoactuator applications with dc motors usually contain two stages of amplification: a first-stage voltage amplified followed by a second-stage power amplifier. Voltage amplifiers are generally quite linear in performance. Power amplifiers are used in two different configurations, type T and type H, as shown in Fig. 18. The type T configuration employs two power sources and only two power transistors. The type H configuration uses only one power source but four transistors. The type T configuration lends itself readily to current feedback schemes. However, the type H is more commonly used because of a single power source requirement. Type H configurations can be operated in two different modes, resulting in bipolar and unipolar drives.

Linear Amplifiers

Linear amplifiers that are dc amplifiers used in the output stage (as H or T configuration) provide gains typically in the range of 2–10. At power levels above 200 W, these amplifiers require external cooling (e.g., fans) to overcome excessive heat generation. This problem is particularly severe in high-performance servo applications that typically have low-impedance rotors operating at low-speed and high-torque conditions. Sometimes this problem is overcome by incorporating a dual-mode current-limiting device which permits high currents for short durations (for overcoming inertia) and then imposes a lower limit for longer durations.

Linear amplifiers are generally configured as voltage amplifiers, but in some applications a current source configuration is employed. Reference 5 discusses the details of voltage and current source configurations, the associated mathematical models, and the influence of the two configurations on the dynamic response of the motor.

Switching amplifiers overcome heat generation problems by switching the voltage on and off at high frequencies. This switching limits the maximum allowable inductance and results in shorter time constants and increased bandwidth. One drawback is that the RFI noise level generated is much higher than with linear amplifiers. There are predominantly two types of switching used in servoactuator applications: (1) PWM and (2) PFM. PWM amplifiers are most commonly used.

PWM Amplifiers

In PWM amplifiers, the voltage applied to the servomotor is varied by changing the pulse width of a high, constant-frequency (typically 10-kHz) train of pulses. Figure 19 shows a schematic of a PWM amplifier. Figure 20*a* shows a photograph of a typical PWM amplifier

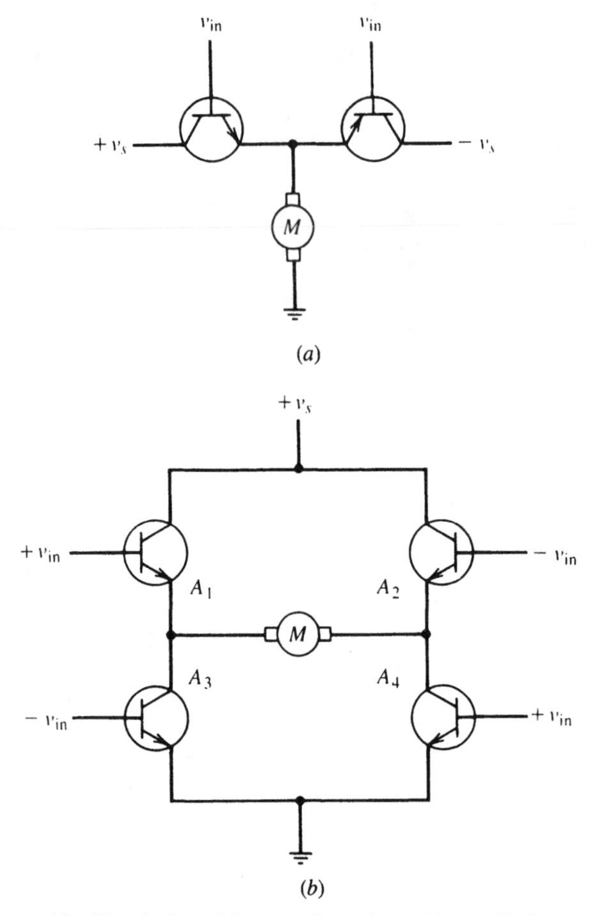

Figure 18 Electrical modulator configurations; (*a*) type T; (*b*) type H.

and Fig. 20*b* shows the associated circuit diagram. Figure 21 shows the firing sequence for unipolar and bipolar configurations.

Mathematical Model—Bipolar Drive PWM Amplifier

The average voltage v_{ma} applied over one cycle of the PWM signal is given by (see Ref. 9, Section 4.5.3)

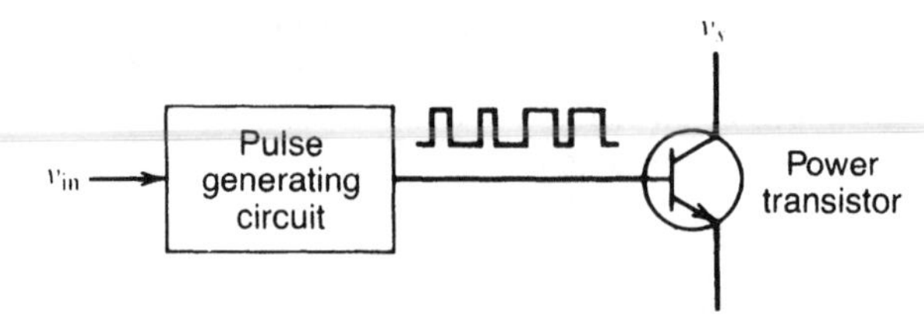

Figure 19 PWM amplifier configuration.

(*a*)

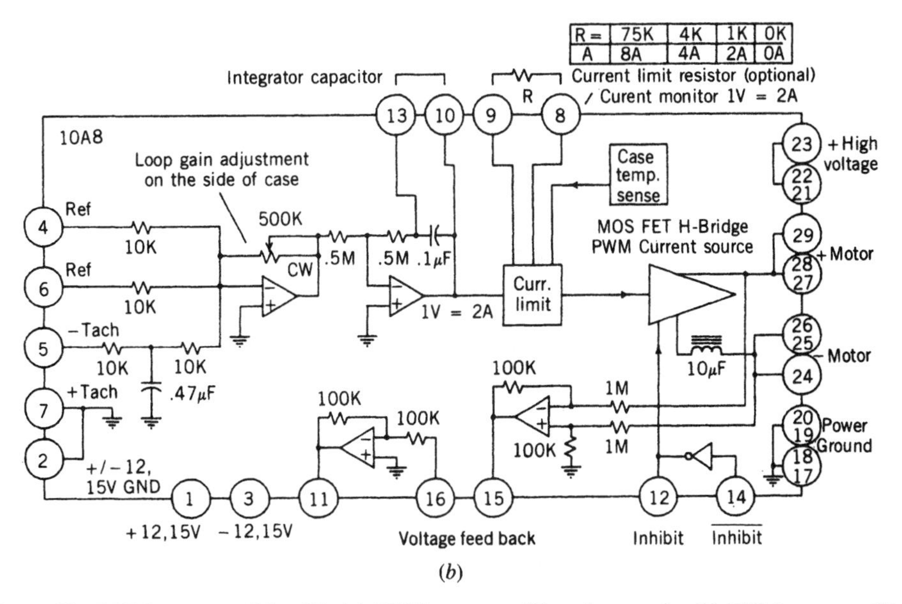

(*b*)

Figure 20 PWM servoamplifier IC: (*a*) PWM servoamplifier photograph; (*b*) PWM servoamplifier circuit diagram. (Courtesy of Advanced Motion Controls, Van Nuys, CA.)

$$v_{ma} = \frac{2v_s t_a}{t_p} - v_s \tag{34}$$

and the change in current over one cycle is given by

$$\Delta i = \frac{2v_s t_a}{L}\left(1 - \frac{t_a}{t_p}\right) \tag{35}$$

where i = current

L = inductance of armature winding

t_p = time period of PWM signal (see Fig. 21)

t_a = actuation time = $(t_p/2) - t_o - t_p v_{in}/v_c$ (see Fig. 21)

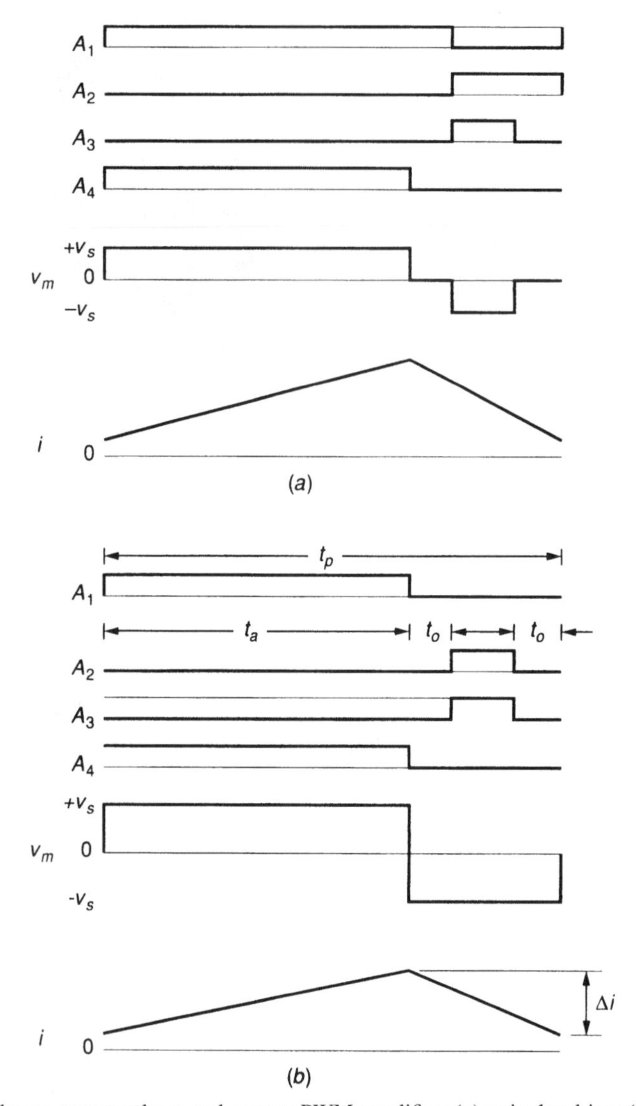

Figure 21 Phases, motor voltage and current PWM amplifier: (*a*) unipolar drive; (*b*) bipolar drive.

t_o = time delay (see Fig. 21)
v_c = peak-to-peak amplitude of a triangular wave voltage signal of time period t_p
v_{in} = input voltage signal applied to amplifier
v_s = supply voltage

Equations (34) and (35) are valid for operating conditions where the instantaneous motor current i remains positive throughout a cycle. The actuation time t_a is determined from the pulse generation circuit characteristics in conjunction with the input signal.[9]

Mathematical Model—Unipolar Drive PWM Amplifier

The average voltage and current amplitudes for a PWM amplifier in a unipolar drive mode are given by

$$v_{ma} = \frac{t_a v_s}{t_p} \tag{36}$$

and

$$\Delta i = \frac{v_s t_a}{L}\left(1 + \frac{t_s}{t_p}\right) \tag{37}$$

For operating conditions where the instantaneous motor current becomes zero for a part of the cycle, the mathematical models can be found in Refs. 5, 20, 23, and 24.

6.2 Stepper Motor Modulators

Modulators for stepper motors are also called "drives." There are three types: oscillator-translators, indexers, and microstepping controls. Oscillators provide a variable-frequency pulse train to the translator. The function of the translator is to convert the pulse train into appropriate signals to operate the power transistors, thereby directing power to the stepper motor. The variable-frequency (3000–15,000 pulses per second) capability of the oscillators provides for accurate manual control of speed. Similarly, a pulse generator in conjunction with a translator provides manual position control. Figure 22 shows a typical oscillator-translator package. Most commercial units also provide for half-stepping and electronic damping.

Indexers are generally programmable microprocessors that perform the functions of the oscillator-translator package, in addition to providing features such as numerical processing, programming, and communications (RS232 and RS274) with computers. The software controls provide such features as setting upper and lower limits of stepper motion and controlling slewing rate. Figure 23 shows a typical indexer. Indexers generally operate on 20–100 V dc power.

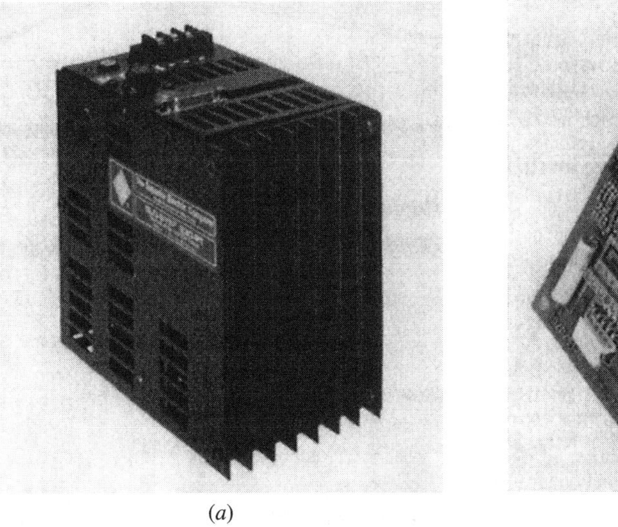

(*a*)

(*b*)

Figure 22 Oscillator-translator modulator for stepper motor control (courtesy of Superior Electric Co., Bristol, CT): (*a*) photograph of modulator; (*b*) Photograph of circuit card.

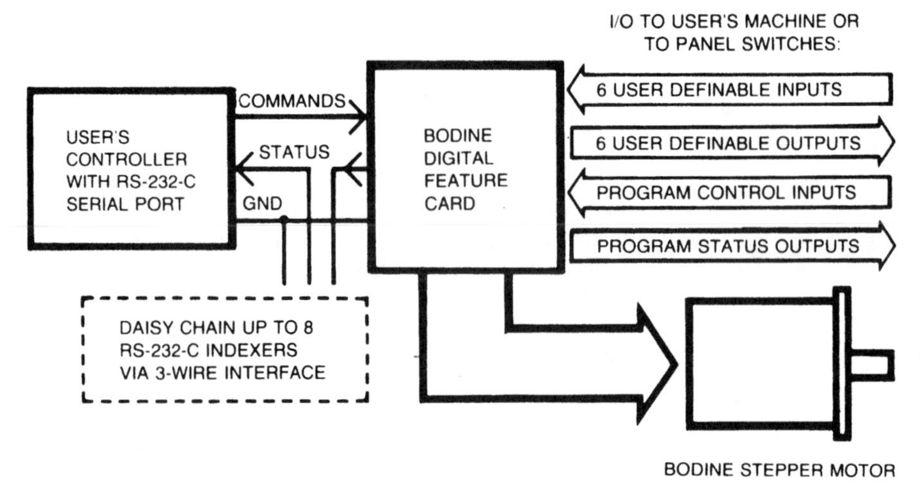

Figure 23 Linear-motion hydraulic servomotors (actuators); indexer-type modulator for stepper motor control. (Courtesy of Bodine Electric Co., Chicago, IL.)

Stepper motors suffer from mechanical resonances (generally in the range of 50–250 steps per minute) due to the excitation resulting from the square shape of the current pulses. Microstepping eliminates this problem by energizing multiple windings simultaneously and by controlling the currents. Hence, modulators for microstepping essentially control the simultaneous currents to the various phases achieving as high as 50,000 microsteps per revolution. Switching frequencies range from 20 to 200 kHz.

7 HYDRAULIC SERVOMOTORS

Hydraulic servomotors or "actuators" convert hydraulic power into mechanical motion. This mechanical motion can be either linear or rotary. Because hydraulic servomotors provide large actuating force or torque capabilities, they are commonly used in heavy-equipment industries. Also, due to their high mechanical stiffness, fast dynamic response, and high power-to-weight ratios, hydraulic servomotors find widespread use in aircraft, missiles, and space vehicles, as well as critical industrial systems where high performance and high-power control are needed.[25]

7.1 Linear-Motion Servomotors

Linear-motion servomotors or actuators provide a translatory motion of a load along a straight line. The motion of the load can be controlled by modulating the flow of hydraulic fluid into or out of the servomotor. There are two basic types of linear-motion servomotor designs: unbalanced and balanced designs, as shown schematically in Fig. 24. The unbalanced design is shorter, while the balanced design has equal extension and return rates when the servomotor is driven by a symmetrical modulator (control valve). High-performance hydraulic and electrohydraulic servosystems require servomotors with low leakage and low friction. Correspondingly, seals are critical elements in servomotor design. Construction details and mounting arrangements are discussed in Refs. 26–28.

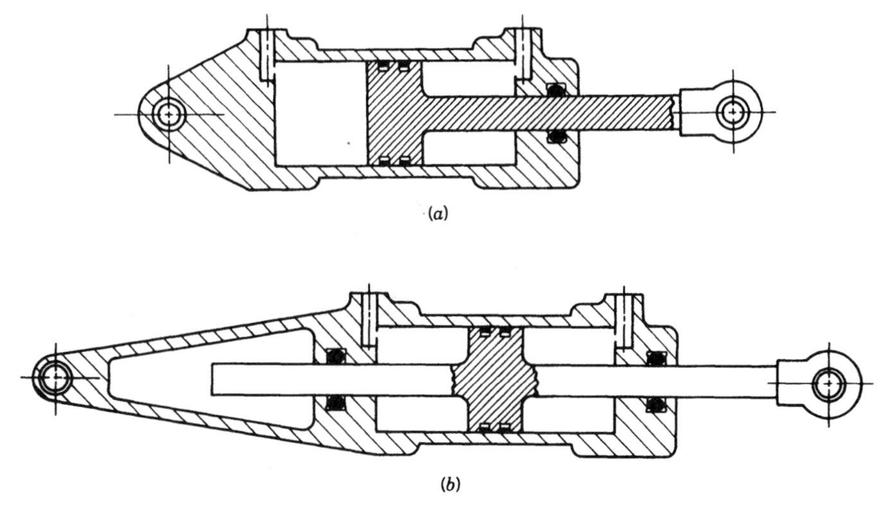

Figure 24 Linear-motion hydraulic servomotors (actuators): (*a*) double-acting unbalanced actuator; (*b*) double-acting balanced actuator.

Since servomotors must move heavy loads quickly and accurately, they should act as very stiff structural members. In a well-designed servomotor, the fluid columns on either side of the piston are the most compliant portions of the structure. The spring rate of one fluid column is given by $K = A\beta/L$ where A is the net column area, L is the column length, and β is the fluid bulk modulus of elasticity. Servomotors used for high-dynamic-performance systems are designed to have the minimum stroke and the shortest permissible connecting passages in order to minimize the volume of fluid under compression. Modulators often are integrated into the servomotor body to minimize lengths of connecting passages (see Section 9.2).

7.2 Rotary-Motion Servomotors

Rotary-motion servomotors are functionally more versatile than their linear counterparts. They are commonly employed even in linear-motion applications through a rack-and-pinion type kinematic conversion. However, linear-motion servomotors also are employed for rotary applications through corresponding kinematic conversion. Rotary actuators can be either simple reversible hydraulic motors of continuous-rotation type or high-performance type with limited angular displacement. Most rotary servomotors used in medium- and high-performance hydraulic and electrohydraulic servosystems are vane (continuous or limited rotation) or piston type (continuous rotation) servomotors.

Rotary-vane servomotors are commonly used in industrial applications where medium to high performance is required and size and weight are not at a high premium. Such applications include earth-moving equipment, agricultural machinery, and materials processing and handling equipment. Figure 25 shows a typical continuous-rotation rotary-vane servomotor capable of operating at speeds up to 4000 rpm (419 rad/s) and pressures up to 2500 psi (1.72×10^7 N·m^2).

In aircraft, missile, and spacecraft applications where high performance and small size and weight are required, piston-type servomotors are commonly used. Figure 26 shows a typical "in-line" piston-type servomotor and Fig. 27 shows a typical "bent-axis" piston-type

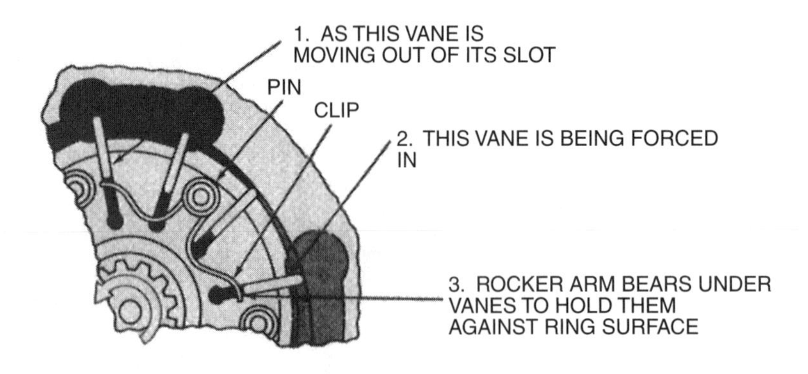

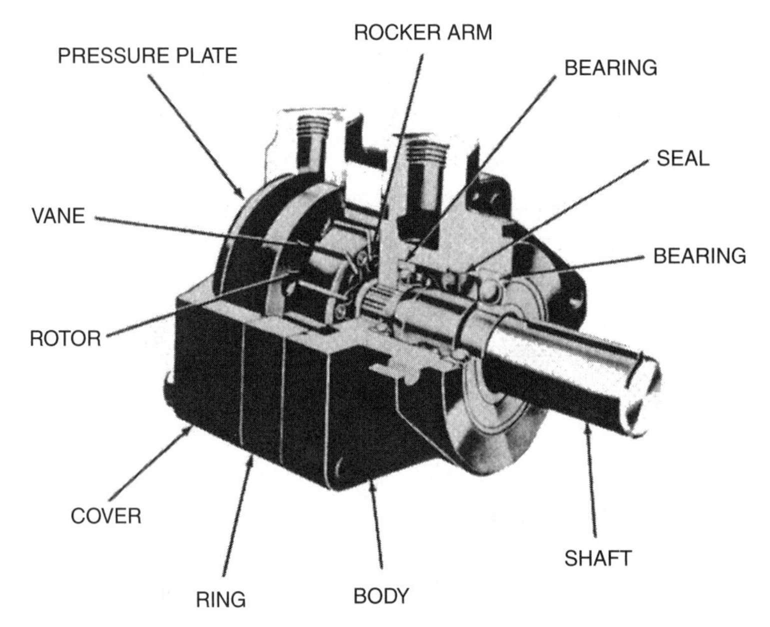

Figure 25 Balanced vane-type hydraulic servomotor. (Courtesy of Vickers, Inc., Troy, MI.)

servomotor. Such servomotors are capable of operating at speeds up to 8000 rpm (838 rad/s) and pressures up to 5000 psi (3.44×10^7 N/m^2).

7.3 Mathematical Models

A hydraulic servomotor (either linear- or rotary-motion type) is a rate-type device. That is, a given flow rate into the servomotor results in a certain velocity (or speed for a rotary servomotor).

Figure 28 shows schematics of linear and rotary servomotors with definitions of the variables. Dynamic models are given in the following descriptions. It is assumed that all structural components are rigid and that external leakage to the environment is negligible. Models that include external leakage effects are presented in Ref. 29.

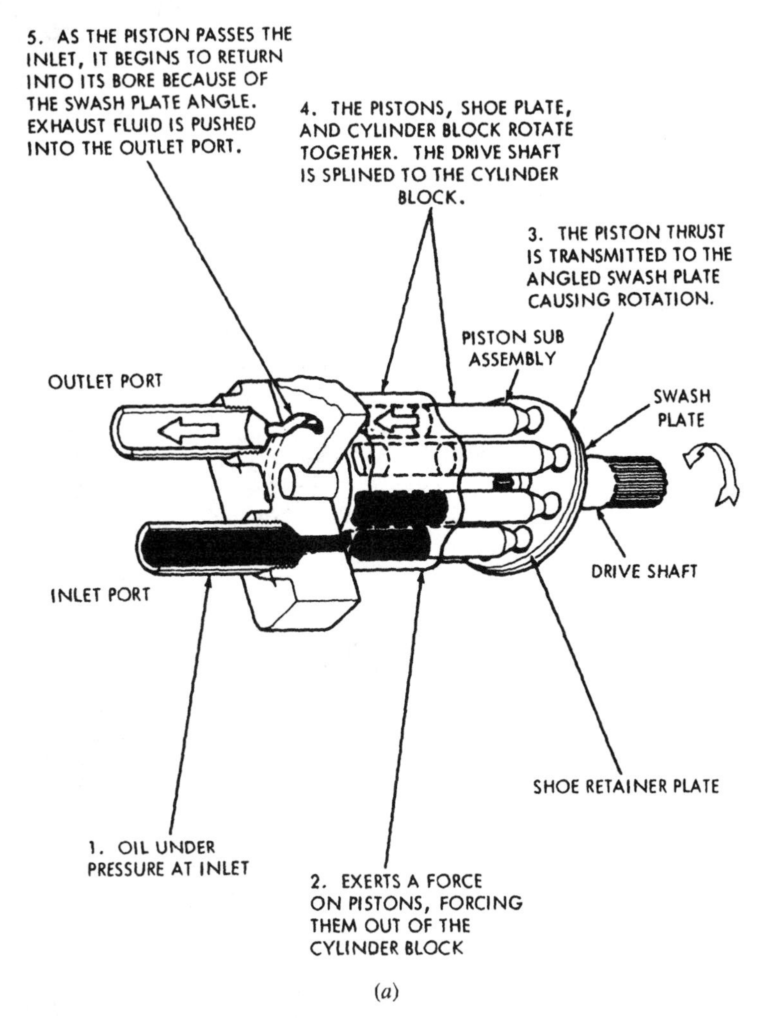

(*a*)

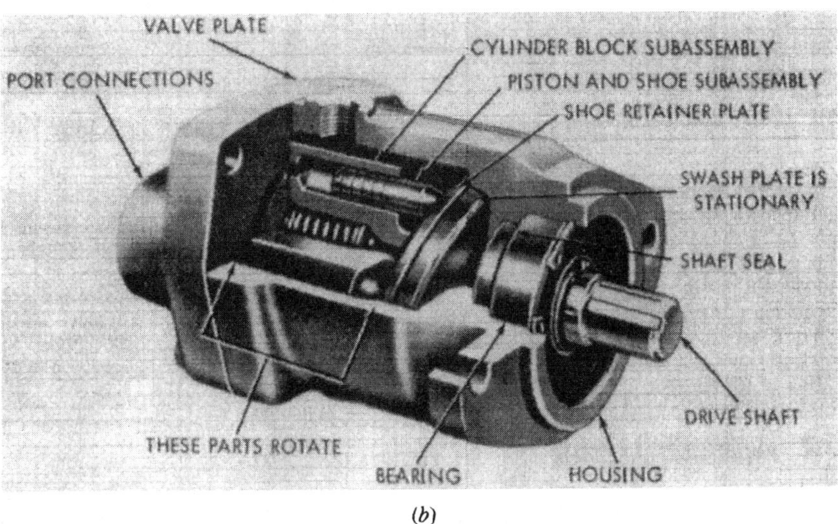

(*b*)

Figure 26 In-line piston servomotor: (*a*) operation of servomotor; (*b*) cutaway of servomotor. (Courtesy of Vickers, Inc., Troy, MI.).

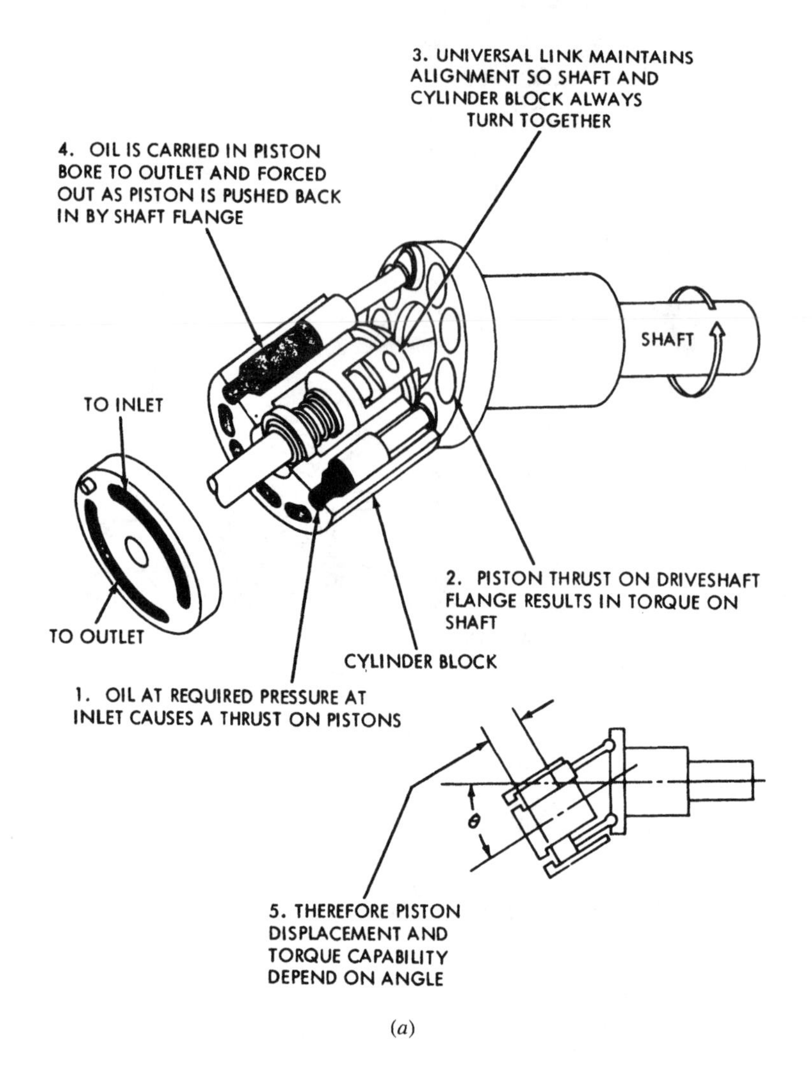

(*a*)

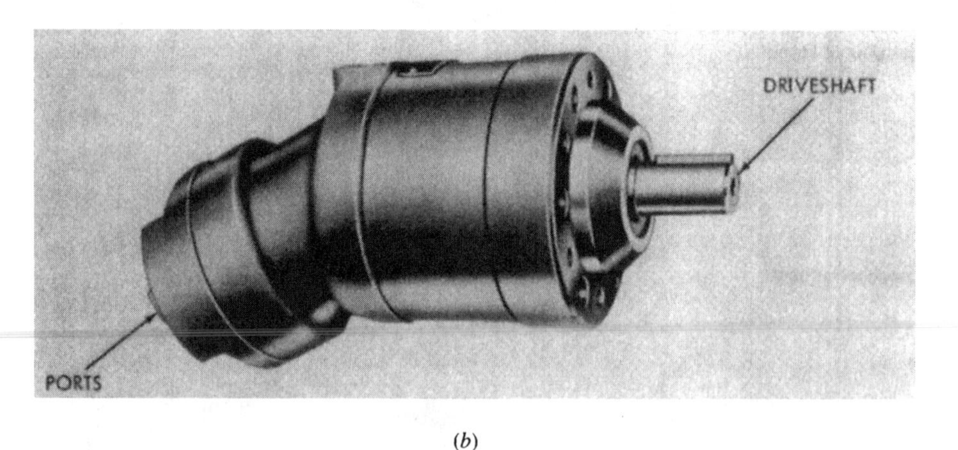

(*b*)

Figure 27 Bent-axis piston servomotor: (*a*) operation of servomotor; (*b*) photograph of servomotor. (Courtesy of Vickers, Inc., Troy, MI.)

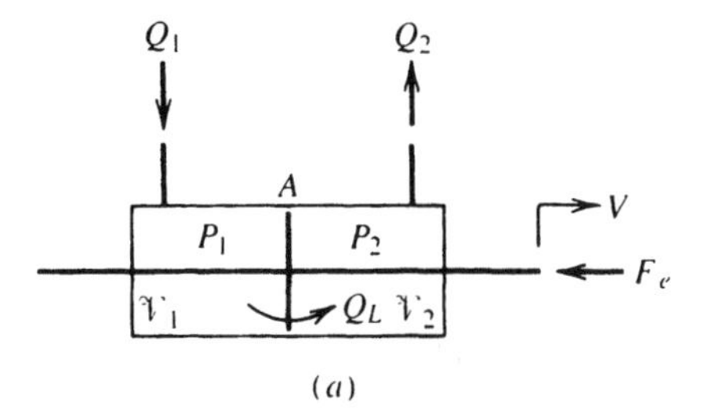

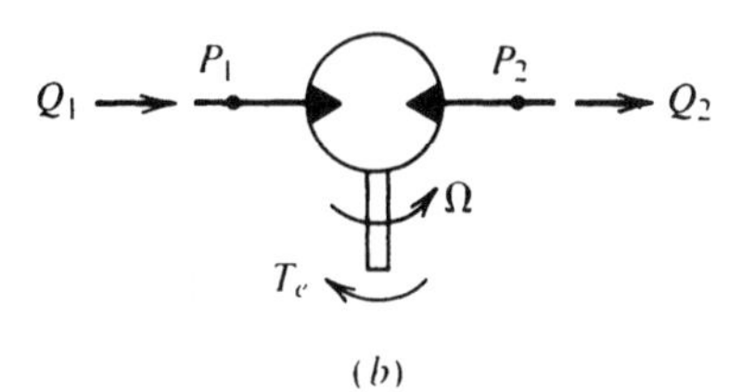

Figure 28 Nomenclature of hydraulic servomotors.

Linear Servomotor (Balanced Design)

Force balance

$$\frac{dV}{dt} = \frac{1}{M_a}(AP_m - B_mV - F_f - F_c - F_e) \tag{38}$$

Chamber continuity

$$\rho_1 Q_1 - \frac{(\rho_1 + \rho_2)}{2} Q_L = \frac{d}{dt}(\rho_1 \mathcal{V}_1) \tag{39}$$

$$\frac{\rho_1 + \rho_2}{2} Q_L - \rho_2 Q_2 = \frac{de}{dt}(\rho_2 \mathcal{V}_2) \tag{40}$$

Equation of state

$$\frac{d\rho_1}{dt} = \frac{\rho_1}{\beta}\frac{dP_1}{dt} \tag{41}$$

$$\frac{d\rho_2}{dt} = \frac{\rho_2}{\beta}\frac{dP_2}{dt} \tag{42}$$

Chamber volume

$$\mathcal{V}_1 = \mathcal{V}_{1i} + AY \tag{43}$$

or

$$\frac{d\mathcal{V}_1}{dt} = AV$$

$$\mathcal{V}_2 = \mathcal{V}_{2i} = -AY \tag{44}$$

or

$$\frac{d\mathcal{V}_2}{dt} = -AV$$

where A = effective area of servomotor piston
B_m = viscous damping coefficient in servomotor
C_2 = leakage coefficient
F_c = force due to Coulomb friction in servomotor
F_e = external load force on servomotor piston
F_f = force due to stiction in servomotor
M_a = mass of moving parts in servomotor
P_1 = pressure at inlet
P_2 = pressure at outlet
$P_m = P_1 - P_2$ = pressure drop across servomotor
Q_1 = flow rate into servomotor
Q_2 = flow rate out of servomotor
$Q_L = C_2 P_m$ = internal leakage flow rate
V = velocity of servomotor piston
$\mathcal{V}_1$ = fluid volume under compression at the inlet
$\mathcal{V}_{1i}$ = fluid volume under compression at inlet for $Y = 0$
$\mathcal{V}_2$ = fluid volume under compression at the outlet
$\mathcal{V}_{2i}$ = fluid volume under compression at outlet for $Y = 0$
Y = linear displacement of servomotor piston from neutral position
β = fluid bulk modulus of elasticity
ρ_1 = fluid mass density at inlet
ρ_2 = fluid mass density at exit

Rotary Servomotor

Torque balance

$$\frac{d\Omega}{dt} = \frac{1}{J_a}(D_m P_m - B_r \Omega - T_f - T_c - T_e) \tag{45}$$

Chamber continuity

$$\rho_1 Q_1 - \frac{(\rho_1 + \rho_2)}{2} Q_L = \frac{d}{dt}(\rho_1 \mathcal{V}_1) \tag{46}$$

$$\frac{(\rho_1 + \rho_2)}{2} Q_L - \rho_2 Q_2 = \frac{d}{dt}(\rho_2 \mathcal{V}_2) \tag{47}$$

Equation of state

$$\frac{d\rho_1}{dt} = \frac{\rho_1}{\beta}\frac{dP_1}{dt} \tag{48}$$

$$\frac{d\rho_2}{dt} = \frac{\rho_2}{\beta}\frac{dP_2}{dt} \tag{49}$$

Chamber volume

$$\mathcal{V}_1 = \mathcal{V}_{1i} + D_m \theta \tag{50}$$

or

$$\frac{d\mathcal{V}_1}{dt} = D_m\Omega$$

$$\mathcal{V}_2 = \mathcal{V}_{2i} - D_m\theta \tag{51}$$

or

$$\frac{d\mathcal{V}_2}{dt} = -D_m\Omega$$

where B_r = viscous damping coefficient in servomotor
C_2 = leakage coefficient
D_m = displacement of servomotor
J_a = polar moment of inertia of servomotor moving parts
P_1 = pressure at inlet
P_2 = pressure at outlet
$P_m = P_1 - P_2$ = pressure drop across servomotor
Q_1 = flow rate into servomotor
Q_2 = flow rate out of servomotor
$Q_L = C_2P_m$ = internal leakage flow rate
T_c = torque due to Coulomb friction in servomotor
T_e = external load torque on servomotor
T_f = torque due to stiction in servomotor
$\mathcal{V}_1$ = fluid volume under compression at inlet
$\mathcal{V}_{1i}$ = fluid volume under compression at inlet for $\theta = 0$
$\mathcal{V}_2$ = fluid volume under compression at outlet
$\mathcal{V}_{2i}$ = fluid volume under compression at outlet for $\theta = 0$
β = fluid bulk modulus of elasticity
ρ_1 = fluid mass density at inlet
ρ_2 = fluid mass density at outlet
θ = angular displacement of servomotor
Ω = angular velocity of servomotor

Simplified Mathematical Model

A simplified mathematical model for hydraulic servomotors can be obtained by assuming that (1) stiction and Coulomb friction effects are negligible, (2) internal leakage is negligible, (3) external load force (or torque) is zero, and (4) the fluid is incompressible. The result is

$$\frac{Z(s)}{P_m(s)} = \frac{K}{\tau s + 1} \tag{52}$$

where the parameters are as defined in Table 7.

Table 7 Hydraulic Servomotor Parameters

Z	V	Ω
K	A/B_m	D_m/B_r
τ	M_a/B_m	J_a/B_r

8 HYDRAULIC MODULATORS

Two basic approaches are used to modulate the flow of a high-pressure fluid to a work-producing device (servomotor). First, modulation may be accomplished by varying the displacement of a rotary pump which supplies fluid directly to a rotary servomotor. A variation on this approach is to use a variable-displacement servomotor and a fixed-displacement pump. Such systems, referred to as pump-displacement-controlled (or motor-displacement-controlled) servoactuators, are not as commonly used in high-performance applications as are valve-controlled servoactuators. The reader is referred to Refs. 29–33 for more detailed discussion of pump-displacement-controlled servoactuators. The discussion here is limited to the second modulation approach (i.e., servoactuators that employ servovalves as modulators).

8.1 Servovalve Design and Operation

Modem servovalves employ one or more of several types of metering devices: flapper-nozzle, poppet, spool, sliding plate, rotary "plate," and jet pipe. Servovalves are typically made in one-, two-, or three-stage configurations. Single- and two-stage configurations are the most common. Generally, the flapper-nozzle valve or the jet-pipe valve is used in the first stage of a two-stage servovalve. The spool-type valve is the most commonly used for single-stage servovalves and for the second stage of two-stage servovalves. Servovalves may be of the two-way, three-way, or four-way type.[30] Four-way types are used in most servosystems where bidirectional motion is required. Three-way types can be used where only unidirectional motion is required; bidirectional control can be achieved if a load biasing scheme is used. Depending on the internal design and application, servovalves may provide flow-rate control or pressure control. Special designs are available which employ flow-rate, pressure, or dynamic pressure feedback within the valve.[34,35]

Servovalves may have mechanical, hydraulic, pneumatic, or electrical input. Most servosystems used in high-performance applications today use electrohydraulic servovalves where an electrical input signal is converted to a mechanical motion through a torque motor. In single-stage valves, the torque motor actuates the control valve, which in turn modulates the flow of hydraulic fluid under pressure from a high-pressure source to a linear- or rotary-motion servomotor. In two-stage valves, the torque motor actuates the first-stage (or pilot) valve, which is typically a flapper-nozzle or jet-pipe valve. The hydraulic output from the first stage drives the second stage (typically a spool-type valve), which in turn modulates the flow from the source to the servomotor. Reference 36 is a detailed history of electrohydraulic servomechanisms with special emphasis on electrohydraulic servovalves.

Figures 29 and 30 show the design features of a modern two-stage electrohydraulic servovalve. The double-coil, double-air-gap torque motor is "dry" (i.e., it is in an environmentally sealed compartment isolated from hydraulic fluid by the flexure tube). The first-stage or pilot valve is a symmetrical, double-nozzle (four-way) flapper-nozzle valve. The flapper is attached to the upper (free) end of the flexure tube. The second stage is a spool valve that slides in a bushing with mating rectangular slots formed by electric discharge machining. The spool-bushing tolerance is held to 0.5 μm. Mechanical force feedback from the second-stage spool to the torque motor is provided by a cantilever spring attached to the flapper at the upper end and to the spool through a ball joint.

Figure 31 is a cross section of a two-stage electrohydraulic servovalve that employs a jet-pipe valve as the first stage. Otherwise the valve is virtually the same in design and operation as the valve shown in Figs. 29 and 30. The jet-pipe valve can pass contamination particles as large as 200 μm, whereas the flapper-nozzle valve can only pass 50-μm particles. Good fluid filtration can negate the importance of these differences.

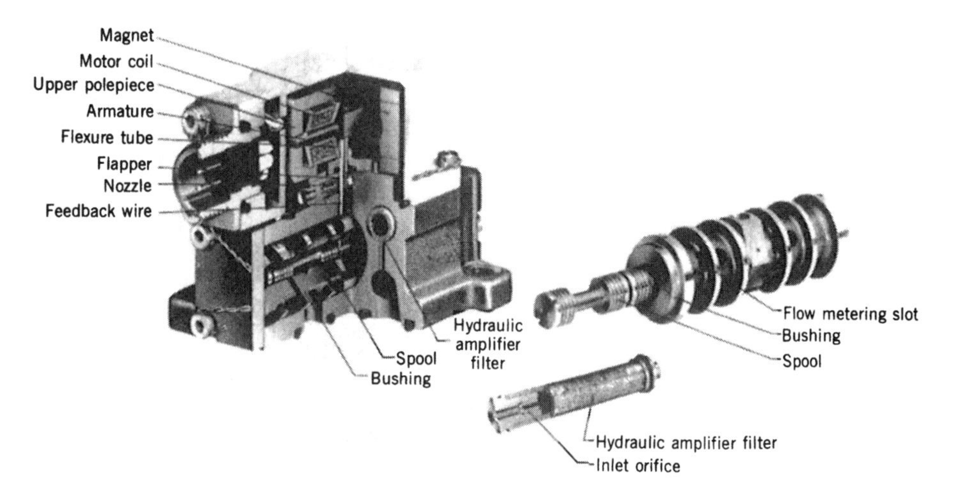

Figure 29 Design features of a two-stage electrohydraulic servovalve. (Courtesy of Moog, Inc., East Aurora, NY.)

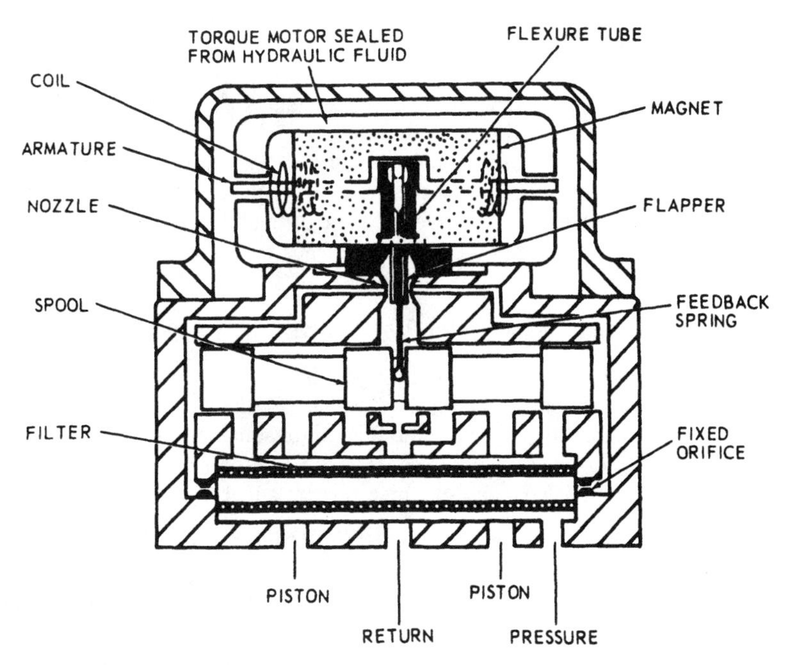

Figure 30 Cross section of two-stage electrohydraulic servovalve with a flapper-nozzle first stage and spool second stage. (Courtesy of Moog, Inc., East Aurora, NY.)

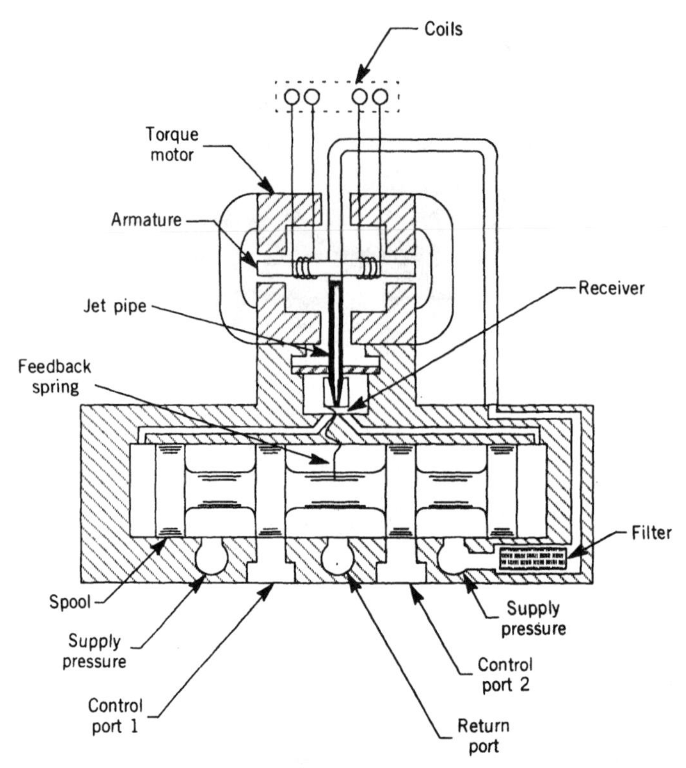

Figure 31 Cross section of a two-stage electrohydraulic servovalve with a jet-pipe first stage and spool second state. (Courtesy of Abex Corporation, Aerospace Division, Oxnard, CA.)

A cutaway diagram of a single-stage swing-plate servovalve is shown in Fig. 16.32*a*; a cross section of the valve is shown in Fig. 16.32*b*. This is an industrial valve that has a high dynamic response (natural frequency about four times that of the two-stage servovalves mentioned earlier). However, the swing-plate valve is considerably heavier than the two-stage aerospace-type valves in Figs. 30 and 31.

Table 8 shows typical performance capabilities of various types of servovalves.

8.2 Mathematical Model of a Spool-Type Valve

Spool-type valves are the most popular due to their ease of construction. They are also easier to analyze than other types of valves. Figure 33 shows a typical spool-valve configuration and defines the important variables and parameters. The valve has three "energy ports" where energy or power flows from or to the environment of the valve. Correspondingly, the valve can be modeled using the three mathematical equations given in functional form as follows:

$$Q_m = f(x, P_m, P_s) \quad (53)$$

$$Q_s = f(x, P_m, P_s) \quad (54)$$

$$F_v = f(x, P_m, P_s) \quad (55)$$

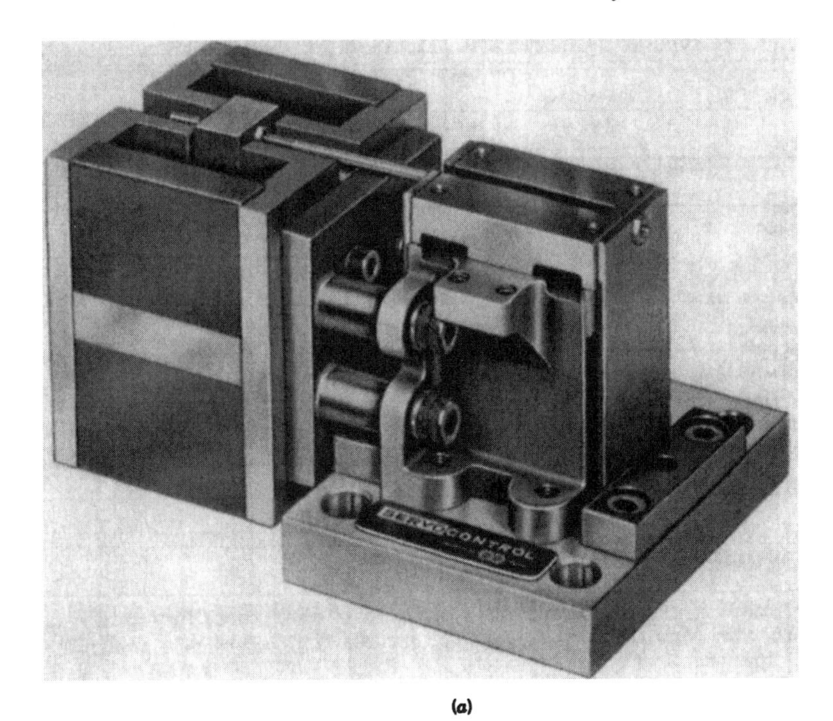

(a)

Connecting Rod
Valve "Swing-Plate" (a)
Return Port #2
Coils
Valve "Swing-Plate" Supports (c)
Valve Body (b)
Permanent Magnets
Electrical Leads
Armature (d)
Armature Torsion Shaft (e)
Cylinder Port #3
Cylinder Port #4
Pressure Port #1

(b)

Figure 32 Cross section of a single-stage, "swing-plate" electrohydraulic servovalve. (Courtesy of The Oilgear Company, Milwaukee, WI.)

Table 8 Performance Specifications of Servovalves

Valve Type	Maximum Working Pressure (psi)	Maximum Flow at 1000 psi Pressure Drop (gpm)	Frequenty at 90° Lag (Hz)	Hysteresis (%)	Resolution (%)
Spool					
One-stage	5000	3500	200	0.1	0.01
Two-stage	7000	1000	200	1	0.01
Three-stage	4500	300	200	1	0.01
Flapper-nozzle/spool					
One- and two-stage	5000	1000	500	0.2	0.01
Three-stage	5000	1000	500	1	0.01
Jet pipe/spool					
One- and two-stage	4500	300	500	2	0.1
Three-stage	4500	300	200	2	0.1
Sliding plate					
One- and two-stage	300	40	150	3	0.1

Source: From Ref. 25.

Equation (53) gives the pressure–flow–displacement characteristics of the valve. These characteristics are needed in the dynamic analysis of a servoactuator which employs the valve. Equation (54) is used to compute the required flow rate from the source and will not be considered further here. Equation (55) is used to calculate the force required to move the spool (e.g., force output requirement of the torque motor in the case of an electrohydraulic servovalve).

The steady-state pressure–flow–displacement characteristics of the spool valve are characterized by the nonlinear orifice equation.[30,37]

$$Q_m = C_d w(x + U)\sqrt{\frac{P_s - P_m}{\rho}} \tag{56}$$

where C_d = effective coefficient of discharge
P_s = supply pressure
P_m = pressure drop across the servomotor (see Fig. 33)

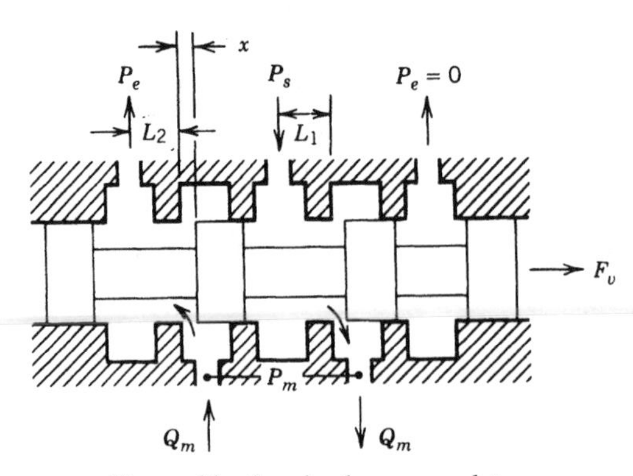

Figure 33 Spool valve nomenclature.

Q_m = flow rate to the servomotor (see Fig. 33)
U = underlap of spool with respect to sleeve (see Ref. 30); $U = 0$ for an "idealized" valve
w = circumferential width of metering ports in the valve
x = displacement of the spool from its neutral position
ρ = mass density of fluid

This model assumes that the flow rates through the metering orifices are steady, the fluid is incompressible, and the valve exhaust pressure $P_e = 0$. A linearized form of this model facilitates the dynamic analysis of a servoactuator containing the valve. The nonlinear model may be linearized by considering small changes of all variables about an initial steady-state operating point, with the result

$$\Delta Q_m = K_1 \, \Delta x - C_1 \, \Delta P_m \tag{57}$$

where

$$K_1 = \left.\frac{\partial Q_m}{\partial x}\right|_{P_{m0},x_0} = C_d w \sqrt{\frac{P_s - P_{m0}}{\rho}} \tag{58}$$

$$C_1 = -\left.\frac{\partial Q_m}{\partial P_m}\right|_{x_0,P_{m0}} = \frac{C_d w x_0}{2\sqrt{\rho(P_s - P_{m0})}} \tag{59}$$

The terms ΔQ_m, Δx, and ΔP_m represent small changes of the corresponding variables about the steady-state operating point x_0, P_{m0}. The constants K_1 and C_1 are evaluated at the operating point. These expressions assume that the valve port shape does not vary with displacement.

The static and dynamic behavior of the valve spool [Eq. (55)] can be modeled by considering the forces which act on the spool.[30] These forces include the externally imposed force (input) as well as steady and unsteady flow forces resulting from flow through the orifices. Additional forces that may be present include the viscous damping between the spool and the sleeve and any mechanical spring forces acting on the spool (not shown in Fig. 33). The force balance equation for the spool shown in Fig. 33 is

$$F_v = m_s \ddot{x} + \frac{\mu A_s}{h} \dot{x} + \left[C_d w \sqrt{2\rho \left(\frac{P_s - P_m}{2} \right)} (L_1 - L_2) \right] \dot{x} + \left[2 C_d C_v w \left(\frac{P_s - P_m}{2} \right) \cos \theta_j \right] x \tag{60}$$

where A_s = net shear area of spool
C_d = metering orifice discharge coefficient
C_v = metering orifice velocity coefficient (see Ref. 30)
F_v = external force on spool (e.g., imposed by torque motor)
h = radial clearance between spool and sleeve (valve body)
L_1 = length of fluid column to be accelerated at inlet (see Fig. 33)
L_2 = length of fluid column to be accelerated at outlet (see Fig. 33)
m_s = mass of spool
P_m = pressure drop across servomotor
P_s = supply pressure
w = circumferential width of metering ports in valve
x = displacement of spool
$\dot{x}$ = velocity of spool

$\ddot{x}$ = acceleration of spool
ρ = mass density of fluid
μ = absolute viscosity of fluid
θ_j = effective angle of fluid jet (see Ref. 30)

The fourth term on the right-hand side of Eq. (60) is the steady flow-induced force and the third term on the right-hand side is the unsteady flow-induced force. The steady flow force is a "springlike" force that always opposes the motion of the spool and hence is a stabilizing force. The unsteady flow force is a "dampinglike" force that changes its direction of action depending on the flow direction, and hence it can be a stabilizing or destabilizing force. The valve is dynamically stable if $L_1 - L_2 > 0$. A more complete discussion of the dynamic modeling of the valve spool is given in Ref. 30.

8.3 Mathematical Models for an Electrohydraulic Servovalve

The steady-state pressure–flow characteristics for an electrohydraulic servovalve of the type shown in Fig. 30 are identical to those of the spool-type valve in the previous section except the input x is replaced by the current I. That is, in the steady state, the motion of the spool in the electrohydraulic servovalve is directly proportional to the current input to the valve. The steady-state pressure–flow–current characteristics for the "idealized" electrohydraulic servovalve (e.g., Fig. 30; spool matched perfectly with sleeve such that effective underlap $U = 0$) are given by the equation

$$Q_m = K_v I \sqrt{\frac{P_s - P_m}{\rho}} \tag{61}$$

where K_v = a size factor
I = current input to servovalve
P_s = supply pressure
P_m = pressure drop across the servomotor
Q_m = control flow rate to the servomotor
ρ = mass density of fluid

This model assumes that the exhaust pressure $P_e = 0$. Equation (61) may be linearized for operation about an initial steady-state operating point with the result

$$\Delta Q_m = K_1 \, \Delta I - C_1 \, \Delta P_m \tag{62}$$

where

$$K_1 = \left.\frac{\partial Q_m}{\partial I}\right|_{P_{m0},I_0} \quad \text{(flow sensitivity)} \tag{63}$$

$$C_1 = -\left.\frac{\partial Q_m}{\partial P_m}\right|_{I_0,P_{m0}} \quad \text{(flow–pressure sensitivity)} \tag{64}$$

Equation (62) is valid for cases when $U \neq 0$ as well. The terms ΔQ_m, ΔI, and ΔP_m represent small changes of the corresponding variables about the initial steady-state operating point I_0, P_{m0}. The constants K_1 and C_1 are evaluated at the operating point.

Typical steady-state pressure–flow–current characteristics for an "idealized" electrohydraulic servovalve that employs a spool-valve second stage [governed by Eq. (61)] are shown in Fig. 34. Characteristics for other types of electrohydraulic servovalves are given in Refs. 34 and 35.

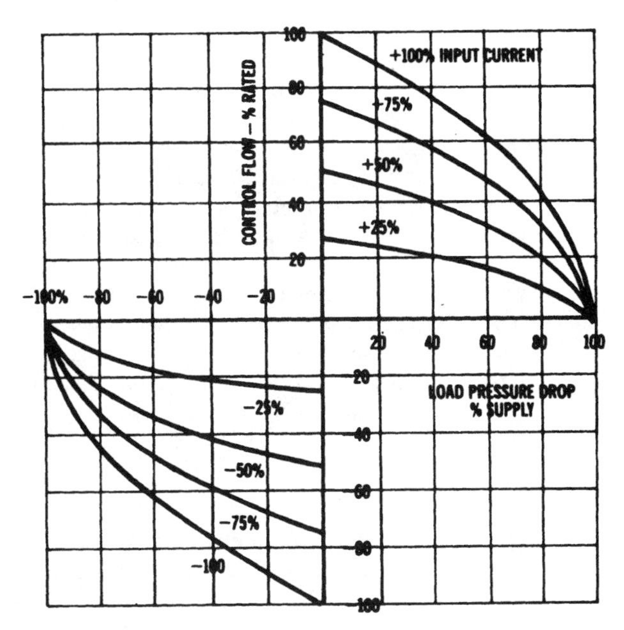

Figure 34 Typical steady-state pressure–flow characteristics of an electrohydraulic servovalve. (Courtesy of Moog Inc., East Aurora, NY.)

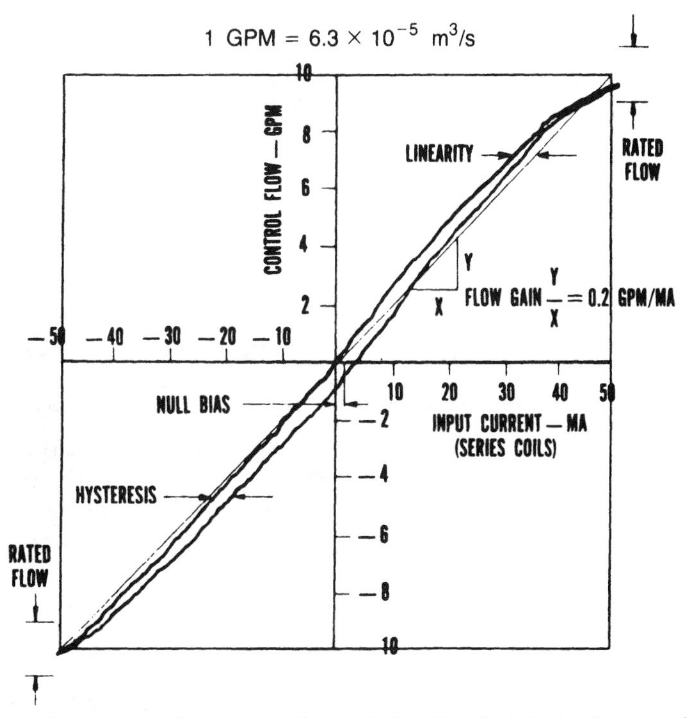

Figure 35 Typical steady-state flow–current characteristics of an electrohydraulic servovalve. (Courtesy of Moog, Inc., East Aurora, NY.)

Another important characteristic of an electrohydraulic servovalve is its hysteresis due to the characteristics of the permanent magnets in the torque motor. The hysteresis characteristic is determined from a measurement of the output flow rate as a function of the input current for a constant (usually zero) pressure drop across the valve (load pressure drop). A typical hysteresis characteristic is shown in Fig. 35. The slope of the flow–current curve is the "flow sensitivity" of the valve [i.e., K_1 in Eq. (63)].

It is often convenient in the dynamic analysis of servoactuators to have an approximate dynamic model for the servovalve. Experience has shown that linearized transfer functions based on empirical approximations from measured servovalve responses are adequate for most system designs. Reference 39 outlines considerations underlying the determination of approximate transfer function models for electrohydraulic servovalves. Figure 36 shows typical frequency response plots for an electrohydraulic flow control servovalve, along with approximate transfer functions. For a frequency range of 0–50 Hz, the following first-order expression has been found to be adequate for two-stage electrohydraulic servovalves:

$$\frac{\Delta Q_m(s)}{\Delta I(s)} = \frac{K_1}{\tau s + 1} \tag{65}$$

where $Q_m(s)$ = Laplace transform of control flow rate to the servomotor
$I(s)$ = Laplace transform of the current input to the servovalve
K_1 = servovalve flow sensitivity at $P_m = 0$, $I = 0$
τ = apparent servovalve time constant (s)

Typical time constants for electrohydraulic flow control servovalves are given in Table 9.

If a good approximation is desired over a wider frequency range, the following second-order model may be preferred:

$$\frac{\Delta Q_m(s)}{\Delta I(s)} = \frac{K_1}{(s/\omega_n)^2 + (2\zeta/\omega_n)s + 1} \tag{66}$$

where $\omega_n = 2\pi f_n$ = apparent natural frequency (rad/s)
ζ = apparent damping ratio (dimensionless)

Typical values of f_n and ζ for two-stage electrohydraulic flow control servovalves are given in Table 9.

9 ELECTROMECHANICAL AND ELECTROHYDRAULIC SERVOSYSTEMS

9.1 Typical Configurations of Electromechanical Servosystems

An electrical servomotor may be combined with an electrical or electronic modulator to form an electromechanical servoactuator. The addition of a feedback transducer forms a servosystem. Figure 2 shows an electromechanical linear-motion servosystem which incorporates a rotary brushless dc servomotor and tachometer feedback; the electronic modulator is not shown.

9.2 Typical Configurations of Electrohydraulic Servosystems

A servoactuator comprising an electrohydraulic servovalve and a servomotor may be combined with an electronic servoamplifier (or modulator) and an appropriate feedback transducer to form a high-performance servosystem. Schematic diagrams of three typical electrohydraulic servosystems are shown in Fig. 37.

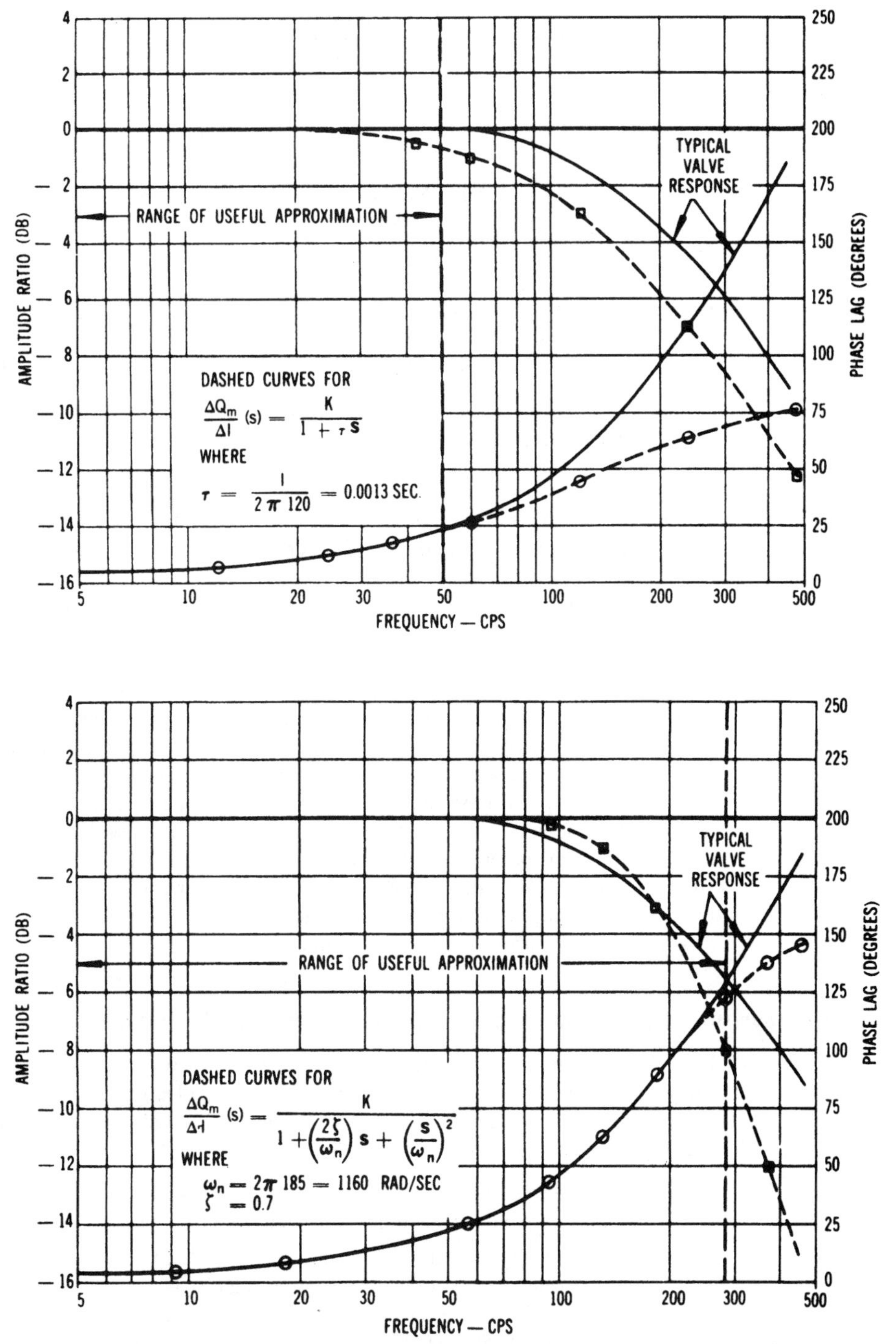

Figure 36 Typical dynamic behavior of electrohydraulic servovalves. (From Ref. 39.)

Table 9 Typical Dynamic Characteristics of Two-Stage Electrohydraulic Flow Control Servovalves

		Approximate Dynamics, 3000 psi, 100°F, Peak-to-Peak Input at 50% Rated Current		
		First Order,	Second Order	
Flow Control Servovalve	Maximum Flow Capacity at 3000 psi (gpm)	τ (s)	f_n (cps)	ζ
A	2	0.0013	240	0.5
B	6	0.0015	200	0.5
C	12	0.0020	160	0.55
D	18	0.0023	140	0.6
E	30	0.0029	110	0.65

Source: From Ref. 39.

Figure 38 shows a photograph and a cross section of a high-performance servosystem which incorporates an electrohydraulic servovalve, an axial-piston servomotor, and a tachometer. The servoamplifier is not shown. Direct manifold mounting of the servovalve results in a small compressed oil volume and therefore high torsional stiffness and fast dynamic response.

Digital control is becoming an important technique for producing near-optimum performance from electrohydraulic servosystems. Figure 39 shows a block diagram and a cutaway of a fully integrated digital electrohydraulic position control system.[40] A two-stage electrohydraulic servovalve drives a linear-motion servomotor. A microcomputer and other electronics integrated within the servovalve housing drive the valve and provide signal conditioning for the position feedback transducer. A ferromagnetic digital position measurement transducer is mounted within the servomotor housing. The feedback management technique uses the microcomputer to model the system and provide digital velocity and acceleration signals without the use of separate sensors (i.e., a digital observer is used). Eight gains are required for accurate and smooth control, and these are calculated within the digital controller to generate the control signal to the torque motor.

9.3 Comparison of Electromechanical and Electrohydraulic Servosystems

Precision motion and force control can be achieved using either electromechanical or electrohydraulic servosystems. The actual choice between electromechanical and electrohydraulic must be based on a number of factors and trade-offs. Studies by Moog, Inc. have produced the following conclusions[3]:

1. Brushless motors with high-energy samarium–cobalt magnets make possible lightweight electromechanical actuators having good response and high efficiency.
2. Electromechanical actuation is currently an alternative to electrohydraulic actuation for applications requiring up to approximately 3–4 hp. Higher power electromechanical actuation systems are limited at the present time by the lack of reliable, compact, lightweight electronics.
3. Electrohydraulic actuation has a proven record in a variety of aerospace and industrial applications requiring high power levels.

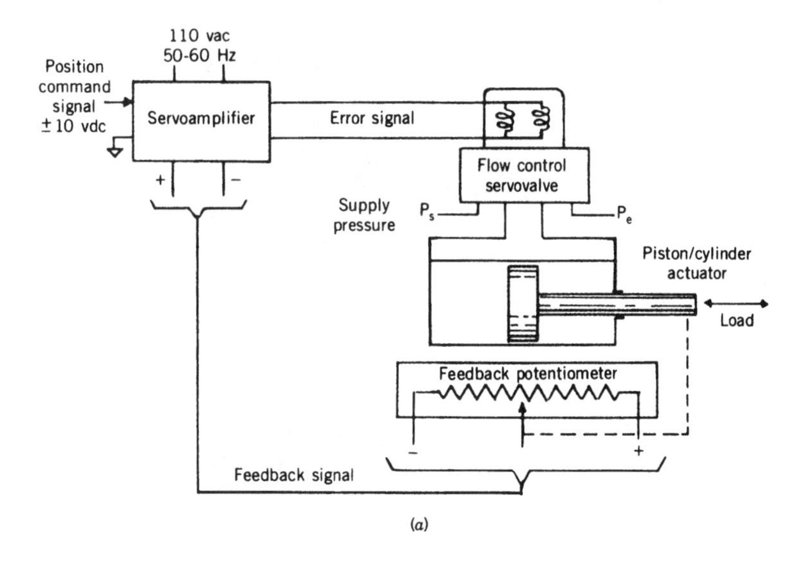

(*a*)

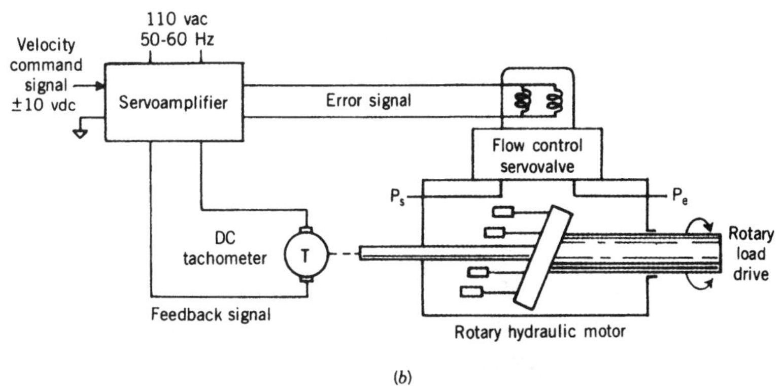

(*b*)

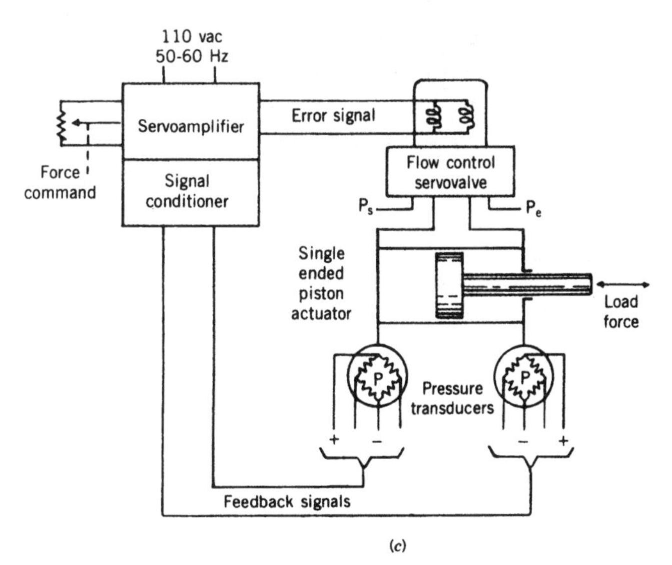

(*c*)

Figure 37 Typical electrohydraulic servosystems: (*a*) linear position servosystem; (*b*) rotary velocity servosystem; (*c*) force servosystem. (Courtesy of Moog, Inc., East Aurora, NY.)

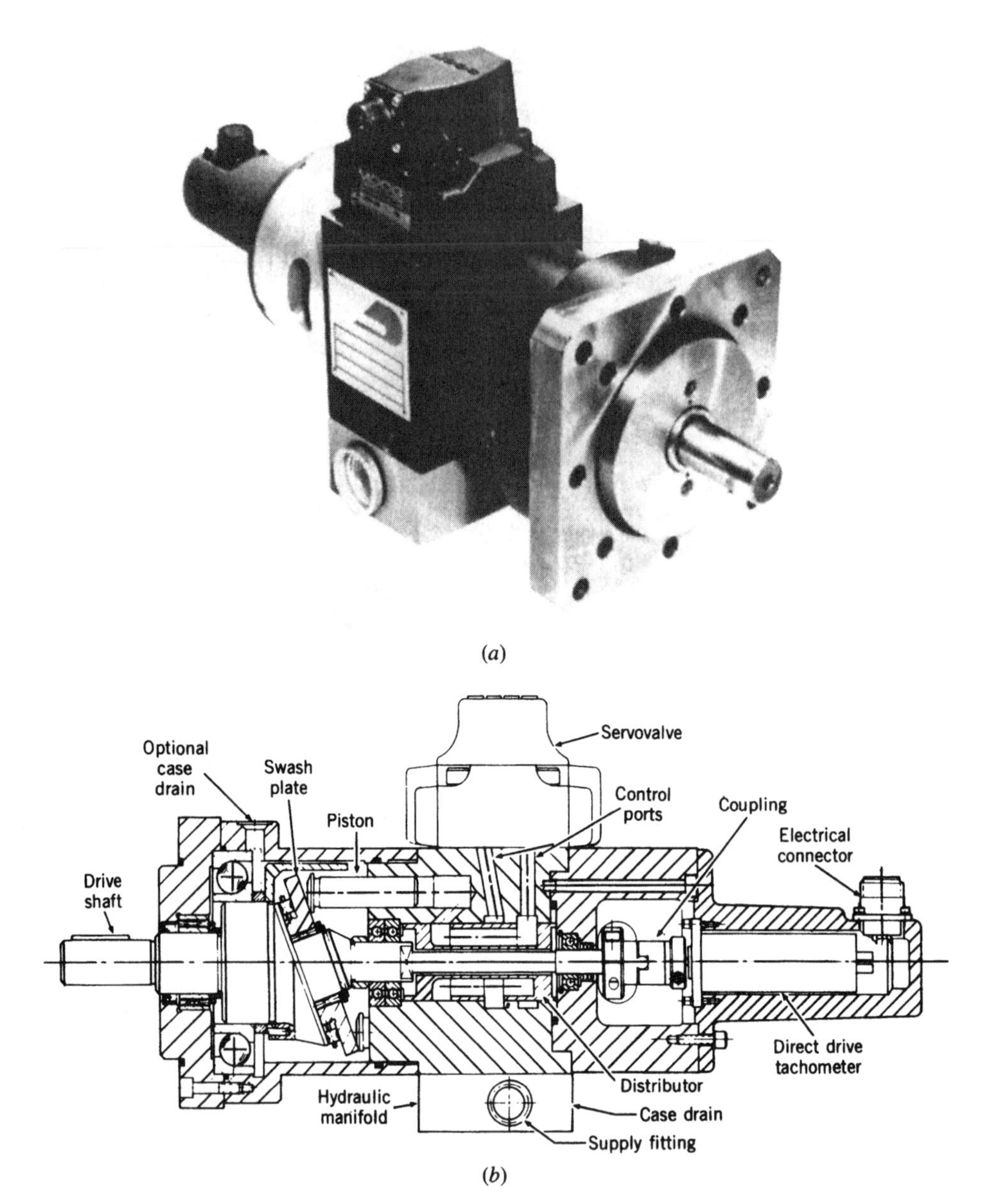

Figure 38 Servosystem with rotary actuator, tachometer, and electrohydraulic servovalve: (*a*) photograph of Moog–Donzelli servosystem; (*b*) cross section of Moog–Donzelli servosystem. (Courtesy of Moog, Inc., East Aurora, NY.)

(*a*)

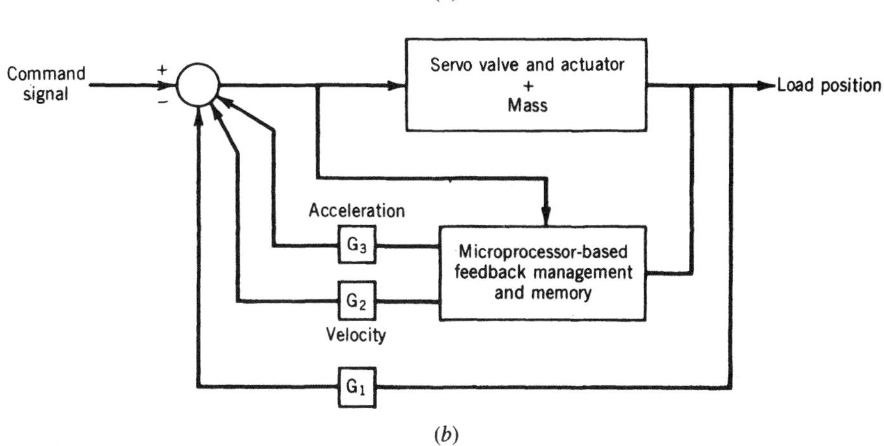

(*b*)

Figure 39 Servosystem with linear actuator, position transducer, and microprocess-based electrohydraulic servovalve: (*a*) cutaway of servosystem; (*b*) block diagram of servosystem. (Courtesy of Vickers, Inc., Troy, MI.)

Table 10 compares advantages and disadvantages of electromechanical and electrohydraulic servosystems.[3]

10 STEADY-STATE AND DYNAMIC BEHAVIOR OF SERVOACTUATORS AND SERVOSYSTEMS

Mathematical models of the components presented in previous sections of this chapter may be combined with a load model to describe the behavior of the servosystem. The system model may be used to study the steady-state and dynamic behavior of the system for various values of system parameters and operating conditions. A number of commercially available digital simulation codes are available for determining the performance in the time or frequency domain.

The combination of a modulator and a servomotor with a load (with or without gearing) forms an open-loop system. Position or velocity feedback may be used to provide a special performance feature (e.g., use of position feedback to convert an open-loop velocity control system to a position control system) or to improve performance.

10.1 Electromechanical Servoactuators

Figure 40 shows schematically a servoactuator comprising a permanent-magnet dc servomotor and an electronic amplifier used to control the velocity of a rotary inertia load. Gearing is used to match the motor torque capability with the load requirements.

Table 10 Comparison of Electromechanical and Electrohydraulic Servosystems

Electromechanical	Electrohydraulic
Advantages	
Lower cost than electrohydraulic	Mature technology
Momentary overdrive capability	Very high reliability
Low quiescent power	Highest actuation performance
Low system weight in low-hp range	Smaller and lighter weight in high-hp range
Packaging flexibility	Continuous power output capability
Conventional or pancake motors	Continuous stall torque capability
Different types of gear reduction	Wide temperature capability
Easy check-out	High vibration and acceleration capability
Single responsibility for servoelectronics and actuators	Proven long-term storability
	Nuclear hardenable
	No EMI generation
	Simple low-power servoelectronics
Disadvantages	
More complex electronics	Usually higher cost
Communication logic for brushless motors	Generally requires more complex power conversion equipment
High-power drive with current limiting	Requires clean hydraulic fluid
Motor inertia-into-stops problems	Quiescent power loss
Overheating with high static loads	
Requires motion reduction/conversion	
Generates EMI	
More difficulty nuclear hardening	
High-power electromechanical actuation not yet proven	

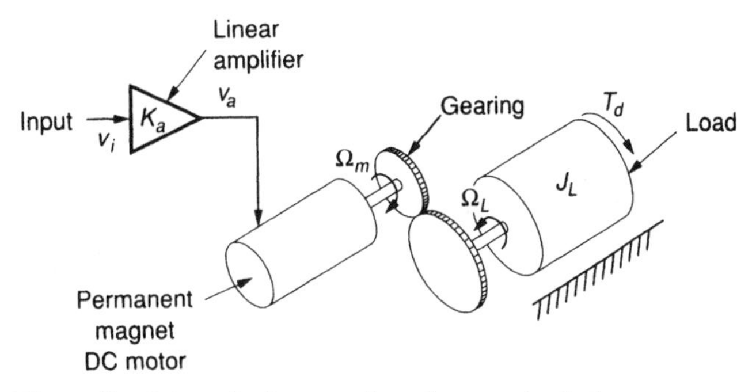

Figure 40 Schematic diagram of an electromechanical servoactuator.

A simplified dynamic model may be derived based on the following assumptions:

1. The amplifier bandwidth is considerably greater than that of the servomotor load portion of the system.
2. The gears have zero backlash and infinite stiffness.
3. The connecting shafts all have infinite stiffness.

Assumptions 2 and 3 eliminate some important dynamic effects that may need to be included in some cases.[41]

Definitions of the critical parameters and variables are as follows:

B_L = viscous damping in the load (N·m·s/rad)
J_L = polar moment of inertia of the load (N·m·s²/rad)
K_a = voltage gain of amplifier (V/V)
K_{ac} = current gain of amplifier (A/V)
n = gear ratio, Ω_m/Ω_L, θ_m/θ_L
T_e = external load torque (N·m)
$T_d(s)$ = Laplace transform of T_d
T_{fL} = Coulomb friction torque in load (N·m)
T_m = total load torque reflected to the motor shaft (N·m)
T_L = total load torque (N·m)
v_a = voltage output of amplifier (V)
v_i = voltage input to amplifier (V)
$V_i(s)$ = Laplace transform of v_i
θ_L = angular displacement of the load (rad)
Ω_L = angular velocity of the load (rad/s)

Other parameters are as defined in Section 3.

Mathematical Model

The basic equations which describe the dynamic behavior of the servoactuator in Fig. 40 are as follows:

Amplifier equation:

$$v_a = K_a v_i \tag{67}$$

Motor equations:

- Electrical:

$$v_a = K_E\Omega_m + L_a \frac{di}{dt} + R_a i \tag{68}$$

- Torque balance:

$$K_T i = J_m \frac{d\Omega_m}{dt} + B_m\Omega_m + T_{fm} + T_m \tag{69}$$

- Gear equation:

$$n = \frac{\Omega_m}{\Omega_L} = \frac{\theta_m}{\theta_L} \tag{70}$$

- Lossless power transfer:

$$T_m\Omega_m = T_L\Omega_L \tag{71}$$

- Load torque balance:

$$T_L = J_L \frac{d\Omega_L}{dt} + B_L\Omega_L + T_{fL} + T_e \tag{72}$$

If the gearing has backlash, the relationship between θ_m and θ_L is as shown in Fig. 41. Similarly, if there is a Coulomb friction load torque, the relationship between the torque and the load velocity is as shown in Fig. 42. These nonlinear effects can be included in the model, but the resulting set of algebraic and differential equations can be studied only through digital simulation. Then the solution must be obtained for a specified input command; a general solution is not possible. It is often sufficient in analysis underlying preliminary design to simplify the model by linearizing the equations and eliminating nonlinear effects such as backlash and Coulomb friction. Such a linearized model can be very useful

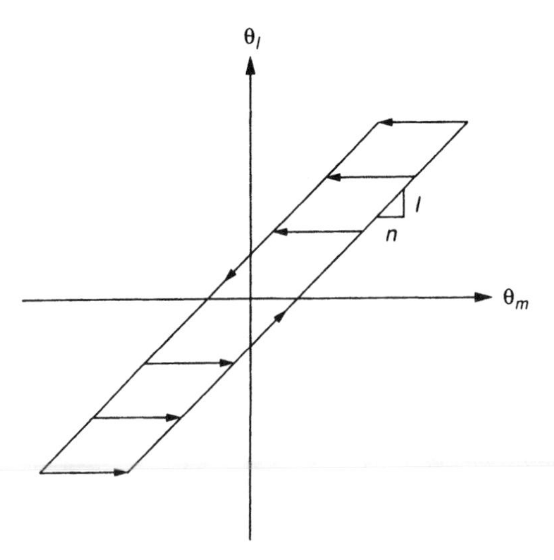

Figure 41 Relationship between θ_m and θ_L due to backlash.

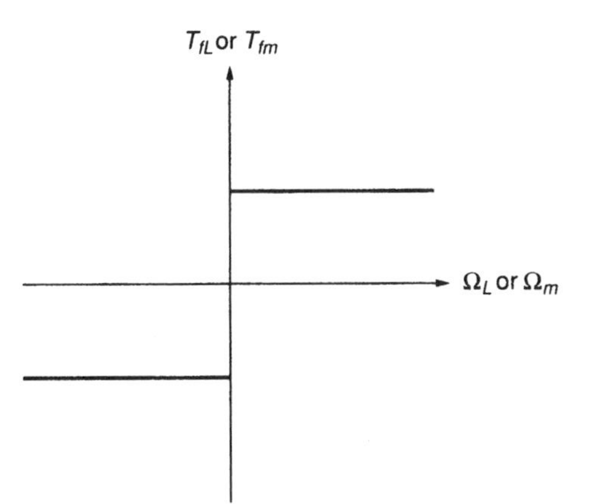

Figure 42 Coulomb friction characteristic.

in gaining an understanding of the general behavior of a system and the sensitivity of the performance to parameter changes.

Equations (67)–(72) can be linearized by assuming small variations of all variables about an initial steady-state operating point. For the system considered here, the linearized dynamic model in the Laplace domain is given by

$$\Delta\Omega_m(s) = G_1(s)\,\Delta V_i(s) - G_2(s)\,\Delta T_e(s) \tag{73}$$

where the transfer functions $G_1(s)$ and $G_2(s)$ are

$$G_1(s) = \frac{K_a K_T}{R_a B_t(\tau_e s + 1)(\tau_m s + 1) + K_T K_E} \tag{74}$$

$$G_2(s) = \frac{(R_a/n(\tau_e s + 1)}{R_a B_t(\tau_e s + 1)(\tau_m s + 1) + K_T K_E} \tag{75}$$

and where

$$\tau_e = \frac{L_a}{R_a} \qquad \tau_m = \frac{J_t}{B_t}$$

$$J_t = J_m + \frac{J_L}{n^2} \qquad B_t = B_m + \frac{B_L}{n^2}$$

A block diagram of Eq. (73) is given in Fig. 43.

In most cases the mechanical time constant (τ_m) is much greater than the electrical time constant (τ_e). In this case, the transfer functions of Eqs. (74) and (75) simplify to first-order forms as follows:

$$G_1'(s) = \frac{K_1}{\tau_1 s + 1} \tag{76}$$

$$G_2'(s) = \frac{K_2}{\tau_2 s + 1} \tag{77}$$

where

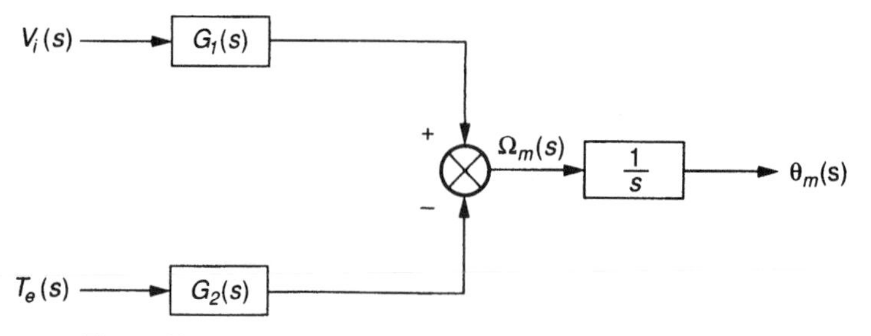

Figure 43 Block diagram of a servoactuator with voltage input.

$$K_1 = \frac{K_a K_T}{R_a B_t + K_T K_E} \tag{78}$$

$$K_2 = \frac{R_a}{n(R_a B_t + K_T K_E)} \tag{79}$$

$$\tau_1 = \tau_2 = \frac{R_a B_t \tau_m}{R_a B_t + K_T K_E}$$

If the mechanical and electrical time constants are comparable, the transfer functions can be expressed in the canonical form as

$$G_1(s) = \frac{K_1}{(s^2/\omega_n^2) + (2\zeta/\omega_n)s + 1} \tag{80}$$

$$G_2(s) = \frac{K_2(\tau_e s + 1)}{(s^2/\omega_n^2) + (2\zeta/\omega_n)s + 1} \tag{81}$$

where K_1 and K_2 are as defined in Eqs. (78) and (79), respectively, and

$$\omega_n = \sqrt{\frac{1}{\tau_e \tau_m}\left(1 + \frac{K_T K_E}{R_a B_t}\right)} \tag{82}$$

$$\zeta = \frac{1}{2}\left[\frac{\tau_e + \tau_m}{\sqrt{\tau_e \tau_m [1 + (K_T K_E / R_a B_t)]}}\right] \tag{83}$$

If a current amplifier is used instead of a voltage amplifier, the resulting linearized dynamic model of the system is given by

$$\Delta\Omega m(s) = G_3'(s)\, \Delta I(s) - G_4'(s)\, \Delta T_e(s) \tag{84}$$

where

$$G_3'(s) = \frac{K_T K_{ac}}{B_t(\tau_m s + 1)} \tag{85}$$

$$G_4'(s) = \frac{1}{B_t n(\tau_m s + 1)} \tag{86}$$

That is, the electrical time constant is not a factor if the current amplifier is used.

Numerical Example

A Magnetic Technology, Inc. dc servomotor (model 3069-237/045; see Ref. 14) is used to rotate an inertia load. System parameters are as follows:

Motor:

$B_m = 1.9$ in.·oz·s/rad (1.34×10^{-2} N·m·s/rad)

$J_m = 0.016$ in.·oz·s^2/rad (1.13×10^{-4} N·m·s^2/rad)

$K_E = 0.236$ V·s/rad

$K_T = 33.4$ in.·oz/A (0.24 N·m/A)

$L_a = 4.8$ mH

$R_a = 4.5\ \Omega$

Gearing:

$n = 3$

Load:

$B_L = 35$ in.·oz·s/rad (0.25 N·m·s/rad)

$J_L = 0.2$ in.·oz·s^2/rad (1.41×10^{-3} N·m·s^2/rad)

Amplifier:

$K_a = 8$ V/V

Substituting these parameters into Eqs. (78), (79), (82), and (83) gives the following gains and performance factors:

$K_1 = 7.87$ rad/s·V

$K_2 = 0.044$ rad/in.·oz·s (6.23 rad/N·m·s)

$\omega_n = 68.5$ Hz

$\zeta = 1.27$

and from the definition for τ_e,

$$\tau_e = \frac{L_a}{R_a} = 1.07 \times 10^{-3}\ \text{s}$$

Since the damping ratio (ζ) is greater than 1, the system will have an overdamped dynamic response. The transfer functions are as follows:

$$G_1(s) = \frac{7.87}{5.40 \times 10^{-6}s^2 + 5.88 \times 10^{-3}s + 1} \tag{87}$$

$$G_2(s) = \frac{0.13(1.07 \times 10^{-3}s + 1)}{5.40 \times 10^{-6}s^2 + 5.88 \times 10^{-3}s + 1} \tag{88}$$

Equations (87) and (88) can be used to determine the transient response or the frequency response.

10.2 Electromechanical Servosystems

Only servosystems including dc servomotors are considered here. The reader should consult Ref. 41 for a more comprehensive treatment of other types of electromechanical servosystems.

Figure 44 shows a closed-loop servosystem that utilizes tachometer feedback around the servoactuator discussed in the previous section. Through the use of velocity feedback, greater accuracy can be achieved than with the open-loop system.

Mathematical Models

An approximate linearized model may be derived for the servosystem by combining the open-loop system model [Eq. (73)] with the following equations:

Summation:

$$v_i = v_{\text{ref}} - v_f \tag{89}$$

Tachometer:

$$v_f = K_t \Delta_m \tag{90}$$

Amplifier:

$$v_a = K_a v_i \tag{91}$$

Equation (90) assumes that the tachometer is mounted directly on the motor shaft, which is normally the case. Reference 18 discusses cases where the tachometer is mounted in other configurations. Also, Ref. 18 presents more accurate tachometer models to account for the magnetic coupling of the tachometer to the motor field. The basic form of the tachometer model in that case is that of a high-pass filter with a phase shift that is dependent on the angular orientation of the tachometer to the motor.

Equations (73), (89), (90), and (91) may be combined to obtain the following closed-loop servosystem model:

$$\Delta\Omega_m(s) = G_3(s)\,\Delta V_i(s) - G_4(s)\Delta T_e(s) \tag{92}$$

where

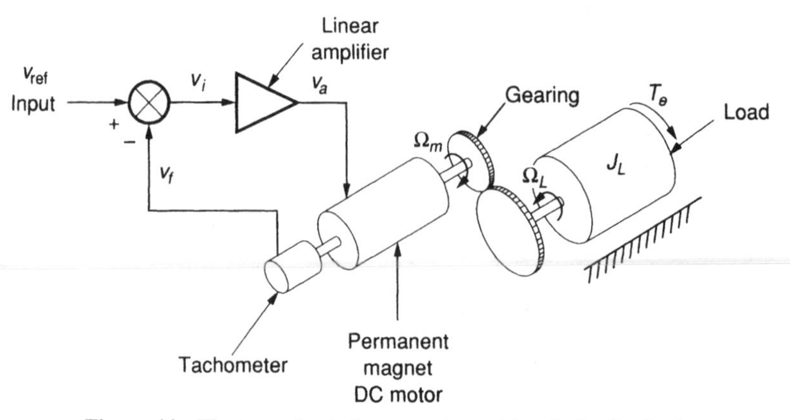

Figure 44 Electromechanical servosystem with velocity feedback.

$$G_3(s) = \frac{K_a K_T}{R_a B_t(\tau_e s + 1)(\tau_m s + 1) + K_T K_E + K_f K_a K_T} \tag{93}$$

$$G_4(s) = \frac{(R_a/n)(\tau_e s + 1)}{R_a B_t(\tau_e s + 1)(\tau_m s + 1) + K_T K_E + K_f K_a K_T} \tag{94}$$

All parameters and variables are defined in the previous section and in Fig. 44. The second-order transfer functions $G_3(s)$ and $G_4(s)$ can be expressed in canonical form as follows:

$$G_3 = \frac{K_3}{(s^2/\omega_{n1}^2) + (2\zeta_1/\omega_{n1})s + 1} \tag{95}$$

$$G_4 = \frac{K_4(\tau_e s + 1)}{(s^2/\omega_{n1}^2) + (2\zeta_1/\omega_{n1})s + 1} \tag{96}$$

where

$$K_3 = \frac{K_a K_T}{R_a B_t + K_T K_E + K_f K_a K_T} \tag{97}$$

$$K_4 = \frac{R_a}{n(R_a B_t + K_T K_E + K_f K_a K_T)} \tag{98}$$

$$\omega_{n1} = \sqrt{\frac{1}{\tau_e \tau_m}\left(1 + \frac{K_T K_E + K_f K_a K_T}{R_a B_t}\right)} \tag{99}$$

$$\zeta_1 = \frac{1}{2}\left[\frac{\tau_e + \tau_m}{\sqrt{\tau_e \tau_m\,[1 + (K_T K_E + K_f K_a K_T)/R_a B_t]}}\right] \tag{100}$$

From the expressions for the closed-loop natural frequency (ω_{n1}) and damping ratio (ζ_1), it is apparent that increasing the feedback gain (K_f) increases the speed of response but decreases the degree of stability.

As in the case with the open-loop system, if the electrical time constant is much smaller than the mechanical time constant, the second-order transfer functions of Eqs. (95) and (96) reduce to the first-order forms given below:

$$G_5(s) = \frac{K_5}{\tau_5 s + 1} \tag{101}$$

$$G_6(s) = \frac{K_6}{\tau_6 s + 1} \tag{102}$$

where

$$K_5 = K_3 \qquad \text{[see Eq. (97)]}$$

$$K_6 = K_4 \qquad \text{[see Eq. (98)]}$$

and

$$\tau_5 = \tau_6 = \frac{R_a B_i \tau_m}{R_a B_t + K_T K_E + K_f K_a K_T}$$

Similar expressions may be derived for the case with a current amplifier instead of a voltage amplifier. It is preferable to use a current amplifier where cost is not prohibitive.

Numerical Example

Consider a closed-loop electromechanical servosystem comprising the open-loop system of Section 10.1 with velocity feedback added. For the same numerical values as in the numerical example of Sec. 10.1 and with K_f = 3 V/krpm, the following results are obtained.

$$K_3 = 6.42 \text{ rad/s·V}$$

$$K_4 = 0.036 \text{ rad/in.·oz·s } (5.10 \text{ rad/N·m·s})$$

$$\omega_{n1} = 75.8 \text{ Hz}$$

$$\zeta_1 = 1.143$$

10.3 Electrohydraulic Servoactuators

Figure 45 is a physical representation of an electrohydraulic servoactuator used to position an inertia load. A two-stage electrohydraulic servovalve modulates the flow of hydraulic fluid to a double-acting hydraulic motor (actuator). The current input to the torque motor of the servovalve, I, is provided by an electronic servoamplifier of gain K_a.

A simplified dynamic model may be derived to relate the voltage input to the servoamplifier, E_s, to the velocity output of the load, $\dot{Y}$. The following basic assumptions simplify the analysis.

1. The amplifier bandwidth is considerably greater than that of the servovalve and the servomotor-load portions of the system.
2. The supply pressure P_s and the exhaust pressure P_e are constant.
3. The bulk modulus of elasticity and viscosity of the hydraulic fluid are constant.
4. The leakage flow rate past the actuator piston is linearly proportional to the pressure drop across the piston; that is, $Q_L = C_2P_m$ [see Fig. 28 and Eqs. (39) and (40)].
5. All connecting passages are rigid and sufficiently short in length and large in diameter to eliminate any resistance or transmission line effects.

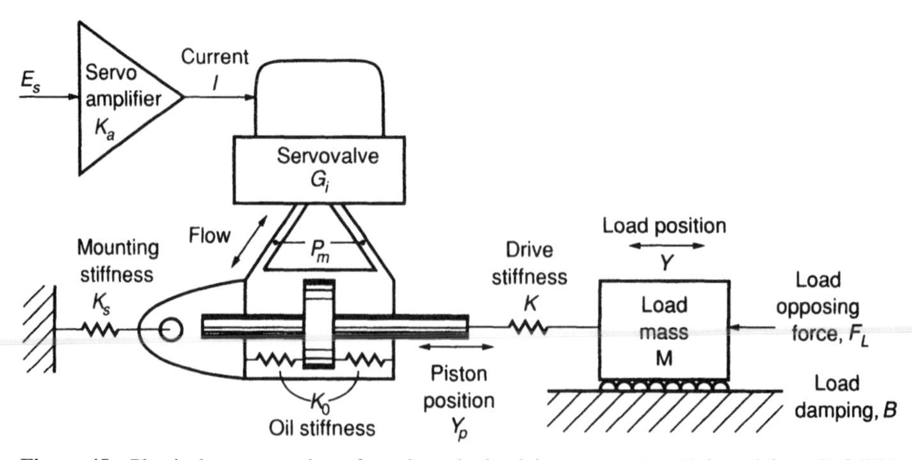

Figure 45 Physical representation of an electrohydraulci servoactuator. (Adapted from Ref. 35.)

6. The mass of the moving parts in the servomotor (M_a) and the viscous damping in the servomotor (B_m) are small compared to the mass (M) and the viscous damping (B) associated with the load.
7. The Coulomb friction and stiction forces in the servomotor and load are negligible.

Mathematical Model

The typical servoamplifier is governed by the equation

$$I = K_a E_s \tag{103}$$

Equation (61) describes the steady-state behavior of the electrohydraulic servovalve. This equation can be used in the dynamic analysis of the servoactuator if the servovalve dynamics are negligible compared to the servomotor-load dynamics. Otherwise, either Eq. (65) or (66) should be used. Equations (38)–(44) describe the dynamic behavior of the servomotor. The external load force in Eq. (38) may be expressed as follows:

$$F_e - F_L = M \frac{d^2Y}{dt^2} + B \frac{dY}{dt} + K(Y_p - Y) \tag{104}$$

The set of algebraic equations outlined above cannot be combined into a single dynamic equation or transfer function because of nonlinearities. Also, since the set of equations is rather complex, conclusions about the performance of the system cannot be drawn without actually solving the equations for a variety of conditions. A computer-based analysis is the only practical method of determining the steady-state and dynamic performance of the system unless the equations are linearized in some fashion.

A linearized model which assumes small perturbations of all variables about an initial steady-state operating condition can be very useful in quickly assessing system dynamic behavior. Such a model is particularly useful for preliminary design and as a reference when an analysis is made using the set of nonlinear describing equations.

It can be shown[45] that for the case when $\mathcal{V}_{1i} = \mathcal{V}_{2i} = \mathcal{V}_i$ (see Fig. 28) and under the assumptions listed above, the linearized model may be expressed as follows:

$$\left[\left(\frac{K_3M}{K_2B + A^2}\right)s^2 + \left(\frac{K_2M + BK_3}{K_2B + A^2}\right)s + 1\right]s(\Delta Y)$$
$$= \left(\frac{K_aA}{K_2B + A^2}\right)G_1(s)(\Delta E_s) - \left(\frac{K_2 + K_3s}{K_2B + A^2}\right)(\Delta F_L) \tag{105}$$

where $K_2 = C_1 + C_2$
C_1 = servovalve pressure–flow sensitivity [see Eq. (64)]
C_2 = internal leakage flow-rate coefficient
$K_3 = A_2/K_t = A(1/K_0 + 1/K_s + 1/K)$
$K_0 = 2\beta A^2/\mathcal{V}_i$ = stiffness of the sealed chamber
K = stiffness of the load drive
K_s = stiffness of the structural mounting

and $G_i(s)$ is the transfer function for the servovalve. When the servovalve dynamics are negligible compared to the servomotor-load dynamics, $G_i(s) = K_1$ [see Eq. (63)]. When the servovalve dynamics are of the same order as the servomotor-load dynamics, $G_i(s)$ is given by the right-hand side of either Eq. (65) or Eq. (66).

Four important measures of performance can be observed from Eq. (105) without actually solving the equation. The steady-state gain or sensitivity of the servoactuator is

$$\left.\frac{\Delta\dot{Y}}{\Delta E_s}\right|_{\text{steady state; } \Delta F_L=0} = \frac{K_a K_1 A}{K_2 B + A^2} \tag{106}$$

The steady-state load sensitivity of the servoactuator is

$$\left.\frac{\Delta\dot{Y}}{\Delta F_L}\right|_{\text{steady state; } \Delta E_s=0} = \frac{-K_2}{K_2 B + A^2} \tag{107}$$

Also observable from Eq. (105) are the natural frequency and the damping ratio associated with the servomotor-load portion of the system. The natural frequency is

$$\omega_{ns} = \sqrt{\frac{K_2 B + A^2}{K_3 M}} \tag{108}$$

and the damping ratio is

$$\zeta_s = \frac{K_2 M + K_3 B}{2\sqrt{K_3 M (K_2 B + A^2)}} \tag{109}$$

The valve dynamics are often negligible compared to the servomotor-load dynamics. In this case, Eq. (105) reduces to a second-order linear differential equation of the generalized form:

$$\left[\left(\frac{1}{\omega_{ns}^2}\right) s^2 + \left(\frac{2\zeta_s}{\omega_{ns}}\right) s + 1\right] s(\Delta Y) = \frac{K_a K_1 A}{K_2 B + A^2} \times \Delta E_s - \frac{K_2 + K_3 s}{K_2 B + A^2} \Delta F_L \tag{110}$$

In most cases, the term $K_2 B$ in Eqs. (105)–(110) is small compared to the term A^2. Simplified forms of these equations result.

The dynamic behavior of the open-loop system may be viewed in terms of the speed of response and the degree of damping. A fast system dynamic response requires a small value of τ and a large value of ω_{ns}. In order for ω_{ns} to be large, $K_3 M$ must be small compared to $K_2 B + A^2$. The value of M is usually fixed, but the designer often has some latitude in varying K_3. The value of K_3 can be minimized by making the volumes within the servomotor chambers small. Increasing the ram area also provides for an increase in ω_{ns}, but the effect is not as great as it first appears since $K_3 = f(\mathcal{V}_i)$ and $\mathcal{V}_i = f(A)$. Also, the actuator area is often set by such practical considerations as maximum load or acceleration requirements.

The degree of damping in the open-loop system is governed by Eq. (109). In most practical systems, $\zeta_s < 1$, and K_3, M, and A are fixed by other considerations. Then the damping may be increased by increasing the load damping B or the value of the effective leakage coefficient, K_L. The value of K_L may be increased by increasing either C_1 or C_2, that is, increasing the valve underlap or the leakage across the actuator piston. Since normally $K_2 B \ll A^2$, Eq. (107) shows that an increase in K_2 results in an increase in sensitivity to load disturbances. Likewise, an increase in the valve underlap (or the use of cross-port leakage) results in an increase in quiescent power dissipation. Clearly, there is a trade-off between degree of damping, steady-state load sensitivity, and quiescent power dissipation.

10.4 Electrohydraulic Servosystems

Figure 46 is a physical representation of a typical electrohydraulic servosystem intended to accurately position an inertia load. The addition of position feedback converts the "rate-type" open-loop system (servoactuator) into a position control system. The linearized model given by Eq. (110) describes the open-loop portion of this position control system, as shown in the block diagram of Fig. 47.

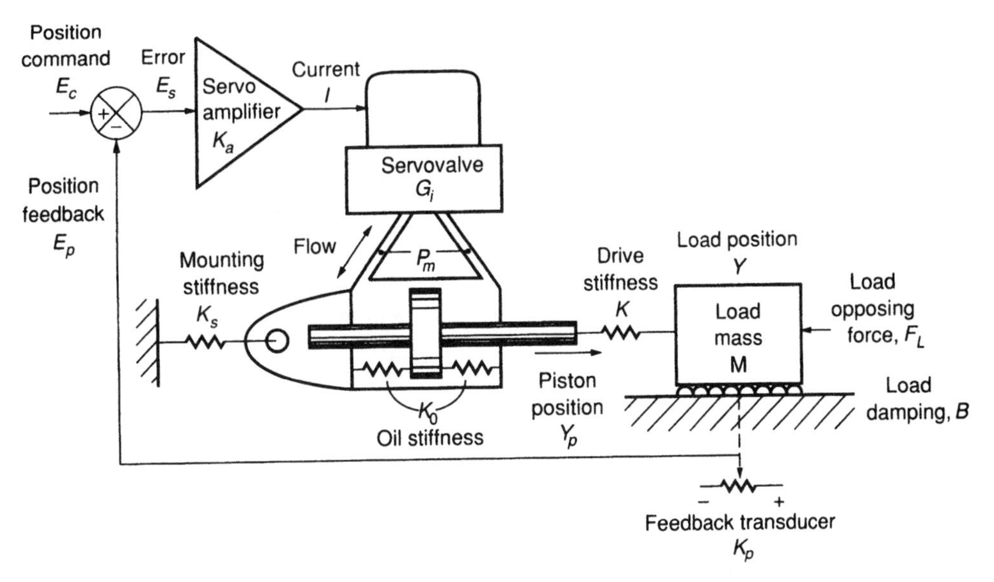

Figure 46 Physical representation of an electrohydraulic servomechanism. (From Ref. 35.)

When the change in the external load force is zero (i.e., $\Delta F_L = 0$), the dynamic model for the closed-loop system may be written as

$$\frac{\Delta Y}{\Delta E_c} = \frac{1}{K_p}\left[\frac{1}{K_{Lp}(1/\omega_{ns}^2)s^2 + (2\zeta_s/\omega_{ns})s + 1)(\tau s + 1)(s) + 1}\right] \tag{111}$$

where

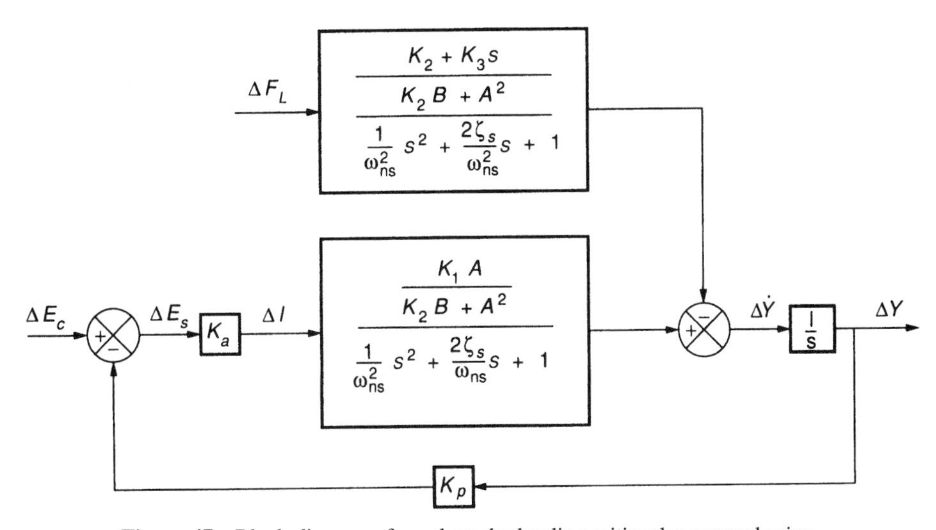

Figure 47 Block diagram of an electrohydraulic positional servomechanism.

$$K_{Lp} = \frac{K_p K_a K_1 A}{K_2 B + A^2} = \text{loop gain}$$

For the case where the servovalve dynamics are negligible compared to the servomotor-load dynamics, Eq. (111) reduces to the following third-order differential equation:

$$\frac{\Delta Y}{\Delta E_c} = \frac{1}{K_p} \left[\frac{1}{(1/K_{Lp})(1/\omega_{ns}^2)s^2 + ((2\zeta_s/\omega_{ns})s + 1s + 1)} \right] \tag{112}$$

In the steady state, Eqs. (111) and (112) both reduce to

$$\left. \frac{\Delta Y}{\Delta E_c} \right|_{\text{steady state}} = \frac{1}{K_p} \tag{113}$$

The integration in the forward loop results in a system with a zero steady-state error for a constant input and a steady-state output that is dependent only on the system input and the position feedback gain. (See Section 7 in Chapter 12 for a discussion of the following errors for systems with nonconstant inputs.) That is, the accuracy with which ΔY follows the input ΔE_c depends only on the accuracy of the feedback measurement device, and not on the accuracy of the forward loop elements.

When the change in the control input is zero (i.e., $\Delta E_c = 0$), the steady-state load sensitivity is

$$\left. \frac{\Delta Y}{\Delta F_L} \right|_{\text{steady state}} = \frac{K_2}{K_p K_a K_1 A} \tag{114}$$

That is, an increase in the feedback gain results in a decrease in the load sensitivity or an increase in the system stiffness. Likewise, an increase in the cross-port leakage in the servomotor or servovalve (i.e., increase in K_2) results in a decrease in the system stiffness.

In most practical cases the roots of the quadratic in Eqs. (111) and (112) are conjugate complex (i.e., $\zeta_s < 1$), or in physical terms, this portion of the system is "underdamped." Consequently, when position feedback is employed, limitations exist in the maximum value of the position feedback gain that can be used while still ensuring system stability. Limitations also exist in the input–output sensitivity and the system stiffness to load disturbances, since these characteristics are dependent on the position feedback gain.

System relative stability can be viewed conveniently by employing the root-locus technique. Figure 48 illustrates root-locus plots for three important cases: (*a*) valve dynamics modeled by a second-order differential equation, (*b*) valve dynamics modeled by a first-order differential equation, and (*c*) negligible valve dynamics. These plots illustrate that the loop gain must be set below some value in order to ensure stability.

When the dynamics of the servovalve are negligible compared to the dynamics of the servomotor-load portion of the system, the system model is third order. Considerable study has been made of third-order dynamic systems. The Routh absolute stability criterion can be employed to determine the maximum value of the feedback gain that can be used and still maintain stability. For the simplified model given by Eq. (112), the maximum value of the feedback gain is

$$K_p = \frac{2\zeta_s \omega_{ns}(K_2 B + A^2)}{K_a K_1 A} \tag{115}$$

and the corresponding maximum value of the loop gain is

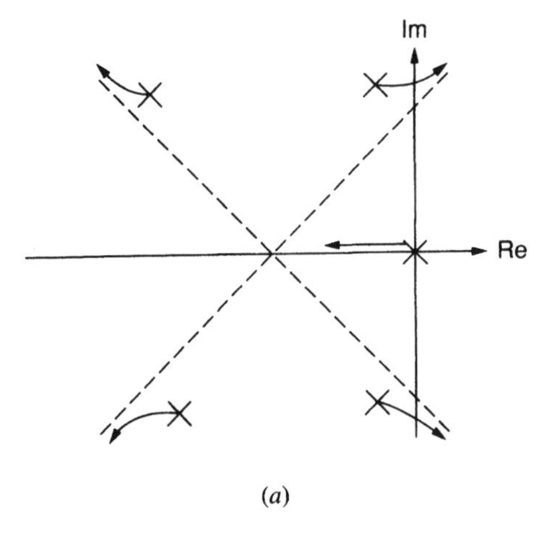

(*a*)

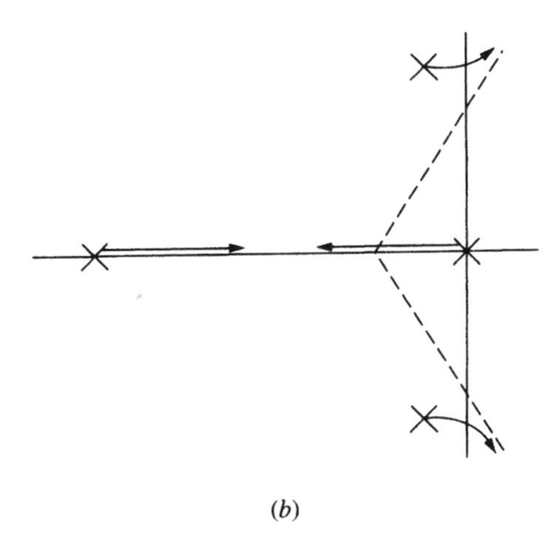

(*b*)

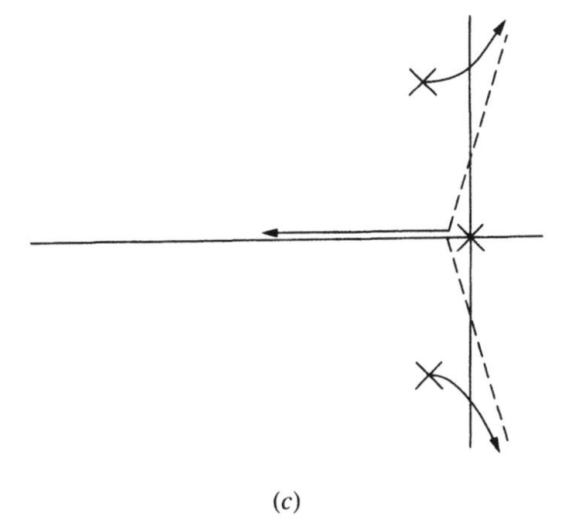

(*c*)

Figure 48 Typical root locus plots for an electrohydraulic positional servomechanism with and without servovalve dynamics: (*a*) with second-order servovalve model [Eq. (66)]; (*b*) with first-order servovalve model [Eq. (65)]; (*c*) with static servovalve model [Eq. (62)].

$$L_{Lp} \equiv \frac{K_p K_a K_1 A}{K_2 B + A^2} = 2\zeta_s \omega_{ns} \tag{116}$$

Equation (116) shows that the maximum value of loop gain depends only on the values of the natural frequency and damping ratio of the open-loop system.

In general, the loop gain is a critical parameter. An increase in loop gain results in a decrease in load sensitivity (or increase in stiffness), an increase in the speed of response, and a decrease in the degree of stability.

For a given electrohydraulic position control system designed to meet certain steady-state performance requirements (e.g., load sensitivity), an optimum or best value of loop gain (and therefore the best feedback gain) exists. This optimum value represents the best compromise between speed of response and degree of stability. Figure 49 illustrates responses of the system output (ΔY) to step changes in the system input (ΔE_c) for four different values of the loop gain [Eq. (112)].

A comprehensive study of Eq. (112) by Meyfarth[46] has shown that the optimum response characteristics to a step input signal are obtained when

$$\zeta_s = 0.5 \tag{117}$$

$$K_{Lp} = 0.34\omega_{ns} \tag{118}$$

In many practical casess, the open-loop system damping ratio (ζ_s) is well below 0.5. The resulting load resonance places severe limitations on the maximum level of loop gain that can be used and therefore limitations on the quality of steady-state and dynamic performance that can be achieved with the closed-loop system. That is, it may not be possible to simultaneously satisfy the steady-state load sensitivity (or stiffness) and dynamic performance requirements without special enhancements to the system.

10.5 Hydraulic Compensation

One of the features of electrohydraulic servosystems is the ease with which electronic feedback and forward-loop compensation networks can be employed to produce improved dy-

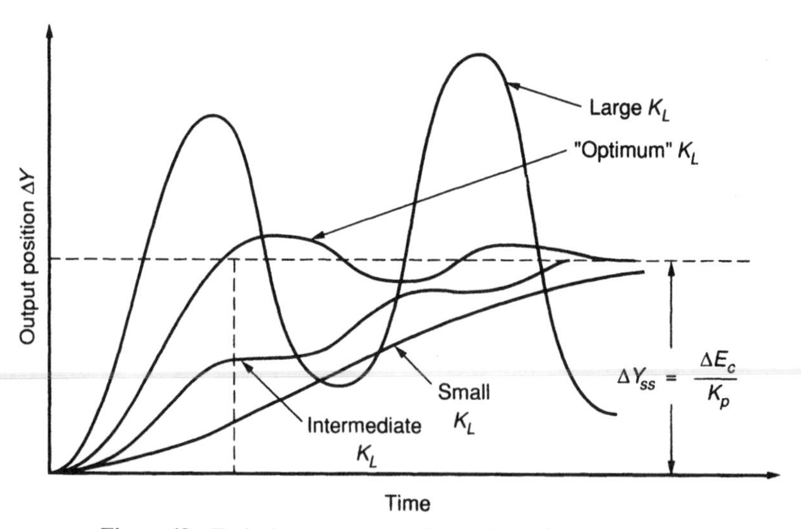

Figure 49 Typical step responses for a third-order linear system.

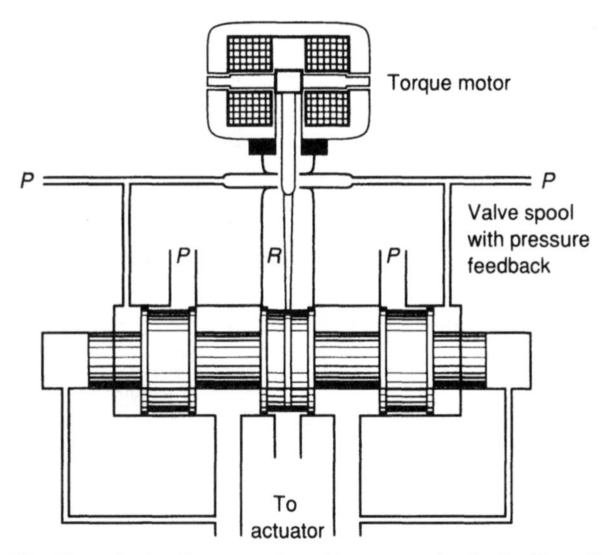

Figure 50 Electrohydraulic servovalve with pressure feedback. (From Ref. 35.)

namic performance. Such techniques are discussed in Chapter 15. The following discussion is limited to techniques for improving the damping within the hydraulic portion of the system; these techniques minimize the need to consider other electronic compensation techniques.

Techniques for improving Damping

Three well-known techniques may be employed to introduce additional damping into the open-loop system. First, *underlap* may be introduced into the servovalve, thereby increasing

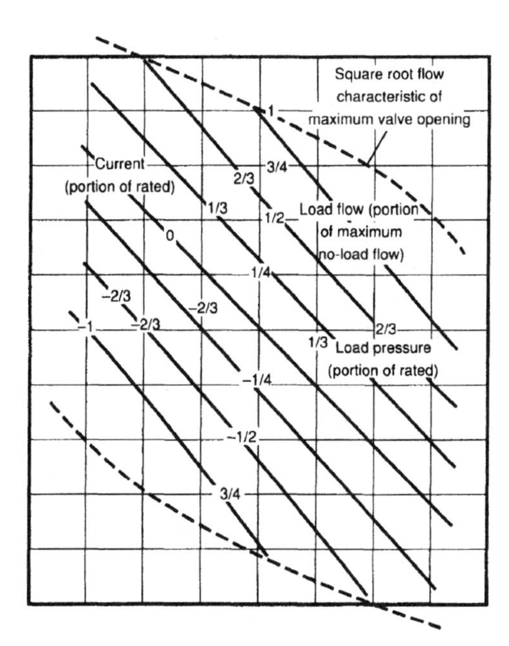

Figure 51 Pressure–flow–current characteristics of an electrohydraulic servovalve with pressure feedback. (From Ref. 35.)

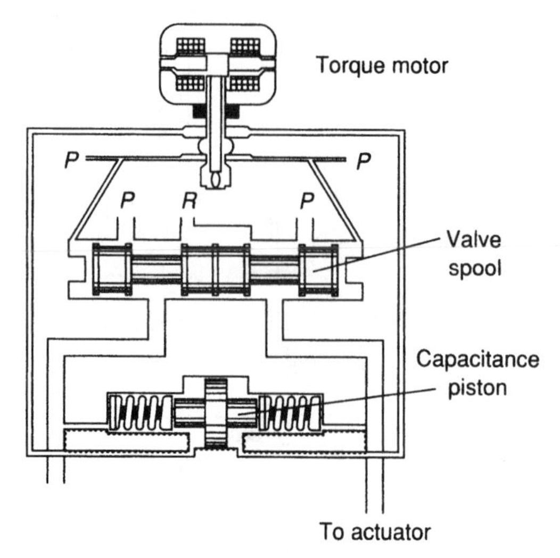

Figure 52 Electrohydraulic servovalve with dynamic pressure feedback. (From Ref. 35.)

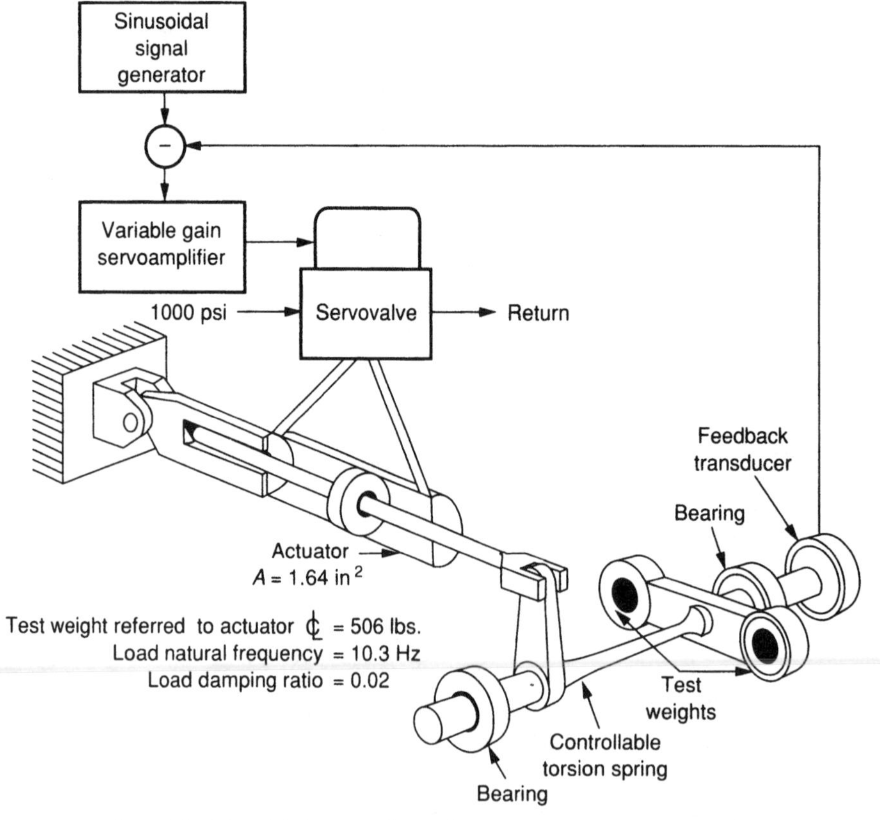

Figure 53 Pictorial diagram of experimental test setup. (From Ref. 35.)

the valve flow–pressure sensitivity [see Eq. (64)]. Second, a leakage path may be provided across the servomotor (i.e., increased value of C_2 in the equation for leakage flow rate across the piston. $Q_L = C_2P_m$). Finally, load force (or load pressure) feedback may be provided around the servovalve–servomotor. The first and second techniques are simple and flexible but often undesirable because they result in decreased steady-state stiffness and increased steady-state power dissipation. The third technique also results in decreased steady-state stiffness but avoids the problem of increased steady-state power dissipation. All three techniques result in an effective modification of the pressure–flow–current characteristics of the servovalve.

Load force (or pressure) feedback is generally preferred in high-performance systems. This feature may be implemented electrically, that is, through feedback of the measured force directly to the servoamplifier. This electrical feedback approach results in a significant increase in system complexity and often a reduction in reliability. These problems can be avoided by direct use of the load pressure itself to reposition the servovalve spool. Figure 50 shows a servovalve in which load pressure is fed back to stub shafts located at the ends of the valve spool.

Experimentally determined steady-state flow–pressure–current characteristics for this "pressure feedback servovalve" are shown in Fig. 51. Clearly, pressure feedback results in

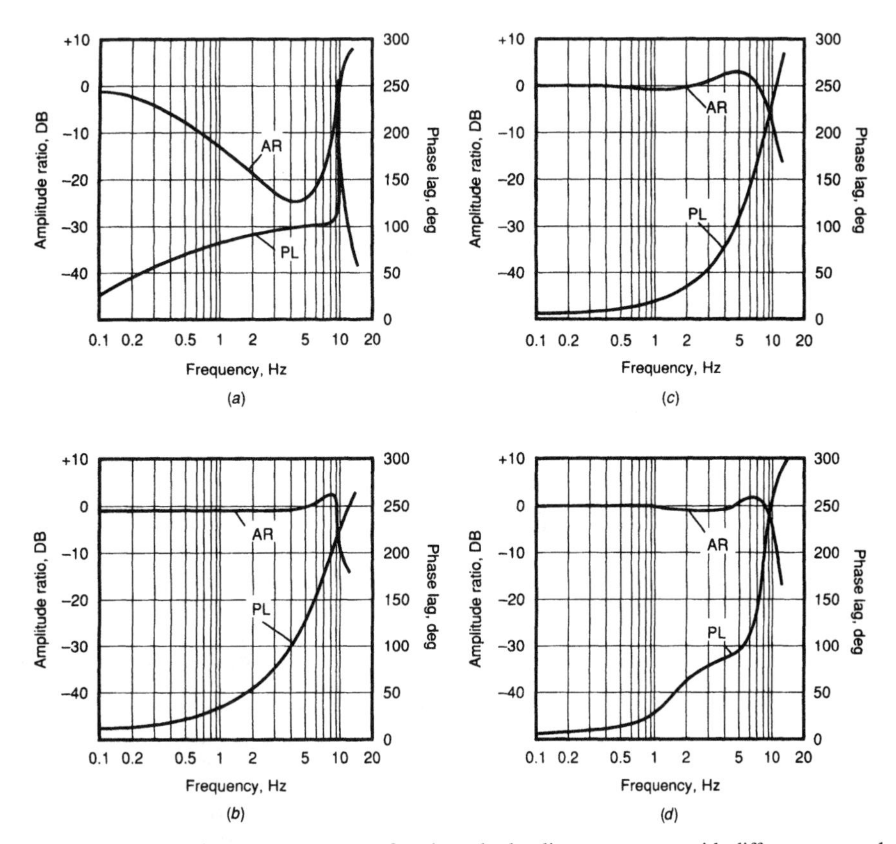

Figure 54 Measured frequency responses for electrohydraulic servosystem with different servovalves: (*a*) measured system response with flow control servovalve; (*b*) measured system response with flow control servovalve and bypass orifice; (*c*) measured system response with pressure feedback servovalve; (*d*) measured system response with servovavle. (From Ref. 35.)

Table 11 Comparative Performance of Various Electrohydraulic Position Servomechanisms

Servo Configuration	Bandwidth (±2 dB), Hz	90° Phase Lag, Hz	Static Load Stiffness, lb/in.
Flow control servovalve	0.15	0.37	9,000
Flow control servovalve with bypass orifice	8.8	5	5,100
Flow control servovalve with pressure feedback	8.8	5	2,500
DPF servovalve	9.2	5	60,000

Source: From Ref. 35.

a reshaping of the characteristics from those of the conventional servovalve (see Fig. 34). In particular, the flow-pressure sensitivity [see Eq. (64)] or slope of the characteristic curves (C_1) is increased in the vicinity of $I = 0$, $P_m = 0$ when pressure feedback is used. But an increase in this slope results in an increase in K_2 and therefore the load sensitivity [see Eq. (114)].

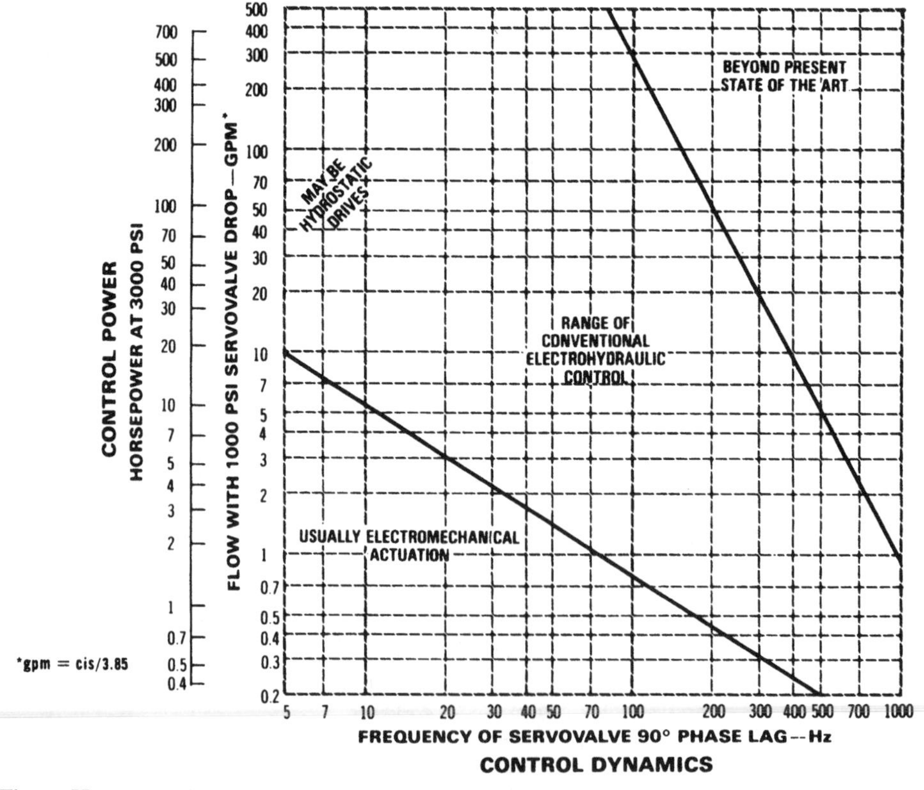

Figure 55 Range of control for electrohydraulic servomechanisms. (From Ref. 36. Courtesy of Moog, Inc., East Aurora, NY.)

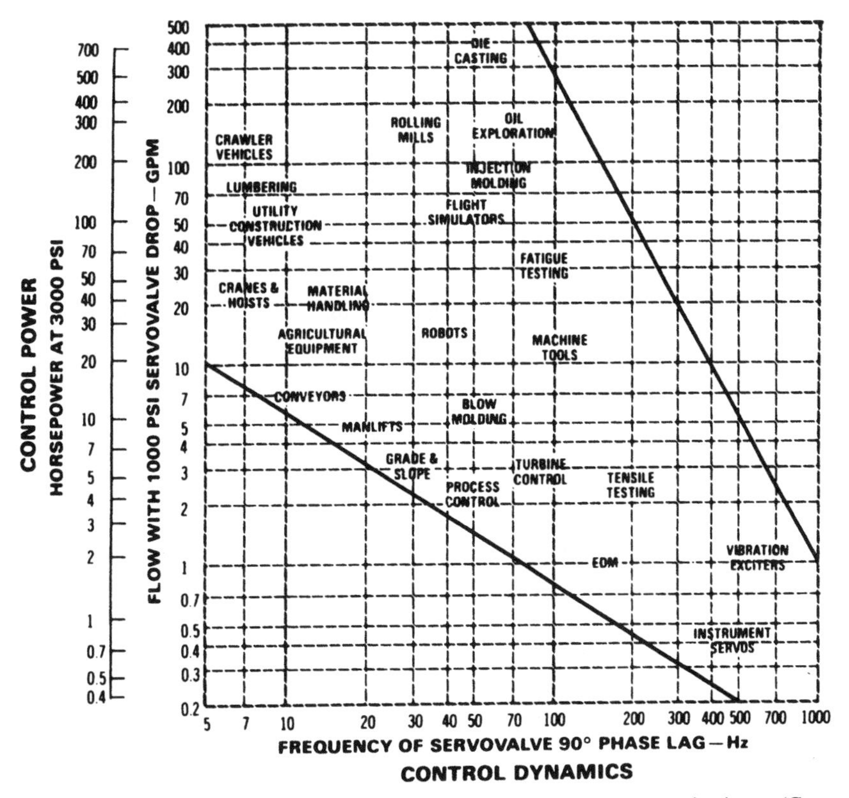

Figure 56 Spectrum of industrial applications of electrohydraulic servomechanisms. (Courtesy of Moog, Inc., East Aurora, NY.)

The *dynamic pressure feedback* (DPF) servovalve combines the best features of the conventional valve (see Fig. 30) and the pressure feedback servovalve (Fig. 50). The DPF servovalve shown in Fig. 52 is similar to the conventional servovalve except for the addition of a high-frequency-pass network (hydraulic resistance and capacitance circuit) to achieve dynamic pressure feedback. Static load pressure feedback is eliminated in this design. The design and application of the DPF servovalve is discussed in detail in Ref. 35.

Examples

A comparative study of the performance of a position control system (see Fig. 53) with different servovalves was conducted by Moog, Inc.[35] The system considered had a load resonant frequency of 10.3 Hz and a damping ratio of 0.02. Tests were conducted with four different servovalves: (a) a conventional flow control servovalve (Fig. 30), (b) a flow control servovalve with a bypass orifice, (c) a flow control servovalve with load pressure feedback (Fig. 50), and (d) a flow control servovalve with dynamic pressure feedback (Fig. 52). In each test the amplifier gain was adjusted such that the peak amplitude ratio was 1.25 (or +2 dB). In cases (b), (c), and (d) the damping was controlled to give an equivalent load damping ratio of 0.6. Measured performance results are given in Fig. 54 and Table 11. When a conventional flow control servovalve is used as in case (a), the low value of loop gain

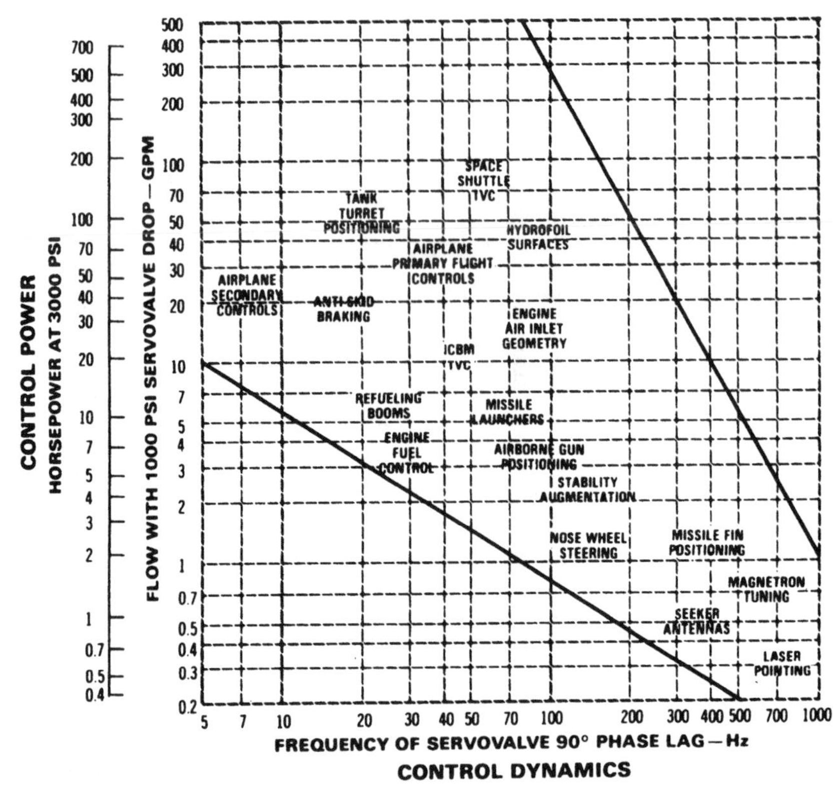

Figure 57 Spectrum of aerospace applications of electrohydraulic servomechanisms. (Courtesy of Moog, Inc., East Aurora, NY.)

required to produce stability results in significantly poorer dynamic performance. But only case (d), with a dynamic pressure feedback servovalve, combines good dynamic performance with a significant improvement in static stiffness to external load disturbances.

10.6 Range of Control for Electrohydraulic and Electromechanical Servosystems

Two principal parameters characterize the range of control for most electrohydraulic servomechanisms[36]: power level and dynamic response. Figure 55 is a graph of control power versus control dynamics showing three dominant ranges of control for electrohydraulic servomechanisms. The middle region of the graph represents the range where electrohydraulic servomechanisms traditionally have dominated applications. Yet developments of high-effeciency and high-speed-of-response rare earth servomotors have led to increased numbers of electromechanical servomechanism applications in parts of this region. For applications requiring low power levels and low dynamic response, electromechanical solutions are generally preferred. There remains a region involving high power level and high dynamic response where neither electrohydraulic nor electromechanical solutions are available.

Areas of typical applications of electrohydraulic servosystems are plotted on the control power versus control dynamics graphs of Figs. 56 and 57. Future needs for both electromechanical and electrohydraulic servomechanisms are lower cost, higher energy efficiency, and higher dynamic response.

REFERENCES

1. G. S. Brown and D. P. Campbell, *Principles of Servomechanisms,* Wiley, New York, 1948.
2. M. F. Marx and T. D. Lewis, "Electromagnetic Force Motor Design Using Rare Earth-Cobalt 1 Permanent Magnets," NAECON 1977 RECORD, National Aerospace & Electronics Conference, pp. 1119–1126.
3. M. A. Davis, "High Performance Electromechanical Servoactuation Using Brushless DC Motors," Technical Bulletin 150, Moog Inc., East Aurora, NY, April 1984.
4. "Electrical and Electronics 1986," *Machine Design* **58**(12), 6–48 (1986).
5. B. C. Kuo and J. Tal, *Incremental Motion Control—DC Motors and Control Systems,* Vol. I, SRL Publishing, Champaign, IL, 1978.
6. J. B. Leonard, "Electromechanical Primary Flight Control Activation Systems for Fighter/Attack Aircraft," Paper No. 821435, Society of Automotive Engineers, Troy, MI, 1982.
7. B. Sawyer and J. T. Edge, "Design of a Somarium Cobalt Brushless DC Motor for Electromechanical Automator Applications," NAECON 1977 RECORD, National Aerospace & Electronics Conference, pp. 1108–1112.
8. M. J. Cronin, "Design Aspects of Systems in All-Electric Aircraft," Paper No. 821436, Society of Automotive Engineers, Troy, MI, 1982.
9. *DC Motors, Speed Controls, Servo Systems,* 5th ed., Electro-Craft Corp., Hopkins, MN, 1980.
10. *Speed and Position Control Systems Distributor Catalog,* Robbins Myers, Inc., Hopkins, MN, Form No. MM-7400-00, January 1988.
11. E. Aha, "Brushless DC Motors—A Tutorial Study," *Motion,* March/April 1987, pp. 20–26.
12. S. Meshkat, "Servo System Design—A Tutorial Study. Vectorial Control of Brushless DC Motors and AC Induction Motors," *Motion,* September/October 1986, pp. 19–23.
13. R. Benzer, "Single-Chip Brushless Motor Controller," *Machine Design,* June 9, 1988, pp. 140–144.
14. *Motion Control Engineering Handbook—DC Servo/Tachometers/Brushless DC,* Magnetic Technology, Inc., Canoga Park, CA, 1985.
15. *Direct Drive Engineering Handbook—DC Torque Motors/Tachometers/Brushless DC,* Magnetic Technology, Inc., Canoga Park, CA, 1985.
16. B. K. Hose, "Adjustable Speed AC Drives—A Technology Status Review," *Proceedings of the IEEE* **70**(2), (1982).
17. A. Kusko and D. G. Galler, "Survey of Microprocessors in Industrial Motor Drive Systems," *IEEE, Industry Applications Society,* 1982, pp. 435–438.
18. J. E. Gibson and F. B. Tuteur, *Control System Components,* McGraw-Hill, New York, 1958, pp. 276–304.
19. B. C. Kuo, *Theory and Applications of Step Motors,* West Publishing, New York, 1974.
20. B. C. Kuo, *Incremental Motion Control—Step Motors and Control Systems,* Vol. II, SRL Publishing, Champaign, IL, 1979.
21. "1.8 deg. PM Hybrid Stepper Motors and Controls," Catalog ST-l, Bodine Electric, Chicago, IL, 1981.
22. B. H. Carlisle, "Stepping Motors: Edging into Servomotor Territory," *Machine Design* **58**(26), 88–100 (1986).
23. M. A. Lewis, "Design Strategies for High-Performance Incremental Servos," in *Proceedings of the Sixth Annual Symposium on Incremental Motion Control Systems and Devices,* B. C. Kuo (ed.), University of Illinois, Urbana, IL, 1977, pp. 141–151.
24. "Motor Selector, DC Motor Selection and Servo System—Design Software for the IBM PC and Compatibles," Par Tech Engineering, Windham, ME.
25. *Machine Design,* Fluid Power Reference Issue, September 17, 1987.
26. R. P. Lambeck, *Hydraulic Pumps and Motors—Selection and Application for Hydraulic Power Control Systems,* Marcel Dekker, New York, 1983.

27. *Fluid Power Design Engineers Handbook,* Parker-Hannifin Corp., Cleveland, OH, 1973.
28. G. B. Keller, *Hydraulic System Analysis,* Industrial Publishing, Cleveland, OH, 1969.
29. H. E. Merritt, *Hydraulic Control Systems,* Wiley, New York, 1967.
30. J. F. Blackburn, G. Reethof, and J. L. Shearer, *Fluid Power Control,* MIT Press, Cambridge, MA, 1960.
31. E. Lewis and H. Stern, *Design of Hydraulic Control Systems,* McGraw-Hill, New York, 1962.
32. M. Guillon, *Hydraulic Servo Systems,* Plenum, New York, 1969.
33. J. Watton, *Fluid Power Systems,* Prentice-Hall International, Hertfordshire, 1989.
34. *Electrohydraulic Servomechanisms in Industry,* Moog, Inc., East Aurora, NY (undated).
35. L. H. Geyer, "Controlled Damping through Dynamic Pressure Feedback," Technical Bulletin 101, Moog, Inc., East Aurora, NY, April 1972.
36. R. H. Maskrey and W. J. Thayer, "A Brief History of Electrohydraulic Servomechanisms," *ASME Transactions, Journal of Dynamic Systems, Measurement and Control* **100** (1978).
37. K. N. Reid, "Fluid Power Control I," Course Notes, School of Mechanical and Aerospace Engineering, Oklahoma State University, Stillwater, OK, 1987.
38. *Industrial Hydraulics Manual,* No. 935100-A, Vickers, Inc., Troy, MI, copyright by Sperry Rand Corp., 1st ed., 1970.
39. W. J. Thayer, "Transfer Functions for Moog Servovalves," Technical Bulletin 103, Moog Inc., East Aurora, NY, January 1965.
40. G. J. Blickley, "Servo Valve Becomes Digital Actuator," *Control Engineering,* June 1986, pp. 76–77.
41. E. B. Canfield, *Electromechanical Control Systems and Devices,* Wiley, New York, 1965, pp. 143–197.
42. T. P. Neal, "Performance Estimation for Electrohydraulic Control Systems," Technical Bulletin 126, Moog, Inc., East Aurora, NY, November 1974.
43. *Stepper Motors: Permanent Magnet 7.5° Step Angle,* Catalog No. STM-REM 987-5M-RL, Crouzet Control, Inc., Schaumburg, IL, 1987.
44. W. E. Wilson, "Performance Criteria for Positive-Displacement Pumps and Fluid Motors," *Transaction of the ASME,* May 1986.
45. J. L. Shearer, "Dynamic Characteristics of Valve-Controlled Hydraulic Servomotors," *ASME Transactions* **76,** 895–903 (1954).
46. P. F. Meyfarth, "Analytical Comparison of the Linear and Bang-Bang Control of Pneumatic Servomechanisms," Thesis, Department of Mechanical Engineering, Massachusetts Institute of Technology, Cambridge, MA, 1958.
47. P. F. Meyfarth, "Dynamic Response Plots and Design Charts for Third-Order Linear Systems," Research Memorandum 7401-3, Massachusetts Institute of Technology, Dynamic Anaysis and Control Laboratory, Cambridge, MA, September 1958.

BIBLIOGRAPHY

Chitayat, A., "Brushless DC Linear Motors," *Motion,* September/October 1987, pp. 22–23.

Fitzgerald, A., and C. Kingley, *Electric Machinery,* McGraw-Hill, New York, 1961.

Humphrey, W. M., *Introduction to Servomechanical System Design,* Prentice-Hall, Englewood Cliffs, NJ, 1973.

Koopman, G., "Operating Characteristics of 2-phase Servomotors," *Transactions of the AIEE* **68,** 319–329 (1949).

Kusko, A., *Solid-State DC Motor Drives,* MIT Press, Cambridge, MA, 1969.

Marinko, J. A.,"PWM Servo-Amplifiers: A Tutorial Study," *Motion,* July/August 1987, pp. 3–8.

Mazurkiewicz, J., "Brushless Motors and Brushless Motor Controllers," *Motion,* November/December 1986, pp. 14–19.

Meshkat, S., "Vectorial Control of Brushless DC Motors and AC Induction Motors," *Motion,* September/October 1986, pp. 19–23, .

Persson, E. K., "Brushless DC Motors in High-Performance Servo Systems," in *Proceedings, Fourth Annual Symposium on Incremental Motion Control Systems and Devices,* B. C. Kuo (ed.), University of Illinois, Urbana, IL, 1975, pp. T1–T16.

Proceedings, Conference on Small Electrical Machines, Institution of Electrical Engineers, London, March 30–31, 1976.

Puchstein, L., and Conrad, *AC Machines,* 3rd ed., Wiley, New York, 1954.

Schept, B., "Servo System Design—A Tutorial Study," Part 3 of 8, in *Motion,* 4th Quarter, 1985, pp. 22–30.

Small and Special Electrical Machines, Publication No. 202, Institution of Electrical Engineers, London, September 1981.

Tal, J., "Quantization Errors in Digital Servo Systems," in *Motion,* September/October 1986, pp. 10–13.

Yeaple, F., *Fluid Power Design Handbook,* Marcel Dekker, New York, 1984.

CHAPTER 15
CONTROLLER DESIGN

Thomas Peter Neal
Consultant
Lake View, NY

1 INTRODUCTION

The purpose of this chapter is to provide a basis for the specification and functional design of electronic servocontrollers. No attempt is made to treat the subject of electronic circuit design. Instead, the goal is to aid the engineer in selecting and applying a suitable off-the-shelf controller or in specifying the controller requirements to a circuit designer. The emphasis is on position, velocity, or force control of mechanical loads, although many of the techniques are applicable to controller design in general. Specialized subjects such as multiaxis control and adaptive control are beyond the scope of this chapter.

As a starting point, it is presumed that a servoactuator has been selected and mounted, together with a suitable power supply, drive amplifier, and mechanical drive mechanism. In addition, the primary feedback transducer has been chosen, and a simple loop closure has been analyzed to determine whether the specified closed-loop performance can be obtained.

Reprinted from *Instrumentation and Control,* Wiley, New York, 1990, by permission of the publisher.

The process of accomplishing these tasks is treated in Chapter 14. If a simple loop closure provides adequate performance, the controller design problem primarily consists of making some basic decisions concerning electronic implementation.

In many applications, a simple loop closure is inadequate, and more elaborate controller functions are required. These latter cases are the primary subject of this chapter. The thrust of the discussion is synthesis of the controller function, rather than analysis of an existing design. For this reason, continuous-time frequency-domain techniques will be used extensively, all starting from block diagrams and transfer functions based on the Laplace operator.[1–3] These techniques, many of which are graphical, are particularly useful in the early design stages. Since most people intuitively relate to time-domain responses, relationships between the frequency-domain results and time histories will be discussed as appropriate.

If the control system is to be implemented in digital hardware or software, it is certainly possible to handle the entire design task using the mathematics of sampled-data systems.[4–7] This approach has not been used here because the so-called classical techniques based on the Laplace transform are more illustrative, and design trade-offs are easier to evaluate. For the control system applications being considered here, it is generally necessary to keep the resolution high and the sampling interval small. In this case, controller characteristics described as transfer functions in Laplace form can be accurately transformed into various mathematical forms appropriate for digital implementation, for example the Z-transform.

2 FUNDAMENTALS OF CLOSED-LOOP PERFORMANCE

To properly design a servocontroller, it is necessary to maintain a clear picture of the desired end result, namely, the achievement of some predetermined performance goals. These performance specifications should be established early in the design. As a minimum, they should define the desired static and dynamic accuracy, bandwidth (response time), and stability. The following sections offer a brief review of these important factors and how they relate to basic loop parameters.

2.1 Accuracy and Loop Gain

The most basic requirement of a servomechanism is probably static accuracy; that is, the controlled variable must accurately hold the command set point. Referring to the generalized block diagram of Fig. 1, several sources of inaccuracy can be described. An external disturbance can cause the load to move without any change in the command signal. The load will continue to move until the resulting error signal causes the actuator to balance the

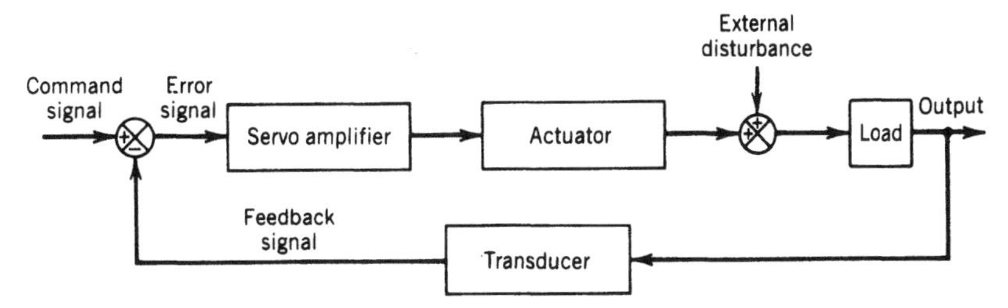

Figure 1 Generalized servomechanism.

disturbance. Anomalies in the actuator and load must also be offset by a finite error signal. Examples are temperature-induced null shifts, hysteresis, threshold, friction, and lost motion. The magnitude of these error signals is minimized if the amplifier gain is high.

Ideally, the amplifier gain would be set high enough that the accuracy of the servo becomes dependent only upon the accuracy of the transducer itself. In practice, however, the amplifier gain is limited by stability considerations. Therefore, it is desirable to provide high gains at low frequencies for accuracy and low gains at high frequencies to minimize stability problems. Since the rate of amplitude roll-off with frequency is directly related to phase lag, excessive roll-off can create more stability problems than it solves. A good compromise is to make the entire forward path look like an integrator over the frequency range of interest (a type I system). This technique is very commonly used to give nearly infinite static gain and a linear gain roll-off with frequency, at the cost of 90° phase lag.

It is important to note that some servoloops contain an inherent integrator, which complicates the accuracy-versus-stability problem. For example, many actuators are inherently rate devices when operated open loop, so that a steady input results in a proportional velocity output. A velocity servo using such an actuator will inherently have a proportional forward loop, and an integrating servoamplifier can be used (Fig. 2). However, the corresponding position servo will inherently have an integration in the forward loop, as shown in Fig. 3. In this latter case, the use of an integrating servoamplifier can cause severe stability problems. From these figures, transfer functions can be written for the closed-loop responses to command and disturbance inputs. Note that the dynamic response characteristics of all elements have been neglected, and it is assumed that the load includes no spring to ground:

$$\frac{V}{e_c} = \frac{1}{K_v} \frac{1}{s/K_{vv} + 1} \tag{1}$$

$$\frac{V}{F_d} = K_4 \frac{s}{s + K_{vv}} \tag{2}$$

$$\frac{X}{e_c} = \frac{1}{K_x} \frac{1}{s/K_{vx} + 1} \tag{3}$$

$$\frac{X}{F_d} = \frac{K_4}{K_{vx}} \frac{1}{s/K_{vx} + 1} \tag{4}$$

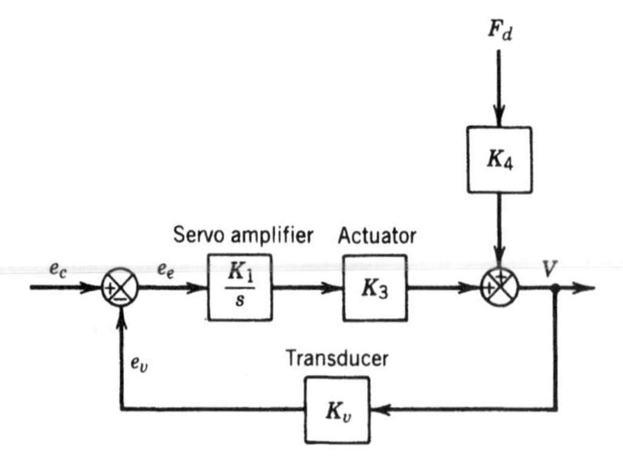

Figure 2 Simplified velocity servo.

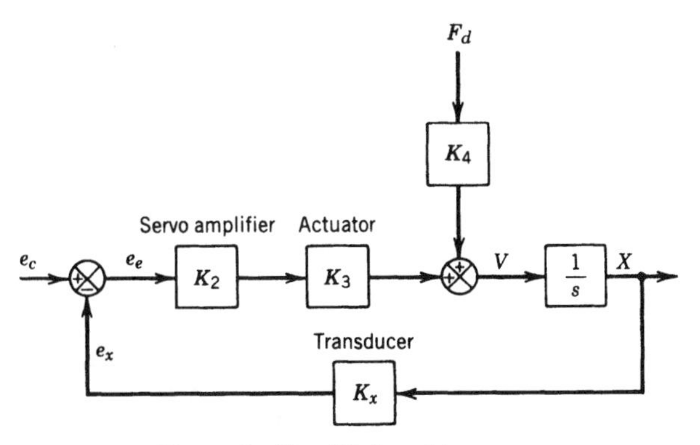

Figure 3 Simplified position servo.

where e_c = command signal V
e_e = error signal, V
e_v = velocity feedback signal, V
e_x = position feedback signal, V
F_d = disturbance force applied to the load, N
K_1 = integrating servoamplifier gain, (V/s)/V
K_2 = proportional servoamplifier gain, V/V
K_3 = actuator gain, (mm/s)/V
K_4 = actuator velocity droop due to force disturbance, (mm/s)/N
K_v = velocity transducer gain, volts/(mm/s)
K_x = position transducer gain, volts/mm
K_{vv} = open-loop gain of velocity servo = $K_1K_3K_v$, s^{-1}
K_{vx} = open-loop gain of position servo = $K_2K_3K_x$, s^{-1}
X = load position, mm
$V = \dot{X}$, mm/s

As shown in Fig. 4, the velocity and position responses to commands are both characterized by a first-order lag having a break frequency equal to the open-loop gain. However, the responses to disturbance forces are quite different in the two cases (Fig. 5). When the disturbance is downstream of the integrator (velocity servo), the servo error is K_4F_d at high frequencies but rolls off at frequencies (in radians per second) below K_{vv} and is zero statically. When the disturbance is upstream of the integrator (position servo), there is a static error inversely proportional to the open-loop gain, which rolls off at frequencies above K_{vx}. Note that Eqs. (1)–(4) remain reasonably valid when the dynamic response characteristics of the various open-loop elements are considered, for those cases in which K_{vv} (or K_{vx}) is well below the lowest break frequencies of those elements. At higher loop gains, the closed-loop dynamics can change considerably, as shown in the following discussion.

The conclusions regarding the effects of disturbance forces on servo accuracy can be generalized to any forward-loop offset or uncertainty. Referring again to Fig. 2, it can be seen that the integrating amplifier will compensate for any forward-loop offset downstream of the integrator, so that the static errors are zero. From Fig. 3, it is apparent that static errors due to offsets upstream of the integration can be quantified as

$$\frac{X_e}{V_0} = \frac{1}{K_{vx}} \tag{5}$$

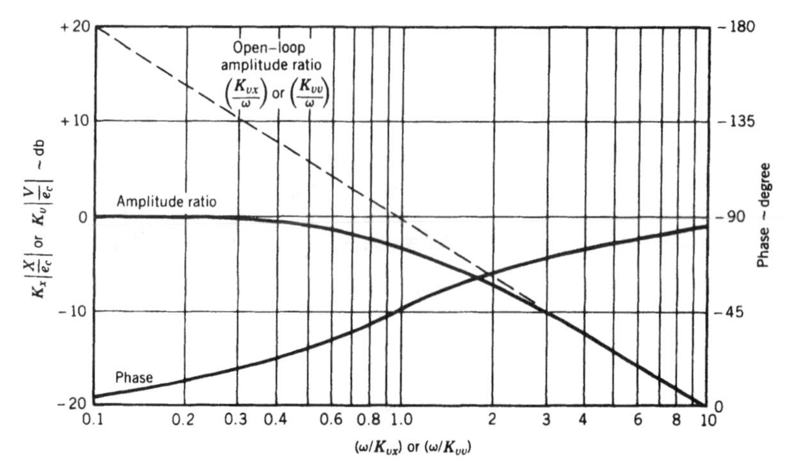

Figure 4 Simplified response to commands.

where X_e = output error = $X_c - X$, mm
V_0 = forward-loop offset, converted to an equivalent open-loop offset in volts, mm/s
X_c = position command = e_c/K_x, mm

Even when no forward-loop offsets or disturbances are present, servos exhibit following errors, sometimes called tracking errors. In servos having forward-loop integrations, these quasi-static errors result whenever the command signal changes at a constant rate, as in a position-tracking servo. For the position servo of Fig. 3, the following error is

$$\frac{e_e}{\dot{e}_c} = \frac{X_e}{\dot{X}_c} = \frac{1}{K_{vx}} \tag{6}$$

For the velocity servo of Fig. 2, the following error is

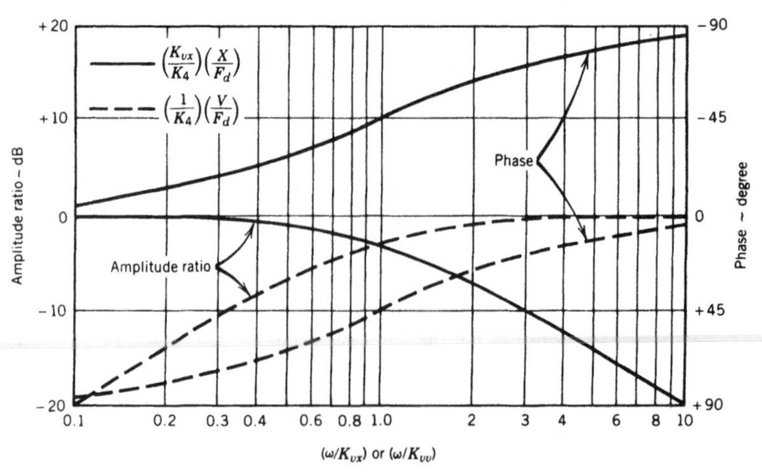

Figure 5 Simplified response to disturbance.

$$\frac{e_e}{\dot{e}_c} = \frac{V_e}{\dot{V}_c} = \frac{1}{K_{vv}} \tag{7}$$

where V_e = output error = $V_c - V$, mm/s
V_c = velocity command = e_c/K_v, mm/s

Note that the following error for the position servo is the position error resulting from a steady rate of change of position command. For the velocity servo, the following error is the velocity error resulting from a steady rate of change of velocity command.

The servo errors discussed thus far have been those that can be minimized by a tight servoloop (high loop gain). To these must be added errors in the transducer mechanism. Even if infinite loop gain were achievable, the servo can be no more accurate than the transducer itself. The most important types of transducer inaccuracies are repeatability, resolution, and linearity. Errors due to transducer location and mounting geometry must also be taken into account.

Many of the foregoing concepts can be applied to servos in general. For example, a force or pressure servo working against a spring load is similar to a position servo in the sense that output force is proportional to actuator position. Also, temperature control servos tend to behave like position servos since the controlling device tends to provide heat flow proportional to temperature error, and thermal loads tend to produce temperature rate of change proportional to heat flow.

2.2 Dynamic Response and Stability

As discussed in Section 2.1, open-loop gain has a strong influence on servo accuracy. High loop gains also provide fast dynamic response in most cases. However, stability considerations will limit the maximum useful loop gain. The dynamic response and stability of a servo are determined by the dynamic characteristics of the various loop components. In many situations, the forward-loop dynamics are dominated by a relatively small number of low-frequency lag elements, and the transducer dynamics are negligible. In these cases, it is often possible to obtain an adequate estimate of servo performance by approximating the combined forward-loop characteristics with an integrator plus a first-order or second-order lag. The adequacy of this approximation can be determined by the match of the frequency response gain and phase for the frequency range in which the phase lag is less than 180°. Using these two rather basic dynamic forms, the relationships among loop gain, stability, and dynamic response are easily seen.

A block diagram using the basic dynamic forms is shown in Fig. 6,

where U = generalized controlled variable
U_c = generalized command input
D = generalized disturbance input
G_d = open-loop response of U to D

The first-order lag is typical of simple temperature control systems and dc servos having short electrical time constants. The second-order lag is often representative of electrohydraulic servos and dc servos having long electrical time constants. The closed-loop responses to command inputs are

$$\frac{U_1}{U_c} = \frac{1}{(\tau_1/K_{u1})s^2 + (1/K_{u1})s + 1} \tag{8}$$

and

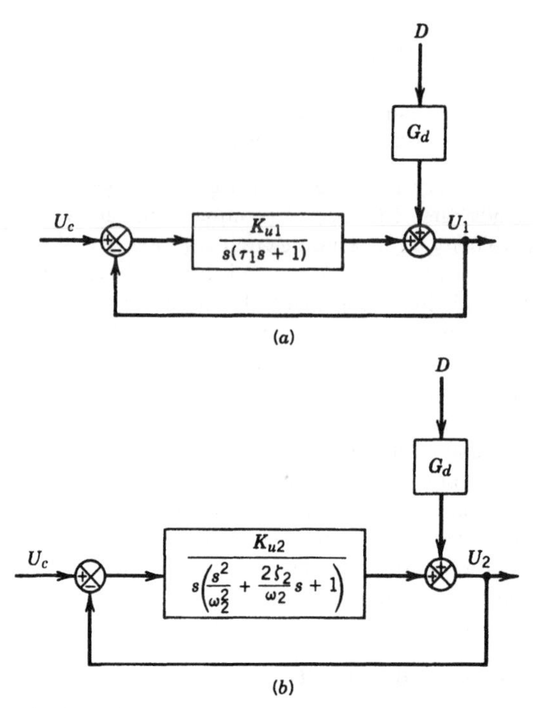

Figure 6 Basic dynamic configurations: (*a*) first-order lag; (*b*) second-order lag.

$$\frac{U_2}{U_c} = \frac{1}{s^3/K_{u2}\omega_2^2 + (2\zeta_2/K_{u2}\omega_2)s^2 + (1/K_{u2})s + 1} \tag{9}$$

Representative root loci are shown in Fig. 7 for both forms. In Fig. 7*a*, the lag combines with the integrator to produce second-order closed-loop poles. The closed-loop natural frequency increases with loop gain, while the damping ratio decreases. In the case of Fig. 7*b*,

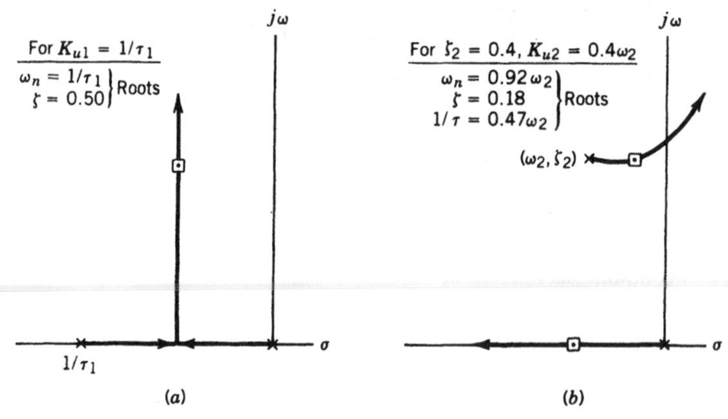

Figure 7 Root loci for basic dynamic configurations: (*a*) first-order lag; (*b*) second-order lag.

the closed-loop transfer function consists of a first-order and a second-order lag. The break frequency of the first-order lag increases with loop gain, while the second-order damping ratio rapidly decreases. In both loop closures, there are clearly trade-offs between closed-loop bandwidth and stability.

There are numerous methods for quantifying the relationships between bandwidth and stability. Closed-loop frequency responses to command inputs are shown for both basic forms in Figs. 8 and 9, while Figs. 10 and 11 present the corresponding step responses. Useful numerical measures of stability are phase margin, gain margin, and damping ratio of the closed-loop complex pair. These are given in Figs. 12 and 13. Note that the gain margin for Fig. 12 is infinite.

Referring to Fig. 6, closed-loop responses to disturbance inputs can be written as

$$\frac{U_1}{D} = \frac{G_d}{K_{u1}} \frac{s(\tau_1 s + 1)}{(\tau_1/K_{u1})s^2 + (1/K_{u1})s + 1} \tag{10}$$

and

$$\frac{U_2}{D} = \frac{G_d}{K_{u2}} \frac{s[s^2/\omega_2^2 + (2\zeta_2/\omega_2)s + 1]}{(s^3/K_{u2}\omega_2^2) + (2\zeta_2/K_{u2}\omega_2)s^2 + (1/K_{u2})s + 1} \tag{11}$$

To determine the final dynamic form of these responses, it is necessary to have a transfer function for G_d. This transfer function can be obtained from the physical model of the system, by deriving the response of the controlled variable to the disturbance input with the controller output equal to zero. As an example, consider a dc motor driving an inertial load and having a short electrical time constant. For an integrating velocity loop, the disturbance transfer function has the form

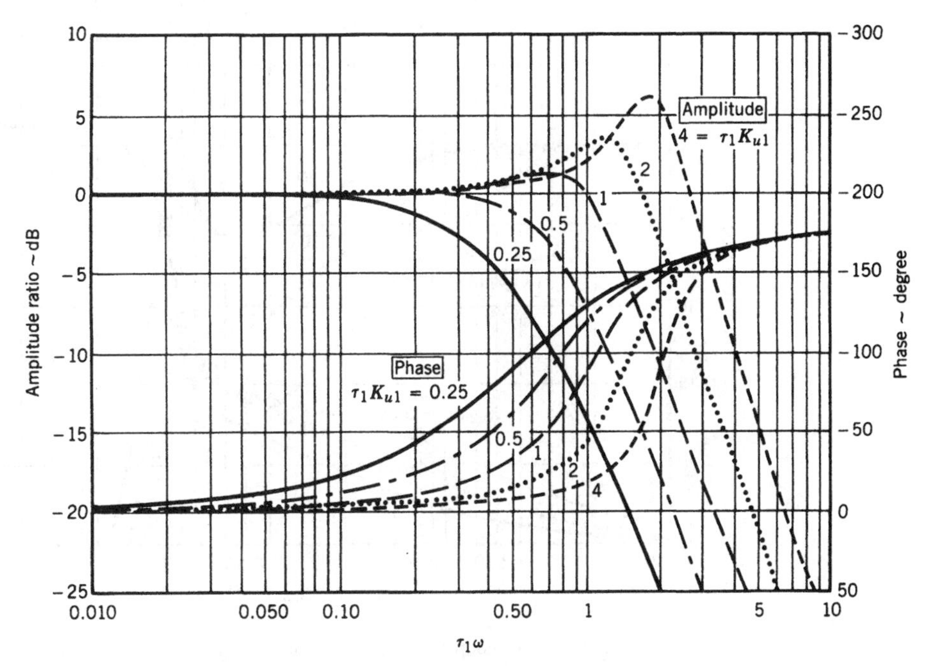

Figure 8 Closed-loop frequency responses for U_1/U_c.

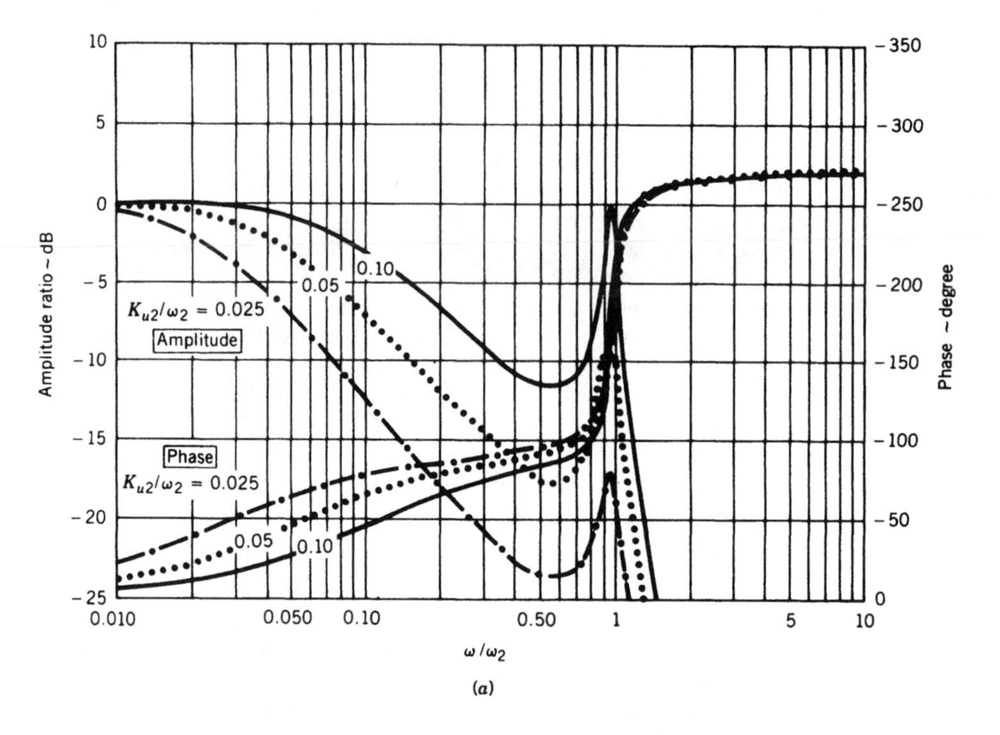

(*a*)

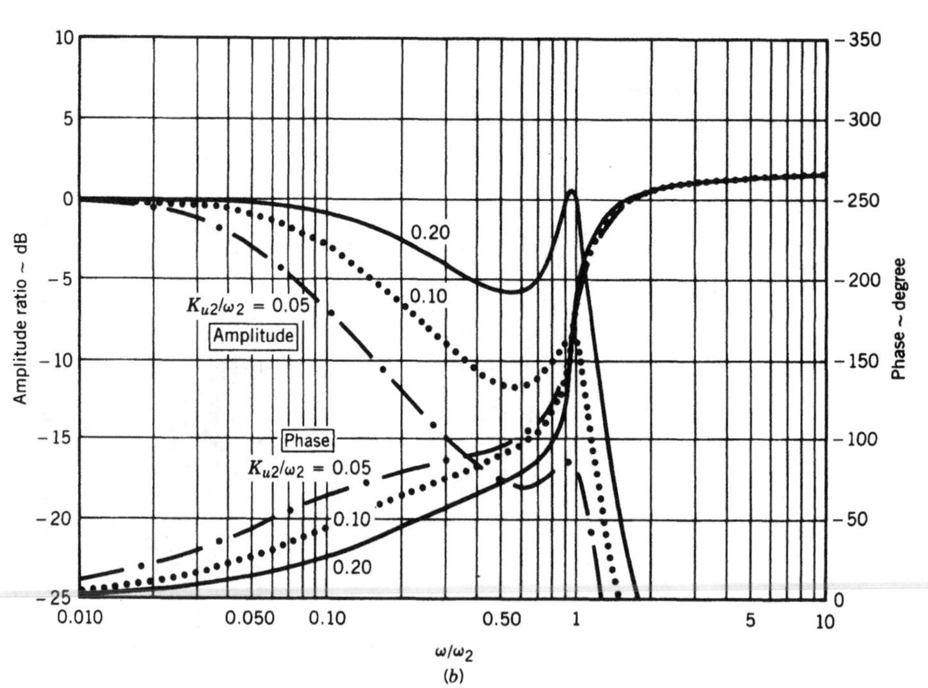

(*b*)

Figure 9 Closed-loop frequency responses for U_2/U_c: (*a*) $\zeta_2 = 0.1$; (*b*) $\zeta_2 = 0.2$.

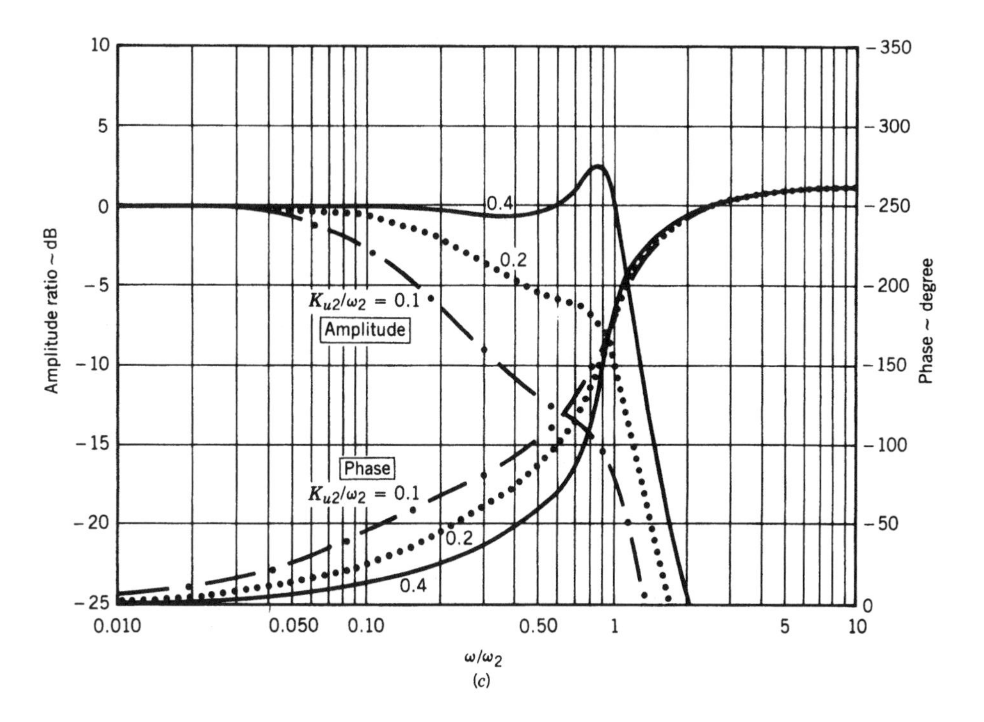

(c)

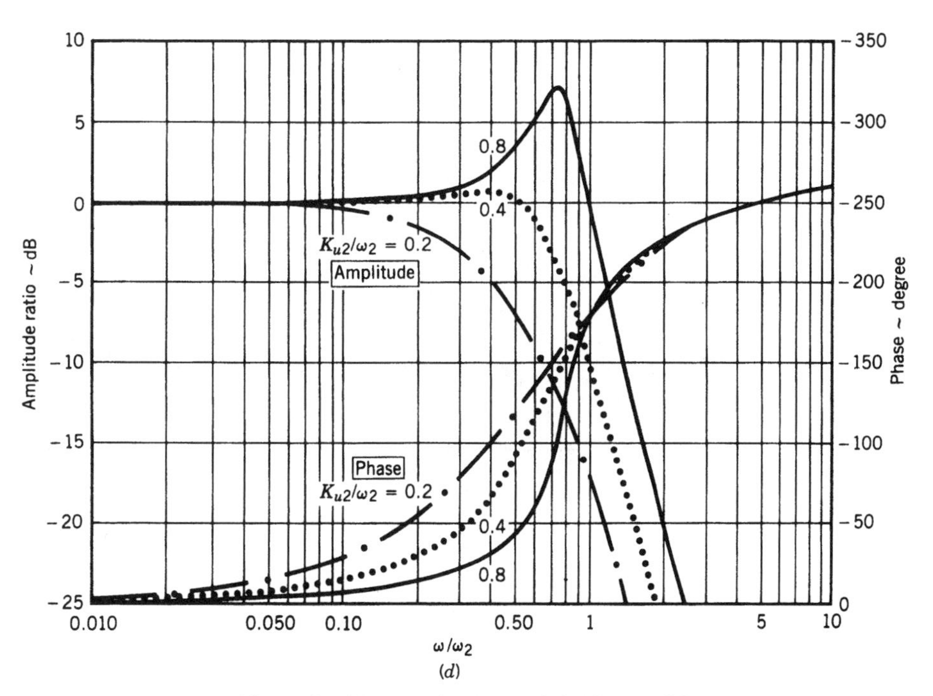

(d)

Figure 9 (*Continued*) (*c*) ζ_2 = 0.4; (*d*) ζ_2 = 0.8.

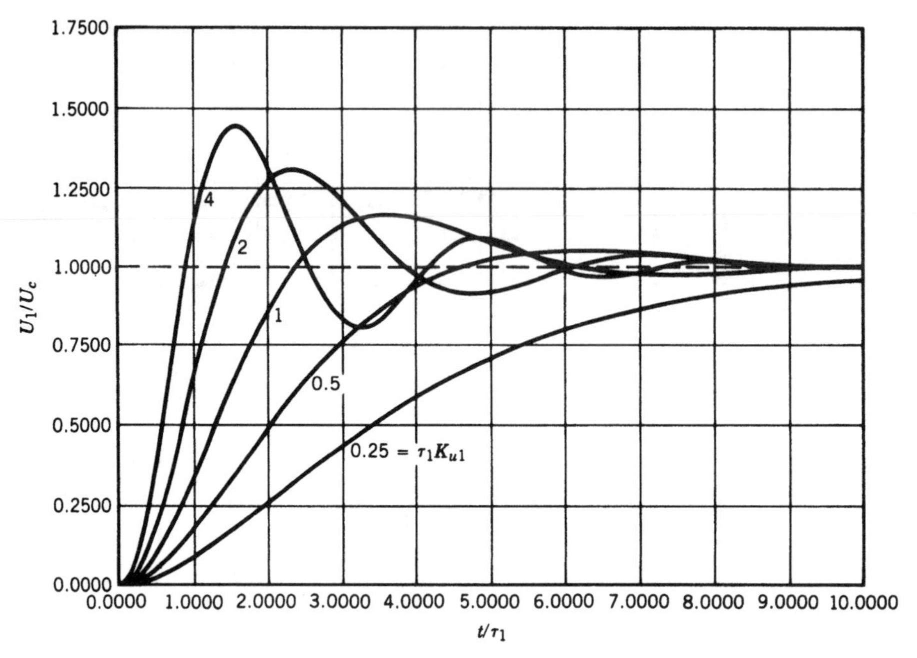

Figure 10 Closed-loop step responses for U_1/U_c.

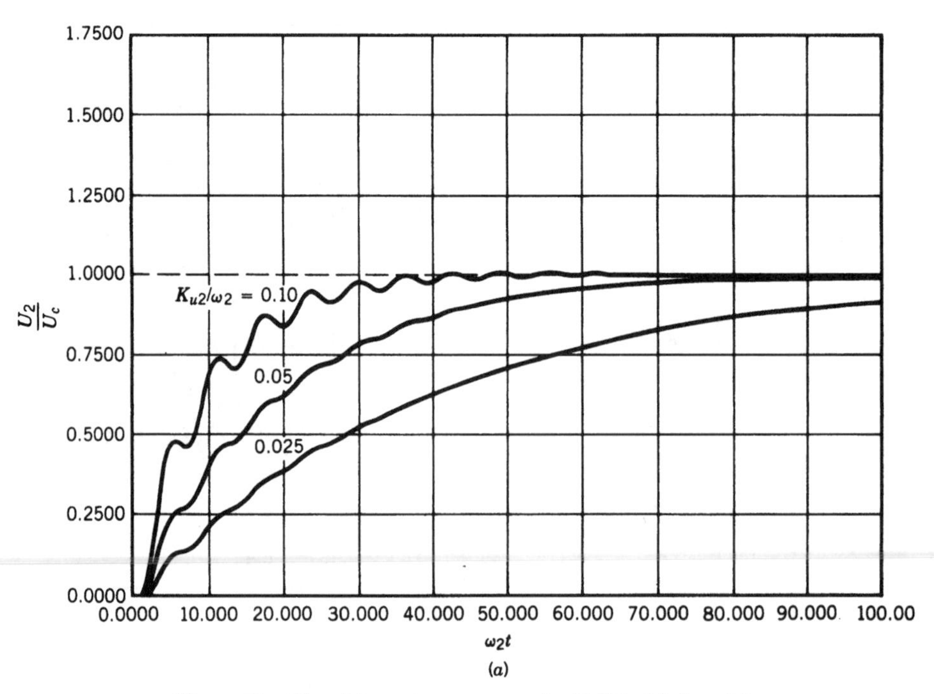

Figure 11 Closed-loop step responses for U_2/U_c: (*a*) $\zeta_2 = 0.1$.

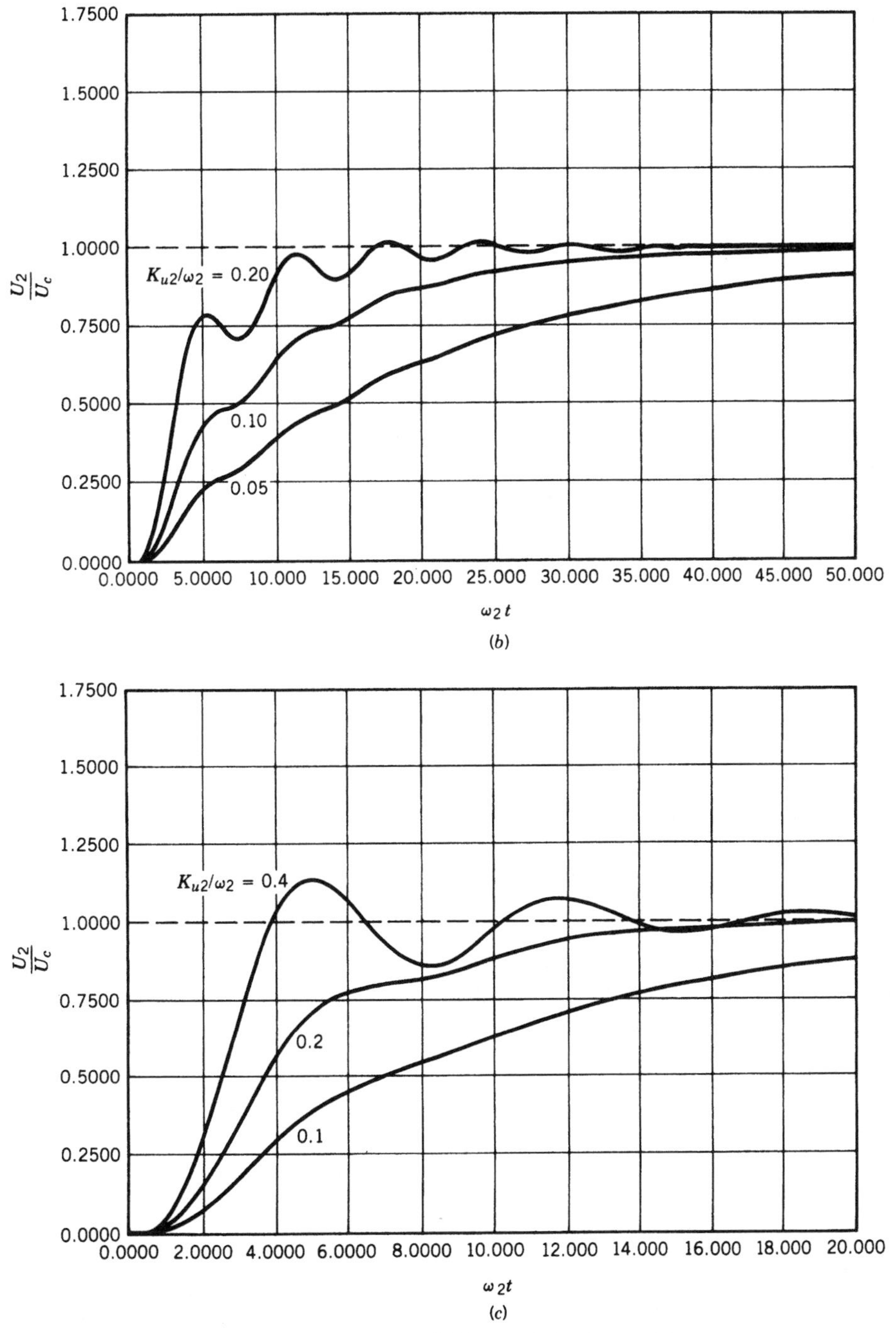

Figure 11 *(Continued)* (*b*) $\zeta_2 = 0.2$; (*c*) $\zeta_2 = 0.4$.

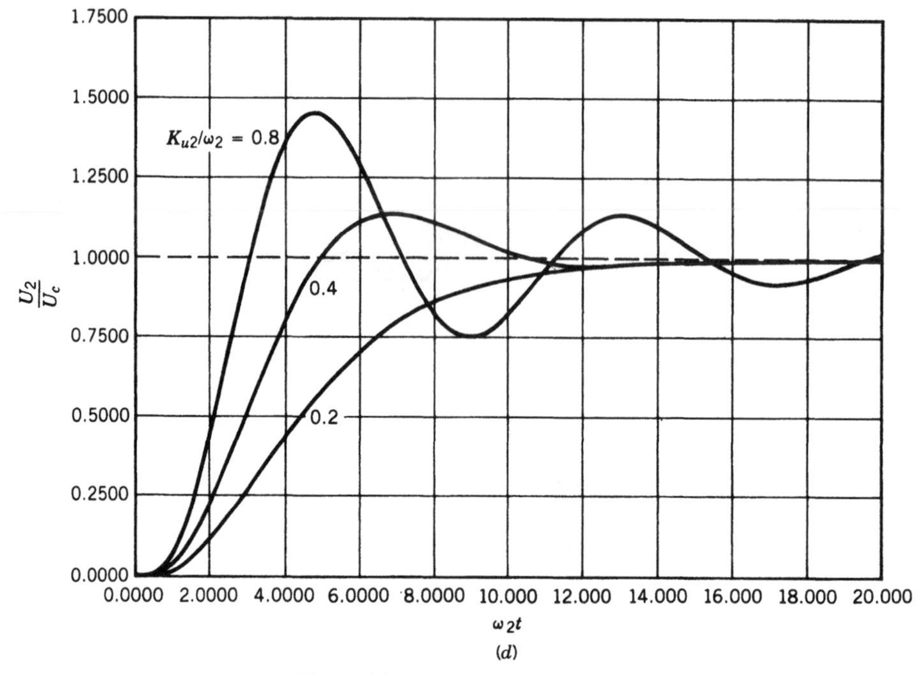

Figure 11 *(Continued)* *(d)* $\zeta_2 = 0.8$.

$$G_d = \frac{K_d}{\tau_1 s + 1} \tag{12}$$

and

$$\frac{U_1}{D} = \frac{K_d}{K_{u1}} \frac{s}{(\tau_1/K_{u1})s^2 + (1/K_{u1})s + 1} \tag{13}$$

For a position loop

$$G_d = \frac{K_d}{s(\tau_1 s + 1)} \tag{14}$$

and

$$\frac{U_1}{D} = \frac{K_d}{K_{u1}} \frac{1}{(\tau_1/K_{u1})s^2 + (1/K_{u1})s + 1} \tag{15}$$

It is instructive to compare Eqs. (13) and (15) with Eqs. (2) and (4), respectively.

3 FREQUENCY COMPENSATION TO IMPROVE OVERALL PERFORMANCE

In Section 2, it is clear that open-loop dynamic characteristics impose profound limitations upon closed-loop performance. However, it is often possible to extend these limitations by

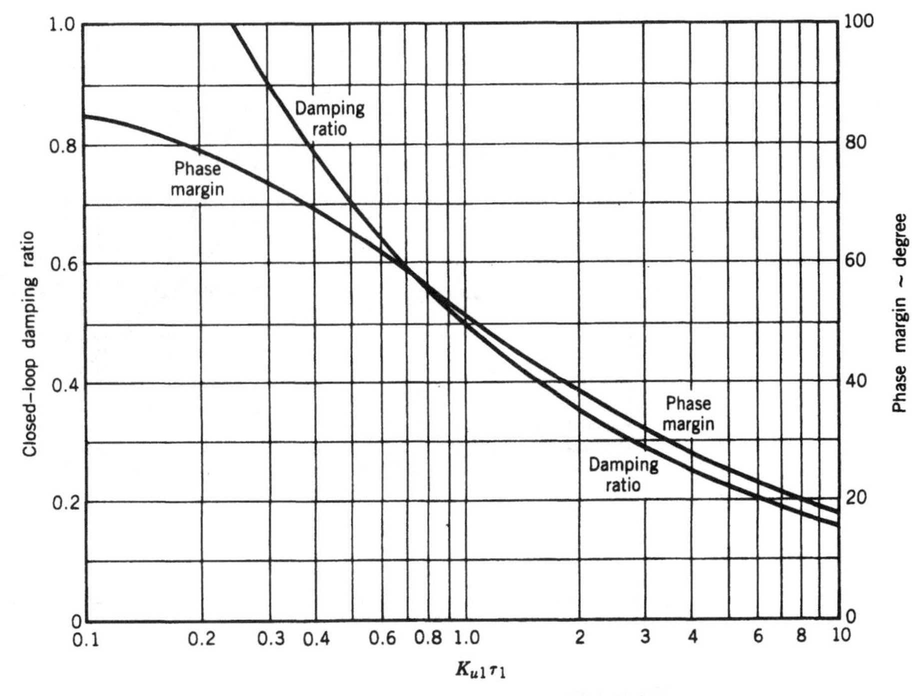

Figure 12 Stability parameters for U_1/U_c.

modifying the inherent open-loop dynamics with frequency compensation (shaping). There are many techniques for designing compensators. The best technique to use is a function of the particular open-loop dynamics under consideration, as well as closed-loop performance goals. The following sections describe some techniques that are useful in various commonly encountered situations.

3.1 Well-Damped Systems

As mentioned in Section 2.1, an ideal form for the combined forward-loop transfer function is an integrator. This ensures very high gains at low frequencies, a linear gain roll-off with frequency, and only 90° of phase lag. For systems in which the dominant open-loop poles are reasonably well damped, loop gain is usually limited by phase lag. This is clearly illustrated by Figs. 12 and 13, in which phase margins deteriorate faster than gain margins (except when ζ_2 is low).

An obvious way to improve phase margins is to make the open-loop transfer function look like an integrator out to higher frequencies. This can be accomplished by using a lead compensator, whose zeros are identical to the dominant forward-loop poles. To make the compensator physically realizable, it must have at least as many poles as zeros, but these poles can be placed at higher frequencies. The net effect of such a lead compensator is to move the break frequencies of the forward-loop poles to higher frequencies. For the example of Fig. 6*a*, the form of the compensator would be

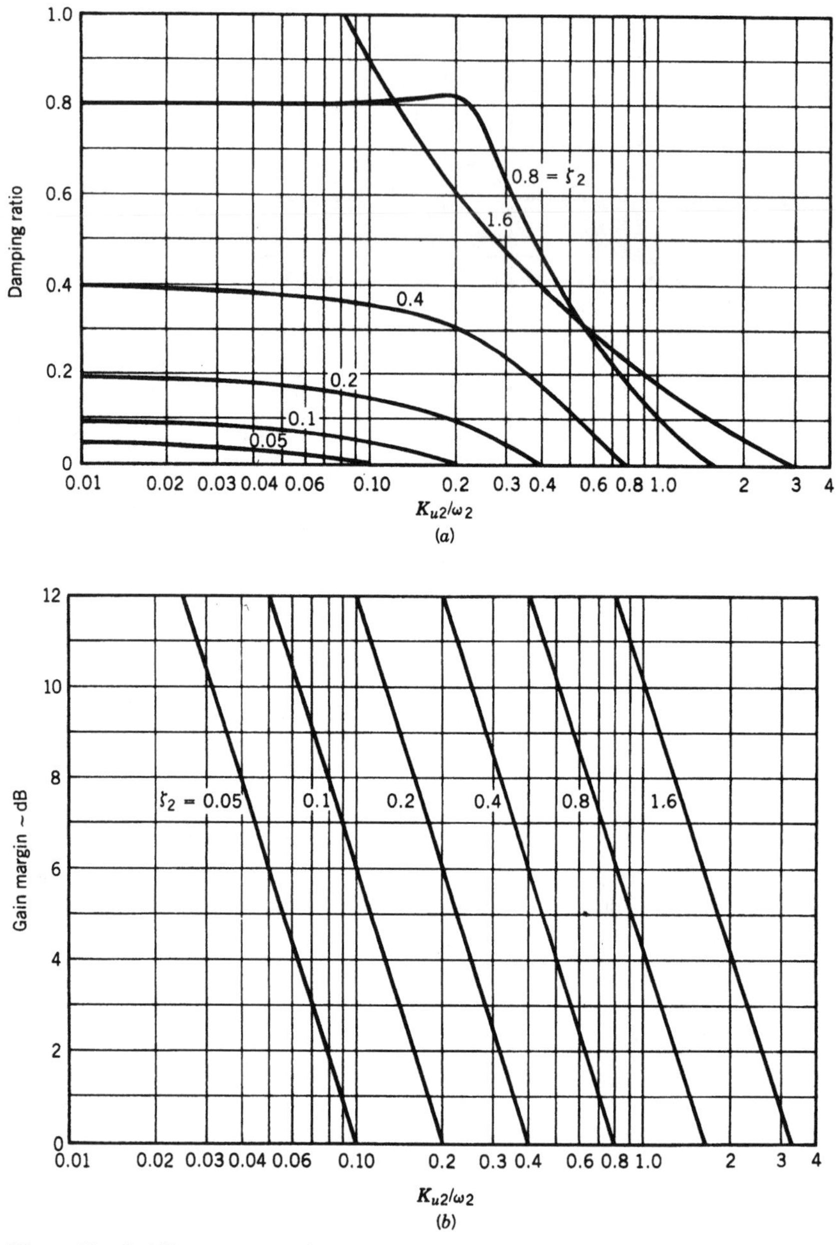

Figure 13 Stability parameters for U_2/U_c: (*a*) closed-loop damping ratio; (*b*) gain margin.

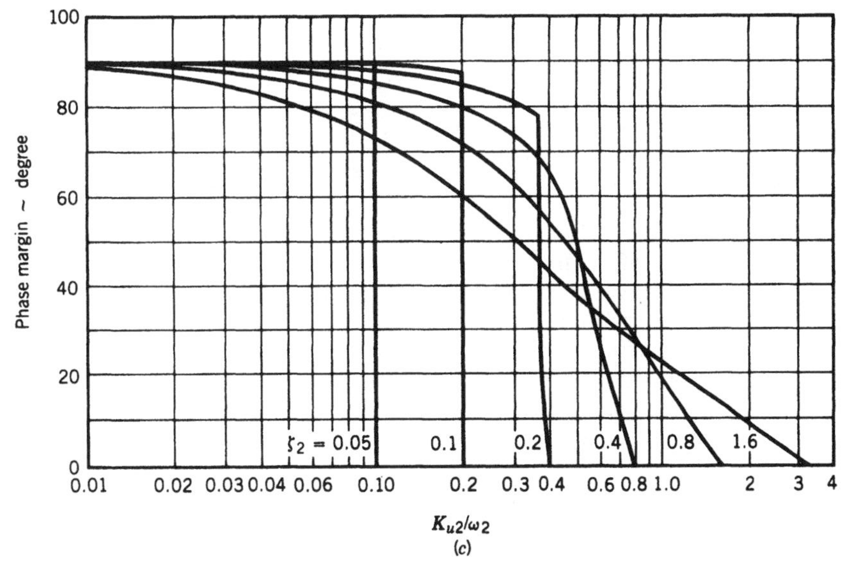

Figure 13 (*Continued*) (*c*) phase margin.

$$G_{c1} = \frac{\tau_{cz}s + 1}{\tau_{cp}s + 1} \tag{16}$$

where $\tau_{cz} = \tau_1$. For the example of Fig. 6*b*, the form would be

$$G_{c2} = \frac{s^2/\omega_{cz}^2 + (2\zeta_{cz}/\omega_{cz})s + 1}{s^2/\omega_{cp}^2 + (2\zeta_{cp}/\omega_{cp})s + 1} \tag{17}$$

where $\omega_{cz} = \omega_2$ and $\zeta_{cz} = \zeta_2$. In both cases, the closed-loop performance is now determined by the compensator poles since the original poles are canceled by the compensator zeros. This augmented performance can be quantified by using Figs. 8–13, with τ_{cp}, ω_{cp}, ζ_{cp} substituted for τ_1, ω_2, ζ_2, respectively.

There are a number of practical limitations on the use of lead compensation, primarily related to the large high-frequency gain of the compensator itself. For the examples of Eqs. (16) and (17), the high-frequency gains are τ_{cp}/τ_{cz} and $(\omega_{cp}/\omega_{cz})^2$, respectively. As a minimum, this characteristic will amplify any high-frequency electrical noise in the system. With reasonable care in the electrical design, high-frequency compensator gains of 10 or more are often practical. For a first-order compensator, this means that the forward-loop break frequencies can be boosted by a factor of 10, while the boost is only the square root of this factor for a second-order compensator. Another problem associated with the gain boost of a lead compensator is that poorly damped high-frequency modes can be excited or even destabilized. This latter effect will be further discussed in Section 3.3. The practicality of lead compensation in any given application can be best determined experimentally (additional high-frequency lags are sometimes needed).

The high-frequency noise situation is improved considerably for systems in which the integrator is electronic (e.g., the velocity servo described in Section 2.1). In this case, the integrator can replace one of the compensator poles so that Eqs. (16) and (17) are replaced by

$$G_{c3} = \frac{\tau_{z3}s + 1}{s} \tag{18}$$

and

$$G_{c4} = \frac{s^2/\omega_{z4}^2 + (2\zeta_{z4}/\omega_{z4})s + 1}{s(\tau_{p4}s + 1)} \tag{19}$$

Because of the noise-attenuating effect of the electronic integrator, $1/\tau_{p4}$ can often be set a factor of 10 greater than ω_{z4}. As previously discussed, the ratio of ω_{cp}/ω_{cz} in Eq. (17) is often limited to $\sqrt{10}$. As shown in the frequency responses of Figs. 14 and 15, the open-loop characteristics of a lead-compensated system will exhibit substantially improved phase characteristics when the integration is electronic rather than inherent. This makes it possible to greatly improve the closed-loop bandwidth of the system for a given level of stability.

In some cases, the open-loop dynamics may be dominated by a low-frequency lead term rather than a lag. When this occurs, a canceling lag compensator can often prevent instabilities due to poorly damped high-frequency modes. Lag compensation is normally well behaved and suffers none of the noise problems that limit the use of lead compensation.

Transfer functions describing the system open-loop characteristics are not always available. Sometimes the only available system description is an experimental frequency response. When this is the case, a compensator can often be designed by graphically subtracting the experimental amplitude ratio (in decibels) and phase from those of an integrator having the same low-frequency gain. This "ideal" compensator can then be approximated with an appropriate transfer function.

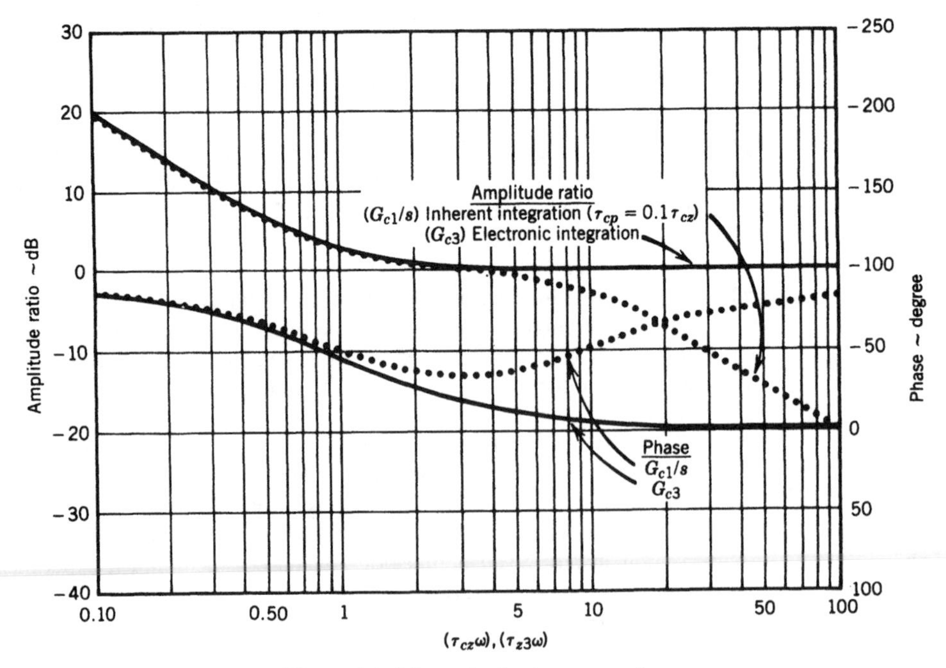

Figure 14 First-order lead compensation.

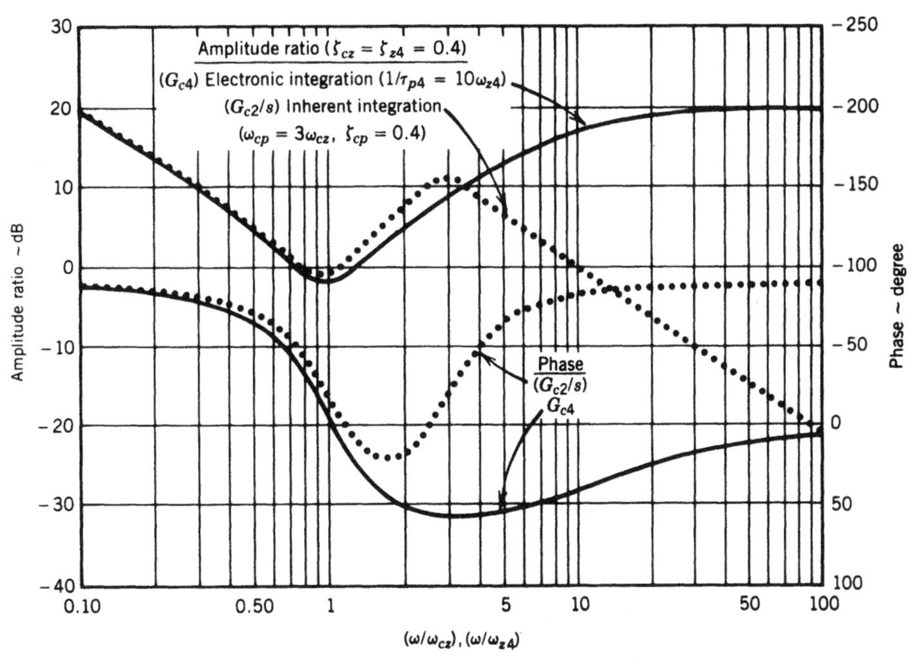

Figure 15 Second-order lead compensation.

3.2 Poorly Damped Systems

Section 3.1 discussed the benefits of lead compensation whose zeros are identical to the dominant forward-loop poles of the system. Theoretically, this technique can be used for any forward-loop transfer function. However, when the dominant forward-loop poles are second order and poorly damped, practical considerations render the technique highly risky in many cases. The basic problem is that the amplitude and phase of a poorly damped pair of poles change very rapidly with frequency in the vicinity of the resonant peak. If the compensator zeros are not precisely matched to the poles, the combined forward-loop transfer function can easily exhibit 180° of phase lag in the vicinity of substantial local peaking.

To illustrate the potential stability problem, consider the system of Fig. 6*b*, with $\zeta_2 = 0.10$. Suppose that the lead compensator of Eq. (7) is added with $\omega_{cz} = \omega_2$, $\zeta_{cz} = 0.10$, $\omega_{cp} = 3\omega_{cz}$, and $\zeta_{cp} = 0.80$. Theoretically, the forward loop is now dominated by the integrator and the compensator poles. Referring to Fig. 9*d*, it can be seen that a well-behaved response can be obtained with a loop gain $K_{u2} = 1.2\omega_{cz}$. However, suppose that ω_2 shifts to a lower value. For example, with an electrohydraulic servoactuator driving an inertial load, the "hydraulic resonance" can change 50% or more over the stroke range of the cylinder. Figure 16 shows how the closed-loop roots and the open-loop frequency response change with variations in ω_2.

Note that a reduction of ω_2 to $0.89\omega_{cz}$ will cause the closed-loop system to become unstable (0 dB at $-180°$ phase). Even if the natural frequency of the forward-loop poles does not change at all, the poles remain poorly damped in the closed-loop transfer function. Although they are masked by the compensator zeros with regard to command inputs, they may be excited by disturbance inputs to the system. The same general comments apply to

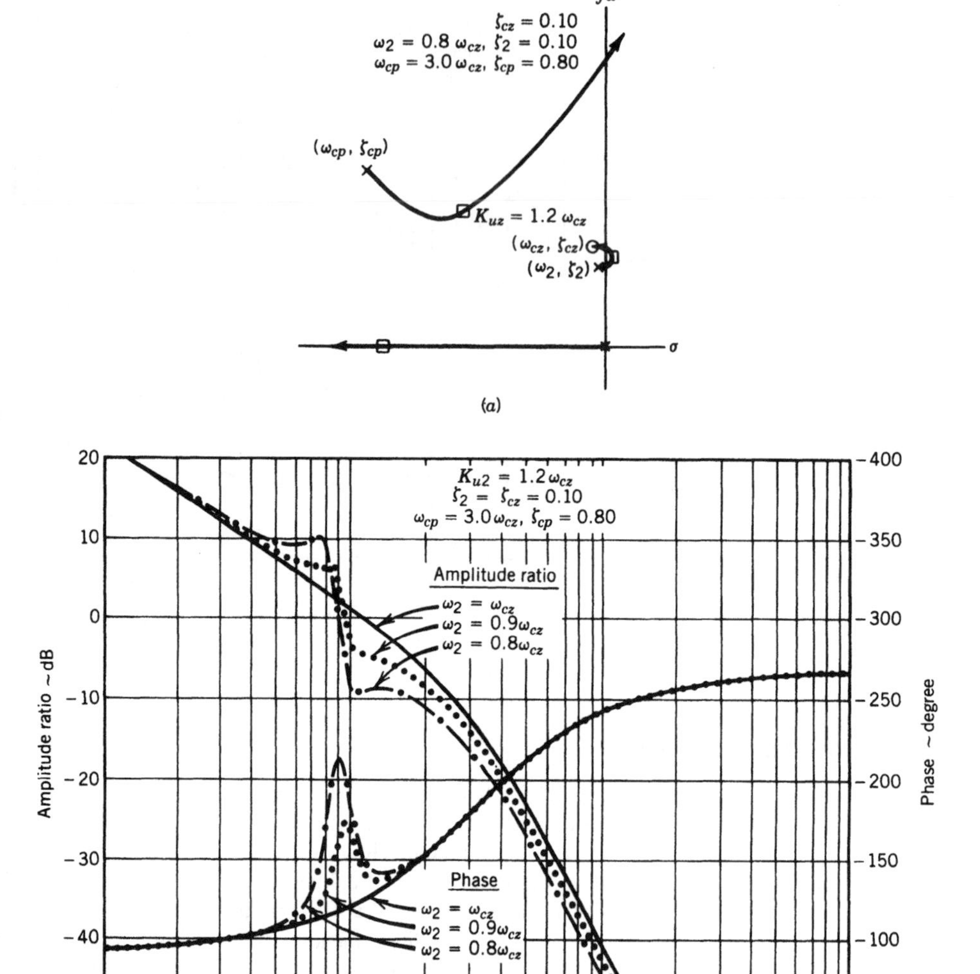

Figure 16 Effects of variations in forward-loop poles on stability of a lead-compensated loop: (*a*) root locus; (*b*) open-loop frequency response.

the use of notch filters. The notch is intended to attenuate the resonant peak, without the bandwidth boost of the lead compensator described in the present example ($\omega_{cp} = \omega_{cz}$, instead of $3\omega_{cz}$).

However, it is possible to design a compensator that will improve the damping of these poles. To accomplish this, it is useful to recall that for any system in which the number of poles exceeds the number of zeros by two or more, the sum of the real parts of all the poles is not changed when the loop is closed.[1] In this case, closing the loop will cause some roots to become more stable and others to become less stable (usually those that were poorly damped to begin with). On the other hand, if the number of poles exceeds the number of zeros by one or less, it is possible for the loop closure to improve the stability of all the roots. Therefore, to be effective in damping a poorly damped dynamic mode, lead compensation of sufficiently high order is required. To illustrate this concept, again consider the system of Fig. 6*b*, with $\zeta_2 = 0.1$. Even if we consider ideal lead compensators having no poles, Fig. 17 makes it clear that damping cannot be improved unless the order of the lead is 2 or more.

As mentioned in Section 3.1, there are important practical constraints on the use of second-order lead compensation. If the forward-loop integrator is electronic in nature, making it practical to achieve a ratio of compensator pole-zero break frequencies on the order of 10, the improvements in damping indicated by Fig. 17*c* are indeed possible. Such a compensator is defined by Eq. (19), with $\omega_{z4} = \omega_2$, $\zeta_{z4} = 0.4$, and $1/\tau_{p4} = 10\omega_{z4}$. The resulting root locus, shown in Fig. 18*a*, indicates that open-loop gain on the order of $8\omega_2$ is possible with good stability (gain margin 6 dB and phase margin 50°). To achieve the same gain margin without compensation, the open-loop gain would be limited to $0.1\omega_2$. Closed-loop frequency responses to commands for both cases are given in Fig. 18*b*.

If the forward-loop integration is inherent rather than electronic, the ratio of compensator pole–zero break frequencies may be limited to values as low as 3 (Section 3.1). Using the compensator of Eq. (17), a good compromise is $\omega_{cz} = \omega_2$, $\zeta_{cz} = 0.4$, $\omega_{cp} = 3\omega_{cz}$, and $\zeta_{cp} = 0.4$. The resulting impact on performance is shown in Fig. 19. To maintain a 6-dB gain margin, the open-loop gain must be reduced to $1.0\omega_2$. Even with this reduced gain, the phase margin is only 30°. If the compensator is located in the forward loop, the closed-loop response to commands exhibits considerable peaking, as shown in Fig. 19*b*. This figure also illustrates that if the compensator is located in the feedback loop, the compensator zeros do not appear in the closed-loop response to commands, and the response exhibits less peaking with more phase lag.

A more straightforward method for improving the performance of poorly damped systems is by the use of lag compensation. A first-order low-pass filter in the forward loop will slow the degradation of closed-loop damping ratio as loop gain is increased. In addition, the lag compensator will attenuate abrupt command or disturbance inputs to the system, thereby reducing their ability to excite the poorly damped mode. Figure 20 shows the effects of a lag compensator optimized for the system of Fig. 6 with $\zeta_2 = 0.10$. The effects of an optimized lag compensator on various stability parameters are illustrated in Fig. 21 for several values of ζ_2. Comparisons with Fig. 13 show that lag compensation can provide a

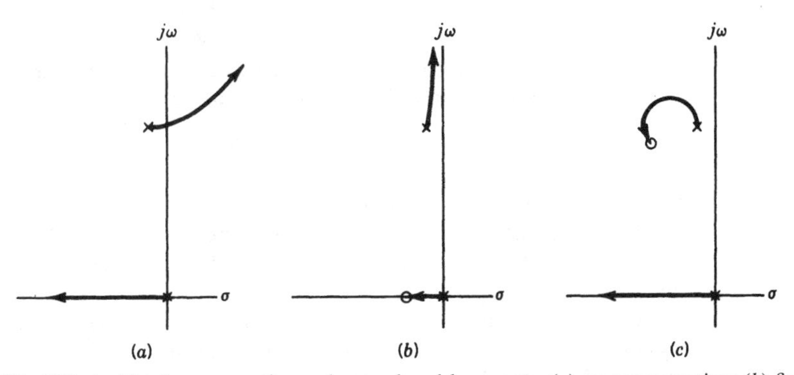

Figure 17 Effect of lead compensation order on closed-loop roots: (*a*) no compensation; (*b*) first-order lead; (*c*) second-order lead.

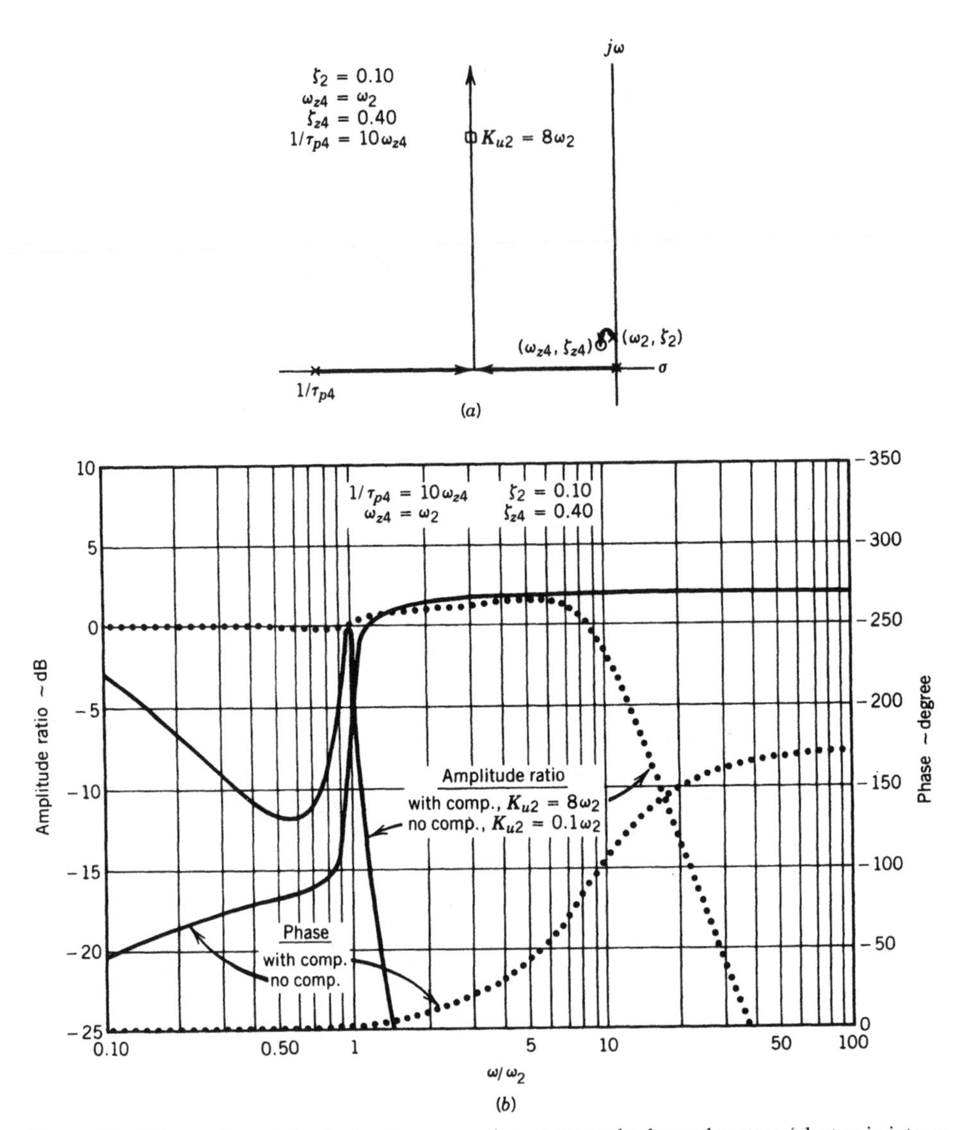

Figure 18 Effects of second-order lead compensation on a poorly damped system (electronic integration): (*a*) root locus; (*b*) closed-loop frequency response.

substantial improvement in loop gain for low ζ_2 but is of little use for $\zeta_2 > 0.3$. For $\zeta_2 = 0.10$, lag compensation allows K_{u2} to be increased from 0.10 to 0.27 s^{-1} with comparable levels of stability (gain margin 6 dB). It should be noted that very little advantage is provided by the use of a higher order low-pass filter instead of the first-order type.

In summary, the use of lead compensation to improve the damping of a poorly damped system is not straightforward, and the end result may be a marginal improvement in closed-loop performance. Lag compensation is more straightforward but offers only a modest improvement in performance. If substantial improvement in system damping and performance

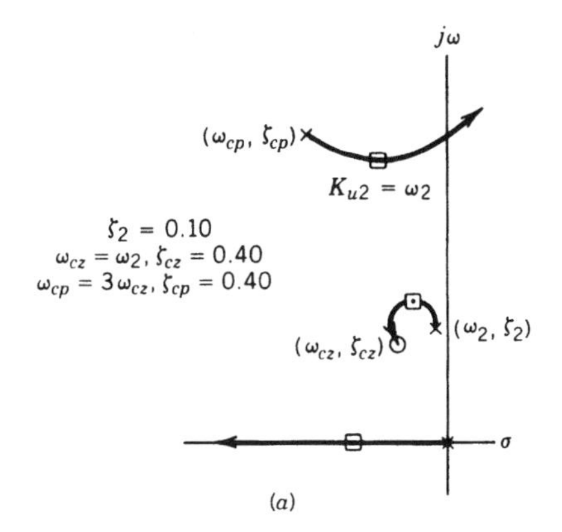

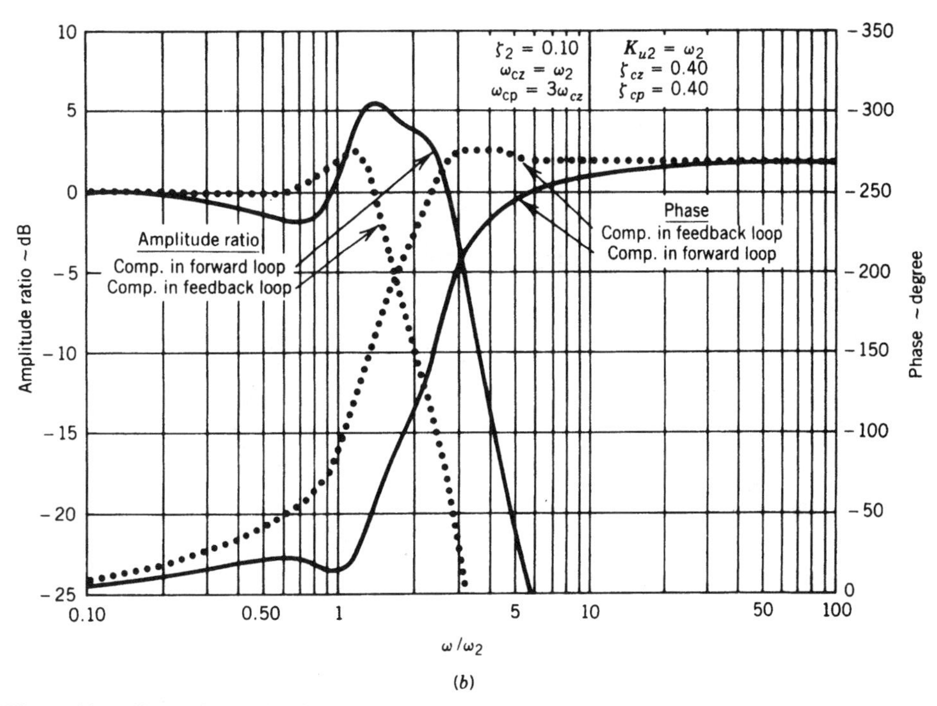

Figure 19 Effects of second-order lead compensation on a poorly damped system (inherent integration): (*a*) root locus; (*b*) closed-loop frequency response.

is required, the use of inner feedback loops is generally more effective, as explained in Section 4.

3.3 Higher Order Effects

In the foregoing sections, the discussion has centered around systems whose forward-loop dynamics can be approximated by relatively simple lag elements. While this approach is often entirely adequate for controller design and performance estimation, the designer should

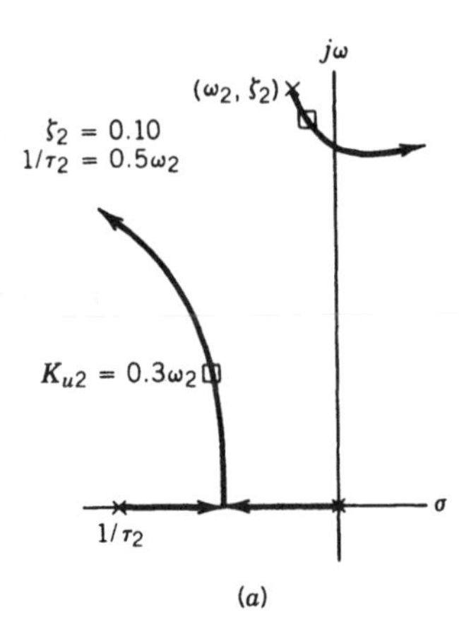

(*a*)

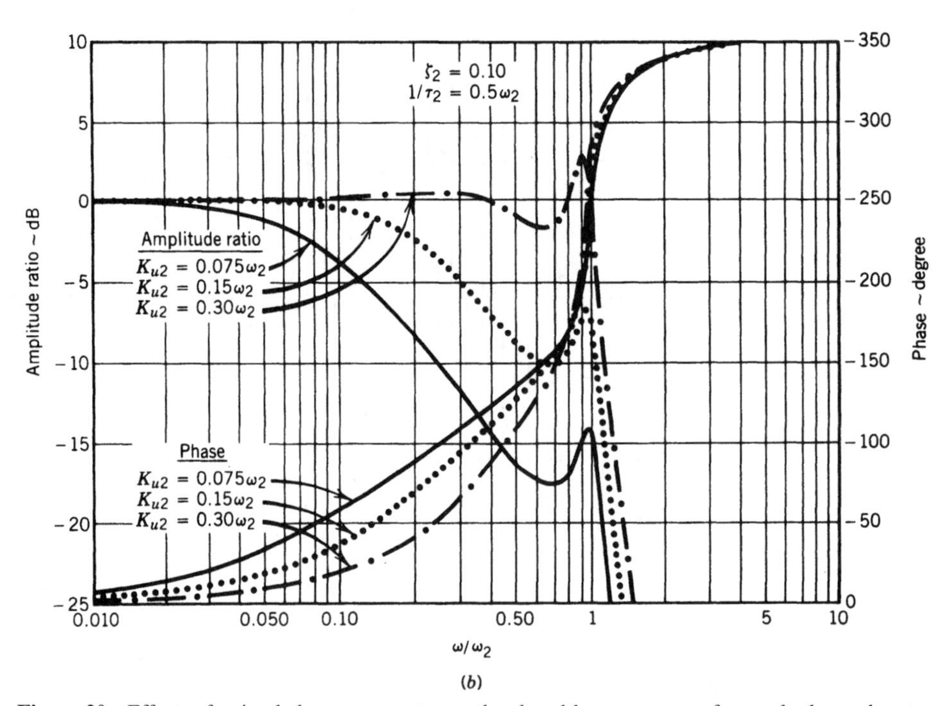

(*b*)

Figure 20 Effects of a simple lag compensator on the closed-loop response of a poorly damped system: (*a*) root locus; (*b*) frequency response.

be aware of several potential limitations. These limitations usually involve the higher order, higher frequency dynamic modes that were unknown or neglected in the early stages of the design. Examples are structural modes, transducer dynamics, and drive amplifier dynamics.

Higher order modes associated with actuator mounting structure, actuator–load mechanical connections, and transducer mounting are usually second order and poorly damped. Often these modes are characterized as poorly damped pole–zero combinations. In either

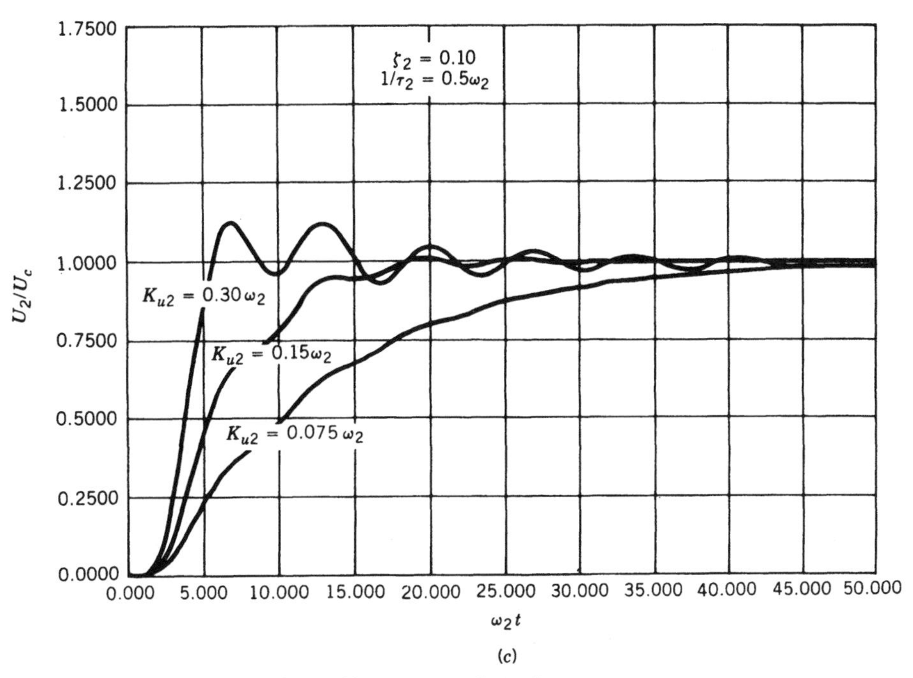

Figure 20 (*Continued*) (*c*) Step response.

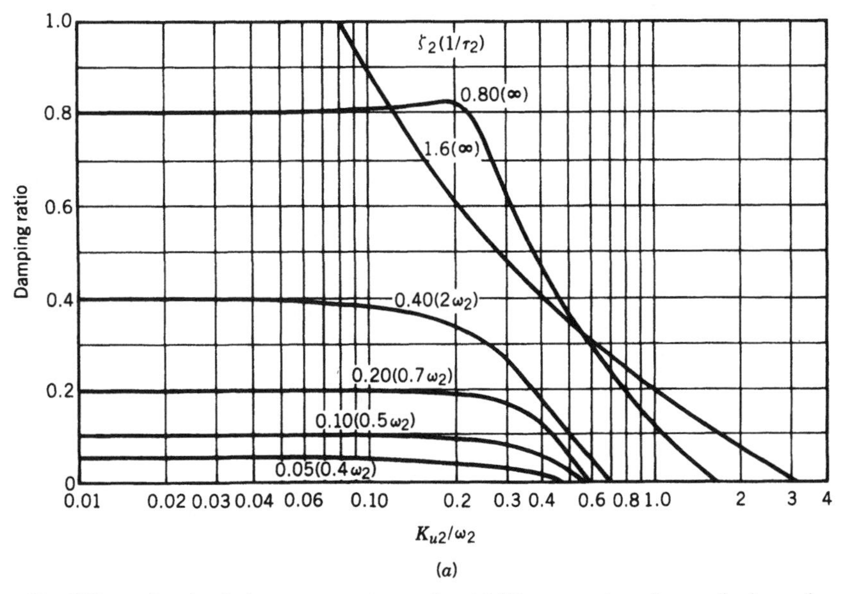

Figure 21 Effects of a simple lag compensator on the stability parameters of a poorly damped system: (*a*) closed-loop damping ratio.

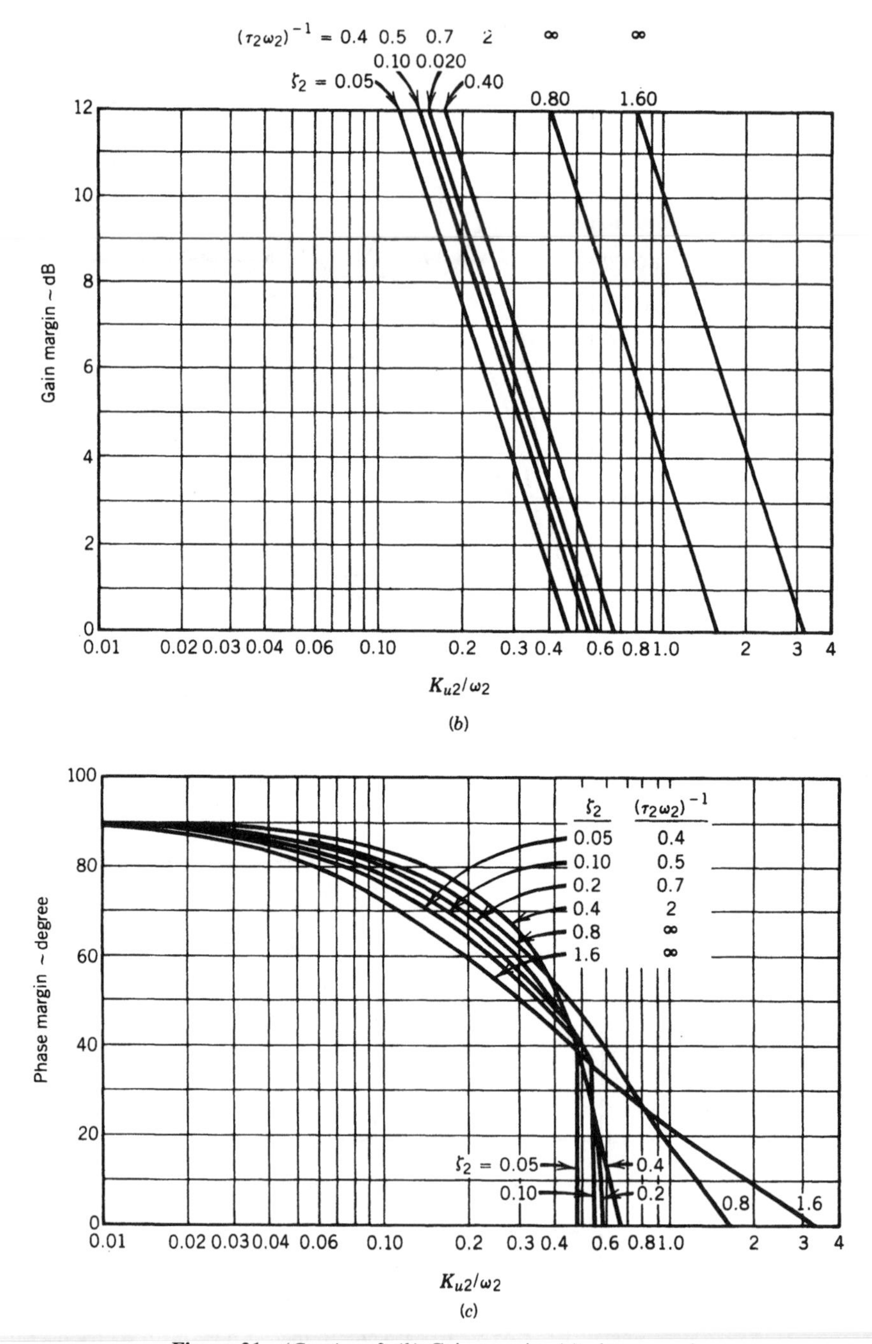

Figure 21 (*Continued*) (*b*) Gain margin; (*c*) phase margin.

case, loop gains selected to obtain adequate performance and stability from the dominant low-frequency modes could be high enough to drive the poorly damped high-frequency modes unstable. This is particularly likely if lead compensation is used to improve the dynamics of the dominant modes. High-frequency stability is probably best assessed by a root-locus plot or Bode plot that includes estimates of all the system dynamic modes. Often, accurate description of the higher order modes is difficult at the design stage. For this reason, it is worthwhile investigating the effects of various high-frequency compensation techniques that can be added when the system is first evaluated experimentally. For example, notch filters are often useful in reducing the effects of high-frequency structural modes at the cost of some low-frequency phase lag.

Many higher order modes are reasonably well damped. The dynamics of drive amplifiers, transducer signal conditioners, and notch filters are typical examples. The primary influence of such modes is that they introduce phase lags that can destabilize the low-frequency dominant modes or limit the effectiveness of lead compensation used to improve low-frequency behavior. A useful approximation of these effects can be made if there is reasonable frequency separation of the higher order modes from the dominant lower frequency modes (a factor of 5 or more). In this case, the system's low-frequency dynamic behavior can usually be assessed by replacing the high-frequency modes with a single first-order lag. The time constant of this first-order approximation, τ_3, can be determined as follows:

$$\tau_3 = \frac{\phi_3}{57.3\omega_3} \tag{20}$$

where ϕ_3 is the net phase lag (degrees) of all the combined high-frequency modes, measured at ω_3 (radians per second). The frequency ω_3 should be approximately equal to the highest natural frequency of the lower frequency dominant modes. It is probably best not to use this simplification if ϕ_3 approaches 30°.

If frequency response data are available, they can be used to estimate ϕ_3. However, at the design stage, it is likely that rough estimates of higher order natural frequencies and damping ratios are the only information available. In this case, ϕ_3 can be determined by adding the phase-lag contributions of the individual high-frequency modes:

$$\phi_3 = \sum \phi_4 + \sum \phi_5 \tag{21}$$

$$\phi_4 = 57.3\tau_4\omega_3 \tag{22}$$

$$\phi_5 = 115\zeta_5 \frac{\omega_3}{\omega_5} \tag{23}$$

where τ_4 is the time constant of a first-order mode, while ω_5 and ζ_5 are the natural frequency and damping ratio of a second-order mode (ϕ_4 and ϕ_5 due to lag terms add to ϕ_3, while lead terms subtract from ϕ_3). The approximations of Eqs. (22) and (23) are accurate to one degree or better for $1/\tau_4 > 3\omega_3$ and $\omega_5 > 3\omega_3$, respectively.

4 INNER FEEDBACK LOOPS

If the desired servo performance cannot be achieved using frequency compensation, as described in Section 3, the addition of inner feedback loops can often provide the needed improvement. This is particularly true when the open-loop damping ratio is poor and the forward loop has an inherent integration. Although inner feedback loops require additional transducers, they offer more flexibility in modifying the servo dynamics than does frequency compensation alone. The following sections discuss the merits of feeding back derivatives of the controlled variable, feeding back variables dynamically different than the controlled variable, and nonelectronic mechanizations of inner loops.

4.1 Derivatives of the Controlled Variable

Section 3 illustrates the benefits of lead compensation but also shows that its effectiveness is limited by the high-frequency gain amplification of the compensator itself. This problem can be alleviated by the use of transducers that directly measure derivatives of the controlled variable. To illustrate the potential benefits, consider again the example of Fig. 6*b* with additional feedback of the first and second derivatives of the controlled variable, as illustrated in Fig. 22*a*. As shown in Fig. 22*b*, the three feedbacks can be mathematically combined into a single loop having a pure second-order pair of zeros, without the added lag normally associated with lead compensators. The result is that much higher forward-loop gains can be achieved for a given level of closed-loop stability. Of course, each transducer will introduce higher frequency dynamic effects that will eventually limit the maximum forward-loop gain, but these effects are usually less restrictive than those imposed by electronic lead compensators. Note that the natural frequency and damping ratio of the feedback zeros in Fig. 22 are determined by the magnitude of the derivative feedback gains relative to the primary feedback gain, which is 1.0 in this case.

There are a variety of uses for derivative feedback loops, including improved closed-loop accuracy, bandwidth, and stability, as well as reduced sensitivity to changes in system parameters. These uses can be illustrated with the aid of Figs. 23 and 24, which show the effects of various sets of feedback zeros on the closed-loop dynamics of Fig. 22*b*. Figure 23 gives root loci for the different zero locations. Because electrical noise and excitation of high-frequency dynamic modes usually limit the gain in the highest derivative loop, a forward-loop gain is selected for each locus that holds $K_{u2}K_{f2}$ constant. Closed-loop frequency responses are then given in Fig. 24. It can be seen that closed-loop damping is improved in all cases. Placement of the feedback zeros at high frequency (low derivative feedback gains) yields high closed-loop bandwidth, but the second-order poles can easily be destabilized by a reduction in forward-loop gain or the presence of the inevitable higher order modes described in Section 3.3. On the other hand, low-frequency feedback zeros (high derivative feedback gains) result in lower closed-loop bandwidth but offer greatly reduced gain sensitivity and are little affected by high-frequency modes. Also in the latter case, the closed-loop response is dominated by the low-frequency second-order poles, which are nearly the same as the feedback zeros. Therefore, the closed-loop characteristics do not vary significantly with large changes in the plant parameters (K_{u2}, ω_2, ζ_2). A good compromise between bandwidth and parameter sensitivity is to set ω_{f2} approximately equal to ω_2.

Comparison of Figs. 23 and 24 with Figs. 18 and 19 show that derivative feedback offers better stability and more flexibility than lead compensation, particularly when the forward loop has an inherent integration. In addition, higher forward-loop gains can generally be used with derivative feedback, which improves the static accuracy of the system. However, forward-loop lead compensation can often produce higher closed-loop bandwidth if the forward-loop integration is electronic in nature, as illustrated in Fig. 18.

A useful way to optimize the system of Fig. 22*a* is to first establish rough estimates of the feedback and forward-loop gains using the combined feedback approach of Figs. 22*b* and 23. It is often helpful to include a first-order approximation of the phase lags caused by the higher order modes [Eq. (20)]. Once the approximate gains are established, the closed-

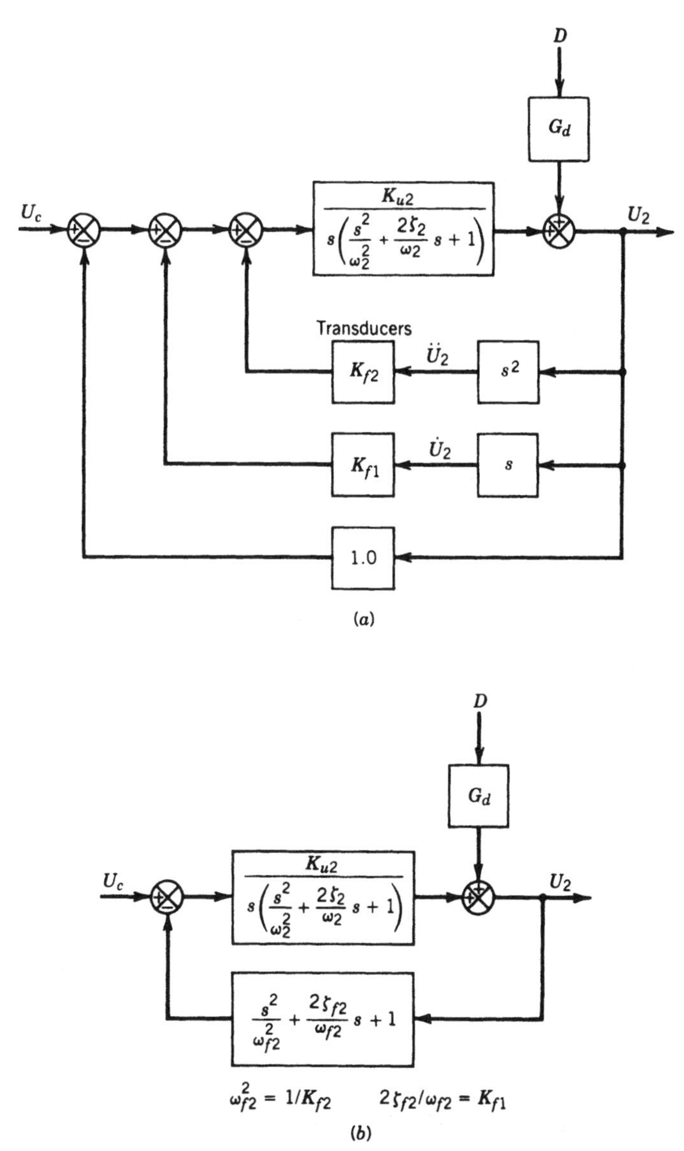

Figure 22 Feedback of controlled-variable derivatives: (*a*) inner loops to improve damping; (*b*) combined feedback loops.

loop response and stability can be checked by analysis of a complete multiloop model, with the higher order modes described more completely and placed in the appropriate loops.

If the forward-loop integrator is electronic in nature, the benefits of derivative feedback can be achieved with one less derivative than if the integrator is inherent. This is accomplished by feeding back the inner 1oops downstream of the integrator, as illustrated in Fig. 25. The static and dynamic characteristics of Fig. 25 are entirely equivalent to those of Fig. 17.22*a* and can also be mathematically reduced to the single-loop configuration of Fig. 22*b*.

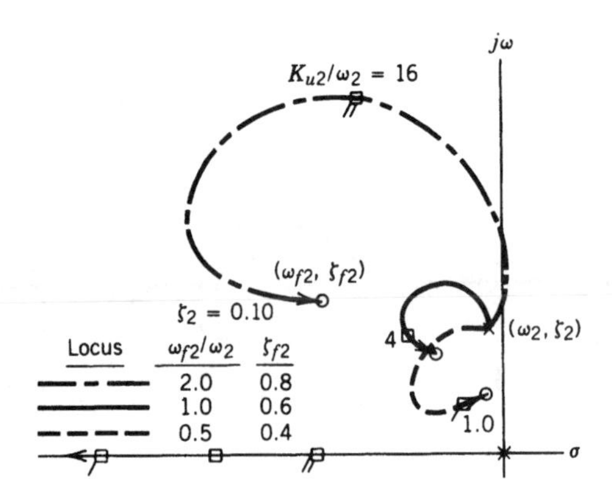

Figure 23 Effects of derivative feedback on closed-loop roots.

Note, however, that the practical implementation of Fig. 25 requires one less transducer than Fig. 22*a*.

Another way in which derivative feedback can be useful is to provide very smooth and repeatable dynamic response and high static accuracy when the primary control loop has an inherent integration. This is particularly useful when closed-loop bandwidth is not a major concern. The technique involves closing a tight integrating loop around the first derivative of the controlled variable, as illustrated in Fig. 26. This loop submerges the effects of forward-loop gain variations, static offsets, and external disturbances. The primary control loop gain can then be set at relatively low levels to ensure smooth, repeatable dynamic

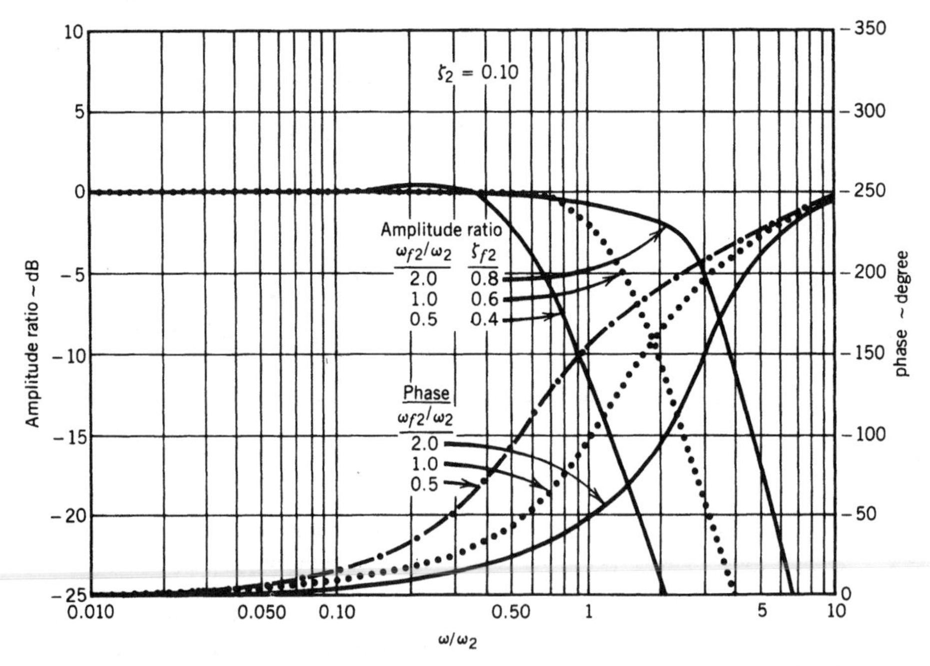

Figure 24 Effects of derivative feedback on closed-loop frequency reqponse of U_2/U_c.

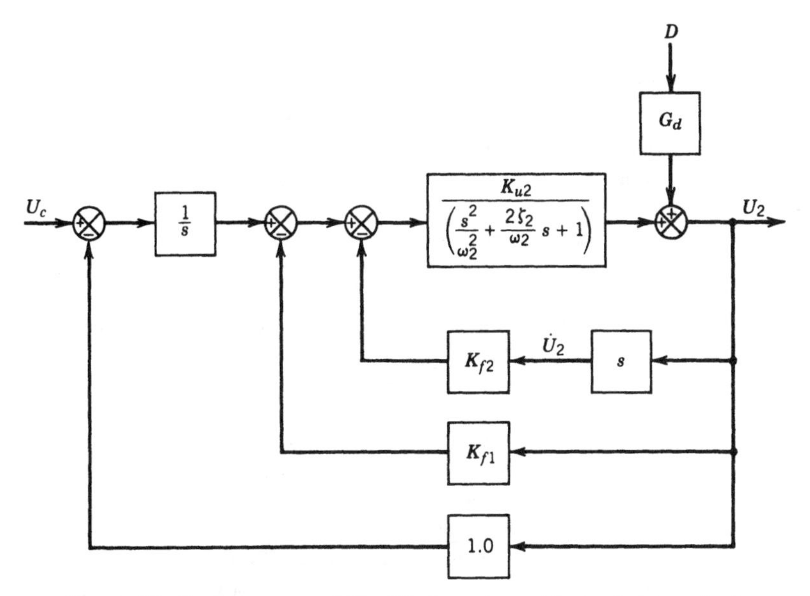

Figure 25 Rearrangement of Fig. 22*a* inner loops (electronic integration).

response without the usual concerns about reduced static accuracy. This technique is also useful if the mounting arrangement of the primary transducer results in gain-limiting higher-order dynamic characteristics in the outer feedback loop.

4.2 Alternative Inner Loop Variables

Sometimes it is advantageous to close inner feedback loops around variables that are dynamically different than derivatives of the primary controlled variable. This might be done because of practical problems related to accurately transducing derivatives of the controlled

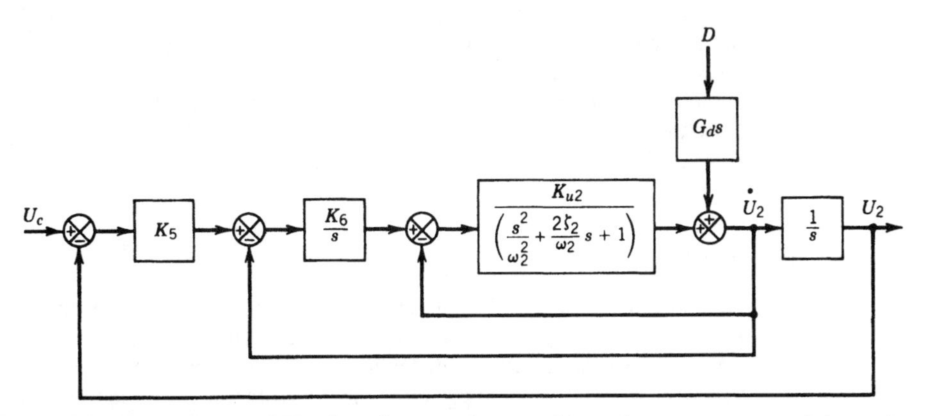

Figure 26 Rearrangement of Fig. 6 to give smooth, repeatable, and accurate response (inherent integration).

variable or because it offers an inherent advantage related to feedback dynamic characteristics. Usually, there are advantages and disadvantages of using alternative feedback loops, as illustrated in the following example.

Consider an electrohydraulic position servo that has rather demanding requirements for dynamic response. Suppose that envelope or environmental constraints make it very difficult to mount velocity and acceleration transducers or that the mounting arrangement itself introduces undesirable higher order dynamics. In this case, consideration might be given to transducing cylinder differential pressure in place of the velocity or acceleration feedbacks. If the load can be represented as primarily a mass, cylinder pressure will be proportional to load acceleration. If cylinder or load friction is large, it may be necessary to use a load cell rather than a pressure transducer. However, the load is often more complex than a simple mass. For example, the load might also have substantial stiffness to ground and viscous damping. In this case, cylinder pressure would have components proportional to load acceleration, velocity, and position. If these components were of the proper relative size and not highly variable, a pressure transducer alone could replace the acceleration and velocity transducers. However, if the load dynamics were complex or highly variable, the use of cylinder pressure as an inner loop might do more harm than good.

Another potential problem associated with the use of alternative inner feedback loops is the influence of external disturbances. Consider again the example of the electrohydraulic position servo. Feedback of either velocity or acceleration will have no effect on closed-loop static stiffness because, by definition, all derivatives of position are zero in the steady state. However, an external force applied to the load will change the cylinder pressure, even in the steady state. Therefore, pressure feedback will reduce closed-loop stiffness unless the pressure feedback signal is high passed, which introduces its own set of dynamic characteristics. Of course, there are applications in which closed-loop stiffness is not critical in the first place.

It should also be mentioned that the mounting of the primary transducer can introduce its own dynamic peculiarities. For example, the controlled variable might be load position relative to ground, and the transducer might be integrally mounted within the servoactuator assembly. If there were substantial compliance in the structure to which the actuator is mounted, the position feedback loop would contain structural zeros. An integral velocity transducer would have the same problem, but a load-mounted accelerometer would not. In this case and in others where the derivative feedback loops are dynamically different from one another, the simplified techniques of Section 4.1 are of limited use, and the complete multiloop model must be analyzed directly. Again it should be noted that these more complex derivative feedback loops may result in better or poorer closed-loop performance than comparable inner loops that feed back pure derivatives of the controlled variable. Proper assessment of these trade-offs requires a good physical model of the system, showing the proper relationships of all the feedback variables being considered.

4.3 Nonelectronic Inner Loops

Occasionally, it is useful to implement inner loop feedbacks by mechanical design rather than electronic means. For example, it may be possible to mount the servoactuator or primary transducer so that structural deflections under load produce favorable feedback zeros that improve closed-loop damping. In an electrohydraulic servo, improved damping can often be obtained by using hydraulic pressure feedback, implemented with a cross-port orifice or laminar leakage path. Both of the schemes will suffer loss of closed-loop stiffness. Sometimes it is possible to mount an electrohydraulic servoactuator so that its rod attaches to the mounting structure and its body attaches to the load. If a mass is then attached to the control valve spool and the spool is aligned with the actuator centerline, a form of acceleration feedback can be achieved (valve porting must be arranged to give proper feedback polarity). Mechanical feedback schemes offer the potential advantages of reduced costs and complexity, as well as alleviating the need for high open-loop bandwidth in the actuator and controller. However, the design of servos using these techniques often requires manipulation of rather complex physical models, which is beyond the scope of this chapter.

5 PREFILTERS AND FEEDFORWARD

As can be seen from Sections 3 and 4, the business of obtaining good closed-loop performance can become rather complex. Sometimes feedback loops and frequency compensation are optimized to achieve the desired stability and accuracy, but the closed-loop response to commands is not particularly desirable. Rather than compromise stability or accuracy by altering the servoloop characteristics, it is often easier to shape the command signal before it enters the servoloop. The following sections discuss some commonly used techniques for accomplishing this, together with their limitations.

5.1 Lag Prefilters

High-gain servoloops are required to achieve static accuracy and rejection of load transient disturbances, but rapid response to commands is often unnecessary or undesirable. In this case, the addition of a simple lag prefilter will often provide the desired result:

$$G_{\text{pf}} = \frac{1}{\tau_{\text{pf}} s + 1} \tag{24}$$

If this prefilter is placed in the command path, as illustrated in Fig. 27, τ_{pf} can be made large enough that G_{pf} dominates the U/U_c response, with an appropriate rise time.

Another way that lag prefilters can be used is to provide a particular set of dynamics for U/U_c, which are easily settable and do not change with variations in the forward-loop parameters. Often called "model following," this technique requires the use of high-bandpass servoloops, so that the dynamics of the prefilter model (G_{pf}) dominate the U/U_c response. In concept, this makes it very easy to obtain any desired U/U_c. transfer function by simply changing the electronic prefilter model. However, to have U/U_c faithfully reflect the model dynamics, it is often necessary for the bandpass of the servoloop to be an order of magnitude higher than the highest frequency singularity in the model transfer function. This is often impractical. If inner feedback loops are needed to achieve the desired servoloop bandpass, it may be more effective to tailor feedback loops to provide a combined feedback transfer

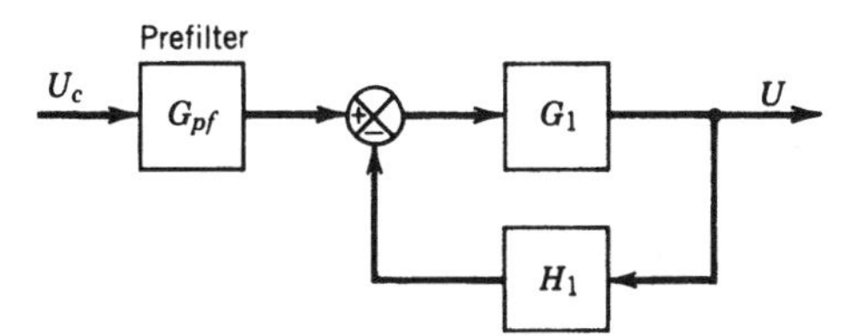

Figure 27 Generalized use of prefilters.

function that is the inverse of the desired U/U_c transfer function. This can be seen with the aid of Fig. 27 by eliminating the prefilter. If the forward-loop gain is high enough, $U/U_c = 1/H_1$. This technique is further discussed in Section 4.1, and an example is given in Fig. 22.

5.2 Lead Prefilters

Since the closed-loop response characteristics of most servoloops are dominated by lag elements, lead prefilters are often used to improve the response to command inputs. Theoretically, this can be accomplished by simply making the prefilter transfer function equal to the reciprocal of the servoloop closed-loop transfer function. For the generalized system illustrated in Fig. 27, the ideal prefilter would be

$$G_{\mathrm{pf}} = \frac{1 + G_1 H_1}{G_1} \tag{25}$$

Unfortunately, the lead required to accomplish this will be limited by its associated poles, which must be selected to prevent excessive electrical noise (as discussed in Section 3.1). Also lead prefilters can accentuate the oscillatory tendencies of a poorly damped servoloop. Even for a well-damped servoloop, overshooting response can occur if the servoloop parameters vary substantially over the operating envelope. In many cases, the lead network may be more effective if it is moved to the forward loop so that higher servoloop gains can be used.

To illustrate the effects of lead prefilters, consider a servoloop of the configuration shown in Fig. 6*a*. Since the closed-loop transfer function will be a second-order lag, as shown by Eq. (8), a second-order prefilter lead would be appropriate. The use of such a prefilter is illustrated in Fig. 28. Assume that the forward-loop gain is selected to give a closed-loop damping ratio $\zeta_6 = 0.50$. This would require $K_{u1} = 1/\tau_1$, as determined from Fig. 12. The closed-loop natural frequency can then be determined from Eq. (8), as follows:

$$\omega_6 = \sqrt{\frac{K_{u1}}{\tau_1}} = \frac{1}{\tau_1} \tag{26}$$

The prefilter zeros can now be set to cancel the closed-loop lag: $\omega_{\mathrm{pz}} = 1/\tau_1$ and $\zeta_{\mathrm{pz}} = 0.50$. As discussed in Section 3.1, a practical upper limit on the high-frequency gain amplification of a lead network is approximately a factor of 10. Therefore, the prefilter poles can be set to $\omega_{\mathrm{pp}} = 3/\tau_1$ and $\zeta_{\mathrm{pp}} = 0.50$. The net effect of the prefilter is an effective boost in closed-loop natural frequency by a factor of 3.

It should be noted that the use of lead prefilters can cause some peculiar effects when variations in the forward-loop characteristics are considered. For the example of Fig. 28, the effects of forward-loop gain variations are illustrated in Figs. 29 and 30. Because the closed-loop peaking at high gains is accentuated by the prefilter, it is usually best to optimize the prefilter for the maximum-gain situation and accept the degraded response at low gains. If

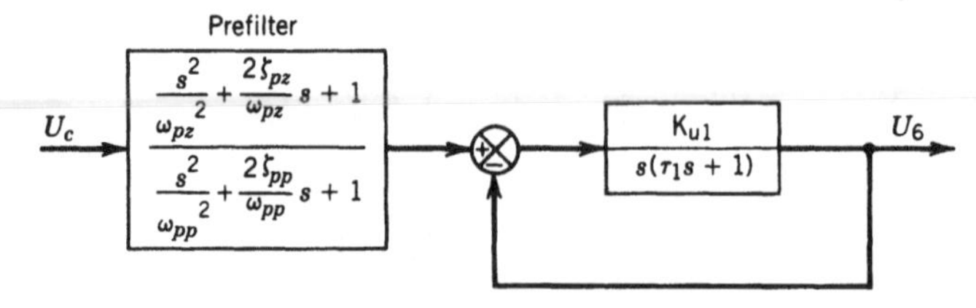

Figure 28 Example of a lead prefilter.

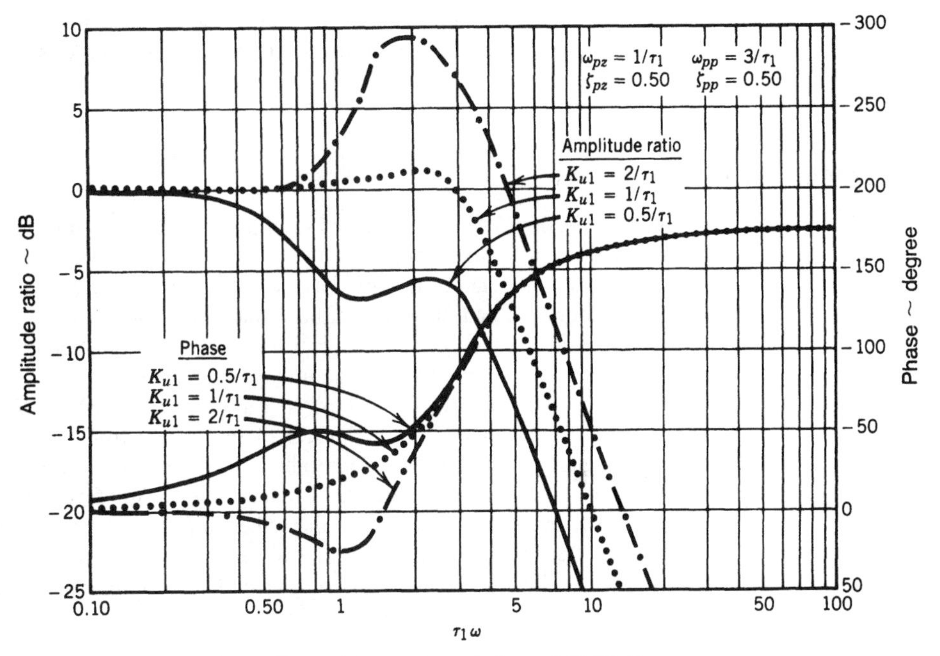

Figure 29 Effects of loop gain variations of U_6/U_c frequency response.

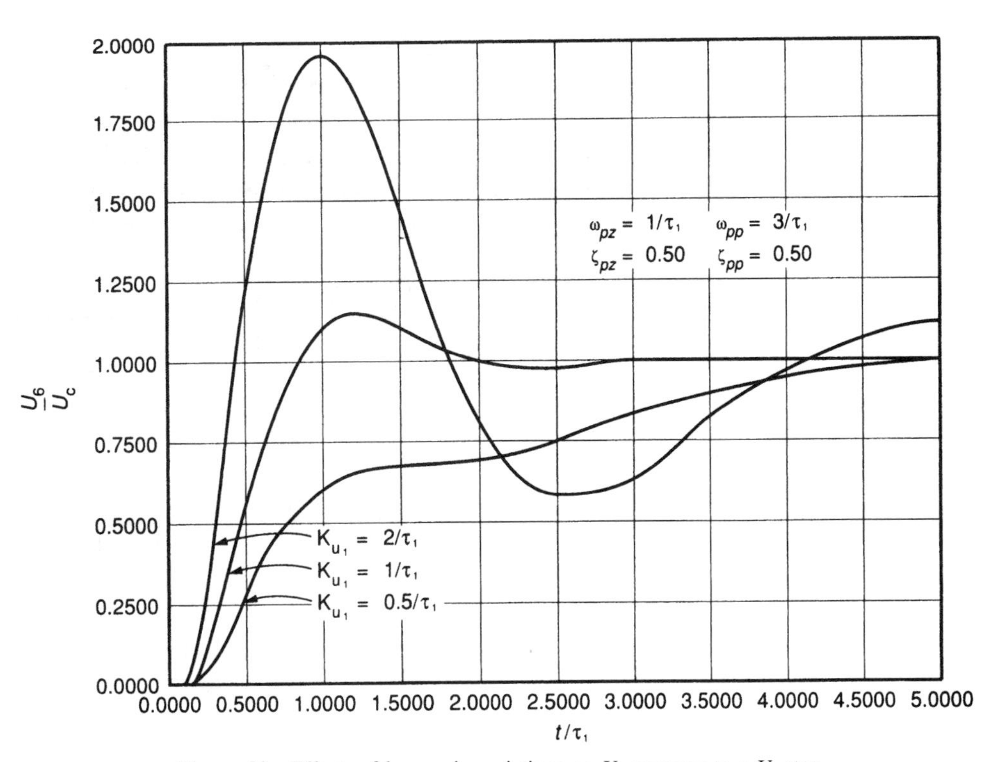

Figure 30 Effects of loop gain variations on U_6 response to a U_c step.

the variations in forward-loop characteristics are large, it may be necessary to use inner feedback loops to minimize the variations in closed-loop response, as explained in Section 4.1.

5.3 Feedforward

As explained in Section 5.2, the effectiveness of lead prefilters is limited by the fact that the command signal must be differentiated. In a system whose forward loop contains electronic lags, the number of command differentiations can often be reduced by the use of feedforward techniques. Feedforward can also be used to reduce the following errors associated with a steady rate of change of the command signal. However, these techniques must be carefully applied to prevent adverse effects on system dynamic response.

The principles of feedforward are illustrated in Fig. 31*a*. If the electronic shaping in the feedforward path approximates the reciprocal of the nonelectronic forward-loop elements, the command signal will nominally be reproduced at the output. Follow-up by the feedback loop will then try to minimize the effects of inaccuracies in the feedforward signal. The use of a prefilter matched to the feedback path further improves the overall response to commands. The net effect of the feedforward configuration can be seen by rearranging the block diagram into an equivalent one that has only a prefilter, as shown in Fig. 31*b*. From this figure, the system response can be written

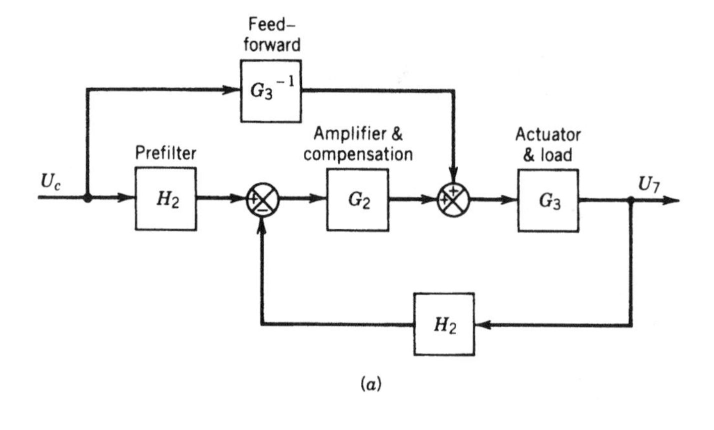

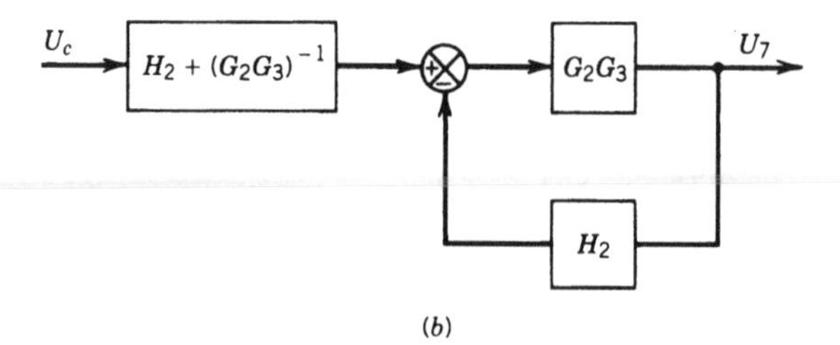

Figure 31 Idealized use of feedforward: (*a*) feedforward model; (*b*) equivalent prefilter model.

$$\frac{U_7}{U_c} = [H_2 + (G_2G_3)^{-1}]\frac{G_2G_3}{1 + G_2G_3H_2}$$

$$= \frac{G_2G_3H_2 + 1}{G_2G_3}\frac{G_2G_3}{1 + G_2G_3H_2} = 1.0 \tag{27}$$

Of course, it is not possible to actually achieve the ideal result given by Eq. (27) because of the lags associated with the lead network in the feedforward path. To illustrate the practical aspects of feedforward, it is useful to reexamine the example of Fig. 28. If the forward-loop integrator is electronic in nature, the use of feedforward offers some advantages over the straight prefilter. Figure 32*a* shows the appropriate form for the feedforward model, and Fig. 32*b* shows it reduced to an equivalent prefilter model. Notice that Fig. 32*a* requires only one differentiation of the command signal, while Fig. 28 requires two. The result is that its equivalent prefilter (Fig. 32*b*) includes a first-order lag rather than a second-order lag (Fig. 28).

To allow direct comparison of the Figs. 28 and 32 examples, the same loop gain is used in each case ($K_{u1} = 1/\tau_1$). Noting that τ_8 will be approximately equal to τ_1 and that the high-frequency gain amplification of the feedforward network should be limited to 10, it is sensible to set $\tau_7 = 0.1\tau_1$. Since the closed-loop poles are given by Eq. (8), the feedforward parameters are selected as follows:

$$\frac{K_8\tau_8}{K_{u1}} = \frac{\tau_1}{K_{u1}} \tag{28}$$

$$\frac{K_8}{K_{u1}} + \tau_7 = \frac{1}{K_{u1}} \tag{29}$$

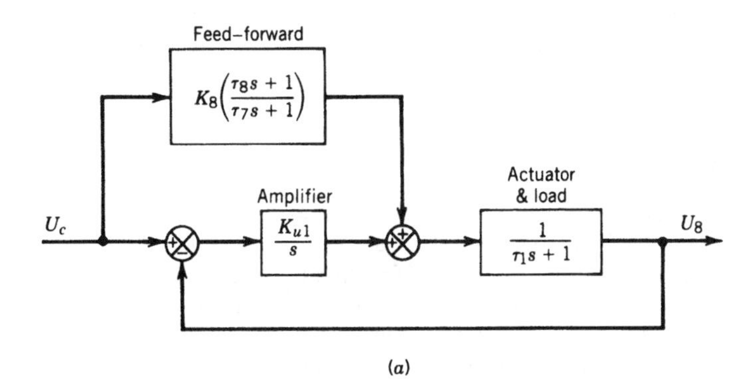

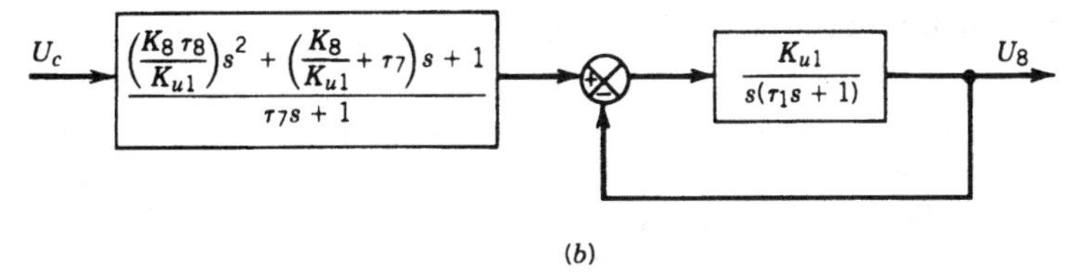

Figure 32 Example of feedforward: (*a*) feedforward model; (*b*) equivalent prefilter model.

Solving these equations after the appropriate substitutions, the feedforward parameters are $K_8 = 0.90$, $\tau_8 = 1.11\tau_1$, $\tau_7 = 0.1\tau_1$. Using these parameters, U_8/U_c frequency and step responses are computed and presented in Figs. 33 and 34, along with the corresponding U_6/U_c responses. In this example, the use of feedforward techniques offers a substantial improvement in system bandwidth, by comparison with a prefilter alone. Of course, the feedforward scheme suffers from sensitivity to forward-loop gain variations similar to those of the prefilter scheme, as illustrated in Figs. 29 and 30.

As previously mentioned, feedforward techniques can also reduce following errors. In general, the following error U_e can be calculated as

$$\frac{U_e}{\dot{U}_c} = \frac{1}{s}\left(1 - \frac{U}{U_c}\right) \tag{30}$$

For the example of Fig. 32, in which the closed-loop transfer function is determined by the feed-forward pole, the following error is determined from

$$\frac{U_{8e}}{\dot{U}_c} = \frac{1}{s}\left(1 - \frac{1}{\tau_7 s + 1}\right) = \frac{\tau_7}{\tau_7 s + 1} \tag{31}$$

For steady command rates, the error is $\tau_7 \dot{U}_c$. Without feedforward, the following error can be determined from Eq. (8):

$$\frac{U_{1e}}{\dot{U}_c} = \frac{1}{s}\left\{1 - \left[\frac{1}{(\tau_1/K_{u1})s^2 + (1/K_{u1})s + 1}\right]\right\}$$

$$= \frac{(\tau_1/K_{u1})s + (1/K_{u1})}{\tau_1/K_{u1})s^2 + (1/K_{u1})s + 1} \tag{32}$$

For steady command rates, the error is $\dot{U}_c/K_{u1}$. For this example, $\tau_7 = 0.1/K_{u1}$. In this case, the proper use of feedforward has reduced the following errors by a factor of 10.

6 PID CONTROLLERS

A very popular form of controller is called PID (proportional-integral-differential). It is very simple in concept and is relatively easy to mechanize. Many essays have been written that describe rules of thumb for "tuning" the controller. Unfortunately these tuning procedures can be rather tedious and are usually applicable for only very simple actuator–load dynamics. The purpose of this section is to offer a unified rationale for applying PID controllers that is useful in synthesizing a control system. This rationale also provides insight for systematically adjusting PID parameters on actual hardware.

6.1 Equivalence to Frequency Compensation

The basis for the ensuing discussion is that a PID controller is simply a particular form of forward-loop frequency compensation. This can be seen from the generalized controller shown in Fig. 35. Note that the differential path is filtered to limit the high-frequency amplitude ratio. Combining the parallel paths of Fig. 35, a single transfer function for the controller can be written:

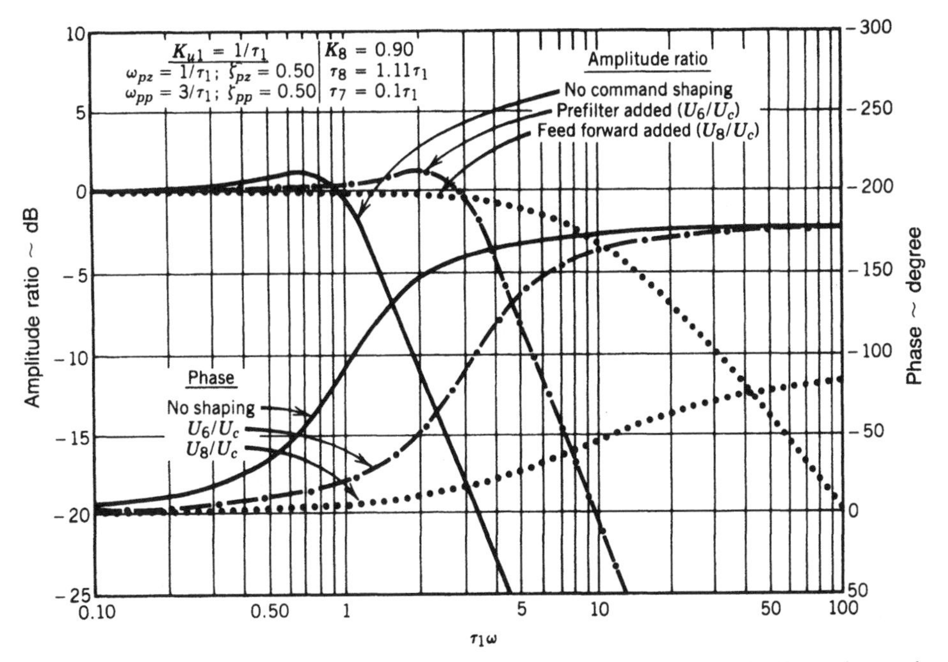

Figure 33 Comparison of closed-loop frequency responses for the prefilter and feedforward examples.

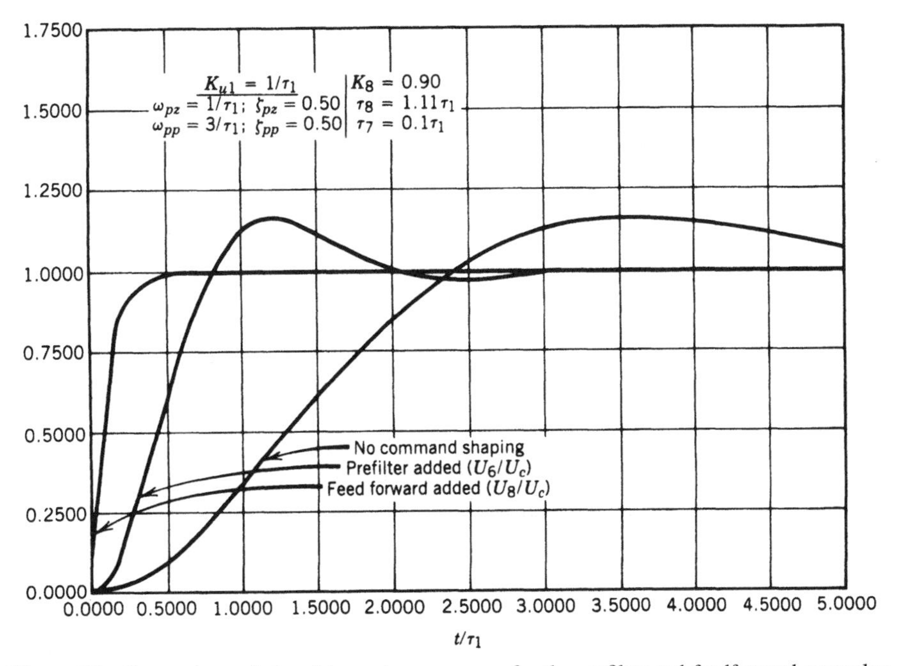

Figure 34 Comparison of closed-loop step responses for the prefilter and feedforward examples.

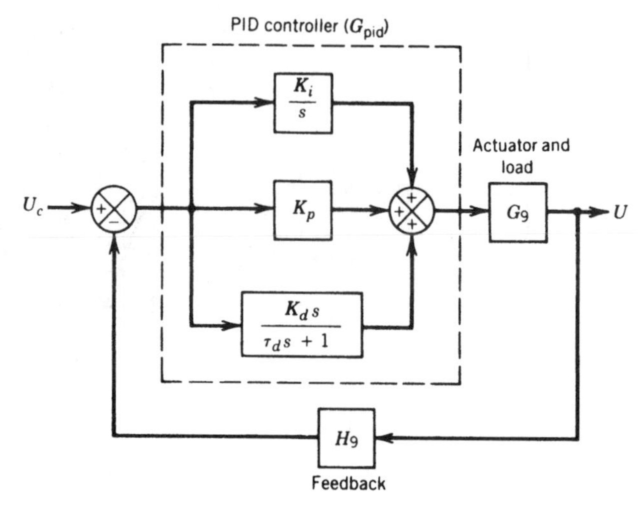

Figure 35 Generalized PID controller.

$$
\begin{aligned}
G_{\text{pid}} &= \left(\frac{K_d s}{\tau_d s + 1}\right) + K_p + \left(\frac{K_i}{s}\right) \\
&= \left(\frac{K_i}{s}\right)\left\{\frac{[K_d/K_i + (K_p/K_i)\tau_d]s^2 + [K_p/K_i + \tau_d]s + 1}{\tau_d s + 1}\right\} \\
&= \left(\frac{K_i}{s}\right)\left[\frac{s^2/\omega_{\text{pid}}^2 + (2\zeta_{\text{pid}}/\omega_{\text{pid}})s + 1}{\tau_d s + 1}\right]
\end{aligned} \tag{33}
$$

Note that Eq. (33) is simply the transfer function of a lead compensator combined with an integrator, as given by Eq. (19). As discussed in Section 3.1, it is usually possible to place the lag break frequency a factor of 10 above the lead natural frequency. In this case, the τ_d terms in the numerator are usually small:

$$
\omega_{\text{pid}} = \sqrt{\frac{K_i}{K_d}} \qquad \text{for } \tau_d \ll \frac{K_d}{K_p} \tag{34}
$$

$$
2\zeta_{\text{pid}}\omega_{\text{pid}} = \left(\frac{K_p}{K_d}\right) \qquad \text{for } \tau_d \ll \frac{K_p}{K_i} \tag{35}
$$

In some applications, second-order lead compensation is not required. In such cases, a simplified version of the PID controller can often be useful. This so-called proportional-integral (PI) controller is formed by setting K_d to zero. The resulting transfer function can then be derived from Eq. (33):

$$
\begin{aligned}
G_{\text{pi}} &= \left(K_p + \frac{K_i}{s}\right) = \left(\frac{K_i}{s}\right)\left(\frac{K_p}{K_i}s + 1\right) \\
&= \left(\frac{K_i}{s}\right)(\tau_{\text{pi}}s + 1)
\end{aligned} \tag{36}
$$

This result is similar to Eq. (18).

Another simplified version of the PID controller is obtained by setting $K_i = 0$. The transfer function of this so-called proportional-differential (PD) controller can also be derived from Eq. (33):

$$G_{\text{pd}} = \left(\frac{K_d s}{\tau_d s + 1}\right) + K_p$$

$$= K_p\left[\frac{(K_d/K_p + \tau_d)s + 1}{\tau_d s + 1}\right]$$

$$= K_p\left[\frac{\tau_{\text{pd}} s + 1}{\tau_d s + 1}\right] \tag{37}$$

This transfer function is the same as the first-order lead compensator of Eq. (16) and is normally used in systems having an inherent integration. Note that the lead break frequency can be a factor of 10 lower than the lag break frequency. In this case the τ_d term in the numerator of Eq. (37) is small:

$$\tau_{\text{pd}} = \left(\frac{K_d}{K_p}\right) \qquad \text{for } \tau_d \ll \frac{K_d}{K_p} \tag{38}$$

Frequency responses of representative PID, PI, and PD controllers are given in Fig. 36, which also shows the effects of the various controller parameters.

6.2 Systems Having No Inherent Integration

Electronic integrators are normally used in the forward loops of systems having no inherent integrations, as explained in Section 2.1. Section 3 explains the various ways in which lead

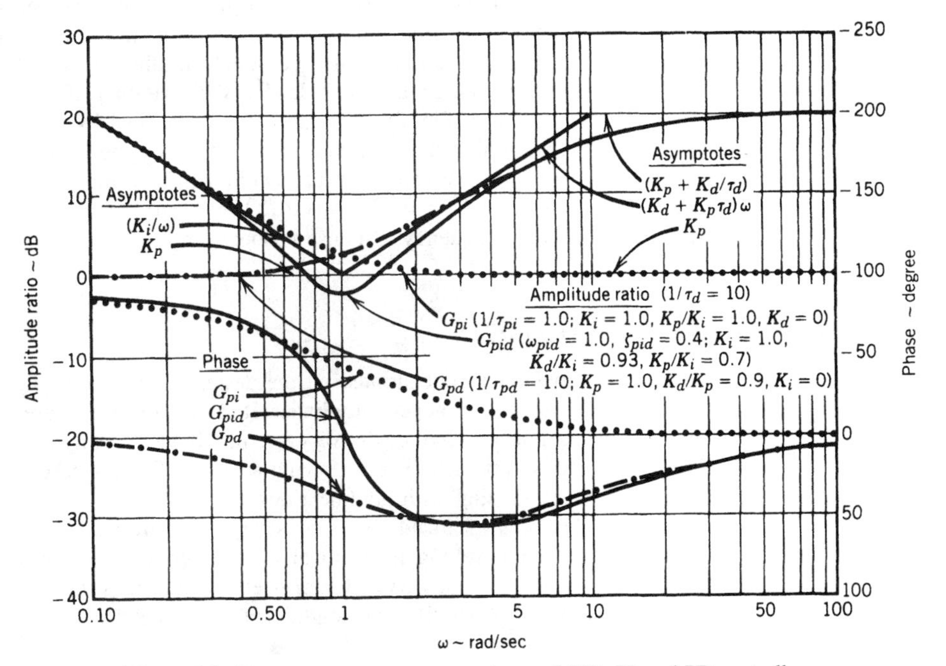

Figure 36 Frequency response comparisons of PID, PI, and PD controllers.

compensation can be usefully applied. A PID controller offers a convenient method for combining the electronic integration with lead compensation. As discussed in Section 6.1, the PI scheme provides first-order lead, while the complete PID scheme offers second-order lead.

6.3 Systems Having an Inherent Integration

Section 2.1 explains that an electronic integrator in the forward loop minimizes static servo errors. However, the addition of an electronic integrator to a system that already has an inherent integration will usually cause dynamic instability. To prevent this type of instability, lead compensation must be combined with the electronic integrator. This can be accomplished with a PI controller.

As shown in Fig. 36, a PI controller contributes nearly 90° of phase lag at low frequencies. If this is added to the 90° of lag already contributed by the integrator inherent in the system, the total low-frequency lag approaches 180°. Since the other system dynamics will add even more phase lag at high frequencies, the PI break frequency must be set low enough to ensure an intermediate frequency range over which the phase lag is reduced. The open-loop frequency responses of Fig. 37*a* illustrates this effect for a system whose inherent characteristics consist of an integrator and a second-order lag. Generally, a PI break frequency greater than 10% of the system's lowest lag frequency will substantially reduce closed-loop stability. This is shown by the phase plots of Fig. 37*a*, by the root loci of Fig. 37*b*, and by the closed-loop frequency responses of 37*c*.

Using the 10% rule of thumb for the PI controller, it is interesting to examine some time histories of the closed-loop system. Figure 38 shows time responses to step and ramp commands for the system of Fig. 37. As illustrated in Fig. 38*a*, the PI controller degrades the step response. Figure 38*b* shows that the PI's double integration at low frequencies eliminates the following error but causes larger overshoot of the steady state.

The 10% rule also applies to systems whose inherent characteristics consist of an integrator and a first-order lag. Frequency responses, root loci, and time histories for such a system are given in Fig. 39. It should also be noted that the effective lag break frequency of the system can be increased by using lead compensation techniques, as described in Section 3. For the system of Fig. 39, this can be accomplished by using a PID, rather than a PI, controller. Reexamining the example of Fig. 39 using the PID characteristics of Fig. 36, a 10-fold increase in effective system bandwidth can be achieved by setting $\omega_{\text{pid}} = 1/\tau_1$ and $\zeta_{\text{pid}} = 1.0$. In this case, one of the two PID lead terms cancels the system lag at $1/\tau_1$. The resulting open-loop frequency response is shown in Fig. 40.

7 EFFECTS OF NONLINEARITIES

The previous sections have concentrated on the design of controllers for linear systems. In practice, physical systems are never truly linear. If the nonlinearities are not large, design of the controller using linear techniques is very useful. However, it is important to understand the limitations of this approach. The following sections discuss these limitations and offer several approaches to dealing with nonlinearities.

Figure 41 illustrates idealized forms of several nonlinearities that are commonly encountered in systems controlling mechanical loads. Saturating nonlinearities can occur in transducers, electronics, and the servoactuator itself. Deadzone is the lack of output for small changes in input and is generally most significant in servoactuators and transducers. Reso-

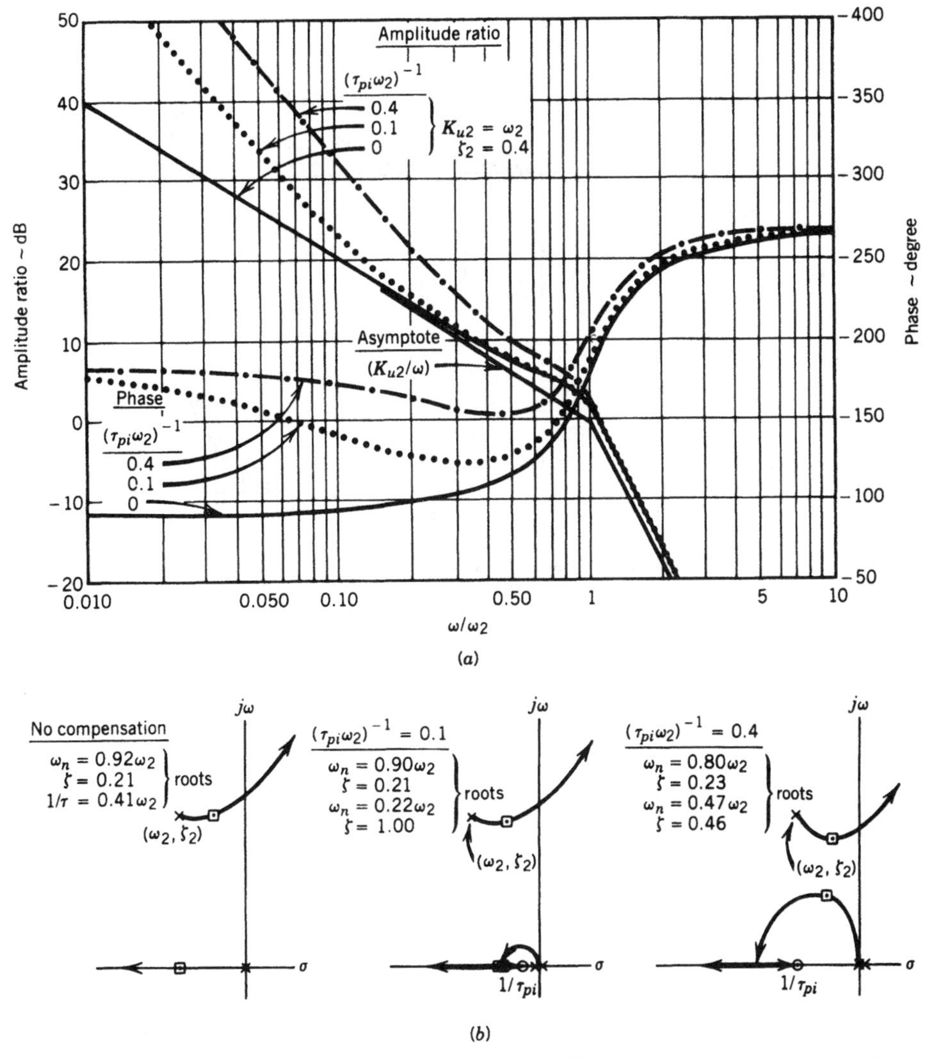

Figure 37 PI controller added to the system of Fig. 6*b*—effects of lead break frequency: (*a*) open-loop frequency response; (*b*) root loci ($\zeta_2 = 0.4$, $K_{u2} = 0.35\omega_2$).

lution is the availability of a limited number of output values and is typical of digital electronics and many types of transducers, including encoders and wire-wound potentiometers. Most servoactuators provide output velocities that are force dependent or output forces that are velocity dependent. The load–velocity curves shown in Fig. 41*d* are typical of an electrohydraulic servoactuator and are highly nonlinear. Coulomb friction is a constant force that always opposes motion. Static friction (stiction) is often larger than Coulomb friction but is very difficult to model.

Mechanical backlash is motion lost when the direction of motion is reversed, as in gear trains and bearings. Since most servoactuators make use of electromagnetic elements, mag-

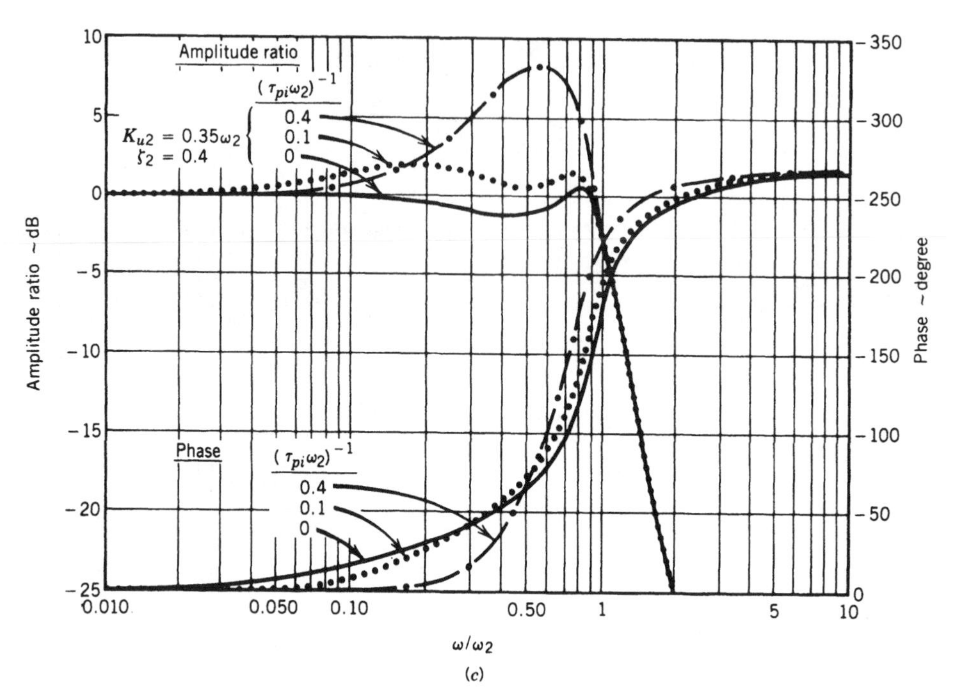

Figure 37 (*Continued*) (*c*) Closed-loop frequency response.

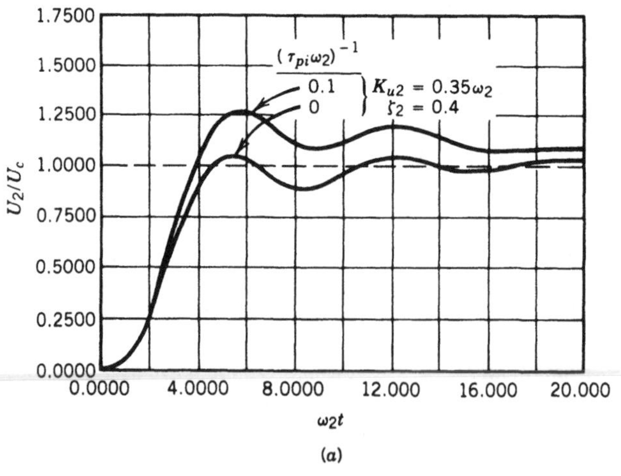

Figure 38 PI controller added to the system of Fig. 6*b*—time histories: (*a*) step response.

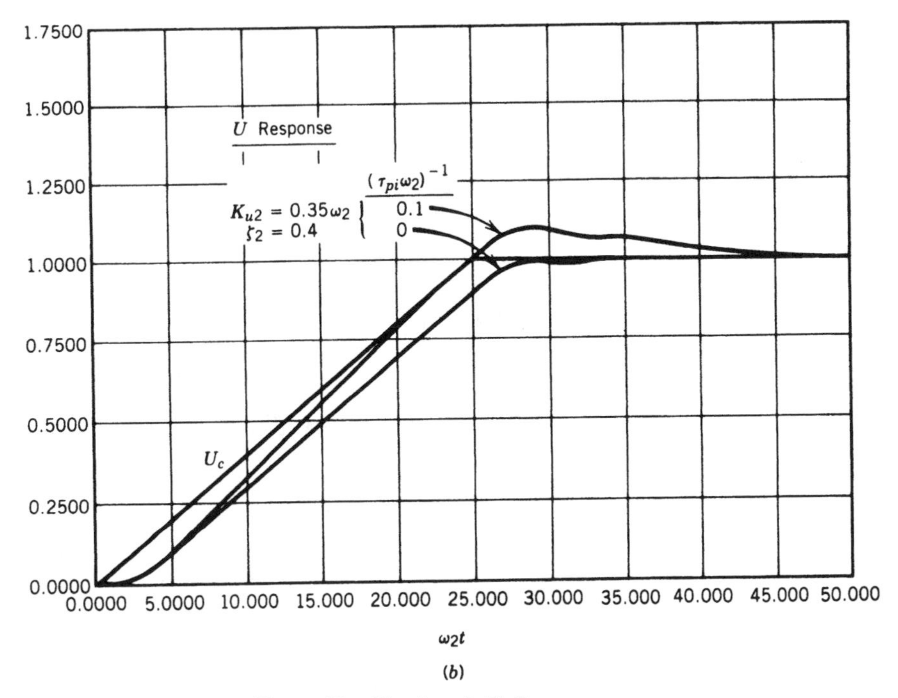

Figure 38 *(Continued)* (*b*) Ramp response.

netic hysteresis effects can cause some system performance anomalies. The width of the hysteresis band is dependent upon the amplitude of the input signal (the output is a function of the input's prior history as well as its present value).

7.1 Simple Nonlinearities

When the output of a nonlinear element depends only upon the present value of its input, the element can often be described by a simple relationship between input and output amplitude. If this function is single valued, it is often possible to assess its effect on the system by using linear approximations. One useful technique is to examine the small-perturbation behavior of the system at a series of operating points along the input–output curve by performing a linear analysis using the local slope at each operating point.

Another technique is describing function analysis, which is useful in estimating the response of nonlinear systems to sinusoidal inputs. In general, a describing function is an amplitude-dependent, frequency-dependent transfer function of a nonlinear element which allows the system to be analyzed by conventional frequency-domain techniques. It is derived from a Fourier analysis of the output of the nonlinear element to a sinusoidal input.[1] For simple nonlinearities that can be described by a single-valued output amplitude versus input amplitude, the describing function is a simple gain that varies with input amplitude. In concept, this gain is the average slope of the input–output curve for the particular input amplitude being considered.

Saturation and deadzone are two of the most common nonlinearities encountered in control of mechanical systems. Referring to Fig. 41*a*, an operating-point analysis would

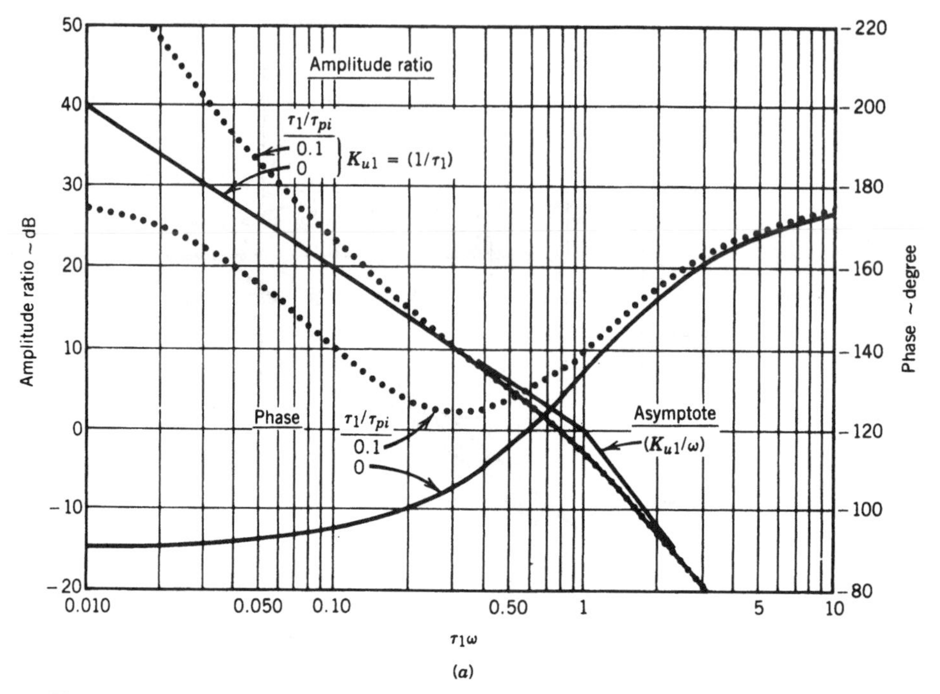

Figure 39 PI controlled added to the system of Fig. 6*a*: (*a*) open-loop frequency response.

idealize the nonlinearity as a simple gain when no saturation takes place and as zero gain when fully in the saturation region. A similar rationale can be applied to Fig. 41*b*. Generally speaking, linear techniques can be used to ensure system stability by analyzing the system with a range of gains determined by the minimum and maximum slopes of the nonlinear amplitude curve. Small-perturbation step response and frequency response of the system around an operating point can also be determined by linear analysis.

Describing-function analysis can provide useful insight into the behavior of the combined deadzone–saturation nonlinearity shown in Fig. 42. Its describing function is a gain that is zero in the deadzone region, increases to a maximum as the input amplitude approaches saturation, then decreases again as the input pushes well into the saturation region. If linear analysis predicts instability at the maximum value of the nonlinear gain, this type of describing function will result in a sustained oscillation at an amplitude corresponding to maximum gain. This behavior is called a stable limit cycle because any tendency of the oscillation to diverge will result in lower gain, which will reduce the tendency to oscillate.

Stable limit cycles can also result from deadzone in a system that is marginally stable at low gains. For example, Section 6.3 explains that a PI compensator used in a system having an inherent integration can exhibit 180° of phase lag at low gains, become stable at intermediate gains, then become unstable at high gains. In this case, a low-frequency oscillation can develop whose amplitude will grow until the describing-function gain is high enough to produce a stable limit cycle.

The effects of saturation and deadzone on system stability are generally straightforward to analyze by operating-point analysis or describing functions, as long as the system consists of single control loops. However, when multiple feedback and feedforward loops are present, the linearized analysis must be performed very carefully. For example, when an inner feed-

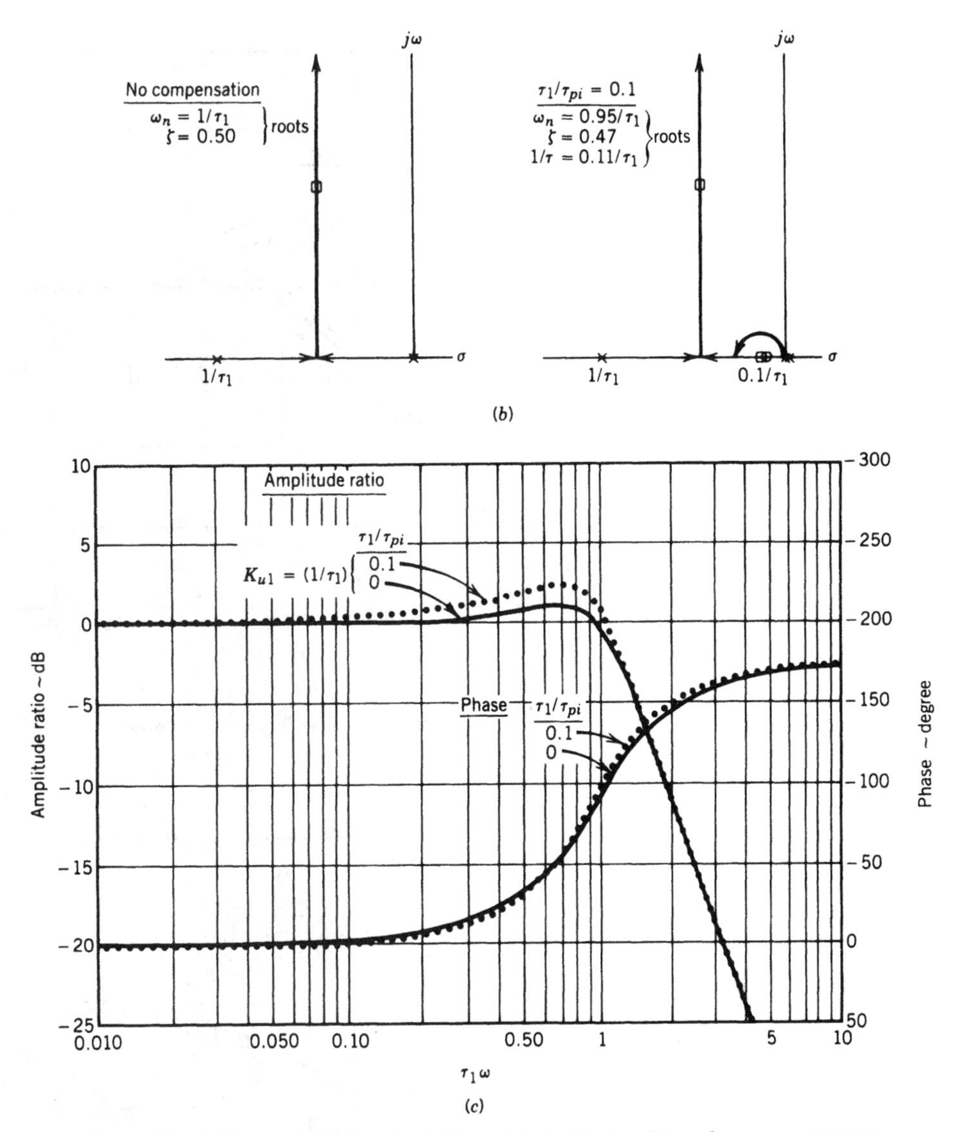

Figure 39 (*Continued*) (*b*) Root loci ($K_{u1} = 1/\tau_1$); (*c*) closed-loop frequency response.

back loop is used to damp an open-loop resonant mode so that higher gains can be achieved in the outer feedback loop, hard saturation or deadzone in the inner feedback path can cause the outer loop to become unstable. Similarly, saturation in a feedforward path or in a lead network can greatly reduce the stabilizing effects they were designed to provide. Another type of saturation is acceleration limiting. Even if it does not create any stability problems, acceleration limiting can cause large overshoots when a position servo decelerates into final position following a large step command.

The resolution of digital systems and certain transducers creates its own set of problems. As shown in Fig. 41*c*, resolution nonlinearities can be described as alternating regions of zero gain and infinite gain. Therefore, resolution will cause most feedback systems to exhibit

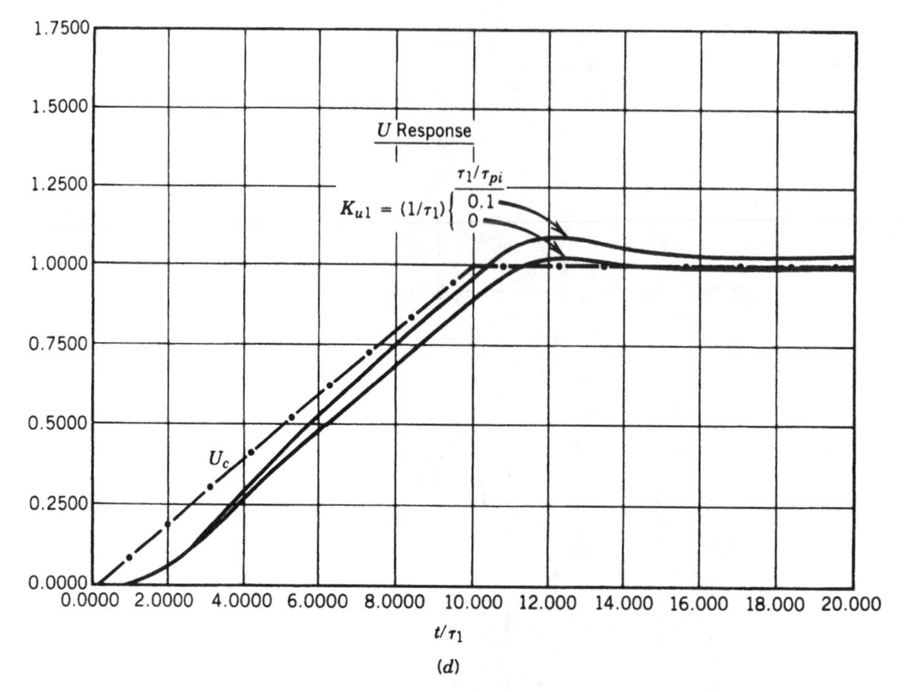

Figure 39 (*Continued*) (*d*) Ramp response.

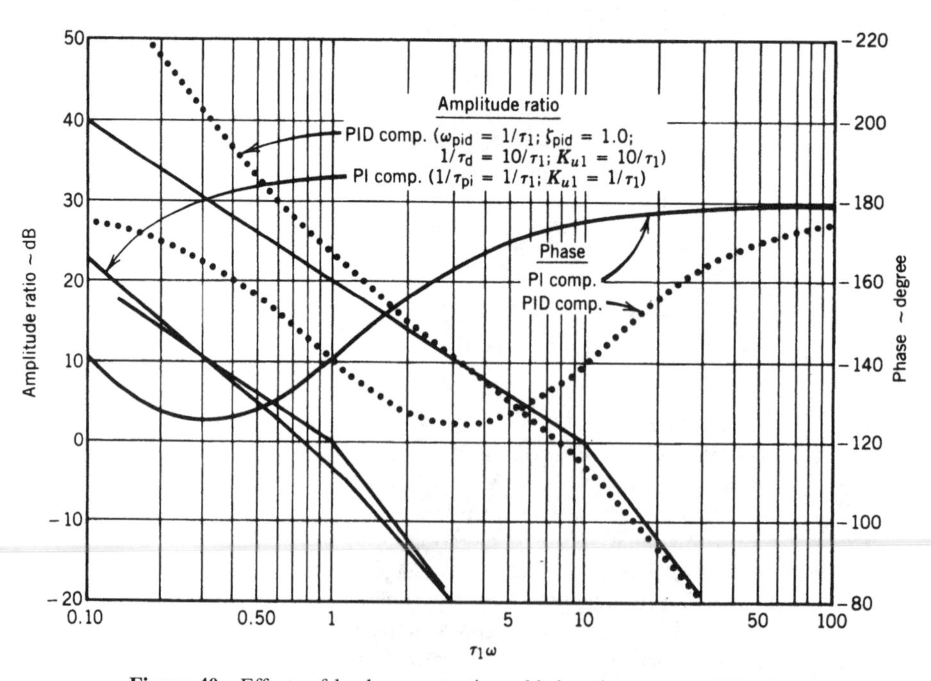

Figure 40 Effects of lead compensation added to the system of Fig. 39.

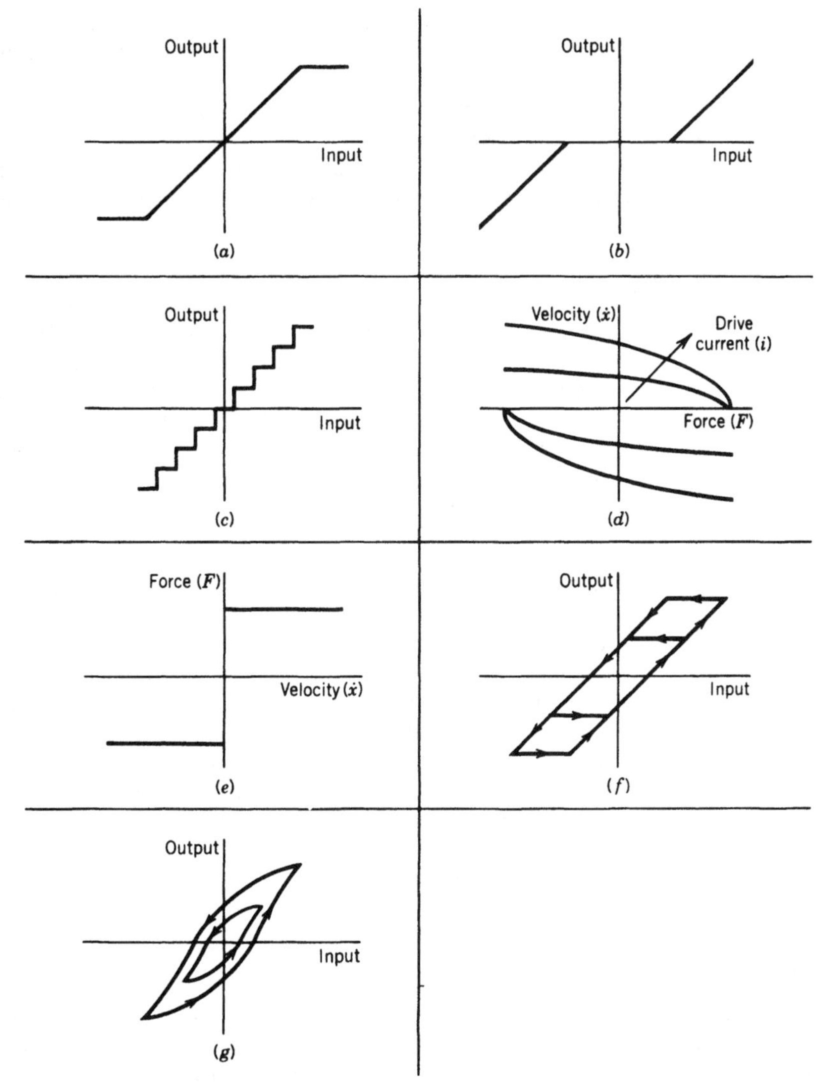

Figure 41 Common nonlinearities: (*a*) saturation; (*b*) deadzone or threshold; (*c*) resolution; (*d*) load–velocity curves; (*e*) Coulomb friction; (*f*) mechanical backlash; (*g*) magnetic hysteresis.

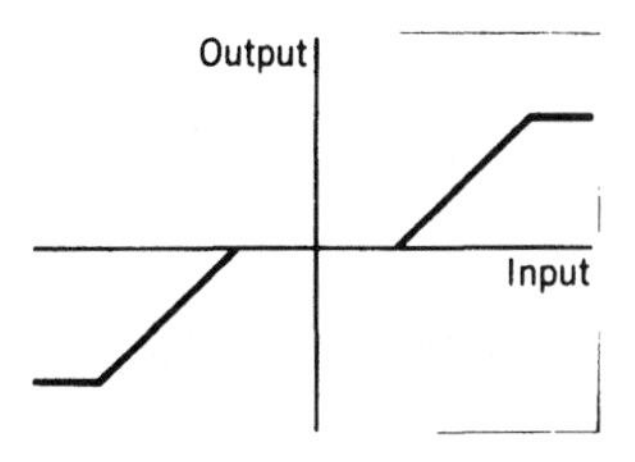

Figure 42 Combined deadzone and saturation.

continuous stable limit cycles with an amplitude corresponding to the resolution increment (least significant bit in a digital system). In a high-resolution system, the magnitude of this limit cycle may be so small that it is not noticeable. Similarly, Coulomb friction (Fig. 41*e*) exhibits infinite gain around zero and a saturating behavior as amplitude increases. In this case, however, the nonlinearity is usually a feedback loop around a mechanical load. When the load is primarily inertial and has no backlash, friction may actually improve system stability rather than decrease it. Of course, friction will also decrease system accuracy.

7.2 Complex Nonlinearities

As nonlinear elements become more complex, linear analysis becomes more complicated and less realistic. However, linear techniques may still be of some use in estimating system stability. For example, a servoactuator's output velocity may be a nonlinear function of output force as well as input drive current. For the example of Fig. 41*d*, system stability can be explored by using linearized characteristics at selected operating points:

$$\Delta\dot{X} = \left(\frac{\partial\dot{X}}{\partial i}\right)\Delta i + \left(\frac{\partial\dot{X}}{\partial F}\right)\Delta F \tag{39}$$

The two derivatives in Eq. (39) can be used in a conventional linear model showing velocity as a function of current input and load force feedback (Fig. 43). Note that the derivative of velocity with respect to force is negative in this case.

Some nonlinear elements such as hysteresis and backlash cannot be approximated by a simple relationship between input and output amplitude. Instead, the output depends upon the history of the input as well as its present value. The describing functions of such elements are typically frequency dependent as well as amplitude dependent.[1] Describing-function analysis with such nonlinearities can become rather complicated and is beyond the scope of this chapter. Also, it can be argued that computer simulation yields more realistic results without much additional effort. This is particularly true if the control system has multiple nonlinearities that are significant.

7.3 Computer Simulation

Unless the control system is extremely complex or highly nonlinear, the use of a simplified linear model is usually the best way to synthesize the basic function of the control system and to perform preliminary performance estimates. The linear techniques described in the body of this chapter are typically faster than simulation, are less prone to major errors, and promote physical understanding of the system's behavior. It is true that simplifying assumptions must be made very carefully, but this process also promotes improved understanding of the system. With the basic system function defined, simulation can then be used to evaluate

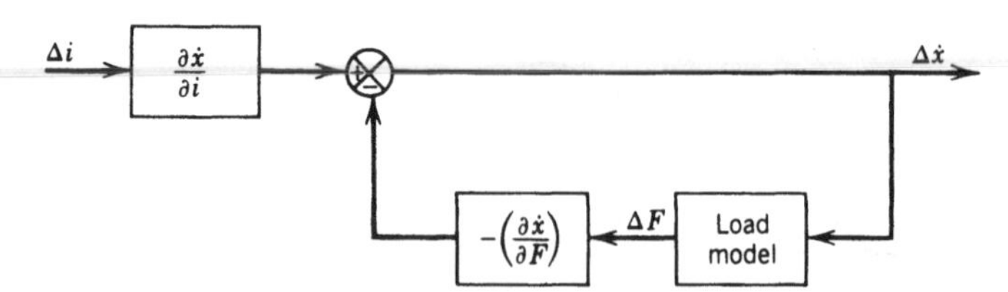

Figure 43 Linearized model of servoactuator load–velocity characteristics (from Fig. 41*d*).

the simplifying assumptions of the linear analysis, "fine tune" the system design, and generate detailed performance data over a wide range of operating conditions.

Early computer simulation was accomplished using analog computers, which offered unlimited opportunity for online operator interaction and real-time operation. Real-time capability meant that there was no waiting for data, and system development work could be accomplished using a combination of simulation and real hardware ("hardware in the loop"). Unfortunately, considerable setup time was required because of the need for patching and scaling, and there were severe restrictions on the size of the simulation that could be handled. These limitations have caused analog computers to become virtually extinct. In their place, a variety of real-time digital and hybrid simulation tools have been developed. These typically offer convenient programming, fast run times, the ability to handle large simulations, and hardware-in-the-loop capability. Their disadvantages include the high cost and maintenance problems associated with specialized computer hardware, inability to time share, inability to use for other kinds of engineering problems, and obsolescence.

To avoid the problems associated with specialized computer hardware, general-purpose simulation software is now readily available which can be run on modern PCs and workstations. If used properly, it can provide realistic results for very complicated, highly nonlinear systems. Programming is typically accomplished quickly and easily. However, such programs can have several drawbacks, such as limited ability for online operator interaction and excessive time required to generate output data, especially frequency responses. Furthermore, most general-purpose programs are carefully designed to minimize limitations on what can be programmed. As a result, they will happily violate the laws of physics without complaint. As with all computer tools, it is good practice to check out all critical program functions prior to generating data and to spot check the early data against the original design calculations wherever possible. An efficient means to accomplish this process is to start with simplified simulation models, then add complexity in layers, as each previous version is checked out.

8 CONTROLLER IMPLEMENTATION

As mentioned in the introduction, it is not the intent of this chapter to address the electronic design of a servocontroller. Nevertheless, some basic understanding of controller implementation is required to properly specify and select a controller. The following discussion describes several basic implementation approaches, together with their relative advantages and disadvantages.

Since control of a mechanical load is inherently a continuous process, the use of a dc analog controller typically provides high servo bandwidth and smooth operation. Furthermore, basic servoloops can be implemented with very simple circuits. Hard-wired digital controllers offer the potential for increased overall accuracy at the cost of degraded resolution and more complex electronic hardware. Microprocessor-based digital controllers offer increased flexibility, versatility, accuracy, and computer interface capability, but maintaining adequate resolution and sampling rates can create throughput problems. The electronic hardware associated with processor-based controllers is typically more complex than a simple analog controller. However, in complex control applications, large amounts of analog circuitry can be replaced with software.

8.1 Analog Controllers

Analog controllers are typically used in servoloops for which high closed-loop bandwidth and smooth operation are required. Modern operational amplifiers have bandwidths of several hundred kilohertz and virtually infinite resolution. Because of this, they are free of the sampling delays, phase lags, and resolution problems associated with microprocessor-based controllers. Furthermore, they are relatively tolerant of electrical noise, and troubleshooting can be accomplished with simple equipment. On the other hand, analog implementation of complex controller functions such as nonlinearities, automatic gain changing, elaborate command processing, and complex failure detection can cause the electronic circuitry to become extremely complicated. In addition, analog controllers typically require periodic adjustment and calibration.

The basic functional elements of a typical analog controller are shown in Fig. 44. The functions of prefilters and compensation networks have been discussed previously in this chapter. Some form of signal conditioning is usually required for transducers. In the case of a simple dc transducer such as a potentiometer, the conditioning may consist of dc excitation together with an output buffer amplifier. Buffer amplifiers can be designed to protect against large voltages erroneously connected to the electronics, to provide consistent loading of the transducer output, to reject electrical cabling noise (electromagnetic interference), and to filter transducer ripple. Low-output transducers, such as those that employ strain gage elements, require high-gain, low-drift amplifiers. Linear variable differential transformers (LVDTs), resolvers, and other ac transducers require ac excitation, demodulation, and filtering to remove ripple. Greatly increased servo accuracy can be obtained with a combination of ac command generation and ac feedback, such as when synchros are utilized. However, this approach has largely been replaced by the use of digital controllers and transducers. In any case, transducer specifications should be carefully studied to determine the proper signal conditioner characteristics. The dynamic characteristics of the signal conditioners, as well as the transducers themselves, can have significant impact on the stability and performance of the servoloops, as explained previously in this chapter.

There are many sources of information concerning the design of analog controllers. References 8 and 9 are excellent sources for the design of operational amplifier circuits for a wide variety of purposes, including compensation and signal conditioning. Furthermore, most manufacturers of operational amplifiers publish useful application handbooks. Also, application literature from transducer manufacturers often discusses signal-conditioning techniques in some detail.

The design of power amplifiers varies widely with the type of servoactuator being driven. In the case of conventional electrohydraulic servoactuators and other types requiring low-power electrical inputs, the power amplifier can be a very simple linear (proportional) circuit. Typically, it utilizes an operational amplifier and a power boost stage consisting of a complementary pair of transistors. Some operational amplifiers have enough output capability to provide the required electrical input directly. In the case of electromechanical devices that must provide a direct electrical-to-mechanical energy conversion, such as brushless servomotors, the power amplifier can become very complex. In this case, its design should be left to an experienced electronics engineer. Linear amplifiers are still the most straightforward and offer the best servo performance but are severely limited in the size of the motor they can control because of the large amount of heat they must dissipate. Since switching transistors typically generate little heat in their full-on or full-off states, various time-modulated on–off power drivers have been developed. The most popular type for servocontrol applications seems to be pulse width modulation (PWM). In this approach, the power devices are switched on and off at a very high fixed frequency. The percent on-time during each cycle is proportional to the dc input voltage from the upstream analog controller circuitry. If the PWM frequency is high enough, only the average cycle voltage will affect the servoactuator output, thereby resulting in nearly proportional control. To accomplish this, the PWM frequency must usually be at least one order of magnitude higher than the bandwidth of the

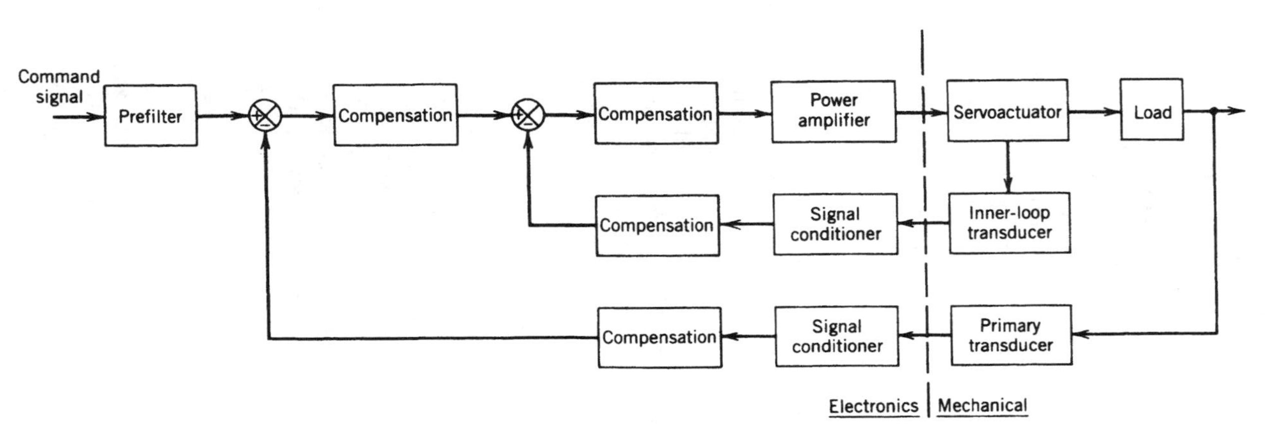

Figure 44 Typical analog controller.

innermost feedback loop (typically a current feedback loop around the power amplifier), or two to three orders of magnitude higher than the bandwidth of the primary control loop. If PWM frequencies of this magnitude are impractical, it may be necessary to reduce bandwidths of the various control loops to prevent unacceptable servo output at the PWM frequency. Design of a PWM amplifier can be difficult, particularly with regard to polarity switching around zero.

It should be noted that servoactuators requiring high-power electrical inputs can create problems related to voltage saturation in the drive electronics. The reason for this is that the input inductive characteristics of many actuators tend to cause long L/R time constants. Current feedback is often used to improve the current response of the actuator, and this leads to large transient voltages for abrupt inputs to the power amplifier. Practical limits on amplifier voltage capability can result in voltage saturation during transients. It is often necessary to design an active voltage-limiting circuit to prevent stability problems during saturation. Careful attention must be devoted to tailoring the current feedback loop and voltage limiter circuits to the servoactuator's electrical dynamics, if amplifier stability and performance problems are to be avoided.

Multiturn potentiometers are usually included in analog controller circuitry to allow proper calibration of transducer scale factors, compensate for electrical and mechanical offsets, and adjust loop gains. The desire for system accuracy often drives the designer to providing many adjustments, but this practice can greatly complicate maintenance procedures. If adequate performance cannot be achieved with a limited number of well-placed adjustments, then serious consideration should be given to the use of a digital controller.

The use of integrating amplifiers in control loops requires some special consideration. First, pure integrators often cause low-frequency oscillations or "hunting" when backlash, deadzone, or friction exists in the system. This behavior can often be controlled by adding a large resistance across the integrator's feedback capacitor. This limits the amplifier's gain and makes it look proportional at low frequencies, while preserving an integrating characteristic in the crossover frequency range. If it is possible for an integrating amplifier to saturate during abrupt commands, it may "latch up" and exhibit a long recovery period, which can result in large servo overshoots. This behavior can be prevented by proper gain distribution in the servoloops or by providing the amplifier with a diode limiter. Of course, an integrating amplifier can drift into saturation if it is powered up before the servoactuator is allowed to move. For example, electronics are often powered up prior to releasing a mechanical brake or applying hydraulic power. In this case, integrator saturation can cause a large engagement transient. This can be prevented by shorting the integrator's feedback capacitor with a relay contact or electronic switch. The short is then opened when the actuator is mechanically or hydraulically engaged.

8.2 Hard-Wired Digital Controllers

The overall accuracy of a servomechanism can be greatly enhanced by the use of a digital transducer and a digital controller. Furthermore, the need for periodic calibration and adjustment can be virtually eliminated. Hard-wired digital electronics can provide this improved accuracy with bandwidths comparable to analog electronics. However, these digital circuits are considerably more complex than comparable analog circuits, are more susceptible to electrical noise, and have finite resolution. For these reasons, hard-wired digital electronics are usually used only in the primary control loop (accuracy is typically not critical in the inner loops). Furthermore, frequency compensation is difficult to implement in digital hardware and is usually left to analog circuitry.

Hard-wired digital electronics are commonly used with high-resolution incremental encoders, as illustrated in Fig. 45*a*. Two pulse trains are generated by the encoder 90° out of phase with one another. Pulse-conditioning circuitry squares up the incoming pulses, determines the transducer's direction of motion, and often increases the resolution by a factor of 4. The asynchronous counter is incremented up or down by each feedback pulse, depending upon the direction of motion. Similarly, command pulses also increment the counter up or down. The net count at any particular time represents the difference between the number of command and feedback pulses since the counter was initialized. After digital-to-analog conversion, this count becomes the error signal transmitted to the analog electronics.

The command pulse train can be generated by additional digital hardware or by a computer. However, if the transducer resolution is high and the desired maximum command rate is high, a computer may be hard pressed to provide the required pulse rates. In this case, hardware comparators and rate multipliers may prove more satisfactory. It should be noted that an incremental system has no inherent knowledge of its absolute position. Therefore, power shutdowns and electrical noise can cause such a system to lose track of where it is. For this reason a "marker pulse" is often provided at some known position to reinitialize the counter periodically. Alternatively, the servo occasionally can be commanded to a mechanical "home" position. As with most transducers having finite resolution, encoders will usually cause limit cycling with an amplitude equal to the least significant bit (Section 7.1).

Absolute digital systems can also be implemented in electronic hardware, as shown in Fig. 45*b*. The encoding transducer outputs a digital word that represents its absolute position at all times. For an optical encoder, the resolution is typically between 12 and 24 bits. After buffering, the feedback word is digitally subtracted from a digital command, and the resulting error is converted to an analog signal that is transmitted to the analog electronics. The digital summing junction can be implemented in a number of ways, including the use of an arithmetic logic unit (ALU), which operates at very high speeds. The digital command and feedback information can be transmitted to the summing junction as parallel digital words or as serial data that must be multiplexed and then decoded. The command information can be generated from a computer or digital thumbwheels.

The absolute system is less susceptible to loss of position information than the incremental system, but the transducers are considerably more expensive and less reliable, and a wire is required for each bit. To improve reliability and reduce cost, resolvers or other sine/cosine output devices are often used, together with a resolver-to-digital (R/D) converter. The penalty is reduced overall accuracy, although this can be improved by using coarse/fine resolvers and appropriate additional hardware logic.

8.3 Computer-Based Digital Controllers

The hard-wired digital controllers of Fig. 45 have limited flexibility and functional capability. The use of a microprocessor may reduce electronic hardware complexity when elaborate system functions are required. There are many such functions that are well suited to microprocessor implementation:

- Command processing (nonlinear functions, limiting, switching, and communication with other computers)
- Redundancy management (fault detection, isolation, and reconfiguration)
- Adaptive control (self-adjustment of control loop parameters as operating conditions or environmental factors change)
- Built-in test (BIT) features

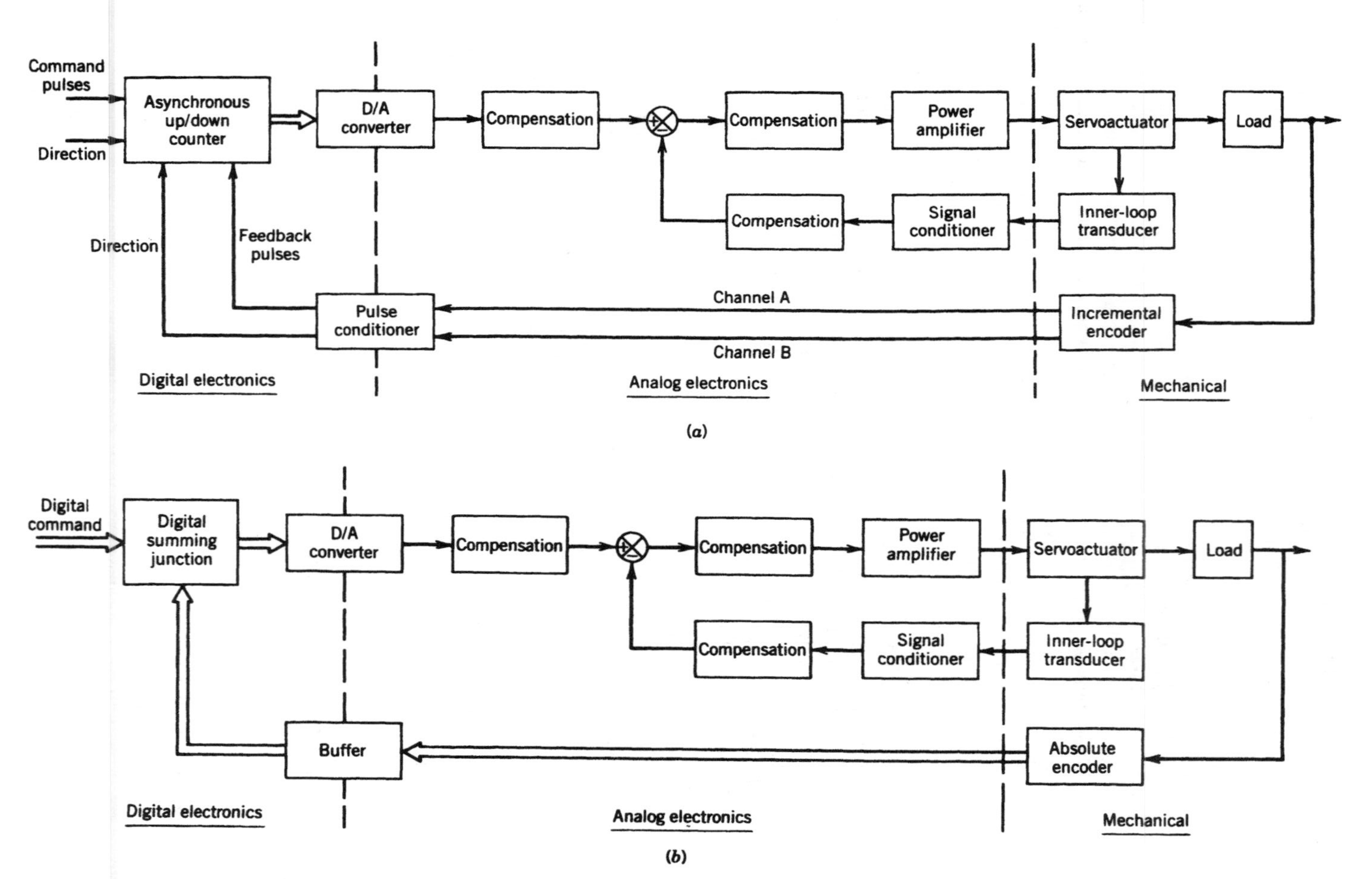

Figure 45 Typical hard-wired digital controller: (*a*) incremental system; (*b*) absolute system.

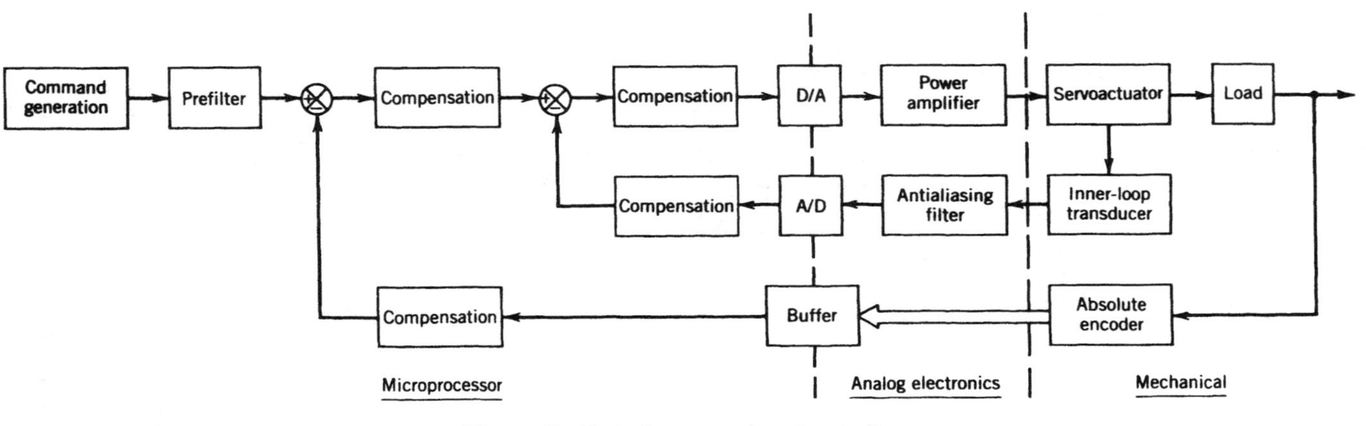

Figure 46 Typical computer-based controller.

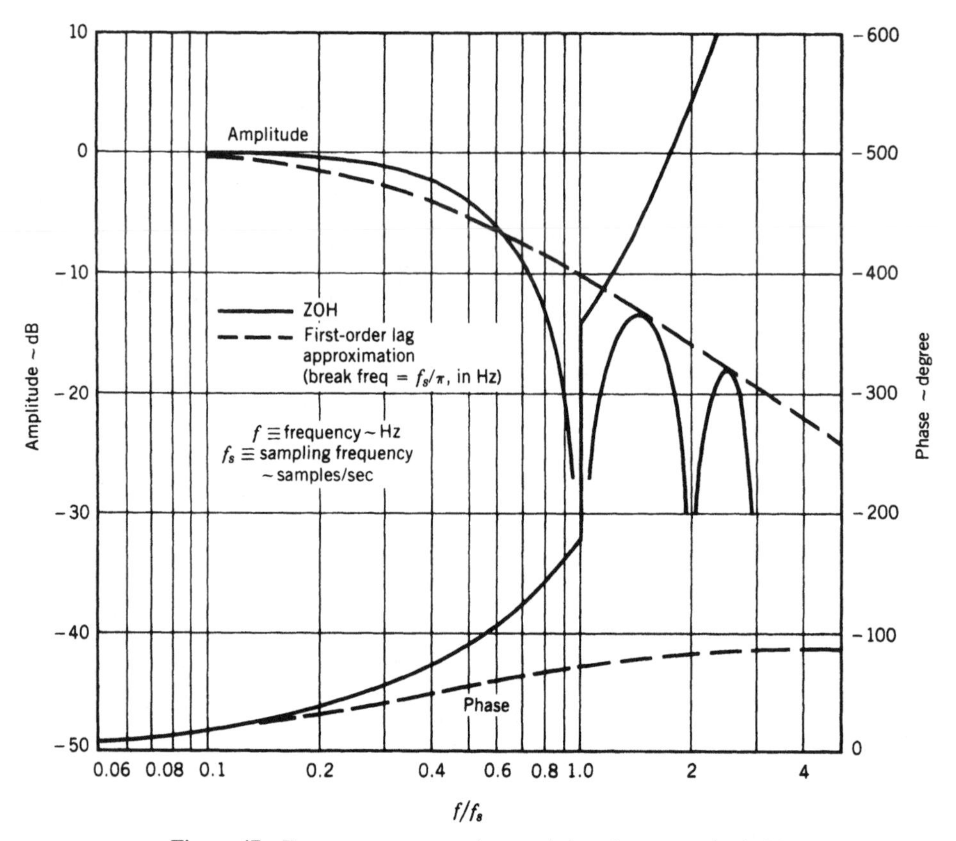

Figure 47 Frequency response characteristics of a zero-order hold.

Once the need for a microprocessor is established, it may also become reasonable to implement the servoloops in software. Note that if the application requires only the closure of simple servoloops without the need for elaborate additional functions, the use of a microprocessor will usually result in more complex hardware than an all-analog system and may be more complex than a hard-wired digital system.

The loop closure architecture of a microprocessor-based controller is illustrated in Fig. 46. This block diagram implements virtually all the loop functions in software, including frequency compensation. Of course, it is not necessary to use a digital outer loop transducer, but use of an analog transducer limits the potential accuracy advantages of the digital controller. Several methods can be used to generate the software compensator designs. Perhaps the most straightforward technique is to construct Laplace transfer functions using the continuous frequency-domain techniques outlined in Sections 1–6. These transfer functions can then be converted to equivalent Z-transforms, from which difference equations can be generated for implementation in software.[7]

It should be noted that a microprocessor-based controller is a sampled-data system, and the digital-to-analog converter usually operates as a zero-order hold (ZOH). The sampling nature of the system, together with computation times, introduces time delays into the control loops which can have a profound influence on system performance and stability. Figure 47 shows a frequency response of a ZOH operated in a sampled-data system. If the system has been designed using frequency-domain techniques, this figure represents the additional phase lag and amplitude that will result from computer implementation of the design. The phase lag is linear with frequency, and an approximate transfer function is a first-order lag with a break frequency equal to f_s/π, where f_s is the sampling frequency of the computer, in hertz. This approximation is very accurate out to a frequency of $f_s/2\pi$. Note that 10° of phase lag exist at a frequency of $f_s/18$. This suggests that the sampling frequency should be at least 20 times the crossover frequency (in hertz) of the loop being implemented if the impact on phase margin is to be minimized. Furthermore, smooth operation of the servo may require heavy filtering at the output of the digital-to-analog converter to reduce sampling-induced ripple. This will add additional phase lag in the servoloops.

The need to minimize phase lags can place severe restrictions on the complexity of computations that the microprocessor can handle in one sampling interval. This problem can be partially overcome by using a separate processor to perform loop closure computations or by implementing compensators in the analog circuitry. Alternatively, the need for high sampling rates can be reduced by implementing high-gain inner feedback loops in analog circuitry, as shown in Fig. 45*b*. If the inner loops utilize analog transducers, the problem of aliasing[4,5,10] adds another reason for using analog electronics. To properly utilize the output of an analog transducer in the computer, an antialiasing filter is required at the input to the analog-to-digital converter. These filters are often first order with a break frequency equal to f_s/π. This doubles the effective phase lag of the computer.

REFERENCES

1. J. J. D'Azzo, and C. H. Houpis, *Feedback Control System Analysis and Synthesis,* McGraw-Hill, New York, 1966.
2. B. C. Kuo, *Automatic Control Systems,* Prentice-Hall, Englewood Cliffs, NJ, 1982.
3. E. O. Doebelin, *Dynamic Analysis and Feedback Control,* McGraw-Hill, New York, 1962.
4. G. F. Franklin, J. D. Powell, and M. L. Workman, *Digital Control of Dynamic Systems,* Addison-Wesley, Reading, MA, 1990.
5. K. J. Astrom and B. Wittenmark, *Computer-Controlled Systems,* Prentice-Hall, Englewood Cliffs, NJ, 1990, 1997.
6. B. C. Kuo, *Digital Control Systems,* Holt, Rinehart and Winston, New York, 1980.
7. J. A. Cadzow, and H. R. Martens, *Discrete-Time and Computer Control Systems,* Prentice-Hall, Englewood Cliffs, NJ, 1970.
8. J. G. Graeme, G. E. Tobey, and L. P. Huelsman, *Operational Amplifiers, Design and Applications,* McGraw-Hill, New York, 1971.
9. J. G. Graeme, *Amplifier Applications,* McGraw-Hill, New York, 1999.
10. E. O. Doebelin, *System Modeling and Response,* Wiley, New York, 1980.

CHAPTER 16

GENERAL-PURPOSE CONTROL DEVICES

James H. Christensen
Holobloc, Inc.
Cleveland Heights, Ohio

Robert J. Kretschmann
Rockwell Automation
Mayfield Heights, Ohio

Sujeet Chand
Rockwell Automation
Milwaukee, Wisconsin

Kazuhiko Yokoyama
Yaskawa Electric Corporation
Tokyo, Japan

1 CHARACTERISTICS OF GENERAL-PURPOSE CONTROL DEVICES

1.1 Hierarchical Control

As shown in Fig. 1, *general-purpose control devices* (GPCDs) occupy a place in the hierarchy of factory automation above the closed-loop control systems described in Chapters 12–15. The responsibility of the GPCD is the coordinated control of one or more machines or processes. Thus, a GPCD may operate at the "station" level, where it controls part or all of a single machine or process, or at the "cell" level, where it coordinates the operation of multiple stations.

In fulfilling its responsibilities, the GPCD must be capable of performing the following functions, as shown in Fig. 2:

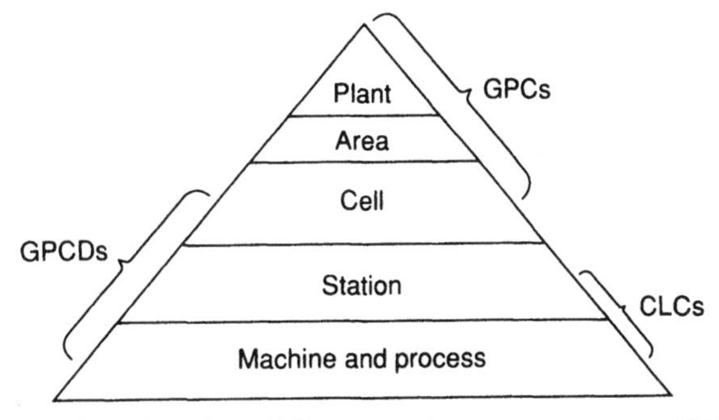

Figure 1 Plant control hierarchy using GPCDs, general-purpose computers (GPCs), and closed-loop controllers (CLCs).

- Issuing commands to and receiving status information from a set of closed-loop controllers that control individual machine and process variables such as velocity, position, and temperature. These closed-loop controllers may be separate devices or integral parts of the GPCD hardware and/or software architecture.
- Issuing commands to and receiving status information from a set of actuators and sensors directly connected to the controlled operation. These actuators and sensors may include signal-processing and transmission elements, as described in Chapter 6.

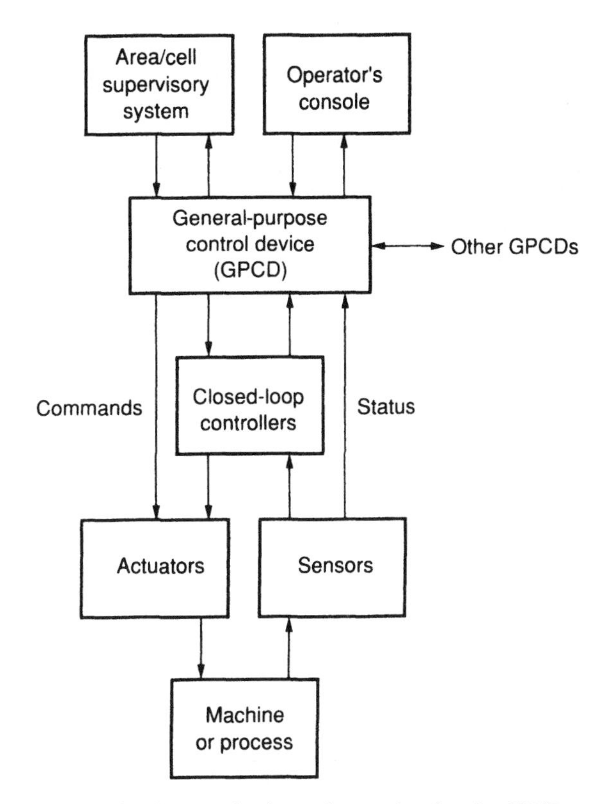

Figure 2 Communication and control paths of a GPCD.

This capability may not be required if all interface to the controlled operation is through the closed-loop controllers described above.

- Receiving commands from and sending information to a control panel or console for the operator of the machine or process.
- Receiving commands from and sending status information to a manual or automated system with the responsibility of supervising the operation of a number of GPCDs within the boundaries of a "cell," for example, a number of coordinated machines or unit operations, or over a wider "area," such as a chemical process or production zone in a factory.
- Interchanging status information with other GPCDs within the same cell or across cell boundaries.

It should be noted that not all of these capabilities are necessarily required for every application of a GPCD. For instance, special closed-loop controllers may not be necessary in an operation requiring only simple on–off or modulating control. Similarly, communication in an automation hierarchy may not be required for simple "stand-alone" applications such as an industrial trash compactor.

However, the capability for expansion into a communicating hierarchy should be inherent in the GPCD architecture if retrofit of stand-alone systems into an integrated production system is considered a future possibility.

1.2 Programmability

To be truly general purpose, a GPCD must be programmable; that is, its operation is controlled by sequences of instructions and data stored in internal memory. The languages used for programming GPCDs are usually *problem oriented:* Programs are expressed in terms directly related to the control to be performed, rather than in a general-purpose programming language such as C++ or BASIC. These languages will be described in appropriate sections for each type of GPCD.

Depending on the application, the responsibility for development and maintenance of GPCD programs may reside with:

- The *original equipment manufacturer* (OEM) of a machine that includes a GPCD as part of its control apparatus
- The *system integrator* who designs and installs an integrated hierarchical control system
- The *end user* who wishes to modify the operation of the installed system

The degree to which the operation of the system can be modified by the end user is a function of:

- The complexity of the system
- The degree to which the end user has been trained in the programming of the system
- The extent to which the operation of the process must be modified over time

For instance, in a high-volume chemical process, only minor modifications of set points may be required over the life of the plant. However, major modifications of the process may be required annually in an automotive assembly plant. In the latter case, complete user programmability of the system is required.

Depending on the complexity of the control program and the degree of reprogrammability required, GPCD programming may be supported by any of several means, including:

- An integral programming panel on the GPCD
- Portable programming and debugging tools
- Minicomputers, personal computers, or engineering workstations that may or may not be connected to the GPCD during control operation
- An online computer system, for example, the area or cell controller shown in Fig. 2.

1.3 Device Architecture

Figure 3 illustrates a GPCD architecture capable of providing the required functional characteristics:

- The *memory* provides storage for the programs and data entered into the system via the *communications processor*.
- The *control processor* performs control actions under the direction of the stored program as well as coordinates the operation of the other functions.
- The *communications processor* provides the means of accepting commands from and providing status information to the supervisory system and operator's console, interchanging status information with other GPCDs, and interacting with program development and configuration tools.

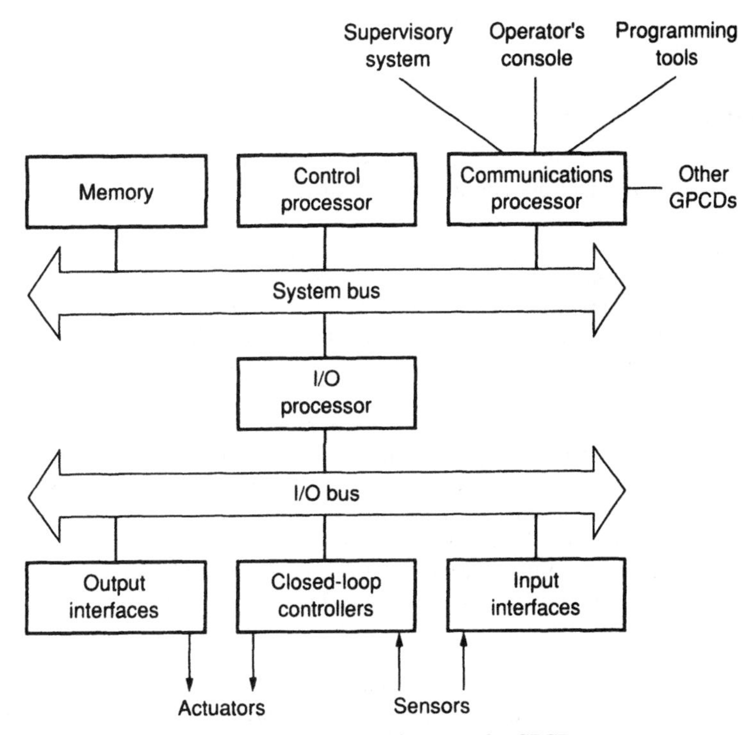

Figure 3 Typical architecture of a GPCD.

- The *I/O (input/output) processor* provides the means by which the control processor can issue commands to and receive status information from the closed-loop controllers as well as interchange information directly with the controlled operation via the *output* and *input interfaces*.
- The *system bus* provides for communication among the functional blocks internal to the GPCD, while the *I/O bus* provides for communication between the internal functional blocks and the "outside world" via the closed-loop controllers and I/O interfaces. As an option, the I/O controller may extend the I/O bus functionality to remote locations using data communications methods such as those described in Section 9.

The architecture shown in Fig. 3 is not the only one possible for GPCDs, nor is it necessarily the most desirable for all applications. For instance:

- Large systems may require multiple control processors on the system bus.
- In applications requiring only a few control loops, it may be more economical to perform the closed-loop control function directly in the software of the control processor.
- A separate I/O processor may not be required if the number of separate I/O interface points is less than a few dozen.
- In small systems, the programmer's console may be interfaced directly to the system or I/O bus.

However, if it is anticipated that control system requirements will grow substantially in the future or if total control system requirements are only partially understood, the use of a flexible, extendable GPCD architecture such as that shown in Fig. 3 is recommended.

1.4 Sequential Control

It is obvious that GPCDs must perform complex sequences of control actions when they are applied to the coordination of material handling and machine operation in the fabrication and assembly of discrete parts or in batch and semibatch processes such as blast furnace operation and pharmaceutical manufacture. However, sequential control is also increasing in importance in "continuous" process control, since no process is truly continuous. At the very least, the process must be started up and shut down for maintenance or emergencies by a predetermined sequence of control actions. In large, integrated processes, these sequences are too complicated to be carried out manually and must be performed automatically by GPCDs.

The increasing importance of sequential control, coupled with the increasing complexity of the controlled processes, have generated the need for graphical programming and documentation techniques for the representation of large, complex sequential control plans. These plans must provide a straightforward representation of the relationship between the operation of the control program in the GPCD and the operation of the controlled machine or process as well as the interrelationships between multiple, simultaneous control sequences.

Recognizing this need, the International Electrotechnical Commission (IEC)[1] has undertaken several efforts to standardize the representation of sequential control plans:

- The IEC 60848 standard[2] defines the GRAFCET specification language for the functional description of the behavior of the sequential part of a control system.

- The IEC 61131-3 standard[3] defines a set of sequential function chart (SFC) constructs, specifically intended to make the IEC 60848 concepts usable for the programming of programmable logic controllers (PLCs).
- The IEC 61512-1 standard[4] defines SFC-like constructs for the description of the sequential operations involved in the control of batch chemical processes.
- The IEC 61499-1 standard[5] defines an execution control chart (ECC) construct which provides for the sequential execution of control algorithms in function blocks for distributed automation systems, similar to the Harel state chart notation of the Unified Modeling Language (UML)[6].

As shown in Fig. 4*a*, an SFC is constructed from three basic types of elements:

- A *step,* representing the current state of the controller and controlled system within the sequential control plan
- A set of associated *actions* at each step
- A *transition condition* that determines when the state of the controller and controlled system is to evolve to another step or set of steps

An SFC consists of a set of independently operating *sequences* of control actions built up out of these basic elements via two mechanisms:

- *Selection* of one of a number of alternate successors to a step based on mutually exclusive transition conditions, as shown in Fig. 4*c*

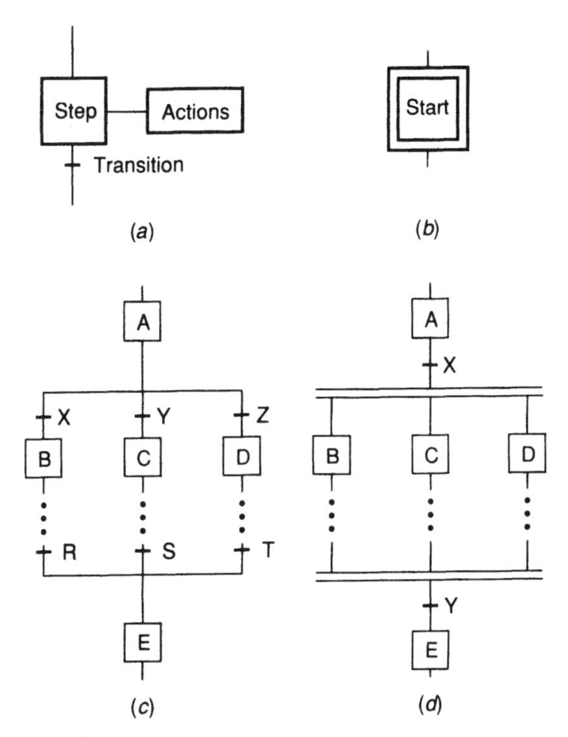

Figure 4 Sequential function chart constructs.

- *Divergence,* that is, initiation of two or more independently executing sequences based on a transitioncondition, as shown in Fig. 4*d*

The mechanisms for representing convergence, that is, resumption of a main sequence after step selection or parallel sequence initiation, are also shown in Fig. 4*c* and Fig. 4*d*, respectively.

Figure 4*b* illustrates the representation of the *initial step* of each sequence. The operation of control sequences can be visualized by placing a *token* in each initial step upon the initiation of system operation. A step is then said to be *active* while it possesses a token and *reset* when it does not possess a token. The actions associated with the step are performed while the step is active and are not performed when it is reset. The resetting of one or more steps and the activation of one or more successor steps can then be envisioned as the processes of token passing, consumption, and generation, as shown in Fig. 5 and Fig. 6. It should be noted that the selection and convergence of alternate paths within a sequence, as shown in Fig. 5, simply involve the passing of a single token. In contrast, divergence to multiple sequences involves the consumption of a single token and the generation of multiple tokens, as shown in Fig. 6*a*, with the converse operation for termination of multiple sequences shown in Fig. 6*b*.

When an action associated with a step is a Boolean variable, its association with the step may be expressed in an *action block,* as shown in Fig. 7. The qualifiers which can be used to specify the duration of the action are listed in Table 1. More complex actions can be specified via one of the programming languages described in Section 2.3; in this case, the action executes continuously while the associated "action control" shown in Table 1 has the Boolean value "1."

An example of the application of SFCs to the control and monitoring of a single motion, for example, in a robot control system, is given in Fig. 8. Here, the system waits until a motion command is received via the Boolean variable CMD_IN. It then initiates the appropriate motion by asserting the Boolean variable CMD. If a feedback signal DONE is not

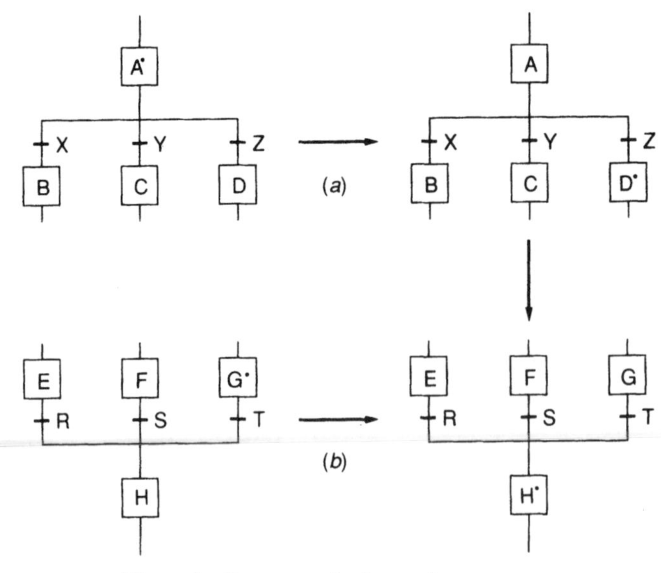

Figure 5 Sequence selection and convergence.

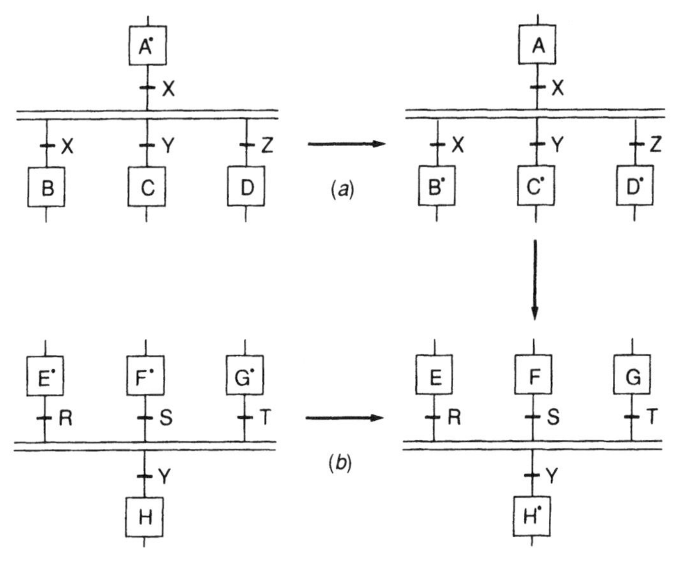

Figure 6 Parallel sequence initiation and convergence.

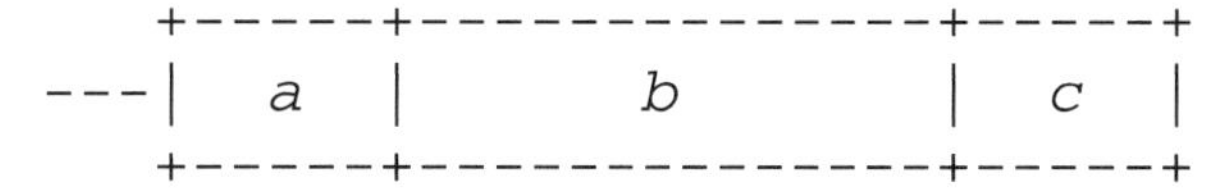

Figure 7 Action block: *a*, action qualifier; *b*, action name; *c*, feedback name.

Table 1 Action Block Qualifiers

Qualifier	Explanation
None	Nonstored (null qualifier)
N	Nonstored
R	Overriding reset
S	Set (stored)
L	Time limited
D	Time delayed
P	Pulse
SD	Stored and time delayed
DS	Delayed and stored
SL	Stored and time limited
P1	Pulse (rising edge)
P0	Pulse (falling edge)

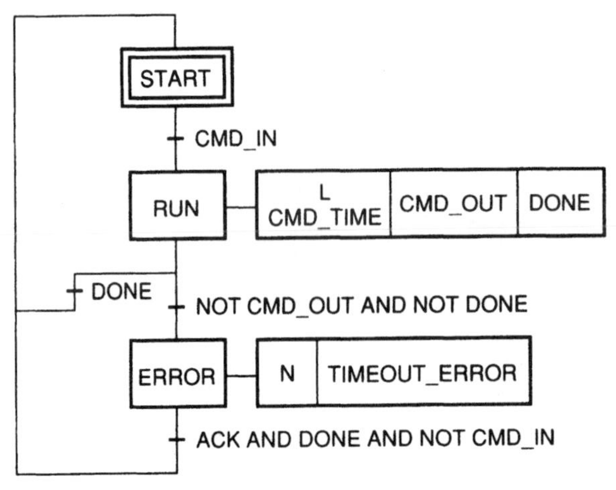

Figure 8 Sequential function chart example.

received within a time interval specified by the variable CMD, the system enters an error step and issues the error message TIMEOUT_ERROR. This error condition is cleared when acknowledged by the signal ACK, for example, from an operator's console, combined with the feedback (DONE) that the motion has been accomplished. The system then reenters the initial step and waits for another command.

1.5 Path Control

General-purpose control devices must often perform *path control,* that is, the coordinated control of several variables at once along a continuous path through time. Typical kinds of path control include:

- The path of a metal-cutting tool or a robot manipulator
- The trajectory of a set of continuous process variables such as temperature, pressure, and composition
- The startup of a set of velocity and tension variables in a paper or steel processing line

The numerical control of metal-cutting tools and robot manipulators is discussed in more detail in Sections 3 and 4, respectively.

A typical application to path control of the general-purpose architecture shown in Fig. 3 has the control processor planning the motions to be accomplished and issuing commands to the closed-loop controllers to perform the required motions. Coordination between the closed-loop controllers may be performed by the control processor, or by direct interaction among the closed-loop controllers, using the I/O bus or special interconnections.

In addition to performing path planning, the control processor also performs sequencing of individual motions and coordination of the motions with other control actions, typically using programming mechanisms such as the SFCs discussed in Section 1.4.

2 PROGRAMMABLE CONTROLLERS (PLCs)

2.1 Principles of Operation

PLC is defined by the IEC as

> A digitally operating electronic system, designed for use in an industrial environment, which uses a programmable memory for the internal storage of user-oriented instructions for implementing specific functions such as logic, sequencing, timing, counting, and arithmetic, to control, through digital or analog inputs and outputs, various types of machines or processes. Both the PLC and its associated peripherals are designed to be easily integrable into an industrial control system and easily used in all their intended functions.[7]

The hardware architecture of almost all programmable controllers is the same as that for the GPCD shown in Fig. 3.

As illustrated in Fig. 9, the operation of most programmable controllers consists of a repeated cycle of four major steps:

1. All inputs from interfaces and closed-loop controllers on the I/O bus, and possibly from other GPCDs, are scanned to provide a consistent "image" of the inputs.
2. One "scan" of the user program is performed to derive a new "image" of the desired outputs, as well as internal program variables, from the image of the inputs and the internal and output variables computed during the previous program scan. Typically, the program scan consists of:
   a. Determining the currently active steps of the SFC (see Section 1.4), if any, contained in the program.
   b. Scanning the program elements or computing the outputs contained in the active actions of the SFC, if any (if the user program does not contain an SFC, then all program elements are scanned). Scanning of program elements in ladder diagrams or function block diagrams (see Section 2.3) typically proceeds from left to right

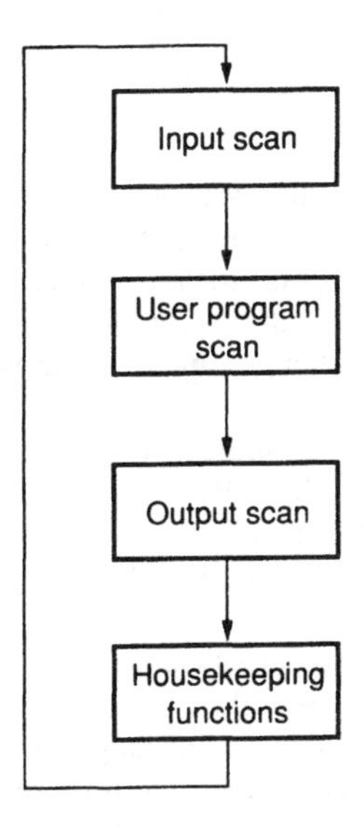

Figure 9 Basic operation cycle of a programmable controller.

and from top to bottom. Programming elements are sometimes provided to enable skipping the evaluation of groups of program elements or to force the outputs of a group of elements to zero.

  c. Evaluating transition conditions of the SFC (if any) at the end of the program scan, in preparation for step 2a in the next program scan.

3. The data from the updated output image are then transferred to the interfaces and closed-loop controllers on the I/O bus and possibly to other GPCDs as well.
4. Finally, "housekeeping" tasks are performed on a time-available basis. These typically include communication with the operator, a supervisory controller, a programming terminal, or other GPCDs.

After the performance of housekeeping tasks, the cyclic operation of the PLC begins again with the input scan. This may follow immediately upon execution of the housekeeping tasks or may be scheduled to repeat at a fixed execution interval.

Some programmable controller systems with separate I/O and/or communications processors provide for overlapping the scanning of the user program with the scanning of the inputs (step 1) and outputs (step 3) and communication functions (step 4). In these cases, special programming mechanisms may be needed to achieve concurrency and synchronization between the program and I/O scans and between the program and communications processing.

A further feature of some PLCs is the incorporation of a mutitasking operating system. As in general computing, the PLC operating system serves to coordinate the multitude of hardware and software resources and capabilities of the PLC. The incorporation of a multitasking operating system serves to allow the PLC to essentially execute multiple instances of the basic operation cycle shown in Fig. 9, rather than a single instance, at the same time. This affords much increased flexibility and capability over a single-tasking operating system PLC.

In PLCs incorporating multitasking operating systems, mechanisms may be needed to achieve appropriate levels of task coordination and synchronization between the multiple instances of the basic operating cycle. As an example, one task's results may affect another task's decisions, especially with regard to I/O.

2.2 Interfaces

The IEC has specified the standard voltage ratings shown in Table 2 for power supplies, digital inputs, and digital outputs of programmable controllers. The IEC standard[8] also defines additional parameters for digital inputs and outputs, shown in Tables 3–5; the parameters specified by the manufacturer should be checked against those defined in the IEC standard in order to assure the suitability of a particular input or output module for its intended use in the control system.

The IEC-specified signal ranges for analog inputs and outputs for programmable controllers are shown in Tables 6 and 7, respectively. The IEC standard lists a number of characteristics whose values are to be provided by the manufacturer, such as input impedance, maximum input error, and conversion time and method, and which must be checked against the requirements of the particular control application.

In addition to simple digital and analog inputs and outputs, closed-loop controllers which can reside on the I/O bus of the programmable controller system may be provided, as illustrated for GPCDs in Fig. 2. In this case, the programming languages for the programmable controller typically provide language elements, in addition to those described in Section 2.3, to support the configuration and supervisory control of these "slave" closed-loop controllers.

Table 2 Rated Values and Operating Ranges for Incoming Power Supplies and Digital I/O Interfaces of Programmable Controllers

	Recommended for	
Rated Voltage	Power Supply	I/O Signals
24 V dc[a]	Yes	Yes
48 V dc[a]	Yes	Yes
120 V rms ac[b]	Yes	Yes
230 V rms ac[b]	Yes	Yes

Note: See the IEC Programmable Controller standard (Ref. 8) for additional notes and rating values.

[a] Voltage tolerance for dc voltage ratings is −15 to +20%.

[b] Voltage tolerance for ac voltage ratings is −15 to +10%.

Table 3 Rated Values for dc and ac Digital Inputs of Programmable Controllers

Rated Voltage	Types
24 V dc	1–3
48 V dc	1–3
120 V ac	1–3
230 V ac	1–3

Note: See the IEC Programmable Controller standard (Ref. 8) for additional notes and rating values.

Table 4 Rated Values for dc Digital Outputs of Programmable Controllers

Rated Current (A)	Maximum Current (A)	Leakage Current (mA)	Input Type Compatibility
0.1	0.12	0.1	1–3
0.25	0.3	0.5	1–3
0.5	0.6	0.5	1–3
1	1.2	1	2, 3
2	2.4	1	2, 3

Note: See the IEC Programmable Controller standard (Ref. 8) for additional notes and rating values.

Table 5 Rated Values for ac Digital Outputs of Programmable Controllers

Rated Current (A)	Maximum Current (A)	Leakage Current (mA)
0.25	0.28	5
0.5	0.55	10
1	1.1	10
2	2.2	10

Note: See the IEC Programmable Controller standard (Ref. 8) for additional notes and rating values.

Table 6 Rated Values for Analog Inputs of Programmable Controllers

Signal Range	Input Impedance
−10 to +10 V	≥ 10 kΩ
0 to +10 V	≥ 10 kΩ
+1 to +5 V	≥ 5 kΩ
4–20m A	≤ 300 Ω

Note: See the IEC Programmable Controller standard (Ref. 8) for additional notes and rating values.

Table 7 Rated Values for Analog Outputs of Programmable Controllers

Signal Range	Load Impedance
−10 to +10 V	≥ 1 kΩ
0 to +10 V	≥ 1 kΩ
+1 to +5 V	≥ 500 Ω
4–20 mA	≤ 600 Ω

Note: See the IEC Programmable Controller standard (Ref. 8) for additional notes and rating values.

Communication interfaces for programmable controllers provide many different combinations of connectors, signal levels, signaling rates, and communication services. The manufacturer's specifications of these characteristics should be checked against applicable standards to assure the achievement of the required levels of system performance and compatibility of all GPCDs in the system.

2.3 Programming

Programmable controllers have historically been used as programmable replacements for relay and solid-state logic control systems. As a result, their programming languages have been oriented around the conventions used to describe the control systems they have replaced, that is, relay ladder logic and function block diagrams. Since these representations are fundamentally graphic in nature, programmable controllers provide one of the first examples of the practical application of graphic programming languages.

Much of the programming of programmable controllers is done in the factory environment while the controlled system is being installed or maintained. Hence, programming was traditionally supported by special-purpose portable programming terminals. In recent years, most of the support for PLC programming has migrated to software packages for personal computers. However, some need still exists for specialized terminals for programming and debugging of programmable controllers in the industrial environment, although most of this functionality can also be supplied by ruggedized, portable personal computers. The selection of support environments is thus an important consideration in the selection and implementation of programmable controller systems.

The standard for PLC programming languages[3] is published by the IEC.[1] This standard specifies a set of mutually compatible programming languages, taking into account the different courses of evolution of programmable controllers in North America, Europe, and Japan and the wide variety of applications of programmable controllers in modern industry. These languages include:

- The SFC elements described in Section 1.4 for sequential control
- Ladder Diagrams (LD) for relay replacement functions
- Function Block Diagrams (FBD) for logic, mathematical, and signal-processing functions
- Structured Text (ST) for data manipulation
- Instruction List (IL) for assembly-language-level programming

The IL language will not be described in this book; for further details the IEC standard[3] should be consulted.

Figure 10 shows the application of the LD, FBD, and ST languages to implement a simple command execution and monitoring function. In general, a desired functionality can be programmed in any one of the IEC languages. Hence, languages can be chosen depending on their suitability for each particular application.

An exception to this portability is the use of iteration and selection constructs (`IF. . .THEN. . .ELSIF`, `CASE`, `FOR`, `WHILE`, and `REPEAT`) in the ST language.

The functionality shown in Fig. 10 can be encapsulated into a reusable *function block* by following the declaration process defined in the IEC language standard. An example of the graphical and textual declaration of this functionality is shown in Fig. 11.

In addition to providing mechanisms for the programming of mathematical functions and function blocks, the standard provides a large set of predefined standardized functions and function blocks, as listed in Tables 8 and 9, respectively. The intent is for these to be used as "building blocks" for user programs.

In addition to being used directly for building functions, function blocks, and programs, the LD, FBD, ST, and IL languages can be used to program the "actions" to be performed under the control of SFCs as described in Section 1.4. These SFCs can then be used to build programs and reusable function blocks using the mechanisms defined in the IEC language standard.[3]

It will be noted in Figure 11 that data types are defined for all variables. The IEC standard provides facilities for strong data typing, with a large set of predefined data types as listed in Table 10. In addition, facilities are provided for user-defined data types as listed in Table 11. The standard allows manufacturers to specify the language features that they support. Users should consult the standard to determine which language features are required by their application and check their language requirements carefully against the manufacturers' specifications when making their choice of a programmable controller system.

2.4 Programmable Controller Standard, IEC 61131

To propagate consistent characteristics and capabilities for PLCs in the marketplace, an international standard describing them has been developed under the auspices of the IEC.[1] It is impossible, in a chapter of this length, to cover all the hardware, software, and programming language characterisitics, features, and scope defined in the IEC standard for programmable controllers. The standard comprises seven parts[3,7–12] under the general title "Programmable Controllers," covering various aspects of PLCs.

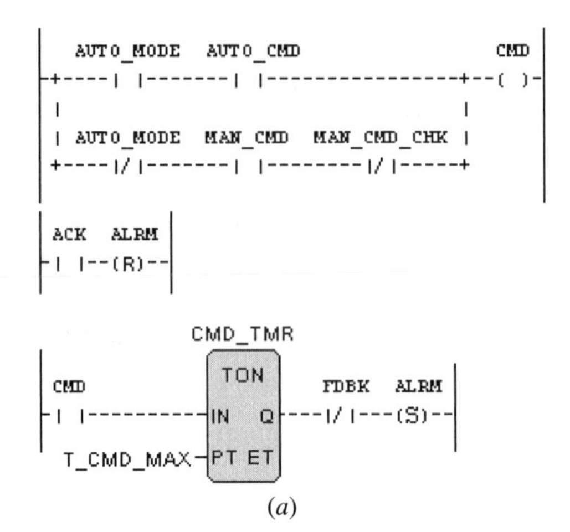

(*a*)

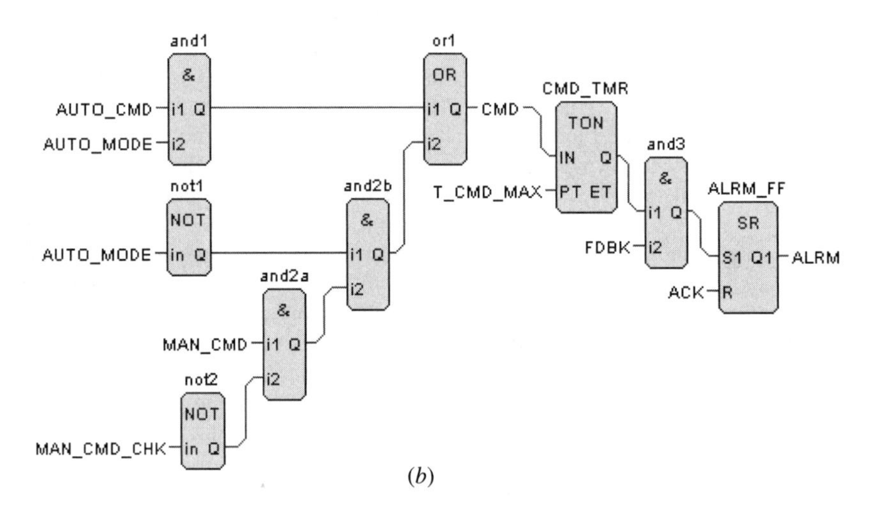

(*b*)

```
CMD := AUTO_CMD & AUTO_MODE OR MAN_CMD & NOT MAN_CMD_CHECK & NOT
AUTO_MODE;
CMD_TMR(IN := CMD, PT := T_CMD_MAX);
ALRM_FF(S1 := CMD_TMR.Q & NOT FBDK, R := ACK);
ALRM := ALRM_FF.Q1;
```

(*c*)

Figure 10 Programmable controller programming example.

```
FUNCTION_BLOCK CMD_MONITOR (* Begin definition of FB CMD_MONITOR *)

  (* Definition of external interface *)

  VAR_INPUT

    AUTO_CMD : BOOL;       (* Automatic Command *)

    AUTO_MODE : BOOL;      (* AUTO_CMD Enable *)

    MAN_CMD : BOOL;        (* Manual Command *)

    MAN_CMD_CHK : BOOL;    (* Negated MAN_CMD for debouncing *)

    T_CMD_MAX : TIME;      (* Maximum time from CMD to FDBK *)

    FDBK : BOOL;           (* Confirmation of CMD completion by operative

    unit *)

    ACK : BOOL;            (* Acknowledgement/Cancel ALRM *)

  END_VAR

  VAR_OUTPUT

    CMD : BOOL;            (* Command to operative unit *)

    ALRM : BOOL;           (* T_CMD_MAX expired without FDBK *)

  END_VAR

  (* Definition of internal state variables *)

  VAR

    CMD_TMR : TON;    (* CMD-to-FDBK timer *)

    ALRM_FF : SR;     (* Note over-riding "S1" input,

                         Command must be cancelled before ACK can cancel

    alarm *)

  END_VAR

(* Definition of Function Block Body per
Figure (a), (b) or (c) *)

END_FUNCTION_BLOCK  (* End definition of FB CMD_MONITOR *)
```

(*a*)

Figure 11 Function block encapsulation: (*a*) textual declaration in ST language; (*b*) graphic representation.

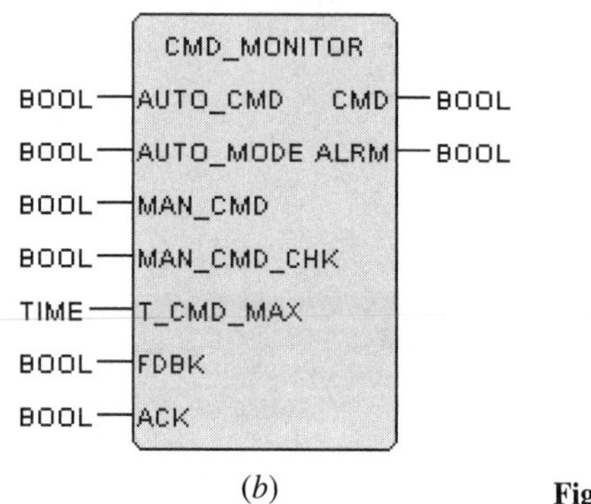

(*b*)

Figure 11 (*Continued*)

Table 8 IEC Standard Functions for Programmable Controllers

Standard Name	Function	Standard Name	Function
Numeric Functions			
ABS	Absolute value	ACOS	Arc cosine
SQRT	Square root	ATAN	Arc tangent
LN	Natural logarithm	ADD or +	Addition
LOG	Logarithm base 10	SUB or −	Subtraction
EXP	Natural exponential	MUL or *	Multiplication
SIN	Sine	DIV or /	Division
COS	Cosine	MOD	Modulo
TAN	Tangent	EXPT or **	Exponentiation
ASIN	Arc sine	MOVE or :=	Assignment
Bit String Functions			
SHL	Shift left, zero filled	SHR	Shift right, zero filled
ROL	Rotate left circular	ROR	Rotate right circular
AND or &	Bitwise Boolean AND	OR or >=1	Bitwise Boolean OR
XOR or =2k+1	Bitwise Boolean exclusive OR	NOT	Bitwise Boolean complement
Selection and Comparison Functions			
SEL	Binary (1 of 2) selection	MUX	Multiplexer (1 of *N*) selection
MIN	Minimum of *N* inputs	MAX	Maximum of *N* inputs
LIM	Hard upper/lower limiter	GT or >	Greater than
GE or >=	Greater than or equal to	EQ or =	Equal to
LE or <=	Less than or equal to	LT or <	Less than
NE or <>	Not equal		
Character String Functions			
CONCAT	Concatenate *N* strings	INSERT	Insert one string into another
DELETE	Delete a portion of a string	REPLACE	Replace a portion of one string with another
NE or <>	Not equal		

Source: From Ref. 3.

Table 9 IEC Standard Function Blocks for Programmable Controllers

Standard Name	Function
Bistable function blocks	
SR	Flip-flop (set dominant)
RS	Flip-flop (reset dominant)
Edge detection function blocks	
R_TRIG	Rising edge detect
R_TRIG	Falling edge detect
Counter function blocks	
CTU	Up counter
CTD	Down counter
Timer function blocks	
TP	One-shot (pulse) timer
TON	On-delay timer
TOF	Off-delay timer
Message transfer and synchronization	
SEND	Messaging requester
RCV	Messaging responder

Source: From Ref. 3

Table 10 IEC Standard Data Types for Programmable Controllers

Keyword	Data Type	Bits
BOOL	Boolean	1
SINT	Short integer	8
INT	Integer	16
DINT	Double integer	32
LINT	Long integer	64
USINT	Unsigned short integer	8
UINT	Unsigned integer	16
UDINT	Unsigned double integer	32
ULINT	Unsigned long integer	64
REAL	Real number	32
LREAL	Long real	64
TIME	Duration	Implementation dependent
DATE	Date (only)	Implementation dependent
TIME_OF_DAY	Time of day (only)	Implementation dependent
DATE_AND_TIME	Date and time of day	Implementation dependent
STRING	String of 8-bit characters	$8n$
WSTRING	String of 16-bit characters	$16n$
BYTE	Bit string of length 8	8
WORD	Bit string of length 16	16
DWORD	Bit string of length 32	32
LWORD	Bit string of length 64	64

Source: From Ref. 3.

Table 11 Examples of User-Defined Data Types for Programmable Controllers

Direct derivation from elementary types, e.g.:
```
TYPE RU_REAL : REAL ; END_TYPE
```
Enumerated data types, e.g.:
```
TYPE ANALOG_SIGNAL_TYPE : (SINGLE_ENDED, DIFFERENTIAL) ; END_TYPE
```
Subrange data types, e.g.:
```
TYPE ANALOG_DATA : INT (-4095..4095) ; END_TYPE
```
Array data types, e.g.:
```
TYPE ANALOG_16_INPUT_DATA : ARRAY [1..16] OF ANALOG_DATA ; END_TYPE
```
Structured data types, e.g.:
```
TYPE
 ANALOG_CHANNEL_CONFIGURATION :
  STRUCT
   RANGE : ANALOG_SIGNAL_RANGE ;
   MIN_SCALE : ANALOG_DATA ;
   MAX_SCALE : ANALOG_DATA ;
  END_STRUCT ;
 ANALOG_16_INPUT_CONFIGURATION :
  STRUCT
   SIGNAL_TYPE : ANALOG_SIGNAL_TYPE ;
   FILTER_PARAMETER : SINT (0..99) ;
   CHANNEL : ARRAY [1..16] OF ANALOG_CHANNEL_CONFIGURATION ;
  END_STRUCT ;
END_TYPE
```

Source: Ref. 3.

Since technology is always advancing, some part(s) of the standard, at any given time, is (are) being updated. Copies of draft standards are normally available for review from the appropriate National Committees for the IEC. Additionally, experts are always welcome to participate in the standards generation effort on the recommendation of National Committees. Information about the IEC, its National Committees, and ordering of the various parts of the IEC 61131 standard for PLCs is available at the IEC website.[1]

3 NUMERICAL CONTROLLERS

3.1 Introduction and Applications

The century from 1760 to 1860 saw the development of a large number of machine tools for shaping cylindrical and flat surfaces, threads, grooves, slots, and holes of many shapes and sizes in metals. Some of the machine tools developed were the lathe, the planer, the shaper, the milling machine, drilling machines, and power saws. With increasing applications for metal machining, the cost in terms of manpower and capital equipment grew rapidly. The attempt at automation of the metal removal process gave birth to numerical controllers.

The history of numerical controllers dates back to the late 1940s, when John T. Parsons proposed a method to automatically guide a milling cutter to generate a smooth curve. Parsons proposed that successive coordinates of the tool be punched on cards and fed into the machine. The idea was to move the machine in small incremental steps to achieve a desired path. In 1952, the U.S. Air Force provided funding for a project at the Massachusetts Institute of Technology (MIT) that developed the Whirlwind computer. In a subsequent project, the Servomechanisms Laboratory at MIT developed the concept of the first workable numerical control (NC) system. The NC architecture was designed to exploit the Whirlwind computer with emphasis on five-axis NC for machining complex aircraft parts.

The MIT NC architecture identified three levels of interaction with the numerical controller.[13–15] At the highest level is a machine-independent language, called APT (Automatically Programmed Tools). APT provides a symbolic description of the part geometry, tools, and cutting parameters. The next level, called the cutter location (CL) level, changes the symbolic specification of cutter path and tool control data to numeric data. The CL level is also machine independent. The lowest level, called the G-code level, contains machine-specific commands for the tool and the NC axis motions.

The conversion from APT to CL data involves the computation of cutter offsets and resolution of symbolic constraints. The conversion from CL data to G-code is called postprocessing.[13,15] Postprocessing transforms the tool center line data to machine motion commands, taking into account the various constraints of the machine tool such as machine kinematics and limits on acceleration and speed. The APT-to-CL data conversion and the compilation of CL data to G-code are computationally intensive; these computationally intensive functions were envisioned to be performed by the Whirlwind computer. The numerical controller works with simple G-codes to keep computational requirements low. The G-codes, punched on perforated paper tape, would be the input medium to the numerical controller.

Since their inception in the 1950s, numerical controllers have followed a similar pattern in the evolution of controller technology as the computer industry in the past 30 years. The first numerical controllers were designed with vacuum tube technology. The controllers were bulky and the logic inside the control was hard wired. The hard-wired nature of the controller made it very difficult to change or modify its functionality. Vacuum tubes were replaced by semiconductors in the early 1960s. In the early 1970s, numerical controllers started using microprocessors for control. The first generation of numerical controllers with microprocessor technology were mostly hybrid, with some hard-wired logic and some control functions in software. Today, most NC functionality is in the software. Microprocessor-based NC is also called computer numerical control (CNC).

The concept of distributed numerical control (DNC) was introduced in the 1960s to provide a single point of programming and interface to a large number of numerical controllers. Most NC users agree that the paper tape reader on a numerical controller suffers the most in reliability. DNC can transfer a program to NC through a direct computer link, bypassing the paper tape reader. DNC has two primary functions: (1) computer-assisted programming and storage of NC programs in computer memory and (2) transfer, storage, and display of status and control information from the numerical controllers. DNC can store and transfer programs to as many as 100 numerical controllers. Distributed numerical controllers commonly connect to numerical controllers through a link called Behind the Tape Reader (BTR). The name BTR comes from the fact that the connection between DNC and CNC is made between the paper tape reader and the control unit.

The use of paper tape is no longer the main form of storing or updating in CNC. With the advent of computer technology that is now incorporated into CNC, the CNC now has memory storage capability such that the programs are stored in the CNC as files (similar to a computer). Thus, loading new programs or storing old programs is now done via electronic connection through RS232, Ethernet, or memory storage devices such as flash memory cards. DNC is still used, especially in die-cutting machines where programs are too long and the programs are transferred through a "drip-feed" technique where the CNC runs the program as the program is downloaded from the DNC.

Numerical controllers are widely used in industry today. The predominant application of NC is still for metal-cutting machine tools. Some of the basic operations performed by machine tools in metal machining are turning, boring, drilling, facing, forming, milling, shaping, and planing.[16] Turning is one of the most common operations in metal cutting. Turning is usually accomplished by lathe machines. The part or workpiece is secured in the chuck of a lathe machine and rotated. The tool, held rigidly in a tool post, is moved at a constant speed along the rotational axis of the workpiece, cutting away a layer of metal to form a cylinder or a surface of a more complex profile.

Applications other than metal machining for numerical controllers include flame cutting, water jet cutting, plasma arc cutting, laser beam cutting, spot and arc welding, and assembly machines.[17]

3.2 Principles of Operation

NC System Components

A block diagram of a NC system is shown in Fig. 12. The three basic components in a NC system are (1) a program input medium; (2) the controller hardware and software, including the feedback transducers and the actuation hardware for moving the tool; and (3) the machine itself.

The controller hardware and software execute programmed commands, compute servo commands to move the tool along the programmed path, read machine feedback, close the servocontrol loops, and drive the actuation hardware for moving the tool. The actuation hardware consists of servomotors and gearing.

The feedback devices on an NC servo system provide information about the instantaneous position and velocity of the NC axes. The servo feedback devices can be linear transducers or rotary transducers. The two most common rotary transducers are *resolvers* and *encoders*.[18] Resolvers consist of an assembly that resembles a small electric motor with a stator–rotor configuration. As the rotor turns, the phase relationship between the stator and rotor voltages corresponds to the shaft angle such that one electrical degree of phase shift corresponds to one mechanical degree of rotation. An optical encoder produces pulsed output that is generated from a disk containing finely etched lines that rotates between an exciter lamp and one or more photodiodes. The total number of pulses generated in a single revolution is a function of the number of lines etched on the disk. Typical disks contain 2000–10,000 lines. A resolver is an analog device with an analog output signal, whereas an encoder is a numerical device that produces a digital output signal.

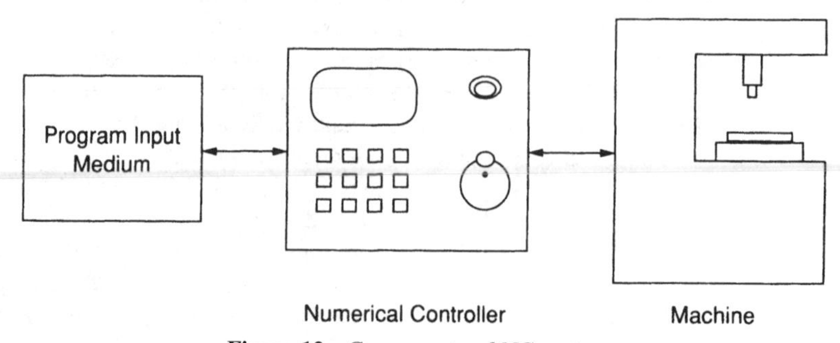

Figure 12 Components of NC system.

Linear transducers may include scales or distance coded markers. Linear transducers have the advantage that they do not introduce backlash errors that a rotary transducer may not be able to detect depending on its mechanical location.

Traditional input media for entering programs into an NC machine include (1) punched cards, (2) punched tape, (3) magnetic tape, and (4) direct entry of the program into the computer memory of the numerical controller.[19] More modern media include flash memory cards.

Originated by Herman Hollerith in 1887, the punched card as an input medium is almost obsolete. The standard "IBM" card's fixed dimensions are 3.25 in. wide, 7.375 in. long, and 0.007 in. thick. Each card contains 12 rows of hole locations with 80 columns across the card. To edit a part program, cards in the deck are replaced with new cards. With a deck of cards, it is easy to lose sequence or have missing blocks due to the loss of a card. Also, punched cards are a low-density storage medium with an input rate that is slower than most other media. As a result, punched cards are now regarded as obsolete for NC program media.

Punched tape was for many years the most popular input medium for a numerical controller. Although punched tape is mostly obsolete, many numerical controllers still provide a punched tape reader. The specifications of the punched tape are standardized by the EIA (Electronic Industries Association) and the AIA (Aerospace Industries Association). Tapes are made of paper, aluminum–plastic laminates, or other materials. Making editorial changes to the punched tape is difficult; only minor editing is possible by splicing new data into the tape. With the advent of online computer editing techniques, rapid editorial changes to a program can be made on a computer screen. At the end of an editing session, a tape can be automatically punched on command from the keyboard.

Magnetic tape is not used as much as punched tape because of its susceptibility to pollutants in the NC environment. Dust, metal filings, and oil can cause read errors on the tape. Sealed magnetic tapes overcome some of these problems.

Along with flash memory cards, direct entry of a part program into the controller memory is a common input medium for today's NC. The programmer can either type in the NC program from a keyboard and a video display terminal or generate the NC program from an interactive graphics environment.[20] Part programming with the aid of interactive graphics is discussed in more detail in Section 3.5.

Operation of a Numerical Controller: Machine Coordinate System

NC requires a point of reference and a coordinate system to express the coordinates of parts, tools, fixtures, and other components in the workspace of the machine tool. The commonly used coordinates are three orthogonal intersecting axes of a right-handed Cartesian coordinate frame, as shown in Fig. 13. The rotations a, b, and c about the x, y, and z axes are used for NC with more than three axes. In most older numerical controllers, a coordinate frame is marked on the machine and all coordinates are with respect to this fixed frame. In the newer numerical controllers, the machine tool user can program, or "teach," a location for the origin of the reference coordinate system. Since such a reference coordinate system is not permanently attached to the machine, it is sometimes called a *floating coordinate system.*

To illustrate the use of an NC coordinate system, let us consider a simple example of drilling a hole in a rectangular plate with a numerical controller. This example will illustrate the steps in the initial setup and operation of an NC machine. Figure 14 shows the drawing of a simple rectangular plate to be drilled by a drilling machine.

The first step in the programming of any NC operation is getting a drawing of the part. The drawing is usually a blueprint with the dimensions and geometrical attributes of the part. Figure 14 shows a rectangular part and the location of the center of the hole to be drilled in the part. Let the lower left-hand corner of the part be the origin of a two-dimensional Cartesian coordinate frame with the x axis and the y axis as shown in Fig. 14.

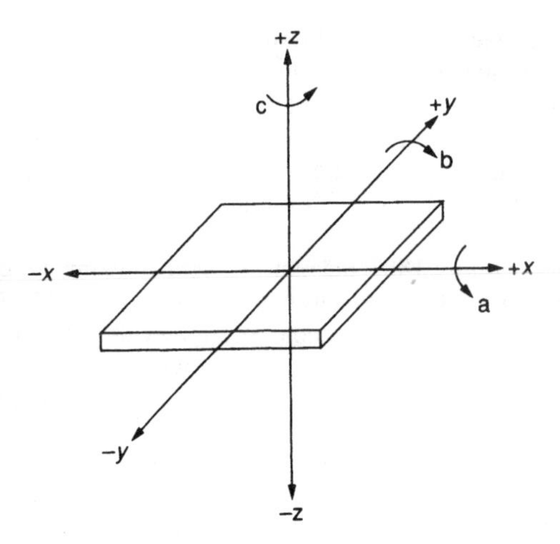

Figure 13 NC coordinate system.

The location of the center of the hole is specified by the coordinates $x = 1.500$ and $y = 2.500$.

The drilling machine must be told the location of the center of the hole. The first step is to establish the location of the NC coordinate system. With the part rigidly held in a fixture, the operator manually moves the machine tool to the lower left-hand corner of the part and presses a button on the machine control panel to teach this point as the origin of the Cartesian coordinate system. Now the machine can locate the center of the hole from the coordinates $x = 1.500$ and $y = 2.500$. In addition to specifying the location of the hole, a programmer can program the rotational speed of the drill, the direction of rotation of the drill (clockwise or counterclockwise), the feed rate, and the depth of cut. The *feed rate* is the distance moved by the tool in an axial direction for each revolution of the workpiece. The *depth of cut* is defined by the thickness of the metal removed from the workpiece, measured in a radial direction.

If the application were milling instead of drilling, the programmer also specifies a cutting speed and a rate of metal removal. The *cutting speed* in a turning or milling operation is the

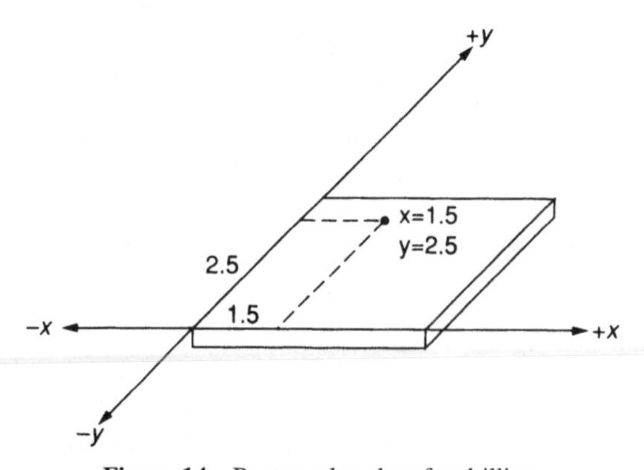

Figure 14 Rectangular plate for drilling.

rate at which the uncut surface of the workpiece passes the cutting edge of the tool, usually expressed in millimeters per minute or inches per minutes. The *rate of metal removal* is given by the product of the cutting speed, the feed rate, and the depth of cut. The cutting speed and the feed rate are the two most important parameters that a machine operator can adjust to achieve optimum cutting conditions.

3.3 Point-to-Point and Contouring Numerical Controllers

In the example of the preceding section, the drill must be moved to the center of the hole before the drilling operation starts. The drill can start from the origin and traverse to the center of the hole. The machine tool may first move along the x axis by 1.5 in., followed by a movement along the y axis by 2.5 in. It may also simultaneously start moving along the x and y axes. A numerical controller that can position the tool at specified locations without control over the path of the tool between locations is called a *point-to-point* or *positioning controller*.[13] A positioning controller may be able to move along a straight line at 45° by simultaneously driving its axes.

A *contouring controller* provides control over the tool path between positions. For instance, the tool can trace the boundary of a complex part with linear and circular segments in one continuous motion without stopping. For this reason, contouring controllers are also called *continuous-path machines*. The controller can direct the tool along straight lines, circular arcs, and several other geometric curves. The user specifies a desired contour which is typically the boundary or shape of a complex part, and the controller performs the appropriate calculations to continuously drive two or more axes at varying rates to follow this contour.

The specification of a contour in point-to-point NC is tedious because it takes a large number of short connected lines to generate a contour such as a circular arc. The programmer must also provide the appropriate feed rates along these short straight-line segments to control the speed along the contour.

NC Interpolators

In multiaxis contoured NC, multiple independently driven axes are moved in a coordinated manner to direct the tool along a desired path. An interpolator generates the signals to drive the servoloops of multiple actuators along a desired tool path.[21] The interpolator generates a large number of intermediate points along the tool path. The spacing between the intermediate points determines the tool accuracy in tracking the desired path. The closer the points are, the better the accuracy. At each interpolated point along the tool path, the positions of the multiple axes are computed by a kinematic transformation. By successively commanding the NC axes to move from one intermediate point to the next, the tool traces a desired contour with respect to the part.

Older NC systems contained hardware interpolators such as integrators, exponential deceleration circuits, linear interpolators, and circular interpolators. Most of today's numerical controllers use a software interpolator. The software interpolator takes as input the type of curve (linear, circular, parabola, spline, etc.), the start and end points, parameters of the curve such as the radius and the center of a circle, and the speed and acceleration along the curve. The interpolator feeds the input data into a computer program that generates the intermediate target points along the desired curve. At each intermediate target point, the interpolator generates the commands to drive the servoloops of the actuators. The servoloops are closed outside of the interpolator. Several excellent references provide details on closed-loop control of servoactuators.[22,23]

3.4 NC Programming

The two methods of programming numerical controllers are *manual programming* and the use of a high-level, computer-assisted *part programming language.*[24] In both cases, the part programmer uses an engineering drawing or a blueprint of the part to program the geometry of the part. Manual programming typically requires an extensive knowledge of the machining process and the capabilities of the machine tool. In manual programming, the part programmer determines the cutting parameters such as the spindle speed and feed rate from the characteristics of the workpiece, tool material, and machine tool.

The most common programming medium for NC machines was formerly perforated tape. EIA standards RS-273-A and RS-274-B define the formats and codes for punching NC programs on paper tape. The RS-273-A standard is called the interchangeable perforated-tape variable-block format for positioning and straight-cut NC systems, while the RS-274-B is the interchangeable perforated-tape variable-block format for contouring and contouring/positioning NC systems. A typical line on a punched tape according to the RS-273-A standard is as follows:

```
n001 g08 x0.0 y1.0 f225 S100 t6322 m03 (EB)
```

where `n` is the sequential block number; `g` is a preparatory code used to prepare the NC for instructions to follow; `x`, `y` are dimensional words or coordinates; `f` is the feed rate; `s` is the code for the spindle rotation speed; `t` is the tool selection code; `m` denotes a miscellaneous function; and `(EB)` is the end-of-block character, typically a carriage return. The dimension words `x` and `y` can be expanded for machines with greater than two axes. The order of the dimension words for machines with more than two axes is given as `x, y, z, u, v, w, p, q, r, i, j, k, a, b, c, d, e.`

Punched tape can be created manually or by means of a high-level computer programming system. Manual programming is more suited to simple point-to-point applications. Complex applications are usually programmed by a computer-assisted part programming language. The most common computer-assisted programming system for NCs is the APT system.[24] APT programming is independent of the machine-tool-specific parameters such as tool dimensions. Some of the other programming systems are ADAPT, SPLIT, EXAPT, AUTOSPOT, and COMPACT II.

NC programming today is getting less specialized and less tedious with the help of interactive graphics programming techniques. The part programmer creates the NC program from a display of the part drawing on a high-resolution graphics monitor in a user-friendly, interactive environment. Graphical programming is discussed in Section 3.5.

Manual Programming

Manual programming of a machine tool is performed in units called *blocks*. Each block represents a machining operation, a machine function, or a combination of both. Each block is separated from the succeeding block by an *end-of-block code*.

The standard tape code characters are shown in Table 12. *Preparatory functions,* or G-codes, are used primarily to direct a machine tool through a machining operation. Examples are linear interpolation and circular interpolation. *Miscellaneous* functions, or M-codes, are generally on–off functions such as coolant-on, end-of-program, and program stop. Details on the G-codes and the M-codes and their corresponding functions are given in several excellent tutorials on NC programming.[19,24]

The *feed-rate function f* is followed by a coded number representing the feed rate of the tool. A common coding scheme for the feed-rate number is the inverse-time code, which is generated by multiplying by 10 the feed rate along the path divided by the length of the path. If FN denotes the feed-rate number in inverse-time code, FN is given as follows:

Table 12 Manual Programming Codes

Character Digit or Code	Description
0	Digit 0
1	Digit 1
2	Digit 2
3	Digit 3
4	Digit 4
5	Digit 5
6	Digit 6
7	Digit 7
8	Digit 8
9	Digit 9
a	Angular dimension around x axis
b	Angular dimension around y axis
c	Angular dimension around z axis
d	Angular dimension around special axis or third feed function
e	Angular dimension around special axis or second feed function
f	Feed function
g	Preparatory function
h	Unassigned
i	Distance to arc center parallel to x
j	Distance to arc center parallel to y
k	Distance to arc center parallel to z
m	Miscellaneous function
n	Sequence number
p	Third rapid traverse direction parallel to x
q	Second rapid traverse direction parallel to y
r	First rapid traverse direction parallel to z
s	Spindle speed
t	Tool function
u	Secondary motion dimension parallel to x
v	Secondary motion dimension parallel to y
w	Secondary motion dimension parallel to z
x	Primary x motion dimension
y	Primary y motion dimension
z	Primary z motion dimension
.	Decimal point
,	Unassigned
Check	Unassigned
+	Positive sign
−	Negative sign
Space	Unassigned
Delete	Error delete
Carriage return	End of block
Tab	Tab
Stop Code	Rewind stop
/	Slash code

```
FN = 10*(feedrate along the path) / (length of the path)
```

For example, if the desired feed rate is 900 mm/min and the length of the path is 180 mm, the feed-rate number is 50 ($f = 0050$). The inverse-time coded feed rate is expressed by a four-digit number ranging from 0001 to 9999. This range of feed-rate numbers corresponds to a minimum interpolation time of 0.06 s and a maximum interpolation time of 10 min.

Spindle speeds are prefixed with the code "s" followed by a coded number denoting the speed. A common code for spindle speed is the *magic-three code*. To illustrate the steps in computing the magic-three code, let us convert a spindle speed of 2345 rpm to magic-three. The first step is to write the spindle speed as 0.2345×10^4. The next step is to round the decimal number to two decimal places and write the spindle speed as 0.23×10^4. The magic-three code can now be derived as 723, where the first digit 7 is given by adding 3 to the power of 10 (4 + 3), and the second two digits are the rounded decimal numbers (23). Similarly, a spindle speed of 754 rpm can be written as 0.75×10^3, and the magic-three code is 675.

Manual programming of the machine tool requires the programmer to compute the dimensions of the part from a fixed reference point, called the origin. Since the NC unit controls the path of the tool center point, the programmer must take into account the dimensions of the machine tool before generating the program. For instance, a cylindrical cutting tool in a milling operation will traverse the periphery of the part at a distance equal to its radius. The programmer also takes into account the limits on acceleration and deceleration of the machine tool in generating a program. For example, the feed rate for straight-line milling can be much higher than the feed rate for milling an inside corner of a part. The tool must be slowed down as it approaches the corner to prevent overshoot. Formulas and graphs are often provided to assist the programmer in the above calculations.

To illustrate manual part programming, let us consider an example of milling a simple part shown in Fig. 15. The programmer initially identifies the origin for the Cartesian co-

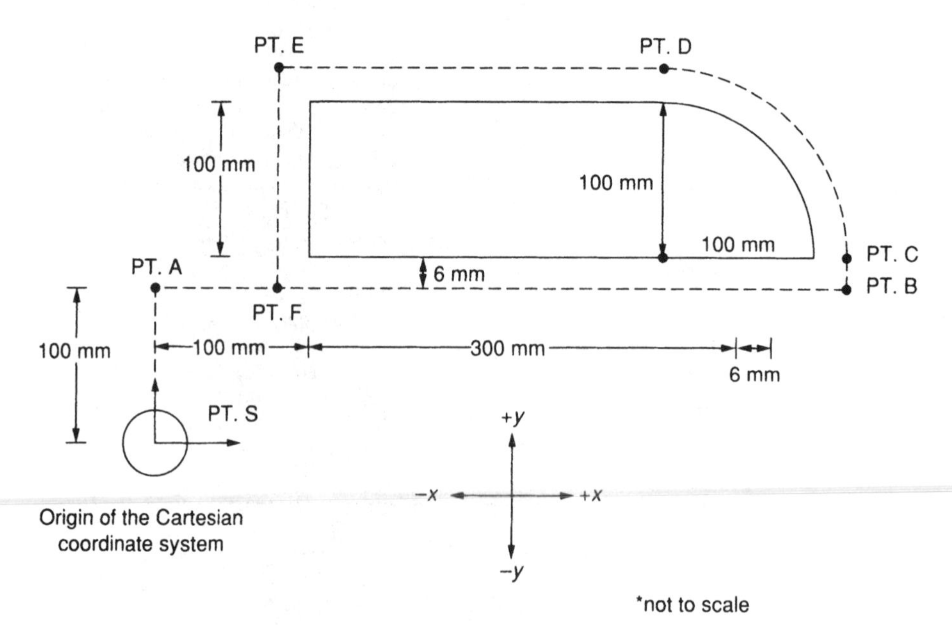

Figure 15 Drawing of a part for machining.

ordinate system and a starting point for the tool. We will assume that the tool starts from the origin (point PT.S). The programmer builds each block of the program, taking into account the tool diameter and the dynamic constraints of the machine tool. Let us assume that the milling tool in our example is cylindrical with a diameter of 12 mm.

In the first block in the program, the tool starts from rest at point PT.S and moves to point PT.A. Let the desired feed rate be 3000 mm/min. Since the distance from PT.S to PT.A is 100 mm, the inverse-time code for the feed rate is 300. Let the spindle speed be 2000 rpm. The magic-three code for 2000 rpm is 720. We will use a preparatory function code, g 08, for exponential acceleration from rest to the desired feed rate. The first block of the program can now be written as follows:

BLOCK 1 (FROM PT.S TO PT.A)

`n`	`0001`	Block 1
`g`	`08`	Automatic acceleration
`x`	`0.0`	No displacement along the x axis
`y`	`100`	A displacement of 100 mm along the y axis
`f`	`300`	Feed rate in inverse-time code
`s`	`720`	Spindle speed in magic-three code
`m`	`03`	Clockwise spindle rotation
`(EB)`		End of block

Similarly, the remaining blocks for moving the tool center point along the dotted line of Fig. 15 can be derived as follows. Note that the feed rate and spindle speed need not be programmed for each block if there is no change.

BLOCK 2 (FROM PT.A TO PT.B)

`n`	`0002`	Block 2
`x`	`406`	Displacement of 406 mm along the x axis
`y`	`0.0`	No displacement along the y axis
`m`	`08`	Turn coolant on
`(EB)`		End of block

BLOCK 3 (FROM PT.B TO PT.C)

`n`	`0003`	Block 3
`x`	`0.0`	No displacement along the x axis
`y`	`6`	Displacement of 6 mm along the y axis
`(EB)`		End of block

BLOCK 4 (FROM PT.C TO PT.D)

`n`	`0004`	Block 4
`g`	`03`	Circular interpolation, counterclockwise
`x`	`106`	Displacement of -106 mm along the x axis
`y`	`106`	Displacement of 106 mm along the y axis
`i`	`300`	The x coordinate of the center of the circle
`j`	`106`	The y coordinate of the center of the circle
`(EB)`		End of block

Block 5 (from PT.D to PT.E)

n	0005	Block 5
x	206	Displacement of −206 mm along the x axis
y	0.0	No displacement along the y axis
(EB)		End of block

Block 6 (from PT.E to PT.F)

n	0006	Block 6
g	09	Automatic deceleration
x	0.0	No displacement along the x axis
y	112	Displacement of −112 mm along the y axis
m	30	Turn off spindle and coolant
(EB)		End of block

These six blocks are entered on a sheet called the *process sheet,* as shown in Fig. 16. The data from the process sheet are punched on a paper tape and input to the NC machine. Notice that each line on the process sheet represents one block of the program. Manual programming as illustrated above can be both tedious and error prone for an inexperienced programmer. In the next section, we illustrate the use of a computer to enter the part program.

Computer-Assisted Programming: Programming in APT
APT was designed to be the common programming language standard for all numerical controllers.[25] APT programs are machine independent. An APT program is a series of English-like statements with a precise set of grammatical rules. Hundreds of keywords embody the huge expanse of NC knowledge into one language. An APT program typically contains process- or part-oriented information. The program does not contain control- and machine-tool-oriented information. APT provides three-dimensional programming for up to five axes.

An APT program containing a description of the part geometry and tool motions is input to a computer program called the *postprocessor*. The postprocessor checks for errors, adds machine-tool-specific control information, transforms the geometric description of the part into tool motion statements, and produces the proper codes for running the machine

<table>
<tr><td colspan="5">PART NUMBER
Sample</td><td colspan="5">TAPE NUMBER
2003A</td><td colspan="3">DATE
6/1/88</td></tr>
<tr><td colspan="5">PART NAME
Example</td><td colspan="5">MACHINE
Allen-Bradley 8200</td><td colspan="3">PROGRAMMER
S. Chand</td></tr>
<tr><td>N</td><td>G</td><td>X</td><td>Y</td><td>Z</td><td>I</td><td>J</td><td>K</td><td>F</td><td>S</td><td>T</td><td>M</td><td>COMMENTS</td></tr>
<tr><td>001</td><td>08</td><td>0.0</td><td>100</td><td></td><td></td><td></td><td></td><td>300</td><td>720</td><td></td><td>03</td><td>From PT. S to PT. A</td></tr>
<tr><td>002</td><td></td><td>406</td><td>0.0</td><td></td><td></td><td></td><td></td><td></td><td></td><td></td><td>08</td><td>From PT. A to PT. B</td></tr>
<tr><td>003</td><td></td><td>0.0</td><td>6</td><td></td><td></td><td></td><td></td><td></td><td></td><td></td><td></td><td>From PT. B to PT. C</td></tr>
<tr><td>004</td><td>0.3</td><td>-106</td><td>106</td><td></td><td>300</td><td>106</td><td></td><td></td><td></td><td></td><td></td><td>From PT. C to PT. D</td></tr>
<tr><td>005</td><td></td><td>-206</td><td>0.0</td><td></td><td></td><td></td><td></td><td></td><td></td><td></td><td></td><td>From PT. D to PT. E</td></tr>
<tr><td>006</td><td>0.9</td><td>0.0</td><td>-112</td><td></td><td></td><td></td><td></td><td></td><td></td><td></td><td>30</td><td>From PT. E to PT. F</td></tr>
</table>

Figure 16 Process sheet for manual programming.

tool. Each special machine has its own postprocessor. Most statements in APT are composed of two parts separated by a slash. The word to the left of the slash is called the major word or the keyword and the one to the right is called the minor word or the modifier. For example, COOLNT/OFF is an APT statement for turning the coolant off. Comments can be inserted anywhere in an APT program following the keyword REMARK.

An APT program is generated in two steps. The first step is to program the geometric description of the part. The second step is to program the sequence of operations that defines the motion of the tool with respect to the part. The programmer defines the geometry of the part in terms of a few basic geometric shapes such as straight lines and circles. The geometric description comprises a sequence of connected shapes that defines the geometry of the part. No matter how complex a part is, it can always be described in terms of a few basic geometric shapes.

Step 1: Programming Part Geometry in APT. The programmer programs the geometry of a part in terms of points, lines, circles, planes, cylinders, ellipses, hyperbolas, cones, and spheres. APT provides 12 ways of defining a line; 2 of the 12 definitions are (1) a line defined by the intersection of two planes and (2) a line defined by two points. There are 10 ways of defining a circle—for instance, by the coordinates of the center and the radius or by the center and a line to which the circle is tangent. The first part of an APT program typically contains the definitions of points, lines, circles, and other curves on the part. For the part shown in Fig. 15, an APT program may start as shown in Fig. 17. Note that the line numbers in the leftmost column are for reference only and are not a part of the APT program.

Line numbers 3 and 4 define the inside and outside tolerances in millimeters.[13,15] Line 5 defines the tool diameter as 12 mm. Lines 6 and 7 define the start point and point PT1. Lines 8–12 define the lines and the circular arc in the part geometry. Note that a temporary line, LIN2, is introduced to guide the tool to the corner at point PT2 (Fig. 17). Line 13 turns the spindle on with a clockwise rotation. Line 14 specifies the feed rate, and line 15 turns the coolant on.

Step 2: Programming the Tool Motion Statements in APT. In programming the motion of the tool, the programmer usually starts at a point on the part that is closest to the origin. The programmer then traverses the geometry of the part as viewed from the tool, indicating directions of turn such as `GOLFT` and `GORGT`. Sometimes the analogy of the programmer "riding" or "straddling" the tool is used to get the sense of direction around the periphery of the part.

An initial sense of direction for the tool is established by the statement INDIRP ("in the direction of a point"). For the part of Fig. 17, let us assume that the tool starts from the point STPT and moves to point PT1. A sense of direction is established by the following statements:

```
FROM / STPT
INDIRP / PT1
```

The statement `GO` is used in startup of the motion. The termination of the motion is controlled by three statements: `TO`, `ON`, and `PAST`. The difference between the statements `GO / TO LIN1`, `GO / ON LIN1`, and `GO / PAST LIN1` is illustrated in Fig. 18.

For the example of Fig. 17, the first GO statement is as follows:

```
GO / TO, LIN1
```

The next statement moves the tool to `PT2`. The sense of direction is set by the previous motion from `STPT` to `PT1`. The tool must turn right at `PT1` to traverse to `PT2`. Also, the

```
1     PARTNO A345, Revision 2
2     REMARK Part machined in 3/4 inch Aluminum
3     INTOL/0.00005
4     OUTTOL/0.00005
5     CUTTER/12
6     STPT   =   POINT/0,0,0
7     PT1   =   POINT/0,106,0
8     LIN1   =   LINE/(POINT/0, 106, 0), (POINT/400, 106,
      0)
9     CIRC1   =   CIRCLE/CENTER, (POINT/300, 106, 0),
      RADIUS, 100
10    LIN2   =   LINE/(POINT/400, 206, 0), LEFT, TANTO,
      CIRC1
11    LIN3   =   LINE/(POINT/400, 206, 0), (POINT/100, 206, 0)
12    LIN4   =   LINE/(POINT/100, 206, 0), (POINT/100, 106, 0)
13    SPINDL/ON, CLW
14    FEDRAT/300
15    COOLNT/ON
```

Figure 17 APT programming of part for machining.

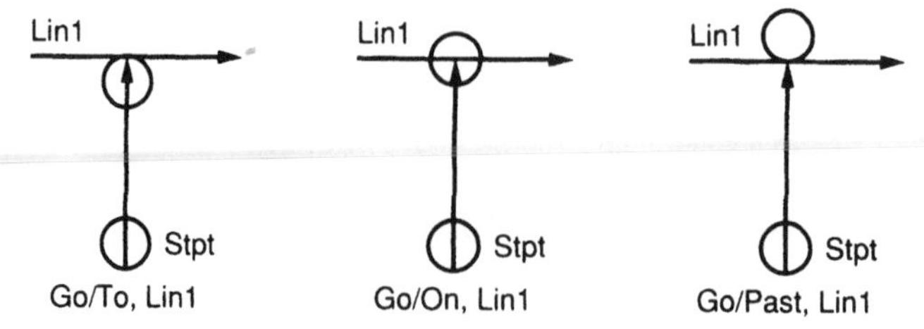

Figure 18 GO statement in APT.

tool must remain to the right of the line LIN1 as seen from the tool in the direction of motion between PT1 and PT2. This move statement is written as follows:

```
TLRGT, GORGT/LIN1, PAST, LIN2
```

The next step is to move the tool from PT2 to PT3 along the circle CIRC1:

```
TLRGT, GOLFT/CIRC1, PAST, LIN3
```

The next two motion statements to complete the motion around the part are as follows:

```
GOFWD/LIN3, PAST, LIN4
TLRGT, GOLFT/LIN4, PAST, LIN1
```

The complete program is shown below:

```
PARTNO A345, Revision 2
REMARK Part machined in 3/4 inch Aluminum
INTOL/0.00005
OUTTOL/0.00005
CUTTER/12
STPT = POINT/0,0,0
PT1 = POINT/0,106,0
LIN1 = LINE/(POINT/0,106,0), (POINT/400, 106, 0)
CIRC1 = CIRCLE/CENTER, (POINT/300, 106, 0), RADIUS, 100
LIN2 = LINE/(POINT/400, 206, 0), LEFT, TANTO, CIRC1
LIN3 = LINE/(POINT/400, 206, 0), (POINT/100, 206, 0)
LIN4 = LINE/(POINT/100, 206, 0), (POINT/100, 106, 0)
SPINDL/ON, CLW
FEDRAT/300
COOLNT/ON
FROM/STPT
INDIRP/PT1
GO/TO, LIN1
TLRGT, GORGT/LIN1, PAST, LIN2
TLRGT, GOLFT/CIRC1, PAST, LIN3
GOFWD/LIN3, PAST, LIN4
TLRGT, GOLFT/LIN4, PAST, LIN1
STOP
FINI
```

3.5 Numerical Controllers and CAD/CAM

A computer-aided design (CAD) system is used for the creation, modification, and analysis of designs. A CAD system typically comprises a graphics display terminal, a computer, and several software packages. The software packages aid the designer in the creation, editing, and analysis of the design. For instance, in the analysis of a part design, the designer can access a library of routines for finite-element analysis, heat transfer study, and dynamic simulation of mechanisms.

Computer-aided manufacturing (CAM) refers to the use of a computer to plan, manage, and control the operations in a factory. CAM typically has two functions: (1) monitoring and control of the data on the factory floor and (2) process planning of operations on the factory floor.

Since the parts manufactured in a factory are usually designed on a CAD system, the design should be integrated with the programming of the machines on the factory floor. The process of integrating design and manufacturing is often referred to as CAD/CAM.

Today, CAD systems are commonly used to design the parts to be machined by NC. The NC programmer uses a CAD drawing of the part in the generation of the part program. Several numerical controllers provide an interactive graphic programming environment that displays the CAD drawing of the part on a graphics terminal. On the screen of the graphics terminal, a programmer can display the cross section of the part, rotate the part in three dimensions, and magnify the part. These operations help the programmer to better visualize the part in three dimensions; a similar visualization of the part from a two-dimensional blueprint is difficult.

In an interactive graphic programming environment, the programmer constructs the tool path from a CAD drawing. In many systems, the tool path is automatically generated by the system following an interactive session with the programmer. The output can be an APT program or a CLFILE, which can be postprocessed to generate the NC punched tape.[20]

There are two basic steps in an interactive graphics programming environment:

1. *Defining the Geometry of the Part.* The geometric definition of the part can be specified during the part design process in a CAD/CAM system. If the geometric definition does not exist, the part programmer must create it on the graphics terminal. This process is interactive, with the part programmer labeling the various edges and surfaces of the part on the graphics screen. After the labeling process is complete, the system automatically generates the APT geometry statements for the part. The definition of the part geometry is much more easily performed on the interactive graphic system than by the time-consuming, error-prone, step-by-step manual process.
2. *Generating the Tool Path.* The programmer starts by defining the starting position of the cutter. The cutter is graphically moved along the geometric surfaces of the part through an interactive environment. Most CAD/CAM systems have built-in software routines for many machining operations, such as surface contouring, profile milling around a part, and point-to-point motion. The programmer can enter feed rates and spindle speeds along each segment of the path. As the program is created, the programmer can visually verify the tool path with respect to the part surface on the graphics screen. The use of color graphics greatly aids in the visual identification of the part, tool, and program parameters.

The advantages of the CAD/CAM approach to NC programming are as follows[26]:

1. The geometric description of the part can be easily derived from the CAD drawing.
2. The programmer can better visualize the part geometry by manipulating the part drawing on the screen, such as rotating the part and viewing cross sections.
3. The interactive approach for the generation of part programs allows the user to visually verify a program on the graphics screen as it is being created.
4. The programmer has access to many machining routines on the CAD system to simplify tool path programming.
5. The CAD environment allows the integration of part design and tool design with programming.

4 ROBOT CONTROLLERS

The robot controller is a device that, via a pretaught software program, controls the operation of the manipulator and peripheral equipment of a robot. This section explains the composition of a typical robot system and the procedure for developing its working program.

4.1 Composition of a Robot System

As shown in Fig. 19, a basic industrial robot system is composed of four components: tools, manipulators, the controller, and the teach pendant. The tool is attached at the end point of the manipulator. The type of tool depends on the work to be done by the manipulator. The manipulator maintains the position and the orientation of the tool. The controller controls the manipulator's motion, tools, and peripherals. The teach pendant is the human–machine interface for the robot controller. The movement of the manipulator and the attached tools are programmed by using the teach pendant.

Tools

The general industrial robot, which is without a dexterous humanlike hand, works by installing a suitable tool for each job in the arm end. Some applications require the use of peripheral equipment, such as power supply or parts supply for the tools. The robot controller can control the operation of peripheral equipment along with the movement of the manipulator. This subject will be described at length later in this section. Table 13 shows the relation between work and the tool. For multiprocess jobs such as assembly tasks, an auto-

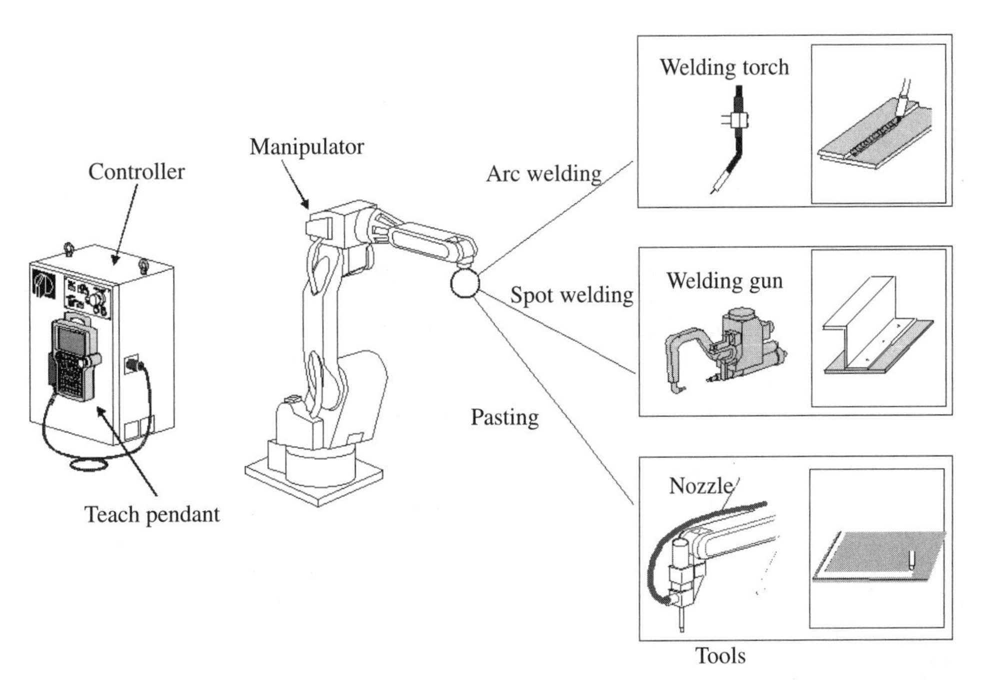

Figure 19 Basic industrial robot system.

Table 13 Relation between Work and Tool

Work	Tool	Peripheral Equipment	Main Controlled Items
Arc welding	Welding torch	Welding power supply, wire supply device	Welding voltage, welding current, arc on/off
Spot welding	Welding gun	Welding power supply	Welding condition, gun on/off, pressurizing power
Pasting/sealing	Nozzle	Paste (seal material) supply device	Gun on/off
Parts transportation	Gripper		Opening and closing
Assembly	Screw driver	Tool exchange device	Tool on/off

matic tool exchange device is necessary to change tools for each individual process. The robot controller also controls the tool exchange device.

Manipulator

The role of the manipulator is to maintain the tool position and orientation. Most industrial robot arms now are very similar to a human arm in that they have a joint. The joint is driven by an actuator in the form of a servomechanism. Actuators include ac servomotors, dc servomotors, and oil pressure motors, for example. In general, industrial robots use six actuators (thereby have six degrees of freedom), but there are some cases where an industrial robot will require less than six. The manipulator needs tools and transfer devices in order to hold and transport payloads.

Controller

As explained earlier, the controller is a device which, using preprogrammed (or "taught") control programs, controls the position and orientation of the manipulator and attached tools. This will be explained in more detail later. A general controller includes the functions to interpret and execute the control program, I/O functions (digital I/O and analog I/O) to control the tools and peripheral equipment, and functions to control servodrives for each actuator.

As shown in Fig. 20, the configuration of a robot controller follows the general scheme of all GPCDs illustrated in Fig. 3, with specialized interfaces as required for the robot control and programming functions. Of particular note is the increasing use of Internet or local-area network (LAN) communication interfaces on the controller, because the robot is frequently working simultaneously with many robots in the factory and also because of the use of remote operation and remote maintenance.

Teach Pendant

The teach pendant is the controller's human–machine interface for making the robot's control program. The screen on which the status of the manipulator and the control program are displayed and the button to operate the manipulator and to teach the position of each joint into the memory of the controller are arranged on the pendant.

4.2 Control Program

The control program is programmed using the teach pendant to make the robot work as described above. The control program is composed of commands, such as those to drive the

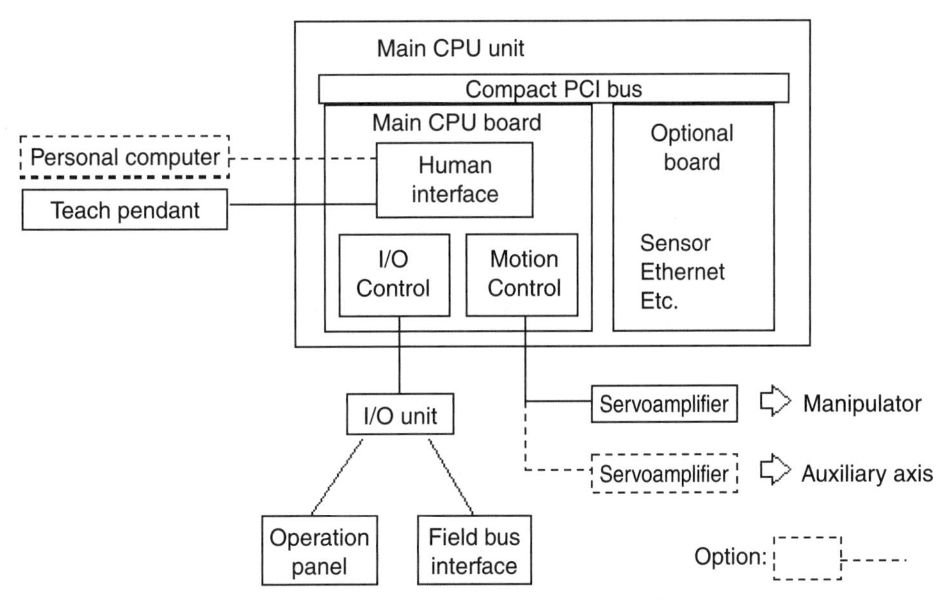

Figure 20 Composition of a robot controller.

controller itself or the manipulator. The program saved in the memory of the controller controls the manipulator and executes the software processing, such as external I/O processing or communication processing, by a set of output signals.

Just as with a computer program, very difficult tasks are made possible by control programs that can be changed through either a change in commands or the creation of subroutines via a series of commands. The basic programming method is to teach the position and the posture of the tool according to the work to be done. For jobs such as arc welding, pasting, and sealing, it is necessary to teach the trajectory of the movement of the tool. In such a case, the controller has a function (interpolation function) to automatically generate any trajectory while the control program is executed by teaching only the starting point and ending point for a straight line and/or the starting point, ending point, and intermediate point for any curve, instead of teaching the entire trajectory directly. Through this function the position and orientation of the tool can be moved. When teaching the trajectory, the speed at which the tool will be moved through the trajectory is very important; therefore, the speed of the tool is taught along with the position and orientation. For example, consider the simple arc welding of two parts shown in Fig. 21. The tool is an arc welding torch, and the peripheral equipment is a welding power supply.

A series of jobs begin with the positioning of the torch at the start point. Next, the arc welding power supply is turned on. To get the best value from the welding, the torch end matches the parts and welds them together while controlling the welding voltage and the current. The arc welding power supply is turned off when reaching the position where the welding is finished. To create this program, the manipulator is moved while the welding torch position and orientation are changed by the teach pendant. The teaching is done as follows:

1. Standby position (P1)
2. Welding start point vicinity (P2)
3. Welding start point (P3)

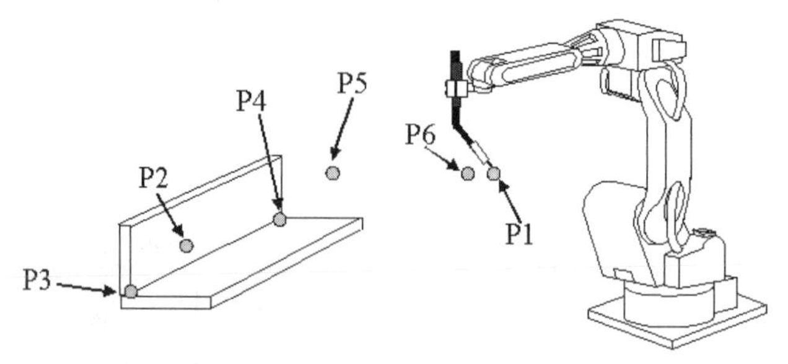

Figure 21 Programming an arc welding robot.

4. Welding end point (P4)
5. Welding end point vicinity (P5)
6. Standby position (P6)

Two teaching points, the start point (P3) and the end point (P4), are necessary for the arc welding. However, supplementary teaching points should be included in order to assure the manipulator does not interfere with the change in the work object or to assure that there is not interference between the work object and the tool. In an actual work situation, it is necessary to teach more supplementary points according to the work environment. The result of the teaching is shown in Table 14 as commands of a robot language. In step 4, the straight-line trajectory is generated automatically, at 1200 mm/min. In an actual situation commands are added using the teach pendant.

After the motion of the manipulator is taught, the control of the peripheral equipment is taught. In this case, when moving from P3 to P4, it is necessary to control the arc welding power supply. Table 15 shows the robot language program.

In the example above, a robot language was used to explain the procedure for creating a robot control program. The teaching method of actually moving the manipulator with the teach pendant is called *online programming*. On the other hand, the method by which the robot is not actually used in making the program but via computer graphics and a robot simulation is called *off-line programming*.

An off-line program can make the control program when an actual manipulator cannot be used, such as the production line design period, and is effective, for example, in the examination of the line arrangement. Online programming and off-line programming are the same as the basic instruction commands shown previously. Figure 22 shows an example screen of a PC-based off-line programming system.

Table 14 Initial Robot Program

```
Step commands
1: MOVEP1
2: MOVEP2
3: MOVEP3
4: MOVE LINEAR P4 SPEED=1200mm/min
5: MOVE P5
6: MOVE P6
```

Table 15 Robot Program with Peripheral Equipment Control

```
Step commands
01: MOVE P1
02: MOVE P2
03: MOVE P3
04: OUT ANALOG PORT=0 VALUE=<some value>
05: OUTANALOGPORT=1VALUE=<some value>
06: OUT DIGITAL PORT=0 VALUE=ON
07: MOVE LINEAR P4 SPEED=600mm/sec
08: OUT DIGITAL PORT=0 VALUE=OFF
09: MOVE P5
10: MOVE P6
```

Acknowledgments

The authors gratefully acknowledge the contributions of Steven Tourangeau of Rockwell Automation to Section 3 as well as the contributions to the previous edition of this chapter by the late Odo J. Struger of Rockwell Automation.

Figure 22 Example screen of an off-line programming system.

REFERENCES

1. http://www.iec.ch.
2. IEC 60848, *GRAFCET Specification Language for Sequential Function Charts,* 2nd ed., International Electrotechnical Commission, Geneva, 2002.
3. IEC 61131-3, *Programmable Controllers—Part 3: Programming Languages,* 2nd ed., International Electrotechnical Commission, Geneva, 2003.
4. IEC 61512-1, *Batch Control—Part 1: Models and Terminology,* 1st ed., International Electrotechnical Commission, Geneva, 1997.
5. IEC 61499-1, *Function Blocks for Industrial-Process Measurement and Control Systems—Part 1: Architecture,* 1st ed., International Electrotechnical Commission, Geneva, 2004.
6. B. P. Douglass, *Real-Time UML,* Addison-Wesley, Reading MA, 1998.
7. IEC 61131-1, *Programmable Controllers—Part 1: General Information,* 2nd ed., International Electrotechnical Commission, Geneva, 2003.
8. IEC 61131-2, *Programmable Controllers—Part 2: Equipment Requirements and Tests,* 2nd ed., International Electrotechnical Commission, Geneva, 2004.
9. IEC 61131-4, *Programmable Controllers—Part 4: User Guidelines,* 2nd ed., International Electrotechnical Commission, Geneva, 2004.
10. IEC 61131-5, *Programmable Controllers—Part 5: Communications,* 1st ed., International Electrotechnical Commission, Geneva, 2000.
11. IEC 61131-7, *Programmable Controllers—Part 7: Fuzzy Control Programming,* 1st ed., International Electrotechnical Commission, Geneva, 2000.
12. IEC 61131-8, *Programmable Controllers—Part 8: Guidelines for the Application and Implementation of Programming Languages,* 2nd ed., International Electrotechnical Commission, Geneva, 2003.
13. R. S. Pressman and J. E. Williams, *Numerical Control and Computer-Aided Manufacturing,* Wiley, New York, 1977.
14. W. Leslie, *Numerical Control Users Handbook,* McGraw-Hill, New York, 1970.
15. Y. Koren, *Computer Control of Manufacturing Systems,* McGraw-Hill, New York, 1983.
16. E. M. Trent, *Metal Cutting,* Butterworths, Woburn, MA, 1977.
17. G. Boothroyd, C. Poli, and L. E. Murch, *Automatic Assembly,* Marcel Dekker, New York, 1982.
18. A. Fitzgerald and C. Kingsley, *Electric Machinery,* 2nd ed., McGraw-Hill, New York, 1958.
19. J. Childs, *Numerical Control Part Programming,* Industrial Press, New York, 1973.
20. M. Groover, and E. Zimmers, *CAD/CAM Computer-Aided Design and Manufacturing,* Prentice-Hall, Englewood Cliffs, NJ, 1984.
21. Y. Koren, A. Shani, and J. Ben-Uri, "Interpolator for a CNC System," *IEEE Transactions on Computers,* **C-25**(1), 32–37 (1976).
22. General Electric, *Pulse Width Modulated Servo Drive,* GEK-36203, March 1973.
23. J. Beckett and G. Mergler, "Analysis of an Incremental Digital Positioning Servosystem with Digital Rate Feedback," *ASME Journal of Dynamic Systems, Measurement and Control,* **87,** March 1965.
24. A. Roberts and R. Prentice, *Programming for Numerical Control Machines,* 2nd ed., McGraw-Hill, New York, 1978.
25. Illinois Institute of Technology Research Institute (IITRI), *APT Part Programming,* McGraw-Hill, New York, 1967.
26. D. Grossman, "Opportunities for Research on Numerical Control Machining," *Communications of the ACM* **29**(6), 515–522 (1986).

BIBLIOGRAPHY

Hughes, T. A., *Programmable Controllers,* 4th ed., ISA, Raleigh, NC, 2004.

Lewis, R., *Modelling Distributed Control Systems Using IEC 61499,* IEE, London, 2001.

Lewis, R. W., *Programming Industrial Control Systems Using IEC 1131-3, rev. ed.,* IEE, London, 1998.

CHAPTER 17

STATE-SPACE METHODS FOR DYNAMIC SYSTEMS ANALYSIS

Krishnaswamy Srinivasan
Department of Mechanical Engineering
The Ohio State University
Columbus, Ohio

1 INTRODUCTION

The use of the state-space approach for the dynamic analysis and control of systems results in analysis and design techniques based in the time domain, as opposed to frequency-domain-based transform techniques. The state-space approach has the following characteristics:

1. It employs a more complete internal representation of dynamic systems as compared to transform methods that use input–output representations. The state of a system represents complete information about the current dynamic condition of the system. It incorporates the effect of all past inputs on the system. When combined with a complete description of the system dynamics in the form of state-space equations and knowledge of all future inputs, the future behavior of the system can be determined. More precise definitions of the notion of state are given in standard textbooks.[1–4]

2. It offers a unified approach to the analysis and synthesis of linear and nonlinear, time-invariant and time-varying, continuous-time and discrete-time, single-input and single-output, and multiple-input and multiple-output systems. Available techniques, however, are more plentiful for some categories of systems.

3. State-space-based methods rely more heavily on digital computers than classical transform-based techniques for dynamic systems analysis and control. In fact, the availability

Reprinted from *Instrumentation and Control,* Wiley, New York, 1990, by permission of the publisher.

of digital computers both for analysis and control synthesis and for implementation of the controllers has been an important factor underlying the growing use of state-space-based methods.

4. State-space-based methods have the potential to improve the performance of controlled systems if such systems can be modeled accurately. They have been less successful in cases where system models are characterized by significant uncertainty. Classical transform-based techniques have been and continue to be widely used in such cases. In fact, one of the more encouraging trends in control systems development has been the establishment of links between state-space-based methods and transform-based methods.[5]

Though the concept of state has been invoked by a number of methods of classical mechanics and is implicit in the phase-plane concept used for nonlinear system stability analysis, the effective application of state-space-based methods for analysis and control of dynamic system behavior has occurred only over the last three decades. Pioneering theoretical work by Kalman[8–11] and others and the availability of digital computers for performing analysis and design computations have been important underlying factors. State-space methods have been most successful in aerospace control applications and less so in a variety of industrial control applications. Among the factors favoring increased emphasis in the future on state-space methods are:

1. The emphasis on controlled system performance improvement resulting from imperatives such as improved efficiency of energy utilization and improved productivity
2. The increasing availability of inexpensive but powerful digital computers for off-line analysis and design computations and online control computations

In Sections 2–7, methods for analysis of dynamic systems using state-space methods are described. Even though most of the results presented in the literature use the continuous-time formulation, the fact that digital computers will be increasingly used for controller implementation implies that discrete-time formulations have significant practical importance. Hence, both continuous-time and discrete-time formulations are presented to the fullest extent possible.

2 STATE-SPACE EQUATIONS FOR CONTINUOUS-TIME AND DISCRETE-TIME SYSTEMS

The differential equations describing the input–output behavior of an nth-order, continuous-time, nonlinear, time-varying, lumped-parameter system can be written in the form of a first-order vector ordinary differential equation and a vector output equation:

$$\mathbf{x}(t) = \mathbf{f}[\mathbf{x}(t), \mathbf{u}(t), t] \qquad t \geq t_0 \tag{1}$$

$$\mathbf{y}(t) = \mathbf{g}[\mathbf{x}(t), \mathbf{u}(t), t] \qquad t \geq t_0 \tag{2}$$

where $\mathbf{x}(t)$ = n-dimensional state vector
$\mathbf{y}(t)$ = p-dimensional output vector
$\mathbf{u}(t)$ = r-dimensional input vector
$\mathbf{f}, \mathbf{g}$ = vectors of appropriate dimension whose elements are single-valued nonlinear functions of the arguments noted

Equation (1) is the state equation and Eq. (2) is the output equation. The state, output, and input vectors are

$$\mathbf{x}(t) = \begin{bmatrix} x_1(t) \\ x_2(t) \\ \vdots \\ x_n(t) \end{bmatrix} \qquad \mathbf{y}(t) = \begin{bmatrix} y_1(t) \\ y_2(t) \\ \vdots \\ y_p(t) \end{bmatrix} \qquad \mathbf{u}(t) = \begin{bmatrix} u_1(t) \\ u_2(t) \\ \vdots \\ u_r(t) \end{bmatrix} \tag{3}$$

The elements $x_1(t), x_2(t), \ldots, x_n(t)$ of the state vector are the state variables of the system.

Formulation of the higher order system differential equations as a set of first-order differential equations has the advantage that the latter are easier to solve by numerical methods than the former. If the functions **f** and **g** are linear functions of $\mathbf{x}(t)$ and $\mathbf{u}(t)$, the system can be described by linear ordinary differential equations. Matrix notation can then be employed to simplify their representation:

$$\dot{\mathbf{x}}(t) = \mathbf{A}(t)\mathbf{x}(t) + \mathbf{B}(t)\mathbf{u}(t) \qquad t \geq t_0 \tag{4}$$

$$\mathbf{y}(t) = \mathbf{C}(t)\mathbf{x}(t) + \mathbf{D}(t)\mathbf{u}(t) \qquad t \geq t_0 \tag{5}$$

where $\mathbf{A}(t) = n \times n$ system matrix
$\mathbf{B}(t) = n \times r$ input–state coupling matrix
$\mathbf{C}(t) = p \times n$ state–output coupling matrix
$\mathbf{D}(t) = p \times r$ input–output transmission matrix

A block diagram representation of Eqs. (4) and (5) is given in Fig. 1 using standard symbols appropriate for simulation diagrams. If the system is linear and time invariant (LTI), the matrices noted become constant matrices, as indicated by the following equations:

$$\dot{\mathbf{x}}(t) = \mathbf{A}\mathbf{x}(t) + \mathbf{B}\mathbf{u}(t) \qquad t \geq t_0 \tag{6}$$

$$\mathbf{y}(t) = \mathbf{C}\mathbf{x}(t) + \mathbf{D}\mathbf{u}(t) \qquad t \geq t_0 \tag{7}$$

If only values of the input and output variables at discrete instants in time are of interest, difference equations are appropriate for describing their relationship. The difference equations describing the input–output behavior of an nth-order, discrete-time, nonlinear, time-varying, lumped-parameter system can be written in the form of a first-order vector difference equation and a vector output equation:

$$\mathbf{x}(t_{k+1}) = \mathbf{f}[\mathbf{x}(t_k), \mathbf{u}(t_k), t_k] \qquad t_k \geq t_0 \tag{8}$$

$$\mathbf{y}(t_k) = \mathbf{g}[\mathbf{x}(t_k), \mathbf{u}(t_k), t_k] \qquad t_k \geq t_0 \tag{9}$$

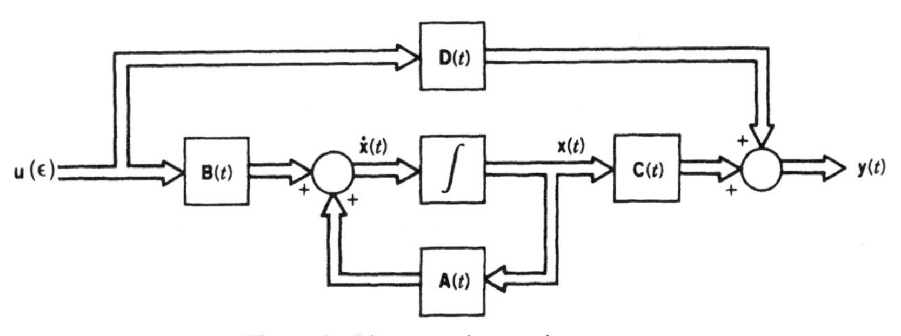

Figure 1 Linear continuous-time system.

where **x**, **y**, **u** = state, output, and input vectors of the same dimensions as noted in connection with Eqs. (1) and (2)
f, **g** = same functions as in the equations already mentioned
t_k, t_{k+1} = the kth and $(k + 1)$th discrete-time instants, respectively

If the interval between consecutive discrete-time instants is constant and equal to T, and if the functions **f** and **g** are linear, the state-space equations become

$$\mathbf{x}(k + 1) = \mathbf{F}(k)\mathbf{x}(k) + \mathbf{G}(k)\mathbf{u}(k) \qquad k \geq k_0 \tag{10}$$

$$\mathbf{y}(k) = \mathbf{C}(k)\mathbf{x}(k) + \mathbf{D}(k)\mathbf{u}(k) \qquad k \geq k_0 \tag{11}$$

where the time instants kT and $(k + 1)T$ are represented by the corresponding sequence numbers k and $k + 1$, for notational convenience. In the preceding equations, **F**, **G**, **C**, and **D** take the place of the matrices **A**, **B**, **C**, and **D** in Eqs. (4) and (5). A block diagram representation of the system equations is given in Fig. 2. The matrices **F**, **G**, **C**, and **D** become constant matrices for time-invariant systems:

$$\mathbf{x}(k + 1) = \mathbf{F}\mathbf{x}(k) + \mathbf{G}\mathbf{u}(k) \qquad k \geq k_0 \tag{12}$$

$$\mathbf{y}(k) = \mathbf{C}\mathbf{x}(k) + \mathbf{D}\mathbf{u}(k) \qquad k \geq k_0 \tag{13}$$

3 STATE-VARIABLE SELECTION AND CANONICAL FORMS

The state vector of a system is comprised of the minimum set of variables necessary to describe the system behavior in the form of the state-space equations already given. It can be shown that the selection of the state vector for a system is not unique.

For linear continuous-time systems, the following development shows that any vector $\mathbf{q}(t)$ related to a valid state vector selection $\mathbf{x}(t)$ by a constant nonsingular transformation matrix **T** is also a valid state vector:

$$\mathbf{q}(t) = \mathbf{T}\mathbf{x}(t) \tag{14}$$

where **T** is a nonsingular $n \times n$ matrix. Equations (4) and (5) may be rewritten in terms of the vector $\mathbf{q}(t)$ as

$$\dot{\mathbf{q}}(t) = [\mathbf{T}\mathbf{A}(t)\mathbf{T}^{-1}]\mathbf{q}(t) + [\mathbf{T}\mathbf{B}(t)]\mathbf{u}(t) \qquad t \geq t_0 \tag{15}$$

$$\mathbf{y}(t) = [\mathbf{C}(t)\mathbf{T}^{-1}]\mathbf{q}(t) + [\mathbf{D}(t)]\mathbf{u}(t) \qquad t \geq t_0 \tag{16}$$

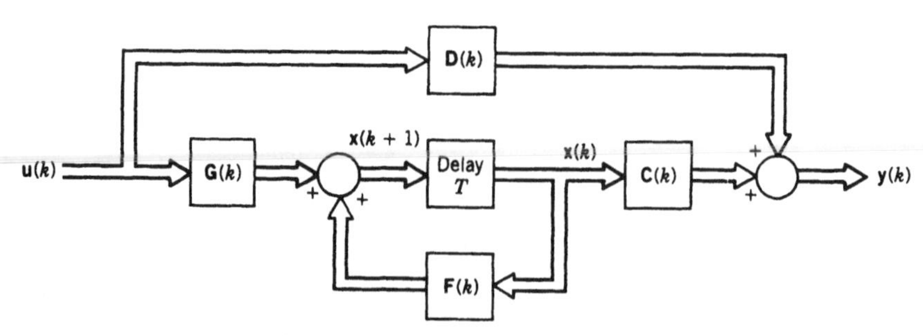

Figure 2 Linear discrete-time system.

Here, $\mathbf{q}(t)$ satisfies the definition of a state vector since it has the same dimension as $\mathbf{x}(t)$. Equations (15) and (16) are the state-space equations in terms of $\mathbf{q}(t)$. The matrices within brackets in these equations are the modified system and coupling matrices.

As for continuous-time systems, the state vector for a given linear, discrete-time system is not unique. Any vector $\mathbf{q}(k)$ related to a valid state vector $\mathbf{x}(k)$ by a constant, nonsingular matrix $\mathbf{T}$ is also a valid state vector:

$$\mathbf{q}(k) = \mathbf{T}\mathbf{x}(k) \tag{17}$$

The corresponding state-space equations are

$$\mathbf{q}(k+1) = [\mathbf{T}\mathbf{F}(k)\mathbf{T}^{-1}]\mathbf{q}(k) + [\mathbf{T}\mathbf{G}(k)]\mathbf{u}(k) \qquad k \geq k_0 \tag{18}$$

$$\mathbf{y}(k) = [\mathbf{C}(k)\mathbf{T}^{-1}]\mathbf{q}(k) + [\mathbf{D}(k)]\mathbf{u}(k) \qquad k \geq k_0 \tag{19}$$

Since the state vector of a system is not unique, the selection of state variables for a given application is governed by considerations such as ease of measurement of state variables or simplification of the resulting state-space equations. If the independent energy storage elements in the system of interest are readily identified, selection of state variables directly related to energy storage in the system is appropriate. An nth-order system has n independent energy storage elements that would enable the selection of n state variables. Examples of energy storage elements are springs and masses in mechanical systems, capacitors and inductors in electrical systems, and capacitance and inertance elements in fluid (hydraulic and pneumatic) systems.

Consider the *RLC* circuit shown in Fig. 3. Let $e_{\text{in}}(t)$ be the input and $e_{\text{out}}(t)$ be the output. The current i_{out} is assumed to be negligible. Kirchhoff's voltage law for the loop yields

$$e_{\text{in}}(t) - Ri_{\text{in}}(t) - L\frac{di_{\text{in}}(t)}{dt} - \frac{1}{C_1}\int i_{\text{in}}(t)\,dt = 0 \tag{20}$$

The current $i_{\text{in}}(t)$ through the inductor and the voltage $e_{\text{out}}(t)$ across the capacitor are directly related to energy storage in the system and are chosen as state variables:

$$x_1(t) = i_{\text{in}}(t) \tag{21}$$

$$x_2(t) = e_{\text{out}}(t) = \frac{1}{C_1}\int i_{\text{in}}(t)\,dt \tag{22}$$

The state equations can be determined from Eqs. (20)–(22) as

$$\dot{x}_1(t) = -\frac{R}{L}x_1(t) - \frac{1}{L}x_2(t) + \frac{e_{\text{in}}(t)}{L} \tag{23}$$

$$\dot{x}_2(t) = \frac{x_1(t)}{C_1} \tag{24}$$

The output equation is

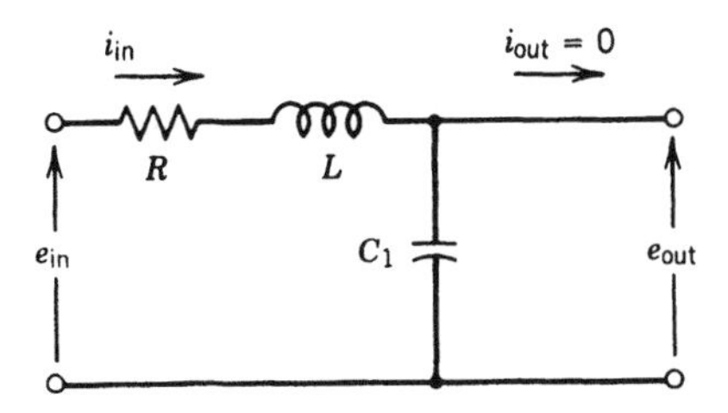

Figure 3 *RLC* circuit.

$$y(t) = e_{out}(t) = x_2(t) \tag{25}$$

The system and coupling matrices for the electrical circuit are

$$\mathbf{A} = \begin{bmatrix} -\frac{R}{L} & -\frac{1}{L} \\ \frac{1}{C_1} & 0 \end{bmatrix} \qquad \mathbf{B} = \begin{bmatrix} \frac{1}{L} \\ 0 \end{bmatrix} \tag{26}$$

$$\mathbf{C} = [0 \quad 1] \qquad \mathbf{D} = 0$$

For sampled-data control applications involving digital control of continuous-time systems, state-variable selection is often based on continuous-time system equations and needs to be retained in the discrete-time formulation. Let the time-invariant, continuous-time system equations be of the form given by Eqs. (6) and (7). The solution for the vector equation of state $\mathbf{x}(t)$ is given by Eqs. (32) and (33) later in this chapter. Applying this solution form over the time interval from kT to $(k + 1)T$, we get

$$\mathbf{x}[(k+1)T] = e^{\mathbf{A}T}\mathbf{x}(kT) + \int_{kT}^{(k+1)T} e^{\mathbf{A}[(k+1)T-t]}\mathbf{B}\mathbf{u}(t)\,dt \tag{27}$$

In sampled-data control applications, the input vector $\mathbf{u}(t)$ is the control input computed by the digital computer and applied to the continuous-time system by a digital–analog converter. Usually, the digital–analog converter has a latch that maintains the output constant between the time instants kT and $(k + 1)T$:

$$\mathbf{u}(t) = \mathbf{u}(kT) \qquad kT < t \le (k+1)T \tag{28}$$

Equation (27) then simplifies to

$$\begin{aligned} \mathbf{x}(k+1) &= (e^{\mathbf{A}T})\mathbf{x}(k) + \left(\int_0^T e^{\mathbf{A}(T-t)}\mathbf{B}\,dt\right)\mathbf{u}(k) \\ &= \mathbf{F}\mathbf{x}(k) + \mathbf{G}\mathbf{u}(k) \end{aligned} \tag{29}$$

Equation (29) establishes the relationship between the system matrices in the discrete-time and continuous-time formulations of the system equations.

3.1 Canonical Forms for Continuous-Time Systems

For high-order, single-input–single-output (SISO) systems or multiple-input–multiple-output (MIMO) systems, the number of elements in the system matrices is large. Selection of state variables that simplify the state-space representation is thus desirable. Such representations of the state-space equations also exhibit significant properties of the system more clearly and are referred to as canonical forms. However, the names and corresponding structures of the canonical forms are not completely standardized.

The controllable canonical form is useful in control system design applications. The controllable canonical form for an nth-order, SISO, LTI, continuous-time system is described in Table 1. The selection of state variables $x_1, x_2, \ldots, x_n$ that results in the controllable canonical form of the state equations is indicated on the simulation diagram in the table. For the case where all the β_i, $i = 1, n$, are zero and $\beta_0 \neq 0$, the transfer function $Y(s)/U(s)$ has no numerator dynamics. The n state variables are then simply the output and $n - 1$ successive derivatives of the output. These are referred to as the phase variables. The phase-variable canonical form is thus a special case of the controllable canonical form. The system

Table 1 State-Space Canonical Forms for SISO Continuous-Time Systems

$$H(s) = \frac{Y(s)}{U(s)} = \frac{\beta_n s^n + \beta_{n-1}s^{n-1} + \cdots + \beta_1 s + \beta_0}{s^n + \alpha_{n-1}s^{n-1} + \cdots + \alpha_1 s + \alpha_0}$$

I. Controllable canonical form
Simulation diagram

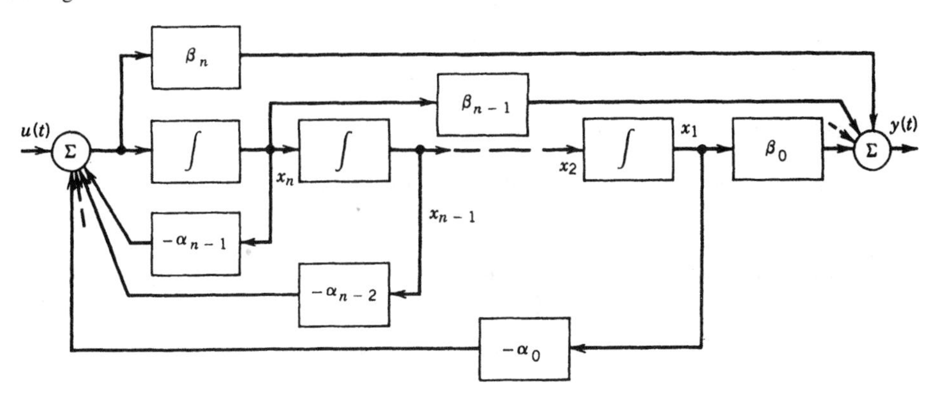

System matrices

$$\mathbf{A} = \left[\begin{array}{c|ccc} 0 & & & \\ 0 & & \mathbf{I}_{n-1} & \\ \vdots & & & \\ 0 & & & \\ \hline -\alpha_0 & -\alpha_1 & \cdots & -\alpha_{n-1} \end{array}\right] \qquad \mathbf{B} = \begin{bmatrix} 0 \\ 0 \\ \vdots \\ 0 \\ 1 \end{bmatrix} \qquad \mathbf{D} = \beta_n$$

$$\mathbf{C} = [\beta_0 - \beta_n\alpha_0 \quad \beta_1 - \beta_n\alpha_1 \quad \ldots \quad \beta_{n-1} - \beta_n\alpha_{n-1}]$$

II. Observable canonical form
Simulation diagram

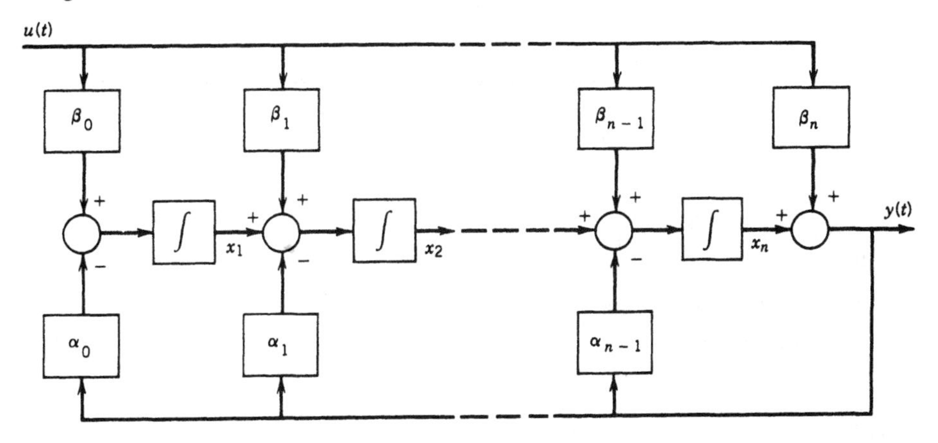

Table 1 (*Continued*)

System matrices

$$\mathbf{A} = \left[\begin{array}{cccc|c} 0 & 0 & \cdots & 0 & -\alpha_0 \\ \hline & & & & -\alpha_1 \\ & \mathbf{I}_{n-1} & & & \vdots \\ & & & & -\alpha_{n-1} \end{array}\right] \qquad \mathbf{B} = \begin{bmatrix} \beta_0 - \beta_n\alpha_0 \\ \beta_1 - \beta_n\alpha_1 \\ \vdots \\ \beta_{n-1} - \beta_n\alpha_{n-1} \end{bmatrix}$$

$$\mathbf{C} = [0 \quad 0 \quad \ldots \quad 0 \quad 1] \qquad \mathbf{D} = \beta_n$$

III. Normal or diagonal Jordan canonical form

Related conditions

(i) Characteristic equation roots s_i, $i = 1, \ldots, n$, are real and distinct

Simulation diagram

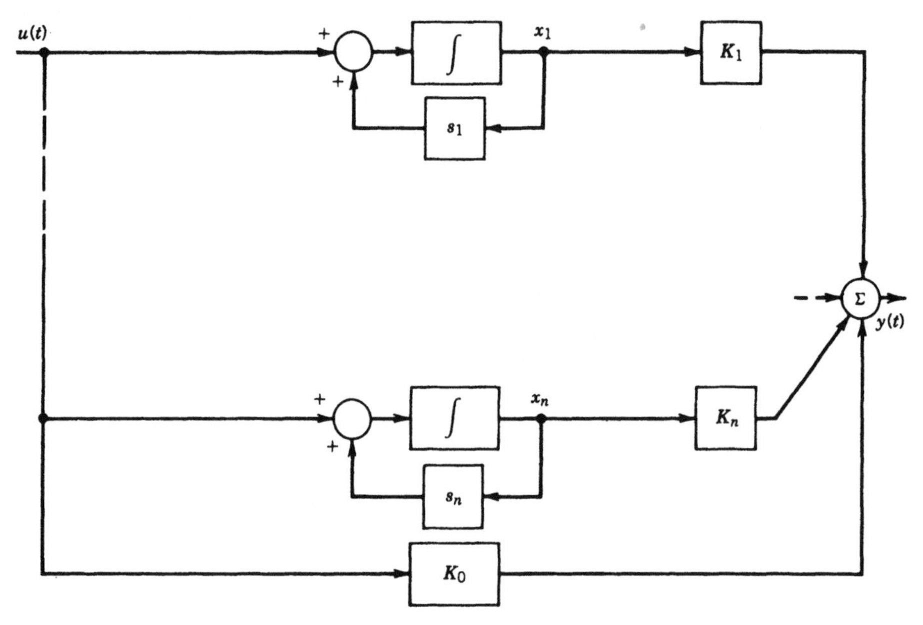

System matrices

$$\mathbf{A} = \begin{bmatrix} s_1 & & & \\ & s_2 & & \mathbf{0} \\ & & \ddots & \\ & \mathbf{0} & & s_n \end{bmatrix} \qquad \mathbf{B} = \begin{bmatrix} 1 \\ 1 \\ \vdots \\ 1 \end{bmatrix}$$

$$\mathbf{C} = [K_1 \quad K_2 \quad \ldots \quad K_n] \qquad \mathbf{D} = K_0$$

where $K_0 = \lim_{s\to\infty} [H(s)]$ and $K_i = \lim_{s\to s_i} [(s - s_i)H(s)] \qquad i = 1, \ldots, n$

IV. Near-normal canonical form

Related conditions

(i) One pair of complex conjugate characteristic equation roots

$$s_k = s_{kr} + js_{ki}$$

$$s_{k+1} = s_{kr} - js_{ki}$$

Table 1 (*Continued*)

(ii) All other roots are real and distinct.

Simulation diagram

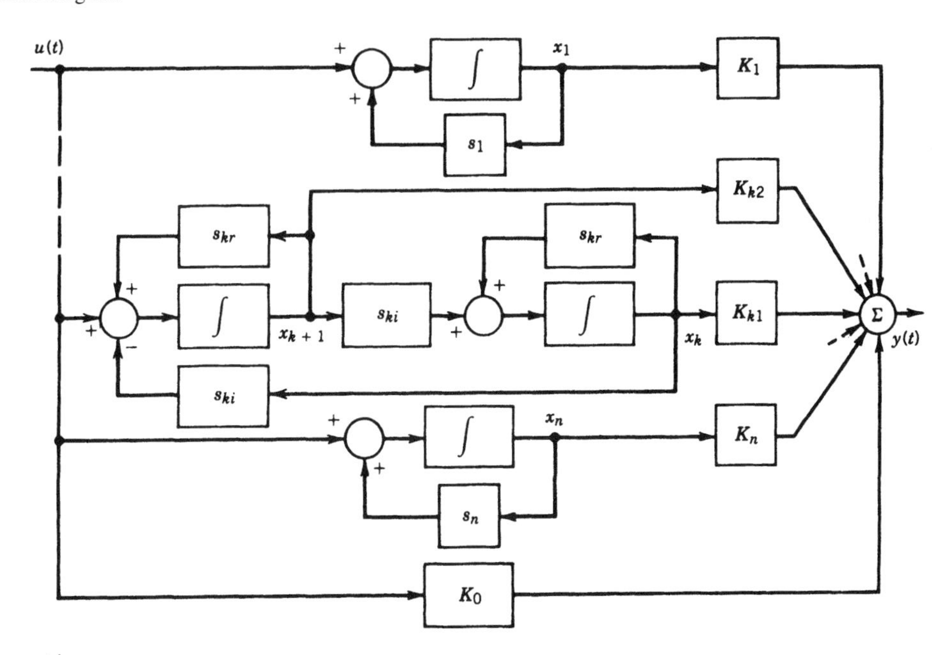

System matrices

$$
\mathbf{A} = \left[\begin{array}{ccc|cc|ccc}
s_1 & & \mathbf{0} & & & & & \\
& \ddots & & & \mathbf{0} & & \mathbf{0} & \\
\mathbf{0} & & s_{k-1} & & & & & \\
\hline
& \mathbf{0} & & s_{kr} & s_{ki} & & \mathbf{0} & \\
& & & -s_{ki} & s_{kr} & & & \\
\hline
& & & & & s_{k-2} & & \mathbf{0} \\
& \mathbf{0} & & & \mathbf{0} & & \ddots & \\
& & & & & \mathbf{0} & & s_n
\end{array}\right] \qquad
\mathbf{B} = \begin{bmatrix} 1 \\ \vdots \\ 1 \\ \hline 0 \\ 1 \\ \hline 1 \\ \vdots \\ 1 \end{bmatrix}
$$

$$
\mathbf{C} = [K_1 \quad . . . \quad K_{k-1} \,\vdots\, K_{k1} \quad K_{k2} \,\vdots\, K_{k+2} \quad . . . \quad K_n] \qquad \mathbf{D} = K_0
$$

where $K_0 = \lim_{s \to \infty} [H(s)] \qquad K_i = \lim_{s \to s_i} [(s - s_i)H(s)]$

$i = 1, k - 1$ and $i = k + 2, \ldots, n$

$$
K_{k1} = -\tfrac{1}{2}\text{Im}[(s - s_k)H(s)]_{s=s_k} \qquad K_{k2} = \tfrac{1}{2}\text{Re}[(s - s_k)H(s)]_{s=s_k}
$$

Table 1 *(Continued)*

V. Nondiagonal Jordan canonical form

Related conditions

(i) One real characteristic equation root s_k repeated m times (i.e., $s_k = s_{k+1} = \cdots = s_{k+m-1}$).

(ii) All other roots are real and distinct.

Simulation diagram

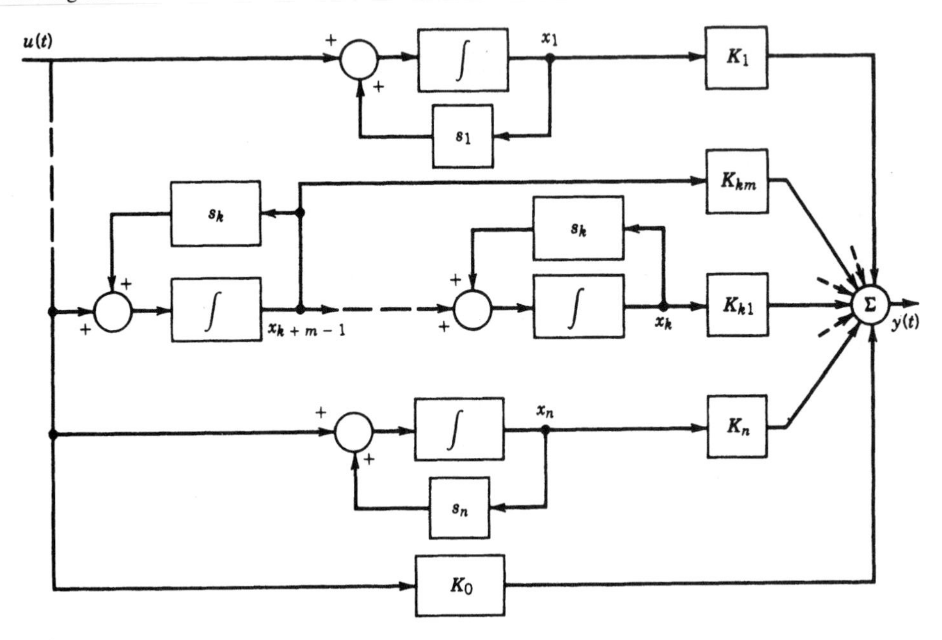

System matrices

$$
\mathbf{A} = \left[\begin{array}{ccc|ccc|ccc}
s_1 & & \mathbf{0} & & & & & & \\
 & \ddots & & & \mathbf{0} & & & \mathbf{0} & \\
\mathbf{0} & & s_{k-1} & & & & & & \\
\hline
 & & & s_k & 1 & \mathbf{0} & & & \\
 & \mathbf{0} & & & \ddots & 1 & & \mathbf{0} & \\
 & & & \mathbf{0} & & s_k & & & \\
\hline
 & & & & & & s_{k+m} & & \mathbf{0} \\
 & \mathbf{0} & & & \mathbf{0} & & & \ddots & \\
 & & & & & & \mathbf{0} & & s_n
\end{array}\right]
\qquad
\mathbf{B} = \left[\begin{array}{c} 1 \\ \vdots \\ 1 \\ \hline 0 \\ 1 \\ \hline 1 \\ \vdots \\ 1 \end{array}\right]
$$

$$
\mathbf{C} = [K_1 \;\; \ldots \;\; K_{k-1} \mid K_{k1} \;\; \ldots \;\; K_{km} \mid K_{k+m} \;\; \ldots \;\; K_n] \qquad \mathbf{D} = K_0
$$

where $K_0 = \lim_{s \to \infty} [H(s)]$ $\quad K_i = \lim_{s \to s_i} [(s - s_i)H(s)]$

$i = 1, k - 1$ and $i = k + m, \ldots, n$

$$
K_{kj} = \frac{1}{(j-1)!} \left\{ \frac{d^{j-1}}{ds^{j-1}} [(s - s_k)^m H(s)] \right\}_{s = s_k} \qquad j = 1, \ldots, m
$$

and coupling matrices for the controllable canonical form are listed in Table 1. The special form of the **A** matrix is referred to as the companion form. Here, $\mathbf{I}_{n-1}$ is the $(n - 1) \times (n - 1)$ identity matrix.

For a SISO, LTI system described by Eqs. (6) and (7), the state transformation matrix **T** in Eq. (14) which transforms the state-space equations into the controllable canonical form exists if the controllability matrix $\mathbf{P}_c$ in Eq. (30) is nonsingular[1]:

$$\mathbf{P}_c = [\mathbf{B} \quad \mathbf{AB} \quad \dots \quad \mathbf{A}^{n-1}\mathbf{B}] \tag{30}$$

The transformation matrix **T** is defined in Table 2.

The observable canonical form is useful in state estimator or observer design applications. The observable canonical form for an nth-order, SISO, LTI system is described in Table 1 in a manner similar to the controllable canonical form. The corresponding **A** matrix is the transpose of the **A** matrix for the controllable canonical form and is also referred to as a companion matrix. The state transformation matrix **T** in Eq. (14), which transforms given state-space Eqs. (6) and (7) into the observable canonical form, exists if the observability matrix $\mathbf{P}_0$ is nonsingular[1]:

$$\mathbf{P}_0 = \begin{bmatrix} \mathbf{C} \\ \mathbf{CA} \\ \vdots \\ \mathbf{CA}^{n-1} \end{bmatrix} \tag{31}$$

State variables can also be chosen to diagonalize or nearly diagonalize the state matrix **A**. The resulting state-space equations are completely or almost completely decoupled from one another and hence show very clearly the effect of initial conditions or forcing inputs on the different characteristic modes of the system response. The resulting physical insight into the system behavior makes the corresponding form of the system equations, called the normal form or diagonal Jordan form, valuable in vibration analysis applications and in control applications involving modal control. The normal form of the state-space equations for a SISO, LTI system with real, distinct characteristic roots is given in Table 1. The diagonal elements of the **A** matrix in the table are the system characteristic roots or eigenvalues. The state variables $x_1, x_2, \dots, x_n$ lie along the eigenvectors of the **A** matrix, in state space. As the corresponding simulation diagram indicates, the behavior of each state variable is governed solely by one eigenvalue, the initial condition on that state variable, and the forcing input.

If some of the distinct characteristic roots of a SISO, LTI system are complex, the matrices **A** and **C** in Eqs. (6) and (7) have complex elements when represented in the normal form just described. Since this could be inconvenient in subsequent matrix manipulations, a nearly diagonal **A** matrix can be obtained for cases where the complex characteristic roots occur in complex-conjugate pairs. This would be the case for system differential equations with only real coefficients. The near-normal form of the system equations for a system with one pair of complex-conjugate characteristic roots is given in Table 1. Extension to the case of multiple complex root pairs is straightforward. The complex characteristic roots result in a few nonzero off-diagonal elements in the **A** matrix; otherwise, the decoupled nature of the system equations is retained.

If one characteristic root of a SISO, LTI system is real and repeated m times, the state equations can only be partially decoupled by appropriate state-variable selection, as shown in Table 1. The resulting state-space equations are said to be in the Jordan canonical form. The corresponding **A** matrix has one submatrix with the repeated eigenvalue at the diagonal positions, ones immediately to the right of the repeated diagonal elements within the submatrix, and zero elements at all other nondiagonal positions.[21] The **A** matrix is then said to

Table 2 Transformation Matrices for Continuous-Time State-Space Canonical Forms

State-space equations (SISO, LTI system)
$\dot{\mathbf{x}}(t) = \mathbf{Ax}(t) + \mathbf{Bu}(t)$
$\mathbf{y}(t) = \mathbf{Cx}(t) + \mathbf{Du}(t)$

Characteristic equation
$\det(s\mathbf{I} - \mathbf{A}) =$
$s^n + \alpha_{n-1}s^{n-1} + \cdots + \alpha_1 s + \alpha_0 = 0$

I. Controllable canonical form

Transformation conditions

(i) $\mathbf{P}_c = [\mathbf{B} \quad \mathbf{AB} \quad \cdots \quad \mathbf{A}^{n-1}\mathbf{B}]$ must be nonsingular

Transformation matrices

(i) $\mathbf{q} = \mathbf{Tx}$, $\mathbf{T} = \mathbf{R}^{-1}\mathbf{P}_c^{-1}$
where

$$\mathbf{R} = \begin{bmatrix} \alpha_1 & \alpha_2 & \cdots & \alpha_{n-1} & 1 \\ \alpha_2 & \alpha_3 & \cdots & 1 & 0 \\ \vdots & \vdots & & \vdots & \vdots \\ \alpha_{n-1} & 1 & \cdots & 0 & 0 \\ 1 & 0 & \cdots & 0 & 0 \end{bmatrix}$$

(ii) New state matrix $= \mathbf{TAT}^{-1}$

$$= \left[\begin{array}{c|ccc} 0 & & & \\ 0 & & & \\ \vdots & & \mathbf{I}_{n-1} & \\ 0 & & & \\ \hline -\alpha_0 & -\alpha_1 & \cdots & -\alpha_{n-1} \end{array}\right]$$

II. Observable canonical form

Transformation conditions

(i) $\mathbf{P}_0 = \begin{bmatrix} \mathbf{C} \\ \mathbf{CA} \\ \vdots \\ \mathbf{CA}^{n-1} \end{bmatrix}$ must be nonsingular

Transformation matrices

(i) $\mathbf{q} = \mathbf{Tx}$, $\mathbf{T} = \mathbf{RP}_0$
where

$$\mathbf{R} = \begin{bmatrix} \alpha_1 & \alpha_2 & \cdots & \alpha_{n-1} & 1 \\ \alpha_2 & \alpha_3 & \cdots & 1 & 0 \\ \vdots & \vdots & & \vdots & \vdots \\ \alpha_{n-1} & 1 & \cdots & 0 & 0 \\ 1 & 0 & \cdots & 0 & 0 \end{bmatrix}$$

(ii) New state matrix $= \mathbf{TAT}^{-1}$

$$= \left[\begin{array}{cccc|c} 0 & 0 & \cdots & 0 & -\alpha_0 \\ \hline & & & & -\alpha_1 \\ & \mathbf{I}_{n-1} & & & \vdots \\ & & & & -\alpha_{n-1} \end{array}\right]$$

III. Normal or diagonal Jordan canonical form

Transformation conditions

(i) $\mathbf{A}$ matrix has only distinct, real eiganvalues s_i, $i = 1, \ldots, n$

Transformation matrices

(i) $\mathbf{q} = \mathbf{Tx}$, $\mathbf{T}^{-1} = \mathbf{M} = [\mathbf{v}_1 \quad \mathbf{v}_2 \quad \cdots \quad \mathbf{v}_n]$

where (a) $\mathbf{v}_i$ are the n linearly independent eigenvectors corresponding to s_i and

(b) $\mathbf{v}_i$ are taken to be equal or proportional to any nonzero columd of $\text{Adj}(s_i\mathbf{I} - \mathbf{A})$

(ii) New state matrix $= \mathbf{TAT}^{-1}$

$$= \begin{bmatrix} s_1 & & & \\ & s_2 & & \mathbf{0} \\ & \mathbf{0} & \ddots & \\ & & & s_n \end{bmatrix}$$

Table 2 (*Continued*)

IV. Normal or diagonal Jordan canonical form

Transformation conditions

(i) **A** matrix has one repeated, real eigenvalue s_k of multiplicity m (i.e., $s_k = s_{k+1} = \cdots = s_{k+m-1}$). All other eigenvalues are real and distinct.

(ii) Degeneracy $d = n - \text{rank}(s_k\mathbf{I} - \mathbf{A}) = m$. Full degeneracy.

Transformation matrices

(i) $\mathbf{q} = \mathbf{Tx}$, $\mathbf{T}^{-1} = \mathbf{M} = [\mathbf{v}_1 \quad \mathbf{v}_2 \quad \cdots \quad \mathbf{v}_n]$

where (a) $\mathbf{v}_i$, $i = 1, k - 1$ nad $i = k + m, n$, are the linearly independent eigenvectors corresponding to the real, distinct eigenvalues;

(b) $\mathbf{v}_i$, $i = 1, k - 1$ and $i = k + m, n$, are taken to be equal or proportional to any nonzero column of $\text{Adj}(s_i\mathbf{I} - \mathbf{A})$; and

(c) $\mathbf{v}_i$, $i = k, k + m - 1$, are the m linearly independent eigenvectors corresponding to the repeated eigenvalue. They are equal or proportional to the nonzero linearly independent columns of

$$\left\{ \frac{d^{m-1}}{ds^{m-1}} [\text{Adj}(s\mathbf{I} - \mathbf{A})] \right\}_{s=s_k}$$

(ii) New state matrix $= \mathbf{TAT}^{-1}$

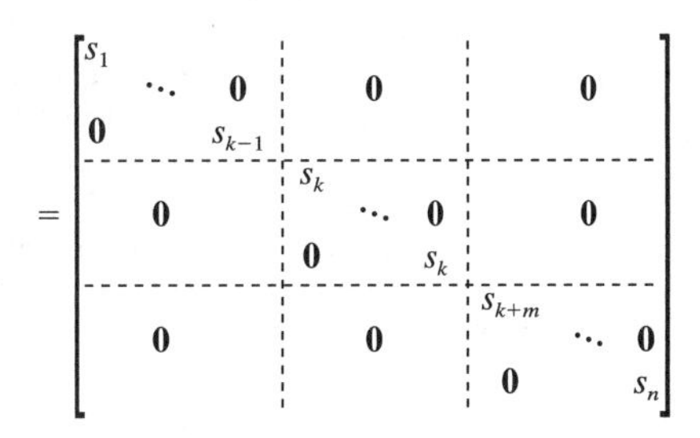

V. Near-normal canonical form

Transformation conditions

(i) **A** matrix has one pair of complex-conjugate eigenvalues, s_k, s_{k+1}

$$s_k = s_{kr} + js_{ki}$$

$$s_{k+1} = s_{kr} - js_{ki}$$

(ii) All other eigenvalues are real and distinct.

Transformation matrices

(i) $\mathbf{q} = \mathbf{Tx}$, $\mathbf{T}^{-1} = [\mathbf{v}_1 \quad \cdots \quad \mathbf{v}_{k-1} \quad \mathbf{v}_{kr} \quad \mathbf{v}_{ki} \quad \mathbf{v}_{k+2} \quad \cdots \quad \mathbf{v}_n]$

where (a) $\mathbf{v}_i$, $i = 1, k - 1$ and $i = k + 2, n$ are the linearly independent eigenvectors corresponding to the real, distinct eigenvalues;

(b) $\mathbf{v}_i$ for $i = 1, \ldots, n$ are taken to be equal or proportional to any nonzero column of $\text{Adj}(s_i\mathbf{I} - \mathbf{A})$; and

(c) $\mathbf{v}_k = \mathbf{v}_{kr} + j\mathbf{v}_{ki}$, $\mathbf{v}_{k+1} = \mathbf{v}_{kr} - j\mathbf{v}_{ki}$ are the complex-conjugate eigenvectors corresponding to s_k and s_{k+1}, respectively.

Table 2 (*Continued*)

(ii) New state matrix = $\mathbf{TAT}^{-1}$

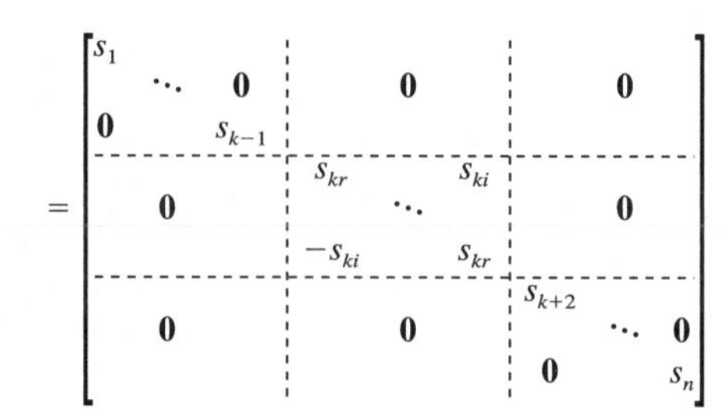

VI. Nondiagonal Jordan canonical form

Transformation conditions

(i) **A** matrix has one repeated, real eigenvalue s_k of multiplicity m (i.e., $s_k = s_{k+1} = \cdots = s_{k+m-1}$). All other eigenvalues are real and distinct.

(ii) Degeneracy $d = n - \text{rank}(s_k\mathbf{I} - \mathbf{A}) = 1$. Simple degeneracy.

Transformation matrices

(i) $\mathbf{q} = \mathbf{Tx}$, $\mathbf{T}^{-1} = [\mathbf{t}_1\ \mathbf{t}_2 \ldots \mathbf{t}_n]$

where (a) $\mathbf{t}_i$, $i = 1, k - 1$ and $i = k + m, n$, are the linearly independent eigenvectors corresponding to the real, distinct eigenvalues;

(b) $\mathbf{t}_i$, for $i = 1, k - 1$ and $i = k + m, n$, are taken to be equal or proportional to any nonzero column of $\text{Adj}(s_i\mathbf{I} - \mathbf{A})$; and

(c) $\mathbf{t}_i$, $i = k, k + m - 1$, are obtained by solution of the equation $\mathbf{AT}^{-1} = \mathbf{T}^{-1}\mathbf{J}$, where **J** is the Jordan canonical matrix given. Each $\mathbf{t}_i$ is determined to within a constant of proportionality.

(ii) New state matrix = $\mathbf{J} = \mathbf{TAT}^{-1}$

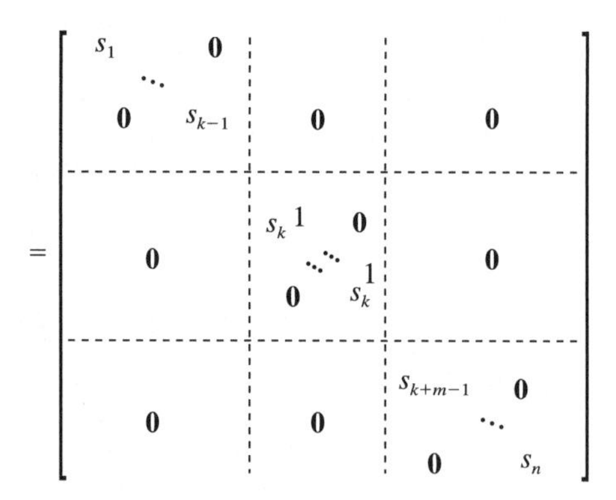

VII. Nondiagonal Jordan canonical form

Transformation conditions

(i) **A** matrix has one repeated, real eigenvalue of multiplicity m (i.e., $s_k = s_{k+1} = \cdots = s_{k+m-1}$). All other eigenvalues are real and distinct.

(ii) Degeneracy $d = n - \text{rank}(s_k\mathbf{I} - \mathbf{A})$ is between 1 and m. General degeneracy.

Table 2 (*Continued*)

Transformation matrices

(i) $\mathbf{q} = \mathbf{Tx}$, $\mathbf{T}^{-1} = [\mathbf{t}_1\ \mathbf{t}_2 \ldots \mathbf{t}_n]$

where (a) $\mathbf{t}_i$, $i = 1, k - 1$ and $i = k + m, n$, are the linearly independent eigenvectors corresponding to the real, distinct eigenvalues;

(b) $\mathbf{t}_i$, $i = 1, k - 1$ and $i = k + m, n$, are taken to be equal or proportional to any nonzero column of $\text{Adj}(s_i\mathbf{I} - \mathbf{A})$;

(c) $\mathbf{t}_i$, $i = k, k + m - 1$, are m linearly independent vectors corresponding to the repeated eigenvalue of multiplicity m, only d of these vectors eigenvectors; and

(d) $\mathbf{t}_i$, $i = k = k, k + m - 1$, obtained by solution of the equation $\mathbf{AT}^{-1} = \mathbf{T}^{-1}\mathbf{J}$, where $\mathbf{J}$ is the Jordan canonical matrix for the problem, with d Jordan blocks. There are d possible choices for $\mathbf{J}$. Each of these choices needs to be tried and the $\mathbf{t}_i$ vectors solved for. Only the correct $\mathbf{J}$ will give the m linearly independent vectors $\mathbf{t}_i$, $i = k, k + m - 1$. Each $\mathbf{t}_i$ is determined to within a constant of proportionality.

(ii) New state matrix $= \mathbf{J} = \mathbf{TAT}^{-1}$. Correct $\mathbf{J}$ determined by trial and error, as previously described.

have one Jordan block. The extension of the result in Table 1 to the case of many different repeated characteristic roots is straightforward.

For SISO or MIMO, LTI systems described by state-space equations (6) and (7), the transformation matrix $\mathbf{T}$ in Eq. (14), which transforms the state-space equations into the diagonal or nondiagonal Jordan form, can be determined. As in Table 1, there are a number of different cases to be considered.[2] If the $\mathbf{A}$ matrix has real, distinct eigenvalues s_i, $i = 1, \ldots, n$, the eigenvectors are linearly independent and can be used to form the modal matrix $\mathbf{M}$ as indicated in Table 2. The transformation matrix $\mathbf{T}$ is then taken to be $\mathbf{M}^{-1}$. If the $\mathbf{A}$ matrix has one pair of complex-conjugate eigenvalues and if system matrices with real elements only are desired, the transformation matrix $\mathbf{T}$ is defined in a slightly different form as indicated in Table 2. The resulting transformed state matrix will have two nonzero off-diagonal elements as indicated.

The transformed state-space equations may be in the nondiagonal Jordan canonical form if the $\mathbf{A}$ matrix has repeated eigenvalues. The procedure for determining the transformation matrix depends on the degeneracy of the matrix $s_k\mathbf{I} - \mathbf{A}$ corresponding to the repeated eigenvalue s_k. If the degeneracy d of $s_k\mathbf{I} - \mathbf{A}$, defined in Table 2, is equal to m, where m is the multiplicity of the repeated eigenvalue s_k, m linearly independent eigenvectors can be found for the repeated eigenvalue. The procedure for doing so is indicated in Table 2. The transformed state matrix is then diagonal. If the degeneracy of $s_k\mathbf{I} - \mathbf{A}$ is 1, only one eigenvector can be determined. Since it can be shown that the degeneracy is equal to the number of Jordan blocks associated with the eigenvector, the transformed state matrix $\mathbf{J}$ has only one Jordan block and is uniquely defined.[2] The nonsingular transformation matrix $\mathbf{T}$ is then determined as indicated in Table 2. If the degeneracy d of $s_k\mathbf{I} - \mathbf{A}$ is greater than 1 but less than m, there are d linearly independent eigenvectors and d Jordan blocks associated with the eigenvalue s_k. In this case, the transformed state matrix $\mathbf{J}$ cannot be uniquely defined but can be one of a finite number of possibilities. A trial-and-error formulation of $\mathbf{J}$ and solution for $\mathbf{T}$, as indicated in Table 2, is necessary until a nonsingular transformation matrix $\mathbf{T}$ is obtained.[2]

3.2 Canonical Forms for Discrete-Time Systems

Canonical forms of discrete-time state-space equations have the same uses that such forms have for continuous-time systems. The development of these canonical forms closely parallels that for continuous-time systems and is therefore summarized here. Table 3 indicates the state-variable selection for SISO, LTI systems described by pulse transfer functions that yield the controllable, observable, and Jordan canonical forms of the state-space equations. For systems already described by discrete-time state-space equations (12) and (13), Table 4 indicates the transformation matrix $\mathbf{T}$ in Eq. (17), which transforms the state-space equations into the canonical forms named previously.

With the exception of the procedures for selecting transformation matrices to convert state-space equations for MIMO, LTI systems to the Jordan canonical form, the procedures and results presented in Tables 1–4 are restricted to SISO, LTI systems. A procedure for representing a SISO, linear time-varying (LTV) system, described by a differential equation, in controllable canonical form has been described by DeRusso, Roy, and Close.[2]

Canonical forms for MIMO, LTI systems cannot, in general, be specified uniquely as is the case for SISO systems. Kailath,[3] Fortmann and Hitz,[1] and Kalman[9] have specified some canonical forms for MIMO systems and have described procedures for representing such systems in these forms, given their transfer function matrix descriptions or state-space equations. The problem of selection of state variables given the transfer function matrix description of MIMO systems is the problem of realization and is considered in Section 7.

4 SOLUTION OF SYSTEM EQUATIONS

4.1 Continuous-Time Systems

The state-space Eqs. (1) and (2) for time-varying, nonlinear systems described by ordinary differential equations can be solved by numerical integration techniques. Such numerical integration would, however, have to be repeated if the initial conditions $\mathbf{x}(t_0)$ or the forcing function $\mathbf{u}(t)$ were to be changed. The computational burden can be reduced for linear systems by using the concept of the state transition matrix.

For LTI systems described by the state equations (6) and (7), the solution is given by

$$\mathbf{x}(t) = \boldsymbol{\phi}(t - t_0)\mathbf{x}(t_0) + \int_{t_0}^{t} \boldsymbol{\phi}(t - \tau)\mathbf{B}\mathbf{u}(\tau)\, d\tau \tag{32}$$

$$\mathbf{y}(t) = \mathbf{C}\boldsymbol{\phi}(t - t_0)\mathbf{x}(t_0) + \int_{t_0}^{t} \mathbf{C}\boldsymbol{\phi}(t - \tau)\mathbf{B}\mathbf{u}(\tau)\, d\tau + \mathbf{D}\mathbf{u}(t) \tag{33}$$

where the $n \times n$ matrix $\boldsymbol{\phi}(t)$ is defined as the state transition matrix of the system. Derivation of this result is available in many standard textbooks on state-space methods.[1,4] The first terms on the right-hand sides of the preceding equations represent the response of the homogeneous system to the initial condition $\mathbf{x}(t_0)$ whereas the second terms represent the forced response of the system. Comparison of the second term in Eq. (33), for the case $\mathbf{D} = 0$, with Eq. (34) for the forced response of a SISO, LTI system indicates that the matrix $\mathbf{C}\boldsymbol{\phi}(t)\mathbf{B}$ is a matrix of impulse responses:

$$y(t_0) = y(t_0) + \int_{t_0}^{t} h(t - \tau)u(\tau)\, d\tau \tag{34}$$

The variable $h(t)$ in Eq. (34) is the impulse response of the system. The interpretation of $\mathbf{C}\boldsymbol{\phi}(t)\mathbf{B}$ as a matrix of impulse responses forms the basis of one of the methods for determining the elements of the transition matrix.[2,4]

The state transition matrix $\boldsymbol{\phi}(t - t_0)$ is the solution of the matrix differential equation

Table 3 State-Space Canonical Forms for SISO Discrete-Time Systems

$$H(z) = \frac{Y(z)}{U(z)} = \frac{\beta_n z^n + \beta_{n-1} z^{n-1} + \cdots + \beta_1 z + \beta_0}{z^n + \alpha_{n-1} z^{n-1} + \cdots + \alpha_1 z + \alpha_0}$$

I. Controllable canonical form
Simulation diagram

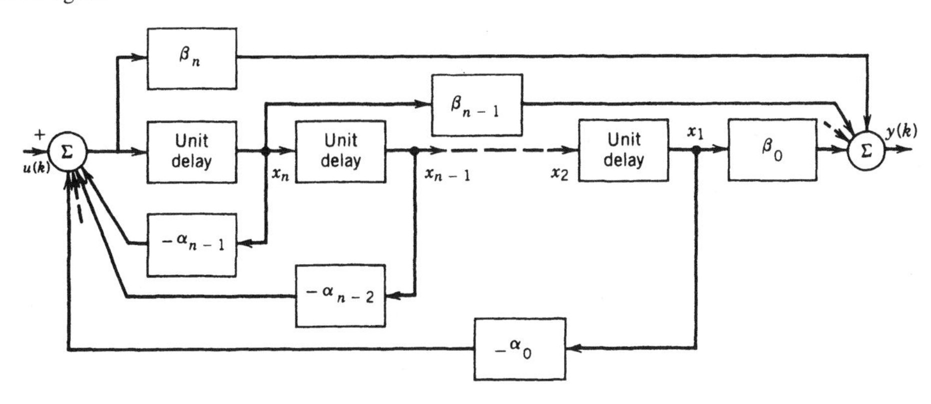

System matrices

$$\mathbf{F} = \left[\begin{array}{c|ccc} 0 & & & \\ 0 & & \mathbf{I}_{n-1} & \\ \vdots & & & \\ 0 & & & \\ \hline -\alpha_0 & -\alpha_1 & \cdots & -\alpha_{n-1} \end{array}\right] \qquad \mathbf{G} = \begin{bmatrix} 0 \\ 0 \\ \vdots \\ 0 \\ 1 \end{bmatrix} \qquad \mathbf{D} = \beta_n$$

$$\mathbf{C} = [\beta_0 - \beta_n\alpha_0 \quad \beta_1 - \beta_n\alpha_1 \quad \ldots \quad \beta_{n-1} - \beta_n\alpha_{n-1}]$$

II. Observable canonical form
Simulation diagram

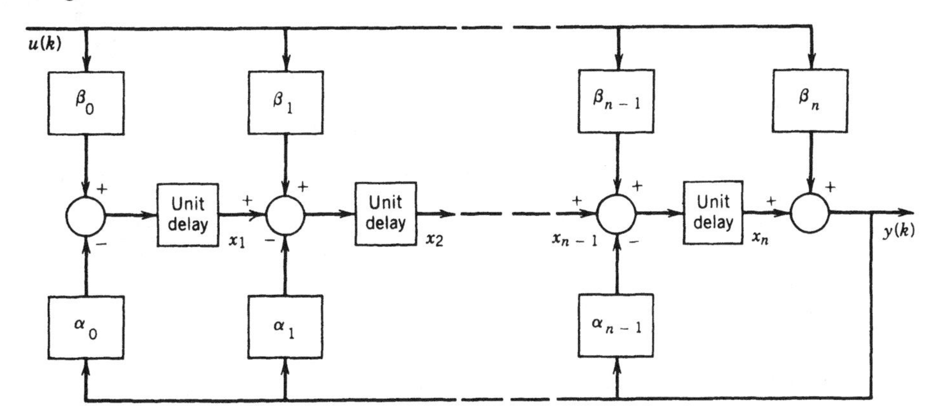

Table 3 (*Continued*)

System matrices

$$\mathbf{F} = \left[\begin{array}{cccc|c} 0 & 0 & \cdots & 0 & -\alpha_0 \\ \hline & & & & -\alpha_1 \\ & \mathbf{I}_{n-1} & & & \vdots \\ & & & & -\alpha_{n-1} \end{array}\right] \qquad \mathbf{G} = \begin{bmatrix} \beta_0 - \beta_n\alpha_0 \\ \beta_1 - \beta_n\alpha_1 \\ \vdots \\ \beta_{n-1} - \beta_n\alpha_{n-1} \end{bmatrix}$$

$$\mathbf{C} = [0 \;\; 0 \;\; \cdots \;\; 0 \;\; 1] \qquad \mathbf{D} = \beta_n$$

III. Normal or diagonal Jordan canonical form
Related conditions
(i) Characteristic equation roots z_i, $i = 1, \ldots, n$, are real and distinct.
Simulation diagram

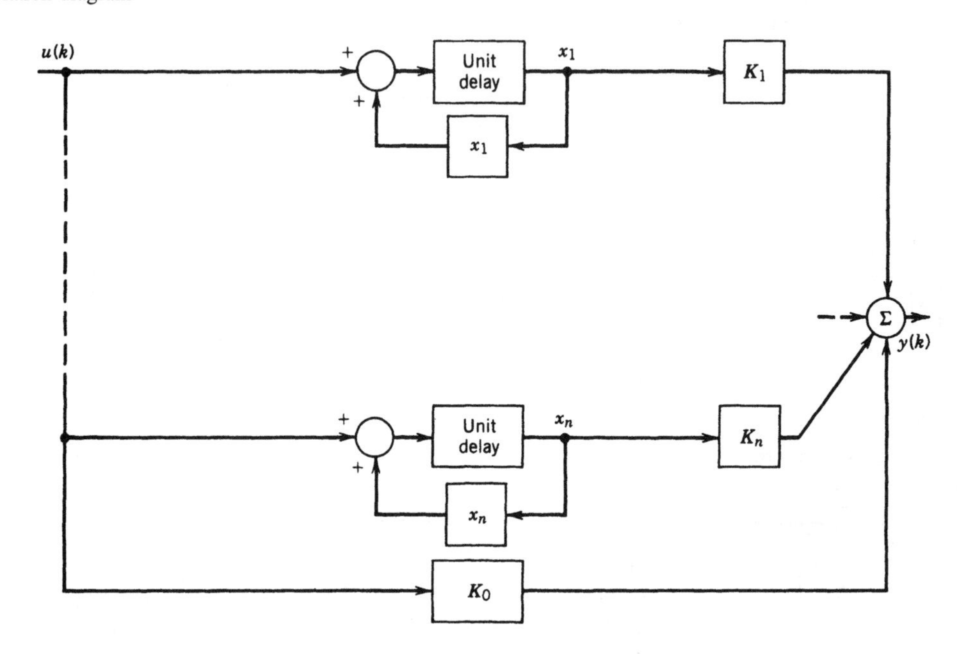

System matrices

$$\mathbf{F} = \begin{bmatrix} z_1 & & & \mathbf{0} \\ & z_2 & & \\ & & \ddots & \\ \mathbf{0} & & & z_n \end{bmatrix} \qquad \mathbf{G} = \begin{bmatrix} 1 \\ 1 \\ \vdots \\ 1 \end{bmatrix}$$

$$\mathbf{C} = [K_1 \;\; K_2 \;\; \cdots \;\; K_n] \qquad \mathbf{D} = K_0$$

where $K_0 = \lim_{z \to \infty} [H(z)]$ and $K_i = \lim_{z \to z_i} [(z - z_i)H(z)] \qquad i = 1, \ldots, n$

Table 3 *(Continued)*

IV. Near-normal canonical form

Related conditions

(i) One pair of complex conjugate characteristic equation roots

$$z_k = z_{kr} + jz_{ki}$$

$$z_{k+1} = z_{kr} - jz_{ki}$$

(ii) All other roots are real and distinct.

Simulation diagram

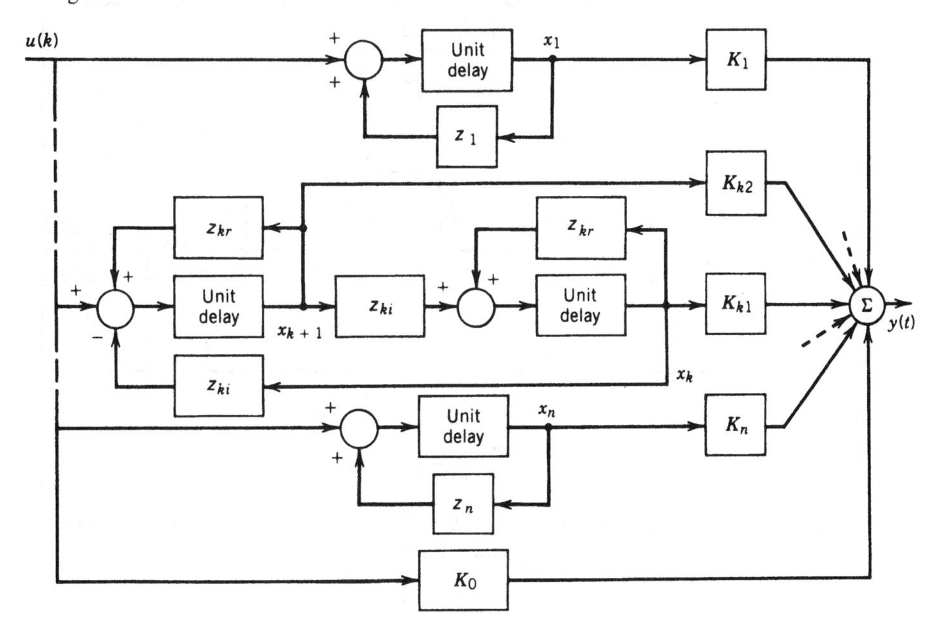

System matrices

$$
\mathbf{F} = \left[\begin{array}{ccc|cc|ccc}
z_1 & & \mathbf{0} & & & & & \\
 & \ddots & & \mathbf{0} & & & \mathbf{0} & \\
\mathbf{0} & & z_{k-1} & & & & & \\
\hline
 & \mathbf{0} & & z_{kr} & z_{ki} & & \mathbf{0} & \\
 & & & -z_{ki} & z_{kr} & & & \\
\hline
 & & & & & z_{k-2} & & \mathbf{0} \\
 & \mathbf{0} & & \mathbf{0} & & & \ddots & \\
 & & & & & \mathbf{0} & & z_n
\end{array}\right]
\qquad
\mathbf{G} = \left[\begin{array}{c}
1 \\ \vdots \\ 1 \\ \hline 0 \\ 1 \\ \hline 1 \\ \vdots \\ 1
\end{array}\right]
$$

where $K_0 = \lim_{z\to\infty} [H(z)] \qquad K_i = \lim_{z\to z_i} [(z - z_i)H(z)]$

$i = 1, k - 1$ and $i = k + 2, \ldots, n$

$$K_{k1} = -\tfrac{1}{2}\mathrm{Im}[(z - z_k)H(z)]_{z=z_k} \qquad K_{k2} = \tfrac{1}{2}\mathrm{Re}[(z - z_k)H(z)]_{z=z_k}$$

Table 3 (*Continued*)

V. Nondiagonal Jordan canonical form

Related conditions

(i) One real characteristic equation root z_k repeated m times (i.e., $z_k = z_{k+1} = \cdots = z_{k+m-1}$).

(ii) All other roots are real and distinct.

Simulation diagram

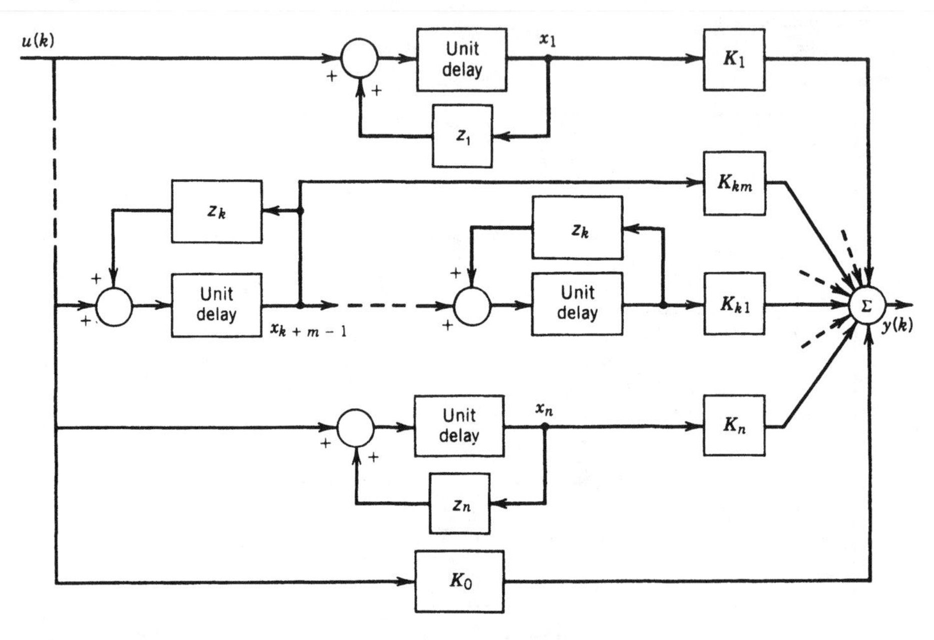

System matrices

$$
\mathbf{F} = \left[\begin{array}{ccc:ccc:ccc}
z_1 & & \mathbf{0} & & & & & & \\
& \ddots & & & \mathbf{0} & & & \mathbf{0} & \\
\mathbf{0} & & z_{k-1} & & & & & & \\
\hdashline
& & & z_k & 1 & \mathbf{0} & & & \\
& \mathbf{0} & & & \ddots & 1 & & \mathbf{0} & \\
& & & \mathbf{0} & & z_k & & & \\
\hdashline
& & & & & & z_{k+m} & & \mathbf{0} \\
& \mathbf{0} & & & \mathbf{0} & & & \ddots & \\
& & & & & & \mathbf{0} & & z_n
\end{array}\right]
\qquad
\mathbf{G} = \left[\begin{array}{c}
1 \\ \vdots \\ 1 \\ \hdashline 0 \\ \\ 1 \\ \hdashline 1 \\ \vdots \\ 1
\end{array}\right]
$$

$$
\mathbf{C} = [K_1 \quad \ldots \quad K_{k-1} \;\vdots\; K_{k1} \quad \ldots \quad K_{km} \;\vdots\; K_{k+m} \quad \ldots \quad K_n] \qquad \mathbf{D} = K_0
$$

where $K_0 = \lim_{z\to\infty} [H(z)]$ $\quad K_i = \lim_{z\to z_i} [(z - z_i)H(z)]$

$$
i = 1, k-1 \text{ and } i = k+m, \ldots, n
$$

and $K_{kj} = \dfrac{1}{(j-1)!} \left\{ \dfrac{d^{j-1}}{js^{j-1}} [(z - z_k)^m H(z)] \right\}_{z=z_k} \qquad j = 1, \ldots, m$

Table 4 Transformation Matrices for Discrete-Time State-Space Canonical Forms

State-space equations (SISO system)
$\mathbf{x}(k+1) = \mathbf{F}\mathbf{x}(k) + \mathbf{G}u(k)$
$y(k) = \mathbf{C}\mathbf{x}(k) + \mathbf{D}u(k)$

Characteristic equation
$\det(z\mathbf{I} - \mathbf{F})$
$= z^n + \alpha_{n-1}z^{n-1} + \cdots + \alpha_1 z + \alpha_0 = 0$

I. Controllable canonical form

Transformation conditions

(i) $\mathbf{P}_c = [\mathbf{G} \quad \mathbf{FG} \quad \cdots \quad \mathbf{F}^{n-1}\mathbf{G}]$ must be nonsingular

Transformation matrices

(i) $\mathbf{q} = \mathbf{T}\mathbf{x}$, $\mathbf{T} = \mathbf{R}^{-1}\mathbf{P}_c^{-1}$
where

$$\mathbf{R} = \begin{bmatrix} \alpha_1 & \alpha_2 & \cdots & \alpha_{n-1} & 1 \\ \alpha_2 & \alpha_3 & \cdots & 1 & 0 \\ \vdots & \vdots & & \vdots & \vdots \\ \alpha_{n-1} & 1 & \cdots & 0 & 0 \\ 1 & 0 & \cdots & 0 & 0 \end{bmatrix}$$

(ii) New state matrix $= \mathbf{TFT}^{-1}$

$$= \left[\begin{array}{c|ccc} 0 & & & \\ 0 & & & \\ \vdots & & \mathbf{I}_{n-1} & \\ 0 & & & \\ \hline -\alpha_0 & -\alpha_1 & \cdots & -\alpha_{n-1} \end{array}\right]$$

II. Observable canonical form

Transformation conditions

(i) $\mathbf{P}_0 = \begin{bmatrix} \mathbf{C} \\ \mathbf{CF} \\ \vdots \\ \mathbf{CF}^{n-1} \end{bmatrix}$ must be nonsingular

Transformation matrices

(i) $\mathbf{q} = \mathbf{T}\mathbf{x}$, $\mathbf{T} = \mathbf{R}\mathbf{P}_0$
where

$$\mathbf{R} = \begin{bmatrix} \alpha_1 & \alpha_2 & \cdots & \alpha_{n-1} & 1 \\ \alpha_2 & \alpha_3 & \cdots & 1 & 0 \\ \vdots & \vdots & & \vdots & \vdots \\ \alpha_{n-1} & 1 & \cdots & 0 & 0 \\ 1 & 0 & \cdots & 0 & 0 \end{bmatrix}$$

(ii) New state matrix $= \mathbf{TFT}^{-1}$

$$= \left[\begin{array}{cccc|c} 0 & 0 & \cdots & 0 & -\alpha_0 \\ \hline & & & & -\alpha_1 \\ & \mathbf{I}_{n-1} & & & \vdots \\ & & & & -\alpha_{n-1} \end{array}\right]$$

III. Normal or diagonal Jordan canonical form

Transformation conditions

(i) $\mathbf{F}$ matrix has only distinct, real eiganvalues z_i, $i = 1, \ldots, n$

Transformation matrices

(i) $\mathbf{q} = \mathbf{T}\mathbf{x}$, $\mathbf{T}^{-1} = \mathbf{M} = [\mathbf{v}_1 \quad \mathbf{v}_2 \quad \cdots \quad \mathbf{v}_n]$
where (a) $\mathbf{v}_i$ are the n linearly independent eigenvectors corresponding to z_i and
(b) $\mathbf{v}_i$ are taken to be equal or proportional to any nonzero column of $\mathrm{Adj}(z_i\mathbf{I} - \mathbf{F})$

(ii) New state matrix $= \mathbf{TFT}^{-1}$

$$= \begin{bmatrix} z_1 & & & \mathbf{0} \\ & z_2 & & \\ & & \ddots & \\ & \mathbf{0} & & z_n \end{bmatrix}$$

IV. Normal or diagonal Jordan canonical form

Transformation conditions

(i) $\mathbf{F}$ matrix has one repeated, real eigenvalue z_k of multiplicity m (i.e., $z_k = z_{k+1} = \ldots = z_{k+m-1}$). All other eigenvalues are real and distinct.

(ii) Degeneracy $d = n - \mathrm{rank}(z_k\mathbf{I} - \mathbf{F}) = m$. Full degeneracy.

Table 4 *(Continued)*

Transformation matrices

(i) $\mathbf{q} = \mathbf{Tx}$, $\mathbf{T}^{-1} = \mathbf{M} = [\mathbf{v}_1\ \mathbf{v}_2 \ldots \mathbf{v}_n]$

where (a) $\mathbf{v}_i$, $i = 1, k - 1$ nad $i = k + m, n$, are the linearly independent eigenvectors corresponding to the real, distinct eigenvalues;

(b) $\mathbf{v}_i$, $i = 1, k - 1$ and $i = k + m, n$, are taken to be equal or proportional to any nonzero column of $\text{Adj}(z_i\mathbf{I} - \mathbf{F})$; and

(c) $\mathbf{v}_i$, $i = k, k + m - 1$, are the m linearly independent eigenvectors corresponding to the repeated eigenvalue. They are equal or proportional to the nonzero linearly independent columns of

$$\left\{\frac{d^{m-1}}{dz^{m-1}}[\text{Adj}(s\mathbf{I} - \mathbf{F})]\right\}_{z=z_k}$$

(ii) New state matrix $= \mathbf{TFT}^{-1}$

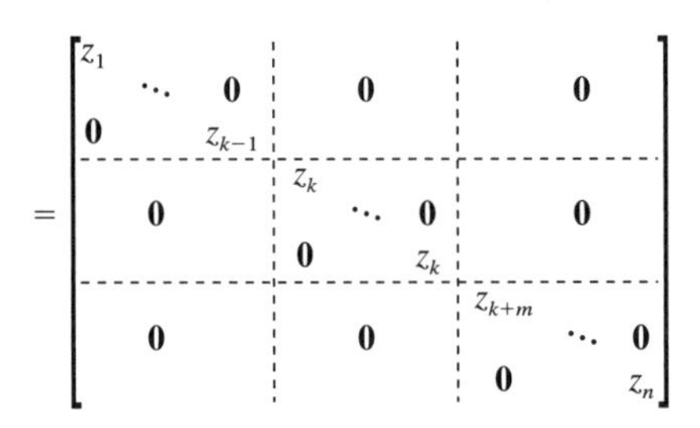

V. Near-normal canonical form

Transformation conditions

(i) $\mathbf{F}$ matrix has one pair of complex-conjugate eigenvalues, z_k, z_{k+1}

$$z_k = z_{kr} + jz_{ki}$$

$$z_{k+1} = z_{kr} - jz_{ki}$$

(ii) All other eigenvalues are real and distinct.

Transformation matrices

(i) $\mathbf{q} = \mathbf{Tx}$, $\mathbf{T}^{-1} = [\mathbf{v}_1\ \cdots\ \mathbf{v}_{k-1}\ \mathbf{v}_{kr}\ \mathbf{v}_{ki}\ \mathbf{v}_{k+2}\ \cdots\ \mathbf{v}_n]$

where (a) $\mathbf{v}_i$, $i = 1, k - 1$ and $i = k + 2, \ldots, n$, are the linearly independent eigenvectors corresponding to the real, distinct eigenvalues;

(b) $\mathbf{v}_i$ for $i = 1, \ldots, n$ are taken to be equal or proportional to any nonzero column of $\text{Adj}(z_i\mathbf{I} - \mathbf{F})$; and

(c) $\mathbf{v}_k = \mathbf{v}_{kr} + j\mathbf{v}_{ki}$, $\mathbf{v}_{k+1} = \mathbf{v}_{kr} - j\mathbf{v}_{ki}$ are the complex-conjugate eigenvectors corresponding to z_k and z_{k+1}, respectively.

(ii) New state matrix $= \mathbf{TFT}^{-1}$

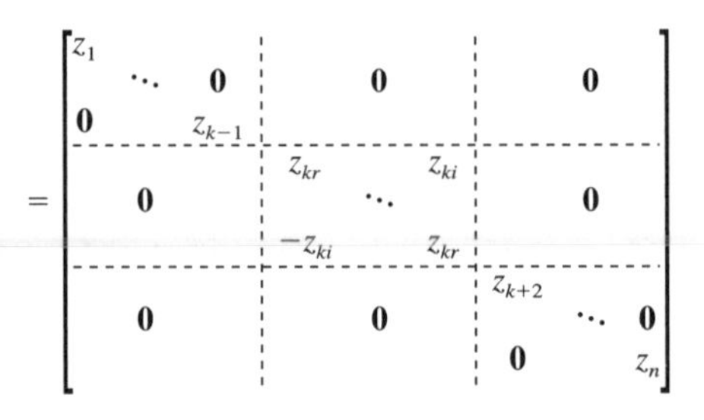

Table 4 (*Continued*)

VI. Nondiagonal Jordan canonical form

Transformation conditions

(i) **F** matrix has one repeated, real eigenvalue z_k of multiplicity m (i.e., $z_k = z_{k+1} = \cdots = z_{k+m-1}$). All other eigenvalues are real and distinct.

(ii) Degeneracy $d = n - \text{rank}(z_k\mathbf{I} - \mathbf{F}) = 1$. Simple degeneracy.

Transformation matrices

(i) $\mathbf{q} = \mathbf{Tx}$, $\mathbf{T}^{-1} = [\mathbf{t}_1\ \mathbf{t}_2 \ldots \mathbf{t}_n]$

where (a) $\mathbf{t}_i$, $i = 1, k - 1$ and $i = k + m, n$, are the linearly independent eigenvectors corresponding to the real, distinct eigenvalues;

(b) $\mathbf{t}_i$, for $i = 1, k - 1$ and $i = k + m, n$, are taken to be equal or proportional to any nonzero column of $\text{Adj}(s_i\mathbf{I} - \mathbf{F})$; and

(c) $\mathbf{t}_i$, $i = k, k + m - 1$, are obtained by solution of the equation $\mathbf{FT}^{-1} = \mathbf{T}^{-1}\mathbf{J}$, where **J** is the Jordan canonical matrix given. Each $\mathbf{t}_i$ is determined to within a constant of proportionality.

(ii) New state matrix = $\mathbf{J} = \mathbf{TFT}^{-1}$

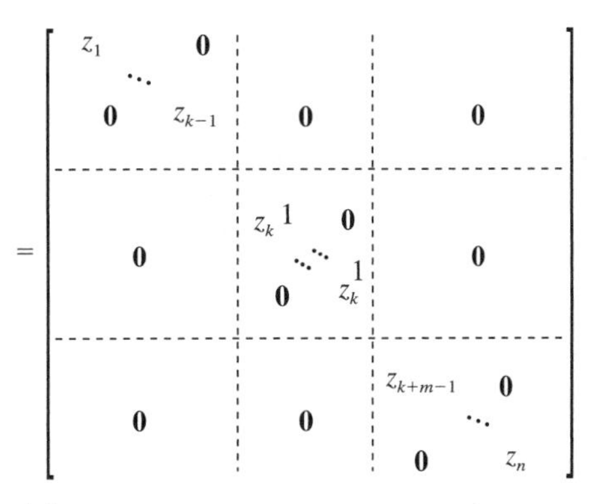

VII. Nondiagonal Jordan canonical form

Transformation conditions

(i) **F** matrix has one repeated, real eigenvalue of multiplicity m (i.e., $z_k = z_{k+1} = \cdots = z_{k+m-1}$). All other eigenvalues are real and distinct.

(ii) Degeneracy $d = n - \text{rank}(z_k\mathbf{I} - \mathbf{F})$ is between 1 and m. Simple degeneracy.

Transformation matrices

(i) $\mathbf{q} = \mathbf{Tx}$, $\mathbf{T}^{-1} = [\mathbf{t}_1\quad \mathbf{t}_2\quad \ldots\quad \mathbf{t}_n]$

where (a) $\mathbf{t}_i$, $i = 1, k - 1$ and $i = k + m, n$, are the linearly independent eigenvectors corresponding to the real, distinct eigenvalues;

(b) $\mathbf{t}_i$, $i = 1, k - 1$ and $i = k + m, n$, are taken to be equal or proportional to any nonzero column of $\text{Adj}(s_i\mathbf{I} - \mathbf{F})$;

(c) $\mathbf{t}_i$, $i = k, k + m - 1$, are m linearly independent vectors corresponding to the repeated eigenvalue of multiplicity m, only d of these vectors eigenvectors; and

(d) $\mathbf{t}_i$, $i = k = k, k + m - 1$, obtained by solution of the equation $\mathbf{FT}^{-1} = \mathbf{T}^{-1}\mathbf{J}$, where **J** is the Jordan canonical matrix for the problem, with d Jordan blocks. There are d possible choices for **J**. Each of these choices needs to be tried and the $\mathbf{t}_i$ vectors solved for. Only the correct **J** will give the m linearly independent vectors $\mathbf{t}_i$, $i = k, k + m - 1$. Each $\mathbf{t}_i$ is determined to within a constant of proportionality.

(ii) New state matrix = $\mathbf{J} = \mathbf{TFT}^{-1}$. Correct **J** determined by trial and error, as previously described.

$$\dot{\boldsymbol{\phi}}(t - t_0) = \mathbf{A}\boldsymbol{\phi}(t - t_0) \qquad t \geq t_0 \tag{35}$$

with the initial condition

$$\boldsymbol{\phi}(t_0 - t_0) = \boldsymbol{\phi}(0) = \mathbf{I} \tag{36}$$

It has the following properties:

$$\boldsymbol{\phi}(t + \tau) = \boldsymbol{\phi}(t)\boldsymbol{\phi}(\tau) = \boldsymbol{\phi}(\tau)\boldsymbol{\phi}(t) \tag{37}$$

$$\boldsymbol{\phi}^{-1}(t) = \boldsymbol{\phi}(-t) \tag{38}$$

The following expressions for $\boldsymbol{\phi}(t)$ can be verified and are useful in its evaluation:

$$\boldsymbol{\phi}(t) = e^{\mathbf{A}t} = \mathbf{I} + \mathbf{A}t + \frac{\mathbf{A}^2t^2}{2!} + \frac{\mathbf{A}^3t^3}{3!} + \cdots \tag{39}$$

and

$$\boldsymbol{\phi}(t) = \mathcal{L}^{-1}[(s\mathbf{I} - \mathbf{A})^{-1}] \tag{40}$$

where $\mathcal{L}^{-1}$ denotes the inverse Laplace transform. Details related to Eqs. (35)–(40) have been described by DeRusso et al.[2] and Brogan.[4]

Knowledge of the state transition matrix for a given system simplifies the task of determining the response of the system to a variety of initial conditions $\mathbf{x}(t_0)$ and forcing functions $\mathbf{u}(t)$. A number of analytical and numerical techniques for its evaluation are available.

Equation (39) forms the basis for a numerical method of determining $\boldsymbol{\phi}(t)$. Closed-form evaluation of $e^{\mathbf{A}t}$ is possible only for special forms of the $\mathbf{A}$ matrix. For example, if $\mathbf{A}$ is a diagonal matrix with diagonal elements equal to the eigenvalues s_i, it can be shown that $\boldsymbol{\phi}(t)$ is also diagonal[4] and is given by

$$\boldsymbol{\phi}(t) = e^{\mathbf{A}t} = \begin{bmatrix} e^{s_1t} & & & \\ & e^{s_2t} & & \mathbf{0} \\ \mathbf{0} & & \ddots & \\ & & & e^{s_nt} \end{bmatrix} \tag{41}$$

Closed-form evaluation of $e^{\mathbf{A}t}$ is only slightly more complex if $\mathbf{A}$ is in the nondiagonal Jordan canonical form.[4] If the transformation matrix $\mathbf{T}$ [Eq. (14)] was used to obtain the diagonal or nondiagonal Jordan matrix $\mathbf{A}$, the transition matrix, for the original state vector $\mathbf{T}^{-1}\mathbf{x}$, is $\mathbf{T}^{-1}e^{\mathbf{A}t}\mathbf{T}$.

Equation (40) provides the basis for an analytical evaluation of $\boldsymbol{\phi}(t)$ that is suitable for low-order dynamic systems. This method requires the inversion of the $n \times n$ matrix $s\mathbf{I} - \mathbf{A}$, followed by the inverse Laplace transformation of the n^2 elements. The matrix inversion is especially cumbersome since the elements of the matrix are functions of s. The matrix inversion can be avoided altogether by using simulation diagrams of the system, in conjunction with block diagram reduction techniques,[2] to determine elements of the matrix $(s\mathbf{I} - \mathbf{A})^{-1}$. Alternative analytical techniques for the evaluation of $\boldsymbol{\phi}(t)$ based on Sylvester's theorem and the Cayley–Hamilton theorem have been described by DeRusso et al.[2] and Brogan.[4]

Numerical evaluation of $\boldsymbol{\phi}(t)$ for a specified value of t can be performed using Eq. (39) and retaining a finite number of terms from the series expansion. The number of terms retained increases with the desired degree of accuracy. An iterative procedure for determining the number of terms to be retained for a specified degree of accuracy has been described by Shinners.[5]

For linear, time-varying systems described by state-space equations (4) and (5), the solution is given by[2,4]

$$\mathbf{x}(t) = \boldsymbol{\phi}(t, t_0)\mathbf{x}(t_0) + \int_{t_0}^{t} \boldsymbol{\phi}(t, \tau)\mathbf{B}(\tau)\mathbf{u}(\tau)\, d\tau \qquad t \geq t_0 \tag{42}$$

$$\mathbf{y}(t) = \mathbf{C}(t)\boldsymbol{\phi}(t, t_0)\mathbf{x}(t_0) + \int_{t_0}^{t} \mathbf{C}(t)\boldsymbol{\phi}(t, \tau)\mathbf{B}(\tau)\mathbf{u}(\tau)\, d\tau + \mathbf{D}(t)\mathbf{u}(t) \tag{43}$$

where the $n \times n$ state transition matrix $\boldsymbol{\phi}(t, t_0)$ for the time-varying system depends on both arguments t and t_0 and not merely on the difference between these two time instants as in the time-invariant system.

The state transition matrix $\boldsymbol{\phi}(t, t_0)$ is the solution of the partial differential equation[2,4]

$$\frac{\partial \boldsymbol{\phi}(t, t_0)}{\partial t} = \mathbf{A}(t)\boldsymbol{\phi}(t, t_0) \tag{44}$$

with the initial condition

$$\boldsymbol{\phi}(t_0, t_0) = \mathbf{I} \tag{45}$$

It has the following properties:

$$\boldsymbol{\phi}(t_2, t_0) = \boldsymbol{\phi}(t_2, t_1)\boldsymbol{\phi}(t_1, t_0) \tag{46}$$

$$\boldsymbol{\phi}(t_1, t_0) = \boldsymbol{\phi}^{-1}(t_0, t_1) \tag{47}$$

Techniques for evaluating the state transition matrix for time-varying systems are considerably more involved than for time-invariant systems and are less widely applicable. A number of analytical methods for determining $\boldsymbol{\phi}(t, t_0)$ for special cases of linear time-varying systems have been described by DeRusso et al.[2]

A simple numerical procedure has been suggested by Palm[6] for computing the transition matrix when analytical determination is not possible. Let the ith column of $\boldsymbol{\phi}(t, t_0)$ be denoted by $\boldsymbol{\psi}_i(t)$ for a specified value of t_0. The matrix partial differential equation (44) becomes n vector ordinary differential equations:

$$\boldsymbol{\psi}_i(t) = \mathbf{A}(t)\boldsymbol{\psi}_i(t) \qquad i = 1, \ldots, n \qquad t \geq t_0 \tag{48}$$

with the initial conditions

$$[\boldsymbol{\psi}_1(t_0) \quad \boldsymbol{\psi}_2(t_0) \quad \cdots \quad \boldsymbol{\psi}_n(t_0)] = \mathbf{I} \tag{49}$$

Numerical solution of the ordinary differential equations gives $\boldsymbol{\psi}_i(t)$ and hence $\boldsymbol{\phi}(t, t_0)$. Note that the computed $\boldsymbol{\phi}(t, t_0)$ would be different for different values of t_0 for time-varying systems.

4.2 Discrete-Time Systems

The solutions of the system equations for linear, discrete-time systems described either by Eqs. (10) and (11) or by Eqs. (12) and (13) are given in Table 5. Expressions for the state transition matrix $\boldsymbol{\phi}(k - k_0)$ for time-invariant systems and $\boldsymbol{\phi}(k, k_0)$ for the time-varying systems and the properties of the state transition matrix are also included in the table. These results can be found in standard textbooks on state-space methods.[2,4]

The state transition matrix $\boldsymbol{\phi}(k - k_0)$ for a time-invariant system depends only on the difference in the sequence number $(k - k_0)$. The transition matrix is given by

Table 5 Solution of the State-Space Equations for Linear, Discrete-Time Systems

	Time-Invariant System [Eqs. (12) and (13)]	Time-Varying System [Eqs. (10) and (11)]
Solution $k > k_0$	$\mathbf{x}(k) = \boldsymbol{\phi}(k - k_0)\mathbf{x}(k_0) + \sum_{m=k_0}^{k-1} \boldsymbol{\phi}(k - m - 1)\mathbf{G}\mathbf{u}(m)$	$\mathbf{x}(k) = \boldsymbol{\phi}(k, k_0)\mathbf{x}(k_0) + \sum_{m=k_0}^{k-1} \boldsymbol{\phi}(k, m + 1)\mathbf{G}(m)\mathbf{u}(m)$
	$\mathbf{y}(k) = \mathbf{C}\boldsymbol{\phi}(k - k_0)\mathbf{x}(k_0) + \sum_{m=k_0}^{k-1} [\mathbf{C}\boldsymbol{\phi}(k - m - 1)\mathbf{G}\mathbf{u}(m)] + \mathbf{D}\mathbf{u}(k)$	$\mathbf{y}(k) = \mathbf{C}(k)\boldsymbol{\phi}(k, k_0)\mathbf{x}(k_0) + \sum_{m=k_0}^{k-1} [\mathbf{C}(k)\boldsymbol{\phi}(k, m + 1)\mathbf{G}(m)\mathbf{u}(m)] + \mathbf{D}(k)\mathbf{u}(k)$
State transition matrix	$\boldsymbol{\phi}(k - k_0) = \mathbf{F}^{k-k_0} \quad k > k_0$ or $\boldsymbol{\phi}(k) = \mathcal{Z}^{-1}[z(zI - F)^{-1}]$	$\boldsymbol{\phi}(k, k_0) = \begin{cases} \prod_{l=k}^{k-1} \mathbf{F}(l) & k > k_0 \\ \mathbf{I} & k = k_0 \end{cases}$
Properties of state transition matrix	$\boldsymbol{\phi}(0) = \mathbf{I}$ $\boldsymbol{\phi}(k_1 + k_2) = \boldsymbol{\phi}(k_1)\boldsymbol{\phi}(k_2)$ $\boldsymbol{\phi}(k) = \boldsymbol{\phi}^{-1}(-k)$ when the inverse exists	$\boldsymbol{\phi}(k_0, k_0) = \mathbf{I}$ $\boldsymbol{\phi}(k_2, k_1)\boldsymbol{\phi}(k_1, k_0) = \boldsymbol{\phi}(k_2, k_0)$ $\boldsymbol{\phi}(k_1, k_2) = \boldsymbol{\phi}^{-1}(k_2, k_1)$ when the inverse exists

$$\boldsymbol{\phi}(k - k_0) = \mathbf{F}^{k-k_0} \tag{50}$$

Numerical computation of $\boldsymbol{\phi}(k - k_0)$ is thus straightforward. Analytical evaluation of $\boldsymbol{\phi}(k - k_0)$ using Eq. (50) is feasible if $\mathbf{F}$ is in the diagonal or nondiagonal Jordan canonical form. If $\mathbf{F}$ is a diagonal matrix with diagonal elements equal to its eigenvalues z_i, $i = 1, \ldots, n$, then

$$\boldsymbol{\phi}(k) = \mathbf{F}^k = \begin{bmatrix} z_1^k & & & \\ & z_2^k & & \mathbf{0} \\ & \mathbf{0} & \ddots & \\ & & & z_n^k \end{bmatrix} \tag{51a}$$

If $\mathbf{F}$ is in the nondiagonal Jordan canonical form, analytical evaluation of $\boldsymbol{\phi}(k)$ is only slightly more complex. If the transformation matrix $\mathbf{T}$ [Eq. (17)] was used to obtain the diagonal or nondiagonal Jordan matrix $\mathbf{F}$, the transition matrix, for the original state vector $\mathbf{T}^{-1}\mathbf{x}$, is $\mathbf{T}^{-1}\mathbf{F}^k\mathbf{T}$.

An alternative analytical evaluation of the transition matrix uses the relationship

$$\boldsymbol{\phi}(k) = \mathcal{Z}^{-1}[z(z\mathbf{I} - \mathbf{F})^{-1}] \tag{51b}$$

where $\mathcal{Z}^{-1}$ denotes the inverse z-transform. This method is useful only for low-order systems because of the need for inverting the matrix $z\mathbf{I} - \mathbf{F}$, which has symbolic elements. The matrix inversion is particularly simple if $\mathbf{F}$ is in the diagonal or Jordan canonical form. As for continuous-time systems, the matrix inversion can be avoided altogether by using simulation diagrams of the discrete-time system and block diagram reduction[2] to directly determine elements of the matrix $(z\mathbf{I} - \mathbf{F})^{-1}$.

The computation of the state transition matrix $\boldsymbol{\phi}(k, k_0)$ for time-varying, discrete-time systems is difficult, as it is for continuous-time systems. For small or moderate values of the order n of the system, numerical evaluation of Eq. (52) is appropriate:

$$\boldsymbol{\phi}(k, k_0) = \begin{cases} \prod_{l=k_0}^{k-1} \mathbf{F}(l) & k > k_0 \\ \mathbf{I} & k = k_0 \end{cases} \tag{52}$$

For larger values of n, analytical and numerical methods for determining $\boldsymbol{\phi}(k, k_0)$ in special cases are available.[2]

5 STABILITY

Since state-space formulation is applicable to a large class of dynamic systems, the question of stability for systems represented in state space is quite a complex one. A more general consideration of stability than that used for SISO, LTI systems would indicate that stability of dynamic systems is not really a property of the systems but is more properly associated with isolated equilibrium points of dynamic systems.[4] A particular point $\mathbf{x}_e$ in state space is an equilibrium point of a dynamic system if, in the absence of inputs, the system state $\mathbf{x}$ is equal to $\mathbf{x}_e$ for time $t \geq t_0$ for continuous-time systems or for $k \geq k_0$ for discrete-time systems. For linear systems described by the state-space equations given in Section 2, the only isolated equilibrium point is at the origin in state space. For nonlinear systems, there may be a number of isolated equilibrium points. Any isolated equilibrium point can be shifted to the origin in state space by a simple change of state variables.[4] The stability definitions to be given assume therefore that the equilibrium point is at the origin in state space and that the system is unforced. Only the more commonly used types of stability will be defined.

The origin is a stable equilibrium point if, for any given value $\epsilon > 0$, there exists a number $\kappa(\epsilon, t_0) > 0$ such that, if the norm $\|\mathbf{x}(t_0)\| < \kappa$, then the norm $\|\mathbf{x}(t)\| < \epsilon$ for all $t > t_0$. The norm of a vector $\mathbf{x}$ may be defined as the Euclidean norm:

$$\|\mathbf{x}(t)\| = \sqrt{\sum_{i=1}^{n} x_i^2(t)} \tag{53}$$

The origin is asymptotically stable if, in addition to being stable, there exists a number $\gamma(t_0) > 0$ such that whenever $\|\mathbf{x}(t_0)\| < \gamma(t_0)$ the following condition is satisfied:

$$\lim_{t \to \infty} \|\mathbf{x}(t)\| = 0 \tag{54}$$

If κ and γ are not functions of t_0 in the previous definitions, the origin is said to be uniformly stable or uniformly asymptotically stable, respectively. If $\gamma(t_0)$ can be arbitrarily large, the origin is said to be globally asymptotically stable. Extension of these stability definitions to discrete-time systems is straightforward and merely requires that the sequence numbers k, k_0 be used instead of the time instants t, t_0, respectively, in the definitions already given. Additional types of stability that depend on the inputs to the system have been defined by Brogan[4] and Kuo.[7]

For LTI systems, the conditions for stability reduce to conditions on the eigenvalues of the system matrix $\mathbf{A}$ or $\mathbf{F}$ and are summarized in Table 6. These eigenvalues are the roots of the system characteristic equation as well, as shown in Section 7. They may be computed

Table 6 Stability Criteria for Linear, Time-Invariant Systems

	Continuous-Time System $\dot{\mathbf{x}}(t) = \mathbf{A}\mathbf{x}(t) + \mathbf{B}\mathbf{u}$ Eigenvalues of $\mathbf{A}$ are $s_i = \alpha_{ic} \pm j\omega_{ic}$	Discrete-Time System $\mathbf{x}(k+1) = \mathbf{F}\mathbf{x}(k) + \mathbf{G}\mathbf{u}(k)$ Eigenvalues of $\mathbf{F}$ are $z_i = \beta_{ic} \pm j\omega_{ic}$
Asymptotically stable	$\alpha_{ic} < 0$ for all roots	$\lvert z_i \rvert < 1$ for all roots
Stable	$\alpha_{ic} < 0$ for all repeated roots and $\alpha_{ic} \le 0$ for all simple roots	$\lvert z_i \rvert < 1$ for all repeated roots and $\lvert z_i \rvert \le 1$ for all simple roots
Unstable	$\alpha_{ic} > 0$ for any simple root or $\alpha_{ic} \ge 0$ for any repeated root	$\lvert z_i \rvert > 1$ for any simple root or $\lvert z_i \rvert \ge 1$ for any repeated root

explicitly by numerical methods. Alternatively, stability criteria such as the Routh–Hurwitz criterion[5] for continuous-time systems or the Jury test[7] for discrete-time systems may be applied. The conditions for asymptotic stability of such systems can also be shown to be sufficient for other types of stability depending on the input, such as bounded-input, bounded-output stability.[4]

For continuous-time, LTV systems, the necessary and sufficient condition for the origin to be a stable equilibrium point is that there exists a number $N(t_0)$ such that the norm of the transition matrix satisfies the following condition:

$$\|\boldsymbol{\phi}(t, t_0)\| \le N(t_0) \qquad \text{for } t \ge t_0 \tag{55}$$

If, in addition, $\|\boldsymbol{\phi}(t, t_0)\| \to 0$ as $t \to \infty$, the system is globally asymptotically stable.[4] The norm of the matrix $\boldsymbol{\phi}$ may be defined as the spectral norm:

$$\|\boldsymbol{\phi}(t, t_0)\| = \sqrt{\max_{\|x\|=1} (\mathbf{x}^T\boldsymbol{\phi}^T\boldsymbol{\phi}\mathbf{x})} \tag{56}$$

The corresponding stability conditions for linear, discrete-time systems are obtained simply by substituting the sequence numbers k and k_0 for time instants t and t_0, respectively, in the development. Time-varying systems that satisfy the property that the state converges exponentially with time to the zero state are said to be exponentially stable.[12] For LTI systems, of course, asymptotic stability is the same as exponential stability.

Stability considerations for nonlinear systems are more complex. For unforced second-order nonlinear systems, the phase-plane method is useful for examining the stability of equilibrium points of the system. The phase plane has the state variables as the coordinates. The state-space equations are used to derive analytical expressions for the trajectories or to draw the trajectories by graphical means. The phase portraits can then be examined to determine the equilibrium points and their stability. Application of the phase-plane method is described by DeRusso et al.[2] for continuous-time systems and by Kuo[7] for discrete-time systems.

Stability analysis of high-order nonlinear systems represented in state space can be done using the second method of Lyapunov. This is a technique requiring considerable ingenuity for effective use and provides sufficient conditions for stability rather than necessary and sufficient conditions.[8]

Lyapunov's method for nonlinear, unforced, time-invariant systems requires the definition of a scalar function of state $V(\mathbf{x})$ called the Lyapunov function. The latter may be thought of as a generalized energy function. The requirement on the Lyapunov function is that it be positive definite in some region about the origin in state space, the origin having been assumed to be an isolated equilibrium point here. A function $V(\mathbf{x})$, which is continuous and has continuous partial derivatives, is said to be positive (negative) definite in some region about the origin if it is zero at the origin and greater than (less than) zero everywhere else in the specified region. If the function is greater than (less than) or equal to zero everywhere in the specified region, it is said to be positive (negative) semidefinite.[4]

Consider the unforced continuous-time system represented by the state equation

$$\dot{\mathbf{x}}(t) = \mathbf{f}[\mathbf{x}(t)] \tag{57}$$

where

$$\mathbf{f}(\mathbf{0}) = \mathbf{0} \tag{58}$$

If a positive-definite function $V(\mathbf{x})$ can be determined in some region Γ about the origin such that its derivative with respect to time is negative semidefinite in Γ, then the origin is a stable equilibrium point. If dV/dt is negative definite, the origin is asymptotically stable. If the region Γ can be arbitrarily large and the conditions for asymptotic stability hold and if, in addition, $V(\mathbf{x}) \to \infty$ as $\|\mathbf{x}\| \to \infty$, the origin is a globally asymptotically stable equilibrium point. Table 7 gives the corresponding stability conditions for nonlinear, time-invariant, discrete-time systems. Extensions of the stability conditions for time-varying systems have been described by Kalman and Bertram[8] and DeRusso et al.[2]

As an example of the application of the second method of Lyapunov, consider the following nonlinear system:

$$\begin{aligned} \dot{x}_1 &= x_2 \\ \dot{x}_2 &= -a_0 x_2 - b_0 x_2^3 - x_1 \end{aligned} \tag{59}$$

where $a_0, b_0 \geq 0$ and both are not zero. The origin is an equilibrium point for this system since, if both x_1 and x_2 are zero,

$$\dot{x}_1 = \dot{x}_2 = 0 \tag{60}$$

Consider the following Lyapunov function:

$$V(x_1, x_2) = x_1^2 + x_2^2 \tag{61}$$

It satisfies the conditions for positive definiteness in an arbitrarily large region about the origin:

Table 7 Application of the Second Method of Lyapunov to Nonlinear, Time-Invariant, Discrete-Time Systems

State equation	$\mathbf{x}(k+1) = \mathbf{f}[\mathbf{x}(k)]$ $\mathbf{f}(\mathbf{0}) = \mathbf{0}$
Lyapunov function	Scalar function $V[\mathbf{x}(k)]$ positive definite in some region about the origin
Condition for stability in Γ	$\Delta V = V[\mathbf{x}(k+1)] - V[\mathbf{x}(k)]$ is negative semidefinite in Γ
Condition for asymptotic stability in Γ	$\Delta V = V[\mathbf{x}(k+1)] - V[\mathbf{x}(k)]$ is negative definite in Γ
Condition for global asymptotic stability	(i) Γ can be arbitrarily large (ii) $V[\mathbf{x}(k)] \to \infty$ as $\|\mathbf{x}(k)\| \to \infty$

$$\frac{dV(x_1, x_2)}{dt} = 2x_1\dot{x}_1 + 2x_2\dot{x}_2$$

$$= -2(a_0x_2^2 + b_0x_2^4) \tag{62}$$

after using the state equations to substitute for $\dot{x}_1$ and $\dot{x}_2$. If a_0, b_0 satisfy the inequalities stated, dV/dt is negative semidefinite in an arbitrarily large region about the origin. The origin is thus a stable equilibrium point. In fact, using a corollary to the main stability theorem provided by Kalman and Bertram,[8] it can be shown that the origin is a global asymptotically stable equilibrium point.

The limitations of Lyapunov' s method are that the Lyapunov function is not unique for a system and there are no systematic procedures for finding a suitable Lyapunov function. Since only sufficient conditions for stability are determined, some choices of Lyapunov functions are better in that they provide more information about system stability than others. Also, appropriate choice of the Lyapunov function can lead to an estimate of the system speed of response.[8] In practice, therefore, the second method of Lyapunov is used primarily to analyze the stability of systems such as high-order, nonlinear systems for which other methods of stability analysis are not available.

6 CONTROLLABILITY AND OBSERVABILITY

The controllability of a linear system is a measure of the coupling between the inputs to the system and the system state. The concept of state controllability was introduced by Kalman[11] in order to clarify conditions for the existence of solutions to specific control problems.

A linear, continuous-time system is said to be state controllable at time t_0 if there exists a finite time $t_1 > t_0$ and a control function $\mathbf{u}(t)$, $t_0 < t < t_1$, that can drive the system state from any initial value to any final value at $t = t_1$. If the system is controllable for all times t_0, the system is completely state controllable.[4] A linear, discrete-time system is said to be state controllable and completely state controllable, respectively, if the sequence numbers k, k_0, k_1 are substituted for the times t, t_0, t_1, respectively, in the two previously given definitions. An additional form of controllability for continuous-time and discrete-time LTV systems is that of uniformly complete state controllability. The mathematical definition of this form of controllability may be found in Kalman.[11] This property implies that the control effort and time interval required to drive the system state to the final value is relatively independent of the initial time. For LTI systems, of course, complete state controllability is the same as uniformly complete state controllability.

Though the control problems formulated above are open-loop control problems, the property of controllability has very significant implications for closed-loop control problems. Section 2 in Chapter 18 indicates that the closed-loop poles of a completely state-controllable time-invariant system can be specified and placed arbitrarily in the complex s-plane (or z-plane for discrete-time systems) by proportional state-variable feedback. Moreover, satisfaction of the controllability conditions to be defined in this section for time-invariant systems ensures that the optimal-control law for a quadratic performance index is a proportional state-variable feedback law and yields an asymptotically stable closed-loop system.[10]

Direct application of the definition of state controllability to LTI systems yields controllability conditions involving the transition matrices. Simple algebraic conditions are usually available for such systems and are used more often in practice to evaluate controllability.

The controllability condition for LTI systems with distinct eigenvalues may be stated very simply if the state equations are transformed to the diagonal Jordan canonical form. Such systems are completely controllable if there are no zero rows in the transformed **B** matrix for continuous-time systems or in the transformed **G** matrix for discrete-time systems.[4] The presence of a zero row in either of these matrices would indicate that the inputs are not coupled to and cannot control the corresponding mode. Algebraic controllability conditions for systems with repeated eigenvalues are given by Palm.[6]

The controllability conditions for LTI systems in general are stated in terms of the matrix $\mathbf{P}_c$, referred to as the controllability matrix in Section 3, and are summarized in Table 8. The $n \times nr$ controllability matrix for a MIMO system is defined by

$$\mathbf{P}_c = [\mathbf{B} \quad \mathbf{AB} \quad \cdots \quad \mathbf{A}^{n-1}\,\mathbf{B}] \tag{63}$$

for continuous-time systems and by

$$\mathbf{P}_c = [\mathbf{G} \quad \mathbf{FG} \quad \cdots \quad \mathbf{F}^{n-1}\mathbf{G}] \tag{64}$$

for discrete-time systems. The condition for complete state controllability is simply that the matrix $\mathbf{P}_c$ has rank n. The controllable canonical form for SISO systems, described in Section 3, derives its name from the fact that transformation to that form is possible if and only if

Table 8 Controllability Conditions for Linear Dynamic Systems

	Continuous Time	Discrete Time
Time-Invariant System		
Necessary and efficient condition for state controllability	(i) $\text{rank}(\mathbf{B} \quad \mathbf{AB} \quad \cdots \quad \mathbf{A}^{n-1}\mathbf{B})$ $= \text{rank}\,(\mathbf{B} \quad \mathbf{AB} \quad \cdots \quad \mathbf{A}^{n-r}\mathbf{B})$ $= n$ or (ii) $\det(\mathbf{P}_c\mathbf{P}_c^{\mathrm{T}}) \neq 0$	(i) $\text{rank}(\mathbf{G} \quad \mathbf{FG} \quad \cdots \quad \mathbf{F}^{n-1}\mathbf{G})$ $= \text{rank}\,(\mathbf{G} \quad \mathbf{FG} \quad \cdots \quad \mathbf{F}^{n-r}\mathbf{G})$ $= n$ or (ii) $\det(\mathbf{P}_c\mathbf{P}_c^{\mathrm{T}}) \neq 0$
Necessary condition for state controllability	$\text{rank}(\mathbf{B}\ \mathbf{A}) = n$	$\text{rank}(\mathbf{G}\ \mathbf{F}) = n$
Necessary and sufficient condition for output controllability	(i) $\text{rank}(\mathbf{CP}_c) = p$ or (ii) $\det(\mathbf{CP}_c\mathbf{P}_c^{\mathrm{T}}\mathbf{C}^{\mathrm{T}}) \neq 0$	(i) $\text{rank}(\mathbf{CP}_c) = p$ or (ii) $\det(\mathbf{CP}_c\mathbf{P}_c^{\mathrm{T}}\mathbf{C}^{\mathrm{T}}) \neq 0$
Time-Varying System: Time Interval of Interest $[t_0, t_1]$ or $[k_0, k_1]$		
Necessary and sufficient condition for state controllability	(i) $\mathbf{W}_c(t_1, t_0)$ is positive definite or (ii) Zero is not an eigenvalue of $\mathbf{W}_c(t_1, t_0)$ or (iii) $\lvert\mathbf{W}_c(t_1, t_0)\rvert \neq 0$ where $\mathbf{W}_c(t_1, t_0)$ $\triangleq \int_{t_0}^{t_1} \boldsymbol{\phi}(t_1, \tau)\mathbf{B}(\tau)\mathbf{B}^{\mathrm{T}}(\tau) \times \boldsymbol{\phi}^{\mathrm{T}}(t_1, \tau)\, d\tau$	(i) $\mathbf{W}_c(k_1, k_0)$ is positive definite or (ii) Zero is not an eigenvalue of $\mathbf{W}_c(k_1, k_0)$ or (iii) $\lvert\mathbf{W}_c(k_1, k_0)\rvert \neq 0$ where $\mathbf{W}_c(k_1, k_0)$ $\triangleq \sum_{k=k_0}^{k_1} \boldsymbol{\phi}(k_1, k)\mathbf{G}(k)\mathbf{G}^{T}(k) \times \boldsymbol{\phi}^{\mathrm{T}}(k_1, k)$
Necessary and sufficient condition for output controllability	$\det \mathbf{W}_y(t_1, t_0) \neq 0$ where $\mathbf{W}_y(t_1, t_0)$ $\triangleq \int_{t_0}^{t_1} \mathbf{C}(\tau)\boldsymbol{\phi}(t_1, \tau)\mathbf{B}(\tau)\mathbf{B}^{\mathrm{T}}(\tau) \times \boldsymbol{\phi}^{\mathrm{T}}(t_1, \tau)\mathbf{C}^{\mathrm{T}}(\tau)\, d\tau$	$\det \mathbf{W}_y(k_1, k_0) \neq 0$ where $\mathbf{W}_y(k_1, k_0)$ $\triangleq \sum_{k=k_0}^{k_1} \mathbf{C}(k)\boldsymbol{\phi}(k_1, k)\mathbf{B}(k) \times \mathbf{B}^{\mathrm{T}}(k)\boldsymbol{\phi}^{\mathrm{T}}(k_1, k)\mathbf{C}^{\mathrm{T}}(k)$

the system is completely state controllable. The transformation matrix **T** in Tables 2 and 4 required to transform the state-space equations for a SISO system to the controllable canonical form exists if and only if the $n \times n$ matrix $\mathbf{P}_c$ is nonsingular; that is, the system is controllable. Equivalent controllability conditions that are simpler to evaluate than the one previously stated are also listed in Table 8 along with a necessary (but not sufficient) condition for complete controllability.

The controllability conditions for LTV systems over a specified time interval are more cumbersome to evaluate in practice as they involve the system transition matrix.[4] These conditions are listed in Table 8. In contrast to time-invariant systems, the controllability of time-varying systems depends on the time interval under consideration.

The concept of output controllability, as opposed to state controllability described earlier, was introduced by Kreindler and Sarachik.[13] A linear, continuous-time system is said to be output controllable at time t_0 if there exists a finite time $t_1 > t_0$ and a control function $\mathbf{u}(t)$, $t_0 < t < t_1$, that drives the system output from any initial value $\mathbf{y}(t_0)$ to any final value $y(t_1)$. If this condition holds true for all times t_0, the system is completely output controllable. Extension of the concept to linear, discrete-time systems is straightforward as before.

Output controllability conditions[4] for linear systems that are purely dynamic [i.e., the matrix $\mathbf{D} = \mathbf{0}$ in Eqs. (5), (7), (11), and (13)] are summarized in Table 8. These conditions are weaker than the corresponding conditions for state controllability if the number of outputs p is less than the number of state variables n. Since this is true in practice, state controllability implies output controllability. On the other hand, output controllability does not imply state controllability in general. It can be shown, however, that for time-invariant systems if the matrix $(\mathbf{CC}^T)$ is nonsingular, output controllability is equivalent to state controllability.

The observability of a linear system is a measure of the coupling between the system state and its outputs. The concept of observability was introduced by Kalman[11] and is relevant to the problem of estimation of system state based on the output vector. The output vector is usually chosen to correspond to measurable variables.

A linear, continuous-time system is said to be observable[4] at time t_0 if there exists a finite time $t_1 > t_0$ such that $\mathbf{x}(t_0)$ can be determined from the history of inputs $\mathbf{u}(t)$ and outputs $\mathbf{y}(t)$ over the time interval $t_0 \leq t \leq t_1$. If the system is observable for all times t_0 and all initial states $\mathbf{x}(t_0)$, the system is completely observable. Extension of the observability concept to discrete-time systems simply requires that the sequence numbers k, k_0, k_1 be substituted for the times t, t_0, t_1, respectively, in the previous definitions. A stronger form of observability for LTV systems is that of uniformly complete observability. The mathematical definition of this form of observability is given by Kalman.[11] This property guarantees that the time interval required to estimate the state is relatively independent of the initial time. For LTI systems, of course, complete observability is the same as uniformly complete observability. A property complementary to observability for LTV systems is that of reconstructibility,[12,14] which concerns the estimation of the state of the system from past measurements of the state. In contrast to this, observability concerns the estimation of the state from future measurements of the output. For time-invariant systems, the two properties of reconstructibility and observability are identical to one another.

As was the case for controllability, direct application of the definition of observability already stated yields conditions involving the transition matrix.[4] Simpler algebraic conditions are available for time-invariant systems. The observability condition for LTI systems with distinct eigenvalues can be stated very simply if the state equations are transformed to the Jordan canonical form. Such systems are completely observable if each column in the transformed **C** matrix has at least one nonzero element.[4] The presence of a column of zeros in this matrix would indicate that the corresponding state variable cannot be estimated from the measured output and input vectors.

More general observability conditions for LTI systems are stated in terms of a matrix $\mathbf{P}_0$ referred to as the observability matrix and are summarized in Table 9. The $np \times n$ observability matrix is defined by

$$\mathbf{P}_0 = \begin{bmatrix} \mathbf{C} \\ \mathbf{CA} \\ \vdots \\ \mathbf{CA}^{n-1} \end{bmatrix} \tag{65}$$

for continuous-time systems and by

$$\mathbf{P}_0 = \begin{bmatrix} \mathbf{C} \\ \mathbf{CF} \\ \vdots \\ \mathbf{CF}^{n-1} \end{bmatrix} \tag{66}$$

for discrete-time systems. The condition for complete observability is simply that the matrix $\mathbf{P}_0$ have rank n. The observable canonical form for SISO systems, described in Section 3, derives its name from the fact that transformation to that form is possible if and only if the system is observable. The transformation matrix $\mathbf{T}$ in Tables 2 and 4, required to transform the state-space equations for a SISO system to the observable canonical form, exists if and only if the $n \times n$ matrix $\mathbf{P}_0$ is nonsingular, that is, the system is observable. Equivalent observability conditions which are simpler to evaluate than the one stated previously are also listed in Table 9 along with a necessary (but not sufficient) condition for complete observability. It should be noted that the observability conditions are independent of time for time-

Table 9 Observability Conditions for Linear Dynamic Systems

	Continuous Time	Discrete Time
Time-Invariant System		
Necessary and sufficient condition for observability	(i) $\operatorname{rank}[\mathbf{C}^T \ \ \mathbf{A}^T\mathbf{C}^T \ \cdots \ (\mathbf{A}^T)^{n-1}\mathbf{C}^T] = \operatorname{rank}[\mathbf{C}^T \ \ \mathbf{A}^T\mathbf{C}^T \ \cdots \ (\mathbf{A}^T)^{n-p}\mathbf{C}^T] = n$ or (ii) $\det(\mathbf{P}_0^T\mathbf{P}_0) \neq 0$	(i) $\operatorname{rank}[\mathbf{C}^T \ \ \mathbf{F}^T\mathbf{C}^T \ \cdots \ (\mathbf{F}^T)^{n-1}\mathbf{C}^T] = \operatorname{rank}[\mathbf{C}^T \ \ \mathbf{F}^T\mathbf{C}^T \ \cdots \ (\mathbf{F}^T)^{n-p}\mathbf{C}^T] = n$ or (ii) $\det(\mathbf{P}_0^T\mathbf{P}_0) \neq 0$
Necessary condition	$\operatorname{rank}(\mathbf{C}^T \ \mathbf{A}^T) = n$	$\operatorname{rank}(\mathbf{C}^T \ \mathbf{F}^T) = n$
Time-Varying Systems:		
Necessary and sufficient condition for observability	Observable at t_0 if and only if there exists a finite time t_1, $t_1 > t_0$ such that (i) $\mathbf{W}_0(t_1, t_0)$ is positive definite or (ii) zero is not an eigenvalue of $\mathbf{W}_0(t_1, t_0)$ or (iii) $\lvert\mathbf{W}_0(t_1, t_0)\rvert \neq 0$ where $\mathbf{W}_0(t_1, t_0) \triangleq \int_{t_0}^{t_1} \boldsymbol{\phi}^T(\tau, t_0)\mathbf{C}^T(\tau)\mathbf{C}(\tau) \times \boldsymbol{\phi}(\tau, t_0)\, d\tau$	Observable at k_0 if and only if there exists a finite time k_1, $k_1 > k_0$ such that (i) $\mathbf{W}_0(k_1, k_0)$ is positive definite or (ii) zero is not an eigenvalue of $\mathbf{W}_0(k_1, k_0)$ or (iii) $\lvert\mathbf{W}_0(k_1, k_0)\rvert \neq 0$ where $\mathbf{W}_0(k_1, k_0) \triangleq \sum_{k=k_0}^{k_1} \boldsymbol{\phi}^T(k_1, k_0)\mathbf{C}^T(k)\mathbf{C}(k)\boldsymbol{\phi}(k_1, k)$

invariant systems. In contrast, the observability conditions for time-varying systems over a specified time interval involve the system transition matrix and hence depend on the time interval.[4] They are also listed in Table 9.

Though the definition of observability above involves an open-loop state estimation problem, the property of observability has important implications for closed-loop realizations of the state estimation problem. It will be shown in Section 5 of Chapter 18 that, if a time-invariant system is completely observable, a closed-loop state estimator can be constructed such that the estimation error transients can be made to decay to zero as rapidly as possible.

The conditions for controllability and observability noted in Tables 8 and 9 have obvious similarities. The two properties can be shown to be duals of each other by formulating the concept of the dual of a dynamic system. Interested readers are referred to Kalman et al.[11,14]

A linear system can, in general, be divided into four subsystems as indicated by Fig. 4. The state vector $\mathbf{x}$ can be written as

$$x^{\mathrm{T}} = x_C^{\mathrm{T}} + x_{CO}^{\mathrm{T}} + x_N^{\mathrm{T}} + x_O^{\mathrm{T}} \tag{67}$$

where the subscripts have the meaning assigned in the figure. The corresponding state-space equations for a time-invariant, continuous-time system are

$$\begin{bmatrix} \dot{\mathbf{x}}_C \\ \dot{\mathbf{x}}_{CO} \\ \dot{\mathbf{x}}_N \\ \dot{\mathbf{x}}_O \end{bmatrix} = \begin{bmatrix} \mathbf{A}_{11} & \mathbf{A}_{12} & \mathbf{A}_{13} & \mathbf{A}_{14} \\ \mathbf{0} & \mathbf{A}_{22} & \mathbf{0} & \mathbf{A}_{24} \\ \mathbf{0} & \mathbf{0} & \mathbf{A}_{33} & \mathbf{A}_{34} \\ \mathbf{0} & \mathbf{0} & \mathbf{0} & \mathbf{A}_{44} \end{bmatrix} \begin{bmatrix} \mathbf{x}_C \\ \mathbf{x}_{CO} \\ \mathbf{x}_N \\ \mathbf{x}_O \end{bmatrix} + \begin{bmatrix} \mathbf{B}_{11} \\ \mathbf{B}_{21} \\ \mathbf{0} \\ \mathbf{0} \end{bmatrix} \mathbf{u} \tag{68}$$

$$\mathbf{y} = [\mathbf{0} \quad \mathbf{C}_{12} \quad \mathbf{0} \quad \mathbf{C}_{14}] \begin{bmatrix} \mathbf{x}_C \\ \mathbf{x}_{CO} \\ \mathbf{x}_N \\ \mathbf{x}_O \end{bmatrix} + \mathbf{D}\mathbf{u}$$

The zero matrices in the **B** matrix correspond to the fact that $\mathbf{x}_N$ and $\mathbf{x}_O$ are not controllable. The zero matrices in the **C** matrix correspond to the fact that $\mathbf{x}_C$ and $\mathbf{x}_N$ are not observable. If the eigenvalues of the **A** matrix are distinct, all off-diagonal elements in the **A** matrix would be zero. The procedure for determining the transformation matrix to convert the state-space equations into the canonical form [Eq. (68)] has been described by Kalman.[9]

The extension of this discussion to time-invariant, discrete-time systems is straightforward. For time-varying systems, the state-space decomposition is a function of time but is similar in structure to that already described.

The significance of the system decomposition as shown in Fig. 4 is that it helps relate the state-space description of linear dynamic systems to transfer function or transfer function matrix descriptions of such systems. The transfer function matrix relating **y** to **u** is a description only of the controllable and observable part of the system and masks other modes that are either not observable or not controllable or neither controllable nor observable. The relationship of the state-space description of dynamic systems to the transfer function matrix description of such systems is discussed in greater detail in Section 7.

Loss of controllability or observability could occur when controllable and observable subsystems are connected together to form composite systems. Gilbert[15] has formulated rules relating the composite system properties to those of the individual open-loop systems. These rules provide greater insight into the conditions leading to loss of controllability or observability than the simple application of the conditions noted in Tables 8 and 9.

The concepts of controllability and observability are obviously very important for MIMO systems since the complexity of such systems frequently masks the nature of the coupling of the system state to the inputs and outputs. For SISO systems, lack of complete

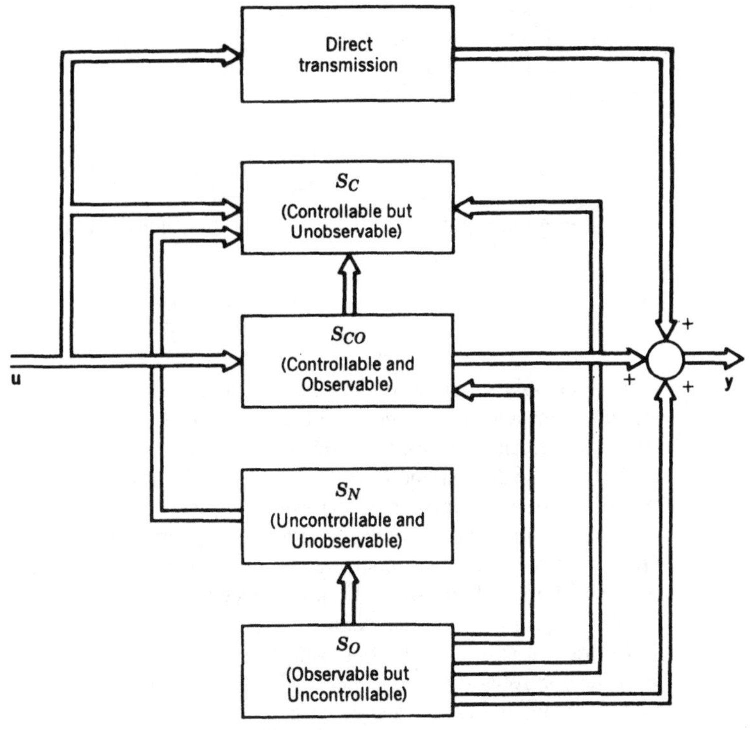

Figure 4 Decomposition of linear system based on controllability and observability.

controllability or observability is a less common occurrence. Conclusions concerning controllability and observability are obvious in many of these cases but less so in others.

Consider the electrical circuit in Fig. 5. The state-space equation for the system is given by

$$\begin{bmatrix} \dot{x}_1 \\ \dot{x}_2 \end{bmatrix} = \begin{bmatrix} -\dfrac{R_1 + R_3}{L_1} & \dfrac{R_3}{L_1} \\ \dfrac{R_3}{L_2} & -\dfrac{R_3 + R_2}{L_2} \end{bmatrix} \begin{bmatrix} x_1 \\ x_2 \end{bmatrix} + \begin{bmatrix} \dfrac{1}{L_1} \\ 0 \end{bmatrix} u(t) \tag{69}$$

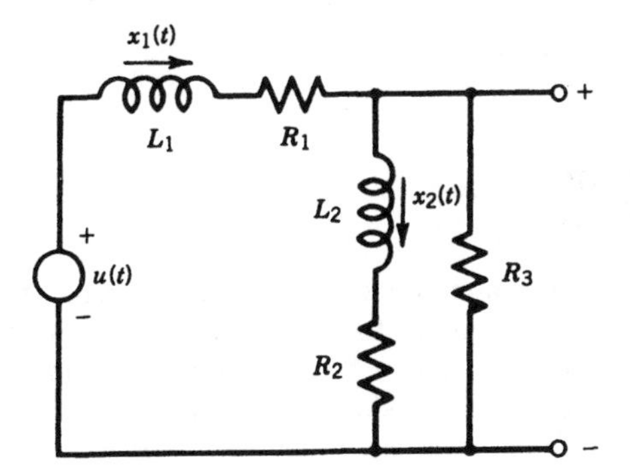

Figure 5 State variables for *RL* circuit.

The system is completely controllable except for the trivial case where R_3 is zero. Similarly, if x_1 or x_2 is chosen as the only output, the system is completely observable as long as R_3 is nonzero. If the voltage across the resistor R_3 is chosen as the output, the corresponding output equation is

$$y(t) = [R_3 \;\; - R_3]\begin{bmatrix} x_1 \\ x_2 \end{bmatrix} \tag{70}$$

The observability matrix $\mathbf{P}_0$ in Eq. (65) is then nonsingular and the system is observable if and only if

$$\frac{R_1}{L_1} \neq \frac{R_2}{L_2} \tag{71}$$

This observability condition is not an obvious one and is equivalent to the requirement that the time constants associated with the two R–L pairs not be identical. However, it should be noted that, given component tolerances in practice, the inequality (71) will be satisfied almost always and the corresponding system will be observable. A discussion on conditions leading to loss of controllability or observability is given by Friedland.[16]

Despite the fact that lack of complete controllability or observability is infrequent for SISO systems, these concepts have practical significance for SISO as well as MIMO systems because of the relationship of these concepts to closed-loop control and state estimation problems. Measures of the degree of controllability and observability can be defined for time-invariant and time-varying systems.

The controllability and observability conditions in Tables 8 and 9 relate these properties to the nonsingularity of square matrices for SISO systems. Measures of the degree of controllability and observability are related to the closeness of these matrices to the singularity condition. Such measures have been defined for time-invariant systems by Johnson[17] and Friedland[18] and are significant for SISO as well as MIMO systems. A system with a better degree of controllability can in general be controlled more effectively. Similarly, a better degree of observability implies that state estimation can be performed more accurately. The proposed measures of the degree of controllability and observability are not in common use but have the potential to quantitatively evaluate proposed control strategies and measurement schemes.[19]

Additional concepts of degrees of controllability and observability for time-varying systems have been described by Silverman and Meadows[20] and for MIMO systems by Kreindler and Sarachik.[13] Properties weaker than state controllability and observability have also been defined[12] and are useful in ensuring that closed-loop control and state estimation problems are well posed. A linear system is said to be stabilizable if the uncontrollable subsystems S_N and S_O in the decomposition of Fig. 4 are stable. Similarly, if the unobservable subsystems S_C and S_N are stable, the system is said to be detectable.[12]

7 RELATIONSHIP BETWEEN STATE-SPACE AND TRANSFER FUNCTION DESCRIPTIONS

The state-space representation of dynamic systems is an accurate representation of the internal structure of a system and its coupling to the system inputs and outputs. For LTI systems, transfer functions (for SISO systems) or transfer function matrices (for MIMO systems) are useful in practice since the dimensions of these matrices are invariably smaller than the dimensions of the corresponding system matrices $\mathbf{A}$ or $\mathbf{F}$ in Eqs. (6) and (12). Analysis and design procedures based on the transfer function matrix descriptions are therefore simpler. The relationship between these two alternative descriptions of LTI systems is described in this section.

Determination of the transfer function matrix from the state-space equations is straightforward. For continuous-time systems, Laplace transformation of Eqs. (6) and (7) with zero initial conditions $\mathbf{x}(0)$ and elimination of $\mathbf{X}(s)$ yields

$$\mathbf{H}(s) = \frac{\mathbf{Y}(s)}{\mathbf{U}(s)} = \mathbf{C}(s\mathbf{I}_n - \mathbf{A})^{-1}\mathbf{B} + \mathbf{D} \tag{72}$$

where $\mathbf{I}_n$ is the $n \times n$ identity matrix. For discrete-time systems, a similar procedure using z-transforms and applied to Eqs. (12) and (13) yields the pulse transfer function matrix

$$\mathbf{H}(z) = \frac{\mathbf{Y}(z)}{\mathbf{U}(z)} = \mathbf{C}(z\mathbf{I}_n - \mathbf{F})^{-1}\mathbf{G} + \mathbf{D} \tag{73}$$

The transfer function matrix corresponding to a given state-space description is therefore unique. However, as indicated in the previous section, the former represents only the controllable and observable part of a system. Unless the entire system is completely controllable and observable, a transfer function matrix description is not a complete characterization of the system dynamic behavior. It can be shown that, for SISO systems, a necessary and sufficient condition for controllability and observability of the system is that there are no pole–zero cancellations between the numerator and denominator of the transfer function matrices in Eqs. (72) and (73). For MIMO systems, this is only a sufficient condition and not a necessary one.[4]

The determination of the state-space description corresponding to a given transfer function matrix description is more complex and is referred to as the problem of realization. Since the transfer function matrix represents only the controllable and observable part of a system, the problem of realization does not have a unique solution. In fact, the transfer function matrix description does not even determine the dimension of the corresponding state vector uniquely. The minimal dimension of the state vector corresponding to a given transfer function matrix is, however, uniquely determined. The associated state-space equations are said to constitute the minimal or irreducible realization of the transfer function matrix. It can be shown that a realization is minimal if and only if it is both controllable and observable.[4] Minimal realizations are not unique. However, any two different minimal realizations of a given transfer function matrix are equivalent in that the corresponding state vectors are related by a nonsingular transformation matrix.

The canonical forms of the state-space equations for SISO systems in Tables 1 and 3 represent minimal realizations if there is no pole–zero cancellation. Techniques for obtaining minimal realizations for MIMO systems are more involved. Brogan[4] has described a procedure for obtaining a Jordan form realization for a given transfer function matrix. When applied to cases where elements of the transfer function matrix have only simple poles, the resulting realization is controllable and observable, as shown in the following example. If one or more elements of the transfer function matrix have repeated poles, the realization that results is controllable but may or may not be observable.

Consider a continuous-time system with two inputs and two outputs and the following transfer function matrix:

$$\mathbf{H}(s) = \begin{bmatrix} \dfrac{1}{s+1} & \dfrac{s}{(s+1)(s+3)} \\ \dfrac{1}{s+3} & \dfrac{1}{s+1} \end{bmatrix} \tag{74}$$

Expand $\mathbf{H}(s)$ using a matrix version of partial-fraction expansion as

$$\mathbf{H}(s) = \frac{\begin{bmatrix} 1 & -\frac{1}{2} \\ 0 & 1 \end{bmatrix}}{s+1} + \frac{\begin{bmatrix} 0 & \frac{3}{2} \\ 1 & 0 \end{bmatrix}}{s+3}$$

$$= \frac{\begin{bmatrix} 1 \\ 0 \end{bmatrix}[1 \quad -\frac{1}{2}] + \begin{bmatrix} 0 \\ 1 \end{bmatrix}[0 \quad 1]}{s+1} + \frac{\begin{bmatrix} 1 \\ 0 \end{bmatrix}[0 \quad \frac{3}{2}] + \begin{bmatrix} 0 \\ 1 \end{bmatrix}[1 \quad 0]}{s+3} \tag{75}$$

It should be noted that the number of vector products each coefficient matrix is factored into is equal to the rank of the matrix. Then $\mathbf{H}(s)$ is written in a form that indicates the matrices $\mathbf{A}$, $\mathbf{B}$, $\mathbf{C}$ clearly, by comparison with $\mathbf{C}(s\mathbf{I} - \mathbf{A})^{-1}\mathbf{B}$:

$$\mathbf{H}(s) = \begin{bmatrix} 1 & 0 & 1 & 0 \\ 0 & 1 & 0 & 1 \end{bmatrix} \begin{bmatrix} \frac{1}{s+1} & & \mathbf{0} & \\ & \frac{1}{s+1} & & \\ & \mathbf{0} & \frac{1}{s+3} & \\ & & & \frac{1}{s+3} \end{bmatrix} \begin{bmatrix} 1 & -\frac{1}{2} \\ 0 & 1 \\ 0 & \frac{3}{2} \\ 1 & 0 \end{bmatrix}$$

$$= \begin{bmatrix} 1 & 0 & 1 & 0 \\ 0 & 1 & 0 & 1 \end{bmatrix} \begin{bmatrix} s+1 & & & \\ & s+1 & \mathbf{0} & \\ \mathbf{0} & & s+3 & \\ & & & s+3 \end{bmatrix}^{-1} \begin{bmatrix} 1 & -\frac{1}{2} \\ 0 & 1 \\ 0 & \frac{3}{2} \\ 1 & 0 \end{bmatrix} \tag{76}$$

Thus, the corresponding realization is

$$\mathbf{A} = \begin{bmatrix} -1 & & & \\ & -1 & \mathbf{0} & \\ \mathbf{0} & & -3 & \\ & & & -3 \end{bmatrix} \qquad \mathbf{B} = \begin{bmatrix} 1 & -\frac{1}{2} \\ 0 & 1 \\ 0 & \frac{3}{2} \\ 1 & 0 \end{bmatrix}$$

$$\mathbf{C} = \begin{bmatrix} 1 & 0 & 1 & 0 \\ 0 & 1 & 0 & 1 \end{bmatrix} \tag{77}$$

The realization is controllable and observable and hence minimal. Modifications of this procedure for cases where $\mathbf{H}(s)$ has elements with repeated poles are described by Brogan.[4] Extensions to discrete-time systems are straightforward.

An alternative two-step procedure for determining a minimal realization for a transfer function matrix involves obtaining a nonminimal realization by any one method as the first step. For example, one of the many realizations in Table 1 (Table 3 for discrete-time systems) can be chosen to represent each of the elements of the transfer function matrix. The state-space descriptions of the elements can then be combined to get the state-space equations for the MIMO system. The resulting realization would, in general, be nonminimal. The second step requires transformation of the state-space equations to the form given by Eq. (68) or an equivalent one for discrete-time systems. Techniques for selecting the transformation matrix are described by Kalman[9] and Fortmann and Hitz.[1] The minimal realization is then given by the controllable and observable subsystem in Fig. 4. The resulting equations for a continuous-time system are

$$\dot{\mathbf{x}}_m = \mathbf{A}_{22}\mathbf{x}_m + \mathbf{B}_{21}\mathbf{u} \tag{78}$$

$$\mathbf{y}_m = \mathbf{C}_{12}\mathbf{x}_m + \mathbf{D}\mathbf{u} \tag{79}$$

where the subscript m indicates a minimal realization. Similar results for discrete-time systems are given by Brogan,[4] Kuo,[7] and Kalman.[9]

8 CONCLUSION

The state-space methods presented in this chapter offer a unifying framework for the dynamic analysis and control of a variety of systems. The primary emphasis in these methods on linear time-invariant systems is a reflection of the state of the literature on the subject and the practice of the art. Results for linear time-varying systems have been given in some of the standard texts[2–4] referred to. The application of state-space methods to nonlinear system analysis and control is treated at some length by Hedrick and Paynter.[21]

Distributed-parameter systems are examples of systems with infinite-dimensional states. Application of state-space methods to these systems has been described by Tzafestas et al.[22] Time-delayed systems are also examples of systems with infinite-dimensional states. The analysis and control of such systems and of many of the other types of systems referred to in this section remains a subject of current research. For current research results in these areas, the reader is referred to journals such as the *ASME Journal of Dynamic Systems, Measurements and Controls; IEEE Transactions on Automatic Control; AIAA Journal of Guidance, Control and Dynamics; SIAM Journal on Control; and Automatica, the Journal of the International Federation of Automatic Control.*

REFERENCES

1. T. E. Fortmann and K. L. Hitz, *An Introduction to Linear Control Systems,* Marcel Dekker, New York, 1977.
2. P. M. DeRusso, R. J. Roy, and C. M. Close, *State Variables for Engineers,* Wiley, New York, 1965.
3. T. Kailath, *Linear Systems,* Prentice-Hall, Englewood Cliffs, NJ, 1980.
4. W. L. Brogan, *Modern Control Theory,* Prentice-Hall, Englewood Cliffs, NJ, 1982.
5. J. C. Doyle and G. Stein, "Multivariable Feedback Design: Concepts for a Classical/Modern Synthesis," *IEEE Transactions on Automatic Control* **AC-26**(1), 4–16 (1981).
6. William J. Palm III, Modeling, *Analysis and Control of Dynamic Systems,* Wiley, New York, 1983.
7. B. C. Kuo, *Digital Control Systems,* SRL Publishing, Champaign, IL, 1977.
8. R. E. Kalman and J. E. Bertram, "Control-System Analysis and Design Via the Second Method of Lyapunov. I—Continuous Time Systems. II—Discrete-Time Systems," *Transactions of the ASME Journal of Basic Engineering* **82D,** 371–400 (1960).
9. R. E. Kalman, "Mathematical Description of Linear Dynamical Systems," *SIAM Journal on Control, Series A* **1**(2), 153–192 (1963).
10. R. E. Kalman, "When Is a Linear Control System Optimal?" *Transactions of the ASME Journal of Basic Engineering* **86D,** 51–60 (1964).
11. R. E. Kalman, "On the General Theory of Control Systems," in *Proceedings of the First International Congress on Automatic Control,* Butterworth's, London, 1960, pp. 481–493.
12. H. Kwakernaak and R. Sivan, *Linear Optimal Control Systems,* Wiley, New York, 1972.
13. E. Kreindler and P. Sarachik, "On the Concepts of Controllability and Observability of Linear Systems," *IEEE Transactions on Automatic Control* **AC-9**(1), 129–136 (1964).
14. R. E. Kalman, P. L. Falb, and M. Arbib, *Topics in Mathematical System Theory,* McGraw-Hill, New York, 1969.
15. E. G. Gilbert, "Controllability and Observability in Multi-Variable Control Systems," *SIAM Journal on Control, Series A* **2**(1), 128–151 (1963).
16. B. Friedland, Control System Design, *An Introduction to State-Space Methods,* McGraw-Hill, New York, 1986.
17. C. D. Johnson, "Optimization of a Certain Quality of Complete Controllability and Observability for Linear Dynamical Systems," *Transactions of the ASME Journal of Basic Engineering* **91D,** 228–238 (1969).
18. B. Friedland, "Controllability Index Based on Conditioning Number," *ASME Transactions, Journal of Dynamic Systems, Measurement and Control* **97,** 444–445 (1975).

19. P. C. Muller and H. I. Weber, "Analysis and Optimization of Certain Qualities of Controllability and Observability for Linear Dynamical Systems," *Automatica* **8,** 237–246 (1972).

20. L. M. Silverman and H. E. Meadows, "Controllability and Observability in Time-Variable Linear Systems," *SIAM Journal on Control* **5**(1), 64–73 (1967).

21. J. K. Hedrick and H. M. Paynter (eds.), *Nonlinear System Analysis and Synthesis:* Vol. 1: *Fundamental Principles,* Workshop/Tutorial Session at the Winter Annual Meeting of ASME, New York, December 1976.

22. S. G. Tzafestas (ed.), *Distributed Parameter Control Systems, Theory and Application,* Vol. 6, International Series on Systems and Control, Pergamon, Oxford, England, 1982.

CHAPTER 18

CONTROL SYSTEM DESIGN USING STATE-SPACE METHODS

Krishnaswamy Srinivasan
Department of Mechanical Engineering
The Ohio State University
Columbus, Ohio

1 INTRODUCTION

The advantages of feedback control in achieving desired input/output relationships are well known. Control system theory based on a frequency-domain approach[1] illustrates clearly that the following aspects of single-input–single-output (SISO) system performance can be improved by feedback: (1) the ability to follow reference inputs accurately in the steady state or under transient conditions and (2) the ability to reject disturbance inputs and reduce sensitivity of the overall controlled system behavior to plant parameter variations and modeling errors. For multiple-input–multiple-output (MIMO) systems, the coupling between individual inputs and outputs can be modified in a desired manner, in addition to the performance features already mentioned, by appropriate control system design.[2]

State-space methods for control system design result in solutions that utilize the state of the system most effectively for feedback. The resulting state-variable feedback control systems improve the same aspects of system performance as previously mentioned. However, the available state-space design procedures accommodate some performance specifications more readily than others. For instance, performance specifications in the form of desired closed-loop pole locations are readily accommodated. Similarly, performance specifications in the form of an index of performance to be optimized can be accommodated by optimal-control theory if the index of performance belongs to a restricted class of performance measures. In fact, recent efforts in control system design using state-space methods have been directed at enhancing the problem formulation to accommodate a greater variety of performance specifications. In spite of these enhancements, performance specifications such as sensitivity of the controlled system performance to plant parameter variations and modeling errors are accommodated more readily by frequency-domain-based design procedures than by state-space or time-domain-based design procedures. Thus, control system design techniques based on frequency-domain and time-domain approaches should be viewed as being complementary to each other in some ways.

2 THE POLE PLACEMENT DESIGN METHOD

2.1 Regulation Problem

It can be shown that, if a linear time-invariant (LTI) system is completely state controllable and if linear instantaneous state-variable feedback is used, the associated feedback gains can be chosen to place the closed-loop poles of the controlled system at any arbitrarily specified locations in the s- or z-plane,[3] depending on whether the system is continuous time or discrete time. Thus, if the continuous-time and discrete-time systems described by Eqs. (6) and (12), respectively, are completely state controllable and the control law is given by (Figs. 1 and 2)

$$\mathbf{u} = -\mathbf{K}\mathbf{x} \tag{1}$$

then the eigenvalues of the matrices $\mathbf{A} - \mathbf{BK}$ and $\mathbf{F} - \mathbf{GK}$ are the closed-loop pole locations and can be assigned any specified locations in the complex plane by appropriate selection of the gain matrix $\mathbf{K}$. If $\mathbf{K}$ is constrained to be a real matrix, the desired eigenvalues should be specified either as real or as complex-conjugate pairs. The resulting design procedure is referred to as the pole placement method and is useful for regulation problems where the objective of the controller is to return the system to equilibrium conditions following an initial disturbance. Specification of the closed-loop poles is equivalent to specification of the damping and speed of response of the closed-loop system transients as the system returns to equilibrium.

For single-input systems, specification of the desired closed-loop pole locations uniquely specifies the gain vector $\mathbf{K}$. A formula for the gain vector $\mathbf{K}$, convenient to evaluate and applicable to both continuous-time and discrete-time systems, is

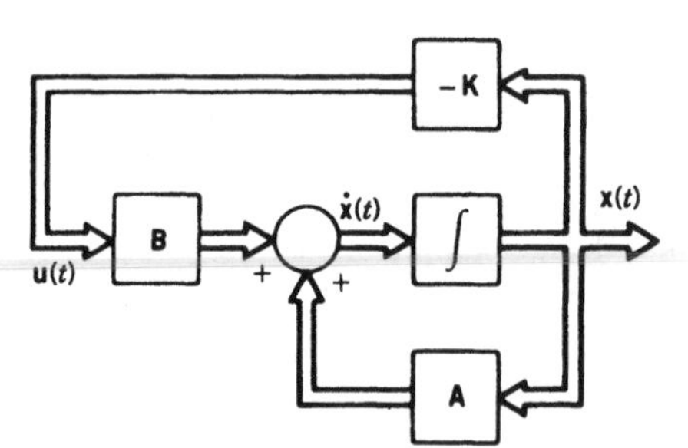

Figure 1 Linear state-variable feedback for continuous-time system.

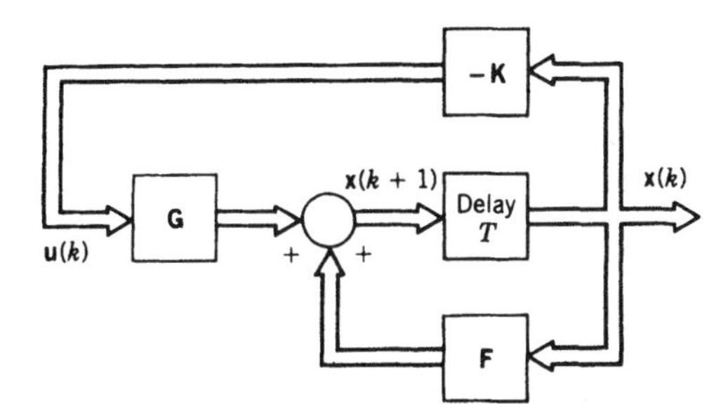

Figure 2 Linear state-variable feedback for discrete-time system.

$$\mathbf{K} = (0 \cdots 0 \; 1)(\mathbf{B} \; \mathbf{AB} \cdots \mathbf{A}^{n-1}\mathbf{B})^{-1}\alpha_c(\mathrm{A}) \tag{2}$$

for continuous-time systems and

$$\mathbf{K} = (0 \cdots 0 \; 1)(\mathbf{G} \; \mathbf{FG} \cdots \mathbf{F}^{n-1}\mathbf{G})^{-1}\alpha_c(\mathbf{F}) \tag{3}$$

for discrete-time systems. In these equations,

$$\alpha_c(\mathbf{A}) = \mathbf{A}^n + \sum_{i=0}^{n-1} \alpha_i \mathbf{A}^i \tag{4}$$

for continuous-time systems and $\alpha_c(\mathbf{F})$ is a similar function of $\mathbf{F}$ for discrete-time systems. The α_i's are the coefficients of the desired characteristic equations of the closed-loop systems. For continuous-time systems we have

$$\begin{aligned} \alpha_c(s) &= \det(s\mathbf{I} - \mathbf{A} + \mathbf{BK}) \\ &= s^n + \sum_{i=0}^{n-1} \alpha_i s^i = 0 \end{aligned} \tag{5}$$

A similar equation describes the discrete-time system characteristic equation. Computer-aided control system design (CACSD) packages supporting state-space methods usually support pole placement designs[4–6] and require only that the designer input information about the system matrices and the desired closed-loop pole locations. The gain vector **K** is then computed and output to the designer.

For multi-input systems. specification of the closed-loop poles does not specify the gain matrix **K** uniquely. The additional freedom in the gain matrix selection can be used to assign eigenvectors (or generalized eigenvectors) or individual transfer function zeros to improve the transient response to nonzero reference inputs.[7] Alternative criteria for gain matrix selection are optimization of feedback gain magnitudes and stability of the closed-loop system in the absence or failure of some of the inputs. Brogan[8] has outlined a procedure for gain matrix selection for multi-input systems, based on closed-loop eigenvector specification in addition to eigenvalue specification. For continuous-time systems described by Eq. (6) in Chapter 17 and Eq. (20) in this chapter, the feedback gain matrix is given by

$$\mathbf{K} = -(\mathbf{e}_{j_1} \; \mathbf{e}_{j_2} \cdots \mathbf{e}_{j_n})[\psi_{j_1}(s_1) \; \psi_{j_2}(s_2) \cdots \psi_{j_n}(s_n)]^{-1} \tag{6}$$

where the desired closed-loop eigenvalues and the corresponding eigenvectors are s_i and $\psi_{j_i}(s_i)$, $i = 1, \ldots, n$, respectively. The eigenvectors are chosen to be n linearly independent columns from the $n \times nr$ matrix $[\psi(s_1) \; \psi(s_2) \cdots \psi(s_n)]$ where

$$\psi(s) = (s\mathbf{I}_n - \mathbf{A})^{-1}\mathbf{B} \tag{7}$$

If the desired s_i are distinct, it will always be possible to find n linearly independent columns as already described. Here $\mathbf{e}_{j_i}$ is defined as the j_ith column of the $r \times r$ identity matrix $\mathbf{I}_r$ and is uniquely determined once j_i is determined. When repeated eigenvalues are desired, the procedure for specifying n linearly independent generalized eigenvectors is different and has been described by Brogan.[8] The results given here are readily applicable to multi-input discrete-time systems described by Eq. (12) in Chapter 17 and Eq. (1) in this chapter. The specified eigenvalues are z_i instead of s_i and the $\mathbf{A}$ and $\mathbf{B}$ matrices are replaced by $\mathbf{F}$ and $\mathbf{G}$, respectively. Also, s is replaced by z in Eq. (7). Alternative methods for gain matrix selection for multi-input systems have been described by Kailath.[7]

If a single-input, linear time-variant (LTV) system is completely state controllable, linear state-variable feedback can be used to ensure that the closed-loop transition matrix corresponds to that of atty desired nth-order linear differential equation with time-varying coefficients, The state-variable feedback gains are time varying in general and can be computed using a procedure described by Wiberg.[9]

If the complete state is not available for feedback, linear instantaneous feedback of the measured output can be used to place some of the closed-loop poles at specified locations in the complex plane. If the continuous-time and discrete-time systems described in Chapter 17 by Eqs. (6), (7), (12), and (13), respectively, satisfy the output controllability conditions listed in Table 8 in Chapter 17, then p of the n eigenvalues of the closed-loop system can, approach arbitrarily specified values to within any degree of accuracy but not always exactly. The control law is

$$\mathbf{u} = -\mathbf{K}\mathbf{y} \tag{8}$$

where $\mathbf{K}$ is a $r \times p$ gain matrix. Brogan[8] has described an algorithm for computing $\mathbf{K}$, given the desired values of p closed-loop eigenvalues. The corresponding characteristic equation is

$$\det[s\mathbf{I}_n - \mathbf{A} + \mathbf{B}\mathbf{K}(\mathbf{I}_p + \mathbf{D}\mathbf{K})^{-1}\mathbf{C}] = 0 \tag{9}$$

for continuous-time systems and

$$\det[z\mathbf{I}_n - \mathbf{F} + \mathbf{G}\mathbf{K}(\mathbf{I}_p + \mathbf{D}\mathbf{K})^{-1}\mathbf{C}] = 0 \tag{10}$$

for discrete-time systems.

An alternative approach to control system design in the case of incomplete state measurement is to use an observer or a Kalman filter for state estimation, The estimated state is then used for feedback. This procedure is discussed in Section 5.

The advantages of the pole placement design method already described are that the controller achieves desired closed-loop pole locations without using pole–zero cancellation and without increasing the order of the system. The desired pole locations can be chosen to ensure a desired degree of stability or damping and speed of response of the closed-loop system. However, there is no convenient way to ensure a priori that the closed-loop system satisfies other important performance specifications such as a desired level of insensitivity to plant parameter variations, acceptable disturbance rejection, and compatibility of control effort with actuator limitations. In addition, for single-input LTI systems, instantaneous state feedback of the form given by Eq. (1) does not affect the locations of zeros of the transfer functions between the system input and system outputs.[3] Thus, the pole placement design method does not afford complete control over the system response to the reference input or disturbance inputs. For multi-input systems, the available freedom in the gain matrix selection can be used to assign individual transfer function zeros, in addition to achieving desired closed-loop pole locations. However, systematic procedures to do this are not available. The consequence of these limitations of the pole placement method is that the design process involves considerable trial and error.

2.2 Modification for Constant Reference and Disturbance Inputs

The pole placement method described is appropriate for regulation problems. For the case of nonzero reference inputs that may be constant or varying with time, the system outputs are required to follow the reference inputs. The control law, Eq. (1), needs to be modified for such problems. If the output vector **y** and the input vector **u** have the same dimension and if the **D** matrix is zero in Eqs. (7) and (13) in Chapter 17, the modified control law has the form

$$\mathbf{u} = -\mathbf{K}\mathbf{x} + \mathbf{N}\mathbf{y}_d \tag{11}$$

where $\mathbf{y}_d$ is a vector of reference inputs. For constant reference inputs $\mathbf{y}_d$, the error $\mathbf{y}_d - \mathbf{y}$ can be reduced to zero under steady-state conditions by selecting

$$\mathbf{N} = [\mathbf{C}(-\mathbf{A} + \mathbf{B}\mathbf{K})^{-1}\mathbf{B}]^{-1} \tag{12}$$

for continuous-time systems and

$$\mathbf{N} = [\mathbf{C}(\mathbf{I}_n - \mathbf{F} + \mathbf{G}\mathbf{K})^{-1}\mathbf{G}]^{-1} \tag{13}$$

for discrete-time systems. The matrices to be inverted on the right-hand sides of the preceding equations exist if and only if the corresponding open-loop transfer matrices $[\mathbf{C}(s\mathbf{I}_n - \mathbf{A})^{-1}\mathbf{B}]$ and $[\mathbf{C}(z\mathbf{I}_n - \mathbf{F})^{-1}\mathbf{G}]$ have no zeros at the origin and at $z = 1$, respectively.[3] The **K** matrix in Eq. (11) is chosen to give the desired closed-loop poles as before. It should be noted that the gain matrix **N** is outside the feedback loop. Hence, the controlled system performance, particularly the steady-state error, would be sensitive to modeling error or error in the elements of the system matrices.

It is well known from classical control theory that integral controller action on the error has the effect of reducing the steady-state error to reference and disturbance inputs. In particular, the steady-state error is reduced to zero for constant reference and disturbance inputs. A similar result can be obtained within the framework of state-variable feedback[3] and will be described for the case where the **y** and **u** vectors have the same dimension and the **D** matrix is zero in Eqs. (7) and (13) in Chapter 17.

Consider the case of constant but unknown disturbance inputs:

$$\dot{\mathbf{x}}(t) = \mathbf{A}\mathbf{x}(t) + \mathbf{B}\mathbf{u}(t) + \mathbf{w}(t) \tag{14}$$

$$\mathbf{y}(t) = \mathbf{C}\mathbf{x}(t) \tag{15}$$

for continuous-time systems and

$$\mathbf{x}(k+1) = \mathbf{F}\mathbf{x}(k) + \mathbf{G}\mathbf{u}(k) + \mathbf{w}(k) \tag{16}$$

$$\mathbf{y}(k) = \mathbf{C}\mathbf{x}(k) \tag{17}$$

for discrete-time systems. The state-space equations are augmented by

$$\dot{\mathbf{q}}_e(t) = \mathbf{y}(t) = \mathbf{C}\mathbf{x}(t) \tag{18}$$

for continuous-time systems and

$$\begin{aligned} \mathbf{q}_e(k+1) &= \mathbf{q}_e(k) + \mathbf{y}(k) \\ &= \mathbf{q}_e(k) + \mathbf{C}\mathbf{x}(k) \end{aligned} \tag{19}$$

for discrete-time systems. The control law is modified to include feedback of the additional states (Figs. 3 and 4):

$$\mathbf{u} = -\mathbf{K}\mathbf{x} - \mathbf{K}_q\mathbf{q}_e \tag{20}$$

If the feedback gains $\mathbf{K}$, $\mathbf{K}_q$ are chosen to ensure asymptotic stability of the resulting closed-loop systems, then

$$\lim_{t\to\infty} \dot{\mathbf{q}}_e(t) = 0 \tag{21}$$

for continuous-time systems and

$$\lim_{k\to\infty} \mathbf{q}_e(k) = \text{const} \tag{22}$$

for discrete-time systems, regardless of the value of the disturbance input. When combined with Eqs. (19) and (20), the preceding equations indicate that the output $\mathbf{y}$ and hence the error go to zero in the steady state for continuous-time and discrete-time systems, respectively. The necessary and sufficient conditions for the existence of an asymptotically stable control law of the form of Eq. (20) are that the continuous-time systems, Eq. (14), and discrete-time systems, Eq. (16), be stabilizable and that the corresponding open-loop transfer matrices $[\mathbf{C}(s\mathbf{I}_n - \mathbf{A})^{-1}\mathbf{B}]$ and $[\mathbf{C}(z\mathbf{I}_n - \mathbf{F})^{-1}\mathbf{G}]$ have no zeros at the origin and at $z = 1$, respectively. It should also be clear that if a constant reference input $\mathbf{y}_d$ is to be included in the problem, the error is $\mathbf{y} - \mathbf{y}_d$ and should be used instead of $\mathbf{y}$ in Eqs. (18) and (19). In this case, the error will go to zero in the steady state while the output $\mathbf{y}$ reaches $\mathbf{y}_d$. The advantage of the integral control action is that it reduces the steady-state error to zero without requiring knowledge of the constant disturbance input or accurate values of the system parameters as in Eqs. (11)–(13). The disadvantage is that it increases the order of the system and, in practice, would degrade system stability or speed of response.

3 THE STANDARD LINEAR QUADRATIC REGULATOR PROBLEM

Controller design in regulation applications using pole placement specifications emphasizes only the transient behavior of the state variables as the system returns to equilibrium. There is no explicit consideration of the required control effort. Control effort can be considered if the controller design problem is formulated as an optimal-control problem with weighting of both control effort and state-variable transients. For regulation applications, the index of

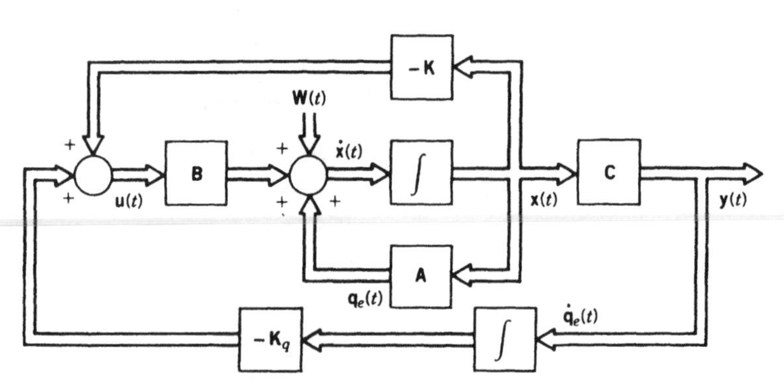

Figure 3 Proportional-plus-integral control, continuous-time system.

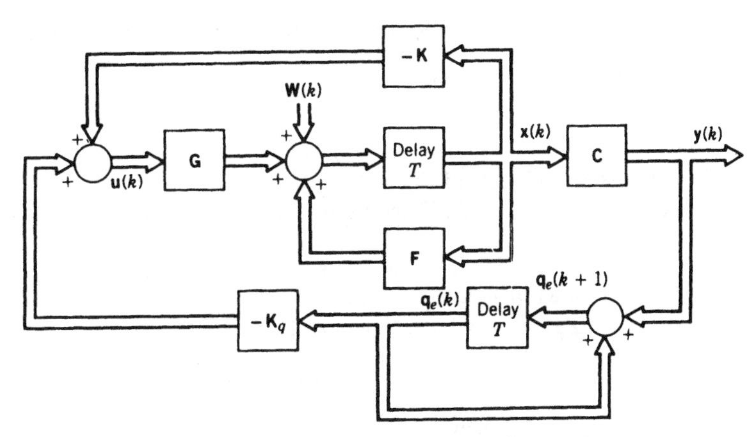

Figure 4 Proportional-plus-integral control, discrete-time system.

performance to be optimized, J, is usually chosen to be a quadratic function of the control inputs and the state variables. The resulting optimal-control law for the control input, when expressed in feedback form, is a linear function of the system state. Hence, this approach to control system design is referred to as the linear quadratic regulator (LQR) problem.

3.1 The Continuous-Time LQR Problem

Consider the continuous-time, LTV system described by Eq. (4) in Chapter 17 and the initial condition, for regulation problems, of

$$\mathbf{x}(t_0) = \mathbf{x}_0 \tag{23}$$

The controller design problem is to choose the control input $\mathbf{u}(t)$ to minimize the quadratic index of performance[3]:

$$J = \int_{t_0}^{t_1} [\mathbf{x}^{\mathrm{T}}(t)\mathbf{R}_1(t)\mathbf{x}(t) + \mathbf{u}^{\mathrm{T}}(t)\mathbf{R}_2(t)\mathbf{u}(t)]\,dt + \mathbf{x}^{\mathrm{T}}(t_1)\mathbf{P}_f\mathbf{x}(t_1) \tag{24}$$

$\mathbf{R}_1(t)$ is a positive-semidefinite symmetric weighting matrix on the state variables, $\mathbf{R}_2(t)$ is a positive-definite symmetric weighting matrix on the control inputs, and $\mathbf{P}_f$ is a positive-semidefinite symmetric weighting matrix on the terminal state. Times t_0, t_1 are initial and terminal time instants. The problem, as previously formulated, is a finite-time, deterministic, LQR problem. Kwakernaak and Sivan[3] have also considered a more general J that includes an additional term of the form $\mathbf{x}^{\mathrm{T}}(t)\mathbf{R}_{12}(t)\mathbf{u}(t)$ within the integral. They have shown that this J can be reduced to the form of Eq. (24) by appropriate redefinition of the weighting matrices and control vector.

The solution of this control problem, using methods from calculus of variations, can be obtained from standard textbooks on optimal control,[3,10] along with conditions for its existence and uniqueness. The optimal-control law, given in feedback form, is a linear, time-varying function of the system state:

$$\mathbf{u}(t) = -\mathbf{R}_2^{-1}(t)\mathbf{B}^{\mathrm{T}}(t)\mathbf{P}(t)\mathbf{x}(t) \tag{25}$$

where $\mathbf{P}$(t) is a $n \times n$ symmetric positive-semidefinite matrix satisfying the matrix Riccati equation

$$-\dot{\mathbf{P}}(t) = \mathbf{R}_1(t) - \mathbf{P}(t)\mathbf{B}(t)\mathbf{R}_2^{-1}(t)\mathbf{B}^{\mathrm{T}}(t)\mathbf{P}(t) + \mathbf{P}(t)\mathbf{A}(t) + \mathbf{A}^{\mathrm{T}}(t)\mathbf{P}(t) \tag{26}$$

and the terminal condition

$$\mathbf{P}(t_1) = \mathbf{P}_f \tag{27}$$

Numerical solution of the matrix Riccati equation is a subject of great importance and of considerable research. Some useful techniques have been briefly described by Kwakernaak and Sivan.[3]

Solution of the LQR problem simplifies as the terminal time t_1 approaches infinity. It can be shown then that the solution of the matrix Riccati equation approaches a steady-state solution $\mathbf{P}_s(t)$ that is independent of $\mathbf{P}_f$. The resulting steady-state control law

$$\mathbf{u}(t) = -\mathbf{R}_2^{-1}\mathbf{B}^{\mathrm{T}}(t)\mathbf{P}_s(t)\mathbf{x}(t) \tag{28}$$

results in an exponentially stable closed-loop system if:

1. The linear system of Eq. (4) in Chapter 17 is uniformly completely state controllable.
2. The pair $\mathbf{A}(t)$, $\mathbf{H}_r^{\mathrm{T}}(t)$ is uniformly completely reconstructible where $\mathbf{H}_r(t)$ is any matrix such that $\mathbf{H}_r(t)\mathbf{H}_r^{\mathrm{T}}(t)$ equals $\mathbf{R}_1(t)$.

The matrix Riccati equation for the steady-state LQR problem simplifies to an algebraic equation and $\mathbf{P}_s$ is a constant if the system matrices and the weighting matrices in the index of performance are constant. The resulting algebraic Riccati equation is

$$\mathbf{R}_1 - \mathbf{P}_s\mathbf{B}\mathbf{R}_2^{-1}\mathbf{B}^{\mathrm{T}}\mathbf{P}_s + \mathbf{A}^{\mathrm{T}}\mathbf{P}_s + \mathbf{P}_s\mathbf{A} = 0 \tag{29}$$

where $\mathbf{P}_s$ is a unique positive-definite solution of Eq. (29) and the resulting time-invariant closed-loop system is asymptotically stable if:

1. The linear system of Eq. (6) in Chapter 17 is completely state controllable.
2. The pair $\mathbf{A}$, $\mathbf{H}_r^{\mathrm{T}}$ is completely observable (reconstructible), where $\mathbf{H}_r$ is any matrix such that $\mathbf{H}_r\mathbf{H}_r^{\mathrm{T}}$ equals $\mathbf{R}_1$.

Another version of the LQR problem involves minimization of the quadratic index of performance for an LTI system over a finite time interval. If the weighting matrices are also time invariant, in many cases the optimal feedback gains are constant over most of the time interval of interest and vary with time only near the terminal time. Since constant feedback gains are easier to implement in practice, implementation of constant gains over the entire time interval would represent a nearly optimal solution that is practically more convenient.[3]

3.2 The Discrete-Time LQR Problem

The results of the LQR problem for discrete-time systems parallel those for continuous-time systems already stated. They are summarized here and described in greater length by Kwakernaak and Sivan.[3] The time-varying, discrete-time system is described by Eq. (10) in Chapter 17 and the initial condition

$$\mathbf{x}(k_0) = \mathbf{x}_0 \tag{30}$$

The index of performance to be minimized by controller design, for the finite-time LQR problem, is

$$J = \sum_{k=k_0}^{k_1-1} [(\mathbf{x}^T)^{-1}(k+1)\mathbf{R}_1(k+1)\mathbf{x}(k+1) + \mathbf{u}^T(k)\mathbf{R}_2(k)\mathbf{u}(k)] + \mathbf{x}^T(k_1)\mathbf{P}_f\mathbf{x}(k_1) \quad (31)$$

where the weighting matrices $\mathbf{R}_1(k)$, $\mathbf{R}_2(k)$, and $\mathbf{P}_f$ serve the same functions as $\mathbf{R}_1(t)$, $\mathbf{R}_2(t)$, and $\mathbf{P}_f$ for continuous-time systems and satisfy the same conditions. The values k_0, k_1 are the initial and final time instants. A more general version of the J, including the term $\mathbf{x}^T(k)\mathbf{R}_{12}(k)\mathbf{u}(k)$ within the summation sign, can be reduced to the form of Eq. (31) by appropriate redefinition of the weighting matrices and control vector.[3]

The solution of this control problem can be obtained using dynamic programming methods and is given by

$$\mathbf{u}(k) = -\mathbf{K}(k)\mathbf{x}(k) \quad (32)$$

where

$$\mathbf{K}(k) = \{\mathbf{R}_2(k) + \mathbf{G}^T(k)[\mathbf{R}_1(k+1) + \mathbf{P}(k+1)]\mathbf{G}(k)\}^{-1} \times \mathbf{G}^T(k)[\mathbf{R}_1(k+1) + \mathbf{P}(k+1)]\mathbf{F}(k) \quad (33)$$

$\mathbf{P}(k)$ is a $n \times n$ symmetric, positive-semidefinite matrix satisfying the matrix difference equation

$$\mathbf{P}(k) = \mathbf{F}^T(k)[\mathbf{R}_1(k+1) + \mathbf{P}(k+1)][\mathbf{F}(k) - \mathbf{G}(k)\mathbf{K}(k)] \qquad k = k_0, k_1 - 1 \quad (34)$$

with the terminal condition

$$\mathbf{P}(k_1) = \mathbf{P}_f \quad (35)$$

Unlike the matrix Riccati equation (26) for continuous-time systems, numerical solution of the preceding matrix difference equations is straightforward for finite-time LQR problems. The procedure involves solution of the difference equations backward in time:

1. Let $k = k_1 - 1$. Then $\mathbf{P}(k+1)$ is equal to $\mathbf{P}_f$ and hence is known.
2. Compute $\mathbf{K}(k)$ using Eq. (33) and the known value of $\mathbf{P}(k+1)$.
3. Compute $\mathbf{P}(k)$ using Eq. (34) and the known values of $\mathbf{K}(k)$ and $\mathbf{P}(k+1)$.
4. Reduce k by 1 and repeat 2 and 3 until $k = k_0$.

The solution to the discrete-time LQR problem also simplifies as the terminal time k_1 approaches infinity. The solutions of the matrix difference equations (33) and (34) converge to steady-state solutions $\mathbf{K}_s(k)$, $\mathbf{P}_s(k)$, which are independent of $\mathbf{P}_f$. The resulting steady-state control law

$$\mathbf{u}(k) = -\mathbf{K}_s(k)\mathbf{x}(k) \quad (36)$$

results in an exponentially stable closed-loop system if:

1. The linear system of Eq. (10) in Chapter 17 is uniformly completely state controllable.
2. The pair $\mathbf{F}(k)$, $\mathbf{H}_r^T(k)$ is uniformly completely reconstructible where $\mathbf{H}_r(k)$ is any matrix such that $\mathbf{H}_r(k)\mathbf{H}_r^T(k)$ equals $\mathbf{R}_1(k)$.

Also, the matrices $\mathbf{K}_s$ and $\mathbf{P}_s$ are constants if the system matrices and the weighting matrices in the index of performance of Eq. (31) are constants. They are given by solution of the following algebraic equations:

$$\mathbf{K}_s = [\mathbf{R}_2 + \mathbf{G}^{\mathrm{T}}(\mathbf{R}_1 + \mathbf{P}_s)\mathbf{G}]^{-1}\mathbf{G}^{\mathrm{T}}(\mathbf{R}_1 + \mathbf{P}_s)\mathbf{F} \tag{37}$$

$$\mathbf{P}_s = \mathbf{F}^{\mathrm{T}}(\mathbf{R}_1 + \mathbf{P}_s)(\mathbf{F} - \mathbf{G}\mathbf{K}_s) \tag{38}$$

The optimal-control law for the infinite-time LQR problem is

$$\mathbf{u}(k) = -\mathbf{K}_s\mathbf{x}(k) \tag{39}$$

and requires only constant state feedback gains. The solution $\mathbf{P}_s$ of Eqs. (37) and (38) is positive definite, and the optimal-control law results in an asymptotically stable closed-loop system if:

1. The linear system of Eq. (12) in Chapter 17 is completely state controllable.
2. The pair $\mathbf{F}$, $\mathbf{H}_r^{\mathrm{T}}$ is completely observable (reconstructible), where $\mathbf{H}_r$ is any matrix such that $\mathbf{H}_r\mathbf{H}_r^{\mathrm{T}}$ equals $\mathbf{R}_1$.

Also, as in the case of continuous-time systems, the optimal feedback gains are nearly constant even for finite-time LQR problems if the system matrices and weighting matrices in J are constant.[11] Finally, a number of techniques for solving the matrix algebraic equations (37) and (38) and the matrix difference equations (33)–(35) are described by Kuo.[12]

3.3 Stability and Robustness of the Optimal-Control Law

An important consideration in the practical usefulness of the optimal-control laws for the LQR problems described is the implication of these laws for performance features of the controlled systems not included in J, such as relative stability and sensitivity of the controlled system to unmodeled dynamics or plant parameter variations. Reference has already been made to the fact that the optimal-control laws for the continuous-time and discrete-time infinite-time LQR problems described result in asymptotically stable closed-loop systems provided that specified controllability and reconstructibility or observability conditions are satisfied. Closed-loop systems with a prescribed degree of stability can be obtained by modifying the performance index J for linear, time-invariant, continuous-time systems:[10]

$$J = \int_0^\infty e^{2\alpha t}(\mathbf{x}^{\mathrm{T}}\mathbf{R}_1\mathbf{x} + \mathbf{u}^{\mathrm{T}}\mathbf{R}_2\mathbf{u})dt \tag{40}$$

where α is a positive scalar constant. If the pair $\mathbf{A}$, $\mathbf{B}$ is completely state controllable and the pair $\mathbf{A}$, $\mathbf{H}_r^{\mathrm{T}}$ is completely observable where $\mathbf{H}_r\mathbf{H}_r^{\mathrm{T}}$ is equal to $\mathbf{R}_1$, the solution to this LQR problem results in a finite value of J. Hence, the transients decay at least as rapidly as $e^{-\alpha t}$. Larger values of α would therefore ensure a more rapid return of the system to equilibrium. The corresponding algebraic Riccati equation is

$$\mathbf{R}_1 - \mathbf{P}_s\mathbf{B}\mathbf{R}_2^{-1}\mathbf{B}^{\mathrm{T}}\mathbf{P}_s + \mathbf{A}^{\mathrm{T}}\mathbf{P}_s + \mathbf{P}_s\mathbf{A} + 2\alpha\mathbf{P}_s = 0 \tag{41}$$

and the optimal-feedback-control law is given by Eq. (28). A similar procedure for discrete-time LTI systems is described by Franklin and Powell.[11]

Additional results concerning the stability properties of the optimal control law for continuous-time, LTI systems described by Eq. (6) in Chapter 17 and employing only constant weighting matrices in the index of performance, Eq. (24), are available and will be summarized. Anderson and Moore[10] have shown that for single-input systems the optimal-control law for the infinite-time LQR problem has $\pm 60°$ phase margin, an infinite gain margin, and 50% gain reduction tolerance before the closed-loop system becomes unstable. These results are best explained with the aid of Fig. 5*a*, where $G_p(s)$ is normalized to be

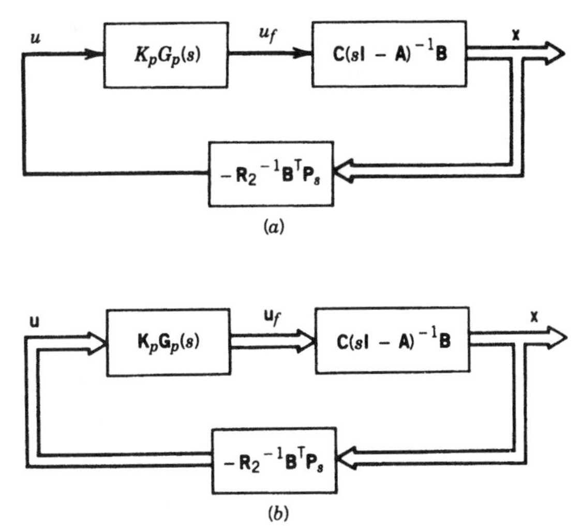

Figure 5 Robustness of optimal LQR control: (*a*) single-input system and (*b*) multi-input system.

unity at $s = 0$. The transfer function $K_pG_p(s)$ characterizes the modeling accuracy and is unity for an exact model. The result stated previously indicates that modeling errors that result either in phase shifts $\underline{/G_p(j\omega)}$ of less than 60° in magnitude for all frequencies or in values of the magnitude ratio K_p greater than one-half would not destabilize the closed-loop system. The result on gain margins extends also to static nonlinear gain relationships between u_f and u.[10]

Safonov[13] has extended these results on gain and phase margins to multi-input infinite-time LQR problems. As shown in Fig. *5b,* the quantities $\mathbf{u}_f(s)$ and $\mathbf{u}(s)$ are vectors and the modeling accuracy, if linear and time invariant, is represented by a transfer matrix $\mathbf{K}_p\mathbf{G}_p^{(s)}$. For an exact model, $\mathbf{K}_p\mathbf{G}_p(s)$ is the identity matrix. The results stated here are special cases of the results derived by Safonov[13] and are valid for the case where $\mathbf{K}_p$ and $\mathbf{G}_p(s)$ are diagonal matrices and $\mathbf{G}_p(s)$ consists of normalized transfer functions $\mathbf{G}_{pj}(s)$ [i.e., $\mathbf{G}_p(0)$ is the identity matrix]. For such a case, as long as all of the phase shifts $\underline{/G_{pj}(j\omega)}$ are less than 60° in magnitude for all frequencies or as long as all of the elements of the $\mathbf{K}_p$ matrix are greater than one-half, the closed-loop system using the optimal-control law is asymptotically stable. More general robustness results, which are more difficult to apply, are also available.[13,14]

Perkins and Cruz[15] have examined frequency-domain characterizations of the infinite-time LQR problem for continuous-time LTI systems and have shown that feedback realization of the optimal-control law results in lower sensitivity to plant parameter variations than an equivalent open-loop realization. Kwakernaak and Sivan[3] have provided other results that are somewhat more useful in relating the sensitivity properties of optimal-control laws to the weighting matrices in J. As the elements of the control effort weighting matrix $\mathbf{R}_2$ are decreased, the control law sensitivity decreases or improves since the optimal feedback gains are higher. However, higher feedback gains naturally imply greater likelihood of actuator saturation. The relative sensitivities of the different state variables depend on the elements of the state weighting matrix $\mathbf{R}_1$. State variables that are weighted more heavily would have lower sensitivity. Also, the sensitivity characteristics of the optimal-control law for non-minimum-phase systems are shown to be inferior to that of minimum-phase systems. Finally, Kwakernaak and Sivan[3] have illustrated that the sensitivity results described do not necessarily extend to discrete-time systems.

4 EXTENSIONS OF THE LINEAR QUADRATIC REGULATOR PROBLEM

The optimal-control law for the LQR problem and the pole placement design method described in the preceding sections have a number of limitations. First, as Horowitz[16] has pointed out, control system design by pole placement or quadratic performance index minimization obscures some practically important aspects and objectives. Among these are sensor noise, loop bandwidths, and sensitivity of system performance to significant plant parameter variations. Rosenbrock and McMorran[17] have pointed out that unconditional stability (i.e., stability for all values of the control gains between zero and design values) is essential for industrial control systems but is not guaranteed by the optimal-control law. Hence, for multivariable optimal-control systems, the failure of a single feedback-measuring instrument could destabilize the closed-loop system. Moreover, the achievement of more modest sensitivity requirements is complicated by the lack of clear guidelines for weighting matrix selection. Available procedures for weighting matrix selection enable the achievement of desired transient response characteristics either by specification of a few dominant closed-loop system poles[10] or by implicit model reference following methods.[18] In the latter case, the reference model is chosen to have desired transient response characteristics. However, there is no available method to ensure a priori that the optimal-control law has other desirable performance characteristics such as low sensitivity. The consequence of the lack of satisfactory guidelines for weighting matrix selection is that practical control system design using the LQR formulation involves considerable trial and error.

Second, the formulation of the standard LQR problem needs to be extended to be able to effectively handle control problems other than regulation. Examples of such problems include regulation in the presence of persistent disturbances, tracking problems, and vibration control problems. Even though some of these problems can be handled by simple extensions of the LQR problem, effective solutions to these problems require significant extensions of the LQR problem formulation.

Extensions of the standard LQR problem formulation addressing some of its limitations will be described here. The extensions involve alternative formulations of the quadratic index of performance to be minimized such that the resulting solutions have desired features. Additionally, the problem formulation utilizes more completely the available information on the systems to be controlled and their environments. One of the measures for evaluating the effectiveness of the resulting problem formulations and solutions is their ability to accommodate a greater variety of problems and performance specifications. Another such measure is their ability to incorporate in the proposed solutions features that are known to be effective in practice. The resulting variety of problem formulations and solutions runs somewhat counter to the unifying nature of the standard LQR problem formulation and constitutes a recognition of its limitations in practice.

4.1 Disturbance Accommodation

Extensions of the standard LQR problem to accommodate unknown disturbance inputs have been proposed by Anderson and Moore,[10] Johnson,[19,20] and Davison and Ferguson.[21] Anderson and Moore[10] consider LTI systems of the following form:

$$\dot{\mathbf{x}}(t) = \mathbf{A}\mathbf{x}(t) + \mathbf{B}[\mathbf{u}(t) + \mathbf{M}_w\mathbf{w}] \tag{42}$$

where $\mathbf{w}$ is a constant, unknown disturbance vector. The restriction that $\mathbf{u}(t)$ and $\mathbf{M}_w\mathbf{w}$ occur additively ensures that the equilibrium $\mathbf{x} = 0$ can be achieved and maintained even if $\mathbf{w}$ is nonzero. The following index of performance with input derivative constraints is to be minimized by the control law:

$$J = \int_0^\infty [\mathbf{x}^T\mathbf{R}_1\mathbf{x} + (\mathbf{u} + \mathbf{M}_w\mathbf{w})^T\mathbf{R}_2(\mathbf{u} + \mathbf{M}_w\mathbf{w}) + \dot{\mathbf{u}}^T\mathbf{R}_3\dot{\mathbf{u}}]\, dt \tag{43}$$

where $\mathbf{R}_1$, $\mathbf{R}_2$ are symmetric positive-semidefinite matrices and $\mathbf{R}_3$ is a symmetric positive-definite matrix. When the system state is augmented to include the vector $\mathbf{u} + \mathbf{M}_w\mathbf{w}$ and the input derivative $\dot{\mathbf{u}}$ is defined to be the new input, the problem reduces to the standard infinite-time LQR problem. The optimal-control law is a proportional-plus-integral state feedback law:

$$\mathbf{u}(t) = -\mathbf{K}\mathbf{x}(t) - \mathbf{K}_I \int \mathbf{x}\, dt \tag{44}$$

where the constant gain matrices $\mathbf{K}$, $\mathbf{K}_I$ are known linear functions of matrices satisfying algebraic Riccati equations. In cases where the complete state $\mathbf{x}$ is not measurable, the state would be estimated from the measured outputs using the methods described in Section 5. The closed-loop system is asymptotically stable provided that certain controllability and observability conditions on the matrices $\mathbf{A}$, $\mathbf{B}$, $\mathbf{R}_1$, and $\mathbf{R}_2$ are satisfied. The gain and phase margin results noted earlier for the standard infinite-time LQR problem are valid here as well.

Johnson[19] considers a more general class of disturbances and state-space equations:

$$\dot{\mathbf{x}}(t) = \mathbf{A}(t)\mathbf{x}(t) + \mathbf{B}(t)\mathbf{u}(t) + \mathbf{B}_w(t)\mathbf{w}(t) \tag{45}$$

$$\mathbf{y}(t) = \mathbf{C}(t)\mathbf{x}(t) + \mathbf{D}(t)\mathbf{u}(t) + \mathbf{D}_w(t)\mathbf{w}(t) \tag{46}$$

where $\mathbf{w}(t)$ is a vector of disturbance inputs. The disturbance inputs are assumed to be described by linear time-varying differential equations that constitute a state-space model for the disturbances:

$$\dot{\mathbf{z}}_\sigma(t) = \mathbf{A}_\sigma(t)\mathbf{z}_\sigma(t) + \mathbf{B}_\sigma(t)\mathbf{x}(t) + \sigma(t) \tag{47}$$

$$\mathbf{w}(t) = \mathbf{C}_\sigma(t)\mathbf{z}_\sigma(t) + \mathbf{D}_\sigma(t)\mathbf{x}(t) \tag{48}$$

where $\mathbf{z}_\sigma(t)$ represents the state of the disturbance and $\boldsymbol{\sigma}(t)$ is a vector of Dirac delta impulses occurring at unknown times. The terms including $\mathbf{x}(t)$ in the preceding equations enable cases of state-dependent disturbances to be considered within this framework. The coefficient matrices are determined experimentally by examination of the records of the disturbances. This type of description of disturbances constitutes a waveform mode description and is applicable to a broad class of disturbances of practical interest that are not described well either by deterministic process models or by stochastic process models. Examples of such disturbances are piecewise linear, piecewise polynomial, or piecewise periodic signals.

The waveform mode description of disturbances is combined with the system state-space equations to provide a rather complete description of the system to be controlled and the inputs affecting its behavior. The exact design of the controller depends on the specific objectives used to govern the design. If one of the objectives of the control system design is to counteract as completely as possible the effects of the disturbance inputs, then the control input is considered to be composed of two parts:

$$\mathbf{u}(t) = \mathbf{u}_m(t) + \mathbf{u}_d(t) \tag{49}$$

where $\mathbf{u}_d(t)$ is the component used to counteract disturbance effects either completely or partially and $\mathbf{u}_m(t)$ is the component used to accomplish other objectives such as closed-loop pole placement. Alternatively, the objective of control system design may be the minimization of a quadratic index of performance. In either of these cases, the control law would require the feedback of the system state $\mathbf{x}$ as well as the disturbance state $\mathbf{z}_\sigma$ (Fig. 6). The state estimation methods described in Section 5 can be used to generate these state estimates from available measurements. The extension of this disturbance accommodation approach to discrete-time systems has also been considered by Johnson.[20]

The formulation of disturbance state models and their incorporation in controller design result in controller features familiar from more classical approaches to disturbance suppression. Examples are integral control for constant disturbances, notch filter control for sinusoidal disturbances, and disturbance feedforward if some components of the disturbance inputs are measurable.[19] Hence, the disturbance accommodation controllers described here may be viewed as generalizations of classical solutions to disturbance suppression. A similar comment may be made concerning the mechanism for disturbance suppression inherent in the robust servomechanism structure described by Davison and Ferguson.[21] The robust controller structure is described later in this section.

4.2 Tracking Applications

Anderson and Moore,[10] Davison and Ferguson,[21] Trankle and Bryson,[22] and Tomizuka et al.[23–26] have considered extensions of the LQR formulation to accommodate tracking applications. Anderson and Moore[10] have considered the servomechanism problem where the linear system state equations are given by Eqs. (4) and (5) in Chapter 17 with $\mathbf{D}(t) = \mathbf{0}$ and the class of desired trajectories $\mathbf{y}_r$ is given by

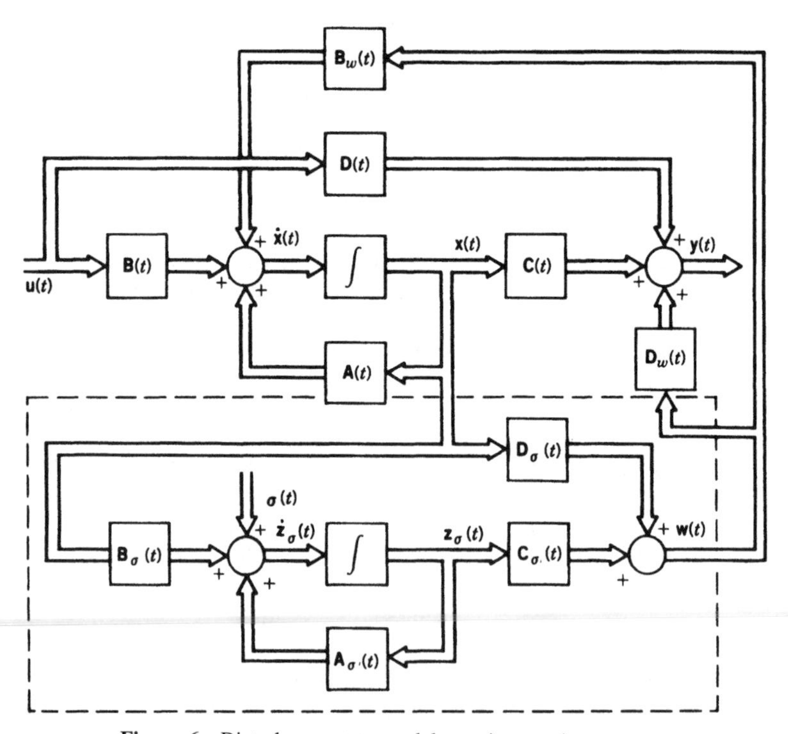

Figure 6 Disturbance state model, continuous-time system.

$$\dot{\mathbf{x}}_r(t) = \mathbf{A}_r(t)\mathbf{x}_r(t) \tag{50}$$

$$\mathbf{y}_r(t) = \mathbf{C}_r(t)\mathbf{x}_r(t) \tag{51}$$

and a specified initial condition $\mathbf{x}_r(t_0)$. The index of performance to be optimized for the finite-time problem is

$$J = \int_{t_0}^{t_1} \{\mathbf{x}^{\mathrm{T}}[\mathbf{I} - \mathbf{C}^{\mathrm{T}}(\mathbf{C}\mathbf{C}^{\mathrm{T}})^{-1}\mathbf{C}]^{\mathrm{T}}\mathbf{Q}_1[\mathbf{I} - \mathbf{C}^{\mathrm{T}}(\mathbf{C}\mathbf{C}^{\mathrm{T}})^{-1}\mathbf{C}]\mathbf{x} + (\mathbf{y} - \mathbf{y}_r)^{\mathrm{T}}\mathbf{Q}_2(\mathbf{y} - \mathbf{y}_r) + \mathbf{u}^{\mathrm{T}}\mathbf{R}_2\mathbf{u}\}dt \tag{52}$$

where the time dependencies of the vectors and matrices have been omitted for convenience. The matrices $\mathbf{Q}_1$ and $\mathbf{Q}_2$ are positive-semidefinite matrices and $\mathbf{R}_2$ is a positive-definite matrix. The weighting on the tracking error $\mathbf{y} - \mathbf{y}_r$ helps reduce it, whereas the weighting on the state $\mathbf{x}$ achieves a smooth response. The optimal-control law involves linear feedback of the system state as well as feedforward of the state of the trajectory model:

$$\mathbf{u} = -\mathbf{K}(t)\mathbf{x}(t) - \mathbf{K}_r(t)\mathbf{x}_r(t) \tag{53}$$

where the gain matrices $\mathbf{K}(t)$ and $\mathbf{K}_r$ are linearly related to solutions of the matrix Riccati differential equations. Conditions for the time-invariant version of this servo problem to reduce to the standard infinite-time LQR problem have also been noted.[10]

Trankle and Bryson[22] have considered the time-invariant servomechanism problem for the case where $\mathbf{y}$, $\mathbf{y}_r$, $\mathbf{u}$ have the same dimension and have proposed the following index of performance:

$$J = \int_0^{\infty} [(\mathbf{y} - \mathbf{y}_r)^{\mathrm{T}}\mathbf{Q}_y(\mathbf{y} - \mathbf{y}_r) + (\mathbf{u} - \mathbf{U}_1\mathbf{x}_r)^{\mathrm{T}}\mathbf{R}_u(\mathbf{u} - \mathbf{U}_1\mathbf{x}_r)]\, dt \tag{54}$$

where $\mathbf{Q}_y$ is positive semidefinite and $\mathbf{R}_u$ is positive definite. A modification of the index of performance to add integral error feedback can also be devised. A matrix $\mathbf{U}_1$ and another matrix $\mathbf{X}$ to occur later in the development are defined by

$$\mathbf{C}\mathbf{X} = \mathbf{C}_r \tag{55}$$

$$\mathbf{A}\mathbf{X} + \mathbf{B}\mathbf{U}_1 = \mathbf{X}\mathbf{A}_r \tag{56}$$

The optimal-control law is asymptotically stable if the pair $\mathbf{A}$, $\mathbf{B}$ is completely state controllable and the pair $\mathbf{A}$, $\mathbf{C}$ is completely observable. The control law is given by

$$\mathbf{u} = (\mathbf{U}_1 + \mathbf{K}\mathbf{X})\mathbf{x}_r(t) - \mathbf{K}\mathbf{x}(t) \tag{57}$$

where $\mathbf{K}$ is related, in the usual manner, to the solution of an algebraic Riccati equation. The first term on the right-hand side represents feedforward control action, and the second term represents feedback control action (Fig. 7). The feedforward action yields faster and more accurate tracking of the desired trajectory than other control schemes that depend more on integral error feedback. Finally, model and system state feedback is required by both Eqs. (53) and (57). If these states are not available for measurement, state estimators such as those described in Section 5 would be needed.

A variation on the servomechanism problem already described is that of tracking where the desired trajectory $\mathbf{y}_r$ is known a priori rather than being defined by a model as in Eqs. (50) and (51). Anderson and Moore[10] have determined the optimal-control law for the index of performance Eq. (52):

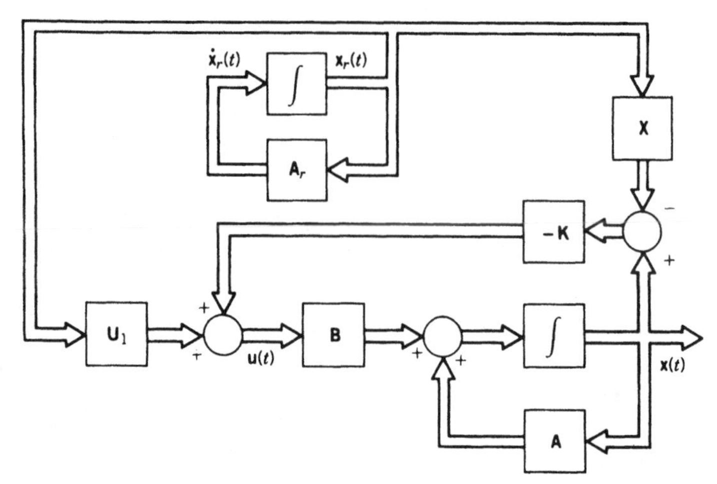

Figure 7 Extension of LAR for time-invariant servomechanism problem.

$$u = -\mathbf{K}(t)\mathbf{x}(t) - \mathbf{R}_2(t)^{-1}\mathbf{B}^{\mathrm{T}}(t)\mathbf{b}(t) \tag{58}$$

where $\mathbf{b}(t)$ is the solution of a linear ordinary differential equation with $\mathbf{y}_r(t)$ as the forcing function and $\mathbf{K}(t)$ is related, in the usual manner, to the solution of a matrix Riccati differential equation.

The control law, Eq. (58), incorporates information about future inputs over the entire interval of interest (t_0, t_1) and is said to have infinite preview control in addition to feedback control. A related problem is one where only finite preview of the desired trajectory is available; that is, at any time τ, $\mathbf{y}_r(t)$ is known for $\tau \leq t \leq \Delta T + \tau$, where ΔT is called the preview length. Tomizuka[23] has examined the continuous-time finite preview problem and determined the optimal-control law for a quadratic index of performance over the entire interval of interest (t_0, t_1) where t_1 is greater than $t_0 + \Delta T$. The desired trajectory, not known from preview at any time t, is assumed to be modeled by a stochastic process. A discrete-time version of the problem is given by Tomizuka and Whitney.[24] Discrete-time finite preview of disturbance inputs in addition to the desired trajectory has also been considered by Tomizuka et al.[25,26] The results indicate that preview control improves the control system performance, especially in the low-frequency range, and that there exists a critical preview length beyond which preview information is less important.

4.3 Frequency Shaping of Cost Functionals

Extensions of the standard LQR problem and the resulting control laws have certain limitations related to the nature of the index of performance used. These limitations become clear when the index of performance is viewed in the frequency domain.[27] The index of performance for the infinite-time, LQR problem for an LTI system

$$J = \int_0^\infty (\mathbf{x}^{\mathrm{T}}\mathbf{R}_1\mathbf{x} + \mathbf{u}^{\mathrm{T}}\mathbf{R}_2\mathbf{u})\, dt \tag{59}$$

can be transformed to the frequency domain using Parseval's theorem:

$$J = \frac{1}{2}\int_{-\infty}^{\infty} [\mathbf{x}^*(j\omega)\mathbf{R}_1\mathbf{x}(j\omega) + \mathbf{u}^*(j\omega)\mathbf{R}_2\mathbf{u}(j\omega)]\, d\omega \tag{60}$$

where the asterisk implies a complex-conjugate transpose of a complex matrix. The index of performance thus weights control and state transients at all frequencies equally despite the fact that, in practice, model accuracy is poorer at high frequencies. Model inaccuracy, sensor noise, and actuator bandwidth limitations are better accommodated by an index of performance that penalizes high-frequency control activity more heavily than low-frequency control activity. In fact, it is recognized good practice in classical control system design to have a steep enough rolloff or attenuation rate of the open-loop transmission functions at high frequencies for adequate noise suppression and robustness to model errors. In contrast, the closed-loop amplitude ratio frequency response curve corresponding to the optimal-control law for an LQR problem may drop off only as slowly as 20 dB/decade.[27] Another limitation of the standard LQR problem formulation is that, in many applications such as vibration control, the objectives of control are stated better in the frequency domain than in the time domain.

Specification of the index of performance in the frequency domain with the weighting matrices being functions of the frequency ω enables many of these limitations to be removed. The frequency-shaped cost functional method described by Gupta[27] is one such method and allows the generalized index of performance:

$$J = \frac{1}{2}\int_{-\infty}^{\infty} [\mathbf{x}^*(j\omega)\mathbf{R}_1(j\omega)\mathbf{x}(j\omega) + \mathbf{u}^*(j\omega)\mathbf{R}_2(j\omega)\mathbf{u}(j\omega)\, d\omega \tag{61}$$

The restrictions on the weighting matrices are:

1. $\mathbf{R}_1(j\omega)$ and $\mathbf{R}_2(j\omega)$ are positive semidefinite and positive definite, respectively, at all frequencies.
2. $\mathbf{R}_1(j\omega)$ and $\mathbf{R}_2(j\omega)$ are Hermitian matrices at all frequencies and, in fact, are rational functions of ω^2
3. $\mathbf{R}_2(j\omega)$ has rank r, where r is the dimension of control vector $\mathbf{u}$.

Two new matrices $\mathbf{P}_1(j\omega)$ and $\mathbf{P}_2(j\omega)$ are defined based on $\mathbf{R}_1(j\omega)$ and $\mathbf{R}_2(j\omega)$:

$$\mathbf{R}_1(j\omega) = \mathbf{P}_1^*(j\omega)\mathbf{P}_1(j\omega) \tag{62}$$

$$\mathbf{R}_2(j\omega) = \mathbf{P}_2^*(j\omega)\mathbf{P}_2(j\omega) \tag{63}$$

where $\mathbf{P}_1$ is $m \times n$, m being the rank of $\mathbf{R}_1$, and $\mathbf{P}_2$ is $r \times r$, r being the rank of $\mathbf{R}_2$ and the dimension of the control vector $\mathbf{u}$. New vectors $\mathbf{x}_p$, and $\mathbf{u}_p$, dynamically related to $\mathbf{x}$ and $\mathbf{u}$, respectively, are then defined:

$$\mathbf{x}_p(s) = \mathbf{P}_1(s)\mathbf{x}(s) \tag{64}$$

$$\mathbf{u}_p(s) = \mathbf{P}_2(s)\mathbf{u}(s) \tag{65}$$

Minimal realizations of $\mathbf{P}_1(s)$ and $\mathbf{P}_2(s)$ are then determined as described in Chapter 18, Section 7, and new states $\mathbf{z}_x$ and $\mathbf{z}_u$ defined:

$$\dot{\mathbf{z}}_x = \mathbf{A}_x\mathbf{z}_x + \mathbf{B}_x\mathbf{x} \tag{66}$$

$$\mathbf{x}_p = \mathbf{C}_x\mathbf{z}_x + \mathbf{D}_x\mathbf{x} \tag{67}$$

$$\dot{\mathbf{z}}_u = \mathbf{A}_u\mathbf{z}_u + \mathbf{B}_u\mathbf{u} \tag{68}$$

$$\mathbf{u}_p = \mathbf{C}_u\mathbf{z}_u + \mathbf{D}_u\mathbf{u} \tag{69}$$

An augmented state $\mathbf{x}_a$ is then defined:

$$\mathbf{x}_a = \begin{bmatrix} \mathbf{x} \\ \mathbf{z}_x \\ \mathbf{z}_u \end{bmatrix} \tag{70}$$

Using Parseval's theorem, the index of performance, Eq. (61), is transformed to the time domain to yield a standard LQR problem formulation with constant weighting matrices:

$$J = \int_0^\infty (\mathbf{x}_a^T \mathbf{R}_{1a} \mathbf{x}_a + 2\mathbf{x}_a^T \mathbf{R}_{12a} \mathbf{u} + \mathbf{u}^T \mathbf{R}_{2a} \mathbf{u})\, dt \tag{71}$$

The optimal-control law is obtained by solving the corresponding algebraic matrix Riccati equation:

$$\mathbf{u} = -(\mathbf{K}\mathbf{x} + \mathbf{K}_x \mathbf{z}_x + \mathbf{K}_u \mathbf{z}_u) \tag{72}$$

The states $\mathbf{z}_x$ and $\mathbf{z}_u$, are dynamically related to $\mathbf{x}$ and $\mathbf{u}$, respectively. Hence, the optimal-control law has dynamic compensators in addition to linear instantaneous state-variable feedback (Fig. 8). If the state $\mathbf{x}$ is not completely measurable, state estimation is required as described in Section 5.

The utility of this design method is that it establishes a clear link between features of the weighting matrices $\mathbf{R}_1(j\omega)$ and $\mathbf{R}_2(j\omega)$ and the resulting controllers. Gupta[27] has shown that the compensator poles and zeros are the same as poles and zeros of the transfer functions $\mathbf{P}_1(s)$ and $\mathbf{P}_2(s)$. For example, if $\mathbf{R}_1(j\omega)$ is singular at $\omega = 0$, integral control results. If $\mathbf{R}_1(j\omega)$ is singular at any other frequency ω_1, the controller has a notch filter at that frequency. This would be desirable if the controlled system has a known resonant frequency at ω_1 and we wish to minimize the excitation of the resonance by disturbances. Finally, if $\mathbf{R}_2(j\omega)$ is chosen to increase with frequency ω, the optimal-control law would have reduced control action at high frequencies. As indicated previously, this is a desirable feature if the controller is to have good noise suppression and robustness to model errors.

4.4 Robust Servomechanism Control

Davison and Ferguson[21] have proposed a controller structure and formulated a design procedure for the robust control of servomechanisms. Robustness is defined here to imply asymptotic stability of the closed-loop system and asymptotic tracking of the desired trajectory for all initial conditions of the controller used and for all variations in the system model parameters that do not cause the controlled system to become unstable. The system equations are the time-invariant versions of Eqs. (45) and (46). The disturbance inputs $\mathbf{w}(t)$ are modeled by time-invariant versions of Eqs. (47) and (48) with no provision for either state-dependent disturbances $[\mathbf{B}_\sigma(t) = \mathbf{0} = \mathbf{D}_\sigma(t)]$ or impulsive inputs $[\sigma(t) = 0]$. The desired trajectory $\mathbf{y}_r(t)$ is described by time-invariant versions of Eqs. (50) and (51).

Under certain specified conditions,[21] the robust servomechanism problem is assured of a solution. The resulting controller structure consists of a servocompensator and stabilizing compensator (Fig. 9), and the robust control input is given by

$$\mathbf{u} = -\mathbf{K}_\eta \eta - \mathbf{K}_\theta \theta \tag{73}$$

where η and θ are the outputs of the servocompensator and stabilizing compensator, respectively, and $\mathbf{K}_\eta$ and $\mathbf{K}_\theta$ are constant-gain matrices. The servocompensator is a dynamic controller with the trajectory error as input and its form and parameters are determined from the state-space models of the system and the disturbance and trajectory inputs. The servo-

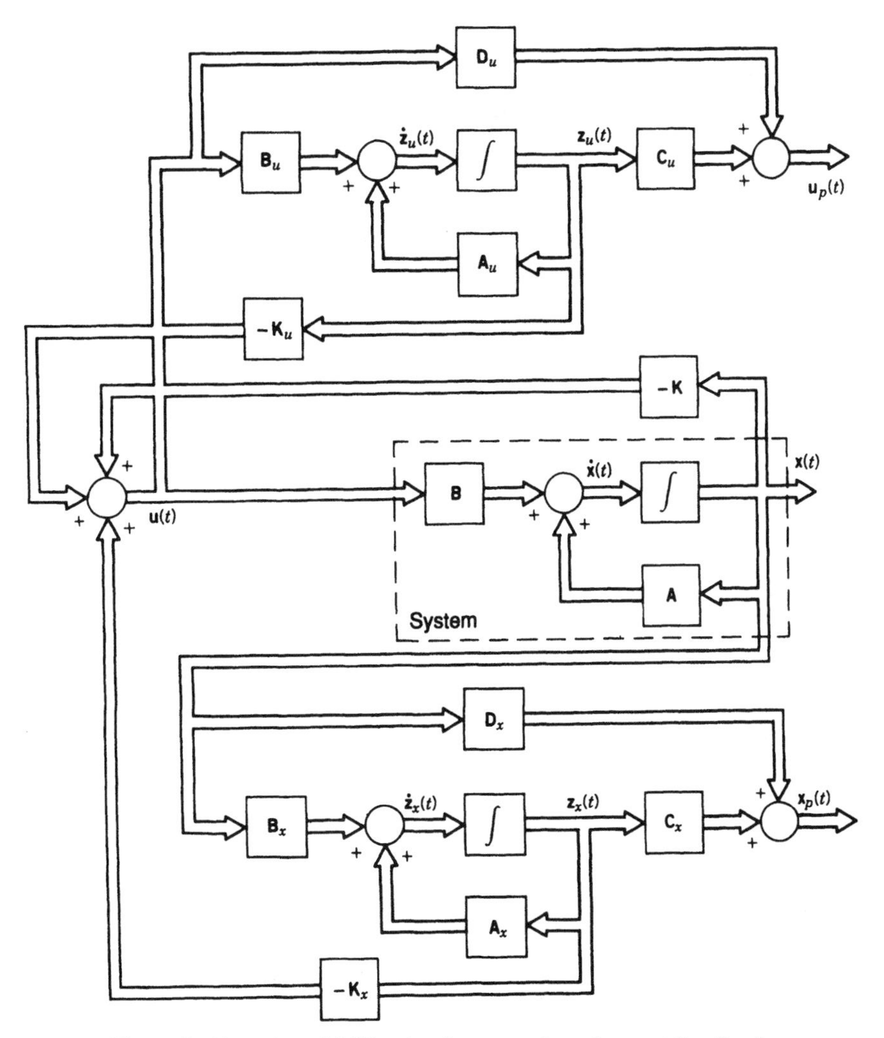

Figure 8 Extension of LQR using frequency-dependent cost functionals.

compensator is a generalization of the integral controller from classical control theory. The stabilizing compensator has the function of stabilizing the augmented system consisting of the servocompensator and the system to be controlled. In general, the stabilizing compensator has a number of inputs as shown in Fig. 9. It is not uniquely defined, however, and is usually chosen to have as simple a form as is possible given the performance requirements on the controlled system. Complete state feedback, if measurable, and observer-based controllers of the type described in Section 6 are among the more elaborate forms of the stabilizing compensator.

Once the structure of the stabilizing compensator is determined, the unknown controller parameters are determined by minimization of a quadratic index of performance. The index of performance is specified such that minimizing it gives a system with fast response and low interaction for MIMO systems. The optimum value of the index of performance is given

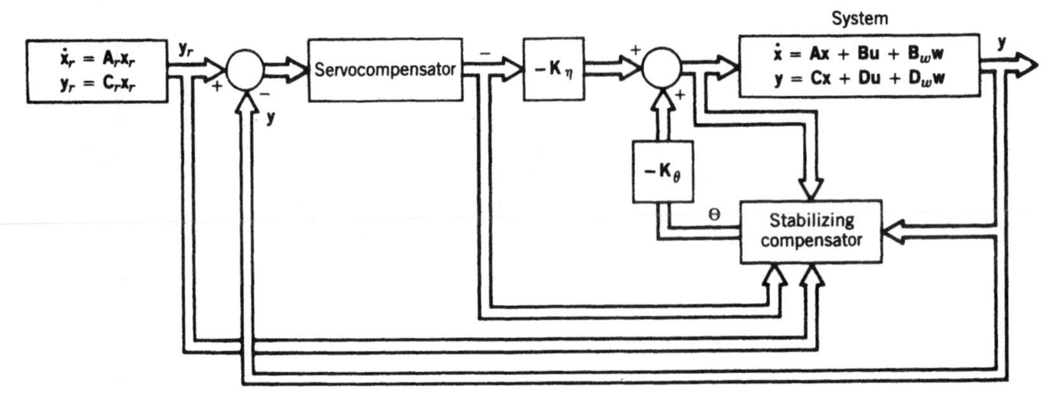

Figure 9 Extension of LQR for the robust LTI servomechanism problem.

in terms of the controller parameters. Since controller parameters are not known a priori, the optimal controller design reduces to a multiparameter optimization problem where the quantity being optimized is the quadratic index of performance. The parameter optimization can be constrained to allow the designer to handle closed-loop system damping constraints, controller gain constraints to avoid saturation effects, controller integrity constraints for sensor and actuator failures, and tolerance constraints to system parameter variations. When applied to example systems,[21] the robust control approach yields controller features commonly obtained from other frequency-domain-based design procedures, such as error integral control, phase-lead compensation, pole–zero cancellation, and low interaction for MIMO systems. Its ability to accommodate a variety of constraints on the controllers to ensure their practical utility makes the robust controller design approach a practically useful approach. Care is needed, however, to keep the resulting controller as simple as possible.

The recent extensions of the LQR problem described have addressed many of the limitations of state-space-based approaches to control system design noted by Horowitz[16] and others. As a result, the state-space approach is expected to be more useful in a greater variety of practical applications. It should be noted, however, that there are still no specific guidelines for the selection of weighting matrices in the quadratic indices of performance used by the extended versions of the LQR problem. Consequently, control system design using these methods would still involve considerable trial and error.

5 DESIGN OF LINEAR STATE ESTIMATORS

The optimal control laws for the standard LQR problem and the extensions described earlier require feedback of the entire system state. Pole placement algorithms require state feedback as well. Complete state measurement is not possible or practical in many instances. Therefore, the state variables often must be estimated from the measured output variables. Closed-loop dynamic realizations of such state estimators are described here. The property of reconstructibility (or equivalently, observability for LTI systems) is critical for the design of such estimators. It plays a role relative to state estimators that is very similar to the role played by state controllability relative to state feedback controller design. This similarity is the consequence of the duality of the two properties, referred to in Chapter 17, Section 6.

Therefore, many of the results presented in this section parallel those of the preceding section on controller design.

5.1 The Observer

The structure of a closed-loop dynamic system, called an observer, to estimate the state $\mathbf{x}(t)$ of an LTV system described by Eqs. (4) and (5) from Chapter 17, from output measurements $\mathbf{y}(t)$ and input measurements $\mathbf{u}(t)$, was proposed by Luenberger[28] (Fig. 10):

$$\dot{\hat{\mathbf{x}}}(t) = \mathbf{A}(t)\,\hat{\mathbf{x}}(t) + \mathbf{B}(t)\mathbf{u}(t) + \mathbf{L}(t)[\mathbf{y}(t) - \mathbf{C}(t)\hat{\mathbf{x}}(t) - \mathbf{D}(t)\mathbf{u}(t)] \tag{74}$$

where $\hat{\mathbf{x}}(t)$ is the estimated state and $\mathbf{L}(t)$ is a time-varying matrix of observer gains. Combination of Eq. (74) and Eqs. (4) and (5) from Chapter 17 yields a dynamic equation for the state estimation error $\mathbf{e}(t)$:

$$\dot{\mathbf{e}}(t) \overset{\Delta}{=} \dot{\mathbf{x}}(t) - \dot{\hat{\mathbf{x}}}(t) = [\mathbf{A}(t) - \mathbf{L}(t)\mathbf{C}(t)]\mathbf{e}(t) \tag{75}$$

with the initial condition

$$\mathbf{e}(t_0) = \mathbf{x}(t_0) - \hat{\mathbf{x}}(t_0) \tag{76}$$

Thus, if the observer is asymptotically stable

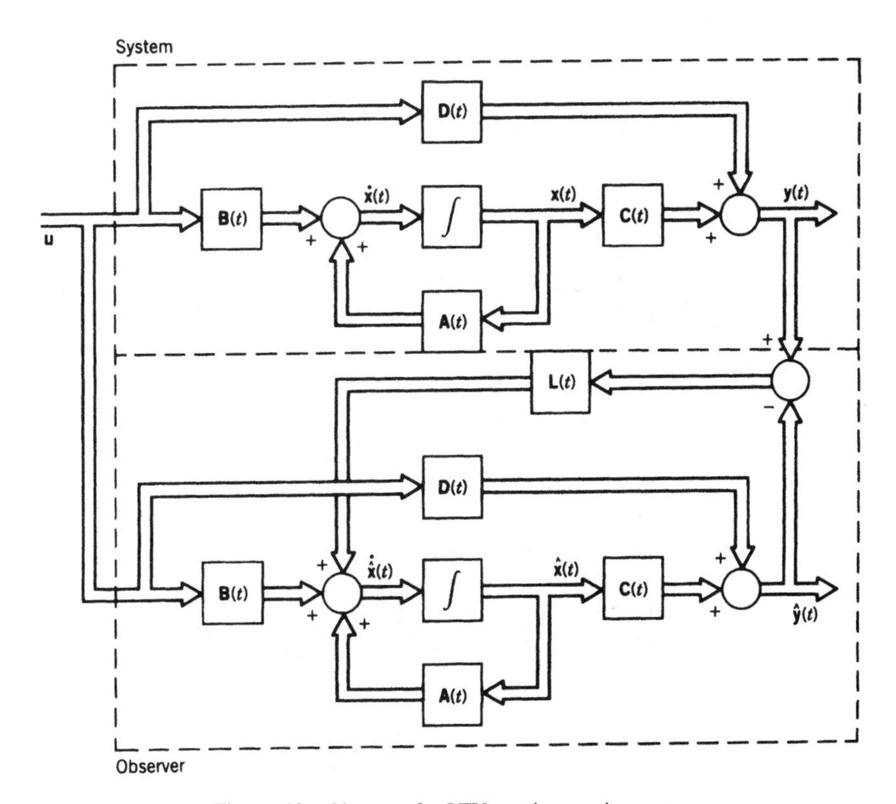

Figure 10 Observer for LTV continuous-time system.

$$\lim_{t \to \infty} \mathbf{e}(t) = 0 \tag{77}$$

for all $\mathbf{e}(t_0)$. If the system under consideration is time invariant, the eigenvalues of the matrix $\mathbf{A} - \mathbf{LC}$ govern the transient behavior of the estimation error. These eigenvalues, referred to as observer poles, can be arbitrarily located in the complex plane by appropriate choice of the constant observer gain matrix $\mathbf{L}$ if and only if the system given by Eqs. (4) and (5) in Chapter 17 is completely reconstructible.[28] For LTI systems, reconstructibility is exactly equivalent to observability. Also, if the $\mathbf{L}$ matrix is to be real, the complex eigenvalues of $\mathbf{A} - \mathbf{LC}$ should be specified as conjugate pairs.

The discrete-time version of the observer is given below.[3] The structure of the observer to estimate the state $\mathbf{x}(k)$ of a linear, time-varying system described by Eqs. (10) and (11) in Chapter 17 is given by (Fig. 11)

$$\hat{\mathbf{x}}(k+1) = \mathbf{F}(k)\,\hat{\mathbf{x}}(k) + \mathbf{G}(k)\mathbf{u}(k) + \mathbf{L}(k)[\mathbf{y}(k) - \mathbf{C}(k)\,\hat{\mathbf{x}}(k) - \mathbf{D}(k)\mathbf{u}(k)] \tag{78}$$

The resulting equation for the state estimation error $\mathbf{e}(k)$ is

$$\mathbf{e}(k+1) \overset{\Delta}{=} \mathbf{x}(k+1) - \hat{\mathbf{x}}(k+1) = [\mathbf{F}(k) - \mathbf{L}(k)\mathbf{C}(k)]\mathbf{e}(k) \tag{79}$$

with the initial condition

$$\mathbf{e}(k_0) = \mathbf{x}(k_0) - \hat{\mathbf{x}}(k_0) \tag{80}$$

If the observer is asymptotically stable

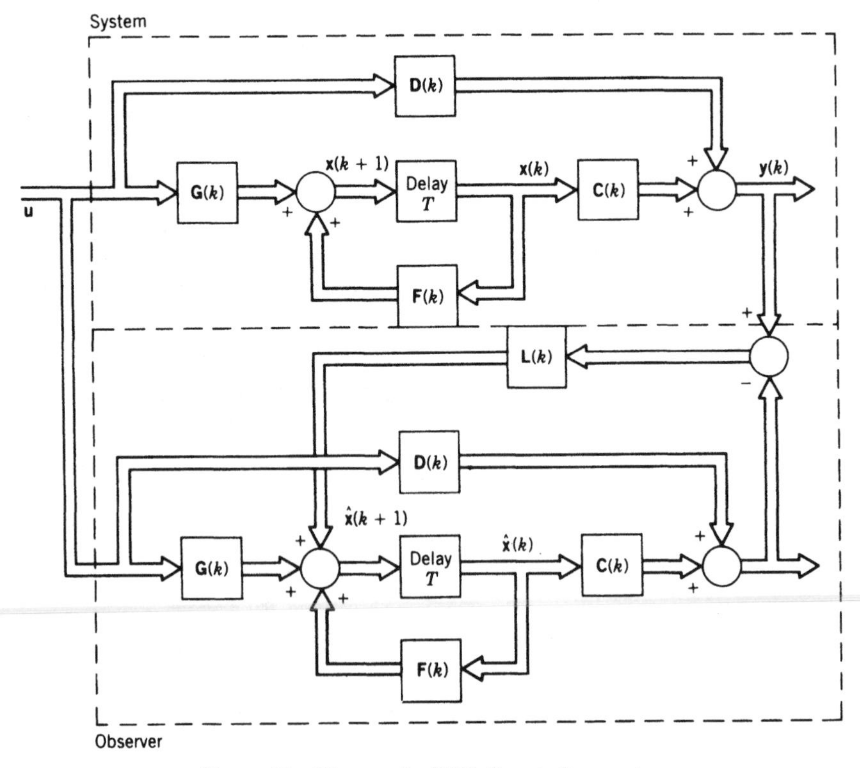

Figure 11 Observer for LTV discrete-time system.

$$\lim_{k \to \infty} \mathbf{e}(k) = 0 \tag{81}$$

for all $\mathbf{e}(k_0)$. If the system under consideration is time invariant also, the eigenvalues of the matrix $\mathbf{F} - \mathbf{LC}$ govern the estimation error transients. Observability of the system given by Eqs. (12) and (13) in Chapter 17 is equivalent to arbitrary pole assignability for the observer.

For single-output systems, specification of the desired observer pole locations uniquely specifies the gain vector $\mathbf{L}$. Formulas for the gain vector $\mathbf{L}$, applicable to both continuous-time and discrete-time systems, will be given here. They are obtained by analogy with Eqs. (2)–(5). The object is to get the observer gain vector $\mathbf{L}$ so that the characteristic equation governing the estimation error transients has specified coefficients:

$$\begin{aligned}\beta_e(s) &= \det(s\mathbf{I} - \mathbf{A} + \mathbf{LC}) \\ &= \det(s\mathbf{I} - \mathbf{A}^{\mathrm{T}} + \mathbf{C}^{\mathrm{T}}\mathbf{L}^{\mathrm{T}}) \\ &= s^n + \sum_{i=0}^{n-1} \beta_{ie} s^i = 0\end{aligned} \tag{82}$$

where the matrix $\mathbf{A} - \mathbf{LC}$ is transposed so that the observer gain selection problem can be reduced to a form similar to the state feedback controller gain selection problem. Comparison of Eqs. (82) and (5) suggests that the following substitutions are necessary to convert algorithms for state feedback controller gain selection to observer gain selection.

CONTROLLER GAIN SELECTION		OBSERVER GAIN SELECTION
$\mathbf{A}$	$\longrightarrow$	$\mathbf{A}^{\mathrm{T}}$
$\mathbf{B}$	$\longrightarrow$	$\mathbf{C}^{\mathrm{T}}$
$\mathbf{K}$	$\longrightarrow$	$\mathbf{L}^{\mathrm{T}}$

(83)

Similarly, for discrete-time systems, we get the following transformations:

CONTROLLER GAIN SELECTION		OBSERVER GAIN SELECTION
$\mathbf{F}$	$\longrightarrow$	$\mathbf{F}^{\mathrm{T}}$
$\mathbf{G}$	$\longrightarrow$	$\mathbf{C}^{\mathrm{T}}$
$\mathbf{K}$	$\longrightarrow$	$\mathbf{L}^{\mathrm{T}}$

(84)

Using the preceding transformations, we get

$$\mathbf{L} = \beta_e(\mathbf{A}) \begin{bmatrix} \mathbf{C} \\ \mathbf{CA} \\ \vdots \\ \mathbf{CA}^{n-1} \end{bmatrix}^{-1} \begin{bmatrix} 0 \\ 0 \\ \vdots \\ 1 \end{bmatrix} \tag{85}$$

for continuous-time systems and

$$\mathbf{L} = \beta_e(\mathbf{F}) \begin{bmatrix} \mathbf{C} \\ \mathbf{CF} \\ \vdots \\ \mathbf{CF}^{n-1} \end{bmatrix}^{-1} \begin{bmatrix} 0 \\ 0 \\ \vdots \\ 1 \end{bmatrix} \tag{86}$$

for discrete-time systems. In the preceding equations,

$$\beta_e(\mathbf{A}) = \mathbf{A}^n + \sum_{i=0}^{n-1} \beta_{ie} \mathbf{A}^i \tag{87}$$

for continuous-time systems and $\beta_e(\mathbf{F})$ is a similar function of $\mathbf{F}$ for discrete-time systems. The β_{ie}'s are the coefficients of the desired characteristic equation (82) for the observer. A similar equation describes the discrete-time observer characteristic equation. If CACSD packages[4–6] supporting only controller pole placement algorithms are available, the transformations of Eqs. (83) and (84) are needed to select observer gains using these algorithms.

The observers of Eqs. (74) and (78) are called full-order observers since their dimensions are the same as those of the systems whose states are being estimated. Reduction of the observer dimension can be achieved by using the fact that the output equation provides us with p linear equations in the unknown state, where p is the number of output variables. Therefore, the observer need only provide $n - p$ additional linear equations and thus need only be of dimension $n - p$. For time-invariant systems, the corresponding observer gain matrix can be chosen to place the $n - p$ observer poles at any desired locations in the complex plane if the original systems of Eqs. (6) and (7) or (12) and (13) in Chapter 17 are completely observable. Equations for reduced-order observers for continuous-time systems are given by Kwakernaak and Sivan[3] and for discrete-time systems by Franklin and Powell.[11] In the presence of measurement noise, the state estimates $\mathbf{x}(t)$ or $\mathbf{x}(k)$ obtained using reduced-order observers are more sensitive than those obtained using full-order observers.

The discrete-time observer of Eq. (78) is also referred to as a prediction estimator since $\mathbf{x}(k + 1)$ is ahead of the last measurements used, $\mathbf{y}(k)$ and $\mathbf{u}(k)$. A variation of this observer, called the current estimator, is useful if the computation time associated with the observer is very short compared to the sampling interval for the sampled-data system. The corresponding observer equation for an LTI system is[11]

$$\hat{\mathbf{x}}(k + 1) = \mathbf{F}\,\hat{\mathbf{x}}(k) + \mathbf{Gu}(k + \mathbf{L}[y(k + 1) - \mathbf{CF}\hat{\mathbf{x}}(k) - \mathbf{CBu}(k) - \mathbf{Du}(k + 1)] \tag{88}$$

The state estimate $\hat{\mathbf{x}}(k + 1)$ therefore depends on the current measurements $\mathbf{y}(k + 1)$ and $\mathbf{u}(k + 1)$. In practice, however, the measurements precede the estimate by a very small computation time. The observer gain formula for a single-output system is then given by

$$\mathbf{L} = \beta_e(\mathbf{F})\begin{bmatrix} \mathbf{CF} \\ \mathbf{CF}^2 \\ \vdots \\ \mathbf{CF}^n \end{bmatrix}^{-1}\begin{bmatrix} 0 \\ 0 \\ \vdots \\ 1 \end{bmatrix} \tag{89}$$

instead of Eq. (86). The term $\beta_e(\mathbf{F})$ has the same meaning as before.

For multioutput LTI systems, specification of the desired observer poles does not specify the gain matrix $\mathbf{L}$ uniquely. The additional freedom in the gain matrix selection can be used to assign the eigenvectors (or generalized eigenvectors) of the matrix $\mathbf{A} - \mathbf{LC}$ or $\mathbf{F} - \mathbf{LC}$ in addition to the eigenvalues. The corresponding procedure would be very similar to that used for gain matrix selection for multi-input systems and described in Section 2. The transformations of Eqs. (83) and (84) can be used to adapt the controller gain selection procedure to observer gain selection. Alternatively, the observer gains can be chosen to design observers whose state estimates have low sensitivity to unmeasured disturbance inputs.[29]

The observer designs described work well in the absence of significant levels of measurement noise or unknown disturbance signals. The sensitivity of the state estimates to measurement noise and unknown disturbance signals depends on the specified location of the observer poles. If these pole locations are too far into the left half of the complex s-plane or too close to the origin in the complex z-plane, the state estimates would be unduly sensitive to measurement noise and disturbance signals. In the limiting case, the observers can be shown to reduce to ideal differentiators or differencing devices. The observer pole locations should therefore be chosen to avoid such high sensitivities of the state estimates but at the same time ensure that the estimation error transients decay more rapidly than the state variables being estimated. Specification of the observer poles in practice involves considerable trial and error in much the same manner as specification of closed-loop poles does for state feedback controller design. If some information is available concerning the disturbance signals affecting the system and the measurement noise, the observer gain matrix selection problem can be formulated as an optimal-estimation problem.

5.2 The Optimal Observer

Optimization of observer design has been primarily performed assuming stochastic models for the disturbance inputs and measurement noise, though the effect of deterministic model errors on observer design is also important.[30] Consider the continuous-time system[3]

$$\dot{\mathbf{x}}(t) = \mathbf{A}(t)\mathbf{x}(t) + \mathbf{B}(t)\mathbf{u}(t) + \mathbf{S}(t)\mathbf{w}_1(t) \qquad t \geq t_0 \tag{90}$$

$$\mathbf{y}(t) = \mathbf{C}(t)\mathbf{x}(t) + \mathbf{w}_2(t) \qquad t \geq t_0 \tag{91}$$

where $\mathbf{w}_1(t)$ is the random disturbance input, $\mathbf{w}_2(t)$ is the random measurement noise, and their joint probabilities are assumed to be known. The column vector $[\mathbf{w}_1^T(t)\ \mathbf{w}_2^T(t)]^T$ is assumed to be a white-noise process with intensity

$$\mathbf{V}(t) = \begin{bmatrix} \mathbf{V}_1(t) & \mathbf{V}_{12}(t) \\ \mathbf{V}_{12}^T(t) & \mathbf{V}_2(t) \end{bmatrix} \tag{92}$$

that is, the expected value

$$E\left\{\begin{bmatrix} \mathbf{w}_1(t) \\ \mathbf{w}_2(t_1) \end{bmatrix}[\mathbf{w}_1^T(t_2) \quad \mathbf{w}_2^T(t_2)]\right\} = \mathbf{V}(t_1)\delta(t_1 - t_2) \tag{93}$$

where $\delta(t_1 - t_2)$ is the Dirac delta function.

The initial state $\mathbf{x}(t_0)$ is assumed to be a random variable uncorrelated with $\mathbf{w}_1$ and $\mathbf{w}_2$ and its probability given by

$$E[\mathbf{x}(t_0)] = \mathbf{x}_0 \quad \text{and} \quad E\{[\mathbf{x}(t_0) - \mathbf{x}_0][\mathbf{x}(t_0 - \mathbf{x}_0]^T\} = \mathbf{Q}_0 \tag{94}$$

The observer form is given by Eq. (74) and Fig. 10. The optimal-observer problem consists of determining $\mathbf{L}(\tau)$, $t_0 \leq \tau \leq t$, and the initial condition on the observer $\hat{\mathbf{x}}(t_0)$ so as to minimize the expected value $E\{[\mathbf{x}(t) - \hat{\mathbf{x}}(t)]^T\mathbf{W}(t)[\mathbf{x}(t) - \hat{\mathbf{x}}(t)]\}$, where $\mathbf{W}(t)$ is a symmetric positive-definite weighting matrix.

If the problem as stated is nonsingular

$$\det[\mathbf{V}_2(t)] > 0 \qquad t \geq t_0 \tag{95}$$

and if the disturbance and measurement noise are uncorrelated

$$\mathbf{V}_{12}(t) = \mathbf{0} \tag{96}$$

the optimal-observer gain matrix is given by Kalman and Bucy[31] as

$$\mathbf{L}(t) = \mathbf{Q}(t)\mathbf{C}^T(t)\mathbf{V}_2^{-1}(t) \qquad t \geq t_0 \tag{97}$$

where $\mathbf{Q}$ is a solution of the matrix Riccati equation:

$$\dot{\mathbf{Q}}(t) = \mathbf{A}(t)\mathbf{Q}(t) + \mathbf{Q}(t)\mathbf{A}^T(t) + \mathbf{S}(t)\mathbf{V}_1(t)\mathbf{S}^T(t) - \mathbf{Q}(t)\mathbf{C}^T(t)\mathbf{V}_2^{-1}(t)\mathbf{C}(t)\mathbf{Q}(t) \qquad t \geq t_0 \tag{98}$$

with the initial condition

$$\mathbf{Q}(t_0) = \mathbf{Q}_0 \tag{99}$$

and the observer initial condition

$$\hat{\mathbf{x}}(t_0) = \mathbf{x}_0 \tag{100}$$

The resulting state estimator is called the Kalman–Bucy filter. The similarity of Eqs. (97) and (98) to the corresponding Eqs. (25) and (26), respectively, for the LQR problem is a result of the duality of the state estimation and state feedback control problems noted earlier. One difference, however, is that the Riccati equation for the optimal observer can be implemented in real time since Eq. (99) is an initial condition for Eq. (98). In contrast, for the finite-time LQR problem, Eq. (27) gives the terminal condition for the Riccati equation (26).

The steady-state properties of the optimal observer for linear time-varying and time-invariant systems parallel those of the optimal controller for the LQR problem and are described by Kwakernaak and Sivan.[3] If the time-varying system

$$\dot{\mathbf{x}}(t) = \mathbf{A}(t)\mathbf{x}(t) + \mathbf{S}(t)\mathbf{w}_1(t) \tag{101}$$

$$\mathbf{y}(t) = \mathbf{C}(t)\mathbf{x}(t) + \mathbf{w}_2(t) \tag{102}$$

is uniformly completely controllable by $\mathbf{w}_1(t)$ and uniformly completely reconstructible, the solution $\mathbf{Q}(t)$ of Eq. (98) converges to a steady-state solution $\mathbf{Q}_s(t)$ as $t_0 \to -\infty$ for any positive-semidefinite $\mathbf{Q}_0$. The corresponding steady-state optimal observer

$$\dot{\hat{\mathbf{x}}}(t) = \mathbf{A}(t)\hat{\mathbf{x}}(t) + \mathbf{L}_s(t)[\mathbf{y}(t) - \mathbf{C}(t)\hat{\mathbf{x}}(t)] \tag{103}$$

where

$$\mathbf{L}_s(t) = \mathbf{Q}_s(t)\mathbf{C}^{\mathrm{T}}(t)\mathbf{V}_2^{-1}(t) \tag{104}$$

is exponentially stable. Also, if the system and the noise statistics are invariant, the matrix Riccati differential equation (98) becomes an algebraic equation as $t_0 \to -\infty$:

$$\mathbf{A}\mathbf{Q}_s + \mathbf{Q}_s\mathbf{A}^{\mathrm{T}} + \mathbf{S}\mathbf{V}_1\mathbf{S}^{\mathrm{T}} - \mathbf{Q}_s\mathbf{C}^{\mathrm{T}}\mathbf{V}_2^{-1}\mathbf{C}\mathbf{Q}_s = \mathbf{0} \tag{105}$$

If the corresponding time-invariant system is completely controllable by the input $\mathbf{w}_1(t)$ and completely observable, Eq. (105) has a unique positive-definite solution $\mathbf{Q}_s$ and the corresponding steady-state optimal observer of Eqs. (103) and (104) is asymptotically stable. Note that the measurable input $\mathbf{u}(t)$ has been omitted from Eqs. (101) and (102) for simplicity but does not change the substance of the results.

The discrete-time version of the optimal linear observer follows.[3] Consider the discrete-time system

$$\mathbf{x}(k+1) = \mathbf{F}(k)\mathbf{x}(k) + \mathbf{G}(k)\mathbf{u}(k) + \mathbf{S}(k)\mathbf{w}_1(k) \tag{106}$$

$$\mathbf{y}(k) = \mathbf{C}(k)\mathbf{x}(k) + \mathbf{D}(k)\mathbf{u}(k) + \mathbf{w}_2(k) \tag{107}$$

where $\mathbf{w}_1(k)$, $\mathbf{w}_2(k)$ are zero-mean, uncorrelated vector random variables representing disturbance and measurement noise, respectively. Their joint probabilities are assumed to be known. The column vector $[\mathbf{w}_1^{\mathrm{T}}(k)\ \mathbf{w}_2^{\mathrm{T}}(k)]^{\mathrm{T}}$ has the variance matrix

$$E\left\{\begin{bmatrix}\mathbf{w}_1(k)\\ \mathbf{w}_2(k)\end{bmatrix}[\mathbf{w}_1^{\mathrm{T}}(k)\quad \mathbf{w}_2^{\mathrm{T}}(k)]\right\} = \begin{bmatrix}\mathbf{V}_1(k) & \mathbf{V}_{12}(k)\\ \mathbf{V}_{12}^{\mathrm{T}}(k) & \mathbf{V}_2(k)\end{bmatrix} \tag{108}$$

The initial state $\mathbf{x}(k_0)$ is considered to be a random variable, uncorrelated with $\mathbf{w}_1$ and $\mathbf{w}_2$ with

$$E[\mathbf{x}(k_0)] = \mathbf{x}_0 \quad \text{and} \quad E\{(\mathbf{x}(k_0) - \mathbf{x}_0)(\mathbf{x}(k_0) - \mathbf{x}_0)^{\mathrm{T}}\} = \mathbf{Q}_0 \tag{109}$$

The observer form is given by Eq. (78) and Fig. 11. The optimal-observer problem consists of determining $\mathbf{L}(k_\tau)$, $k_0 \le k_\tau \le k$, and initial condition on the observer $\hat{\mathbf{x}}(k_0)$ so as to minimize the expected value $E\{[\mathbf{x}(k) - \hat{\mathbf{x}}(k)]\mathbf{W}(k)[\mathbf{x}(k) - \hat{\mathbf{x}}(k)^{\mathrm{T}}]\}$, where $\mathbf{W}(k)$ is a symmetric positive-definite weighting matrix.

If the problem as stated is nonsingular

$$\det[\mathbf{V}_2(k)] > 0 \qquad k \geq k_0 \tag{110}$$

the optimal-observer gain matrix is given by the recurrence relations

$$\mathbf{L}(k) = [\mathbf{F}(k)\mathbf{Q}(k)\mathbf{C}^{\mathrm{T}}(k) + \mathbf{V}_{12}(k)][\mathbf{V}_2(k) + \mathbf{C}(k)\mathbf{Q}(k)\mathbf{C}^{\mathrm{T}}(k)] \tag{111}$$

$$\mathbf{Q}(k+1) = [\mathbf{F}(k) - \mathbf{L}(k)\mathbf{C}(k)]\mathbf{Q}(k)\mathbf{F}^{\mathrm{T}}(k) + \mathbf{V}_1(k) - \mathbf{L}(k)\mathbf{V}_{12}^{\mathrm{T}}(k) \tag{112}$$

with the initial condition

$$\mathbf{Q}(k_0) = \mathbf{Q}_0 \tag{113}$$

and $k \geq k_0$. The initial condition on the observer state should be

$$\hat{\mathbf{x}}(k_0) = \mathbf{x}_0 \tag{114}$$

Again, the similarity of the optimal-observer equations (111)–(113) to the optimal-controller equations (33)–(35) for the LQR problem results from the duality of state estimation and state feedback control problems.

The similarity extends to the steady-state behavior of the optimal observer and the optimal controller.[3] If the time-varying system

$$\mathbf{x}(k+1) = \mathbf{F}(k)\mathbf{x}(k) + \mathbf{S}(k)\mathbf{w}_1(k) \tag{115}$$

$$\mathbf{y}(k) = \mathbf{C}(k)\mathbf{x}(k) + \mathbf{w}_2(k) \tag{116}$$

is uniformly completely controllable by $\mathbf{w}_1(k)$ and uniformly completely reconstructible, the solution $\mathbf{Q}(k)$ of Eqs. (111) and (112) converges to a steady-state solution $\mathbf{Q}_s(k)$ as $k_0 \to -\infty$ for any positive-semidefinite $\mathbf{Q}_0$. The corresponding steady-state observer

$$\hat{\mathbf{x}}(k+1) = \mathbf{F}(k)\,\hat{\mathbf{x}}(k) + \mathbf{L}_s(k)[\mathbf{y}(k) - \mathbf{C}(k)\,\hat{\mathbf{x}}(k)] \tag{117}$$

where $\mathbf{L}_s(k)$ is obtained using $\mathbf{Q}_s(k)$ for $\mathbf{Q}(k)$ in Eq. (111) is exponentially stable. Also, if the system and the noise statistics are time invariant, the matrix difference equations (111) and (112) become algebraic equations as $k_0 \to -\infty$. If the corresponding time-invariant system is completely controllable by the input $\mathbf{w}_1(k)$ and completely observable, the resulting steady-state optimal observer is asymptotically stable. Again, note that the measurable input $\mathbf{u}(k)$ has been omitted from Eqs. (115) and (116) for simplicity but does not change the substance of the results.

Interested readers are referred to Kwakernaak and Sivan[32] for a more complete consideration of the optimal observer. There is also extensive literature available on Kalman filters.[31,33,34]

6 OBSERVER-BASED CONTROLLERS

The observers described in the preceding section can be used to provide estimates of system state that, in turn, can be used to provide state feedback as described in Sections 2–4. The resulting controllers are dynamic compensators and are referred to as observer-based controllers.

The design of such observer-based controllers for LTI systems is simplified somewhat by the fact that their modes or eigenvalues satisfy the separation property, that is, the eigenvalues of the observer-controller are the same as the eigenvalues of the observer and the eigenvalues of the controller, the latter evaluated assuming perfect state measurement. For continuous-time LTI systems described by Eqs. (6) and (7) in Chapter 17, the control law given by Eq. (1), and the observer given by Eq. (74) with constant coefficient matrices, the observer-controller is given by Fig. 12 and the characteristic equation of the corresponding closed-loop system is[7]

$$\det(s\mathbf{I} - \mathbf{A} + \mathbf{BK})\det(s\mathbf{I} - \mathbf{A} + \mathbf{LC}) = 0 \tag{118}$$

A similar result can be shown to be true for discrete-time LTI systems.[11] The corresponding closed-loop system is shown in Fig. 13 and has the characteristic equation

$$\det(z\mathbf{I} - \mathbf{F} + \mathbf{GK})\det(z\mathbf{I} - \mathbf{F} + \mathbf{LC}) = 0 \tag{119}$$

For LTI systems subjected to unmeasured randomly varying disturbance inputs and measurement errors, if the statistics of these signals are known, state estimators of the Kalman–Bucy type will be used. The resulting estimator-based controllers have eigenvalues that also satisfy a separation property. As a result of the separation property for controllers based on observers or Kalman–Bucy filters, the design of the controllers can be treated independently of the observer.

The use of observers or Kalman filters to provide state estimates for state feedback controllers does, however, impair overall controller performance. For instance, the transient

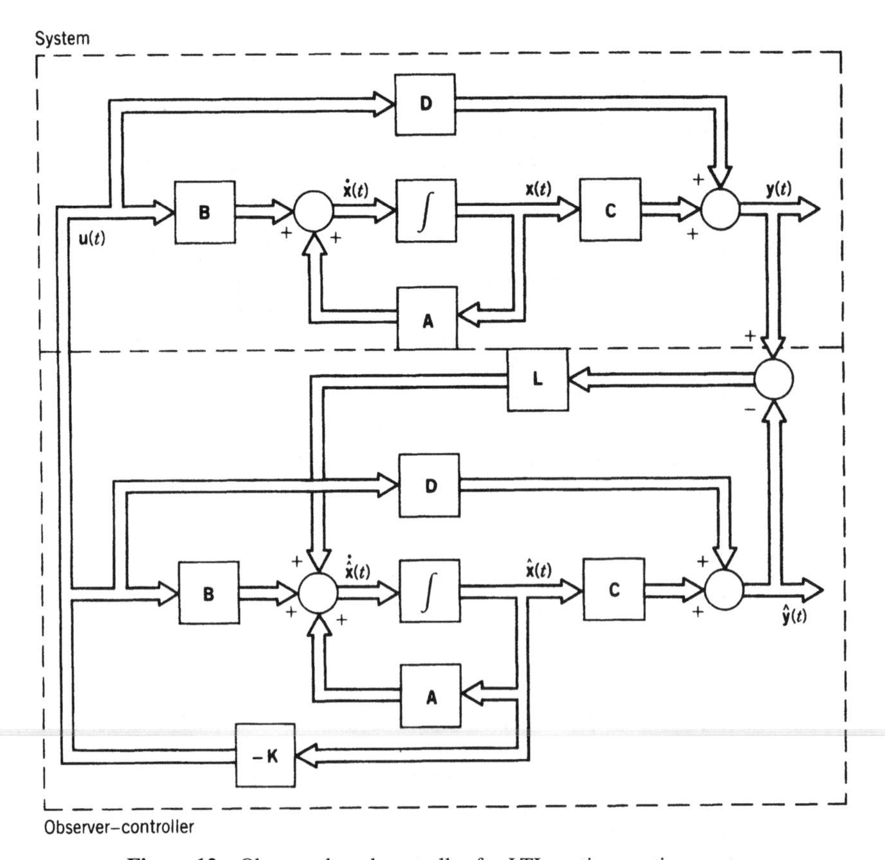

Figure 12 Observer-based controller for LTI continuous-time system.

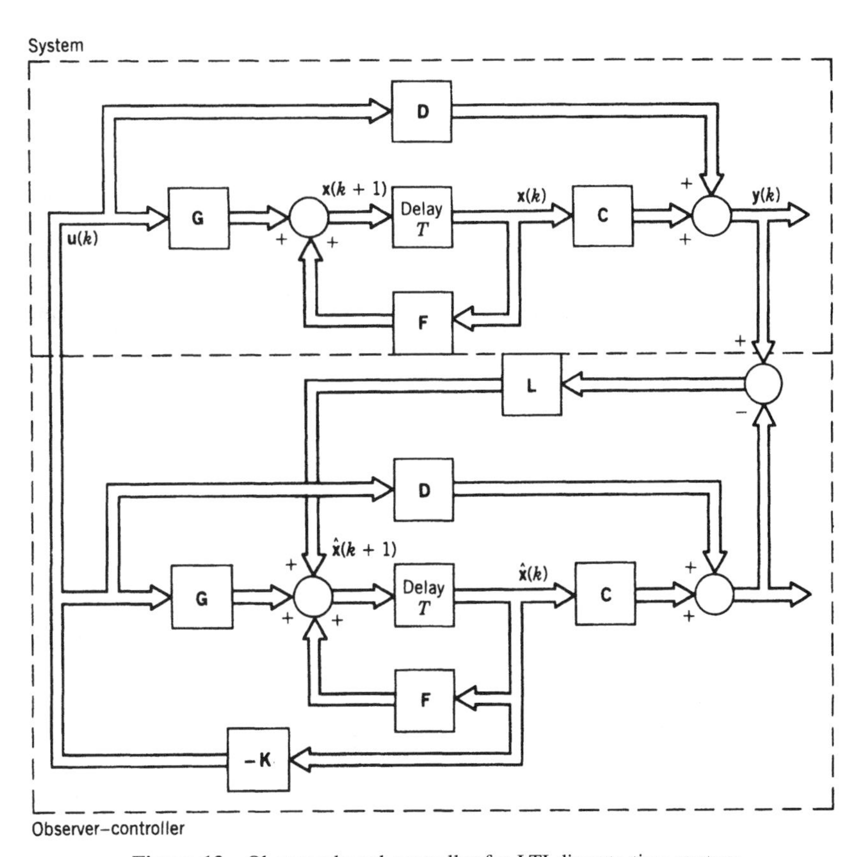

Figure 13 Observer-based controller for LTI discrete-time system.

response of such controllers is poorer than that of controllers using complete state feedback. Moreover, the properties relating to the gain and phase margins of optimal controllers for the LQR problem are not applicable if the controllers use estimated states for feedback rather than measured states.[13] There is a definite loss of controller robustness associated with the use of state estimators. The robustness of such controllers is more properly evaluated by considering them as dynamic compensators and using methods common to the frequency-domain approach.[35]

Observer-based controllers for LTI systems can naturally be examined using transfer function or frequency-domain-based methods. Such a linking of time-domain- and frequency-domain-based controller designs offers a number of advantages. The state-space-based design approach leads to consideration of controller structures that are not obvious from a transfer function approach. In addition, the state-space approach alerts the designer to the problem of loss of controllability or observability via pole–zero cancellation. On the other hand, the transfer function approach can result in controllers that cannot be obtained by using observer-based controllers.[7] In addition, the consideration of observer-based controllers from a transfer function perspective helps evaluate the controllers to see whether proven and practical design guidelines are violated. Such guidelines invariably use transfer function terminology since they have evolved from years of experience using classical control techniques. If such guidelines are violated, the state-space-based controller design procedure can be modified appropriately.[36]

A recent example of a MIMO controller design method that has evolved from a combination of frequency-domain methods and state-space methods is the linear-quadratic-Gaussian method with loop-transfer-recovery (LQG/LTR), developed by Athans.[37] The procedure relies on the fact that results and requirements relating to control system robustness to modeling errors are best presented in the frequency domain. The powerful controller and estimator structures resulting from LQR formulations of the control and state estimation problem are used. These structures are useful in this design method because their robustness and performance have been well studied, using frequency-domain measures.[14] The resulting method relies upon designer expertise in formulating good performance specifications at the outset. The design method therefore avoids the main weakness of state-space methods—namely, the weak connection between performance measures used by these methods and performance measures of engineering significance. The computation of the controller is, however, straightforward in this method, since the controller structures are derived from well-established state-space methods and are well supported by commercial CACSD packages.[4–6]

The first step in the design method is the definition of the design plant model. This model includes not only the nominal model of the system to be controlled but also scaling of the variables and augmentation of the dynamics, such as the inclusion of integrators dictated by control objectives. The number of control inputs r and the number of outputs p are assumed to be equal in the following development. The model is also linear and time invariant and strictly proper, that is, the transmission matrix $\mathbf{D}$ in the system equations (6) and (7) in Chapter 17 is zero. The transfer function matrix of the system is

$$\mathbf{H}(s) = \mathbf{C}(s\mathbf{I}_n - \mathbf{A})^{-1}\mathbf{B} \tag{120}$$

Modeling inaccuracy is treated as follows. The actual transfer function matrix is given by

$$\mathbf{H}_A(s) = [\mathbf{I}_n + \mathbf{E}(s)]\mathbf{H}(s) \tag{121}$$

where $\mathbf{E}(s)$ characterizes the modeling error. The maximum singular value $\sigma_{\max}$ of the matrix $\mathbf{E}(j\omega)$ is assumed to be bounded by a known bound $e_m(\omega)$:

$$\sigma_{\max}[\mathbf{E}(i\omega)] < e_m(\omega) \tag{122}$$

The second step in the design procedure is the specification of a target feedback loop that has satisfactory robustness, stability, and performance specifications. The target feedback loop is shown in Fig. 14. It is obviously not directly implementable since the control inputs $\mathbf{u}$ do not appear in the system. The matrix $\mathbf{L}$ is a constant matrix and is chosen as described later. It is the designer's task to experiment with different choices of $\mathbf{L}$ and evaluate whether

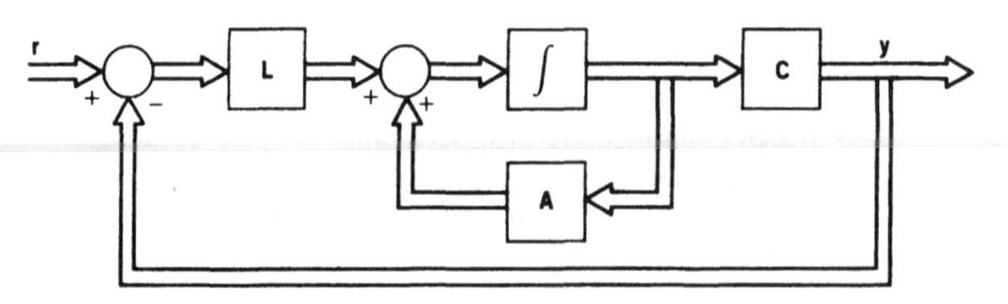

Figure 14 Target feedback loop block diagram.

the resulting system has satisfactory performance. This stage of the design therefore requires considerable trial and error.

Athans[36] has suggested using the steady-state Kalman–Bucy filter formulation with stationary system and noise characteristics. It should be noted, however, that the formulation is not used here to perform an estimation task. It is being used here to help calculate the matrix **L** because the resulting target feedback loop has well-known performance and robustness characteristics. At the minimum, selection of **L** as described by Athans[36] guarantees that the target feedback loop will not amplify disturbances entering the system at the output. In addition, the target feedback loop will not go unstable as long as the modeling uncertainty $e_m(\omega)$ in Eq. (122) is below 0.5.

Once the target feedback loop is chosen, the final step in the design procedure is the design of a compensator to enable the controlled system to approximate the behavior of the target feedback loop closely. Athans[36] has proposed the compensator structure shown in Fig. 15. The similarity of the controller to the observer-based controller in Fig. 12 is clear. The gain matrix **K** is computed via the solution of a version of the LQR problem. The loop transfer recovery (LTR) result, credited by Athans to other researchers, guarantees that for minimum-phase systems the procedure described yields a controlled system behavior that approximates the target feedback loop behavior as closely as desired.

For non-minimum-phase systems, the design procedure remains the same. The only difference is that the final design may not approximate the behavior of the target feedback loop closely. In effect, this would result in additional design iterations to arrive at a satisfactory final design.

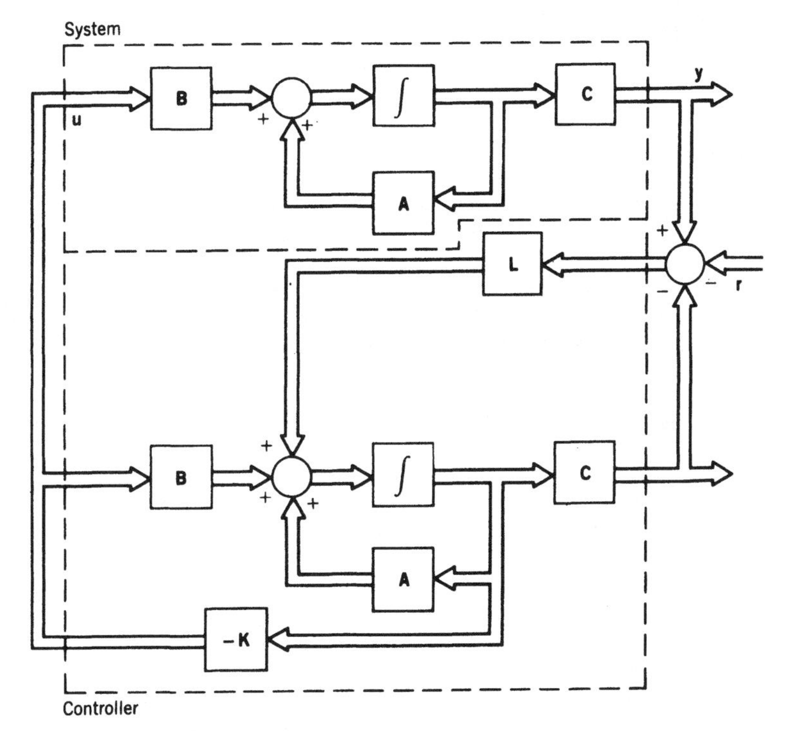

Figure 15 Compensator structure for LQR/LTR method.

The LQG/LTR design procedure has been applied successfully to evaluate the feasibility of MIMO control for aircraft and helicopter flight control, jet engine control, and submersible control.

7 CONCLUSION

The primary emphasis in this chapter on linear time-invariant finite-dimensional systems is a reflection of the state of the literature on the subject and the practice of the art. The reader is referred to the following works for more exhaustive treatment of some of the topics not covered at great length here. The subject of multivariable control using state-space methods has been addressed at much greater length by Kailath[7] and others.[38] The application of state-space methods to nonlinear system analysis and control is treated at some length by Ramnath et al.[32] Optimal-control problems other than the LQR formulation have been described in detail in a number of textbooks.[39,40] The subject of adaptive control refers to control situations where the controller parameters are adapted or adjusted as the behavior of the system being controlled changes. One approach to adaptive control, termed the Model Reference Approach and employing state-space description, has been described at length by Landau.[41] Distributed-parameter systems are examples of systems with infinite-dimensional states. Application of state-space methods to these systems has been described by Tzafestas et al.[42] Time-delayed systems are also examples of systems with infinite-dimensional states. The analysis and control of such systems and of many of the other types of systems referred to in this section remains a subject of current research. For current research results in these areas, the reader is referred to journals such as the *ASME Journal of Dynamic Systems, Measurement and Control, IEEE Transactions on Automatic Control, AIAA Journal of Guidance, Control and Dynamics, SIAM Journal on Control, and Automatica, The Journal of the International Federation of Automatic Control.*

REFERENCES

1. I. M. Horowitz, *Synthesis of Feedback Control Systems,* Academic Press, New York, 1963.
2. A. G. J. MacFarlane, *Frequency Response Methods in Control Systems,* IEEE Reprint Series, IEEE, New York, 1979.
3. H. Kwakernaak and R. Sivan, *Linear Optimal Control Systems,* Wiley-Interscience, New York, 1972.
4. Anonymous, *CTRL-C, A Language for the Computer-Aided Design of Multivariable Control Systems,* Systems Control Technology, Palo Alto, CA, 1983.
5. R. Walker, C. Gregory, Jr., and S. Shah, "MATRIX$_x$: A Data Analysis, System Identification, Control Design and Simulation Package," *IEEE Control Systems Magazine,* December 1982, pp. 30–36.
6. K. J. Astrom, "Computer Aided Modeling, Analysis and Design of Control Systems—A Perspective," *IEEE Control Systems Magazine,* May 1983, pp. 4–16.
7. T. Kailath, *Linear Systems,* Prentice-Hall, Englewood Cliffs, NJ, 1980.
8. W. L. Brogan, *Modern Control Theory,* Prentice-Hall, Englewood Cliffs, NJ, 1982.
9. D. M. Wiberg, *State Space and Linear Systems,* Schaum's Outline Series, McGraw-Hill, New York, 1971.
10. B. D. O. Anderson and J. B. Moore, *Linear Optimal Control,* Prentice-Hall, Englewood Cliffs, NJ, 1971.
11. G. F. Franklin and J. D. Powell, *Digital Control of Dynamic Systems,* Addison-Wesley, Reading, MA, 1980.
12. B. C. Kuo, *Digital Control Systems,* SRL Publishing, Champaign, IL, 1977.
13. M. G. Safonov, *Stability and Robustness of Multivariable Feedback Systems,* MIT Press, Cambridge, MA, 1980.

14. N. A. Lehtomaki, N. R. Sandell Jr., and M. Althaus, "Robustness Results in Linear-Quadratic-Gaussian Based Multivariable Control Designs," *IEEE Transactions on Automatic Control* **AC-26**(1), 75–93 (1981).

15. W. R. Perkins and J. B. Cruz, Jr., "Feedback Properties of Linear Regulators," *IEEE Transactions on Automatic Control* **AC-16**(6), 659–664 (1971).

16. I. M. Horowitz and U. Shaked, "Superiority of Transfer Function over State-Variable Methods in Linear Time-Invariant Feedback System Design," *IEEE Transactions on Automatic Control* **AC-20**(1), 84–97 (1975).

17. H. H. Rosenbrock and P. D. McMorran, "Good, Bad or Optimal?" *IEEE Transactions on Automatic Control* **AC-16**(6), 552–554 (1971).

18. J. S. Tyler, Jr., "The Characteristics of Model Following Systems as Synthesized by Optimal Control," *IEEE Transactions on Automatic Control* **AC-9**(5), 485–498 (1964).

19. C. D. Johnson, "Theory of Disturbance-Accommodating Controllers," in *Control and Dynamic Systems, Advances in Theory and Applications,* Vol. 12, C. T. Leondes (ed.), Academic Press, New York, 1976, pp. 387–489.

20. C. D. Johnson, "A Discrete-Time Disturbance-Accommodating Control Theory for Digital Control of Dynamical Systems," in *Control and Dynamic Systems, Advances in Theory and Applications,* Vol. 18, C. T. Leondes (ed.), Academic Press, New York, 1982, pp. 223–315.

21. E. J. Davison and I. J. Ferguson, "The Design of Controllers for the Multivariable Robust Servomechanism Problem Using Parameter Optimization Methods," *IEEE Transactions on Automatic Control* **AC-26**(1), 93–110 (1981).

22. 1. L. Trankle and A. E. Bryson, Jr., "Control Logic to Track Outputs of a Command Generator," *AIAA Journal of Guidance and Control* **1**(2), 130–135 (1978).

23. M. Tomizuka, "Optimal Continuous Finite Preview Problem," *IEEE Transactions on Automatic Control* **AC-20**(3), 362–365 (1975).

24. M. Tomizuka and D. E. Whitney, "Optimal Finite Preview Problems (Why and How Is Future Information Important?)" *ASME Transactions, Journal of Dynamic Systems, Measurement and Control* **97**(4), 319–325 (1975).

25. M. Tomizuka and D. E. Rosenthal, "On the Optimal Digital State Vector Feedback Controller with Integral and Preview Actions," *ASME Transactions, Journal of Dynamic Systems, Measurement and Control* **101**(2), 172–178 (1979).

26. M. Tomizuka and D. H. Fung, "Design of Digital Feedforward/Preview Controllers for Processes with Predetermined Feedback Controllers," *ASME Transactions, Journal of Dynamic Systems, Measurement and Control* **102**(4), 218–225 (1980).

27. N. K. Gupta, "Frequency Shaped Cost Functionals: Extension of Linear-Quadratic Gaussian Design Methods," *AIAA Journal of Guidance and Control* **3**(6), 529–535 (1980).

28. D. G. Luenberger, "Observing the State of a Linear System," *IEEE Transactions on Military Electronics* **8**, 74–80 (1964).

29. S. L. Shah, D. E. Seborg, and D. G. Fisher, "Design and Application of Controllers and Observers for Disturbance Minimization and Pole Assignment," *ASME Transactions, Journal of Dynamic Systems, Measurement and Control* **102**(1), 21–27 (1980).

30. F. E. Thau and A. Kestenbaum, "The Effect of Modeling Errors on Linear State Reconstructors and Regulators," *ASME Transactions, Journal of Dynamic Systems, Measurement and Control* **46**(4), 454–459 (1974).

31. R. E. Kalman and R. J. Bucy, "New Results in Linear Filtering and Prediction Theory," *ASME Transactions, Journal of Basic Engineering, Series D* **83**, 45–108 (1961).

32. R. V. Ramnath and H. M. Paynter (eds.), *Nonlinear System Analysis and Synthesis:* Vol. 2—*Techniques and Applications,* Workshop/Tutorial Session at the Winter Annual Meeting of ASME, New York, December 1980.

33. J. M. Mendel and D. L. Gieseking, "Bibliography on the Linear-Quadratic-Gaussian Problem," *IEEE Transactions on Automatic Control* **AC-16**(6), 847–869 (1971).

34. B. O. Anderson and J. B. Moore, *Optimal Filtering,* Prentice-Hall, Englewood-Cliffs, NJ, 1979.

35. J. C. Doyle and G. Stein, "Robustness with Observers," *IEEE Transactions on Automatic Control* **AC-24**(4), 607–611 (1979).

36. A. E. Bryson, Jr., "Some Connections between Modern and Classical Control Concepts," *ASME Transactions, Journal of Dynamic Systems, Measurement and Control* **101**(3), 91–98 (1979).

37. M. Athans, "A Tutorial on the LQG/LTR Method," in *Proceedings of the 1986 American Control Conference,* Seattle, WA, June 1986, pp. 1289–1296.
38. M. Sain (ed.), "Special Issue on Linear Multivariable Control," *IEEE Transactions on Automatic Control* **AC-26**(6), 1–295 (1981).
39. M. Athans and P. Falb, *Optimal Control,* McGraw-Hill, New York, 1966.
40. A. P. Sage, *Optimum Systems Control,* Prentice-Hall, Englewood Cliffs, NJ, 1968.
41. Y. D. Landau, *Adaptive Control. The Model Reference Approach,* Marcel-Dekker, New York, 1979.
42. S. G. Tzafestas (ed.), *Distributed Parameter Control Systems, Theory and Application,* Vol. 6, International Series on Systems and Control, Pergamon Press, Oxford, England, 1982.

CHAPTER 19

NEURAL NETWORKS IN FEEDBACK CONTROL SYSTEMS

F. L. Lewis
Automation and Robotics Research Institute
University of Texas at Arlington
Fort Worth, Texas

Shuzhi Sam Ge
Department of Electrical and Computer Engineering
National University of Singapore
Singapore

1 INTRODUCTION

Dynamical systems are ubiquitous in nature and include naturally occurring systems such as the cell and more complex biological organisms, the interactions of populations, and so on, as well as man-made systems such as aircraft, satellites, and interacting global economies. Von Bertalanffy[1] were among the first to provide a modern theory of systems at the beginning of the century. Systems are characterized as having outputs that can be measured, inputs that can be manipulated, and internal dynamics. *Feedback control* involves computing suitable control inputs, based on the difference between observed and desired behavior, for a dynamical system such that the observed behavior coincides with a desired behavior prescribed by the user. All biological systems are based on feedback for survival, with even the simplest of cells using chemical diffusion based on feedback to create a potential difference across the membrane to maintain its *homeostasis,* or required equilibrium condition for survival. Volterra was the first to show that feedback is responsible for the balance of two populations of fish in a pond, and Darwin showed that feedback over extended time periods provides the subtle pressures that cause the evolution of species.

There is a large and well-established body of design and analysis techniques for feedback control systems which has been responsible for successes in the industrial revolution, ship and aircraft design, and the space age. Design approaches include classical design methods for linear systems, multivariable control, nonlinear control, optimal control, robust control, H_∞ control, adaptive control, and others. Many systems one desires to control have unknown dynamics, modeling errors, and various sorts of disturbances, uncertainties, and noise. This, coupled with the increasing complexity of today's dynamical systems, creates a need for advanced control design techniques that overcome limitations on traditional feedback control techniques.

In recent years, there has been a great deal of effort to design feedback control systems that mimic the functions of living biological systems. There has been great interest recently in "universal model-free controllers" that do not need a mathematical model of the controlled plant but mimic the functions of biological processes to learn about the systems they are controlling online, so that performance improves automatically. Techniques include fuzzy logic control, which mimics linguistic and reasoning functions, and artificial neural networks (NNs), which are based on biological neuronal structures of interconnected nodes, as shown in Fig. 1. By now, the theory and applications of these nonlinear network structures in feedback control have been well documented. It is generally understood that NNs provide an elegant extension of adaptive control techniques to nonlinearly parameterized learning systems.

This chapter shows how NNs fulfill the promise of providing *model-free learning controllers* for a class of nonlinear systems, in the sense that a structural or parameterized model of the system dynamics is not needed. The control structures discussed are *multiloop controllers* with NNs in some of the loops and an outer tracking unity-gain feedback loop. Throughout, there are repeatable design algorithms and guarantees of system performance, including both small tracking errors and bounded NN weights. It is shown that as uncertainty about the controlled system increases or as one desires to consider human user inputs at higher levels of abstraction, the NN controllers acquire more and more structure, eventually acquiring a hierarchical structure that resembles some of the elegant architectures proposed by computer science engineers using high-level design approaches based on cognitive linguistics, reinforcement learning, psychological theories, adaptive critics, or optimal dynamic programming techniques.

Many researchers have contributed to the development of a firm foundation for analysis and design of NNs in control system applications. See Section 11 on historical development and further study.

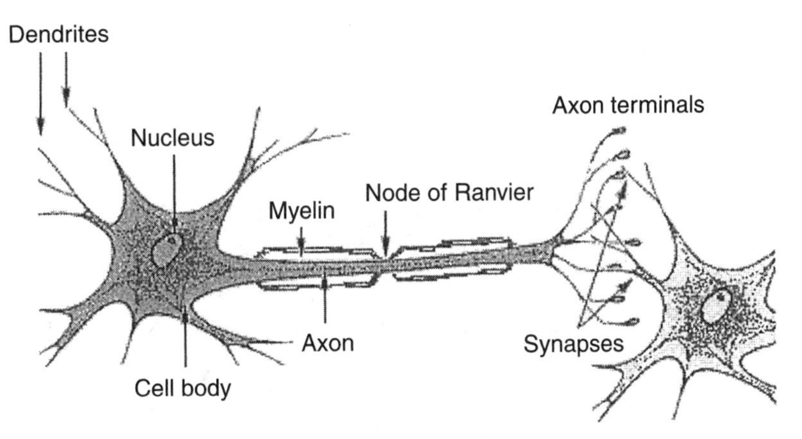

Figure 1 Nervous system cell. (With permission from http://www.sirinet.net/~jgjohnso/index.html.)

2 BACKGROUND

2.1 Neural Networks

The multilayer NN is modeled based on the structure of biological nervous systems (see Fig. 1) and provides a nonlinear mapping from an input space $\Re^n$ into an output space $\Re^m$. Its properties include function approximation, learning, generalization, classification, and so on. It is known that the two-layer NN has sufficient generality for closed-loop control purposes. The two-layer NN shown in Fig. 2 consists of two layers of weights and thresholds and has a hidden layer and an output layer. The input function $x(t)$ has n components, the hidden layer has L neurons, and the output layer has m neurons.

One may describe the NN mathematically as

$$y = W^{\mathrm{T}}\sigma(V^{\mathrm{T}}x)$$

where V is a matrix of first-layer weights and W is a matrix of second-layer weights. The second-layer thresholds are included as the first column of the matrix W^{T} by augmenting the vector activation function $\sigma(\cdot)$ by 1 in the first position. Similarly, the first-layer thresholds are included as the first column of the matrix V^{T} by augmenting vector x by 1 in the first position.

The main property of NNs we are concerned with for control and estimation purposes is the *function approximation property.*[2,3] Let $f(x)$ be a smooth function from $\Re^n \rightarrow \Re^m$. Then, it can be shown that if the activation functions are suitably selected and is restricted to a compact set $S \in \Re^n$, then for some sufficiently large number L of hidden-layer neurons, there exist weights and thresholds such that one has

$$f(x) = W^{\mathrm{T}}\sigma(V^{\mathrm{T}}x) + \varepsilon(x)$$

with $\varepsilon(x)$ suitably small. Here, $\varepsilon(x)$ is called *the neural network functional approximation error*. In fact, for any choice of a positive number ε_N, one can find a NN of large enough size L such that $\varepsilon(x) \leq \varepsilon_N$ for all $x \in S$.

Finding a suitable NN for approximation involves adjusting the parameters V and W to obtain a good fit to $f(x)$. Note that tuning of the weights includes tuning of the thresholds as well. The neural net is *nonlinear in the parameters V,* which makes adjustment of these parameters difficult and was initially one of the major hurdles to be overcome in closed-

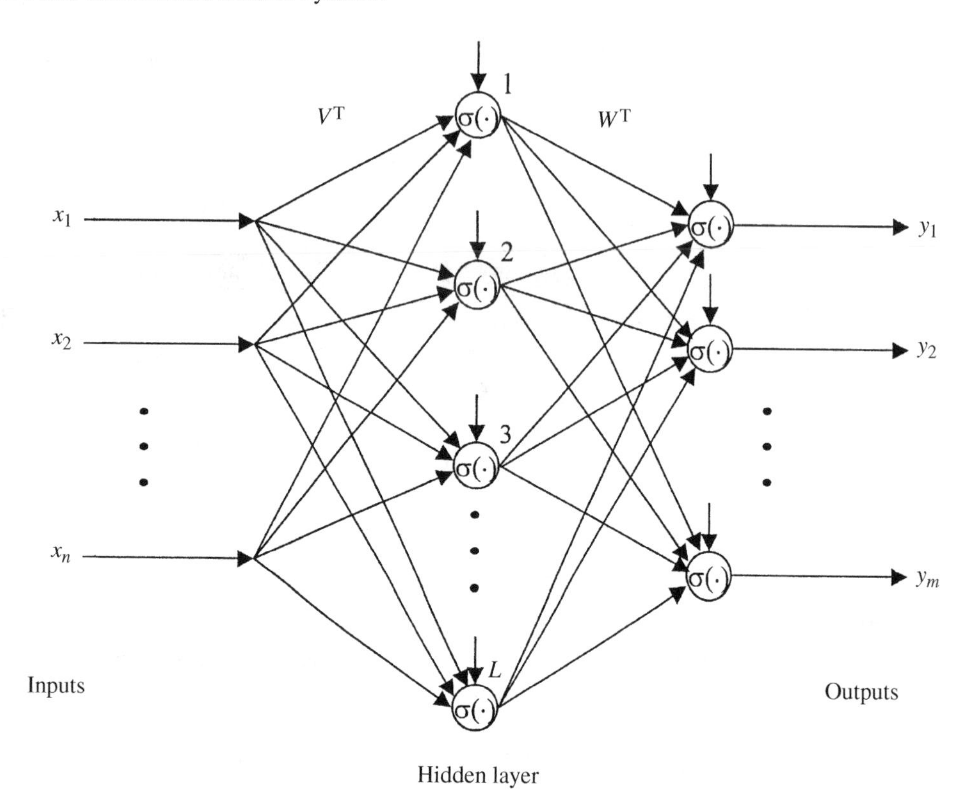

Figure 2 Two-layer NN.

loop feedback control applications. If the first-layer weights V are fixed, then the NN is linear in the adjustable parameters W (LIP). It has been shown that, if the first-layer weights V are suitably fixed, then the approximation property can be satisfied by selecting only the output weights W for good approximation. For this to occur, $\sigma(V^T x)$ must provide a *basis*. It is not always straightforward to pick a basis $\sigma(V^T x)$. It has been shown that the cerebellar model articulation controller (CMAC),[4] radial basis function (RBF),[5] fuzzy logic,[6] and other structured NN approaches allow one to choose a basis by suitably partitioning the compact set S. However, this can be tedious. If one selects the activation functions suitably (e.g., as sigmoids), then it was shown in Ref. 7 that $\sigma(V^T x)$ is almost always a basis if is selected randomly.

2.2 NN Control Topologies

Feedback control involves the measurement of output signals from a dynamical system or *plant* and the use of the *difference* between the measured values and certain prescribed *desired values* to compute system inputs that cause the measured values to follow, or *track*, the desired values. In feedback control design it is crucial to guarantee by rigorous means both the tracking performance and the internal stability or boundedness of all variables. Failure to do so can cause serious problems in the closed-loop system, including instability and unboundedness of signals that can result in system failure or destruction.

The use of NNs in control systems was first proposed by Werbos[8] and Narendra and Parthasarathy.[9] NN control has had two major thrusts: approximate dynamic programming, which uses NNs to approximately solve the optimal control problem, and NNs in closed-loop feedback control. Many researchers have contributed to the development of these fields. See Section 11 and the References and Bibliography.

Several NN feedback control topologies are illustrated in Fig. 3,[10] some of which are derived from standard topologies in adaptive control.[11] Solid lines denote control signal flow loops while dashed lines denote tuning loops. There are basically two sorts of feedback control topologies: indirect and direct techniques. In *indirect* NN control there are two functions; in an identifier block, the NN is tuned to learn the dynamics of the unknown plant, and the controller block then uses this information to control the plant. *Direct* control is more efficient and involves directly tuning the parameters of an adjustable NN controller.

The challenge in using NNs for feedback control purposes is to select a suitable control system structure and then to demonstrate using mathematically acceptable techniques how the NN weights can be tuned so that closed-loop stability and performance are guaranteed. In this chapter, we shall show different methods of NN controller design that yield guaranteed performance for systems of different structure and complexity. Many researchers have participated in the development of the theoretical foundation for NNs in control applications. See Section 11.

3 FEEDBACK LINEARIZATION DESIGN OF NN TRACKING CONTROLLERS

In this section, the objective is to design an NN feedback controller that causes a robotic system to follow, or track, a prescribed trajectory or path. The dynamics of the robot are unknown, and there are unknown disturbances. The dynamics of an n-link robot manipulator may be expressed as[12]

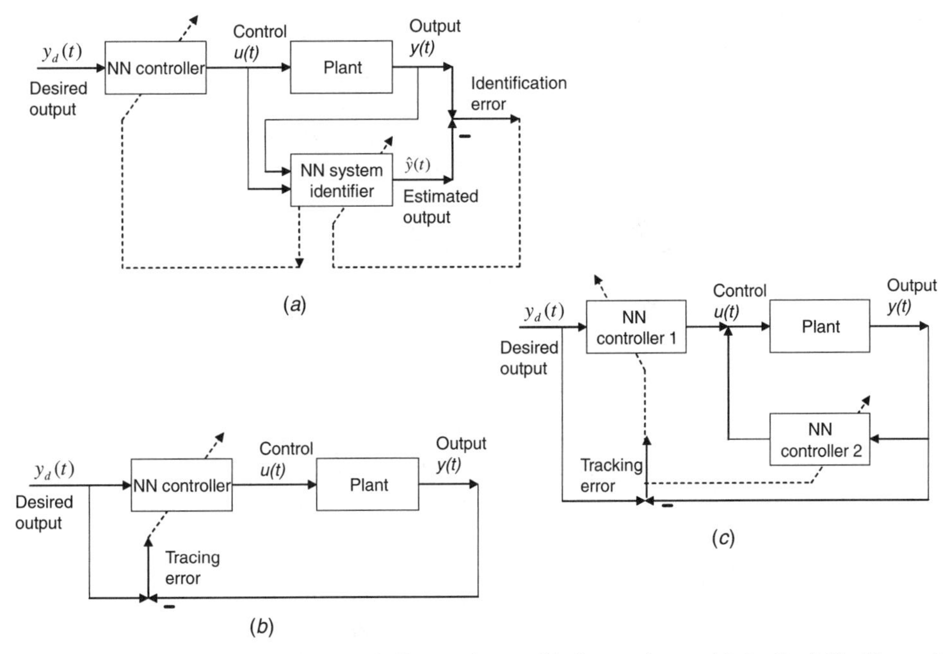

Figure 3 NN control topologies: (*a*) indirect scheme; (*b*) direct scheme; (*c*) feedback/feedforward scheme.

$$M(q)\ddot{q} + V_m(q,\dot{q})(\dot{q} + G(q) + F(\dot{q}) + \tau_d = \tau \tag{1}$$

with $q(t) \in R^n$ the joint variable vector, $M(q)$ an inertia matrix, V_m a centripetal/Coriolis matrix, $G(q)$ a gravity vector, and $F(\cdot)$ representing friction terms. Bounded unknown disturbances and modeling errors are denoted by τ_d and the control input torque is $\tau(t)$.

The sliding-mode control approach of Slotine[13,14] can be generalized to NN control systems. Given a desired arm trajectory $q_d(t) \in R^n$, define the tracking error $e(t) = q_d(t) - q(t)$ and the sliding variable error $r = \dot{e} + \Lambda e$, where $\Lambda = \Lambda^T > 0$. A *sliding-mode manifold* is defined by $r(t) = 0$. The NN tracking controller is designed using a feedback linearization approach to guarantee that $r(t)$ is forced into a neighborhood of this manifold. Define the nonlinear robot function

$$f(x) = M(q)(\ddot{q}_d + \Lambda\dot{e}) + V_m(q,\dot{q})(\dot{q}_d + \Lambda e) + G(q) + F(\dot{q}) \tag{2}$$

with the known vector $x(t)$ of measured signals suitably defined in terms of $e(t)$, $q_d(t)$. The NN input vector x can be selected, for instance, as

$$x = [e^T \quad \dot{e}^T \quad q_d^T \quad \dot{q}_d^T \quad \ddot{q}_d^T]^T \tag{3}$$

3.1 Multilayer NN Controller

A NN controller may be designed based on the *functional approximation properties* of NNs, as shown in Ref. 15. Thus, assume that $f(x)$ is unknown and given approximately as the output of a NN with unknown "ideal" weights W, V so that $f(x) = W^T\sigma(V^Tx) + \varepsilon$ with ε an approximation error. The key is now to approximate $f(x)$ by the NN functional estimate $\hat{f}(x) = \hat{W}^T\sigma(\hat{V}^Tx)$, with $\hat{V}$, $\hat{W}$ the current (estimated) NN weights as provided by the tuning algorithms. This is *nonlinear in the tunable parameters* $\hat{V}$. Standard adaptive control approaches only allow LIP controllers.

Now select the control input

$$\tau = \hat{W}^T\sigma(\hat{V}^Tx) + K_v r - v \tag{4}$$

with K_v a symmetric positive-definite (PD) gain and $v(t)$ a certain robustifying function detailed in Ref. 15. This NN control structure is shown in Fig. 4. The outer PD tracking loop guarantees robust behavior. The inner loop containing the NN is known as a *feedback linearization loop,*[16] and the NN effectively learns the unknown dynamics online to cancel the nonlinearities of the system.

Let the estimated sigmoid Jacobian be $\hat{\sigma}' \equiv d\sigma(z)/dz|_{z=\hat{V}^Tx}$. Note that this jacobian is *easily computed in terms of the current NN weights*. Then, the next result is representative of the sort of theorems that occur in NN feedback control design. It shows how to tune or train the NN weights to obtain guaranteed closed-loop stability.

Theorem (NN Weight Tuning for Stability) Let the desired trajectory $q_d(t)$ and its derivatives be bounded. Take the control input for (1) as (4). Let NN weight tuning be provided by

$$\dot{\hat{W}} = F\hat{\sigma}r^T - F\hat{\sigma}'\hat{V}^Txr^T - \kappa F\|r\|\hat{W} \qquad \dot{\hat{V}} = Gx(\hat{\sigma}'^T\hat{W}r)^T - \kappa G\|r\|\hat{V} \tag{5}$$

with any constant matrices $F = F^T > 0$, $G = G^T > 0$, and scalar tuning parameter $\kappa > 0$. Initialize the weight estimates as $\hat{W} = 0$, $\hat{V} = $ random. Then the sliding error $r(t)$ and NN weight estimates $\hat{W}$, $\hat{V}$ are uniformly ultimately bounded.

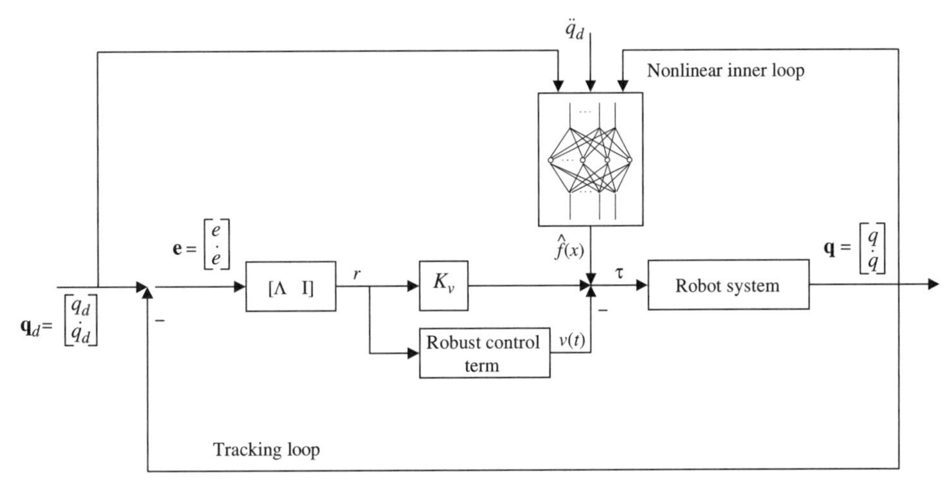

Figure 4 NN robot controller.

A proof of stability is always needed in control systems design to guarantee performance. Here, the stability is proven using nonlinear stability theory (e.g., an extension of Lyapunov's theorem). A Lyapunov energy function is defined as

$$L = \tfrac{1}{2}r^{\mathrm{T}}M(q)r + \tfrac{1}{2}\,\mathrm{tr}\{\tilde{W}^{\mathrm{T}}F^{-1}\tilde{W}\} + \tfrac{1}{2}\,\mathrm{tr}\{\tilde{V}^{\mathrm{T}}F^{-1}\tilde{V}\}$$

where the weight estimation errors are $\tilde{V} = V - \hat{V}$, $\tilde{W} = W - \hat{W}$, with $\mathrm{tr}\{\cdot\}$ the trace operator so that the Frobenius norm of the weight errors is used. In the proof, it is shown that the Lyapunov function derivative is negative outside a compact set. This guarantees the boundedness of the sliding variable error $r(t)$ as well as the NN weights. Specific bounds on $r(t)$ and the NN weights are given in Ref. 15. The first terms of (4) are very close to the (continuous-time) backpropagation algorithm.[17] The last terms correspond to Narendra's e-modification[18] extended to nonlinear-in-the-parameters adaptive control.

Robust adaptive tuning methods for nonlinear-in-the-parameters NN controllers have been derived based on the adaptive control approaches of e-modification, Ioannou's σ-modification, or projection methods. These techniques are compared by Ioannou and Sun[19] for standard adaptive control systems.

Robustness and Passivity of the NN When Tuned Online

Though the NN in Fig. 4 is static, since it is tuned online, it becomes a dynamic system with its own internal states (e.g., the weights). It can be shown that the tuning algorithms given in the theorem make the NN *strictly passive* in a certain novel strong sense known as "state-strict passivity," so that the energy in the internal states is bounded above by the power delivered to the system. This makes the closed-loop system *robust* to bounded unknown disturbances. This strict passivity accounts for the fact that no persistence of excitation condition is needed.

Standard adaptive control approaches assume that the unknown function $f(x)$ is linear in the unknown parameters and a certain regression matrix must be computed. By contrast, the NN design approach allows for nonlinearity in the parameters, and in effect the NN learns its own basis set online to approximate the unknown function $f(x)$. It is not required to find a regression matrix. This is a consequence of the NN universal approximation property.

3.2 Single-Layer NN Controller

If the first-layer weights V are fixed so that $\hat{f}(x) = \hat{W}^{\mathrm{T}}\sigma(V^{\mathrm{T}}x) \equiv \hat{W}^{\mathrm{T}}\phi(x)$, with $\phi(x)$ selected as a basis, then one has the simplified tuning algorithm for the output layer weights given by

$$\dot{\hat{W}} = F\phi(x)r^{\mathrm{T}} - \kappa F\|r\|\hat{W}$$

Then, the NN is LIP and the tuning algorithm resembles those used in adaptive control. However, NN design still offers an advantage in that the NN provides a universal basis for a class of systems, while adaptive control requires one to find a regression matrix, which serves as a basis for each particular system.

3.3 Feedback Linearization of Nonlinear Systems Using NNs

Many systems of interest in industrial, aerospace, and U.S. Department of Defense (DoD) applications are in the affine form $\dot{x} = f(x) + g(x)u + d$, with $d(t)$ a bounded unknown disturbance, nonlinear functions $f(x)$ unknown, and $g(x)$ unknown but bounded below by a known positive value g_b. Using nonlinear stability proof techniques such as those above, one can design a control input of the form

$$u = \frac{-\hat{f}(x) + v}{\hat{g}(x)} + u_r \equiv u_c + u_r$$

that has two parts, a *feedback linearization* part $u_c(t)$ and an *extra robustifying part* $u_r(t)$. Now, *two NNs* are required to manufacture the two estimates $\hat{f}(x)$, $\hat{g}(x)$ of the unknown functions. This controller is shown in Fig. 5. The weight updates for the $\hat{f}(\mathbf{x})$ NN are given exactly as in (5). To tune the $\hat{g}$ NN, a formula similar to (5) is needed, but it must be modified to ensure that the output $\hat{g}(x)$ of the second NN is bounded away from zero, to

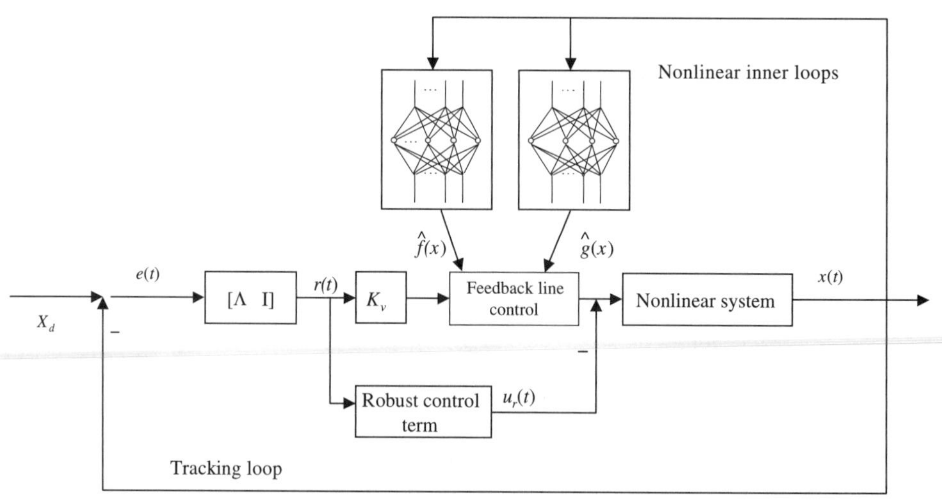

Figure 5 Feedback linearization NN controller.

keep the control $u(t)$ finite. It is called a controller singularity problem if $u(t)$ becomes infinity. More advanced control is possible using novel techniques. One good example is the use of integral Lyapunov functions in Refs. 20 and 21.

3.4 Partitioned NNs and Input Preprocessing

In this section we show how NN controller implementation may be streamlined by partitioning the NN into several smaller subnets to obtain more efficient computation. Also discussed in this section is preprocessing of input signals for the NN to improve the efficiency and accuracy of the approximation.

Partitioned NNs

A major advantage of the NN approach is that it allows one to partition the controller in terms of partitioned NN or neural subnets. This (i) simplifies the design, (ii) gives added controller structure, and (iii) makes for faster weight-tuning algorithms.

The unknown nonlinear robot function (2) can be written as

$$f(x) = M(q)\zeta_1(x) + V_m(q,\dot{q})\zeta_2(x) + G(q) + F(\dot{q})$$

with $\zeta_1(x) = \ddot{q}_d + \Lambda\dot{e}$, $\zeta_2(x) = \dot{q}_d + \Lambda e$. Taking the four terms one at a time,[22] one can use a small NN to approximate each term, as depicted in Fig. 6. This procedure results in four neural subnets, which we term a *structured or partitioned NN*. This approach can also utilize the properties of the physical systems conveniently for control system design and implementation. It can be directly shown that the individual partitioned NNs can be separately tuned exactly as in (5), making for a faster weight update procedure.

An advantage of this structured NN is that if some terms in the robot dynamics are well known [e.g., inertia matrix $M(q)$ and gravity $G(q)$], then their NNs can be replaced by equations that explicitly compute these terms. NNs can be used to reconstruct only the unknown terms or those too complicated to compute, which will probably include the friction $F(\dot{q})$ and the Coriolis/centripetal terms $V_m(q,\dot{q})$.

Preprocessing of Neural Net Inputs

The selection of a suitable NN input vector $x(t)$ for computation should be addressed. Some preprocessing of signals yields a more advantageous choice than (3) since it can explicitly

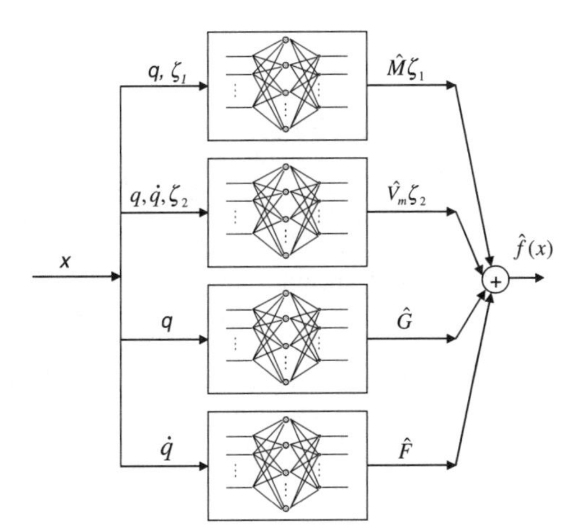

Figure 6 Partitioned NN.

introduce some of the nonlinearities inherent to robot arm dynamics. This reduces the burden of expectation on the NN and, in fact, also reduces the functional reconstruction error.

Consider an n-link robot having all revolute joints with joint variable vector $q(t)$. In revolute joint dynamics, the only occurrences of the joint variables are as sines and cosines,[23] so that the vector x can be taken as

$$x = [\zeta_1^T \quad \zeta_2^T \quad (\cos q)^T \quad (\sin q)^T \quad \dot{q}^T \quad \mathrm{sgn}(q)^T]^T$$

where the signum function is needed in the friction terms.

4 NN CONTROL FOR DISCRETE-TIME SYSTEMS

Most feedback controllers today are implemented on digital computers. This requires the specification of control algorithms in *discrete-time* or digital form.[22] To design such controllers, one may consider the discrete-time dynamics $x(k + 1) = f(x(k)) + g(x(k))u(k)$, with functions $f(\cdot)$ and $g(\cdot)$ unknown. The digital NN controller derived in this situation still has the form of the feedback linearization controller shown in Fig. 4.

One can derive tuning algorithms for a discrete-time NN controller with N layers that guarantee system stability and robustness.[15] For the ith layer the weight updates are of the form

$$\hat{W}_i(k + 1) = \hat{W}_i(k) - \alpha_i \hat{\phi}_i(k)\hat{y}_i^T(k) - \Gamma\|I - \alpha_i \hat{\varphi}_i(k)\hat{\varphi}_i^T(k)\|\hat{W}_i(k)$$

where $\hat{\varphi}_i(k)$ are the output functions of layer i, $0 < \Gamma < 1$ is a design parameter, and

$$\hat{y}_i(k) \equiv \hat{W}_i^T(k)\hat{\varphi}_i(k) + K_v r(k) \quad \text{for } i = 1, \ldots, N - 1$$

and

$$\hat{y}_N(k) \equiv r(k + 1) \quad \text{for last layer}$$

with $r(k)$ a filtered error. This tuning algorithm has two parts: The first two terms correspond to a gradient algorithm often used in the NN literature. The last term is a discrete-time robustifying term that guarantees that the NN weights remain bounded. The latter has been called a ''forgetting term'' in NN terminology and has been used to avoid the problem of ''NN weight overtraining.''

Recently, NN control has been successfully extended to systems in strict-feedback form with a modified tuning law.[24]

5 MULTILOOP NN FEEDBACK CONTROL STRUCTURES

Actual industrial or military mechanical systems may have *additional dynamical complications* such as vibratory modes, high-frequency electrical actuator dynamics, compliant couplings or gears, and so on. Practical systems may also have *additional performance requirements* such as requirements to exert specific forces or torques as well as perform position trajectory following (e.g., robotic grinding or milling). In such cases, the NN in Fig. 4 still works if it is modified to include *additional inner feedback loops* to deal with the additional plant or performance complexities. Using Lyapunov energy-based techniques, it can be shown that, if each loop is state-strict passive, then the overall multiloop NN controller provides stability, performance, and bounded NN weights. Details appear in Ref. 15.

5.1 Backstepping Neurocontroller for Electrically Driven Robot

Many industrial systems have high-frequency dynamics in addition to the basic system dynamics being controlled. An example of such systems is the n-link rigid robot arm with motor electrical dynamics given by

$$M(q)(\ddot{q}) + V_m(q,\dot{q})\dot{q} + F(\dot{q}) + G(q) + \tau_d = K_T i$$

$$L\dot{i} + R(i,\dot{q}) + \tau_e = u_e$$

with $q(t) \in R^n$ the joint variable, $i(t) \in R^n$ the motor armature currents, $\tau_d(t)$ and $\tau_e(t)$ the mechanical and electrical disturbances, and motor terminal voltage vector $u_e(t) \in R^n$ the control input. This plant has unknown dynamics in both the robot subsystem and the motor subsystem.

The problem with designing a feedback controller for this system is that one desires to control the behavior of the robot joint vector $q(t)$; however, the available control inputs are the motor voltages $u_e(t)$, which only affect the motor torques. As a second-order effect, the torques affect the joint angles.

Backstepping NN Design

The NN tracking controller in Fig. 7 may be designed using the *backstepping* technique.[25] This controller has *two neural networks,* one (NN 1) to estimate the unknown robot dynamics and an additional NN in an inner feedback loop (NN 2) to estimate the unknown motor dynamics. This multiloop controller is typical of control systems designed using rigorous system-theoretic techniques. It can be shown that by selecting suitable weight-tuning algorithms for both NNs, one can guarantee closed-loop stability as well as tracking performance in spite of the additional high-frequency motor dynamics. Both NN loops are state-strict passive. Proofs are given in terms of a modified Lyapunov approach. The NN tuning algorithms are similar to the ones presented above, but with some extra terms.

In standard backstepping, one must find several regression matrices, which can be complicated. By contrast, NN backstepping design does not require regression matrices since the NNs provide a universal basis for the unknown functions encountered.

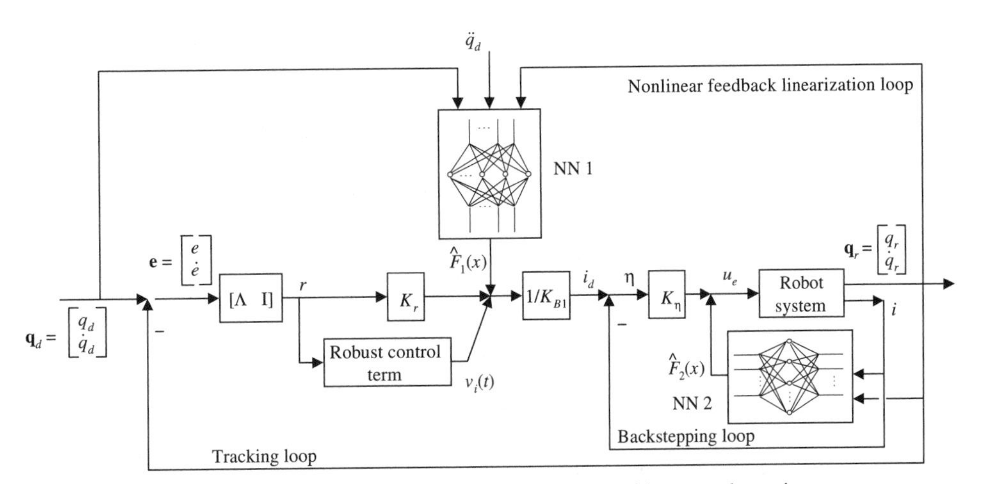

Figure 7 Backstepping NN controller for robot with motor dynamics.

5.2 Compensation of Flexible Modes and High-Frequency Dynamics Using NNs

Actual industrial or military mechanical systems may have additional dynamical complications such as vibratory modes, compliant couplings or gears, and so on. Such systems are characterized by having more degrees of freedom than control inputs, which compounds the difficulty of designing feedback controllers with good performance. In such cases, the NN controller in Fig. 4 still works if it is modified to include additional inner feedback loops to deal with the additional plant complexities.

Using the Bernoulli–Euler equation, infinite series expansion, and the assumed mode shapes method, the dynamics of flexible-link robotic systems can be expressed in the form

$$\begin{bmatrix} M_{rr} & M_{rf} \\ M_{fr} & M_{ff} \end{bmatrix}\begin{bmatrix} \ddot{q}_r \\ \ddot{q}_f \end{bmatrix} + \begin{bmatrix} V_{rr} & V_{rf} \\ V_{fr} & V_{ff} \end{bmatrix}\begin{bmatrix} \dot{q}_r \\ \dot{q}_f \end{bmatrix} + \begin{bmatrix} 0 & 0 \\ 0 & K_{ff} \end{bmatrix}\begin{bmatrix} q_r \\ q_f \end{bmatrix} + \begin{bmatrix} F_r \\ 0 \end{bmatrix} + \begin{bmatrix} G_r \\ 0 \end{bmatrix} = \begin{bmatrix} B_r \\ B_f \end{bmatrix}\tau$$

where $q_r(t)$ is the vector of rigid variables (e.g., joint angles), $q_f(t)$ the vector of flexible mode amplitudes, M an inertia matrix, V a Coriolis/centripetal matrix, and matrix partitioning is represented according to subscript r for the rigid modes and subscript f for the flexible modes. Friction F and gravity G apply only for the rigid modes. Stiffness matrix K_{ff} describes the vibratory frequencies of the flexible modes.

The problem in controlling such systems is that the input matrix $B = [B_r^{\mathrm{T}} \quad B_f^{\mathrm{T}}]^{\mathrm{T}}$ is not square but has more rows than columns. This means that while one is attempting to control the rigid-mode variable $q_r(t)$, one is also affecting $q_f(t)$. This causes undesirable vibrations. Moreover, the zero dynamics of such systems is non–minimum phase, which results in unstable flexible modes if care is not taken in choosing a suitable controller.

Singular Perturbations NN Design

To overcome this problem, an additional *inner feedback loop* based on singular perturbation theory[26] may be designed. The resulting multiloop controller is shown in Fig. 8, where a NN compensates for friction, unknown nonlinearities, and gravity and the inner loop manages the flexible modes. The internal dynamics controller in the inner loop may be designed using a variety of techniques, including H_∞ robust control and linear quadratic Gaussian/loop transfer recovery (LQG/LTR). Such controllers are capable of compensating for the effects

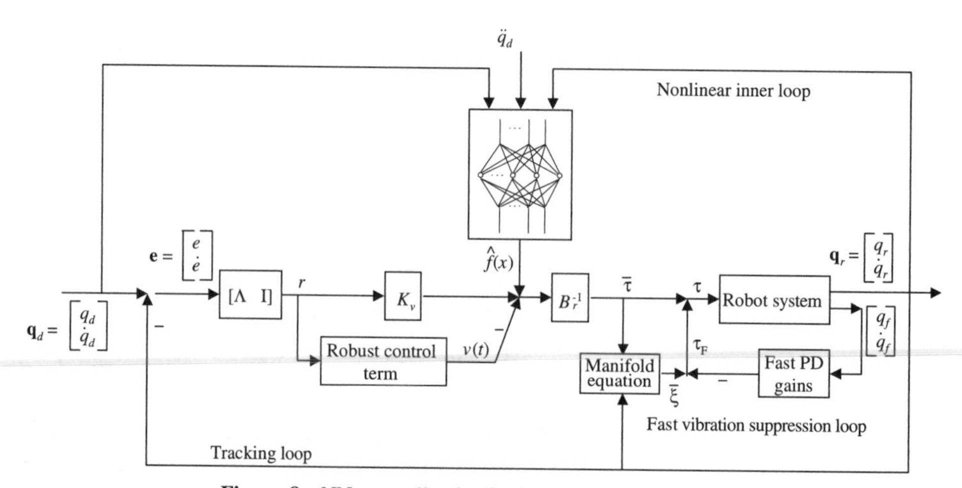

Figure 8 NN controller for flexible-link robotic system.

of inexactly known or changing flexible mode frequencies. An observer can be used to avoid strain rate measurements.

In many industrial or aerospace designs, flexibility effects are limited by restricting the speed of motion of the system. This limits performance. By contrast, using the singular perturbation NN controller, *a flexible system can far outperform a rigid system in terms of speed of response*. The key is to use the flexibility effects to speed up the response in much the same manner as the cracking of a whip. That is, the flexibility effects of advanced structures are not merely a debility that must be overcome, but they offer the possibility of *improved performance* over rigid structures if they are suitably controlled. By exploiting recent advances in materials, such as piezoelectric materials, further improved performance is attainable for the so-called smart material flexible robots.[20b]

5.3 Force Control with Neural Nets

Many practical robot applications require the control of the force exerted by the manipulator normal to a surface along with position control in the plane of the surface. This is the case in milling and grinding, surface finishing, and so on. In applications such as MEMS assembly, where highly nonlinear forces, including van der Waals, surface tension, and electrostatics dominate gravity, advanced control schemes such as NNs are especially required.

In such cases, the NN force/position controller in Fig. 9 can be derived using rigorous Lyapunov-based techniques. It has guaranteed performance in that both the position-tracking error $r(t)$ and the force error $\tilde{\lambda}(t)$ are kept small while all the NN weights are kept bounded. The figure has an *additional inner force control loop*. The control input is now given by

$$\tau(t) = \hat{W}^{\mathrm{T}}\sigma(\hat{V}^{\mathrm{T}}x) + K_v(Lr) - J^{\mathrm{T}}(\lambda_d - K_f\tilde{\lambda}) - v$$

where the selection matrix L and Jacobian J are computed based on the decomposition of the joint variable $q(t)$ into two components—the component $q_1(t)$ (e.g., tangential to the given surface) in which position tracking is desired and the component $q_2(t)$ (e.g., normal to the surface) in which force exertion is desired. This is achieved using holonomic constraint techniques based on the prescribed surface that are standard in robotics (e.g., work by

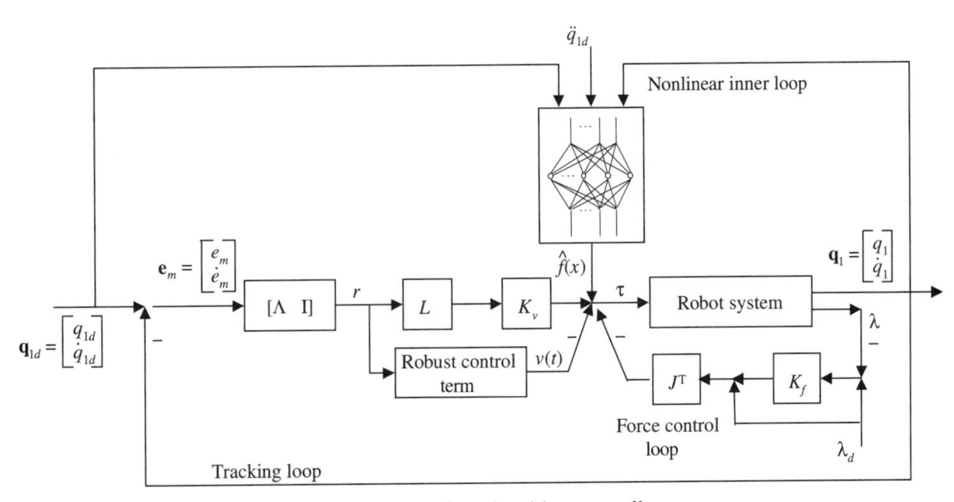

Figure 9 NN force/position controller.

McClamroch[27] and others). The filtered position-tracking error in $q_1(t)$ is $r(t)$, that is, $r(t) = q_{1d} - q_1$ with $q_{1d}(t)$ the desired trajectory in the plane of the surface. The desired force is described by $\lambda_d(t)$ and the force exertion error is captured in $\tilde{\lambda}(t) = \lambda(t) - \lambda_d(t)$ with $\lambda(t)$ describing the actual measured force exerted by the manipulator. The position-tracking gain is K_v and the force-tracking gain is K_f.

6 FEEDFORWARD CONTROL STRUCTURES FOR ACTUATOR COMPENSATION

Industrial, aerospace, DoD, and MEMS assembly systems have actuators that generally contain deadzone, backlash, and hysteresis. Since these actuator nonlinearities appear in the *feedforward* loop, the NN compensator must also appear in the feedforward loop. The design problem for neurocontrollers where the NN appears in the feedforward loop is significantly more complex than for feedback NN controllers. Details are given in Ref. 28.

6.1 Feedforward Neurocontroller for Systems with Unknown Deadzone

Most industrial, vehicle, and aircraft actuators have deadzones. The deadzone characteristic appears in Fig. 10 and causes motion control problems when the control signal takes on small values or passes through zero, since only values greater than a certain threshold can influence the system.

Feedforward controllers can offset the effects of deadzones if properly designed. It can be shown that a NN deadzone compensator has the structure shown in Fig. 11. The NN compensator consists of *two* NNs: NN II is in the direct feedforward control loop, and NN I is not directly in the control loop but serves as an observer to estimate the (unmeasured) applied torque $\tau(t)$. The feedback stability and performance of the NN deadzone compensator have been rigorously proven using nonlinear stability proof techniques.

The two NNs were each selected as having one tunable layer, namely the output weights. The activation functions were set as a basis by selecting fixed random values for the first-layer weights.[7] To guarantee stability, the output weights of the inversion NN II and the estimator NN I should be tuned respectively as

$$\dot{\hat{W}}_i = T\sigma_i(U_i^{\mathrm{T}} w)r^{\mathrm{T}}\hat{W}^{\mathrm{T}}\sigma'(U^{\mathrm{T}}u)U^{\mathrm{T}} - k_1T\|r\|\hat{W}_i - k_2T\|r\|\|\hat{W}_i\|\hat{W}_i$$

$$\dot{\hat{W}} = -S\sigma'(U^{\mathrm{T}}u)U^{\mathrm{T}}\hat{W}_i\sigma_i(U_i^{\mathrm{T}}w)r^{\mathrm{T}} - k_iS\|r\|\hat{W}$$

where subscript i denotes weights and sigmoids of the inversion NN II and variables without subscripts correspond to NN I. Note that σ' denotes the Jacobian. Design parameters are the positive-definite matrices T and S and tuning gains k_1, k_2. The form of these tuning laws is

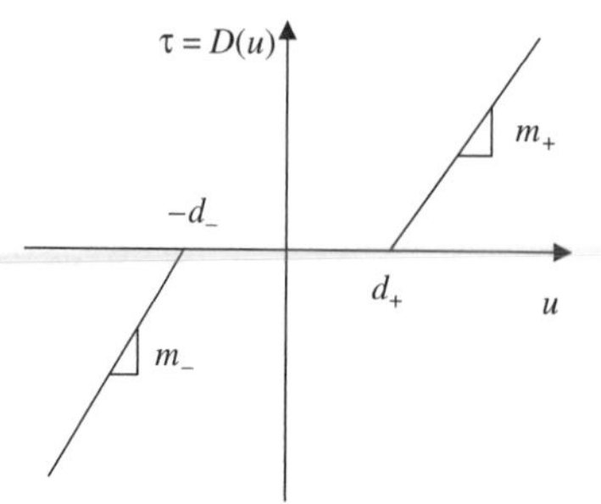

Figure 10 Deadzone response characteristic.

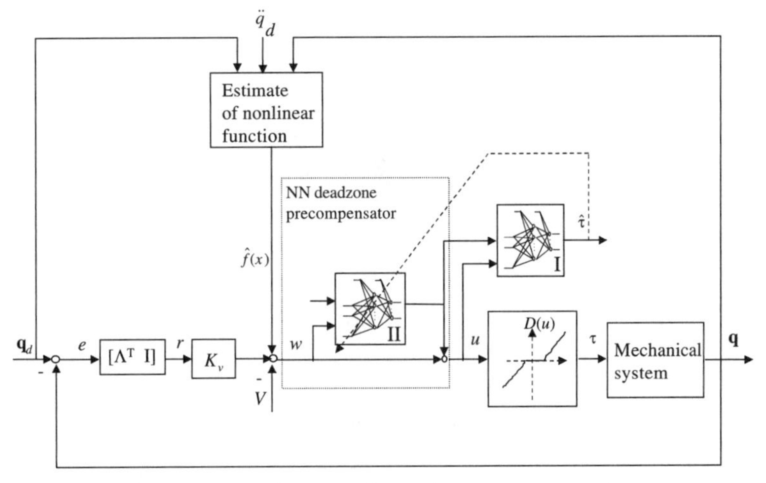

Figure 11 Feedforward NN for deadzone compensation.

intriguing. They form a coupled nonlinear system with each NN helping to tune itself and the other NN. Moreover, signals *are backpropagated through NN I* to tune NN II. That is, the two NNs function as *a single NN with two layers,* first NN II, then NN I, but with the second layer not in the direct control path. Note the additional terms, which are a combination of Narendra's e-modification and Ioannou's σ-modification.

Reinforcement Learning Structure
Neural network I is not in the control path but serves as a *higher level critic* for tuning NN II, the action-generating net. The critic NN I actually functions to provide an estimate of the torque supplied to the system in the absence of deadlock, which is a *target torque.* It is intriguing that this use of NN in the feedforward loop (as opposed to the feedback loop) requires such a *reinforcement learning* structure. Reinforcement learning techniques generally have the critic NN outside the main feedback loop, on a higher level of the control hierarchy.

6.2 Dynamic Inversion Neurocontroller for Systems with Backlash

Backlash is a common problem in actuators with gearing. The backlash characteristic is shown in Fig. 12 and causes motion control problems when the control signal reverses in direction, often due to dead space between gear teeth.

Dynamic inversion is a popular controller design technique in aircraft control and elsewhere.[29] Dynamic inversion by NNs has been used by Calise and co-workers[30] in aircraft control using NNs. Using dynamic inversion, a NN controller for systems with backlash is designed in Ref. 28. The neurocontroller appears in the feedforward loop, as in Fig. 13, and is a *dynamic or recurrent NN.* In this neurocontroller, a desired torque $\tau_{\text{des}}(t)$ to be applied is determined; then, using a backstepping type of approach,[25] the neurocontroller structure shown in Fig. 13 is derived. A NN is used to approximate certain nonlinear functions appearing in the derivation. Unlike backstepping, dynamic inversion lets the required derivative appear explicitly in the controller. In the design, a filtered derivative $\xi(t)$ is used to allow implementation in actual systems.

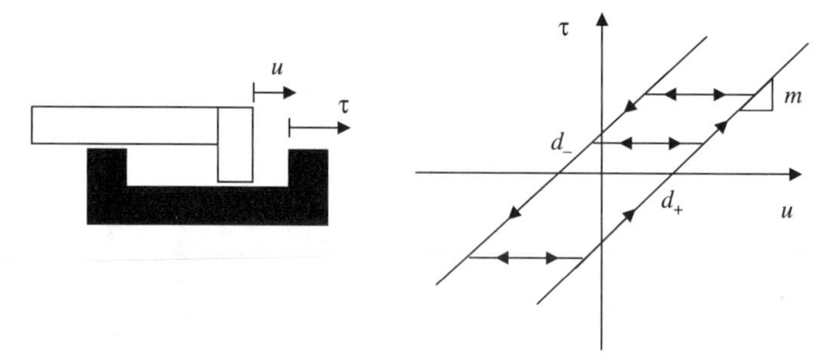

Figure 12 Backlash response characteristic.

The NN precompensator shown in Fig. 13 effectively adds control energy to invert the dynamical backlash function. The control input into the backlash element is given by

$$\hat{u}(t) = K_b\tilde{\tau} + \xi - y_{nn} + v_2$$

where $\tilde{\tau}(t) = \tau_{des}(t) - \tau(t)$ is the torque error, $y_{nn}(t)$ is the NN output, and $v_2(t)$ is a certain robust control term detailed in Ref. 28. Weight-tuning algorithms given there guarantee closed-loop stability and effective backlash compensation.

7 NN OBSERVERS FOR OUTPUT FEEDBACK CONTROL

Thus far, we have described NN controllers in the case of full state feedback, where all internal system information is available for feedback. However, in actual industrial and com-

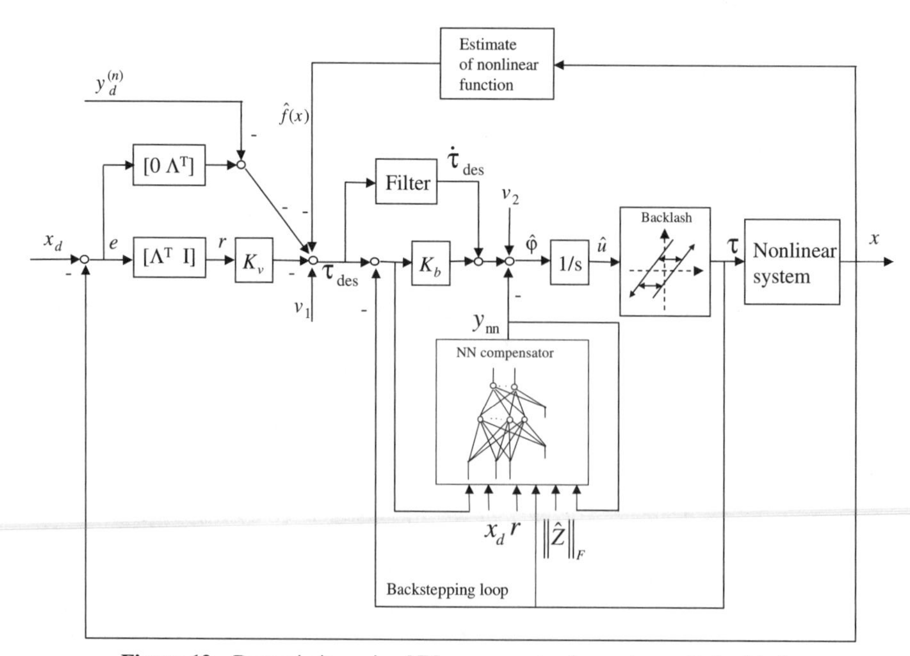

Figure 13 Dynamic inversion NN compensator for system with backlash.

mercial systems, there are usually available only certain restricted measurements of the plant. In this output feedback case one may use an *additional dynamic NN* with its own internal dynamics in the controller. The function of this additional NN is effectively to provide estimates of the unmeasurable plant states, so that the dynamic NN functions as what is known as an *observer* in control system theory.

The issues of observer design using NNs can be appreciated with rigid robotic systems.[12] For these systems, the dynamics can be written in state-variable form as

$$\dot{x}_1 = x_2$$

$$\dot{x}_2 = M^{-1}(x_1)[-N(x_1, x_2) + \tau]$$

where $x \equiv q$, $x_2 \equiv \dot{q}$ and the nonlinear function $N(x_1, x_2) = V_m(x_1, x_2)x_2 + G(x_1) + F(x_2)$ is assumed to be unknown. It can be shown[31] that the following dynamic NN observer can provide estimates of the entire state $x = [x_1^{\mathrm{T}} \quad x_2^{\mathrm{T}}]^{\mathrm{T}} \equiv [q^{\mathrm{T}} \quad \dot{q}^{\mathrm{T}}]^{\mathrm{T}}$ given measurements of only $x_1(t) = q(t)$:

$$\dot{\hat{x}}_1 = \hat{x}_2 + k_D x_1$$

$$\dot{\hat{z}}_2 = M^{-1}(x_1)[-\hat{W}_o^{\mathrm{T}}\sigma_o(\hat{x}) + k_P \tilde{x}_1 + \tau].$$

$$\hat{x}_2 = \hat{z}_2 + k_{P2}\tilde{x}_1$$

In this system, the hat denotes estimates and the tilde denotes estimation errors. It is assumed that the inertia matrix $M(q)$ is known, but all other nonlinearities are estimated by the observer NN $\hat{W}_o^{\mathrm{T}}\sigma_o(\hat{x})$, which has output layer weights $\hat{W}_o$ and activation functions $\sigma_o(\cdot)$. Signal $v_o(t)$ is a certain observer robustifying term, and the observer gains k_P, k_D, k_{P2} are positive design constants detailed in Ref. 31.

The NN output feedback tracking controller shown in Fig. 14 uses the dynamic NN observer to reconstruct the missing measurements $x_2(t) = \dot{q}(t)$ and then employs a second static NN for tracking control, exactly as in Fig. 4. Note that the outer tracking PD loop structure has been retained but an additional dynamic NN loop is needed. In Ref. 31, weight-tuning algorithms that guarantee stability are given for both the dynamic estimator NN and the static control NN.

8 REINFORCEMENT LEARNING CONTROL USING NNs

Reinforcement learning techniques are based on psychological precepts of reward and punishment as used by I. P. Pavlov in the training of dogs at the turn of the century. The key tenet here is that the performance indicators of the controlled system should be simple, for instance, $+1$ for a successful trial and -1 for a failure, and that these simple signals should tune or adapt a NN controller so that its performance improves over time. This gives a learning feature driven by the basic success or failure record of the controlled system. Reinforcement learning has been studied by many researchers, including Refs. 32 and 33.

It is difficult to provide rigorous designs and analysis for reinforcement learning in the framework of standard control system theory since the reinforcement signal has reduced information, which makes study, including Lyapunov techniques, very complicated. Reinforcement learning is related to the so-called sign error tuning in adaptive control[34] which has not been proven to yield stability.

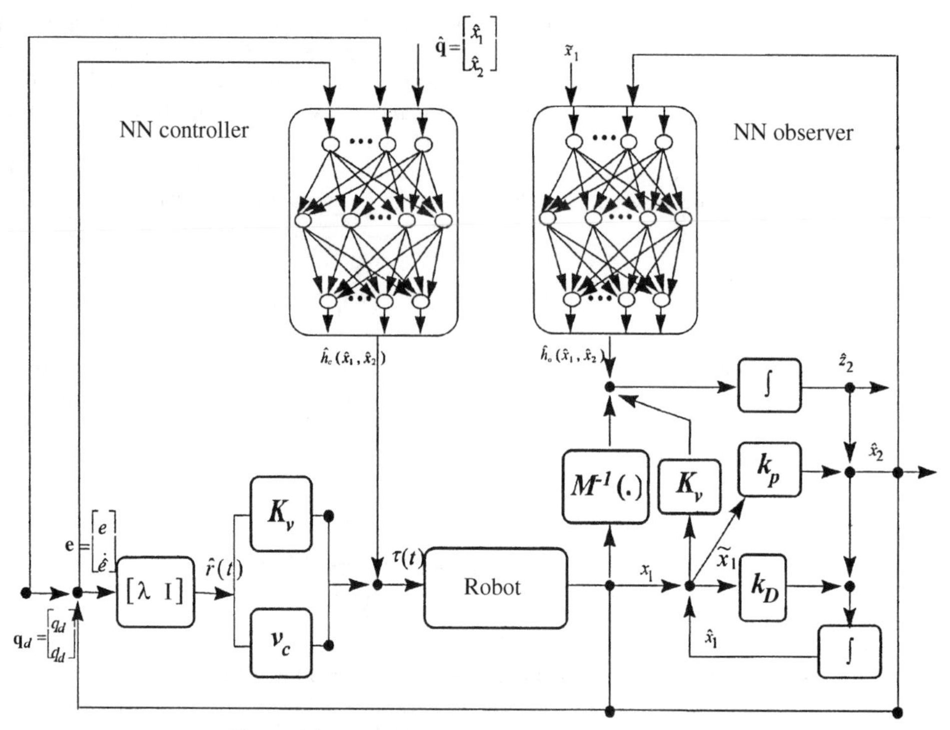

Figure 14 NN observer for output feedback control.

8.1 NN Reinforcement Learning Controller

A simple signal related to the performance of a robotic system is the signum of the sliding variable error $R(t) = \mathrm{sgn}(r(t))$, with the sliding variable error given by $r = \dot{e} + \Lambda e$, where $e = q_d - q$ is the tracking error and matrix Λ is positive definite. Signal $R(t)$ satisfies the criteria required in reinforcement learning control: (i) It is simple, having values of only 0, ± 1 and (ii) the value of zero corresponds to a reward for good performance, while nonzero values correspond to a punishment signal. Therefore, $R(t)$ may be taken as a suitable reinforcement learning signal.

Rigorous proofs of closed-loop stability and performance for reinforcement learning may be provided[31] by (i) using *nonstandard Lyapunov functions,* (ii) deriving novel modified NN tuning algorithms, and (iii) selection of a suitable multiloop control structure. The architecture of the reinforcement adaptive learning NN controller derived is shown in Fig. 15. A performance evaluation loop has the desired trajectory $q_d(t)$ as the user input; this loop manufactures $r(t)$, which may be considered as the *instantaneous utility*. The critic element evaluates the signum function and so provides the reinforcement signal $r(t)$ which critiques the performance of the system.

It is not easy to show how to tune the action-generating NN using only the reinforcement signal $r(t)$, which contains significantly less information than the full error signal $r(t)$. A successful proof can be based on the *Lyapunov energy function*

$$L(t) = \sum_{i=1}^{n} |r_i| + \frac{1}{2}\,\mathrm{tr}(\tilde{W}^{\mathrm{T}} F^{-1} \tilde{W})$$

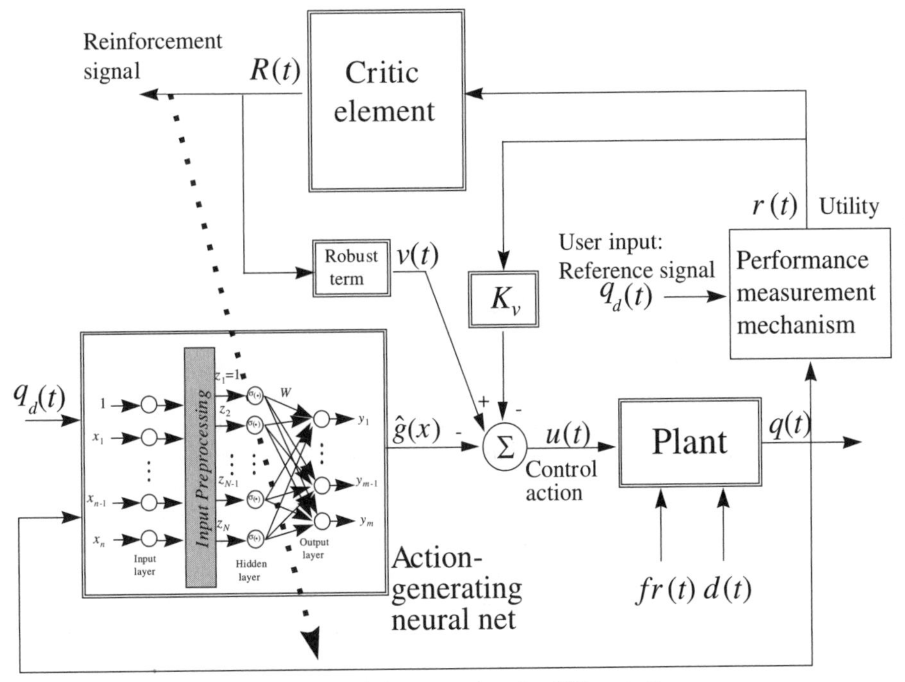

Figure 15 Reinforcement learning NN controller.

where $r(t) \in R^n$. This is not a standard Lyapunov function in feedback system theory but is similar to energy functions used in some NN convergence proofs (e.g., by Hopfield). Using this Lyapunov function, one can derive NN tuning algorithms that guarantee closed-loop stability and tracking. The NN weights are tuned using *only the reinforcement signal* $R(t)$ according to

$$\dot{\hat{W}} = F\sigma(x)R^{\mathrm{T}} - \kappa F\hat{W}$$

This is similar to what has been called *sign error tuning* in adaptive control, which has usually been proposed without giving any proof of stability or performance.

8.2 Adaptive Reinforcement Learning Using Fuzzy Logic Critic

Fuzzy logic systems are based on the higher level linguistic and reasoning abilities of humans and offer intriguing possibilities for use in feedback control systems. The idea of using backpropagation tuning to tune fuzzy logic systems was proposed by Werbos.[33] Through the work of Wang,[6] K. Passino, and S. Yurkovich,[35] and others, it is now known how to tune fuzzy logic systems so that they learn online to yield very good performance in closed-loop control applications.

A fuzzy logic (FL) system with product inferencing, centroid defuzzification, and singleton output membership functions has output vector $y(t)$ whose components are given in terms of the input vector $x(t) \in R^n$ by

$$y_k = \sum_{j=1}^{L} w_{kj}\sigma_j(x,U) \qquad \text{or} \qquad y = W^{\mathrm{T}}\sigma(x,U)$$

where $W^{\mathrm{T}} = [w_{kj}]$ is a matrix of output representative values and the FL basis functions $\sigma_j(\cdot)$ play the role of NN activation functions. Using product inferencing, the basis functions are given in terms of the one-dimensional membership functions (MFs) $\mu_{ij}(x,U_{ij})$ by

$$\sigma_j(x,U) = \frac{\prod_{i=1}^{n} \mu_{ij}(x_i,U_{ij})}{\sum_{j=1}^{L} \prod_{i=1}^{n} \mu_{ij}(x_i,U_{ij})}$$

where U_{ij} is a vector of parameters of the MFs including the centroids and spreads. The number of rules is L. The standard choice for the MFs is triangle functions. However, other choices have been used, including splines (c.f. Ref. 4, CMAC NN), second- or third-degree polynomials, or the RBF functions.[36]

FL systems have the connotation of higher level supervisors since they are rule based. The fuzzy-neural reinforcement learning scheme shown in Fig. 16 has been developed, where a FL system serves as a critic and a NN serves as an action-generating network that controls the system. The reinforcement controller is adaptive in the sense that the FL critic is tuned as well as the NN action-generating network to improve system performance through online learning. Stability and convergence proofs have been provided and depend on using certain specialized tuning schemes for the FL critic membership functions and the NN weights. Tuning the membership functions has the effect of modifying them so they converge onto the region in with highest state trajectory activity, a form of *dynamic focusing of awareness.*

The advantage of the FL/NN adaptive reinforcement learning structure is that the critic can be initialized using linguistic/heuristic notions by the human user. Finally, for FL systems one can look at the final MFs and interpret what information has been stored in the system through learning.

9 OPTIMAL CONTROL USING NNs

Heretofore we have discussed the design of NN controllers for tracking and stabilization based on control theory techniques including feedback linearization, backstepping, singular perturbations, force control, dynamic inversion, and observer design. The point was made

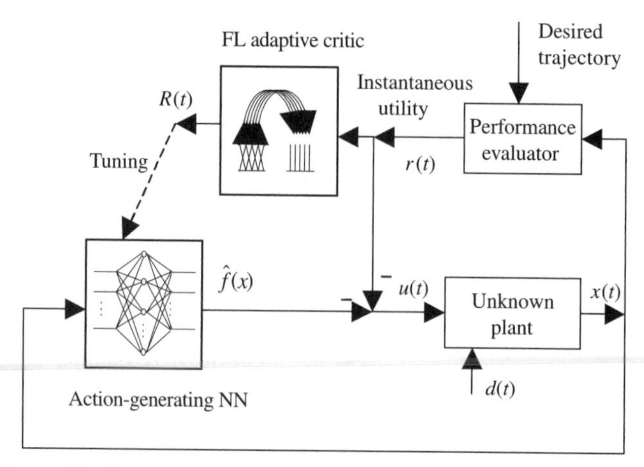

Figure 16 Fuzzy logic adaptive reinforcement learning NN controller.

that as the system dynamical structure becomes more complex or the performance requirements become more stringent, it is necessary to add more feedback loops. Rigorous neurocontroller design algorithms may be given in terms of Lyapunov energy-based techniques, passivity, and so on.

Nonlinear optimal control design provides a very powerful theory that is applicable for systems in any form. Solution of the so-called Hamilton–Jacobi (HJ) equations will directly yield a controller with guaranteed properties in terms of stability and performance for any sort of nonlinear system. Unfortunately, the HJ equations are difficult to solve and may not even have analytic solutions for general nonlinear systems. In the special case of *linear* optimal control,[37] solution techniques are available based on Riccati equation techniques, and that theory provides a cornerstone of control design for aerospace systems, vehicles, and industrial plants. It would be very valuable to have tractable controller design techniques for general nonlinear systems. In fact, it has been shown that NNs afford computationally effective techniques for solving general HJ equations, and so for designing closed-loop controllers for general nonlinear systems.

9.1 NN H_2 Control Using the Hamilton–Jacobi–Bellman Equation

In work by Abu-Khalaf and Lewis[38] it has been shown how to solve the Hamilton–Jacobi–Bellman (HJB) equation that appears in optimal control for general nonlinear systems by a successive approximation (SA) technique based on NNs. Rigorous results have been proven, and a computationally effective scheme for nearly optimal controller design was provided based on NNs.

This technique allows one to consider general affine nonlinear systems of the form

$$\dot{x} = f(x) + g(x)u(x) \tag{6}$$

To give internal stability and good closed-loop performance, one may select the L_2 norm performance index

$$V(x(0)) = \int_0^\infty [Q(x) + u^T Ru]\, dt \tag{7}$$

with matrix R positive definite and $Q(x)$ generally selected as a norm. It is desired to select the control input $u(t)$ to minimize the cost $V(x)$. Under suitable assumptions of detectability, this guarantees that the states and controls are bounded and hence that the closed-loop systems is stable.

An infinitesimal equivalent to the cost is given by

$$0 = \frac{\partial V^T}{\partial x}(f + gu) + Q + u^T Ru \equiv H\left(x, \frac{\partial V}{\partial x}, u\right) \tag{8}$$

which defines the Hamiltonian function $H(\cdot)$ and the costate as the cost gradient $\partial V/\partial x$. This is a nonlinear Lyapunov equation. It has been called a *generalized HJB equation* by Saridis and Lee.[39]

Differentiating with respect to the control input $u(t)$ to find a minimum yields the control in the form

$$u(x) = -\frac{1}{2} R^{-1} g^T(x) \frac{\partial V(x)}{\partial x} \tag{9}$$

Substituting this into the previous equation yields the HJB equation of optimal control

$$0 = \frac{\partial V^{\mathrm{T}}}{\partial x} f + Q - \frac{1}{4} \frac{\partial V^{\mathrm{T}}}{\partial x} g(x) R^{-1} g(x) \frac{\partial V}{\partial x} \tag{10}$$

The boundary condition for this equation is $V(0) = 0$. Solving this equation yields the optimal value function $V(x)$, whence the optimal control may be computed from the cost gradient using (4).

This procedure will give the optimal control in feedback form for any nonlinear system. Unfortunately, the HJB equation cannot be solved for most nonlinear systems. In the linear system case, the HJB equation yields the Riccati equation, for which efficient solution techniques are available. However, most systems of interest today in aerospace, vehicles, and industry are nonlinear.

Therefore, one may use a SA approach wherein (8) and (9) are iterated to determine sequences $V^{(i)}$, $u^{(i)}$. The initial stabilizing control $u^{(0)}$ used in (8) to find $V^{(0)}$ is easily determined using, for example, the linear quadratic regulator (LQR) for the linearization of (6). It has been shown by Saridis and Lee[39] that the SA converges to the optimal solution V^*, u^* of the HJB equation. Let the region of asymptotic stability of the optimal solution be Ω^* and the region with asymptotic stability (RAS) at iteration i be $\Omega^{(i)}$. Then, in fact, it has been shown that

$u^{(i)}$ is stabilizing for all i;

$V^{(i)} \to V^*$, $u^{(i)} \to u^*$, $\Omega^{(i)} \to \Omega^*$ uniformly;

$V^{(i)}(x) \geq V^{(i+1)}(x)$, that is, the value function decreases; and

$\Omega^{(i)} \leq \Omega^{(i+1)}$, that is, the RAS increases.

In fact, Ω^* is the largest RAS of any other admissible control law.

NNs for Computation of Successive Approximation Solution

It is difficult to solve Eqs. (8) and (9) as required for the SA method just given. Beard et al.[40] showed how to implement the SA algorithm using the Galerkin approximation to solve the nonlinear Lyapunov equation. This method is computationally intensive, since it requires the evaluation of numerous integrals. It was shown in Ref. 38 how to use NNs to compute the SA solution at each iteration. This yields a computationally effective method for determining nearly optimal controls for a general class of nonlinear constrained input systems. The value function at each iteration is approximated using a NN by

$$V(x) \approx V(x,w_j) = w^{(i)\mathrm{T}} \sigma(x)$$

with w_j the NN weights and $\sigma(x)$ a basis set of activation functions. To satisfy the initial condition $V^{(i)}(0) = 0$ and the symmetry requirements on $V(x)$, the activation functions were selected as a basis of even polynomials in x. Then the parameterized nonlinear Lyapunov equation becomes

$$0 = w^{(i)\mathrm{T}} \nabla\sigma(x)(f(x) + g(x)u^{(i)}) + Q + u^{(i)\mathrm{T}} R u^{(i)}$$

with $u^{(i)}$ the current control value. Evaluating this equation at enough sample values of x, it can easily be solved for the weights using, for example, least squares. The sample values of x must satisfy a condition known as *persistence of excitation* in order to obtain a unique least-squares solution for the weights. The number of samples selected must be greater than the number of NN weights. Then, the next iteration value of the control is given by

$$u^{(i+1)}(x) = -\tfrac{1}{2} R^{-1} g^{\mathrm{T}}(x) \nabla\sigma^{\mathrm{T}}(x) w^{(i)}$$

Using a Sobolev space setting, it was shown that under certain mild assumptions the NN solution converges in the mean to a suitably close approximation of the optimal solution.

Moreover, if the initial NN weights are selected to yield an admissible control, then the control is admissible (which implies stability) at each iteration.

The control given by this approach is shown in Fig. 17. It is a *feedback control* in terms of a nonlinear NN. This approach has also been given for constrained input systems, such as industrial and aircraft actuator systems.

9.2 NN H_∞ Control Using the Hamilton–Jacobi–Isaacs Equation

Many systems contain unknown disturbances, and the optimal control approach just given may not be effective. In this case, one may use the H_∞ design procedure.

Consider the dynamical system in Fig. 18, where $u(t)$ is an action or control input, $d(t)$ is a disturbance or opponent, $y(t)$ is the measured output, and $z(t)$ is a performance output with $\|z\|^2 = h^T h + \|u\|^2$. Here we take full state feedback $y=x$ and desire to determine the action or control $u(t) = u(x(t))$ such that, under the worst disturbance, one has the L_2 gain bounded by a prescribed γ so that

$$\frac{\int_0^\infty \|z(t)\|^2\,dt}{\int_0^\infty \|d(t)\|^2\,dt} = \frac{\int_0^\infty (h^T h + \|u\|^2)\,dt}{\int_0^\infty \|d(t)\|^2\,dt} \le \gamma^2$$

This is a differential game with two players[41,42] and can be confronted by defining the utility

$$r(x,u,d) = h^T(x)h(x) + \|u(t)\|^2 - \gamma^2\|d(t)\|^2$$

and the long-term value (cost-to-go)

$$V(x(t)) = \int_t^\infty r(x,u,d)\,dt = \int_t^\infty (h^T(x)h(x) + \|u(t)\|^2 - \gamma^2\|d(t)\|^2)\,dt \tag{11}$$

The optimal value is given by

$$V^*(x(t)) = \min_{u(t)} \max_{d(t)} \int_t^\infty r(x,u,d)\,dt$$

The optimal control and worst-case disturbance are given by the stationarity conditions as

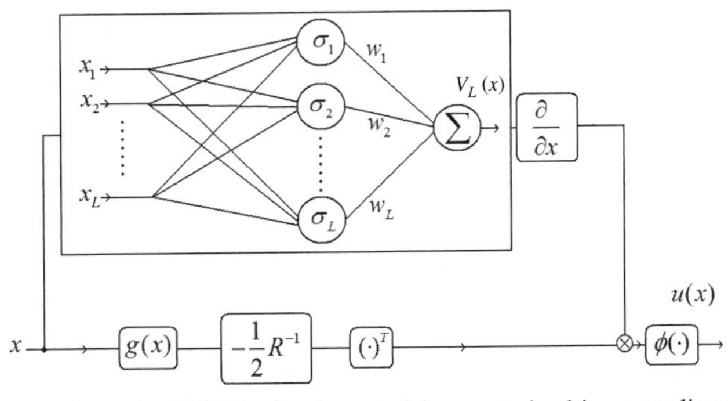

Figure 17 Nearly optimal NN feedback control for constrained input nonlinear systems.

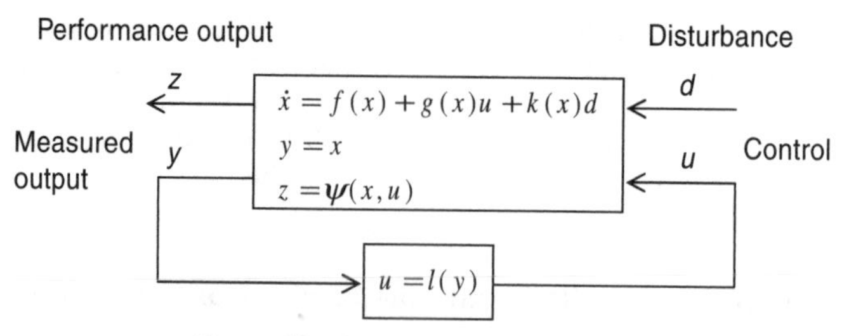

Figure 18 Bounded L_2 gain problem.

$$u^*(x(t)) = -\frac{1}{2}g^{\mathrm{T}}(x)\frac{\partial V^*}{\partial x} \tag{12}$$

$$d^*(x(t)) = \frac{1}{2\gamma^2}k^{\mathrm{T}}(x)\frac{\partial V^*}{\partial x} \tag{13}$$

If the *min–max* and *max–min* solutions are the same, then a saddle point exists and the game has a unique solution. Otherwise, we consider the min–max solution, which confers a slight advantage to the action input $u(t)$.

The infinitesimal equivalent to (11) is found using Leibniz's formula to be

$$0 = \dot{V} + r(x,u,d) = \left(\frac{\partial V}{\partial x}\right)^{\mathrm{T}}\dot{x} + r(x,u,d) = \left(\frac{\partial V}{\partial x}\right)^{\mathrm{T}}F(x,u,d) + r(x,u,d) \equiv H\left(x,\frac{\partial V}{\partial x},u,d\right) \tag{14}$$

with $V(0)=0$, where $H(x,\lambda,u,d)$ is the Hamiltonian with $\lambda(t)$ the costate and $\dot{x} = F(x,u,d) \equiv f(x) + g(x)u + k(x)d$. This is a nonlinear Lyapunov equation.

Substituting u^* and d^* into (14) yields the nonlinear HJI equation

$$0 = \left(\frac{dV^*}{dx}\right)^{\mathrm{T}} f + h^{\mathrm{T}}h - \frac{1}{4}\left(\frac{dV^*}{dx}\right)^{\mathrm{T}} gg^{\mathrm{T}}\frac{dV^*}{dx} + \frac{1}{4\gamma^2}\left(\frac{dV^*}{dx}\right)^{\mathrm{T}} kk^{\mathrm{T}}\frac{dV^*}{dx} \tag{15}$$

whose solution provides the optimal value V^* and hence the solution to the min–max differential game. Unfortunately, this equation cannot generally be solved.

In Ref. 43 it has been shown that the following two-loop successive approximation policy iteration algorithm has very desirable properties like those delineated above for the H_2 case. First one finds a stabilizing control for zero disturbance. Then one iterates Eqs. (13) and (14) until there is convergence with respect to the disturbance. Now one selects an improved control using (12). The procedure repeats until there is convergence of both loops. Note that it is easy to select the initial stabilizing control u_0 by setting $d(t)=0$ and using LQR design[37] on the linearized system dynamics.

NN Solution of HJI Equation for H_∞ Control

To implement this algorithm practically one may approximate the value at each step using a one-tunable-layer NN as

$$V(x) \approx V(x,w^i_j) = w_j^{i^{\mathrm{T}}}\sigma(x)$$

with $\sigma(x)$ a basis set of activation functions. The disturbance iteration is in index i and the control iteration is in index j. Then the parameterized nonlinear Lyapunov equation (14) becomes

$$0 = w_j^{iT} \nabla\sigma(x)\dot{x} + r(x,u_j,d^i) = w_j^i \nabla\sigma(x)F(x,u_j,d^i) + h^T h + \|u_j\|^2 - \gamma^2\|d^i\|^2$$

which can easily be solved for the weights using, for example, least squares. Then, on disturbance iterations the next disturbance is given by

$$d^{i+1}(x) = -\tfrac{1}{2}R^{-1}k^T(x)\,\nabla\sigma^T(x)w_j^i$$

and on control iterations the improved control is given by

$$u_{j+1}(x) = -\tfrac{1}{2}R^{-1}g^T(x)\,\nabla\sigma^T(x)w_j^i$$

This algorithm is shown to converge to the approximately optimal H_∞ solution. This yields a NN feedback controller as shown in Fig. 17 for the H_2 case.

10 APPROXIMATE DYNAMIC PROGRAMMING AND ADAPTIVE CRITICS

Approximate dynamic programming (ADP) is based on the optimal formulation of the feedback control problem. For discrete-time systems, the optimal control problem may be solved using dynamic programming,[37] which is a backward-in-time procedure and so unsuitable for online implementation. ADP is based on using nonlinear approximators to solve the HJ equations forward in time and was first suggested by Werbos.[44] See the Section 11 for cited works of major researchers in the area of ADP. The current status of work in ADP is given in Ref. 45.

The previous section presented the continuous-time formulation of the optimal control problem. For discrete-time systems of the form

$$x_{k+1} = f(x_k,u_k)$$

with k the time index, one may select the cost or performance measure

$$V(x_k) = \sum_{i=k}^{\infty} \gamma^{i-k} r(x_i,u_i)$$

with γ a discount factor and $r(x_k,u_k)$ known as the instantaneous utility. A first-difference equivalent to this yields a recursion for the value function given by

$$V(x_k) = r(x_k,u_k) + \gamma V(x_{k+1})$$

One may invoke Bellman's principle to find the optimal cost as

$$V^*(x_k) = \min_{u_k} (r(x_k,u_k) + \gamma V^*(x_{k+1}))$$

and the optimal control as

$$u^*(x_k) = \arg\min_{u_k} (r(x_k,u_k) + \gamma V^*(x_{k+1}))$$

Determining the optimal controller using these equations requires an iterative procedure known as dynamic programming that progresses backward in time. This is unsuitable for real-time implementation and is computationally complex.

The goal of ADP is to provide approximate techniques for evaluating the optimal value and optimal control using techniques that progress forward in time, so that they can be implemented in actual control systems. Howard[46] showed that the following successive iteration scheme, known as *policy iteration,* converges to the optimal solution:

1. Find the value for the prescribed policy $u_j(x_k)$:

$$V_j(x_k) = r(x_k, u_j(x_k)) + \gamma V_j(x_{k+1})$$

2. Policy improvement:

$$u_{j+1}(x_k) = \arg\min_{u_k} (r(x_k, u_k) + \gamma V_j(x_{k+1}))$$

Werbos[33] and others (see Section 11) showed how to implement ADP controllers by four basic techniques—HDP, DHP, and their action-dependent forms ADHDP, and ADDHP—to be described next.

Heuristic Dynamic Programming (*HDP*). In HDP, one approximates the value by a *critic* NN with tunable parameters w_j and the control by an *action-generating* NN with tunable parameters v_j so that

$$\text{Critic NN:} \quad V_j(x_k) \approx V(x_k, w_j)$$

$$\text{Action NN:} \quad u_j(x_k) \approx u(x_k, v_j)$$

HDP then proceeds as follows:

Critic Update

Find the desired target value using

$$V^D{}_{k,j+1} = r(x_k, u_j(x_k)) + \gamma V(x_{k+1}, w_j)$$

Update critic weights using recursive least squares (RLS), backprop, and so on:

$$w_{j+1} = w_j + \alpha_j \frac{\partial V}{\partial w_j} [V^D{}_{k,j+1} - V(x_k, w_j)]$$

Action Update

Find the desired target action using

$$u^D{}_{k,j+1} = \arg\min_{u_k} (r(x_k, u_k) + \gamma V(x_{k+1}, w_{j+1}))$$

Update critic weights using RLS, backprop, and so on:

$$v_{j+1} = v_j + \alpha_k \frac{\partial u}{\partial v_j} [u^D{}_{k,j+1} - u(x_k, v_j)]$$

This procedure is straightforward to implement given today's software (e.g., MATLAB). The value required for the next state x_{k+1} may be found either using the dynamics equation (1) or the next state can be observed from the actual system.

Dual Heuristic Programming (*DHP*). Noting that the control only depends on the value function gradient [e.g., see (9)], it is advantageous to approximate not the value but its gradient using a NN. This yields a more complex algorithm, but DHP converges faster than HDP. Details are in Ref. 33.

Q-Learning or Action-Dependent HDP. A function that is more advantageous than the value function for ADP is the Q function, defined by Watkins[47] and Werbos[8] as

$$Q(x_k, u_k) = r(x_k, u_k) + \gamma V(x_{k+1})$$

Note that Q is a function of both x_k and the control action u_k and that

$$Q_h(x_k,h(x_k)) = V_h(x_k)$$

where subscript h denotes a prescribed control or policy sequence $u_k = h(x_k)$. A recursion for Q is given by

$$Q_h(x_k,u_k) = r(x_k,u_k) + \gamma Q_h(x_{k+1},h(x_{k+1}))$$

In terms of Q, Bellman's principle is particularly easy to write; in fact, defining the optimal Q value as

$$Q^*(x_k,u_k) = r(x_k,u_k) + \gamma V^*(x_{k+1}))$$

one has the optimal value as

$$V^*(x_k) = \min_{u_k}(Q^*(x_k,u_k))$$

The optimal control policy is given by

$$h^*(x_k) = \arg\min_{u_k} (Q^*(x_k,u_k))$$

Watkins showed that the following successive iteration scheme, known as *Q learning,* converges to the optimal solution:

1. Find the Q value for the prescribed policy $h_j(x_k)$:

$$Q_j(x_k,u_k) = r(x_k,u_k) + \gamma Q_j(x_{k+1},h_j(x_{k+1}))$$

2. Policy improvement:

$$h_{j+1}(x_k) = \arg\min_{u_k} (Q_j(x_k,u_k))$$

Using NN to approximate the Q function and the policy, one can write the ADHDP algorithm in a very straightforward manner. Since the control input action u_k is now explicitly an input to the critic NN, this is known as action-dependent HDP. Q learning converges faster than HDP and can be used in the case of unknown system dynamics.[48]

An action-dependent version of DHP is also available wherein the gradients of the Q function are approximated using NNs. Note that two NNs are needed, since there are two gradients, as Q is a function of both x_k and u_k.

11 HISTORICAL DEVELOPMENT, REFERENCED WORK, AND FURTHER STUDY

A firm foundation for the use of NNs in feedback control systems has been developed over the years by many researchers. Included here is a historical development and references to the body of work in neurocontrol.

11.1 NN for Feedback Control

The use of NNs in feedback control systems was first proposed by Werbos.[8] Since then, NNs control has been studied by many researchers. Recently, NNs have entered the mainstream of control theory as a natural extension of adaptive control to systems that are nonlinear in the tunable parameters. The state of NN control is well illustrated by papers in the Automatica Special issue on NN control.[49] Overviews of the initial work in NN control are provided by Miller et al.[50] and the *Handbook of Intelligent Control,*[51] which highlighted a host of difficulties to be addressed for closed-loop control applications. Neural network applications in closed-loop control are fundamentally different from open-loop applications such as classification and image processing. The basic multilayer NN tuning strategy is backpropagation.[17] Basic problems that had to be addressed for closed-loop NN control[33,44] included weight initialization for feedback stability, determining the gradients needed for backpropagation tuning, determining what to backpropagate, obviating the need for preliminary off-line tuning, modifying backprop so that it tunes the weights forward through time, and providing efficient computer code for implementation. These issues have since been addressed by many approaches.

Initial work in NN was for system identification and identification-based indirect control. In closed-loop control applications, it is necessary to show the stability of the tracking error as well as boundedness of the NN weight estimation errors. Proofs for internal stability, bounded NN weights (e.g., bounded control signals), guaranteed tracking performance, and robustness were absent in early works. Uncertainty as to how to initialize the NN weights led to the necessity for "preliminary off-line tuning." Work on off-line learning was formalized by Kawato.[52] Off-line learning can yield important structural information.

Subsequent work in NNs for control addressed closed-loop system structure and stability issues. Work by Sussmann[53] and Albertini and Sontag[54] was important in determining system properties of NNs (e.g., minimality and uniqueness of the ideal NN weights, observability of dynamic NNs). The seminal work of Narendra and Parthasarathy[9,10] had an emphasis on finding the gradients needed for backprop tuning in feedback systems, which, when the plant dynamics are included, become recurrent nets. In recurrent nets, these gradients themselves satisfy difference or differential equations, so they are difficult to find. Sadegh[55] showed that knowing an approximate plant Jacobian is often good enough to guarantee suitable closed-loop performance.

The approximation properties of NN[2,3] are basic to their feedback controls applications. Based on this and analysis of the error dynamics, various modifications to backprop were presented that guaranteed closed-loop stability as well as weight error boundedness. These are akin to terms added in adaptive control to make algorithms robust to high-frequency unmodeled dynamics. Sanner and Slotine[5] used radial basis functions in control and showed how to select the NN basis functions, Polycarpou and Ioannou[56,57] used a projection method for weight updates, and Lewis and Syrmos[37] used backprop with an e-modification term.[18] All this work used NNs that are *linear* in the unknown parameter. In linear NNs, the problem is relegated to determining activation functions that form a *basis set* (e.g., RBF[5] and functional link programmable network (FLPN)[55]). It was shown by Sanner and Slotine[5] how to systematically derive stable NN controllers using approximation theory and basis functions. Barron[58] has shown that using NNs that are linear in the tunable parameters gives a fundamental limitation of the approximation accuracy to the order of $1/L^{2/n}$, where L is the number of hidden layer neurons and n is the number of inputs. Nonlinear-in-the-parameters NNs overcome this difficulty and were first used by Chen and Khalil,[59] who used backprop with deadzone weight tuning, and Lewis et al.,[60] who used Narendra's e-modification term in the backprop. In nonlinear-in-the-parameters NNs, the basis is automatically selected online by tuning the first-layer weights and thresholds. Multilayer NNs were rigorously used for discrete-time control by Jagannathan and Lewis.[61] Polycarpou[62] derived NN controllers that do not assume known bounds on the ideal weights. Dynamic/recurrent NNs were used for control by Rovithakis and Christodoulou,[63] Poznyak,[64] Rovithakis,[65] who considered multiplicative disturbances; Zhang and Wang,[66] and others.

Most stability results on NN control have been local in nature, and global stability has been treated by Kwan et al.,[67] and others. Recently, NN control has been used in conjunction with other control approaches to extend the class of systems that yields to nonparametric control. Calise and coworkers[30,68,69] used NNs in conjunction with dynamic inversion to control aircraft and missiles. Feedback linearization using NNs has been addressed by Chen and Khalil,[59] Yesildirek and Lewis,[70] Ge et al.,[22] and others. NNs were used with backstepping[25] by Lewis, et al.,[15] Arslan and Basar,[71] Wang and Huang,[72] Ge et al.,[20,21] and others.

NNs have been used in conjunction with the Isidori–Byrnes regulator equations for output-tracking control by Wang and Huang.[72] A multimodel NN control approach has been given by Narendra and Balakrishnan.[73] Applications of NN control have been extended to partial differential equation systems by Padhi et al.[74] NNs have been used for control of stochastic systems by Poznyak and Ljung.[75] Parisini and co-workers have developed receding horizon controllers based on NNs[76] and hybrid discrete-event NN controllers.[77]

In practical implementations of NN controllers there remain problems to overcome. Weight initialization still remains an issue, and one may also find that the NN weights become unbounded despite proofs to the contrary. Practical implementation issues were addressed by Chen and Chang,[78] Gutierrez and Lewis,[79] and others. Random initialization of the first-layer NN weights often works in practice, and work by Igelnik and Pao[7] shows that it is theoretically defensible. Computational complexity makes NNs with many hidden layer neurons difficult to implement. Recently, work has intensified in wavelets, NNs that have localized basis functions, and NNs that are self-organizing in the sense of adding or deleting neurons automatically.[36,80,81]

By now it is understood that NNs offer an elegant extension of adaptive control and other techniques to systems that are nonlinear in the unknown parameters. The universal approximation properties of NNs[2,3] avoid the use of specialized basis sets, including regression matrices. Formalized improved proofs avoid the use of assumptions such as certainty equivalence. Robustifying terms avoid the need for persistency of excitation. Recent books on NN feedback control include Refs. 15, 20, 22, 28, 31, and 82.

11.2 Approximate Dynamic Programming

Adaptive critics are reinforcement learning designs that attempt to approximate dynamic programming.[83,84] They approach the optimal solution through forward approximate dynamic programming. Initially, they were proposed by Werbos.[44] Overviews of the initial work in NN control are provided by Miller et al.[50] and the *Handbook of Intelligent Control*.[52] Howard[46] showed the convergence of an algorithm relying on the successive policy iteration solution of a nonlinear Lyapunov equation for the cost (value) and an optimizing equation for the control (action). This algorithm relied on perfect knowledge of the system dynamics and is an off-line technique. Later, various online dynamic-programming-based reinforcement learning algorithms emerged and were mainly based on Werbos's HDP,[33] Sutton's temporal differences (TDs) learning methods,[85] and Q-learning, which was introduced by Watkins[47] and Werbos[8] (called *action-dependent* critic schemes there). Critic and action network tuning was provided by RLS, gradient techniques, or the backpropagation algorithm.[17] Early work on dynamic-programming-based reinforcement learning focused on discrete finite-state and action spaces. These depended on lookup tables or linear function approximators. Convergence results were shown in this case, such as Dayan.[86]

For continuous-state and action spaces, convergence results are more challenging as adaptive critics require the use of nonlinear function approximators. Four schemes for approximate dynamic programming were given in Ref. 33, the HDP and DHP algorithms and their action-dependent versions (ADHDP and ADDHP). The linear quadratic regulation (LQR) problem[37] served as a testbed for much of these studies. Solid convergence results were obtained for various adaptive critic designs for the LQR problem. We mention the work of Bradtke et al.,[48] where Q learning was shown to converge when using nonlinear function approximators. An important persistence of excitation notion was included. Further work was done by Landelius,[87] who studied the four adaptive critic architectures. He demonstrated convergence results for all four cases in the LQR case and discussed when the design is model free. Hagen and Krose[88] discussed the effect of model noise and exploration noise when the adaptive critic is viewed as a stochastic approximation technique. Prokhorov and Feldkamp[89] looked at Lyapunov stability analysis. Other convergence results are due to Balakrishnan and co-workers,[74,90] who have also studied the optimal control of aircraft and distributed-parameter systems governed by partial differential equations. Anderson et al.[91] showed convergence and stability for a reinforcement learning scheme. All of these results were done for the discrete-time case.

A thorough treatment of neurodynamic programming is given in the seminal book by Bertsekas and Tsitsiklis.[92] Various successful practical implementations have been reported, including aircraft control examples by Ferrari and Stengel,[93] an Auto Lander by Murray et al.,[94] state estimation using dynamic NNs by Feldkamp and Prokhorov,[95] and Neuro-Observers by Liu and Balakrishnan.[96] Si et al. have provided analysis[97] and applied ADP to aircraft control.[46] An account of adaptive critic designs is found in Prokhorov and Wunsch.

Applications of adaptive critics in the continuous-time domain were mainly done through discretization and the application of well-established discrete-time results (e.g., Ref. 99). Various continuous-time nondynamic reinforcement learning strategies were discussed by Campos and Lewis[100] and Rovithakis,[101] who approximated a Lyapunov function derivative. In Kim and Lewis[31] the HJB equation of dynamic programming is approximated by a Riccati equation, and a suboptimal controller based on NN feedback linearization is implemented with full stability and convergence proofs. Murray et al.[102] prove convergence of an algorithm that uses system state measurements to find the cost to go. An array of initial conditions is needed. Unknown plant dynamics in the linear case is confronted by estimating a matrix of state derivatives. The cost functional is shown to be a Lyapunov function and is approximated using either quadratic functions or an RBF neural network. Saridis and Lee[39] showed the convergence of an off-line algorithm relying on the successive iteration solution of a nonlinear Lyapunov equation for the cost (value) and an optimizing equation for the control (action). This is the continuous-time equivalent of Howard's work. Beard et al.,[40] showed how to actually solve these equations using Galerkin integral approximations, which require much computational effort.

Q-learning is not well posed when sampling times become small and so is not useful for extension to continuous-time systems. Continuous-time dynamic-programming-based reinforcement learning is reformulated using the so-called advantage learning by Baird,[103] who defines a differential increment from the optimal solution and explicitly takes into account the sampling interval Δt. Doya[104] derives results for online updating of the critic using techniques from continuous-time nonlinear optimal control. The advantage function follows naturally from this approach and in fact coincides with the continuous-time Hamiltonian function. Doya gives relations with the TD(0) and TD(λ) techniques of Sutton.[85] Lyshevski[105] has focused on a general parametrized form for the value function and obtained a set of algebraic equations that can be solved for an approximate value function.

Acknowledgments

The referenced work of Lewis and co-workers was sponsored by National Science Foundation grant ECS-01-40490 and Army Research Office grant DAAD19-02-1-0366.

REFERENCES

1. von Bertalanffy, L., *General System Theory,* Braziller, New York, 1968.
2. G. Cybenko, "Approximation by Superpositions of a Sigmoidal Function," *Mathematics of Control, Signals and Systems* **2**(4), 303–314 (1989).
3. J. Park and I. W. Sandberg, "Universal Approximation Using %Radial-Basis-Function Networks," *Neural Computation* **3,** 246–257 (1991).
4. J. S. Albus, "A New Approach to Manipulator Control: The Cerebellar Model Articulation Controller Equations (CMAC)," *Transactions ASME Journal of Dynamics, Systems, Measurement, and Control,* **97,** 220–227 (1975).
5. R. M. Sanner and J.-J. E. Slotine, "Gaussian Networks for Direct Adaptive Control," *IEEE Transactions on Neural Networks* **3**(6), 837–863 (1992).
6. L.-X. Wang, *Adaptive Fuzzy Systems and Control: Design and Stability Analysis,* Prentice-Hall, Englewood Cliffs, NJ, 1994.
7. B. Igelnik and Y.-H. Pao, "Stochastic Choice of Basis Functions in Adaptive Function Approximation and Functional-Link Net," *IEEE Transaction on Neural Networks* **6**(6), 1320–1329 (1995).
8. P. J. Werbos, "Neural Networks for Control and System Identification," in *Proceedings of the IEEE Conference on Decision and Control,* FL, 1989.
9. K. S. Narendra and K. Parthasarathy, "Identification and Control of Dynamical Systems Using Neural Networks," *IEEE Transactions on Neural Networks,* **1,** 4–27 (1990).
10. K. S. Narendra and K. Parthasarathy, "Gradient Methods for the Optimization of Dynamical Systems Containing Neural Networks," *IEEE Transactions on Neural Networks* **2**(2), 252–262 (1991).
11. Y. D. Landau, *Adaptive Control,* Marcel Dekker, New York, 1979.
12. F. L. Lewis, D. M. Dawson, and C. T. Abdallah, *Robot Manipulator Control,* Marcel Dekker, New York, 2004.
13. J. J. E. Slotine and J. A. Coetsee, "Adaptive Sliding Controller Synthesis for Nonlinear Systems," *International Journal of Control* **43**(4), 1631–1651 (1986).
14. J. J. E. Slotine and W. Li, "On the Adaptive Control of Robot Manipulators," *International Journal of Robotics Research* **6**(3), 49–59 (1987).
15. F. L. Lewis, S. Jagannathan, and A. Yesildirek, *Neural Network Control of Robot Manipulators and Nonlinear Systems,* Taylor and Francis, London, 1999.
16. L. R. Hunt, R. Su, and G. Meyer, "Global Transformations of Nonlinear Systems," *IEEE Transactions on Automatic Control* **28,** 24–31 (1983).
17. P. J. Werbos, "Beyond Regression: New Tools for Prediction and Analysis in the Behavior Sciences," Ph.D. Thesis, Committee on Applied Mathemathics, Harvard University, 1974.
18. K. S. Narendra and A. M. Annaswamy, "A New Adaptive Law for Robust Adaptation without Persistent Excitation," *IEEE Transactions on Automatic Control,* **AC-32**(2), 134–145 (1987).
19. P. Ioannou and J. Sun, *Robust Adaptive Control,* Prentice-Hall, Englewood Cliffs, NJ, 1996; electronic copy available at http://www-rcf.usc.edu/~ioannou/Robust_Adaptive_Control.htm
20. S. S. Ge, C. C. Hang, T. H. Lee, and T. Zhang, *Stable Adaptive Neural Network Control, Kluwer, Boston, MA, 2001.*
21. S. S. Ge, T. H. Lee, and Z. P. Wang, "Adaptive Neural Network Control for Smart Materials Robots Using Singular Perturbation Technique," *Asian Journal of Control* **3**(2), 143–155 (2001).
22. S. S. Ge, T. H. Lee, and C. J. Harris, *Adaptive Neural Network Control of Robotic Manipulators,* World Scientific, Singapore, 1998.
23. F. L. Lewis, *Applied Optimal Control and Estimation: Digital Design and Implementation,* TI Series, Prentice-Hall, Englewood Cliffs, NJ, 1992.
24. S. S. Ge, G. Y. Li, and T. H. Lee, "Adaptive NN Control for a Class of Strict Feedback Discrete-Time Nonlinear Systems," *Automatica* **39,** 807–819 (2003).
25. M. Krstic, I. Kanellakopoulos, and P. Kokotovic, *Nonlinear and Adaptive Control Design,* Wiley, New York, 1995.
26. P. V. Kokotovic, "Applications of Singular Perturbation Techniques to Control Problems," *SIAM Review,* **26**(4), 501–550 (1984).

27. N. H. McClamroch and D. Wang, "Feedback Stabilization and Tracing of Constrained Robots," *IEEE Transactions on Automatic Control,* **33,** 419–426 (1988).

28. F. L. Lewis, J. Campos, and R. Selmic, *Neuro-Fuzzy Control of Industrial Systems with Actuator Nonlinearities,* Society of Industrial and Applied Mathematics Press, Philadelphia, PA, 2002.

29. B. L. Stevens and F. L. Lewis, *Aircraft Control and Simulation,* 2nd ed., Wiley, New York, 2003.

30. A. J. Calise, N. Hovakimyan, and H. Lee, "Adaptive Output Feedback Control of Nonlinear Systems Using Neural Networks," *Automatica* **37**(8), 1201–1211 (2001).

31. Y. H. Kim and F. L. Lewis, *High-Level Feedback Control with Neural Networks,* World Scientific, Singapore, 1998.

32. A. G. Barto, R. S. Sutton, and C. W. Anderson, "Neuron-like Elements That Can Solve Difficult Learning," *IEEE Transactions on Systems, Man, and Cybernetics,* **13**(5), 634–646 (1983).

33. P. J. Werbos, "Approximate Dynamic Programming for Real-Time Control and Neural Modeling," in *Handbook of Intelligent Control,* D. A. White and D. A. Sofge (eds.), Van Nostrand Reinhold, New York, 1992.

34. C. R. Johnson, Jr., *Lectures on Adaptive Parameter Estimation,* Prentice-Hall, Englewood Cliffs, NJ, 1988.

35. K. M. Passino and S. Yurkovich, *Fuzzy Control,* Addison-Wesley, Menlo Park, NJ, 1998.

36. R. M. Sanner and J.-J. E. Slotine, "Structurally Dynamic Wavelet Networks for Adaptive Control of Robotic Systems," *International Journal of Control* **70**(3), 405–421 (1998).

37. F. L. Lewis and V. Syrmos, *Optimal Control,* 2nd ed., Wiley, New York, 1995.

38. M. Abu-Khalaf and F. L. Lewis, "Nearly Optimal State Feedback Control of Constrained Nonlinear Systems Using a Neural Networks HJB Approach," *IFAC Annual Reviews in Control* **28,** 239–251 (2004).

39. G. Saridis and C. S. Lee, "An Approximation Theory of Optimal Control for Trainable Manipulators," *IEEE Transactions on Systems, Man, and Cybernetics,* **9**(3), 152–159 (1979).

40. R. Beard, G. Saridis, and J. Wen, "Galerkin Approximations of the Generalized Hamilton-Jacobi-Bellman Equation," *Automatics,* **33**(12), 2159–2177 (1997).

41. H. Knobloch, A. Isidori, and D. Flockerzi, *Topics in Control Theory,* Springer Verlag, Boston, 1993.

42. T. Başar and P. Bernard, *H_∞ Optimal Control and Related Minimax Design Problems,* Birkhäuser, 1995.

43. M. Abu-Khalaf, F. L. Lewis, and J. Huang, "Computational Techniques for Constrained Nonlinear State Feedback *H*-Infinity Optimal Control Using Neural Networks," paper 1141, presented at the Mediterranean Conference on Control and Automation, Kusadasi, Turkey, June 2004.

44. P. J. Werbos., "A Menu of Designs for Reinforcement Learning Over Time," in *Neural Networks for Control,* W. T. Miller, R. S. Sutton, and P. J. Werbos (eds.), MIT Press, Cambridge, MA, 1991, pp. 67–95.

45. J. Si, A. Barto, W. Powell, and D. Wunsch, *Handbook of Learning and Approximate Dynamic Programming,* IEEE Press, West Conshohocken, PA, 2004.

46. R. Howard, *Dynamic Programming and Markov Processes,* MIT Press, Cambridge, MA, 1960.

47. C. Watkins, "Learning from Delayed Rewards," Ph.D. Thesis, Cambridge University, Cambridge, England, 1989.

48. S. Bradtke, B. Ydstie, and A. Barto, "Adaptive Linear Quadratic Control Using Policy Iteration," CMPSCI-94-49, University of Massachusetts, Amherst, MA, June 1994.

49. K. S. Narendra and F. L. Lewis, Special Issue on Neural Network Feedback Control, *Automatica* **37**(8) (2001).

50. W. T. Miller, R. S. Sutton, and P. J. Werbos (eds.), *Neural Networks for Control,* MIT Press, Cambridge, MA, 1991.

51. D. A. White and D. A. Sofge (eds.), *Handbook of Intelligent Control,* Van Nostrand Reinhold, New York, 1992.

52. M. Kawato, "Computational Schemes and Neural Network Models for Formation and Control of Multijoint Arm Trajectory," in *Neural Networks for Control,* W. T. Miller, R. S. Sutton, and P. J. Werbos (eds.), MIT Press, Cambridge, MA, 1991, pp. 197–228.

53. H. J. Sussmann, "Uniqueness of the Weights for Minimal Feedforward Nets with a Given Input-Output Map," *Neural Networks* **5,** 589–593 (1992).

54. F. Albertini and E. D. Sontag, "For Neural Nets, Function Determines Form," *Proceedings of the IEEE Conference Decision and Control,* December 1992, pp. 26–31.

55. N. Sadegh, "A Perceptron Network for Functional Identification and Control of Nonlinear Systems," *IEEE Transactions on Neural Networks* **4**(6), 982–988 (1993).

56. M. M. Polycarpou and P. A. Ioannou, "Identification and Control Using Neural Network Models: Design and Stability Analysis," Technical Report 91-09-01, Dept. Elect. Eng. Sys., University of Southern California, Los Angeles, CA, September 1991.

57. M. M. Polycarpou and P. A. Ioannou, "Neural Networks as On-Line Approximators of Nonlinear Systems," in *Proceedings of the IEEE Conference on Decision and Control,* Tucson, December 1992, pp. 7–12.

58. A. R. Barron, "Universal Approximation Bounds for Superpositions of a Sigmoidal Function," *IEEE Transactions on Information Theory* **39**(3), 930–945 (1993).

59. F.-C. Chen and H. K. Khalil, "Adaptive Control of Nonlinear Systems Using Neural Networks," *International Journal of Control,* **55**(6), 1299–1317 (1992).

60. F. L. Lewis, A. Yesildirek, and K. Liu, "Multilayer Neural Net Robot Controller with Guaranteed Tracking Performance," *IEEE Transactions on Neural Networks,* **7**(2), 388–399 (1996).

61. S. Jagannathan and F. L. Lewis, "Multilayer Discrete-Time Neural Net Controller with Guaranteed Performance," *IEEE Transactions on Neural Networks* **7**(1), 107–130 (1996).

62. M. M. Polycarpou, "Stable Adaptive Neural Control Scheme for Nonlinear Systems," *IEEE Transactions on Automatic Control* **41**(3), 447–451 (1996).

63. G. A. Rovithakis and M. A. Christodoulou, "Adaptive Control of Unknown Plants Using Dynamical Neural Networks," *IEEE Transactions on Systems, Man, and Cybernetics* **24**(3), 400–412 (1994).

64. A. S. Poznyak, E. N. Sanchez, and W. Yu, *Differential Neural Networks for Robust Nonlinear Control,* World. Scientific, Singapore, 2001.

65. G. A. Rovithakis, "Performance of a Neural Adaptive Tracking Controller for Multi-Input Nonlinear Dynamical Systems," *IEEE Transactions on Systems, Man, and Cybernetic, Part A* **30**(6), 720–730, (2000).

66. Y. Zhang and J. Wang, "Recurrent Neural Networks for Nonlinear Output Regulation," *Automatica* **37**(8), 1161–1173 (2001).

67. C. Kwan, D. M. Dawson, and F. L. Lewis, "Robust Adaptive Control of Robots Using Neural Network: Global Stability," *Asian Journal of Control* **3**(2), 111–121 (2001).

68. J. Leitner, A. J. Calise, and J. V. R. Prasad, "Analysis of Adaptive Neural Networks for Helicopter Flight Control," *Journal of Guidance, Control, and Dynamics* **20**(5), 972–979 (1997).

69. M. B. McFarland and A. J. Calise, "Adaptive Nonlinear Control of Agile Anti-Air Missiles Using Neural Networks," *IEEE Transactions on Control Systems Technology* **8**(5), 749–756 (2000).

70. A. Yesildirek and F. L. Lewis, "Feedback Linearization Using Neural Networks," *Automatica* **31**(11), 1659–1664 (1995).

71. G. Arslan and T. Basar, "Disturbance Attenuating Controller Design for Strict-Feedback Systems with Structurally Unknown Dynamics," *Automatica* **37**(8), 1175–1188 (2001).

72. D. Wang and J. Huang, "Neural Network Based Adaptive Tracking of Uncertain Nonlinear Systems in Triangular Form," *Automatica* **38,** 1365–1372 (2002).

73. K. S. Narendra and J. Balakrishnan, "Adaptive Control Using Multiple Models," *IEEE Transactions on Automatic Control* **42**(2) 171–187 (1997).

74. R. Padhi, S. N. Balakrishnan, and T. Randolph, "Adaptive-Critic Based Optimal Neuro Control Synthesis for Distributed Parameter Systems," *Automatica* **37**(8), 1223–1234 (2001).

75. A. S. Poznyak and L. Ljung, "On-Line Identification and Adaptive Trajectory Tracking for Nonlinear Stochastic Continuous Time Systems Using Differential Neural Networks," *Automatica* **37**(8), 1257–1268 (2001).

76. T. Parisini, M. Sanguineti, and R. Zoppoli, "Nonlinear Stabilization by Receding-Horizon Neural Regulators," *International Journal of Control* **70,** 341–362 (1998).

77. T. Parisini and S. Sacone, "Stable Hybrid Control Based on Discrete-Event Automata and Receding-Horizon Neural Regulators," *Automatica* **37**(8), 1279–1292 (2001).

78. F.-C. Chen and C.-H. Chang, "Practical Stability Issues in CMAC Neural Network Control Systems," *IEEE Transactions on Control Systems Technology,* **4**(1), 86–91 (1996).

79. L. B. Gutierrez and F. L. Lewis, "Implementation of a Neural Net Tracking Controller for a Single Flexible Link: Comparison with PD and PID Controllers," *IEEE Transactions on Industrial Electronics* **45**(2), 307–318 (1998).

80. J. A. Farrell, "Stability and Approximator Convergence in Nonparametric Nonlinear Adaptive Control," *IEEE Transactions on Neural Networks* **9**(5), 1008–1020 (1998).
81. J. Y. Choi and J. A. Farrell, "Nonlinear Adaptive Control Using Networks of Piecewise Linear Approximators," *IEEE Transactions on Neural Networks* **11**(2), 390–401 (2000).
82. R. Zbikowski and K. J. Hunt, *Neural Adaptive Control Technology,* World Scientific, Singapore, 1996.
83. A. G. Barto, "Connectionist Learning for Control," in *Neural Networks for Control,* W. T. Miller, R. S. Sutton, P. J. Werbos (eds.), MIT Press, Cambridge, MA, 1991.
84. A. G. Barto and T. G. Dietterich, "Reinforcement Learning and Its Relationship to Supervised Learning," in *Handbook of Learning and Approximate Dynamic Programming,* J. Si, A. Barto, W. Powell, and D. Wunsch (eds.), IEEE Press, West Conshohocken, PA, 2004.
85. R. Sutton, "Learning to Predict by the Method of Temporal Differences," *Machine Learning* **3,** 9–44 (1988).
86. P. Dayan, "The Convergence of TD(λ) for General λ," *Machine Learning* **8**(3–4), 341–362 (1992).
87. T. Landelius, "Reinforcement Learning and Distributed Local Model Synthesis," Ph.D. Dissertation, Linköping University, 1997.
88. S. T. Hagen and B. Krose, "Linear Quadratic Regulation Using Reinforcement Learning," in *Proceedings of the Eighth Belgian-Dutch Conference on Machine Learning, BENELEARN'98,* F. Verdenius and W. van den Broek (eds.), Wageningen, October 1998, pp. 39–46.
89. D. V. Prokhorov amd L. A. Feldkamp, "Analyzing for Lyapunov Stability with Adaptive Critics," in *Proceedings of the International Conference on Systems, Man, Cybernetics,* Dearborn, MI, 1998, pp. 1658–1661.
90. X. Liu and S. N. Balakrishnan, "Convergence Analysis of Adaptive Critic Based Optimal Control," in *Proceedings of the American Control Conference,* Chicago, IL, 2000, pp. 1929–1933.
91. C. Anderson, R. M. Kretchner, P. M. Young, and D. C. Hittle, "Robust Reinforcement Learning Control with Static and Dynamic Stability," *International Journal of Robust and Nonlinear Control,* **11** (2001).
92. D. P. Bertsekas and J. N. Tsitsiklis, *Neuro-Dynamic Programming,* Athena Scientific, MA, 1996.
93. S. Ferrari and R. Stengel, "An Adaptive Critic Global Controller," in *Proceedings of the American Control Conference,* Anchorage, AK, 2002, pp. 2665–2670.
94. J. Murray, C. Cox, R. Saeks, and G. Lendaris, "Globally Convergent Approximate Dynamic Programming Applied to an Autolander," in *Proc. ACC,* Arlington, VA, 2001, pp. 2901–2906.
95. L. Feldkamp and D. Prokhorov, "Recurrent Neural Networks for State Estimation," paper presented at the Twelfth Yale Workshop on Adaptive and Learning Systems, New Haven, CT, 2003, pp. 17–22.
96. X. Liu and S. N. Balakrishnan, "Adaptive Critic Based Neuro-Observer," in *Proceedings of the American Control Conference,* Arlington, VA, 2001, pp. 1616–1621.
97. J. Si and Y.-T. Wang, "On-Line Control by Association and Reinforcement," *IEEE Transactions on Neural Networks* **12**(2), 264–276 (2001).
98. D. Prokhorov and D. Wunsch, "Adaptive Critic Designs," *IEEE Transactions on Neural Networks* **8**(5) (1997).
99. J. N. Tsitsiklis, "Efficient Algorithms for Globally Optimal Trajectories," *IEEE Transactions on Automatic Control* **40**(9), 1528–1538 (1995).
100. J. Campos and F. L. Lewis, "Adaptive Critic Neural Network for Feedforward Compensation," in *Proceedings of the American Control Conference,* San Diego, CA, June 1999.
101. G. A. Rovithakis, "Stable Adaptive Neuro-Control Via Lyapunov Function Derivative Estimation," *Automatica* **37**(8), 1213–1221 (2001).
102. J. Murray, C. Cox, G. Lendaris, and R. Saeks, "Adaptive Dynamic Programming," *IEEE Transactions on Systems, Man, and Cybernetics* **32**(2) (2002).
103. L. Baird, "Reinforcement Learning in Continuous Time: Advantage Updating," in *Proceedings of the International Conference on Neural Networks,* Orlando, FL, June 1994.
104. K. Doya, "Reinforcement Learning in Continuous Time and Space," *Neural Computation* **12,** 219–245 (2000).
105. S. E. Lyshevski, *Control Systems Theory with Engineering Applications,* Birkhauser, Berlin, 2001.

BIBLIOGRAPHY

Beard, R., G. Saridis, and J. Wen, "Approximate Solutions to the Time-Invariant Hamilton-Jacobi-Bellman Equation," *Automatica* **33**(12), 2159–2177 (1997).

Gong, J. Q., and B. Yao, "Neural Network Adaptive Robust Control of Nonlinear Systems in Semi-Strict Feedback Form," *Automatica* **37**(8), 1149–1160 (2001).

Lewis, F. L., "Nonlinear Network Structures for Feedback Control," *Asian Journal of Control* **1**(4), 205–228 (1999).

Lewis, F. L., K. Liu, and A. Yesildirek, "Neural Net Robot Controller with Guaranteed Tracking Performance," *IEEE Transactions on Neural Networks* **6**(3), 703–715 (1995).

Li, Y., N. Sundararajan, and P. Saratchandran, "Neuro-Controller Design for Nonlinear Fighter Aircraft Maneuver Using Fully Tuned Neural Networks," *Automatica* **37**(8), 1293–1301 (2001).

Mendel, J. M., and R. W. MacLaren, "Reinforcement Learning Control and Pattern Recognition Systems," in *Adaptive, Learning, and Pattern Recognition Systems: Theory and Applications,* J. M. Mendel and K. S. Fu (eds.), Academic, New York, 1970, pp. 287–318.

Miyamoto, H., M. Kawato, T. Setoyama, and R. Suzuki, "Feedback-Error-Learning Neural Network for Trajectory Control of a Robotic Manipulator," *Neural Networks,* **1,** 251–265 (1988).

Narendra, K. S., "Adaptive Control of Dynamical Systems Using Neural Networks," in *Handbook of Intelligent Control,* D. A. White and D. A. Sofge (eds.), Van Nostrand Reinhold, New York, 1992, pp. 141–183.

Poznyak, A. S., W. Yu, E. N. Sanchez, and J. P. Perez, "Nonlinear Adaptive Trajectory Tracking Using Dynamic Neural Networks," *IEEE Transactions on Neural Networks* **10**(6), 1402–1411 (1999).

Selmic, R., F. L. Lewis, A. J. Calise, and M. B. McFarland, "Backlash Compensation Using Neural Network," U.S. Patent 6,611,823, August 26, 2003.

Wang, J., and J. Huang, "Neural Network Enhanced Output Regulation in Nonlinear Systems," *Automatica* **37**(8) 1189–1200 (2001).

Zhang, T., S. S. Ge, and C. C. Hang, "Adaptive Neural Network Control for Strict Feedback Nonlinear Systems Using Backstepping Design," *Automatica* **36**(12), 1835–1846 (2000).

CHAPTER 20
MECHATRONICS

Shane Farritor
University of Nebraska–Lincoln
Lincoln, Nebraska

Mechatronics is the integration of computers, electronics, and information sciences into mechanical engineering. The advent of inexpensive and small microcomputers and electronics has led to the widespread integration into mechanical systems. The goal of this integration is to improve the performance of mechanical systems. Today's mechanical engineers cannot design state-of-the art systems without considering the cross-disciplinary field of mechatronics.

A typical mechatronic system is depicted in Fig. 1. It shows a mechanical system whose performance is monitored by some type of sensor. The sensor output is conditioned so it can be input into a microcomputer usually embedded in the product. The microcomputer determines an input to the mechanical system that will improve performance. The microcomputer outputs this information, and this output is converted and amplified to cause an actuator to alter the behavior of the mechanical system. This sense–plan–act cycle is repeated (usually several times per second) to improve the performance of the mechanical system.

Knowledge of mechatronics first and foremost requires a solid understanding of mechanical engineering. Mechanical engineering and control systems are covered throughout this book so this chapter will give only a basic overview of electrical engineering and computer engineering.

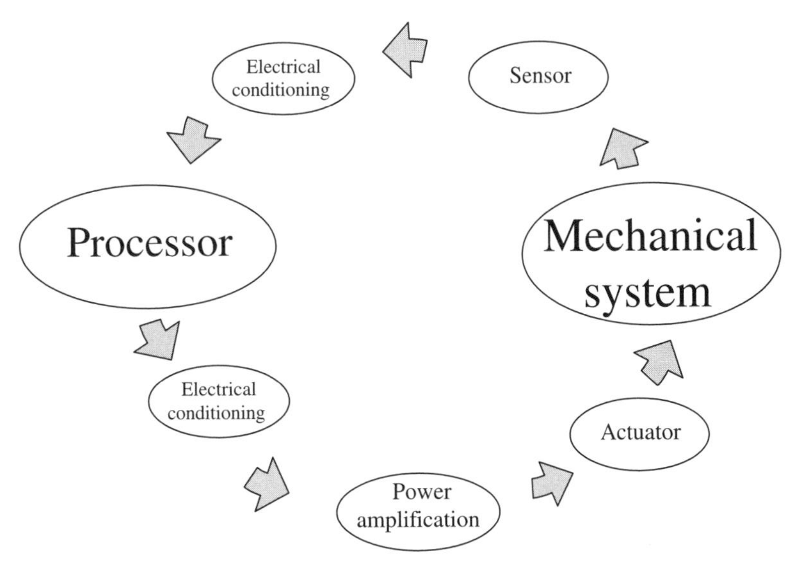

Figure 1 Mechatronic system.

1 BASIC ANALOG ELECTRONICS

Basic circuit analysis and design are critical in the field of mechatronics. With improving computer technology it is tempting to treat all systems as digital; however, this is an analog world. Analog circuits are often the best choice to condition and modify signals to and from analog actuators and sensors.

1.1 Definitions

Engineering modeling and analysis are often described as the analysis of the flow of power between various elements. The state of an engineering element can be defined by two variables—the effort variable and the flow variable. The instantaneous power for that element is defined as the product of the effort variable and the flow variable. For example, in traditional linear mechanical systems the power is defined as the product of the velocity of a mass and the net applied force to that mass. In fluid systems this can be written as the product of the volumetric flow rate and the pressure.

These analogies are often helpful to mechanical engineers trying to understand electrical circuits. In electrical systems the power is defined as the product of the electrical current (amperage) and the electrical potential (voltage).

Definitions of the variables used in circuit analysis are given below:

1. *Current.* Electrical current is usually designated with the symbol i (or I). It is defined as the time rate of change of the charge (designated q with units of coulombs). The electrical current has the units of the ampere (A).

$$i = \text{time rate of change of charge} = \frac{dq}{dt} \qquad q = \text{charge (C)}$$

2. *Voltage.* Voltage is the electromotive force or potential difference and is designated with the symbol V (or e, E, v). It is a measure of the potential of an electric field. The electrical voltage has the units of volts and is generally written

$$V_{AB} = \text{potential of } A \text{ with respect to } B$$

or more commonly with a (plus sign) at point A and a (minus sign) at point B.

3. *Power.* Power is defined as the ability to do work. The flow of power is generally what is described by engineering models. Power has units of watts and is the product of the voltage and current.
4. *Work.* Work (or energy) has units of joules and is the integral of power over time:

$$W = \int_{t_0}^{t_f} P(t)\,\partial t \qquad (1)$$

1.2 Basic Electrical Elements

Several elements and their constitutive relationships need to be understood to analyze basic electrical circuits. The constitutive relationship for each element describes the relationship between the voltage (effort variable) and the current (flow variable):

1. *Ideal Voltage Source.* The ideal voltage source produces a voltage, or electrical potential, as a function of time. This voltage produced by the source is an input to the system and the voltage source adds power to the system. The current passing through the ideal voltage source is not generally known. This current is determined by the other elements connected to the voltage source. The ideal voltage source is capable of producing infinite current and therefore infinite power. See Fig. 2.
2. *Ideal Current Source.* The ideal current source produces a current as a function of time. This current produced by the source is an input to the system and the current source adds power to the system. The voltage across the ideal current source is not generally known. This voltage is determined by the other elements connected to the current source. The ideal current source is capable of producing infinite voltage and therefore infinite power. See Fig. 3.
3. *Resistor.* The resistor restricts the flow of current and dissipates power. The magnitude of this resistance is denoted R and has the units of ohms (Ω). The resistor is a passive element in that it does not add energy to the system. The resistor is governed by a passive sign convention, as shown in Fig. 4, in that the current flows across the resistor from the side of higher potential (voltage) to the side of lower potential (voltage). If the current and the voltage are negative, this indicates that current flow is in the opposite direction and the potential is higher on the opposite side. In linear

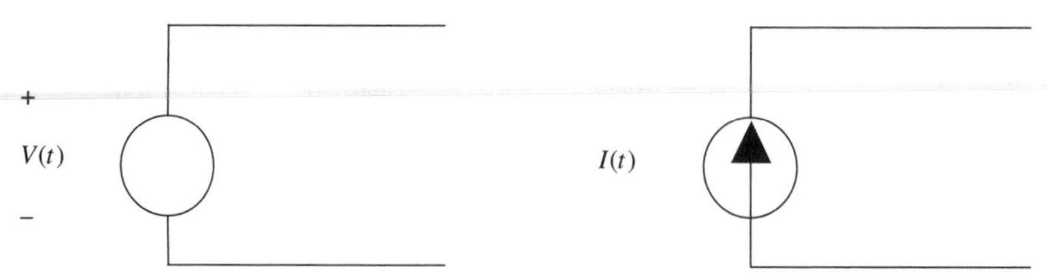

Figure 2 Ideal voltage source.

Figure 3 Ideal current source.

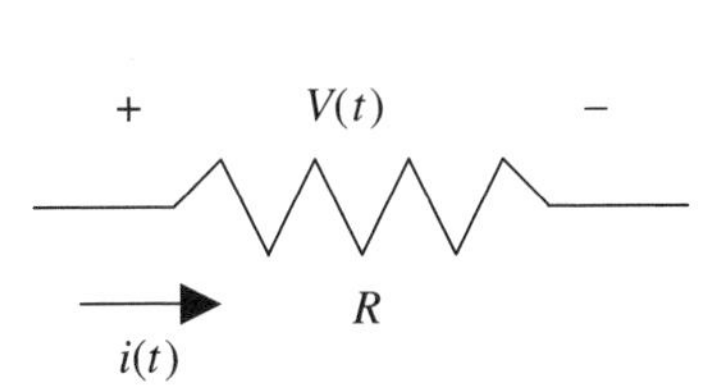

Figure 4 Ideal resistor.

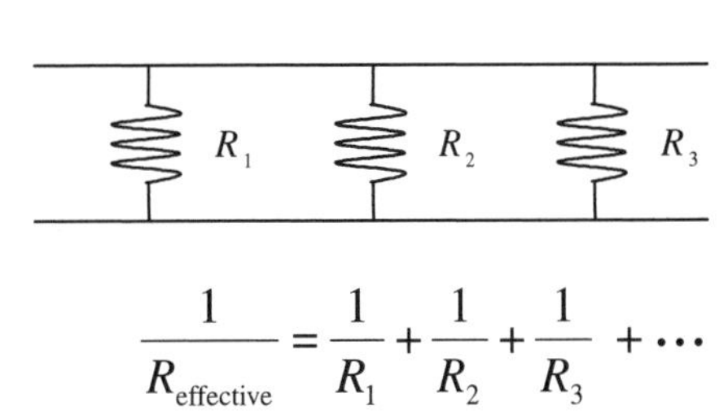

Figure 5 Resistors.

resistors the relationship between voltage and current is linear. Linear resistors are governed by Ohm's law, given by $V(t) = i(t)R$.

The magnitude of the resistance of a long cylindrical wire is related to the resistivity of the material the wire is made of (copper and gold have low resistivity) and the length of the wire. It is inversely related to the cross-sectional area of the wire:

$$R = \frac{\rho \ell}{A} = \frac{(\text{resistivity})(\text{length})}{\text{cross-sectional area}} \tag{2}$$

Series and parallel combinations of resistors can be thought of as an effective resistance. Resistors connected in series (share common current) create an effective resistance equal to the sum of the resistors. Resistors connected in parallel create an effective resistance where the inverse of the effective resistance is related to the sum of the inverse of each of the resistors. See Fig. 5.

If there are two parallel resistors, this expression becomes

$$R_E = \frac{R_1 R_2}{R_1 + R_2} \tag{3}$$

The resistance is sometimes described by the conductance G, defined as the inverse of the resistance:

$$G = \frac{1}{R} \tag{4}$$

4. *Capacitor.* The capacitor collects electrical charge and stores energy. The magnitude of the capacitance is denoted C and has the units of farads (F, or coulombs per volt). The capacitor is an energy storage element. The capacitor, like the resistor, is gov-

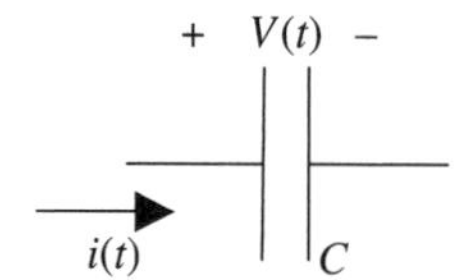

Figure 6 Ideal capacitor.

erned by a passive sign convention as shown in the figure in that the current flows "across" the capacitor from the side of higher potential (voltage) to the side of lower potential (voltage). If the current (and the voltage) are negative, this indicates that current flow is in the opposite direction and the potential is higher on the opposite side. See Fig. 6. In linear capacitors, the voltage is linearly related to the time integral of the current. Linear capacitors are governed by the equation

$$V(t) = \frac{1}{C}\int_0^t i(t)\,\partial t \tag{5}$$

or

$$i(t) = C\,\frac{\partial V(t)}{\partial t} \tag{6}$$

The simplest capacitor is a parallel-plate capacitor that consists of two plates separated by a nonconducting material called a dielectric. The magnitude of the capacitance for a parallel-plate capacitor is proportional to the dielectric constant of the material property between the plates and the area of the plates. It is inversely related to the distance between the plates. Such capacitors are often used as position sensors by changing either the distance between the plates or the area of overlap between the plates (see Fig. 7):

$$C = \frac{K\epsilon_0 A}{d} \tag{7}$$

where K = dielectric constant
ϵ = permeability constant
A = area of plates
d = distance between plates

Series and parallel combinations of capacitors can be thought of as an effective capacitance. Capacitors connected in series (share common current) create an effective capacitance where the inverse of the effective resistance is related to the sum of the inverse of each of the capacitances. Capacitors connected in parallel create an effective capacitance equal to the sum of the capacitances. See Fig. 8.

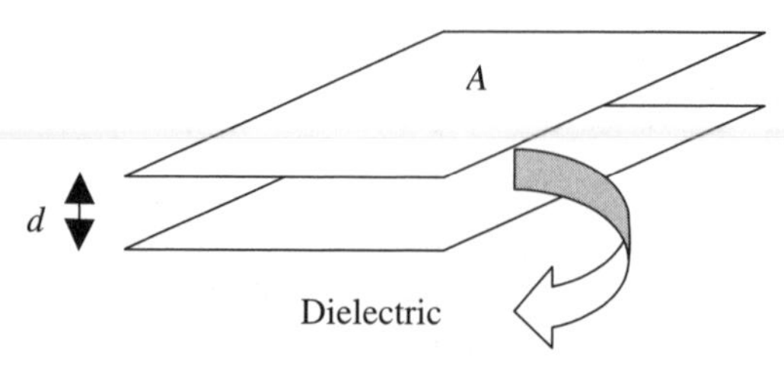

Figure 7 Parallel-plate capacitor.

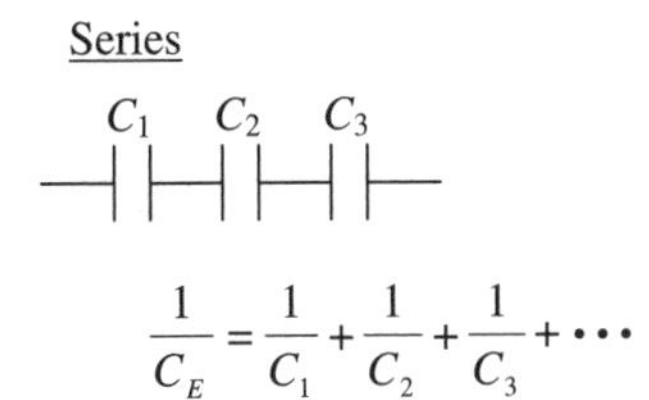

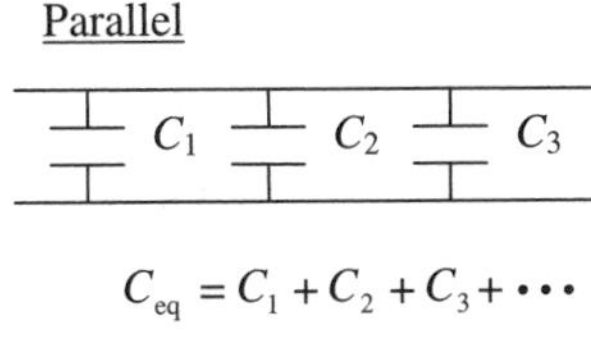

Figure 8 Capacitors.

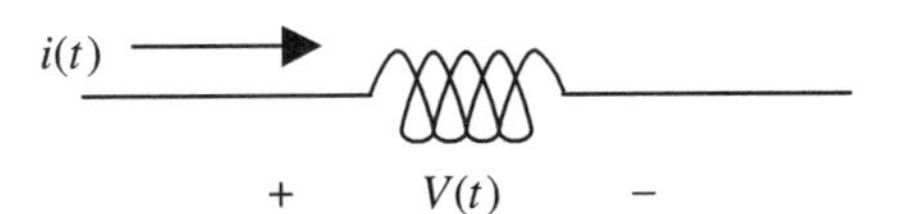

Figure 9 Ideal inductor.

5. *Inductors.* The inductor stores magnetic energy by creating a magnetic field. The magnitude of the inductor is denoted L and has the units of henrys (H). The inductor is an energy storage element. The inductor, like the resistor, is governed by a passive sign convention, as shown Fig. 9, in that the current flows through the inductor from the side of higher potential (voltage) to the side of lower potential (voltage). If the current and the voltage are negative, this indicates that current flow is in the opposite direction and the potential is higher on the opposite side. In linear inductors, the drop across the inductor, $V(t)$, is linearly related (by L) to the time rate of change of the current passing through the inductor.

Linear inductors are governed by the equations

$$V(t) = L\frac{dI(t)}{dt} \tag{8}$$

$$I(t) = \frac{1}{L}\int_0^t V(t)\,\partial t \tag{9}$$

Series and parallel combinations of inductors can be thought of as an effective inductance. Inductors connected in series (share common current) create effective inductance equal to the sum of the inductors. Inductors connected in parallel create an effective inductance where the inverse of the effective inductance is related to the sum of the inverse of each of the inductors. See Fig. 10.

6. *Ideal Diode.* The ideal diode is a nonlinear element. It does not store, add, or dissipate energy. The ideal diode can be compared to an ideal check valve in a fluidic system. The diode allows current to flow unimpeded (no voltage drop) in the positive direction. The diode allows no current to flow in the negative direction. Therefore, I can only be positive and V can only be negative, as shown in Fig. 11.

1.3 Circuit Analysis

Basic circuit analysis relies on two laws: (1) Kirchhoff's voltage law and (2) Kirchhoff's current law. There are several approaches and techniques to circuit analysis, but a straight-

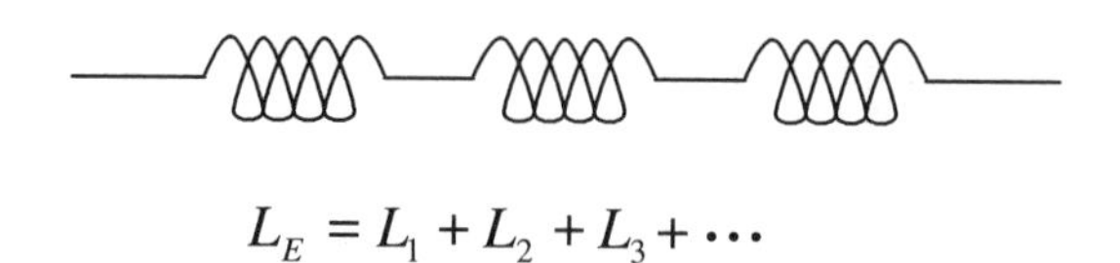

Parallel

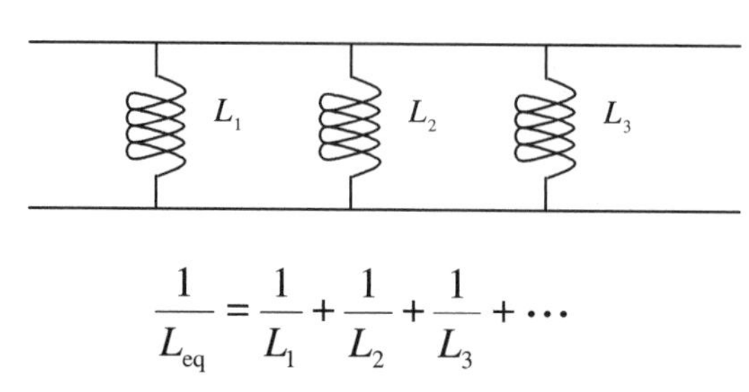

Figure 10 Inductors.

forward application of these two laws as presented here will be effective in all situations. This straightforward application, however, may lead to some additional algebra. Experience with these problems can lead to shortcuts.

Kirchhoff's Voltage Law

$$\sum_{i=1}^{n} V_i = 0 \tag{10}$$

Kirchhoff's voltage law states that the sum of voltages around a closed loop is zero. To a mechanical engineer, this is analogous to the movement of a mass in a potential field (such as gravity). For example, it is clear that if a bowling ball is picked up, rolled down the lane, and then returned to the original position, there is no net change in the ball's potential energy

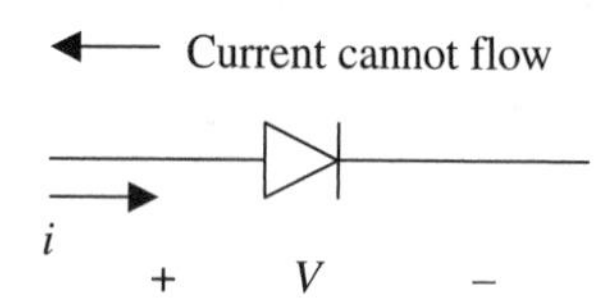

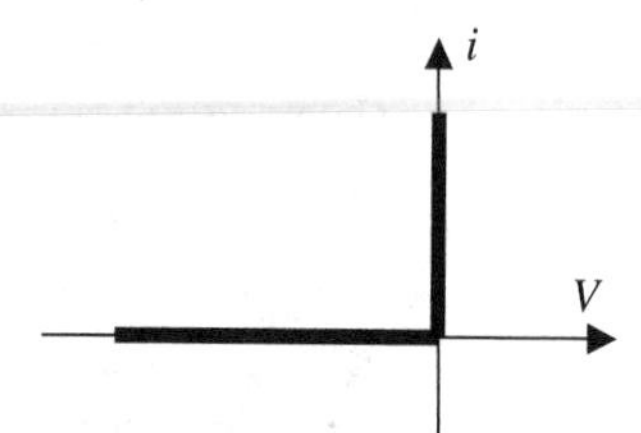

Figure 11 Ideal diode.

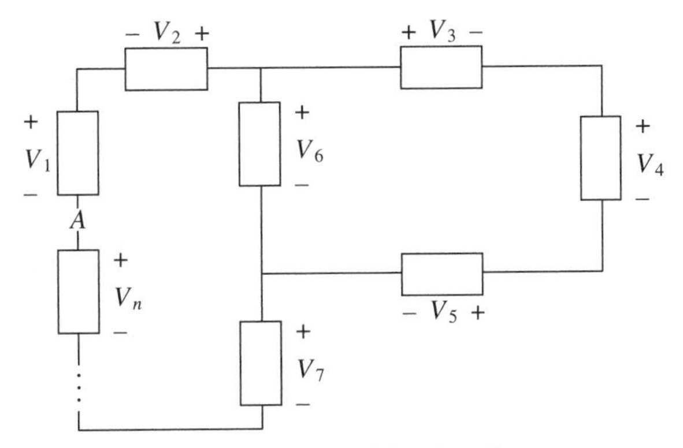

Figure 12 Kirchhoff's voltage law.

as it moved through this closed path. Similarly, Kirchhoff's voltage law states that the sum of the potential changes (voltages) around a closed loop is zero. The closed loops that can be used to formulate these equations are not unique. However, they will not all be independent (see Fig. 12):

1. Pick a starting point (e.g., *A*).
2. Choose a loop direction [clockwise (CW)].
3. Write the equation as you go around the loop:

$$V_1 + V_2 - V_3 - V_4 - V_5 - V_7 + \cdots + V_n = 0 \tag{11}$$

Kirchhoff's Current Law

$$\sum_{i=1}^{N} I_i = 0 \tag{12}$$

Kirchhoff's current law states that the net sum of the current flowing into a node is zero. To a mechanical engineer, this is analogous to fluid flow into and out of a node of a pipe network. If no fluid is stored in the node, then mass conservation requires that the mass flow rates into (positive flow) and out of (negative flow) the node must be zero; (that is, what goes in must come out) (see Fig. 13):

1. Assume current directions and draw arrows.
2. Assume current directions must be consistent with assumed voltage drops.
3. Apply Kirchhoff's current law (assume current into the node is positive):

$$i_1 - i_2 + i_3 - i_4 + \cdots + i_n = 0 \tag{13}$$

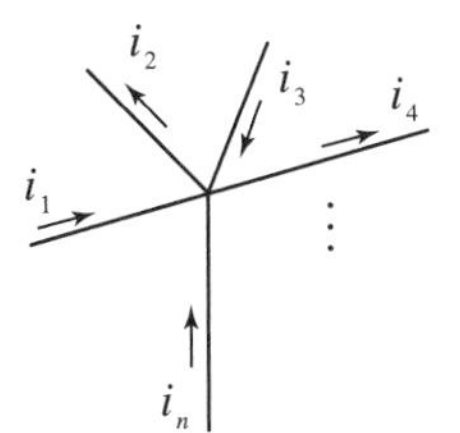

Figure 13 Kirchhoff's current law.

1.4 Sources and Meters

Some practical considerations are important when building mechatronic systems. Sources and meters (and all elements) do not behave as the ideal representation described above. Some of the limitations are described here.

1. *Voltage Source.* An ideal voltage source as described above does not have any output resistance (no resistor shown in the above representation) and can supply infinite current. Obviously, a real voltage source does have an output resistance. The output resistance (usually $<1\ \Omega$) can be represented as the model given below. Now, as current is output from the source, the output voltage (V_{out}) is no longer the same voltage as the source voltage (V_s). The magnitude of this output resistance (and the corresponding voltage drop) changes with many factors. For example, the output resistance of a rechargeable battery (such as a car battery) increases as the battery ages. The most significant difference between an ideal voltage source and a real source is that the real source can provide only a limited current. This is a limitation on both the instantaneous current that can be produced as well as the total amount (time integral) of current that can be produced. For example, the power supply limitation on instantaneous current is often a constraint for battery-operated devices such as cell phones or robots. See Fig. 14.

2. *Current Source.* An ideal current source as described above does not have any output resistance (no resistor shown in the above representation) and can supply infinite voltage. Obviously, a real current source does have an output resistance. The output resistance (usually $>1\ \text{M}\Omega$) can be represented as the model given below. Now, as current is output from the source, the output current (I_{out}) is no longer the same current as the source current (I_s). The magnitude of this output resistance (and the corresponding voltage drop) changes with many factors. The most significant difference between an ideal current source and a real source is that the real source can provide only a limited voltage. This is a limitation on both the instantaneous voltage that can be produced as well as the total amount (time integral) of voltage that can be produced. See Fig. 15.

3. *Voltmeter.* An ideal voltmeter has infinite input resistance and draws no current from the voltage being measured. A real voltmeter has a finite input resistance as modeled below and does draw some current from the source which can change the voltage being measured. However, the input resistance of most real voltmeters is very large (usually several mega ohms) and this makes the voltmeter a very safe device as it does not draw "significant" current. See Fig. 16.

4. *Ammeter.* An ideal ammeter has zero input resistance and does not produce a voltage drop. A real ammeter has a finite input resistance as modeled below and does have a small voltage drop across the leads of the meter. However, the input resistance of most real ammeters is very small (a few ohms), and this makes the ammeter a device that requires careful consideration before use. If the ammeter leads are placed between two points with a potential difference, a short circuit will occur and very large current will be produced. Therefore, most

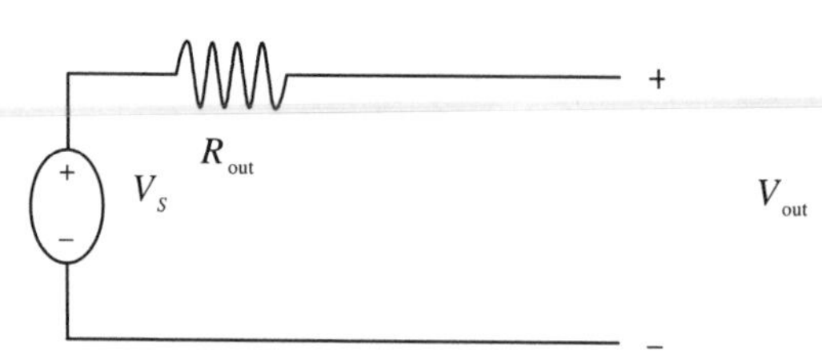

Figure 14 Real voltage source.

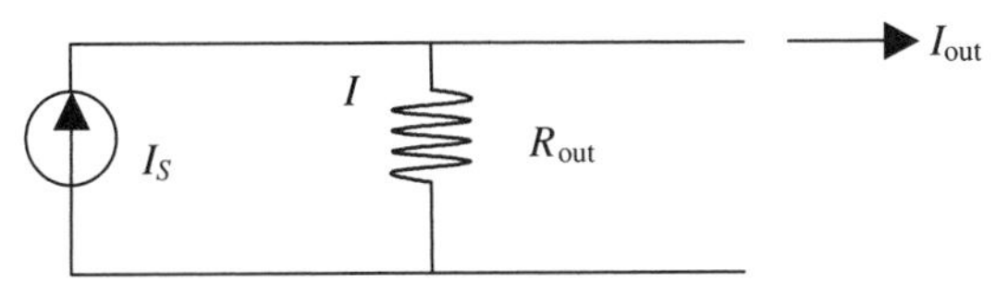

Figure 15 Real current source.

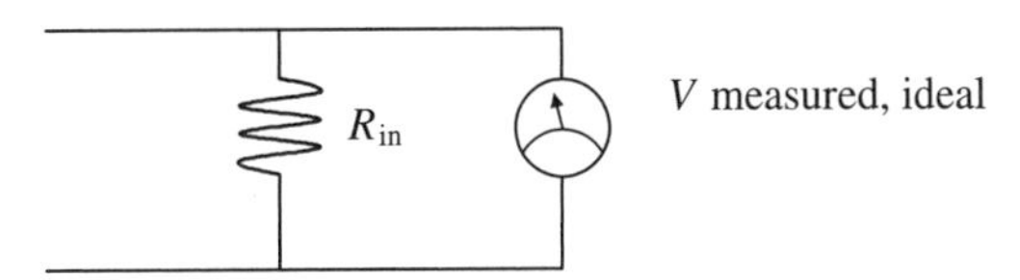

Figure 16 Real voltage meter.

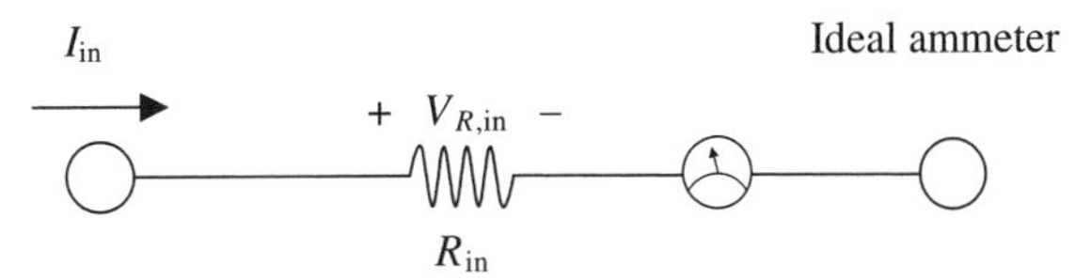

Figure 17 Real current meter.

devices that function as ammeters and voltmeters (multimeters) require that the user remove the leads and place them into new terminals before the devices are used as ammeters. This requires the user to be very deliberate about using the ammeter. See Fig. 17.

5. *Ohmmeter.* An ohmmeter is used to measure the resistance between the two terminals. The ohmmeter consists of a voltage source and an ammeter. A voltage is applied between the leads, the current is measured, and the resulting resistance is found using Ohm's law.

1.5 RL and RC Transient Response

To develop a basic understanding of the effects of inductors and capacitors in circuits, it is useful to understand *RL* and *RC* circuits. These circuits contain one resistor and one inductor (*RL*) or one resistor and one capacitor (*RC*). The equations of these circuits (developed with the laws above) are first-order differential equations with constant coefficients of the form

$$X(t) = B\frac{dy}{dt} + Cy \tag{14}$$

If we assume step inputs, a solution of the forms in Figs. 18 and 19 is assumed as solution to this equation. These first-order responses to step inputs rise or decay according to a time constant *T*:

1. First-order rise:

$$y(t) = Ae^{-(1/T)t} \tag{15}$$

2. First-order decay:

$$y(t) = A(1 - e^{-(1/T)t}) \tag{16}$$

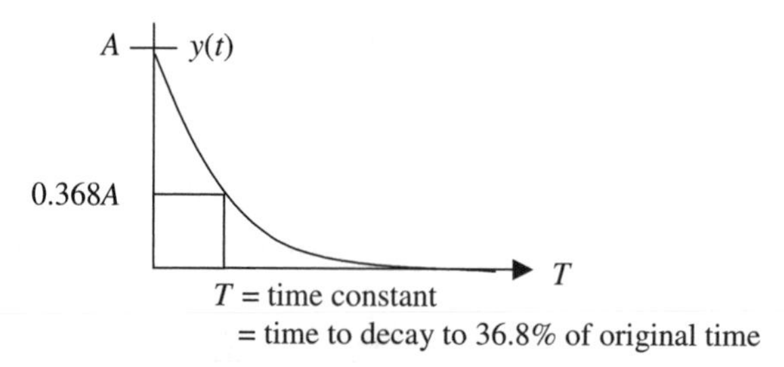

Figure 18 First-order response.

Consider the following circuits and the resulting equations, solutions, and time responses to step inputs:

1. *RL* circuits:

$$T = \frac{L}{R} \tag{17}$$

a. Differential equation:

$$Ri + L\frac{di}{dt} = V \tag{18a}$$

Solution:

$$i = \frac{V}{R}(1 - e^{-t/T}) \quad \text{(see Fig. 20)} \tag{18b}$$

b. Differential equation:

$$Ri + L\frac{di}{dt} = 0 \tag{19a}$$

Solution:

$$i = \frac{V}{R}(1 - e^{-t/T}) \quad \text{(see Fig. 21)} \tag{19b}$$

2. *RC* circuits:

$$T = RC \tag{20}$$

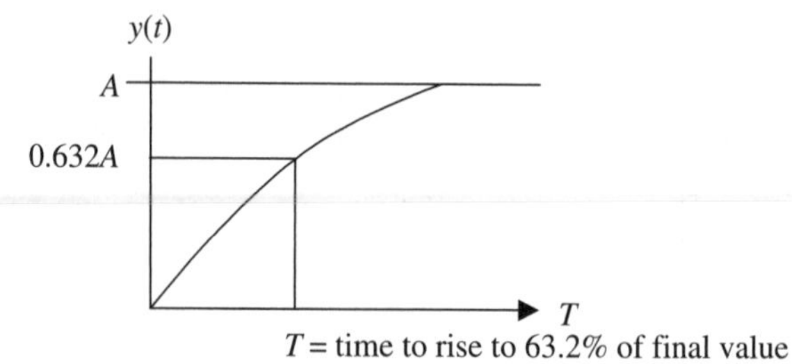

Figure 19 First-order response.

Figure 20 An *RL* circuit (see Eq. 18b).

Figure 21 An *RL* circuit (see Eq. 19b).

a. Differential equation:

$$\frac{V}{R} + C\frac{dv}{dt} = I \tag{21a}$$

Solution:

$$v = RI(1 - e^{-t/T}) \quad \text{(see Fig. 22)} \tag{21b}$$

b. Differential equation:

$$\frac{V}{R} + C\frac{dv}{dt} = 0 \tag{22a}$$

Solution:

$$v = RIe^{-t/T} \quad \text{(see Fig. 23)} \tag{22b}$$

The resulting observations from these circuits are that the voltage does not change instantly across a capacitor, but the current can change instantly. Also, the voltage can change instantly across an inductor, but the current cannot change instantly:

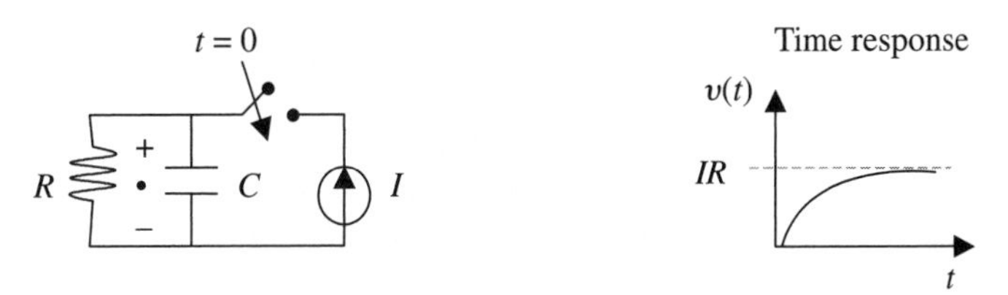

Figure 22 An *RC* circuit (see Eq. 21b).

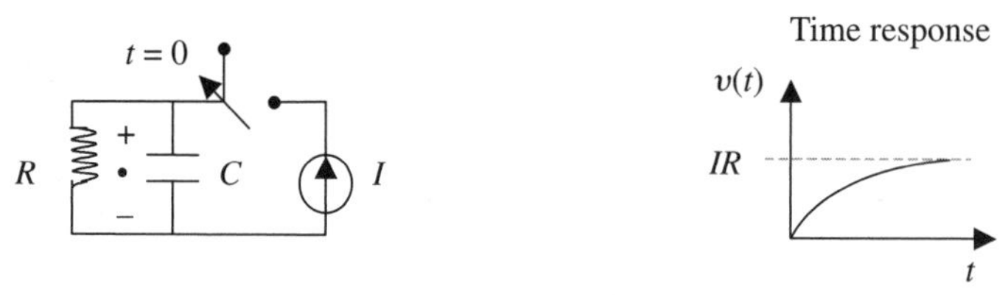

Figure 23 An RC circuit (see Eq. 22b).

	V	I
C	No	Yes
L	Yes	No

The instantaneous behavior (t is small) steady-state behavior (t is large) is also interesting. This shows that a capacitor begins like a short circuit and becomes an open circuit. Also, an inductor begins as an open circuit and becomes a short circuit:

	Capacitor	Inductor
$t = 0$	Short	Open
$t = \infty$	Open	Short

1.6 Basic Electronics Examples

Consider the basic circuit in Fig. 24. Applying Kirchhoff's current law to node A reveals

$$i_1 + i_2 = 1 \tag{23}$$

Now Kirchhoff's voltage law can be written as

$$500i_1 + 100i_2 - 100 - 200i_2 = 0 \tag{24}$$

and it is possible to substitute for

$$i_1 \ (i_1 = 1 - i_2) \tag{25}$$

Thus

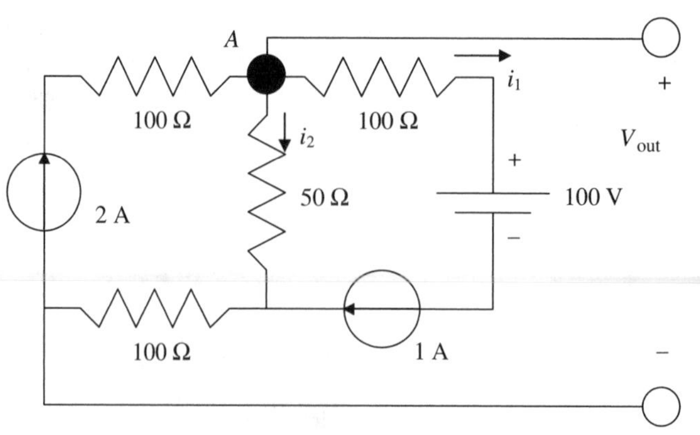

Figure 24 Example electrical circuit.

$$500 - 500i_2 - 100i_2 - 100 - 200i_2 = 0 \tag{26}$$

$$800i_2 = 400 \tag{27}$$

where $i_2 = ½$ and $i_1 = 1$. Now Kirchhoff's voltage law can be written for the outside loop:

$$V_{\text{out}} - 200 - 100i_2 - 100 = 0 \tag{28}$$

Therefore,

$$V_{\text{out}} = 350 \text{ V} \tag{29}$$

Transfer Function for an RL Circuit

Consider the circuit in Fig. 25. Kirchhoff's current law can be written as

$$I_{\text{in}} = i_1 + i_2 \tag{30}$$

Kirchhoff's voltage law can be written as

$$V_{\text{out}} + L\frac{di_2}{dt} - R_1 i_1 = 0 \tag{31}$$

and as

$$\frac{V_{\text{out}}}{R_L} = i_2 \tag{32}$$

Substituting (32) into (31) gives

$$V_{\text{out}} + \frac{L}{R_L}\frac{dV_{\text{out}}}{dt} - R_1\left(I_{\text{in}} - \frac{V_{\text{out}}}{R_L}\right) = 0 \tag{33}$$

The transfer function can be found using La Place transformation:

$$V_{\text{out}}(s) + \frac{L}{R_L} sV_{\text{out}}(s) - R_1 I_{\text{in}}(s) + -\frac{R_1}{R_L} V_{\text{out}}(s) = 0 \tag{34}$$

Rearranging gives

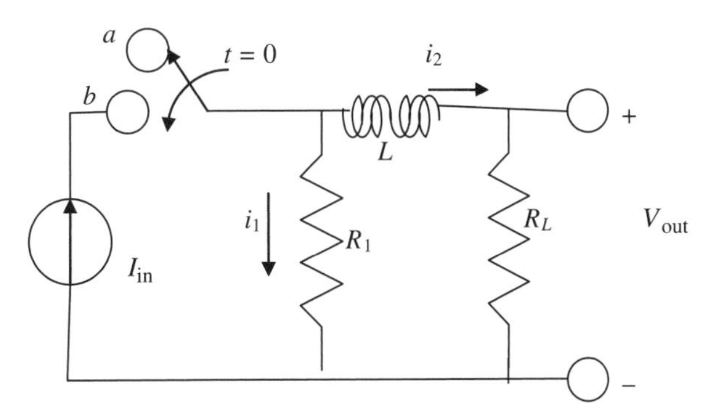

Figure 25 Example electrical circuit.

$$\frac{V_{out}(s)}{I_{in}(s)} = \frac{R_1}{(L/R_L)\, s + (1 + R_1/R_L)} \tag{35}$$

2 OPERATIONAL AMPLIFIERS

Operational amplifiers, or op amps, are extremely useful in mechatronics. They are primarily used to modify an analog signal, such as increasing the voltage output of a strain gauge circuit or removing noise from a signal. They are often used to amplify, sum, subtract, integrate, or differentiate analog signals. They are also used to create active filters. They are active devices in that they add energy to the circuit.

Consider the ideal op amp model in Fig. 26. All voltages shown in the figure are referenced to the same ground. As active elements, the op amp adds energy to the circuit through the power supplied by $+V_s$, $-V_s$ (V_s is often but does not have to be 15 V). The supply voltages are often left off Fig. 26 for convenience.

Circuit analysis of op amps requires only two additional observations. For ideal op amp behavior it must be remembered that

$$\Delta V = (V_+ - V_-) = 0 \tag{36}$$

or

$$V_- = V_+ \tag{37}$$

and

$$i_- = i_+ = 0 \tag{38}$$

It should also be noted that, in general,

$$i_{out} \neq 0 \tag{39}$$

Op amps produce an output voltage (V_{out}) that is related to an input voltage (V_{in}, not necessarily V_+ or V_-). The relationship can frequently be viewed as a gain defined by

$$A_v = \frac{V_{out}}{V_{in}} \tag{40}$$

The op amp generally has a very high input impedance (ideally infinite) defined by

$$Z_{in} = \frac{V_{in}}{i_{in}} \tag{41}$$

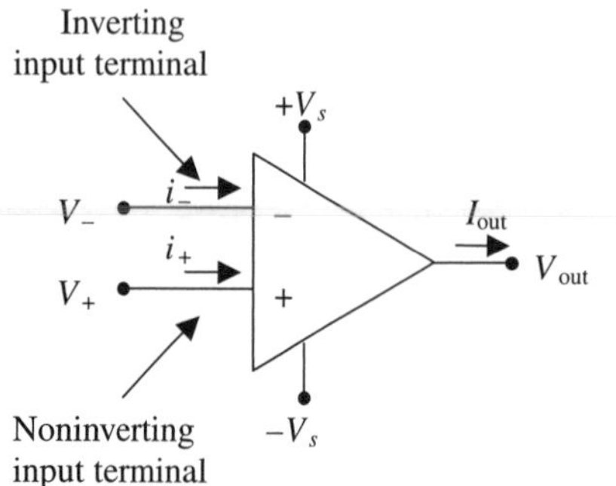

Figure 26 Ideal operational amplifier.

For real op amps the input impedance is usually >100 kΩ. Op amps also have a very low output impedance (ideally zero) defined by

$$Z_{\text{out}} = \frac{V_{\text{out}}}{i_{\text{out}}} \tag{42}$$

For real op amps this is usually less than a few ohms.

Circuits for two of the most common op amps, the inverting op amp and the noninverting op amp, are analyzed in Figs. 27 and 28.

2.1 Inverting Amplifier

Apply op amp rules:

$$\Delta V = 0 \tag{43}$$

or

$$V_{-} = V_{+} \tag{44}$$

$$i_{-} = i_{+} = 0 \tag{45}$$

Kirchhoff's current law gives

$$i_R = i_{Rf} = i \tag{46}$$

Kirchhoff's voltage law gives

$$V_{\text{in}} - i_{\text{in}}R = 0 \tag{47}$$

$$\frac{V_{\text{in}}}{R} = i_{\text{in}} \tag{48}$$

$$V_{\text{out}} + i_{\text{out}}R_f = 0 \tag{49}$$

$$-\frac{V_{\text{out}}}{R_f} = i_{\text{out}} \tag{50}$$

Setting $i_{\text{in}} = i_{\text{out}}$ gives

$$\frac{V_{\text{in}}}{R} = \frac{-V}{R_f} \tag{51}$$

or

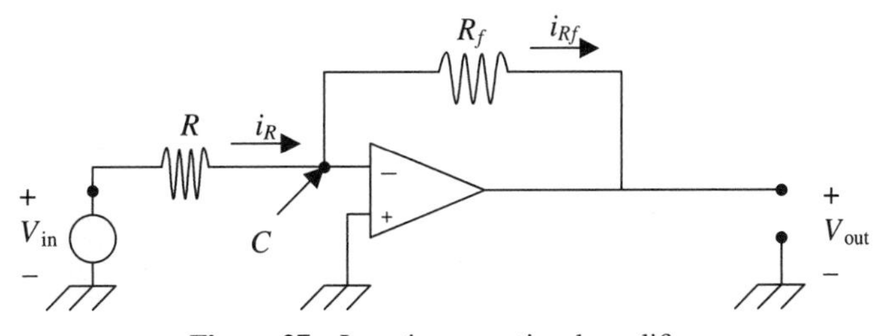

Figure 27 Inverting operational amplifier.

$$\frac{V_{\text{out}}}{V_{\text{in}}} = -\frac{R_f}{R} \tag{52}$$

Notice that for the inverting amplifier the gain is negative—hence the name "inverting." This means that for periodic signals there would be a 180° phase shift between the output voltage and the input voltage. The gain for this circuit is

$$A = -\left(\frac{R_f}{R}\right) \tag{53}$$

$$V_{\text{out}} = AV_{\text{in}} \tag{54}$$

2.2 Noninverting Amplifier

Kirchhoff's current law yields

$$i_{\text{in}} = i_{\text{out}} \tag{55}$$

since $i_+ = i_- = 0$. Kirchhoff's voltage law yields

$$V_{\text{in}} + Ri = 0 \tag{56}$$

$$V_{\text{out}} + R_f i - V_{\text{in}} = 0 \tag{57}$$

$$i = -\frac{V_{\text{in}}}{R} \tag{58}$$

$$i = \frac{V_{\text{in}} - V_{\text{out}}}{R_f} \tag{59}$$

Setting the i's equal gives

$$-\frac{V_{\text{in}}}{R} = \frac{V_{\text{in}} - V_{\text{out}}}{R_f} \tag{60}$$

Dividing by V_{in} yields

$$-\frac{R_f}{R} = 1 - \frac{V_{\text{out}}}{V_{\text{in}}} \tag{61}$$

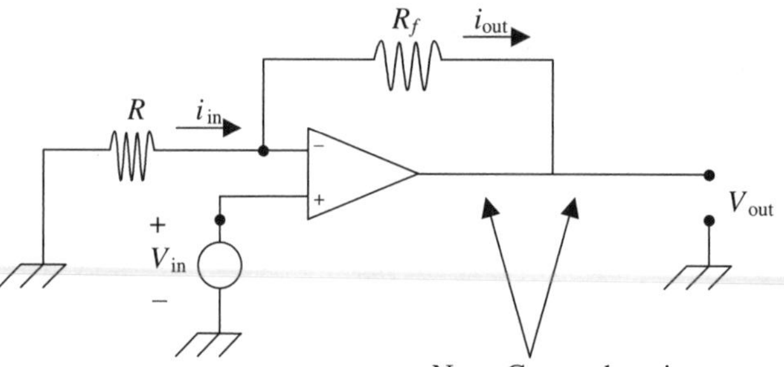

Figure 28 Noninverting operational amplifier.

$$\frac{V_{\text{out}}}{V_{\text{in}}} = \underbrace{1 + \frac{R_f}{R}}_{\text{Gain} \geq 1} \tag{62}$$

Here V_{out} and V_{in} have the same sign so the amplifier is noninverting. This circuit is often used as a buffer to "isolate" one portion of a circuit from another.

Type	Schematic	Input/Output Equation
Summer	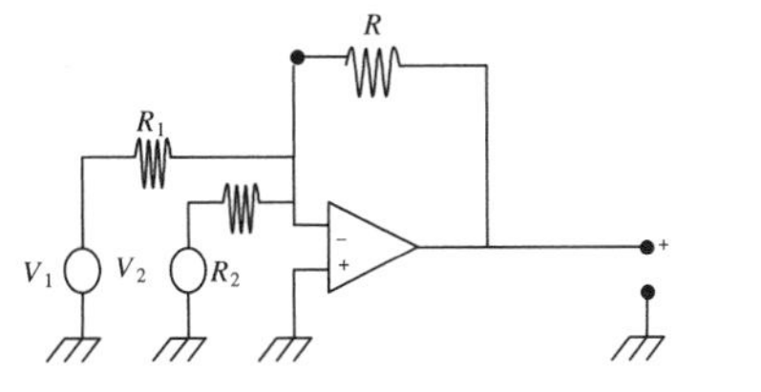	$V_{\text{out}} = -(V_1 + V_2)$
Difference amplifier	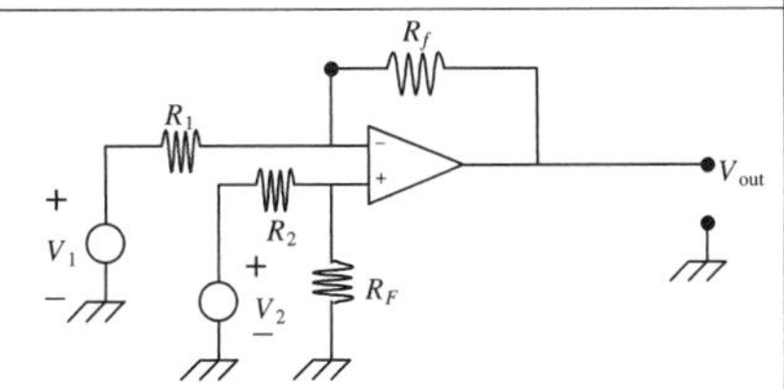	$V_{\text{out}} = \frac{R_f}{R}(V_2 - V_1)$
Integrator	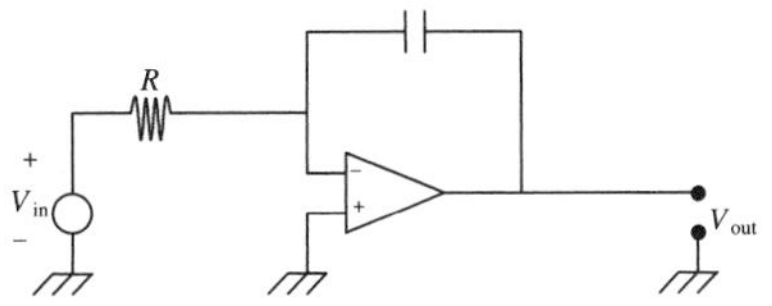	$V_{\text{out}}(t) = -\frac{1}{RC}\int_0^t V_{\text{in}}(x)\, dx$
Differentiator	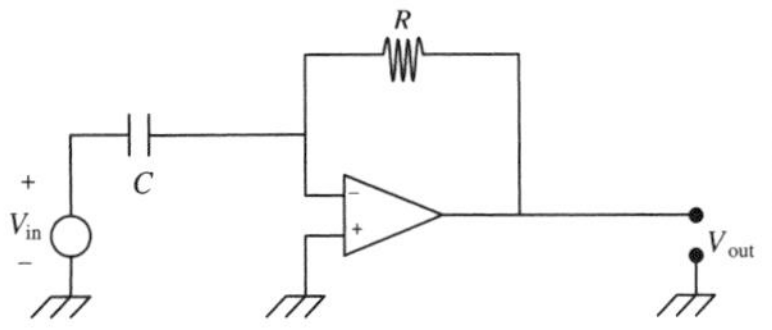	$-RC\left(\frac{dV_{\text{in}}}{dt}\right) = V_{\text{out}}(t)$
Gain and shift amplifier	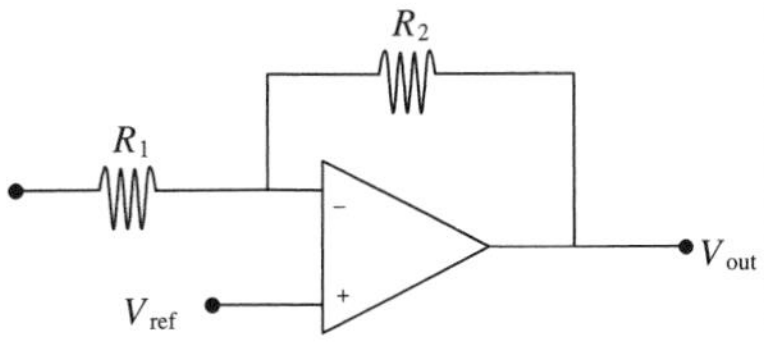	$V_{\text{out}} = -\frac{R_2}{R_1}V_1 + \left(1 + \frac{R_2}{R_1}\right)V_{\text{ref}}$ if $R_1 = R_2 = R$ $V_{\text{out}} = \frac{R_f}{R}(V_2 - V_1)$

Figure 29 Useful op amp circuits.

2.3 Other Common Op Amp Circuits

Some other useful op amp circuits are described in Fig. 29 (page 843).

3 BINARY NUMBERS

Computers and digital electronics used in mechatronic systems are described by binary arithmetic. Therefore, it is important to understand binary numbers to fully understand the function of computers. First, consider a base-10 number as in standard mathematics. Base-10 numbers have 10 possible digits (0, 1, 2, 3, 4, 5, 6, 7, 8, 9). Consider a number such as 234. Here the 4 represents the 1's digit, the 3 represents the 10's digit, and the 2 represents the 100's digit. The number 234 represents four 1's and three 10's and two 100's ($4 * 1 + 3 * 10 + 2 * 100 = 234$). The value of each digit increases by a factor of 10 as you move to the left and decreases by a factor of 10 as you move to the right. Consider a number such as 234:

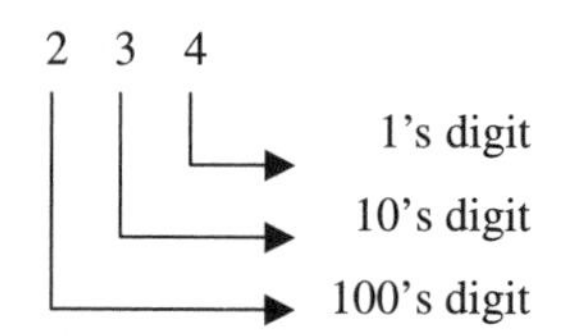

4 represents the sum of nine 1's

3 represents the sum of seven 10's

2 represents the sum of one 100

Now, consider a base-2 number, or a binary number. Base-2 numbers have two possible digits (0, 1). Again, consider the number 234. With binary numbers the value of each digit increases by a factor of 2 as you move to the left and decreases by a factor of 2 as you move to the right. To write this number as a binary number requires zero 1's and one 2's and zero 4's and one 8's and zero 16's and one 32's and one 64's and one 128's ($234 \equiv 11101010 = 0 * 1 + 1 * 3 + 0 * 4 + 1 * 8 + 0 * 16 + 1 * 32 + 1 * 64 + 1 * 128$):

Binary base 2: two possible digits (0, 1)

$$234 \equiv 11101010$$

0	1's
1	2's
0	4's
1	8's
0	16's
1	32's
1	64's
1	128's

3.1 Binary Numbers of Different Size

Each digit in a binary number (either a 0 or a 1) is called a *bit*—"binary digit." A nibble is a group of 4 bits, a byte is a group of 8 bits, a word is a group of 16 bits, and a double word (dword) is a group of 32 bits. Each bit is numbered starting with zero and moving to the left. The Kth bit represents the 2^K place holder. The zeroth bit is called the least significant bit (LSB) and the highest bit (e.g., seventh in illustration) is called the most significant bit (MSB):

Identifying individual bits (see Fig. 30):

- Starting from the right
- The Kth bit represents the 2^K slot

Least significant bit: bit furthest to the right

Most significant bit: bit furthest to the left

3.2 Hexadecimal Numbers

Hexadecimal numbers are binary numbers that are easier for humans to work with. Hexadecimal numbers have 16 unique digits (0, 1, 2, 3, 4, 5, 6, 7, 8, 9, A, B, C, D, E, F). Since hexadecimal numbers sometimes use both letters and numbers an "h" is usually placed at the end of the number to distinguish it as a hexadecimal number (e.g., ACEh). Again, the value of each digit increases by a factor of 16 as you move to the left and decreases by a factor of 16 as you move to the right. The advantage of hexadecimal numbers is that each hexadecimal digit represents a 4 bit binary number. This makes it very simple to convert from hexadecimal to binary (see Fig. 31).

Example 1 34h = 52:

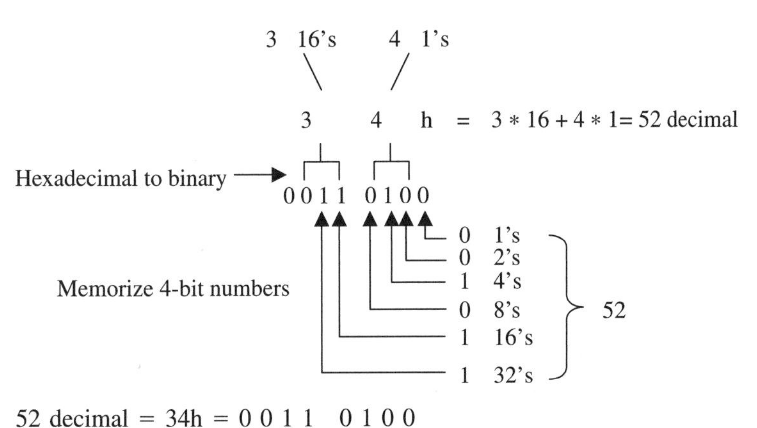

Example 2 Binary to hexadecimal:

0 0 1 0 1 0 0 1 0 0 1 0

1. Break into 4-bit segments (add two 0's to left).
2. Convert each 4-bit segment into one hexadecimal digit using Fig. 31:

0 0 1 0 1 0 0 1 0 0 1 0

2 9 2h = 2 (256) + 9 (16) + 2 (1) = 658

7th bit			*K*th bit		2nd bit	1st bit	0th bit
1	0	1	1	0	0	1	1

Figure 30 A byte.

Hexadecimal	Binary
0	0000
1	0001
2	0010
3	0011
4	0100
5	0101
6	0110
7	0111
8	1000
9	1001
A	1010
B	1011
C	1100
D	1101
E	1110
F	1111

Figure 31 Hexadecimal-to-binary conversion.

3.3 Binary Addition

Binary addition is straightforward and similar to decimal addition. The two numbers are first aligned by their respective columns. Then the 1's digits are added. If the result is greater than the maximum allowable number for that column (i.e., 1), the remainder is carried into the next column. This process is repeated until all columns are added and no remainder is left:

$$\begin{array}{r} 3 \\ +\ 1 \\ \hline 4 \end{array} \qquad \begin{array}{r} 011 \\ +\ 001 \\ \hline 100 \end{array}$$

3.4 Two's Complement Binary Numbers

Straight (regular) binary notation cannot represent a negative numbers. A new convention called two's complement binary numbers is used. Then binary arithmetic can be used to add a positive number to a negative number and in this way it is possible to perform binary subtraction. Consider a 3-bit two's complement example:

(3-Bit Example)		For N bits
011	3	2^{N-1}
010	2	↑
001	1	
000	0	0
111	−1	
110	−2	
101	−3	↓
100	−4	-2^{N-1}

The procedure to convert from a positive number (e.g., 3) to the negative of that number (e.g., −3) is:

(i) Invert all bits.

(ii) Add 1 to the result.

The procedure to convert from a negative number (e.g., −3) to the positive of that number (e.g., 3) is:

(i) Subtract 1 (add −1).

(ii) Invert all bits.

Consider the following 3-bit examples:

Example 3 3 bits

To change a positive 3 (written as 011 in 3-bit binary) to a negative 3

(i) Invert all bits:

$$100$$

(ii) Add 1:

$$\begin{array}{l} 100 \\ \underline{001} \\ 101 \Rightarrow -3 \end{array}$$

Example 4 To convert a 3-bit two's compliment negative 2 to a positive 2.

(i) Subtract 1:

$$\begin{array}{l} 110 \\ \underline{111} \\ 101 \end{array}$$

(ii) Invert all bits:

$$010 \Rightarrow 2$$

Now two two's complement numbers can be added and this is like binary subtraction. Consider the following example:

$$\begin{array}{rl} (-3) + (2) \longleftrightarrow & 101 \\ & \underline{010} \longleftrightarrow -1 \\ & 111 \end{array}$$

4 DIGITAL COMPUTERS

Mechanical engineers should be familiar with the basic low-level operations of microcontrollers and computers so as to (1) understand what computers can and cannot do in mechatronic systems, (2) understand how computers are used in mechatronic systems (what their job is), (3) be able to communicate with electrical and computer engineers on interdisciplinary design teams, and (4) be able to select a microcontroller when designing a mechatronic system. With these stated goals only the general concepts are presented here. The details of low-level computer operation are numerous and beyond the scope of this chapter.

Bit #	7 (MSB)	6	5	4	3	2	1	0 (LSB)
	X	X	X	X	X	X	X	X

where x can be a 0 or a 1.

Figure 32 Register.

4.1 Most Basic Computer

The most basic computer includes a central processing unit (CPU) and some amount of memory. The memory holds numbers while the CPU, sometimes called the microprocessor, holds numbers and processes these numbers. The CPU sends some control information (including addresses) to the memory and the CPU and memory exchange data. The CPU holds and processes data in registers. A register is a set of flip-flops (a digital electronic device with two states—0, 1) whose contents are read or written as a group. Registers come in several sizes such as 8, 16, and 32 bits. Consider the 8-bit register shown in Fig. 32, where x can be a 0 or a 1.

Registers are used for temporary storage of numbers and manipulation of numbers (addition, subtraction, etc.).

4.2 Memory

In almost all computers, memory is arranged as a series of bytes. Each memory location can hold a single 8-bit binary number (00000000 to 11111111). More than one memory location is needed to store numbers larger than 255. Memory can be thought of as an 8-bit register, as shown above. Memory can come in many forms, such as solid-state electronic chips [random-access memory (RAM)], magnetic storage devices such as floppy and hard disks, and optical devices such as CDs and DVDs. A computer usually has thousands, millions, or many, many more memory locations. For example:

- 1 kb of memory is $2^{10} = 1024$ memory locations
- 64 kb of memory is $64 \times 2^{10} = 2^{16} = 65{,}563$ memory locations
- 1 Mb of memory is 1024 kb $= 2^{20} = 1{,}048{,}576$ memory locations
- 1 GB of memory is 1024 Mb $= 2^{30} = 1{,}073{,}741{,}824$ memory locations

One of the most important concepts in understanding computer memory is that *memory has an address and contents.* The address is a fixed number that is a unique identifier for a specific memory location. The contents (a single byte of data) of a specific address can be changed (see Fig. 33).

To get data from memory, the CPU loads the memory address on a special register, then sends that address to memory. Memory returns the contents of that address to another register on the CPU. The size of the number needed for the address can often be a limitation of the amount of memory a computer can have. For example, most modern desktop computers have 32-bit registers. A single register can hold a number up to 2^{32}, or 4 Gb, or 4,294,967,296. This is the number of memory locations that can be addressed with a single register (although computers often use more than one register for a memory address).

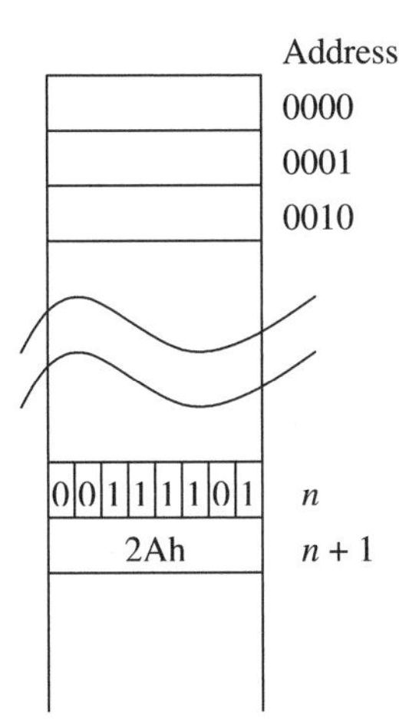

Figure 33 Schematic of computer memory.

Storing Information In Memory

Each memory location always contains a number between 0 and 255. This number (along with the contents of other memory locations) can be interpreted in many ways to mean many different things:

1. *Binary Numbers.* Memory can store only unsigned binary numbers.

2. *Two's Complement Binary Numbers.* Memory can store signed binary numbers in the form of two's complement binary numbers.

3. *ASCII.* ASCII stands for the American Standard Code for Information Interchange. ASCII code is a format for characters such as letters (both upper- and lowercase), digits 0–9, and punctuation symbols. ASCII only has 128 possible symbols, so only a 7-bit binary number is needed. Since memory comes in 8-bit locations, an extra bit is added to the MSB. Consider the following example:

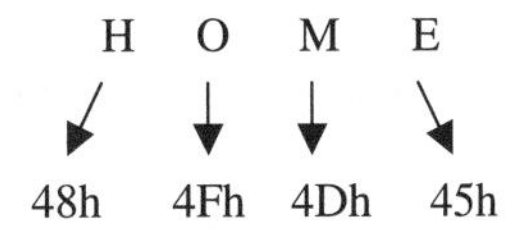

Since ASCII requires only 7 bits, the eighth bit is wasted. For this and several reasons there has been other formats similar to ASCII that encode more information. For example, a format used by the American National Standards Institute (ANSI) includes accented characters and a few Greek letters. Another is UNICODE, which uses 16 bits and has 65,536 different symbols.

4. *Floating-Point Numbers.* Floating-point numbers are stored in several memory locations in scientific notation. For example, 1749 must first be written in scientific notation as 1.749×10^3. There are several different types of floating-point numbers. Single-precision floating-point numbers use 32 bits (23 bits for the significant figures 1.749, 1 bit for the sign on these figures, and 8 bits for the exponent and sign on the power of ten). Single-precision numbers can range from 10^{-38} to 10^{38} with only 6–8 digits of accuracy. Double-precision floating-point numbers use 64 bits (53 bits for the significant figures with the sign and 11 bits for the exponent and sign). Double-precision numbers can range from 10^{-308} to 10^{308} with 13–16 digits of accuracy.

4.3 Microcomputers

Microprocessors often refer to the CPU described above. Most microprocessors contain some memory in the form of registers. Microcomputers contain microprocessors connected to other devices by a common data bus. The data bus is a series of conductors that allow information to flow between each subsystem. The other devices can include input/output (I/O) devices such as monitors and keyboards or additional memory devices or I/O devices such as analog-to-digital (A/D) converters. A microcontroller is just enough microcomputer on a chip to do a specific control job. For example, a microcontroller could be a microprocessor with some digital I/O to control a home security system.

5 TRANSFER OF DIGITAL DATA

Digital data can be transferred in several ways. The most common methods are parallel and serial data transfer. Parallel communication transfers several bits at the same time along separate conductors. In serial communication data bits are sent one after the other along a single conductor. There are two types of serial communication: synchronous and asynchronous.

5.1 Parallel Data Transfer

In parallel data transfer all bits occur simultaneously on a set of data lines. Each bit is placed on each data line by the transmitting digital system and then can be read by the receiving digital system. The advantage of parallel communication is that it is relatively fast (relative to serial communication) because several bits of data can be transferred at the same time. The disadvantage is that several conductors are required. See Fig. 34.

5.2 Serial Data Transfer

In serial data transfer a sequence of bits, or train of pulses, occur on a single data line, as shown in Fig. 35.

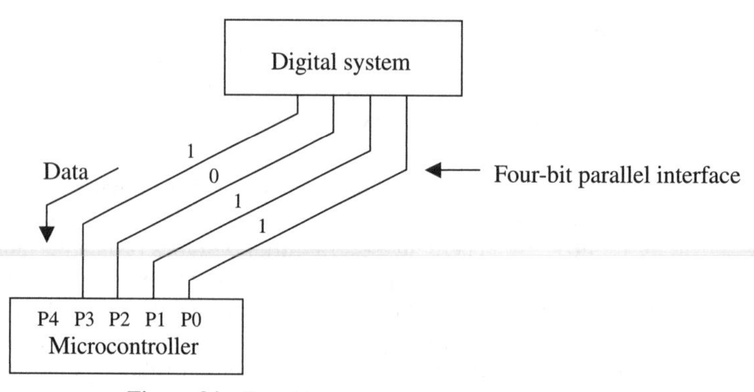

Figure 34 Four-bit parallel transfer of digital data.

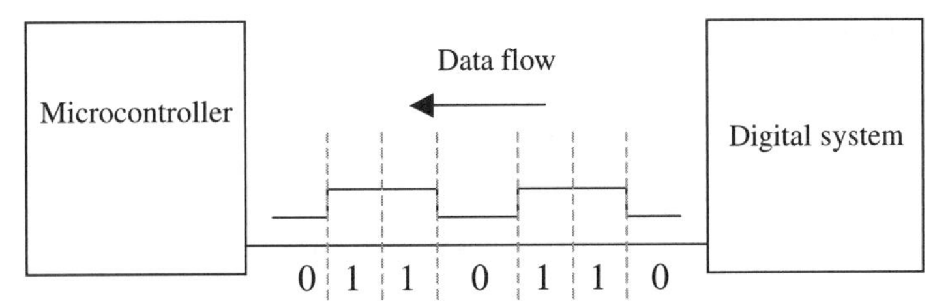

Figure 35 Serial transfer of digital data.

Asynchronous Serial Data Transfer

In asynchronous serial communication the data are written to the serial line at a predefined rate. Both the transmitter and receiver must be set for identical timing. This timing rate is called bits per second (bps) or baud rate (e.g., 4800 baud or 9600 baud). The advantage of asynchronous serial communication is that it only requires one wire (and ground) to communicate the data. This method is used in RS-232 (com ports on PCs) communication. The microprocessor detects the first "edge" of the first data bit and then (because of the baud rate) it knows how long to wait for the next bit. The baud rate is not exactly the rate the data will be transferred because there are some overhead bits required for the transfer (e.g., parity). See Fig. 36.

Synchronous Serial Data Transfer

In synchronous serial communication the data are written to the serial line and a separate line is used as a clock, or signal, to indicate the data are ready to be transferred. In this case the rate of data transfer is controlled by the digital device that provides the clock. The advantage is that the rate of transfer can be directly controlled and the transmitter and receiver do not require precise coordination. The disadvantage is that an extra line (clock) is required. See Fig. 37.

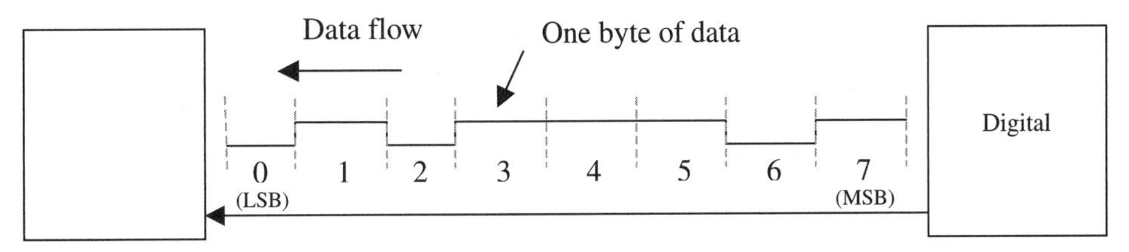

Figure 36 Asynchronous serial data transfer.

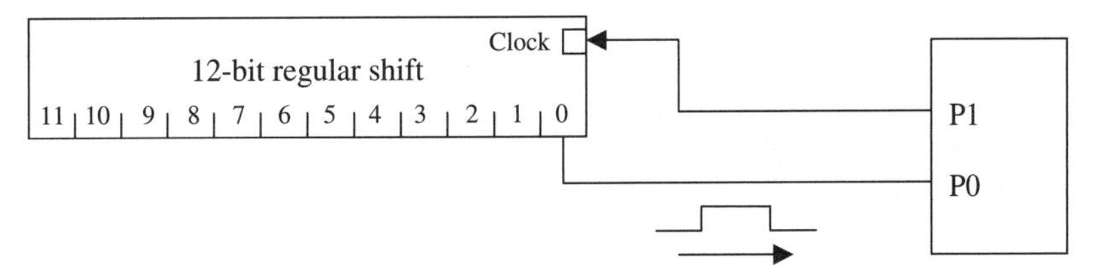

Figure 37 Synchronous serial data transfer.

6 ANALOG-TO-DIGITAL CONVERSION

The world we live in is analog. To interface digital computers to an analog mechanical system, an A/D conversion process is required. Converting from an analog signal to a digital number is a two-step process that involves (1) quantization and (2) coding. The process of quantization is where the range of the analog signal is broken down into a discrete number of bins. In coding, the bin "location" of the analog signal is converted into a digital number that can be understood by the computer. There are literally hundreds of A/D converter types.

Several definitions are required to discuss A/D converters:

Resolution n refers to the number of bits used to digitally approximate the analog value of the input.

Number of possible states $N = 2^n$.

Analog quantization size Q is a measure of the analog change that can be resolved (minimum error):

$$Q = \frac{V_{\max} - V_{\min}}{N}$$

6.1 Four-Bit A/D Converter

To understand the terms explained above, the example in Fig. 38 will be used. In the example, a tachometer is used to measure the speed of an electric motor. The tachometer produces a voltage between -12 and $+12$ V that is linearly proportional to the speed of the motor. The analog voltage output of the motor will be read with a 4-bit A/D converter. Therefore, the A/D converter has a resolution of 4 bits and can have 16 (2^4) possible states (0000–1111 or 0–15). The analog quantization size will be 1.5 V [(12 − (−12))/16]. This means, for example, that all voltages between 10.5 and 12 V will be represented by the encoded digital number 1111.

7 DIGITAL-TO-ANALOG (D/A) CONVERTER

A D/A converter has many of the same issues as an A/D converter. The resolution is given in the number of bits used to create the output analog voltage. For example, a 4-bit D/A

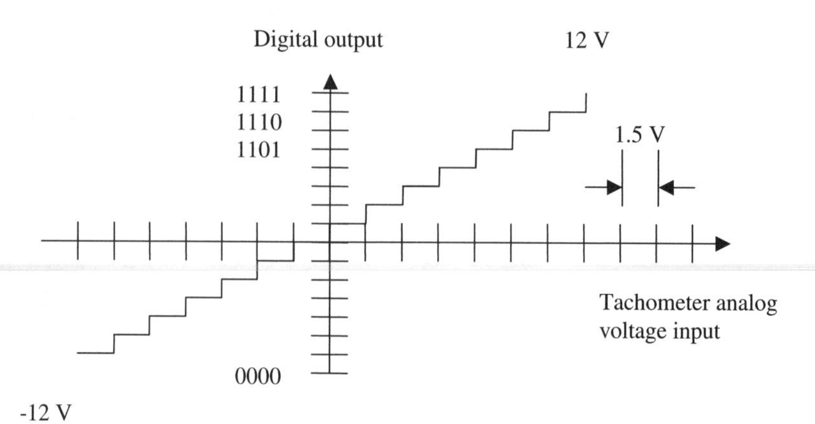

Figure 38 Analog-to-digital conversion.

converter can produce 16 different output states. Issues to consider closely when selecting a D/A converter include the resolution and the amount of current that can be supplied.

8 SENSORS

8.1 Position Sensors

Potentiometers

One of the most basic position sensors is the potentiometer. A potentiometer is a device that generally has a mechanical input in the form of a rotary shaft or a linear slide. The mechanical input moves a wiper across a resistor. The potentiometer also has an electrical input that places a fixed voltage across the resistor. As the position of the mechanical input is changed, the location of the wiper and the voltage of an output electrical lead attached to the wiper change. The change in voltage is linearly related to the angular change of the mechanical input. Linear potentiometers that use a slide input instead of a shaft are also common.

Potentiometers are traditionally used in many applications, such as the volume knob on a radio. See Fig. 39.

Digital Optical Encoder

A digital optical encoder is another common position sensor (both rotary and linear are available) that provides a digital output. The encoder consists of a disk attached to a shaft that is the mechanical input to be measured (θ). The disk contains several "tracks" of alternating slots. One side of the disk has a light source and the opposite side of the disk has a photosensor. As the disk rotates, the slots alternately either block the light or allow it to pass to the photosensor. In this way the digital encoder "encodes" the position of the disk as either a 0 (no light) or a 1 (light). See Fig. 40.

In this way it is possible to count the alternating regions of light and dark and determine the angular change in the position of the disk/shaft. Several observations can be made from this principle. First, with only one row (track) of slits it is not possible to determine the direction the disk is rotating. Also, all measurements are made relative to a starting point. Finally, the encoder has a finite number of slits, limiting the resolution of the sensor.

These limitations are eliminated in different ways with different types of encoders. A quadrature optical encoder (described in the next section) uses two rows of slits to determine the direction of travel. Other encoders use many tracks so that each position of the shaft is unique, allowing for an absolute measurement of shaft position.

Quadrature Optical Encoder

A quadrature optical encoder includes two rows of slits that are placed 90° (a quarter cycle) out of phase. This allows the encoder to measures relative position and direction. Consider the diagram in Fig. 41, which presents the two tracks (A and B) with the encoder shaft

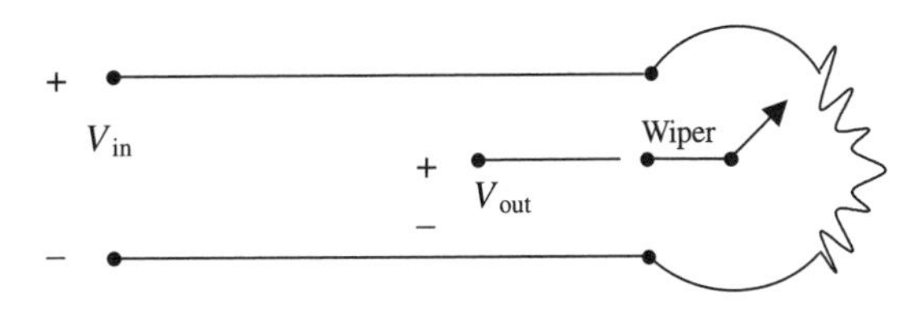

Figure 39 Potentiometer.

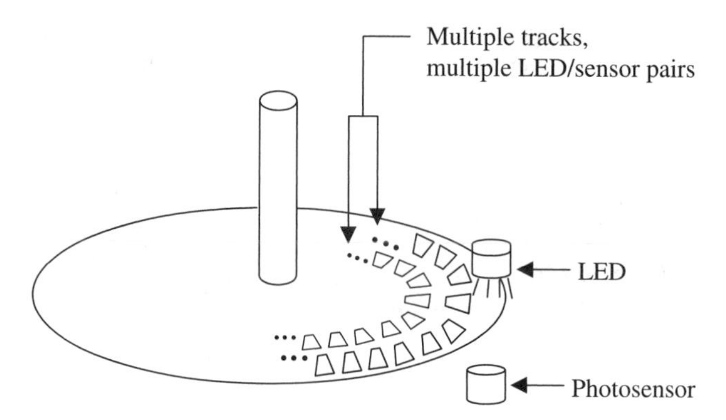

Figure 40 Rotary optical encoder.

rotating at constant velocity. Figure 41 presents the alternating regions of light and dark created by the two rows of slits. Since the slits are one-quarter cycle out of phase, there are four states created: (1) *A* on, *B* off; (2) *A* on, *B* on; (3) *A* off, *B* on; and (4) *A* off, *B* off. The direction of rotation can be determined by the order that these states appear:

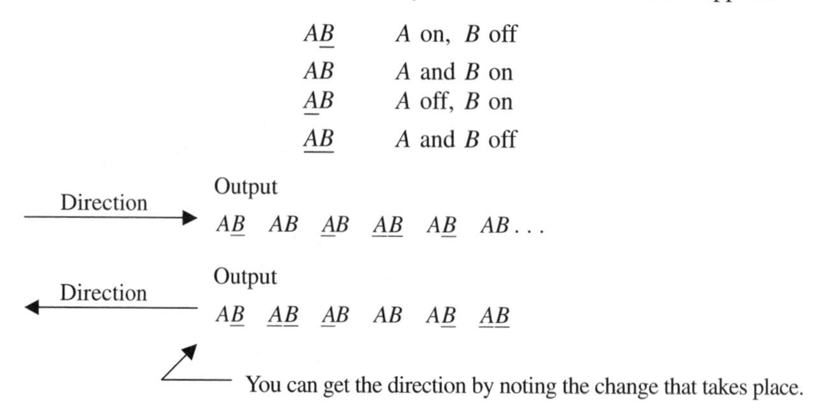

Linear Variable Differential Transformer

Another common position sensor is the linear variable differential transformer (LVDT). This sensor is based on an electrical transformer depicted in Fig. 42.

The transformer consists of a primary coil of wire wrapped around an iron core with a secondary coil also wrapped around the same core. As an alternating current is passed through the primary core, a changing magnetic field is created by the coil. The magnetic

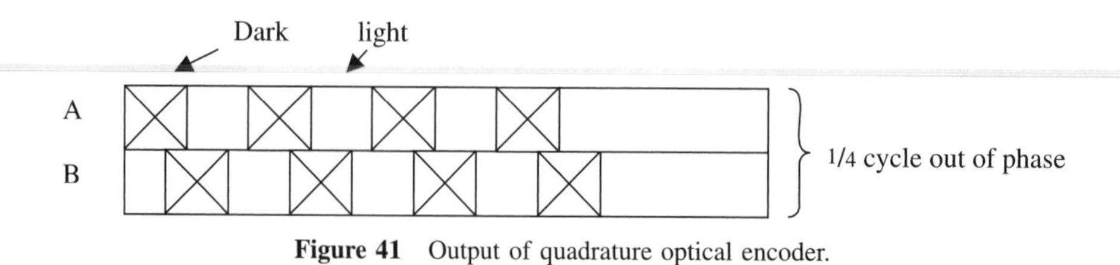

Figure 41 Output of quadrature optical encoder.

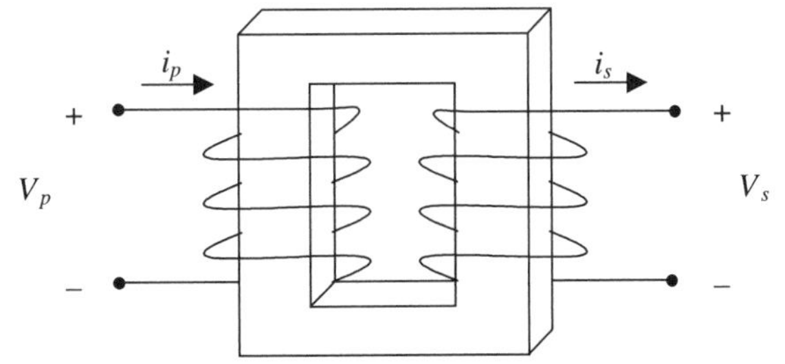

Figure 42 Electrical transformer.

field is then transferred through the high-permeability iron core. The changing magnetic filed induces a current in the secondary coil. The ideal inductor neither adds nor dissipates energy and power is conserved between the primary and secondary coils ($P_{in} = P_{out}$). The current induced (and hence the voltage) in the secondary coil is related to the current in the primary coil by the ratio of turns of each coil ($V_p/n_p = V_s/n_s$ or $n_p i_p = n_s i_s$). This principle is used (in a slightly different arrangement) to produce a position sensor.

The LVDT arrangement in Fig. 43 shows how moving the iron core (x direction) will change the coupling between the primary and secondary coils as described above. This arrangement produces a linear region where the magnitude of the ac output voltage ($=V_s$) is linearly related to the position of the movable iron core.

8.2 Force Sensors

Strain Gauge

It is impossible to directly measure a force due to the principle of causality. The effect of a force is often measured and then the force is estimated based on its effect. Strain gauges are frequently used to measure the surface stress on a structure and then estimate the force.

Strain gauges are generally a foil material where resistance changes as the foil is deformed. Strain gauges are generally glued to the surface of a structure and force is measured along a single measuring axis.

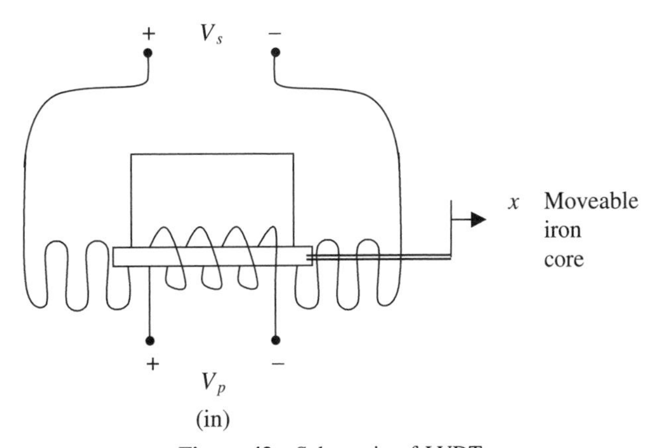

Figure 43 Schematic of LVDT.

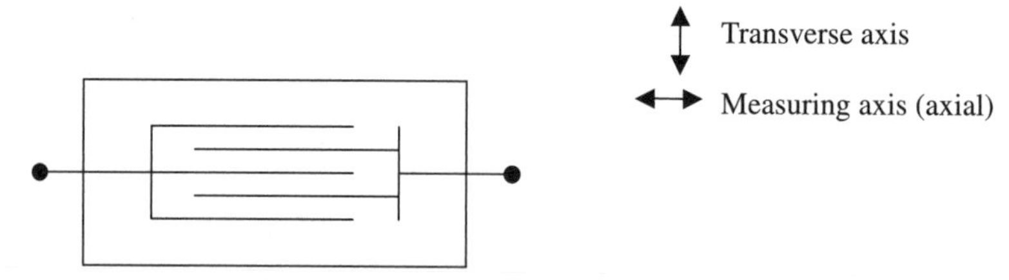

Figure 44 Schematic of strain gauge.

Strain gauges can be shown electrically as variable resistors where the resistance is related to the force. The principle is based on the physics of a long square rod, as shown in Fig. 44. The electrical resistance of the rod (R) is related to the resistivity of the material and the length of the rod and inversely related to the cross-sectional area of the rod. As a force is applied along the axis of the rod, its length (and cross-sectional area, related by Poisson's ratio) will change and thereby change the resistance. Strain gauges are designed to take maximum effect of this basic principle. It should also be noted that resistivity can change depending on many factors, including temperature. Much effort is put into advanced strain gauge applications to make the gauges independent of factors such as temperature variation. See Fig. 45.

Strain gauges are described by the gauge factor, which relates the change in resistance normalized by the zero load resistance of the gauge to the strain along the axis of the gauge:

$$F = \frac{\Delta R/R}{\epsilon_{\text{axial}}} \quad \text{(often} \approx 2\text{)} \tag{63}$$

Consider the cantilevered aluminum beam example given below. The strain gauge is mounted in the center of the beam 1 in. from the end and 1 in. from the wall. The beam is 2 in. long, 1 in. wide, and $\frac{1}{8}$ in. thick. A 5-lb load is applied at the end of the beam and the gauge factor is 2. Figures 46 and 47 show a mathematical model of the beam and the strain is estimated. This is then translated into the expected change in resistance of the gauge through the gauge factor.

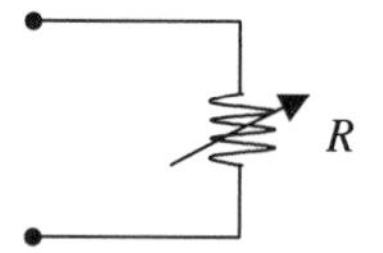

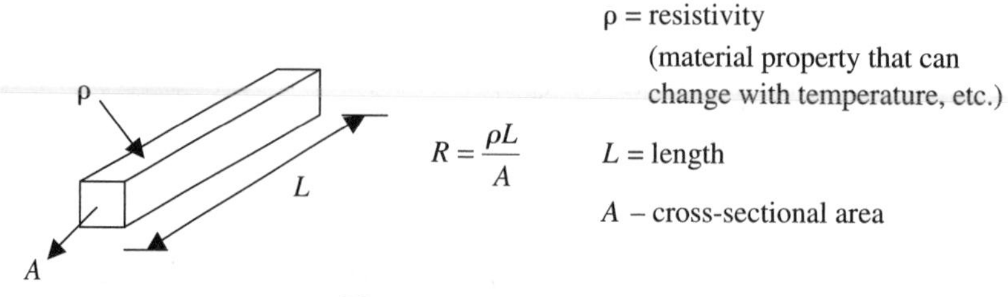

Figure 45 Basic resistance.

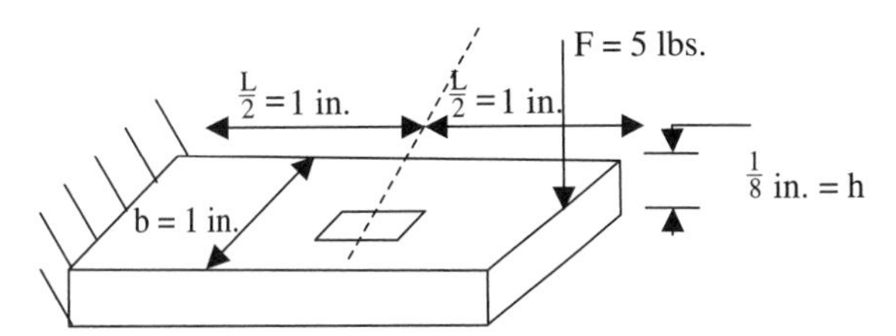

Figure 46 Strain gauge example.

Stress Modulus

$$\sigma = E\epsilon = \text{stress} = (\text{modulus of elasticity})(\text{strain}) \tag{64}$$

$$\sigma_{\text{bending}} = \frac{MC}{I} = \frac{(\text{bending moment})(\text{distance from neutral axis})}{(\text{moment of inertia})} = \text{bending stress} \tag{65}$$

$$E_{\text{Al}} = 10 \times 10^6 \text{ lb/in.}^2 \text{ (psi)} = \text{modulus of elasticity for aluminum} \tag{66}$$

$$M = F\frac{L}{2} = (5 \text{ lbs})(1 \text{ in.}) = 5 \text{ in.}\cdot\text{lb} = \text{bending moment} = \text{force}\,\frac{\text{length}}{2} \tag{67}$$

$$c = \tfrac{1}{16} \text{ in.} \tag{68}$$

$$I = \frac{1}{12}bh^3 = \frac{1}{12}\,1\,\frac{1^3}{8} = 1.6 \times 10^{-4} \text{ in.}^4 = \frac{1}{12}(\text{base})(\text{height})^3 \tag{69}$$

$$\epsilon = \frac{(5 \text{ in.}\cdot\text{lb} \times 0.0625 \text{ in.})/1.6 \times 10^{-4} \text{ in.}^4}{10 \times 10^6 \text{ lb/in.}^2} = .000192$$

$$= 192\mu\epsilon \text{ (unitless)} \tag{70}$$

$$\text{Strain} = \epsilon = 1.92\mu\epsilon \tag{71}$$

$$\text{Gauge factor} = \frac{\Delta R/R}{\epsilon_{\text{axial}}} = Z \tag{72}$$

$$\Delta R = RFE = (120)(2)(0.000192) = 0.046\ \Omega \tag{73}$$

This change in resistance is very small. It must be amplified in some way to be able to practically measure the change. If the above strain gauge is used in a voltage divider, (1)

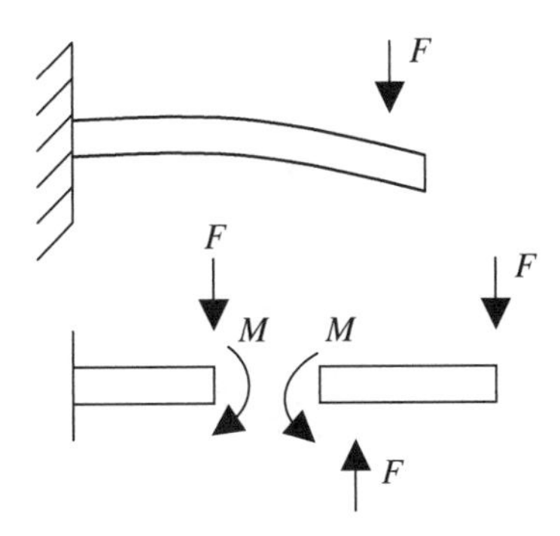

Figure 47 Free-body diagram.

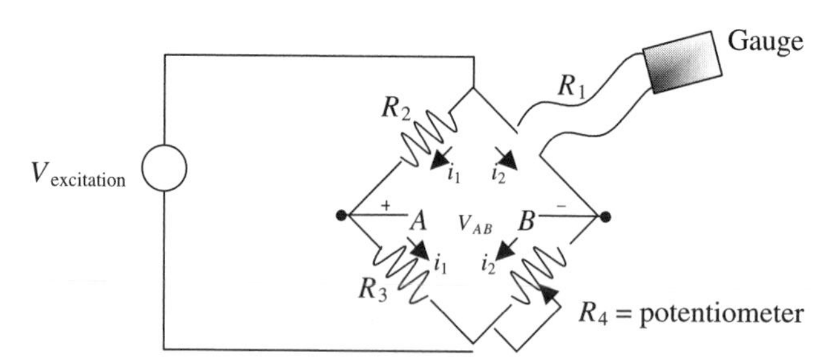

Figure 48 Strain gauge circuit.

the change in voltage will be small and (2) the voltage will be centered about a larger unchanged voltage.

For these reasons a Wheatstone bridge is used. It is shown in the analysis below that the change in voltage (V_{AB}) resulting from the change in the gauge is centered about 0 V and can then be amplified using a differential amplifier (described earlier in this chapter).

- used to measure small changes in resistance (Fig. 48).

Turn potentiometer until

$$V_A = V_B \tag{74}$$

Kirchhoff's voltage law gives

$$V_{AB} + i_1 R_2 - i_2 R_1 = 0 \tag{75}$$

$$V_{AB} - i_1 R_3 + i_2 R_4 = 0 \tag{76}$$

and

$$V_{ex} - i_1 R_2 - i_1 R_3 = 0 \tag{77}$$

$$V_{ex} - i_1(R_2 + R_3) = 0 \tag{78}$$

where

$$i_1 = \frac{V_{ex}}{R_2 + R_3} \tag{79}$$

and

$$V_{ex} - i_2 R_1 - i_2 R_4 = 0 \tag{80}$$

$$V_{ex} - i_2(R_1 + R_4) = 0 \tag{81}$$

where

$$i_2 = \frac{V_{ex}}{R_1 + R_4} \tag{82}$$

Substituting (79) and (82) into (75) yields

$$V_{AB} + \left(\frac{V_{ex}}{R_2 + R_3}\right)(R_2) - \left(\frac{V_{ex}}{R_1 + R_4}\right)R_1 = 0 \tag{83}$$

Substituting (79) and (82) into (76) gives

$$V_{AB} - \left(\frac{V_{\text{ex}}}{R_2 + R_3}\right)(R_3) + \left(\frac{V_{ex}}{R_1 + R_4}\right)R_4 = 0 \tag{84}$$

and

$$V_{AB} = -V_{\text{ex}}\left(\frac{R_2}{R_2 + R_3} - \frac{R_1}{R_1 + R_4}\right) = V_{\text{ex}}\left(\frac{R_1}{R_1R_4} - \frac{R_2}{R_2 + R_3}\right) \tag{85}$$

We now deform the strain gauge (R_1 becomes $R_1 + \Delta R_1$):

$$\frac{V_{AB}}{V_{\text{ex}}} = \frac{R_1 + \Delta R_1}{(R_1 + \Delta R_1) + R_4} - \frac{R_2}{R_2 + R_3} \tag{86}$$

Equation (86) relates V_{AB} to ΔR where ΔR is related to strain.

Shuffle (86):

$$\Delta R_1 = R_1\left[\frac{\dfrac{R_4}{R_1}\left(\dfrac{V_{AB}}{V_{\text{ex}}} + \dfrac{R_2}{R_2 + R_3}\right)}{\left(1 - \dfrac{V_{AB}}{V_{\text{ex}}} - \dfrac{R_2}{R_2 + R_3}\right)} - 1\right] \tag{87}$$

Use (87) and

$$F = \frac{\Delta R/R}{\epsilon_{\text{axial}}} \tag{88}$$

This analysis [(63)–(88)] gives an example of how to design a system with strain sensors.

9 ELECTROMECHANICAL MODELING EXAMPLE

Consider the mechatronic system shown in Fig. 49. Here a permanent-magnet dc motor is used to drive a one-degree-of-freedom robotic arm in a gravity field. The motor is connected to the arm with a pair of spur gears. The arm is used to actuate a spring (which could be used to represent interaction with the environment, i.e., applying a force).

The motor can be modeled electrically as an inductor, a resistor, and a voltage source connected in parallel; see Fig. 50. If the gears are assumed to be ideal, the following assumptions may be appropriate:

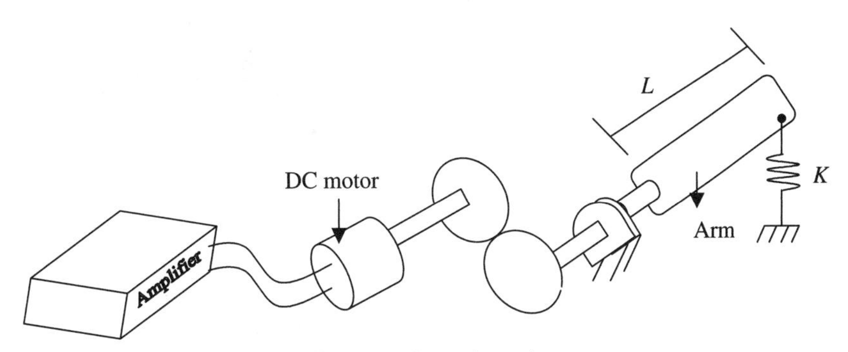

Figure 49 Example mechatronic system.

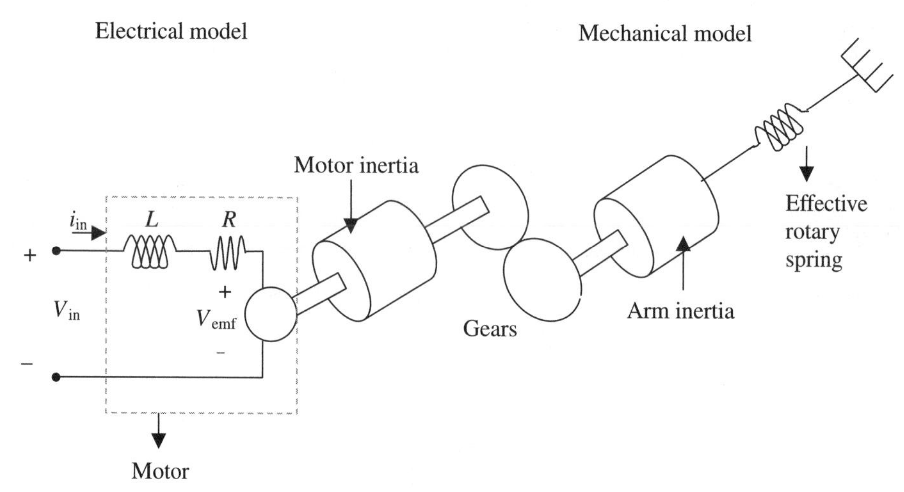

Figure 50 Model of mechatronic system.

1. No friction
2. Motor rotor and arm are the only significant masses
3. Small motions

Now the constitutive relationships for the motor and gears can be written and Kirchhoff's voltage law can be applied to the electrical circuit. Then, free-body diagrams can be created for each mechanical element; see Fig. 51:

- Motor relations:

$$\tau = K_t i_{in} = \text{torque of the motor} = (\text{torque constant})(\text{input current}) \quad (89)$$

where K_t is the torque constant.

$$V_{emf} = \omega K_e = K_e \dot{\theta}_{motor} \quad (90)$$

where K_e is the electromotive force (emf) constant.

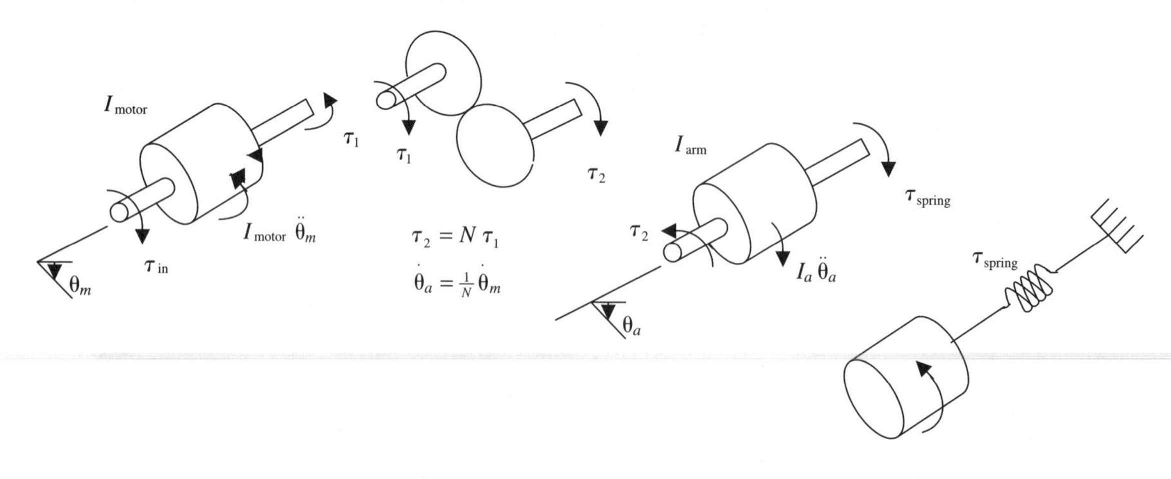

Figure 51 Free-body diagrams.

- Electrical equations:

$$V_{\text{in}} = L\frac{di_{\text{in}}}{dt} + Ri_{\text{in}} + K_{\text{emf}}\theta_{\text{motor}} \tag{91}$$

$$\tau_{\text{in}} - \tau_1 = I_{\text{motor}}\,\ddot{\theta}_{\text{motor}} \tag{92}$$

$$\tau_2 - \tau_{\text{spring}} = I_a\ddot{\theta}_a \tag{93}$$

where $\tau_1 = (1/N)\tau_2$ and $\tau_2 = \tau_{\text{spring}} + I_a\,\ddot{\theta}_a$. Substituting into (92) yields

$$\begin{aligned} \tau_{\text{in}} - \frac{1}{N}(\tau_{\text{spring}} + I_a\ddot{\theta}_a) &= I_{\text{motor}}\,\ddot{\theta}_{\text{motor}} \\ \ddot{\theta}_a &= \frac{1}{N}\,\ddot{\theta}_m \end{aligned} \tag{94}$$

The arm can then be modeled with an effective spring constant:

$$\tau_{\text{in}} - \frac{\tau_{\text{spring}}}{N} - \frac{I_a}{N}\left(\frac{1}{N}\,\ddot{\theta}_m\right) = I_{\text{motor}}\,\ddot{\theta}_m \tag{95}$$

Then τ_{spring} can be derived as

$$\tau_{\text{spring}} = K_{\text{effective}}\theta_a \tag{96}$$

$$= LF = \underbrace{KL^2}_{K_{\text{effective}}}\ \theta_a \quad \text{(see Fig. 52)} \tag{97}$$

Substituting (95)–(97) into (94) yields

$$\tau_{\text{in}} - \frac{KL^2}{N^2}\,\theta_{\text{arm}} - \frac{I_a}{N^2}\,\ddot{\theta}_m = I_{\text{motor}}\,\ddot{\theta}_{\text{motor}} \tag{98}$$

Inertial forces Spring force reflected to motor axis

$$\tau_{\text{in}} = \left(\frac{I_a}{N^2} + I_{\text{motor}}\right)\ddot{\theta}_m + \frac{KL^2}{N^2}\,\theta_m \tag{99}$$

Arm inertia reflected to motor axis Motor inertia

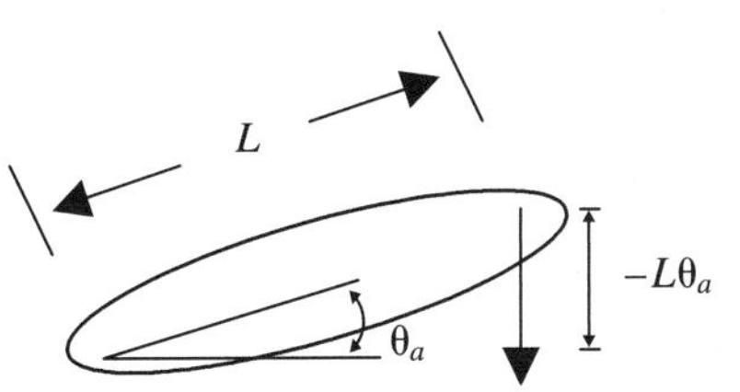

Figure 52 Free-body diagram of the arm.

Now, since

$$\tau_{\text{in}} = K_t i_{\text{in}} \tag{100}$$

substituting (100) into (99) yields

$$K_t\, i_{\text{in}} = \left(\frac{I_a}{N^2} + I_m\right)\ddot{\theta}_m + \frac{KL^2}{N^2}\,\theta_m \tag{101}$$

Substituting (101) into (91) yields

$$V_{\text{in}} = L\,\frac{d}{dt}\left\{\frac{1}{K_t}\left[\left(\frac{I_a}{N^2} + I_m\right)\ddot{\theta}_m + \frac{KL^2}{N^2}\,\theta_m\right]\right\} + R\left\{\frac{1}{K_t}\left[\left(\frac{I_a}{N^2} + I_m\right)\ddot{\theta}_m + \frac{KL^2}{N^2}\,\ddot{\theta}_m\right]\right\}$$

$$+ K_e\dot{\theta}_{\text{motor}} \tag{102}$$

This differential equation relates θ_m to V_{in}. The Laplace transform can be used to obtain the transfer function:

$$\frac{\text{Output}}{\text{Input}} = \frac{\theta_m(s)}{V_{\text{in}}(s)} \tag{103}$$

Also, substituting $\theta_a = (1/N)\,\theta_m$ yields $T(s) = \theta_a(s)/V_{\text{in}}(s)$.

BIBLIOGRAPHY

Asulander, D. M., and C. J. Kempf, *Mechatronics: Mechanical System Interfacing,* Prentice-Hall, Englewood Cliffs, NJ, 1996.

Cogdell, J. R., *Foundations of Electrical Engineering,* 2nd ed., Prentice-Hall, Englewood Cliffs, NJ, 1996.

Fraser, R. E., *Process Measurement and Control: Introduction to Sensors, Communication, Adjustment, and Control,* Prentice-Hall, Englewood Cliffs, NJ, 2001.

Histand, M. B., and D. G. Alciatore, *Introduction to Mechatronics and Measurement Systems,* 2nd ed., WCB/McGraw-Hill, New York, 2003.

Horowitz, P., and W. Hill, *The Art of Electronics,* 2nd ed., Press Syndicate of the University of Cambridge, Cambridge, 1989.

Necsulescu, D., *Mechatronics,* Prentice-Hall, Englewood Cliffs, NJ, 2002.

Paul, C. R., S. A. Nasar, and L. E. Unnewehr, *Introduction to Electrical Engineering,* 2nd ed., McGraw-Hill, New York, 1992.

Rizzoni, G., *Principles and Applications of Electrical Engineering,* 3rd ed., McGraw-Hill, New York, 2000.

Rizzoni, G., *Principles and Applications of Electrical Engineering,* 4th ed., McGraw-Hill, New York, 2003.

Sargent, M., and R. Shoemaker, *The Personal Computer from the Inside Out,* 3rd ed., Addison-Wesley, Reading, MA, 1995.

Shetty, D., and R. A. Kolk, *Mechatronics System Design,* PWS Publishing Company, Boston, MA, 1997.

CHAPTER 21

INTRODUCTION TO MICROELECTROMECHANICAL SYSTEMS (MEMS): DESIGN AND APPLICATION

M. E. Zaghloul
Department of Electrical and Computer Engineering
The George Washington University
Washington, D.C.

1 INTRODUCTION

In general, microelectromechanical systems have features in the micrometer- and, increasingly, nanometer-size range. Often, they are miniaturized systems that combine sensors and actuators with high-performance embedded processors on a single integrated chip. The word *electromechanical* implies the transfer of technology from mechanical to electrical and vice versa. Those devices embedded in functional systems are some times referred to as microsystems. This field is increasingly leading to devices and material systems whose size is on the order of a nanometer, that is, the size of molecules. Microsystems and nanotechnology enable the building of very complex systems with high performance at a fraction of the cost and size of ordinary systems. As such, these systems are the enabling technology for today's explosive growth in computer, biomedical, communication, magnetic storage, transportation, and many other technologies and industries. Microsystems and nanotechnology challenges range from the deeply intellectual to the explicitly commercial. This field is by its very nature a link between academic research and commercial applications in the aforementioned and other disciplines. Indeed, these disciplines span a very broad range of industries that are at the forefront of current technological growth.

The integration of microelectronics and micromechanics is a historic advance in the technology of small-scale systems and is very challenging for designers and producers of MEMS. The addition of micromachined parts to microelectronics opens up a large and very important parameter space to technological development and exploitation.

The MEMS structures and devices result from the sequence of design, simulation, fabrication, packaging, and testing. There are varieties of devices that can be classified as MEMS. There are passive devices, that is, nonmoving structures. There are devices that involve sensors and devices that involve actuators, which have micromechanical components. These are conceptually reciprocal in that sensors respond to the world and provide information and actuators use information to influence something in the world. Another class includes systems that integrate both sensors and actuators to provide some useful function. This classification, like most, is imperfect. For example, some devices that are dominantly sensors have actuators built into them for self-testing. Airbag triggers are an example. However, the framework provides a simple but quite comprehensive framework for considering MEMS devices.

In this chapter we will discuss some aspects of the design of these devices and introduce the reader to the technology used. In addition, we will discuss the structure of some of those devices.

2 MICROFABRICATION PROCEDURES

The fact that the field of MEMS largely grew out of the integrated circuit (IC) industry has been noted often. There is no doubt that the use of fabrication processes and associated equipment that were developed initially for semiconductor industry has given the MEMS industry the impetus it needed to overcome the massive infrastructure requirements. However, it is noted that the field of MEMS has gone far beyond the materials and processes used for IC production. The situation is indicated schematically in Fig. 1. About a half dozen materials, notably silicon and its oxide and nitride, and standard microfabrication processes, such as lithography and ion implantation, oxidation, deposition, and etching, have generally been employed to make ICs. The set of materials used in IC devices is expanding to include, for example, low-dielectric-constant materials, polymers, and other nonconventional IC materials.

Many MEMS can be made with the same set of materials and processes as used for microelectronics. However, one of the hallmarks of the emerging MEMS industry is the use of numerous other materials and processes. Most basically, substrates other than silicon are being employed for MEMS. Silicon carbide has been demonstrated to be a good basis for many mechanisms that can stand higher temperature service than silicon. Diverse materials can be used within MEMS devices. While aluminum and, recently, copper are the metals used in IC devices, micromachining of many other metals and alloys has been demonstrated. Magnetic materials have been incorporated into some MEMS devices. Piezoelectric materials are especially attractive for MEMS because of their electrical–mechanical reciprocity. That is, application of a voltage to a piezoelectric material deforms it, and application of a strain produces a voltage. Zinc oxide and lead zirconium titinate (PZT) are important piezoelectric materials for MEMS. Many other examples of materials employed for MEMS could be given. However, the point is clear. Micromechanics are made of many more kinds of materials than microelectronics.

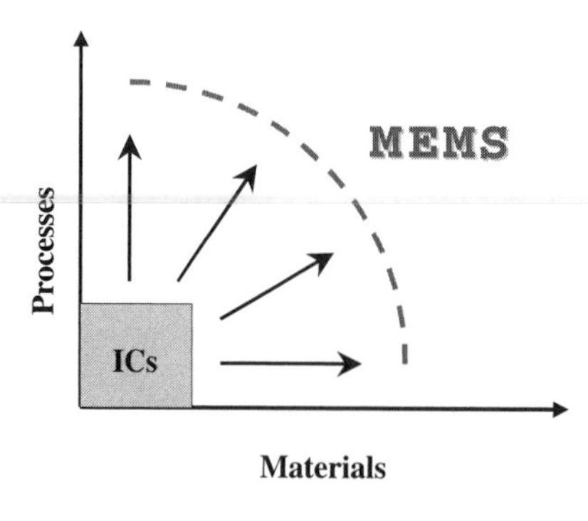

Figure 1 The number of materials and processes employed to make MEMS greatly exceeds those used to manufacture integrated circuits.

Because of the varieties of materials used in MEMS fabrication, the processes for producing and modifying them widened far beyond those found in the IC industry. However, something more fundamental is at work when it comes to processes for making MEMS. Integrated circuits are monolithic and, despite up to 30 layers in some cases, are made by largely two-dimensional thin-film processes that yield what some call 2.5-dimensional structures. By contrast, micromechanical devices must have space between their parts so they can move, and the dimension perpendicular to the substrate is often very fundamentally necessary for their performance. Development of processes to make micrometer-scale parts that can move relative to each other was the breakthrough that enabled MEMS. Such micromachining processes fall into three major categories, which will now be reviewed briefly.

Surface micromachining involves the buildup of micromechanical structures on the surface of a substrate by deposition, patterning, and etching processes. The key step is the etching away of an earlier deposited and patterned sacrificial layer in order to free the mechanism. Figure 2 shows the steps needed for such processing.[1,2]

This process was first demonstrated about 35 years ago, when a metal–oxide–semiconductor field effect transistor (MOSFET) with a cantilever mechanical gate was produced.[3] The most common sacrificial material now is silicon dioxide, which is conveniently

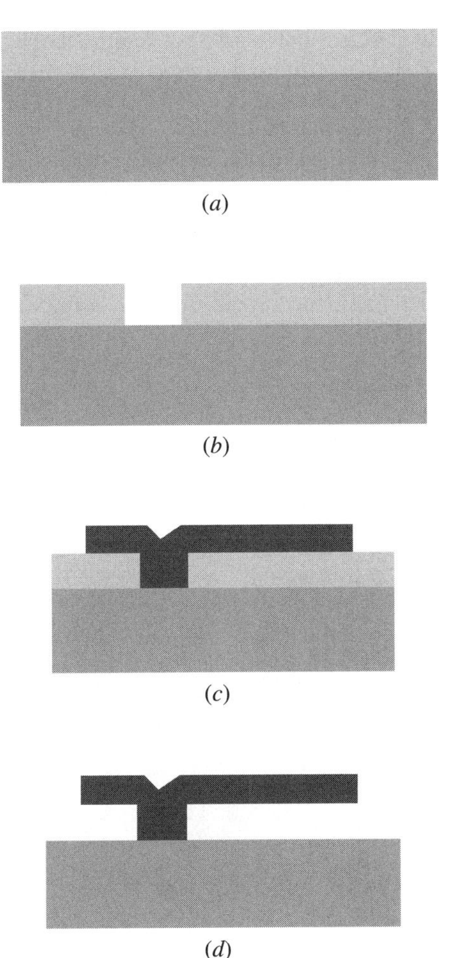

Figure 2 Surface micromachining steps: (*a*) Step 1: Deposit sacrifical layer. (*b*) Step 2: Pattern layers. (*c*) Step 3: Deposit structure layer. (*d*) Step 4: Etch sacrifical layer.

dissolved from under a movable part using hydrofluoric acid. Surface micromachining has been used to produce an amazing variety of micromechanical devices, some of which are now in large-scale production. Microaccelerometers and MEMS angle rate sensors are examples.

Figure 3 shows examples of mechanical structures built by surface micromachining.

Bulk micromachining, as the name implies, involves etching into the substrate to produce structures of interest. It can be done with either wet or "dry," that is, plasma, processes, either of which can attack the substrate in any direction (isotropically) or in preferred directions (anisotropically). Bulk micromachining has two primary variants. The first depends on the remarkable property of some wet chemical etches to attack single-crystal silicon as much as 600 times faster along some crystallographic directions compared to others. This anisotropic process is called orientation-dependent etching (ODE). It was known long before the emergence of MEMS technologies and has become a mainstay of the industry. ODE is especially useful for producing thin membranes that serve as the sensitive element in micropressure sensors. It is employed for production of these and other commercial MEMS devices. The second approach to bulk micromachining is to use plasma-based etching processes that attack the substrate, usually silicon, in preferential directions. Deep reactive ion etching (DRIE) is a plasma process that is used increasingly to make MEMS. It can produce structures that are over 10 times as deep as they are wide. Bulk micromachining steps are shown in Fig. 4*a*. Examples of devices developed using bulk micromachining are shown in Fig. 4*b*.

The third general class of micromachining processes is a collection of the numerous and varied techniques that can produce structures and mechanisms on the micrometer scale. Laser-induced etching and deposition of materials, electroetching and electroplating, ultrasonic and electron discharge milling, ink jetting, molding, and embossing are all available to the MEMS designer.

Similar to ICs, MEMS devices are made using creative combinations of the materials and processes noted above. Some remarkable micromechanisms have been demonstrated, largely in academic fabrication facilities, and commercialized using diverse foundries.

3 DESIGN AND SIMULATIONS

To verify that the devices function, the designer has to model the MEMS device. The modeling involves writing the equation of motion or physical modeling of the performance of

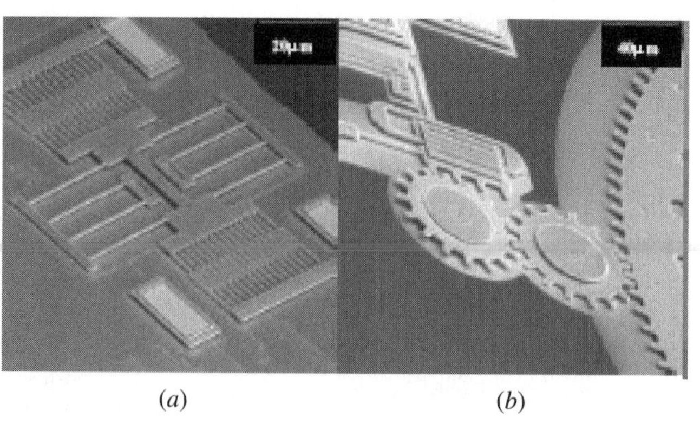

(*a*) (*b*)

Figure 3 Examples of surface micromachining: (*a*) simple sensors and actuators; (*b*) gear train.

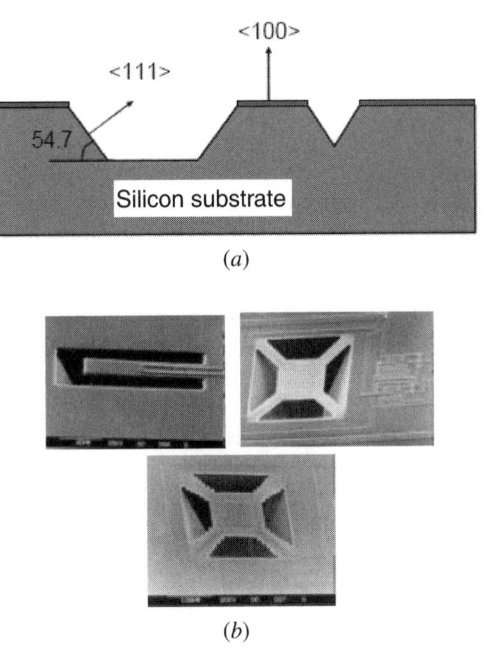

Figure 4 (*a*) Bulk micromachining anisotropic etch. (*b*) Examples of bulk micromachining devices.

the device. Finite-element techniques are used to solve these modeling equations. There are a variety of computer aided design (CAD) tools to aid the designer in the simulation and modeling of the device. In a very fundamental way, these tools are more complicated than the software for design of either solely ICs or solely mechanical devices. This is due to the close coupling of both electrical and mechanical effects within many MEMS. Consider a microcantilever that is pulled down by electrostatic forces. Its simulation has to take into account both the flow of electrical charge and mechanical elasticity in an iterative and self-consistent fashion.

Thermal, optical, magnetic, fluidic, and other mechanisms are also active in some MEMS and have to be handled self-consistently in the simulation phase.

Two basic approaches have been taken in the past decade to the need for specialized software for the design and simulation of MEMS. In the first approach, CAD design, tools and available software from electronic design were modified to accommodate the requirement for MEMS design. In the second approach finite-element modeling was applied to MEMS. Software from the Tanner Tools very large scale integrated (VLSI) design suite were used for MEMS, for example MEMS-PRO,[4] which was recently acquired by MEMSCAP,[5] as was the popular mechanical engineering software from ANSYS.[6] Recently, new suites of software specifically developed for MEMS were marketed. Most of them include electronic, mechanical, and thermal simulation, and some have other physical mechanisms as well as processing simulation tools. Such software is available from CFD Research Corporation, Coventor (formerly called CRONOS Technologies), IntelliSense Corporation, and Integrated Systems Engineering. These tools vary widely in the mechanisms and material parameters that they include, the details of design and simulation of devices, and the fabrication facilities with which they interface. The choice of suitable software to use for MEMS design is still challenging. MEMSCAP is based on CADENCE (which is the most popular IC design tool[7]). It consists of a set of tools which enable the design flow either bottom up or top down. It incorporates the MEMS design environment into existing and well-known environments with easy intellectual property (IP) cells or design reuse and the ability to exchange data between multidisplinary teams. The MEMSCAP simulator is based on the CADENCE environment, so the designer can simulate MEMS devices with the IC schematics and simulation. Models can be generated from the ANSYS finite-element model or from written analytical equations. Behavior models/scalable symbolic view can be generated. The Verilog-A model can also be generated. The generated model can be used to perform optimization simulations inside the environment or to realize a system simulation. In addition, emulators are available for etching, cross section projection of the different material layers.

The design of MEMS devices involves knowledge of the sequence of materials to be used to realize the device. The sequence of materials used could be the standard sequence, in which case a standard technology process may be used in conjunction with other processing steps, for example, postprocessing. The sequence of materials to be used could be custom designed by the designer, which requires knowledge of the materials and their thin-film properties. Designers usually design the device and identify the material to be used and then use CAD tools to verify the performance. Iteration procedures are part of the design until the required performance is reached. After satisfactory simulation performance, the device is sent to fabrication foundries.

4 FABRICATION FOUNDRIES

After designing and simulating MEMS and deciding on the materials they will contain and the processes needed to make them, the next concern is which fabrication facility to employ. Sometimes, a standard IC facility can be used with postprocessing steps. Postprocessing involves adding or removing materials from the standard fabricated device. For example, using a standard complementary metal–oxide–semiconductor (CMOS) fabrication facility, we could realize the suspended-plate structure shown in Fig. 5. In this case the CMOS is removed from the bulk substrate to create the suspended structure on top of the etched pit.

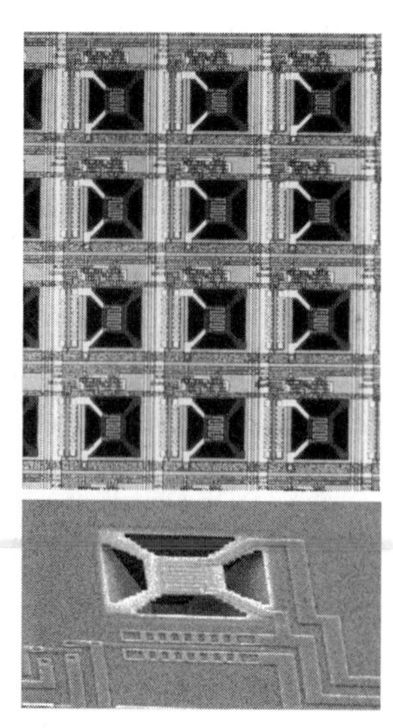

Figure 5 Micro-hot-plate array and scanning electron micrograph of one of the elements.

For example, the micro-hot-plate shown in Fig. 5 was realized in CMOS technology in a standard foundry with a postprocessing step of bulk micromachining to produce the suspended thin film with a resistive heater. The small mass of the heated element permits temperature changes of 300°C in a few milliseconds.

Many MEMS structures and devices have been produced by such postprocessing of CMOS chips.[8–10] Techniques which are not compatible with CMOS have also been used, in which case surface micromachining techniques produce mechanical structures on top of the substrate. A micromirror fabricated using surface micromachining is an example of such a device.

Figure 6 is a schematic of two pixels of the Digital Mirror Device manufactured by Texas Instruments. The torsion hinges are 5×1 μm in area and about 100 nm thick. The individual mirrors are 16 μm square. Over 500,000 of them are found in a single device, making this the system with the most moving parts produced in the history of mankind. The inventor, Larry Hornbeck, and the company received Emmy Awards in 1998 for outstanding achievement in engineering development.

There are now several foundries specifically for the production of MEMS. The fact that design rules in MEMS are roughly two generations behind those in ICs is significant. This enables MEMS foundries to buy used equipment from the microelectronics industry. Mass production of many MEMS now is done using 100- and 150-mm wafers. Several companies and organizations in the United States and abroad offer fabrication services for MEMS somewhat analogous to those in IC foundries. They include BFGoodrich, Advanced MicroMachines, CMP (France), Institute of Microelectronics (Singapore), IntelliSense, ISSYS, Surface Technology Systems (U.K.), and many more.

Most of these foundries have all the facilities in-house to produce complete MEMS devices. However, the wide variety of materials and processes that can be designed into MEMS means that it is not always possible to find all the needed tools under one roof. Hence, the Defense Advanced Research Projects Agency instituted a new type of foundry service several years ago. It is called the MEMS-Exchange.[11] This organization contracts with diverse industrial and academic fabrication facilities for a wide range of services. The MEMS designer can draw from any of them. A completed design is sent to MEMS-Exchange, which handles scheduling, production, billing, and other factors, such as the protection of proprietary designs.

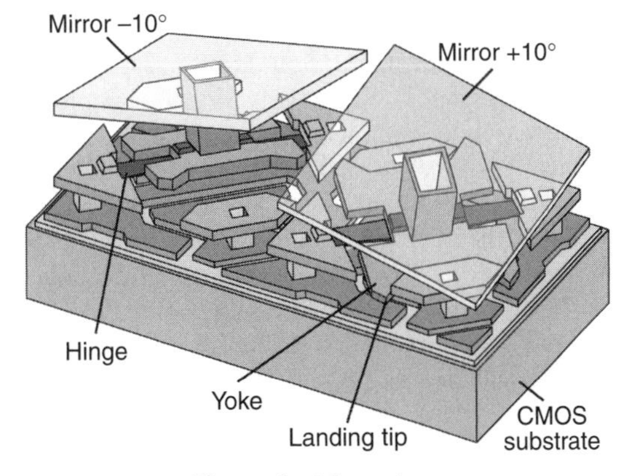

Figure 6 Micromirror.

5 EXAMPLES OF MEMS DEVICES AND THEIR APPLICATIONS

There are varieties of applications of MEMS. This section gives a brief overview of MEMS applications with reference to commercial devices. There are many fields in which MEMS devices have been introduced. Table 1 shows examples of MEMS applications.

Table 1 summarizes some of the applications of MEMS and shows the air bag accelerometer developed by Analog Devices in which the structure of the sensors is based on a variable-capacitor device. Figure 7 shows the surface micromachining of the Analog Devices accelerometer. Mechanical structures were studied to develop miroresonators, such as the fixed–fixed beam of Fig. 8, and circular resonators.

Researchers are using MEMS techniques to produce an array of nanoresonators that can be integrated with other components. Figure 8 shows a working radial contour-mode disk resonator with 10 μm radius and quality factor $Q = 1595$ at atmospheric pressure.[12] Work is aimed at coupling such resonators together to make large arrays. These devices were used in the design of electric filters for high-frequency communication systems. The benefit of using MEMS device is the high quality factor, which implies a high-efficiency circuit. Figure 9 shows an example of a mechanical switch which is an electrostatic switch. It is used in high-frequency circuits as a small-loss switch. The radio-frequency switch is small, on the order of 50 μm $\times$ 50 μm.

Figure 9 shows a micrograph (top) and schematic of the Raytheon MEMS microwave switch. The electrode under the flexible membrane is the actuator. The capacitance of the switch varies from near zero (open) to 3.4 pF (closed). The signal path is about 50 μm wide.

Figure 10 shows a gas sensor developed using CMOS technology and the associated circuits that make it a smart gas sensor.[13]

The above examples illustrate the variety of applications of MEMS devices as well as the variety of materials used. Other examples can be found in Ref. 14.

The last letter in MEMS stands for systems. This is due partly to the fact that a MEMS device is quite complex. However, MEMS devices are "only" components which are used in larger and more complex systems. That is, individual sensors or actuators can be used as components and incorporated into subsystems or systems in order to perform some useful function. The accelerometer in the air bag subsystem of an automobile and the DMD in a projection system in a theater are examples. However, it is also possible to closely couple both MEMS sensors and actuators into miniature systems all on one substrate. These are called "systems on a chip." Microfluidics with all the needed functionality on a substrate, including pumps and valves, as well as channels, mixers, separators, and detectors, are under development for compact analyzers. These will be relatively evolutionary advances over current microfluidic chips. High-density data storage systems with both actuation and sensing functions represent a more revolutionary example of integrated microsystems.

The variety of commercially available MEMS and their applications have both increased dramatically in recent years. The production of MEMS is now more than a $20 billion industry worldwide, with about 100 million devices marketed annually. While this industry grew out of the microelectronics industry, it is more complex in many important ways. Most fundamentally, it requires the integration of both microelectronics and micromechanics. Many MEMS involve several closely coupled mechanisms, some of which behave differently on the micrometer spatial scale than on familiar macroscopic scales. This complicates both the design and simulation of MEMS. So also does the much wider variety of materials and processes used to make MEMS compared to microelectronics. Because many MEMS have to be open to the atmosphere, their packaging, calibration, and testing are complex. Questions about the long-term reliability of MEMS are being answered as MEMS devices spend more years in use by consumers and industries.

Table 1 Examples of MEMS Applications

Pressure Sensors	Inertia Sensors	Optical Devices	Data Storage	RF-MEMS	Acoustic MEMS	MicroFluidic	Chemical Sensors
Aeronautical	Air bag accelerometer	Optical beam steering	Hard disk component	Miniature antenna	Microphone	Micropumps	Polymer gas sensors
Blood pressure	Motion control sensors	Microlasers	Miniature read/write	RF-switch	Acoustic vibratos	Microvalves, microchannels	Tin oxide gas sensors
Auto tire	Automobile suspension	Optical switch	Magnetic devices	Filters and resonators		Lab on chip	Preconcentarors
Touch pressure	Vibration sensor	Micromirrors	Optical storage	Inductors and capacitors		Ink bubble jet nozzle	Smart gas sensors

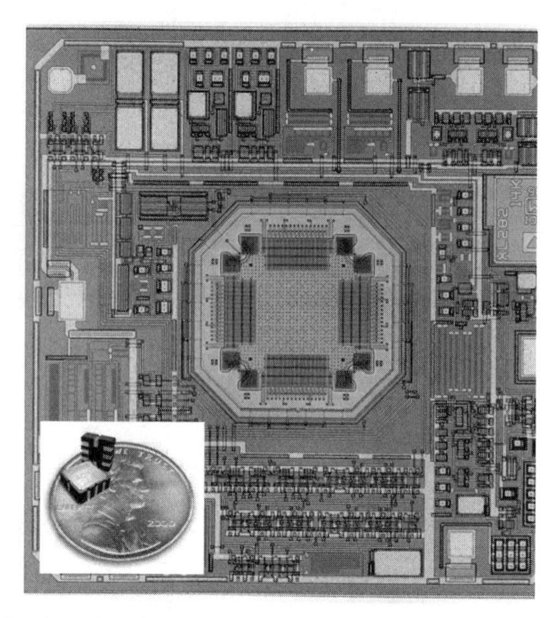

Figure 7 Photographs of exterior of new two-axis microaccelerometers in leadless packages on a penny and micrograph of chip from Analog Devices. In this device, the microelectronic and micromechanical components are tightly integrated on the silicon substrate.

6 CONCLUSIONS

We have discussed the advances in the microfabrication structures that allowed the realization of three-dimensional structures at the micrometer scale. As listed above, the materials and microfabrication processes used are unlimited to realize MEMS devices. Many MEMS involve several mechanical structures which behave differently on the micrometer spatial scale. Several CAD tools were developed to aid in the design of such devices. The design of such miniature devices is challenging and complicated as compared to devices at the macroscale level. Despite such engineering challenges, MEMS offer high performance and are small, have low power, and are relatively inexpensive. They both improve on some existing applications and enable entirely new systems. Some applications are targeted from the outset of

Figure 8 Circular resonators.

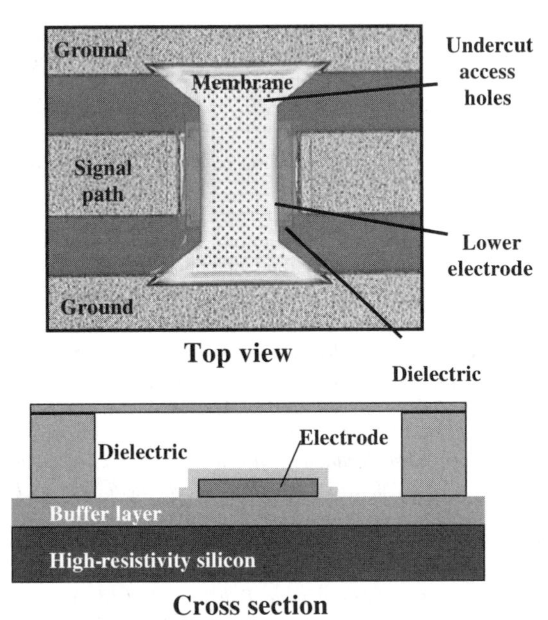

Figure 9 Microwave switch.

design, but others are opportunistic. That is, the large number of MEMS components on the market makes them available to design engineers for a very wide variety of uses.

APPENDIX: BOOKS ON MEMS

A great deal of information on the design, simulation, fabrication, packaging, testing, and application of MEMS is available. This information is presented as journal articles, confer-

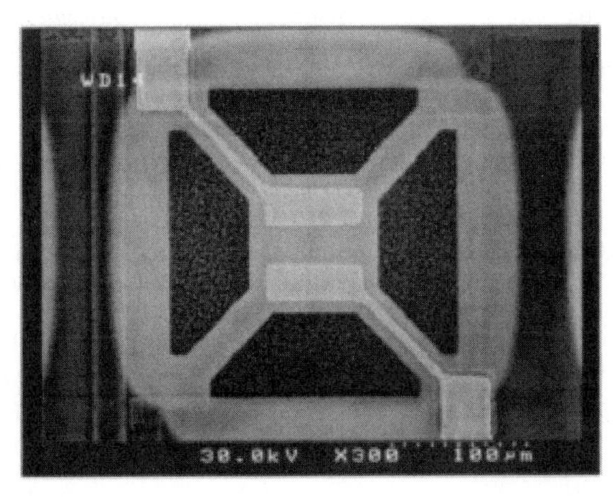

Figure 10 Scanning electron micrograph of CMOS gas sensor.

ence proceedings, and books as well as being available from the World Wide Web. In the past several years many books on MEMS have been published:

1995: *Integrated Optics, Microstructures and Sensors,* M. Tabib-Azar (ed.), Kluwer Academic.

1996: *Micromachines* (*A New Era in Mechanical Engineering*), I. Fujimasa, Oxford University Press.

1997: *Micromechanics and MEMS* (*Classical and Seminal Papers to 1990*), W. S. Trimmer (ed.), IEEE Press, New York.

1997: *Fundamentals of Microfabrication,* M. Madou, CRC Press, Boca Raton, FL.

1997: *Handbook of Microlothography, Micromachining and Microfabrication,* Vol. 2: Micromachining and Microfabrication, P. Rai-Choudhury (ed.), SPIE Press.

1998: *Micromachined Transducers Sourcebook,* G. T. A. Kovacs, WCB McGraw-Hill, New York.

1998: *Microactuators,* M. Tabib-Azar, Kluwer Academic.

1998: *Modern Inertial Technology,* 2nd ed., A. Lawrence, Springer.

1998: *Methodology for Modeling and Simulation of Microsystems,* B. F. Romanowicz, Kluwer Academic.

1999: *Selected Papers on Optical MEMS,* V. M. Bright and B. J. Thompson (eds.), SPIE Milestone Series, Vol. MS 153, SPIE Press.

1999: *Microsystem Technology in Chemistry and Life Sciences,* A. Manz and H. Becker (eds.), Springer.

2000: *An Introduction to Microelectromechanical Systems Engineering,* N. Maluf, Artech House.

2000: *MEMS and MOEMS Technology and Applications,* Vol PM85, P. Rai-Choudhury (ed.), SPIE Press.

2000: *Electromechanical Systems, Electric Machines and Applied Mechatronics,* S. E. Lyshevski, CRC Press, Boca Raton, FL.

2000: *Handbook of Micro/nano Tribology,* 2nd ed., B. Bhushan, CRC Press, Boca Raton, FL.

2001: *MEMS Handbook,* Mohamed Gad-El-Hak (Editor in Chief), CRC Press, Boca Raton, FL.

2001: *Microsystem Design,* S. D. Senturia, Kluwer Academic.

2001: *MEMS and Microsystems: Design and Manufacture,* T.-R. Hsu, McGraw-Hill College Division, New York.

2001: *Mechanical Microsensors,* M. Elwenspoek and R. Wiegerink, Springer.

2001: *Nano- and Microelectromechanical Systems,* S. E. Lyshevski, CRC Press, Boca Raton, FL.

2001: *Microflows: Fundamentals and Simulations,* 2nd ed., G. E. Karniadakis and A. Berskok, Springer Verlag.

2001: *Microsensors, MEMS and Smart Devices,* J. W. Gardner, V. K. Varadan, and O. O. Awadelkarim, Wiley, New York.

2001: *Microstereolithography and Other Fabrication Techniques for 3D MEMS,* V. K. Varadan, X. Jiang, and V. V. Varadan, Wiley, New York.

2002: *Fundamentals of Microfabrication* (*The Science of Miniaturization*), 2nd ed., M. Madou, CRC Press, Boca Raton, FL.

2002: *MEMS and NEMS: Systems, Devices and Structures,* S. E. Lyshevsky, CRC Press, Boca Raton, FL.

2002: *Microfluidic Technology and Applications,* M. Koch, A. Evans, and A. Brunnschweiler, Research Studies Press.

2002: *Fundamentals and Applications of Microfluidics,* N.-T. Nguyen and S. T. Wereley, Artech House.

2002: *Microelectrofluidic Systems Modeling and Simulation,* T. Zhang, K. Chakrabarty, R. B. Fair, and S. E. Lyshevsky, CRC Press, Boca Raton, FL.

2002: *Modeling MEMS and NEMS,* J. A. Pelesko and D. H. Bernstein, CRC Press, Boca Raton, FL.

2002: *Nanoelectromechanics in Engineering and Biology,* M. P. Hughes, CRC Press, Boca Raton, FL.

2002: *Optical Microscanners and Microspectrometers Using Thermal Bimorph Acutators,* G. Lammel, S. Schweizer, and P. Renaud, Kluwer.

2003: *RF MEMS Theory, Design and Technology,* G. M. Rebeiz, Wiley-Interscience, New York.

2003: *MEMS and Their Applications,* V. K. Varadan, K. J. Vinoy, and K. A. Jose, Wiley, New York.

REFERENCES

1. M. Madou, *Fundamental of Microfabrication,* CRC, Boca Raton, FL, 1779.
2. M. Madou, *Fundamental of Microfabrication, The Science of Miniaturization,* CRC Press, Boca Raton, FL, 2002.

3. H. C. Nathanson et al., "The Resonant Gate Transistor," *IEEE Transactions on Electron Devices* **ED-14**(3), 117–133 (1967).

4. www.tanner.com.

5. www.memscap.com.

6. www.ansys.com.

7. www.CADENCE.com.

8. V. Milanovic, M. Gaitan, E. Bowen, N. Tea, and M. E. Zaghloul, "Thermoelectric Power Sensor for Microwave Applications by Commercial CMOS Fabrication," *Transactions of IEEE Electron Device Letters* **18**(9), 450–452 (1997).

9. V. Milanovic, M. Gaitan, E. Bowen, and M. E. Zaghloul, "Micromachining Microwave Transmission Lines in CMOS Technology," *IEEE Transactions on Microwave and Theory Techniques* **45**(5), 630–635 (1997).

10. M. Ozgur, M. E. Zaghloul, and M. Gaitan, "High Q Backside Micromachined CMOS Inductors," paper presented at the IEEE International Symposium on Circuits and Systems, Orlando, FL, May 1999, pp. II-577–II-580.

11. www.MEMS_EXCHANGE.com.

12. J. Wang, Z. Ren, and C. T. C. Nguyen, "1.14 GHz Self Aligned Vibrating Micromechanical Disk Resonator," paper presented at the IEEE Radio Frequency Integrated Circuits (RFIC) Symposium, June 2003, pp. 335–338.

13. M. Afridi, J. S. Suehle, M. E. Zaghloul, D. W. Berning, A. R. Hefner, R. E. Cavicchi, S. Semacik, C. B. Montgomery, and C. J. Taylor, "A Monolithic CMOS Microhotplate-Based Gas Sensor System," *IEEE SENSORS Journal* **2**(6), 644–655 (2002).

14. *Proceedings of the IEEE Special Issue on Integrated Sensors, Microactuators, and Microsystems* [*MEMS*], IEEE, New York, August 1998.

INDEX

C

E

F

G

H

I

J

K

L

M

N

O

P

Q

R

S

T